Combustion Technologies for a Clean Environment

Energy, Combustion and the Environment

A series edited by Scott Samuelsen
Professor of Mechanical, Aerospace, and Environmental Engineering and Director, UCI Combustion Laboratory, University of California, Irvine.

The subject of this series is the production of energy and the relationship of this production to environmental impact. The scope focuses on: (1) fossil fuels, since these are projected to serve as the principal energy sources for more than 100 years; (2) three levels of environmental air quality impact, including urban air quality, tropospheric global warming and acid rain, and stratospheric ozone degradation; and (3) both stationary and aeroengine sources of energy production.

Volume 1
Combustion Technologies for a Clean Environment
Edited by Maria da Graça Carvalho, Woodrow A. Fiveland, F. C. Lockwood, and Christos Papadopoulos

This book is part of a series. The publisher will accept continuation orders which may be cancelled at any time and which provide for automatic billing and shipping of each title in the series upon publication. Please write for details.

Combustion Technologies for a Clean Environment

Selected Papers from the Proceedings of the First International Conference
Vilamoura, Portugal, September 3–6, 1991

Edited by

Maria da Graça Carvalho
Instituto Superior Técnico
Lisbon, Portugal

F. C. Lockwood
Imperial College of Science, Technology and Medicine
London, UK

Woodrow A. Fiveland
Babcock & Wilcox
Alliance, Ohio, USA

Christos Papadopoulos
CINAR SA
Athens, Greece

Gordon and Breach Publishers
Australia • Austria • Belgium • China • France • Germany • India • Japan
Malaysia • Netherlands • Russia • Singapore • Switzerland
Thailand • United Kingdom • United States

Gordon and Breach Science Publishers SA
Postfach
4004 Basel
Switzerland

Many of the articles appearing in this book were originally published in *Combustion Science and Technology*, volume 93, numbers 1–6.

Library of Congress Cataloging-in-Publication Data

International Conference on Combustion Technologies for a Clean
Environment (1st : 1991 : Vila Moura, Portugal)
Combustion technologies for a clean environment :
selected proceedings of the First International Conference,
Vilamoura, Portugal, September 3–6, 1991 / edited by Maria da Graça
Carvalho ... [et al.].
p. cm.
-- (Energy, combustion and the environment, ISSN 1073-7804; v1)
ISBN 2-88124-619-2
1. Combustion engineering--Environmental aspects--Congresses.
I. Carvalho, M. G. (Maria da Graça), 1955– . II. Title. III. Series.
TJ254.5.I547 1991
628.5'32--dc20 93-46756
CIP

CONTENTS

PART 6 — BIOMASS COMBUSTION

PART 7 — FLUIDISED BED COMBUSTION

PART 8 — INCINERATORS

PART 9 — INTERNAL COMBUSTION ENGINES

PART 10 — REACTION KINETICS

INTRODUCTION TO THE SERIES

The world is in transition toward acknowledging, and accepting, that a basic conflict exists between the provision of energy to support the world's economies and standard of living, and the protection of the air resource that sustains the world's life. The *Energy, Combustion and the Environment* book series is designed to place this conflict into perspective, and to provide readers — students, consulting engineers, regulators, legislative staff, and environmental action groups — with the information and tools required to understand the challenges and conduct analyses to address these challenges.

PREFACE

The existence and development of civilization owes as much to the exploitation of combustion processes as to any other single cultural or scientific influence. However, it is increasingly recognized that the unrelenting pursuit of current economic growth rates, with ever-increasing material prosperity, cannot continue indefinitely on the basis of established and traditional combustion processes. Although the consequences of unrestricted growth are not easily quantified by even the best available scientific models, all projections show that the results would be environmentally unacceptable and/or unsustainable.

Against this background, the aim of *Combustion Technologies for a Clean Environment* is to describe, compare, and evaluate new combustion technologies which maximize energy conversion while minimizing undesirable emissions. This volume contains selected papers presented at the First International Conference on Combustion Technologies for a Clean Environment held in Vilamoura, Algarve, Portugal, September 3–6, 1991. A range of interesting contributions are presented both at academic and applied levels.

The papers were selected following review by members of the Scientific Committee, whose names are listed on the following page. The editors are grateful to the members of this committee, to the authors, and to all participants who helped to make the conference a success. We also wish to acknowledge the critical financial support of the sponsoring organizations listed a page later. The assistance of Maria Fernanda Afonso, Cristina Gomes, and Drs. Phil Costen and Joao Toste Azevedo in the proofreading and typing of this book is greatly appreciated.

The Second International Conference on Combustion Technologies for a Clean Environment was held in Lisbon, Portugal, in July 1993, and the proceedings will be published through this series.

Maria da Graça Carvalho — *F. C. Lockwood*

Woodrow A. Fiveland — *Christos Papadopoulos*

SCIENTIFIC COMMITTEE

- **H. Barklage-Hilgefort**, Deutsche Glastechnische Gesellschaft, Frankfurt, Germany
- **C. Bertrand**, Lafarge Coppee Recherche, Viviers-sur-Rhone, France
- **R. B. Collin**, Royal Institute of Technology, Stockholm, Sweden
- **I. Gulyurtlu**, LNETI, Lisboa, Portugal
- **K. Hein**, Delft University of Technology, Delft, The Netherlands
- **J. D. Hickerson**, U.S. Department of Energy, Pittsburgh, Pennsylvania, U.S.A.
- **P. Hughes**, Combustion and Carbon Research Laboratory, Ottawa, Canada
- **D. Lalas**, Centre for Renewable Energy Sources-CRES, Athens, Greece
- **T. Miura**, Tohoku University, Sendai, Japan
- **T. Oliver**, National Power Plc Technology and Environmental Centre, Leatherhead, U.K.
- **K. van Ommen**, Shell Laboratorium, Amsterdam, The Netherlands
- **P. Roberts**, International Flame Research Foundation, Ijmuiden, The Netherlands
- **J.-L. Roth**, IRSID, Maizierres-les-Metz, France
- **M. Sala**, Ansaldo Componenti, Milano, Italy
- **P. Sens**, Commission of the European Communities, Brussels, Belgium
- **N. Shah**, Sandstrand Power Systems, San Diego, California, U.S.A.
- **I. W. Smith**, CSIRO Division of Coal and Energy Technology, North Ryde, Australia
- **L. D. Smoot**, Brigham Young University, Provo, Utah, U.S.A.
- **M. Souza-Santos**, IPT, Instituto de Pesquisas Tecnológicas do Estado de São Paulo, Brasil
- **T. Suzuki**, Kobe Steel Co., Kobe, Japan
- **T. F. Wall**, The University of Newcastle, Australia
- **M. C. Welbourne**, NEI-International Combustion Ltd., Derby, U.K.
- **J. Wendt**, University of Arizona, Tucson, U.S.A.
- **T. Yamada**, Idemitsu Kosan Co. Limited, Kimitsu Chiba, Japan

SPONSORS

ACERC — Advanced Combustion Engineering Research Center
ANSALDO Componenti
Babcock & Wilcox
Banco Comercial Português
Banco Nacional Ultramarino
Banco Totta & Açores
The British Council
Câmara Municipal de Loulé
Centre for Renewable Energy Sources
Centro de Termodinâmica Aplicada e Mecânica dos Fluidos da U.T.L.
Comissão das Comunidades Europeias
Comission of the European Communities
CSIRO
Direcção Geral de Energia
EDP-Electricidade de Portugal
European Office of Aerospace Research and Development
Fundação Calouste Gulbenkian
Fundação Luso Americana para o Desenvolvimento
Fundação Oriente
Idemitsu Kosan Co. Ltd.
Instituto Nacional de Investigação Científica
Instituto de Promoção Turistica
Instituto Superior Técnico
INTERG-Instituto da Energia
International Flame Research Foundation
MAGUE
Ministério do Ambiente e Recursos Naturais
NEI-International Combustion Ltd.
NESTE

1.

STRATEGIES: NOW AND IN THE FUTURE

Strategies for Future R & D in Combustion Processes

Invited lecture by P. F. SENS *Commission of the European Communities, DG XII*

INTRODUCTION

Designing an energy R & D strategy is a very speculative and risky matter. The development of a new technology to a commercial product apparently takes some 30 years and another 20 years may be required to implement the technology on a large scale. An example of this can be found in the development of the jet engine which was invented in 1929, but it has lasted until the end of the fifties, with several interruptions in the development, explosions of test engines and bankruptcy, that jet aircraft was introduced in regular air traffic. It has also taken about thirty years between the discovery of uranium fission and the first civilian application of nuclear energy, but further introduction of nuclear power suffers now from lack of public support in many countries.

The prototype fast breeder reactor in Kalkar (D) will not be taken into operation and will be dismantled after more than 25 years of development. It is to be hoped that the development of fusion technology can obtain timely assurance of continued public support in order to avoid a fate similar to the Kalkar reactor.

Because of the long lead times, today's decision makers will very likely no longer be around by the time the full effect of their decision, or lack of decision, becomes apparent. Their decision will affect their children much more, but it is difficult to predict in which way.

The conditions under which a specific technology may have to compete with others may change drastically during the development period. The last two decades have given examples of significant changes, caused not only by market forces but also by geopolitical factors and public acceptance. The fluctuations in oil price have caused fluctuations in coal price as well. The upward instabilities have stimulated developments from offshore technology to renewable energy while the downward instabilities have made their economy doubtful.

Uncertainties of this kind will remain and require a flexible R & D planning. On the one hand one should be careful not to drag on R & D activities on a new energy technology which has reached the phase of demonstration but whose application is postponed due to a change in the economic environment. Such a technology should be mothballed and kept in reserve as an option for "better" times.

On the other hand, a decision not to develop an energy option because it will never become economic under the present conditions may, in retrospect, appear to be unjustified if looked at under the circumstances of the year 2010 or beyond, when oil prices, according to present scenarios, have increased considerably due to exhausting resources.

CRITERIA

It may be evident from the above that energy R & D strategy has to be directed to long term targets and that economic prospects should not be the only guiding principle. In judging potential energy sources one should apply criteria such as:

— contribution to the energy supply;
— contribution to supply security and diversification;
— R & D requirement;
— environmental impact;

— economic prospects;
— societal acceptance.

As indicated above, the last two criteria imply such large uncertainties, that they can hardly be applied for long term R & D strategy. Another question is: is long term equally long for all the actors in the field?

For the R & D programmes, supported by the European Communities there are a number of specific criteria such as:
— industry oriented;
— multinational and multipartner;
— innovative;
— positive for the environment;
— precompetitive;
— subsidiary;
— in support of SME's;
— reflecting the European dimension.

It must be evident from these criteria that, although the EC may recognize the need for R & D in a specific area, it does not necessarily have to be supported by the EC. The subject may already be too close to competition or no longer suited for a multipartner arrangement.

Subsidiary means that the EC should preferably support those activities which can be better performed at the Community level than at the national level.

The overall effect of these criteria will be that in general the EC-supported R & D aims at longer term objectives to the benefit of the whole Community.

THE SCENERY

The abatement of the environmental effects of the utilisation of fossil fuels is, in particular, a subject which can be better studied at the Community level. The environmental impact does not stop at the borders. It is an international phenomenon, requiring an international approach.

Fossil fuels cover at present some 84% of the European energy consumption and contribute to a large extent to the emissions of nitrous oxides, sulphur oxides, carbon dioxide and particulates. Liquid fuels are particularly important in the transport sector, while coal generates about 40% of the electricity consumed in the EC. It is evident from scenarios on the future energy situation in the EC that the dependence on fossil fuels will remain for a very long time. However, the mix within the fossil fuel sector has changed in the past and may do so in the future. Substitution of oil and gas by coal in the generation of electricity has been a distinct policy of a number of countries as a reaction to the high oil prices due to the energy crises of 1973 and 1979. In response to the concern about global warming, there is now the opposite tendency, in particular to strengthen the share of natural gas. A recent study (1) shows that, besides energy conservation measures and a marginal contribution of renewable energies, this switch in fuel mix is a major element in obtaining the anticipated CO_2 emission reduction target in the year 2010. But at the same time, the study concludes that:

"an increased reliance on natural gas in EC countries would address the inter-related questions of security of supply, significant gas price increase and exhaustion of resources. Major shifts from coal to gas make the contribution of clean coal technologies to CO_2 reduction less significant. Nevertheless, one might expect this decreasing trend to be reversed in the longer term with the development of carbon removal techniques. To maintain a well balanced range of supply options, it is therefore recommended to direct R & D efforts to these abatement technologies."

Because this switching to natural gas cannot be a long term policy, it may be considered as an interim action, which provides the time for active development of clean coal technologies. Despite the fact that mankind has been burning coal for centuries, there is no reason to be satisfied with the present state of the technology. Apart from coal being dirty and inconvenient to transport and handle, its efficiency in the conversion to electricity, where more than 70% of the coal is used, is low and there is a large scope for improvement.

The main argument, however, for intensified R & D in clean coal utilisation is given in the following table:

TABLE I
RESOURCES AND CONSUMPTION

	Resources Gtce	Consumption 1988, Gtce/yr
Coal	11.000	3.5
Conventional Oil	270	4.4
Heavy Oils & Natural Bitumen	200	≈ 0
Shale Oil	2.700	≈ 0
Natural Gas	350	2.4

1988 Data from: IEA-Panel on "Clean Use of Coal"
1 tce = 29.3 GJ

It is evident that mankind will continue to use the enormous coal resource long beyond the point at which oil and gas reserves will be exhausted. In the future oil and gas will become more expensive but gradually they will lose their role as price leader. Energy prices will be determined by coal prices, because coal will remain the only fossil fuel for electricity generation. As for the EC one may expect that the future electricity production will come for 40% from solid fuels and nuclear each and for 20% from all the other sources together of which more than three quarter from hydropower and natural gas (2).

It is about time that this reality becomes accepted and that more means are made available to the development of a clean coal technology. One should realize that already a small reduction in the emissions from coal combustion will give an enormous benefit to the environment because of the large volume of coal involved.

These required means for clean coal technology development should be seen in relation to the World Energy Conference prospective that by the year 2000 worldwide 600 to 1100×10^9 US dollar will be spent per year for developments and investments in electric power generation.

Continued coal utilisation is not a subject of great public and political support and "green" technologies are far more popular. Nevertheless, it is encouraging to learn that the power industry remains more realistic and that coal import into the European Community has increased with 11.9% from 1989 to 1990 to more than compensate a reduction in domestic production of 4.6%. Over the first few months of 1991 the production has increased again with 2.3% compared to the same period of 1990.

FUTURE R & D

Considerable and successful efforts have already been undertaken to reduce the emissions to the environment of coal fired power stations. In the recent past, systems have been developed for reducing the SO_2, NO_x and dust content of combustion gases before they leave the stack. However, they are all tail-end systems: first the problem is created and then solved at the expense of the overall conversion efficiency from coal to electricity.

It seems more sensible to try to avoid the formation of the pollutants during the combustion. Combustion research may make an important contribution by developing proper aerodynamic designs for the near-burner area of coal combustors and by optimizing staged combustion in fluidised beds. The EC supports these developments in the current JOULE programme, as well as a number of projects relating to improved capture of particulates.

Cleaning coal prior to its use is another interesting option. Microbial desulphurisation is a potential process for which at present a pilot plant is under construction in the JOULE programme. It is a rather slow and bulky process. There are doubts about its economy, but as argued above this should not be the only consideration. The pilot plant will provide the data on which the economy can be better evaluated. This option for cleaning coal is closely connected to the application of coal-water slurries. It would be a great disadvantage in cost and energy if the clean slurry from the microbial process has to be reconverted to powder coal. Combustion of coal-water slurries remains therefore of interest and it is positive to learn that Tohoku Electric Power Company of Japan has recently ordered a 600 MW power plant, to operate on a coal-water mixture with 70% coal content.

Besides the awareness about the emissions mentioned earlier, there is the increasing concern about the generation of the "greenhouse" gases CO_2 and N_2O from fossil fuels combustion. Although N_2O formation may probably be manipulated in a similar way as NO_x, CO_2 is inherent in fossil fuel combustion and solutions to absorb or store CO_2 to any reasonable level are presently not available. This is the main reason why continued use of coal is being criticized by antagonists with claims such as:

— it will never be possible to capture CO_2 from flue gases at a reasonable price;
— capture and disposal of CO_2 will require so much energy that the net production of CO_2 will only increase;
— it will never be possible to store CO_2 permanently. In the long run it will always be released to the environment.

In fact these claims are only assumptions and it is the task of the coal community to provide the scientific, technical and economic evidence in these matters on which politicians and the public can base their opinions and decisions. The objective of future R & D should therefore be the assessment and development of coal fired power production systems with minimum emissions into the environment. Reduction of CO_2 emissions is presently the main concern and can be obtained by:

— increasing the efficiency of the conversion of coal to electricity in order to generate less CO_2 per kWh and to compensate for the additional energy requirements and cost of CO_2 capture and disposal;
— adapting the combustion conditions in order to facilitate CO_2 capture and disposal.

Although the efficiency of the present day steam cycle technology may be stretched from the conventional 38% to 42%, to increase the conversion efficiency significantly one has to apply combined cycle technology. The predominant option here is coal gasification combined cycle, but when defining the R & D needs in this area one should take into account that a 250 MWe demonstration plant is under construction in the Netherlands to start operation in 1993. This plant will have a conversion efficiency of 43.2% (LHV).

It may be speculated that an increase to 46% is possible by further development of gas turbine blade material to allow higher inlet temperatures and by development of hot gas clean-up systems to protect the blades against corrosion and erosion. However, this 46% will very likely be the upper limit. Coal gasification is in the hands of the industry and is receiving large interest from the utilities. Publicly supported R & D should therefore be restricted to more basic and new aspects such as improved process control by on-line analysis of the product gas and the control of CO_2 emissions by application of the shift reaction and subsequent CO_2 capture.

Another promising system but still less developed is pressurized fluidized bed combustion with topping cycle and gas and steam turbines. This system also has the potential to reach an efficiency of 45-46% with good environmental properties. In particular the N_2O formation may be reduced in the topping burner. Further R & D work is required to develop this option to demonstration.

Atmospheric bubbling and circulating fluidized beds as components of a combined cycle are also well in hands of the industry and it is therefore up to the manufacturers to perform further R & D.

In order to have a better chance for CO_2 capture in these systems it would be attractive to have the CO_2 in concentrated form and preferably under pressure. It is therefore appropriate to investigate coal gasification and pressurized combustion, albeit as pulverized coal or in a PFBC, with recirculation of the flue gases and oxygen injection. It is evident that such a system requires additional investments, e.g. in an air separation plant, but it is not evident whether this cannot be compensated for by using multiple cycle electricity production with increased efficiency.

To investigate this and to indicate the combined cycle systems with a large potential for efficiency increase on which future research should be focused, a study has been initiated by the University of Ulster under contract with the EC.

The study "Techno-Economic Evaluation of Zero Emission Fossil Fuel Power Plants" (3) has started with calculating the efficiency and kWh price of nine combined cycle systems, giving due account of all the additional energy and cost penalties connected to CO_2 capture and liquefaction.

The nine systems are summarized in the table below:

TABLE II
COMBUSTION AND CO_2 REDUCTION OPTIONS

	Conventional Combustion	with MEA Scrubber and CO_2 Liquefaction	with Oxygen Firing and CO_2 Recirculation & Liquefaction
Pulverised Coal with FGD.	System 1	System 2	System 3
Pressurised Fluidised Bed Combined Cycle.	System 4	System 5	System 6
Magnetohydrodynamics with steam cycle.	System 7	System 8	System 9

An overview of the results is presented in table III. The calculations were performed with the ECLIPSE process simulator, developed under the previous non-nuclear energy R & D programme of the EC.

It should be emphasized that these results are only the first. Presently a coal gasification combined cycle is being calculated and more systems will follow.

These calculations will set the path for future experimental work to develop details of the most promising systems and to provide data to improve the input into the ECLIPSE process simulator.

TABLE III
EVALUATION RESULTS

	Conventional			Combined Cycle			MHD		
	Bas. 1	MEA 2	O_2 3	Bas. 4	MEA 5	O_2 6	Bas. 7	MEA 8	O_2 9
Thermal Energy Input (MW)	1324	1324	1324	1306	1306	1306	1261	1261	1261
Electricity in MW									
Total Generated	524	461	540	847	785	751	720	671	668
To FGD	13	13							
To MEA Scrubbers		19			19			14	
To O_2 Sep.			101			103			90
To CO_2 Liquef.		36	47		36	49		32	40
To Other Utilities	19	20	20	286	287	189	116	117	77
Net Production	492	373	372	561	443	410	604	508	461
Overall Gen Eff.(%)	37.2	28.2	28.1	43.0	33.9	31.4	47.9	40.3	36.6
Gaseous Emissions									
CO_2(g/kWh)	880	120	0	780	100	0	630	75	0
NO_2(g/kWh)	1.8	2.4	0	0.9	1.1	0	0	0	0
SO_2(g/kWh)	1.1	0.2	0	0.3	<0.1	0	0.8	<0.1	0
Greenhouse Effect Total as (gCO_2/kWh)*	1180	510	0	920	280	0	630	75	0
Capital Costs** (10^6 UK£)									
Basic System	279	264	299	380	366	310	422	424	343
FGD Plant	88	88							
MEA Scrubbers		46			46			43	
O_2 Separation			120			123			100
CO_2 Liquefaction		31	46		31	46		28	39
Utilities	5	6	6	4	5	5	4	5	5
Total Capital Cost	372	435	471	384	448	484	426	500	487
Annual Income and Costs (10^6 UK£)									
Income from Sales	157	180	182	165	186	192	166	191	183
Cost of Materials	58	65	57	63	67	64	53	59	54
Net Annual Income	90	105	114	93	109	117	103	121	118
Prod. Cost (UK£ / MWH)	19.4	28.4	25.8	18.2	24.8	25.9	14.8	19.6	20
Electricity Selling Price for NPV=O UK£ / MWH	45.6	68.6	69.6	42.0	59.8	66.8	39.2	53.6	56.5

Assumes NO_x is all N_2O with greenhouse potential 160 CO_2

** Raw Materials costs include labour and maintenance. Production cost does not include capital charges.

CONCLUSIONS

The results of the ECLIPSE calculations, although far from complete, indicate already that high efficiency combined cycles have the potential of providing conditions under which zero emissions occur and give the penalty on the kWh price that one has to pay for this advantage. This is exactly the kind of information required to determine in which direction future activities should go and to take decisions concerning future power generation strategies.

Some of the systems considered up to now, such as systems 2 and 3, are particularly suited for retrofitting existing plant with additional installation of equipment for the MEA (mono-ethanol amine) scrubbers and the CO_2 liquefaction. Although this equipment is not revolutionary, it is a considerable investment weighing heavily on the kWh price due to the reduced conversion efficiency. Other systems like 5 and 6 are more appropriate for new plant which due to the application of combined cycles suffer less from decrease in efficiency.

The most promising option up to now is the MHD combined cycle, which provides such a step upwards in efficiency that the penalty for "no emission" is limited to about 20% of the kWh price. The systems 7, 8 and 9 can be considered for future installations though it must be emphasized that MHD is still a long way from commercial viability. MHD development programmes are in progress in the US, the USSR, Japan, Italy, the Netherlands, Australia, India, China and a number of other countries. If MHD can be successfully developed it offers the prospect of "clean" power at the same price as today's "dirty" power and at much higher fuel conversion efficiencies.

None of the evaluations have been optimized. In the MHD case only a linear MHD generator has been considered, while a disk generator may be more efficient and cost effective. No attempt has been made up to now to study triple cycles such as the combination of MHD, a gas turbine and a steam turbine.

In summary, future R & D on clean combustion and gasification technologies should address the following points:

— pre-cleaning of coal;
— coal water slurry preparation and combustion;
— burner development to reduce pollutant formation in the flame;
— combustion with CO_2 recirculation and oxygen injection;
— development of pressurized burners with oxygen injection and CO_2 recirculation;
— behaviour of SO_2, NO_x and N_2O in systems with CO_2 capture;
— investigation of combined cycles, specially designed and integrated for minimum CO_2 production and maximum CO_2 capture;
— application of the shift reaction in coal gasification.

In addition, the long term perspectives for MHD and its potential for zero emission power generation at acceptable cost should justify the launching of a European development in this area. Such a programme should address CO_2 recirculation with oxygen injection, -which would eliminate the technically difficult air preheater-, the feasibility of a disk generator and the performance of high interaction experiments. In the world scene such a European activity would be a welcome contribution to the programmes in, mainly, the US, the USSR and Japan. Because such a development will have a long term objective, will be precompetitive and could be made subsidiary by merging with the present activities in two of the Member States, it would constitute an excellent opportunity to create a major multidisciplinary European project, responding to the criteria mentioned earlier and clearly targeted on zero emission fossil fuel power generation.

In conclusion it may be said that there is no single solution to the greenhouse problem but that as far as coal combustion is concerned there are prospects to limit CO_2 emissions to a level which will finally be determined by the amount mankind is prepared to pay for energy, the development of clean conversion systems and thus for a clean environment.

REFERENCES

Cost-effectiveness Analysis of CO_2 Reduction Options, prepared for the Commission of the European Communities, DG XII, by Coherence s.c. (Belgium), under contracts JOUM-0002, -0006, -0007 and -0010.

Energy in Europe; Energy for a New Century: The European Perspective, Commission of the European Communities, Directorate-General for Energy, Special issue July 1990.

Techno-economic Evaluation of Zero Emission Fossil Fuel Power Plants, prepared for the Commission of the European Communities D.G. XII by the Centre for Energy Research, University of Ulster, Coleraine, Northern Ireland, U.K.BT52 1SA, under contract nr PSS*0205/UK.

Clean and Efficient Electric Power Generation for the Next Century: The British Coal Topping Cycle

M. St. J. ARNOLD, G. J. KELSALL AND D. M. HUDSON *Coal Research Establishment, Cheltenham, Gloucestershire, GL52 4RZ, UK*

Abstract — An advanced coal-fired electricity generating system is being developed by British Coal. Known as the British Coal Topping Cycle, the system offers the potential for:

— 20% reduction in electricity generating costs compared to conventional pulverised fuel power stations equipped with flue gas desulphurisation.
— 20% reduction in coal burnt per unit of electricity produced and a corresponding reduction in CO_2 emissions.
— Low emissions of particulates and oxides of sulphur and nitrogen.

Because of the potential advantages, British Coal, with support currently from the UK Department of Trade and Industry, PowerGen plc and GEC ALSTHOM, have embarked upon a substantial development programme. The paper describes the system and focuses upon the current developments in the areas of coal gasification and desulphurisation, char combustion, gas filtration and gas combustion with particular reference to nitrogen oxide control.

Key Words: Topping Cycle, Gasification, Char Combustion, Gas Cleaning, Gas Combustion, Gas Turbine

INTRODUCTION

Against the background of rising worldwide energy demand, coupled with increasing environmental concerns, attention has focused on the development of cleaner and more efficient power stations. One immediate commercial response has been to turn to natural gas-fired combined cycles. However, coal is very abundant and there is a longer-term desirability to preserve premium fuels for applications in which their natural advantages can be most appropriately exploited. These factors have led to a resurgence of interest in advanced coal-fired systems.Several clean coal technologies are available (Dawes *et al.* 1990). These include circulating fluidised bed combustion (CFBC), pressurised fluidised bed combustion (PFBC) and integrated gasification combined cycles (IGCC). Such technologies must offer a number of advantages compared with conventional technology:

— Low capital cost and short construction periods.
— Low cost of electricity.
— Good environmental performance for SO_x and NO_x.
— High efficiency with a consequent reduction in CO_2 emissions.
— Good availability, turndown and response characteristics.
— Ability to handle a wide range of coals with different size ranges and ash characteristics.

These requirements have been taken into account by British Coal in examining the various options, assessing factors such as efficiency, cost, environmental factors and development risk. It has concluded that a Topping Cycle based on a partial fluidised bed gasifier with fluidised bed combustion of the residue offers the potential for a 20% or more reduction in specific capital cost and cost of electricity compared with conventional pulverised fuel stations equipped with environmental control equipment. Improved efficiency will result in a 20% or more reduction in coal burnt per unit of electricity, with corresponding reduction in CO_2 emissions. Inherent features of the process will ensure low emissions of dust, sulphur and nitrogen oxides.

Because of these potential advantages, British Coal with financial and technical support from PowerGen plc and GEC ALSTHOM have embarked upon a substantial development programme. The objective is to establish a commercially available technology, when required, early in the next century.

THE BRITISH COAL TOPPING CYCLE

System Description

The system is based on the partial gasification of coal with separate combustion of residual char in a PFBC or CFBC. Initial development will focus on the CFBC variant, with fuel gas cleaning using ceramic candle filters at 400–600°C. Such temperatures offer a lower risk option whilst maintaining high cycle efficiency. The cleaned gas will be burned in a new generation, high temperature gas turbine fitted with a waste heat boiler. Figure 1 shows a schematic of the Topping Cycle selected by British Coal for initial development. Coal is gasified in an air-blown spouted bed gasifier at elevated pressure and temperatures up to 1000°C. Sorbent is injected into the gasifier to retain sulphur. The fuel gas is precleaned in a cyclone and then cooled; it still contains some fine particulate material which is removed using ceramic candle filters. Virtually all the volatile alkali salts present condense onto the particulate matter collected by the filter. Between 70 and 80% of the coal is converted into a low calorific value fuel gas. The residue is removed mainly as fines collected by the cyclones and hot gas filters. These fines are transferred to the CFBC where, together with gasifier char, they are burned to raise steam. The sulphided sorbent is converted to calcium sulphate. Additional sorbent can be fed to the combustor if required.

Particular advantages of the British Coal spouted fluidised bed gasifier are that it can operate using a wide range of hard coals, including high swelling bituminous coals, and operates in a non-agglomerating mode. It is thus not subject to problems due to variation in ash properties. Advantage is taken of the ability to capture sulphur within the bed, thus allowing dry dust removal at high temperature to be utilised with consequent increase in thermal efficiency. The gasifier is air blown; the cost and efficiency loss arising from oxygen production are thus avoided. The use of a separate char combustor means that high temperature heat is available in an oxidising atmosphere to ensure that state of the art steam turbines can be used, including supercritical conditions.

Efficiency

The efficiency of the Topping Cycle has been calculated using the British Coal ARACHNE flowsheeting package (Topper *et al.* 1989). Comparison has been made with a number of other power generation systems and are shown on a lower heating value (LHV) basis in Table I. Calculations have been carried out on a strictly comparable basis using state of the art gas turbines (entry temperature 1260°C) and subcritical steam conditions of 538°C, 160 bar, and reheat to 538°C. The technologies employing gas turbines, as expected, show an improvement in efficiency over steam only plants. The advantage indicated for the Topping Cycle is mainly due to use of an air-blown system and the use of hot gas filtration rather than aqueous scrubbing to clean the gas.

There is scope for significant improvement on these efficiencies in due course. The development of advanced supercritical steam turbines (for which the CFBC system is particularly well suited) will mean that they can readily be introduced to the Topping Cycle configuration. IGCC systems have no additional source of high temperature heat and may be less able to take advantage of advanced supercritical steam conditions. There should also

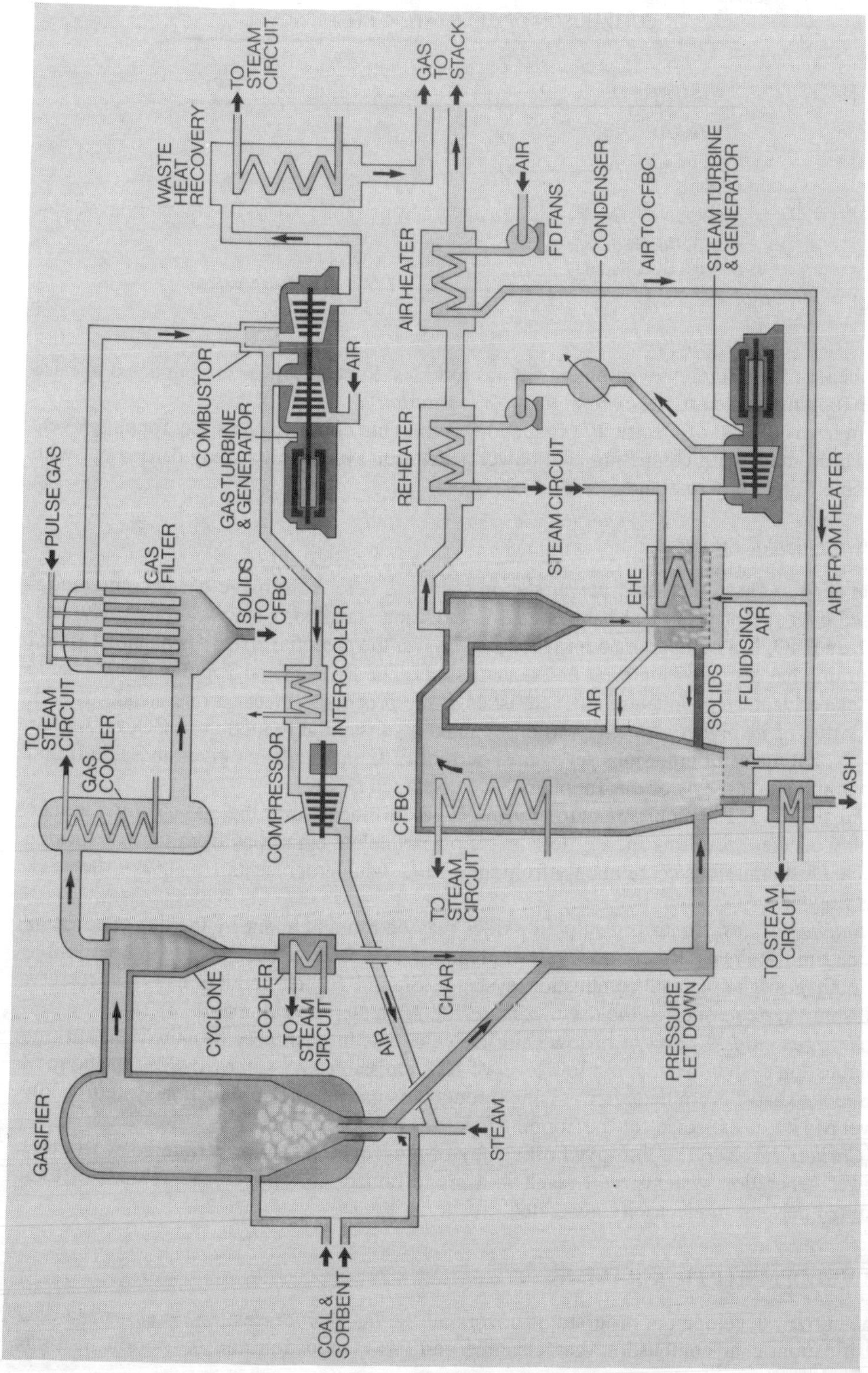

FIGURE 1 The British coal topping cycle.

TABLE I

COMPARISON OF GENERATING EFFICIENCY

Technology	Net efficiency (%LHV)
PF+FGD	38.3
CFBC	39.1
PFBC	40.8
IGCC	41.7
BCC Topping Cycle (CFBC Option)	46.4

be benefit from future developments in gas turbines. Such advances are predicted to raise the Topping Cycle efficiency to around 52% or more.

In terms of cost of electricity production, favourable comparison of the Topping Cycle has been made with competing coal power generation systems in a study carried out with Bechtel Limited (Cross and Sheikh, 1990).

Environmental Effects

The need to clean coal-fired power station stack gas before discharge has become recognised over recent years and has resulted in emission standards being set for particulates, SO_x and NO_x in a number of countries. Concern over the potential effect of enhanced global warming has also led to interest in CO_2 emissions. The British Coal Topping Cycle is able to take advantage of inherent characteristics of the process to meet current standards and, by virtue of its higher efficiency, to bring about a substantial reduction in CO_2 emissions levels. Estimates of emissions set against current EEC standards are given in Table II and environmental aspects of the Topping Cycle discussed below.

Particulates. The requirement to protect the gas turbine ensures that gas cleaning will be highly efficient resulting in very low levels of particulate emissions from the gas turbine stack. Dust emissions come mainly from the CFBC, which represents 10–30% of the stack gas flow.

Sulphur oxides. Control of sulphur oxides may be brought about by the use of sorbents in the fluidised beds. Retention levels of greater than 90% have been shown to be possible in both gasification and combustion systems. Should further reductions be necessary, polishing systems such as the use of zinc-ferrite are under development.

Nitrogen oxides. The use of low calorific value gas in a suitably designed gas turbine combustion system results in a low level of NO_x emissions. As some 70–90% of the total gaseous exhaust is emitted after passing through this gas turbine combustion system, a low level of NO_x is expected for the Topping Cycle as a whole.

Carbon dioxide. The improved efficiency of the Topping Cycle, compared with other power generation systems, will result in a proportionate decrease in the amount of CO_2 emitted per unit of electricity generated.

DEVELOPMENT PROGRAMME

The current development programme covers all the main processes, i.e., pressurised coal gasification, char combustion, gas cleaning and gas combustion and utilisation in a gas

TABLE II
ENVIRONMENTAL IMPACT DATA

	Environmental impact			
Technology	CO_2 kg/kWh	SO_x retention %	NO_x mg/m^3 @ 6% O_2	Particulates mg/m^3 @ 6% O_2
PF+FGD	0.87	90	500–650	50*
CFBC	0.86	90	100–300	~30**
PFBC	0.82	90	150–300	~1–2***
IGCC	0.78	99	120–300	Negligible emission
BCC Topping Cycle (CFBC Option)	0.72	90+	200–300	~5–10**
EEC emission standard		90++	650	50

*Plant designed to meet limit, could be less depending upon FGD process. **Filter Bag House on CFBC. ***Ceramic filters. +Without sulphur polishing. ++Based on a 500 MW_{th} plant burning a typical UK sulphur coal.

turbine. The programme also covers a number of other important issues such as control, process optimisation and availability. Development work on the main process items is outlined below.

Gasification

The British Coal pressurised fluidised bed gasifier is based on an atmospheric pressure spouted bed design which has been developed previously for industrial fuel gas applications (Arnold *et al.* 1992, Gale *et al.* 1991). The gasifier was developed specifically to handle a wide range of bituminous coals in terms of swelling characteristics and ash properties.

Current development is centred on a 12 t/d pressurised pilot plant which can be operated at pressures up to 20 bar. The gasifier, illustrated in Figure 2, consists of a 10.8m high refractory lined pressure vessel with a 0.32m internal diameter, which expands to 0.45m at 6 m above the base flange. The coal is crushed and injected into the base through the central air spout. This method of feeding breaks up any agglomerates that may form, particularly with high swelling coals. Limestone (–1 mm) is also fed pneumatically through the spout. The fluidising air is preheated to 250°C to simulate a commercial plant. This is injected together with steam from a package boiler into the central spout and additional nozzles in the cone to even out bed temperature gradients.

Commissioning tests on the pressurised gasifier were completed in March 1991. The subsequent test programme, using three typical UK bituminous coals with swelling numbers up to 7½, have provided confidence that gasifier performance will be suitable for the requirements of the Topping Cycle.

In addition to pressurised gasifier operation, a substantial supporting programme is being carried out, aimed at understanding the chemical and physical process in the gasification and desulphurisation processes and at providing a basis for future scale-up.

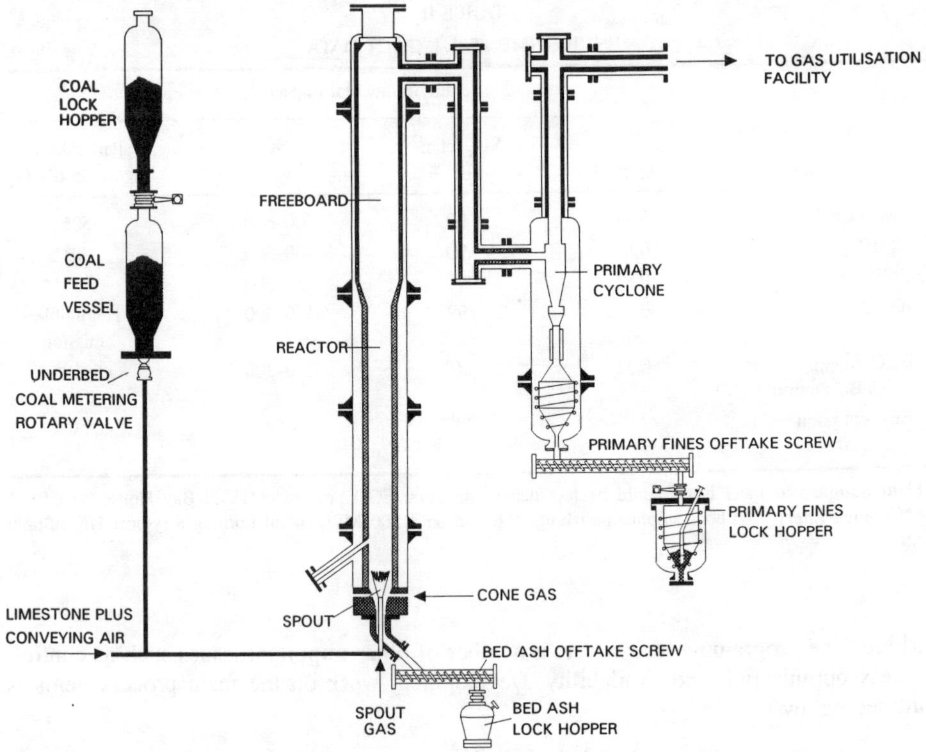

FIGURE 2 Schematic of the pressurised fluidised bed gasifier pilot plant.

Char Combustion

An atmospheric pressure CFBC char combustor was selected for the initial development of the Topping Cycle because CFBC is proven in many commercial plants worldwide. Also, the relatively low degree of cycle integration in the CFBC Topping Cycle should result in simpler control and higher operating availability. The main uncertainties which need to be addressed are char cooling and transfer, combustion efficiency, operation with a very fine material, sulphation of calcium sulphide and disposal of ash. The possible issue of nitrous oxide (N_2O) and potential control measures must also be addressed. Combustion tests will be undertaken on a 2MW CFBC test facility at CRE (see Figure 3) using residual char and fines, produced from the gasifier, to address these issues.

Gas Filtration

Critical to the success of the system is a reliable flow of clean fuel gas to the gas turbine. The gas must contain low levels of dust if acceptable lifetimes of the turbine blades are to be achieved. This has the additional advantage that the gas turbine exhaust can be released direct to atmosphere.

British Coal has considerable experience in the filtration of hot gas in both a combustion environment (Morrel *et al.* 1990; Tassicker *et al.* 1989) and in a fuel gas environment (Bower et al. 1989). A joint British Coal/Central Electricity Generating Board experimental

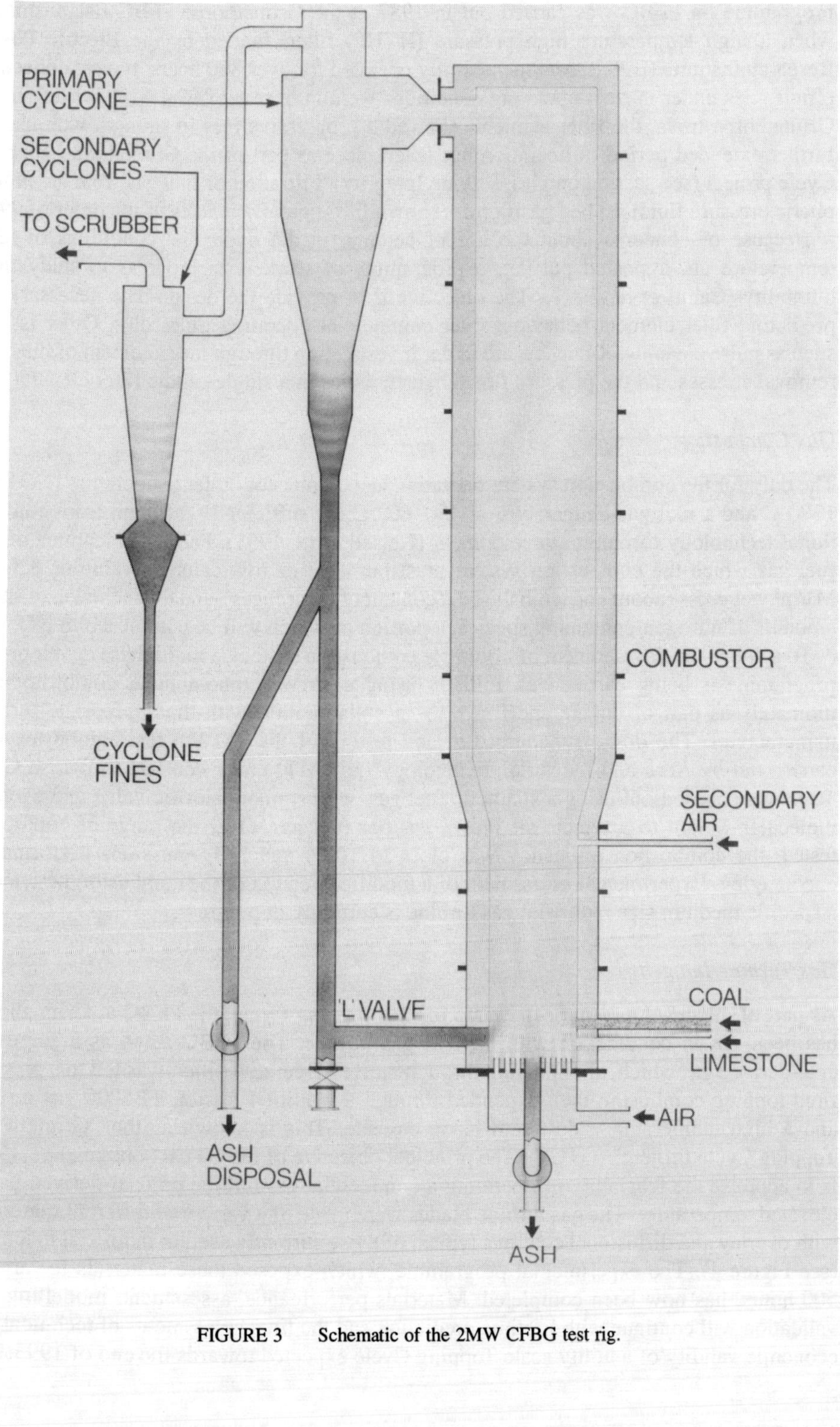

FIGURE 3 Schematic of the 2MW CFBG test rig.

programme on PFBC was carried out in 1987 at the Grimethorpe PFBC Establishment when a high temperature high pressure (HTHP) filter, funded by the Electric Power Research Institute (EPRI), was successfully operated for over 860 hours; typical collection efficiencies under normal operating conditions were in excess of 99.4%. However, in the Grimethorpe trials, the filter elements showed a progressive loss in strength with use. A further extended period of operation has taken place as part of the Grimethorpe Topping Cycle project (see section on Gas Turbine Integrity). Filtration of fuel gas from an atmospheric pressure fluidised bed gasifier has shown filtration efficiencies of the order of 99%.

Because of concerns about the life of ceramic media under the conditions of high temperature and repeated pulsing, a programme of work is in progress to study their durability (Cahill *et al.* 1991). The objective is to provide the design data necessary for predicting filter element behaviour over commercial operating timescales. Other issues, such as pulse cleaning efficiency, are under investigation through measurement of the cake removal stresses and the pressure/flow characteristics in a single candle filter element.

Gas Combustion

The duty for the combustion system operating at a combustor outlet temperature (COT) of 1380°C and a fuel gas temperature of 400–600°C are sufficiently different from conventional technology to require development (Kelsall *et al.* 1991). Particular features of the fuel gas which the combustion system must burn are its low calorific value of 3.5–4.0 MJ/m^3 wet gross (about one tenth the heat content of natural gas), and the inclusion of small amounts of nitrogen-containing species, a portion of which will be combusted to NO_x.

To achieve the development of a suitable combustion system, a multiphase development programme is being carried out, initially using a 1MWth tubo-annular combustor and ultimately as part of engine specific tests in collaboration with the appropriate turbine manufacturer. The design, manufacture and testing of the 1MWth test combustor was carried out by Aero and Industrial Technology Ltd (AIT) under contract to British Coal. Tests were carried out using a synthetic fuel gas, with similar calorific value and average molecular weight to a typical air blown gasifier fuel gas. Over the range of conditions tested, the combustion efficiency was close to 100% and NO_x emissions performance encouraging. Experimental assessment of a modified version of the combustor, targeted at a specific medium size industrial gas turbine is currently in progress.

Gas Turbine Integrity

As part of the development programme for the Topping Cycle, the PFBC at Grimethorpe has been run in conjunction with a small gas turbine. The PFBC acted as a source of coal-derived gas which, after passing through an HTHP ceramic filter, was fed to a propane fired topping combustor, then expanded through a modified Ruston TB5000 gas turbine and a high temperature sidestream blade cascade. This is known as the "Grimethorpe Topping Cycle Project" (GTCP). The principal objective of the GTCP (Clarke *et al.* 1991) is to appraise the feasibility of operating a commercial gas turbine on coal-derived gas at elevated temperature. The gas turbine blades were made of a variety of different materials, with overlay and diffusional coatings typical of those currently used in industrial machines (see Figure 4). The experimental programme, which exposed these materials for around 500 hours, has now been completed. Materials performance, assessment, modelling and validation will continue with lifetime prediction and the final assessments of technical and economic validity of a utility scale Topping Cycle expected towards the end of 1993.

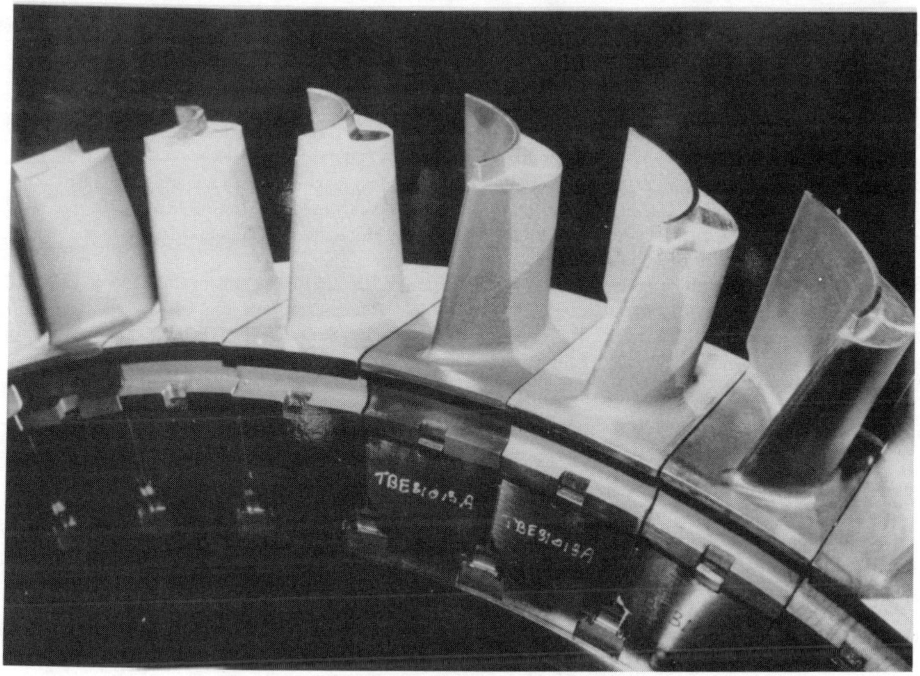

FIGURE 4 Section of the Grimethorpe TCP turbine rotor.

SUMMARY

The British Coal Topping Cycle has the potential to provide a cost effective, high efficiency combined cycle system for power generation with low environmental impact. Such a system should offer significant economic benefit over both conventional coal-fired systems and emerging advanced technologies such as CFBC, PFBC and IGCC. Of the Topping Cycle options available, studies have identified that the preferred initial development choice is the CFBC Topping Cycle with intermediate temperature (400–600°C) gas cleaning, since this has the lowest technical risk. The programme established by British Coal has been directed towards the development of this system. It is intended that the current work will be followed by scale-up and cycle integration, leading to the demonstration of a commercial prototype.The environmental performance of the system has been designed to conform to the current UK and EEC emissions standards. Due regard has been given to possible future tightening of these standards and to the possible future need to limit carbon dioxide.

ACKNOWLEDGEMENTS

The studies described in this paper form part of the British Coal Topping Cycle programme, being developed by British Coal in collaboration with the UK Department of Trade and Industry, Power Gen plc and GEC Alsthom. Further funding is being provided by the Commission of European Economic Communities, the European Coal and Steel Community and EPRI. The contributions of colleagues at both British Coal and other participating organisations are gratefully acknowledged. However, any views expressed are those of the authors and not necessarily those of the supporting organisations.

REFERENCES

Arnold M. St. J., Gale J.J. and Laughlin K. M. (1992), "The British Coal Spouted Fluidised Bed Gasification Process," 41st Canadian Chemical Engineering Conference, Vancouver, *Can J. Chem. Eng.*, **70**.

Bower C. J., Arnold M. St. J., Oakey J. E. and Cross P. J. I. (1989), "Hot Particulate Removal for Coal Gasification Systems," *Proc. Filtech 1989*, Karlsruhe.

Cahill P., Hudson D. M. and Minchener A. J. (1991), "Hot Gas Cleaning for the British Coal Topping Cycle," *EPRI 10th International Conference on Gasification Power Plants*, San Francisco, USA.

Clark R. K., Arnold M. St. J., Fackrell J. E., Mordlecai M. and Dawes S. G. (1991) "The Grimethorpe Topping Cycle Project, Institute of Energy Conference on Fluidised Beds," London.

Cross P. J. I., and Sheikh K. A. (1990), "Development of Advanced Coal-Fired Power Generation Technologies in the UK," *Power Gen '90 Conference*, Orlando, Florida, USA.

Dawes S. G., Brown D., Hyde J. A. C., Bower C. J. and Henderson C. (1990), "Options for Advanced Power Generation From Coal," *Inst. of Mech. Eng. Conference*, Power Generation and the Environment, London.

Gale J. J., Steel J. G., Laughlin K. M., and Reed G. P., (1991), "Development of a Pressurised Fluidised Bed Gasifier for use in an Advanced Coal Fired Power Generation System," *VGB Coal Gasification Conference*, Dortmund, Germany.

Kelsall G. J., Smith M. A., Todd H. and Burrows M. J. (1991), "Combustion of LCV Coal Derived Fuel Gas for High Temperature, Low Emissions Gas Turbines in the British Coal Topping Cycle," *36th Gas Turbine and Aeroengine Conference and Exposition*, Orlando, ASME 91-GT-384.

Morrel R., Butterfield D. M., Clinton D. J., Barrat P. G., Oakey J. E. O., Durst M. and Burnard G. K. (1990), "The Mechanical Performance of Ceramic Dust Filter of The Grimethorpe Pressurised Fluidised Bed Combustor (PFBC)," *Proceedings 1st International Conference Ceramics in Energy Applications*, Sheffield.

Tassicker O. J., Burnard G. K., Leitch A. J. and Reed G. P., (1989), "Performance of a Large Filter Module Utilising Porous Ceramics on a Pressurised Fluidised Bed Combustor," *Tenth International Conference on Fluidised Bed Combustion*, San Francisco, USA.

Topper J. M., Cross P. J. I., Davison J. E. and Goldthorpe S. H. (1989), "Process Assessment with the ARACHNE Process Simulation Package," *I. Chem. E Symposium Series No. 114*, EFCE Publication No. 77.

Low-NO_x Combustion in Steelworks

A. MILANI and D. NELLI *Centro Sviluppo Materiali, Stazione Sperimentale Combustione, Genova, Italy*

Abstract — Integrated steelworks based on coke/iron-ore cycle make large use of by-product fuel gases (coke oven-, blast furnace- and oxygen converter-gas). The nitrogen oxide is mainly formed in the flame "near field" of the burner with a mechanism strictly dependent on the local mixing pattern and therefore on the burner design. Typical cases, based on R&D trials carried out on CSM's test facilities with industrial size burners (flat flame and high velocity burners) for steel slab reheating furnaces for strip mills are presented and NO_x emissions from conventional and from low-NO_x equipment are compared. A brief review of the nitric oxide pollution from other steel works plants based on free-flame combustion of fluid fuels is also reported.

Key Words: nitric oxide, lean fuel gases, reheating furnaces, air pollution in steelworks

INTRODUCTION

The present contribution concerns the case of NO_x formation and emission from free turbulent flames encountered in the steel-making industry and is mainly based on experimental research and development work carried out by Centro Sviluppo Materiali (CSM) on behalf of Italimpianti and Ilva, an engineering and a manufacturing company respectively, with the financial support of the European Coal and Steel Community (Milani *et al.* 1990).

Although more than 50% of the total man-made NO_x stems from traffic and internal combustion engines, and a minor share only from industrial stationary appliances, large integrated steelworks are a concentrated NO_x source and are facing this problem in the framework of planned programs for the reduction of air pollution (SO_x, CO, solids etc). Within the works, *almost half of the NO_x* is produced by the sinter plant and is mainly derived from solid bed combustion processes. This problem deserves special attention, but is beyond the scope of the present paper which is restricted to gas and oil based combustion processes.

In other plants (Furnaces, Coke Ovens, Power Plant, Hot Stoves, utilities, ...), NO_x is formed in free turbulent flames, fired with by-product gases such as coke-oven, blast-furnace or converter-gas (COG, BFG, LDG). Fuel oil and natural gas are added to cope with the overall energy balance required by the integrated steelworks, based on raw coal and iron ore. Natural gas tends to be utilised for its intrinsic clean-fuel properties and case of use, as it is made available at high pressure.

NO_x FORMATION IN TURBULENT FLAMES

Accumulated NO_x Emission Data

In Figure 1 accumulated data from different sources and plants fired with a variety of typical fuels, burned in free flames, are reported in mg of NO_2 emitted per unit of thermal input, based upon the lower calorific value (LCV), as a function of the theoretical flame temperature.

Even for the same process and/or the same fuel, field data are scattered depending on differences in operating conditions, in fuel composition and in combustion equipment design that affect in-flame NO_x formation mechanisms: thermal NO_x, according to the Zeldovich scheme (that will produce exponentially increasing emissions for flame tempera-

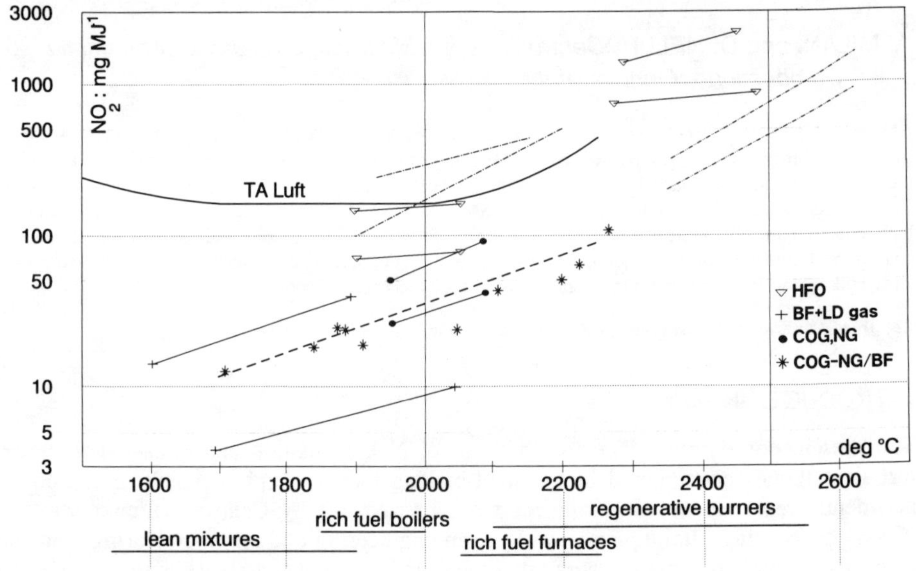

FIGURE 1 Accumulated data vs. flame temperature.

ture above ~1550°C), and fuel and "prompt" NO_x for fuels containing nitrogen bound species, in which case NO_x formation depends on the local stoichiometry pattern rather than on temperature (Kremer, 1986).

Figure 1 suggests a subdivision of NO_x emissions data from plants in three groups based on typical fuels/processes:

i) For the family of lean gases (LCV < ~12.5 MJ/Nm³) that may be obtained by mixing blast furnace and/or oxygen converter gas with rich gases (COG or NG), the adiabatic temperature seems a good criterion to predict if the NO_x emissions are to be maintained within the desired limits; it should be easy to keep the level below 100 mg/Nm³ (equivalent to ~30 mg/MJ for LCV = 12.5 MJ/Nm³) if the flame temperature is <~1900°C; for this class of fuels, NO_x emission is mainly determined by the average flame temperature.

ii) The second class includes all rich fuel-gases (NG, COG, LPG) that allow high process temperatures to be achieved. In this case, the local peak temperature and its turbulent fluctuations (instead of the time-averaged value) are the key factors in determining NO_x emissions. The NO_x formation rate is physically controlled by the turbulent mixing pattern, temperature and residence time history within the flame: as a consequence, the combustion system performance is greatly influenced by the burner and hence the fluid-dynamic design; practical examples of burners of this class will be shown in greater detail later.

iii) With heavy fuel oil (HFO), a large proportion of the NO_x emission arises from fuel-NO_x because of the organic nitrogen compounds in HFO (as it is the case with coal and sometimes with coke oven gas containing traces of ammonia or —CN groups). This fuel NO_x depends primarily on the nitrogen content of the fuel and may be controlled to a certain extent by means of a fuel-rich followed by a fuel-lean zone within the combustion region ("staging"). Inevitably, NO_x emissions are normally higher than for gasoil (which is almost nitrogen free) and with gaseous fuels, because fuel-NO_x is added to thermal-NO_x.

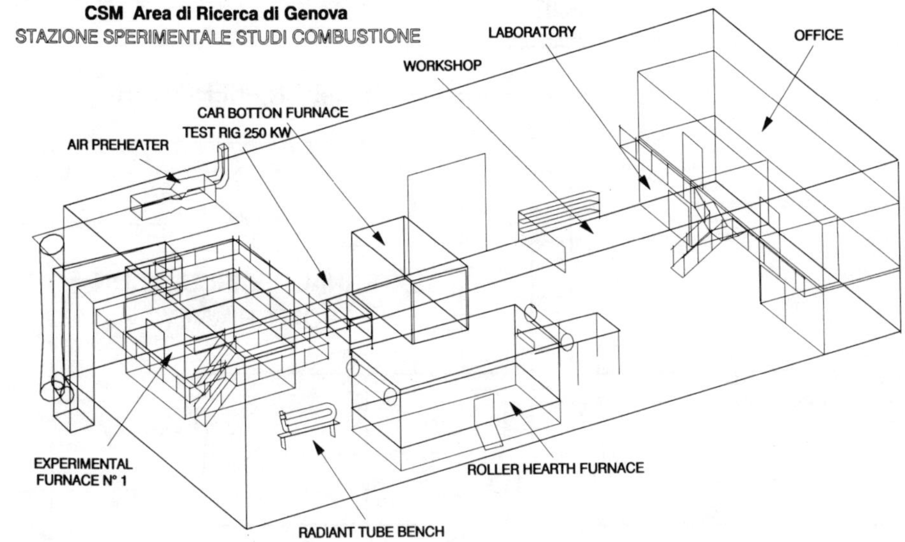

FIGURE 2 CSM's experimental station and pilot plants.

Low-NO_x Burner Design

Gas fired equipment accounts for a large number of high temperature applications in steelworks, that fall into the ii) class above (flame temperature ~1900–2300°C). The observed NO_x emissions easily span one entire decade, in the range from ~40 to 400 mg NO_2/MJ correspondent to ~150 to 1500 mg NO_2/Nm^3, which is just across the commonly accepted reference limit known as "TA Luft" (the German regulation). Such burners are often suitable for retrofitting to "low-NO_x" design, if the observed emissions are close to the upper boundary reported in Figure 1.

Burner manufacturers have expended considerable effort in attempts to reduce thermal NO_x emissions by means of appropriate aerodynamic adjustment: these attempts are regarded as "primary measures," as distinct from expensive waste gas treatment or fuel beneficiation techniques ("secondary measures").

The aim of the new low-NO_x design is to avoid local temperature peaks, to reduce the excess air and the firing density as far as possible, without impairing the general performance of the plant. The designer has the following interrelated options:

— modify the port and nozzle geometry of the burner;
— reduce the average flame temperature, by introducing a diluting ballast into the combustion air (external recirculation, water or steam injection);
— apply fuel and/or air staging to the burner.

The above options may require minor burner modifications only or merely replacement of the burners, but retrofitting to low NO_x performance may also involve the combustion supply and control system or even the design of the combustion chamber. In general low-NO_x burners tend to adversely affect the plant performance from the point of view of flame stability, of fuel burnout (CO and unburnt hydrocarbon emissions) and of fuel consumption. Solutions to the problem should obviously take all these points into account and should be "friendly" to the process performance, for example uniformity of heating, that may affect product quality.

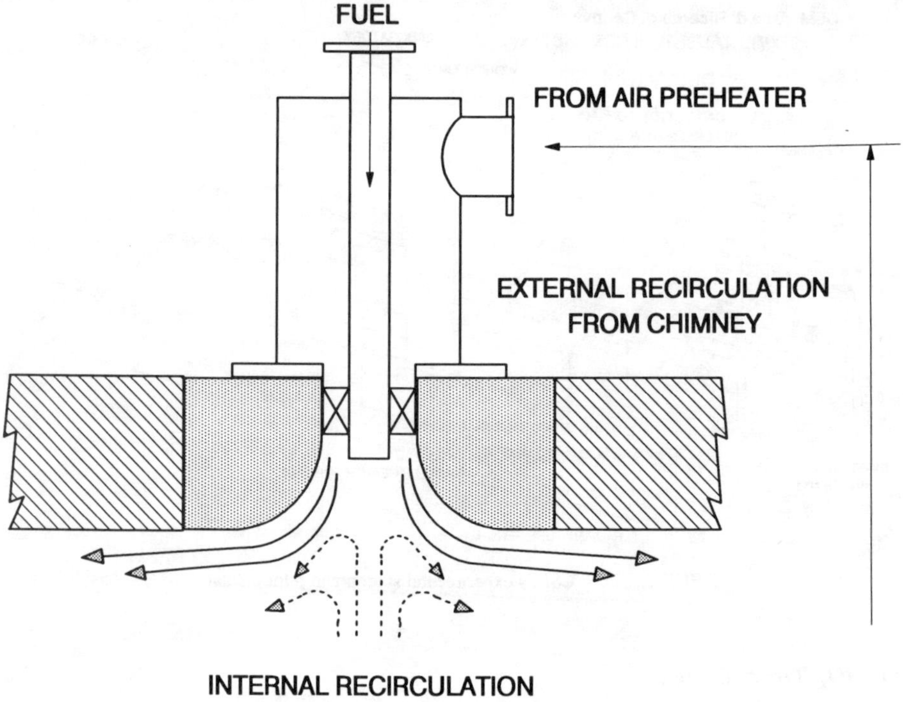

FIGURE 3 Italimpianti "radiant-roof" burner.

CSM's Experimental Station

In spite of the recent encouraging development of "CFD" (Computational Fluid Dynamics) implemented as software packages for design purposes (Swithenbank, 1991; Weber *et al.* 1986), burners cannot be designed purely on the basis of desk work and require an experimental development, based on a check and trial approach on a test furnace, of a suitable scale, close to the industrial reference plant.

Centro Sviluppo Materiali (CSM) has for many years exploited a large pilot plant within the Steelworks of Cornigliano (Genoa), connected to the existing fluid and energy supply (Cavelli *et al.* 1978). Some test facilities are listed in the 3D sketch of Figure 2 and are mainly oriented around the steel industry. The roller-hearth furnace and the car-bottom furnace, both fired with natural gas up to ~3.5 MW, are pilot units meant for process purposes (steel reheating, heat treatment, sintering etc.); the burner development is mainly carried out on bench scale or on small rigs for preliminary reduced scale tests: an example is the "radiant tube," that has also been retrofitted to low-NO_x performance (Mattarini, 1989). Large, full scale flames are investigated on the bigger burner test furnace, suitable for typical full-size burners and connected to fuel supply equipment, that includes natural gas, blast furnace gas, coke oven gas and liquid fuels up to some 3.5 MW, with combustion air preheating, flue gas recirculation, vitiated air supply and ancillary facilities. The experimental furnace is instrumented for combustion monitoring and in-flame intrusive sampling.

The furnace has been exploited in particular for the design of burners for steel reheating furnaces, mainly walking-beam or pusher-furnaces for steel slabs and semi-products, on behalf of the engineering company Italimpiant. An extensive work program was carried out

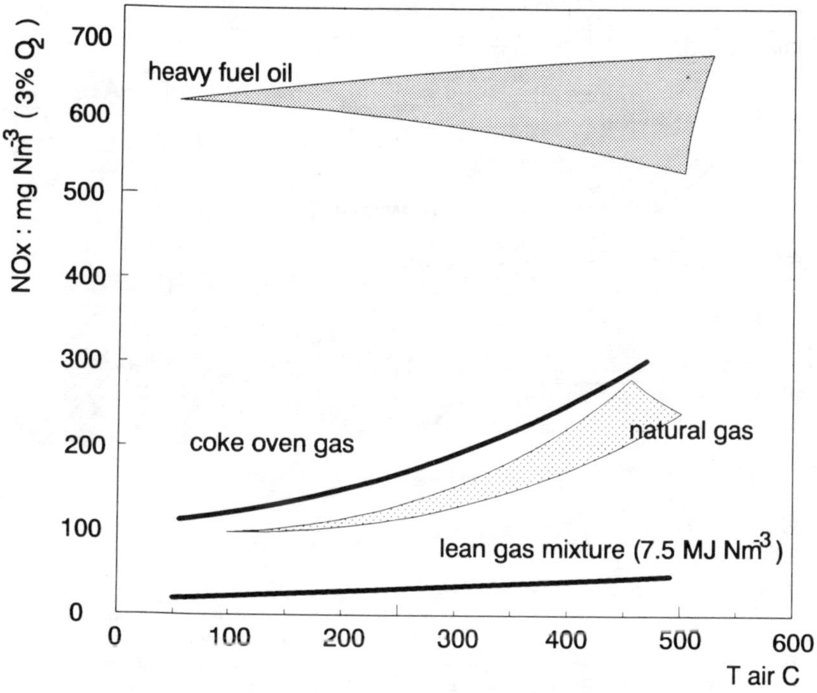

FIGURE 4 NO_x emissions from roof burners.

to check and to retrofit the burners for low-NO_x performance: the present paper describes how primary measures can be applied to gas burners for reheating furnaces and a few results concerning other free flame plants.

REHEATING FURNACES

Roof Burners

Typical burners employed in upper zones of reheating furnaces and installed on the roof in a regular pattern are the "flat flame" or "radiant roof" burners. In the flat flame burner, sketched in Figure 3, combustion takes place within a narrow layer attached to the burner refractory quarl with an open "central recirculation zone": flue gases are entrained by this recirculation from the combustion chamber back into the flame root. This mechanism dilutes the local concentration of the reactants (fuel and oxygen) and thereby reduces local oxygen concentration and temperature peaks, conducive to high NO_x.

The NO_x emission levels shown in Figure 4 are quite satisfactory for gas firing, though they appear high for oil. According to conventional design, flame stability was obtained by creating local hot spots, but this was at the expense of high NO_x emissions, as confirmed by literature results (Klug *et al.* 1991).

A way of NO_x reduction without alteration of the burner design is by external flue gas recirculation into the combustion air; driving up to ~10% of the flue gases back from the exhaust duct at a temperature of ~300°C, via a booster fan, can decrease the NO_x emission by as much as 50% (Figure 5). This system is effective for rich fuels (e.g., COG, NG gas

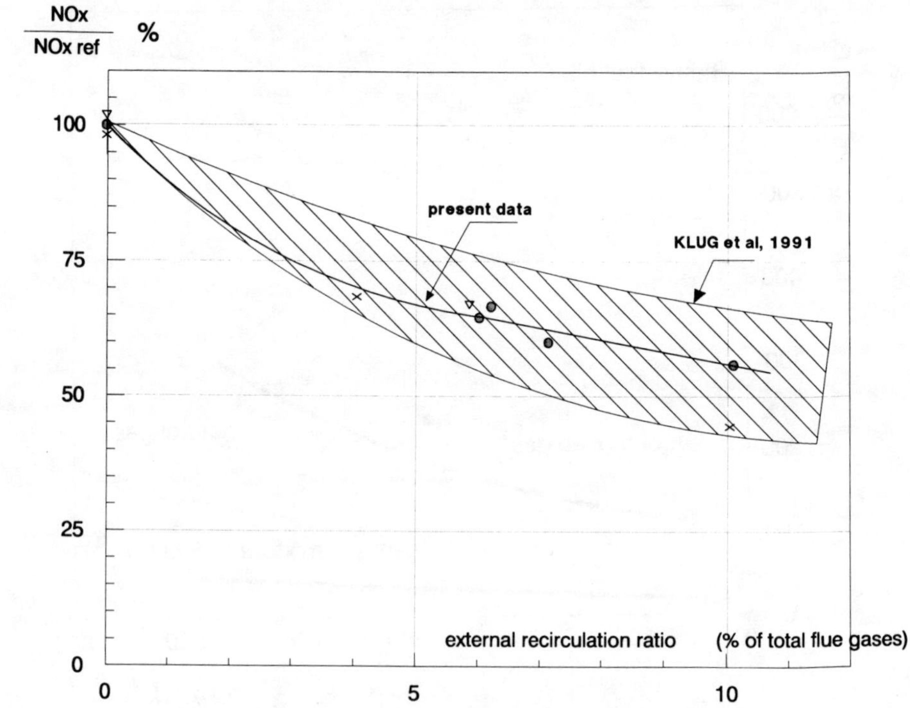

FIGURE 5 Effect of external flue gas recirculation.

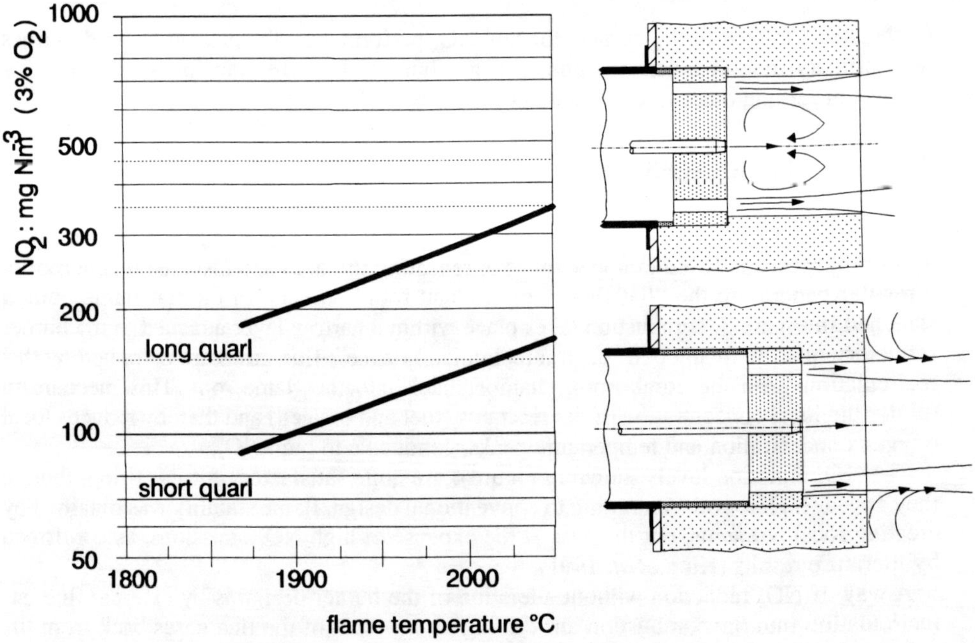

FIGURE 6 NO_x emissions from wall firing burners.

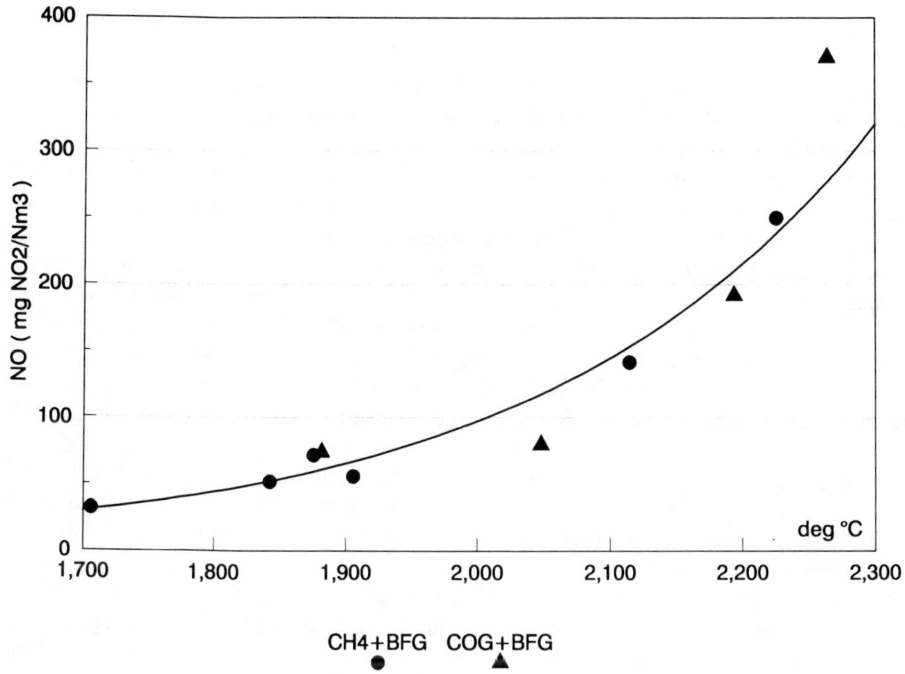

FIGURE 7 NO$_x$ emissions from lean gas mixtures.

or HFO) in spite of some penalty in consumption; water injection (either steam or liquid) into the combustion air would give similar results.

Wall Burners

Development work has been carried out on wall burners that produce a long straight flame and are frequently employed in steel reheating furnaces for side- and front-wall firing. According to the common design scheme in Figure 6, the combustion air is injected at high velocity through individual circular holes provided in a refractory "baffle"; a minor proportion of the combustion air (primary air) may be injected closer to the fuel supply (generally at the axis). At the outlet of this baffle, the air and the fuel jets end up in a cylindrical port, built into a refractory casting within the thickness of the furnace wall. This port acts as a "quarl" for the burner and improves combustion stability and flame pattern considerably.

A long quarl (i.e., large values of the aspect ratio *L/D*) increases NO$_x$ emissions: faster mixing of air and fuel within the quarl obtained with long cylindrical ports, high swirl effects, particular jet geometry etc., increases flame stability, but should be avoided for low-NO$_x$ purposes. A value of the aspect-ratio *L/D* not greater than 0.6 is recommended and may be adopted safely also in retrofitting existing plants (Figure 6); this is a good example of a primary measure that may be implemented in a relatively non expensive way. Similar results have been found by Japanese investigators (Narazaki *et al.* 1983; Takashima, 1988).

The NO$_x$ emission is also sensitive to the ratio of fuel/air velocities and to other parameters affecting the near field of the burner. For a fixed geometry, NO$_x$ data relevant to different gas mixtures of varying lower calorific value (LCV) fall on a single curve as a function of the flame temperature (Figure 7).

Variable Flame Burners

In modern walking beam reheating furnaces, some heating zones are equipped with variable flame burners; the flame shape control aims at improving the temperature uniformity of the slabs delivered to the mill. The heat release pattern is controlled, according to the production rate and to the arrangement of the slabs, by controlling the ratio of secondary to primary air at the burner. This air staging effect is not meant for NO_x reduction, but may affect the emission level very much, as it may be expected.

An impressive example of NO_x emission sensitivity to the adjustment of primary / (primary + secondary) air flow is reported in Figure 8. With primary air forming only 20% of the total ($R\% = 80$), a long stable flame with very low NO_x is produced (<~100 mg/Nm³). As soon as primary air is increased, combustion intensity increases while the flame gets shorter, as it is evident from Figure 9, where detailed in-flame measurements with the experimental burner have been reported in order to compare the long flame and the short flame.

Going back to Figure 8, it should be noted that intermediate values of *R%* (~60 to ~80) involve a relatively moderate variation of the furnace temperature profile: a variation of ~60°C of this maximum is amplified to ~300°C in peak flame temperature and results in a 3.5-fold magnification of NO_x.

A further point worth mentioning is that low-NO_x configurations are only slightly sensitive to operating variables, opposite to what happens with high-NO_x configurations: the two above reference cases are compared in Figure 10 to show this very important effect.

Application to Industrial Scale

Observed NO_x ranges as those presented in Figures 8 and 10 explain why any general correlation of NO_x data cannot be established: a reliable prediction of the stack emissions in operating conditions can be set up for a well tested design only. In the examples shown above, the main trick adopted for optimizing the burner consisted in allowing the premixing of flue gases from the *internal recirculation zone* into the combustion air upstream of the flame front. The experimental burners were tested on a semi-industrial scale in thermal and stoichiometric boundary conditions typical of an industrial plant. No wonder, therefore, that NO_x emissions observed on working walking-beam furnaces exactly reproduced values expected from the test furnace measurements (obviously using the same fuel mixture and the same burner design). Furthermore it has been experimentally proved that scaling from a smaller prototype to a larger size unit follows the square-root criterion (constant velocity), as far as NO_x and the flame length are concerned.

Several new and retrofitted plants, in Italian steelworks and all over the world, have been equipped with Italimpianti low-NO_x burners that were kept up-to-date with the ongoing R&D work.

In spite of efforts still going on (Weber *et al.* 1986), a quantitative NO_x prediction based on a mathematical fluid dynamic model was not yet successful and therefore nozzle design (and, even more difficult, NO_x prediction) has to be carried out experimentally.

The situation for the designer is, however, not always as clearcut and fortunate as with gas burners in furnaces: in other plants of the steel industry, where the combustion head is "embebbed" in the intricacies of traditional process plants, modifications of existing equipment are by far more difficult.

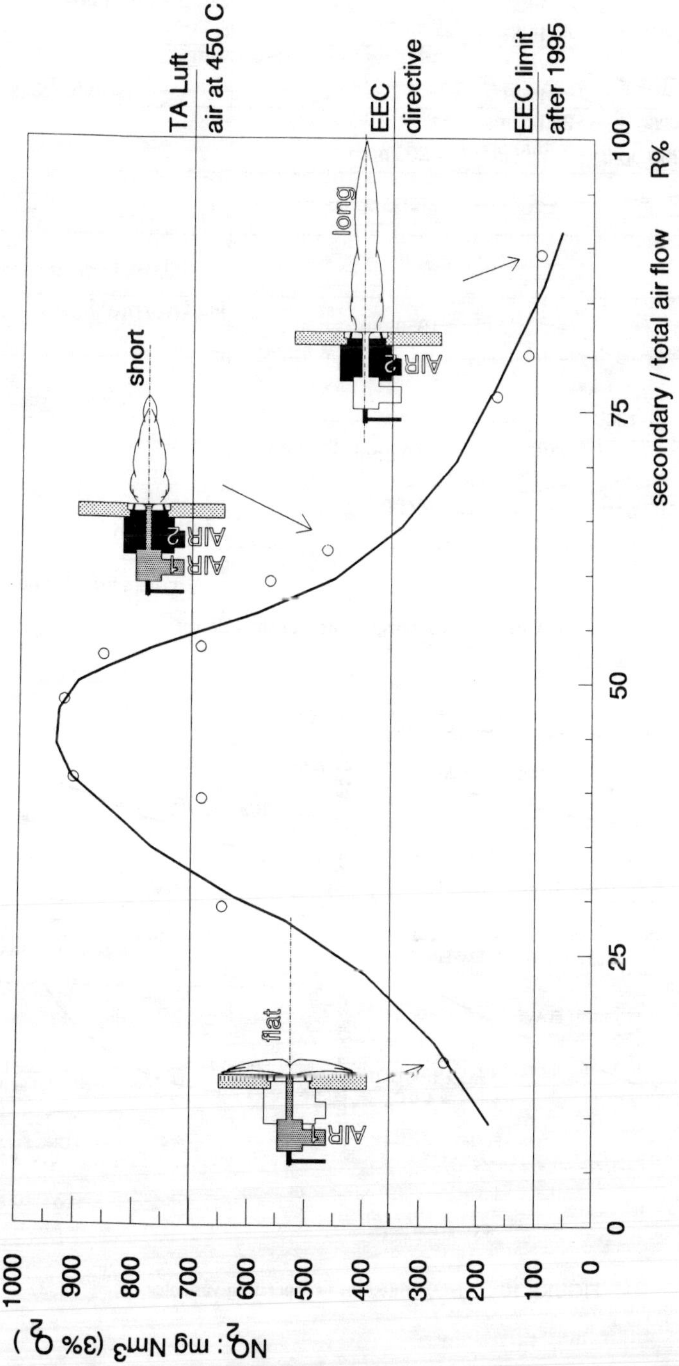

FIGURE 8 NO$_x$ emissions vs. air setting control.

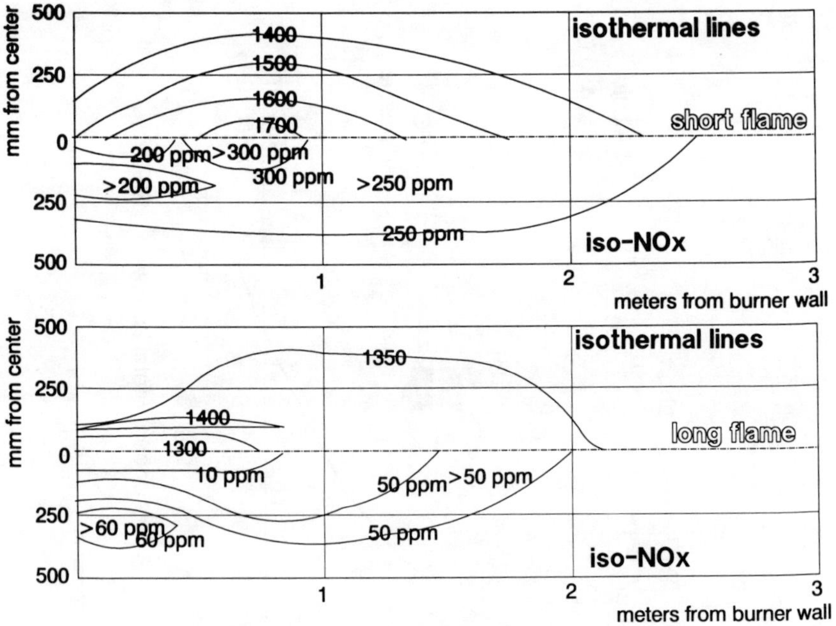

FIGURE 9 Patterns of long and short flames.

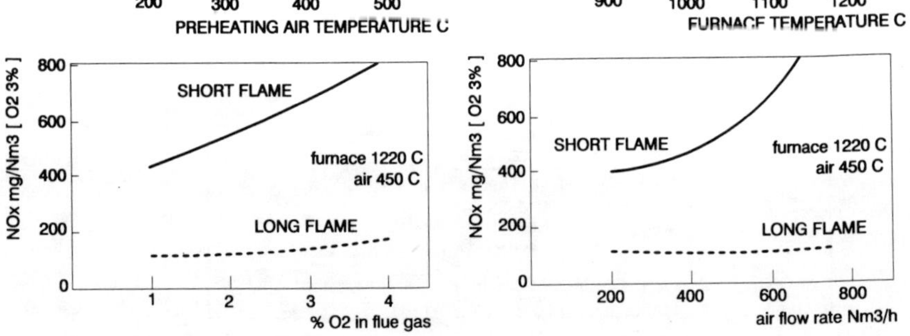

FIGURE 10 NO_x emissions vs. operating variables.

OTHER FLAME BASED PLANTS

Large plants typical of integrated steelworks are the coke ovens, the hot blast stoves ("cowpers") and the power plant (often in combined cycle heat/process steam and power).

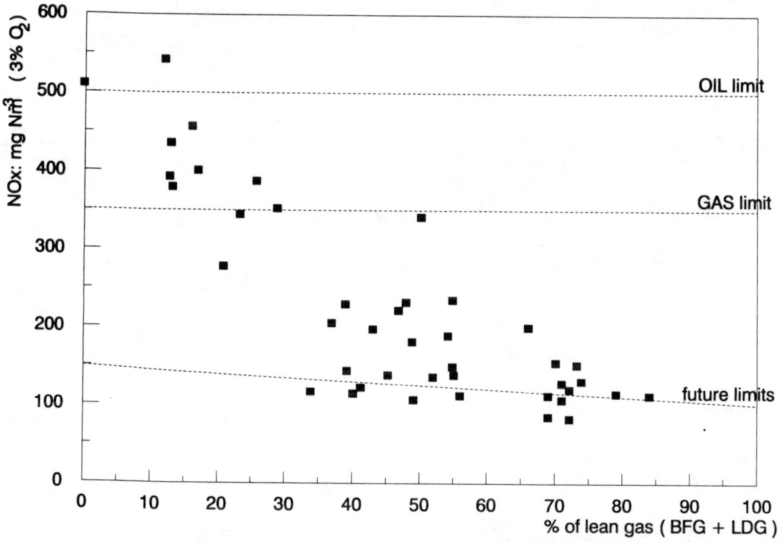

FIGURE 11 NO_x from multi-fuel power plants.

All these plants make use of large amounts of gas mixtures, that is of lean gas consisting of blast furnace gas (LCV ~ 3 MJ/Nm^3) and of basic oxygen converter gas (LCV ~ 8 MJ/Nm^3). With lean fuel gas mixtures, the thermal NO_x emission is automatically limited, as we have already seen in Figure 1 (see the class i) in paragraph "Accumulated NO_x Emission Data" above). None of the above processes, however, can afford to use lean fuel gas alone, because of flame stability, process control and incomplete combustion problems. Addition of rich fuel is therefore necessary and this may result in NO_x emissions beyond the allowed limits.

The *hot blast stove* is not a strong NO_x emitter in general and can experience severe plant consequences from combustion equipment modification; a revamping project should therefore be carefully designed and tested before being applied in practice.

The *coke oven* heating process usually produces too much NO_x; this problem has been investigated in specialized works and is difficult to solve for old oven batteries, due to complicancy and limited possibilities of retrofitting the combustion chamber and the flow distribution masonry (Eisenhut *et al.* 1990).

There may be a NO_x problem for large boilers and steam generators within integrated steelworks, because these power plants are subject to strict mandatory regulations (Nelli, 1991). If the contribution of the lean gas (BFG + LDG) to the thermal power required for steam raising falls below ~50% the remaining portion being covered by coke oven gas (which is a stronger NO_x emitter than natural gas), the emission may be higher than the required limit as shown in Figure 11, that reproduces several experimental data relevant to a power plant. If the lean by-product gases are not sufficient and natural gas is not available to cope with the required load (total electric power + process steam), then heavy fuel oil is burned and obviously NO_x increases, because of the "fuel-NO_x." The problems arising from these particular low-NO_x multifuel burners, may be tackled by looking at solutions and techniques proposed or adopted by the power industry for large, heavy fuel oil fired steam generators (Pasini *et al.* 1991).

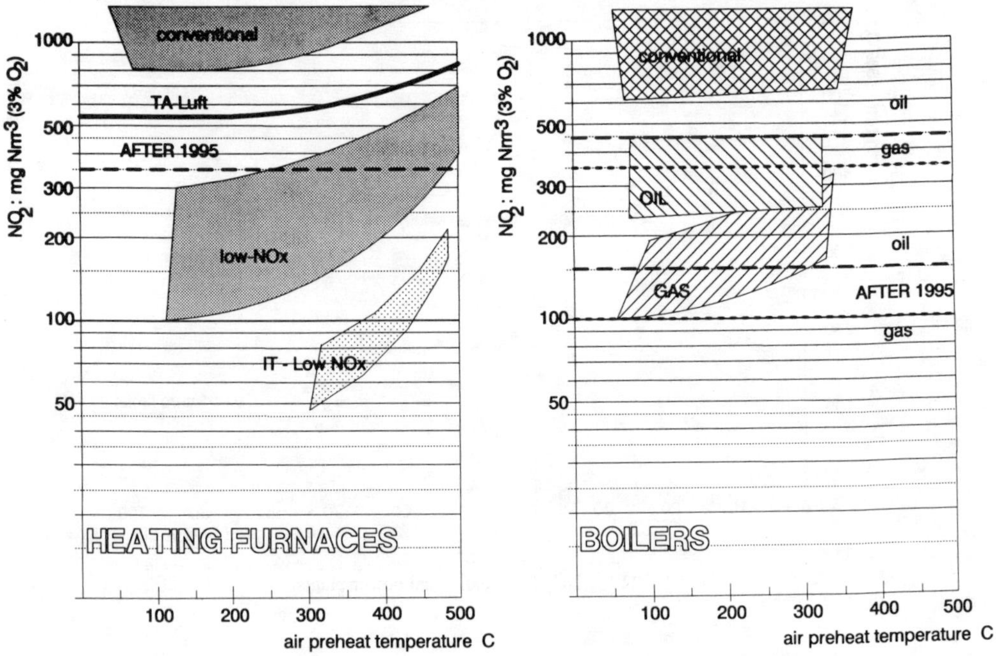

FIGURE 12 Low-NO_x and conventional burners.

CONCLUSIONS

The situation of NO_x emissions from steel reheating furnaces is summarized in Figure 12 where the German TA-Luft regulation with envisaged European limits are reported as a function of the air preheat temperature. Conventional equipment based on old designs where NO_x emission was not considered at all, can be strong emitters, even when process and operating variables allow for the possibility of attaining lower limits (Hovis, 1986; Kremer, 1986).

All major manufacturers have developed new low-NO_x burners in the range ≈100–350 mg/Nm³ (gas firing); therefore many plants may be retrofitted to low-NO_x performance on the basis of the present state of the art. Higher emission values are to be expected for oil firing and for "regenerative" burners. The latter case is particularly interesting because of its large energy saving potential; coordinated R&D effort is going on in order to improve the emission performance of these systems (Hibberd, 1988).

For the case of large steam generators, there is a considerable pressure toward further reductions. There will be difficulties in keeping NO_x emissions below 100 mg/Nm³ for gas and 150mg/Nm³ for oil firing, without producing unacceptable CO and/or hydrocarbon emission. The case of recovery gas boilers is fortunate as long as the proportion of rich and "dirty" fuels (fuel oil, non depurated coke oven gas etc) may be kept to a minimum. Otherwise, sophisticated technologies should be applied to power plants based on the experience gained by intensive R&D efforts by the power industry for solid, liquid and gaseous fuels (Pasini *et al.* 1990).

Development and validation of mathematical models and of available software for computational fluid dynamics, in the field of flames and of reacting flows, will greatly enhance and help both the basic understanding and practical design criteria for future plants

(Swithenbank, 1991; Weber *et al.* 1986). The strategic interest of the steel industry in advanced background knowledge and knowhow in this domain, both theoretical (mathematical modelling) and experimental (advanced measuring and control techniques) goes far beyond the described examples of existing plants and will be more and more extensively applied to the development and the design of new processes in steel and iron making.

REFERENCES

Cavelli M., Facco G., Milani A. Forno sperimentale per ricerche sulla combustione. BTF Nov-Dic 1978, 335–342.

Eisenhut W. NO_x Emissionen und Immissionen in Zusammenhang mit der Beheizung von Koksoefen. Bergbau Forschung, ECSC Contract 7261-01/411/01.

Hibberd D. The Application of Regenerative Burners to a Continuous Heat Treatment Furnace, Energy Efficiency Office Rep.ED/166/191 Aug 1988.

Hovis J. Low NO_x burners for metallurgical furnaces, *Iron & Steel Engineer*, June 1986.

Klug K. *et al.* Reduzierung der NO_x-Emission von Deckenstrahlbrennern, *Gas Waerme International*, 40, 1991, H.1.

Kremer H. Grundlagen der NO_x-Entstehung und -Minderung, Gas Waerme International Band 36 (1986), H.4.

Mattarini A. Ottimizzazione delle Caratteristiche di un Tubo Radiante, CSM Rapporto 6150R-CSM, Genova Feb. 1989.

Milani A., Nelli D. Studio su forno pilota degli ossidi di azoto emessi dai forni di riscaldo, Draft Report, ECSC Contract 7261 01/407/04, CSM Report 6828R, Dic. 1990.

Narazaki S., Matsuzaki K. Scaling up of low-NO_x Gas Burners, IFRF 6th Members Conference, IJmuiden, 1983.

Nelli D. Riduzione delle emissioni di ossidi di azoto dalle centrali termiche tipiche degli stabilimenti siderurgici. CSM-ECSC Contract 7261-01/451/04, 1991.

Pasini S., Trebbi G. Aspetti innovativi della combustione in caldaie per la produzione di energia elettrica. ENEL/DSR/CRTN, Oct. 1990.

Swithenbank J. Mathematical Modelling as a tool for waste incinerator design and optimisation, IFRF 3day "ToTeM 5" Meeting on Waste Incineration Technology, Bari, Oct. 1991.

Takashima H. The Development of High Efficiency, Low NO_x Burners, Sumitomo Search, 36, May 1988.

Weber R., Boysan F., Swithenbank and P.A. Roberts. Computations of near field aerodynamics of swirling, expanding flows, *21st Symposium (Int) on Combustion*, 1986, 1435.

Optimization of a Cement Plant Preheating Tower, Equipped with a Precalciner, Burning Pulverized Coal

C. BELOT, D. GOFFE, D. GROUSET *(Sté Bertin et Cie—F40220 TARNOS)*
C. BERTRAND, B. HOMASSEL, J. L. PHILIPPE *(Lafarge Coppée—F38291 LA VERPILLIERE)*

INTRODUCTION

An optimization of a cement plant precalciner has been undertaken with modelling techniques and on-site experiments. The aim of that project was to enhance the performance of the preheating tower, mainly by optimizing the combustion in the precalciner in order to reduce the emission of unburnt products.

This work has been achieved by two different modelling approaches. A global modelling of the complete tower has been executed, followed by a detailed modelling of the processes occurring inside the precalciner. The first model intends to describe the effect of the complete set of operating parameters on the process performance in order to be able to find the optimal solution. The detailed model is specific to the precalciner; its aim is to determine the optimal design and setting.

PROCESS OVERVIEW

Portland cement is produced from a mixture of lime-stone and clay which is introduced in a rotary kiln where the sintering reactions take place. From a chemical point of view the following steps are realized: drying of the material, decarbonation of lime-stone, melting of the alumina components, combination of silica and lime. The product resulting from the combination of silica and lime is called clinker, this material is then finely ground and mixed with additives, mainly gypsum, to give the final cement.

In the very earliest rotary kilns for cement manufacturing, the crude material was mixed with water and that paste was introduced into the kiln so that all the above described processes took place in the kiln. This process, called the wet process, was very energy consuming because of all the water that needed to be evaporated. A development realized in the 50's consisted of equipping the front end of the kiln with a preheating tower. No more water is used to prepare the material, but it is finely ground and introduced at the top of a cyclone tower. The material which is introduced in the preheater is called raw meal.

The function of the tower is to exchange heat from the hot gases coming from the kiln to the crude material, thereby working like a countercurrent exchanger (figure 1). The heat exchange takes places in a series of cyclones. On one stage of cyclones the following steps are realized. the material coming from the upper stage cyclone underflow is picked up by gases coming from the lower stage cyclone overflow, the material is heated up by the hot gases in the entrance duct of the cyclone in the current stage, inside the cyclone the powder is separated from the gas and flows then to the stage underneath.

Typically, these kind of towers are equipped with four stages of cyclones. The separation efficiency of the cyclones is in the order of 80 to 90% and the overall pressure loss of the complete tower is in the order of 500mm WG. The overall exchange efficiency of that process depends on the separation efficiency of the cyclones. When bad separation occurs in the cyclones, the temperature of the top levels rises due to the fact that hot solid material is carried to the upper stage.

In this process, called the dry process, drying of the raw material and part of the decarbonation is achieved in the cyclone tower. Typically, one can reach decarbonation yields in the order of 50% at the kiln inlet. Energy consumption of a wet process kiln is in the order of 1200 Kcal per kg of clinker; with the dry process this figure drops to 800.

A more recent development of this process was to equip the preheating tower with a precalciner. In principle this modification consists of bringing the energy for decarbonation to where it is needed. Decarbonation is the most energy consuming reaction in the process.

A precalciner is a combustion chamber placed at a lower stage in the cyclone tower. About 60% of the fuel supply is fired in that chamber, with the combustion air being derived from the clinker cooler. The use of precalcining systems allows the rotary kiln to be shortened because a high degree of decarbonation can be obtained at the kiln entrance, typically of the order of 90%. On the other hand, the production capacity of existing kilns can be enhanced by adding a precalciner. When one compares the situation of an existing kiln with a given diameter, one can see that the production can be roughly multiplied by two when a cyclone tower with a precalciner is added to that kiln.

Several types of precalciners are used through the world, differing from the flow in the devices (up- or down-flow, with or without a vortex flow), the design of the air, fuel and material supply to the device. In the case of the Ciment Lafarge plant at Lexos, which is studied in this paper, it is a RSP (Reinforced Suspension Preheater) precalciner which is used.

In that precalciner, the flow structure is that of a down-flowing vortex. The parts described below make up the components of the reactor (from top to bottom):

— a cylindrical part equipped with one scroll entrance for air and a coal injector placed on the axis,
— a conical part allowing the vortex flow to develop,
— two tangential air entrances where the raw meal from the upper cyclone flows in,
— a main cylindrical part which joins a square duct

In principle, the device separates the combustion and decarbonation functions. The fuels ignite at the top where no raw meal is present. The flame then develops along the axis. The raw meal is introduced tangentially into the main cylindrical part and flows down, along the side walls under the effect of centrifugal forces. The result of this is that complete combustion can be achieved before the mixing of the fuel and the raw material. Also, the powder material flowing along the walls protects the wall from the heat transmited by radiation in that sense that all of the heat transferred is taken over by the endothermal decarbonation reaction.

This type of precalciner was initially developed for liquid fuel. When it was switched to solid fuel firing, some difficulties in obtaining complete combustion inside the device appeared. The work described in this paper is a part of the actions taken to understand the phenomena and determine ways of optimisation.

GLOBAL MODELLING OF THE PREHEATING TOWER

Purpose:

The aim of this study is to obtain an overall description of the operating conditions within the preheating tower. This description will be used to analyze those mechanisms which lead to the emission of unburnt products. Then a sensitivity analysis will allow the potential reduction of emissions by modifying operating conditions to be quantified.

Model Description

Basic Features

The model that has been set up is a zone model. The tower is discretized into 9 elements in which balance equations for mass, species and energy are solved in order to obtain average concentrations and temperatures. Figure 1 shows the arrangement of the elements:

* elements 1 to 4 stand for cyclone exchangers
* elements 5 and 6 stand for the upper and lower part of the precalciner.
* elements 7 and 8 stand for connecting pipes
* element 9 stands for the dust chamber (at the back end of the kiln).

The parameter $\lambda 2$ is introduced to simulate the fraction of the preheated combustion air that is delivered to cyclone n°3, due to an air leakage at the raw material input of the precalciner.

The parameter ν describes the "mixing" of raw material with the gas flow in the precalciner. It is assumed that, over the entire flow rate of raw meal from cyclone n°3 to the precalciner, only the fraction ν is mixed with the gas flow. The fraction $(1 - \nu)$ flows along the walls of the precalciner, so that effective mixing with the gas flow only occurs in the pipe downstream.

Each element is assumed to behave like a perfect stirred reactor at steady state. The following variables are computed in each reactor:
- gas and solid mass flow rates
- temperature, assumed to be the same for both gas and solid matter
- gas composition, described by the mass fractions of 7 species: H_2O, CO_2, CO, CH_4, O_2, SO_2 and inert (N_2)
- solid matter composition: bound H_2O and CO_2, CaO, coal (d.a.f. basis), char, SiO_2, Al_2O_3, Fe_2O_3, MgO.

The above variables are obtained by solving balance equations related to each element of the tower, accounting for a number of chemical reactions:
- dehydration and decarbonation of mineral matter
- coal devolatilization
- combustion of volatile products, assumed to be CH_4 and CO
- char oxidation to CO

Balance equations

For each type of variable, we define a vector containing the values of the variable within the 9 elements of the tower:

Qg: vector of gas flow rates leaving the elements
Qm: solid mass flow rates leaving the elements
Yj: mass fractions of species in the elements
T: temperatures within the elements.

We can now derive the balance equations for these variables;
Mass conservation within the gas phase can be written as:

$$A \times Qg = Qg_{feed} + Sg$$

where Qg feed gathers, for each element, the gas flow rates entering the tower, Sg is the mass transferred from the solid phase due to chemical reactions, and A is a 9×9 matrix describing the connexions between elements.

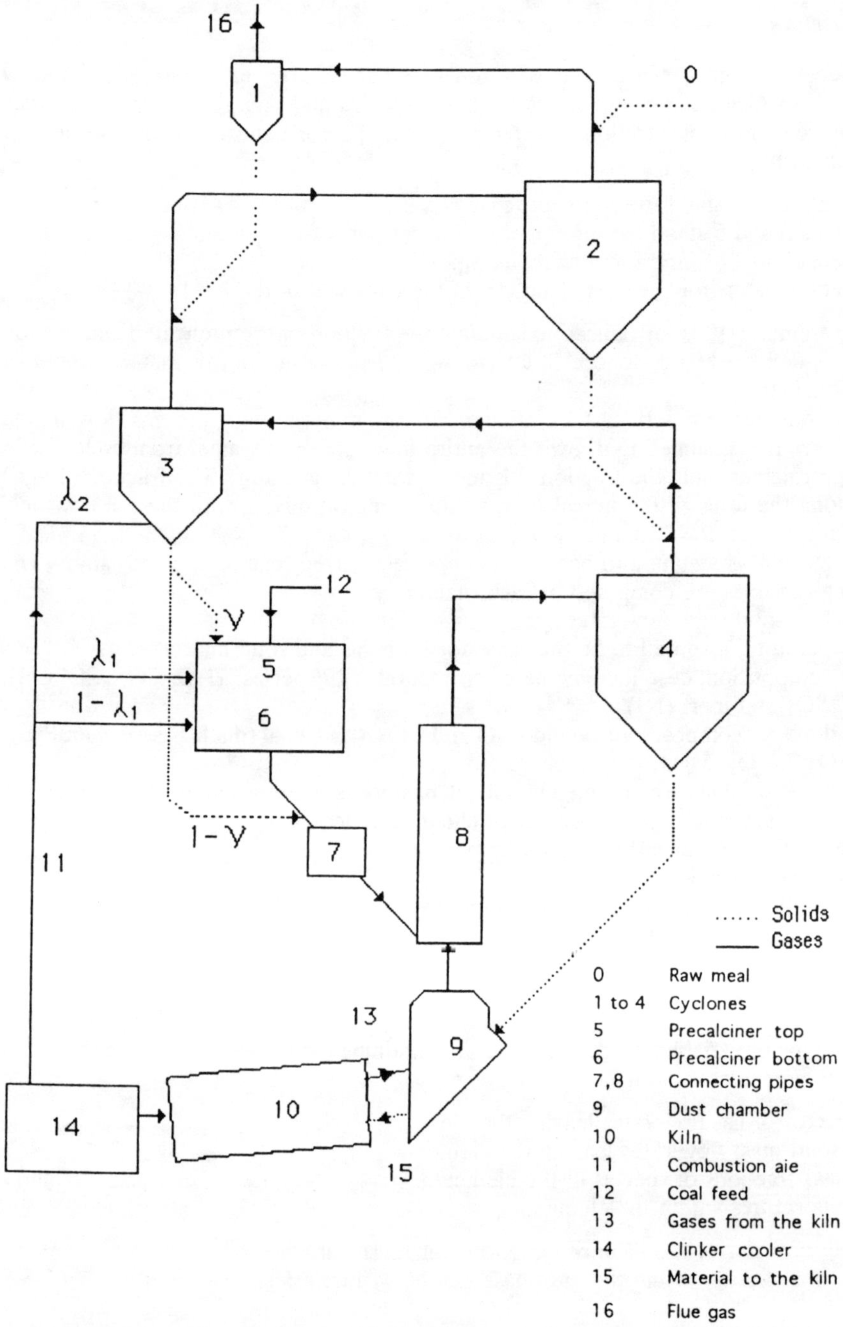

FIGURE 1 Flowsheet of the preheater.

From figure 1 it can be seen that:

$$
A=\begin{bmatrix}
1 & -1 & 0 & 0 & 0 & 0 & 0 & 0 & 0 \\
0 & 1 & -1 & 0 & 0 & 0 & 0 & 0 & 0 \\
0 & 0 & 1 & -1 & 0 & 0 & 0 & 0 & 0 \\
0 & 0 & 0 & 1 & 0 & 0 & 0 & -1 & 0 \\
0 & 0 & 0 & 0 & 1 & 0 & 0 & 0 & 0 \\
0 & 0 & 0 & 0 & 0 & 0 & 0 & 0 & 0 \\
0 & 0 & 0 & 0 & 0 & -1 & 1 & 0 & 0 \\
0 & 0 & 0 & 0 & 0 & 0 & -1 & 1 & -1 \\
0 & 0 & 0 & 0 & 0 & 0 & 0 & 0 & 1
\end{bmatrix}
$$

In the same way, the conservation of the mass of solid particles leads to:

$$\mathrm{B} \times \mathrm{Qm} = \mathrm{Qm}_{\mathrm{feed}} - \mathrm{Sg}$$

where B is a matrix of coefficient accounting for both the geometrical arrangement, collection efficiencies $\rho 1, \rho 2, \rho 3, \rho 4$ of the cyclones, and raw materials mixing inside the precalciner:

$$
B=\begin{bmatrix}
1 & \rho 2 - 1 & 0 & 0 & 0 & 0 & 0 & 0 & 0 \\
-\rho 1 & 1 & \rho 3\text{-}1 & 0 & 0 & 0 & 0 & 0 & 0 \\
0 & -\rho 2 & 1 & \rho 4\text{-}1 & 0 & 0 & 0 & 0 & 0 \\
0 & 0 & 0 & 1 & 0 & 0 & 0 & -1 & 0 \\
0 & 0 & 0 & 0 & 1 & 0 & 0 & 0 & 0 \\
0 & 0 & 0 & 0 & 0 & 0 & 0 & 0 & 0 \\
0 & 0 & \rho 3.\nu & 0 & -1 & 1 & 0 & 0 & 0 \\
0 & 0 & \rho 3(1-\nu) & 0 & 0 & -1 & 1 & 0 & 0 \\
0 & 0 & 0 & 0 & 0 & 0 & -1 & 1 & -1
\end{bmatrix}
$$

Species and energy balances are used in a similar way to derive equations for Yj and T.

Source terms

The balance equations contain source terms due to chemical reactions. The rates of these reactions are modelled as follows. Dehydration and coal devolatilisation are assumed to occur instantaneously and completely, as soon as raw meal and coal particles are fed into the tower. This assumption is supported by the high temperature levels inside the tower. Consequently, the reaction rates are taken to be the input mass flow rate of bounded water and coal volatiles in elements n°1 and 5 respectively. The amount of volatile products generated by coal is deduced from the ASTM value. The volatile products are assumed to be CH_4 and CO, the ratio of which is obtained from the coal elemental analysis. The resulting reaction is:

$$1\text{kg coal (daf)} = 0.235\text{kg } CH_4 + 0.115\text{kg CO} + 0.650\text{kg char.}$$

Decarbonation is also assumed to be instantaneous and irreversible, but the amount of CO_2 released from mineral $CaCO_3$ is dependant on temperature level. It is well known that decarbonation does not take place below 700°C; and is completed above approximately 860°C. To evaluate the decarbonated part of the mineral matter (i.e. the percentage of calcium found in the form of CaO), we propose to use a linear function between these two temperatures. The decarbonation rate is then computed as the flow rate of bounded CO_2 entering the element and exceeding the maximum value compatible with the decarbonated fraction at the temperature of the element.

Gaseous combustion of volatile matter is described by a two reaction global mechanism, as proposed by Dryer and Glassman (1972):

I $CH_4 + 1{,}5\ O_2 \rightarrow CO + 2H_2O$

II $CO + 0{,}5\ O_2 \rightarrow CO_2$

The kinetic rates of these reactions are expressed as:

$WI = 10^{13.2}\ \exp(\text{ -24400/T })\ [CH4\ ]^{0,7}\ [O2\ ]^{0,8}$

$WII = \alpha\ 10^{14.75}\ \exp(\text{ -21600/T })\ [CO\ ][H2O\ ]^{0,5}\ [O2\ ]^{0,25}$ (with W in $mol.cm^{-3}s^{-1}$)

The correction coefficient a has been introduced to account for the specific operating conditions of the precalciner (lack of oxygen and high concentrations of CO_2). The value $\alpha = 0,1$ has been found to correlate satisfactorily from experiments by Gadiou and Prado (1987) on coal samples used at the LEXOS plant. In these experiments a drop tube furnace was used characterize the combustion of single coal particles under controlled temperature and partial pressure of oxygen. The experimental data are the delays taken by each step in the combustion process. By a data regression procedure, relations can be developped to correlate the combustion time with burner temperature and oxygen concentration. In our case, one obtains:

$$\tau_c = 0.29 + 3,22.\ 10^{-2}/PO2 - 6,25.\ 10^{-4}(T\ - 1273)$$

where τ_c is the combustion time in seconds, PO_2 is the ambient partial pressure of O_2 in atm, T is the ambient gas temperature in Kelvin.
Char oxidation rate is then obtained by:

$$W_{char} = Q_{char} \cdot \min(1, \tau_s/\tau_c)$$

where Q_{char} is the mass flow rate of char entering the element and τ_s is the residence time.

Application to the LEXOS Plant.

The model described above was used to simulate the operating conditions of the LEXOS plant in march 1986 with the data given in table 1.
Agreement was searched between numerical results and measurements made at the LEXOS plant in March 1986. Table 2 (case n°1) displays the results finally obtained with $\lambda 2 = 4\%$ and $\nu = 30\%$. It can be seen that the model is able to reproduce experimental measurements with an acceptable accuracy:

— good prediction of mass flow rates and temperatures
— good prediction of CO_2 content of mineral matter at the exhaust box (the discrepancy between computed and measured value is compatible with the uncertainty affecting the measurements).
— fair prediction of CO and CH_4 within the elements, due to the large fluctuations of concentrations inside the tower and to the simplicity of the model used.

Moreover, it is interesting to note that the model can provide complementary information, such as the rate of unburnt char at the exhaust box and the rate of decarbonation in each element, that are useful for the understanding of the process and its optimization.

TABLE I
Global Modelling Operating Conditions (flowrates in Kg/s)

	Raw material	Kiln outlet	Combustion air	Coal + air	Fresh air
Temp.K	348.00	1277.00	1073.00	298.00	288.00
Gas flowrate		1.67E-01	1.72E+01	4.35E-01	2.00E+00
H2O% weight.		3.86E-02	9.30E-03	9.30E-03	9.30E-03
CO2		3.49E-01			
CO					
CH4					
O2		1.94E-02	2.26E-01	2.26E-01	2.26E-01
SO2					
N2		5.93E-01	7.65E-01	7.65E-01	7.65E-01
Solid flow rate	4.50E-01	3.13E+00		2.65E+00	
H2O % weight	2.26E-02				
CO2	3.49E-01	5.00E-03			
CaO	4.44E-01	5.72E-01		1.94E-02	
Coal(daf)				6.82E-01	
Si O2	1.19E-01			1.62E-01	
Al2 O3	2.81E-02			7.82E-02	
Fe2 O3	1.86E-02			2.24E-02	
N2O K2O	1.56E-02	1.62E-01		1.21E-02	

Optimization Study

The zone model was used to explore some possibilities of improvement of the tower operation. In order to reduce unburnt emissions from the tower, three configurations were tested:

* suppression of air leakage between cyclone n°3 and the precalciner $\lambda 2 = 0$, case n°2)

* staging of mineral matter input in the precalciner ($\nu = 0,1$ instead of 0.3, case n°3)

* increase of primary combustion air in the precalciner (case n°4).

Table 2 shows the results of the calculations and exhibits the differences with respect to the basic configuration.

The suppression of air by-pass is an efficient way of improving the process, since it cuts the CO emission by a factor of 2. Moreover, char combustion and decarbonation are enhanced, so that mineral matter enters the main kiln with a lower content of bounded CO_2.

The staging of mineral input is not very valuable: it only has little affect on unburnt emissions, but it increases the temperature level within the precalciner. This is likely to be harmful, both for the reactor (risk of hot spots) and the process (risk of early clinkerisation).

Increasing primary air in the precalciner is a sound way of reducing CO emissions. However, this enhancement of combustion (char and CO) does not lead to any increase in decarbonation rate at the exhaust box. In other words, the energetic efficiency of the process has been lowered.

TABLE II

Global Modelling Results

		Measured	Mathematical Model			
			Case n°1	Case n°2	Case n°3	Case n°4
Tower outlet						
T	K	638	622	625	622	632
O2	% vol.	1.9	0.29	0	0.27	0.8
CO2	% vol.	35.5	32.3	33	32.4	31.4
CO	% vol.	0.1	0.06	0.03	0.05	0
CH4	% vol.	0.15	0	0	0	0
η			97.4%	100%	97.8%	100%
Cyclone 4 overflow						
T	K	1123	1109	1117	1109	1116
O2	% vol.	0.3	0	0.01	0.01	0.54
CO2	% vol.	38.5	33	33	33	31.9
CO	% vol.	0.3	0.11	0.04	0.09	0.01
CH4	% vol.	0.3	0	0	0	0
Precalciner outlet						
T	K	1390	1352	1367	2195	1316
O2	% vol.	2.2	3.7	4.1	1.7	5
CO2	% vol.	21	24	23	18.4	22
CO	% vol.	6.8	0.09	0.08	0	0.11
CH4	% vol.	10	0	0	0	0
Kiln inlet						
Loss on Ignition		8%	10.9%	9.9%	10.8%	10.9%
Residual coke			3.4%	0%	3%	0%

DETAILED MODELLING OF THE PRECALCINER

Purpose

The global model has proven useful in analysing the operation of the tower and identifying ways of achieving clean and efficient combustion. However, this analysis is based on the assumption of perfect mixing within each element of the tower. This assumption is worth checking at least in the precalciner where both combustion and decarbonation occur at high rates. Three dimensional simulations of the flowfield inside the precalciner were undertaken in order to assess the "mixing level" of gas and particules, and also to evaluate temperature and concentration gradients inside the burner.

A commercially available computer code, FLUENT*, was used to perform the calculations. The detailed description of the model has been published elsewhere (see for example WILD et al. 1987) and only the main features will be recalled here. We will concentrate then on the case description and on the results obtained.

Model Description

FLUENT is a general purpose 3D code which solves the Navier-Stokes equations over

* FLUENT is a registered trademark of Creare Inc. and is available from FLUENT Europe Ltd, Hutton's building, 146 West Street Sheffield 51 4ES (GB).

TABLE IIIa
Air flow rates

			Transport air	Primary air	Secondary air
CASE 1	Flow rate	(kg/s)	0,450	4,106	13,150
	Temperature	(K)	298	1 075	1 123
	Density	(kg/m3)	1,186	0,329	0,315
	Velocity	axial (m/s)	11,170	—	—
	radial	(m/s)	—	—	—
	tangential	(m/s)	5,585	35,660	43,490
CASE 2	Flow rate	(kg/s)	0,270	2,770	13,050
	Temperature	(K)	298	1 075	1 123
	Density	(kg/m3)	1,186	0,329	0,315
	Velocity	axial (m/s)	6,700	—	—
	radial	(m/s)	—	—	—
	tangential (m/s)	3,350	24,060	43,160	

a structured cartesian grid. Models are available for the description of turbulence (k-ε or ASM model), combustion (eddy break-up model) and radiation (6 flux model). A stochastic model for particle tracking is included, with coupling terms of mass, momentum and heat transfer between the particles and the continuous phase. Some modifications were brought to the source code in order to treat the specific case of the precalciner, i.e. presence of two types of particles within the flowfield (coal and raw mineral). The particle tracking module was extended in the following way:

— each class of particles, defined by the injection parameters (position, velocity, flowrate, temperature, diameter) is provided with a specific set of physico-chemical properties: density, specific heat, composition (volatile matter, char, inerts)
— devolatilisation is described by a number of chemical reactions, which may take place simultaneously and may be different from one particle class to the another.
— volatile matter may result in several gaseous species, depending on the stoichiometry and kinetic rate of devolatilisation reactions.
— the kinetics of char combustion is described by the classical model of Field *et al.* (1967)

This treatment allows coal pyrolysis and raw material decarbonation to be modelled in a similar way.

Case Description

Two calculations were performed, corresponding respectively to the operating conditions of the LEXOS plant precalciner in March 1986 and June 1990. The former intended to complement the results obtained with the global model (see chapter 3), whereas the latter was performed in order to compare numerical results and detailed measurements made in the precalciner.

Table 3 summarises gas and particle flow rate characteristics, physical properties and chemical data used for the calculations. The main differences between the two sets of operating conditions lie in coal properties (flow rate, calorific value, grinding) and raw material flow rate.

An Arrhenius law was used to model coal pyrolysis

$$d/_{dt}\, m_p = k.\ \exp(-E/RT).m_p,$$

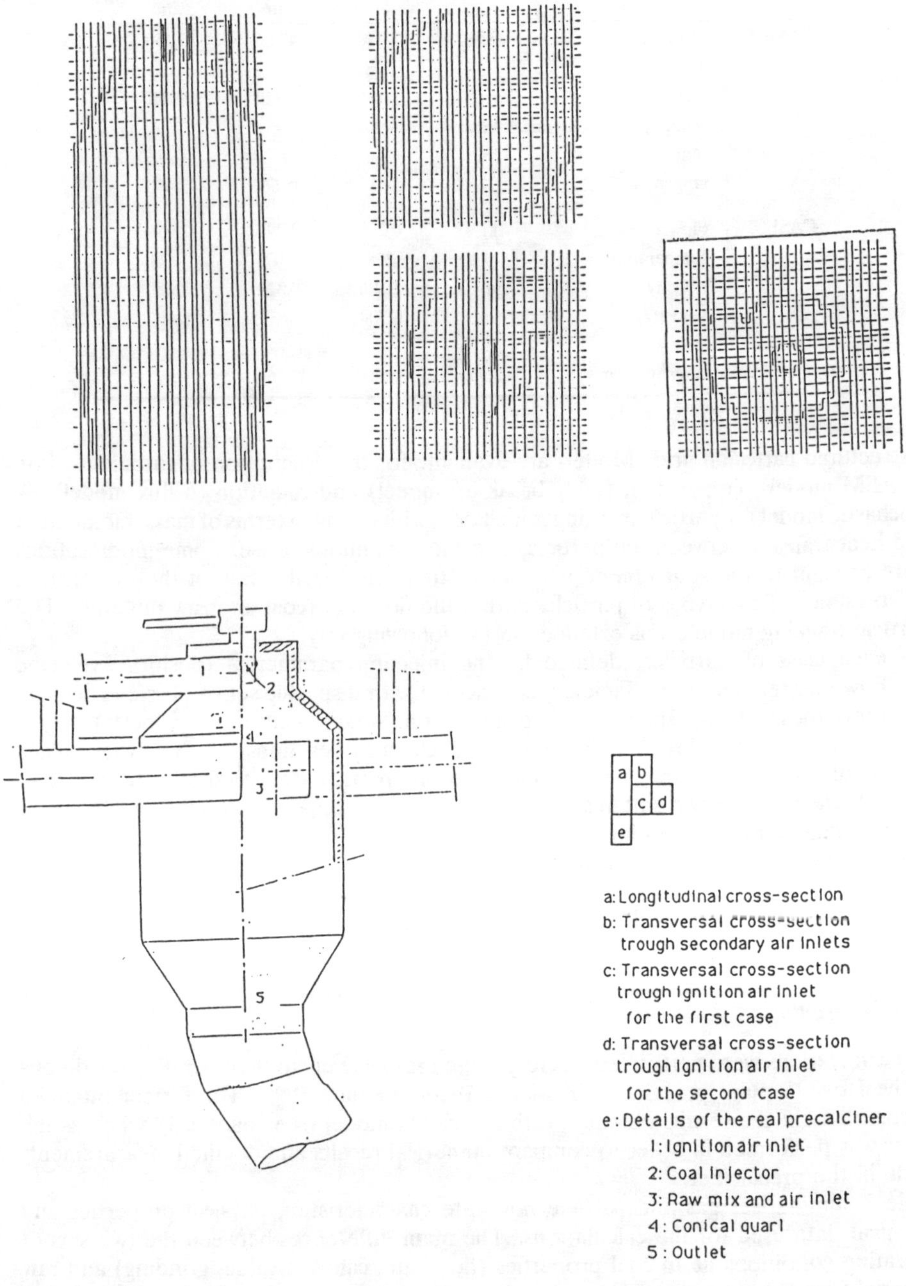

FIGURE 2 Geometry of the precalciner and grid generation.

TABLE IIIb
Solid phase properties

	COAL		RAW MATERIAL	
	Case 1	Case 2	Case 1	Case 2
Flow rate (kg/s)	2.67	1.91	52	66.3
T (K)	300	300	1000	1000
Density (kg/m3)	2 000	2 000	2000	2000
Part. diam. (μ m)	100	1.8 to 115 (8 classes)	200	1.7 to 92 (10 classes)
Composition	(% weight)			
volatile matter	23.9	28.7		
ash	31.8	18.3		
char	44.3	53.0		
loss on ignition			29.5	22.9
Inerts			70.5	77.2

where m_p denotes the mass of volatile matter remaining in the coal particle at time t. The values of k and E were obtained from the drop tube experiments described above. The precalciner is discretized on a 26 × 24 × 24 grid in x, y and z direction respectively, the former being the axis of the reactor. Figure 2 shows some views of the grid in a meridian plan and two cross-sections passing through air inlets. Note that for the second calculation, the grid was slightly modified to account for the relative orientation of tangential air inlets (the primary air inlet lies at 45° with respect to the secondary inlets) and for the eccentricity of the coal nozzle with respect to the burner axis.

Results of the First Simulation

Figures 3a to 3c depict some views of the velocity field and particle trajectories obtained from the first run.

The flowfield exhibits a general rotating motion due to the tangential air inlets. Particles are entrained in this motion, but coal and raw material do not mix together: coal particles remain in the center, whereas raw material is centrifugated and falls along the wall.

The temperature field (figure 3d) indicates that combustion takes place downstream of the reactor, with some "hot spots" in the upper part due to volatile combustion. This situation is attributed to rather slow combustion of char particles. The presence of raw material along the wall maintains the latter at a moderate temperature (around 750°C). This results in a very high temperature gradient in the outlet section. The contours of CO mass fraction (figure 3e) show relatively high concentration, around 10%, in the upstream zone, whereas output concentration falls down to approximately 2%.

Overall, general agreement is found between averaged quantities computed by FLUENT at the exit of the precalciner and the values obtained by experimental measurements and global simulation of the tower, as shown in table 4. Yet the FLUENT computation shows that in this particular case, only poor mixing is achieved in the precalciner, due to the delayed combustion of char particles and the centrifuging of raw material.

Results of Second Simulation

The results of the second run are displayed in figure 4. The velocity field is not very different from the previous one, except that coal nozzle eccentricity causes a small recirculation zone to occur in the primary air volute. Coal particles remain along the axis

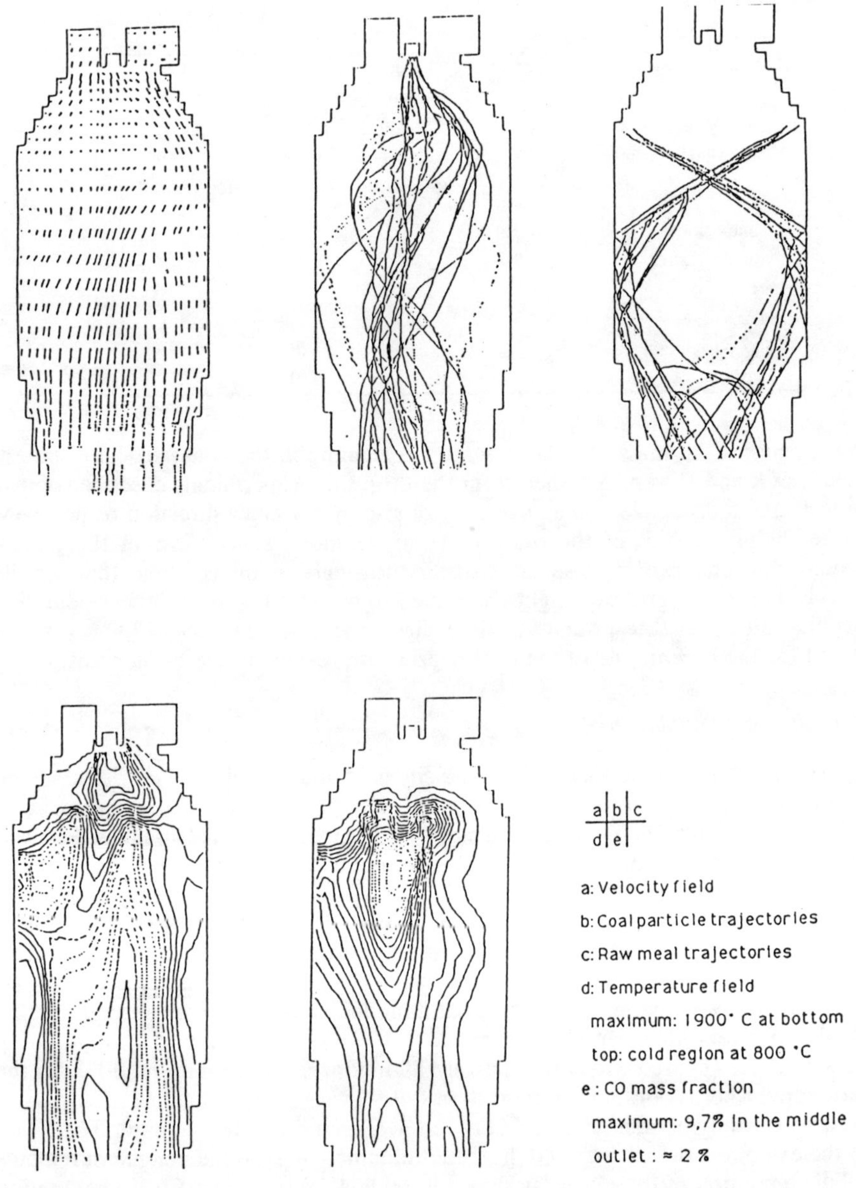

FIGURE 3 Results of the first 3D simulation.

TABLE IV
Comparison of precalciner output conditions: experimental, global model, 3D model

	T (K)	O2 (% vol)	CO2 (% vol)	CO (% vol)	CH4 (% vol)
Measurements	1390	2.2	21	6.8	10
Global model	1352	3.7	24	0.09	0
3D model (average)	1500	5	19	1.5	0.5

of the burner and are less dispersed than in the first case, due to their smaller diameter. On the contrary, raw material particles undergo a larger dispersion, spreading over the whole wall of the precalciner.

The temperature field is quite different from the previous one. Hot zones are always present near the coal nozzle, indicating combustion of volatile matter. But the maximum temperature is obtained in the middle of the reactor, owing to rapid combustion of small char particles. Downstream, combustion ceases due to lack of oxygen, and temperature decreases.

Contours of CO mass fraction exhibit high concentrations in the upper part of the burner. This is attributed to the high content of CO from pyrolysis products and to the beginning of char oxidation.

From a global point of view, this second case results in a much more homogeneous flowfield than the first one: two zones can clearly be identified, a "hot" one located in the conical part and a "cold" one downstream of the precalciner, where the temperature is driven by raw material decarbonation.

Comparison With Detailed Measurements

Measurements were taken through 5 holes in the precalciner wall at 3 different levels: one port was placed at the top on the conical part at the opposite side of the ignition air entrance; two ports were placed just below the tertiary air entrance; and two ports were placed at the bottom of the precalciner where the cylindrical part becomes rectangular. The holes at the two lower levels were positioned in order to be able to achieve orthogonal probe passages at the same level. This was done to get information about eventual asymmetries inside the combustion chamber.

The experiments were carried out by means of two types of water cooled probe: a suction pyrometer for the measurement of temperature profile and a gas sampling probe for gas analysis. The back end of that probe was equipped with a filter to retain dust, a cooling system to condense water included in the sampled gas stream and was then connected to gas analyzers. The following species have been measured: CO, CO_2, O_2 and unburnt gases given in equivalent CH_4.

The overall campaign lasted for three days and the passages on each hole were repeated twice. The process parameters in real industrial conditions are always very fluctuating. In the precalciner, measurement conditions were tough, mainly because of the high dust concentration which introduced complications in the sampling system. These two facts explain the wide spread in the experimental data collected.
However, some interesting features of the process inside the precalciner can be drawn from the experiments. Table 5 shows the mean values obtained at each level. In that table some preliminary measurements undertaken at an another period are also given. This set of data corresponds to a different production level.

One can see that there are some difficulties in obtaining complete combustion inside the combustion chamber. This is based on the following observations: in the first set

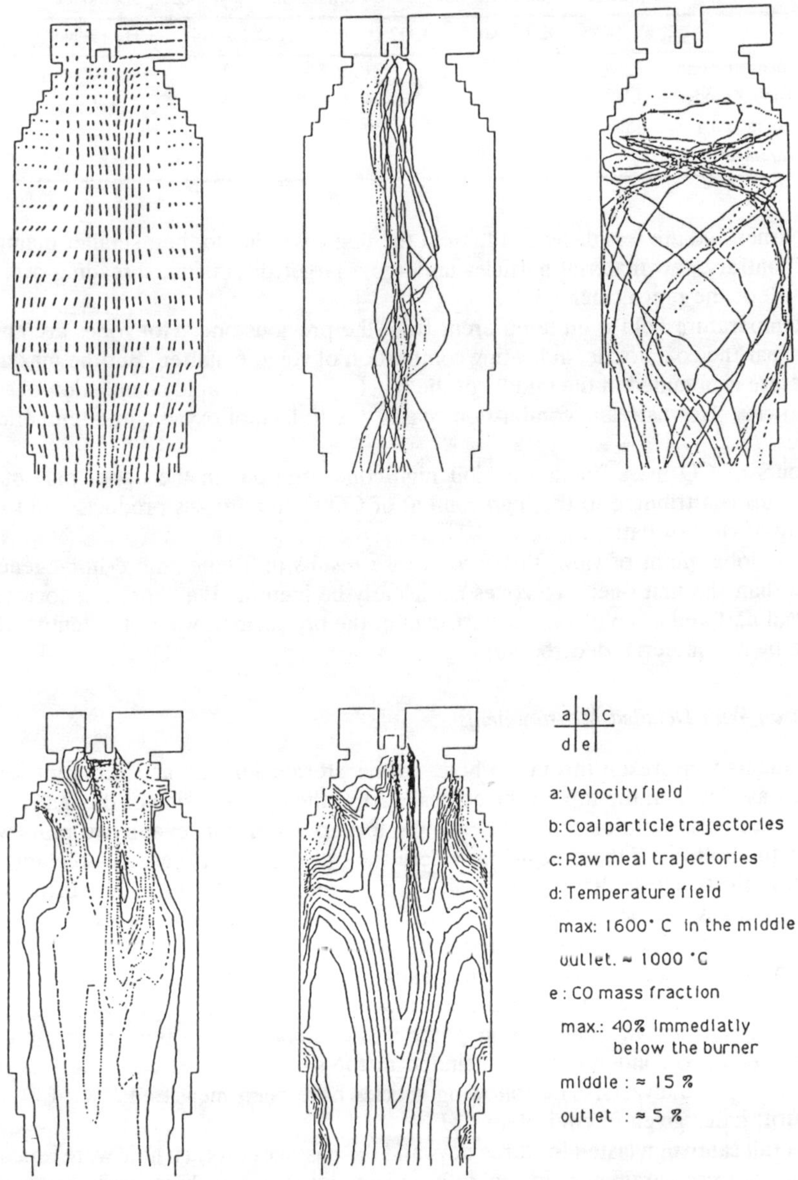

FIGURE 4 Results of the second 3D simulation.

TABLE V
Mean values (Measurements and 3D model)

		october 89	march 90	model
Production	(T/d)	2230	2090	2230
Thermal input	(MW)	53	50	53
Tertiary air	(Nm3/h)	44043	41278	44043
Temperature	(°C)	820	≈820	820
Top level				
T	(°C)	1200	—	1500
Intermediate level				
T	(°C)	1200	1300	1100
O2	(%vol)	5	10	≈0 (axis)
CO2	(%vol)	25	15	25
CO	(%vol)	10	4	12
CH4	(%vol)	9	1	17
Bottom level				
T	(°C)	1100	1200	900
O2	(%vol)	1	5	1
CO2	(%vol)	35	25	35
CO	(%vol)	9	2	5
CH4	(%vol)	9	1	2

of data one remarks on a high level of CO at the bottom of the furnace, and in both sets of data the temperature in the lower part remains high, sometimes higher than in the upper part, indicating that the combustion is still progressing after the outlet of the precalciner. Due to residence time effects, this is amplified when the production level is raised. When more fuel is used, one can see that the level of unburnt gases rises at the precalciner outlet.

Figure 5 shows a comparison between experimental (dots) and calculated (solid lines) radial profiles for a path placed below the two tangential air entrances.

Qualitatively, the model is coherent with the measurements. The radial profile shows the same tendencies and no asymmetries can be seen. Temperature profiles have the same shape: the maximum is on the axis and temperature drops near the wall. The CO and CH_4 concentration is highest at the top level where the coal devolatilizes while there is still a high concentration at the outlet. The radial oxygen profile also indicates the same shape: the concentration is low in the middle, rises when moving to the wall and then drops again very close to the wall. The CO_2 profile is the opposite (reverse of the oxygen profile). This profile is observed in both the model and the experiments. It can be explained by the combination of combustion, which takes place along the axis, and decarbonation, which takes place at some distance from the wall with a rise in the concentration of CO_2 therein.

The detailed modelling explains the behaviour of the processes inside the precalciner. At first, one can confirm by the calculation that the combustion and decarbonation are effectively separated. A second point is that with the modelling, the combustion problems inside the chamber are also confirmed.

When one looks at the absolute values, the comparison is not good, as can be seen in table 5. This can easily be explained when one looks at the rate of fuel burnt. During the measurement campaign the rate of production was below its nominal value for which the calculation was done. Although the absolute values given by the model are in the

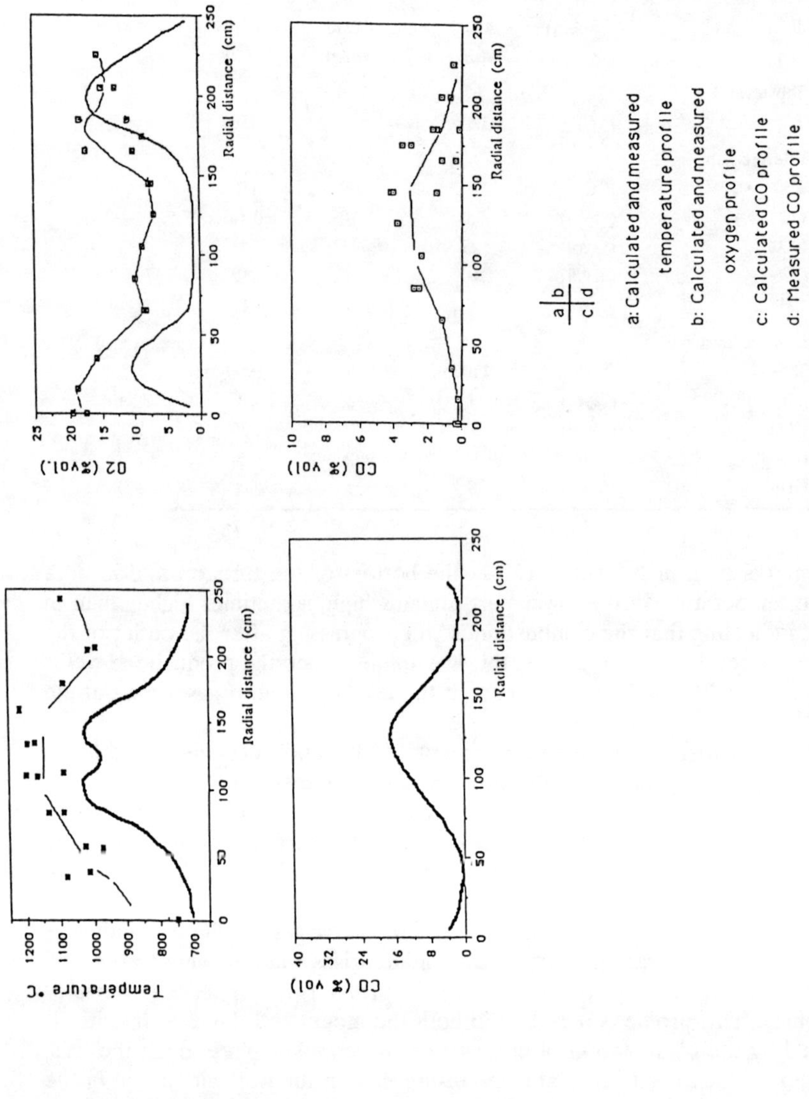

FIGURE 5 Comparison between the second 3D simulation and measurements.

order of the preliminary measurements, unfortunately no profiles have been measured during that period.

CONCLUSION

Useful information for optimizing the combustion in the precalciner was derived from both modelling approaches. The global modelling gave interesting information on the operating conditions of the plant and produced guidelines for optimisation. The detailed calculation of the flow inside the combustion chamber should be interesting for optimizing the precalciner itself. With such a tool, it becomes possible to evaluate the impact of geometrical modifications on the efficiency of the reactor. The analysis of the fields of temperature and gaseous species inside the precalciner calculated by the model is also useful for understanding the phenomena in the precalciner. However, this software is very time-consuming and cannot be systematically used for this application on currently available workstations.

REFERENCES

Dryer, F. L. and Glassman, I (1972). High Temperature Oxidation of CO and CH_4. 14th symp. (int.) on Combustion, pp 987-1001

Field, M. A., Gill P. W., Morgan B. B. and Hawksley P. G. W. (1967) Combustion of Pulverized Coal, Combustion and Flame, 14, pp 237-248

Gadiou R., and Prado G. (1987). Rapport d'étude de la combustion du charbon LEXOS 200 á 250μm. Private communication.

Wild, P. N., Boysan, F., Swithenbank, J. and Lu, X. (1987). 3-Dimensional gas turbine combustion modelling. AGARD CP 422, pp 27.1-27.14

IGCC Plants As Alternatives to Pulverized Coal Plants in Western Countries

P.J. DECHAMPS and PH. MATHIEU *Department of Nuclear Engineering and Power Plants University of Liège, Belgium*

Abstract — After the Chernobyl accident in 1986, most western countries have slowed down, suspended or even sometimes stopped their nuclear programme. In many cases, natural gas has been regarded as the best alternative to nuclear fuel and many countries have opted for combined cycle plants. Because combined cycle plants are too dependent on natural gas, the generation mix planners of Belgium intend to balance the share of coal and gas in fuel supplies. Combined cycle plants featuring coal gasification (IGCC) appear as environmentally sound alternatives to pulverized coal plants. In this paper, IGCC plant performances for four commercially available gasification processes (Texaco-Shell-Dow-British Gas/Lurgi) are compared when associated with similar CC plants.

Key Words: Coal gasification, Integration, IGCC, phased construction.

NOMENCLATURE AND ABBREVIATIONS

BGL	British Gas/Lurgi
CA/S	Limestone over Sulphur Weight Ratio
CC	Combined Cycle
GT	Gas Turbine
HP	High Pressure
HRSG	Heat Recovery Steam Generator
IGCC	Integrated Coal Gasification in Combined Cycle
kWh	KiloWatt.Hour (3600000 J)
LHV	Low Heating Value (MJ/kg)
LP	Low Pressure
$\dot{m}$	Mass Flow (kg/s)
p	Pressure (bar)
P	Power Output (MW)
PFBC	Pressurized Fluidized Bed
PC	Pulverised Coal
ST	Steam Turbine
t/d	Ton per Day (of dry coal)
TET	Turbine Entry Temperature (°C)

INTRODUCTION

In its last electricity equipment plan for the decade 1988–98, the Belgian Committee for Electricity Management predicted the installation of 4000 MW of capacity in the form of GT, CC and PC plants. One PC plant rated at 600 MW has already been decided but from now on the IGCC alternative is open. Indeed, the Belgian operators want to ensure a good security of their fuel supply and a good diversification of fuel sources. They don't want to be too dependent on natural gas as they previously were on oil. In that respect, the contributions of gas and coal should be shared properly and the operators should be able to switch from natural gas to coal if gas happened to be lacking or too expensive.

More generally speaking, a safe fuel supply and environmental requirements are bringing new technology options for coal-based power plants into the competition. For the time being, natural gas is abundant and rather cheap. Its demand is however getting higher and higher with the danger that its price might substantially increase.

As the kWh generated by CC plants is very sensitive to fuel price, a return to coal is predicted in the near future but with clean and efficient combustion technologies. Now decisionmakers would like that an integrated process be available taking into account technical performances, economy, impacts on human health and on local and global environment and last, but not least public acceptance. Unfortunately these parameters cannot be compacted in a single index. Hence the choice of technology will be based on tradeoffs among efficiency, emission levels and costs.

In this paper we don't take the fluidized beds into account in a comparison of technologies. Their chances are indeed decreasing as IGCC technology is developing and going into industrial maturity. Even PFBCs are generally considered as limited in efficiency by the upper temperature, which has to remain under the ash softening temperature, and also in their ability to comply with the most stringent environmental requirements.

We don't however claim that this technology has no future, especially when hot gas clean up will be efficiently operating. Moreover hybrid systems combining IGCC and PFBC units are already now thought of as possible improvements for a longer term future. In this paper, we decide to stay in the framework of the Belgian generation mix in order to compare the IGCC technique, generally considered as the future coal-based technology complying with stringent emission regulations, and PC plants equipped with $deNO_x$ and $deSO_x$ devices. To be fair, we have to mention that advances in clean combustion (by multistage combustion) and in efficiency when operating with supercritical or even hypercritical cycles are in progress for PC plants, but so far are not broadly implemented.

SELECTED GASIFIERS

Four gasification technologies have been selected for this study with regard to their commercial availability and proven experience. They represent what utility planners are faced with when trying to select a gasifier for an IGCC project. They hopefully fall within basically different technological options, so that they may exhibit the same overall performance level with different drawbacks and advantages. We have restrained the study to oxygen fed gasifiers, providing medium LHV gases.

The Texaco Process

The Texaco gasification process is based on a slurry fed entrained flow gasifier (EPRI, 1988) (Gluckman and Spencer,1987). This latter delivers a high temperature (≈1400°C) crude syngas to the gas clean-up units. The gas cooling produces high pressure saturated steam. When treated and desulphurised, the syngas is usually saturated before it goes to the gas turbines in order to meet the most severe NO_x legislations.

The Shell Process

The Shell gasification process features a dry fed entrained flow gasifier (EPRI, 1989). The water required by the gasification reactions is brought to the reaction zone as saturated steam. Among the considered processes, Shell has the unique feature of producing superheated steam (540°C) in the syngas cooling and treatment unit.

The Dow Process

Like the Texaco process, the Dow gasification system is based on a slurry fed entrained flow gasifier (EPRI, 1989). Unlike the Texaco unit, the gasifier comprises two stages. The first one receives all of the oxygen together with 80% of the slurry, whilst the remaining

TABLE I
SELECTED GASIFICATION TECHNOLOGIES

	TEXACO	SHELL	DOW	BGL
Gasifier type	Entrained bed	Entrained bed	Entrained 2 stages	Fixed bed
Feeding method	Slurry	Dry coal	Slurry	Dry coal
Reactants (kg per kg of coal)				
Oxygen	0.87	0.80	0.78	0.54
Liquid water	0.5	–	1.0	–
Steam	–	0.13	–	0.31
Carbon conversion rate	>99	99.8	>99	>99.5
Averaged syngas properties (possibly moisturized or saturated), composition, vol%:				
H_2	25	33	38	27
CO	35	61	32	56
CH_4	<0.1	<0.1	<0.1	6
CO_2	6.5	1	22	1.5
H_2O	33	0.5	6	6
N_2, misc.	<0.5	4.5	2	3.5
LHV (MJ/kg)	8.16	12.2	8.54	13.75
Adiabatic combustion temperature, K	2315	2620	2380	2520
Syngas delivery temperature (°C)	1400	1400	1000	400
Fraction of coal LHV contained in cooled and treated syngas (%)	76	78	74	90
Fraction of coal LHV available as sensible heat in syngas (%)	8	16.8	18.2	2
Fraction of coal LHV required by the process as heat input (%)	11.5	2.4	4.7	3.5

20% goes to the second stage. This peculiar arrangement helps in recovering some of the sensible heat of the syngas before it leaves the gasifier. Hence, the syngas leaves the gasifier at 1000°C and its cooling produces saturated high pressure steam.

The British Gas/Lurgi Process

The BGL process is basically different from the three above described techniques because it is based on a fixed bed gasifier (sometimes called a moving bed as well) (EPRI, 1988) (Herster, 1990). The fixed bed intrinsically behaves like a counterflow heat exchanger and delivers a low temperature syngas (approximately 400°C). This process produces the highest heating value syngas (13 to 14 MJ/kg) with a considerable methane content (5 to 6% in volume).

Table I summarises the main characteristics of the four considered processes. Part two of this table gathers the information that is relevant for the assessment of the integration required by each process. Let us point out that the Texaco and Dow processes provide a syngas with an LHV around 8 MJ/kg while the Shell and BGL processes generate a richer syngas with an LHV around 12 to 13 MJ/kg.

IGCC PERFORMANCE ASSESSMENT

Model Assumptions and Design Criteria

In Belgium, two CC plants rated at 450 MW with a 2-2-1 arrangement will be commissioned in 1993 and successive ones are currently in the decision process. Therefore we have selected the components of this CC type as reference for our calculations and comparisons.

The gas turbine. All IGCC plants are supposed to be equipped with 150 MW-class gas turbines adapted so that they are capable of operating with the different syngases at their design pressure ratio and TET.

The ambient conditions for these gas turbines are 288 K and 1 bar, with typical pressure drops of 10 mbar at the inlet and 25 mbar at outlet. It is assumed that there is no capacity matching problem between the gasifiers and the gas turbines. The syngas might hence be produced in excess. When IGCC plants are part of an electricity generation mix, their best performances are obtained when the CC unit operates in load following mode and the gasification unit in base load. In a previous paper (Mathieu and Pilidis, 1991), we proposed to design integrated sites comprising of GTs or aeroderivative engines operating in open cycle and using the excess syngas to cover the peak load.

Heat recovery steam generators. HRSGs are of a conventional dual pressure design with assisted circulation.

Basic assumptions typically include:

— HP drum pressure = 72 bar
— LP drum pressure = 6.25 bar
— pinch point temperature difference = 10°C
— HP superheat temperature = 515°C
— BP superheat temperature = 200°C (possibly more if the temperature difference with the gas stream is more than 40°C)
— deaerator pressure = 3 bar (fitted with a feedwater heater)
— the stack temperature has to be higher than 100°C.

Rankine cycle. The steam turbines are state-of-the-art machines with classical isentropic efficiencies. They expand the steam down to the condenser pressure (50 mbar).

Discussion of the Results

Introduction. In order to help with the performance comparison, all the results are presented for an hypothetical plant comprising of only one gas turbine, one HRSG and one steam turbine set.

The Texaco based plant. Figure 1 schematically presents the CC part of the Texaco based IGCC plant and illustrates the main connections with the gasification side (links 1 to 5).

Significant characteristics include a gross output equal to 292.5 MW in which the steam turbine contributes 37%. The net output of 259.5 MW is achieved with an efficiency of 42.4% (based on coal LHV).

Singular features are the very high capacity required at the water treatment plant by both the slurry preparation (link 2: $\dot{m} = 9.1$ kg/s) and the syngas saturation (link 5: $\dot{m} = 19.5$ kg/s in saturation), and the bypass of the HP evaporator section by more than 50% of the HP flow.

The Shell based plant. Figure 2 illustrates the CC part of the Shell based IGCC plant and shows its main connections with the gasification side (links 1 to 4).

Significant features include a gross output of 284.9 MW in which the steam turbine contributes 45.4%. The net output of 248.5 MW is achieved with an efficiency of 42.15% (based on coal LHV).

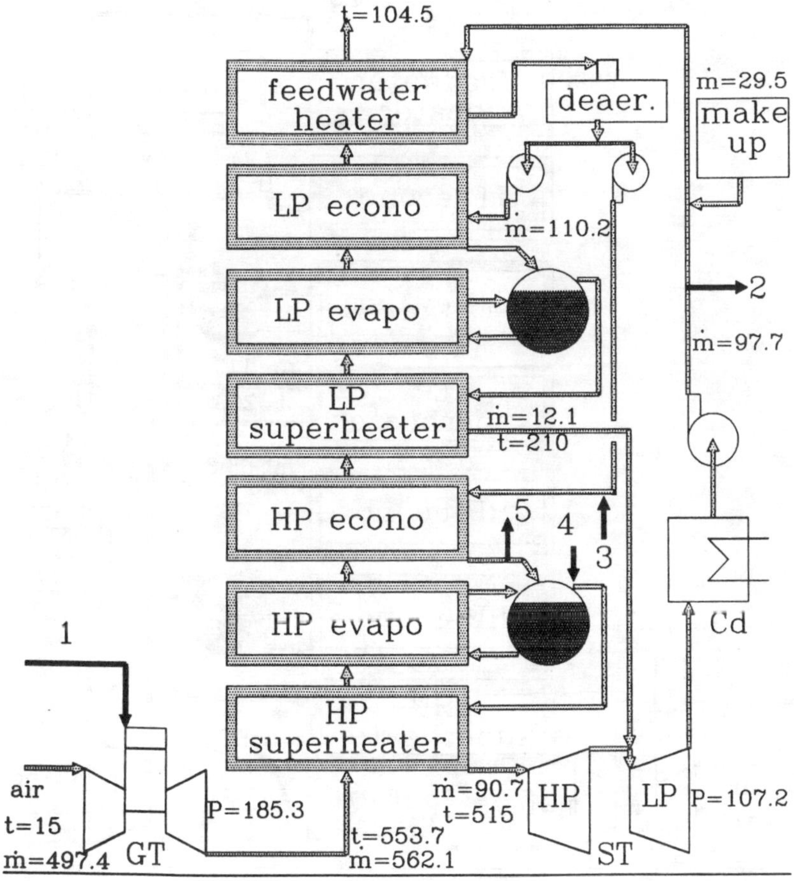

1 : syngas ṁ=64.7 LHV=8.15 MJ/kg
2 : water to slurry preparation ṁ=9.1
3 : water from saturator ṁ=21.5 t=200
4 : saturated steam from gasifier ṁ=46.7 p=75
5 : water to gasifier and saturator ṁ=87.7 t=286.5

FIGURE 1 Texaco-based IGCC plant CC arrangement.

The most singular characteristic is the production of HP superheated steam (link 4: $\dot{m}$ = 40.2 kg/s) by the gasification process, starting with feedwater ($\dot{m}$ = 41.2 kg/s). This allows for a quite conventional HRSG design as the superheated steam directly goes to the inlet of the HP steam turbine.

The Dow based plant. The CC part of the Dow based IGCC plant is schematically presented in figure 3, together with its principal links with the corresponding gasification installations (links 1 to 6).

The gross output amounts to 306.6 MW in which the steam turbine contributes 43.7%, whilst the net output of 289.8 MW is achieved at 42.2% efficiency (based on coal LHV).

Singular features are the high water treatment capacity required by the slurry preparation (links 3 and 4: $\dot{m}$ = 10 kg/s) and the bypass of the HP evaporator by 63% of the HP steam

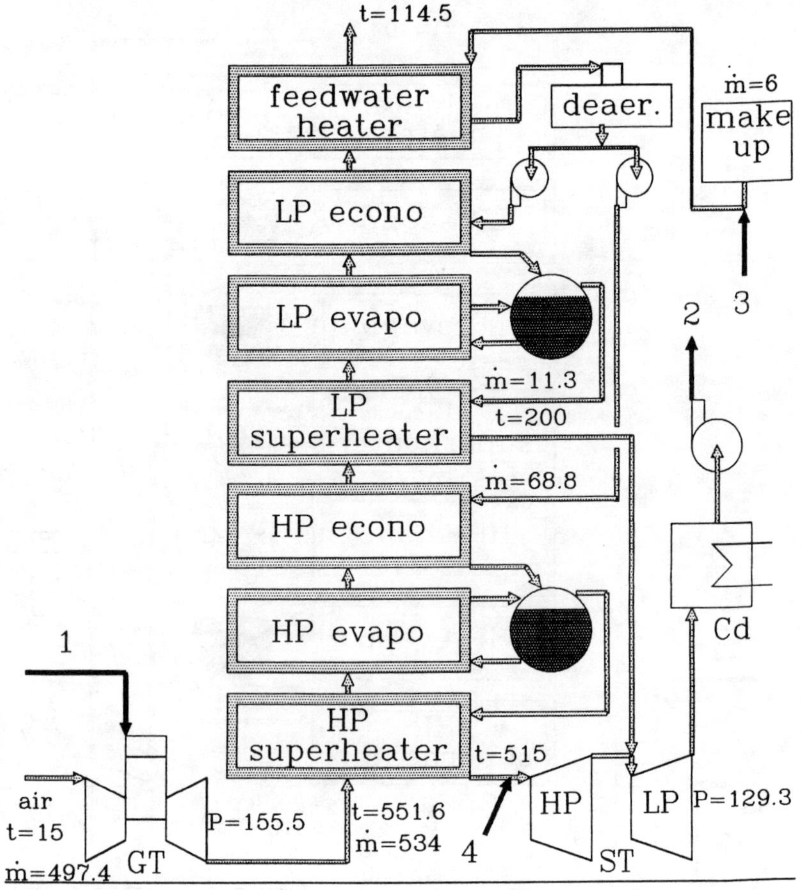

1 : syngas ṁ=36.6 LHV=12.8 MJ/kg
2 : water to syngas treatment and to gasifier ṁ=115.2 t=30.7
3 : water from syngas treatment ṁ=74 t=45
4 : steam from gasifier ṁ=40.2 t=540 p=72

FIGURE 2 Shell-based IGCC plant CC arrangement.

mass flow (links 5 and 6: ṁ = 77 kg/s). It is very remarkable that all the LP saturated steam is used by the process, so that there is no need for an LP superheater (link 2: ṁ = 5.6 kg/s).

The British Gas/Lurgi based plant. The CC part of the BGL based IGCC plant is described by figure 4, with the connections to and from the gasification plant (links 1 to 5).

The gross output amounts to 236.8 MW, with a 34.2% contribution from the steam turbine, as it would be in a genuine combined cycle. The net output of 213.2 MW corresponds to an efficiency of 42.0% (coal LHV).

The syngas treatment section acts like a feedwater heater (for 78.7% of the feedwater mass flow) while the syngas heater is fed by HP warm water picked up at connection 4 and sent back at connection 5 with a very low mass flow. BGL is obviously the less integrated arrangement among the four ones examined here, while the Dow is the most integrated one.

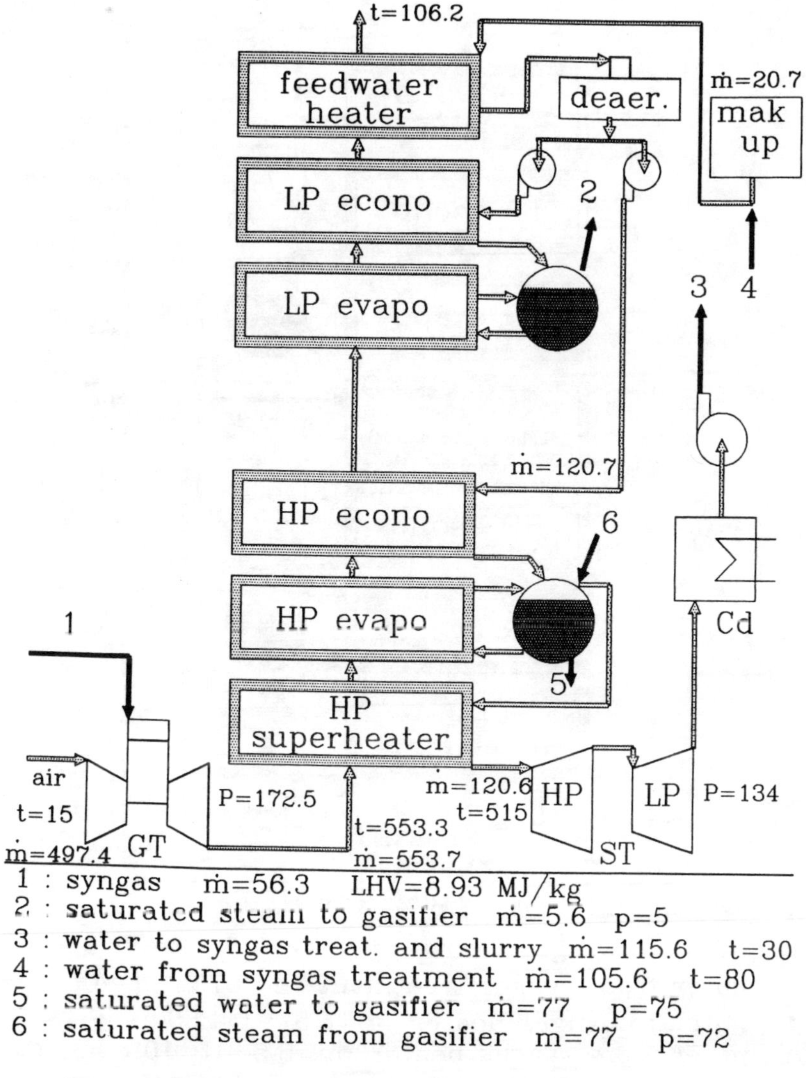

FIGURE 3 Dow-based IGCC plant CC arrangement.

Comparison and Discussion

Figure 5 shows how the energy cascades between the different units of the four considered IGCC power plants. All figures are quoted according to the reference [coal LHV * coal mass flow = 100] and represent energy flows.

Table II gathers the main design point characteristics for the same four power plants, but in absolute terms (and again, for one gas turbine).

From these two inserts, it can be observed that:

— the four considered gasification technologies achieve the same overall net efficiency, i.e., around 42%;
— the share of the steam turbine power ranges from 34% (as ingenuine CC plants) to 45% for BGL and Shell, respectively;

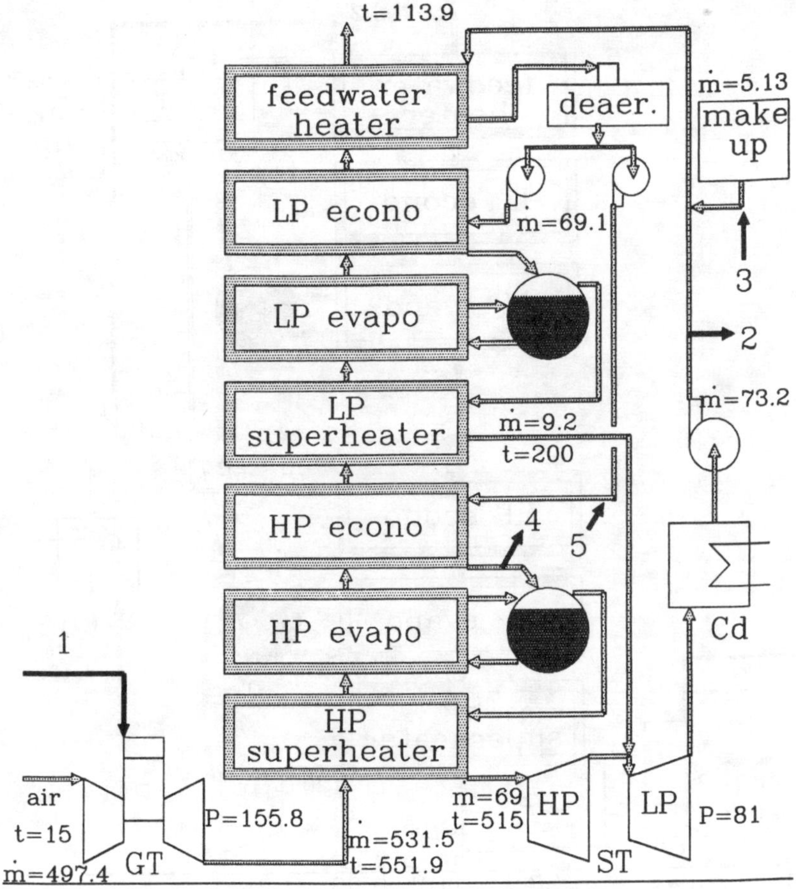

1 : syngas ṁ=34.1 LHV=13.75 MJ/kg
2 : water to syngas treatment ṁ=57.6 t=30.7
3 : water from syngas treatment ṁ=57.6 t=49
4 : water to syngas heater ṁ=6.5 t=286.5
5 : water from syngas heater ṁ=6.5 t=108

FIGURE 4 BGL-based IGCC plant CC arrangement.

- the net power outputs and coal inputs vary from 213 to 290 MW and 1570 to 2040 t/d for BGL and for Dow, respectively, basically depending on the syngas LHV and on the resulting move of the gas turbine operating point in the compressor map;
- the auxiliary requirements, chiefly for the air separation plant, are quite different, depending on the gasifiers' needs, and going from 23 to 36 MW for BGL and Dow respectively;
- cold gas efficiencies and integration levels are quite different, especially between moving bed technology and entrained beds. Based on energy flows, the BGL is the less integrated process, while the Dow is the most integrated one.

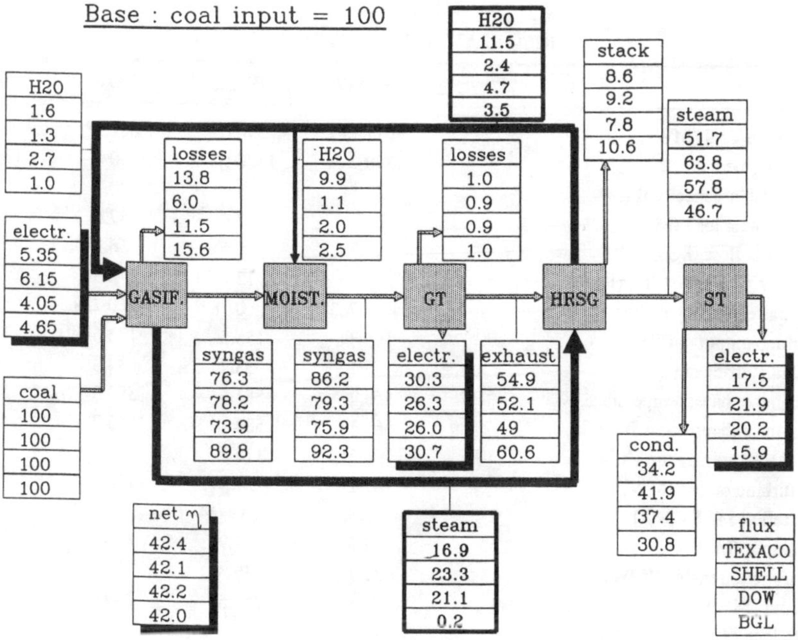

FIGURE 5 The four IGCC plants: summary of energy flows.

ENVIRONMENTAL CONSIDERATIONS

Introduction

If coal has to be used for power generation, the pressure of environmental considerations and the stringency of legislation could make the difference, so that the respective impacts of IGCCs and PCs must be addressed. Emissions of SO_2 and NO_x are responsible for acid rain and emissions of CO_2 are partly responsible for the global warming of the earth's atmosphere. If the scientists agree on the effects of the former emissions and on global warming, there is no consensus on the effects of this warming. However some western countries have already decided to limit the CO_2 emissions growth and even opted for a zero CO_2 growth from now to the year 2000. Although the debates are still in process, technicians have to anticipate the regulations and promote clean and efficient fossil fuel based technologies.

Emissions and Air Toxics

CO_2. For a given fuel, CO_2 emissions only depend on the efficiency of the process. Because IGCCs have an efficiency advantage (41 to 43%) over PCs (38%), CO_2 emissions are less (typically 10%) with IGCCs. Moreover, CO_2 emissions will still be reduced further as advanced GTs are used, the efficiency typically going up to 46% with advanced technologies.

SO_2. Current state-of-the-art technology for deSO_x devices in PC plants can achieve a 95% desulphurization (Conley, 1990), whilst more than 99% is achievable with IGCC technology (Curtis and Isles, 1990). In PCs, the increase of the sulphur retention from the

TABLE II
IGCC PLANT CHARACTERISTICS

	TEXACO	SHELL	DOW	BGL
Heat input as coal (MW)	611.97	590.65	662.13	508.11
Coal requirements (t/d)	1890	1820	2040	1570
Make-up water capacity (kg/s)	29.5	6	20.7	5.13
Condenser heat load (MW)	209.4	247.4	247.6	156.6
Syngas mass flow (kg/s)	64.7	36.6	56.3	34.1
Treated syngas LHV (MJ/kg)	8.16	12.2	8.54	13.75
Cooled gas efficiency (0–1)	0.763	0.782	0.739	0.898
Gas turbine output (MW)	185.3	155.5	172.5	155.8
Gas turbine efficiency (%)	35.1	33.2	34.3	33.2
Gas turbine exhaust temperature, K	553.7	551.6	553.3	551.9
Gas turbine exhaust mass flow (kg/s)	562.1	534	553.7	531.5
HRSG stack temperature (°C)	104.5	114.5	106.2	113.9
Steam turbine output (MW)	107.2	129.3	134	81
Output ratio [ST/(ST + GT)]	0.366	0.454	0.437	0.342
Gross output (MW)	292.5	284.9	306.6	236.8
Auxiliary requirements (MW)	32.7	36.3	26.8	23.6
Net output (MW)	259.5	248.5	279.8	213.2
Net efficiency (%)	42.4	42.1	42.2	42.0

classical 90 to 95% requires very efficient scrubbers and produces a global efficiency reduction and hence an increase in CO_2 emissions. Similarly, obtaining 95% sulphur retention with PFBCs requires a much higher CA/S ratio than for 90% and produces more solid wastes.

NO_x. Although synthesis gases have heating values far less than that of natural gas, adiabatic combustion temperatures are usually 50 to 200°C higher. Therefore, if nothing is done about it, the rate of thermal NO_x formation is dramatically higher. The mixing of syngas with N_2 from the air separation plant or better with steam can bring the adiabatic combustion temperature to the level experienced with natural gas. It is even better when mixing syngas with CO_2. NO_x emissions are then comparable to those of current gas turbine standards (20 ppm), well below those achievable with PC plants (Alpert and Gluckman, 1986) (Mueller and Karg, 1987). For the latter, the NO_x emissions are typically 5 times higher and to reduce them, a Selective Catalytic Reactor is needed in the flue gases. This decreases the efficiency and hence increases CO_2 emissions. Moreover, NO_x emissions from IGCC could be reduced further when advanced GTs with dry low NO_x burners come into operation. Also when mixing syngas with N_2 or steam, the LHV goes down and when approaching about 12% of the natural gas LHV, NO_x emissions come down to 10 ppm (which is the target of the Clean Air Act)

Cooling Water

Cooling water requirements for IGCC are less than half those of PCs because the steam cycle contributes less than 50% of the overall IGCC output. This is a considerable advantage in places where rivers are already highly thermally loaded.

Wastes and By-Products

Ash versus slag. The slag produced by the gasifiers (a vitreous material) is produced in a smaller amount (typically 50% less) and is easier to handle than the fly-ash produced by PCs. The slag may even be used in road works and buildings. Slag is more easily stored on, site because, unlike ash, it is not leachable.

Gypsum versus elemental sulphur. The desulphurization in PCs has gypsum as a waste. Gypsum has no commercial value. Desulphurization technologies used in IGCCs produce pure elemental sulphur, which is less in volume and weight and has a non-negligible commercial value. The revenue of selling this sulphur must even be taken into account when evaluating the economic viability of IGCCs.

Costs

Investment costs are currently 100 to 200 $/kW higher for IGCC than PC but typical screening curves show that the total cost in $/kW comes under that of a PC for a high capacity factor, hence in base load.

Environmental Conclusions

Figure 6 presents the comparison of environmental performances for a PC and an IGCC plant of the same capacity. The above discussion showed that the environmental impacts for the four considered gasification technologies don't considerably differ. For this reason, this figure presents average values for IGCCs in order to illustrate their huge advantage over PCs. The heat rejection would be the only variable parameter among the gasification technologies as they have different contributions of the steam cycle into the overall power output.

CONCLUSIONS

In this paper, we show that, despite the differences among the various gasifiers and more particularly with respect to their degree of integration into the CC plant, the resulting IGCC plants using Texaco, Shell, Dow and BGL processes exhibit a fairly similar global efficiency, amounting to 42%.

This efficiency, if slightly higher than that of a classical PC plant, is however at the same level of advanced PC plants operating at supercritical pressures and high temperatures thanks to new materials.

So, in a first stage, the decision-maker has to make his choice between a PC and an IGCC plant. The two important points are that the IGCC investment cost is slightly higher but its emissions (air toxics, SO_2, NO_x, CO_2, particulates, solid wastes and heat rejection) are all smaller.

In a second stage, if the decision-maker has opted for the IGCC technique, its choice of the gasification unit to be added or associated with a given CC unit, can be also made on other criteria than performance, i.e., on economy or any other consideration.

As pointed out in this study, the different IGCC plants strongly differ in their degree of integration and if the plant had to be build in phases, a careful analysis of mass and energy fluxes should be carried out. This is the subject of a future publication (Dechamps and Mathieu, 1992).

This makes the decision-maker free with respect to environmental impacts, the costs and any other considerations.

In any case, the IGCC technology exhibits huge environmental advantages over classical PC technology. Stringent legislation could make IGCC the only acceptable solution when coal has to be used for power generation. Even if the cost per installed kW is a little bit higher for IGCC than for PC plants, cleanness has to be paid for, this cost going expected

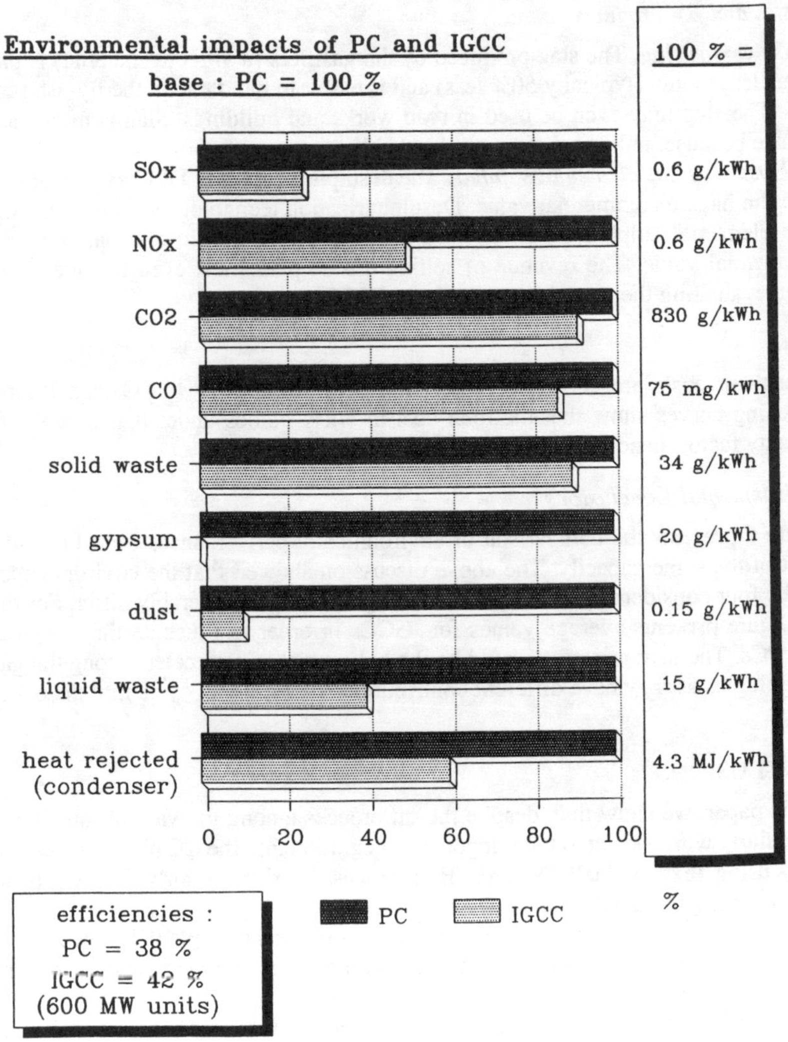

FIGURE 6 Environmental impacts: comparison of PC and IGCC plants.

to decrease as the technology comes to maturity and the number of installed units increases. On the other hand, it is generally admitted from now on that in base load, the cost of an IGCC is below the cost of a PC plant. As the Netherlands, Germany and the USA are already on the way of IGCC, we believe that it is the coal technology for the future.

REFERENCES

Alpert, S.B., and Gluckman, M.J. (1986), Coal Gasification Systems for Power Generations, *Ann. Rev. Energy*, 1986, **11**, pp. 315–55.

Conley B. (1990), Advanced Limestone Scrubbers Demonstration, *Modern Power Systems*, April 1990, pp. 61–65.

Curtis, R.H. and Isles, W.J.H (1990), Control of Flue Gas Desulphurisation in Coal Burning Power Plants. Paper presented at the International Conference on Coal and Power Technology '90, May 1990, Amsterdam.

Dechamps and Mathieu (1992), Phasing the Construction of an IGCC Plant for Fuel Flexibility, ASME paper 92-GT-144, presented at the ASME Turbo-Expo Conference, Koln, Germany.

EPRI (1988), Evaluation of a Texaco Gasification-Combined-Cycle Plant With KWU Gas Turbines, Report GS-6160.

EPRI (1989), Baltimore Gas and Electric Company's Study of a Shell-Based IGCC Power Plant, Report GS-6283.

EPRI (1989), Evaluation of a Dow-Based Gasification-Combined Cycle Plant Using Low-Rank Coals, Report GS-6318.

EPRI (1988), A 180-MWe British Gas/Lurgi-Based IGCC Power Plant, Report AP-6011.

EPRI (1984), Economics of Phased Gasification-Combined-Cycle Plants: Utility Results, Report AP-5466-SR.

Gluckman, M.J., and Spencer, D.F. (1987), A Reconciliation of the Cool Water IGCC Plant Performance and Cost Data with Equivalent Projections for a Mature 360 MW Commercial IGCC Facility, Cool Water Coal Gasification Conference, 1987.

Hester, J.C. (1990), Florida Site for Air Blown Clean Coal IGCC Project, *Modern Power Systems*, May 1990, pp. 32–38.

Mathieu, Ph., and Pilidis, P. (1991), The Use of Gaseous Fuels Aero-Derivative Gas Turbine Engines, ASME paper 91-GT-44, presented at the International Gas Turbine and Aeroengine Congress and Exposition, Orlando, June 3–6, 1991.

Mueller, R., and Karg, J. (1987), Assessment of Coal Gasification Systems for Combined Cycle Power Plants, ASME paper 87-JPGC-PWr-17, presented at the ASME/JEEE Power Generation Conference, Florida, Oct. 1987.

2.

PULVERIZED COAL COMBUSTION

Fossil Fuel Utilisation

K. HEIN *Delft University of Technology, The Netherlands*

Energy is not only the base of life, but—in its usable forms for the customers—the driving force behind industrial development, the major source of welfare and, hence, the backbone of our present state of living. These impacts of energy are clearly mirrored in the world's energy consumption (e.g. IZE, 1990a). The majority of the primary energy is utilized in highly industrialized regions (fig. 1, left) and there are distinct differences in the state of development and/or the degree of secondary energy application, which is expressed as the specific per capita utilization (fig. 1, right).

The vast majority of the present days' energy is based on fossil fuels (fig. 2, left), of which oil, the most versatile source, plays the major role, followed by natural gas, which has been growing in demand since the 60's (e.g. Coal Resources and, IZE 1990b) (fig. 3).

Coal, being the most abundant fossil source (fig. 2, right), has been the most important source of primary energy since the start of industrialization and has gained renewed interest during the last 15 years due to the two major oil crises in the early seventies. Following the development of coal utilization in industrialized countries, as in the Federal Republic of Germany (fig. 4), it can be noticed that the traditional markets, like the application for domestic purposes and the use in the steel industry, lost importance in absolute and relative terms. In contrast, the utilization of coal for direct heat generation in industry—although having a small share—increased. However, the major coal application was, and still is, electricity generation. Because of the fact that electricity production has been related to primary energy consumption with increasing tendency for almost 30 years now (e.g. Klätte 1986) (fig. 5), its future production is assumed to be strongly linked with the development of coal utilization worldwide.

The utilization of coal for electricity generation is a well developed technique which provides a most versatile form of secondary energy available just about anywhere, at anytime, to the customer. Electricity is produced in a reliable, safe, and economical way, which is generally well accepted by the public.

The major steps of development of fossil fuel utilization for the production of electricity are shown in fig. 6. Whilst during the earlier stages the development of the steam generation process, the fuel combustion, and the upscaling to more economical larger units were the main areas of concern, cost efficiency, availability, and the utilization of inferior fuel qualities took on a high importance during the fifties and sixties.

At present, utility boilers are under operation in single unit sizes up to a maximum of 600 MW_e for brown coal and 700–1300 MW_e for bituminous coals. The high technical standard is mirrored by the fact that availabilities of base load stations are typically above 90%, with overall efficiencies close to 40% for bituminous coal and 37% for low calorie fuels, such as brown coal.

For about one decade now, the installation of emission control processes in compliance with regulations in many countries has formed a major task for plant operators and equipment manufacturers. For emission control, flue gas treatment processes are predominantly installed (e.g. Hildebrand, 1990). Also alterations or additions to the combustion process are used for SO_2– and NO_x– reduction (e.g. Schwirten, *et al.* 1988, Hein and Kallmeyer, 1989). Apart from these techniques being installed mainly in retrofit situations, new combustion methods with incorporated emission control, such as fluidized bed systems, are well-developed and already available for large scale steam production (e.g. Plass 1991).

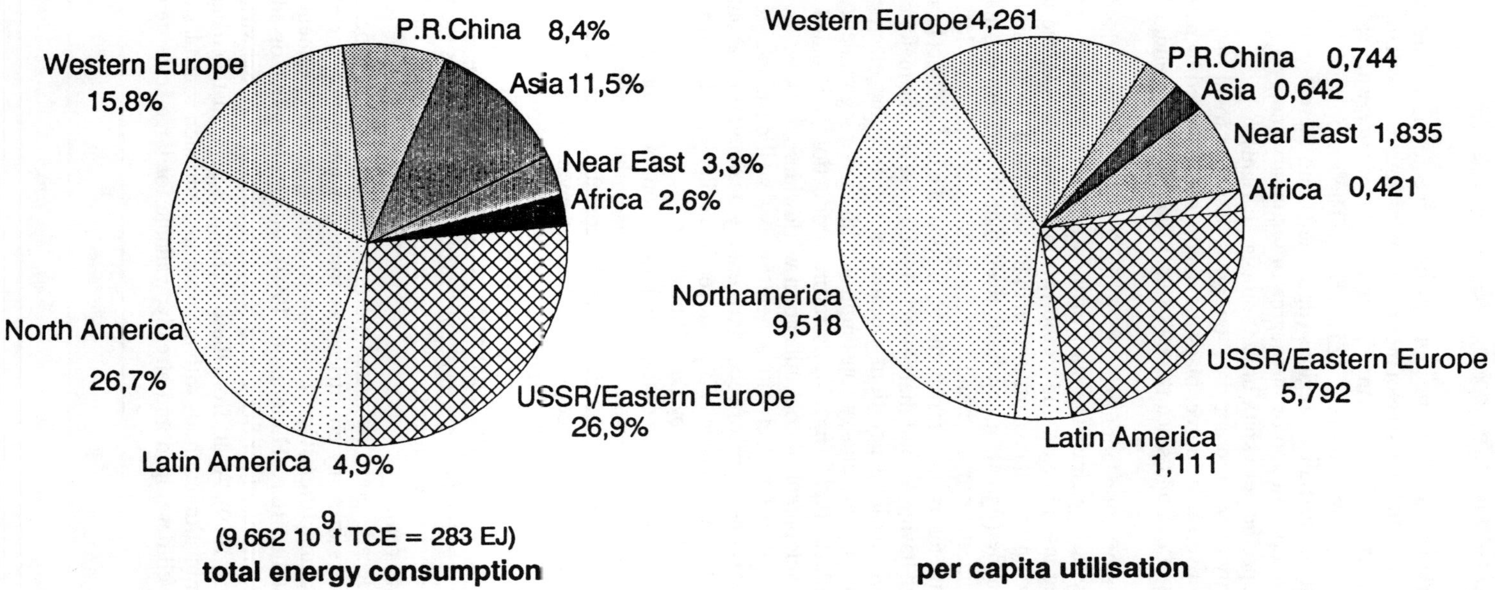

FIGURE 1 World primary energy consumption 1987 (from IZE 1990a).

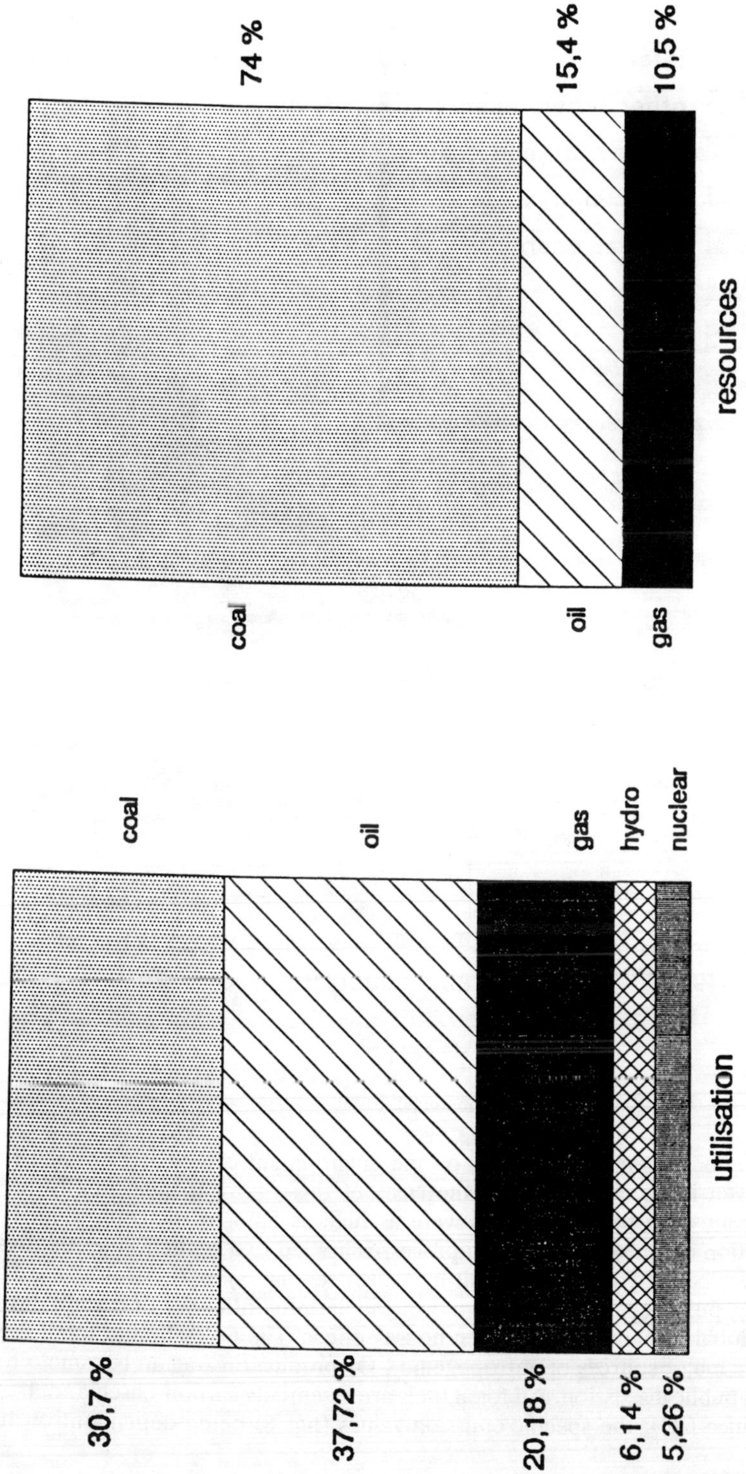

FIGURE 2 Primary Energy Sources (from IZE 1990a).

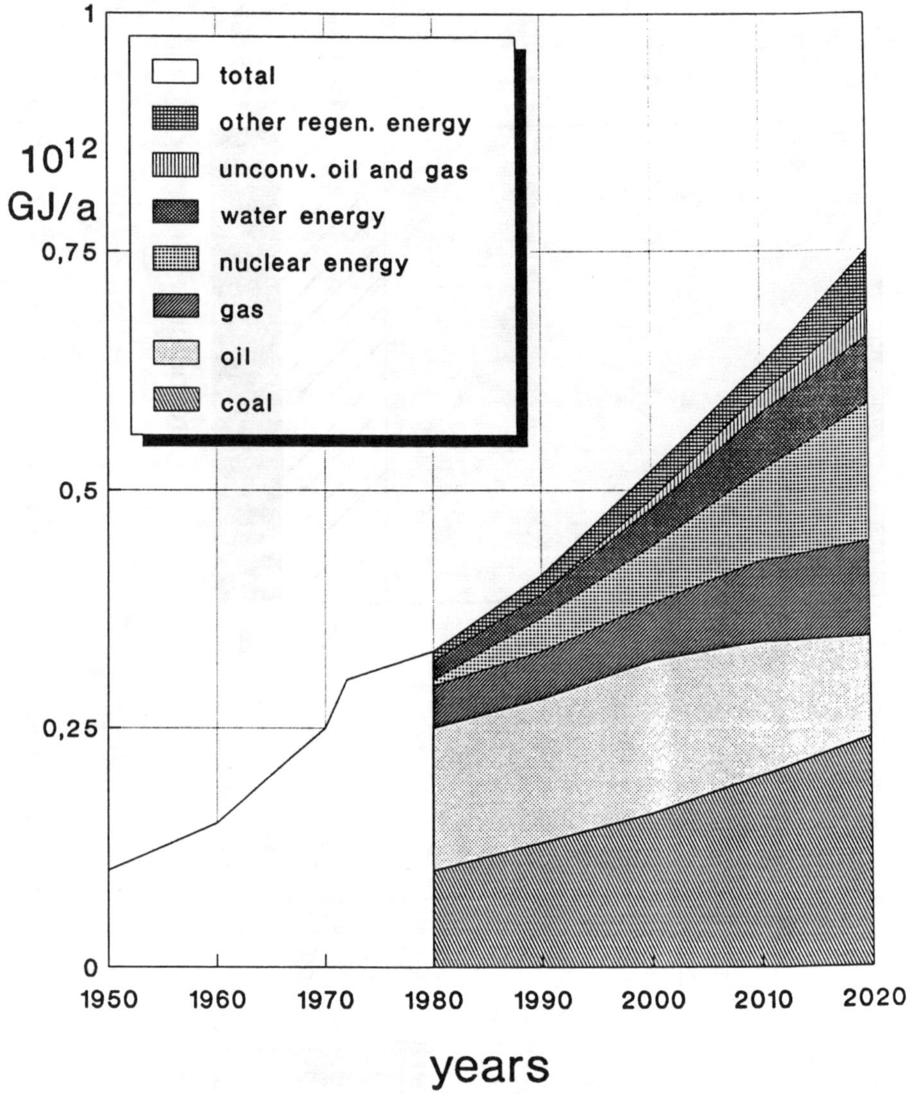

FIGURE 3 World energy supply (from Coal Resources).

The positive effect of emission control on the environment can clearly be seen from the example given in fig. 7. However, the costs of these retrofit measures were not cheap—investment costs for single units were as high as 50% of the total plant costs and the installation of emission reduction processes has led to a reduction in the overall efficiency of the electricity generation unit by 1–2% (see fig. 8).

More recently, public concern has grown over the harmful influence of carbon dioxide (CO_2) on the global climate via the "green house" effect. The utilisation of fossil fuel—being one of the major sources of antropogenic CO_2-production—again became a topic of controversial public discussion. All fossil fuels are essentially carbon-based and, hence, eventually produce CO_2; the specific emission values (fig. 8) being dependent on their composition.

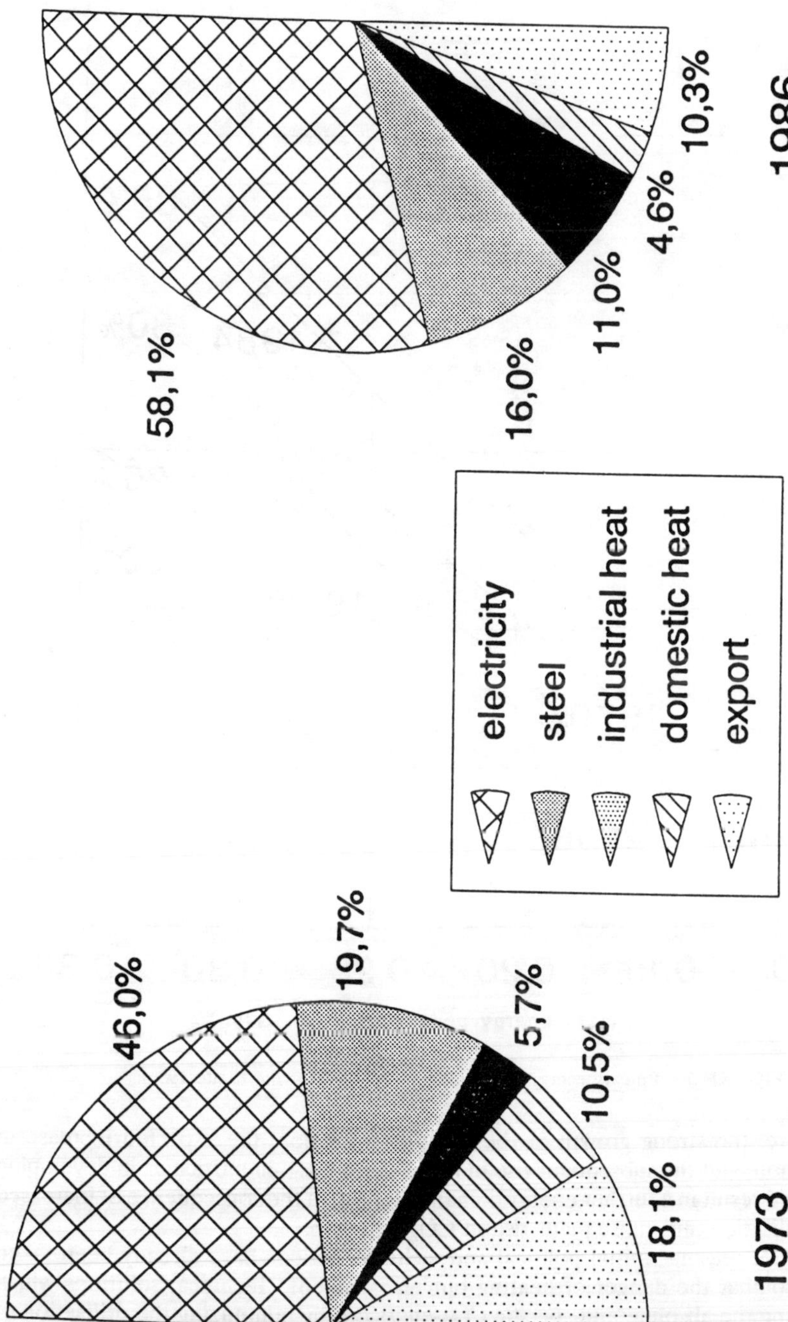

FIGURE 4 Development of coal utilization—Federal Republic of Germany (from IZE 1990b).

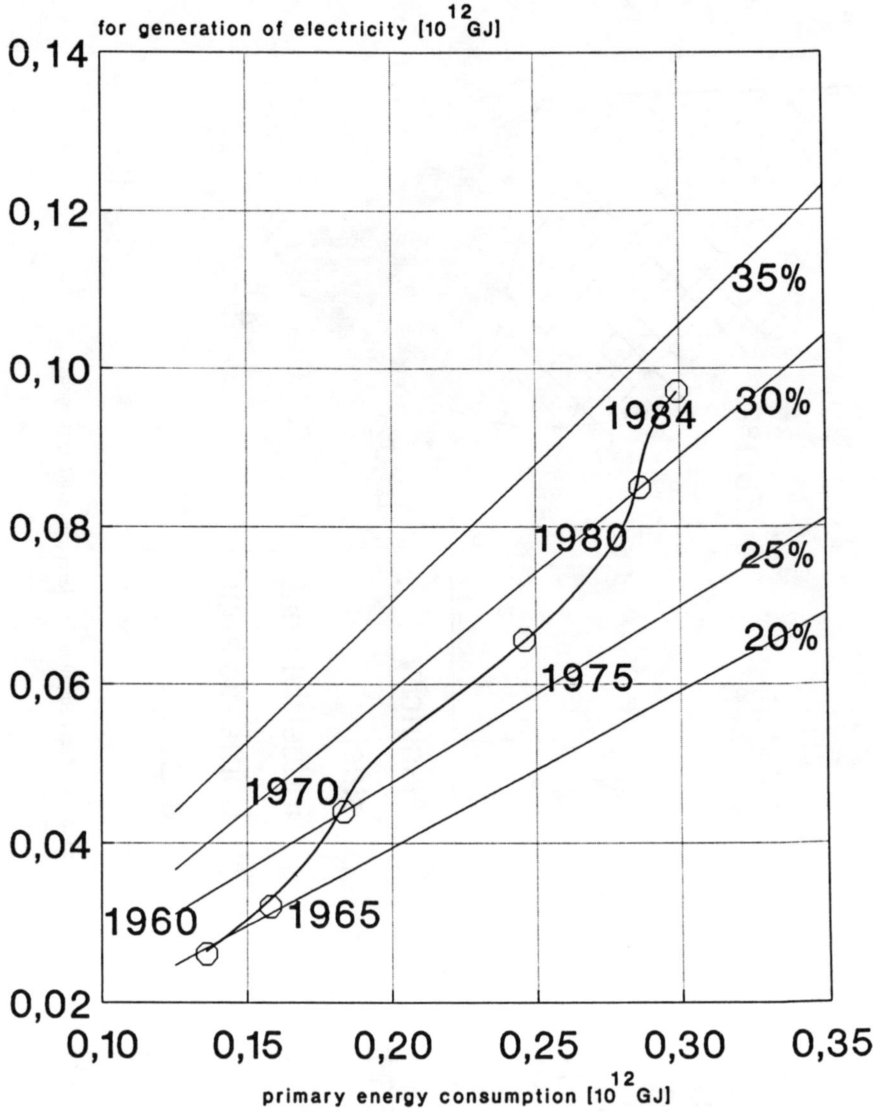

FIGURE 5 Primary energy consumption, total world (from Klätte 1986).

Furthermore, the strong growth of the world's population, the expected increase in industrialization, and the subsequent rise in living standards—in particular in developing countries—will result in a further energy demand along with the consequence of increased fossil fuel utilisation and subsequent rise in CO_2-emission.

For the above reasons, initial political and technical actions have already been taken in order to combat the danger of heating our environment. Technical solutions which aim at reducing the absolute and specific CO_2-emissions by minimizing the utilisation of C-containing fuels are under investigation.

One of the first alternatives is the optimization of the conventional steam generation process. However, the technical possibilities (multiple reheat, optimization or preheat,

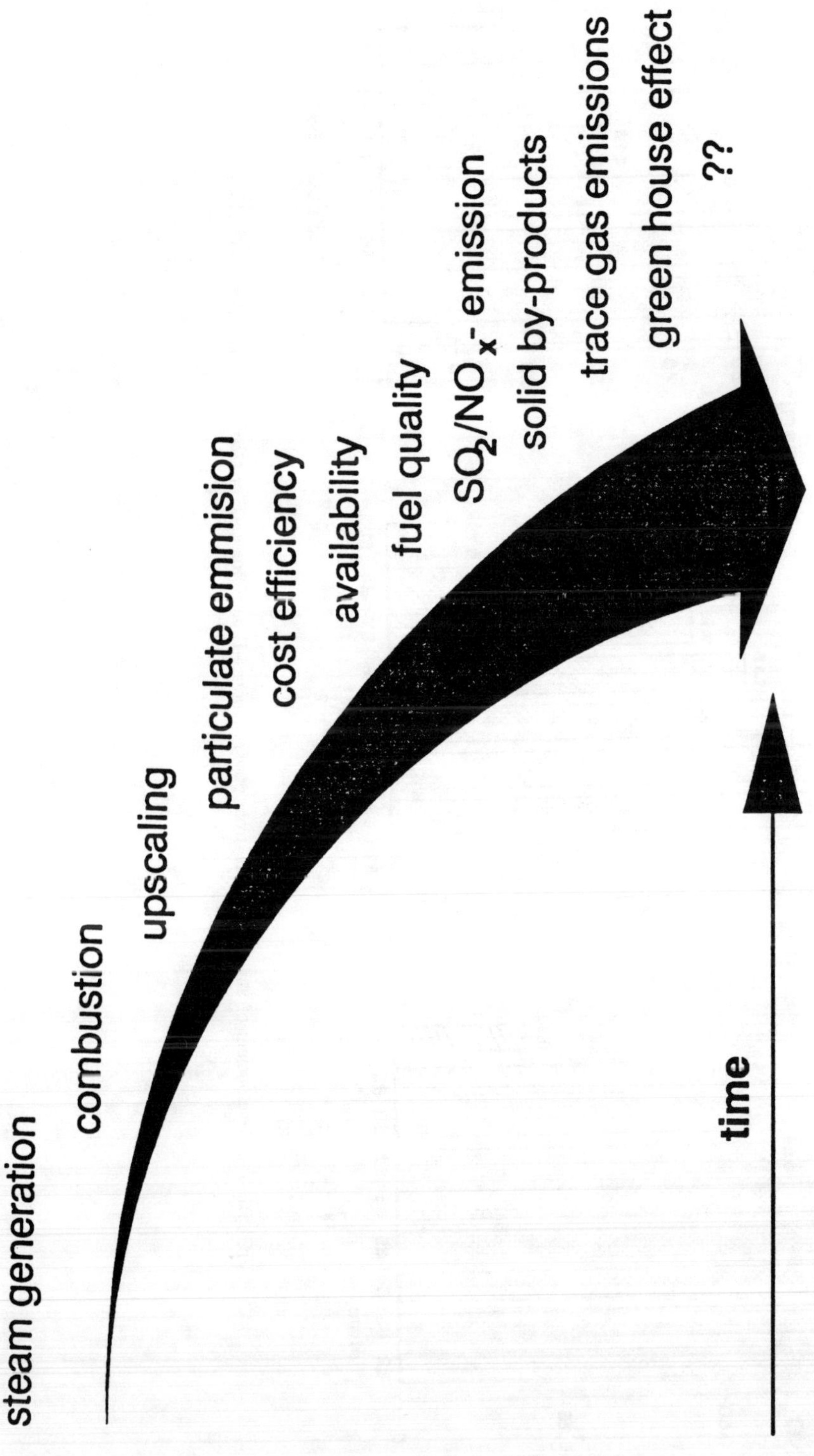

FIGURE 6 Electricity from fossil primary energy area of concern.

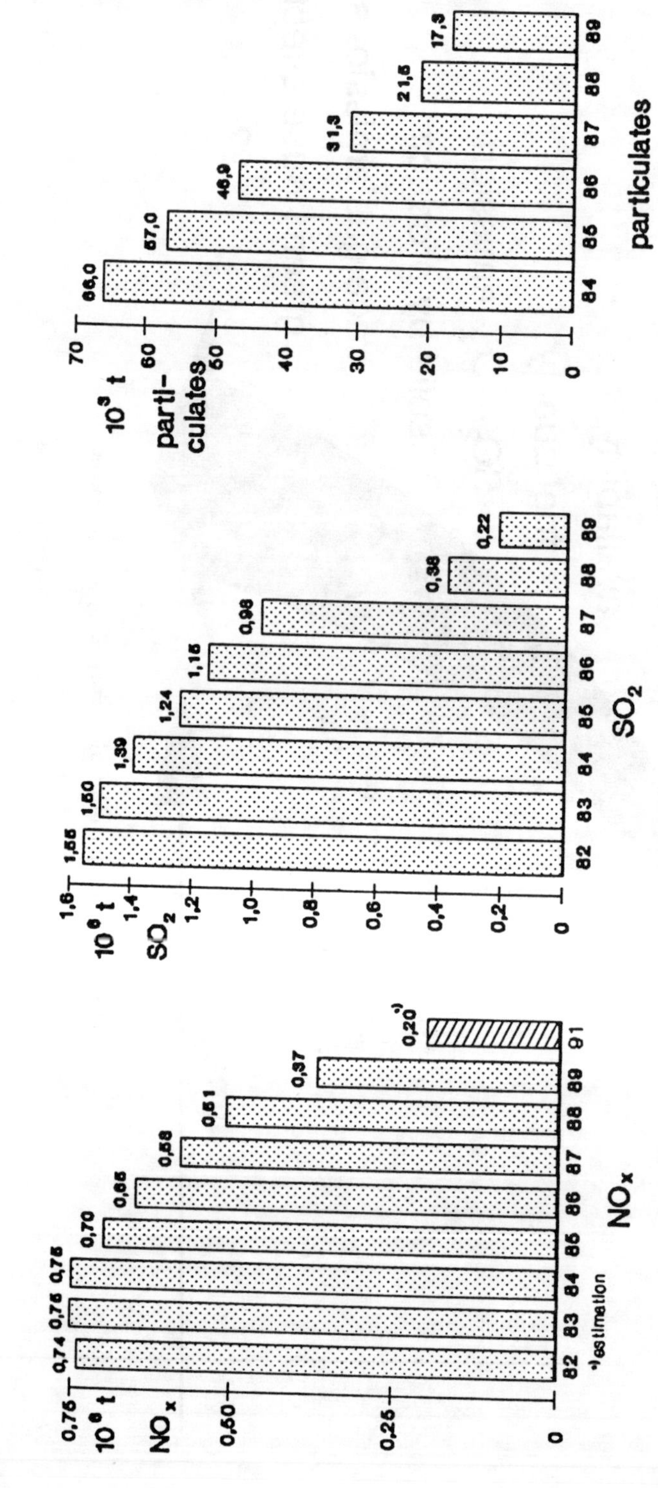

FIGURE 7 Development of annual emissions from power stations—Federal Republic of Germany (from Hildebrand 1990).

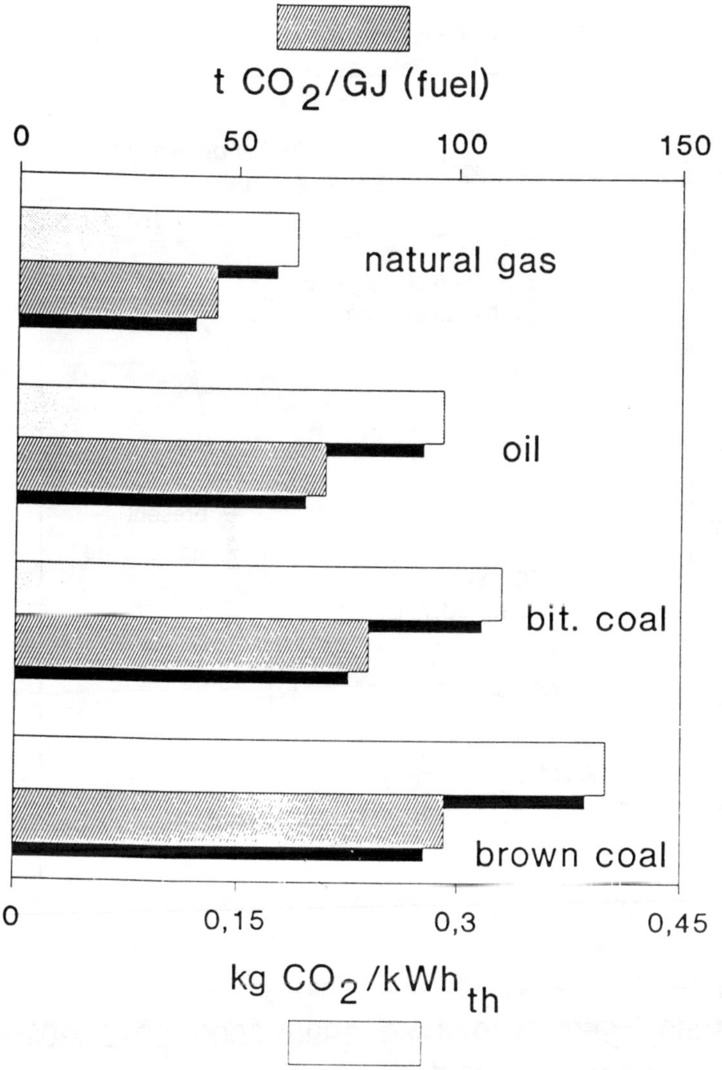

FIGURE 8 Specific CO_2—emissions from fossil fuels.

and "cold end" as well as supercritical operation) are limited and will only lead to marginal improvements in the total process efficiency of about 1%.

Also, a replacement within the fossil fuel group by changing to higher H/C-ratios is very limited for economical reasons and may offer only a solution for a restricted time period. Further improvements are expected from a thorough screening of heat and energy sinks, the reduction of losses within the electricity production process and—even more important—an improvement in the efficiency of secondary electricity consuming systems.

Another option is offered by the part change to non-carbon energy sources. However, the utilisation of such sources is either only locally accessible with small quantities

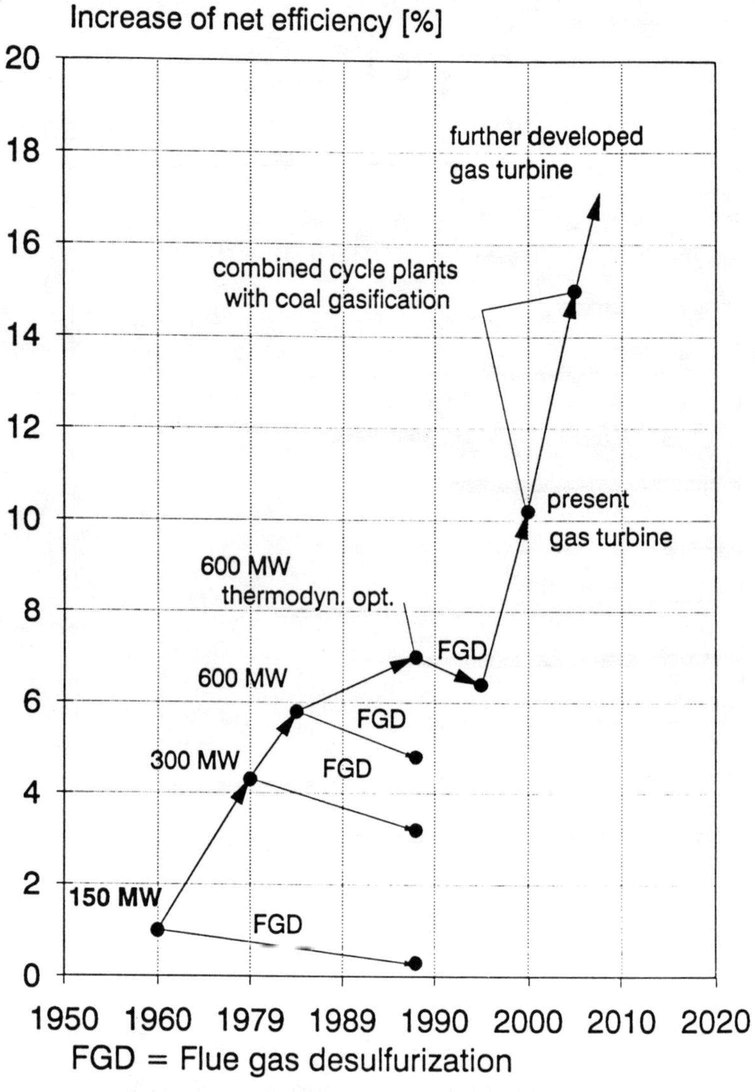

FIGURE 9 Development of efficiency for brown coal based electricity generation (from Klätte 1986).

available, and is, thus, economically impractical (regenerative energies) or suffers, at present, from only a limited public acceptance (nuclear energy).

For the above given reasons new concepts of coal utilisation in electricity production are under consideration. Substantial improvements in fuel conversion efficiencies can only be expected by making better use of the thermodynamics by providing for energy conversion at elevated temperatures above the ones of the water/steam cycle (e.g. Klätte, 1986) (fig.9). Therefore, an attractive solution is offered by the combined cycle arrangement, i.e. the combination of a gas turbine with a steam turbine, due to an increase of the usable temperature difference.

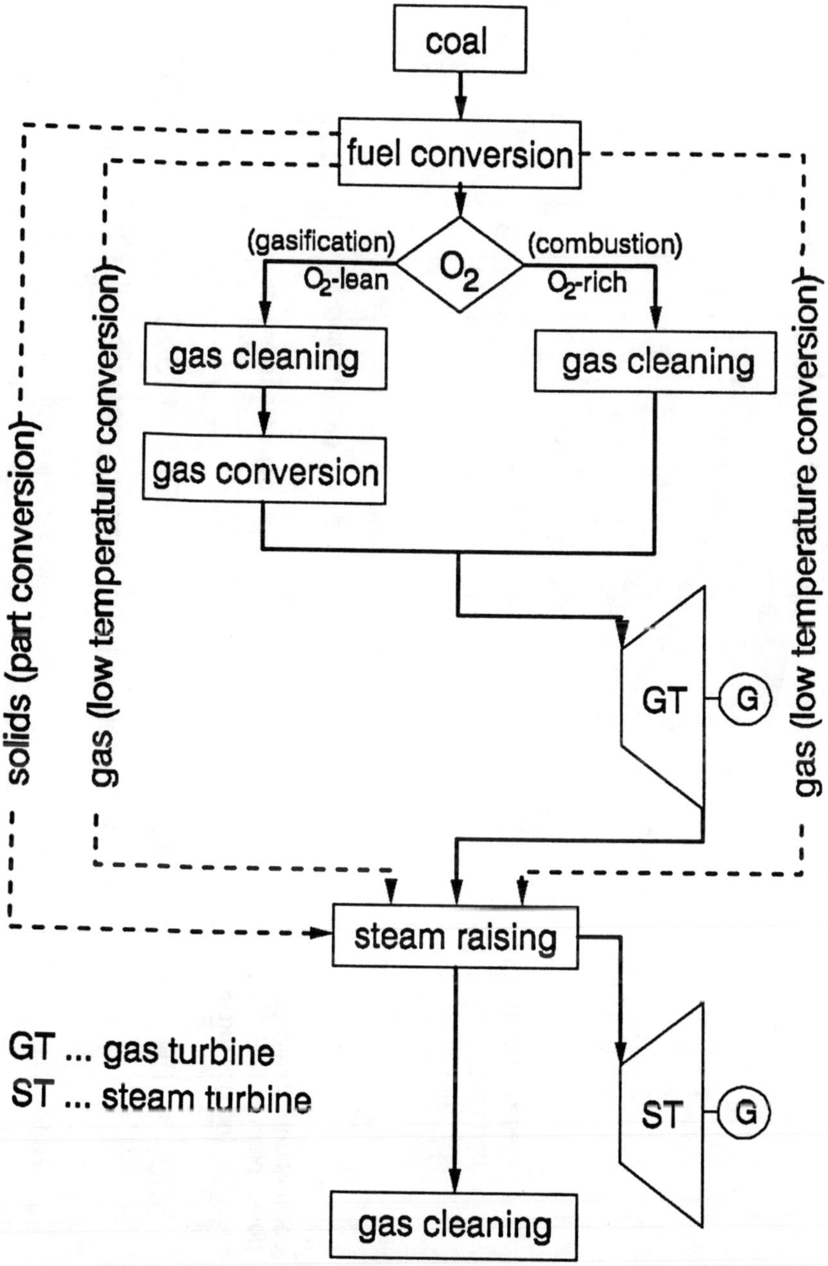

FIGURE 10 Gas/steam combined cycles for electricity generation.

With regard to coal, the discussion concerning CO_2-emissions brought about a renewed interest in the known principles of pressurized fuel conversion for integration into such a combined cycle. For combined cycles using coal, and based on both pressurized combustion and gasification, schematically given in fig. 10, technical solutions are in

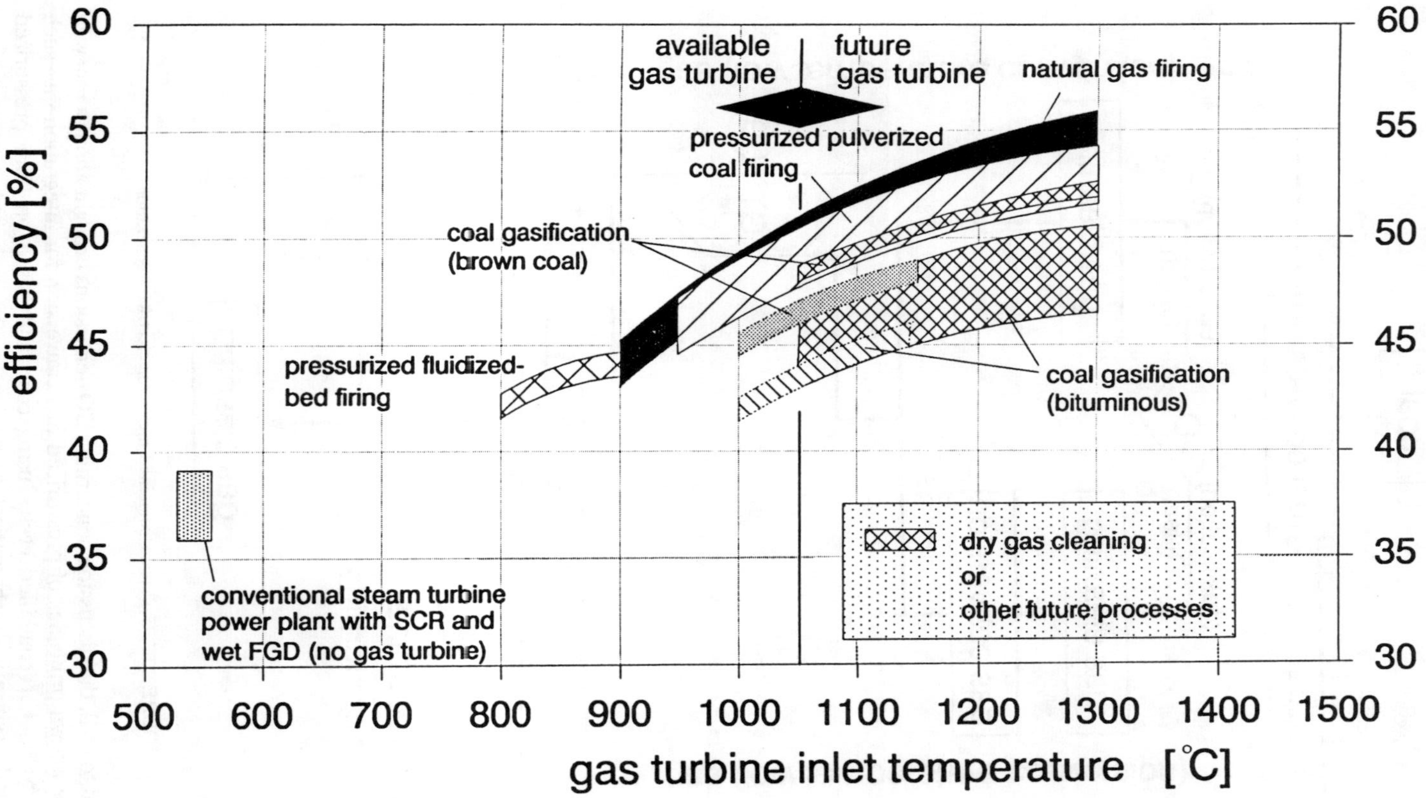

FIGURE 11 Combined cycle plants; Comparison of efficiencies (from Pruschek et al 1990).

different stages of development (concept studies, research investigations, planning, construction and large-scale demonstration respectively).

The first two large-scale, combined cycles based on coal gasification are at present under construction in the Netherlands for bituminous coal (e.g. de Winter and Busschen, 1991) and in Germany for brown coal (e.g. Bergmann and Ewers, 1990). Successful demonstration of the combined cycle technique is expected to be available by the end of this decade. These projects are assumed to improve the efficiency of coal utilization (e.g. Pruschek, *et al.* (1990) as given in fig. 11.

In conclusion it could be stated that the future energy demand can only be satisfied by an economically and ecologically sound symbiosis of:

— a continuing utilisation of further advanced nuclear energy generation,
— an increased use of regenerative fuels, in particular, solar and wind systems, and
— high efficiency fossil fuel conversion systems using predominantly coal with a minimum impact on the environment, the last of which is now, and also expected to be in the foreseeable future, the predominant mode of electricity generation.

REFERENCES

Bergmann, H., and Ewers, J. (1990). New Power Station Concepts for Lignite. VGB Conference "Power Plant Engineering 2000", February.

Coal Resources 1985–2020. Reports to the Conservation Commission of the World Energy Conference, IPL Science and Technology Press.

De Winter, H. J. M., ten Busschen, A. (1991). ICGCC Demonstration Project Buggenun. VGB Conference "Coal Gasification 1991", May.

Hein, K., and Kallmeyer, D. (1989). Reduction of NO_x-Emission of Brown Coal Fired Boilers. VGB Kraftwerkstechnik, 69. pp. 515–520.

Hildebrand, M. (1990). Stand der Rauchgasreinigung bei EVU-Kraftwerken SO_2- und NO_x-Minderung Sonderdruck Elektrizitätswirtschaft Jg. 89 (9).

IZE (1990a).—Energiewirtschaft, Kurz und Bündig, Ausgabe.

IZE (1990b).—Energie in der Europäischen Gemeinschaft. Westermann Schulbuchverlag GmbH; Braunschweig.

Klätte, G. (1986). Die Elektrizitätswirtschaft in den 80er und 90er Jahren—Trends, Aufgaben und Erfordernisse. Jahrbuch Bergbau, Öl und Gas, Elektrizität, Chemie 85/86, Verlag Glückauf.

Plass, L. (1991). Wirbelschichttechnik für Vergasung, Verbrennung, Trocknung und Gasreinigung und ihre Anwendung in Kraftwerken. VGB-Kraftwerkstechnik 71, No. 3 pp. 184–191.

Pruschek, R., Renz U., and Weber E. (1990). Kohlekraftwerk der Zukunft. Minister für Wirtschaft, Mittelstand und Technologie des Landes Nordrhein Westfalen, Düsseldorf.

Schwirten, D., Böcker, D., and Jüssen P. (1988). Versuche zur Entschwefelung von braunkohlebefeuerten Industriekesseln nach dem trocken-Additiv-Verfahren (TAV) VGB-Kraftwerkstechnik 68. No. 9, p. 935.

Influence of the Quality of Greek Brown Coals on NO_x–SO_2 Emissions from Thermal Power Stations

E. KAKARAS, Z. NTOUROS, M. FOUNTI and N. PAPAGEORGIOU *National Technical University of Athens, Mechanical Engineering Department, Thermal Engineering Section, Athens, Greece*

Abstract — In the paper, measurements of NO_x and SO_2 emissions from the Greek Public Power Corporation power stations at Megalopolis and Ag. Dimitrios are presented. The adopted measuring techniques are described and the effects of fuel and firing system on NO_x emissions and on the natural SO_2 retention in the flue gases are discussed. The Megalopolis power station does not emit considerable levels of NO_x because of the low calorific value of the Megalopolis-lignite. The same does not hold true for the Northern Greece (Ag. Dimitrios) power stations. The different ash composition of the two fuels considerably affects the natural desulphurization of the flue gases.

Key Words: Brown coal, pollutants, emissions, desulphurization, concentrations, measurements

INTRODUCTION

A major incentive for environmental protection is the ability to control the formation of nitrogen oxides from either atmospheric nitrogen ("thermal" NO_x) or from fuel-bound nitrogen ("fuel" NO_x) under operating conditions. It is, by now, well known that the rate of formation of nitrogen oxides in flames is highly dependent on temperature (Zeldovich, 1946, Thomson *et al.* 1971, Williams *et al.* 1971) which renders the requirements for high combustion efficiency and low NO_x formation contradictory (Beer, 1972).

On the other hand, in the case of sulphur, only the form of the sulphur oxide (SO_2 or SO_3) can be affected by the combustion process as the total emission will effectively depend upon the amount of sulphur contained in the fuel.

The total amount of NO_x and SO_2 emitted to the atmosphere depends strongly on the chemical analysis of the ash and of the pulverized fuel as well as on its calorific value.

In Greece, the Public Power Corporation (PPC) is exploiting soft brown coals for energy production. For example, in 1989 (Daily Press, "Kathimerini," 1991) 77.8% of the total electricity production, 29391 GWh, stemmed from burning lignite whereas only 13.5% came from oil and 7.3% from Hydroelectric power stations. Greece has to import 1.4% of its annual energy demand.

The aim of this work is to present the various relations which combine the NO_x/SO_2 emission levels from the Greek power stations with different characteristics of pulverized fuel. It is essential to distinguish between the different types of pulverized coal used at each thermal power plant in order to suggest the relevant primary measures to reduce the NO_x and SO_2 emissions, according to the fuel (and furnace) specifications, in order to be able to meet the EC emission limits without any further secondary "post-combustion" techniques (flue gas desulphurisation, catalysts etc.).

Greek brown coals feature a high humidity percentage, namely 40–65%, low to average ash percentage, i.e., 7–17%, and a net calorific value 3.0–10 MJ/kg.

Measurements have been obtained for the purposes of this work from the two above named power stations, at Megalopolis and Ptolemais, and the performance data were analysed. In the following section the experimental set-up and the measuring techniques used in the two cases are described. The measurements of NO_x emissions are presented in the sequence and the effects of fuel and firing system on NO_x emissions are discussed.

Subsequently, the measurements of the SO_2 emissions and the natural SO_2 retention in the flue gases is discussed. The paper closes with conclusions and remarks for the future environmental protection of the areas in question.

EXPERIMENTAL TECHNIQUES

Measurements of O_2, CO_2, NO_x and SO_2 concentrations, as well as of gas temperature were performed at two different heights above the burners inside the furnace of each of the two power stations and in the flue gases at each furnace outlet.

Description of Megalopolis Power Station, Unit I

After the evaluation of several operational parameters, such as maintenance, flexibility and fouling, it was found for the Megalopolis power plant, that boiler No. I was the more suitable for the purposes of this investigation.

Unit I of Megalopolis was commissioned in 1970. The rated power is 125 MW_{el} and with a maximum continous steam output of 420 ton/h.

The boiler was erected by VKW, has natural circulation, two passages of gases and a steam reheater from the outlet of the HP-Turbine.

The boiler has six mills of the DGS BABCOCK type. The capacity of each mill is about 69 ton/h. Inside the mill drying of the lignite takes place by hot gases (ca. 800°C) and pulverization.

The furnace cross-section is 12×12 m. Openings for the measuring probe were available at heights of 20 and 26 m, respectively. The soot blowers had to be removed as a consequence. At each of these heights three openings on one side of the furnace were available.

Measurements at the Ag. Dimitrios III Power Plant

The power plant Ag. Dimitrios is located in the Ptolemais area. Boiler No. III has a forced circulation, and a rated power of 300 MW_{el}. The furnace has a 15×15 m cross-section. There are several openings at various heights.

Instrumentation and Measuring Techniques

Table I tabulates the instruments used for measuring species concentrations and the expected accuracies. Figure 1 shows the measuring arrangement used for the determination of species concentration inside the furnace. The setup shown in the figure has been used for the measurements at Megalopolis power station. For the measurements at Ag. Dimitrios the construction of the probe was different and the sample was quenched after suction.

The high H_2O and SO_2 content of the flue gas has a considerable influence on the measurement accuracy due to the increase of the dew point DP and the acid dew point ADP temperatures.

The DP and ADP can be calculated by a method given, for instance, by J.T. Banchero, F.H. Verhoff (1975) of D. Lambert (1978).

The high DP and ADP temperature led to the conclusion that the intense cooling of the gas sample inside the water cooled pyrometer suction probe would inevitably cause a considerable condensation of water vapour of the flue gas inside the probe which would influence the accuracy of any SO_2 measurement.

To overcome this problem a new and improved probe for SO_2 measurements was designed and constructed by PPC.

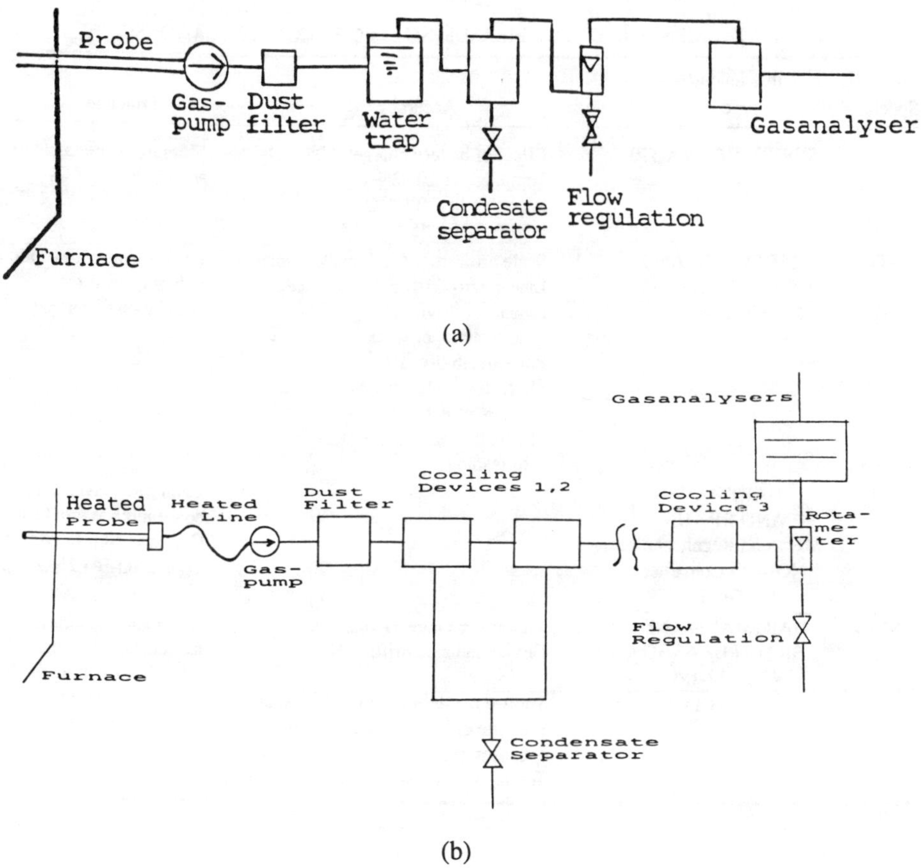

FIGURE 1 Measuring setup for Megalopolis I and Agios Dimitrios III.

Gas-temperature measurements were taken with the use of a suction pyrometer which was approximately 4m in length.

The measurements have been performed on a 3×3 grid of measuring points (3 locations on the one boiler side at 3 different depths) for two furnace cross-sections between the level of the burners and the exit of the furnace. The species concentration before the air preheater was also measured at the same time. Additional measurements took place at the exit of the mills in order to estimate the inlet conditions of the fuel and air.

The methods used for the fuel analysis and the inaccuracies introduced due to the instrumentation used are shown in Table II.

NO_X EMISSIONS

Several series of measurements have been performed at the power stations of Megalopolis and Ag. Dimitrios in order to determine not only the NO_x emission levels but also their distribution inside the furnace and at various heights. The distribution of NO_x concentration at a furnace cross-section of 26 m for the Megalopolis boiler is shown in Figure 2a. Due to operational reasons, the excess air was kept very high which resulted in the respectively

TABLE I
INSTRUMENTS USED FOR MEASURING SPECIES CONCENTRATIONS

Species	Instrument/ type	Accuracy	Principle
O_2	SERVOMEX O.A.570	Effect of battery charge: 0.05% O_2 Temp. effect ±0.02% O_2/°C Repeatability: 0.1% O_2 ≈0.5% O_2 for 45°C	Measures paramagnetic susceptibility of O_2
NO	HARTMANN AND BRAUN RADAS 1G 42/20-22-1 EN	Scale error <0.3°C of smallest span Linear zero drift <±2%/per week Linear sensitivity change: <±2% of deflection per week Reproducibility: 1% Short-time stability 1% Pressure effects: 1% Fluctuation causes 1% change in leading	UV resonance absorption utilising gas filter correlation technique
NO_2	HARTMANN AND BRAUN Mess-und Regeltechnik NO_2-NO-Converter CGO-K		Catalytic converter of coal/molybdenum basis Converts NO_2 to NO and $O_2 + 2NO_2 \leftrightarrow 2NO + O_2$
SO_2	HARTMANN AND BRAUN URAS 3G 42/20-12-1×L	Measuring range: 0–0.03 vol% Device-induced drift: ≤1% of span Short-term stability: ±0.2% of span Pressure effects: 1% fluctuation causes 1% change in reading Temp. effects: 1% of span per 10°C	Resonance absorption associated with specific vibration-rotation bands of heteroatomic gases in the infrared at $2.5 < \lambda < 12$ μm

high O_2 concentrations shown in Figure 2b. The high excess air percentage influences the measured NO_x concentration both quantitatively and qualitatively. After converting NO_x concentrations for 6% O_2, concentrations between 50 and 80 ppm NO_x occur at furnace heights of 26 or 20 m when the electrical power produced by the unit varies between 75 and 85 MW_{el} (Figure 3). The measured NO_x concentrations of Figure 2a and Figure 3 clearly indicate that for the Megalopolis power station no considerable NO_x emission is detected.

The low NO_x emission level for the Megalopolis power plant has also been confirmed from measurements at the stack exit (before the I-D Fans) which have shown 80–100 ppm NO_x concentrations with 8.5% O_2.

The NO_x emission level from the Ag. Dimitrios power station (Ptolemais area, Northern Greece) was found to be significantly higher than the one measured at Megalopolis. The data shown in Figure 4a represent the NO_x concentration at a furnace cross section at a height of 36m and the respective O_2 concentration is presented in Figure 4b.

The distribution of NO_x concentration along the furnace height as measured at three different locations at the left side of the furnace is shown in Figure 5.

A simple comparison between Figures 3 and 5 shows that the Ag. Dimitrios power plant emits up to three times more NO_x than the one in Megalopolis, although both of them are lignite fired boilers. Overall emission measurement performed at the stack of Ag. Dimitrios boiler III (before IDF's) has shown NO_x emissions of 200–230 ppm (at 6,5% O_2) when the unit's load was 290 MW_{el}.

TABLE II
METHOD USED FOR THE ANALYSIS OF LIGNITE AND FLUE GASES

Analysis	Method	Accuracy
Sample preparation	ASTM D2013/72 (78)	—
Fly-liquid ash	ASTM D2795/69 (80)	±0.05% up to 1%
SO_3	ASTM D1757/80	±0.2%
Lignite analysis:		
Total C	ASTM D3178/73 (79)	±0.3%
H_2	ASTM D3178/73 (79)	±0.07%
CO_2	ASTM D1756/62 (79)	±0.1%
Total S	ASTM D3177/82	±0.05%
Total N_2	ASTM D3179/73	±0.05%
Calorific value	ASTM D3286/77	±(0.15%) or 10 cal/gr
Total moisture	ASTM D3302/82	±0.2%
Ash	ASTM D3174/82	±0.3%
Volatiles	ASTM D3175/82	±1.0%

TABLE III
FUEL ANALYSIS

Fuel analysis "as received"	Megalopolis I			Ag. Dimitrios III		
	Test 1	Test 2	Test 3	Test 1	Test 2	Test 3
Proximate analysis						
Water content %	56.87	60.18	52.50	53.4	51.9	51.19
Ash %	16.71	18.62	24.32	16.68	17.85	18.45
Vol. mater. %	16.2	12.8	14.3	18.69	21.40	*
Lower cal. value kcal/kg	1105	781	908	1173	1227	1254
Stochastic analysis						
C %	16.22	13.05	14.04	17.04	17.28	18.82
H_2 %	1.33	1.03	1.39	1.27	1.41	1.42
S %	1.72	1.02	1.25	0.42	0.59	0.6
N_2 %	0.54	0.35	0.41	0.44	0.55	0.45
CO_2 %	0.49	1.03	1.47	3.43	4.47	*
O_2 %	6.12	4.72	4.62	7.32	5.95	8.36

*Not measured.

INFLUENCE OF THE FUEL QUALITY AND THE FIRING SYSTEM ON NO_x FORMATION

The different operating parameters which are indicated by the different O_2 concentrations for the two boilers (Figures 2a and 4) as well as the different NO_x emissions level can be explained by the different fuel characteristics from the two lignite types which are used at the power stations in the Megalopolis and Ptolemais areas. Several fuel and ash analyses have been performed during the measurements at both power stations, which resulted in the fuel data shown in Table III.

From Table III it becomes obvious that the Ag. Dimitrios lignite features less moisture and a higher calorific value than the one from Megalopolis.

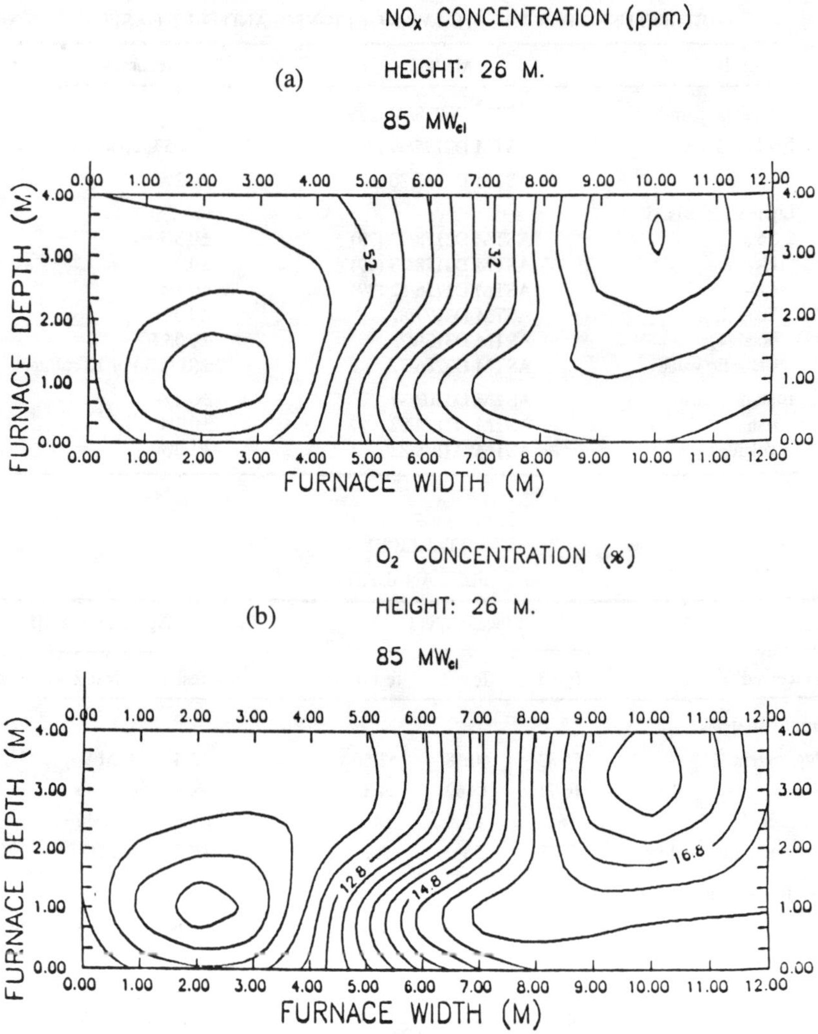

FIGURE 2 NO_x concentration (a) and O_2 concentration (b) for the Megalopolis I furnace (as measured).

The different fuel characteristics lead to a different firing system for the two boilers, which is shown in Figure 6a,b.

Due to the different firing system used, the unburnt fuel content in the liquid and the fly ash of Megalopolis power station is considerably lower than that at Ag. Dimitrios (see Table IV).

The poor fuel quality of the Megalopolis lignite leads to lower temperatures inside the furnace, as shown in Figure 7, in comparison to the temperature distribution at a vertical cross section of the Ag. Dimitrios furnace, which is shown in Figure 8. The higher temperature inside the furnace of Ag. Dimitrios and the higher volatiles content of the lignite are considered to be the reasons for the higher NO_x emission.

Generally the NO_x emission is considered to be low. This is especially so in the case of Megalopolis as the design of the firing system should have resulted in higher NO_x emis-

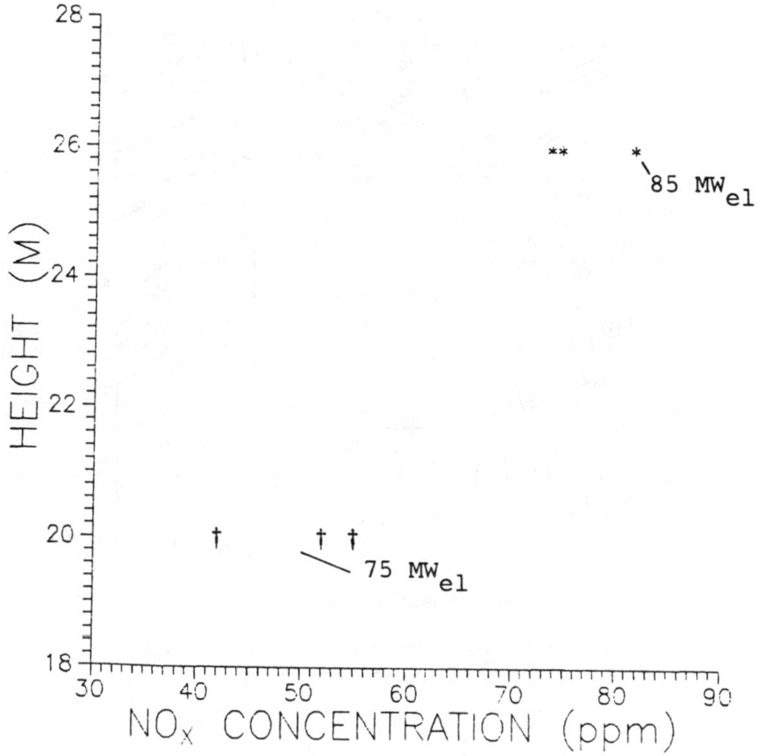

FIGURE 3 NO_x emissions (6% O_2) for Megalopolis I at various furnace heights.

TABLE IV

UNBURNT FUEL CONTENT OF THE LIQUID AND FLY ASH

Unburnt fuel % in	Megalopolis I			Ag. Dimitrios III		
	Test 1	Test 2	Test 3	Test 1	Test 2	Test 3
Liquid ash	24.25	27.46	36.47	42.8	41.9	37.5
Fly ash	1.3	1.5	1.6	2.4	2.5	2.4

TABLE V

CONSUMPTION OF BROWN COAL AND OIL AT THE MEGALOPOLIS POWER STATION

Date	Oil (Lit/h)	(tons/h)	Brown coal Power (MW)
10/01/91	5500	189.06	65
11/01/91	6500	182.80	85
12/01/91	5000	131.10	75
13/01/91	5500	191.13	62
14/01/91	0000	193.20	110

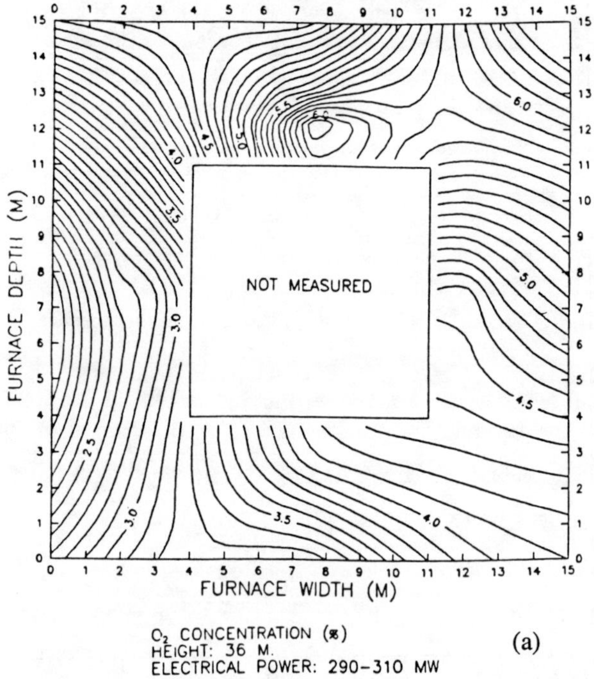

(a)

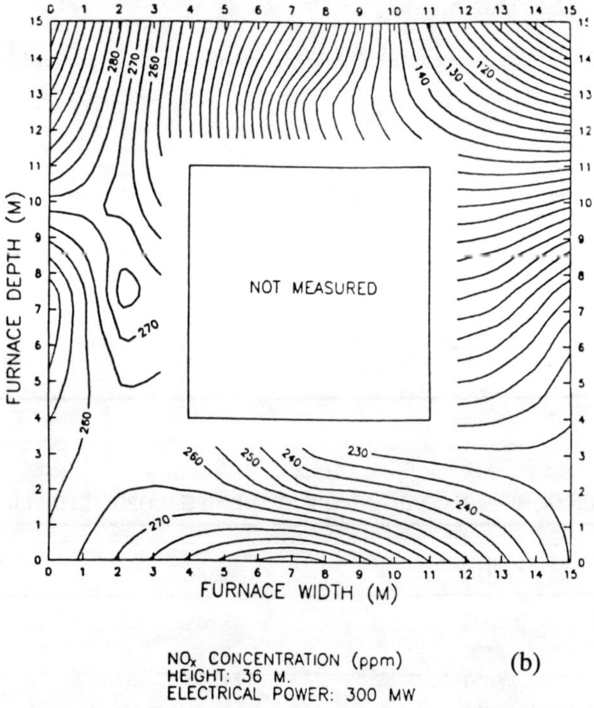

(b)

FIGURE 4 NO_x (a) and O_2 (b) concentrations for Ag. Dimitrios III furnace (as measured).

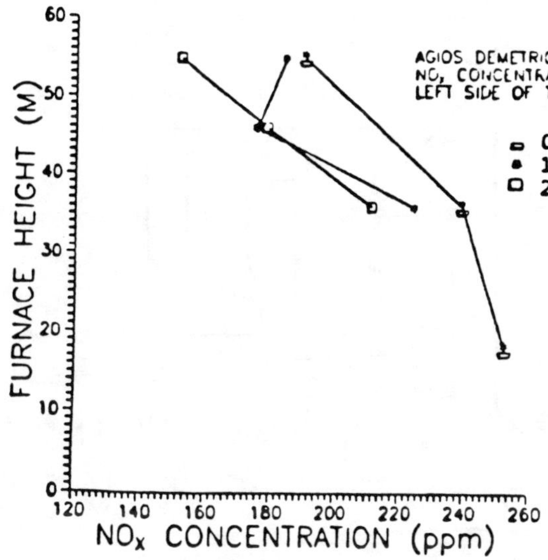

FIGURE 5 NO_x emissions (6% O_2) for Ag. Dimitrios III at various furnace heights.

TABLE VI
THEORETICAL SO_2 CONTENT OF FLUE GAS (DRY)

	Megalopolis			Ag. Dimitrios		
	Test 1	Test 2	Test 3	Test 1	Test 2	Test 3
Fuel S (%) (as received)	1.72	1.02	1.25	0.42	0.59	0.6
Y_{SO_2T} (ppm) (dry)	4288	3194	3430	1473	2061	2110

sions than the ones actually measured, considering also that, due to the poor fuel quality, some diesel oil had to be simultaneously burnt to assure the combustion stability as shown from Table V. The firing system in Ag. Dimitrios, as with every power station in Northern Greece, features only two main air streams (primary and secondary air).

Therefore, it should be expected that with the proper burner arrangement and air staging, the boilers can meet the international standards for the NO_x emissions (Hein *et al.* 1986, 1988, Reidick 1987, Reidick *et al.* 1990).

The influence on the NO_x formation from the simultaneous oil burning is shown in Figure 9, which represents the emission of NO_x in relation to the boiler load. From this figure it is obvious that the production of 85 MW_{el} with simultaneous oil combustion produces more NO_x than the production of 110 MW_{el} with pure lignite combustion.

SO_2 EMISSIONS

Because of the different sulphur content of the lignite used in the Megalopolis and in the Ptolemais area, the SO_2 emissions from the two power stations are expected to be complete-

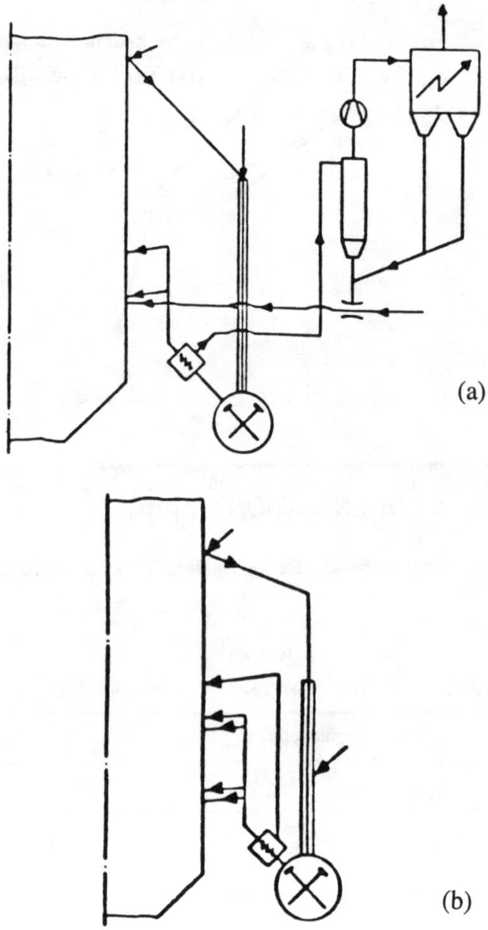

FIGURE 6 Pulverization and firing of Megalopolis (a) and Ptolemais (b) lignite.

ly different. Taking into account the fuel analysis shown in Table III and considering the excess air conditions during the measurements, the maximum equivalent SO_2-content of the flue gases which represents the total conversion of fuel sulphur to SO_2 has been estimated and the results are shown in Table VI.

The flue gas analysis at the stack (before the IDFs) has resulted in the following values for the SO_2 concentration: Megalopolis I: 2300 ppm SO_2, 8,6% O_2, S = 1,5%; Ag. Dimitrios III: 120–170 ppm SO_2, 7,1% O_2, S = 0,42–0,6%.

It is therefore obvious that not only does the different sulphur content of the two fuels affect the SO_2 emissions but also the different ash composition. A more detailed set of measurements has been performed at Ag. Dimitrios III concerning the SO_2 formation. Figure 10 shows the measured SO_2 concentrations at a furnace cross section, at a height of 36 m.

The distribution of SO_2 concentration along the furnace height, which is represented in Figure 11, provides qualitative information on the progress of natural SO_2 retention in relation to the flue gas residence time inside the furnace.

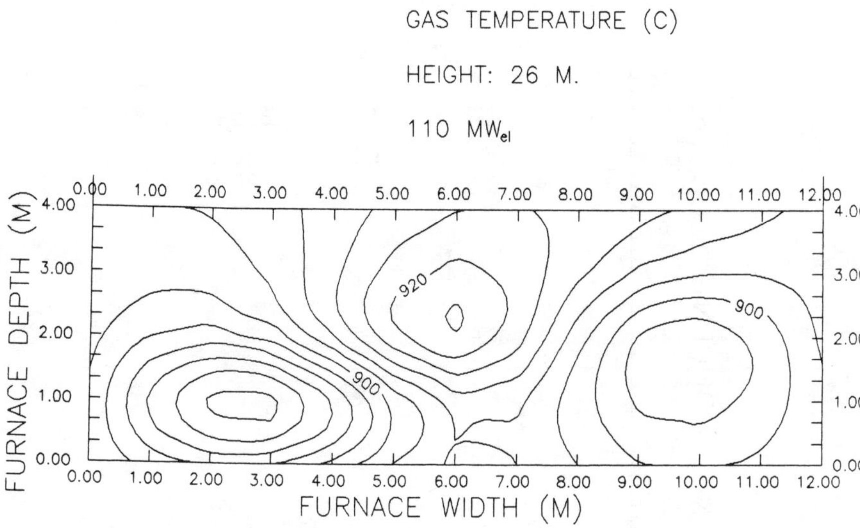

FIGURE 7 Temperature distribution at a cross-section of Megalopolis I boiler.

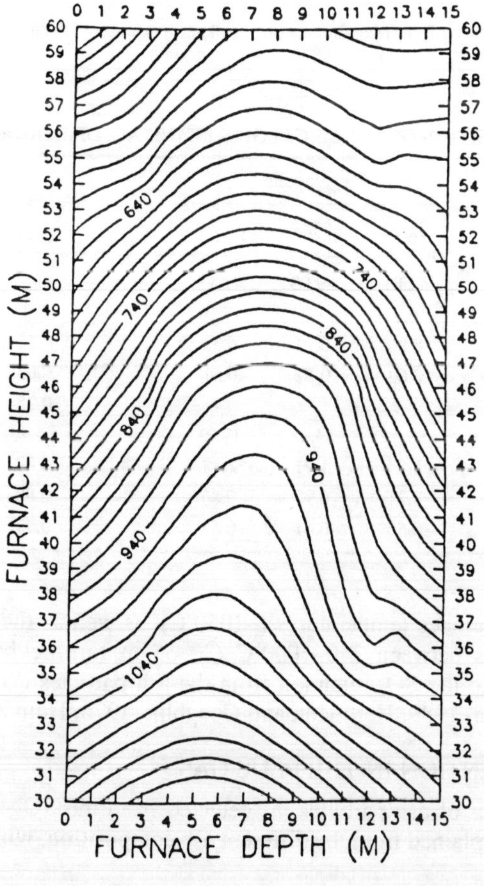

FIGURE 8 Temperature distribution at a vertical furnace cross-section for the Ag. Dimitrios III boiler.

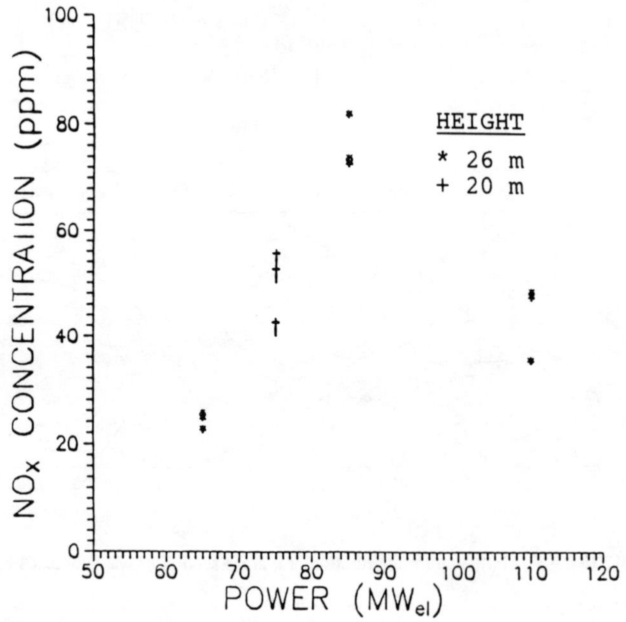

FIGURE 9 NO_x emission from Megalopolis power plant in relation to the produced electric power.

TABLE VII
ASH COMPOSITION OF THE MEGALOPOLIS AND AG. DIMITRIOS LIGNITE

Ash analysis %	Megalopolis			Ag. Dimitrios		
	Test 1	Test 2	Test 3	Test 1	Test 2	Test 3
SiO_2	42.01	49.03	47.13	36.53	37.95	34.63
Al_2O_3	16.71	14.74	17.14	10.88	12.37	11.84
Fe_2O_3	8.86	6.13	6.39	4.58	4.80	5.00
TiO_2	0.63	0.52	0.51	0.63	0.69	0.53
P_2O_5	–	–	–	0.25	0.33	0.38
CaO	13.96	14.43	17.01	34.98	33.2	35.72
MgO	1.54	1.91	1.48	6.82	4.0	4.40
K_2O	1.31	0.92	0.86	0.80	0.72	0.94
Na_2O	0.58	0.61	0.53	0.77	0.52	0.83

The relatively low furnace temperatures (~1050°C), as well as the existance of a large area with temperatures between 740–1000°C (30–55 m furnace height, see Figure 5) contributes favorably to the SO_2 retention from the ash particles ("natural desulphurization") (Bort and Roman, 1989, Hartmann and Coughlin, 1976, Hein and Schiffer, 1979).

NATURAL SO_2 RETENTION IN THE FLUE GAS

The different behaviour of ash particles, as detected from the SO_2-retention for the two types of fuel, can be explained from the different ash composition, which is shown in Table VII.

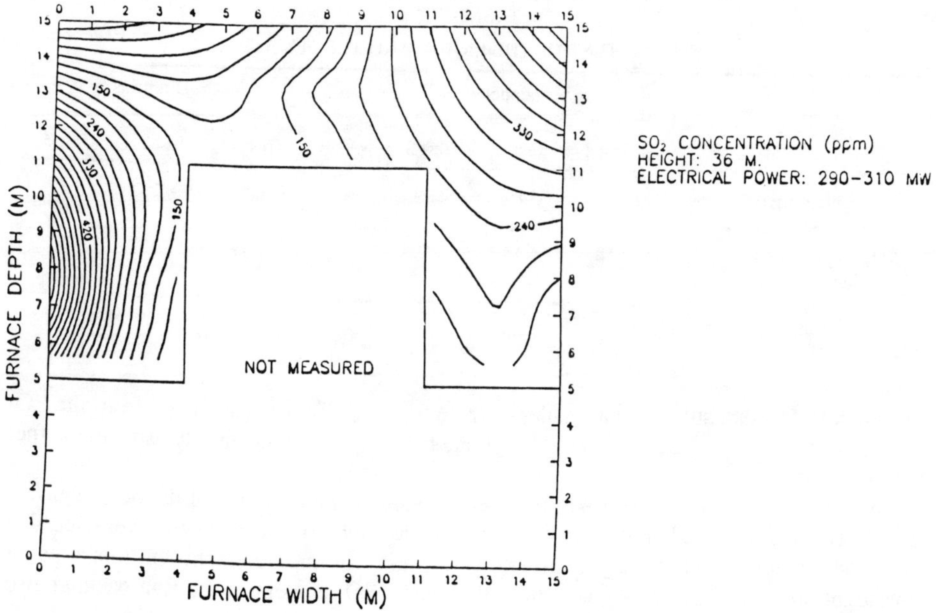

FIGURE 10 SO_2 concentration inside the furnace of the Ag. Dimitrios III boiler (36 m).

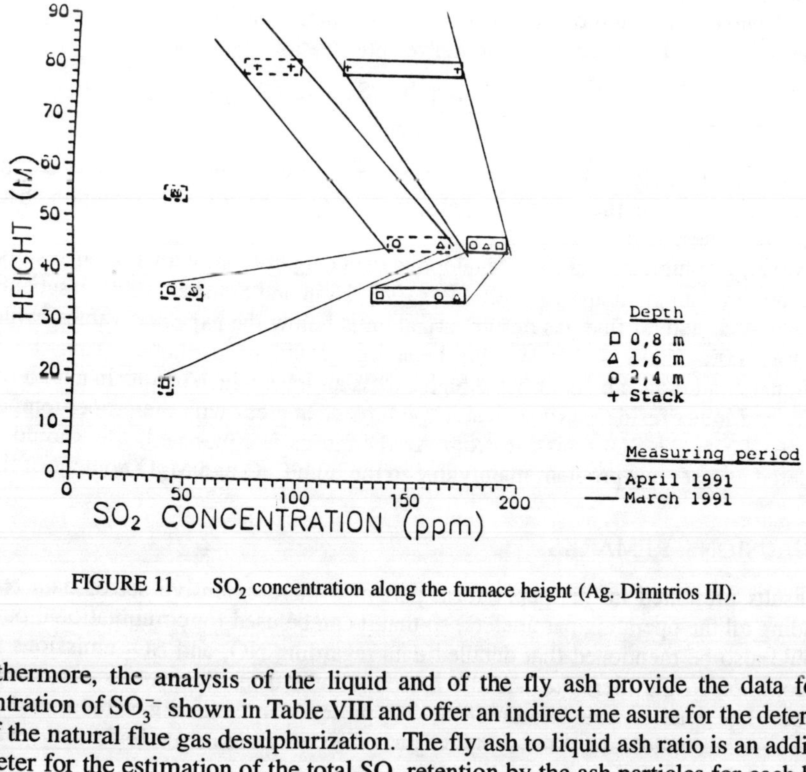

FIGURE 11 SO_2 concentration along the furnace height (Ag. Dimitrios III).

Furthermore, the analysis of the liquid and of the fly ash provide the data for the concentration of SO_3^- shown in Table VIII and offer an indirect me asure for the determination of the natural flue gas desulphurization. The fly ash to liquid ash ratio is an additional parameter for the estimation of the total SO_2 retention by the ash particles for each boiler.

TABLE VIII

SO_3 CONTENT OF THE LIQUID AND THE FLY ASH

	Megalopolis			Ag. Dimitrios		
	Test 1	Test 2	Test 3	Test 1	Test 2	Test 3
SO_3 content in fly ash %	4.4	3.89	4.73	6.23	6.23	5.53
SO_3^{--} content in in liquid ash %	3.58	4.48	3.52	2.79	3.45	3.57

For each test case a sulphur mass balance (Stratmann 1958) was calculated from the SO_2 content of the flue gas, the equivalent SO_2 content of the liquid and the fly ash and the fuel sulphur content.

The mass balance was performed in order to evaluate the accuracy of the desulphurization determination based on the ash analysis. The deviation of this balance never exceeded 10%, which represents a very satisfactory accuracy of measurement for this type of investigation. This discrepancy is mainly due to the SO_3 content of the flue gas and also due to the deposition of a part of the sulphur containing fly ash on the boiler surfaces.

With the use of a "DIONEX" ionchromatography, the measured SO_3 concentration in the condensate was found to represent an equivalent inaccuracy in the SO_2 measurement of less than 1%. It should be noted that this accuracy has been achieved after the construction of the new heated sampling probe and the special treatment for the flue gas cooling.

Figures 12a and 12b present the sulphur retention ratio

$$\eta_{DES} = \frac{\text{S fly ash} + \text{S liquid ash}}{\text{total S fuel}}$$

in relation to the Ca/S mole ratio of the fuel for the two power plants. The completely different behaviour of the two fuel types, with respect to the SO_2 retention, becomes obvious from these two figures.

However, a comparison of the Megalopolis data (Figure 10a) with similar results from the literature (Chunghtai and Michelfelder, 1983, Hein and Schiffer, 1979, Jeschar, 1986) has shown that natural flue gas desulphurization is below the expected values, which are from 30% 60% (Ca/S = 0.6–2). This behaviour could be explained from the Units' operational conditions, i.e., mainly the high excess air level which results in a high flue gas velocity and lower residence times inside the furnace at areas with the proper temperature levels. As far as the SO_2 emission from Ag. Dimitrios is concerned, this should not be considered as a serious problem mainly due to the high CaO and MgO content of the ash.

CONCLUSIONS: REMARKS

The results presented in this paper form part of a more extensive set of data recorded (including all the operating parameters) so that it can be used for computational purposes. It should also be mentioned that detailed data regarding NO_x and SO_2 emissions are not very common in the literature and the present results offer themselves for comparison purposes. The detailed examination of the SO_2 presence in the flue gas, the liquid and the fly ash and the condensate of the sampled gas, offers an accurate method for investigating the fuel sulphur content.

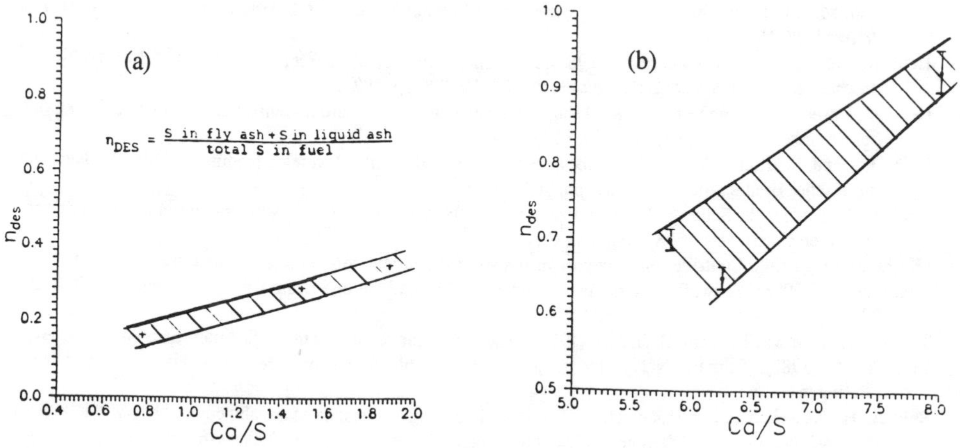

FIGURE 12 Natural desulphurization ratio for Megalopolis I (fly ash/total ash = 0.8–0.95) (a) and Ag. Dimitrios III power plants (fly ash/total ash = 0.8–0.95) (b).

The evaluation of the experimental results shows that the power stations of Megalopolis and Ptolemais (Ag. Dimitrios) exhibit a completely different behaviour regarding their NO_x and SO_2 emissions. Therefore the primary measures for the NO_x and the SO_2 retention which are suitable for the first type of power station cannot be applicable to the second. Furthermore, it is found that the main concern regarding environmental protection for the Megalopolis power plant should be focussed on the SO_2 emission levels, whereas for the Ptolemais area this should be on the NO_x emissions. However, based on international experience, it is estimated that the NO_x emissions can be reduced if the air supply system is modified according to a proper airstaging technique.

In conclusion, it can be stated that in order to control the emission levels of NO_x and SO_2 from browncoalburning power stations, the fuel quality, firing system and operating parameters should be taken into account in order to assess the proper technique for the emissions reduction according to international standards.

ACKNOWLEDGMENTS

This work has been sponsored by the EEC, within the frame of the JOULE programme, contract No. JOUF0046. The authors would also like to acknowledge the support of the PPC for the performance of measurements.

REFERENCES

Beér, J.M. (1972). *Journal Inst. Fuel*, **45**, p. 377.

Bortz, S.J., and Roman, V. (1989). Hydrate and process parameters controlling SO_2 removal during hydroxide injection near 1000 F, in: *Proc. 1st Combined Flue Gas Desulfurization and Dry SO_2* Control Symposium, Palo Alto.

Chughtai, M.Y., and Michelfelder, S. (1983). Schadstoffeinbindung durch Additiveinblasung um die Flamme, BWK 35, pp. 75–83.

Frank B., and Stahl, H. (1962). Schwefelbilanzuntersuchungen an Dampferzeugern mit Heiz und Steinkohlenstaubfeuerungen, VGB Heft 81, pp. 411–413.

Hartmann, M., and Coughlin, R.W. (1976). Reaction of Sulphurdioxide with limestone and gein model, *AICHE Journal*, **22**, No 3.

Hein, K., König, J., and Rüsenberg, D. (1988). Erprobung von NO_x Minderungsmaßnahmen an Braunkohlefeuerungen, etc., *Elektrotechnische Zeitschrift*, **109**, Nr. 7/8, pp. 292–296.

Hein, K., Rüsenberg, D., and Meuzel, K. (1986). Untersuchungen zur feuerungsseitigen Minderung der Stickoxide bei Braunkohlefeuerungen, *VGB*, **10**, pp. 957–959.

Hein, K., and Schiffers, A. (1979). Verbesserung der natürlichen Schwefeleinbindung bei der Verfeuerung rheinischer Braunkohlen, *BWK*, **31**, pp. 389–391.

Jeschar, R., Scholz, R., Schopf, N., and Klôppner, G. (1986). Schadstoffarme Verbrennung in einen Drallbrennkammersystem, *5e DVV Kolloquium*, Clausthal.

"Kathimerini" (1991). Contribution of energy resources to the production of electricity in Greece, January.

Stratmann, H. (1958). Schwefelbilanzuntersuchungen bei Steinkohlen und Braunkohlenfeuerungen, *VGB*, **2**, pp. 23–30.

Thomson, B., Brown, T.D., and Beér, J.M. (1971). Proc. American Instit. Chem. Eng., Atlantic City Meeting, Aug.

Reidick, H. (1987). Primäre NO_x Minderung bei Braunkohlefeuerungen, Deutscher Flammentag (DVV), Göttingen.

Reidick, H., Kessel, W., Fritz, P., and Hein, K. (1990). Feuerungstechnische Massnahmen zur NO_x Minderung an ausgewählten braunkohlebefeuerten Dampferzeugern, VGB Kraftwerke 1990, VGB TB 016.

Williams, G.C., Sarofim, A.F., and Lamber, N. (1971). Emission from Continuous Combustion Systems, Symp. G.J. Detroit.

Zeldovich, J. (1946). *Acta Physiochim, USSR*, **21**, 4.

Measurements of NO_x Emissions from Coal Boilers

RUI AFONSO, *New England, Power Westborough, MA, USA*
GEORGE C. DUSATKO, *Energy System Associates, Pittsburgh, PA, USA*
JOHN H. POHL, *Energy International, Mission Viejo, CA, USA*

Abstract —This paper presents and refines semi-empirical techniques to predict the NO_x emissions and compares the results with measurements from coal-fired power plants. The predictions are made by assuming independent effects of fuel properties, boiler design, and boiler operating parameters. The main effects are described by

$$\frac{NO}{NO_r} = \left(\frac{NO}{NO_r}\right)_{Coal} \left(\frac{NO}{NO_r}\right)_{Burner} \left(\frac{NO}{NO_r}\right)_{Boiler} \left(\frac{NO}{NO_r}\right)_{Fineness}$$

Each of the individual effects are further divided into more detailed, but also independent effects. The above techniques have been presented in previous work. This paper extends that work by comparing the predictions made using the above techniques with measurements made in 4 additional power plants firing 9 different coals. In addition, this paper presents new evidence showing that increasing the fineness of the coals increased NO_x emissions by up to 40 ppm (d, 3% O_2). The new information on the effect of fineness on NO_x emissions is integrated into the previous estimating techniques.

Key Words: NO_x, Coal, Burner, Coal Fineness

INTRODUCTION

Utility plants are changing the coals which they fire to reduce pollution and/or to acquire a less expensive or more reliable source of fuel. The changes in coal properties of the fuels fired will result in changes in the performance of and emissions from the plant. The operators of the utility plants need techniques to predict these changes in performance prior to committing to fuel purchases. This paper presents semi-empirical techniques to predict the emission of NO_x when coals are changed in power plants. These techniques have been presented previously by Pohl *et al.* (1983), Pohl *et al.* (1987), Pohl and Dusatko (1989a), Pohl and Dusatko (1989b), and Juniper and Pohl (1990). This paper extends the previous work by comparing the predicted NO_x emissions with measured emissions from 4 different plants firing 9 different coals. In addition, this paper presents new evidence on the effect of coals fineness on NO_x emissions and integrates that effect into the previously developed estimating techniques.

APPROACH

The approach used to develop empirical techniques to predict nitrogen oxide formation is to divide nitrogen oxide production into three main independent causes of nitrogen oxide emissions:

1. coal properties
2. burner design and operation
3. boiler design and operation

Each influence is assumed to be independent of the others. This is not strictly true for two reasons. First, nitrogen oxide produced by one influence will reduce the potential for production by the others. Second, the results of one influence will enter and alter the other regions of influence. However, the assumption of independent influence is a

TABLE I

Symbols

4 Boilers	9 Coals
Brayton Point 1	
	○ Baseline
	• Columbian
	◒ 5-Blend
Brayton Point 2	
	△ Baseline
	▲ Pocahontas
Salem Harbor 1	
	□ Baseline
	■ Alpine
Salem Harbor 3	
	◇ Baseline
	◆ Mingo-Logan

reasonable first approximation and is partially compensated for by the use of empirical data to derive the expression for prediction of nitrogen oxide emissions.

The total relative effects of all the influences are approximated by multiplying the independent effects. Each independent influence is taken as a change in nitrogen oxides from a reference condition as shown in Equation 1.

$$\frac{NO}{NO_r} = \left(\frac{NO}{NO_r}\right)_{Coal} \left(\frac{NO}{NO_r}\right)_{Burner} \left(\frac{NO}{NO_r}\right)_{Boiler} \left(\frac{NO}{NO_r}\right)_{Fineness} \tag{1}$$

Each of the three major influences are divided further into sub-influences which are handled in the same multiplicative fashion.

The absolute value of nitrogen oxide emissions is best approximated from measured nitrogen oxide emissions at a known condition. In the absence of a measured level of nitrogen oxide, the absolute nitrogen oxide can be estimated from the calculated nitrogen oxide at the reference conditions used in equation (1).

Coal Properties

The relative influences of coal properties under a range of fixed mixing conditions has been established by Pohl *et al.* (1983), Schultz (1985), Schultz and Kremer (1986). Emissions of nitrogen oxides are influenced first by the fuel nitrogen content and second by the volatile matter of the coal. Increases in fuel nitrogen always result in increases in nitrogen oxide emission. However, increases in the volatile content of the coal results in increases in nitrogen oxides emissions under normal operating conditions but decreases in emission of nitrogen oxide emission under staged conditions as shown in Figures 1 and 2. These figures are prepared for mixing conditions which approximate those of utility burners.

Data points shown in Figures 1 and 2 represent calculated points for the coals of this study. Symbols for and the properties of the coals of this study are shown in Tables I and II. It should be noted that the plots do not show that $NO/NO_R = 1.0$ at all reference conditions. We think this is a result of selecting average reference conditions which do not correspond to actual coals.

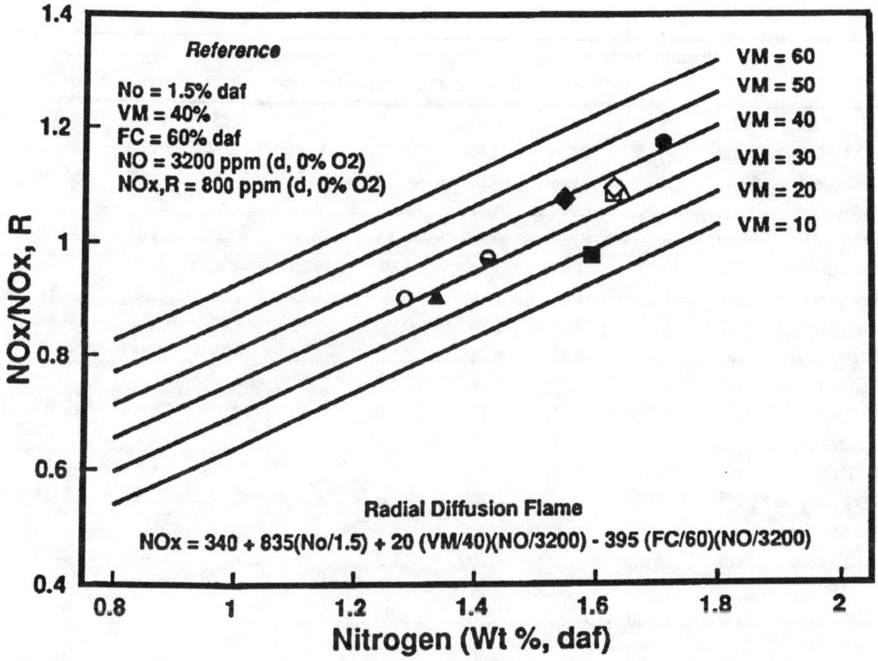

FIGURE 1. Approximate Influence of Coal Properties on NO_X Formation

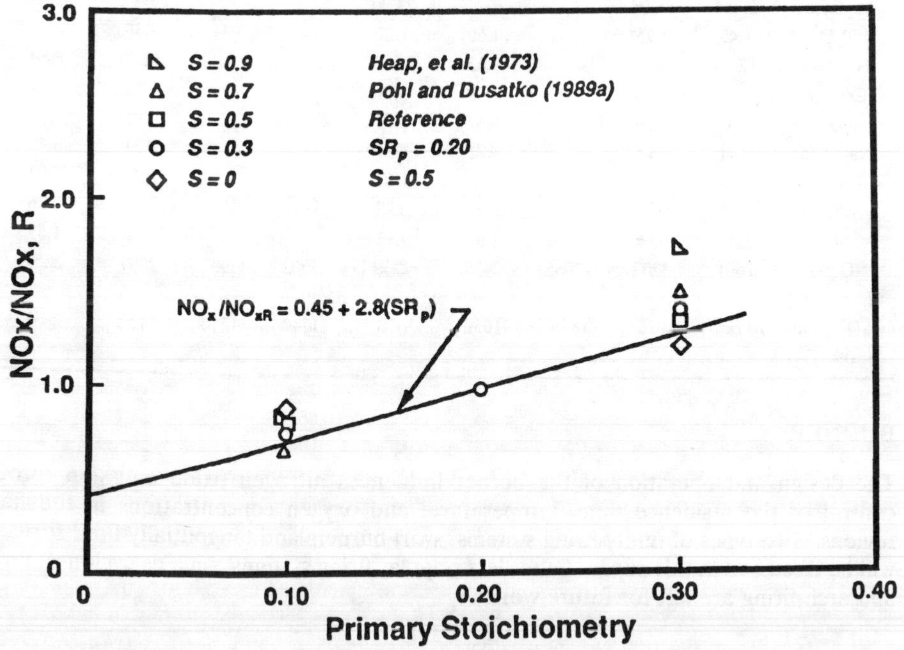

FIGURE 2. Influence of Primary Stoichiometry on Relative NO_X Emissions

TABLE II
Coal Analysis

	Brayton Point 1			Brayton Point 2		Salem Harbor 1		Salem Harbor 3	
	Baseline	Columbia	5-Blend	Baseline	Pocahontas	Baseline	Alpine	Baseline	Mingologan
H_2O (ar)	8.26	6.61		5.56	6.13	5.77	6.55	7.91	8.38
VM (daf)	31.68	41.99	28.51	31.95	19.22	33.84	22.29	34.10	34.72
FC (daf)	68.32	58.01	71.49	68.05	80.78	66.16	77.71	65.80	65.29
ASH (ar)	9.47	9.06	10.39 (a)	8.84	4.83	7.88	9.20	8.85	7.74
C (daf)	86.99	84.03	87.96	87.02	91.55	85.34	87.47	85.04	84.29
H (daf)	5.12	5.82	4.93	5.13	4.80	5.28	4.74	5.54	5.55
N (daf)	1.57	1.73	1.42	1.64	1.35	1.63	1.60	1.63	1.54
S (daf)	1.65	0.94	0.87	1.33	0.88	1.55	1.62	1.44	0.99
O (by diff) (daf)	4.66	7.48	4.82	4.86	1.57	6.18	4.28	6.34	6.83
HHV Btu/lb (ar)	12,656	12,682	13,802	13,238	14,079	13,167	13,017	12,598	12,696
HGI	76	59		76	98			65	61
IDT (red) °F	2577	2692		>2700	2330			2255	2434
AFT (red) °F	2648	2698		>2700	2370			2468	2633
HT (red) °F	2694	>2700		>2700	2400			2569	2696
FT (red) °F	>2700	>2700		>2700	2470			2653	2700
S_1O_2(a)	48.46	60.80		51.40	37.65			50.86	52.50
A_2O_2(a)	28.84	26.04		30.68	24.90			26.92	28.19
T_1O_2	1.47	1.25		1.60	1.27			1.30	1.34
Fe_2O_3(a)	12.74	6.74		9.63	14.47			12.40	8.80
CeO (a)	1.81	0.98		1.49	8.01			1.15	1.71
MgO (a)	0.68	0.50		0.61	1.73			0.96	0.94
Na_2O (a)	0.41	0.28		0.21	1.04			0.51	0.48
K_2O (a)	2.25	1.64		1.88	1.67			3.16	3.03
P_2O_3 (a)	0.55	0.24		0.47	0.22			0.34	0.26
SO_3	1.33	0.66		1.65	7.91			1.06	1.59
NO ppm (d, 0% O_2)	2601	3202	2565	2978	2339	3002	2907	2997	2857
SO_2 ppm (d, 0% O_2)	1186	938	686	1059	670	1242	1290	1162	810

BURNER

The design and operation of the burner influences nitrogen oxide emission through control of the residence time, temperature, and oxygen concentrations in the flame regions. Two types of utility firing systems, swirl burners and tangentially fired furnaces, will be discussed in this paper. Other less popular firing systems, such as cyclone-, turbo- and arch-firing are left for future work.

Swirl Burner

The influence of design and operation of swirl burners on nitrogen oxide production is

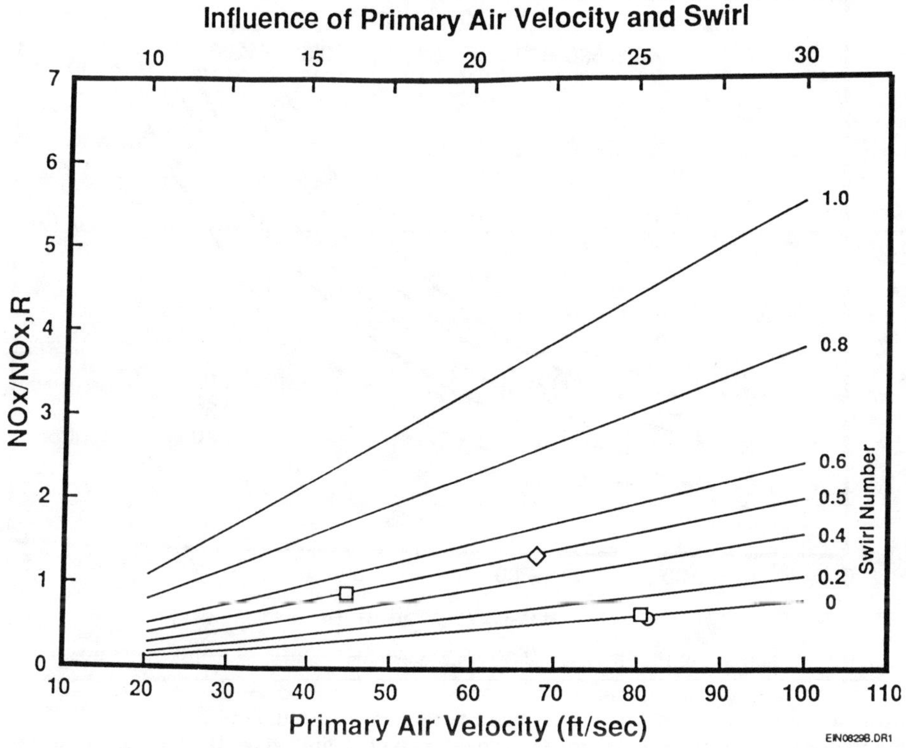

FIGURE 3. Influence of Primary Air Velocity and Swirl Primary Velocity (m/sec)

difficult to predict because design and operation of such burners alters the flame shape and the flame shape has a large influence on the nitrogen oxide formation, Payne, *et al.* (1980), Heap, *et al.* (1973), Michelfelder, *et al.* (1977), and Weber, *et al.* (1987). The influence of swirl burner design and operation is divided into three categories:

1. primary air stoichiometry as it influences flame stoichiometry
2. primary air velocity as it influences flame residence time
3. swirl number as it influences mixing and flame residence time

The influence of primary air stoichiometry on nitrogen oxide emissions can be approximated as shown in Figure 2, Pohl and Dusatko (1989a).

The primary air velocity affects nitrogen oxide emission in two ways. First, the primary air velocity, the secondary air velocity and the swirl number determine the geometry and type of flame developed. Second, the primary air velocity determines the flame residence time. The first effect will be discussion with the influence of swirl number.

The second effect, residence time, can be approximated by estimating the flame residence time as a function of the primary air velocity and the burner geometry, Pohl, *et al.* (1987) and Payne (1980). The proposed relationship between primary velocity, swirl number, and relative nitrogen oxide emissions is shown in Figure 3.

Nitrogen oxide emissions have been shown to be a complicated function of swirl number and the types of flames, Heap, *et al.* (1973), Michelfelder, *et al.* (1977) and Weber, *et al.* (1987). Most swirl flames currently used in utility boilers are Type II flames

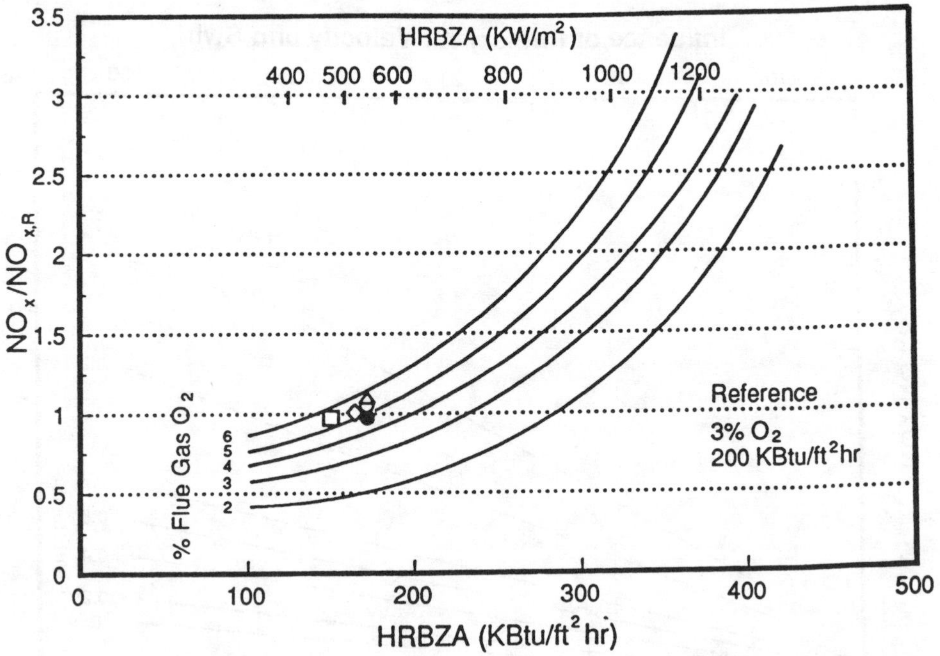

FIGURE 4. Influence of Heat Release and Oxygen concentration on Relative Emission of Nitrogen Oxides

with closed recirculation patterns which operate at swirl numbers of 0.3-1.0. Pilot scale data shows an increase in nitrogen oxide for Type I and Type II coal flames at swirl numbers between 0.3 and approximately 1.0, Heap, *et al.* (1973) and Michelfelder, *et al.* (1977). Decreases in nitrogen oxides emissions can occur at swirl numbers larger than 1 for these types of coal flames. A tentative relationship between swirl numbers and nitrogen oxide emissions from utility coal flames is proposed in Figure 3. It should be noted that while the relationship was derived from pilot scale data, Figure 3 accurately accounts for the difference in normal wall swirl burners and the high swirl cell burners of nitrogen oxides measured in this study.

Tangential-Firing

The mixing and combustion region for tangentially-fired boilers are different from those of the boilers using swirl burners, and may need to be handled differently. In tangentially-fired boilers, the fuel is injected in each corner of the boiler at a tangent to a small circle in the center of the boiler. Air is injected above and below the fuel jet without swirl. This injection system creates a large stable fireball in the burner region.

For normally-fired tangential boilers, the nitrogen oxide levels are controlled by the volume and excess air level in the burner zone. The flame residence time can be approximated as the burner volume divided by the amount of flue gas at the flame temperature. The flame temperature can be normalized by the ratio of the Burner Zone Heat Release Area (BZHRA) to a standard BZHRA of 200,000 Btu/ft^2. Such calculations yield a relative residence time in the flame zone. At this time, data available (Pohl and Dusatko 1989b) is too limited to develop a correlation for the influence of tangential-firing on NO_x emissions.

Correlations for the nitrogen oxide emissions from tangentially-fired will be developed in future work. For the purposes of this study, the influence of primary air stoichiometry

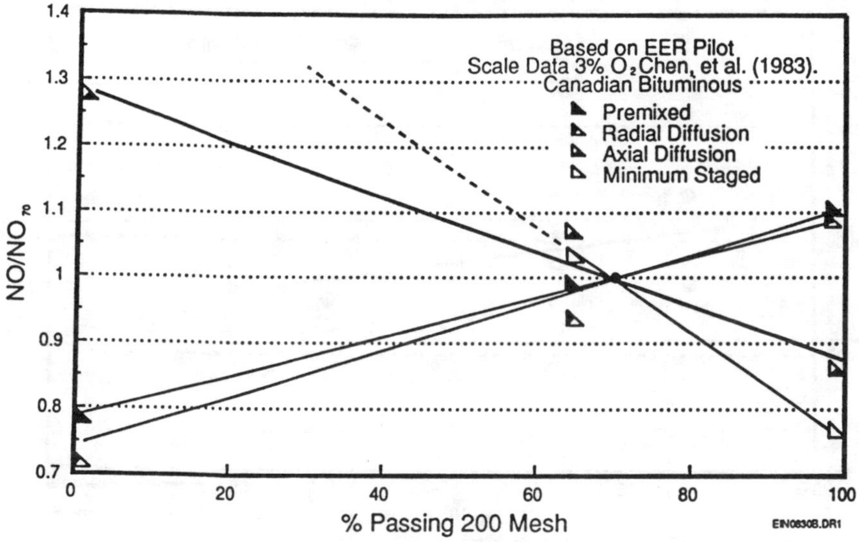

FIGURE 5. Influence of Particle Fineness on NO_X Emissions

and velocity on nitrogen oxide emissions will be approximated using the correlations developed for swirl burners. None of the boilers of this study were staged. However, the influence of staging (OFA, LNCF) on nitrogen oxide emissions can be estimated from the relationship for staged combustion (Pohl and Dusatko 1989b) which agrees with the general rule of thumb that staging combustion can reduce the emissions of nitrogen oxide from 40-60 percent.

BOILER

The influence of boiler design and operation on nitrogen oxide emission can be separated into two parts. The first is the effect of oxygen concentration outside the flame region and the second is the time-temperature profile outside the flame region. Boiler characteristics are shown in Table III. The influence of oxygen concentration has been correlated with nitrogen oxide emission as shown in Figure 4. The influence of the time-temperature history in the boiler on nitrogen oxide formation has been correlated with the Heat Release in the Burner Zone Area (HRBZA = heat release in Btu divided by the area of walls in the burner zone plus 10 feet above the burner zone plus the the area of planes across both ends of the burner zone) as shown in Figure 4.

COAL FINENESS

Limited information is available on the effect of coal fineness on NO_X emissions. Early work (Chen, *et al.* 1983) indicated that under fuel lean conditions, increased coal fineness increased NO_X emissions, but under staged conditions, increased fineness reduced NO_X emissions. This result was explained by the increased exposure and higher conversion of fuel nitrogen compounds to oxidization under lean combustion and increased exposure to reducing conditions under rich combustion. This explanation is supported by evidence that the conversion of volatile nitrogen is higher than that of char nitrogen (Pohl and Sarofim 1977 and Song, *et al.* 1982).

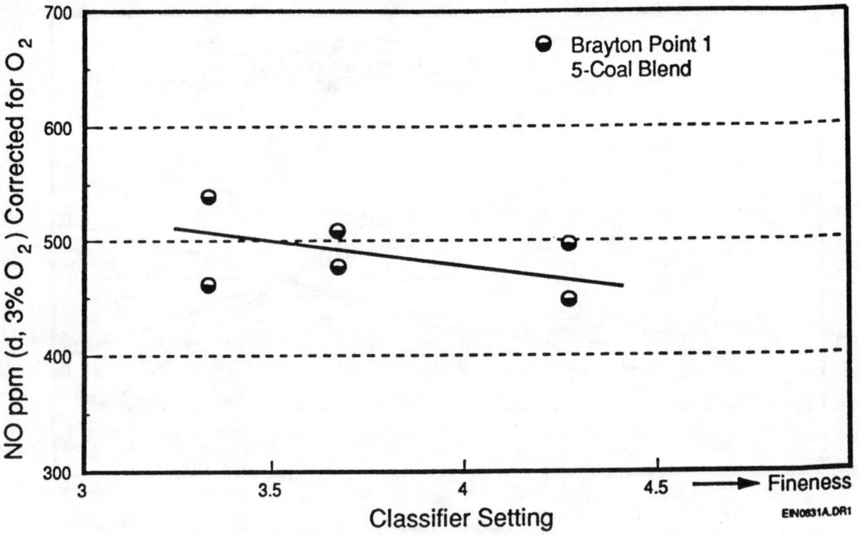

FIGURE 6. Influence of coal Fineness on NO_x Emission from Brayton Point 1

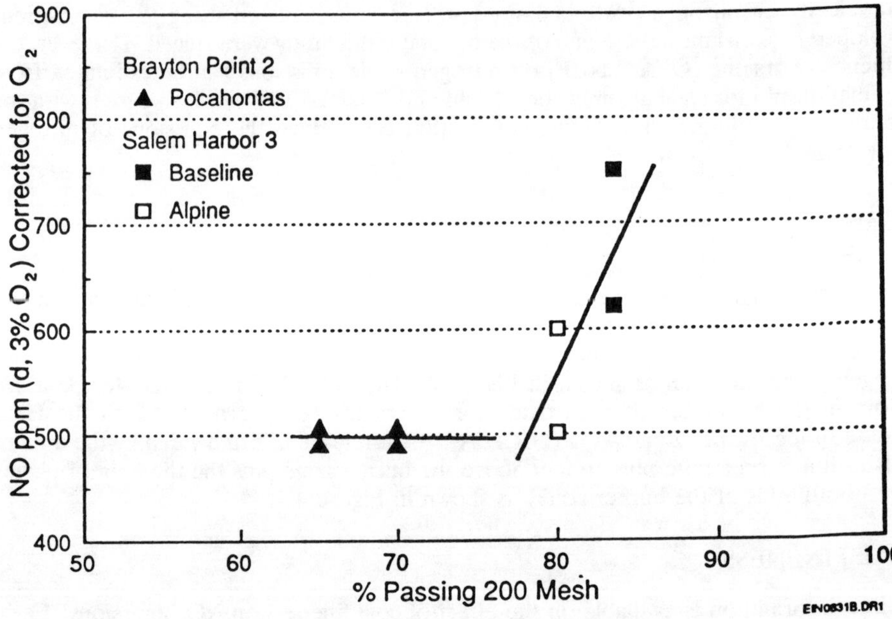

FIGURE 7. Influence of Coal Fineness on NO_x Emissions from Brayton Point 2 and Salem Harbor 3

The tentative conclusion from these studies is that coal fineness can effect NO_x emissions. The direction and extent of this influence depends on the stoichiometry and mixing conditions in the flame regions. Figure 5 shows the tentative relationship between coal fineness and NO_x emission for several limiting flame mixing conditions. The figure

TABLE III
Boilers

	Brayton Point	Salem Harbor	
	Units 1 & 2	Unit 1	Unit 3
Manufacturer	CE	B & W	B & W
MCR MW	250	85	150
Firing Mode	TDW (a)	FW (b)	FW
Number of Burners	4 × 8	4 × 3	4 × 4
Height (ft)	124.6	730	84
Width (ft)	27.7	22.5	36
Depth (ft)	48.3	30.4	22
Volume (Kft^3)	166.6	49.9	66.5
Projected SA (Kft^2)	28.5	9.1	11.3
BZA (Kft^2)	14.1	5.0	7.6
Coal Rate 10^6 (Btu/hr)	2429	748	1260
VHR ($KBtu/ft^3$-hr)	14.8	13.0	18.9
HRPSA (Btu/ft^2-hr)	85.2	82.2	111.5
BZAHR ($KBtu/ft^2$-hr)	172	150	166
Coal Pipe (in.)	10	14.75	11.75
Primary Air (Klbm/hr)	365	114	200
PA/Fuel (lb/lb)	1.98/1	2/1	2/1
PA Excess Air (%)	0.20	0.18	0.18
Primary Temp (°F)	160	150	150
PA Velocity (ft/sec)	89.4	33.6	68.8
Burner throat in.	—	33	33
Secondary Air Flow (Klbm/hr	2,450	858	1,520
SA Temp (°F)	560	540	500
SA Velocity (ft/sec)	211	72	99
Swirl Number	0	1.0 (c)	1.0 (c)
Air/Fuel (lb/lb)	14.6/1	15.1/1	15.2/1
Excess Air	1.37	1.35	1.35
Stack O_2 %	5.5	5.3	5.3

a. Tangential Divided Wall; b. Front Wall; c. Estimated

was developed based on pilot scale data for a Canadian bituminous coal (Chen, *et al.* 1983).

The influence of fineness on NO_x measured in this study is shown in Figures 6 and 7. The data in Figure 6 for Brayton Point 1 firing the blend of 5 coals shows a slight decrease in NO_x emissions with increasing classifier setting (increasing fineness, actual fineness was not measured). Figure 7 shows that increasing the fineness from 65 to 70 percent through 200 mesh (74 microns) for firing the Pocahontas coal in Brayton Point 2 had little influence on NO_x emissions. However, increasing the fineness from 80 percent passing 200 mesh to 87 percent passing 200 mesh caused the NO_x emissions to increase from 550 ppm to 630 ppm. The NO_x emissions reported here have been corrected for dilution and normalized for the influence of oxygen using Figure 4.

The results are consistent with those reported in Figure 5. Brayton Point 1 and 2 are tangentially-fired boilers with longer, less rapidly mixed flames. Based on the results, these tangentially-fired furnaces would have flames which have mixing patterns between those of the Radial Diffusion and Axial Diffusion Flames. Whereas the swirl stabilized

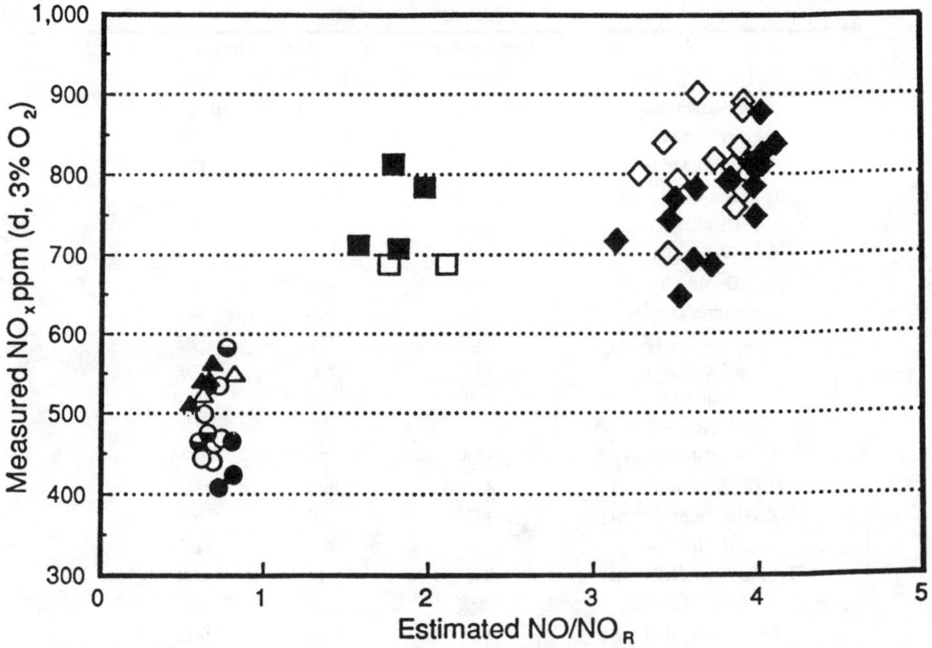

FIGURE 8. Comparison Estimated and Measured NO_x Emissions

flames of Salem Harbor approach the mixing conditions between the Radial Diffusion Flames and the Premixed Flames.

With the limited data available at present, we recommend that the influence of coal fineness be approximated as follows:

- Tangential-Fired Boilers—No influence
- Swirl Burners—

$$\frac{NO}{NO_{70}} = 0.8 + 0.003(\% < 200m)$$

Use of this relationship for Salem Harbor 3 firing the Alpine coal with an increase in the percent passing 200 mesh from 80 to 87 yields an increase in NO_x emissions from 550 to 561 whereas the measured increase was much larger from 550 to 630. Additional data is needed before a relationship can be established to account for the influence of fineness on NO_x emissions from coal fired boilers.

Estimated NO_x

The NO_x emissions were estimated using the techniques described in this paper. The influence of the properties of the coals were estimated from Figure 1, the influence of the primary stoichiometry from Figure 2, the influence of primary air velocity and stoichiometry from Figure 3, the influence of heat release in the burner zone and oxygen from Figure 4, and the influence of coal fineness from Figure 5. These effects were combined and the comparison is shown in Figure 8. The comparison is made by comparing the combined influences with the actually measured NO_x emissions.

CONCLUSION

This paper has presented further comparison of the semi-empirical techniques previously developed for estimating NO_x emissions and has extended the techniques to include the influence of coal fineness in NO_x emissions. The predicted NO_x emissions have been compared with measurements made in 4 boilers firing 9 coals. More data and refinement are required before these techniques can be fully accepted to estimate NO_x emissions.

REFERENCES

Chen, S. L., Heap, M. P. and Pershing, D. W., (1983). "Bench-Scale NO Emissions Testing of World Coals' Influence of Particle Size and Temperature," *Proceedings of the 1982 Joint Symposium on Stationary Combustion NO_x Control,* EPRI CS-3182, July.

Heap, M. P., T. M. Lowes, R. Walmesley, and H. Bartelds, (1973). Burner Design Principles for Minimum NO_x Emissions. In: Proc. Coal Combustion Seminar, EPA-650/2-73-021 (NTIS PB 224-210). U.S. Environmental Protection Agency, Research Triangle Park, NC, p. 141.

Michelfelder, S., M. P. Heap, and T. M. Lowes, (1977). "The Effect of Burner Construction and Burner Parameters on the Emission of NO_x from Natural Gas, Heavy Fuel Oil, and Pulverized Coal Flames," International Flame Research Document Number K20/a/90, Ijmuiden, The Netherlands, April.

Payne, R., (1981). Private communication. Energy and Environmental Research Corp., Irvine, CA.

Payne, R., T. Akiyama, and J. G. Witkamp, (1980). Studies of the Pollutant Emission Characteristics of a Number of Oil-Fired Industrial Burners and the Relationship to Firing Conditions. IFRF Doc. Nr. F3T/2/6, International Flame Research Foundation, Ijmuiden, Netherlands.

Pohl, J. H. and G. C. Dusatko, (1989). "The Influence of Coal Properties, Boiler Design and Boiler Operating Conditions on Emission of Nitrogen Oxides," International Flame Research Foundation 9th Members Conference, Noordwijkerhout, The Netherlands, May.

Pohl, J. H., S. L. Chen, M. P. Heap, and D. W. Pershing, (1983). Correlation of NO_x Emissions with Basic Physical and Chemical Characteristics of the Coal. In: Proc. Joint Symposium on Stationary Combustion NO_x Control. EPRI CS-3182, Electric Power Research Institute, Palo Alto, CA, p. 36-1.

Pohl, J. H., and G. C. Dusatko, (1989). "Estimates of NO_x Formation in Utility Boilers", ASME Fossil Fuels Combustion Symposium, Houston, TX, February.

Pohl, J. H., G. C. Dusatko, P. C. Orban, and R. W. McGraw, (1987). "The Influence of Fuel Properties and Boiler Design and Operation on NO_x Emissions", Joint Symposium on Stationary Combustion NO_x Control, New Orleans, L.A., March.

Schultz, W., (1985). Experimentelle Untersuchung der Bildung von Stickstoffoxiden bei der Kohlenstaubuer-brennung. Doctorarbeit, Ruhr-Universitut Bochum, West Germany, p. 218.

Schultz, W., and Kremer, H., (1986). Influence on Temperature of the Formation of NO_x During Pulverized Coal Combustion. Paper presented at Twenty-First Symposium (Intl) on Combustion, Munich, West Germany. The Combustion Institute, Pittsburgh, PA, August 3-8.

Weber, R., J. P. Smart, and W. J. Phelan, (1987). "NO_x Reduction with Coal Firing by Application of Both Internal Air Staging and Fuel Rich Precombustors: Report on the AP 14 Trials, International Flame Research Foundation Document Number F03712/16, Ijmuiden, The Netherlands, February.

Use of Numerical Modeling in the Design of a Low-NO_x Burner for Utility Boilers

W. A. FIVELAND and C. E. LATHAM *The Babcock & Wilcox Company Research and Development Division Alliance, Ohio 44601*

Abstract—This paper describes the role of numerical modeling in the design of a low-NO_X burner for utility boilers.

Environmental regulations for boiler systems are becoming more stringent. With an increasing emphasis on the use of lower-grade fuels, emission controls, and reliability under more complex operating conditions, conventional, empirically-derived design criteria for fossil fuel boilers cannot be adequately extended. As a result, Babcock & Wilcox has developed mathematical modeling techniques to enhance design capability. The role of numerical modeling in an ongoing burner development program is discussed. A review of pilot-scale studies demonstrates the feasibility of using the novel low-NO_X burner in a commercial unit. Recent full-scale modeling of the baseline unit and the retrofit condition demonstrates that unit performance, e.g., combustion efficiency, is not degraded with the new burner. Modeling at the utility scale confirms trends predicted at the pilot scale.

Ongoing work will collect data for the retrofit conditions, and models will be compared with data and improved. As feasibility is shown for use of the burner in a utility boiler and experience is gained with the models, they will provide valuable tools for use on other "clean air" projects.

INTRODUCTION

NO_X emissions from stationary sources like boilers are regulated by the United States Federal government, and in some cases emissions are regulated by the state or local authorities. Federal New Source Performance Standards (NSPS) have set NO_X limits for gas-, oil-, and coal-fired boilers sold since 1971. Many pre-NSPS units will face NO_X reductions by Acid Rain legislation enacted in the early 90s through the Clean Air Act. In regions of non-compliance, very strict NO_X emission levels on existing or new units are placed. Suppliers of combustion systems are being required to reduce NO_X to lower and lower levels for new boilers and for a diverse population of existing units. In the population of existing boilers, units with standard cell burners account for 26,000 MW of electric power generating capacity. These burners were designed for rapid mixing of the fuel and oxidant, producing good combustion efficiency and large quantities of NO_X making them excellent candidates for low-NO_X burner retrofits. Low-NO_X burners incorporate features to restrict formation of thermal and fuel NO_X. Low-NO_X burners are designed to provide controlled mixing of the air with fuel. Combustion is generally delayed, producing a longer, but cooler flame that limits the production of NO_X. Air staging ports (NO_X ports) can be used to divert combustion air from the burners to further reduce NO_X. The design and overall arrangement of the burners, NO_X ports, gas recirculation system, and furnace, in combination with the fuel properties, will dictate the resulting NO_X emissions.

Combustion efficiency benefits from rapid, complete mixing of air and fuel. Limiting the rate at which the air and coal mix, particularly during early stages of combustion, can effectively control NO_X formation, but these measures for coal-fired units have a tendency to reduce combustion efficiency. Consequently, one aspect or the other can be controlled fairly easily, but controlling both can prove difficult and trade-offs are necessary. This is primarily true for U.S. eastern bituminous coals, and is less of a concern with western free-burning coals.

Low-NO_X combustion systems will eventually be utilized on a wide variety of existing units, including single wall-fired, opposed-fired, and roof-fired boilers. These boilers range in size from 12.6 kg/s to 1260 kg/s (10^5 lb/hr to 10^7 lb/hr) main steam. The small units have one or two burners; and the largest units have over 100 burners. Furnace sizes and configurations vary considerably, firing coals ranging from low-volatile bituminous and sub-bituminous to lignite. Controlling the NO_x in the various boilers will result in a variety of combustion system/furnace configurations.

Traditionally, combustion systems have been designed based upon empirical indices and correlations derived from full-scale performance data. The indices and correlations would be used to arrange equipment, predict combustion efficiency, and predict NO_x emission levels. Although the empirical methods work well for designing systems similar to those already in service, they are often unreliable for new configurations outside of the "design envelop" of the data base.

With the development of high-speed engineering workstations, numerical simulation of combustion systems is now feasible and provides a new engineering tool that is ideal for evaluating design options. Since the mid-1970s, Babcock & Wilcox (B&W) has been developing flow, combustion, and heat transfer models that can be used together with conventional methods to optimize system performance. The objective of the model development has been to produce computational models applicable to a wide variety of coal-fired furnaces. The models are based on a fundamental description of the various interacting processes that occur during flow, combustion, and heat transfer, and contain a minimum of empiricism.

In this paper, the role of numerical modeling in an ongoing burner development program is discussed. A review of pilot-scale results is presented demonstrating the feasibility of using a novel low-NO_x burner concept in a commercial unit. Full-scale boiler modeling of the baseline and the retrofit configurations demonstrated that the unit performance is not significantly affected with the new burners and confirmed the effects observed in the pilot-scale tests. Velocity, temperature, and species concentrations for the baseline condition are compared to the field data. Future work which will lead to the proof of feasibility of the low-NO_x cell burner in the field is discussed.

BURNER DEVELOPMENT PHILOSOPHY

The large number of existing units in the field with cell burners makes these units good retrofit candidates. The standard two-nozzle cell burner configuration is shown in Figure 1. Close spacing between the nozzles and rapid mixing of the fuel and the air produced by the cell burner create a short, stubby flame that maximizes the heat release rate. These burners are desirable for their high combustion efficiency; however, they promote the formation of NO_x.

The low-NO_x cell burner was conceived and then designed as the retrofit burner for cell units. It was designed to fit into the existing boiler openings without any modification to boiler pressure parts, providing an economical alternative to reduce NO_x. The novel design of the burner necessitated characterizing the burner at several scales showing feasibility at each scale to settle concerns about maintaining combustion performance. An integrated numerical and experimental program was designed (LaRue and Rogers, 1985) to fully characterize the burner at several scales: 1.76 MW, 30 MW, and utility scale. Numerical modeling was done before testing to project burner performance and locate instrumentation. After the tests, predictions were compared to data, models were modified if required, and performance was scaled to the next level. The low-NO_x burner configuration is shown in Figure 2. All of the fuel and about 50% of the air flows through the lower throat. The balance of the air is directed through the upper throat. The burner throats control the mixing of the fuel and oxidant, creating a long flame. The lower throat produces a fuel-rich region suitable for reducing NO_x, while the upper throat provides the balance of air to complete the combustion.

Utilities are concerned that new retrofits do not significantly affect plant performance. For example, the controlled mixing which produces longer flames may also increase the amount of unburned fuel and gas temperatures at the furnace exit causing overheating of the pendant surfaces.

MODELING APPROACH

Computational techniques have been developed to solve the governing, partial-differential equations. In general, the models solve the fully elliptic, three-dimensional, finite-difference approximations of the conservation equations for mass, momentum, turbulence, combustion, and heat transfer. The details of the models are documented in Fiveland and Wessel (1988), and they are similar in many respects to models developed by Lockwood, et al. (1980), and Smith and Smoot (1989). The furnace geometry is divided into control volumes (52,000 active) as shown in Figure 3, depicting the grid layout used to analyze a 605-MW boiler. All features of the furnace are simulated such as the burners, NO_x ports, hopper, arch, and pendant surfaces. The governing equations are integrated over each control volume and discretized using finite-difference techniques described by Patankar (1980). The modeled equations for flow, combustion, and heat transfer are solved for each control volume in the figure.

Detailed information for each control volume is produced for the flow characteristics, major chemical species (CH_4, C, CO_2, CO_4, SO_2, H_2O), temperature, and heat flux distributions throughout the boiler—information which is difficult or impractical to obtain experimentally. The model must be sufficiently complex to describe flow, combustion, and heat transfer, and yet simple enough for application to practical systems.

Figure 4 shows the sequence of reactions modeled. Coal devolatilization and subsequent char oxidation are characterized by a total fuel evolution rate (Fiveland and Wessel, 1988). The first stage of fuel evolution represents devolatilization and char formation processes due to pyrolysis. A second stage of fuel evolution represents the rate of char conversion to fuel before it forms products. Ash remains after the fuel evolves from the char.

Combustion of the fuel is characterized by a two-step process: In the first step, fuel is assumed to react with oxygen to form carbon monoxide and products; in the second step, there is a kinetic reduction of the carbon monoxide to carbon dioxide.

This two-step reaction sequence is important because of its effect on heat transfer and NO pollutant processes. In many practical combustion devices operating with fuel-rich conditions like the low-NO_x cell burner, CO concentrations are about 10^5 ppm. Neglecting this intermediate combustion product could result in large inaccuracies in the distribution of gas temperature and oxygen concentration. Consequently, this has an effect on NO formation and reduction mechanisms, since the reaction rates depend to a large extent on temperature and oxygen concentration.

Because the pollutant species have low concentrations (10^2 ppm) relative to the major species (10^5 ppm), the formation and reduction of NO can be considered a perturbation in the flow and combustion processes; therefore, the flow and combustion calculations can be performed first and the pollutant calculations done subsequently.

As discussed in Fiveland and Wessel (1991), fuel NO is a major contributor to overall NO inventory for all pulverized-coal applications. Formation of fuel NO can occur by homogeneous or heterogeneous reactions. Nitrogen is released from the coal during devolatilization of the coal and oxidation of the char. There is still widespread controversy over the rate at which the nitrogen is released; however, following the recommendations of Smith, et al. (1982):

- Nitrogen is released at a rate equal to the mean rate of coal weight loss.
- Fuel nitrogen evolved during char oxidation follows the same reaction sequence as nitrogen that evolves during devolatilization.

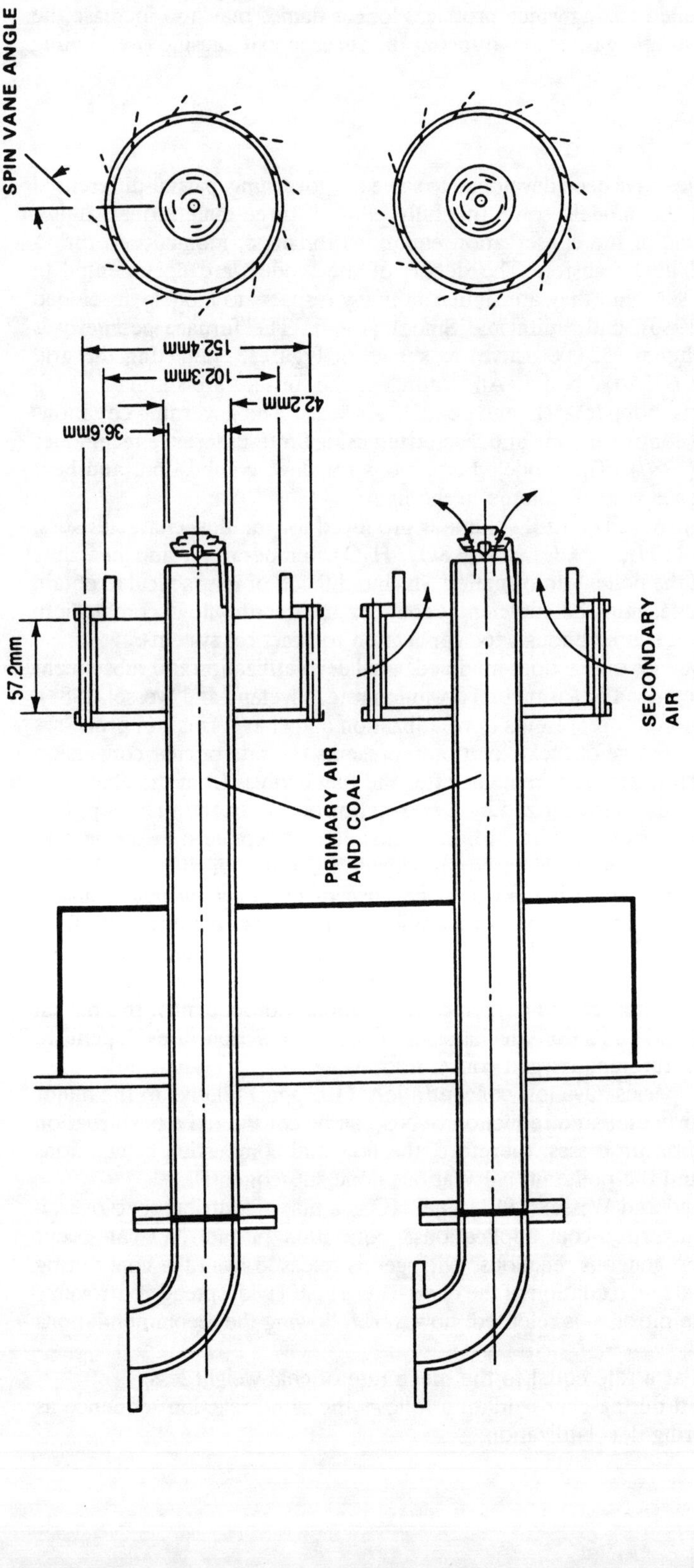

FIGURE 1. Standard cell burner for the combustion and fuel preparation facility.

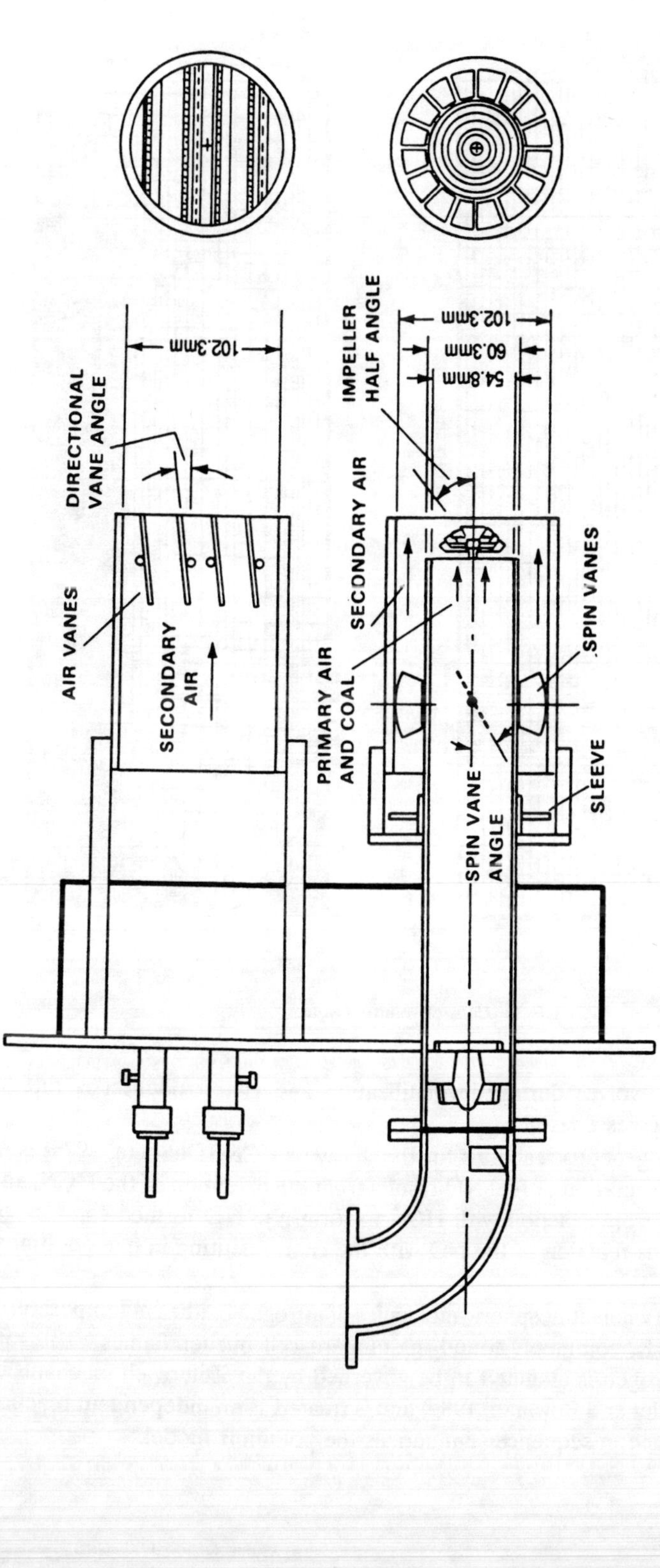

FIGURE 2. Low-NO_x cell burner for the combustion and fuel preparation facility.

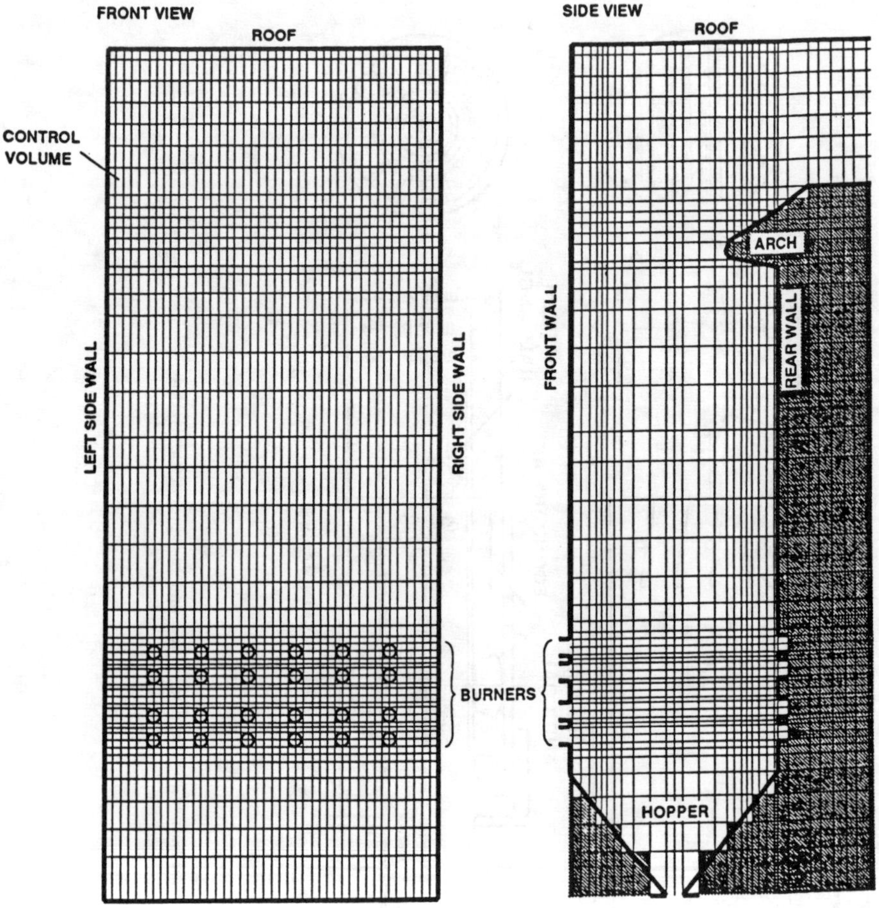

FIGURE 3. Control volume layout.

• Fuel nitrogen that evolves during devolatilization and char oxidation is rapidly converted to HCN in the gas phase.

The sequence of homogeneous reactions for the decay of HCN (DeSoete, 1975) is as shown in Figure 4. The formation of fuel NO results from the oxidation of the HCN, and depletion of NO results from reaction with HCN to form N_2. NO formed can also be reduced by heterogeneous reaction of the NO with the char, resulting in one additional path.

Thermal NO is formed when atmospheric molecules of nitrogen oxidize at temperatures in the range 1600-1800 K, commonly found in standard cell burner flames. The rate of conversion of the nitrogen is assumed to be governed by the Zeldovich mechanisms (Bowman, 1975, and Miller and Bowman, 1989) and is treated as an independent reaction path, as shown. This reaction sequence summarizes the pollutant model.

$$Coal \xrightarrow{T} Fuel + Char \qquad HCN + O_2 \rightarrow NO + \text{---}$$

$$Char \xrightarrow{K_{ch}} Fuel + Ash \qquad HCN + NO \rightarrow N_2 + \text{---}$$

$$Fuel + O_2 \rightarrow yCO + ZH_2O + \text{---} \qquad NO + Char \rightarrow CO + \text{---}$$

$$CO + 1/2\,O_2 \xrightarrow{K_{CO}} CO_2 \qquad O + N_2 \xrightarrow{T} NO + N$$

$$Coal \xrightarrow{T} HCN \qquad N + O_2 \rightarrow NO + O$$

$$N + OH \rightarrow NO + H$$

FIGURE 4. Reaction sequence for coal combustion and pollutant formation.

Pilot-Scale Modeling

Pulverized-coal combustion was investigated in a pilot-scale unit using the standard cell burner and the low-NO_x cell burner. Tests were conducted in the combustion and fuel preparation facility (CFPF) at the Babcock & Wilcox Alliance Research Center. This facility's furnace was designed for a heat input of 1.76 MW, with coal firing. Figure 5 shows this wall-fired furnace, which is geometrically similar to many commercial boilers. The surfaces in the furnace are cooled by natural circulation of water at atmospheric pressure, and the walls are insulated with refractory and fibrous insulation to reduce heat absorption and to maintain furnace-gas temperatures that are typical of commercial-scale furnaces.

The furnace is equipped with two cell-type burners mounted on the front wall and arranged vertically on the furnace centerline. Tests were conducted with Ohio No. 6 coal, and three firing configurations were used. Pulverized coal was supplied to the burners by an MPS-32 pulverizer with a rotating classifier and primary air heating and supply system. Table I summarizes the characteristics of the pulverized coal and the operating conditions for the three firing configurations.

In tests with the standard cell burner, the rapid mixing of the primary and secondary flow streams promotes high combustion temperatures and complete carbon utilization within 1 m of the throat. This was confirmed by experimental observations of flame length and by local measurements of CO, CO_2, and O_2 concentrations. NO concentrations for the standard cell burner rose rapidly to the final exit value within 0.5 m of the burner throat. The predicted concentration of NO at the exit agreed well with the measurements as shown in Table II.

In the low-NO_x burner tests, swirling fuel and air were directed through the lower throat (all of the fuel and approximately 56% of the secondary air); the balance of the air is directed through the upper throat.

The low-NO_x cell burner heat release is illustrated by the complex temperature distribution shown in Figure 6. The highest temperatures (approximately 2100 K)

TABLE I

CHARACTERISTICS OF PULVERIZED COAL AND FURNACE OPERATING CONDITIONS FOR THE COMBUSTION AND FUEL PREPARATION FACILITY

Furnace Configuration Coal Type	Std. Cell Ohio #6	Low-NO_x Cell Cleaned Ohio #6	Overfire Air Cleaned Ohio #6
Ultimate Analysis, %			
Moisture	4.60	8.08	6.84
Carbon	63.80	67.84	68.75
Hydrogen	4.50	4.82	4.88
Nitrogen	1.22	1.35	1.37
Sulphur	4.14	2.26	2.29
Ash	15.50	7.66	7.77
Oxygen	6.24	7.99	8.10
Higher Heating Value, kJ h^{-1}	26851.3	28674.9	29062.0
Primary Air			
Flow Rate, kg h^{-1}	396.9	369.7	374.7
Temperature, K	334	333	333
Moisture, kg/kg dry air	0.003 0	0.018 0	0.011 3
Secondary Air			
Flow Rate, kg h^{-1}	1187	1795	1842
Temperature, K	599	593	588
Moisture, kg/kg dry air	0.003 0	0.018 0	0.011 3
Flow Rate of Pulverized Coal, kg h^{-1}	222	198	201
Furnace			
Stoichiometry	1.174	1.174	1.174
Load, MW	1.66	1.58	1.62
Static Pressure, atm	1.0	1.0	1.0

TABLE II

SUMMARY OF NO_x PREDICTIONS AND MEASUREMENTS

		NO_x Emissions (ppm)	
Firing Configuration	Coal Type	Measured	Predicted
Standard Cell	Ohio No. 6	973	898
Standard Cell	Lower Kittaning	796	799
Low NO_x Cell	Ohio No. 6	310	292
Overfire Air	Ohio No. 6	120	155

are within 0.6 m of the burner throats, where the combustion is fuel-rich or nearly stoichiometric and where heat-release rates are the greatest. Near the air ports of each low-NO_x cell burner, the temperatures are much lower because the combustion in these regions is fuel-lean. Steep temperature gradients exist between the hot and cool regions of the alternating fuel and air ports of the two burners. In the middle and upper furnace regions, the temperatures are highest near the walls, where the flow is upward. Temperatures are lower in the center of the furnace, where the downward recirculating gases have had a chance to cool. Predicted temperatures in the lower furnace compared well with the data, in view of the steep temperature gradients in those regions. The predicted average furnace exit gas temperature of 1482 K is 33 K lower than the measured value.

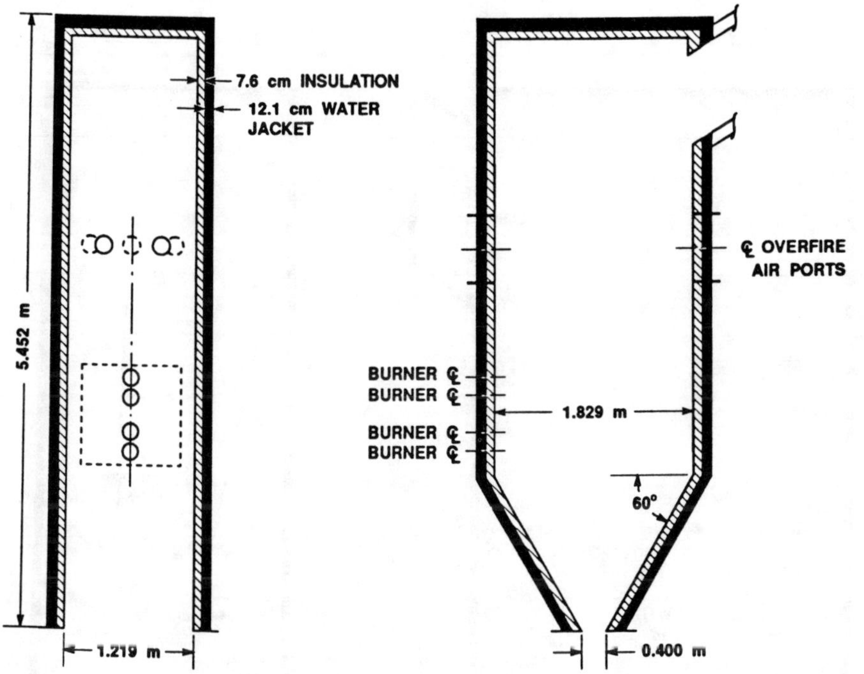

FIGURE 5. Pilot-scale furnace.

The long flame produced by this burner arrangement controls the mixing of the fuel and air, and it delays combustion. This produces regions of high-temperature, fuel-rich combustion, which is better suited for the reduction of fuel NO. Carbon monoxide concentrations were predicted as large as 17,000 ppm in the combustion zone and decreased to values less than 100 ppm at the furnace exit. Figure 7 shows the measured and predicted NO_x values. Table II shows the exit values, which agree well with the measured values.

Overfire air injection was used with the low NO_x cell burner to produce staged combustion in the furnace; 38% and 28% of the air was injected through the lower and upper throats, respectively, and the balance through the overfire air port. Predictions of combustion and heat transfer were similar to the low NO_x cell burner configuration, except that combustion was delayed further by lack of oxygen in the lower furnace, resulting in a longer flame.

Although the flame was less distinguishable during pilot-scale tests, it was predicted to extend across the furnace and up the back wall by about 1 m. This was confirmed by local measurements of CO, CO_2 and O_2 concentrations in the furnace. Complete utilization of the carbon at the furnace exit was measured and also predicted. Values of CO were predicted in the burner zone near 160,000 ppm and 80 ppm at the furnace exit. Experimental data confirmed these predictions. Concentrations of NO are as high as 600 ppm in fuel-rich regions of the flame, and as low as 50 ppm in fuel-lean regions of the furnace. Mixing within the furnace results in an average NO value of 149 ppm at the furnace exit. An experimental value of 125 ppm was measured.

Utility-Scale Modeling

Pulverized coal combustion was investigated for a 605-MW B&W universal pressure utility boiler. The configuration of the furnace is shown in Figure 8. The dry ash

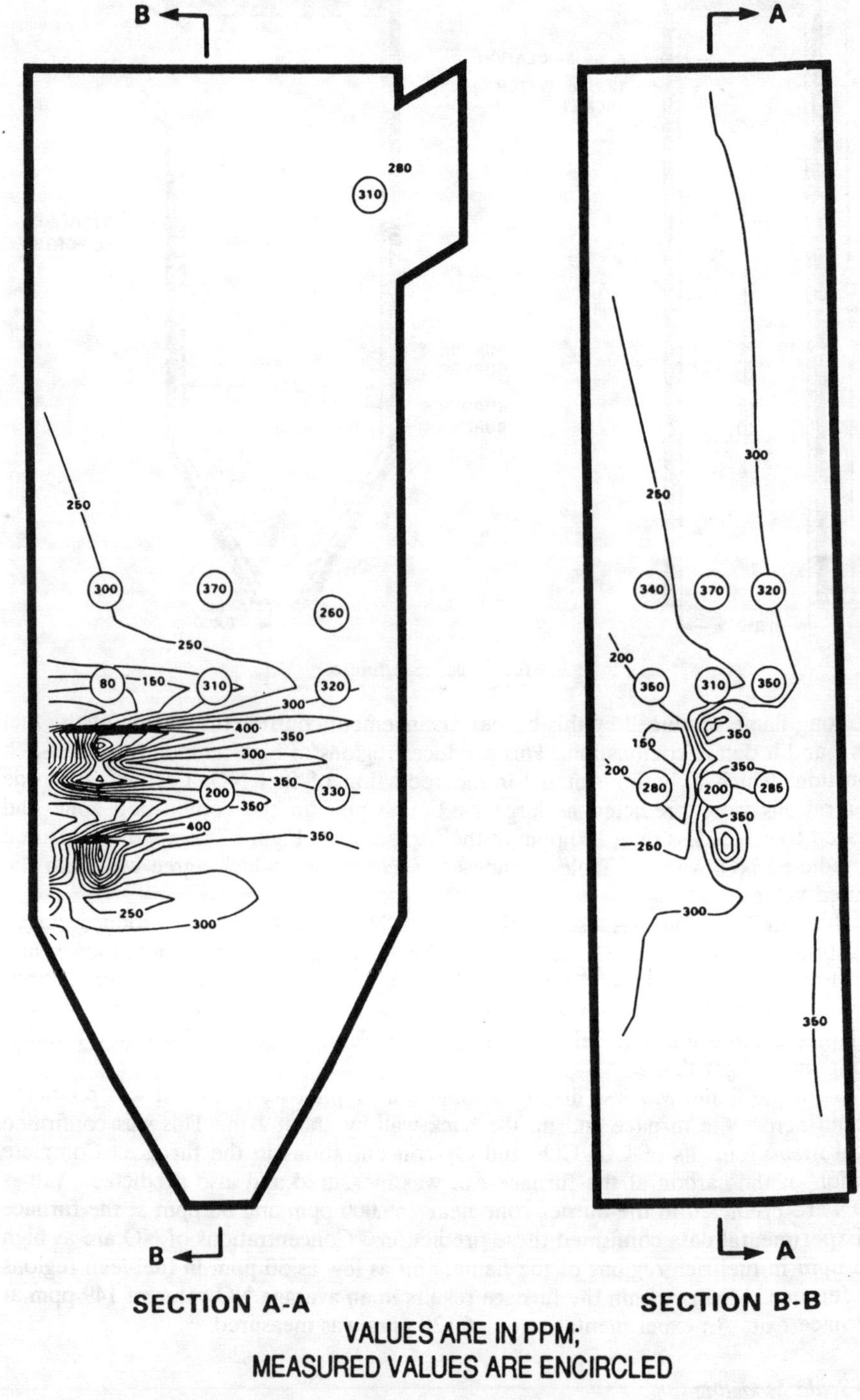

FIGURE 6. Gas temperature contours in the pilot-scale furnace.

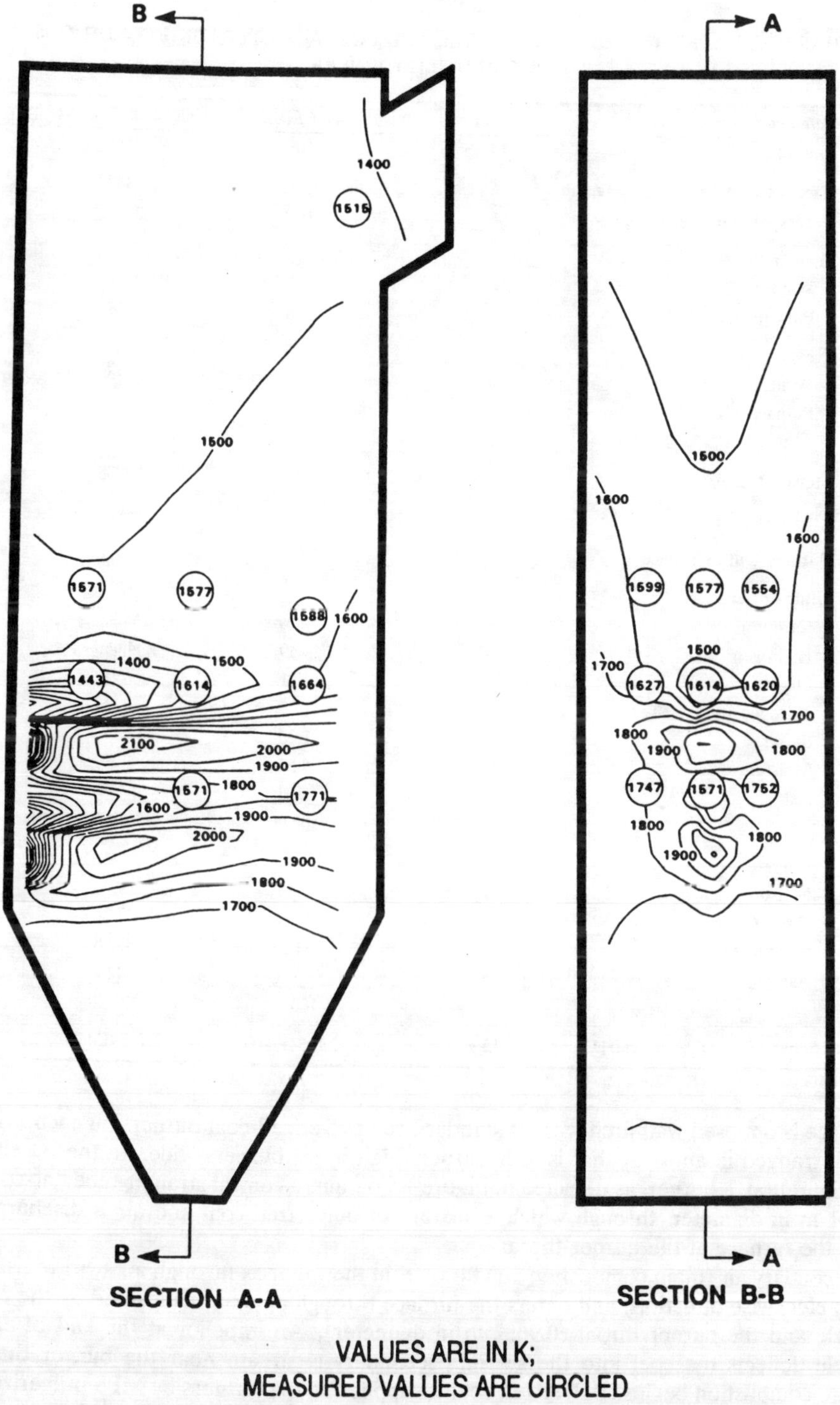

FIGURE 7. Concentration of nitric oxide in the pilot-scale furnace.

TABLE III

CHARACTERISTICS OF PULVERIZED COAL AND FURNACE OPERATING CONDITIONS FOR THE UTILITY BOILER

Parameter	*Units*	*Full Load*	*Intermediate Load*
Output	MWe	605	458
Excess O_2 at Economizer Outlet	%	3.739	5.348
Excess Air Leaving Economizer	%	21.2	33.3
Temperatures			
Windbox	K	599	583
Pulverizer Outlet	K	338	338
Pressures			
Windbox	Pa	5960	3334
Furnace	Pa	5462	3160
Atmospheric	Pa	99903	100343
Moisture in Air	kg/kg-dry air	0.0073	0.0038
Coal Flow Rate	kg/sec	57.2	44.2
Burners Out-of-Service	—	None	8
Ultimate Analysis			
Carbon	%	68.52	65.95
Hydrogen	%	4.49	4.48
Oxygen	%	7.51	7.91
Nitrogen	%	1.23	1.16
Sulfur	%	1.00	1.12
Moisture	%	5.55	5.39
Ash	%	13.70	13.99
		100.00	100.00
Proximate Analysis			
Moisture	%	5.55	5.39
Fixed Carbon	%	47.61	47.24
Volatile Matter	%	33.14	33.38
Ash	%	13.70	13.99
		100.00	100.00
Higher Heating Value (HHV)	J/kg	2.7654E+07	2.7559E+07

furnace is opposed-wall fired with 24 standard cell pulverized-coal burners on each firing wall arranged in an array that is four burners high by six burners wide. In the baseline configuration, a burner, as depicted in Figure 1, contains a central straight-tube "nozzle", 0.381 m in diameter, through which a mixture of pulverized coal and air is discharged into the furnace at the burner throat.

Secondary air (main combustion air) flows from the windbox through air swirl registers (in a clockwise direction) and enters the furnace through an annulus formed by the coal nozzle and the burner throat (0.9652 m in diameter). An impeller at the end of each nozzle deflects the coal into the swirling secondary airstream near the burner throat where combustion begins. Pulverized coal is supplied to the burners by MPS pulverizers. At full load, a high-volatile eastern bituminous coal is supplied at a rate of 57.10 kg/s and 21.2% excess air.

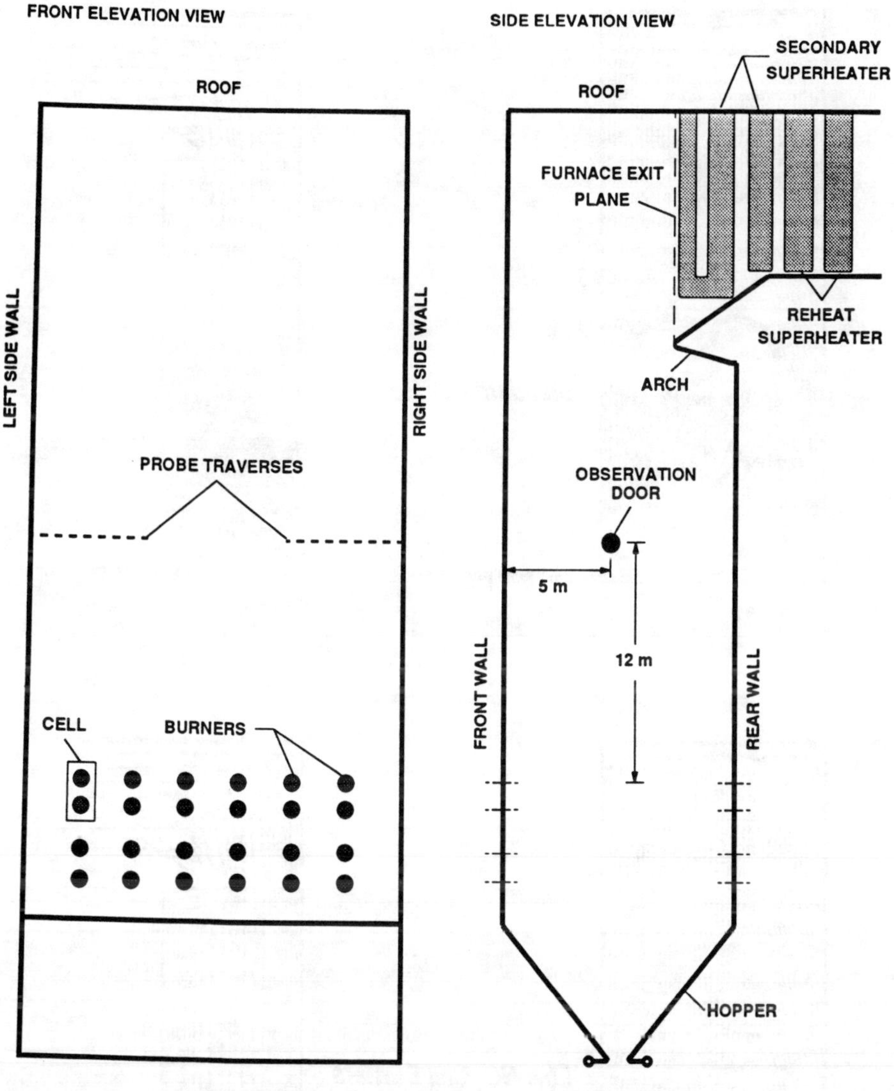

FIGURE 8. 605-MW utility boiler.

The furnace is enclosed by rows of vertically-oriented, water-cooled tubes in a membrane wall construction. Water at supercritical pressure is circulated sequentially through all the heated surfaces, the membrane walls, and convection pass elements. At full load, the boiler produces 554 kg/s of steam at 814 K and 26.34 MPa. The operating conditions modeled are listed in Table III.

Velocity vectors predicted by the model for both the standard cell and low-NO_x cell configurations are shown in Figure 9. The arrows indicate the direction of the flow in the plane and the length of the arrows are directly proportional to the magnitude of the

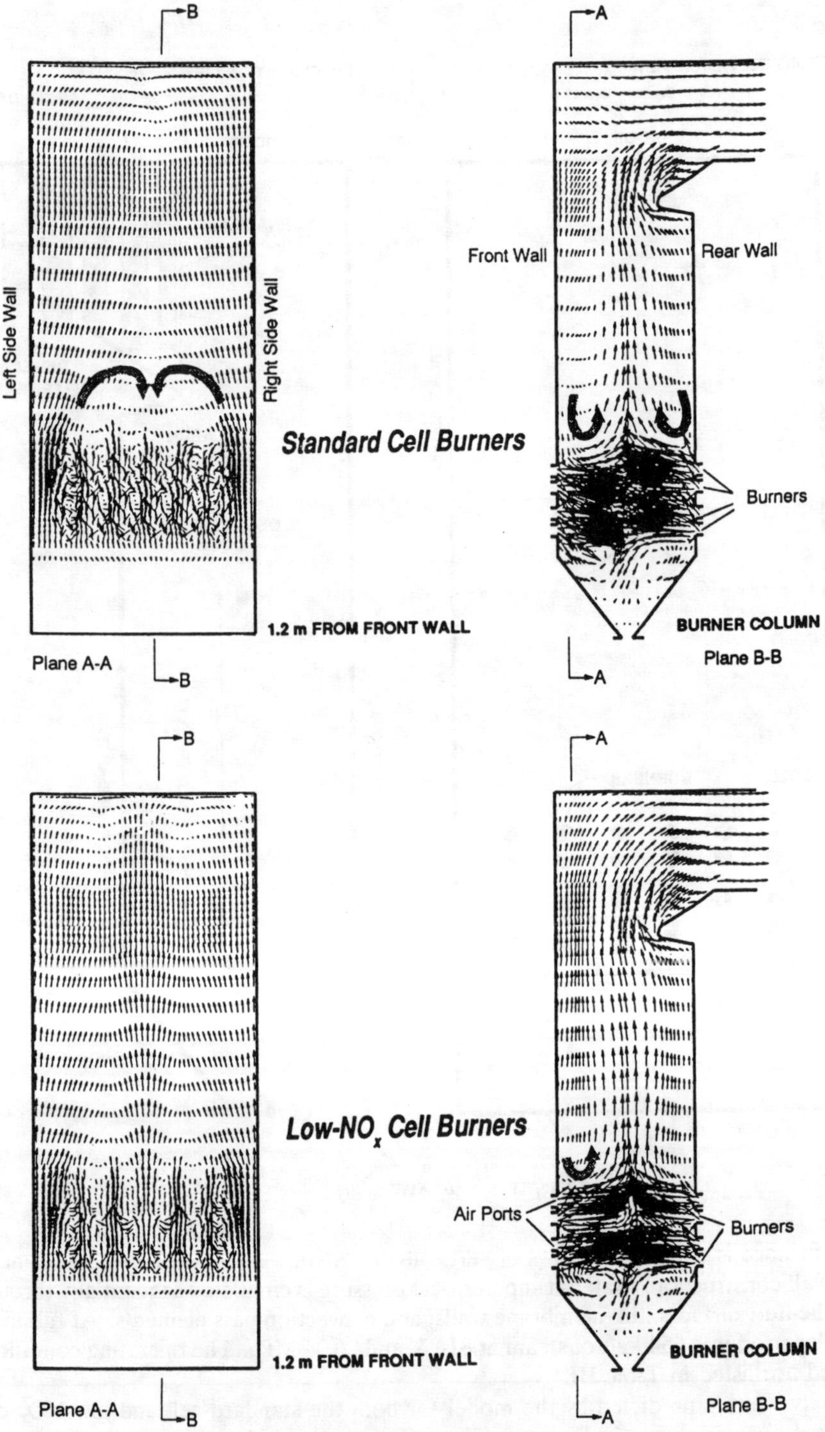

FIGURE 9. Velocity distributions in the utility boiler.

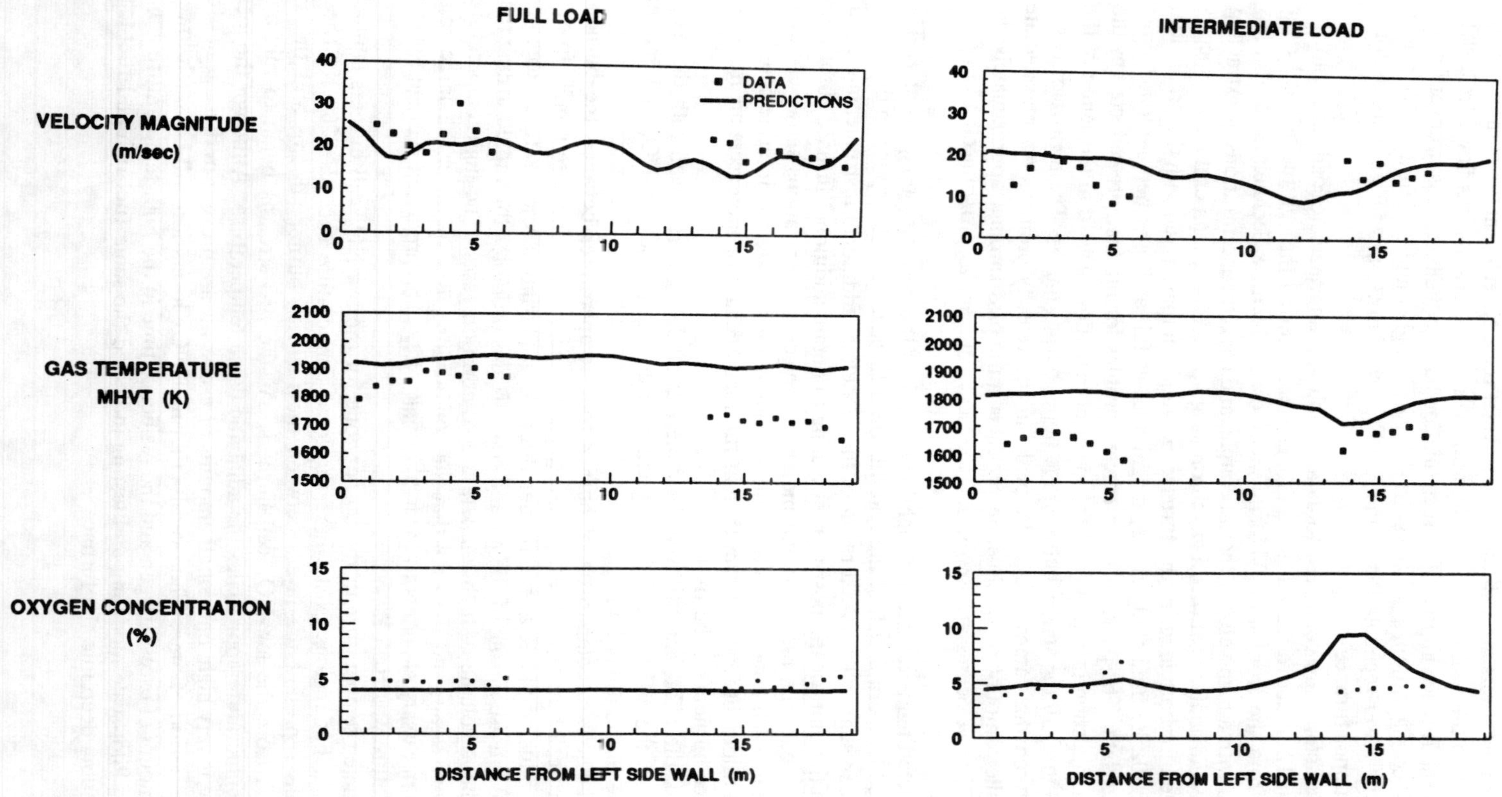

FIGURE 10. Gas velocity magnitude, gas temperature, and oxygen concentration at a plane 12m above the burners for full and intermediate load.

velocity. The flow enters through the burners. Inlet velocity vectors at the burners are not perpendicular to the wall due to swirling flow. For both cases, the gas flow is relatively quiescent in the furnace hopper. The recirculation back to the burners is stronger for the standard cell configuration, since the swirl intensity is higher and the burner entrainment is higher. The lower recirculation of the low-NO_x cell burner provides a more uniform flow in the upper furnace.

Model predictions are compared to data for full and intermediate loads in Figure 10 at a location 12 m above the burners. Probes were inserted through the left and right walls, 5 m from the front wall. Velocity was measured with a Fechheimer probe, gas temperature with a high-velocity thermocouple, and oxygen concentrations by extracting a gas sample and analyzing it with a gas chromatograph. Velocity temperature and species were collected for both loads at the furnace exit, and compared well with the prediction.

In Figure 10, the predicted velocity magnitudes at full and intermediate loads agree well with the data. The peaks in the predicted profiles for full load are reflected by the data. The combustion products from the lower burners flow upward in the center of the furnace and through the spaces between the upper burner flames. The high-velocity gas flowing between the flames causes the peaks in the velocity profiles. At intermediate load, a smaller amount of combustion products must pass through approximately the same area. The velocities are lower, and the peaks are not as pronounced 12 m from the burners.

The predicted temperatures are slightly higher than the measured temperatures. The overpredicted temperatures indicate that the lower furnace heat absorption in the model may have been slightly low. As a result of the modeling, additional instrumentation will be added to the unit for the post-retrofit testing to obtain additional furnace absorption data. At full load, the predicted and measured oxygen concentrations show very little variation along each traverse and from side-to-side in the furnace. At intermediate load, the O_2 is considerably less uniform than at full load because of the excess air from the out-of-service burners on the front wall.

In general, the pilot- and utility-scale numerical modeling agrees well with the data. Consequently, the models will be a good tool to assess the low-NO_x cell burner performance in the utility boiler.

Figure 11 shows a comparison of burner zone temperatures between the burners. The standard cell burner combustion causes gas temperature to increase rapidly from a mean inlet value of 517 K at the burner throat to a maximum temperature exceeding 2090 K. The intense mixing of the air and coal streams promotes the rapid burning that causes the large volume of high-temperature combustion gases. The high temperature, the 21.2% excess air, and the time that the combustion gases spend in this large zone provide the necessary elements to form "thermal" NO_x. The predicted average furnace exit temperature was 1655 K.

The predicted temperature in the burner zone for the low-NO_x cell burners is much lower overall, but regions exist where the local temperature exceeds 2200 K. These small regions represent the layers between the fuel-rich burner streams and the air port streams from the low-NO_x cell burners. As the two streams begin to mix and react, the temperature approaches the adiabatic flame temperature. Although the local temperature is very high, only small amounts of excess air are present in these zones for a very short period of time. The overall "thermal" NOx formation will be much lower, since the amount of excess oxygen and the residence time in the high-temperature zones are limited. Pilot-scale modeling and testing indicated the same fuel-rich and fuel-lean zones extending deep into the furnace.

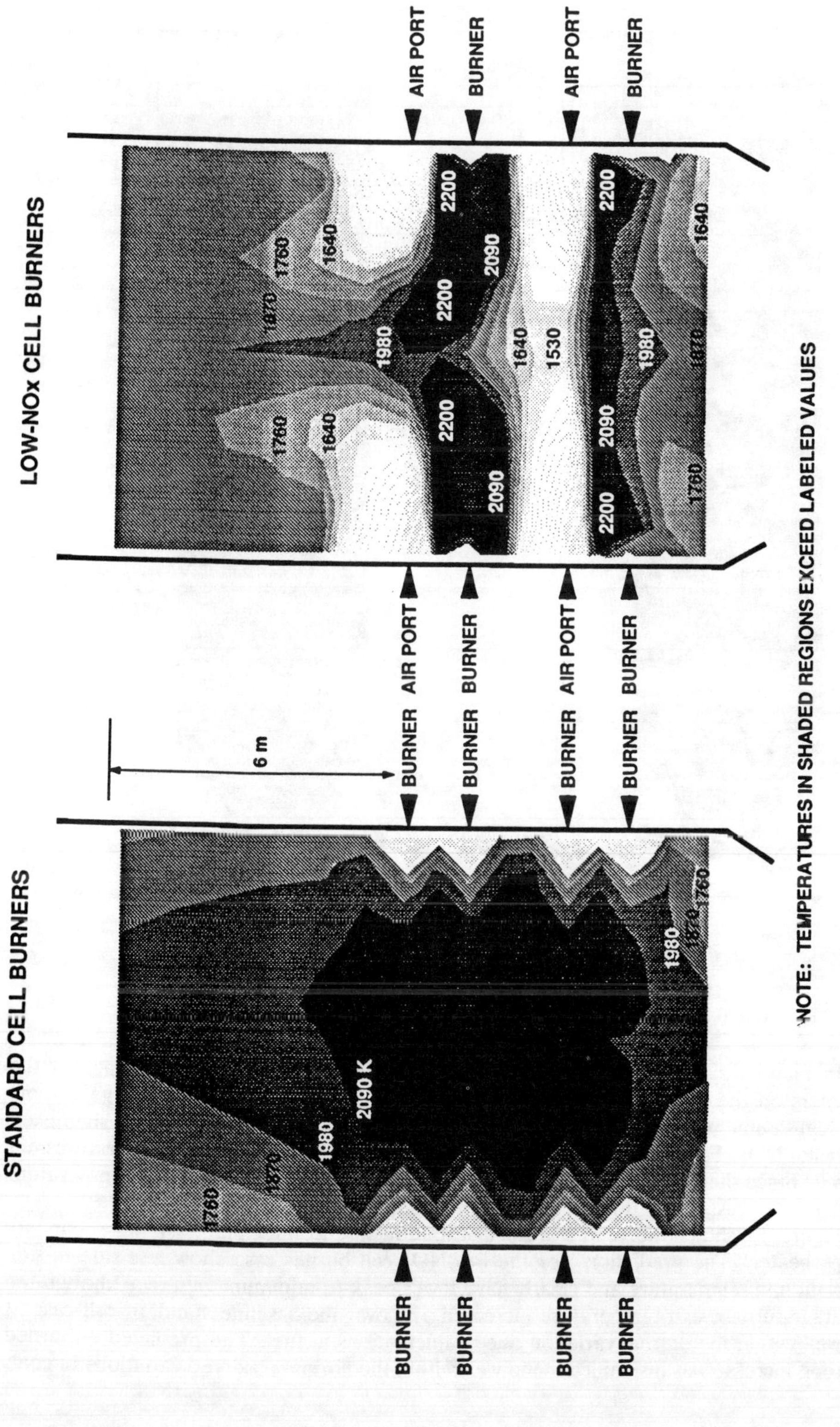

FIGURE 11. Predicted burner zone temperatures.

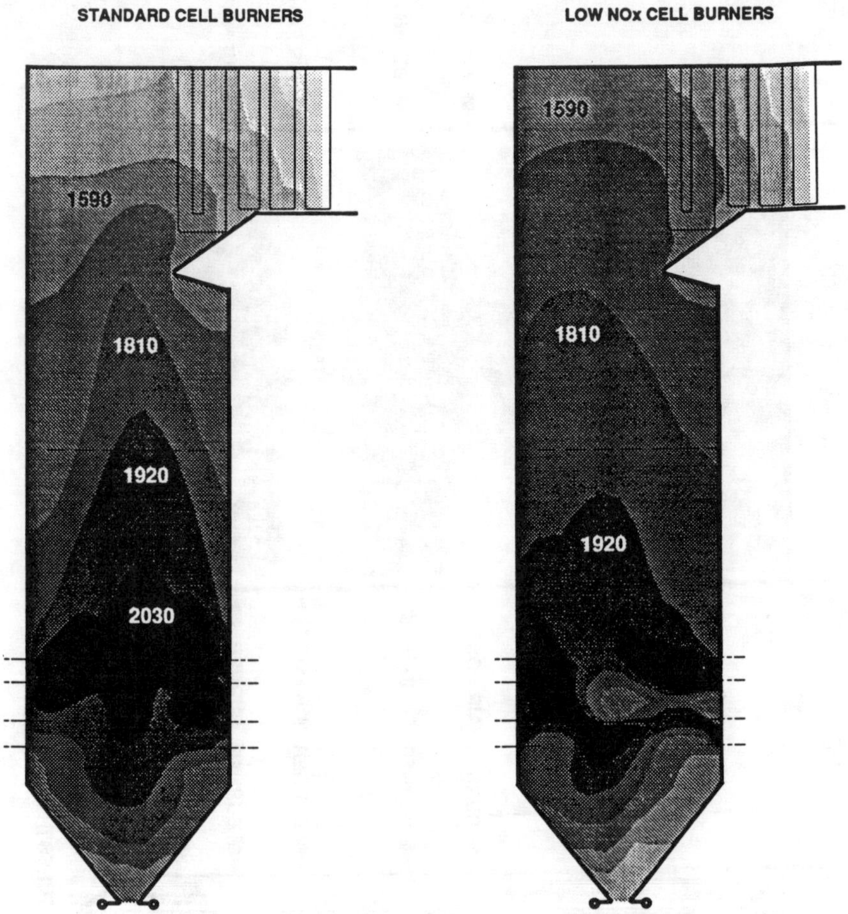

FIGURE 12. Gas temperatures at the furnace centerline.

In Figure 12, the temperature distributions for the unit are shown for each of the burners on the furnace centerline. For the standard cell burner case, a large volume of high-temperature combustion products is predicted, which will produce significant thermal NO_x. For the low-NO_x cell burner case, the combustion zone temperatures are lower. Since the lower furnace gas temperatures are more uniform, smaller temperature differences exist at the inlet of the secondary superheater.

Figure 13 shows the temperature distribution at the inlet plane of the secondary superheater. The predictions for the low-NO_x cell burner case show less side-to-side variation in temperature and clearly have lower peak-to-minimum values. Although the average furnace exit temperature increased 5 K over the baseline standard cell case, it is well within the diurnal variation due to fluctuations in fuel. The predicted unburned carbon increase was insignificant and well within the normal observed variations in coals

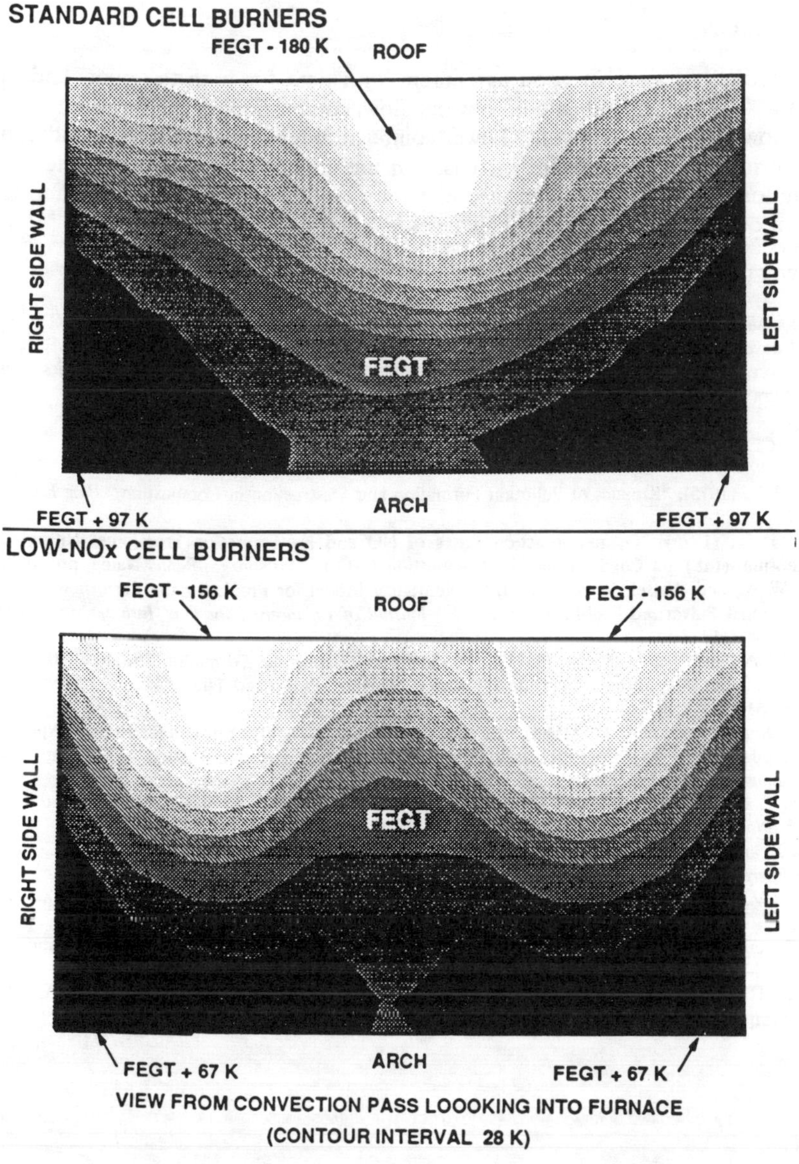

FIGURE 13. Predicted furnace exit gas temperature distributions.

used at the plant. The modeling indicates no obvious flow, combustion, or heat transfer problems in using the low-NO_x cell burner in the boiler.

The low-NO_x cell burner was shown to provide low NO_x levels in the pilot-scale facility. Modeling the two burner configurations in the utility unit showed that there was no significant change in boiler performance. Future unit testing will be done to provide data quantifying the retrofit operation. Data and model predictions will be compared, and with the successful use of the model, it will provide a tool for future applications.

Numerical modeling can be used as a tool to quantify flow, combustion, and heat transfer in combustion equipment. As experience is gained with the models and validation efforts continue, the role of models in the development of "clean air" equipment will increase.

CONCLUSION

Model predictions were in good agreement with data for both the pilot- and utility-scale data. The models indicate no obvious flow, combustion, or heat transfer problems with the low-NO_x burner retrofit: Lower combustion temperatures were found implying lower thermal NO; insignificant increases in unburned carbon were predicted; and a more uniform distribution of furnace exit temperature was predicted, leading to less chance of overheating the pendant surfaces.

ACKNOWLEDGMENT

The Babcock and Wilcox Company with support from the Electric Power Research Institute (EPRI) conducted the initial development of the low-NO_x cell burner. The U.S. Department of Energy (DOE), EPRI, B&W, Dayton Power & Light (DP&L) (host site), and a consortium of cell burner unit owner utilities sponsored the field demonstration.

REFERENCES

Bowman, C. T., (1975), "Kinetics of Pollutant Formation and Destruction in Combustion," *Prog Energy Comb Sci,* **1,** pp. 33-45.

DeSoete, G. G., (1975), "Overall Reaction Rates of NO and N_2 Formation From Fuel Nitrogen," 15th Symposium (Int.) on Combustion, The Combustion Institute, Pittsburgh, Pennsylvania, pp. 1093-1102.

Fiveland, W. A., and Wessel, R. A., (1988), "A Numerical Model for Predicting the Performance of Three-Dimensional Pulverized-Fuel-Fired Furnaces," *Journal of Engineering for Gas Turbines and Power,* **110,** No. 1, pg. 117.

Fiveland, W. A., and Wessel, R. A., (1991), "Model for Predicting the Formation and Reduction of Nitric Oxide Pollutants in Three-Dimensional Furnaces Burning Pulverized Fuel, "*Journal of the Institute of Energy,* **64,** No. 458.

LaRue, A. D., and Rodgers, L. W., (1985), "Development of Low-NO_x Cell Burners for Retrofit Applications," presented at the 1985 EPA/EPRI Joint Symposium on Stationary Combustion NO_x Control, Boston, Massachusetts, May 6-9, 1985.

Lockwood, F. C., Salooja, A. P., and Syed, S. A., (1980), "A Prediction Method for Coal-Fired Furnaces," *Combustion and Flame,* **38,** pp. 1-15.

Miller, A. J., and Bowman, C. T., (1989), "Mechanism and Modeling of Nitrogen Chemistry in Combustion," *Prog Energy Comb Sci,* **715,** No. 4, pp. 287-338.

Patankar, S. V., (1980), Numerical Heat Transfer and Fluid Flow, Hemisphere Publishing, McGraw-Hill, New York.

Smith, P. J., and Smoot, L. D., (1989), "Detailed Model for Practical Pulverized Coal Furnaces and Gasifiers, Volume I—General Technical Report," Brigham Young University, Provo, Utah.

Smith, P. J., Hill, S. C., and Smoot, L. D., (1982), "Theory for NO Formation on Turbulent Coal Flames," 19th Symposium (Int.) on Combustion, The Combustion Institute, pp. 1263-1270.

The Effect of the Near Burner Aerodynamics on Pollution, Stability and Combustion in a PF-Fired Furnace.

T. ABBAS*, P. COSTEN*, M. A. HASSAN**, and F. C. LOCKWOOD* * *Department of Mechanical Engineering, Thermofluids Section, Imperial College of Science, Technology and Medicine, London, UK* ** *Mechanical Power Engineering Department, Faculty of Engineering, Cairo University, Giza, Egypt*

Abstract—An investigation into the effect of a change in the near burner region aerodynamics was undertaken in a large-scale laboratory furnace firing a high-volatile bituminous pulverised fuel. To this end two variations of primary air/fuel jet design were installed in a swirl stabilized burner of the type used in wall-fired boilers. Significant differences in flame shape, intensity, stability performance and pollutant emissions were observed and recorded when a burner gun with a single annular fuel injection orifice (SAO), appropriate to industrial practice, wherein the central bluff body provides a location for the light-up burner, was replaced with a simplified version with the bluff body removed. The replacement, with a single central orifice (SCO), maintained the same cross-sectional area to that of the original.

A dramatic reduction in NO emissions (30%) was observed for this burner with no discernible deterioration in flame efficiency brought about by a reduction in particle burnout. Under normal operating conditions the SCO burner exhibits a similar stability performance to that of the SAO.

Key Words: Pollution, PF, Diffusion Flame, Aerodynamics, Stability, Burnout.

NOMENCLATURE

D_s	Diameter of the secondary air tube (burner throat)=56 mm
X	Axial distance from the burner throat (mm)
M_p	Mass of primary jet (coal +air) (kg/hr)
V_p	Velocity of the primary air (m/s)
M_s	Mass of secondary air (kg/hr)
V_s	Velocity of secondary air (m/s)
NBR	Near burner region
IRZ	Internal recirculation zone
ERZ	External recirculation zone
R_m	Momentum ratio
S_s	Swirl number
λ	Excess air (%)

INTRODUCTION

The industrial use of pulverised fuel (PF) in combustors faces the challenge of not only maintaining high combustion efficiency but also in establishing high safety standards along with minimum pollutant emissions. To meet these demands a burner designer needs to address criteria regarding combustor exit burnout, flame stability and pollutant emission limits. Among the pollutants, nitric oxide is of great interest to researchers because the resultant fuel nitrogen conversion to NO can be significantly reduced by combustion modifications.

The last decade has seen an increased awareness of the all important fuel nitrogen reactions that take place during the combustion of fossil fuels. As a result numerous methods have been developed to either reduce the promotion of N to NO or promote the reduction of NO to N_2. These techniques incorporate either one or a combination of low excess air, low temperature operation, ammonia or urea injection, with or without the catalyst, and air or fuel staging (Sarofim and Flauan 1976; Pershing and Wendt 1979; Lyon and Cole 1990; Raymant 1990).

These methods have been adopted by industry to a greater or lesser extent but in most cases are either expensive or not suitable for retrofitting. Some cause a deterioration in the combustion efficiency or increase the likelihood of the conversion of NO to N_2O (another pollutant) instead of N_2. To overcome these drawbacks, near burner region (NBR) aerodynamics can be manipulated to achieve low NO_x emission values from existing combustion facilities. The key factor in designing a low-NO_x burner using such an approach is to increase the proportion of coal particle devolatilization and ignition in an oxygen-lean internal recirculation zone (IRZ) in order to convert a higher proportion of the volatile nitrogen to N_2. Recently Smart *et al.* (1988) have shown that such a regime, wherein NO_x emissions are significantly reduced, can be established simply through a physical downstream displacement of the burner jet position inside the quarl.

In comparison with NO_x reduction research, little work has been carried out on the stability performance of PF flames. Some experimental studies include Beer (1964), Hein and Leuckel (1969), Wolanski and Nojcicki (1974), Godoy *et al.* (1986), Hassan *et al.* (1991), and Costa *et al.* (1991a). All these studies emphasize the importance of the internal recirculation zone and the environment in which volatile release takes place. In the present work, comparative stability performance, particle burnout and NO_x emissions for a wide range of secondary air swirl numbers, excess air levels, and also primary jet and secondary jet momentum ratios for two types of burner jet designs have been investigated in order to optimise the operating conditions for both burners.

The two burner designs chosen here are shown in the Figure 1, the first having a centrally located bluff-body creating a single annular orifice (SAO) jet. This design conforms to industrial practice where the central bluff-body would normally house the gas or oil light-up burner. The second design, of identical primary jet cross-sectional area, was achieved by removing the bluff-body to provide a single central orifice (SCO). This latter burner design has been the subject of a series of investigations involving stability (Hassan *et al.*, 1991; Costa *et al.*, 1991a), particle burnout (Godoy *et al.* 1991) and NO_x emission values (Hirji, 1986). The relative measurement locations and an approximate interpretation of the size and shape of the IRZ boundary and NBR, along with the trajectories of a 'typical' coal particle (Yehia 1991) for each configuration is shown in Figure 1.

EXPERIMENTAL METHODS

A complete description of the large-scale PF-fired cylindrical furnace used in this study is available elsewhere (Costa, 1991b), therefore only its most important features are reiterated here.

The combustion chamber consists of ten individual water-cooled sections of internal diameter 0.6 m and height 0.3 m. The furnace roof and the upper five segments are refractory lined. The burner geometry is typical of that used in wall-fired power station boilers. It consists of a burner jet and a secondary air supply in a conventional double concentric configuration, terminating in a castable refractory quarl of half angle 24°C and a length to diameter ratio of unity. The preheated secondary air (300°C) enters via a plenum situated above the burner in which it encounters a moveable block swirl generator. The PF is delivered by a twin-screw feeder with homogeniser to a balanced Venturi injector in the primary air line. Once the flame is established, the feed rate is controlled with an accuracy of $\pm$ 1%.

Steady state operating conditions were adversely affected by a build-up of coal particles on the SAO burner tip. A high proportion of these particles were entrained by the

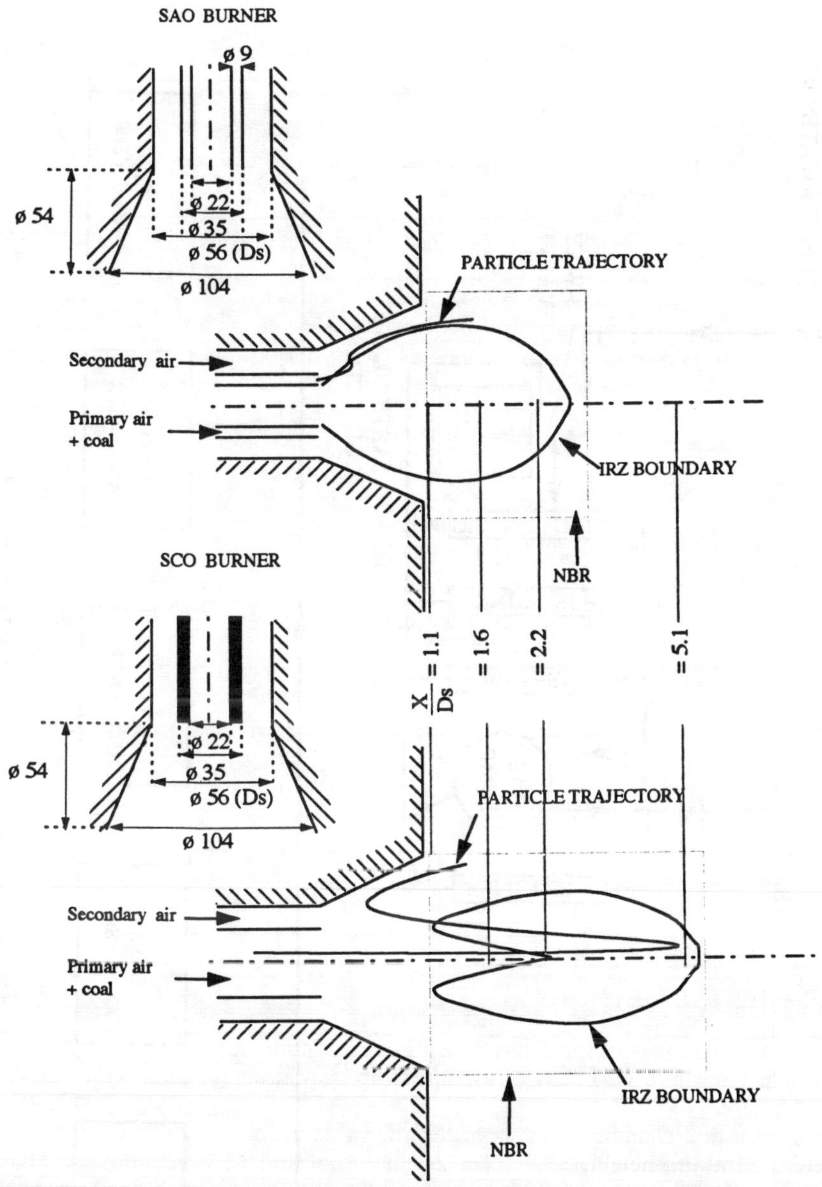

FIGURE 1 Details of SAO and SCO burner jets (all dimensions are in mm) with NBR aerodynamics.

recirculating gas attached to the central bluff-body (Costa *et al.,* 1991a). Under the most adverse operating conditions some 10% of the total fuel input to the furnace ended up as semi-combusted 'particle-pellets', each of almost identical size and shape and falling through the furnace at regular time intervals as the result of their formation and detachment from the bluff-body tip. Several bluff-body designs were installed in an attempt to eliminate this particle build-up. However, all failed to address the problem.

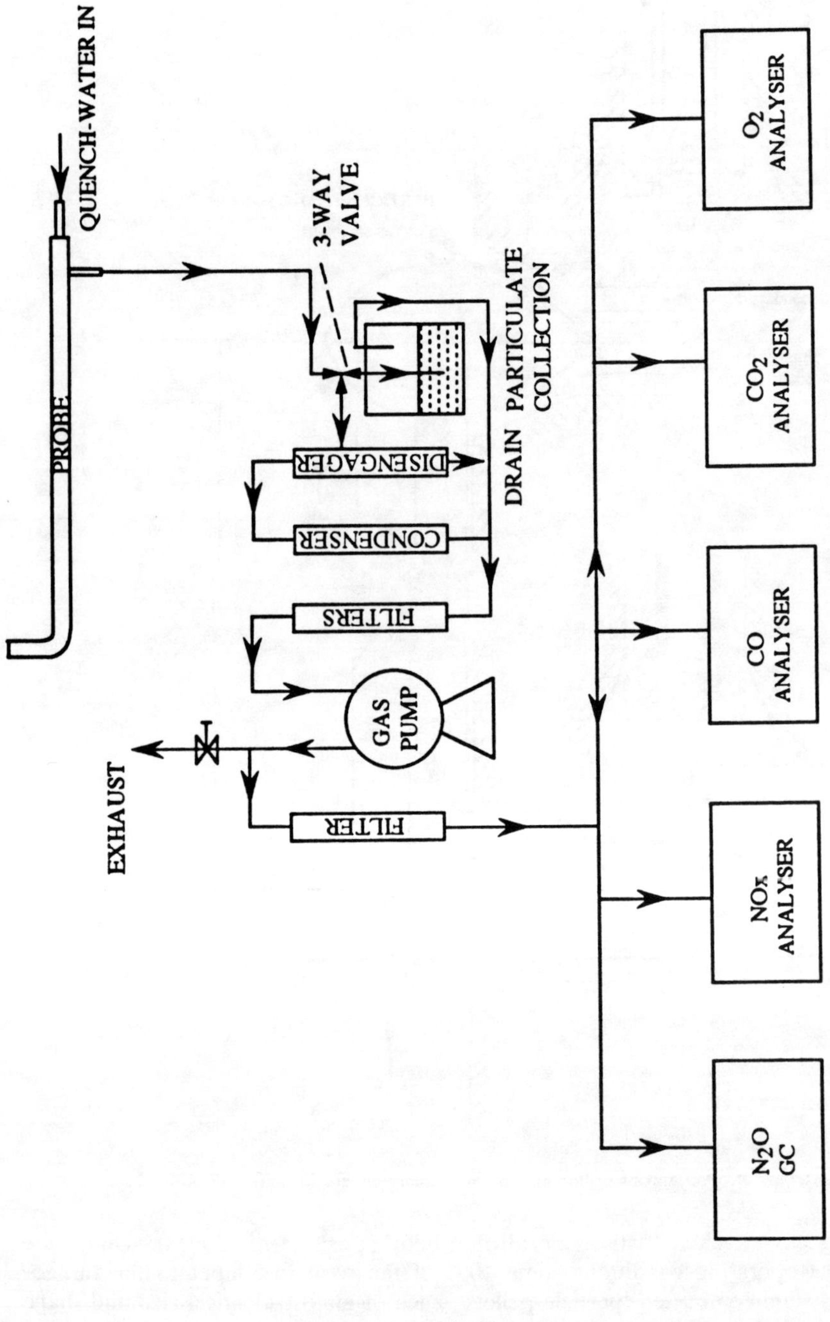

FIGURE 2a Schematic of the gas and particulate sampling system.

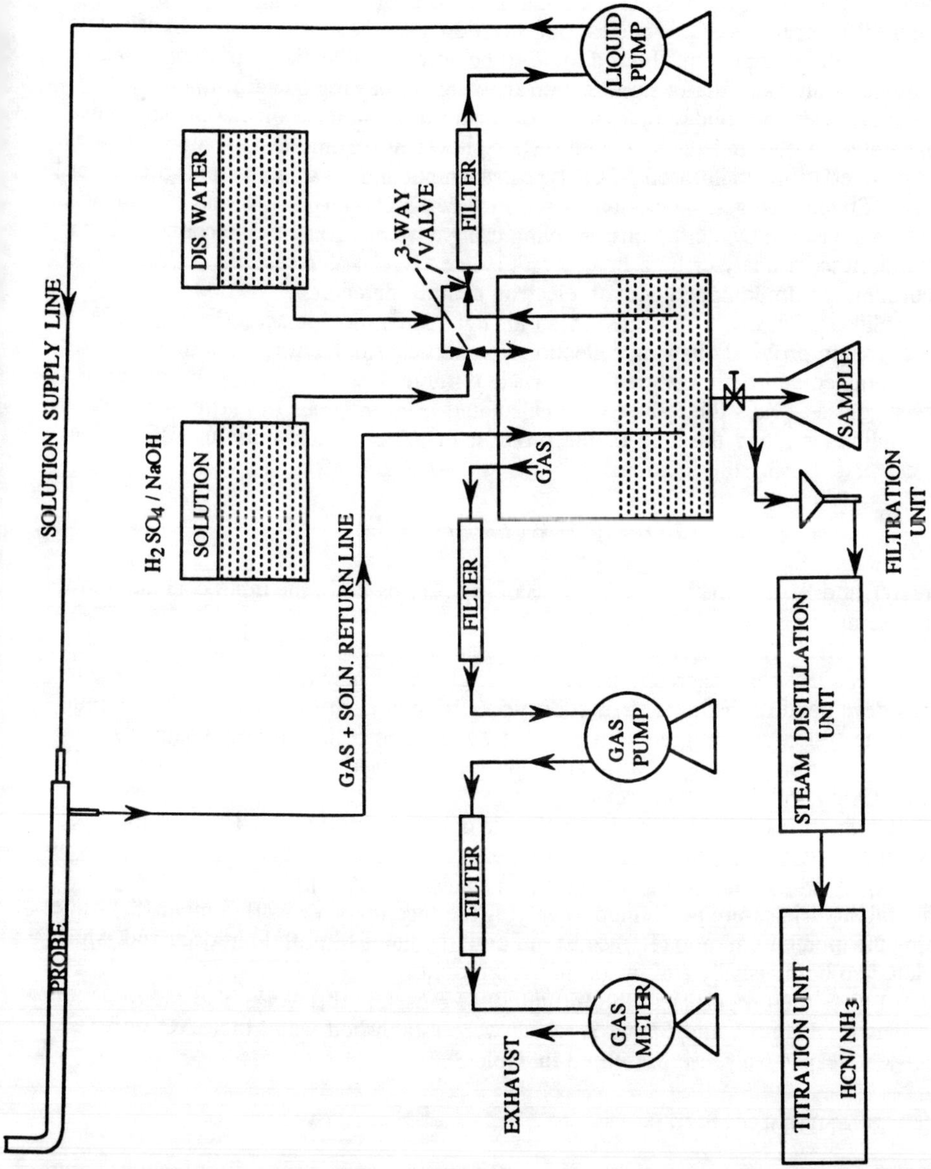

FIGURE 2b Schematic of the sampling system for HCN and NH_3.

The present successful arrangement incorporates a thin walled tube, its outlet chamfered to give the minimum possible surface area. This design still led to the generation of coal deposits, although their size, frequency and density were significantly reduced. The problem was finally solved by introducing some 4% of the primary air contribution through the central tube. Gas species concentration measurements at the quarl exit, with and without this component, showed there to be no detectable difference, implying that this source of air had little or no effect on the near burner region aerodynamics.

The gas species and solid samples were collected using a water-cooled, water-quenched stainless steel probe and the sampling system shown in Figure 2a. CO concentrations were recorded using an infrared NDIR type instrument and O_2 with a paramagnetic type analyser. Stable nitrogen species measured included NO, NH_3, HCN, and N_2O. Nitric oxide was measured by continuous sampling using a water cooled probe connected to a chemiluminescent analyser by a heated PTFE line. N_2O was measured on-line with a gas chromatograph equipped with an electron capture detector.

The solid particles were collected in an air-tight separation vessel following retrieval using the same probe. Following collection, the particle sample was dried in an oven at 110°C, weighed and then analyzed for particle burnout as well as hydrogen, carbon and nitrogen release. Mass flux values were calculated from the mass of particles collected, the sampling time and the area of the probe inlet. The particle burnout (BO) values were calculated using the relationship:

$$BO = [(1 - W_k/W_x)/1 - W_k)]$$

where W_k and W_x are the weight fractions of ash (dry basis) in the input coal and burnt particle sample respectively (Ismail 1989).

NH_3 and HCN gases extracted from the flame were first dissolved in 0.05M H_2SO_4 and 0.1M NaOH solutions, respectively. A special sampling system, illustrated in Figure 2b, was developed to facilitate their collection. Prior to each solid and liquid sampling session, the system was purged with distilled water. The collected liquid samples were then filtered prior to a steam distillation process followed by titrating for HCN/NH_3 using an Orion 960 Autotitrant device.

Gas temperatures were recorded with the aid of a 40 μm Pt/Pt-13% Rh thermocouple. The temperature uncertainties due to radiation losses were about 10% in the regions of highest temperatures and are discussed by Costa *et al.* (1991b).

The stability trials were performed by varying the secondary air swirl number (S_s) while keeping the momentum ratio (R_m) constant, until the flame 'lift off' limit occurred, which was detected both visually and by an infrared monitor.

Table 1 gives the properties of the Bituminous PF used in this work. The experimental conditions for Flame 1 and Flame 2, which were established using the SAO and SCO burner jets, respectively, are presented in Table 2.

RESULTS AND DISCUSSION

The NBR aerodynamics play an important role in the formation of nitrogen oxides, flame stability, and particle burnout. An assessment of their influence is quantified through the collection of axial data along the combustor centre line for two flames established by the SAO and SCO burner jets. Data reported herein includes flame temperature, concentrations of CO, O_2, NO, N_2O, HCN, NH_3, particle burnout (BO), mass flux (MF), and nitrogen release (NR).

Axial profiles for the two flames are shown in Figure 3 where Flame 1 and Flame 2 are denoted by the symbols ◇ and ■, respectively. For Flame 1, the BO, NR, MF and

TABLE I
Analysis of Bituminous Coal: Gedling, Nottinghamshire, UK.

Proximate analysis	
(Weight % as received)	
Moisture	6.3
Volatile matter,	35.8
Fixed carbon	53.7
Ash	4.2
Ultimate analysis	
(Weight % as received)	
Carbon	72.6
Hydrogen	5.05
Nitrogen	1.29
Sulphur	1.55
Oxygen (by difference)	15.31

TABLE II
Furnace operating conditions for Flame 1 and Flame 2.

Coal feed rate, kg/hr	14.0
Excess air, %	15
Momentum ratio*	0.25
Secondary air swirl number	1.03
Primary air flow rate, kg/hr	32.8
Secondary air flow rate, kg/hr	120.4
Primary air temperature, °C	80
Secondary air temperature, °C	300

* Momentum ratio, $R_m=M_p\ V_p/M_s\ V_s$

temperature profiles, at the quarl exit (axial position of X/D_s)=1.1), show that a significant proportion of particle combustion and primary jet dispersion has already occurred. Very high CO, NO, NH_3, HCN and low O_2 concentrations indicate that combustion has been initiated inside the burner quarl along with intense combustion within the primary coal jet itself. Further downstream and between $X/D_s=1.6$ and 2.2, a slight increase in NO values and a decrease in CO, HCN and NH_3 values reveals continuing mixing and volatile nitrogen conversion to NO. At $X/D_s=5.1$, MF and CO values are approaching their minima while BO and NR have nearly reached their maxima. The leveling-off of the concentrations of O_2, NO, HCN and NH_3 indicate the end of the IRZ, with the slight variation due to the entrainment of the vitiated combustion gases. With increasing axial displacement HCN, NH_3, N_2O, CO, O_2 and temperature values decrease while NO, BO and NR values increase slightly due to mixing and the final heterogeneous reactions associated with char combustion.

For Flame 2, at the quarl exit, the MF profile, unlike its counterpart, indicates the penetration of the well-defined primary air/coal jet into the IRZ. CO, NO and O_2 values suggest that, also unlike Flame 1, little or no chemical reaction has been initiated at this stage. However, burnout and nitrogen release has commenced while high concentrations of the NH_3 and HCN from the evolution of volatile-N are in evidence as the result of coal pyrolysis reactions brought about by the high temperatures therein. From a further data archive of radial profiles for the two flames (Abbas *et al.* 1992), coal particle combustion for the SCO burner is not initiated on the combustor axis, but in a relatively

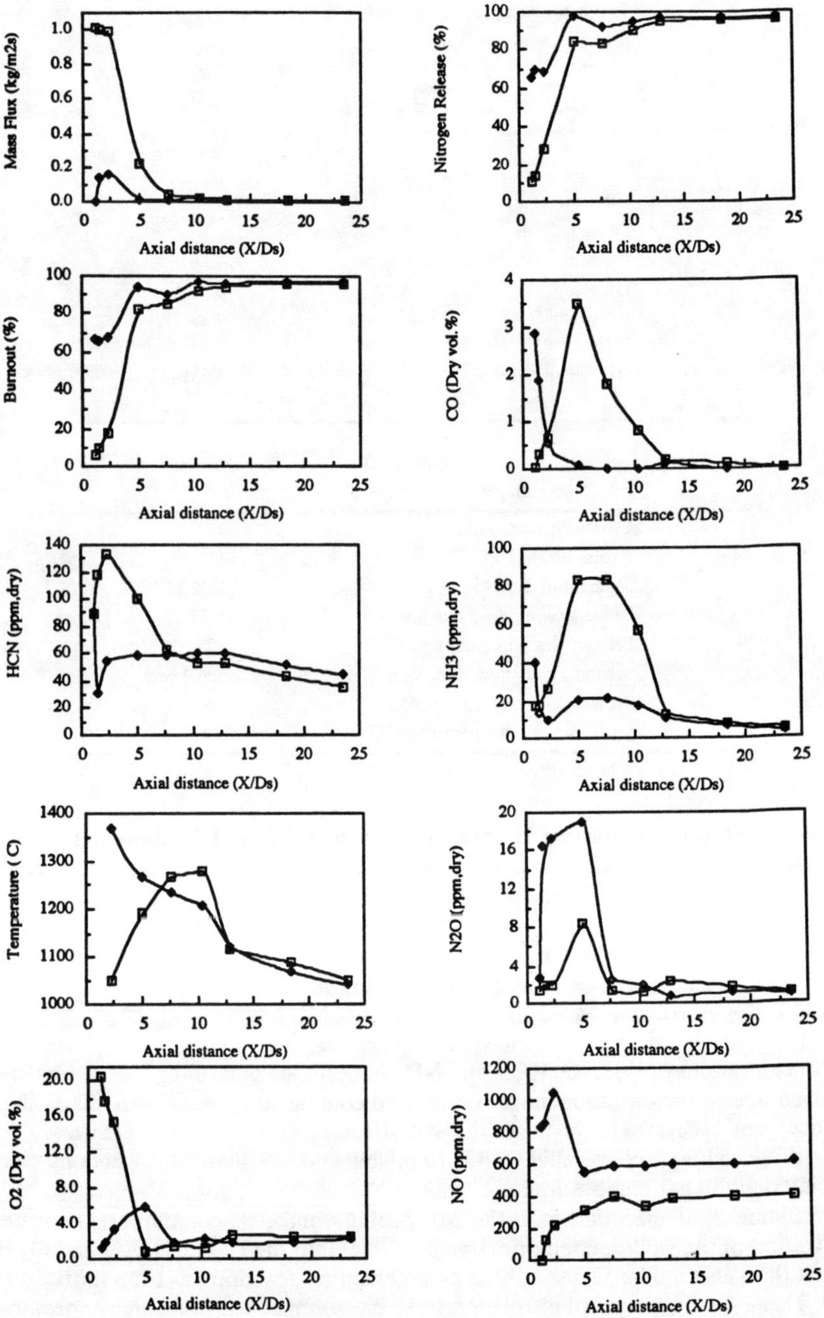

FIGURE 3 Axial profiles for Flame 1 and Flame 2.

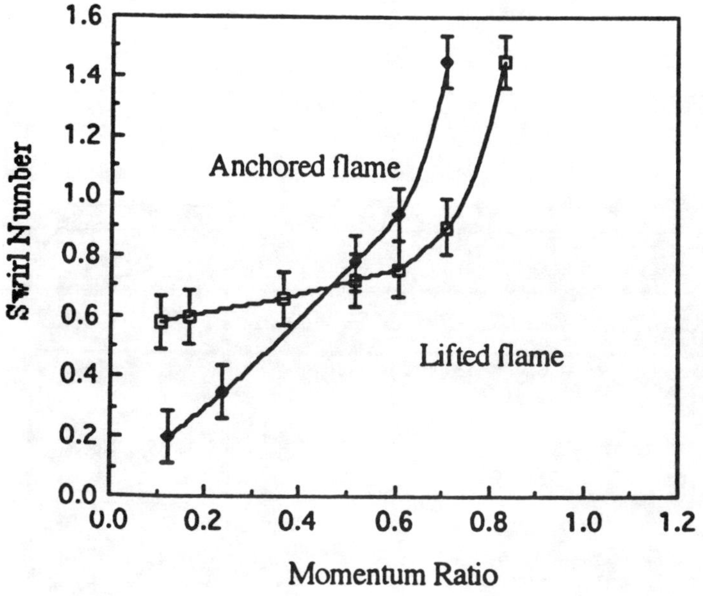

FIGURE 4 Stability of SAO and SCO burner jets at 15% excess air.

fuel-rich region between the periphery of the central jet and the inner boundary of the IRZ. In this flame coal particle combustion is not initiated on the combustor axis at this position. Further downstream at $X/D_s=1.6$ and 2.2, increases in NO, CO and HCN concentrations, along with a further increase in particle burnout and nitrogen release, are now accompanied by a decrease in O_2 concentration. This phenomenon reveals that not only pyrolysis but also combustion reactions have been initiated. The mass flux values are similar to those occurring in the previous axial position, indicating that the primary coal jet is still intact, with very little divergence being evident.

The intense combustion continues between X/D_s of 2.2 and 5.1. O_2, NO, HCN, NH_3 concentrations and flame temperature profiles reveal that, although most of the combustion has been completed by this stage, the mixing between the relatively ‘cooler’ external recirculation zone (ERZ) gases and the ‘hotter’ air stream continues. High CO values indicate the existence of a region of hot-fuel-rich gas close to the centre line. N_2O values have almost doubled compared with the previous stations. The presence of CO reveals the continuation of heterogeneous reactions and mixing with further increases in axial displacement, until the furnace exit ($X/D_s=43.3$), where these reactions appear to have been completed. However, for N_2O, the concentration values remain fairly constant, (2 ppm) throughout the combustor beyond $X/D_s=5.1$.

Comparative photographic and illustrative images of both flames are shown in Plates 1 and 2 and Figure 1, respectively. Flame 1 is seen to be shorter and wider with intense devolatilization and combustion being initiated inside the quarl. The very high reaction rates are a result of the burner jet design which brings the primary jet immediately under the influence of the secondary air stream, thereby shifting the initial volatile liberation and ignition point well upstream. In contrast to its counterpart, Flame 2 is longer and narrower with the combustion being initiated on the centre line, but well downstream of the quarl.

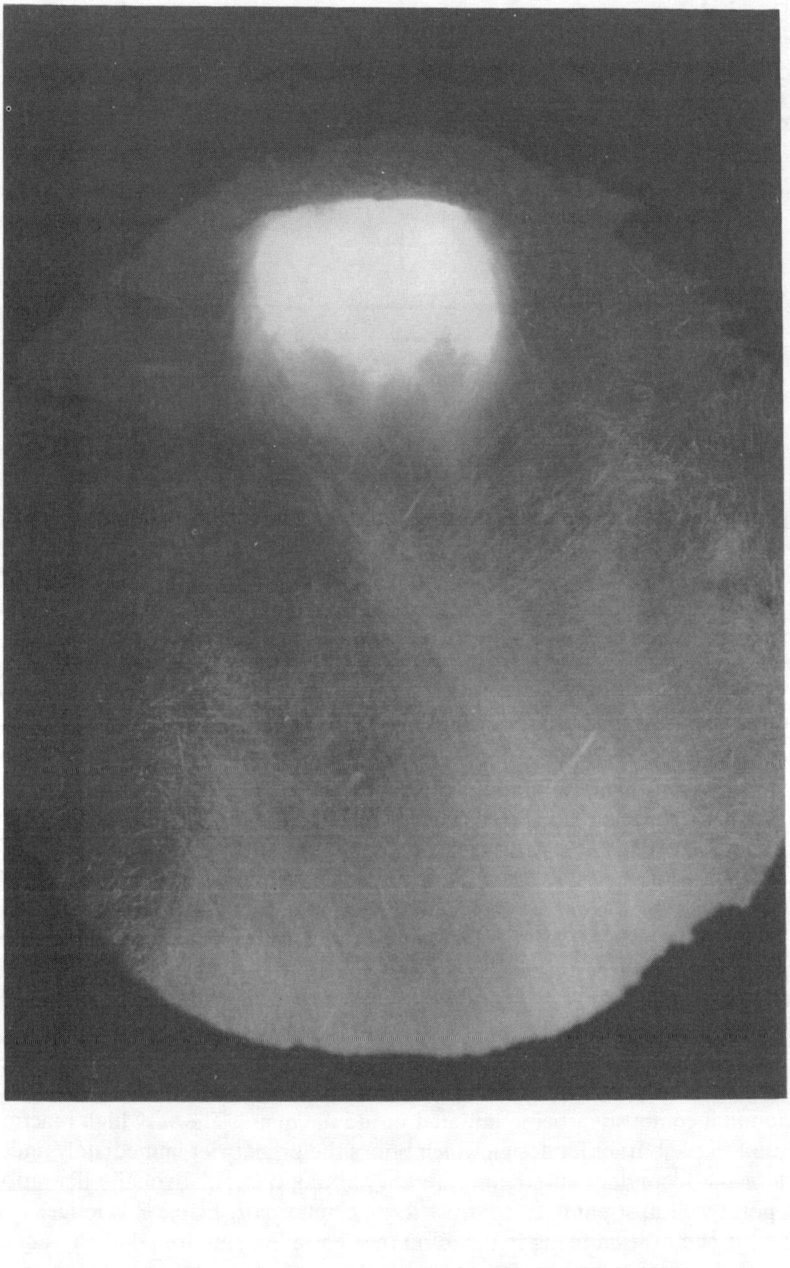

PLATE 1 A photograph of Flame 1 (SAO burner). SEE COLOR PLATE I.

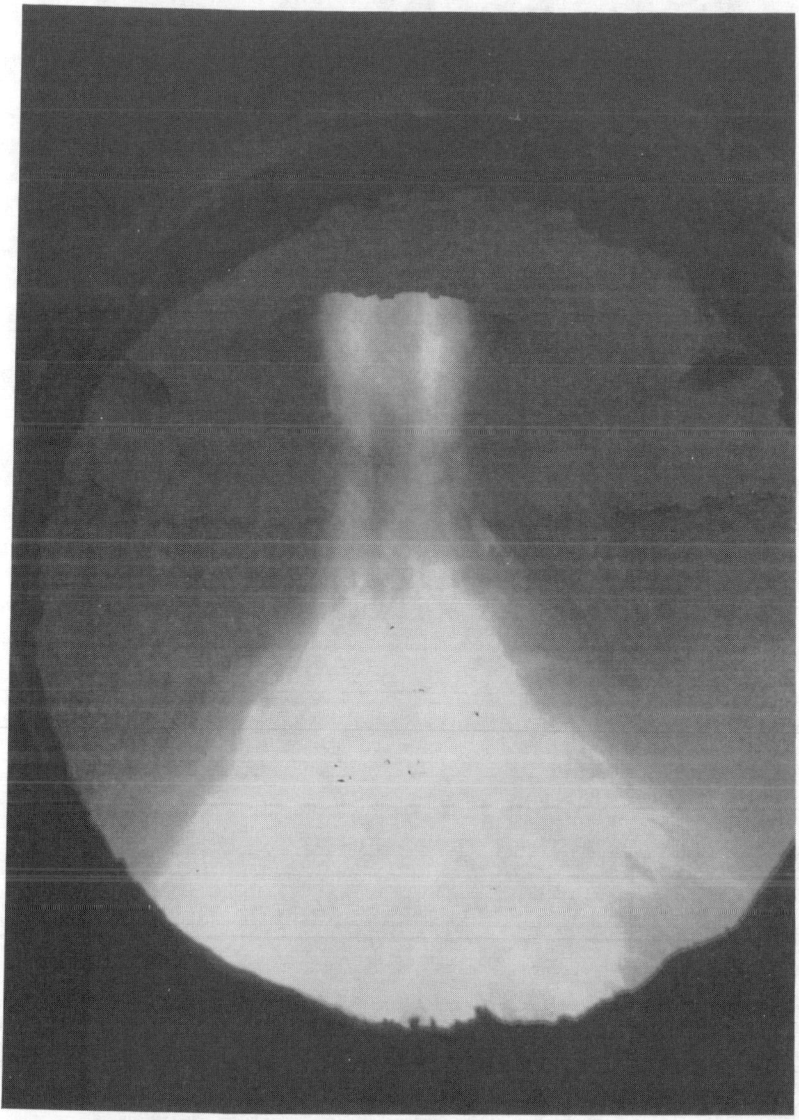

PLATE 2 A photograph of Flame 2 (SCO burner). SEE COLOR PLATE II.

PLATE 3 A photograph of a typical unstable flame. SEE COLOR PLATE III.

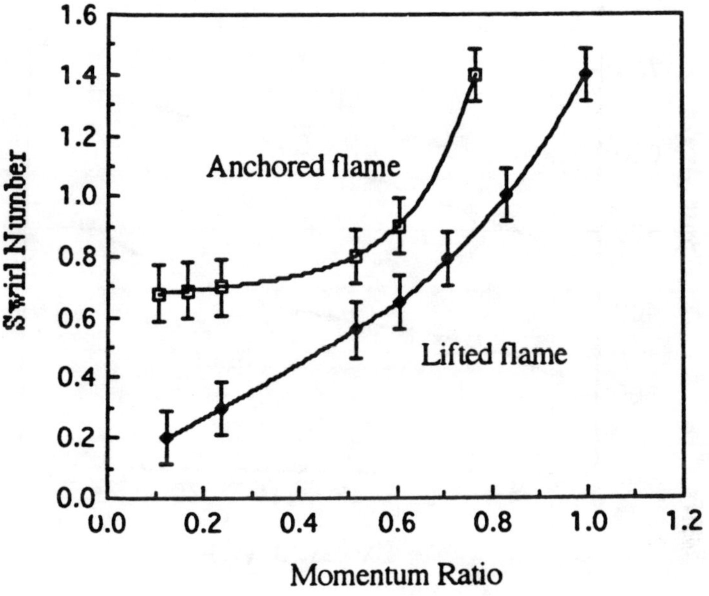

FIGURE 5 Stability of SAO and SCO burner jets at 2% excess air.

Stability Performance

The flame stability criteria adopted herein regards a stable flame as one which is lit-back and ignites well inside the quarl. Plate 3 is an example of a lifted flame and, although acceptable in other applications, e.g. the cement industry, does not satisfy our stability criteria designed for wall-fired boilers. Flame stability is greatly influenced by the particle residence times in the IRZ along with a sufficient volatile release within that region to provide an ignition source for the incoming coal particles. For a particular burner jet design and coal type and condition, the burner stability will be a function of secondary air swirl level as well as the secondary and primary jet velocities. Therefore stability curves plotted here are in the form of secondary air swirl number versus the ratio of axial momentum of the primary stream (coal +air) to that of the secondary air. The error-bars on the curve are indicative of the experimental uncertainties and hysterisis incurred due to the vague transition boundary between the lit-back and lifted flame. It is also important to mention that in these stability curves there is a lower limit of the momentum ratio, R_{mmin}, below which a stable flame cannot be maintained. This lower R_m value is a physical constraint on the coal feed system; below this value there is an insufficient amount of primary air available to entrain the PF in the Venturi injector and transport it to the burner.

Figures 4 and 5 show the stability limits for two levels of excess air (15 and 2%) for the SAO and SCO burner jets where the symbols ◇ and ■ represent these burner jets, respectively. For 15% excess air (Figure 4), the SAO burner produces a stable flame at lower swirl numbers for a lower momentum ratio as compared with the SCO burner. However, the stability advantage of the SAO burner is eroded as R_m is increased. In the case of 2% excess air, the stability margins at the lower R_m for the SAO burner remain the same, while the upper stability range is well extended up to a momentum ratio of unity. The stability curve for the SCO burner remains unaffected by the decrease in the

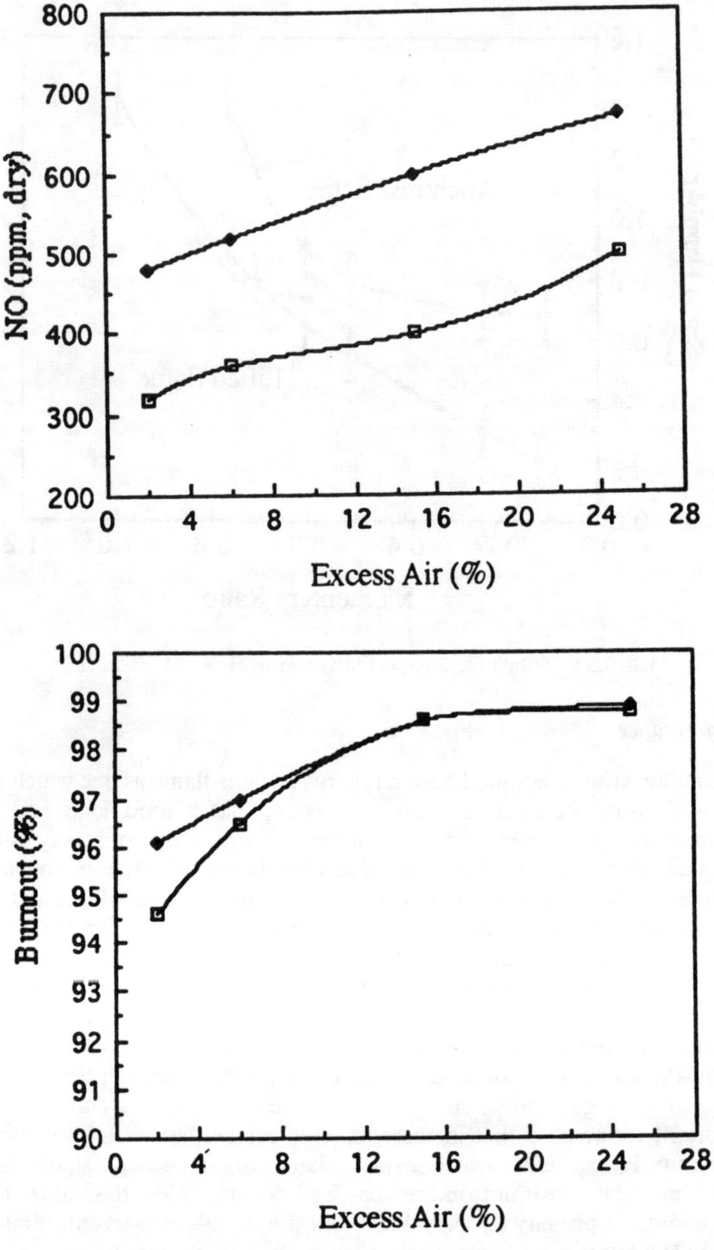

FIGURE 6 Effect of stoichiometric ratio on NO and BO values for two burner designs for R_m=.24 and S_s=1.

excess air level-an observation consistent with a previous stability study of this burner, Hirji (1986).

The difference in the stability ranges for the two burners and the effect of a change in stability at the higher range of R_m with the excess air of the SAO burner stems from the particles' trajectories and their residence time in the IRZ. The position of the IRZ

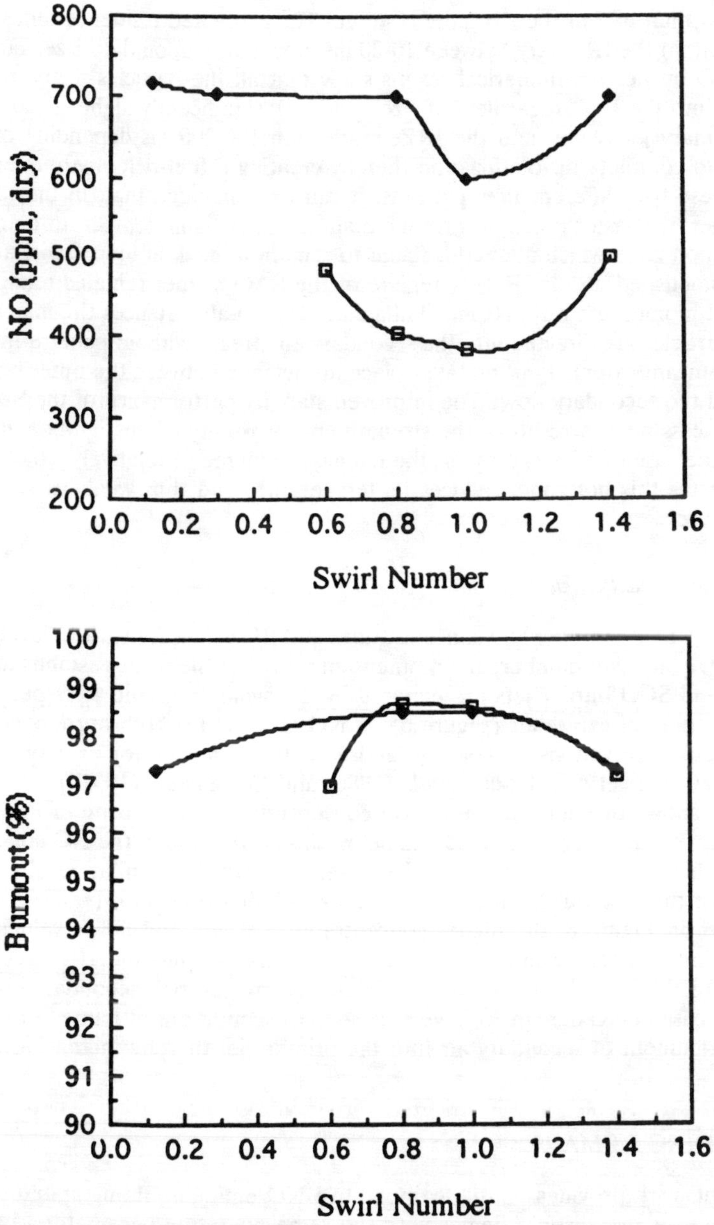

FIGURE 7 Effect of swirl number on NO and BO values for two burner designs for $R_m=1$ and $\lambda=15$.

itself relative to the burner tip will also be a crucial factor affecting flame stability. The numerical study (see Figure 1) suggests that for the case of the SAO burner, an IRZ is established very close to the burner tip. The predicted particles' history shows that the smaller particles fail to penetrate the IRZ and burn in the oxygen-rich shear layer adjacent to it. However, only the larger particles partially penetrate the IRZ, where they release less than half of their volatile content. The estimated residence times for these particles within the IRZ vary between 10-20 ms, depending upon their size. For the case of the SCO burner the numerical results show that all the particles in the primary jet penetrate into the IRZ, regardless of size. The now significantly higher residence times in this primary jet region and the IRZ, more than 100-200 ms depending on particle size, lead to complete devolatilization thereby creating a fuel-rich environment. In the light of these two different flow patterns, it can be concluded that, in the case of the SAO burner, the faster reaction rates maintain a higher flame temperature and a more active ignition zone which allow this flame to remain lit back at lower swirl numbers.

Once established, the NBR flow regime of the SAO burner remains relatively independent of momentum ratio (Romo-Millares 1991). In all instances the majority of the injected particles are directed into the secondary air stream without penetrating the IRZ to any significant extent. Ignition takes place in the zone between the outer boundary of the jet and the secondary flow. The improved stability performance of the SAO burner at lower excess air is a result of the strength and proximity of this ignition zone to the quarl. In the case of 15% excess air, the increased volume of relatively 'cold' secondary air, attenuates this hot zone, reduces its temperature and thus weakens its role as an ignition source.

NO Emissions and Particle Burnout

Figures 6, 7, and 8 show the NO concentrations and BO values for various levels of excess air, secondary air swirl number, and momentum ratio. All the flames established by using the SAO and SCO burner jets are denoted by the symbols △ and ■, respectively. The significant effect of excess air (Figure 6) on NO and BO for both burners is consistent with previous observations on coal, oil and gaseous fuels. See for example Rees *et al.* (1981), Asay *et al.* (1983), Godoy *et al.* (1986) and Costa *et al.* (1991b).

Figure 7 shows that the minimum NO concentrations and maximum BO values for both burners occurs at $S_s=1$. Above and below this swirl number, the NO concentrations and BO values start to increase and decrease, respectively. Similar results have been reported for the Burnout for the SCO burner by Godoy *et al.* (1991).

As shown in Figure 8, the higher concentrations of NO and lower values of BO at lower R_m are due to the limited primary jet penetration into the IRZ, resulting in a tendency for the particles to be flung out into the oxygen-rich secondary stream. The similar increase in NO at high R_m levels, close to the conditions of flame 'lift-off' are due to the entrainment of secondary air into the primary jet thereby increasing the oxygen availability.

CONCLUDING REMARKS

The present work provides unique comparative NO emission, flame stability and combustion data for two distinct flames with the same operating parameters, being useful for the validation of mathematical models, and the prediction of NO concentrations associated with burner design.

The resulting flame changes were a manifestation of the influence of primary air/fuel jet design on the aerodynamics of the IRZ. This region is created immediately downstream of the jet as a result of the swirling combustion air flow, quarl design and the blockage ratio of the burner jet to the secondary air inlet. The SAO burner produced a strong recirculation zone as a consequence, not only of the swirl level, but also of the effect of the bluff body in the centre of the burner jet. This resulted in an IRZ that was attached to the gun and filled a high proportion of the divergent quarl. Primary air penetration into

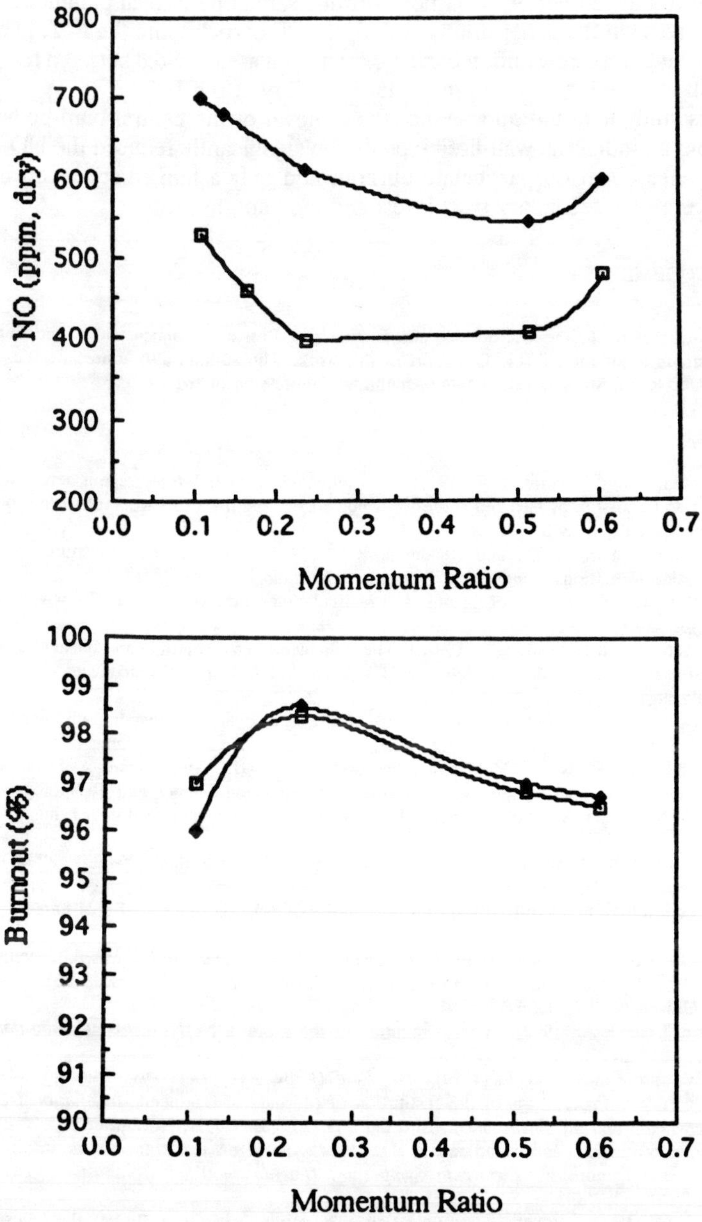

FIGURE 8 Effect of momentum on NO and BO values for two burner designs for S_S=1 and λ=15.

this strong recirculation zone was limited, with the fuel particles being heavily influenced by the high swirling forces at their point of injection. As a consequence, a high proportion of the particle trajectories followed the outer boundary of the IRZ, with devolatilization and burning being initiated in areas of high turbulence, flame temperature and relatively high oxygen concentration.

A change to the simpler SCO burner, with the same operational parameters, brought about an increase in the penetration depth of the PF particles into the IRZ. Devolatilization and burning were now initiated in an oxygen lean area located between the periphery of the a still well-defined jet and the inner boundary of the IRZ.

From this study it would appear that the removal of the central bluff-body (light up burner) from an industrial wall-fired type design significantly reduced the NO emissions with no adverse effect on particulate burnout and only a limited reduction in stability performance at low secondary swirl levels and momentum ratios.

ACKNOWLEDGEMENTS

The authors are grateful to the UK Science and Engineering Research Council and to the European Coal and Steel community for their financial support of this work. The authors also appreciate the assistance of M. Yehia and C. Romo-Millares in creating a qualitative impression of the IRZ.

REFERENCES

Abbas, T., Costen, P. and Lockwood, F. (1992). The influence of near burner region aerodynamics on the formation and emission of nitrogen oxides in a pulverized coal-fired furnace. Submitted to Combustion and Flame. Vol 91, p. 346.

Asay, B. W., D. P., Smoot, L. D., and Headman, P. O. (1983). Effect of coal moisture on burnout and nitrogen oxide formation. *Combustion Science and Technology,* **15,** 35.

Beer, J. M. (1964). IFRF: the effect of fineness and recirculation on the combustion of low volatile pulverized coal. *J. Inst. Fuel,* **37**:286.

Costa, M., Costen, P., and Lockwood, F. (1991a). The influence of coal volatility and burner aerodynamics on pulverized-coal flame stability *2nd European Conference on Industrial Furnaces and Boilers,* Vilamoura-Algarve-Portugal.

Costa, M., Costen, P., And Lockwood, F. (1991b). Pulverised-coal and heavy-fuel-oil flames: large-scale experimental studies at Imperial College, London. *Journal of Institute of Energy,* **64**:64.

Godoy, S., Hassan, M. A., Ismail, M.A. and Lockwood, F.C. (1991). Measurements of char burnout in a laboratory pulverized coal combustor. To appear in Combustion Science and Technology.

Godoy, S., Hirji, K., Ismail, M. and Lockwood, F.C. Stability limits of pulverized coal firing. *J. Inst. Energy,* **59**:38.

Hassan, M. A., Ismail, M. A. and Lockwood, F.C. (1991). Pulverized coal flame stability: effect of the burner quarl cooling. *Chemical Eng. Sci.,* **49.**

Hein, K. and Leuckel, W. (1969). Trial C-14-further studies of swirl on pulverized coal flames. IFRF report, F32/a/40.

Hirji, Kaa (1986). Combustion measurements in Pulverised Coal Flames PhD Thesis. University of London.

Ismail, M. (1989). Char burnout and flame stability in a pulverized fuel furnace. PhD Thesis Imperial College, University of London.

Lyon, R. K. and Cole, J. A. (1990). A reexamination of the rapre ± NO_x process. *Combustion and Flame,* **82**:435.

Pershing, D. W., and Wendt, J. O. L. (1979). *Ind. Eng. Chem. Proc. Des. Dev.,* **18**:60.

Raymant, A. P. (1990). Proceedings of the Institution of Mechanical Engineers, *International Conference on Power Generation and the Environment,* Birdcage Walk, London, P.37, November.

Rees, D. P., Smoot, L. D., and Headman, P. O. (1981). Nitrogen oxide formation inside a laboratory pulverized coal combustor. *Eighteenth Symposium (International) on Combustion,* The Combustion Institute, Pittsburgh, 1305.

Romo-Millares, C. (1991). Private communication, Thermofluids Section, Imperial College of Science Technology and Medicine, London, U.K.

Sarofim, A. F. and Flauan, R. C. (1976). *Prog. Energy Combust. Sci.,* **2**:1.

Smart, J. P., Knill, K. J., Visser, B. M., and Weber, R. (1988). *Twenty-Second Symposium (International) on Combustion*, The Combustion Institute, 1117.

Wendt, J. O. L., Pershing, D. W., Lee, J. W., and Glass, J. W. Z. (1979). *Seventeenth Symposium (International) on Combustion*, The Combustion Institute, Pittsburgh, 77.

Wolanski, P. and Wojcicki, S. (1974). Stabilization of coal dust-air mixture by bluff bodies. *15th Symposium (International) on Combustion*, The Combustion Institute, 1295.

Yehia, M. (1991). Private communication, Thermofluids Section, Imperial College of Science Technology and Medicine, London, U.K.

Experimental and Numerical Investigation of NO_x Formation and its Basic Interdependencies on Pulverized Coal Flame Characteristics

U. SCHNELL, M. KAESS, H. BRODBEK *University of Stuttgart, Institut fuer Verfahrenstechnik und Dampfkesselwesen (IVD), Stuttgart, Germany*

Abstract—The formation of nitrogen·oxide in pulverized coal combustion is analysed by experiments in a 500 kW test rig and by numerical predictions. This study extracts some of the dozens of parameters that were investigated during the trials. It turned out that the huge number of process variables that have an impact on NO_x formation can be reduced to a smaller figure by an appropriate data analysis. Major parameters presented in this contribution are the fineness of the coal dust and the amount of swirl. Special emphasis is put on the fact that the level of NO_x concentration must necessarily be monitored together with the degree of burnout of the coal dust.

Key Words: Coal dust combustion, NO_x emissions, Swirl flames, Experiments in coal flames, Mathematical flame model

INTRODUCTION

In the past decade a strong effort was made in Germany to reduce pollutant emissions from power station combustors. It has been demonstrated that the emission of nitrogen oxides can effectively be reduced by primary measures. This results from the strong dependence of NO_x emissions from the combustion process. Most of the basic parameters that influence NO_x formation are subject of experimental investigations. While some parameters can be studied in laboratory scale experiments or in semi-industrial test facilities, the vast majority of process variables in practical combustion systems and their impact on NO_x emission are normally unknown.

Mathematical models of flames and furnaces have proved to be a useful tool in understanding the complex correlations between flame characteristics and pollutant formation pathways. These models were developed by the help of experiments that were conducted in order to test the predictive qualities of the numerical approach. After validation of the modelling assumptions the knowledge of the complex NO_x formation process can be enhanced by numerical simulation.

Internal dependencies which are not amenable to experimental investigations can be studied in detail and concrete pollutant abatement techniques can be conceived.

EXPERIMENTAL INVESTIGATIONS

Experimental Set-up

Experimental research work concerning the deeper understanding of the pulverized coal combustion process demands a test rig with definite and reproducible experimental conditions. For this purpose, a coal combustion facility was built up in semi-industrial scale. In its basic construction, it reproduces the combustion conditions in industrial furnaces. Additionally, it allows for the variation of a broad range of process parameters (Kaess, 1991).

The installation of a coal preparation rig for drying, milling and sifting facilitates the use of different ranks (chemical components) and granulations of the raw coal. The

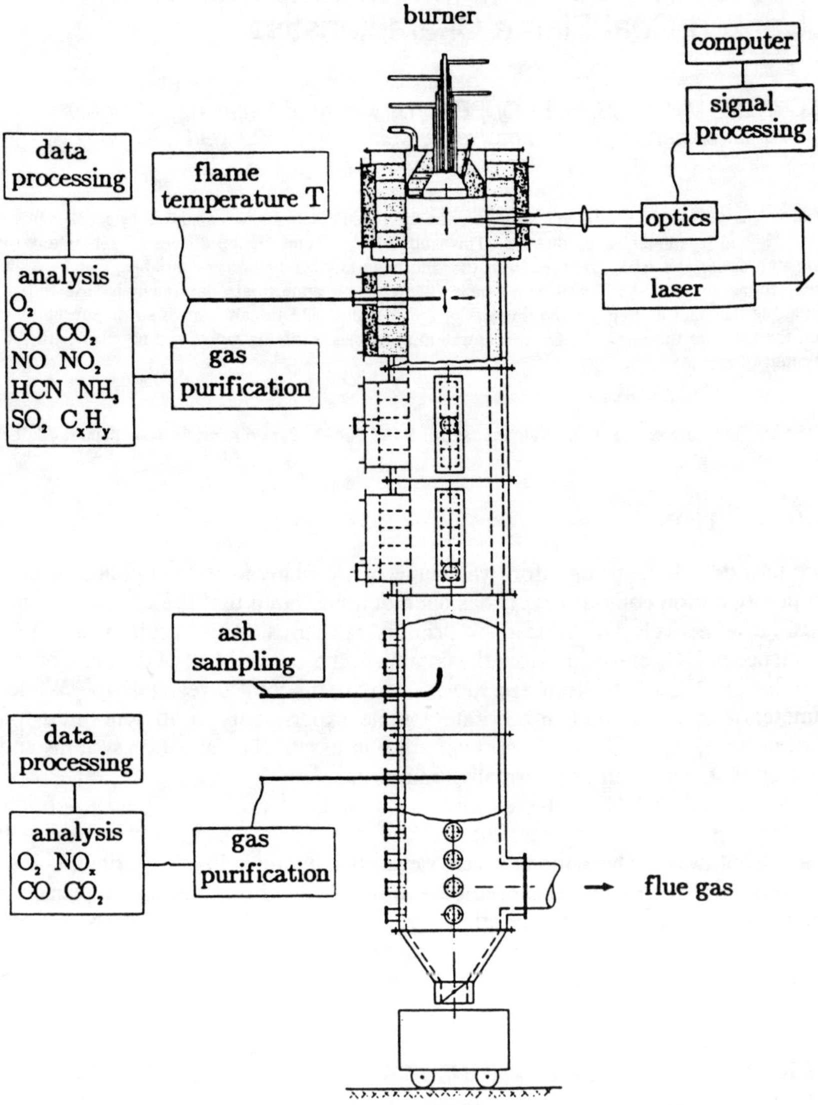

FIGURE 1 Coal combustion chamber with measurement equipment

coal mill with an aerodynamic classifier permits the adjustment of different particle size distributions of the pulverized coal.

The vertical combustion chamber with a maximal thermal input of 500 kW and a length of 7 m is the core of the test facility (fig. 1).

Especially investigations in mixing properties necessitate full turbulent combustion conditions, which exist in this combustion chamber. The presence of rotationally symmetrical flames in the cylindrical round chamber reduces the expenditure of measurements. Also the exchange of data with two-dimensional mathematical flame models is more efficient (Schnell and Kaess, 1989; Schnell, 1991). The disturbing influence of gravity force on the symmetry of the flame is only avoidable with a vertical furnace. Investigations into char burnout by flue dust sampling necessitate a furnace with symmetric sedimentation and buoyancy. So differences in density and gravity forces do not cause an asymmetric segregation of gas and particle phase. Slagging and fouling of combustion residues at furnace walls becomes fully symmetric by using a vertical arrangement, where the burner is installed on top.

The combustion chamber was constructed in order that temperature and residence time conditions during the combustion of pulverized coal particles correspond with the characteristics of power station combustors. The first section of 4 m, refractory lined and water cooled, simulates the radiative part of a steam generator furnace. The following section of 3.5 m, which is only water cooled, simulates the zone of convective heat transfer. The diameter of the furnace is 0.75 m in the refractory lined part and 0.85 m in the non-refractory lined part. Three rows of accesses at the vertical combustion chamber allow the use of probes and non-intrusive measurement techniques.

Measurements in Coal Flames

Field measurements of temperature and gas concentrations in the flames are performed using suction probes. Sampling of flue dust is realized by an isokinetic probe. For measurements of velocities and turbulence intensities Laser Doppler Anemometry (LDA) is used. In the past years, this measurement technique has been applied to large flames with various fuels (Brodbek, 1990). It has been optimized for radiating flames as heavy fuel oil flames and pulverized coal flames.

The two-component LDA system consists of a 15 W Argon Laser and a backscatter LDA-optics with a Bragg cell and two photomultipliers.

The geometry of the probe volume is minimized by two beam expanders with an expansion ratio of 1:7. Due to the thick refractory walls of the furnace the focal length of the front lens must be 1200 mm.

Two FFT-processors especially designed for LDA measurements ensure high data rates in spite of the poor signal quality in flames. A PC-AT computer is used for statistical evaluation, graphics presentation and data storage.

To ensure the usefulness of LDA measurements in pulverized coal flames, several limitations of this measurement technique have to be considered. One of these items is the particle slip. Coal flames are two-phase flows with partially high local particle concentrations. The fraction of large particles is not able to follow the motion of the gas phase without any slip. In a theoretical analysis it was shown that the high viscosity of the hot flue gas and the particle number distribution of the coal (approx. $95\% < 10\ \mu m$) result in a very high probability of detected signals sent out from small particles which follow the gas flow nearly without any slip (Brodbek, 1991). A higher particle slip even for smaller particles is only expected in the near-burner region where high velocity gradients prevail in order to intensify the mixing process of coal and air jets. LDA measurements in pulverized coal flames can be affected by refractive index fluctuations. High temperature and density gradients in reactive flows induce refractive index fluctuations along the beams.

The consequence is that the beams do not intersect at each time and thus a decrease of the data rate is perceived. This effect limits the maximum measuring depth in diffusion flames to about 1 m. Errors in velocity determination due to phase fluctuations can be neglected because the expected error is less than 1% (Bouis, 1975).

Another item to be considered when judging LDA applicability in flames is light absorption. The high particle concentration in the flame and in the flue gas causes an increase of light absorption in deep measuring positions. This leads to a poor contrast, a bad signal quality and a low data rate. A limited compensation of this effect is possible by using high laser power.

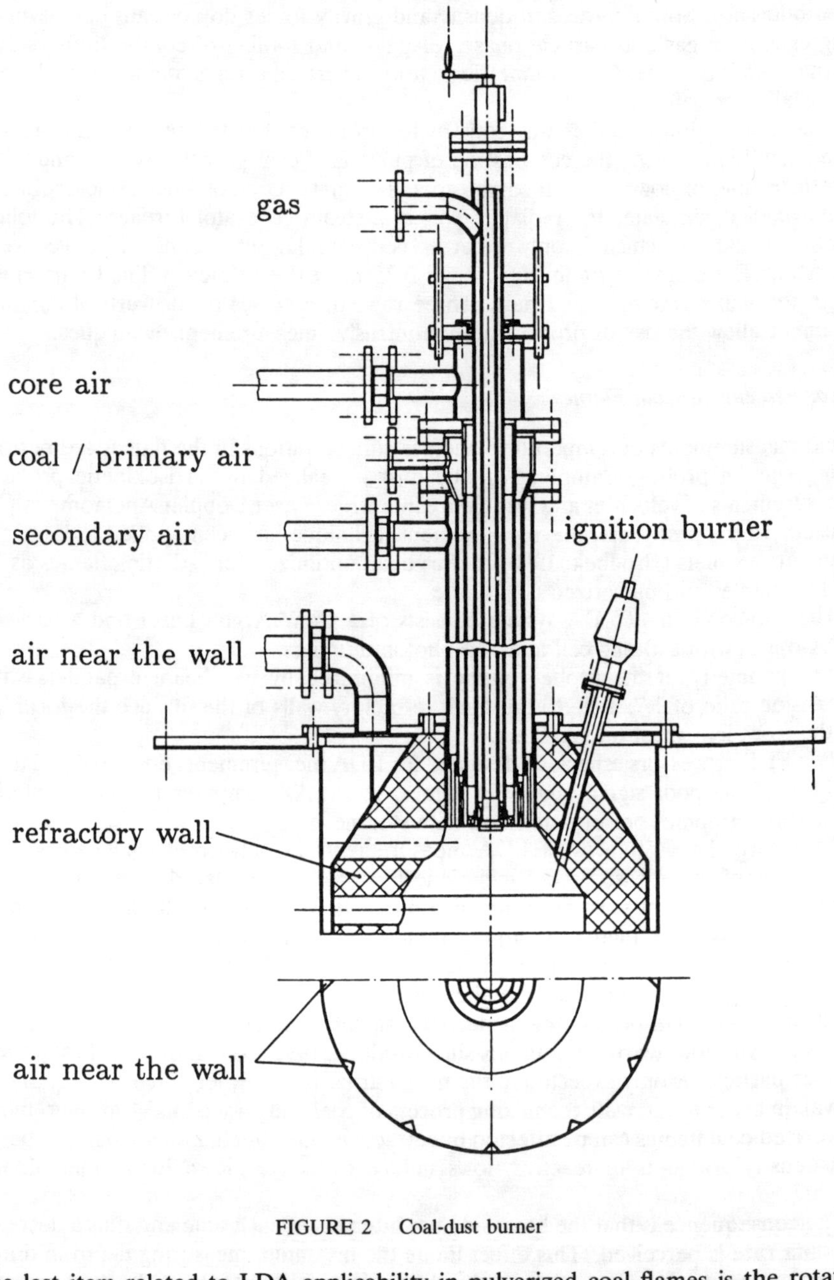

FIGURE 2 Coal-dust burner

The last item related to LDA applicability in pulverized coal flames is the rotation of coal particles. Tangentially issuing jets of volatile matter impart rotation to burning coal particles during their devolatilization (Beér, 1987). LDA measurements at rotating particles show that due to the non-spherical surface the measured velocities differ from the real particle velocities (Kale *et al*. 1988). With increasing particle rotation rate the measured velocity range becomes broader. The mean value is not influenced but the measured turbulence intensity may be erroneous. This effect causes only a very small error since the maximum deviation from the real particle velocity was found to be below approx. 1 cm/s.

General Combustion Parameters

The combustion conditions can be varied by altering several process parameters, e.g. burner type and nature of the fuel. This study extracts some of the dozen of parameters that were investigated during the trials. There was no change in geometry, thermal power and heat extraction. The nature of the fuel, a German hard coal (high-volatile bituminous coal) with a volatile matter content of 34%, was varied by using different comminution rates.

Influence of Burner Type and Swirl Intensity on NO_x Emission

The burner (fig. 2) allows for a wide range of different flow patterns by using different settings for core air, primary and secondary air and air near the wall. The flow patterns can be influenced by the mass flow of each inlet.

Additionally, it is possible to choose between swirled or unswirled core air, primary air and secondary air. Two alternative configurations of the burner are shown in fig. 3. Burner type 1 is characterized by swirled core air and unswirled secondary air while type 2 has unswirled core air and swirled secondary air. Both types use swirled primary air. The different burner configurations result in different flow patterns (fig. 4) and thus in different combustion characteristics. Burner type 1 has a large recirculation zone produced by the swirled core air. Fuel and air are heated up very fast by recirculated hot flue gases. This burner arrangement is characterized by a very intense reaction process with high temperatures. It is quite insensitive to coal type and comminution rate.

For burner type 2, swirled secondary air is responsible for a pressure drop in a small area that is filled up with unswirled core air. Thus, almost no hot flue gases can be recirculated to heat up the incoming air and coal. The combustion progress and the burnout of the coal is strongly dependent on coal fineness and coal type. Fig. 5 shows NO_x emission and burnout as a function of swirl number for burner type 2, using coarse grinding of the coal.

Low swirl numbers lead to low NO_x emissions at the expense of a bad burnout. An increasing swirl number is necessary for a sufficient burnout when a coarse grinding of the coal with 49% <90 μm is used.

When the coal particle size distribution was set to 92% <90 μm, an improved burnout can be achieved even for low swirl numbers (fig. 6). The NO_x emission, however, increases drastically with increasing swirl number. High combustion intensity that is unnecessary for a good burnout in this case of fine grinded coal, coincides with high NO_x values.

The results of burner type 1 and coarse grinding of the coal (fig. 7) also show that NO_x emission is strongly correlated with the degree of burnout. In this case, only limited influence of swirl number on NO_x emissions and burnout can be found. Similar trends with respect to the effect of coal fineness on NO_x emissions and burnout are observed for burner type 1. Using a fine coal particle size distribution, higher NO_x emissions and a better degree of burnout are found (Kaess, 1991).

The different behaviour of both burner types with respect to NO_x emissions can be attributed to the different flow patterns which were detected by LDA measurements. For burner type 1, the intenseness of the heat-up and the early devolatilization process leads to a rapid conversion of evolved fuel-nitrogen species to NO_x in a high-temperature environment which additionally favours thermal NO_x formation. Besides, the recirculation of flue gases does not neutralize the oxygen surplus in this area.

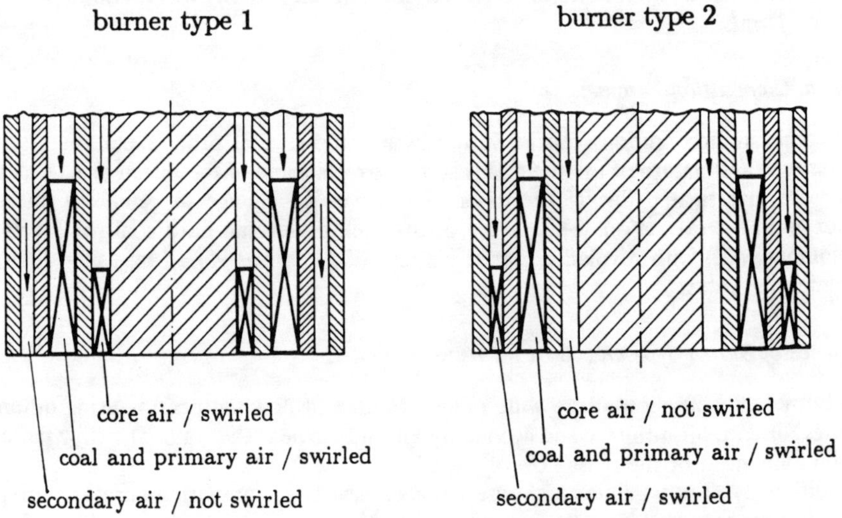

FIGURE 3 Different configurations of the coal dust burner

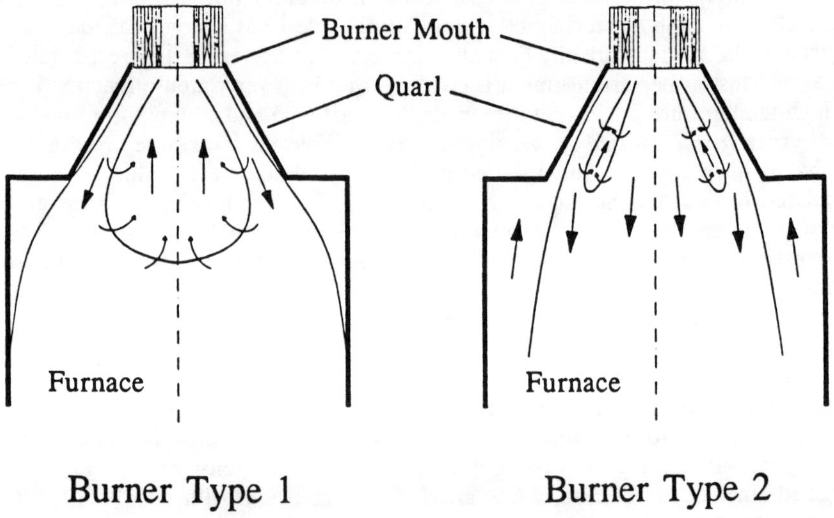

FIGURE 4 Flow patterns of different burner configurations

NUMERICAL INVESTIGATIONS

Merits and Fundamentals of Mathematical Flame Modelling

In-flame measurements and mathematical flame models are a useful tool to detect global correlations between NO_x emissions and process parameters. To study these dependencies in more detail, a "complete" model of NO_x formation in pulverized coal flames is applied.

"Complete" in this context means that the mathematical model of the particle-loaden fluid flow, heat transfer by radiation and the reactions of the chemical species are

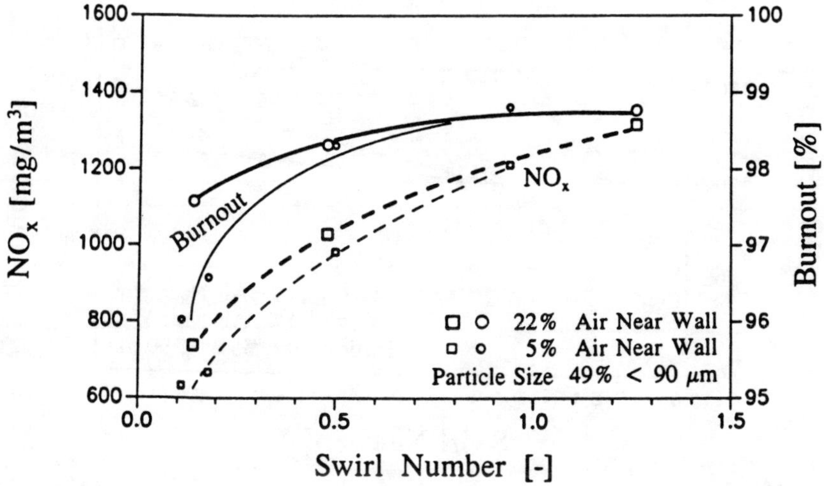

FIGURE 5 NO_X emission and burnout as a function of swirl number for burner type 2 with coarse grinding of thc coal

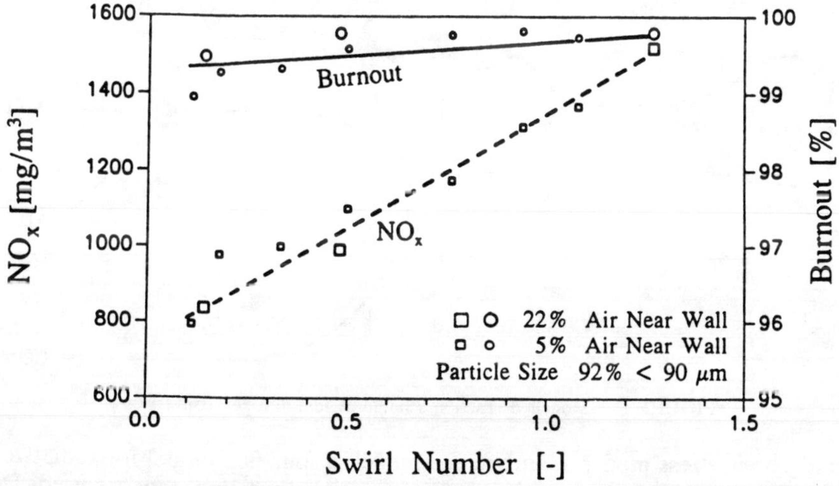

FIGURE 6 NO_X emission and burnout as a function of swirl number for burner type 2 with fine grinding of the coal

modelled in a comparable level of complexity as the kinetics of nitrogen oxide formation. The turbulent flow field, heat release and composition of the mixture is calculated from the numerical solution of the transport equations for momentum, enthalpy and species concentration with a finite difference code in Eulerian formulation. The particle cloud is approximated as a continuum and equal mean velocities of gas and solid phase are presumed.

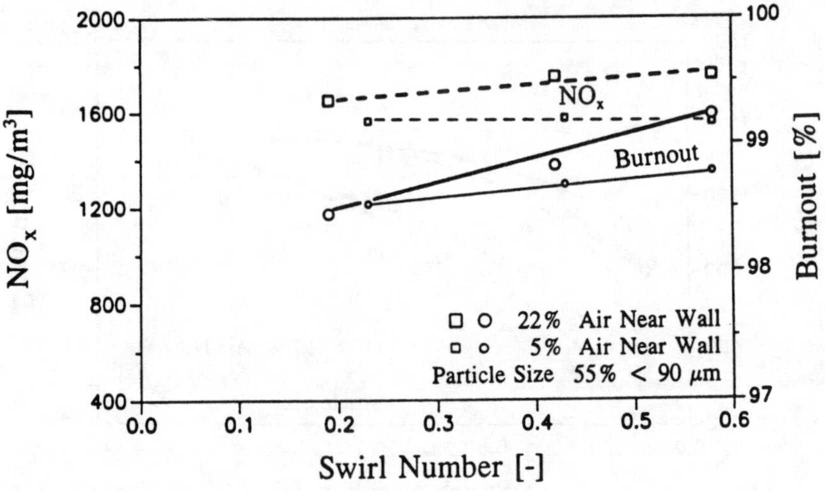

FIGURE 7 NO_x emission and burnout as a function of swirl number for burner type 1 with coarse grinding of the coal

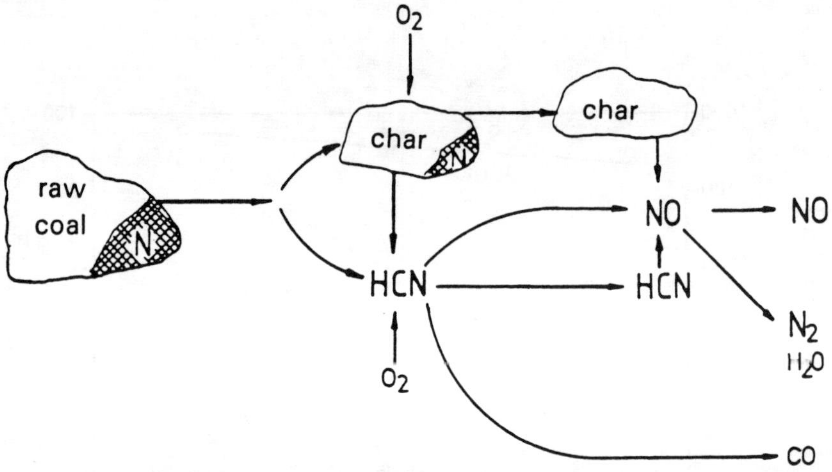

FIGURE 8 Modelled reaction scheme of fuel-nitrogen conversion

An algebraic stress model as turbulence model, a four flux model for radiative heat transfer and a reaction model consisting of several global reaction steps are used. Details of the model of pulverized coal combustion applied to the prediction of NO_x emissions in axisymmetric furnaces are presented in Schnell (1991), Schnell et al. (1991).

NO_x Reaction Model

The NO_x model comprises the fixation of atmospheric nitrogen according to the Zeldovich mechanism and a model of fuel NO_x formation. Usually, the latter mechanism is the major source of NO_x emissions in pulverized coal combustion. Fuel-NO formation and destruction mechanisms are modelled by two reaction schemes (Schnell, 1991). The simpler one is depicted in fig. 8.

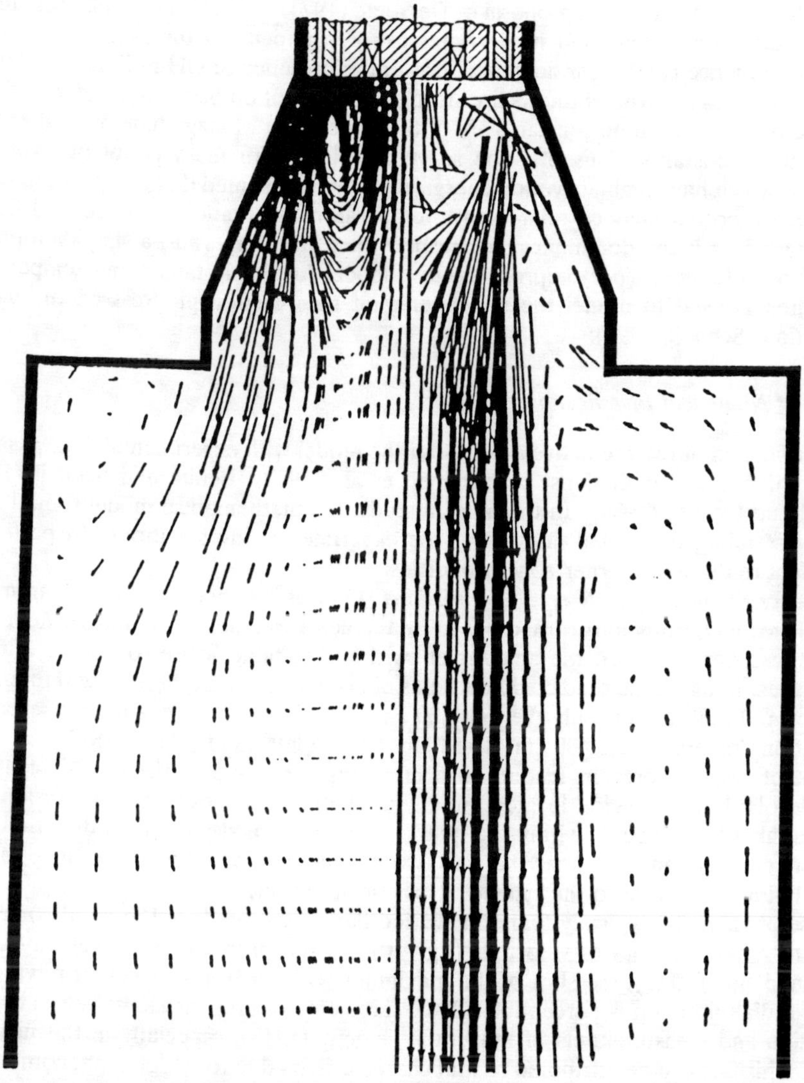

FIGURE 9 Predicted flow field of a medium-intensity swirl flame (LHS) and a low-intensity swirl flame (RHS)

It is assumed that the release of coal nitrogen as cyanide and ammonia during pyrolysis and char burnout are the initial steps in fuel-nitrogen conversion. As experimental evidence under laboratory conditions showed, it is safe to assume that a relatively high portion of the fuel-bound nitrogen is devolatized as HCN with a rate proportional to the volatiles release rate. The same consideration holds for the evolution of nitrogen during char burnout that is modelled analogously with the equivalent rate of carbon conversion. Both heterogeneous reactions are modelled with regard to a potential diffusional limitation in the porous char structure.

Depending on local stoichiometry and temperature conditions, the intermediate nitrogenous species can either be oxidized to NO or they may react with earlier formed

NO yielding N^2. Following a proposal of De Soete (1974), the intermediate N-containing species such as ammonia and its derivates can be lumped together, and the complex reaction sequence can be considered as fast in the presence of OH radicals.

Reaction rates in turbulent flows are strongly dependent on fluctuations of local flame conditions if reaction time constants are of the same order of magnitude or smaller than mixing time constants. This situation is typically found for many pollutant formation reactions which have high activation energies. It can be assumed that concentration fluctuations are not strongly correlated with temperature fluctuations in large coal flames where radiation is the dominant mechanism of heat transfer. Thus, a simplified probability density function (pdf) approach with a model for the instantaneous temperature fluctuation is used to model the dependency of the reaction progress on turbulence interaction (Schnell, 1991).

Results of Numerical Investigations

The verification of the predictive qualities of the model with experimental data was done by several furnace predictions, e.g. Schnell et al. (1987), Benim and Schnell (1989), Schnell and Görner (1989). In the present study, the mathematical model should assist in understanding the global findings of the experimental investigations. In particular, this refers to the near burner zone properties.

Influence of swirl intensity on NO_x emissions. One of the most important parameters to influence NO_x emissions turned out to be the aerodynamics of the burner near field.

As it was stated earlier, the amount of swirl is a suitable figure to classify different flame types. In fig. 9, the predicted flow fields of two flames with different swirl intensities are shown. In all figures, only the first 100 cm after the burner are shown whereas the calculation domain covers 400 cm. The flame on the left-hand side (LHS) has a higher amount of swirled core air, leading to a more rapid heat-up of the coal dust and an early devolatilization region located directly after the burner exit. The expansion after ignition takes place mainly within the quarl and predominantly in radial direction. The exterior recirculation zone near the wall is very small compared with the huge interior recirculation zone close to the symmetry axis of the furnace.

These phenomena typically found in swirled flames are predicted in good correspondence to measured data of velocities (fig. 10). In the profile of axial velocity on the right-hand side (RHS), the shift of the maximum position from the center towards the wall in radial direction is perceptible. There are some discrepancies, however, between prediction and measurements of tangential velocity (LHS), especially in the near wall region. This fact can be attributed to wall characteristics due to slagging phenomena and subsequent eddy dissipation on the rough surface. Due to the axisymmetrical formulation of the code with uniform circumferential velocity, these phenomena are not reconcilable with the numerical results.

The low-intensity swirl flame on the RHS of fig. 9 shows different near burner zone properties. Most of the energy to establish ignition stems from recirculating flue gases in the exterior wall region. The huge external recirculation zone expands upstream into the quarl whereas the interior recirculation zone is very small. Further downstream and obviously detached from the burner, the axial expansion of the flame after ignition takes place.

To judge the implication of the near burner aerodynamics on NO_x formation, not only the flow pattern of the gas phase has to be considered. Another valuable tool is the prediction of particle tracks in the furnace. Based on the converged flow field solution of the Eulerian treatment, this is accomplished by a Lagrangian calculation of particle motion. In this approach, the momentum equations of particles are numerically

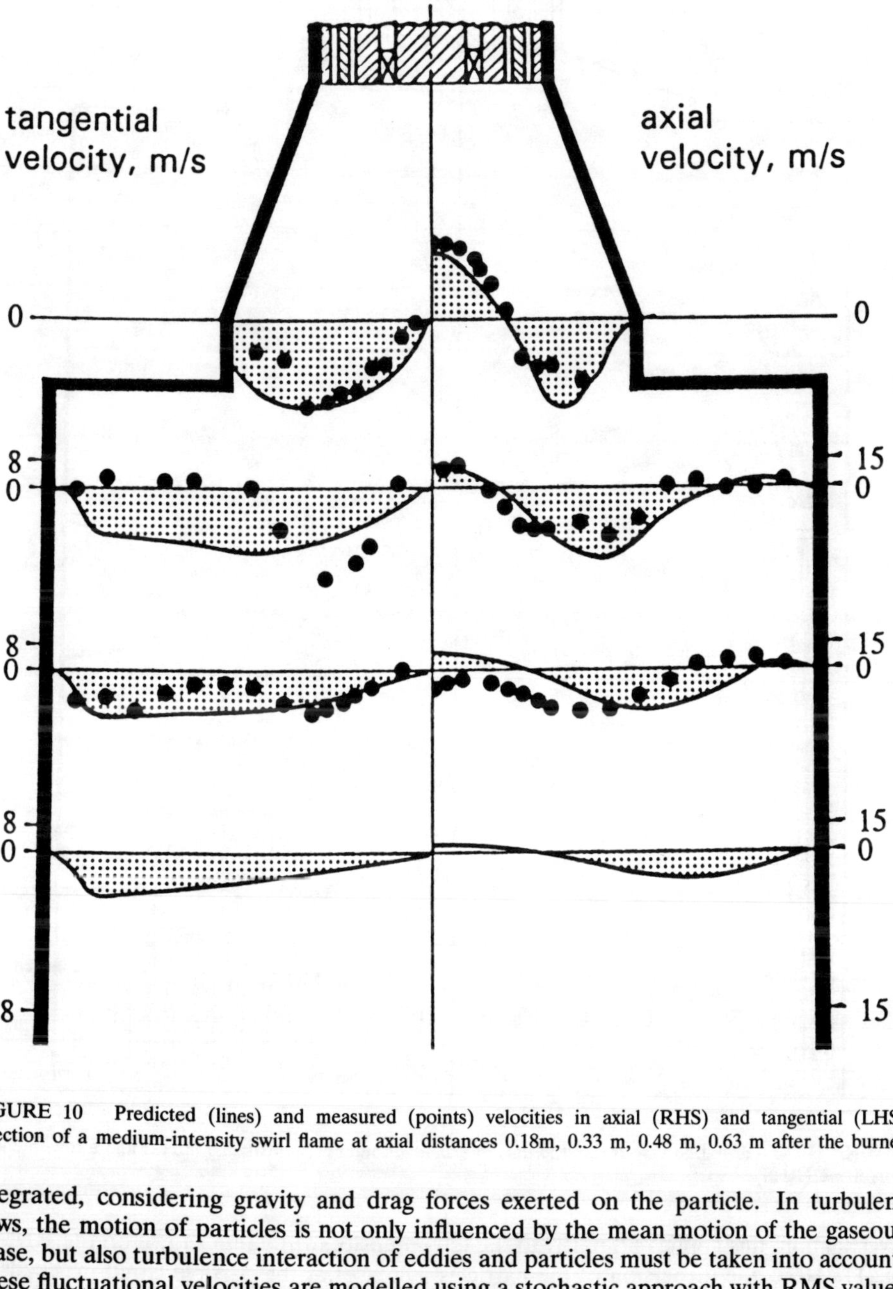

FIGURE 10 Predicted (lines) and measured (points) velocities in axial (RHS) and tangential (LHS) direction of a medium-intensity swirl flame at axial distances 0.18m, 0.33 m, 0.48 m, 0.63 m after the burner

integrated, considering gravity and drag forces exerted on the particle. In turbulent flows, the motion of particles is not only influenced by the mean motion of the gaseous phase, but also turbulence interaction of eddies and particles must be taken into account. These fluctuational velocities are modelled using a stochastic approach with RMS values provided by the turbulence model (Schnell, 1991).

In fig. 11, particles with a diameter of 30 μm equivalent to the mean particle size in the experiments were "injected" in the primary duct of the burner, assuming equal inlet velocity as the gas flow. In earlier work of Benim (1990) and Schnell and Görner (1989) it was demonstrated that the particle paths computed with turbulence effects differ considerably from those neglecting turbulence. For the two flame types under

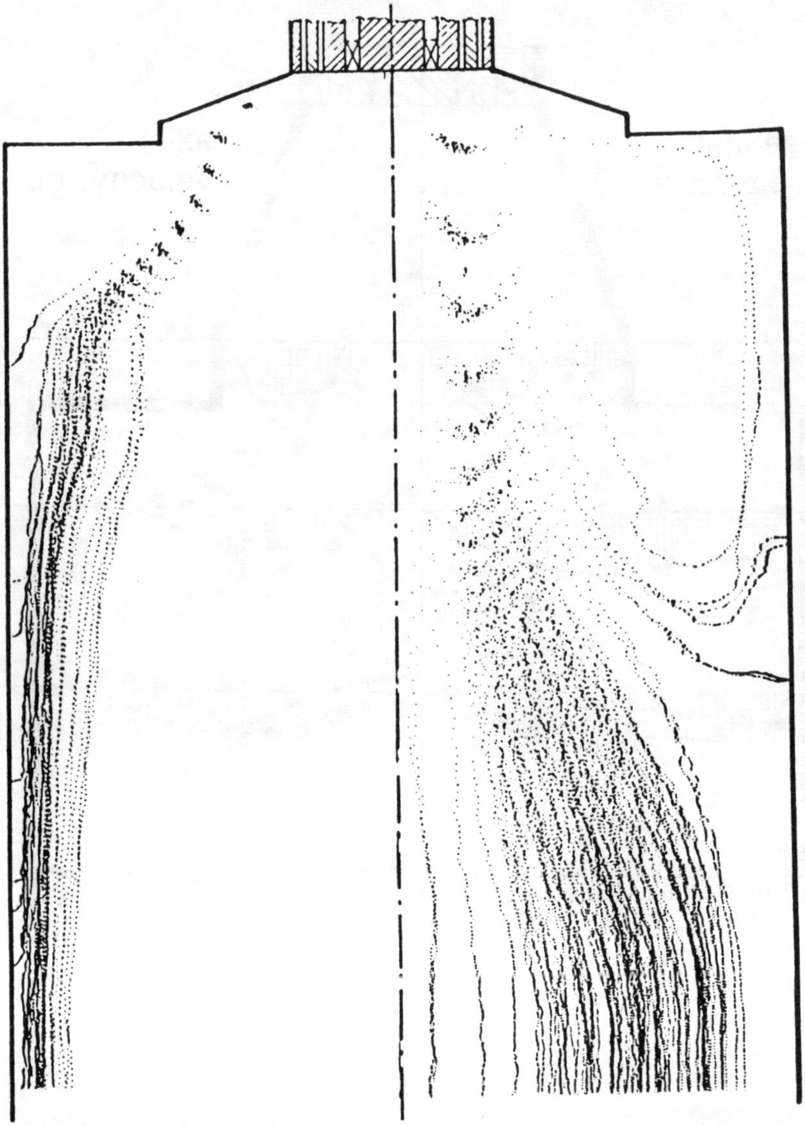

FIGURE 11 Predicted particle tracks (30 μm) in a medium-intensity swirl flame (LHS) and a low-intensity swirl flame (RHS)

investigation, this applies primarily to the low-intensity swirl flame shown on the RHS of fig. 11. Here, the dispersion of the particles increases as the particle cloud proceeds in axial direction. For judging the impact of particle motion on NO_x emissions it is essential to notice that most of the particles undergo devolatilization in the interior zone close to the symmetry axis. Only few particles are likely to penetrate the exterior recirculation zone.

In the medium-intensity swirl flame (fig. 11, LHS), the situation is completely different. The spread of the particle cloud is smaller, and as a result of the swirl, most particles

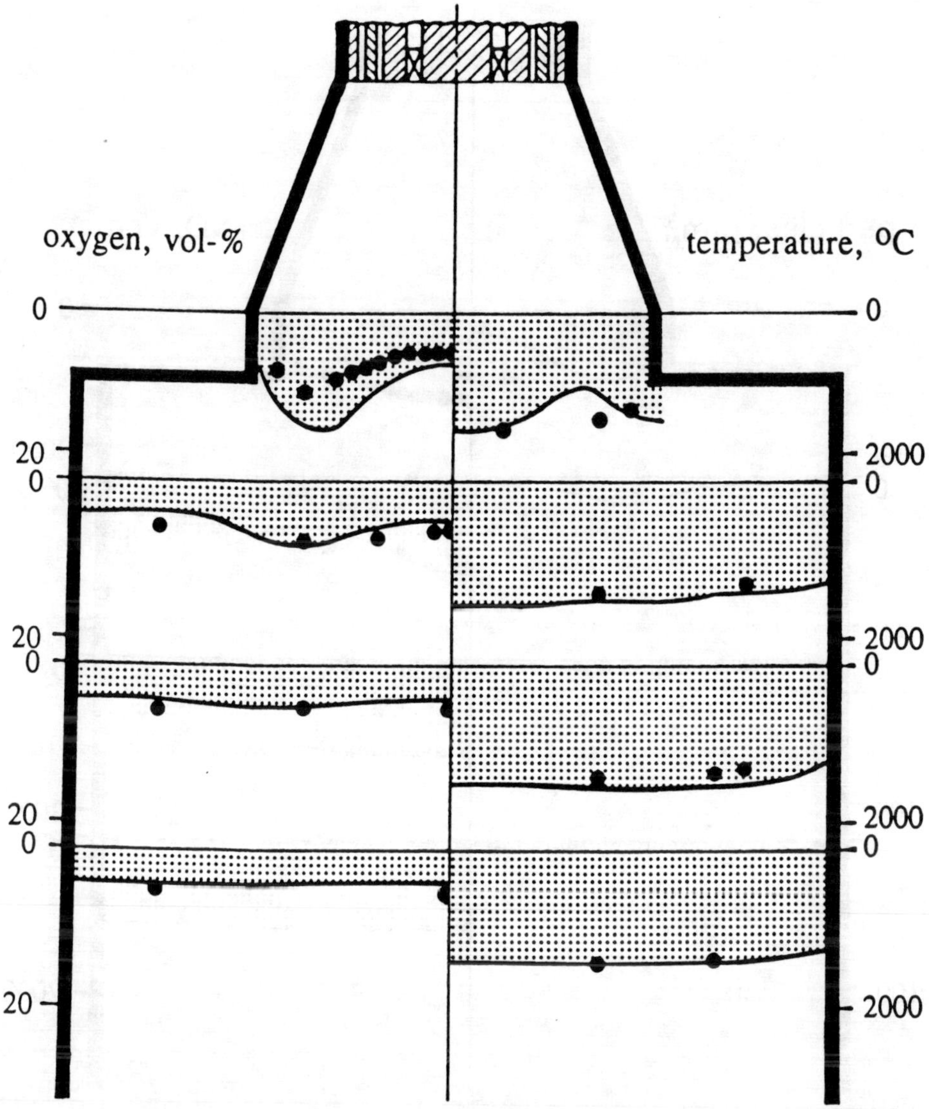

FIGURE 12 Predicted (lines) and measured (points) temperature (RHS) and oxygen concentration (LHS) in a medium-intensity swirl flame at axial distances 0.18m, 0.33 m, 0.48 m, 0.63 m after the burner

are thrown towards the quarl wall respectively the furnace wall. The devolatilization process starts directly after the burner exit and proceeds along the outer boundary of the rotating flame gases in the quarl. There is only a slight probability for re-entry into the interior recirculation zone although the radial displacement of the particle cloud seems to be overpredicted.

In the experiments, the different wall slagging behaviour of the two flame types could be reconciled with the different wall striking probability of particles.

To continue with the discussion of the medium-intensity swirl flame first, the predicted

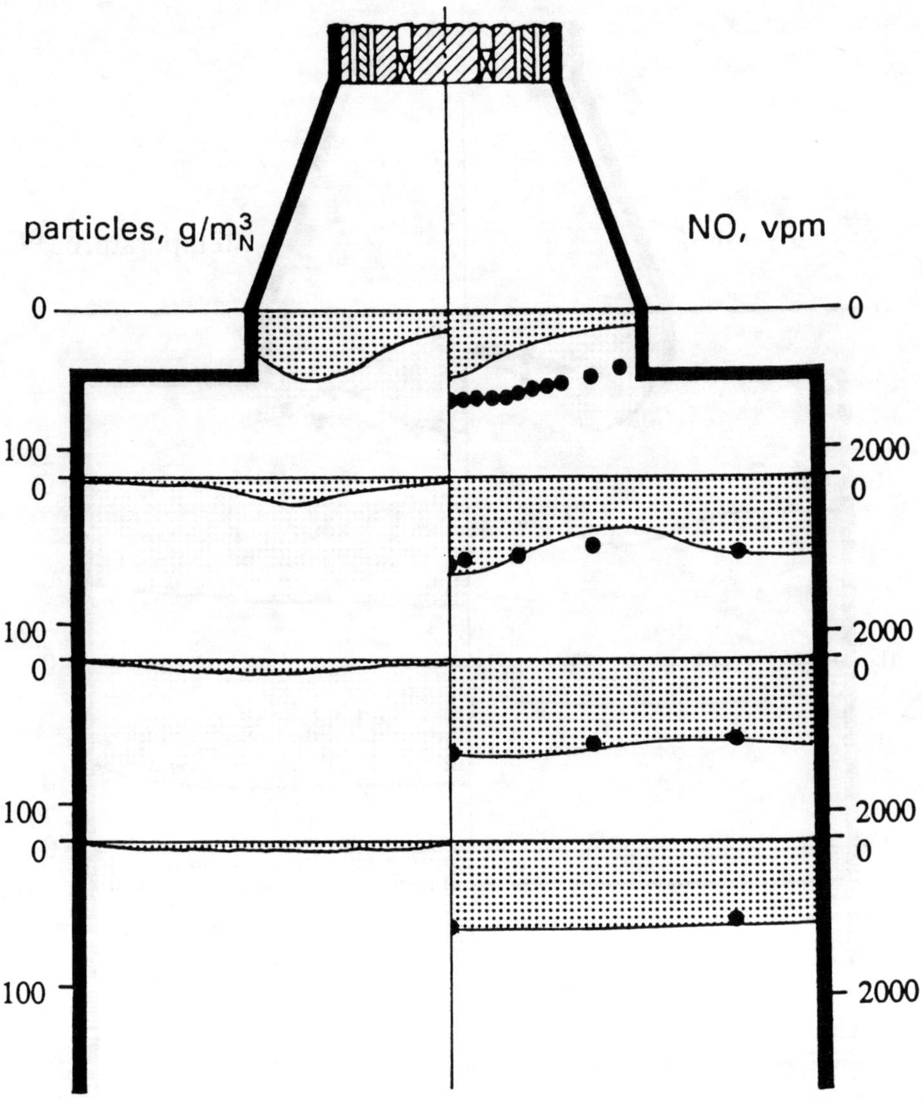

FIGURE 13 Predicted (lines) and measured (points) NO concentration (RHS) and particle concentration (LHS) in a medium-intensity swirl flame at axial distances 0.18m, 0.33 m, 0.48 m, 0.63 m after the burner

temperature and oxygen profiles have to be considered next. As a result of the strong turbulent exchange, the temperature rise following the very early ignition of the coal dust jet is very intense, with maximum values of nearly 1500°C (fig. 12, RHS). This result coincides with measured data, and it corresponds with the phenomena discussed in the context of the flow field.

The profile of oxygen concentration (fig. 12, LHS) remains nearly uniform after the axial position of 0.48 m. The very fine grinding of the coal results in a very rapid burnout, and hence the oxygen consumption is completed thus early. On the other hand, the

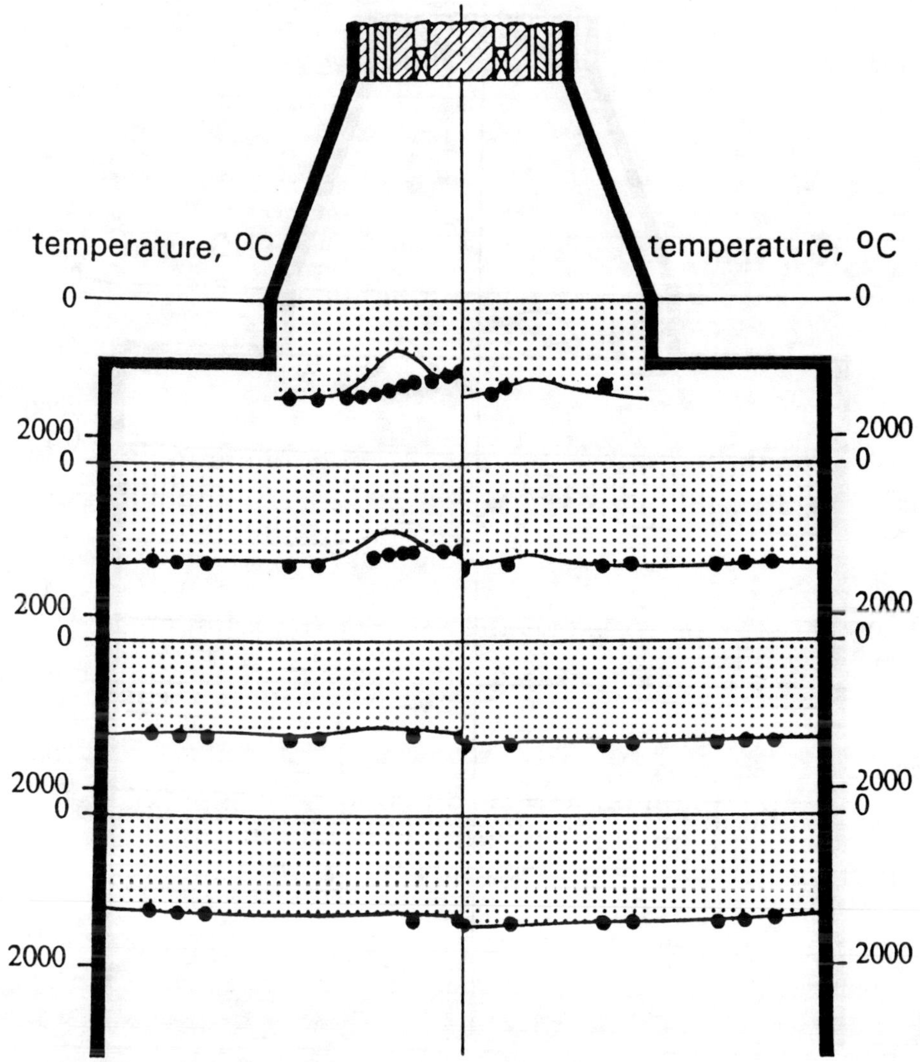

FIGURE 14 Predicted (lines) and measured (points) temperature distribution in a low-intensity swirl flame with fine grinding (RHS) and coarse grinding (LHS) of the coal at axial distances 0.18m, 0.33 m, 0.48 m, 0.63 m after the burner

high amount of oxygen surplus in this unstaged flame does not favour any reduction of previously formed NO_x.

These findings are confirmed in the NO_x profile (fig. 13, RHS). The relatively high amount of swirl, the intense mixture of coal dust with air, the rapid heat-up and the availability of surplus oxygen lead to a nearly unlimited conversion of fuel-bound nitrogen to NO_x.

In addition to this source of NO_x, there is a relatively high portion of NO_x formed out of atmospheric nitrogen according to the Zeldovich mechanism in this high-temperature flame. In the first slab at 0.18 m after the burner, the discrepancies between measurement and prediction can be attributed to some deficiencies in the formulation of the pyrolysis

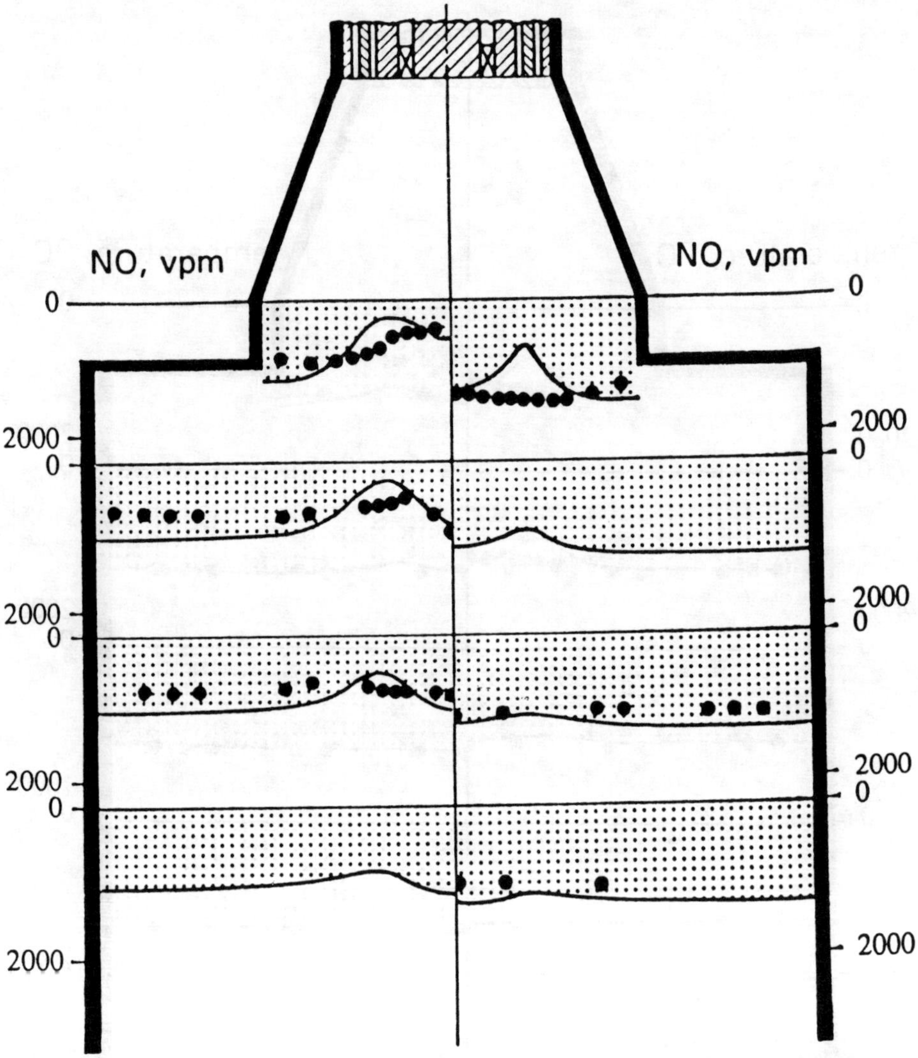

FIGURE 15 Predicted (lines) and measured (points) NO concentration in a low-intensity swirl flame with fine grinding (RHS) and coarse grinding (LHS) of the coal at axial distances 0.18m, 0.33 m, 0.48 m, 0.63 m after the burner

reaction. The single one-step reaction does not account for a staged evolution of nitrogeneous intermediates. For that reason the conversion of HCN and ammonia species to NO_x is underestimated.

Further downstream, the measured and predicted NO_x concentration profiles show good agreement. The predicted outlet emission of NO_x totals to 1150 vpm (measurement: 1170 vpm).

If the particle concentration profile in fig. 13, LHS is compared with the Lagrangian particle tracks in fig. 11, LHS, there is considerable correspondence between both numerical treatments. The swirl-induced radial displacement of the particle cloud, visualized in the particle track simulation, and the subsequent rapid burnout may also

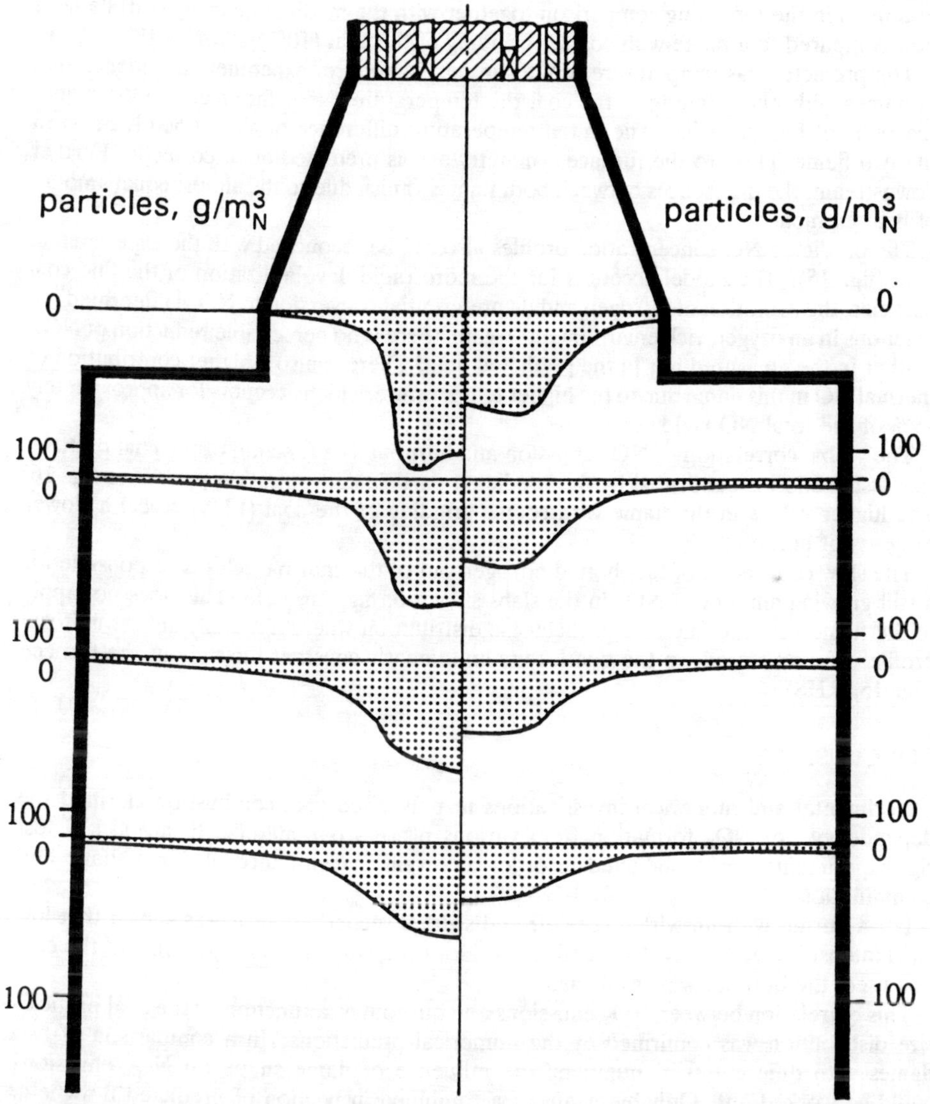

FIGURE 16 Predicted particle concentration in a low-intensity swirl flame with fine grinding (RHS) and coarse grinding (LHS) of the coal at axial distances 0.18m, 0.33 m, 0.48 m, 0.63 m after the burner

be identified in the predicted concentration profile. Nevertheless, the rate of coal and char weight loss even seems to be slightly underestimated. The reactivity of the very fine grinded coal particles with a high content of fuel-bound nitrogen (1.38% daf) provides more nitrogen oxide inside the quarl than it is anticipated from the numerical solution.

This deficiency is comparable to the overprediction of radial particle displacement in the near burner quarl zone discussed earlier.

Influence of coal particle size distribution on NO_x emissions. The second parameter with a strong impact on NO_x emissions worked out in the experiments is the coal particle size distribution, prepared in the coal mill. The flame with low swirl intensity examined in the foregoing comparison together with the medium-intensity swirl flame is now compared to a flame with coarser grinding of the coal (10%>90μm vs 1%>90μm).

The predicted gas temperature profiles in fig. 14 confirm experimental evidence that in flames with fine grinding of the coal the temperature rise is faster due to the higher reactivity of fine particles. The initial temperature difference of almost 500 K between the two flames close to the furnace symmetry axis is predicted quite correctly. Further downstream, the differences between both flames vanish due to the almost equal amount of heat extraction.

The predicted NO concentration profiles also can be reconciled with the experimental data (fig. 15). The model accounts for the more rapid devolatilization of the fine coal particles, the evolution of nitrogen and its preferential conversion to NO via intermediate reactions in an oxygen-rich environment. Again, there is no perceptible reduction of NO_x, neither in the measured nor in the predicted data. There is also a higher contribution of thermal NO in this flame due to the higher temperature, which accounts for approximately 15% of the total NO yield.

The global correlation of NO_x emission and burnout as a function of the coal particle-size distribution is confirmed by the predicted particle concentration pattern in fig. 16. The higher values in the flame with coarse grinding of the coal (LHS) reveal a slower progress of burnout.

The slow conversion of fuel-bound nitrogen out of the char particles is responsible for a still growing amount of NO_x in the slabs shown on fig. 16, LHS. This does not apply to the flame with the fine coal particle-size distribution where a radially almost uniform profile is found already in the quarl, remaining nearly constant throughout the furnace (fig. 15, RHS).

CONCLUSIONS

Experimental and numerical investigations in pulverized coal combustion clarified the dependencies of NO_x formation from various process parameters. In a 500 kW test rig, experiments were conducted in order to reveal the influence of flame shape and comminution rate.

For a burner with unswirled core air and swirled secondary air it was shown that low swirl intensities lead to low NO_x emissions. However, when a coarse grinding of the coal was used, the burnout was insufficient.

This correlation between NO_x emissions and burnout as a function of the coal particle-size distribution was confirmed by the numerical predictions. In a comparison of two flames with different swirl numbers, the influence of flame shape on NO_x emissions could be worked out. Only by means of a combined inspection of predicted flow field, particle tracks, temperature and gas species concentration the formation and destruction process of NO_x in pulverized coal flames can be understood. By these predictions, the usefulness and reliability of mathematical furnace simulations and its predictive qualities with respect to NO_x emissions have been demonstrated.

The implication of these results helped in understanding the fundamental factors that influence the formation of NO_x. The optimization of primary NO_x reduction measures in pulverized coal combustion is currently being investigated by staged combustion and reburning trials.

ACKNOWLEDGEMENTS

This work was carried out within the TECFLAM project on Mathematical Modelling and Laser Diagnostics of Technical Flames. Funding by the German Ministry of Research and Technology (BMFT) and the Federal Government of Baden-Württemberg is gratefully acknowledged.

REFERENCES

Beér, J. M. (1987). 1986 BCURA Lectures: 'Combustion of Coal—A New Look at an Old Problem', *J. Inst. Energy* **143,** pp. 143-151.

Benim, A. C., and Schnell, U. (1989). Finite Element Simulation of Turbulent Reacting Flows with Emphasis on Pulverized Coal Combustion and Nitrogen Oxide Formation. *5th Int. Symp. Num. Meth. Eng. (INME),* Lausanne (Switzerland).

Benim, A. C. (1990). Finite Element Analysis of Confined Swirling Flows. *Int. J. Num. Meth. in Fluids* **11,** pp. 697-717.

Brodbek, H. (1990). Geschwindigkeits- und Turbulenzmessungen in industriellen Oel- und Kohlenstaubflammen mittels der Laser-Doppler-Anemometrie. *LASER—Technologie und Anwendungen, Jahrbuch 2.* Ausgabe, Vulkan Verlag, Essen (Germany).

Brodbek, H. (1991). Messung von Geschwindigkeits- und Turbulenzfeldern in Kohlenstaubflammen mit dem Laser-Doppler-Verfahren *VDI-Fortschritt-Berichte Reihe* 6, No. 258, VDI-Verlag, Düsseldorf (Germany).

Bouis, X. (1975). Applications récentes à l' ISL, de la velocimetrie laser aux measures dans des écoulements turbulentes. AGARD Conference Preprint 193 *"Application of Non-Intrusion Instrumentation in Fluid-Flow-Research".*

De Soete, G. (1974). Overall Reaction Rates of NO and N_2 Formation from Fuel Nitrogen. 15th Symp. (Int.) on Combustion, The Combustion Institute, pp. 1093-1102.

Kale, S. R., Ramezan, M., and Anderson, R. J. (1988). Measurement of Particle Rotational Velocity Using a Laser Doppler Anemometer. *Proc. of the 4th International Symposium on Application of Laser Anemometry to Fluid Mechanics,* No. 2.10, Lisbon (Portugal).

Kaess, M. (1991). Experimentelle Untersuchungen von Stickstoffoxidemissionen und Ausbrand an einer Versuchsanlage zur Kohlenstaubverbrennung. *VDI-Fortschritt-Berichte Reihe* 6, No. 261, VDI-Verlag, Düsseldorf (Germany).

Schnell, U., Görner, K., and Benim, A. C. (1987). Mathematische Modellierung von Kohlenstaubflammen—Methoden und Teilmodelle. Proceedings of the *3rd TECFLAM Seminar,* Karlsruhe (Germany), pp. 149-164.

Schnell, U., and (Kaess, (1989). Vergleich von Rechnung und Messung an der TECFLAM-Kohlenstaubversuchs-brennkammer. *Proceedings of the 5th TECFLAM Seminar,* Stuttgart (Germany), pp. 25-39.

Schnell, U., and Görner, K. (1989). Interaction of Pyrolysis Kinetics with Heat Transfer and Fluid Flow in Brown Coal Flames. *Intern. Conf. Coal Science,* Tokyo (Japan).

Schnell, U. (1991). Berechnung der Stickoxidemissionen von Kohlenstaubfeuerungen. *VDI-Fortschritt-Berichte Reihe* 6, No. 250, VDI-Verlag, Düsseldorf (Germany).

Schnell, U., Epple, B., and Spliethoff, H. (1991). Modelling of Pulverized Coal Combustion with Special Emphasis on NO_X Formation and Mixing Phenomena Related to NO_X Reduction. *Proceedings of the 2nd Int. Symp. on Coal Combustion,* Beijing (China), pp. 98-107.

MODELLING THE NEAR REGION OF A SWIRLING PULVERISED COAL BURNER

J. L.T. AZEVEDO and M.G. CARVALHO *Instituto Superior Técnico, Mechanical Engineering Department, Av. Rovisco Pais, 1096 Lisboa Codex, Portugal*

Abstract — A numerical model for pulverised coal combustion is applied to operating conditions for which detailed experimental data on the near burner region exists. Comparisons are made between the model predictions and the experimental radial profiles of gas composition and temperature. The model used is based on an Eulerian set of equations for the gas phase and a stochastic Lagrangian representation of the particulate phase. Particle drying, is considered and the devolatilization is simulated by a single or two competing reaction models. The subsequent combustion of volatiles is controlled by turbulent mixing and chemical kinetics. The char burning rate is calculated using kinetic and diffusional contributions. The influence of char and volatiles combustion rates are analysed with special emphasis on the near burner region. The results show the importance of the volatiles' combustion rate close to the burner. The influence of char combustion rate and particle dispersion is important in the recirculation zones.

INTRODUCTION

The prediction of pulverised coal combustion has been the subject of a large amount of research in the last twenty years. Since the late seventies most of the models used for axisymmetric geometries are based on Eulerian equations for the gas phase and Lagrangian models for the coal particles (see e.g., Lockwood and Mahmud, 1988, Wennerberg, 1988, Fiveland *et al.*, 1984, Smith *et al.* 1981, Truelove and Holcombe, 1990, and Visser *et al.* 1990). The comparison of model predictions with experimental results has been limited to temperature and heat fluxes in some cases.

The calculations presented by Visser *et al.* (1990) present a good agreement with experimental results, even within the quarl. They have used a reduced mixing rate and the combustion model was modified considering carbon monoxide formation when the local conditions are sub-stoichiometric. This procedure was also adopted by Truelove and Holcombe (1990) yielding a better prediction of the gas concentrations close to the burner. The predictions presented by Costa *et al.* (1990) for a different kind of burner show deficiencies close to the quarl. Possible causes of these deficiencies are the motivation for the present work.

The geometrical configuration of the burner considered in the present work consists of a central pipe for primary air and entrained coal, surrounded by the coaxial secondary air, with swirl imparted to promote a recirculation flow pattern. In this type of arrangement the location of the central recirculation zone depends on the momentum ratio of the primary and secondary streams and swirl intensity. The exact location of the stagnation points of the recirculation zones is difficult to predict and this in turn affects the near burner region.

In power station boilers, the presence of a fuel gun inside the primary air delivery tube may act as a stabilizer to the central recirculation zone. To simulate this type of burner, the primary air and coal are fed through an annular coal injector, in most pulverised coal experimental combustors. The central core acts as a bluff body and the recirculation zone may be appended to the burner, being more easily predicted. This is the case of flame I presented by Visser *et al.* (1990), which compares more favourably with experimental results than flame II where the recirculation zone is penetrated by the primary air stream.

Lockwood *et al.* (1991) have compared the two types of burner geometries and obtained better results for the prediction of the annular burner gun.

The kinetic rates of devolatilization and char combustion used by different authors are in some cases a result of the optimization of model predictions based on gas temperatures and wall heat fluxes. This procedure is adopted due to the absence of kinetic data for the specific coals under consideration. However the calibration of combustion model constants from comparisons with experiments, should be based on char burnout and gas composition which is the most relevant experimental information for this submodel.

Chemical kinetics of volatiles combustion are neglected in some of the models or are considered through the carbon monoxide reaction rate. Recently Shaw *et al.* (1990) derived global kinetic reaction rates for volatiles combustion as a single-step reaction. Kinetic rates for intermediate steps, such as carbon monoxide combustion or the decomposition and combustion of tar, can also be used to represent the limiting step of volatiles combustion. The consideration of a transport equation for carbon monoxide can be used to consider the incomplete oxidation of char. The consideration of tar as an intermediate specie introduces the need for a detailed devolatilization and combustion model in the gas phase.

The main objective of this work was to compare gas concentrations in the near burner region and analyse the effect of the coal and volatiles combustion model. The model used is a modified version of the model of Lockwood and Mahmud (1988). Particle drying and chemical kinetics of volatiles combustion are now considered. A stochastic model for particle tracking (see e.g., Azevedo and Pereira, 1991) and a two competing reaction model for devolatilization were also implemented. The model was used to predict a flame studied by Lockwood and Mahmud (1988) and Costa *et al.* (1990) which was initially documented by Godoy *et al.* (1988). After this introduction, the following section presents the physical and numerical model used with reference to the modifications performed. The model results are presented in the next section showing the effects of char and volatile's combustion. The conclusions from the present work are summarized in the final section.

PHYSICAL AND NUMERICAL MODEL

The numerical model is based on two different modules corresponding to the gas and coal particle phases, respectively, which are coupled and interact during the numerical calculation.

Gas Phase

The gas phase model comprises transport equations for the three momentum components, enthalpy, gas concentration of oxygen, volatiles and products and for the turbulent quantities, k and ε. The general form of these equations is:

$$\frac{\partial}{\partial x}(\rho U\phi) + \frac{1}{r}\frac{\partial}{\partial r}(\rho r V\phi) = \frac{\partial}{\partial x}\left(\frac{\mu_e}{\sigma_\phi}\frac{\partial \phi}{\partial x}\right) + \frac{1}{r}\frac{\partial}{\partial r}\left(r\frac{\mu_e}{\sigma_\phi}\frac{\partial \phi}{\partial r}\right) + S_\phi + S_\phi^p, \quad (1)$$

where S_ϕ and S_ϕ^p are the source terms due to the gas and particle phase. The form of these source terms for the momentum are similar to the isothermal case (see e.g., Azevedo and Pereira, 1991) and thus are omitted here. The gas mass fractions and the enthalpy sources are presented in Table 1.

The gas source terms in the gas concentration equations are derived from the combustion model of the volatiles. The volatiles are released to the gas phase leading to S_{vol}^p and the subsequent combustion of the volatiles (considered as a single specie) is controlled by

TABLE I

SOURCE TERMS OF GAS EQUATIONS DUE TO COMBUSTION

ϕ	S_ϕ	S_ϕ^P
C_{O_2}	$-R_{VC} \times O_{2\,STOC}$	$-\sum K_{ch} A_p n_p \frac{32}{12}$
C_{VOL}	$-R_{VC}$	$\sum K_v V_p n_p$
C_{CO_2}	$R_{VC} \times CO_{2\,STOC}$	$\sum K_{ch} A_p n_p \frac{44}{12}$
h	$R_{VC} H_{VC} + 4\sigma\varepsilon(T_R^4 - T^4)$	$\sum A_p \left[h(T_p - T) + \varepsilon_p \sigma (T_p^4 - T_p^4)\right] n_p$

$R_{VC} = R_k R_m/(R_k + R_m)$ = volatile combustion rate, kg/m^{3}s
$O_{2\,STOC}$, $CO_{2\,STOC}$ = stoichiometric coefficient of volatile reaction

mixing and chemical kinetics. The eddy-dissipation model of Magnussen and Hjertager (1977) is used to estimate the mixing rate using a timescale derived from the k-ε turbulence model:

$$R_m = A\rho \frac{\varepsilon}{k} \min\left\{C_{VOL}, \frac{C_{O_2}}{O_{2\,STOC}}, \frac{C_{O_2}}{CO_{2\,STOC} + B}\right\}, \tag{2}$$

where the constants A = 3.5 and B = 2 were taken. The stoichiometric coefficients of volatiles combustion are derived from coal composition and volatile yield. The volatiles chemical kinetic rate of Shaw *et al.* (1990) is considered:

$$R_k = A_{vc} e^{-E_{vc}/RT} \rho^2 C_{VOL} C_{O_2}, \tag{3}$$

where the parameters A_{VC} and E_{VC} were taken as 380800 m^3/kg s and 48 KJ/mol.

The combustion of volatiles to form products (CO_2, H_2O) in a single-step reaction is the simplest model for volatiles combustion. The consideration of intermediate reactions such as the formation of tar and decomposition into lighter gas components would require the solution of the corresponding transport equations and the knowledge of the volatiles composition and kinetic rates for all possible reactions. Although such a detailed description is not attempted here, the heavy hydrocarbon combustion rate is considered to represent the combustion of tar and the importance of this step is analysed. The combustion rate of long-chain hydrocarbon molecules from Siminski *et al.* (1972) is taken as representative of tar combustion:

$$\frac{dC_{TAR}}{dt} = -59.8 T P^{0.3} C_{TAR}^{0.5} C_{O_2}^{0.5} \exp(-12200/T). \tag{4}$$

The tar molecular weight was assumed to be 120.

The heterogeneous combustion is assumed to be complete within the boundary layer of the coal particle leading to two source terms in the oxygen and carbon dioxide equations. The enthalpy source includes the heat exchanged with the coal particles while the contribution from the gas phase includes the heat released by volatiles combustion and the radiative heat transfer contribution. A nonequilibrium diffusion model is used for radiation (Arscott *et al.*, 1973), which leads to an algorithm of the same form as for the general transport equations without convective terms. The pressure field is calculated from the continuity equation using the PISO algorithm.

Particle Phase

Coal combustion is simulated by tracking representative particles in the flow domain. The particle momentum equation

$$\frac{\partial \vec{U}_p}{\partial t} = \frac{\vec{U} - \vec{U}_p}{\tau} + \vec{g} \tag{5}$$

is analytically integrated in time steps dictated by the grid dimensions and by turbulence. τ is the particle characteristic time and $\vec{U}$ is the instantaneous gas velocity. This value is obtained as the sum of the mean value with a fluctuating component with Gaussian distribution and standard deviation of $\sqrt{2/3\,k}$. The interaction of a given eddy with the particle trajectory is limited in time by the eddy lifetime and the time for a particle to cross an eddy (see e.g., Azevedo and Pereira, 1991).

The temperature of the particle is calculated from the energy balance equation:

$$\begin{aligned} m_p c_p \frac{dT_p}{dt} &= A_p \left[h(T - T_p) + \sigma\varepsilon_p (T^4 - T_p^4)\right] \\ &+ K_{ch} A_p P_{O_2} H_C - H_{DV} \frac{dV}{dt} V_p - H_{H_2O} \frac{dm_{H_2O}}{dt}. \end{aligned} \tag{6}$$

The variation in particle temperature is the result of heat exchanged with the gas and particle heat source terms. These sources are the heat generated by char combustion (H_c) and the heat necessary to release moisture (H_{H_2O}) and volatiles (H_{DV} taken as zero in the present work). Particle drying is considered in the model by setting the particle temperature at a constant value of 100°C and decreasing the moisture content in the coal. The specific enthalpies associated with combustion processes are the carbon calorific value (H_C) and the volatiles calorific value (H_{VC}) taken from a global balance using the gross calorific value of coal.

Two models of devolatilization are used, namely, the single first-order and the two competing reaction model. According to the first-order model the devolatilization rate is expressed as:

$$\frac{dV}{dt} = K_v (V^* - V), \tag{7}$$

where the kinetic rate K_v is expressed in Arrhenius form as $A_v \exp(-E_v/RT)$. The kinetic parameters used are those recommended by Badziok and Hewksley (1970) ($A_v = 134000$ s^{-1}; $E_v = 74$ kJ/mol). The ultimate yield of volatiles, V*, exceeds the proximate volatile matter by a specified factor V*/VM taken as 1.35.

The two-competing reaction model is described by two parallel first order reactions

$$\text{coal} \begin{cases} \xrightarrow{K_{v1} = A_{v1} e^{-E_{v1}/RT}} \alpha_1 \text{ vol} + (1 - \alpha_1) \text{ char} \\ \xrightarrow{K_{v2} = A_{v2} e^{-E_{v2}/RT}} \alpha_2 \text{ vol} + (2 - \alpha_2) \text{ char} \end{cases} \tag{8}$$

where the second reaction has a larger activation energy ($E_{v2} > E_{v1}$). The devolatilization rate is given by:

$$\frac{dV}{dt} = (\alpha_1 K_{v1} + \alpha_2 K_{v2})\, C_{unr}, \tag{9}$$

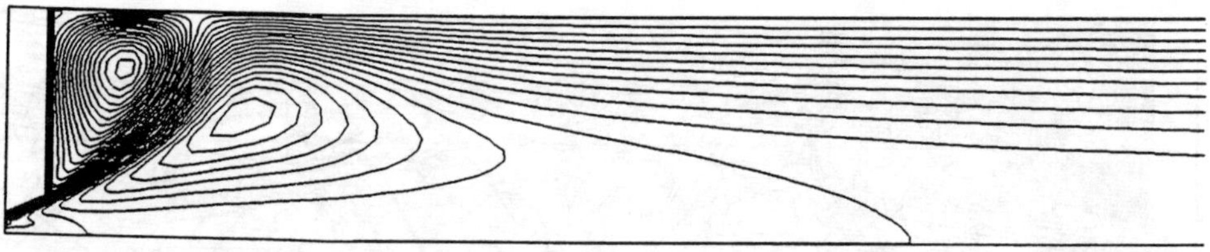

FIGURE 1 Predicted gas streamlines.

where C_{unr} is the unreacted coal fraction. In this model the yield of volatiles varies between α_1 and α_2, depending on the particle temperature history and thus is not a parameter as in the single-reaction model. The kinetic rates of Ubhayakar *et al.* (1977) were used (A_{v1} = 134000 s^{-1}, E_{v1} = 74 kJ/mol, $A_{v2} = 1.46 \times 10^{13}$, E_{v2} = 251 kJ/mol, α_1 = VM, α_2 = 1). The preexponential of the first reaction was modified to the one of Badziok and Hewksley (1970), following the recommendation of Jamaluddin *et al.* (1985).

The heterogeneous reaction rate is expressed by a combination of diffusional and kinetic rates as usual for coal combustion. Two different char kinetic rates presented by Field (1970) were used in the present work: higher (A_c = 0.005 kg/m²s; E_c = 74 kJ/mol) and lower (A_c = 0.086 kg/m²s; E_c = 150 kJ/mol).

RESULTS

The experimental results, used in the present work, were obtained in the Imperial College vertical cylindrical furnace, 0.6 m in diameter and 3 m long. The burner consists of an annular inlet for the swirled secondary air and a central pipe where the primary air and coal is delivered. The test case considered in this work had primary and secondary air mass flow rates of 27.5 and 103.4 kg/h respectively and the coal mass flow rate was 11.7 kg/h, corresponding to 15% excess air. The inlet temperatures were 80°C and 300°C for the primary and secondary air flow, respectively. The tangential velocity profile considered was of a forced vortex with swirl number 0.78. The axial velocity was specified with a uniform value at each stream. The fluctuating velocity was specified as 10% of the mean velocity and the characteristic turbulent length scale, taken as 6% of the passage width, was used to specify the inlet turbulence dissipation rate. The inlet particle mean velocities and fluctuating velocities were assumed equal to the gas velocities. The wall temperature was taken from the experimental results as 1020 ± 30°C. Coal characteristics can be found in Costa el at. (1990) and the particle size distribution was discretized in ten fractions ranging from 4 to 150 μm with a mean diameter of 45 μm (see e.g., Godoy *et al.* 1988).

Figure 1 represents the gas streamlines. The gas flow patterns are in agreement with the flow observations of Lockwood and Mahmud (1988). The primary air jet is decelerated and reversed by the central recirculation zone, which extends up to about one metre. Figure 2 shows sets of 100 trajectories of burning coal particles for three conditions considered in the present work, namely: a) considering turbulence dispersion and using the lower kinetic rate for char burnout; b) neglecting turbulent dispersion and; c) considering the higher kinetic rate for char burnout. Turbulent dispersion has to be considered to account for the particles' spreading into the recirculation zones (see figures 2a) and 2b)). The two kinetic rates of char burnout chosen are limit values of the rates presented by Field (1970). Using the lower kinetic rate some of the particles are not completely burned in the furnace, while using the higher combustion rate the coal particles burn close to the quarl. In the present calculations the coal particles' trajectories were reflected when they hit the wall. This

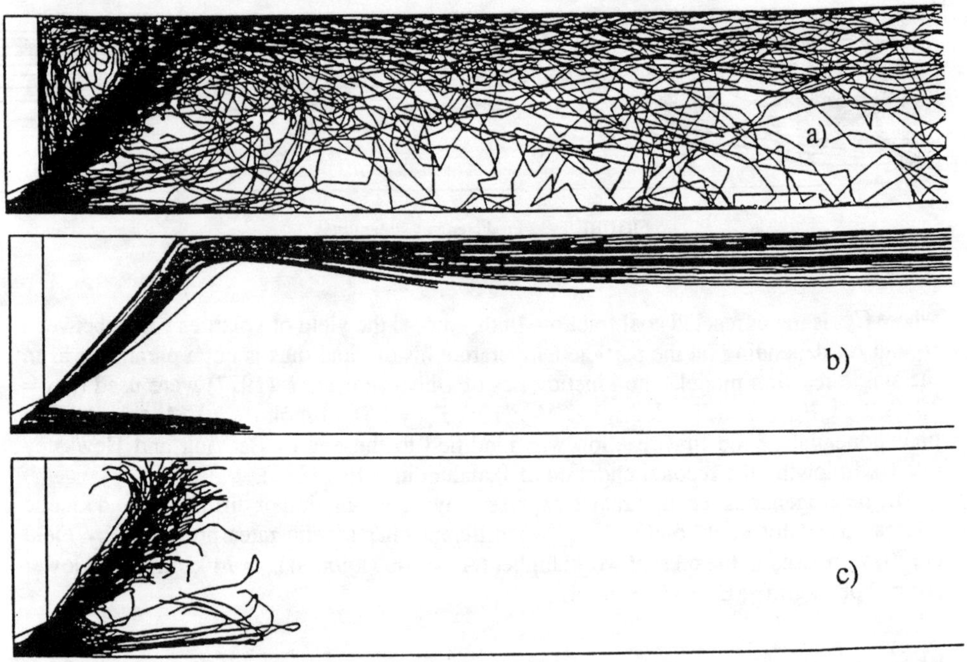

FIGURE 2 Predicted particle trajectories: a) standard case, b) neglecting particle dispersion, c) higher reaction rate.

procedure is open to criticism because coal particles may stick at the refractory walls whether burning or not. This condition is now considered in the model, but was not used for the present results.

The location of volatiles' release is a result of the dispersion and devolatilization models. Using the lower kinetic rate for the char burnout model, contour maps of the volatiles' sources are presented in figure 3. Figure 3b) compared with figure 3a) shows the effect of neglecting turbulent dispersion. Figure 3c) shows the results obtained by neglecting particle drying and considering moisture evolved with the volatiles. Although the devolatilization of each particle begins sooner, only minor modifications in the location of volatiles sources is observed due to the reversing characteristic of the flow and the small amount of moisture (6.3%). Figure 3d) shows the volatile sources using the two-competing reactions model of Ubhayakar *et al.* (1977), where the first reaction rate is equal to the one of the single reaction model. The predicted volatile yield in the present case is close to the one assumed in the single reaction model (1.35 VM). The devolatilization reaction rate of the two-competing reaction model is a bit higher than in the single reaction model, leading to more concentrated volatile sources. However this minor difference does not significantly affect the results.

Figure 4 shows equivalent contour maps for the char burnout. Figure 4a) shows the influence of turbulent dispersion when compared with figure 4b). The consideration of velocity fluctuations produces a more uniform distribution of particles in the flow, with some char combustion inside the recirculation zones. The predicted char burnout in this case is not complete, as observed in the experiments. Using the higher kinetic rate, char burnout is complete close to the top of the furnace. Lockwood and Mahmud (1988)

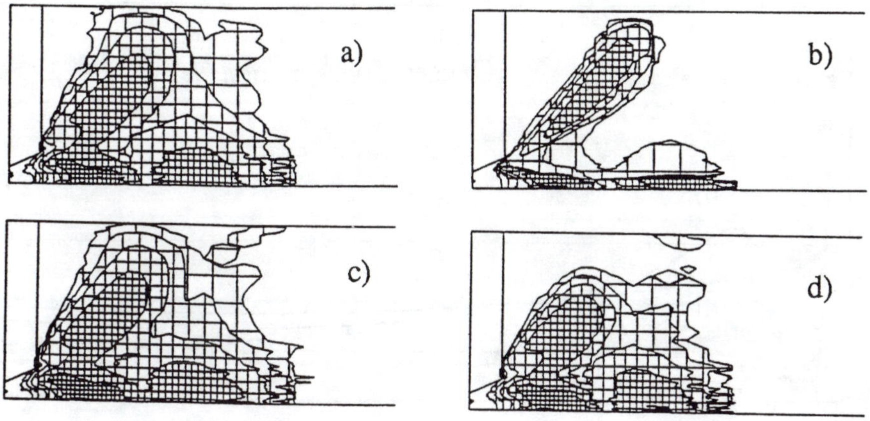

FIGURE 3 Contour maps of volatiles distribution: a) standard case, b) neglecting particle dispersion, c) neglecting particle drying, d) parallel reaction model.

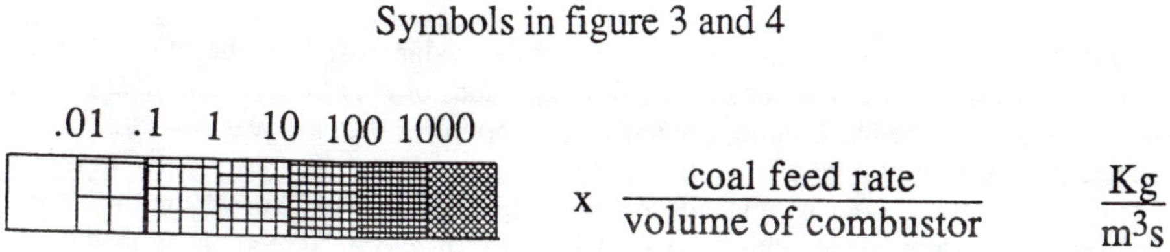

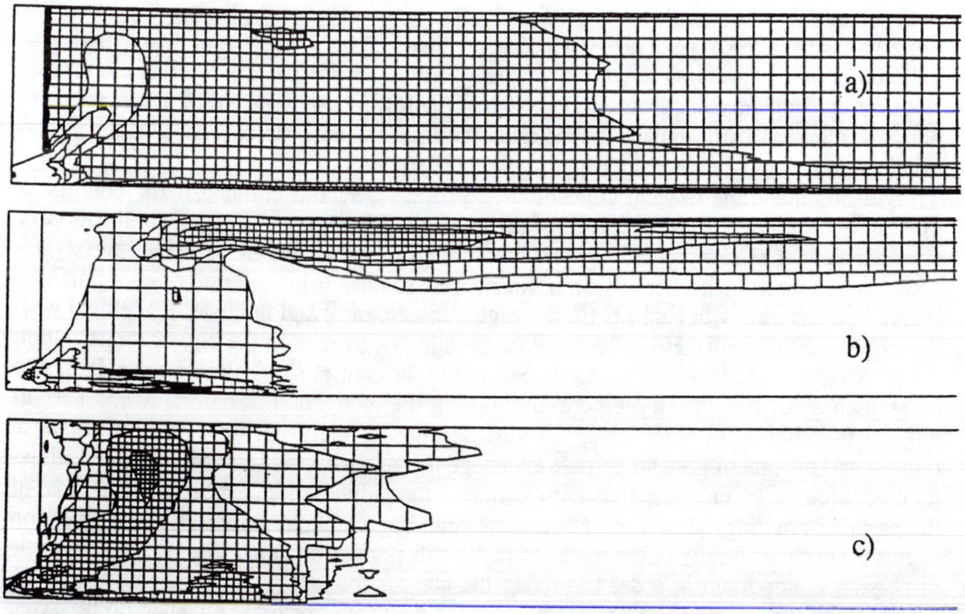

FIGURE 4 Contour maps of char combustion: a) standard case, b) neglecting particle dispersion, c) higher reaction rate.

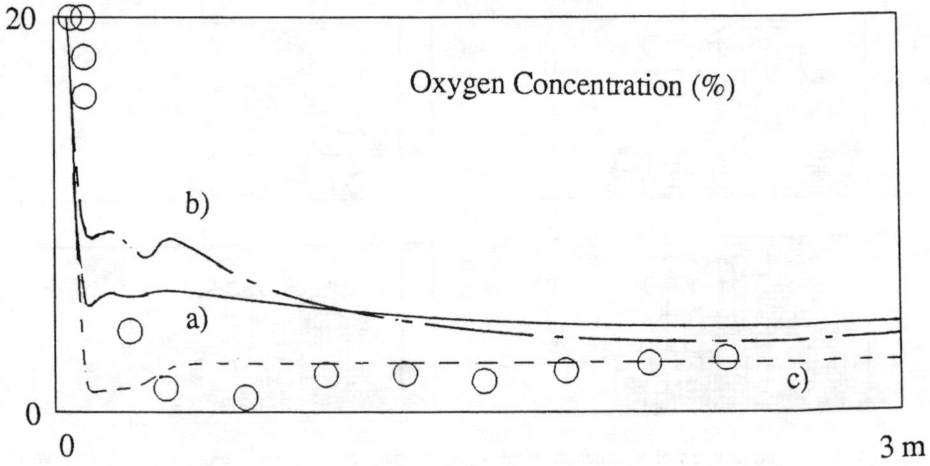

FIGURE 5 Centreline oxygen profiles: a) standard case, b) neglecting particle dispersion, c) higher reaction rate.

recommended a char burnout law close to the lower value based on the burnout comparison. However that study was based on few (2) values of char burnout and thus further comparisons with more local values are required to optimize the char reaction rate.

The centreline evolution of oxygen for the cases discussed above is represented in figure 5. The lower kinetic rate (curve A) leads to a larger level of oxygen inside the recirculation zone compared to the higher rate (curve C) which is in closer agreement with the experimental results. This cannot justify the choice of a given char reaction rate as the position of the central recirculation zone is predicted as being too close to the burner. This observation follows from the oxygen decay which is predicted closer to the burner than experimentally observed, suggesting an error in the position of the central recirculation zone. Neglecting turbulence dispersion in the case of the lower combustion rate, fewer particles burn and release volatiles inside the recirculation zone and the predicted oxygen concentration is higher than when considering turbulent dispersion.

Figure 6 shows the oxygen concentration profiles near the burner for the conditions discussed above (a,b,c) and neglecting the chemical kinetics of volatiles combustion or considering that rate expressed by the kinetic rate of Siminski *et al.* (1972). The oxygen consumption near the burner is slower considering the lower kinetic rate of char particles (curve A compared with curve C for the higher kinetic rate) and in closer agreement with the experimental results. However the concentration of oxygen in the corner recirculation zone is overpredicted, as was the case observed in the central recirculation zone. It should be reemphasised that the oxygen consumption in the recirculation regions is not a result only of the char combustion rate, but also depends strongly on the extension of these regions and the rate of gas and particle exchange through their boundaries. The deterministic model (curve B) underpredicts combustion in the recirculation region because most of the particles are dragged by the air stream surrounding the recirculation. This comparison shows the high sensitivity of the model to particle dispersion and char kinetic rates. Particle dispersion is important in order to predict the correct concentrations in the recirculation zones. No attempt is made here to optimize char kinetic parameters which should be based on more comparisons of local char burnout results or simpler flow situations. Nevertheless, the smaller kinetic rate considered here improves the overall results.

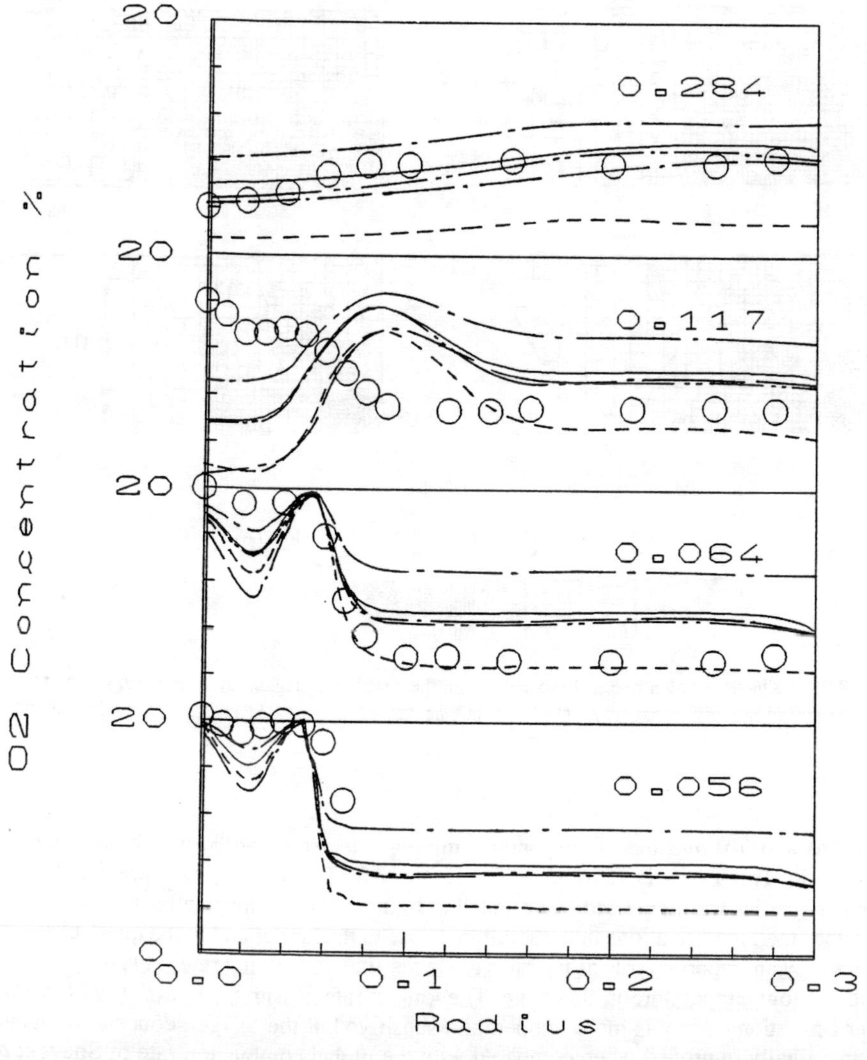

FIGURE 6 Near burner oxygen concentrations: a) standard case, b) neglecting particle dispersion, c) higher reaction rate, d) neglecting chemical kinetics, e) kinetic rate of Siminski *et al.* (1972).

The oxygen concentrations predicted near the burner, neglecting chemical kinetics of volatiles' combustion, are represented in figure 6 by curve D, while curve E shows the results obtained using the kinetic rate representative of tar combustion. The figure shows that near the burner, some improvements are obtained by considering chemical kinetic rates of volatiles combustion. However, their consideration does not explain the earlier oxygen consumption.The proximity of the central recirculation zone associated with the predicted larger spreading of the incoming jet, promotes a fast deceleration of the particles which are allowed to burn close to the quarl outlet.

Figure 7 shows the kinetic and mixing rates on the near burner region. The kinetic rates of Shaw *et al.* (1990) (equation 3) and of Siminski *et al.* (1972) (equation 4) are shown in

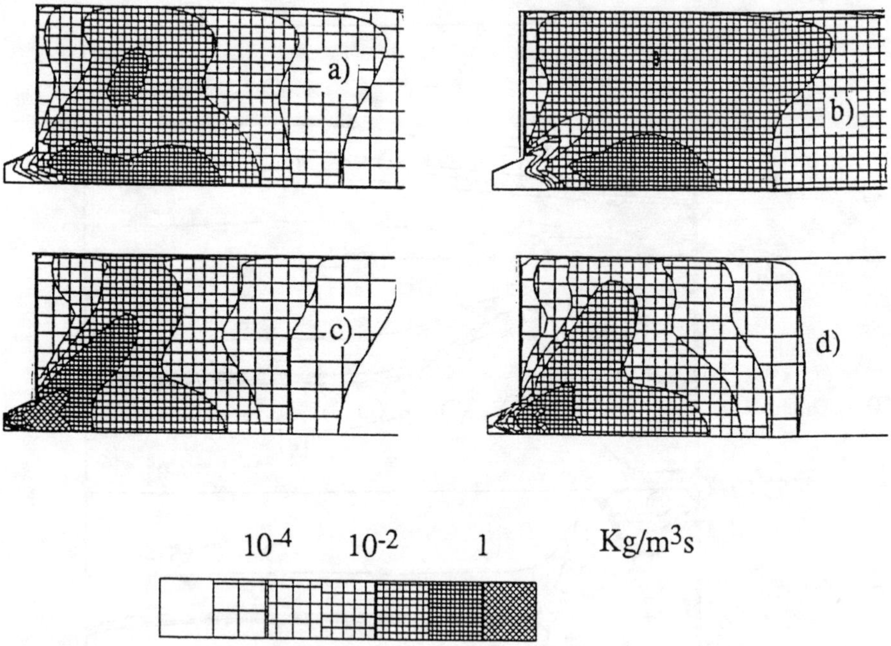

FIGURE 7 Kinetic reaction rate and mixing rate on the near burner region: a) kinetic rate of Shaw *et al.* (1990), b) kinetic rate of Siminski *et al.* (1972), c) mixing rate when using a kinetic rate, d) mixing rate when neglecting the kinetic rate.

figure 7a) and 7b) and the corresponding mixing rate (equation 2) which were similar between the two cases is represented in figure 7c). The mixing rate predicted when chemical kinetics was neglected is presented in figure 7d) showing smaller values than 7c) due to the small modifications in the solution. Close to the burner and in the quarl, chemical kinetics play an important role as the rates are lower than the mixing rate. This is essentially due to the low temperature in this zone. The kinetic rate of Siminski *et al.* (1972) with a higher activation energy is more temperature sensitive but the oxygen concentrations are only marginally improved when compared with the global combustion rate of Shaw *et al.* (1990). The use of the rate of Siminski *et al.* (1972) is only an approximation undertaken to investigate the importance of kinetics of large molecules such as tar. The content of tar in the volatiles depends on the coal and particle temperature history and its consideration requires a more complete model for the gas phase reaction. The comparison of these results shows the importance of chemical kinetics which can be considered by a single global reaction rate for volatiles. A possible further development of this model is the consideration of the carbon monoxide equation to allow for partial combustion of both volatiles and char.

Figure 8 shows the predicted gas temperatures. Using the higher combustion rate (curve C) the temperature is overpredicted outside the central recirculation zone in the first half meter of the combustor where combustion is completed. Neglecting turbulence dispersion (curve B) the temperature is also overpredicted outside the recirculation zone at $x = 0.5$ m due to the concentration of burning particles in this region. The first temperature profile shows the overprediction of the incoming jet spreading. The large gradients predicted are a consequence of the proximity of the recirculation zone and the heat transfer model.

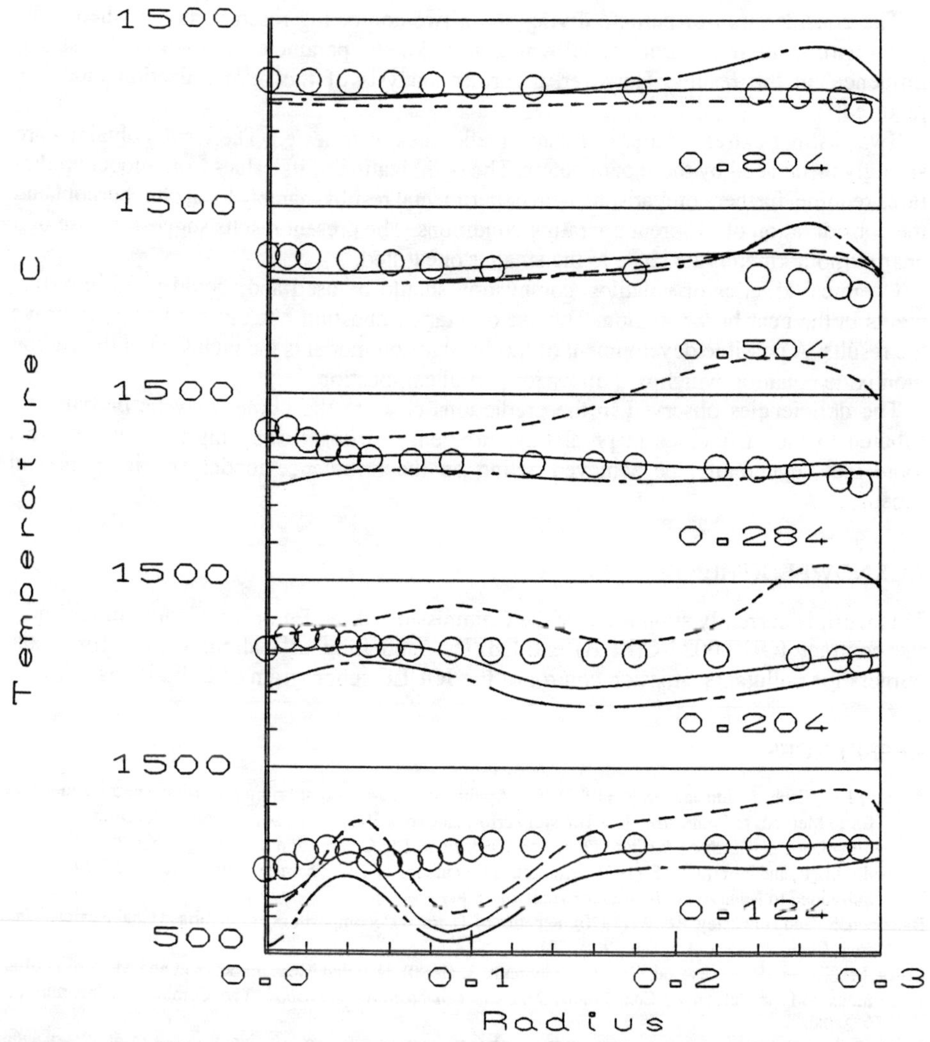

FIGURE 8 Temperature profile: a) standard case, b) neglecting particle dispersion, c) higher reaction rate ($A_c = 0.005$, $E_c - 74$).

The comparison of the gas concentration profiles at different locations in the combustor are not presented here. They show the results already discussed concerning the oxygen consumption in the stream surrounding the recirculation zones and the concentration inside these zones.

CONCLUSIONS

A stochastic Lagrangian model was applied to predict a swirling pulverised coal flame. The consideration of turbulence dispersion leads to a better prediction of the gas concentrations and temperature when compared with a deterministic model.

The consideration of particle drying and a two-competing reaction model when compared with a single reaction model with similar kinetic parameters did not have a strong influence on the results. However, a further analysis of the devolatilization model is required.

Two distinct extreme values of char kinetic rates were tested. The results obtained are strongly influenced by these parameters. The optimization of its values from model predictions requires further comparisons with experimental results, namely local char burnout and the consideration of different operating conditions. The present results suggest the use of a char burnout kinetic rate close to the smaller one tested.

Chemical kinetics of volatiles' combustion should be used and should yield improvements in the near burner region. The use of a tar combustion rate did not further improve the results. A possible development of the combustion model is the inclusion of the carbon monoxide equation which will allow for partial combustion.

The deficiencies observed in the predictions close to the burner may be partially attributed to the difficulties in predicting turbulent swirling flows. Improvements on this topic may, eventually, be achieved acting on the turbulence model and in numerical schemes.

ACKNOWLEDGMENTS

This work is currently supported by the Commission of the European Communities under the contract JOUF-0023-C(EDB) entitled The modelling of wall fired pulverised coal burners for pollutants emission control of the JOULE subprogram of Solid Fuels.

REFERENCES

Arscott, J.A., Gibb, J., and Jenner, R. (1973), The Application of Non-Equilibrium Diffusion Theory and Monte Carlo Methods to Predict the Heat Transfer Performance of a 500 MW Power Station Boiler from Isothermal Flow Data, *Combustion Institute European Symposium*, The Combustion Institute, pp. 674–679.

Azevedo, J.L.T., and Pereira J.C.F. (1991), Numerical Calculation of Two-Phase Turbulent Confined Swirling Jets, Submitted to *Numerical Methods for Heat and Fluid Flow*.

Badzioch, S., and Hawksley, P.G.W. (1970), Kinetics of Thermal Decomposition of Pulverised Coal Particles, *Ind. Eng. Chem. Process Des. Dev.*, **9**(4): 521.

Costa, M., Costen, P., Lockwood, F.C., and Mahmud, T. (1990), Detailed Measurements in and Modelling of an Industry-Type Pulverised Coal Flame, *23rd Int. Combustion Symposium*, The Combustion Institute, pp. 973–980.

Field, M.A. (1970), Measurements of the Effect of Rank on Combustion Rates of Pulverised Coal, *Combustion and Flame*, **14**: 237–248.

Fiveland, W.A., Cornelius, D.K. and Oberjohn, W.J. (1984), A Numerical Model for Predicting Furnace Performance in Axisymmetric Geometries, ASME paper 84-HT-103.

Godoy, S., Hirji, K.A., and Lockwood, F.C. (1988), Combustion Measurements in a Pulverised Coal-Fired Furnace, *Comb. Sci. and Tech.*, **59**: 165–182.

Jamaluddin, A.S., Truelove, J.S., and Wall, T.F. (1985), Modelling of Coal Devolatilization and Its Effect on Combustion Calculations, *Combustion and Flame*, **62**: 85–89.

Lockwood, F.C., and Mahmud, T. (1988), The Prediction of Swirl Burner Pulverised Coal Flames, *22nd Symp. (Int.) on Combustion*, pp. 165–173.

Lockwood, F.C., Mahmud, T., and Yehia, M.A. (1991), Simulation of Pulverised-Coal Experimental Furnace Performance, IC internal report, Submitted for publication.

Magnussen, B.F., and Hjertager, B.H. (1977), On Mathematical Modelling of Turbulent Combustion with Special Emphasis on Soot Formation and Combustion, *16th Symp. (Int.) on Combustion*, pp. 719–729.

Shaw, D.W., Zhu, X., Misra, M.K., and Essenhigh, R.H. (1990), Determination of Global Kinetics of Coal Volatiles Combustion, *23rd Symp. (Int.) on Combustion*, The Combustion Institute, pp. 1152–1162.

Siminski, V.J., Wright, F.J., Edelman, R.B., Economos, C., and Fortune, O.F. (1972), Research on Methods of Improving the Combustion Characteristics of Liquid Hydrocarbon Fuels, *AFAPTR 72–74*, Vol. I and II. Air Force Aeropropulsion Lab. Wright Patterson Air Force Base, Ohio.

Smith, P.J., Fletcher, T.H., and Smoot, L.D. (1981), Model for Pulverised Coal-Fired Reactors, *18th Symp. (Int.) on Combustion*, The Combustion Institute, pp. 1285–1293.

Truelove, J.S., and Holcombe (1990), Measurements and Modelling of Coal Flame Stability in a Pilot-Scale Combustor, *23rd Symp. (Int.) on Combustion*, The Combustion Institute, pp. 963–971.

Ubhayakar, S.K., Stickler, D.B., Rosenberg, C.W., and Gannon, R.E. (1977), Rapid Devolatilization of Pulverised Coal in Hot Combustion Gases, *16th Symp. (Int) on Combustion*, The Combustion Institute, pp 427–436.

Visser, B.M., Smart, J.P., Van den Kamp, W.L., and Weber, R. (1990), Predictions of Quarl Zone Properties of Swirling Pulverised Measurements and Coal Flames, *23rd Symp. (Int.) on Combustion*, The Combustion Institute, pp. 949–955.

Wennerberg, D. (1988), Prediction of Pulverised Coal and Peat Flames, *Combust. Sci. and Tech.*, **58**: 25–41.

NOMENCLATURE

A	Preexponential term in kinetic rate
A_p	Coal particle area
C	Mass concentration of gas specie in index
E	Activation energy
I	Enthalpy
k	Turbulent kinetic energy
K_v	Kinetic rate of devolatilization
m_p	Particle mass
n_p	Particle number
R	Reaction rate of gas phase
P	Pressure
T	Temperature
U, U_p	Velocity of gas and particle
V	Volatile yield or velocity
V_p	Coal particle volume

GREEK SYMBOLS

ε	Dissipation rate of turbulent kinetic energy
ε_p	Particle emissivity
ρ	Density
σ	Stephan–Boltzmann constant
σ_ϕ	Schmidt number of ϕ

INDICES

DV	Devolatilization
c	coal or char
VC	Volatile combustion
VOL	Volatiles
STOC	Stoichiometric conditions

Development of a Low-NO_x Emission Slagging Combustor

TOMIO SUZUKI, TAKEO YOSHIGAE, D. C. PALTER, TOSHIO KASAI and SHUZO ITOH *Kobe Steel, Ltd., Wakinohamacho 1-chome, Chuo-ku, Kobe, Japan*

Abstract—A vertical slagging combustor has been developed for use with pulverized coal, coal/water mixtures, coal ash and sewage sludge ash. Primary air is supplied from the top of the combustor through a swirler and secondary air is tangentially injected into the combustor from the upper side. The secondary air produces a stronger swirl pattern than the primary air. The flow pattern in the combustor is optimized for high slag capture and combustion efficiency by controlling the ratio of primary air to secondary air. Two-stage combustion is employed to produce low NO_x emissions. Slag capture and combustion efficiencies for pulverized coal and coal/water mixtures are 90-99%. The minimum value of NO_x emissions is approximately 150-200 ppm, corrected to 6% O_2.

This slagging combustor has many possible applications including use as a gasifier for a scrap melter. The reducing gas produced by the slagging combustor is injected into the scrap melter, along with oxygen, in order to melt the scrap. It is also possible to use this slagging combustor for the slagging of coal ash and sewage sludge ash in order to safely and economically dispose of the ash. The slag capture rate is over 95% for ash. The slag shows no leaching of harmful substances such as mercury, chromium, lead, cadmium, arsenic, cyanogen or PCB.

Key Words: NO_x, Slagging Combustor, Coal Water Mixture, Pulverized Coal, Sewage Sludge, Heavy Metals

INTRODUCTION

Slagging combustors or cyclone combustors have been developed for the combustion of coal in MHD systems and boilers by Seidl (1955), Braswell et al. (1984), Mattsson et al. (1985), Grunert et al. (1947) and Fond et al. (1985). Conventional slagging combustors must keep the wall temperature a few hundred degrees higher than the ash fusion temperature to continuously drain the slag away through the slag tap hole. This results in a very high slag capture and combustion efficiency. There are, however, disadvantages to the conventional slagging combustor, these being high NO_x and excessive refractory wear due to the high temperature. A precombustor or a high temperature preheater is also required to increase the temperature of the combustion air supplied to the slagging combustor.

In the present study, a new vertical slagging combustor has been developed for pulverized coal, coal/water mixtures, coal ash and sewage sludge ash in order to eliminate the disadvantages of the conventional design. This vertical slagging combustor has been applied to a commercial plant for a melting-after-carbonizing sludge process.

VERTICAL SLAGGING COMBUSTOR

Figure 1 shows a schematic drawing of the vertical slagging combustor. The combustion chamber is 500 mm in diameter and 1100 mm in height. Primary air is supplied from the top of the combustion chamber through a swirler, and secondary air is tangentially injected into the combustion chamber from the upper side. The secondary air produces a stronger swirl flow than the primary air. The flow pattern in the combustion chamber is controlled by manipulating the ratio of primary to secondary air in order to produce a high combustion efficiency, high slag capture and low NO_x emissions. A two-stage combustion process is then employed to further lower the NO_x emissions.

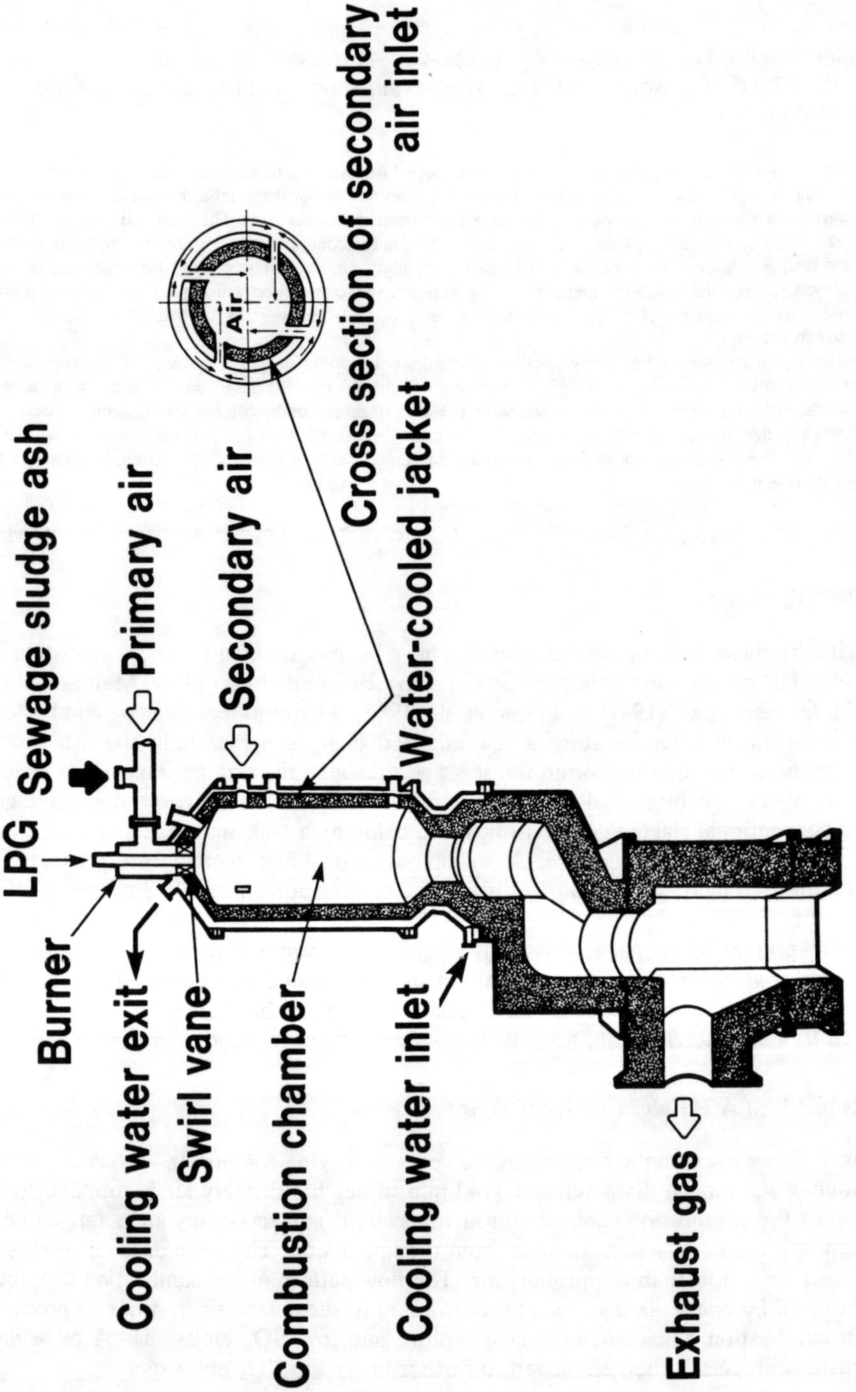

FIGURE 1 Schematic of vertical slagging combustor

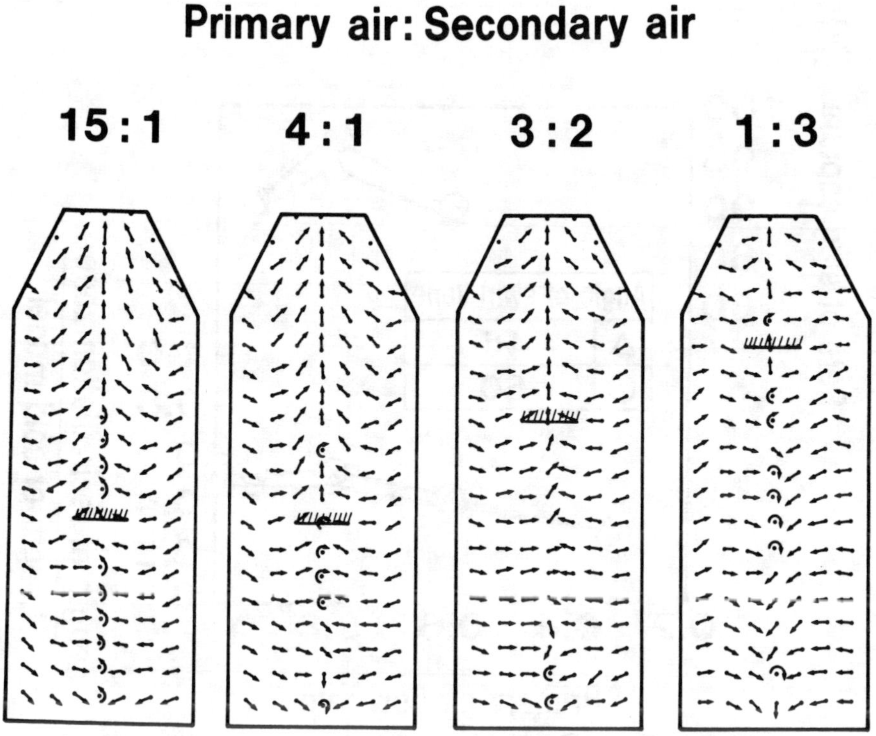

FIGURE 2 Effect of ratio of primary air to secondary air on flow pattern

Fuel is injected into the strongly turbulent swirling flow and then quickly mixed with combustion air and burnt. Particles such as ash collide with the wall of the combustion chamber due to centrifugal force. When the wall temperature is higher than the ash fusion temperature, the ash becomes molten slag and flows down into the slag pot. However, if the wall temperature is lower than the ash fusion temperature, the ash particles fall into the ash pot as they do in a non-slagging cyclone combustor.

The vertical combustion chamber exploits two phenomena: both high temperature gas and natural gravitation to promote the flow of molten slag into the slag pot.

COLD MODEL TEST

Experiment

The cold flow scale replica of the combustion chamber is 130 mm in diameter and 330 mm in height, and built of transparent acrylic resin. The experiments were conducted under the same air momentum and swirl number as the slagging combustor. The flow patterns in the model are measured using a tuft method. The air velocities are measured by a hot anemometer.

Flow Pattern

Figure 2 shows the effect of the ratio of the primary air to secondary air on the flow pattern

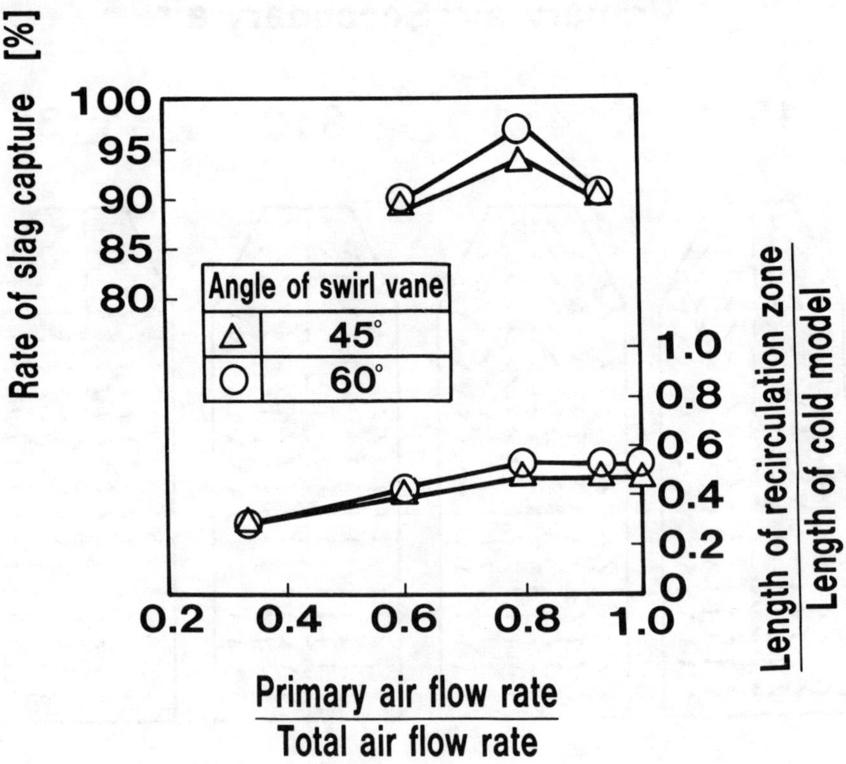

FIGURE 3 Effect of ratio of primary air to secondary air on slag capture in combustion tests and internal recirculation zone length in cold model test

when the primary air swirler vane angle is 60°. The end of the internal recirculation zone is indicated by a shaded line. The internal recirculation zone is lengthened by increasing the ratio of the primary air to the secondary air. The maximum length of the internal recirculation zone is about half that of the chamber model length.

The swirling flow velocity near the wall of the combustion chamber model is increased by decreasing the ratio of primary to secondary air. The location of the maximum velocity moves towards the center of the model as the ratio of primary to secondary air increases. The swirling flow velocity in the internal recirculation zone is increased as the ratio of primary to secondary air increases. The maximum value of the flow swirl velocity in the internal recirculation is obtained with a 4:1 primary to secondary ratio.

Relation Between Flow and Slag Capture

Figure 3 shows a comparison of the effect of the primary to secondary air ratio on the slag capture in combustion tests of sewage sludge ash with the effect on internal recirculation zone length in the cold model test. Maximum slag capture is obtained at a 4:1 ratio of primary air to secondary air for both swirler vane angles.

The internal recirculation zone length and slag capture are increased with an increase in swirler vane angle. The larger the swirler vane angle and the higher the ratio of

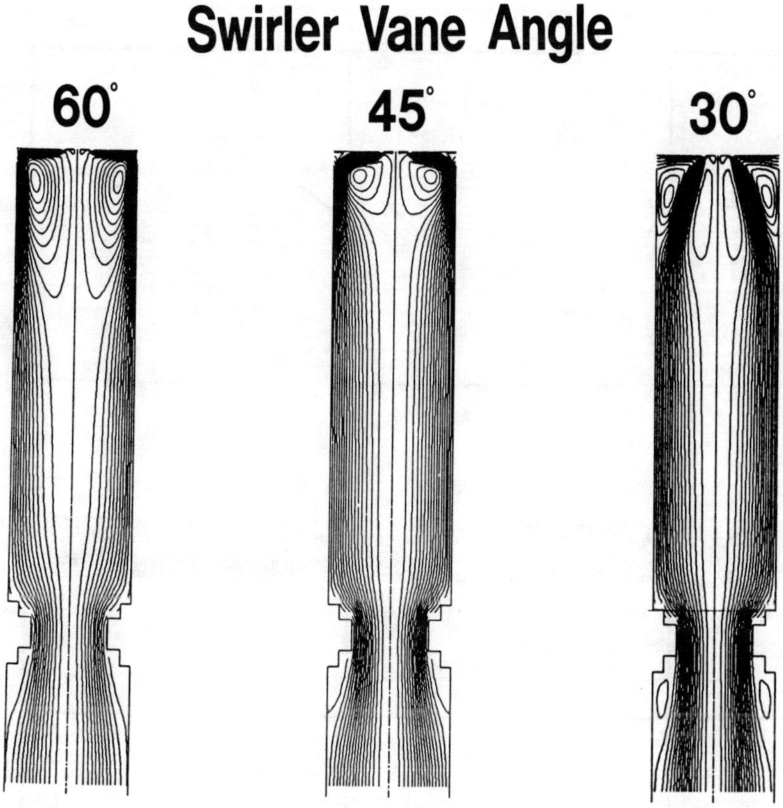

FIGURE 4 Combustor flow pattern as a function of swirler vane angle (Calculated)

primary to secondary air flowrate, the stronger the swirl and the more primary air that is pressed against the wall. This creates a large area of negative pressure in the center of the combustor and therefore a larger area of recirculation in the center of the combustor. On the other hand, the lower the ratio of primary to secondary air flowrate, the larger the amount of secondary air flow that adheres to the wall and prevents the primary air from being pressed against it, and thus the smaller the area of recirculation that is generated in the center of the combustor.

These results indicate that the optimization of the flow pattern and velocity in the slagging combustor is critical for improving the slag capture and combustion characteristics.

SIMULATION

A computer simulation was conducted in order to help visualize the flow pattern inside the combustor, predict the slag capture rate, and optimize the design.

The computer model was developed using FLUENT, a general purpose computational fluid dynamics package. Both the flow pattern and the combustion of butane were calculated. A ratio of primary to secondary air of 4:1 was assumed. Particle tracking was calculated using a stochastic, Lagrangian model that considered the effect of turbulence

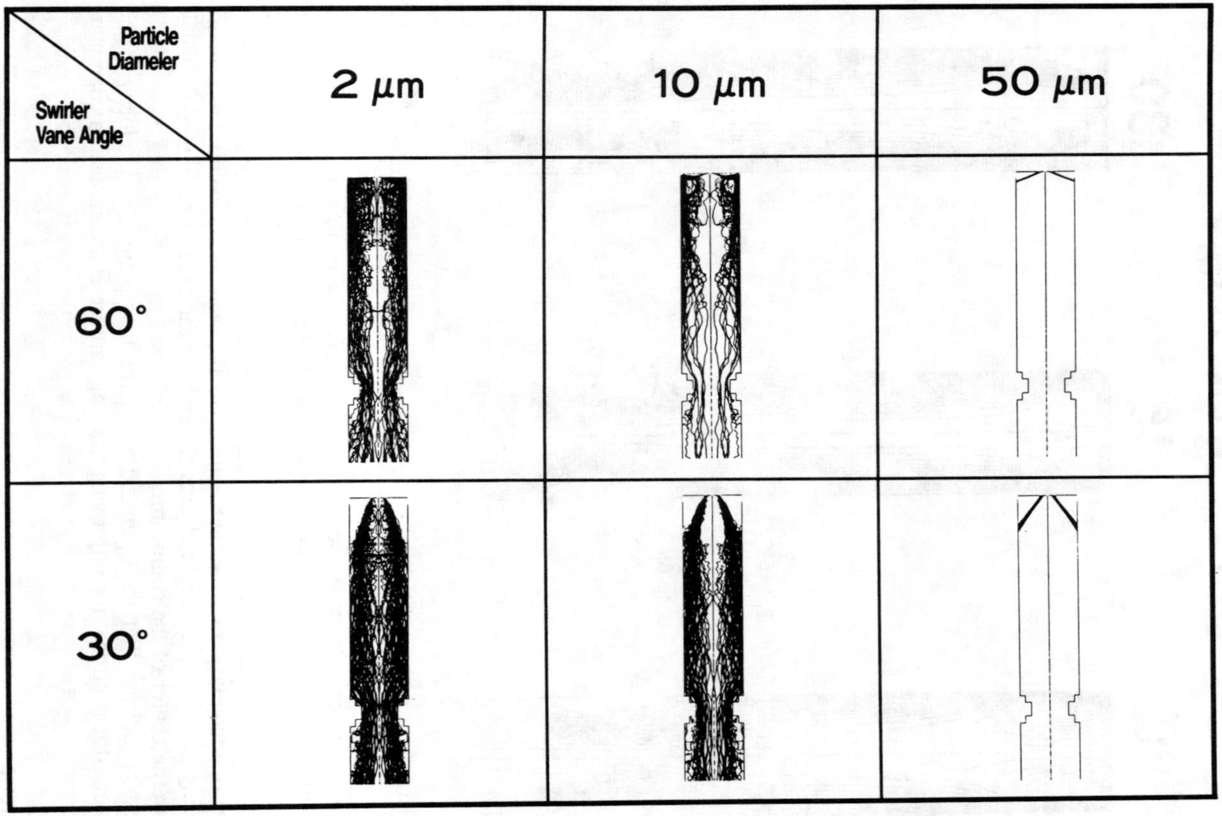

FIGURE 5 Particle tracks for 2 swirler vane angles and 3 particle diameters

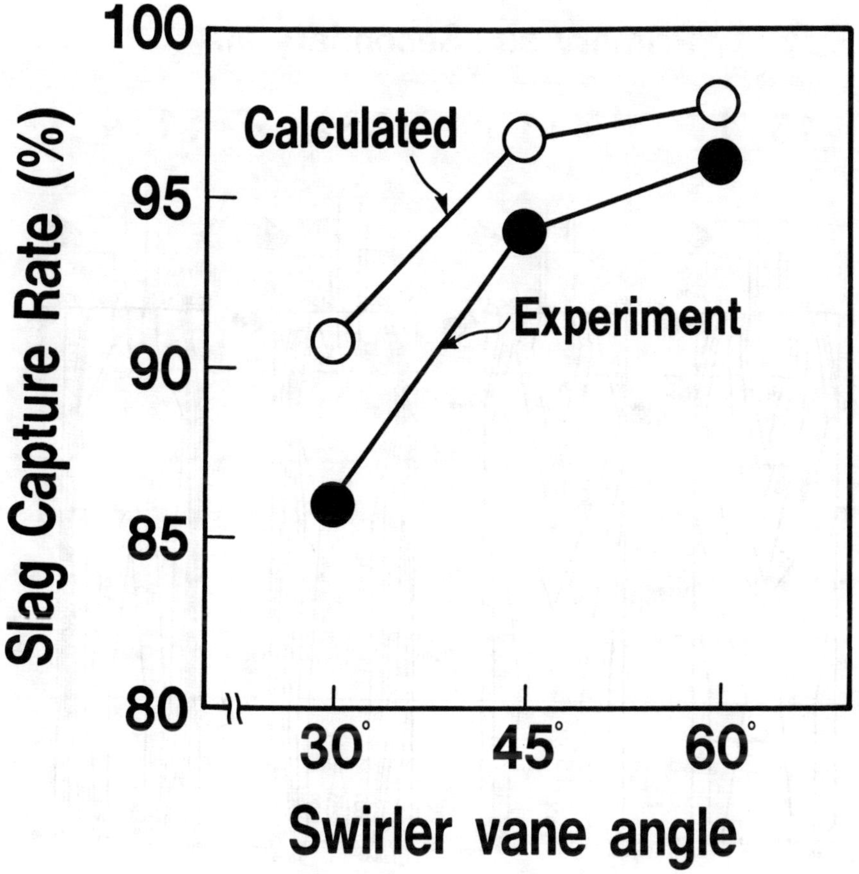

FIGURE 6 Slag capture rate vs. swirler vane angle

in the gas flow on the motion of the particles. All particles that contacted the wall were assumed captured. Particle-particle collisions and the effect of the particles on the gas flow were not considered.

Figure 4 shows the streamlines inside the combustor for three swirler vane angles. This figure shows that at a swirler vane angle of 60°, a strong recirculation zone is formed in the center of the combustor that forces the primary air against the wall. At an angle of 45°, a weaker recirculation zone is formed. At an angle of 30°, there is almost no recirculation in the center of the combustor, but instead, a recirculation zone is formed in the upper corner which prevents the primary air from contacting the wall.

Particle tracks for 2 μm, 10 μm, and 50 μm particles, for swirler vane angles of 30° and 60°, are shown in Figure 5. Large particles are quickly forced against the wall, regardless of swirler vane angle, by the centrifugal force created by the swirling flow. Fine particles closely follow the gas flow pattern and are thus more difficult to capture. This figure shows that the swirler vane angle of 60°, which produces the strong recirculation zone in the center of the combustor, is much more effective at capturing the particles than a swirler vane angle of 30°.

The comparison between the calculated and experimental values of slag capture versus swirler vane angle in Figure 6 shows that, while the calculation shows slag capture rates

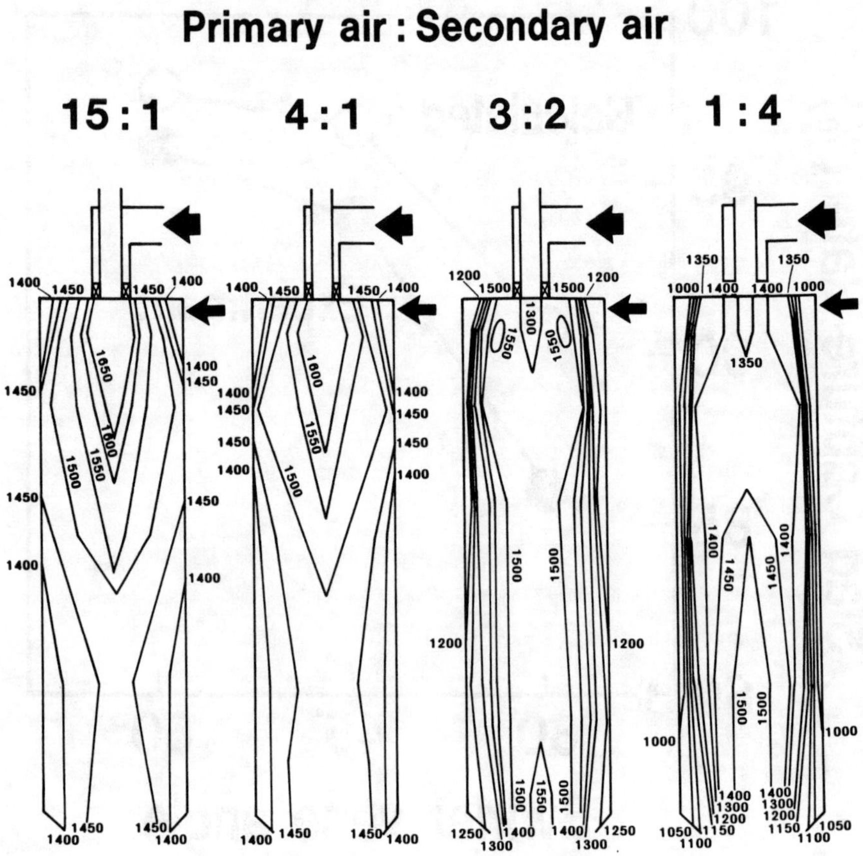

FIGURE 7 Effect of ratio of primary air to secondary air on temperature distribution in slagging combustor

slightly higher than the experimental values, the trends match very closely. This verifies that the simulation produces results that can be used to forecast the change in slag capture rate due to changes in combustor design or operating conditions.

COMBUSTION TESTS

Experimental

Pulverized coal, coal ash or sewage sludge ash up to 250 kg/hr, is supplied with primary air to the burner of the slagging combustor. Combustion air heated by a hot stove to 150° - 400° is fed to the slagging combustor at O_2=21%-100% by the addition of O_2 into the exhaust gas (vitiated air) of the stove. In the case of coal ash or sewage sludge ash slagging, butane gas is used as the fuel. The coal used has a volatiles content of 35.4% and ash of 10.8%, pulverized to a size distribution of 75% less than 200 mesh.

In this experiment, solid and gas samples of both the slagging combustor and the secondary combustion furnace are extracted at the exit by using water-cooled probes in order to evaluate its performance. The carbon burnout and slag capture rate are calculated by the ash tracer method and by the weight of the sampled ash at the exit of the combustor.

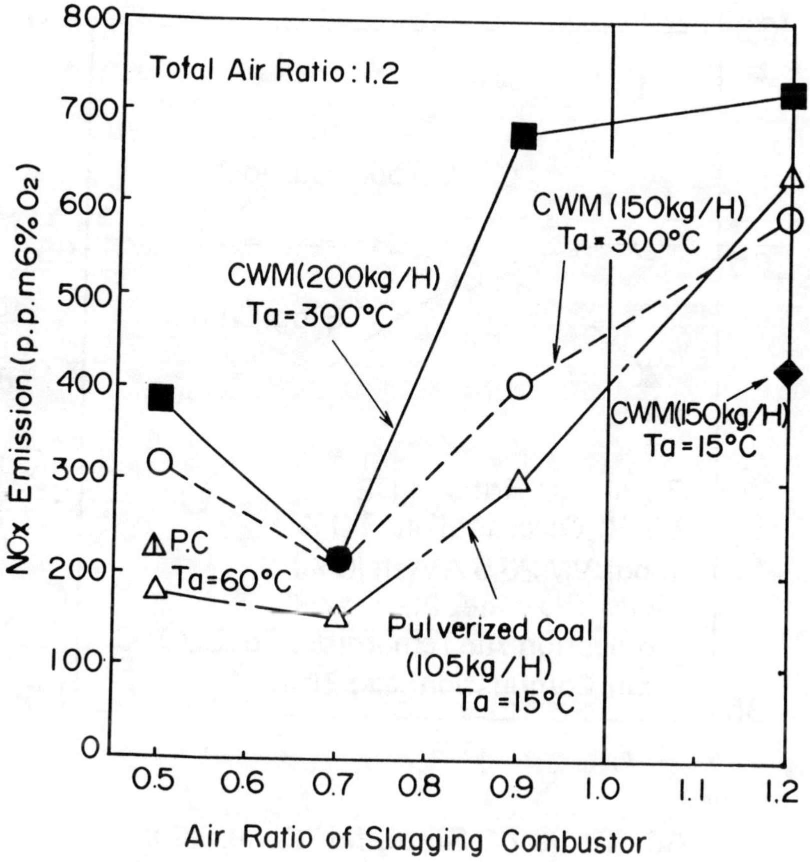

FIGURE 8 Effect of combustion air ratio of slagging combustor on NO_X emissions in two-stage combustion

Temperature Distribution

Figure 7 shows the effect of the primary air to secondary air ratio on the temperature distribution in the combustor. At high ratios of primary to secondary air flowrate, the temperature generated in the upper portion of the combustor is higher, and the temperature of the combustor is kept more uniform. The higher the ratio of primary to secondary air flowrate, the larger the area of the recirculation zone generated in the center of the combustor. This recirculation zone stabilizes the flame in the upper portion of the combustor.

On the other hand, the lower the ratio of primary to secondary air flowrate, the smaller the recirculation zone. Fuel and air are therefore mixed and burned in the lower portion of the combustor.

NO_x Emissions

Figure 8 shows the effect of the combustion air ratio of the slagging combustor (primary and secondary air) on NO_X emissions in the case of two-stage combustion. NO_X emissions are analyzed in the sampled gas at the exit of the secondary combustion furnace under a

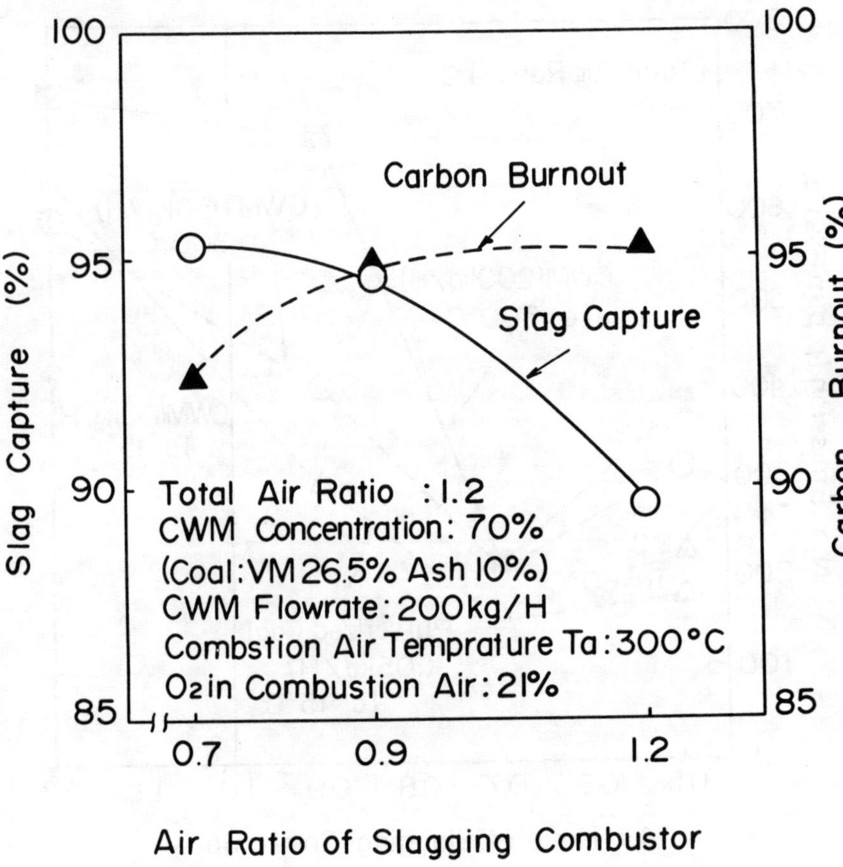

FIGURE 9 Effect of combustion air ratio of slagging combustor on slag capture and carbon burnout

condition of 1.2 times the total combustion air ratio (primary, secondary, and two-stage air). The air ratio is defined as the ratio of the air flow rate to the stoichiometrically correct concentration. The NO_x concentration decreases greatly with a decrease in the combustion air ratio of the slagging combustor. The lowest NO_x concentration occurs at a primary air ratio of 0.7. The reduction percentages are 75% for pulverized coal and 65-70% for CWM compared to non-two-stage combustion. Similarly, Stansel et al. (1984) found that low NO_x levels could be obtained at an air ratio of 0.7. The NO_x concentration of CWM is below that of pulverized coal when the air temperature is ambient and the air ratio is 1.2.

Combustion Efficiency

Figure 9 shows the effect of the combustion air ratio of the slagging combustor on carbon burnout and slag capture in CWM combustion. Carbon burnout is increased with an increase in the air ratio. Because flame temperature and oxygen concentration in the combustor are increased with an increase in the air ratio, the combustion is improved. This carbon burnout is higher than that of a cyclone combustor and lower than that of the MHD slagging combustor.

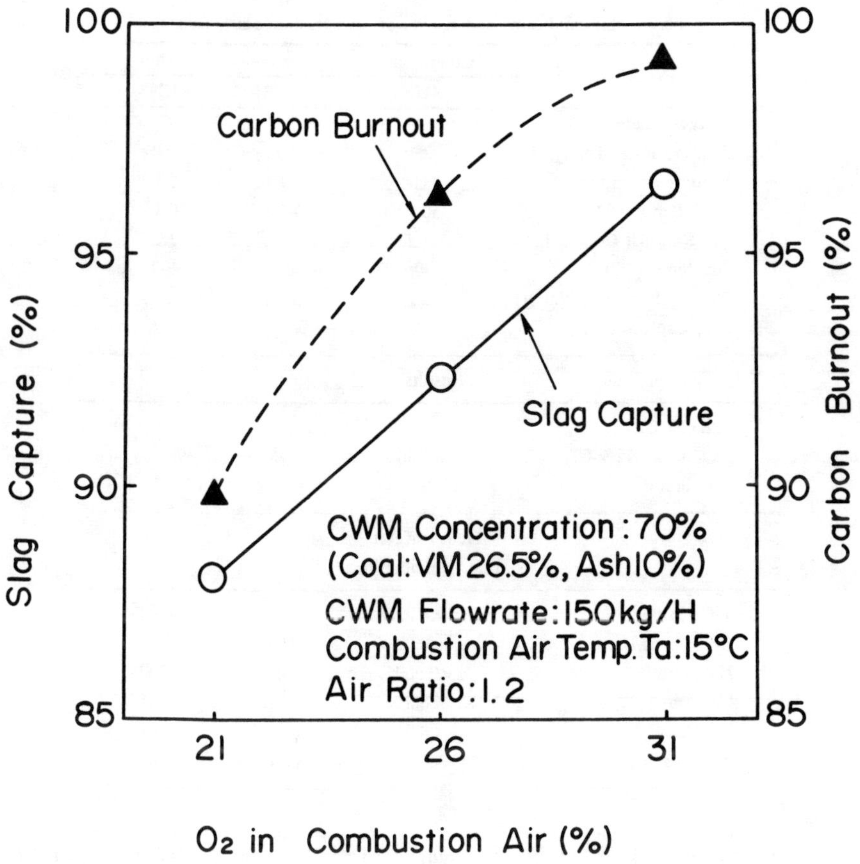

FIGURE 10 Effect of O_2 in combustion air on slag capture and carbon burnout

Figure 10 shows the effect of the O_2 concentration in the combustion air on carbon burnout and slag capture in CWM combustion. Carbon burnout is greatly increased with an increase in O_2 concentration. Oxygen enrichment promotes combustion due to the higher flame temperature and higher overall O_2 concentration in the combustor.

Slag Capture

Slag capture and efficiencies are shown in Figures 9 and 10. The slag capture decreases with an increase in the air ratio and a decrease in O_2 concentration. High air ratios and low O_2 concentrations increase the flow rate of combustion air. Therefore, the residence time in the slagging combustor decreases, and the carry-over of fine particles may be significantly increased.

A higher flame temperature, however, gives a higher slag capture. Therefore, it is very important to control the flow rate of primary and secondary air in the combustor to improve slag capture, especially under low temperature conditions.

Table I shows experimental conditions and results of sewage sludge ash slagging. Sewage sludge ash (fly ash) is emitted from the sewage sludge fluidized incinerator. The sewage sludge ash is mixed with primary air before entering the burner and injected uniformly

TABLE I
Experimental conditions and results of sewage sludge ash slagging

Conditions			
	RUN1	RUN2	RUN3
Sewage Sludge Ash	A	B	C
Feeding Rate (kg/H)	250	250	250
Feeding Temp (°C)	20	20	20
Wall Temp (°C)	1350-1450	1350-1500	1350-1500
Air Ratio of Combustor	1.2	1.2	1.2
Air Temp (°C)	400	400	400
C_4 H_{10} Flowrate (Nm^3/H)	25	25	25
Slag Capture (%)	99.0	98.1	96.6
Results			
Exhaust Gas			
NO_x (ppm 12%-02)	43-80	25-75	90-95
Temp(°C)	1550-1650	1500-1600	1500-1550
Heavy Metal Water-Miscible Test			
T-Cr (mg/l)	<0.02	<0.02	<0.02
Cr^{+6} (mg/l)	<0.02	<0.02	<0.02
Hg (mg/l)	<0.0005	<0.0005	<0.0005
Cd (mg/l)	<0.01	<0.01	<0.01
Pd (mg/l)	<0.01	<0.01	<0.01
As (mg/l)	<0.01	<0.01	<0.01

into the slagging combustor through the swirler of the burner. The wall temperature of the combustor is kept over one hundred degrees higher than the ash fusion temperature, by means of butane gas combustion.

The sewage ash melts quickly after injection into the combustor. The slag capture rates are higher than those of coal or CWM slagging. A heavy metal water-miscible test indicates no problem with slag disposal.

APPLICATION

Sewage Sludge Melting Process

Process flowchart Figure 11 shows a flowchart of the Melting After Carbonizing Sludge (MACS) process.

A 50 ton-cake per day demonstration plant for the MACS process was constructed in 1987 and Kobe Steel has since conducted full-scale technical demonstrations. The sludge slagging combustor has been applied to a 40 ton-cake per day MACS process commercial plant which has been operating since March 1990. Dewatered cake is stored in a conical-silo type hopper and delivered to the fluidized-bed carbonizing furnace at the desired feed rate. In the furnace, the dewatered cake is carbonized at a combustion air ratio of 1.0. Using such a ratio can suppress the generation of Cr^{+6} and NO_x. The residue is separated from the exhaust gas, collected by the cyclone, and discharged into the slagging combustor without lowering the temperature. The residue is then melted to

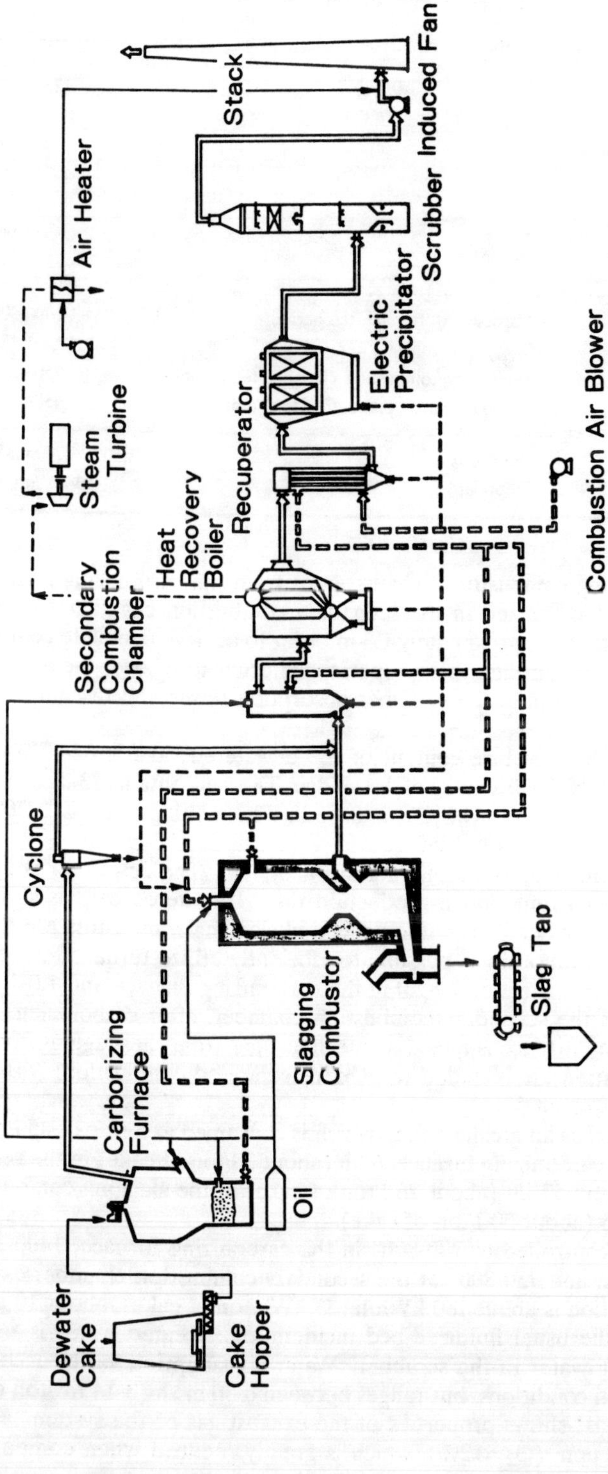

FIGURE 11 Schematic of Melting after Carbonizing Sludge (MACS) process

TABLE II
Operation results of MACS process

Item		Experimental value
Cake feed rate		2.0~2.2 t/h
Carbonizing furnace		
Consumption of heavy oil A		.41~45 *l*/h
Secondary combustion chamber		6 *l*/h
Slagging combustor		55~57 *l*/h
Total		102~108 *l*/h
per unit cake		50~51 *l*/t. cake
Temperature	Carbonizing furnace temperature	700~800°C
	Slagging combustor temperature	1,400~1,450°C
	Secondary combustion chamber temperature	750~800°C
	Combustion air preheating temperature	350~400°C
Pressure	Wind box	1,500~1,700mmH_2O
	Freeboard	-10~-20mmH_2O

make slag in the slagging combustor. Exhaust gases from the carbonizing furnace and the slagging combustor are mixed in the secondary combustion chamber to bring the total combustion air ratio to approximately 1.3 in order to achieve complete combustion. Exhaust gas then proceeds through the recuperator in order to recover the heat energy. It is then treated in the EP and the scrubber absorption tower and discharged to the atmosphere.

Operation results The moisture content of the dewatered cake is 62-67% and the lower calorific value is 2300-2600 kcal/kg of dry sludge. The ash contains 23-31% SiO_2 and 26-28% CaO while its basicity (CaO/SiO_2) is 0.9-1.3. The melting point is 1220°-1250° and the melt-flowing point is 1240°-1280°.

Table II shows the principal operation data of the system. The dewaterer is operated 6 days per week at the current sludge production rate. Dewatered cake is stored in a 1000m^3 conical storage hopper. The carbonizing and slagging combustors are operated 3 days per week—as continuously as possible to efficiently utilize furnace temperature.

The combustion air ratio is set at 0.9-1.0 in the carbonizing furnace and 1.0-1.1 in the slagging combustor. In the secondary combustion chamber, after carbonizing, exhaust gas from the carbonizing furnace is mixed with exhaust gas from the slagging combustor and secondary combustion air is added to achieve complete combustion, with a total combustion air ratio of about 1.3.

No. 6 heavy oil is used as an auxiliary fuel, which is consumed at a rate of 45 l/h (about 21 l/ton of cake) in the carbonizing furnace, 6 l/h (about 3 l/ton of cake) in the secondary combustion chamber, and 57 l/h (about 26 l/ton of cake) in the slagging combustor, for a total of about 108 l/h (about 50 l/ton of cake).

The temperature is controlled to 700-800° in the carbonizing furnace, 1400-1450° in the slagging combustor, and 750-800° at the secondary combustion chamber exit.

The power consumption is about 180 kWh/h (83 kWh/ton of cake), about 10 kWh/ton of cake greater than the usual fluidized-bed incineration. Treated water is used as a coolant and as washing water in the scrubber. Water consumption varies according to white-smoke prevention conditions, but ranges between 6-30 m^3/h (3-14 m^3/ton of cake).

Exhaust gas Table III shows properties of the exhaust gas of the system. The NO_x concentration is 20-30 ppm (O_2=12%), which is greatly reduced when compared with

TABLE III
Properties of exhaust gas of MACS process

Item		Measured value			Specified value
Sampling position		Carbonizing furnace	Secondary combustion chamber	stack	—
Exhaust gas temperature (°C)		730	790	124	—
Amount (Nm^3/h)	Humidity	4,240	5,440	11,900	—
	Dryness	2,320	3,440	9,210	—
Dust concentration (g/Nm^3)		—	—	0.019	—
NO_x $O_2$12%	(ppm)	30	25	23	<250
CO_2	(%)	16.4	8.9	3.0	—
O_2	(%)	0.0	6.0	17.0	—
CO	(%)	0.5	<0.01	<0.01	—
H_2	(%)	0.77	<0.01	<0.01	—
SO_x	(ppm)	61	46	<5	—
HCl	(mg/Nm^3)	1,129	550	10	<700

NO_x concentrations of 100-200 ppm for conventional fluidized-bed incineration. This is thought to result from the effect of the carbonizing secondary combustion system which combines the carbonizing furnace with the secondary combustion chamber (two-stage combustion). Unburnt gas generated from carbonizing primarily consists of H_2, CO, and CH_4. Complete combustion takes place in the secondary combustion chamber at a combustion air ratio of about 1.3 and at temperatures over 750°.

Slag None of the slag showed any leaching of harmful substances over detection limits. The properties of the hardened slag differed markedly according to the cooling rate.

(1) Water-cooled slag forms fine amorphous particles. It might, in some cases, be rejected due, for example, to abrasion loss of standard road crushed stone. For possible applications, it can be used for backfilling soil for buried pipes.
(2) In rapidly-cooled slag (cooled at more than 10° /min, for example) crystalization is scarcely observed through X-ray diffraction.
(3) X-ray diffraction results show that development of crystalization proceeds significantly in slowly-cooled slag (cooled at less than 5° /min). Slowly-cooled slag has a strength equivalent to natural stone, and this would hardly be lowered even if it were used as a concrete aggregate. Slowly-cooled slag has a wide variety of direct applications, and the manufacturing technique has been established.

Scrap Melting Process

Process flowchart Figure 12 shows a schematic of the scrap melting process. This melting process directly uses coal energy in the system. The pulverized coal used as the heat source is gasified in the slagging combustor, which is connected to the scrap melter by a duct, with most of ash and sulphur in the coal being simultaneously removed in this combustor.

The primary gas which contains a 60 to 70% CO and H_2 gas mixture is directly supplied into the melting furnace in order to melt the cold iron resources. The cold iron

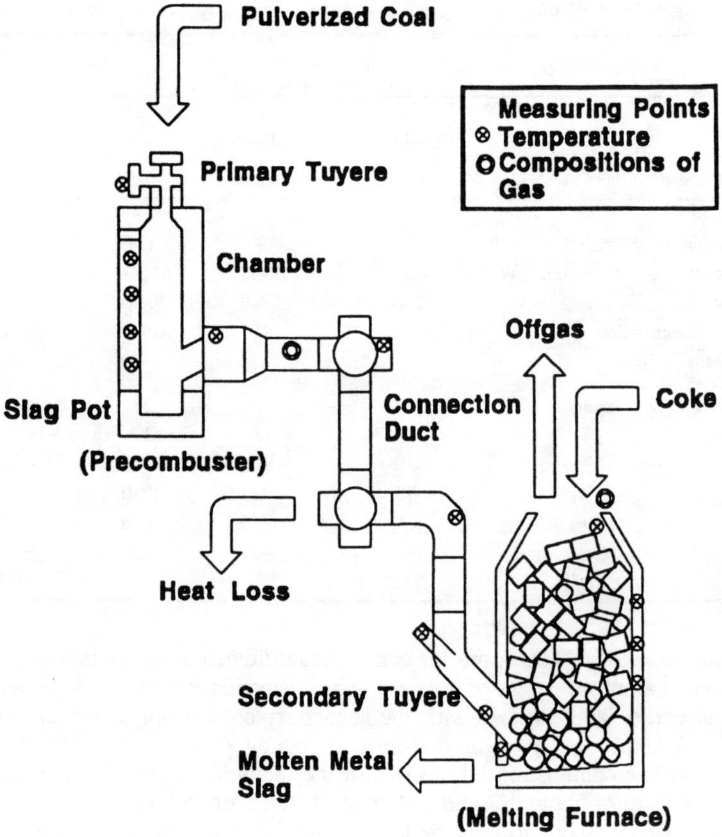

FIGURE 12 Schematic of scrap melting process

resources, such as the scrap and pig iron, are added continuously from the top of the melting furnace and preheated by the combustion gas during the decent to the melting zone located in the lower part of melting furnace.

Operation results The total heat balance for a 1.5 ton/hr scale bench plant is given in Figure 13. Total energy consumption of the scrap melter is about 3390 MJ/metric ton. The chemical heat of coal provides approximately 56% of the total input energy, and the chemical heat of coke provides about 42%. Pulverized coal is sufficiently gasified in the slagging combustor and supplied to the melting furnace. The major role of coke in the melting furnace is as the source of carbon in the molten iron. The sensible heat of the molten iron is 37% of the total output energy despite the small scale of the plant. The sensible heat of the off-gas is 11% and the chemical heat of the off-gas is 35% of the total output energy. The heat loss is 17.6% of the total output energy. Heat efficiency mainly depends on the heat transfer between the cold iron resources and the off-gas, as well as the post-combustion ratio. Further development will be expected to increase the total energy efficiency of this process by applying additional post-combustion techniques and optimizing the melting furnace shape.

Coal Gasification

The slagging combustor can also be used for coal gasification. Steam is added to promote

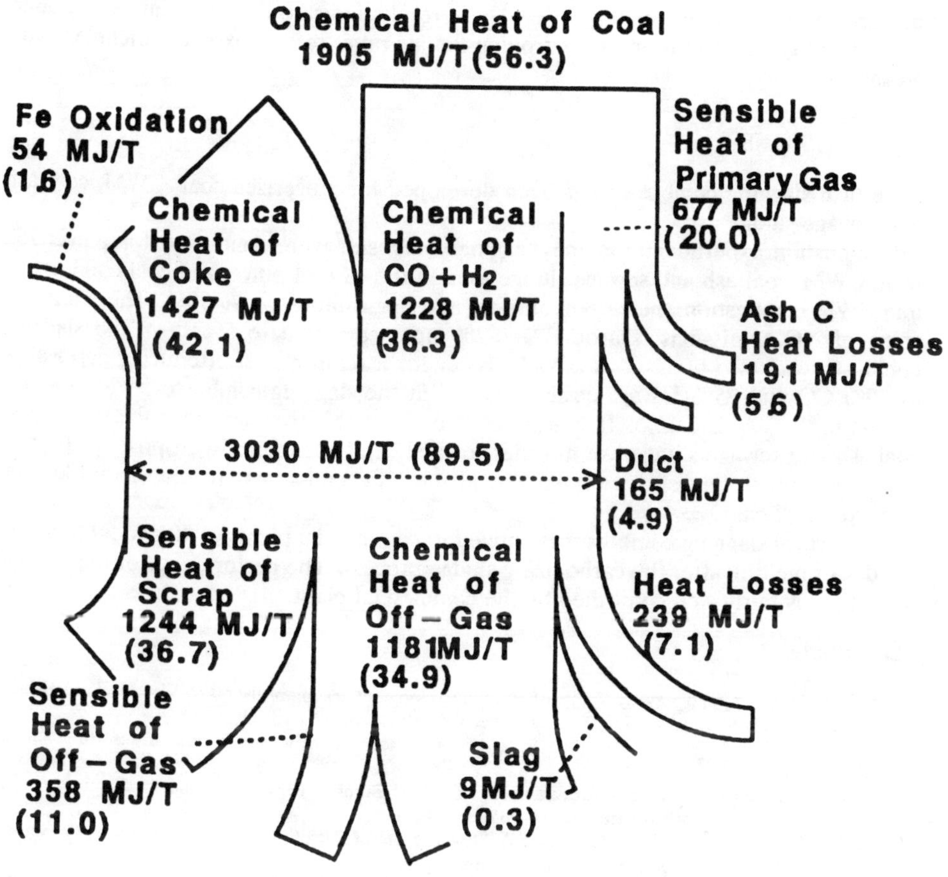

FIGURE 13 Total heat balance of scrap melting process

TABLE IV
Experimental conditions and gas composition of coal gasification (Carrier gas : N_2)

	Experimental Conditions				Gas Composition (Dry)				Gas	
	Coal (kg/H)	Air Ratio	O_2 in Air(%)	H_2O (Nm^3/H)	H_2 (%)	CO (%)	CO_2 (%)	N_2 (%)	NO_x (ppm)	Temp (°C)
1	102	0.41	100.0	47	25.7	52.5	13.9	7.9	74	1422
2	235	0.40	46.5	130	24.2	24.7	17.4	33.7	110	1315
3	225	0.57	34.5	50	7.6	21.9	12.5	58.0	306	1450
4	200	0.70	34.5	50	6.6	17.8	17.5	58.1	410	1675

this gasification. The concentration of the reducing gases, such as CO and H_2, increases with a decrease in the air ratio. CO_2, however, increases with an increase in the air ratio. Table IV lists gasified coal gas compositions. Gas 2 is composed of $CO/H_2 \doteq 1$. Gas 1 is another gasified coal gas which is composed of $CO/H_2 \doteq 2$ and has high CO and H_2 concentrations. The unburnt carbon in the slag is 0.02%. The carbon burnout and slag capture rates are approximately 95%. These results indicate that it is simple to obtain the desired gas composition by controlling the air ratio, rate of oxygen enrichment and steam ratio.

CONCLUSIONS

A vertical slagging combustor has been developed for pulverized coal, CWM, coal ash and sewage sludge ash.

Combustion experiments on the slagging combustor were conducted for pulverized coal, CWM, coal ash and sewage sludge ash. In the case of pulverized coal combustion and CWM combustion, the carbon burnout and slag capture are 90-99%. The minimum value of NO_x emissions is around 150-200 ppm corrected to 6% O_2. The slagging combustor can also be used as a coal gasifier for a scrap melter. Reducing gases such as 50% CO and 25% H_2 are easily obtained in the slagging combustor. The unburnt carbon in the slag is 0.02%. It is also possible to use this combustor for the slagging of coal ash and sewage sludge ash in order to safely and economically dispose of the ash. The slag capture for ash is over 95%, and the NO_x emissions are less than 100 ppm corrected to 12% O_2.

The vertical slagging combustor is applied to commercial plants of 40 and 50 ton-cake per day of melting after the carbonizing sludge process. The performance of the vertical combustor is fairly well confirmed by the commercial plant.

REFERENCES

Braswell, R., Iwata, H., and Stewart, J. (1984). Early Development Testing of the 50 MW_T MHD Coal-Fired Combustor, 22nd Symp. on Engineering Aspects of Magnetohydrodynamics.

Grunert, A. E., Skog, L., and Wilcoxson, L. S. (1947). The Horizontal Cyclone Burner. Transactions of The A.S.M.E., pp. 613.

LaFond, J. F., and Payne, R. (1985). Development of a Coal Burning Vortex Containment Combustor. 2nd Annual Pittsburgh Coal Conference, pp. 785.

Mattsson, A. C. J., and Stankevics, J. O. A. (1985). Development of Retrofit External Slagging Coal Combustor System. 2nd Annual Pittsburgh Coal Conference, pp. 776.

Seidl, H. (1955). The Development and Practice of Cyclone Firing in Germany. Joint Conf. on Comb. A.S.M.E. 137, pp. 92.

Stancel, J., Kuenzly, J., Sheppard, D., and Ubhayakars (1984). TRW's Slagging Combustor System Progress, 7th Int. Coal & Lignite Utilization Exhibition & Conference, Houston, Texas.

3.

OIL COMBUSTION

Improving Availability and Economy of Oil Fired Boilers by Reducing Deposition and Corrosion Caused by Flue Gases — Part II

SAADIA REES *Apollo Chemical and Technologies Ltd., Haifa, Israel*

Abstract — Chemical technology has been implemented as a significant means of energy saving and environmental reduction of emissions in oil-fired steam generation. Applications include combustion improvement, slagging modifications, fouling prevention, corrosion inhibition and particulate emission control. Chemical technology offers savings in steam generation through: (a) elimination of equipment replacement; (b) reduction of fuel costs and (c) reduction of operating expenses at the same time that particulate and acidic stack emissions, referred to as "extreme air pollution," can be achieved to meet current requirements of the regulatory agencies. The operating experience at a utility with a cold-end neutralizing additive to lower the acid dew point is detailed.

Key Words: Air Heaters (A/Hs), Corrosion, Boilers, Environmental Regulations, Catalyst, Particulate, Emission

INTRODUCTION

A major objective in every steam generating plant is to obtain maximum boiler efficiency at minimum operational cost. However, fouling and corrosion caused by low melting ash compounds in the hot section of the boiler and condensation of sulfuric acid in the cold section of the boiler, limit efficient and reliable operation.

With the strict environmental regulations which apply to steam generating units on the one hand and the constant search for low cost fuel-oils on the open market on the other, the steam generation operator finds himself caught between the clean air acts and the resultant high costs of generating megawatts to satisfy the emission requirement of the regulatory agencies. As a result, a conflict occurs between operating an economical steam plant and keeping a clean stack.

A method (Lev Er 1977) was developed that has enabled the power generation management to burn relatively low cost fuels and meet the environmental standards. This paper reports on the 15 years of operational experience, the modifications made in the system employed during the past 15 years and the results obtained with this method, "Cold End Neutralization," trade name of "Coaltrol M" (to Apollo Technologies International, Livingston, New Jersey 07039) (Kukin 1989).

IMPROVEMENT OF COMBUSTION

Combustion may be defined as the rapid chemical combination of oxygen with the combustible elements of the fuel. The objective of efficient combustion is to release the maximum heat possible, while minimizing losses from combustion imperfections such as high excess air and high exit gas temperatures (Kober 1984).

As the stoichiometric relationship during combustion cannot practically be achieved, any amount of excess air will cause adverse oxidation of the impurities in the fuel as well as loss of thermodynamic efficiency.

The immediate effect of incomplete combustion will be emissions of unburned carbon and high temperature of the exit gases. In either case, and definitely in both cases, a reduction in the boiler's efficiency is experienced.

In oil burning boilers, a method that can be employed to achieve a minimum oxygen operation involves the use of combustion catalysts. Certain metallic compounds, notably those containing manganese, iron or cerium have the capability of reducing carbon carryover in the flue gases (Giammar *et al.* 1975). When properly applied, these metallic containing materials as finely dispersed metallic oxide slurries, permit boiler operation at reduced excess air levels without adversely affecting plume opacity, often reducing the latter (Goldberg 1978).

With the use of specific metallic catalyst, excess air reduction in the 40–50% range has been achieved with proportional SO_3 reduction.

During the past 10 years data were collected showing the effect of a catalyst added to the fuel oil. The catalyst containing slurry consisted of a finely micronized and dispersed oil slurry of manganese oxide (Kukin, 1989). The latter was added to the fuel-oil to provide 15–40 parts per million by weight of manganese. In Figure 1, the composite results of treating the fuel oils with catalyst are shown. Particulate concentrations and particulate emissions were measured before and after treatment with boilers averaging in size from 75 to 220 Megawatts.

The particulate emissions were analysed by isokinetic sampling utilizing EPA method 17, as well as Loss on Ignition (LOI), by standard test procedures. The overall reduction in particulate concentration, as well as reduction in emissions, were found to be, on the average, 45%.

The reduction in carbon particulates that resulted from the catalytic oxidation provides for increased combustion efficiency at the same time that the reduced emissions helped the operators to meet emissions standards.

REDUCTION OF DEPOSIT BUILDUP ON BOILER SURFACES

There has been a plethora of published information on the use of magnesium oxide added to the fuel in order to minimize vanadium corrosion of superheater and reheater tubes and tube supports, much of it controversial for the following reasons:

(1) The addition of magnesium oxide to raise the fusion point of the low melting vanadium oxides or sodium vanadates adds mass to the boiler, often reducing the "cleanliness" in the hot section as well as in the air heaters. The net result could be an increase in particulate emissions unless the air heaters are washed frequently.

(2) Adding adequate quantities of MgO to the fuel-oil to provide a minimum ratio of 1 part Mg to 0.72 parts vanadium, considered by many operators to be optimum, could result in boiler fouling particularly when used in oil-designed boilers.

(3) The use of MgO is particularly useful in coal-designed units that were converted to oil, where the primary purpose of the MgO is to increase superheat and reheat temperatures.

(4) The added MgO could result in a reflective coating on the furnace walls which in turn could necessitate additional use of an attemperating spray with decreased boiler efficiency.

Properly applied, the MgO addition may provide the following benefits:

a. In boilers where high temperature vanadium oxide corrosion has been documented, the magnesium oxide can form a light high melting point eutectic with the vanadium oxide, V_2O_5. The V_2O_5 is a low melting compound of about 670°C. In its liquid phase, it will coat the tubes, dissolving the protective coating on them, and promoting corrosion.

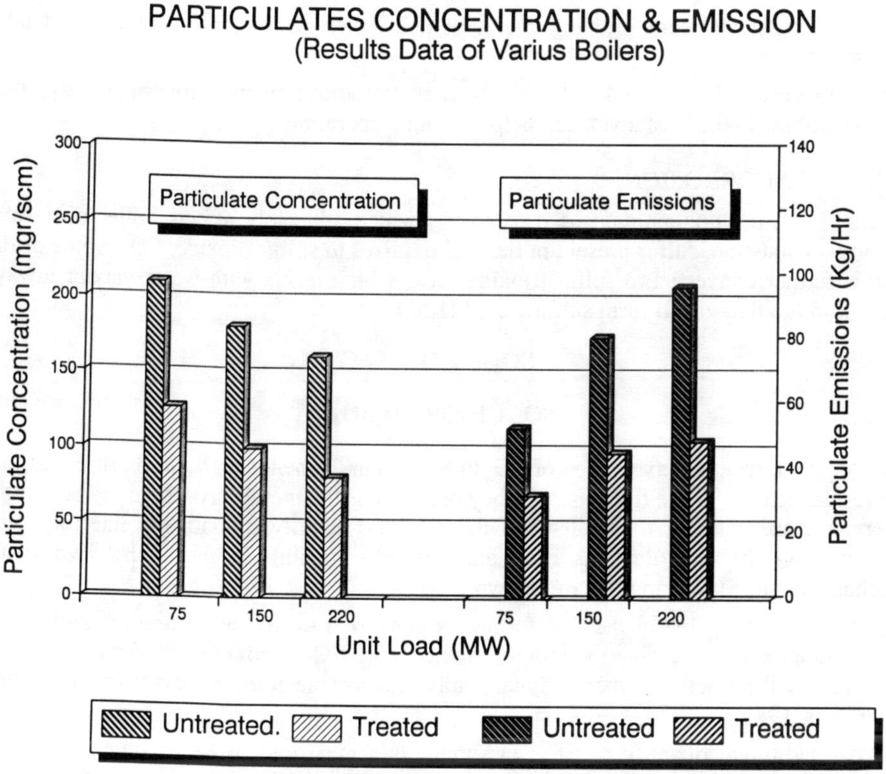

FIGURE 1 Improvement of combustion.

b. By means of the MgO-based products added either to the fuel or to the boiler, the slagging, or impinging characteristics of the vanadium oxide is reduced. The melting point of the latter will be raised. With sufficient MgO, the liquid phase is converted to a higher melting solid $(MgO)_3V_2O_5$. This modified compound has less tendency to form molten deposits on the tubes:

$$3MgO + V_2O_5 \rightarrow (MgO)_3V_2O_5$$

The $(MgO)_3V_2O_5$ is characterized as noncorrosive and has a significantly higher melting point (1240°C) than the V_2O_5.

c. An additional benefit of the conversion of the V_2O_5 to $(MgO)_3V_2O_5$ is that the molten vanadium pentoxide is not available to catalyze SO_2 to SO_3, albeit the temperatures in the furnace proper are too high for SO_3 to exist because the latter is unstable at the high temperatures in the boiler.

d. Where the tube spacings in the superheat and reheat sections are wide enough to accommodate the addition of MgO, as for example either in coal-to-oil converted boilers or a boiler that was designed to burn oil but where the manufacturer specifically designed the boiler with widely spaced tubes, then the MgO will react with the vanadium oxides, as well as with other impurities in the fuel without any serious detriment, because of additional ash buildup. The physical properties of the magnesium vanadates are frail and flake-like deposits that are more easily removed by the gas stream and soot blowing cycles, which assists in keeping the tubes and tube spaces cleaner.

Again, in a properly designed and managed fireside additive-treated program, the MgO, or MgO combined with a catalyst, can help the unit's operations.

COLD END CORROSION

Sulfur is another contaminant of heavy fuels which adversely affects boiler operation. During combustion, sulfur present in the fuel oxidizes to sulfur dioxide, SO_2. Some of the SO_2 is further converted to sulfur trioxide, SO_3, which reacts with water vapour always present in the flue gas to form sulfuric acid H_2SO_4:

$$SO_2 + \tfrac{1}{2}O_2 \leftrightarrow SO_3$$

$$SO_3 + H_2O \rightarrow H_2SO_4$$

The degree to which conversation of SO_2 to SO_3 occurs depends on the level of excess air, the residence time of the flue gases as they pass across the metal surfaces in the boiler, the internal temperature of the boiler and the catalytic activity of both the interior boiler surfaces and the slag deposits originating from vanadium in the fuels. Two major mechanisms for SO_3 formation are known:

(1) Approximately 1% of the SO_2 formed is converted to SO_3 by a direct reaction with atomic oxygen contained within the flame. More SO_2 is not converted by this means because the reaction is thermodynamically unfavorable at these elevated temperatures (Reid, 1971).

(2) An additional quantity of SO_2, in some cases measured to be as much as 5%, is converted to SO_3 as the flue gases passes over catalytically active surfaces in the boiler (Jenkinson, 1963).

Both iron and vanadium are effective catalysts for this reaction. At the boiler's cold end temperatures, equilibrium would favour almost complete conversion of SO_2 to SO_3.

The rate of reaction, however, is slow and equilibrium is never reached due to the relatively rapid flue gases velocities. As a consequence, a maximum of about 6% of the sulfur present in the fuel converts to SO_3. This was also proven to be correct by actual measurements of SO_3 concentration in the flue gases of various size boilers. Upstream of the air heater (A/H), the SO_3 remains in the flue gases as relatively harmless SO_3 vapour. As the flue gases cool through the A/H, a second equilibrium reaction takes place between SO_3 vapour and H_2O vapour resulting in conversion to H_2SO_4 (sulfuric acid) vapour. At a temperature of about 200°C, the sulfuric acid is still in the vapour state and is relatively harmless. However, as the temperature drops below 200°C, the H_2SO_4 vapour eventually reaches its dew point and condenses on the relatively cooler metal surfaces of the A/H and duct work, causing severe corrosion, commonly referred to as Cold End Corrosion. It also results in acid emissions from the stacks, referred to as "extreme air pollution."

Normal boiler operating procedure is for the exit gases temperatures (A/H outlet gases temperature) to be at least 10 to 20°C above the acid dew point to prevent condensation on the cold-end surfaces. However, the cold-end average temperature is always below the acid dew point.

The problems caused by acid condensation will occur despite the 10 to 20°C "safety margin" due to uneven duct temperatures and air leaks.

Disregarding the exact quantity of sulfuric acid that will condense in the A/H or duct work, all the H_2SO_4 will eventually condense once it is out of the stack and will cause adverse damage to the environment.

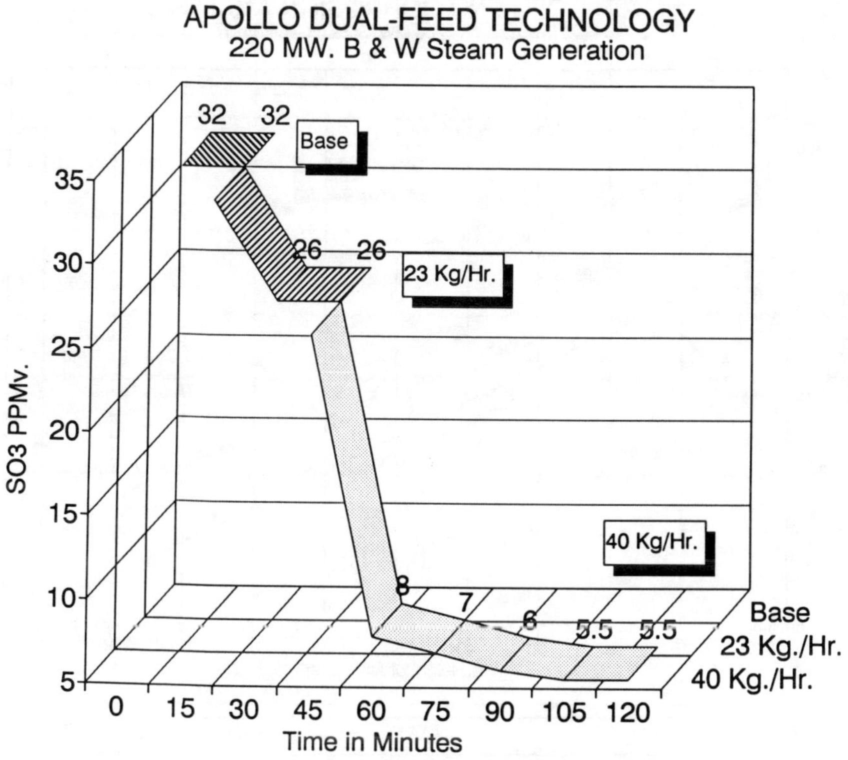

FIGURE 2 SO_3 reduction.

If for any level of sulfur in the fuel, the SO_3 formed in the critical areas in the boiler, as mentioned above, can be successfully neutralized, the boilers will be able to operate more efficiently at lower exit gases temperatures without concern from the highly acidic emissions, or acid condensation and corrosion in the boiler, or pollution of the environment.

The reduction of SO_3 can be accomplished by the use of additives in two ways:

(1) Minimize the conversion of SO_2 to SO_3 in the furnace. This accounts for about 1% of the conversion that takes place in the untreated fuel.

(2) Neutralize the SO_3 just prior to the location in the boiler where the SO_3 condensation begins to occur, e.g., within the air heaters.

The two effective means mentioned above will minimize the condensation of H_2SO_4 on the air heaters efficiently and economically.

The reduction of the SO_3 in the flue gases as a function of time in minutes after addition of the neutralizing, as well as the influence of the feed rate of the additive is shown in Figure 2. By the use of Coaltrol M, it was possible to maintain the acidity below 10 PPMv which effectively prevents acid corrosion of the air heaters.

AIR HEATER ANALYSES

The A/H in a steam generating plant serves as an efficient heat recovery unit from the thermodynamic point of view but, in practice, the A/H and the downstream ductworks are subjected to adverse fouling and corrosion.

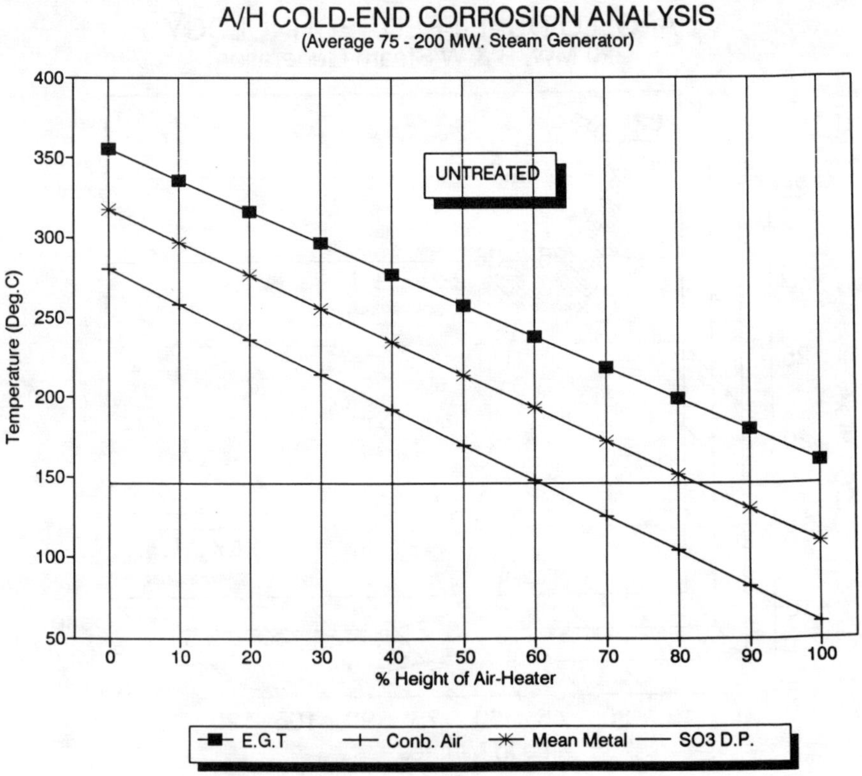

FIGURE 3 Air heater analysis untreated.

In the A/H the boiler operators are recovering as much of the heat as possible which is stored in the flue gases, otherwise very hot gases will exit through the stack and result in a tremendous heat loss.

In the A/H the heat from the hot gases is transferred to the combustion air prior to combustion air entrance to the combustion chamber.

Combustion air enters the A/H either at the boiler's house ambient temperature or, every so often, it is heated, with steam coils to 80–90°C — well bellow the SO_3 dew point. The A/H, while rotating, is at all times half-subjected to the hot gases passing through on their way to the stack and thus heating the A/H elements, and the other half is subjected to the combustion air passing through the heated elements, absorbing part of the heat prior to entering the combustion chamber.

Due to the A/H rotating movement, a certain quantity of flue gases are trapped whenever the A/H elements are moving from the gas side to the air side and immediately subjected to the relative low temperature of the air, thus lowering the temperature of the gases and the SO_3 within the gases to a temperature below the SO_3 dew point.

Also, due to the deterioration of the A/H mechanical seals (between the air and gases side), more and more quantities of ambient air, referred to as parasite-air, enters the gases half of the A/H, causing an addition reduction of the flue gas temperature to below the SO_3 dew point.

From a large number of analyses of rotating air heaters, it was found that on average, 40% of the A/H height of the individual sections of the A/H baskets will be subjected to

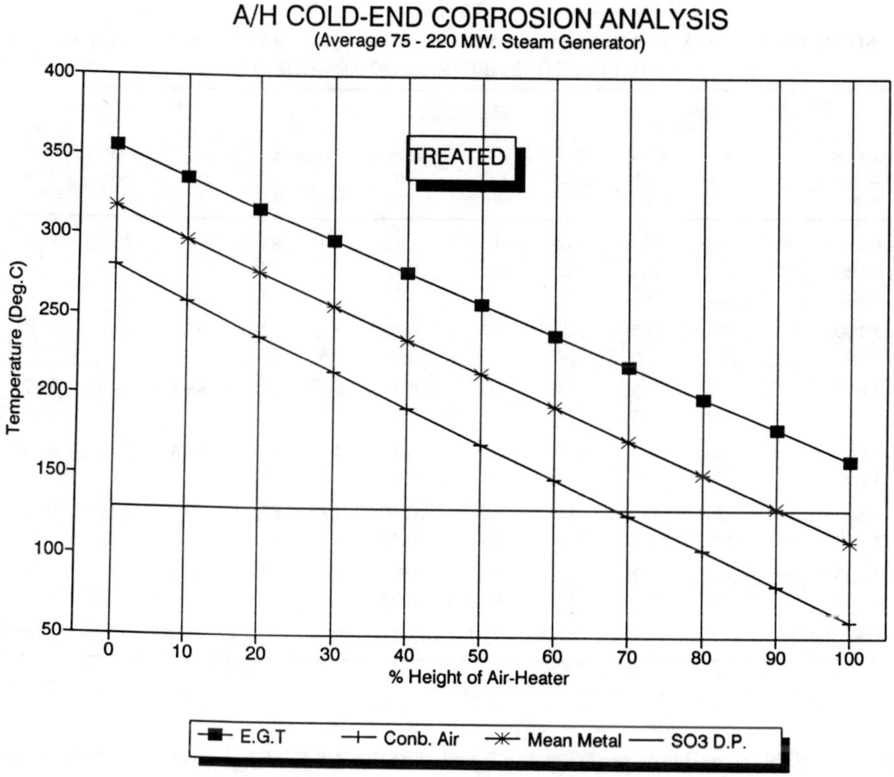

FIGURE 4 Air heater analysis treated.

temperatures below the SO_3 dew point and 15% to 20% of the A/H height will be subjected to temperatures below the SO_3 dew point on every cycle. The Coaltrol-M neutralizing system enables the reduction of the above mentioned values to 30% and 9% respectively, as shown in Figures 3 and 4.

TREATMENT PERFORMANCE AT ISRAEL ELECTRIC CO.

At the Israel Electric Co., cold-end neutralization with Coaltrol M has been in continuous used for over 15 years. Currently, 12-oil fired boilers with a total capacity of 1800 MW. are being treated (Lev Er *et al.* 1977), ranging in size from 75 MW. to 220 MW.

The injection equipment for accurately feeding the neutralizing additive, a powder, was designed and manufactured by the additive supplier, Apollo Technologies (Kukin *et al.* 1988). The additive is aspirated into the gas stream prior to entering the A/Hs at temperatures ranging between 365–385°C.

The average sulfur content in the fuel was 2.7%. In all the units, SO_3 in the treated boilers is capable of being maintained at all times below a value of 10 PPMv. As a result, the excess air ratio is kept close to stoichiometric, the exit gas temperature is low, generally only 20°C above the acid dew point, and the total boiler efficiency is increased. Also, as the A/Hs and ductwork to the stack are constantly protected against any potential condensation of sulfuric acid and pluggage, the boilers on-line period is increased. This results in greater

TABLE I
SUMMARY OF DATA AND CALCULATIONS OF AIR HEATER PARAMETERS BY TREATMENT WITH FLUE GAS NEUTRALIZATION ADDITIVE

Additive to fuel ratio	Exit gas Hot end °C	Exit gas Cold end °C	Acid dew point °C	CO ppm	Excess air $\%O_2$	Deposit pH	Deposit %C	Gas side eff. %	dP A/H %	Gas/air heat ratio %
Base	385	165	142	70	1.10	2.4	88	68	40.0	91.0
Base	375	165	142	100	0.95	3.2	77			
Base	372	167	143	60	1.10	3.1	78			
1:1000	376	165	140	84	0.90	3.5	74	68.8	38.2	91.8
1:1000	365	165	139	75	0.95	3.4	70			
1:650	385	162	135	60	0.85	4.2	67	68.1	38.2	91.7
1:650	370	165	135	60	0.85	4.2	70			
1:600	380	166	133	80	0.85	4.4	68	69.4	39.6	87.9
1:600	370	167	133	75	0.90	4.5	67			
1:550	380	165	131	80	0.80	4.8	62	69.7	39.0	93.1
1:550	370	167	132	80	0.80	4.7	63			
1:400	379	162	131	75	0.90	5.3	61	70.3	37.9	93.4
1:400	360	165	132	80	0.90	4.8	63			
Improvements			−7.0%	12.5	−18.2%	50.0	−28.4%	3.3%	−5.3%	1.6%

boiler availability and also prolongs the interval of the A/H wash periods by a minimum of 150%.

Over a regular period (average every 18 months) of internal inspection of the A/Hs, no corrosion damage was reported. A summation of the data obtained by treatment with flue gas neutralization additive is shown in Table I (Lev Er *et al.* 1977).

SUMMARY

The combination of a catalyst added to the fuel oil and a neutralizing additive injected into the air heaters upstream gas achieves the following main objectives:

1. Enables the boiler operator to operate as close as possible to the stoichiometric combustion relation.
2. Reduction of the unburned carbon in the flue gases.
3. Reduction of 40 to 60% in the stack particulate concentration.
4. Reduction of the SO_3 concentration to less than 10 PPMv.
5. Keeps the pressure differential across the air heater within design values for a prolonged period.
6. Increases boiler availability.
7. Increases boiler efficiency.

REFERENCES

Lev Er, J., Peer, A., and Frindland, A.C. (1977). Improving Availability and Economy of Oil Fired Boilers by Reducing Deposition and Corrosion Caused by Gases. *Proc. Tenth World Energy Conference*, Istanbul, Turkey, September.

Kukin, I. (1989). U.S. Patent No. 4,842,617, June 27.

Kober, A.E., and Galanis, C.A. (1984). Dual Feed Treatment to Protect Hot and Cold Sections of Oil Fired Generation Units, *Proc. 13th National Meeting of Chemical Engineers of Comission Federal de Electricidad*, Mexico, D.F., December.

Giammar, R.D., and Krause, H.H., D.W. ASME, 75-WA/CD-75.

Goldberg, H.J., and Lewis, D. (1978). Prescription for a Successful Fuel Treatment Program, *Combustion*, **50**, No. 6, December.

Reid, W.T. (1971). *External Corrosion and Deposits*, American Elsevier, New York.

Jenkinson, J.R. (1963). *Sulphur Oxide Reactions, The Mechanism of Corrosion by Fuel Impurities*, Butterworth Publications, London.

Kukin, I. (1989). U.S. Patent No. 4,804,388, June 27.

Kukin, I., and Pepe, W.C. (1988). U.S. Patent No. 4,793,268, December.

Emission Characteristics of the Shell Low-NO_x Burner

B. DE JONG *Koninklijke/Shell-Laboratorium, Amsterdam (Shell Research B.V.)*
Badhuisweg 3 1031 CM Amsterdam The Netherlands

Abstract—At the Koninklijke/Shell-Laboratorium in Amsterdam a low-NO_x burner is being developed for the combustion of heavy fuel oil, asphalt, natural gas and refinery gas, as well as for combinations of liquid and gaseous fuels in dual fuel firing mode. The NO_x reduction is accomplished by air staging, while the particulate emission is simultaneously kept low using intense steam atomisation (40%w steam on fuel intake). The emission characteristics have been determined experimentally in a vertical test furnace, which can accommodate a maximum thermal input of 15 MW. The experiments so far indicate that the NO_x emission limits for new equipment, as proposed in the amendment to the draft decree of the Dutch clean air act, cannot be met with any of the burners considered. As the projected NO_x emission limits for existing equipment are not exceeded, the application of the Shell low-NO_x burner should primarily be regarded as a solution for retrofit situations.

Key Words: Oil combustion; NO_x reduction; low-NO_x burner technology

1. INTRODUCTION

1.1 Pollutant Legislation

Dutch industry operating stationary combustion equipment such as boilers and furnaces is confronted with tightening pollutant legislation to limit the detrimental effects of combustion products on the environment. Figure 1 shows a recently proposed amendment to the Dutch clean air act. It is obvious that the progressive reduction of the maximum allowable NO_x emission will require a considerable effort from industry. The oil refineries in particular are confronted with a major task, as their fuel diets cannot be arbitrarily chosen: large amounts of heavy residual fuels have to be dealt with, especially in complex refineries with a high conversion capacity. This has also been recognised by Kelly et al. (1989), who reviewed work carried out at the burner test rig at the BP Research Centre. More recently, a comprehensive review on the control of combustion-generated NO_x emissions has been given by Bowman (1992).

In view of the relevance of this subject, the atmospheric combustion group of the equipment engineering department of the Koninklijke/Shell-Laboratorium in Amsterdam has been working since 1985 on the development of a low-NO_x burner for the Shell type vertical process heater. This paper reports on the current position of this ongoing burner development and compares the emission characteristics with pending Dutch pollutant legislation. Attention will be focused on the NO_x- and particulates emission limits for existing as well as new process furnaces after 1994.

These limits are depicted in Figure 2 for dual fuel firing.

1.2 Principle of the Shell Low-NO_x Burner

In general various options exist to reduce the NO_x emission from a combustion process: air staging, fuel staging and flue gas recirculation.

With air staging the burner is fired sub-stoichiometrically, i.e. it is operated with less than the theoretical amount of combustion air, while the remainder of the combustion air is supplied further downstream via staging air nozzles. In the case of gas firing this results in a reduction of the peak flame temperature and consequently in a lower thermal NO_x formation. When firing liquid fuel containing organically bound nitrogen

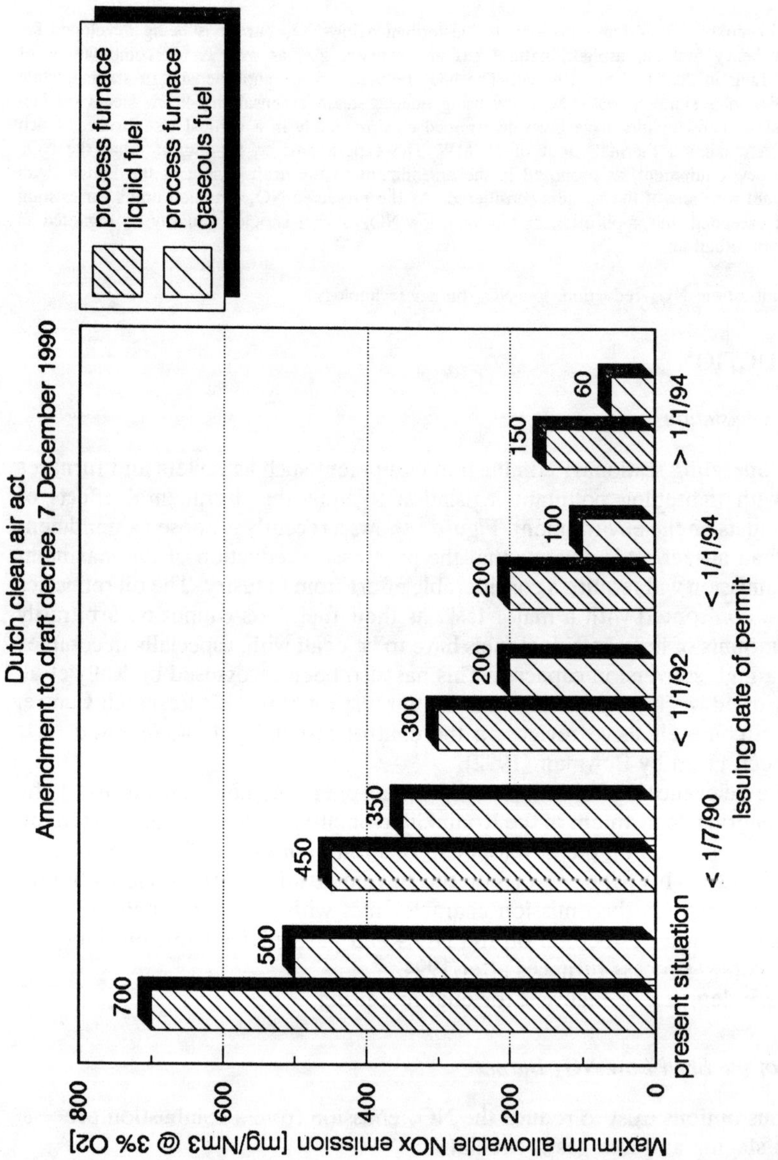

FIGURE 1 Emission requirements for combustion facilities

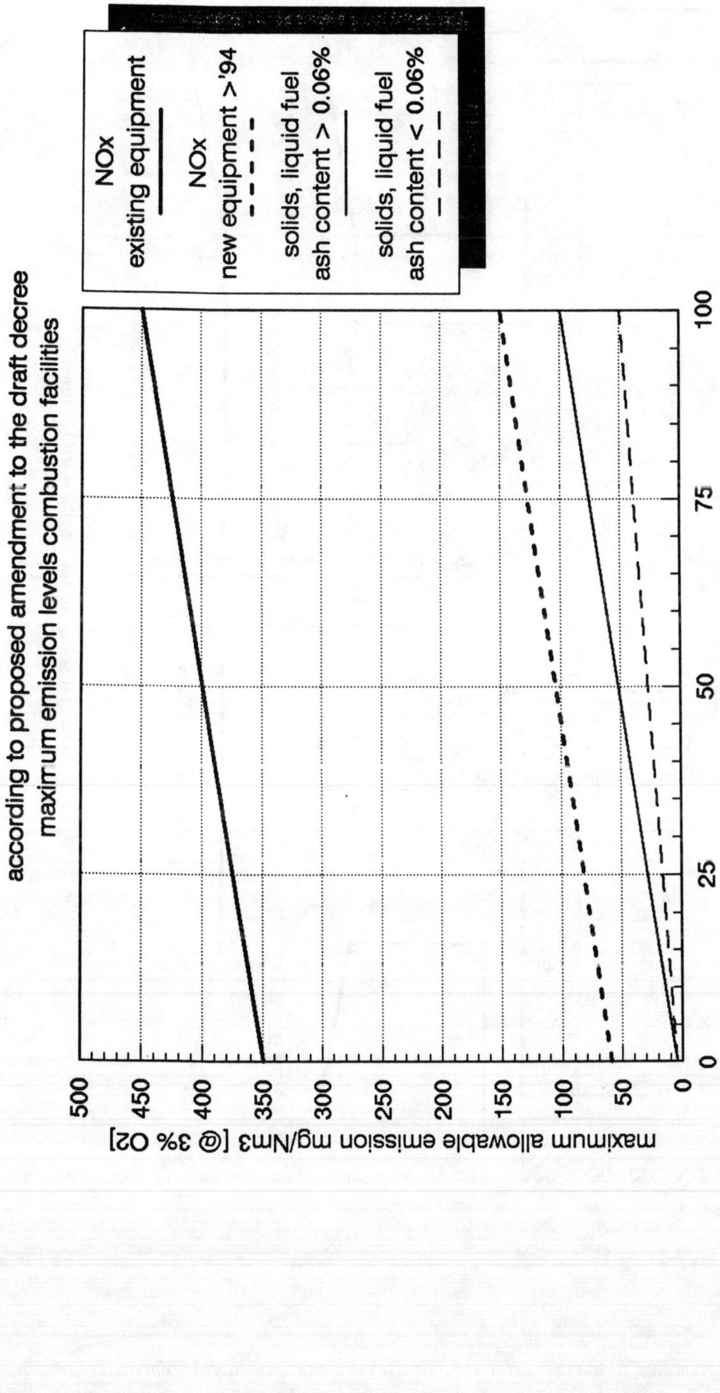

FIGURE 2 NO_X/solids emission limits fcr dual fuel firing

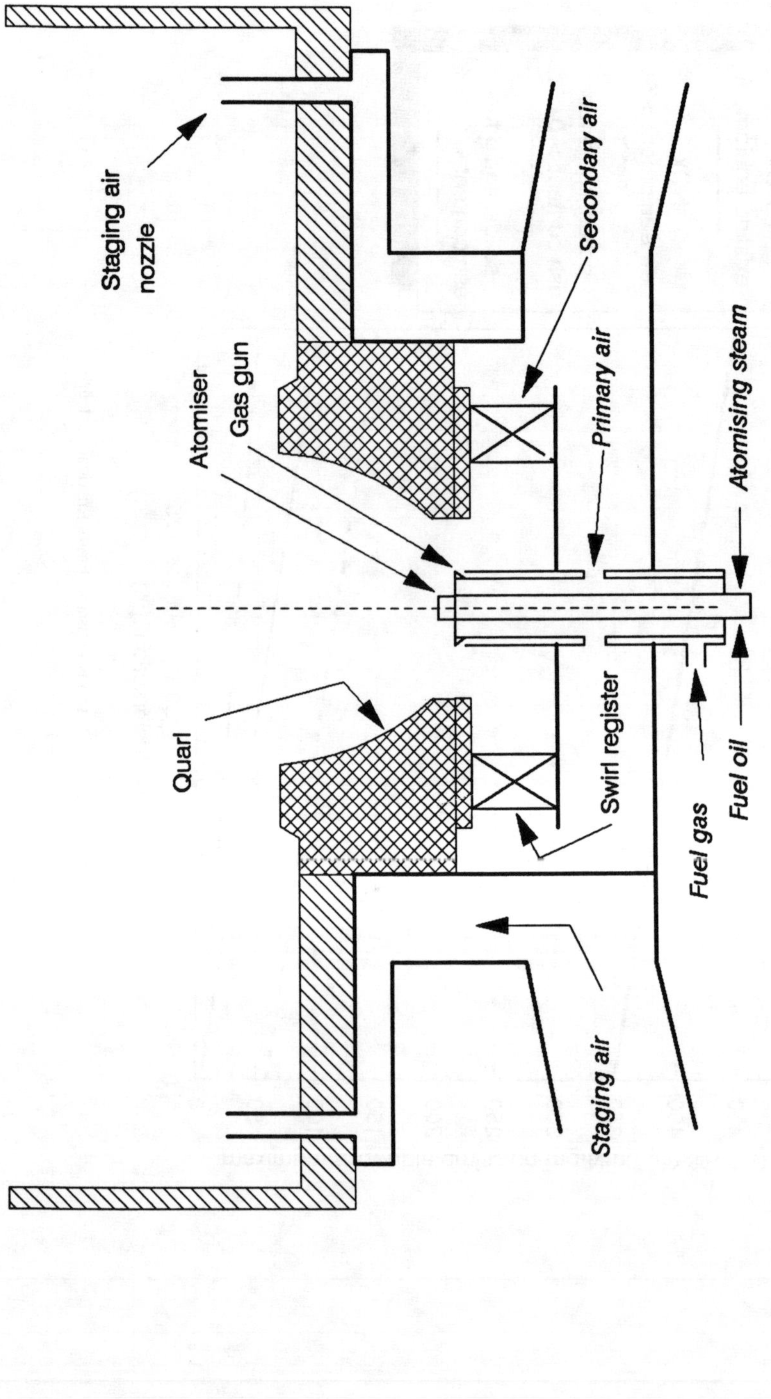

FIGURE 3 General set-up of the Shell low-NO_x burner

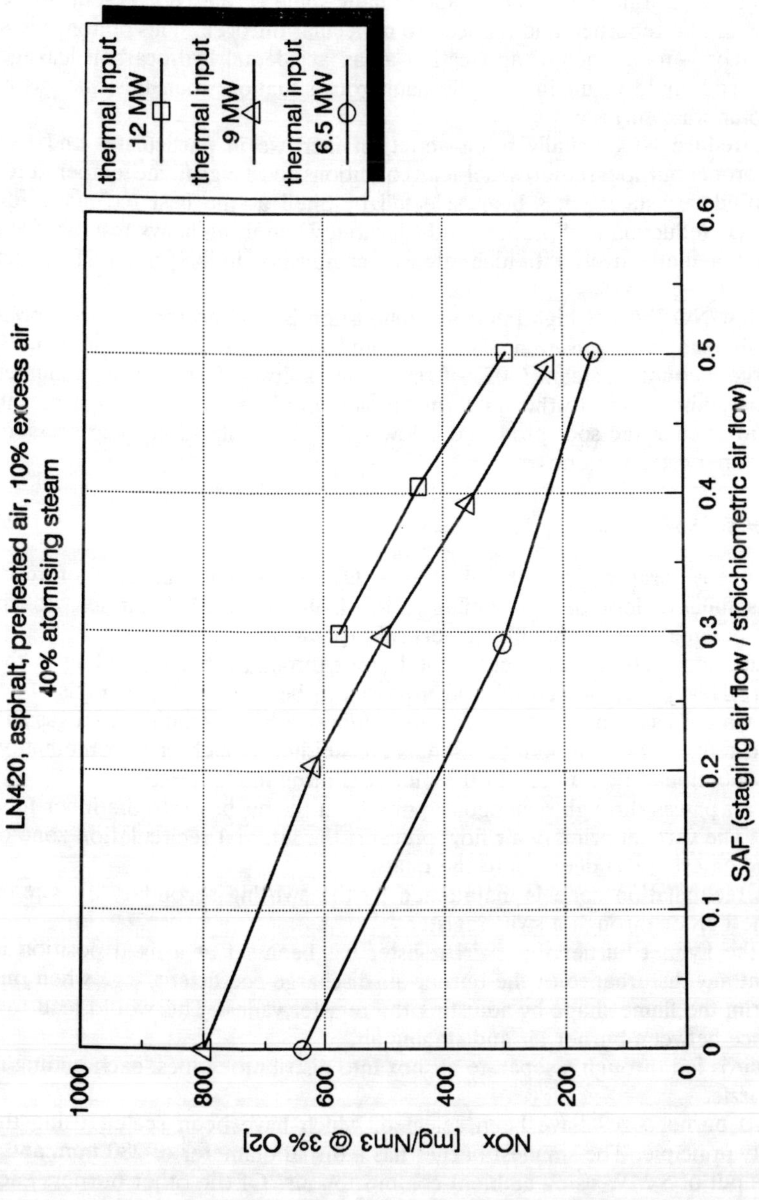

FIGURE 4 NO_x as a function of SAF and thermal input

compounds, these compounds are released in the fuel-rich zone, where they are converted to molecular nitrogen, rather than oxidised to NO.

Fuel staging can be realised by injecting fuel through spuds downstream of the main burner. Thus a second fuel-rich zone is created where initially formed NO can be further reduced through the reburning principle.

Finally, flue gas recirculation involves diluting the combustion air with oxygen-deficient combustion products, so that again the peak flame temperature is reduced, resulting in a lower thermal NO_x formation, whereas additionally some of the NO present in the recirculated gas can be reburned and reduced to molecular nitrogen. This option is less suitable for refinery process heater application as an accidental hydrocarbon leakage from a furnace coil could result in a highly dangerous situation when this flue gas is re-injected into another furnace.

Measures to reduce NO_x usually bring about an increase in particulates and CO emissions, as a proper burnout requires fuel-lean conditions and a high flame temperature. With this in mind, air staging has been selected by Shell as the best technique for simultaneous NO_x reduction and proper solids burnout when firing heavy residual fuel, as it can be combined effectively with high intensity combustion in the primary, fuel-rich flame zone.

In the Shell low-NO_x burner high intensity combustion is realised through the application of very fine steam atomisation. The preheated oil is mixed at a pressure of 10 barg with a large amount of steam (40% of the oil mass flow) in the mixing chamber of a twin fluid atomiser. Due to the small oil droplets produced and the short mixing time the amount of char and soot produced is lower than in combustion using pressure atomisers or ordinary steam atomisers.

1.3 Outline of the Shell Low-NO_x Burner

Figure 3 shows the general set-up of the Shell low-NO_x burner. This burner is a forced-draught, swirl-stabilised floor burner similar to the Lyunet-type Shell burner. It has especially been designed for the Shell-type vertical process heater.

Liquid fuel and atomisation steam enter the burner through a barrel, ending in the steam atomiser. Heavy fuel oil (HFO) and asphalt can be used as liquid fuels. Fuel gas is supplied via the gas gun: a set of co-annular tubes with a circumferential gas slit. Natural gas and refinery gas can both be used as fuels either as such or in combination with the above-mentioned liquid fuels in the dual fuel firing mode.

The primary air passes through rectangular ports from the air box into the inner tube of the gas gun. The vertical primary air flow prevents the internal recirculation zone of the burner from entering too deeply into the quarl.

The internal recirculation zone is maintained by the swirling secondary air stream which enters the burner through a swirl register.

Contrary to the Lyunet burner this swirl register has been set at a fixed position in order to prevent any disturbance of the burner air discharge coefficient, e.g. when one would like to trim the flame shape by adjusting the register vanes. This would lead to a disturbed balance between burner air and staging air.

The staging air is fed through a separate air box into distributor tubes, each ending in a staging air nozzle.

Four standard burner sizes have been selected, which have been scaled using the constant velocity principle. The smallest burner has a throat diameter of 280 mm, and a rated thermal input of 5 MW using ambient combustion air. Of the other burners each has a 50% higher rated thermal input than the preceding one, resulting in quarl throat diameters of 340, 420 and 520 mm, covering a firing range between 5 and 16 MW. The burners have been designated LN280, LN340 etc., indicating the quarl throat diameter.

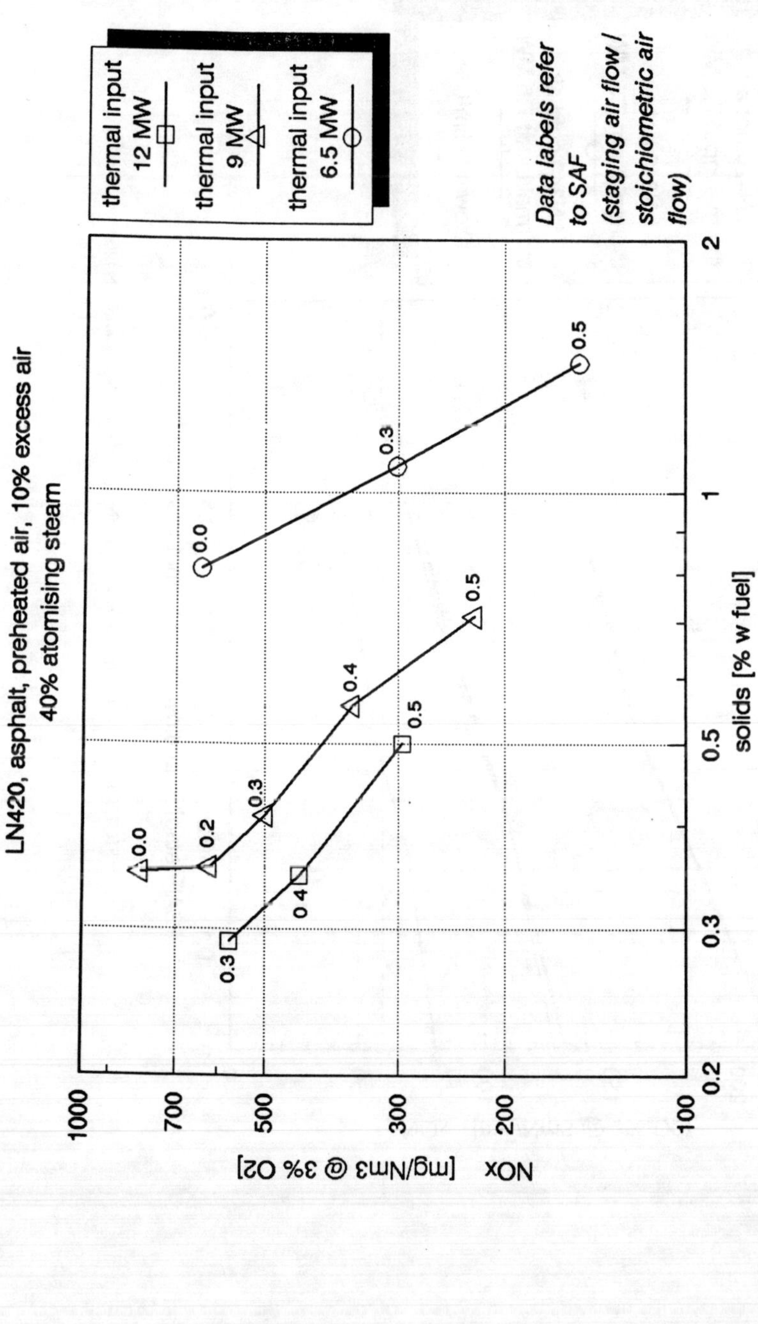

FIGURE 5 NO$_X$ versus solids emissions

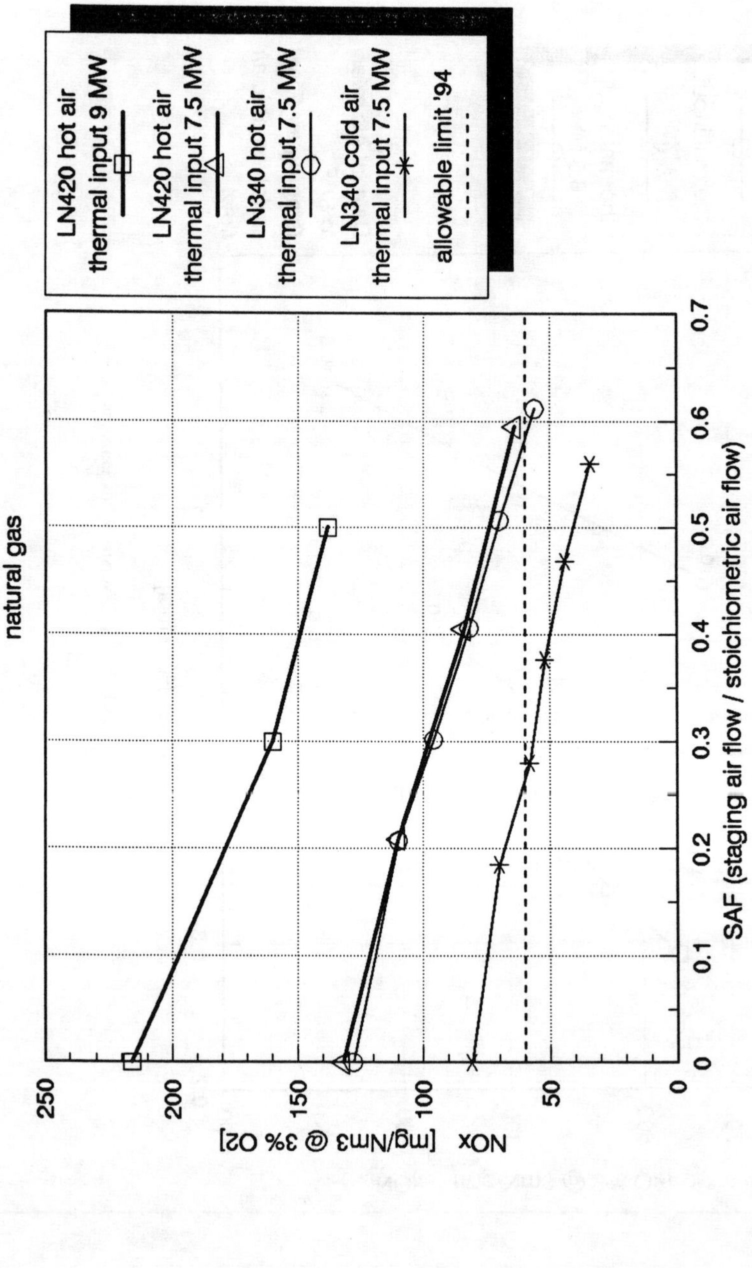

FIGURE 6 NO_x emissions for gas firing on the LN340 and LN420

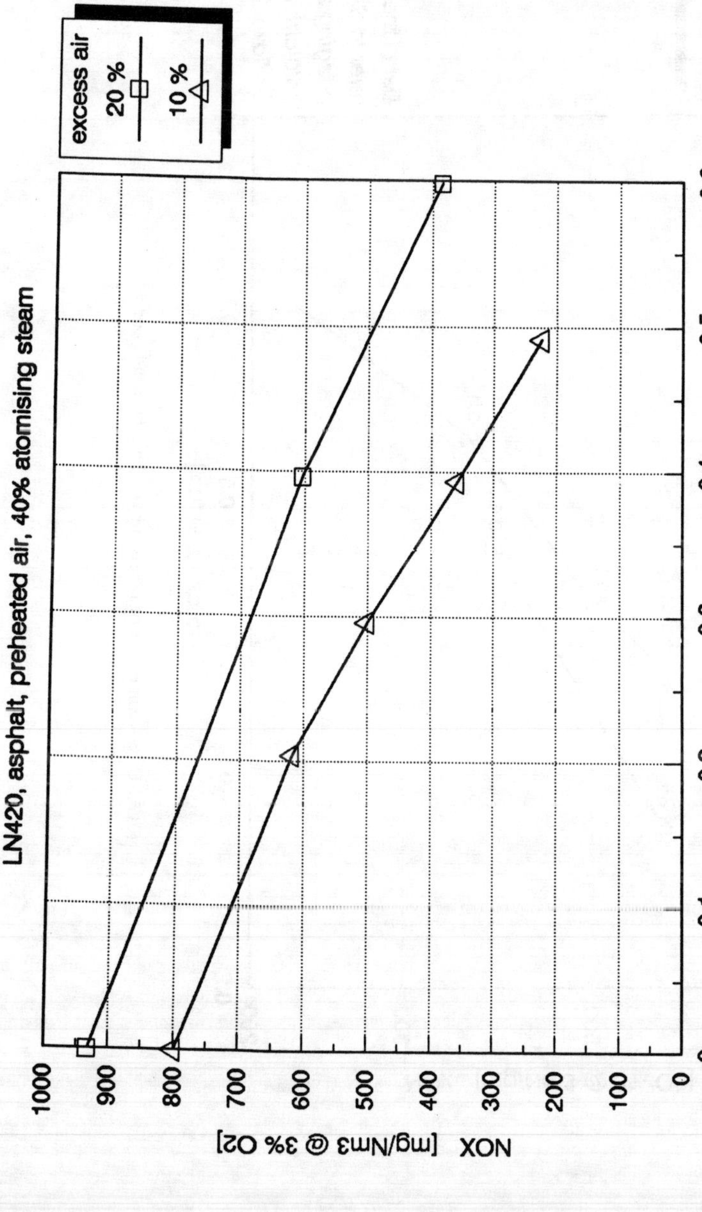

FIGURE 7 Influence of excess air on $N0_x$

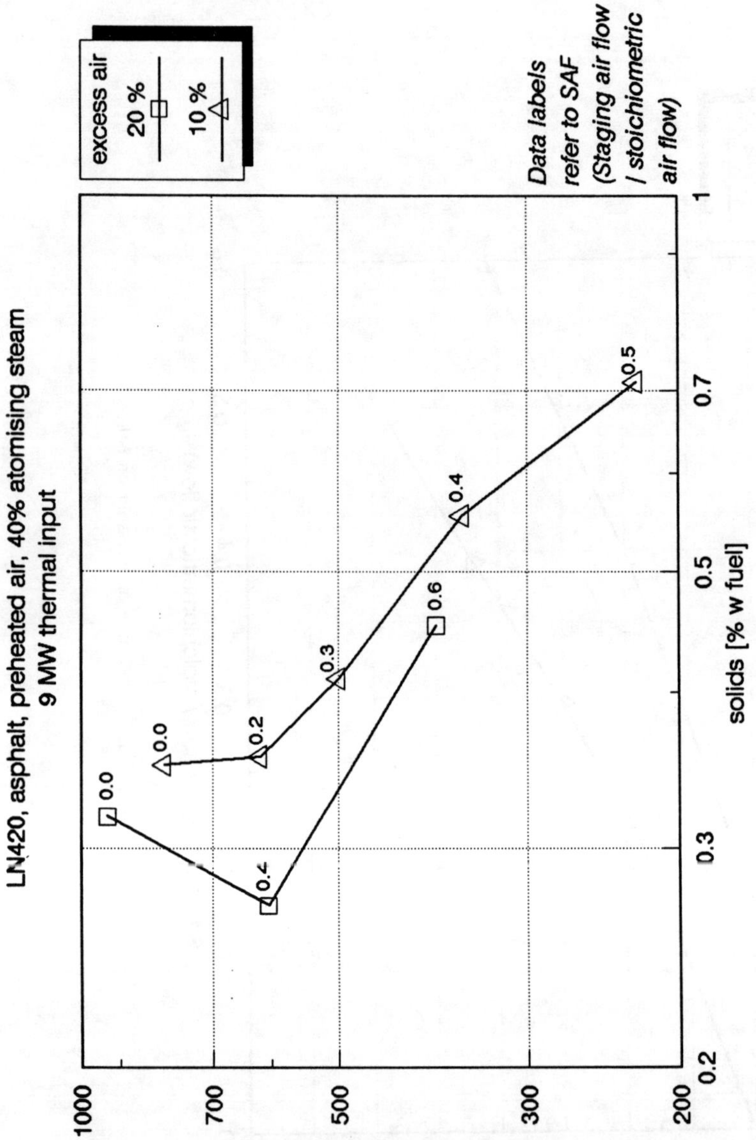

FIGURE 8 Influence of excess air on the NO_x-solids curve

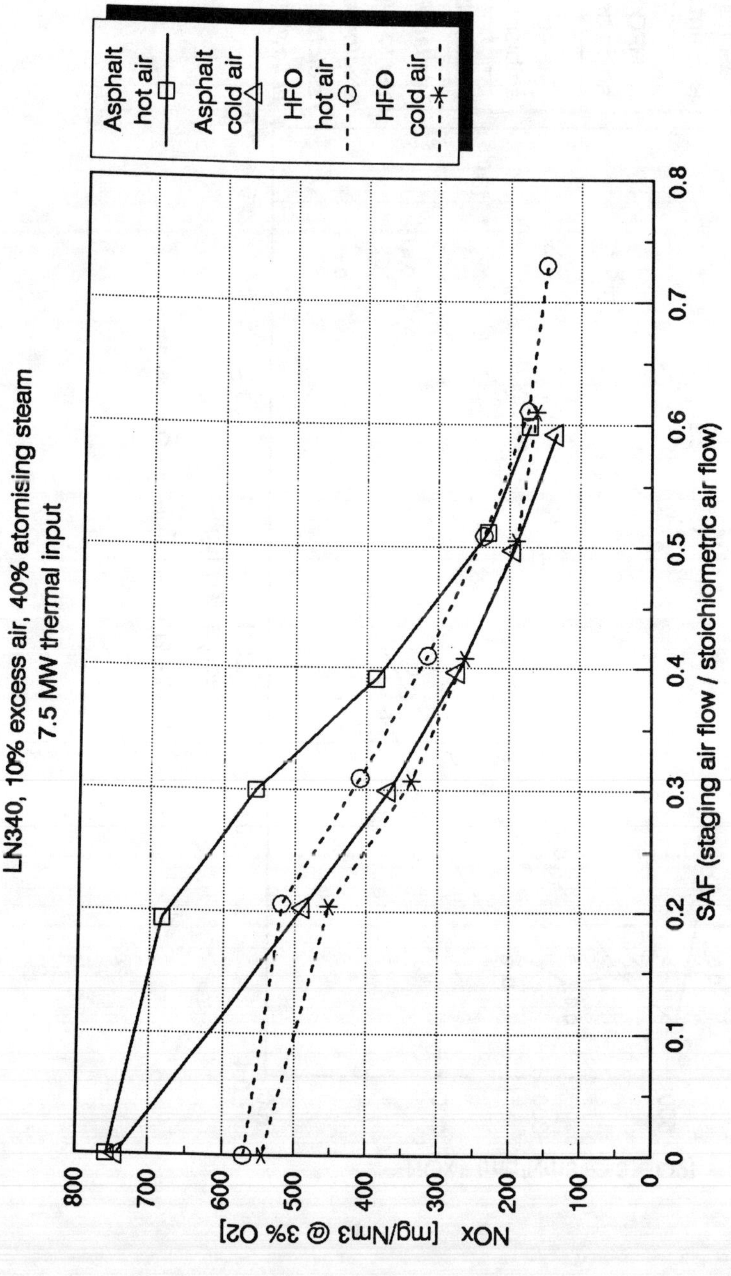

FIGURE 9 NO_x as a function of SAF and air temperature

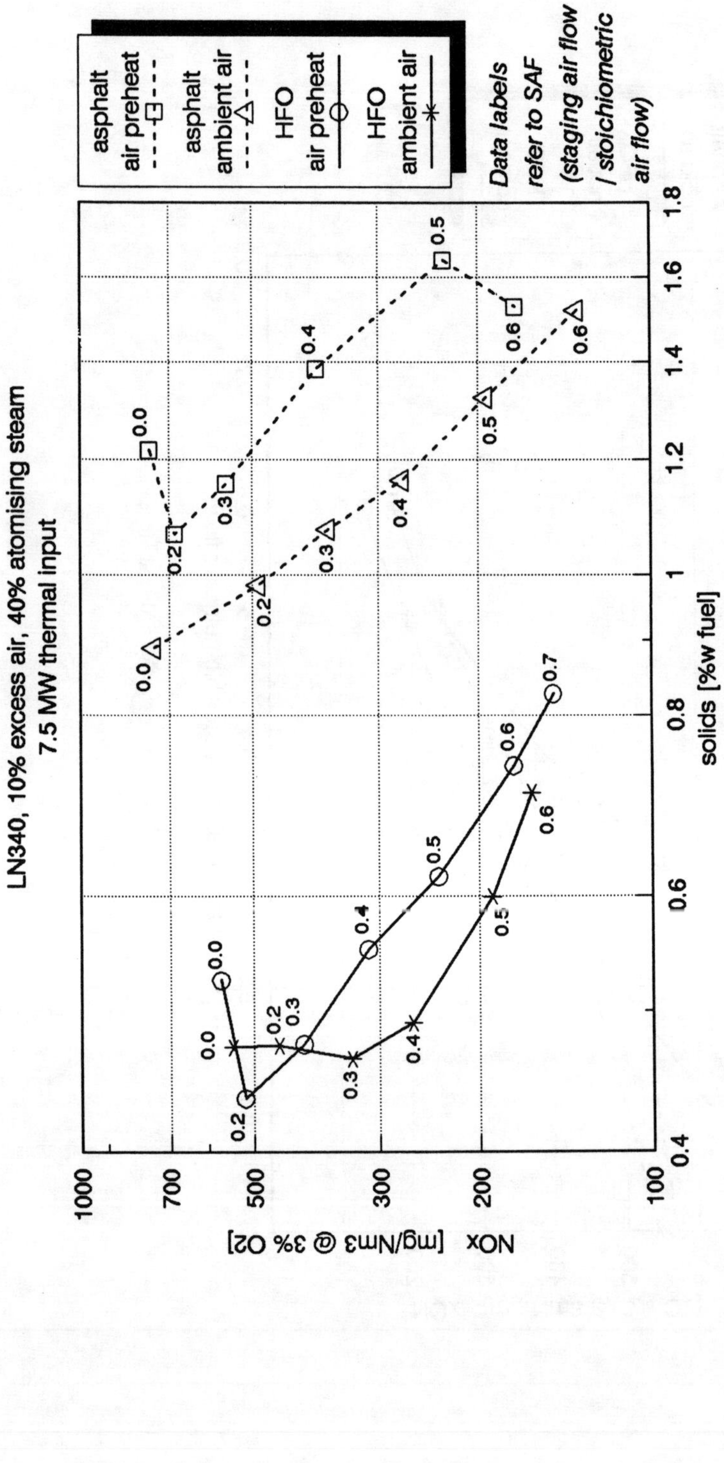

FIGURE 10 Influence of air preheating on NO_x-solids curves

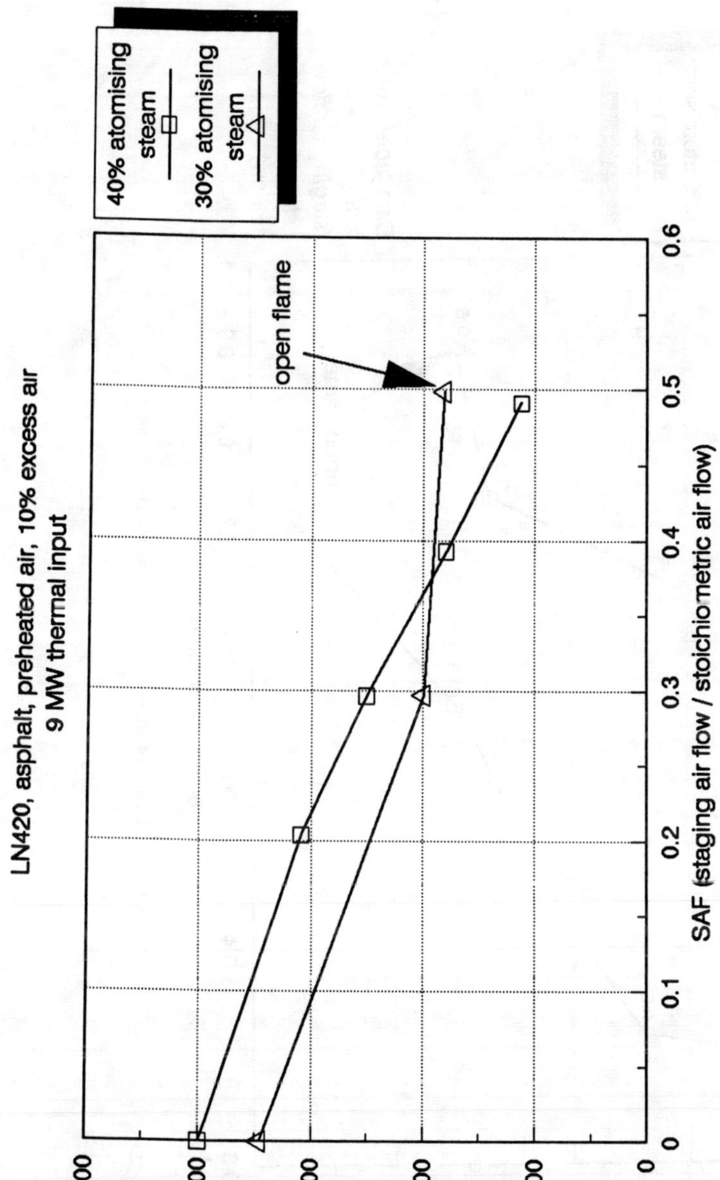

FIGURE 11 Influence of atomising steam to oil ratio on NO_x

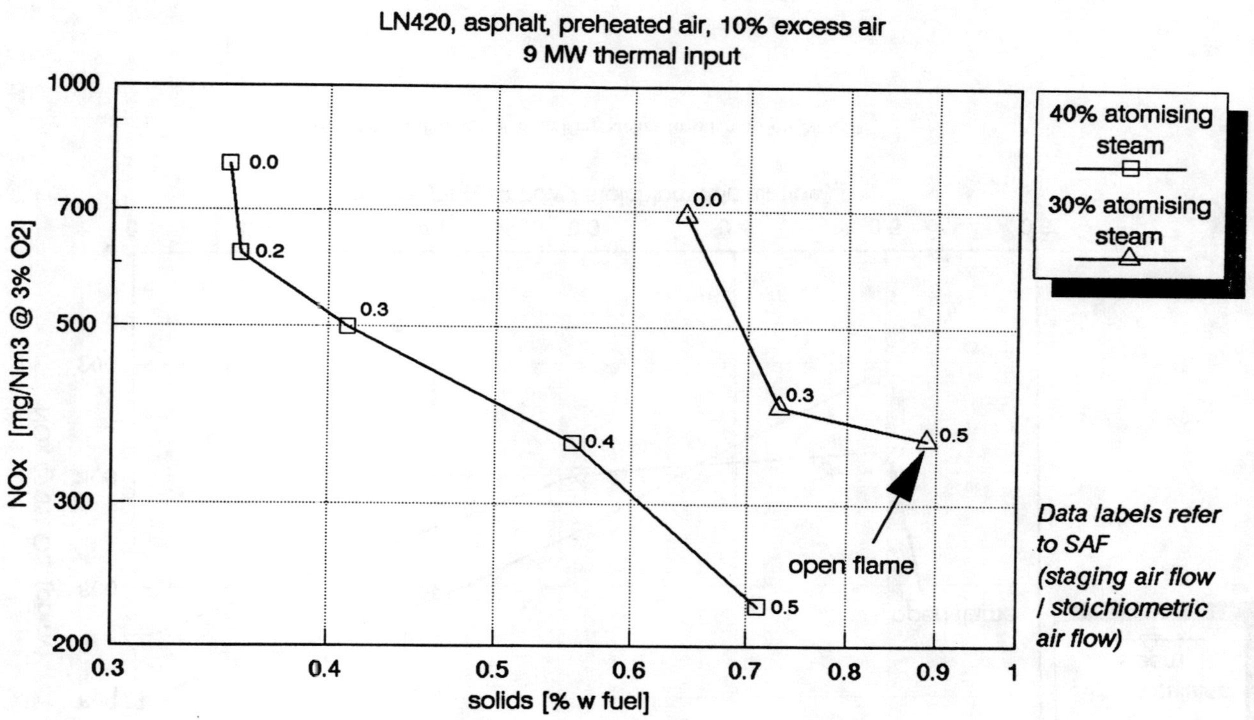

FIGURE 12 Influence of steam to oil ratio on NO_x-solids curve

2. EXPERIMENTAL

2.1 *The Vertical Test Furnace*

The emission characteristics have been determined experimentally in a vertical test furnace (VTF) which can accommodate a maximum thermal input of 15 MW. The furnace contains one radiant cell which measures $4 \times 4 \times 11$ m. The heat sink consists of 40 wall tubes of 6″ diameter grouped in two coils, creating a total heat exchanging area of 232 m^2. The oil circulating in these coils is heated from 150 to 240°C, which means that the radiant cell temperature is relatively low in comparison with that in furnaces for practical applications. The primary and secondary combustion air can be preheated to 200°C by means of an air preheater. The staging air temperature was kept constant at 20°C.

2.2 *Measuring Equipment*

The composition of the flue gas has been measured using the standard on-line instrumentation of the VTF. The instruments used were:

for NO_x : Beckman chemiluminiscence analyser type 951

for CO : Beckman infrared analyser type 864

for O_2 : Servomex paramagnetic analyser type 1100

The solids measurements were performed, for a limited number of test conditions, using a gravimetric technique based on an isokinetic extraction of a sample flow through a glass-fibre filter. The solids concentration was derived from the filter weight before and after the measurement and the dry flue gas volume as measured with a gas meter. For all test runs the stack opacity was determined using a DURAG opacity meter; it can be correlated with the gravimetric solids measurements, provided that the atomiser dimensions, the oil flow, the oil/gas ratio for dual fuel firing and the atomising steam flow are kept constant.

2.3 *Test Conditions*

The influences of the following process variables have been investigated:

1. relative amount of staging air
2. thermal input
3. excess air
4. air preheating temperature
5. atomising steam to oil ratio
6. oil/gas ratio when firing in dual fuel mode.

The fuels used during the experiments were heavy fuel oil, asphalt and natural gas. The typical compositions of these fuels can be found in Tables I and II.

The burners tested were the three smallest burners, of which the rated thermal input could be accommodated by the VTF.

TABLE I
Typical composition and properties of fuel oil

Property	Unit	HFO	Asphalt
Specific density (15/4°C)	kg/dm^3	0.97	1.04
Viscosity at 100°C	mm^2/s	33.2	925
Conradson carbon residue	%wt	6.2	23.6
Asphaltenes	%wt	3.0	15.7
Nitrogen content	%wt	0.37	0.74
Ash content	%wt	0.01	0.07
Lower heating value	MJ/kg	41.3	38.7

TABLE II
Typical composition and properties of natural gas

component/property	unit	
He	%v	0.055
N_2	%v	5.160
CO_2	%v	7.750
CH_4	%v	82.135
C_2H_6	%v	3.770
C_3H_8	%v	0.680
C_4H_{10}	%v	0.240
C_5H_{12}	%v	0.070
C_6H_{14}	%v	0.039
C_7H_{16}	%v	0.076
C_8H_{18}	%v	0.025
Lower heating value	MJ/kg	37.37
St. air requirement	Nm^3/kg gas	9.97
St. air requirement	kg/kg gas	12.89
Density	kg/Nm^3	0.88
Av. molecular weight	kg/kmole	19.76

3. RESULTS AND DISCUSSION

3.1 *Influence of Air Staging and Thermal Input for Liquid Fuel Firing.*

The amount of air staging can be expressed in terms of the staging air factor (SAF), being the staging air flow divided by the stoichiometric air requirement. Similarly, the burner air factor (BAF) refers to the burner air flow, whereas the total air factor (TAF) refers to the total air flow, and therefore equals SAF + BAF.

Figure 4 gives the NO_x as a function of the SAF for various thermal inputs for asphalt firing in the LN420. The NO_x decreases almost linearly with the SAF from 800 mg/Nm^3 to 224 mg/NM^3 at the design thermal input for air preheating (9MW). The NO_x emission increases with the thermal input, which can be explained by the increase in thermal NO_x emission due to the higher flame temperature.

The solids emissions measured during these tests are given in Figure 5. Obviously, the higher flame temperature has a beneficial effect on the solids emission, which decreases with increasing thermal input. The difference between the 12 MW and 9 MW curves is smaller than that between the 6.5 MW and 9 MW curves, as the 6.5 MW tests had been performed with the 9 MW atomiser, resulting in a lower mixing chamber pressure and consequently a lower atomisation quality.

3.2 *Influence of Air Staging and Thermal Input for gas firing*

Figure 6 gives the NO_x as a function of the SAF for various thermal inputs for natural gas firing. It is clear that the NO_x emission for 9 MW and no staging is considerably lower than for oil firing, but on the other hand the influence of staging is rather limited: the NO_x emission goes down by only 78 mg/Nm^3.

The thermal input has a significant influence: reducing the thermal input from 9 to 7.5 MW results in a 50% reduction of the NO_x emission for deep-staged conditions. Furthermore, it can be observed that the NO_x emission is independent of the burner size, as the 7.5 MW test results on the LN340 exactly match those on the LN420. This can be explained as follows. The heat transfer of a gas flame mainly occurs through convection and not so much by radiation. This convective heat transfer is independent of the burner size, but does depend on the radiant cell geometry, which remained unchanged during the tests.

3.3 *Influence of Excess Air*

Figure 7 demonstrates that increasing the excess air from 10 to 20% results in a considerable upward shift of the NO_x emission of the order of 130 mg/Nm^3 for asphalt firing. Figure 8 shows that this simultaneously leads to a better solids burnout, as the NO_x-solids curve shifts to lower solids levels. The improvement in solids burnout increases from 13% at zero staging to 40% at deep staging (SAF = 0.5).

3.4 *Influence of Air Preheating*

The majority of the experiments have been performed using preheated air with temperatures varying between 170 to 200°C. Air preheating is a common situation in process heaters, as it increases the combustion performance and overall efficiency. As a consequence, the flame temperature will be higher, leading to higher NO_x emissions.

Figure 9 shows the influence of air preheating on the NO_x emission from HFO and asphalt flames. It is apparent that the increase in NO_x emission due to the application of preheated air is almost constant for mildly staged conditions, whereas at deep-staged conditions there is hardly any influence at all. Here the increase in flame temperature with the resulting higher NO_x emission is counterbalanced by the improved fuel nitrogen release, which leads to quick conversion to molecular nitrogen in the fuel-rich zone.

Figure 10 gives the corresponding NO_x-solids curves. Contrary to expectation, the solids burnout deteriorates with increased combustion air temperature. A possible explanation is the decreased difference in velocity between the fuel spray jets and the surrounding burner air, due to the lower air density. The air entrainment in the primary combustion zone will therefore be smaller, leading to increased soot formation and poorer solids burnout.

Looking back at Figure 6 it can be observed from the LN340 gas firing tests that the burner air temperature also has a significant influence on the NO_x formation in gas flames. This is due to the fact that this NO_x is entirely of thermal origin and will therefore be directly influenced by the air preheating temperature.

3.5 *Influence of Atomising Steam to Oil Ratio*

The standard steam to oil ratio (STOR) for which the low-NO_x burner has been designed is 40%. However, for economic reasons it is desirable to limit steam consumption to as low a level as technically possible. Therefore the influence of the STOR on the various emissions has been considered by executing a number of tests with a STOR of 30%.

In Figure 11 the NO_x emission has been plotted versus the SAF. The atomisers used were the same as for a STOR of 40%. It appears that the NO_x emission generally decreases, apparently due to a less intense mixing of fuel spray and burner air. For deep staging the NO_x emission increases, which is probably due to the occurrence of an open flame.

Figure 12 gives the STOR influence on the NO_x-solids curve for the LN420. Decreasing the STOR from 40 to 30% leads to a less intense mixing of fuel and burner air, as the jet momentum decreases. This results in higher solids emissions. The influence of STOR decreases with increasing SAF.

It is concluded that decreasing the STOR at deep-staged conditions has a negative influence on the solids emission. Furthermore it may lead to open flames and should be avoided.

3.6 *Influence of Oil/Gas Ratio when Firing in Dual Fuel Mode*

Oil firing results in higher NO_x emissions than gas firing, due to the fuel bound nitrogen present in fuel oil. Dual fuel firing will therefore have a beneficial effect on the NO_x emissions as compared to 100% liquid fuel firing.

The total NO_x produced cannot simply be calculated from the NO_x produced by gas and oil flames separately. This can be concluded from Figure 13, which shows that there is no linear relation between NO_x and oil fraction. Instead the curves are S-shaped, indicating that interaction between the flames occurs.

At low oil fractions the injection of gas in the primary, substoichiometric zone leads to quick oxygen consumption, thus creating a more highly reducing environment, which is beneficial for fuel NO_x reduction. Furthermore the extra heat release in this zone leads to a quick fuel-bound nitrogen release, which also adds to the NO_x reducing effect.

At high oil fractions the effect of a lower atomiser mixing chamber pressure predominates, resulting in higher NO_x emissions.

The same effect is observed for the solids emissions, which have been given as a function of the oil fraction in Figure 14. Apparently, the effect of a deteriorated atomisation quality resulting from a smaller fuel oil flow is stronger than the smaller solids production due to the use of natural gas. The increase in solids production, expressed in weight percentage of (liquid) fuel intake is clearly demonstrated in Figure 15, showing the NO_x-solids curves for various oil/gas ratios. Figure 14 also reveals that the measured solids emissions exceed the allowable limits considerably. This may be attributed to the relatively low radiant cell temperature of the VTF. Further research is needed in order to determine the effect of a higher radiant cell temperature on the solids- as well as the NO_x emissions.

It is concluded that if dual fuel firing is the normal mode of operation, and if the oil/gas ratio is fairly constant, the atomiser choice should be based on the design liquid fuel flow. If the choice is based on the maximum thermal input at 100% liquid fuel consumption, the solids emission will be higher than necessary, possibly even exceeding the solids emission for 100% liquid fuel firing.

Figure 16 shows the NO_x emission curves for deep-staged dual fuel firing on the three burners tested. From this figure it can be concluded that the NO_x emission limits for new equipment as proposed in the amendment to the draft decree of the Dutch clean air act cannot be met with any of the burners considered. Only for natural gas firing could the NO_x emission be kept below 60 mg/Nm^3 if the thermal input was kept below 7.5 MW.

The projected NO_x emission limits for existing equipment, however, are not exceeded. Therefore the application of the Shell low-NO_x burner should primarily be regarded as a solution for retrofit situations.

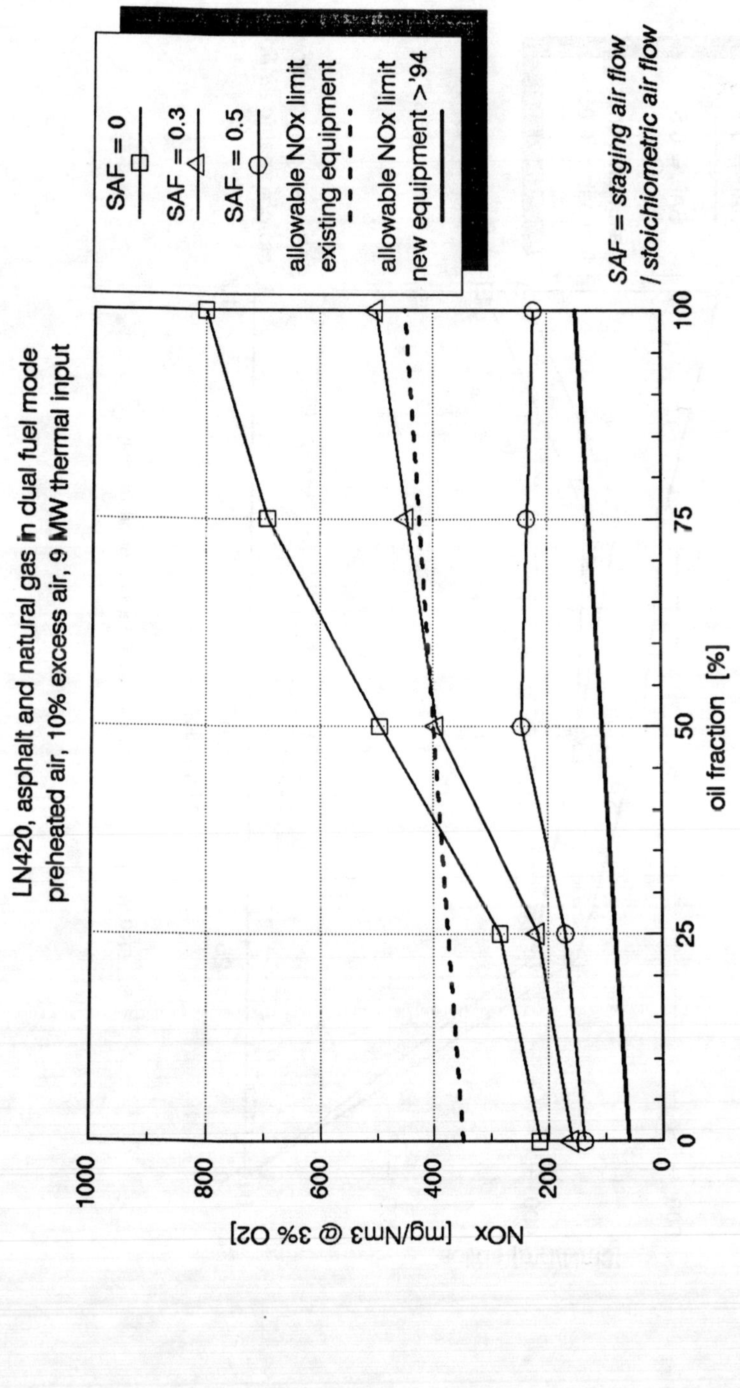

FIGURE 13 NO_x emission as a function of oil fraction.

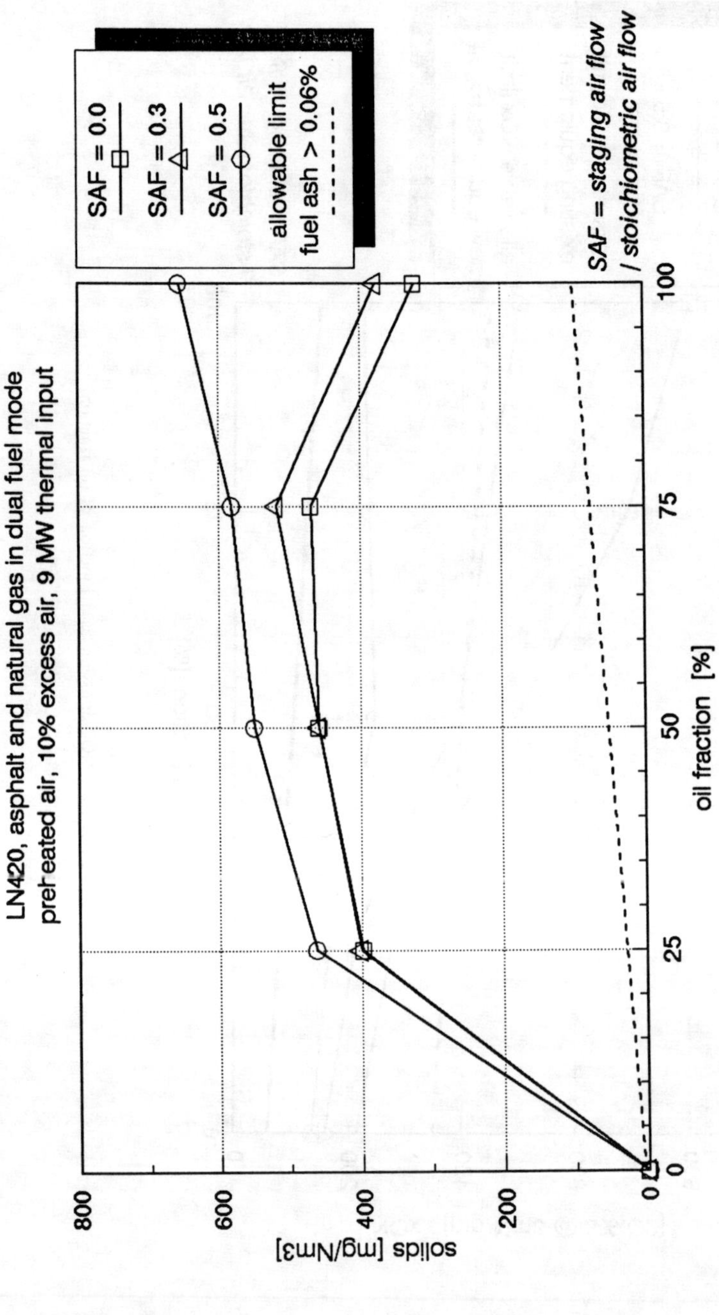

FIGURE 14 Solids emission as a function of oil fraction.

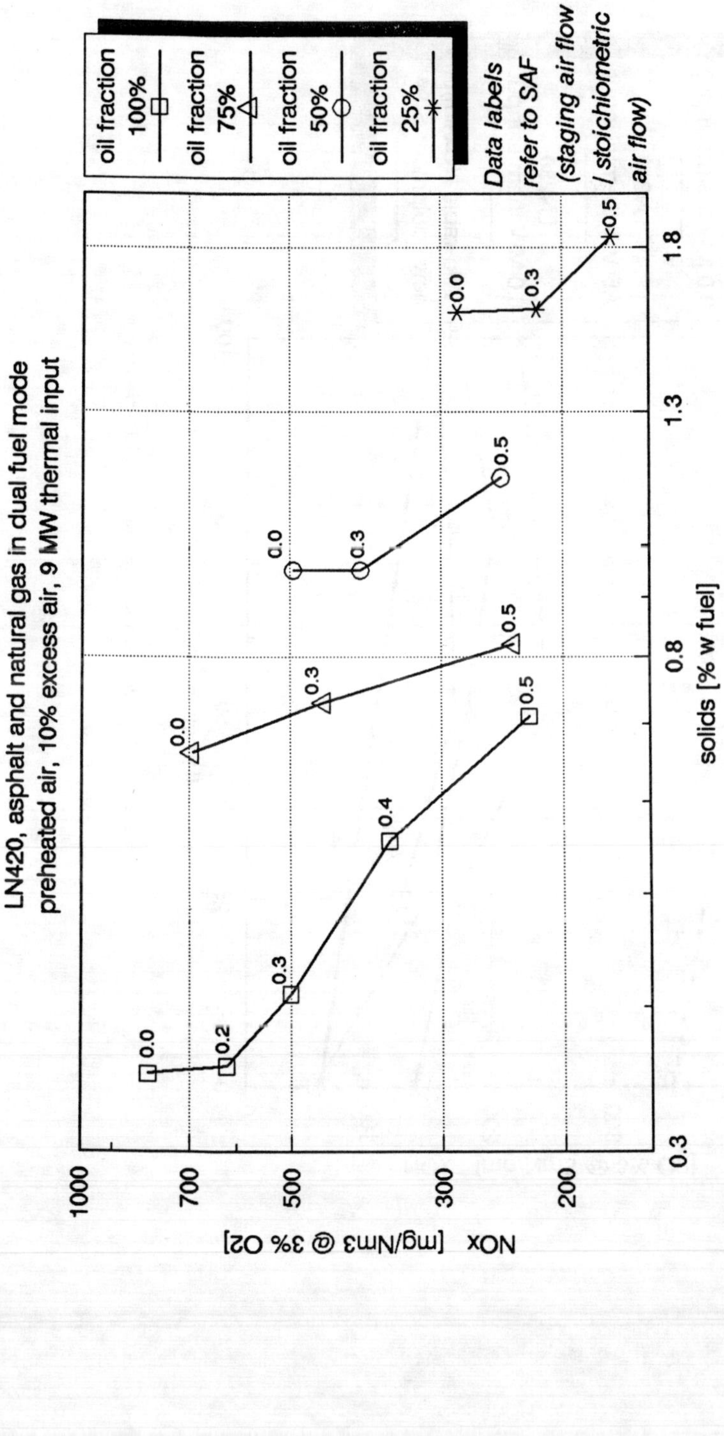

FIGURE 15 Influence of oil fraction on NO_X-solids curve.

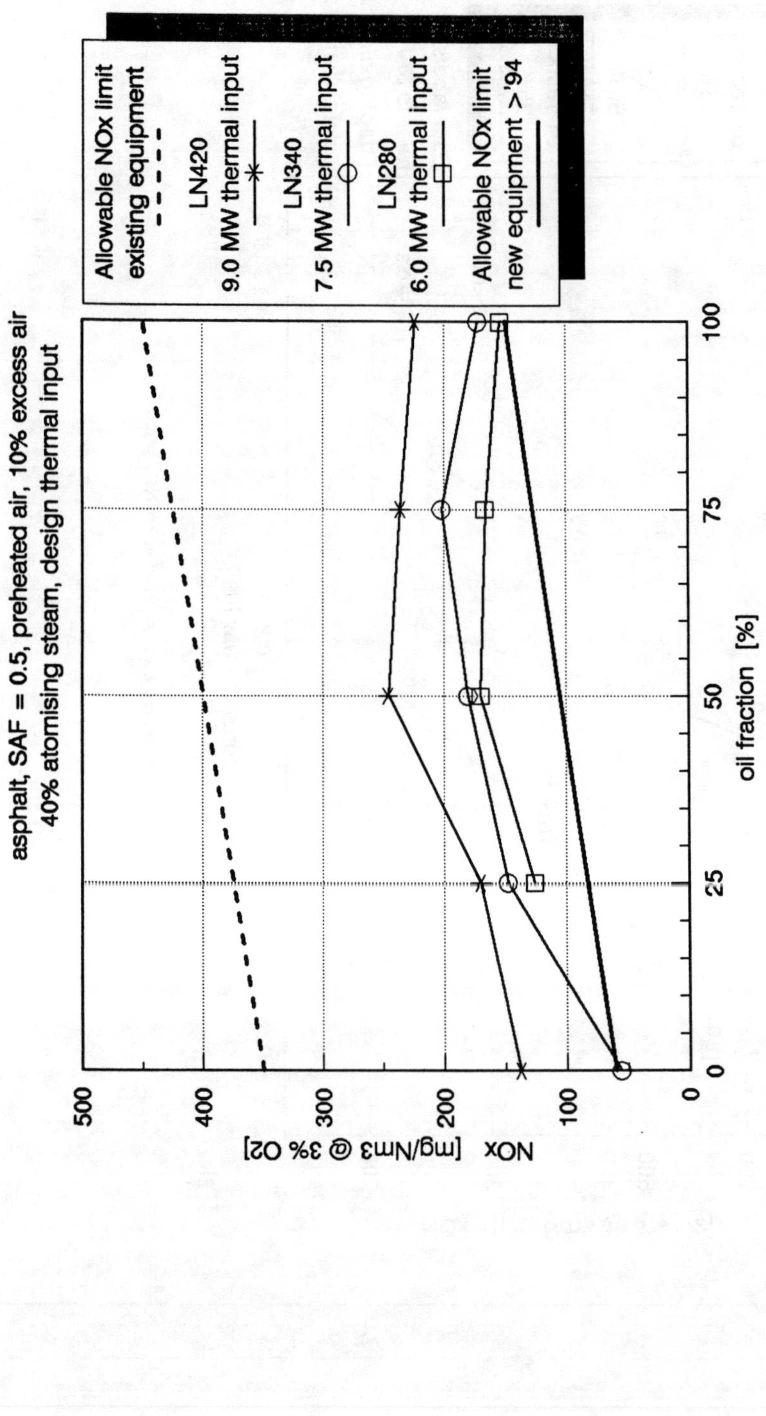

FIGURE 16 NO_x emission as a function of oil fraction.

4. CONCLUSIONS

The test runs performed at the experimental vertical test furnace give rise to the following conclusions:

1. The NO_X emissions for liquid fuel firing can be reduced by a factor of 3, resulting in NO_X emissions between 150 and 225 mg/Nm^3 (@ 3% O2). The maximum practical amount of staging air was found to be 60% of the stoichiometric air requirement, though this may have to be limited to smaller values for the larger burners in order to prevent flame impingement.

2. The solids emissions found for asphalt firing are very high (up to 650 mg/Nm^3). Further investigations are required to determine whether this can be reduced if the radiant cell temperature is increased to more practical figures.

3. Dual fuel firing gives only limited improvement in the solids emissions, as the atomiser mixing chamber pressure drops if the liquid fuel flow is reduced below its design value. The use of another atomiser, suitable for the actual fuel flow during dual fuel firing would improve this situation.

4. The NO_X emissions for gas firing appear to be primarily dependent on the thermal load and amount of staging air and not so much on the burner size. For natural gas firing the NO_X emission is generally reduced by a factor 2.

5. The NO_X emission limits for new equipment as proposed in the amendment to the draft decree of the Dutch clean air act cannot be met with any of the burners considered. Only for natural gas firing can the NO_X emission be kept below 60 mg/Nm^3 if the thermal input is kept below 7.5 MW.

6. The projected NO_X emission limits for existing equipment are not exceeded. Therefore the application of the Shell low-NO_X burner should primarily be considered as a solution for retrofit situations.

REFERENCES

Bowman, C. T., Control of combustion-generated nitrogen oxide emission: technology driven by regulation. *Proceedings Twenty-Fourth Symposium (Intern.) on Combustion* (1992), to appear.

Kelly, F. T., Liddy, J. P., Sutherland, H., Sutton, G. W., Whitehead, D. M. (1989). Clean combustion of heavy fuels. *Erdoel Kohle, Erdgas, Petrochem. Brenst. Chem.* **42,** 408

The Prediction of NO_x Emissions from Spray Combustion

E. HAMPARTSOUMIAN, W. NIMMO, M. POURKASHANIAN, and A. WILLIAMS
Department of Fuel and Energy, The University of Leeds, Leeds, UK
M. MISSAGHI *FLUENT Inc., Hanover, New Hampshire, USA*

Abstract—The study of in-flame chemical processes involving nitrogenous species forms an important part of the understanding of the design parameters which lead to lower NO_x emissions from combustion systems. This paper presents data from experimental and modelling studies on the nitrogenous emissions from an oil fired furnace using staged combustion for the control of NO_x emissions. Measurements of the in-flame NO concentration profiles are compared for the same burner operating in the unstaged and staged mode. The exit gas NO concentration was reduced by 30% during staged combustion with 35% secondary air. A post-processing NO model linked to the output from a commercial computational code was used to predict the rates of formation and concentrations of thermal, fuel and prompt-NO from the experimental system. The predictions showed that NO emission is in good qualitative agreement with the experimentally observed values when the effects of superequilibrium radical formation and turbulence/chemistry interaction are considered.

Key Words: NO_x Modelling, spray Combustion, oil.

INTRODUCTION

Increasing awareness over environmental pollution, coupled with a need to implement more stringent pollution control legislation, has accelerated the development and installation of NO_x reduction and control systems. Low NO_x burners making use of staged combustion are one of the most widely used techniques and they have now been installed in many fossil fuel fired combustion systems.

The present study is concerned with modelling NO_x emissions from a pilot-scale oil fired combustor operating under normal and air-staged conditions. Whilst modelling NO_x formation from coal flames have been well studied (eg. Fiveland and Wessel, 1991) little work has been undertaken on the modelling of NO_x emissions from oil sprays. Since most practical combustors operate as turbulent, non-premixed flames, computer models for the prediction of NO_x concentrations must be capable of describing the turbulence/chemistry interactions. Considerable progress has been made in this area using joint probability density functions (pdf's) (Missaghi et al. 1990). The work described in this paper is concerned with the development of a post-processing model for the prediction of thermal, prompt and fuel NO_x from oil spray combustion which incorporates improved, decoupled NO formation kinetics by the imposition of superequilibrium O atoms and OH radicals concentration.

NO_x *Formation and Model Kinetics*

There are two clear aims to the application of the technique of air-staging to reduce NO_x emissions, namely (i) to achieve more even distribution of fuel and oxygen and hence less intense combustion down the centre of the flame and (ii) to obtain a fuel rich primary (central) zone where high concentrations of HCN and NH_3 are produced. The effect of staging on the NO formation chemistry is considered below together with a brief description of the reaction scheme employed in the post-processing model.

Thermal NO

The formation of thermal NO via the extended Zeldovich mechanism is well known and is represented by:

$$O + N_2 \underset{k_{-1}}{\overset{k_1}{\rightleftharpoons}} NO + N \tag{I}$$

$$N + O_2 \underset{k_{-2}}{\overset{k_2}{\rightleftharpoons}} NO + O \tag{II}$$

$$N + OH \underset{k_{-3}}{\overset{k_3}{\rightleftharpoons}} NO + H \tag{III}$$

Reaction (III) becomes important for rich flames. The rate of thermal NO formation is strongly temperature dependent so that reductions in the local flame temperature and oxygen concentration will be clearly beneficial.

Assuming a quasi-steady state assumption for the concentration of nitrogen atoms, the following expression was used for the rate of thermal NO formation:

$$\frac{d[NO]_{Th}}{d_t} = 2k_1[O][N_2]\left[\frac{1 - [NO]^2/K[O_2][N_2]}{1 + k_{-1}[NO]/k_2[O_2] + k_3[OH]}\right] \tag{1}$$

where the equilibrium constant $K = (k_1/k_{-1})(k_2/k_{-2})$

The rate constants given by Hanson and Salimian (1984) were used for reactions (I-III). However, thermal NO formation is also strongly dependent on the concentration of O and OH atoms. From Baulch et al. (1973) assuming partial equilibrium

$$O_2 + M \underset{k_{-4}}{\overset{k_4}{\rightleftharpoons}} O + O + M \tag{IV}$$

so that

$$[O] = \gamma[K_O[O_2]]^{0.5} \tag{2}$$

The coefficient γ determines the deviation from equilibrium and $K_O = k_4/k_{-4}$. However O and OH radical concentrations influence reactions (I-III), and therefore assuming partial equilibrium values of O and OH species can lead to an error on the predicted NO formation rate. Therefore, in the developed model, allowance for super-equilibrium concentration of radicals in and near the primary reaction zone were made by imposing the data from a detailed investigation given by Dixon-Lewis (1991), on O and OH concentrations in CH_4 air flame.

Fuel-NO

The formation of NO via fuel nitrogen is strongly dependent on flame stoichiometry. Under fuel lean conditions [OH] ≅ [O] ≅ [H]. The reactions (V-VII) below of HCN with O radicals are regarded as rate controlling reactions:

$$HCN + O \rightleftharpoons NCO + H \tag{V}$$

$$HCN + O \rightleftharpoons NH + CO \tag{VI}$$

$$HCN + O \rightleftharpoons CN + OH \tag{VII}$$

followed by

$$NCO + O \rightleftharpoons NO + CO \tag{VIII}$$

$$NH + OH \rightleftharpoons NO + H_2O \tag{IX}$$

$$CN + O \rightleftharpoons N + CO \underset{O_2}{\overset{OH}{=}} NO \tag{X}$$

Under fuel rich conditions [H] > [OH] ≫ [O]. This leads to the formation of hydrogenated nitrogenous species, NH_i, which subsequently produce species such as NH_3 and N_2.

$$HCN + OH \rightarrow NH_2 + CO \tag{XI}$$

$$NH_2 + RH \rightarrow NH_3 + R \tag{XII}$$

$$NH_2 + NO \rightarrow N_2 + H + OII \tag{XIII}$$

There is also a greater probability of hydrocarbon radicals and carbon particles in the fuel rich flame depleting NO by the following reactions:

$$CH + NO \rightarrow HCN \tag{XIV}$$

$$C + NO \rightarrow N_2 + CO \tag{XV}$$

The chemical reaction scheme involved in the formation of fuel-NO is highly complex involving many reactions which have been investigated in detail (e.g. Chen et al, 1989). The correct prediction of fuel-NO formation within the flame requires coupling the NO formation kinetics to the hydrocarbon combustion mechanism in which the number of reactions to be considered can involve long computer processing times. A number of investigators (e.g. Smoot et al, 1985) have consequently adopted a simplified model based on the scheme below:

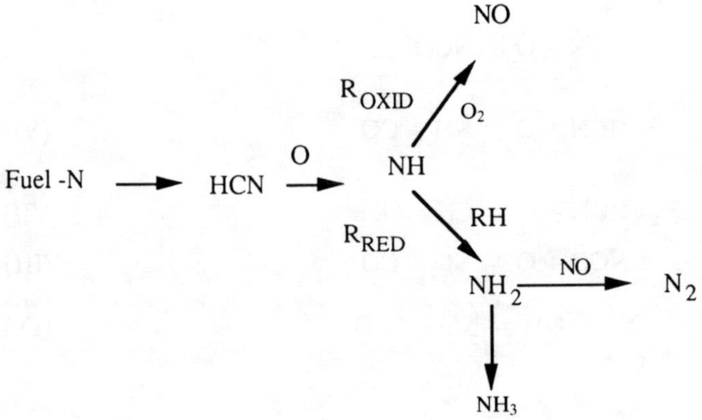

(XVI)

where the rate of formation of fuel-N is determined by the relative rates of oxidation and reduction so that

$$\frac{d[\mathrm{NO}]_F}{dt} = R_{\mathrm{OXID}} - R_{\mathrm{RED}} \quad (3)$$

In the model the fuel-N is assumed to enter the flame at a rate proportional to the rate of droplet evaporation and in turn is fully converted to HCN. The rate of fuel-NO formation is controlled by reaction (VI) and depends on the concentration of the relevant reactants.

$$d[\mathrm{NO}]_{\frac{F}{dt}} \alpha[\mathrm{Fuel} - \mathrm{N}][O_2] \quad (4)$$

Similarly the rate of HCN reduction to nitrogen is given by

$$\frac{d[\mathrm{N}]_2}{dt} \alpha[\mathrm{NO}][R_i] \quad (5)$$

The fuel-NO yield in the model is calculated by solving the time dependent differential equations (4) and (5), which, after simplification, lead to an expression similar to that derived by De Soete (1975).

$$[\mathrm{NO}]_F\mathrm{yield} \cong [A_a[O_2]^{\alpha}/A_b[\mathrm{Fuel} - \mathrm{N}i]]^{0.5} \exp(E_{\mathrm{OX}} - E_{\mathrm{RED}})/2T \quad (6)$$

where A_a and A_b are the pre-exponential factors, E the activation energy, α the reaction order and [Fuel-N] the fuel nitrogen concentration which is time dependent. Values for the pre-exponential factor and activation energy were based on the assumption that all the fuel nitrogen is initially converted to HCN. Equation (6) shows that the rate of NO formation increases with oxygen concentration but reduces with increasing fuel nitrogen content. These instantaneous laminar rates were time-averaged for the turbulent system employed in the post-processor model. However the dependence of fuel-NO formation on temperature is weak in turbulent diffusion oil type flames in comparison to air to fuel ratio and mixing.

Prompt-NO

In the early stages of the flame prompt-NO can form as a result of reaction of molecular nitrogen with fuel fragments such as CH principally by reaction (XVII):

$$CH + N_2 \rightarrow HCN + N \quad \text{(XVII)}$$

$$C_2 + N_2 \rightarrow 2CN \quad \text{(XVII)}$$

If it is assumed that rate of formation of prompt-NO, [NO] p, is given by

$$\frac{d[\mathrm{NO}]_p}{dt} \cong \frac{d[\mathrm{HCN}]}{dt} = A \exp -E_p/RT[\mathrm{CH}][\mathrm{N_2}] \quad (7)$$

then, based on a similar reaction scheme to the formation of fuel-NO from HCN, the formation of prompt-NO under these conditions can be represented by an equation analogous to the empirical equation derived by De Soete (1975).

$$\frac{d[\mathrm{NO}]_p}{dt} = f[\mathrm{O2}]^{\alpha}[\mathrm{N_2}][\mathrm{Fuel}]AT^{\beta}\exp(-E_p/RT) \quad (8)$$

where α is the reaction order and f a correction factor dependent on the air to fuel ratio and fuel type (Missaghi et al. 1990).

Post Processing Model

The kinetics of NO formation and destruction described above are essentially based on data obtained from laboratory laminar premixed flames or shock tube studies where the molecular diffusion conditions are well defined. However, the flow in any practical combustion system is highly turbulent and flames are characterised by temporal fluctuations in temperature, velocity and species concentration. In order to account for these factors, a post processing computer model was developed at Leeds incorporating a commercial computer code. In the model, predictions of the NO level in the combustion chamber are obtained from the solution of an extra partial differential equation for the conservation of NO. Equations (1), (6) and (8) describing the rate of formation of NO are effectively time integrated along the stream line solutions produced by the FLUENT code. In cylindrical-polar co-ordinates, the Eulerian equation for mass fraction of NO (m_{NO}) can be given as:

$$\frac{\partial}{\partial x}\left[\rho\, U m_{NO} - \frac{\gamma\,\mathrm{eff}}{\sigma_{NO,x}}\frac{\partial m_{NO}}{\partial x}\right] + \frac{1}{r}\frac{\partial}{\partial r}\left[r\,\rho\, V m_{NO} - r\frac{\gamma\,\mathrm{eff}}{\sigma_{NO,r}}\frac{\partial m_{NO}}{\partial r}\right] = S_{NO}W_{NO} \quad (9)$$

where U and V are the average velocities in the axial (x) and radial (r) directions respectively, ρ is the mean density, W_{NO} is the molecular weight of NO, γ_{eff} is the effective combined laminar and turbulent viscosity, and $\sigma_{NO,x}$, and $\sigma_{NO,r}$ are the Prandtl numbers in the x and r directions for turbulent diffusion of NO based on eddy diffusivity hypothesis, S_{NO} is the mean turbulent molar rate of NO production per unit volume.

TABLE I
Characteristics of fuel oil and spray size distribution

Composition % mass	Gross CV MJ/kg	Distillation Temperature	Radial Distance mm	SMD (micron)
C 86.8	42.9	°C	0	48
H 10.6	Viscosity mm^2/s	Initial bp 200	20	65
N 0.4	80°C 33.6	5% volume 234	40	70
S 2.1	100°C 17.8	10% volume 254	60	86
Ash 0.06	Density g/cm^3	15% volume 271	70	89
O 0.04*	0.98 @ 15.6°C	Final bp 273	80	90

* by difference

In order to ensure numerically stable solutions, equation (9) was discretised by the upwind-difference technique. The mean turbulent reaction rate was directly determined using probability density function (pdf) techniques described in detail in a previous publication (Hand et al. 1989) and reviewed here briefly. The performance of the turbulent model employed in the post processor was further improved by describing the reacting mixture with two dependent variables to account for the increased number of fluctuating field variables such as reacting species and temperature involved in the prediction of NO (Janika and Kollmann, 1979). The model expresses the mean turbulent reaction rate in terms of the instantaneous rate, w, and a joint pdf of several variables, $(P\nu_1, \nu_2, ..)$ which is given as

$$\bar{w} = \int w(\nu_1, \nu_2 \ldots) P(\nu_1, \nu_2 \ldots) d\nu_1, d\nu_2 \tag{10}$$

In the present study a two variable joint pdf was used, and the mean turbulent rate of production of NO based on equation (7) was calculated from:

$$\bar{S}_{NO} = \int\int S_{NO}(\nu_1, \nu_2) P(\nu_1, \nu_2) d\nu_1, d\nu_2 \tag{11}$$

where $\bar{S}_{NO}$ is the mean turbulent rate of NO production, S_{NO} the instantaneous rate of production and $P(\nu_1, \nu_2)$ the joint pdf of the statistically independent variables ν_1 and ν_2. The variables ν_1 and ν_2 are selected depending on the source of NO. For thermal and fuel-NO, the temperature and mean radical oxygen mass fraction were used. For prompt-NO, the mean CH mass fraction is more critical than oxygen.

EXPERIMENTAL FACILITY

The experimental furnace used in this work was a ceramic lined tunnel furnace 3m long × 0.8m in diameter having a 16 degree smooth tapered refractory quarl at the burner end to reduce dead space and minimise recirculation. This configuration was chosen as being representative of some fire-tube boiler combustion chambers. Throughout the length of the furnace a umber of ports allowed access for water cooled gas sampling and temperature measurement probes. The burner used was a typical utility boiler pressure jet package oil burner (250 kW rating) incorporating radial swirl on the combustion air. The physical and chemical properties of the medium fuel oil used are shown in Table I. The radial variation of the SMD droplet size at a distance of 15cm from the atomiser is also shown in Table I for the 45° angle fuel jet nozzle and typical experimental fuel pressure, fuel preheat and flow rate conditions.

When operating in the unstaged mode, all the combustion air was introduced through the swirler at the burner nozzle. For staged operation, secondary air was introduced through four circular ports equally spaced at a radial distance of 15cm from the fuel inlet jet. The flow of secondary air through each port was equalised and axially directed to be parallel to the burner horizontal axis. Up to 40% of the total combustion air could be introduced via the four secondary air ports.

For NO and NO_2, gaseous samples were obtained using water-cooled quartz lined probes and measured by chemiluminescent analyser. For the purposes of this investigation a standard unstaged and staged flame operating at 16 to 17% excess air overall was investigated. This gave good combustion efficiencies but also a relatively high NO_x concentration in the exhaust gases (ca 280 vpm) in the unstaged form.

Figure 1 shows additional burner details. Grid generation in the model accurately reflected both the burner and furnace geometry (Fig. 2) for staged and unstaged operation. The burner head was modelled with a centrally located fuel injection nozzle together with the swirl plate which separated the main combustion air flow into three parts, namely a primary flow immediately surrounding the nozzle, a secondary swirling flow through the swirl plate itself and finally a tertiary flow through the annulus between the swirl plate and the burner wall. Staging was modelled by further considering the air flow through the staging ports.

RESULTS AND DISCUSSION

Experimental

Experimental data are given for the concentrations of NO in a spray flame of medium fuel oil for 3% oxygen in the flue gases for both staged and unstaged operation. Figure 3 shows the effect of increasing the proportion of secondary air during staged combustion. The NO emission was reduced by around 45% compared to an unstaged flame at 40% secondary air injection through the four burner wall ports. At this level of air staging no reduction in carbon combustion efficiency was apparent and visual flame stability was essentially unchanged. Further increases of secondary air were not considered practical with the existing burner in order to prevent overheating of the burner head which is cooled by the incoming primary air flow. To minimize this risk, a flame with 35% secondary air was chosen for detailed in-flame measurements presented here and for modelling purposes.

The variations in the concentration of NO along the centre axis of the flame are shown in Fig. 4 with radial profiles shown at various axial distances from the burner fuel inlet jet. The data points are the average of at least two separate measurements. The exit NO concentrations were approximately 280 and 185 ppm for unstaged and staged (at 35% secondary air) combustion respectively. The staged flame was also characterised by a central zone much lower in O_2 and higher in CO. Figure 5 shows axial oxygen profiles for the unstaged and staged flames at three different radial distances (0cm, 10cm and 15cm from the central axis). The O_2 concentration profile of the staged flame showed a rapid consumption of O_2 down the central axis reaching extremely low levels ($< 0.5\%$ for 10cm and almost zero for 30cm). In contrast at a radius of 15cm during staging much higher O_2 levels were measured due to the location of one of the secondary air injection ports directly upstream from this point. At a position between the two previously discussed, at 10cm radius, evidence of the recirculation of combustion products can be seen characterised by an increase in O_2 concentration from a low level of $\sim$2.5% to $\sim$5% over an axial distance of 27cm. It is likely that recirculation of combustion products from the flame is partially induced by the relatively high velocity of the secondary air flows and

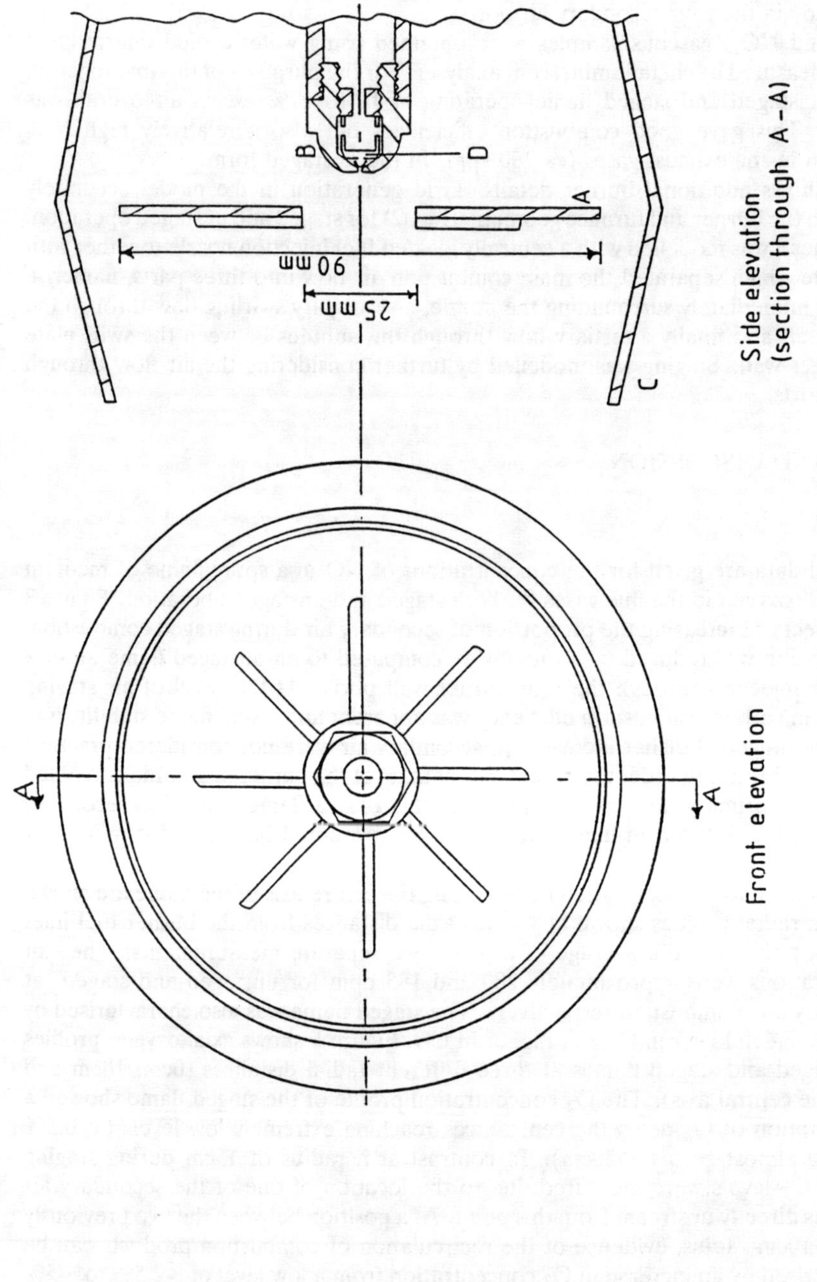

FIGURE 1 Oil burner schematic.

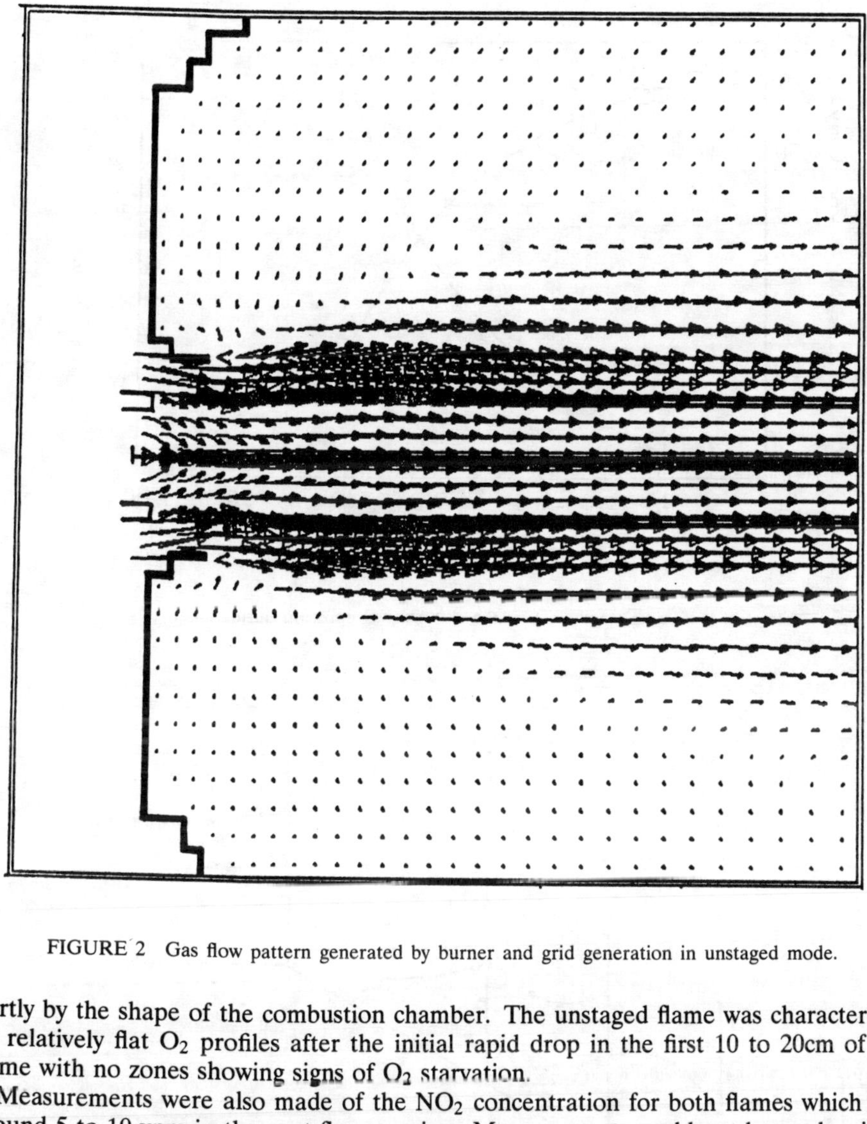

FIGURE 2 Gas flow pattern generated by burner and grid generation in unstaged mode.

partly by the shape of the combustion chamber. The unstaged flame was characterised by relatively flat O_2 profiles after the initial rapid drop in the first 10 to 20cm of the flame with no zones showing signs of O_2 starvation.

Measurements were also made of the NO_2 concentration for both flames which was around 5 to 10 vpm in the post-flame region. Measurements could not be made within the flame due to interference from NH_3 and HCN in the catalytic NO_2 converter of the chemiluminescent analyser.

Computational studies

The combustion model employed in this investigation uses a two-step reaction mechanism. Residual fuel oil is considered to consist of C_5H_{12} as volatile component coupled with coke combustion representing 0.3% asphaltene present in the fuel. The gas phase combustion is considered as

$$C_5H_{12} + 5\frac{1}{2}O_2 \rightarrow 5CO + 6H_2O \qquad (A)$$

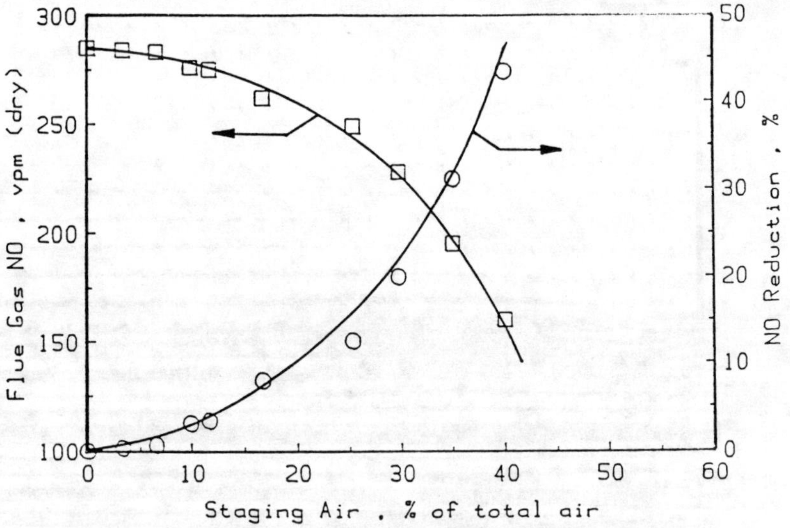

FIGURE 3 Effect of increasing staging air on the exit gas NO emission during air-staging. (3% O_2 in flue gas).

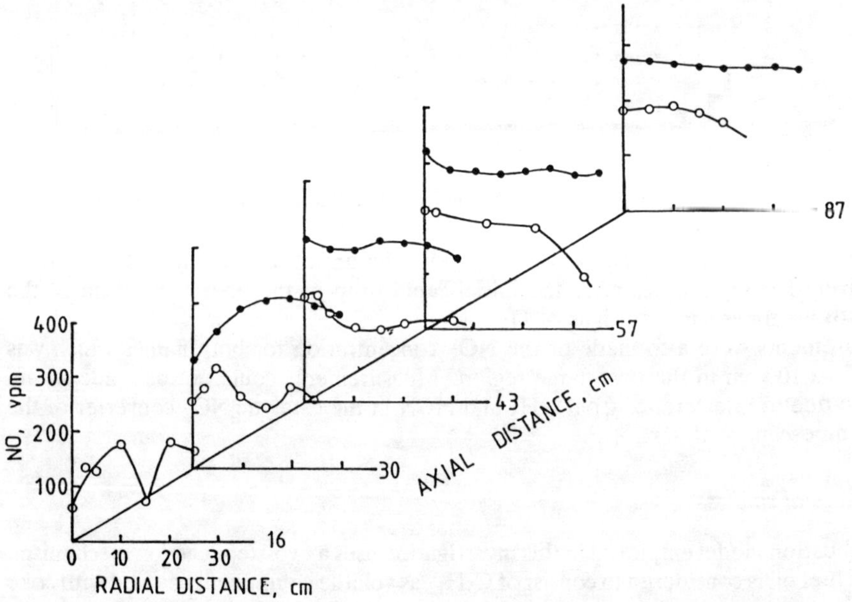

FIGURE 4 Radial in-flame NO concentration profiles (closed symbols; unstaged: open symbols; staged).

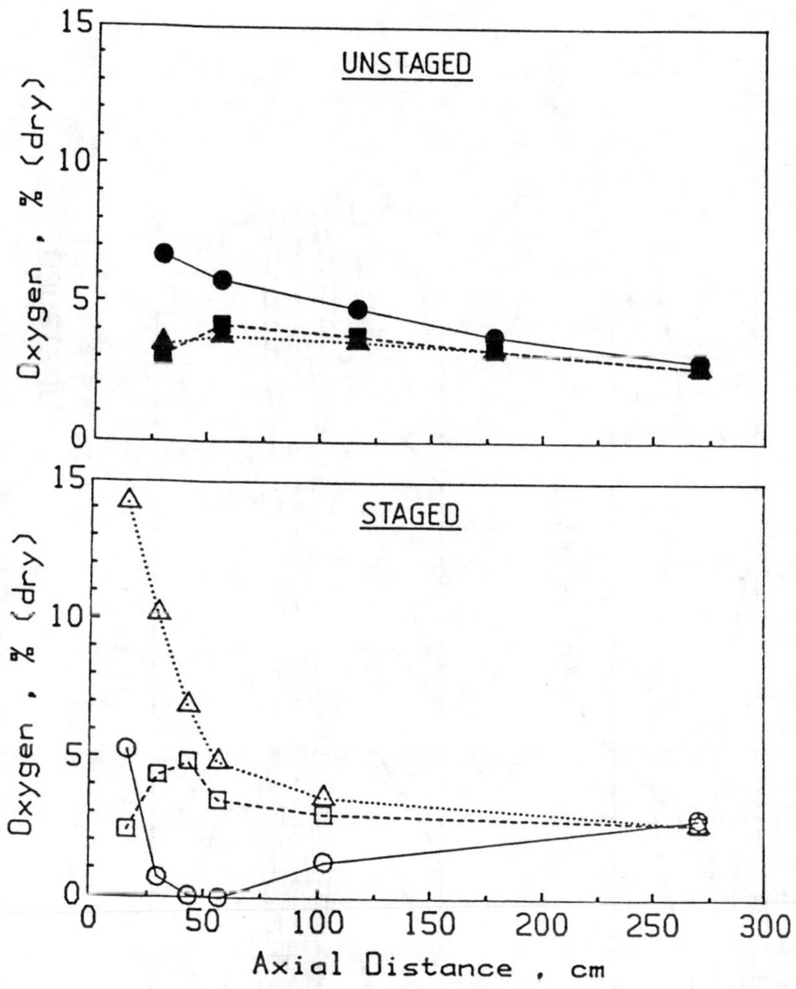

FIGURE 5 Axial in-flame oxygen concentration profiles (closed symbols unstaged, open symbols staged, radius: O and O centre, □ and □ 10cm, Δ and Δ 15cm).

$$CO + \frac{1}{2}O_2 \rightarrow CO_2 \tag{B}$$

By employing a two-step reaction mechanism, the flame temperature is predicted with greater accuracy, particularly for staged combustion which produces significantly higher levels of CO in the near flame region (> 5%) due to the fuel rich operating conditions. If this intermediate species were neglected, the predicted flame temperature and species concentrations would be subject to significant error.

Figure 6a presents the normalized contours of predicted temperature for the unstaged experimental conditions investigated.

The theoretical predictions for thermal prompt-NO and fuel-NO concentrations considered separately are shown in Figures 6b, 7a and 7b, respectively.

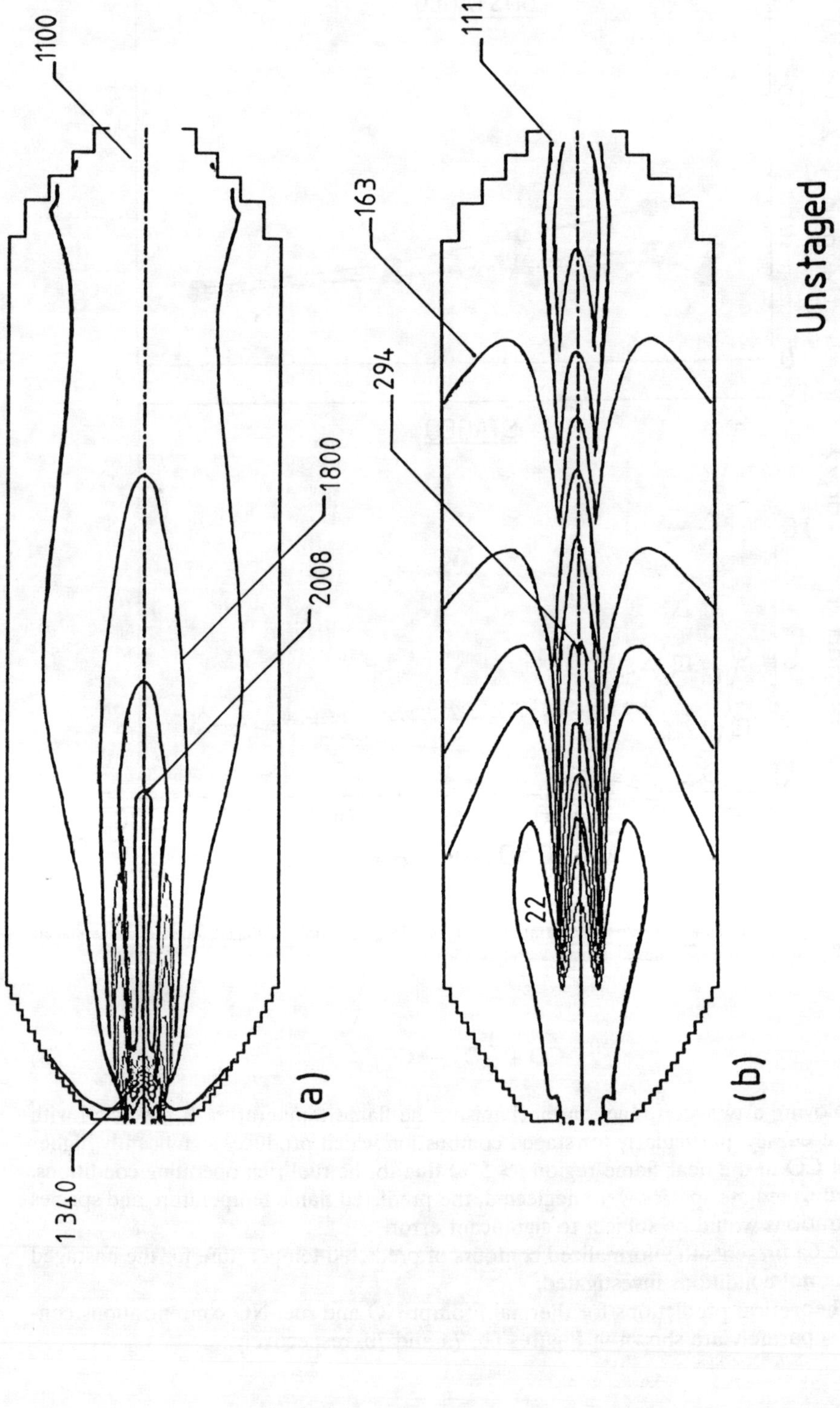

FIGURE 6 (a) Predicted temperature, K, and (b) thermal NO, concentration (ppm) inside the furnace for unstaged combustion.

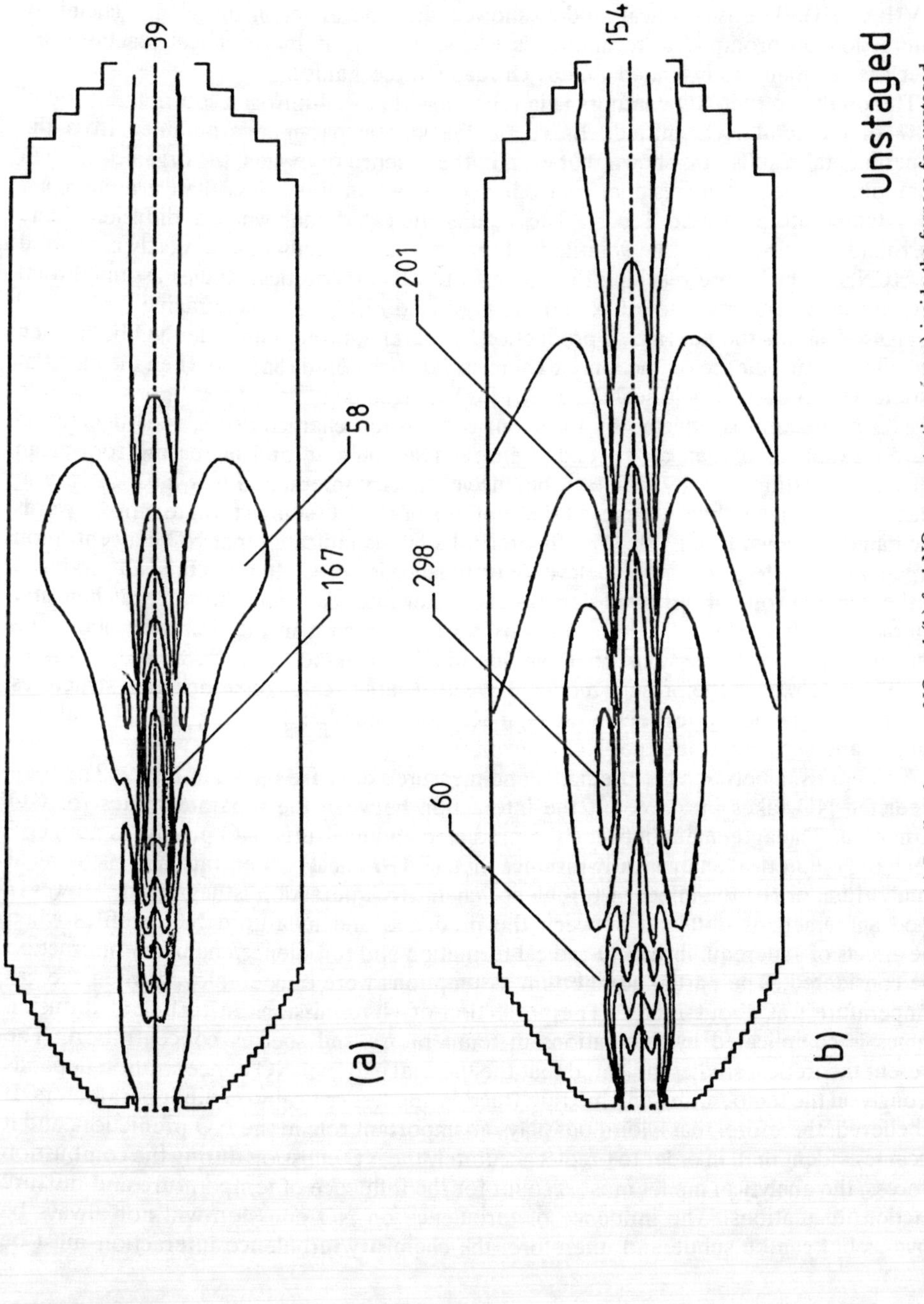

FIGURE 7 (a) Predicted prompt-NO and (b) fuel-NO, concentrations (ppm) inside the furnace for unstaged combustion.

Computer predictions indicate that thermal NO is produced mainly in the initial 160cm from the burner port and it reaches its maximum concentration on 130cm from the burner. The thermal NO concentration follows closely the temperature profile, and in general the temperature and NO concentrations diminish considerably in the region just downstream of the flame front. Figure (7a) indicates that the prompt-NO contribution to the total NO emission is around 10% in fuel-lean conditions. This could be due to the fact that the fuel fragments oxidize more rapidly and therefore are not available to react with N_2 to form CN species that are apparently the precursors to prompt-NO. (Reaction XVII-XVIII). The theoretical studies showed that the effect of turbulence chemistry interaction on prompt-NO formation is less significant as the chemical reactions are much faster than the extended Zeldovich reaction mechanism.

The predicted fuel-NO contours (Fig. 7b) show that fuel-nitrogen contributes around 52% of the total NO emitted. Based on the kinetic parameters obtained from the experimental results, it is shown that even in the absence of oxygen, the organic nitrogen compounds, present in residual fuel oil, decompose in much less than a millisecond at a temperature around 1200K. The results indicated that when a diffusion flame surrounds an oil droplet, the volatile fuel nitrogen compounds will be mostly converted to HCN species before reaching the flame front. The theoretical studies assumed total fuel-nitrogen conversion to HCN and the predicted results are encouraging.

Figure 8 shows the net rate of production of thermal, prompt and fuel-NO both when the effect of turbulence on the kinetic chemistry is considered (8a) and when the effect of turbulence on chemistry is neglected (8b) for unstaged combustion conditions. The rate of NO production is affected by turbulent combustion characteristics, (i.e. decreasing concentration of oxygen atoms and chemical reaction rate and increasing root-mean square fluctuating velocity). The turbulence/chemistry interaction was most intense in the flame region and hence it caused the contours of the NO source term to move towards the flame as shown in Fig. 8b. The theoretical analysis indicated that NO concentration in general tends to go down as the level of turbulence is raised although the NO produced in the reaction zone of turbulent flame is at a faster rate than when turbulence/chemistry interaction was neglected. There are variations of opinion in the literature about the dependency of NO emissions on Reynolds number (Drake et al. 1987 and Peters et al. 1981). However the present model shows that large scale coherent flame structures quench the reaction rates locally at high Reynolds number lowering the thermal NO and, to a lesser extent, the fuel-NO concentration.

A comparison between the predicted and measured data are shown in Fig. 9. The total predicted NO takes into account the interaction between the separate routes for NO formation. The agreement between the predicted and measured NO profiles is not good when the theoretical analysis only assumes that, (i) O-radical concentration is in chemical equilibrium, or (ii) the effect of turbulence/chemistry interaction is neglected. However good agreement is obtained between the predicted and measured NO profiles when the affects of superequilibrium O-radical formation and turbulence/chemistry interaction are considered. The partial equilibrium assumptions were only employed when the gas temperature was about 1200K. The prediction of NO emissions in turbulent diffusion flames is complicated by fluctuations in temperature and species concentration. The present theoretical studies indicated that the thermal and fuel-NO concentration depends strongly on the temperature and mixture fraction and less critically on other parameters. It is believed, therefore, that a joint pdf plays an important role in the NO prediction, and it became evident that, in order to predict accurately the NO emission during the combustion process, the analytical model must account for the influence of temperature and mixture fraction fluctuations. The influence of turbulence on NO emission will not always be obvious but can be subtle and, therefore, the chemistry/turbulence interaction must be

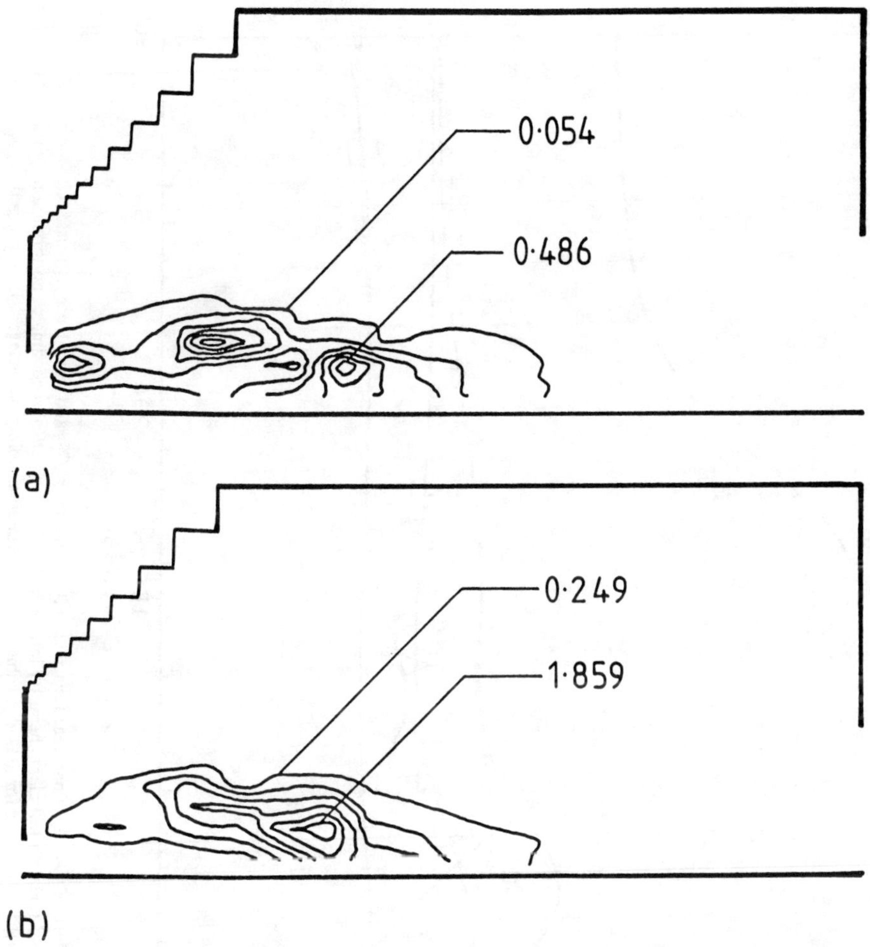

FIGURE 8 The effect of turbulence on theoretical rates of formation of NO (a) with turbulence, (b) laminar ($kg/m^3 \times 10^7$).

carefully designed and independently evaluated and the question of dependency of the NO emission on the Reynolds number should be investigated further. From comparison between predicted and experimental results it is clear that although theoretical prediction is in good agreement with measured values down the stream of the flame, however in the near burner zone there is a difference between predicted and measured values.

Figure 10 shows the predicted temperature (10a) and thermal NO (10b) and Fig. 11 the prompt-NO (11a) and fuel-NO (11b) concentrations for the staged combustion conditions investigated. Compared to unstaged combustion, the flame shape is changed due to the presence of secondary air and the heat transfer to surrounding wall is increased due to the recirculation resulting from the staged air inlet ports. The predicted flame shape is also slightly wider at its maximum diameter and slightly shorter than for unstaged condition, this being confirmed by visual observations. The concentration of thermal NO

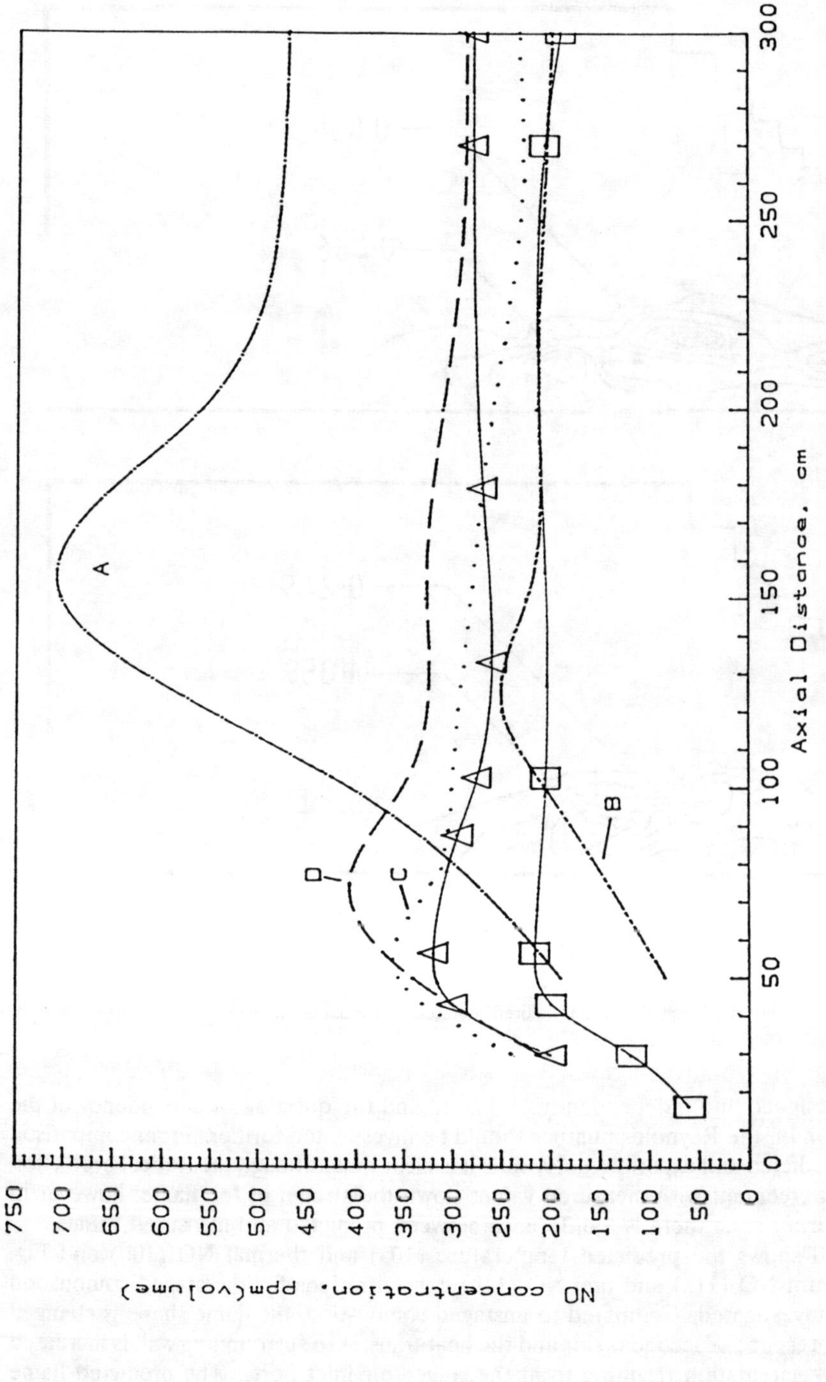

FIGURE 9 The effect of (A) Laminar reaction rates, (B) equilibrium O_2 and turbulence effects. (C) superequilibrium O and OH with turbulent chemistry interaction on the theoretical modelling of no formation (unstaged mode). (D) is as (C) but for staged combustion. Experimental points, □—staged, Δ—unstaged.

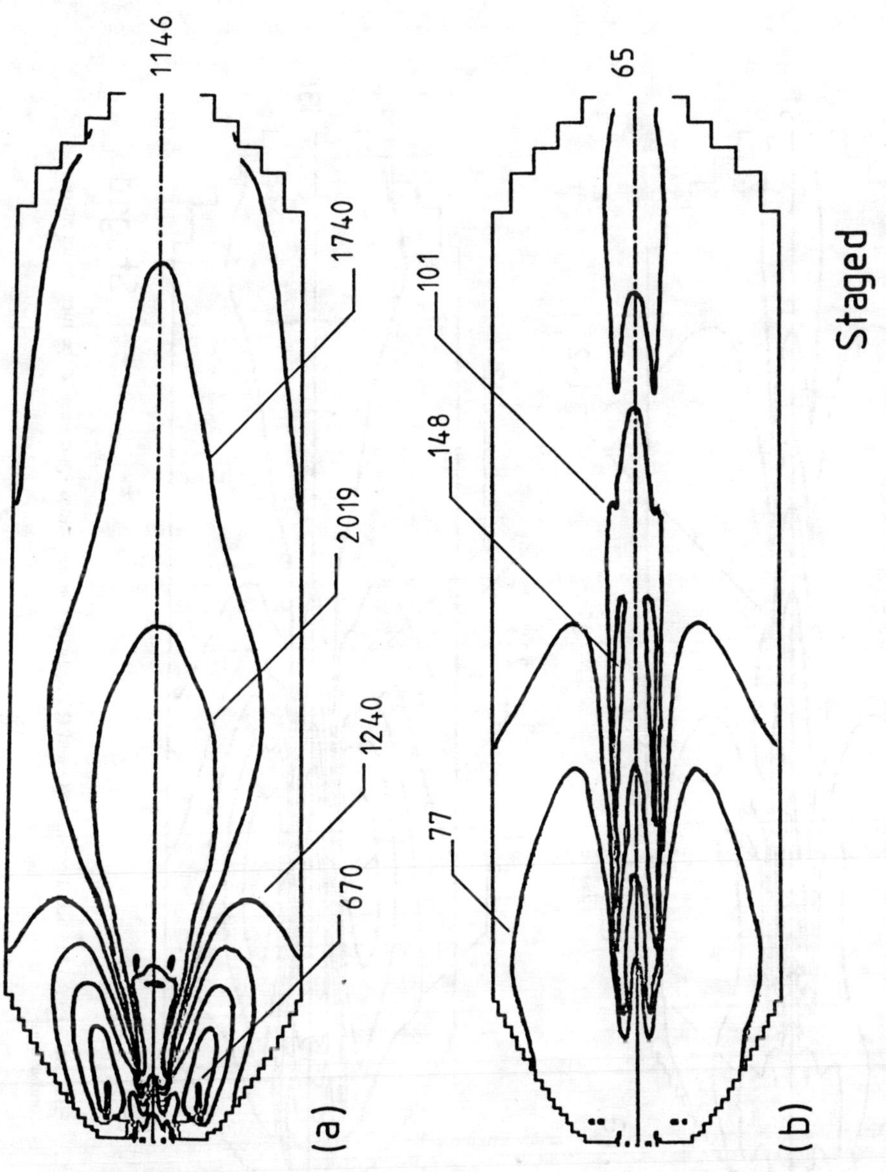

FIGURE 10 (a) Predicted temperature, K, and (b) thermal NO, concentration (ppm) inside the furnace for staged combustion.

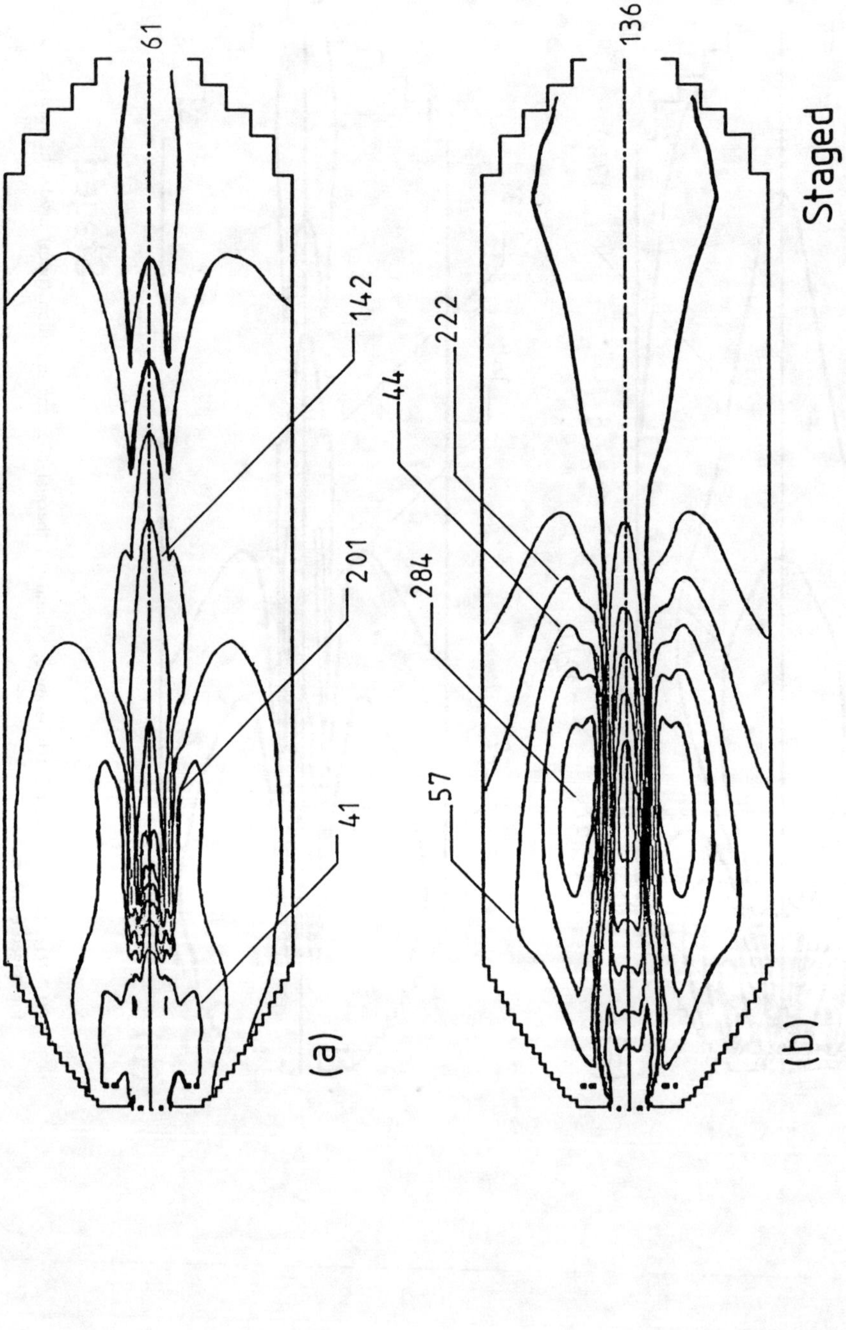

FIGURE 11 (a) Predicted prompt-NO and (b) fuel-NO concentrations (ppm) inside the furnace for staged combustion.

is reduced considerably by around 37% and formed initially at 70cm from the burner. The predicted reduction in the in-flame thermal NO is a major contributing factor to the lower NO levels observed for the staged flame and is due to the extremely low oxygen availability compared to the unstaged flame. It is interesting to note that the predicted contribution from prompt-NO is increased significantly accounting for around 24% of the total NO formed. The contribution from fuel-NO is also increased, however this increase is less than 3%. Overall, the theoretical results indicated a reduction of 18% in the total NO concentration in the flue gas for staged combustion which is in reasonable qualitative agreement with the experimental results.

CONCLUSIONS

The use of air-staging as a means of NO_x emission control in fuel oil combustion, has been studied in terms of experimental measurements of nitrogenous compounds and O_2. A model has been tested in application to certain experimental conditions. It has been shown that:

(a) By staging the combustion to a level of 35% staged air, a 30% reduction in NO_x emission can be achieved.

(b) The implications for the production of NO_x are that air-staging, in the manner described in this paper, generates a fuel-rich centre to the flame where extremely low levels of O_2 can be found where the nitrogenous, fuel-derived intermediates, HCN and NH_3, are highest in concentration resulting in a reduced likelihood of being oxidised to NO.

(c) The location of the formation of prompt-NO, thermal NO and fuel-NO, has been determined theoretically for the experimental conditions of the unstaged flame by the application of commercial computational code and a post processing model.

(d) The accuracy of the predicted level of NO emission is found to be dependent on predicted flame temperature, species concentration and gas velocity.

(e) By comparing experimental results with computational results the importance of incorporating turbulence/chemistry interaction and superequilibrium O and OH concentrations has been shown in modelling practical combustion systems such as that described here.

ACKNOWLEDGEMENTS

The authors would like to thank the UK SERC for a grant in aid of the experimental work presented in this paper.

REFERENCES

Basevich, V. Y. A., (1987). Prog. Energy Combust. Sci., pp. 199.

Baulch, D. L., Drysdall, D. D., Horne, D. G., and Lloyd, A. C. (1973) Evaluated Kinetic Data for High Temperature Reactions, vol. 1, 2, 3, Butterworths.

Chen, S. L., McCarthy, J. M., Clark, W. D., Heap, M. P., Seeker, W. R., and Pershing, D. W. (1987). *Twenty-First Symposium (International) on Combustion, The Combustion Institute,* pp. 1159.

De Soete, G. G. (1975). *Fifteenth Symposium (International) on Combustion, The Combustion Institute,* pp. 1093.

Dixon-Lewis, G. (1991). Private communications.

Drake, M. C., Correa, S. M., Pitz, R. W., Shyy, W., and Fennimore, C. P. (1987). *Combustion and Flame,* **69,** pp.347.

Fiveland, W. A., and Wessel, R. A. (1991). A model for predicting formation and reduction of NO_x in three-dimensional furnaces burning pulverised fuel. *J.Inst.E.* **64,** pp. 458.

Hand, G., Missaghi, M., Pourkashanian, M., and Williams, A., (1989). Proceedings of the 9th Members Conference, 2, IFRF Doc. No. K21/g/30.

Hanson, R. K., and Salimian, S. (1984). Survey of Rate Constants in N/H/O System, pp. 361, *Combustion Chemistry* (Ed. W. C. Gardiner).

Jawika, J., and Kollmann, W. (1979). *Seventeenth Symposium (International) on Combustion,* pp. 421, the Combustion Institute.

Jones, W. P. (1975). *Combustion Science and Technology,* **10,** pp. 93.

Kolb, T., Jansohn, P., and Leuckel, W. (1989). *Twenty-Second Symposium (International) on Combustion, The Combustion Institute,* pp. 1193.

Lyon, R. K. (1975). U.S. Patent No. 3,900,554.

Missaghi, M., Pourkashanian, M., Williams, A., and Yap, L. (1990), Proc. of American Flame Foundation, USA.

Muzio, L. J., Arand, J. K., and Teixeira, D. P. (1977). *Sixteenth Symposium (International) on Combustion, The Combustion Institute,* pp. 199.

Peters, N., and Donnerhask, S. (1981). *Eighteenth Symposium (International) on Combustion, The Combustion Institute,* Pittsburgh, pp. 33.

Smoot, L. D., and Smith, P. J. (1985). *Coal Combustion and Gasification,* Plenum Press, New York.

4.

GAS COMBUSTION

Comparison of Measurements and Predictions of Flame Structure and Thermal NO_x, in a Swirling, Natural Gas Diffusion Flame

R. D. BOARDMAN[1], C. N. EATOUGH, G. J. GERMANE, and L. D. SMOOT
Departments of Mechanical and Chemical Engineering, Brigham Young University, Provo, UT 84602

Abstract—A combined thermal and fuel NO_X submodel has recently been added to a generalized, 2-dimensional pulverized coal gasification and combustion model (PCGC-2). This model is applicable to reacting and non-reacting gaseous and particle-laden flows. The thermal NO model is based on the extended Zel'dovich mechanism. To perform an evaluation of the NO_X submodel, combustion measurements of gas velocities, temperatures, and species concentrations were made in a laboratory-scale, experimental reactor with a 150 kW natural gas flame at an equivalence ratio of 1.05 and a secondary-air swirl number of 1.5. Combustion measurements of velocities and major species concentrations show generally good agreement with predicted values. Gas temperature measurements closely match predictions in the recovery region but fail to show predicted high temperature in the annular region. This study provides an evaluation of a comprehensive combustion model and the NO_X submodel which can be useful as a design tool to provide pollutant formation trends in applied systems as combustion parameters are varied.

Key Words: Combustion measurements, Combustion modeling, Natural Gas, NO_X.

INTRODUCTION

Background

The world's basic and advanced-technology industries are highly reliant upon an adequate supply of high-quality energy, the production of which depends upon combustion technology. The international competitiveness of these industries hinges on their ability to more efficiently and cleanly use, through advanced combustion technology, available fuel resources. The design of advanced combustion systems and processes for the clean use of these fuels can be greatly enhanced by the utilization of verified predictive and interpretive combustion models.

The development and inclusion of a combined thermal and fuel NO predictive method into a generalized pulverized coal combustion and gasification model provides a tool for predicting NO_X formation during gaseous and/or heterogeneous combustion (Boardman, 1990). The NO_X submodel couples the homogeneous reactions with turbulence interactions. This feature allows the submodel to be used for predicting the impact of combustor design and operating parameters. For example, the formation of NO_X in practical combustors can be reduced significantly by using low-NO_X burners. The amount of NO_X formed by a particular burner depends on the local mixing pattern and equivalence ratio. In order for the NO_X submodel to be used as a tool to aid in burner design, it must be capable of predicting the effects of equivalence ratio and swirl number, the primary variables manipulated in developing low-NO_X burners.

Development and evaluation of a comprehensive combustion model requires accurate combustion measurements and control of system parameters during testing. Control of various combustion parameters is essential in determining their effects on flame dynamics and chemistry.

---

[1] Currently at Westinghouse, Idaho Nuclear Company, Idaho Falls, ID.

Because of the unavailability of adequate combustion measurements for model evaluation, a new, highly controllable, experimental, laboratory-scale Controlled Profile Reactor (CPR) has been designed and constructed (see Figure 1) (Eatough, 1991). The CPR was designed to obtain complete combustion data sets for combustion code validation as well as to evaluate process strategies and develop diagnostic instrumentation. The facility has unique features which fill many identified needs to adequately perform advanced combustion research and provide reliable combustion measurements for code evaluation. These features include the ability to carefully control and define inlet and boundary conditions, as well as to specify wall temperature and control of flame heat loss, which are important controllable parameters in laboratory-scale facilities.

Objectives

The objectives of this study were to make laboratory furnace measurements under carefully controlled natural gas combustion conditions in the CPR and to evaluate a thermal NO_x submodel, in combination with PCGC-2, by comparison with these data. The tests included the acquisition of spatially resolved gas species, temperature and velocity measurements, while the comparisons of measurements to code predictions included interpretation of the results in terms of experimental or model deficiencies.

PCGC-2 MODEL DESCRIPTION

Basic Combustion Model

The combustion model used to predict the turbulent flame structure of the CPR reactor was PCGC-2 ("Pulverized Coal Gasification and Combustion-2 dimensional"), a computer code applicable to axi-symmetric combustors. PCGC-2 solves a set of general conservation equations for gaseous or gas/particle-laden (coal) reacting flows, according to specified initial and boundary conditions for mass, momentum, and energy. Auxiliary equations are used to describe the chemical and physical processes while the two equation k-ϵ turbulence model is used to predict turbulence quantities. Simulations provide predictions for the mean gas field properties and local particle properties for steady-state, turbulent flames. A complete description of PCGC-2 has been published by Smoot and Smith (1985).

The solution of the major field variables for a combustion case without coal particles is based on the "fast" chemistry assumption where all gaseous reactions go to equilibrium locally, as soon as the reactants are released to the gas phase and mixed at a molecular level. Specification of the local instantaneous pressure, enthalpy, and elemental composition is sufficient to determine the instantaneous equilibrium composition, temperature, and fluid density of the gas mixture. For turbulent flames, it can be reasonably assumed that all species have equal turbulent diffusion coefficients. Thus the local elemental composition is determined throughout the system by tracking the mixing of primary gas, secondary gas, and coal off-gas. Two conserved scalars, f (primary gas mixture fraction) and η (coal off-gas mixture fraction), are used in PCGC-2 to track this mixing process. The energy equation is then solved to determine the local enthalpy (h). With knowledge of the instantaneous values of f, η, and h, the instantaneous values of species concentrations and temperature are uniquely determined.

Conservation equations, which depend on turbulence quantities, are solved to obtain both the mean and variance of the mixture fraction variables. Then, given an assumed shape for the statistical probability of f and η, mean gas phase properties (*i.e.*, density, temperature and species concentrations) can be obtained by convolving over the probability of the mixture fraction variables. The statistical distributions of f and η are characterized by the degree of turbulence at each spatial point throughout the reactor. This approach requires that all variables, including enthalpy, have statistical similitude with the mixture fractions. Statistical similarity exists between any property and the mixture fraction when it can be shown that the two are uniquely correlated.

All of the gas-phase conservation equations are Eulerian, steady-state, second-order, non-linear, elliptical, partial differential equations. Particle trajectories are calculated using a Lagrangian framework which provides source terms (mass, momentum, and energy) for the gas equations. The resulting model describes the response of a coal particle to its thermal, chemical, and physical environment.

The measurements and predictions of the current study were for combustion of natural gas diffusion flames. This reduced the model predictions to a Eulerian reference frame except for predictions that were made to simulate the effects of soot formation on the thermal environment in the reactor by seeding the inlet flow with inert, non-reacting, microscopic soot particles.

NO_x Submodel

The foundation for the NO_x submodel was laid by Smith *et al.,* (1982). The theory for the combined thermal and fuel NO model is discussed in this paper; however, the NO_x predictions presented herein are limited to natural gas flames where fuel nitrogen is not present. Comparisons between predicted and measured coal combustion and gasification cases, where the fuel NO mechanism typically dominates NO formation, have been previously published (Hill *et al.,* 1984, Smith *et al.,* 1986, Boardman and Smoot, 1988). Joint thermal NO and fuel NO predictions of coal combustors have also been made and have been reported elsewhere (Solomon et al., 1990).

In comparison to the moderately fast species reaction rates of light hydrocarbons, nitrogen fixation reactions are slower. The instantaneous reaction assumption is not valid for the formation of many minor species in coal flames, including the formation and destruction of nitrogen oxides. In order to successfully predict the slower formation of trace species, the reaction kinetics and chemistry/turbulence interactions must be adequately described. Fortunately, the formation of trace species does not affect the fast fuel-oxidation kinetics or temperature significantly. This allows the NO_x submodel calculations to be completed following prediction of the steady-state, reacting flow structure.

Nitrogen oxide formation mechanisms. Most of the NO_x emitted to the atmosphere by combustion systems is in the form of nitric oxide (NO), with only a small fraction appearing as nitrogen dioxide (NO_2). Other nitrogenous pollutants generated during combustion include ammonia, hydrogen cyanide, amine compounds, and nitrous oxide (N_2O). Because N_2O and NO_2 typically only exist as transient species due to the simultaneous presence of radical species, they are not thought to be significant NO_x species in oil and pulverized coal flames nor in unquenched effluent of most gas combustors. This leaves NO as the significant nitrogen oxide species emitted from most combustors.

It is generally agreed that the reaction network for thermal NO is correctly described by the extended Zel'dovich mechanism based on research originally carried out by Zel'dovich (1947).

$$N_2 + O \leftrightarrow NO + N \tag{1}$$

$$N + O_2 \leftrightarrow NO + O \tag{2}$$

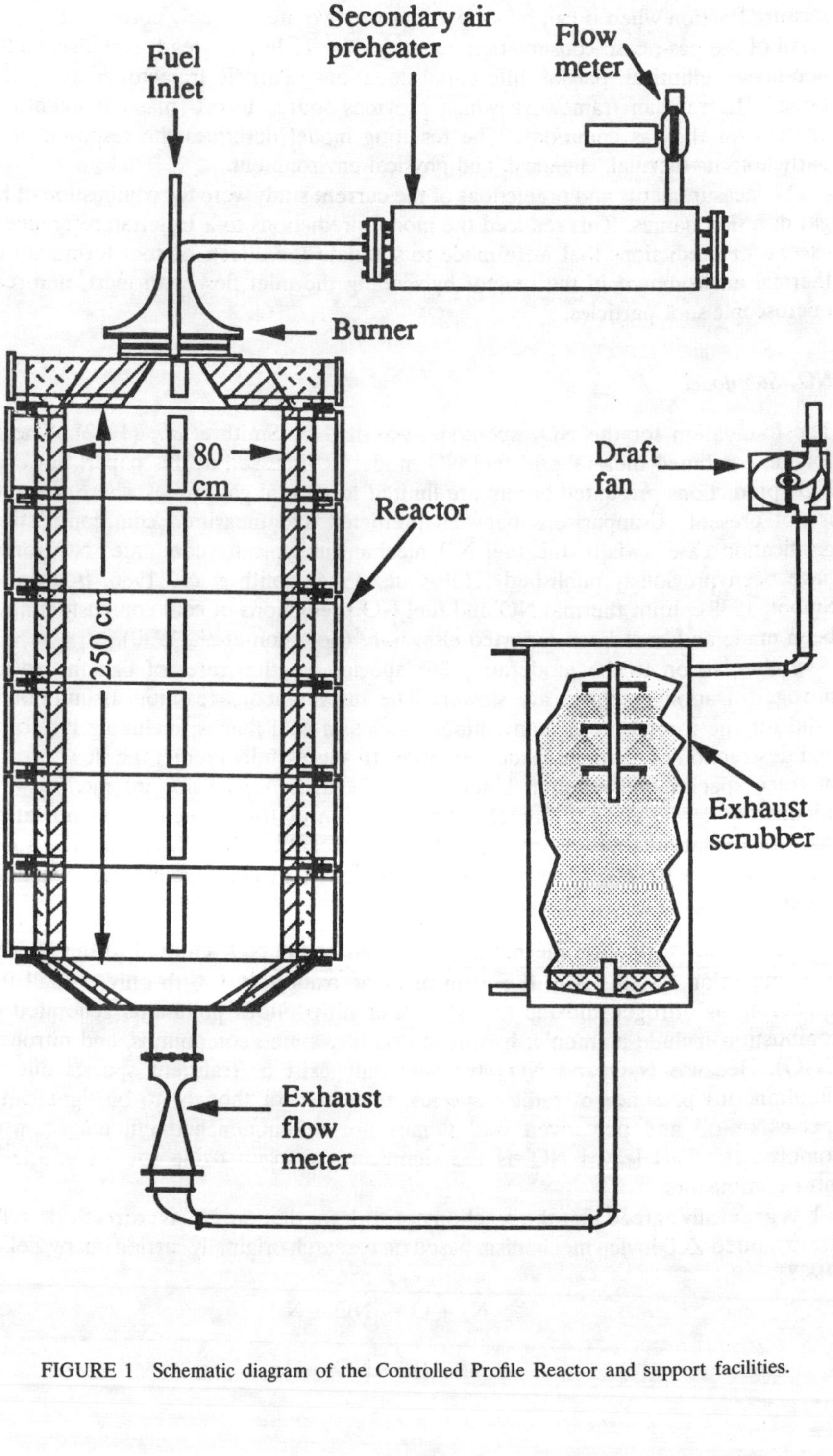

FIGURE 1 Schematic diagram of the Controlled Profile Reactor and support facilities.

In fuel-rich environments (stoichiometric ratios of air to fuel < 1.0), at least one additional step should be included in this mechanism (Lavoie *et al.,* 1970).

$$N + OH \leftrightarrow NO + H \tag{3}$$

The general rate expression, which includes the forward and reverse reactions in the three-step extended Zel'dovich mechanism, is:

$$\frac{d[NO]}{dt} = 2[O]\left(\frac{k_1[N_2] - \frac{k_{-1}k_{-2}[NO]^2}{k_2[O_2]}}{1 + \frac{k_{-1}[NO]}{k_2[O_2]+k_3[OH]}}\right) gmole\ cm^{-3}s^{-1} \tag{4}$$

Rate parameters for Reactions 1-3 of the form, $k = AT^b \exp(-E/RT)$, were obtained from Miller and Bowman (1989).

The thermal NO rate equation is coupled to the fuel oxidation process through competition for the oxygen atom. In fuel-lean, secondary combustion zones, oxygen atoms have been assumed to be in equilibrium with molecular oxygen.

$$[O] = K_{eq}^{1/2}[O_2]^{1/2} \tag{5}$$

Since the measurements in this study were limited to near stoichiometric fuel-air equivalence ratios, Expression 5 was used to predict atomic oxygen concentrations (Boardman, 1990). A separate partial equilibrium expression, derived from the assumption of reaction equilibrium among fast-shuffle oxy-hydrogen reactions was also investigated but was found to overpredict thermal NO concentrations for these cases. All of the predictions presented in this paper assumed partial equilibrium between atomic and molecular oxygen was valid.

Basic NO_x submodel equations. The governing equation used in the NO_x submodel for calculating reaction species concentrations throughout the combusting flow field is species continuity.

$$\bar{W}_i = \bar{\rho}\tilde{u}\left(\frac{\partial \tilde{Y}_i}{\partial x}\right) + \bar{\rho}\tilde{v}\left(\frac{\partial \tilde{Y}_i}{\partial r}\right) - \left(\frac{\partial}{\partial x}\right)\left(\tilde{D}y\frac{\partial \tilde{Y}_i}{\partial x}\right) - \frac{1}{r}\left(\frac{\partial}{\partial r}\right)\left(r\tilde{D}y\frac{\partial \tilde{Y}_i}{\partial r}\right) \tag{6}$$

$\bar{W}_i$ is the overall mean chemical reaction source or sink term calculated by summing all individual (global or elementary-step) mean reaction rates, ω_j involving formation and destruction of species i.

$$\bar{W}_i = \sum_j \bar{\omega}(X_i, X_k, T)_j \tag{7}$$

Mean reaction rates for various species are obtained by convolving instantaneous rates over the probability of primary-stream and coal off-gas mixture fraction progress variables.

$$\bar{\omega}_i = \bar{\rho}\int_f\int_\eta (\omega_i(f,\eta,h)/\rho(f,\eta,h))\tilde{P}(f)\tilde{P}(\eta)d\eta df \tag{8}$$

This integration accounts for the effects of turbulence interactions on the kinetics. Several assumptions are made when using Equation 8 and, although none of these assumptions

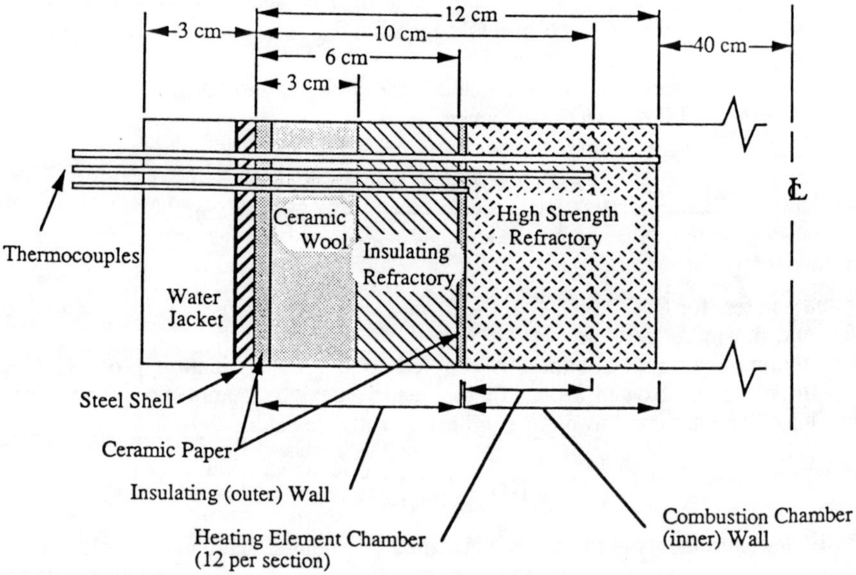

FIGURE 2 Controlled Profile Reactor combustion chamber wall design.

have been exactly proved, experimental evidence suggests that they are acceptable. The main assumption required is that the instantaneous reaction rates are uniquely correlated to the mixture fractions, f and η. A clipped Gaussian distribution with intermittency for pure primary-stream, pure secondary-stream, and pure coal off-gas material is assumed to be adequate for describing the statistics of compositional fluctuations. Fluctuations in the residual enthalpy (enthalpy not associated with chemical enthalpy) are also assumed to be small and negligible.

EXPERIMENTAL FACILITY

Configuration

The CPR is cylindrical and down-fired with an internal diameter of 80 cm and a length of 250 cm (Figure 1) (Eatough, 1991). Swirl is imparted to the secondary air stream using a moveable-block swirl generator. The design is an improvement upon previous reactors in that it is highly instrumented so that more extensive and more accurate combustion measurements as well as flow and thermal boundary control and measurement are possible. The reactor body is sectional with an interchangeable top section so that different burner types may be used. The reactor provides ample optical accessibility for the use of non-intrusive laser diagnostic systems, visual observations, radiometer access and photography. Accurate combustion zone measurements such as radiation, species concentration, temperatures and velocities can be made. Wall heat flux and temperature are measured with thermocouple systems and controlled by means of electrical heaters inbedded in the reactor walls (Butler, 1988). Both inlet and outlet flow rates are measured for use in model computations as well as calculating mass balances.

The multi-layered furnace wall system is composed, from inside out, of a high density refractory layer, electric wall heaters, a low density refractory, a refractory fiber blanket, an annular space for cooling water, and the structural steel shell. Final design of the combustion chamber wall is shown in Figure 2.

Instrumentation

In order to manage the large number of signals requiring measurement for the purpose of data collection, system control, material and thermal balances, and operating condition specification, a computerized data acquisition and process control system was developed and installed on the CPR. A suction pyrometer was used to measureL gas temperatures and a five-hole pitot tube was used for measuring gas velocities. A water-quenched sample probe was utilized to collect gas samples for species determination.

Combustion gas samples were removed from the combustion zone and water-quenched with a stainless steel, water-cooled probe. The combustion sample was subsequently dried using an ice bath. The resulting dry gas sample was analyzed for CO_2, CO, CH_4, O_2, and N_2 with a gas chromatograph. NO_x, O_2, CO and CO_2 were also measured using on-line instrumentation. NH_3 and HCN were collected in the sample gas condensate and measured using ion-selective electrodes.

Reactor Characterization

CPR characterization tests, including flame symmetry determination and mass and thermal balances, were performed to demonstrate to operational ability of the reactor. Flame symmetry was demonstrated in the CPR by measuring and comparing radial flame temperature profiles in four quadrants of the reactor. Mass and energy balances across the reactor were performed for a 150 kW natural gas flame. Mass balance closures within about 2% were typical while energy balance closures of around 5% were typical.

COMBUSTION MEASUREMENTS

Velocities

Comparisons of predicted and measured axial, radial, and tangential velocity profiles are shown in Figure 3a-c. Negative axial velocities indicate regions of recirculation. Axial velocity measurements of Figure 3a show a large central recirculation zone extending from about the 0.13 m axial position to about 0.95 m. This recirculation zone has the effect of stabilizing the flame by enhancing mixing and recirculating hot combustion gases toward the burner, preheating the incoming fuel and air. An outer recirculation zone is located in the annular corner created by the reactor wall and top.

Axial velocity measurements show generally good agreement with model predictions outside the central recirculation zone. Within the central recirculation zone, there is some disparity between measured and predicted values. The predicted axial velocity profile is skewed slightly toward the wall in accordance with the data trend which is indicative of the higher predicted radial velocities immediately beyond the quarl expansion region (Figure 3b). PCGC-2 model predictions show a more rapid jet expansion in this region than measurements do. The prediction of negative centerline axial velocities and positive peak velocities higher than measured is the result of a larger and more intense predicted central recirculation zone. The inability of the code, with a linear k-ϵ model, to correctly predict velocity component decay and complex recirculation zone details in swirling flow has previously been reported by Sloan *et al.* (1986). Beyond the initial jet expansion

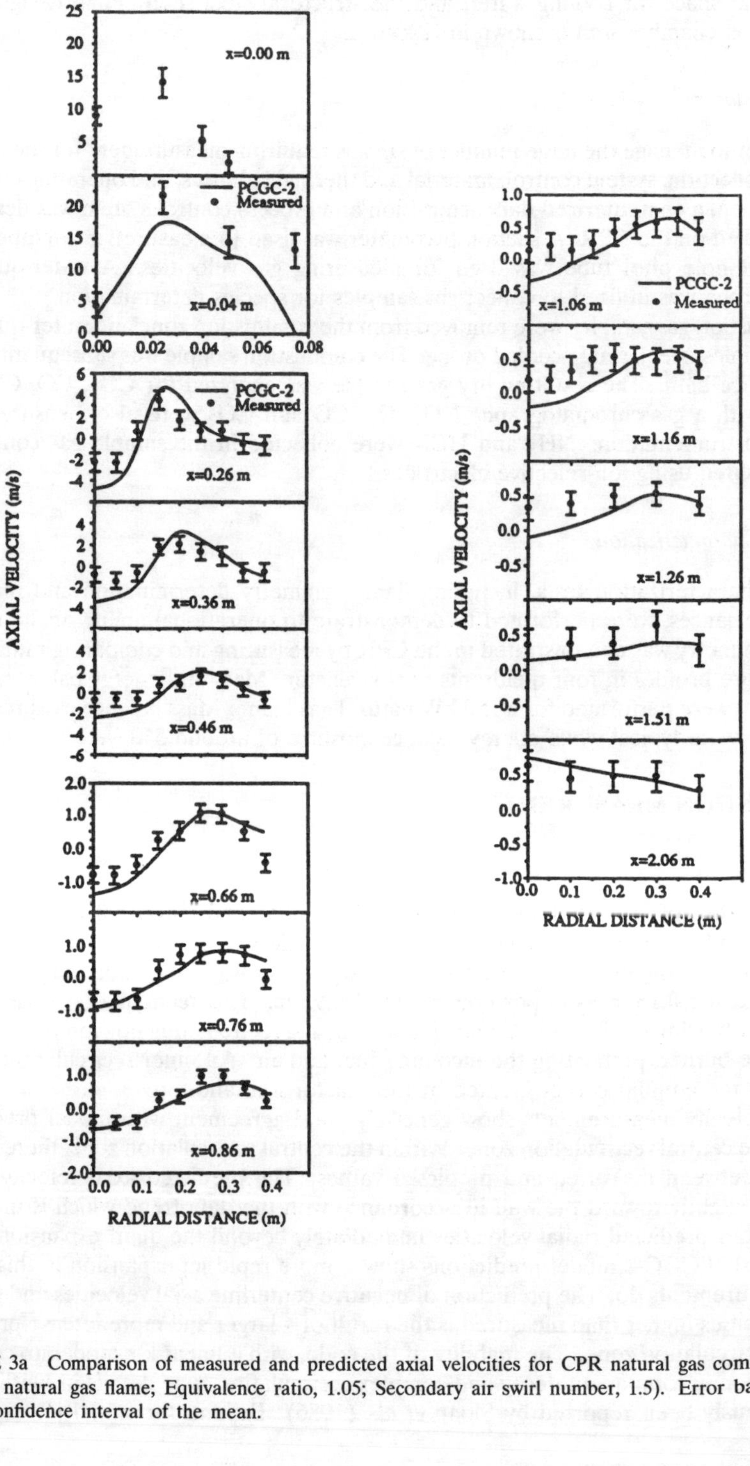

FIGURE 3a Comparison of measured and predicted axial velocities for CPR natural gas combustion tests (150 kW natural gas flame; Equivalence ratio, 1.05; Secondary air swirl number, 1.5). Error bars represent a 90% confidence interval of the mean.

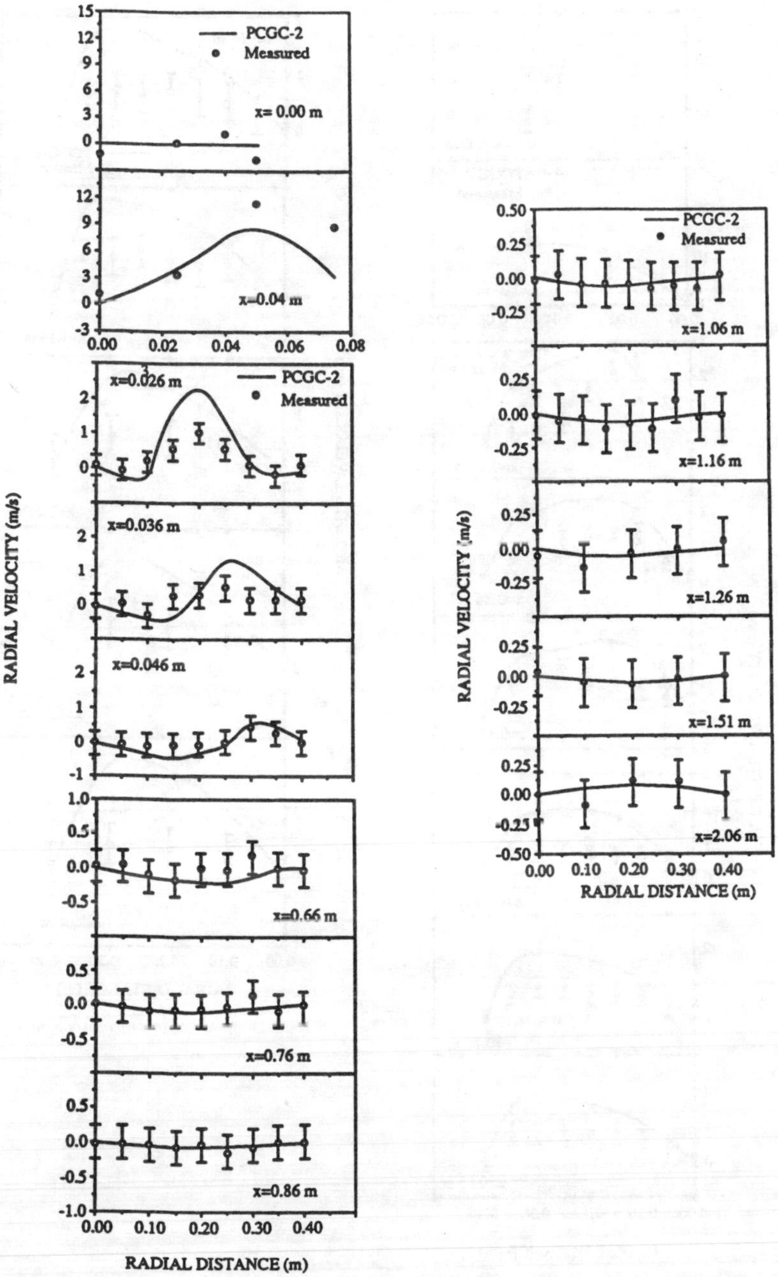

FIGURE 3b Comparison of measured and predicted radial velocities for CPR natural gas combustion tests (150 kW natural gas flame; Equivalence ratio, 1.05; Secondary air swirl number, 1.5). Error bars represent a 90% confidence interval of the mean.

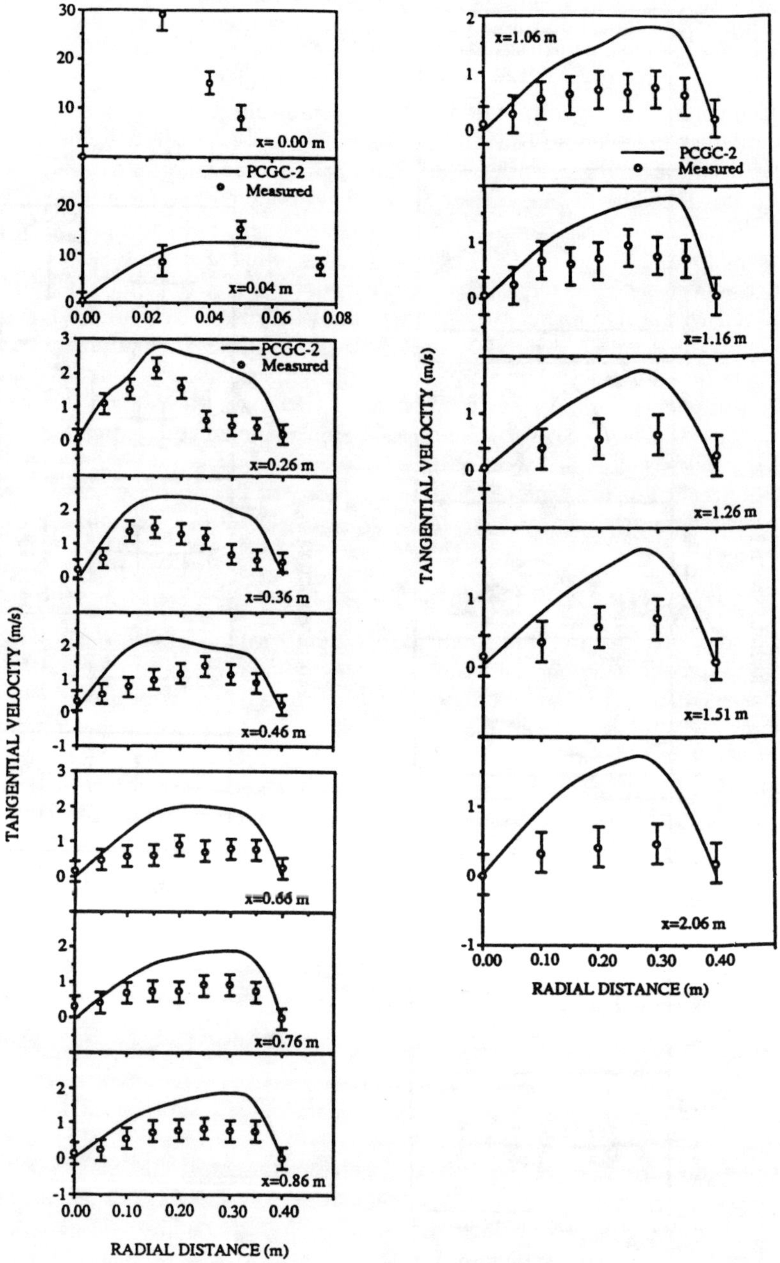

FIGURE 3c Comparison of measured and predicted tangential velocities for CPR natural gas combustion tests (150 kW natural gas flame; Equivalence ratio, 1.05; Secondary air swirl number, 1.5). Error bars represent a 90% confidence interval of the mean.

into the annular corner recirculation zone, radial velocity predictions agree closely with measurements (Figure 3b).

Comparisons of predicted and measured tangential velocities (Figure 3c) exhibit the greatest disparity of the three velocity components. Measured velocities are consistently lower than those predicted by the model. PCGC-2 model predictions of tangential velocity profiles show the flow decaying to a forced vortex distribution while measurements show a combined vortex velocity distribution with an overall velocity decay more pronounced than indicated by predictions. The tendency of the linear k-ϵ model to predict tangential velocity to decay into a forced vortex type of flow has also been reported by Sloan (1986), who attributes the disparity to deficiencies in the kinetic energy dissipation equation. However, aerodynamic drag produced from the rough quarl and reactor walls, which are not accounted for in the model, may also be a contributing factor.

Improvement in the agreement between axial velocity predictions and measurements was achieved with a non-linear turbulence model which is utilized in a 3-dimensional combustion code based on PCGC-2. The application of the 3-D comprehensive model with the non-linear k-ϵ turbulence model to this geometry simulated use of 2-dimensional non-linear turbulence formulations (Smoot *et al.,* 1991). An explanation for the closer agreement observed with the 3-D version of the model is thought to be related to one or more of the following: 1) inclusion of the non-linear turbulence model, 2) greater grid refinement possible with the new model, or 3) effects of the newer solution technique. This work is continuing as the 3-D version of the code is further developed.

Gas Temperatures

Comparisons of measured and predicted gas temperatures are shown in Figure 4. Measured values show the highest temperature region near the reactor centerline while predictions indicate a high temperature annulus surrounding the reactor centerline. Similar predictions made with the developing 3-D model referenced above retained the high temperature annulus, but the general temperature profile tended to be lower. This resulted in slightly better comparisons with data along selected temperature profiles but slightly worse along others.

Flame heat loss predictions early in the reactor may be considered suspect due to the fact that, since the model's primary application is to coal combustion and gasification, it does not consider soot formation in gaseous combustion. The localized formation of soot could significantly lower the flame temperature through two separate mechanisms: First, the formation of soot particles in a gaseous flame results in less heat release than if complete combustion (formation of CO_2) occurred; Second, the high emissivity of soot particles can significantly reduce flame temperature through radiant heat loss to the walls. Neither of these two mechanisms are accounted for in PCGC-2 model predictions of gaseous combustion. The fact that luminosity, resulting from soot radiation, was visually detected in the flame suggests a need for the inclusion of a soot forming mechanism during gaseous combustion in the model.

To estimate the effect of soot radiation on predicted gas temperatures, a prediction was made by introducing non-reacting, 1 mm, simulated soot particles to the reactor through the primary inlet tube. These particles were specified to have radiative properties similar to soot. The concentration of simulated soot particles was 10^{-6} volume fraction in the primary tube which diluted to a typical experimental value of about 10^{-7} volume fraction when mixed with secondary air (Syed *et al.,* 1990). Comparisons of gas temperature predictions from this case with measurements are also shown in Figure 4. These comparisons can only be considered qualitative since the soot particle size and number density present in the flame were unknown and the heat release effects due to soot

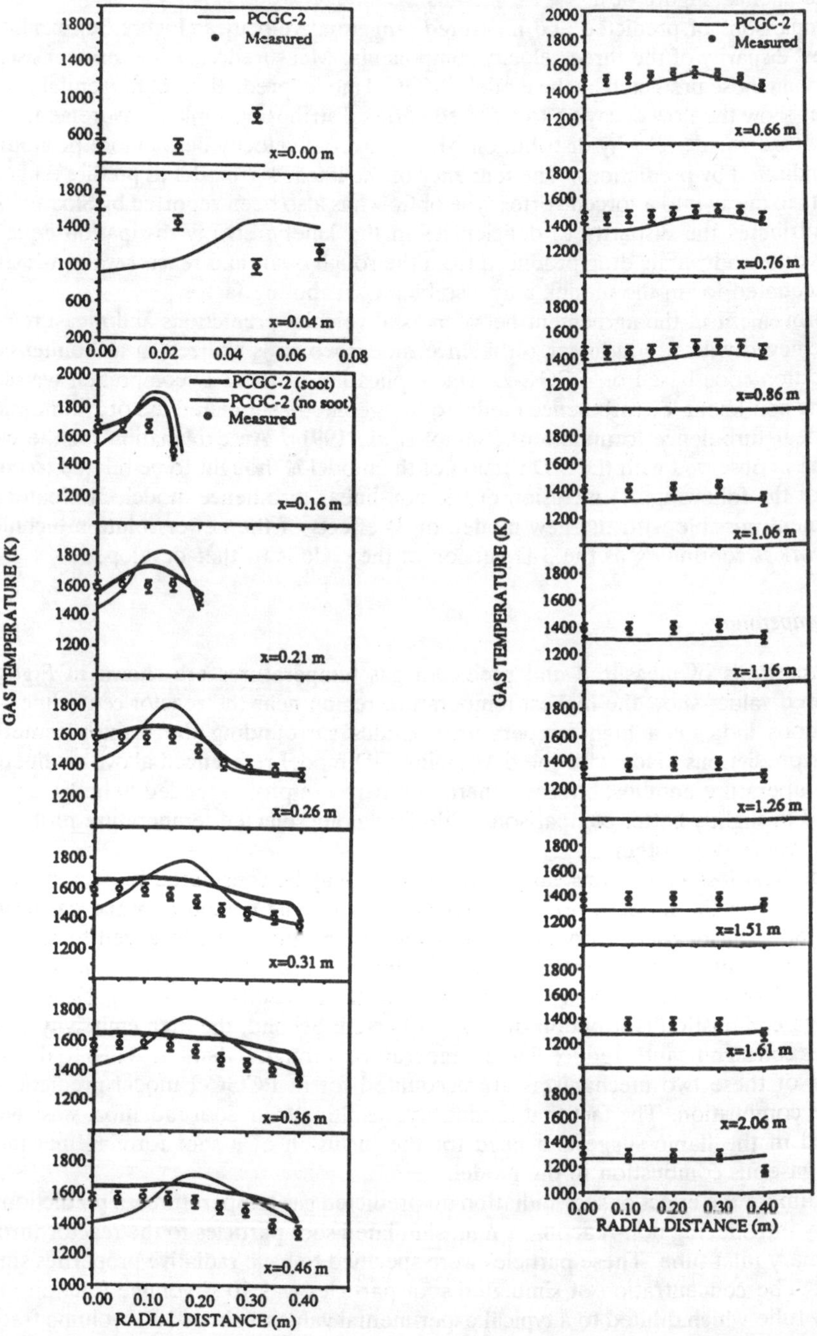

FIGURE 4 Comparison of measured and predicted gas temperatures for CPR natural gas combustion tests (150 kW natural gas flame; Equivalence ratio, 1.05; Secondary air swirl number, 1.5). Error bars represent a 90% confidence interval of the mean.

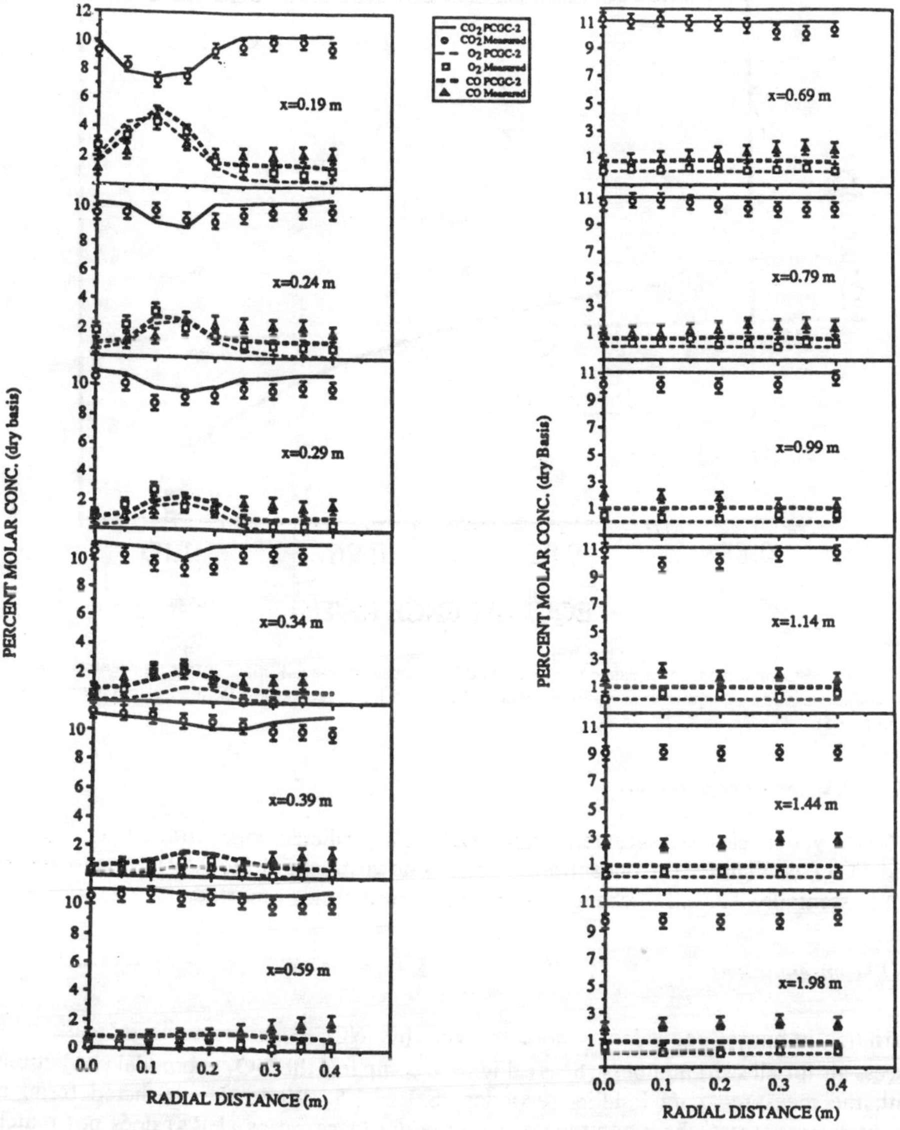

FIGURE 5 Comparison of measured and predicted major species concentrations for CPR natural gas combustion tests (150 kW natural gas flame; Equivalence ratio, 1.05; Secondary air swirl number, 1.5). Error bars represent a 90% confidence interval of the mean.

formation and destruction are not taken into account in this estimation. Nevertheless, the comparisons show significantly better agreement between predicted and measured gas temperatures than does the non-sooting case, particularly with regard to the predicted high temperature ridge which was significantly decreased by the addition of soot particles.

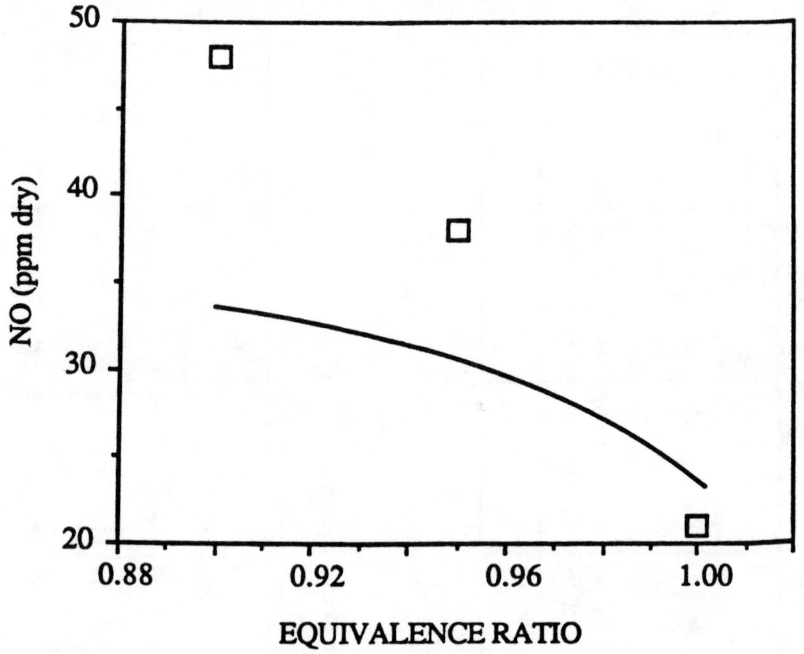

FIGURE 6 Comparison of predicted (curve) and measured (symbols) nitric oxide concentrations for systematic variation in equivalence ratio number holding swirl constant (150 kW natural gas flame; Swirl number 1.5).

Major Species Concentrations

Major gas species measurements agree well with predicted concentrations of O_2, CO and CO_2 (see Figure 5). Concentration profiles downstream of the 0.59 m axial position were essentially flat and unchanged and were therefore not reported.

NO_x Concentrations

Variation in burner equivalence ratio. Typically, NO_x concentrations increase with excess air for all swirl numbers (SN's). Figure 6 compares the NO_x submodel predictions with the measured data holding secondary SN at 1.5. While the predicted trend is qualitatively correct, the magnitude at lower equivalence ratios (ER's) does not match the measured magnitude. This suggests: 1) There might be an effect of flame structure that is not accounted for, 2) The O_2-O equilibrium assumption is invalid, or 3) Thermal NO may be formed at lower ER's through other mechanisms, such as the possibility of the formation of nitrous oxide in cooler flame regions which is subsequently converted into NO (Miller and Fisk, 1987).

Variation in burner swirl number. The formation of NO_x is impacted by the degree of swirl due to the effects of enhanced turbulent mixing and recirculation characteristics of the flow field. The observed effects of swirl on final NO emissions vary and depend on parameters other than secondary swirl number. NO_x emissions either increase or decrease with swirl number and often pass through a local maximum or minimum.

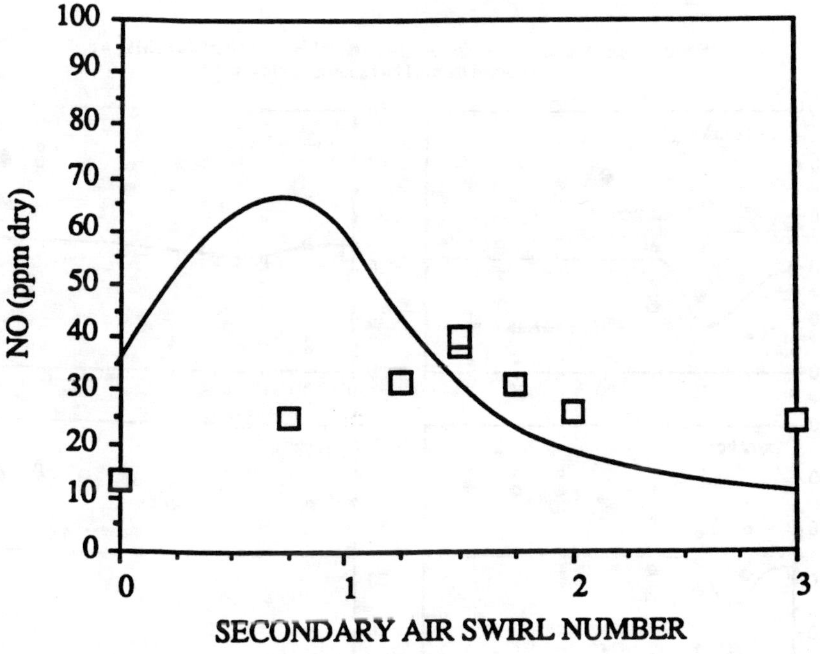

FIGURE 7 Comparison of predicted (curve) and measured (symbols) nitric oxide concentrations for systematic variation in secondary swirl number holding E.R. constant (150 kW natural gas flame; Equivalence ratio 0.95).

Figure 7 shows the measured data and NO submodel predictions for one burner geometry. The predicted concentrations shown by the curve follow the trend of the measured data although the predicted local maximum is higher and occurs at an earlier SN. Other predictions also revealed some disparities between the model results and the data. The lack of agreement between predicted and measured NO_x concentrations was attributed to deficiencies in the NO_x submodel chemistry and possibly to inadequate prediction of the governing flame structure at the lower swirl numbers.

NO_x profiles. Predicted NO profiles are compared to the measured nitrogen pollutant data in Figure 8. The model qualitatively predicted the nitric oxide profiles at each axial location in the reactor. The difference between predicted and measured effluent concentrations is about 30 percent.

The detection of significant concentrations of cyanide and amine components infers that some NO is likely formed from HCN and NH_3 oxidation on the fuel-lean side of the flame front. The Zel'dovich mechanism only predicts the formation of NO by oxidation of the nitrogen in the combustion air. In hydrocarbon flames, nitrogen can also be fixed by "prompt" NO reactions; for example:

$$CH + N_2 \rightarrow HCN + N \tag{9}$$

HCN can also be formed by reduction of NO by reactions following the global sequence:

$$NO + CH_i \rightarrow HCN + \ldots \tag{10}$$

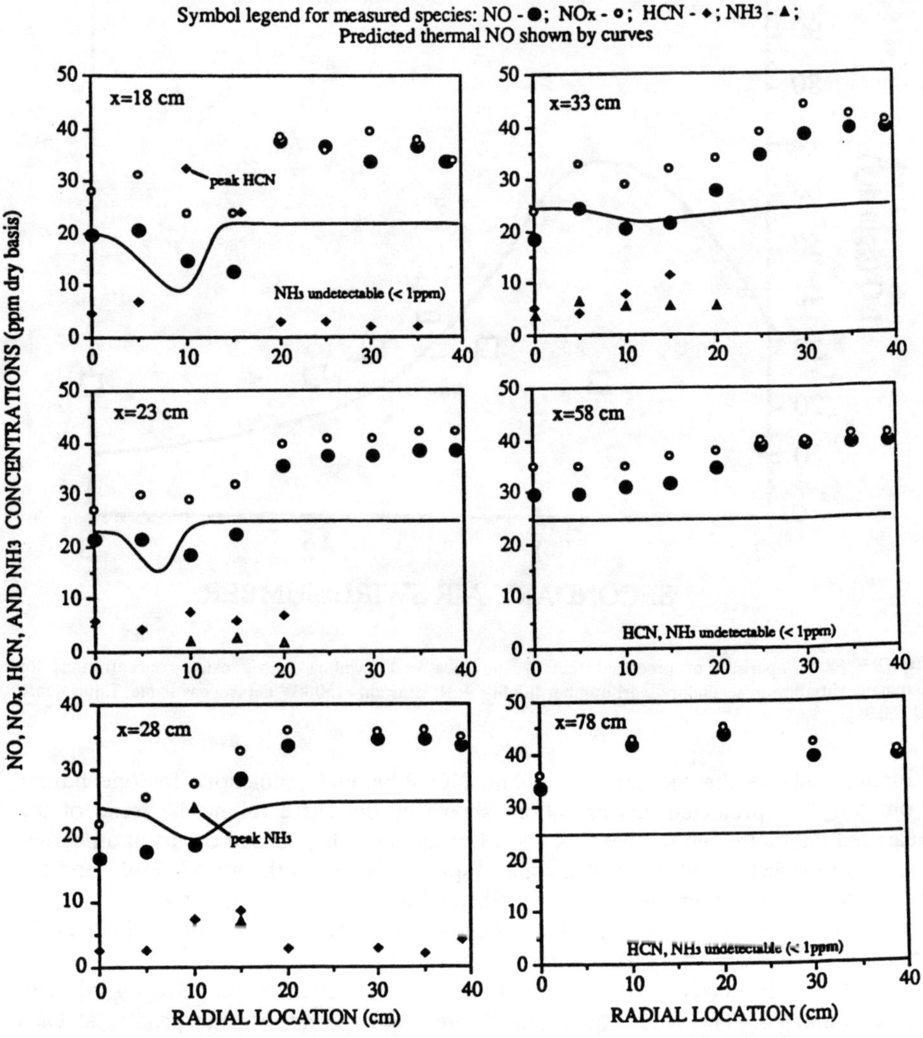

FIGURE 8 Comparison of predicted thermal nitric oxide with measured nitric oxide concentrations. Other measured species (NO_x, HCN, and NH3) shown to demonstrate complexity of the chemistry (150 kW natural gas flame; Equivalence ratio = 1.05; Secondary air swirl = 1.5).

Reaction 10 is called a "reburning" reaction and does not change the total amount of fixed nitrogen while Reaction 9 increases total fixed nitrogen concentrations. Once HCN is formed, it decays quickly to NH_3, NO and N_2 by the following global sequence:

$$HCN \rightarrow NH_3 \rightarrow NO \rightarrow N_2 \tag{11}$$

The detection of NO_2 (taken as the difference between the NO_x and NO profiles) further demonstrates the complexity of the nitrogen chemistry occurring, even in gas flames. NO_2

is formed by oxidation of NO by HO_2 in fuel-rich regions. It is subsequently converted back to NO by reaction with H atoms (Miller and Bowman, 1989). The measured data show that NO_2 is formed in the fuel-rich recirculating zone and then is essentially all converted back to NO shortly following uniform mixing in the reactor.

CONCLUSIONS

Reactor Facility

The new CPR facility provides for detailed and complete optical access to the flame zone in the furnace environment and makes possible intrusive and non-intrusive measurements of inlet turbulence intensities, gas and particle temperatures, major gas and pollutant species and velocities. By allowing for both increased flame access and control of combustion conditions, the facility expands the current ability to perform important parametric studies in complex, combusting systems. The reactor is a key laboratory tool for ongoing instrumentation development required for more detailed combustion zone studies.

Combustion and NO_x Pollutant Tests

A natural gas-air combustion data set for 2-D and 3-D comprehensive model evaluation was obtained, including gas velocities, gas temperatures, and species concentrations. Combustion measurements of velocities and major species concentrations show variable agreement with predicted values. There is some disparity between measured and predicted velocities in the central recirculation zone. Gas temperature measurements closely match predictions in the recovery region but do not show the substantial high temperature annular region predicted by the model. This may be partly attributed to the code not predicting soot formation in gaseous combustion, and possibly also to turbulence model or numerical solution method.

A goal of the experimental work was to operate in a region of stoichiometries where prompt NO and NO_2 formation would be minimized. This attempt was thwarted by the locally fuel-rich zone inherent in diffusion flames. The formation of HCN in the fuel-lean zone confirms that it was not possible to completely avoid prompt NO formation and decay in the hydrocarbon diffusion flame. Whether or not prompt NO formation is compensated or exceeded by the destruction of NO by nitrogen intermediates in the fuel-rich core cannot be determined from these data and predictions.

The purpose of the prediction and measurement comparisons was to demonstrate the ability of PCGC-2 with the Zel'dovich mechanism to predict thermal NO formation. The detection of HCN, NH_3 and NO_2 reveals that the Zel'dovich mechanism cannot precisely predict total NO concentrations in swirling, natural gas diffusion flames. Despite the complexity of the chemistry, the Zel'dovich mechanism approximately predicts the measured profile trends. At least for this case, thermal NO appears to be the major contribution to NO_x concentrations.

BIBLIOGRAPHY

Boardman, R. D. (1990). "Development and Evaluation of a Combined Thermal and Fuel Nitric Oxide Model," Ph.D. Dissertation, Department of Chemical Engineering, Brigham Young University, Provo, UT.

Boardman, R. B. and Smoot, L. D. (1988). "Prediction of Nitric Oxide in Advanced Combustion Systems," *AIChE Journal*, **34,** 1573.

Butler, B. W. (1988). "Design of a Wall Heating System for a Laboratory-Scale Furnace," M.S. Thesis, Department of Mechanical Engineering, Brigham Young University, Provo, UT.

Eatough, C. N. (1991). "Controlled Profile Reactor Design and Combustion Measurements," Ph.D. Dissertation, Department of Mechanical Engineering, Brigham Young University, Provo, UT.

Hill, S. C., Smoot, L. D., and Smith, P. J. (1984). "Prediction of Nitrogen Oxide Formation in Turbulent Coal Flames," *Twentieth Symposium (International) on Combustion,* The Combustion Institute, Pittsburgh, PA, 1391.

Lavoie, G. A., Heywood, J. B., and Keck, J. C. (1970). "Experimental and Theoretical Study of Nitric Oxide Formation in Internal Combustion Engines," *Combustion Science and Technology,* **1,** 313.

Miller, J. A. and Bowman, C. T. (1989). "Mechanism and Modeling of Nitrogen Chemistry in Combustion," *Progress in Energy and Combustion Science,* **15,** 287.

Miller, J. A. and Fisk, G. A. (1987). "Combustion Chemistry," *C&EN* Special Report, 22, August 31.

Sloan, D. G., Smith, P. J., and Smoot, L. D. (1986). "Modeling of Swirl in Turbulent Flow Systems," *Progress in Energy and Combustion Science,* **12,** No. 3, 163.

Smith, P. J., Hill, S. C. and Smoot, L. D. (1982). "Theory for NO Formation in Turbulent Coal Flames," *Nineteenth Symposium (International) on Combustion,* The Combustion Institute, Pittsburgh, PA, 1263.

Smoot, L. D. and Smith P. J. (1985). Coal Combustion and Gasification, Plenum Press, New York, NY.

Smoot, L. D., Bartholomew, C. H., and Pershing, D. W. (1 Feb. 1991). Fifth-Year Progress Report, Volume II, Research Program Volume, NSF Advanced Combustion Engineering Research Center, Brigham Young University, Provo, UT.

Solomon, P. R., Hamblen, D. G., Serio, M. A., Smoot, L. D., and Brewster, B. S. (1991). "Measurement and Modeling of Advanced Coal Conversion," U.S. DOE contract No. DE-AC21-86MC23075.

Syed, K. J., Stewart, C. D., and Moss J. B. (1990). "Modelling Soot Formation and Thermal Radiation in Buoyant Turbulent Diffusion Flames," *Twenty-third Symposium (International) on Combustion,* The Combustion Institute, Pittsburgh, PA, 1533-1541.

Zel'dovich, Y. B., Sadovnikov, P. Y., and Frank-Kamentskii, D. A. (1947). "Oxidation of Nitrogen in Combustion", (translated by Shelef), Academy of Sciences of USSR.

Prediction of Flame Structure and NO Emission by Numerical Simulation

D. AIULFI AND J. GASS *Swiss Federal Institute of Technology Laboratory for Energy Systems, ETH-Zentrum, Zürich, Switzerland*

Abstract — The main aim of this work is the development of a numerical simulation tool for industrial flames. The present phase of the work concentrated on turbulent diffusion flames, mainly of natural gas. As a model, the standard k-ε model was used, with a one step chemistry and a representation of the fuel/air mixture fraction by a probability density function as closure. A large number of flames were simulated and the results compared with measurements, either found in the literature or collected in the own test facility. Concerning the prediction of NO emission, it was found that the results based on a pure Zeldovich mechanism for thermal NO production were satisfactory for lean flame conditions, but for fuel-rich flames a more detailed chemistry should be used.

Key Words: Computer Simulation, natural gas combustion, NO prediction, comparison with experiment

INTRODUCTION

Nearly 50% of the overall Swiss energy consumption is used for space heating, of which the major part is accomplished using light fuel oil. In the near future the majority of private motor cars will be equipped with a catalyst, therefore space heating will be the biggest producer of NO. Although the models under consideration are of more general use, this particular field of application comes under scrutiny, which means looking at flames in the range of 20 kW up to 1 MW.

If one starts modelling combustion processes in turbulent flames one very quickly ends up at the limits of the available computational resources. As the model is intended to be a tool used for the development of industrial combustion facilities, the search is mainly for "simplified methods," where the need for computer time stays within acceptable limits for industrial purposes. One question that can be posed is, how much simplification is allowed in order to obtain still reliable results concerning pollutant formation, mainly nitrogen oxide?

As it is known that the reaction within diffusion flames is mainly controlled by the mixing process, the work concentrated on modelling the turbulent flow field, with the reaction only being taken into account by a one-step chemistry. With the knowledge of temperatures and oxygen concentrations it is possible to calculate the formation of NO by a pure Zeldovich mechanism. If a more detailed chemistry is needed to describe the NO formation, it can no longer be achieved using a one-step model as there is no information available about intermediate products, such as the CH- or OH-radicals (Miller and Bowman 1989) which are involved in the formation of prompt or fuel NO and the destruction of already formed NO.

In modern low NO_x combustion technology more and more premixed flames are in use, so in future it will be imperative to focus on the problem as to how the necessary chemistry can be described simply enough to be combined with fluid dynamics calculations. The first phase of this work concentrated on one-phase flow only, mainly the combustion of natural gas, but it is clear that future work has to also include two-phase flows (oil sprays). As the computational resources of modern computers are increasing nearly every year, the restrictions that have to put on the models can be eased.

THE MODEL

Combustion modelling can be undertaken at various levels of sophistication. Also the necessary simplifications that have to be applied, depend on the type of flame that is going to be modelled. The models for a candle flame, an explosion in a cylinder of a car engine or, in this case, a combustor for space heating need to have different properties.

The model used for this work can be described briefly as follows: the turbulent flow field, including the mixing process of fuel and air, is simulated by the time-averaged fluid dynamics equations, where the influence of turbulence is taken into account by the k-ε model. The modelling of the combustion process involves the assumption of an infinitely fast reaction (mixed = burned) equal to the simulation of the mixing process of fuel and air. It is convenient to describe this process by the so called mixture fraction, f, which is defined as follows:

$$f = \frac{\Phi - \Phi_A}{\Phi_F - \Phi_A}, \qquad \Phi = m_{fu} - mO_2/s,$$

where Φ_A = pure air, Φ_F = pure fuel, s = stoichiometric factor; f is a conserved scalar, which means it does not change its value when reaction takes place.

To take into account the fluctuations of f at a certain place due to turbulence, its mean value is represented by a probability density function (PDF):

$$f = \int_0^1 f \cdot P(f) \cdot df.$$

This allows a definition of a measure g for the fluctuations of the mixture fraction:

$$g = \int_0^1 (f - \bar{f})^2 \cdot P(f) \cdot df.$$

For both quantities, f and g, a transport equation can be formulated. The shape of the PDF has to be presumed. In this case rectangular-shaped PDFs are compared with clipped Gaussian shaped PDFs and with PDFs of the shape of so-called β functions.

Another approach for modelling the combustion process is given by Magnusson (1976) with his eddy-dissipation model. This model assumes that the mixture averaged on a macroscale is inhomogeneous on a microscale level and reaction takes place according to a mixing process which is proportional to the eddy dissipation. The equation he uses is:

$$R_f = C \cdot \bar{\rho} \frac{\varepsilon}{k} \min\left(m_f, \frac{m_{O_2}}{\lambda}, \frac{m_p}{1+\lambda}\right),$$

where p stands for product; the proposed value for C is 4. This method was also taken into consideration and comparisons are included in the results.

For the formation of NO only the rate limiting reaction of the Zeldovich mechanism was taken into account:

$$O + N_2 \rightarrow NO + N.$$

Immediate oxidation of the free N-radical is assumed. The reaction constant, dependent on temperature as well as concentrations of O_2 and N_2, was calculated with the following formula:

$$\dot{m}_{NO} = C \cdot \frac{M_{NO}}{M_{N_2}} \cdot m_{N_2}(f) \cdot \rho \cdot \left(\frac{m_{O_2}(f) \cdot \rho}{1000 \cdot M_{O_2}}\right)^{0.5} \cdot e^{-67337/T},$$

where M is the molecular weight and $C = 5.74 \times 10^{14}$.

Although the method is fully three-dimensional, all the work was performed in a quasi two-dimensional geometry, which means a flow field with rotational symmetry (swirl-free flow with no tangential components of velocity).

The complete model for this geometry consists of equations for (for more details see Aiulfi 1991):

— conservation of mass
— transport of momentum in the axial and radial direction
— conservation of energy
— transport of kinetic energy k of turbulence and its dissipation rate ε
— transport of mixture fraction f and its fluctuation
— formation and transport of thermal NO
— radiant flux in the axial and radial direction

This model is applicable only to diffusion flames. Premixed or partially premixed flames, where the combustion process is controlled by the chemical reaction, cannot be represented.

All the simulation work was carried out with the PHOENICS-Program. A typical case used between 1 and 2 hours of CPU time on a CRAY XMP machine.

THE TEST FACILITY

The ETH natural gas test facility was specially designed to produce a flow-field that could easily be simulated. The burner consists of coaxial flows of fuel gas, primary and secondary air. The flame is stabilized on a bluff body (see fig. 1).

The combustion chamber itself is a tube 2.5 m long with a diameter of 0.8 m. This size allows flames up to nearly 1 MW and is too large for the 300 kW which was the limit from the natural gas supply. The usable size was reduced by a movable wall (see fig. 2) to a size matching the enclosure for the flame. The flow-field established under these conditions consists of a small intense recirculation zone (stabilizing the flame) just behind the bluff body and a larger one filling the whole combustion chamber.

A second large recirculation zone is established in the space behind the partitioning wall. The outside wall consists of eight water-cooled segments with a relatively high water flow rate, establishing well-defined wall temperatures as boundary conditions. Each segment is provided with a pair of windows, giving easy access for measurements. For the data acquisition a computerized system was used which was able to automatically measure profiles through the flame. Thermocouples with a radiation shield were used to measure temperatures. The flow field was investigated with a water-cooled five-hole probe. For the evaluation of the gas composition (in the chimney as well as in the flame) a flue gas analyser was used which had the capability of measuring O_2, CO_2, CO, NO and NO_2.

MEASUREMENTS AND COMPARISONS WITH COMPUTATIONAL RESULTS

A major ability of a computational tool for the purpose of combustion engineering is to predict flow fields and temperature distributions. Many simulations and comparisons have been made to this end and also to evaluate the influence of certain assumptions of the model.

Methane Jet Flame

Experimental situation. For one set of comparisons, measurements of a methane jet flame by Sislian *et al.* were used (1988). The main reason was the availability of very

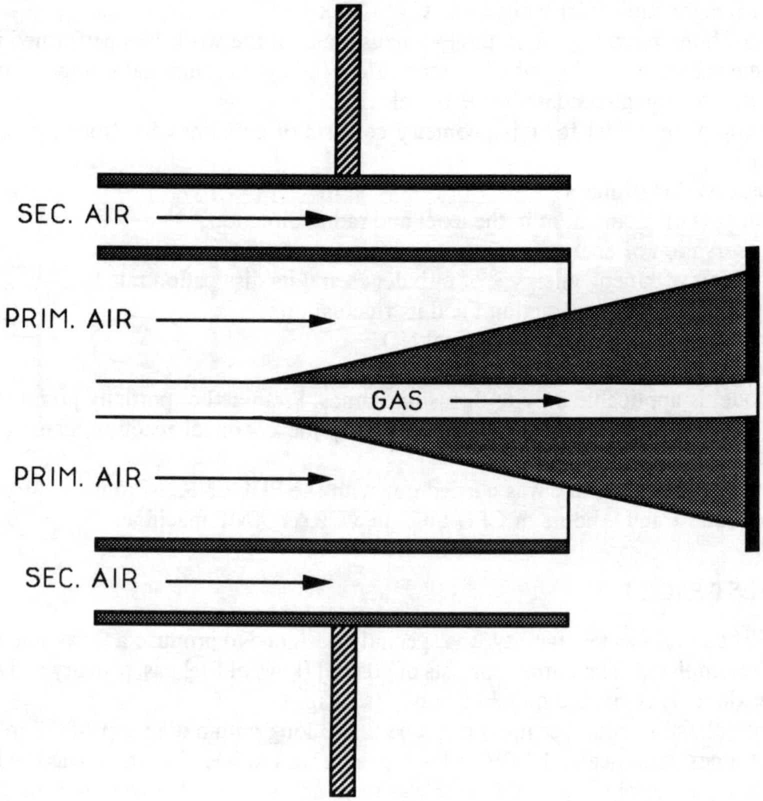

FIGURE 1 Burner head.

detailed LDA measurements for this flame, so that a comparison of measurements concerning turbulence parameters with modelled results could be made.

The experimental arrangement consisted of a simple axisymmetric diffusion flame of 96% methane and 4% argon, burning vertically upwards into stagnant air. This air volume was shielded against room draughts by a mesh screen. The average gas velocity at the nozzle exit was 17.2 m/sec, corresponding to a Reynolds number — based on the nozzle diameter of 10 mm — of 7850.

Model variations under consideration. In the following, comparisons of three variations of the model were included:

Model 1: The mixture fraction is represented by a rectangular shaped probability density function. This approach has the advantage that integration over the PDF is very simple, but it does not represent physical reality very well.

Model 2: A so called β-function is used to represent the PDF for the mixture fraction. This function is defined as follows:

$$P(f) = \frac{f^{a-1}(1 - f^{b-1}}{\int_0^1 f^{a-1}(1 - f)^{b-1}\, df},$$

where $a = f[(f(1 - f) - g)/g]$, $b = (1 - f)[(f(1 - f) - g)/g]$.

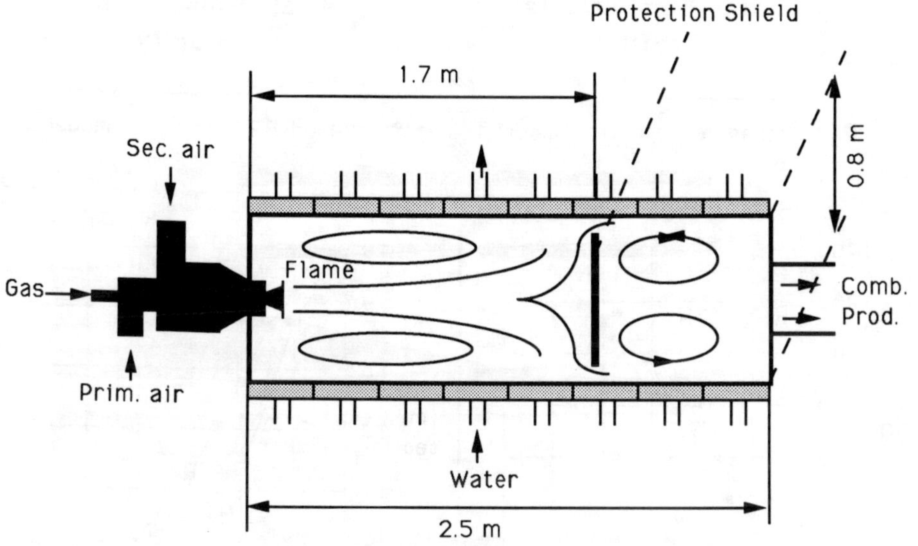

FIGURE 2 Cross-section through the test facility.

This function has the advantage of being a better representation of physical reality. The β-function looks Gaussian-like in the middle of the definition interval and becomes more and more asymmetric towards the ends (0 or 1). Integration over the PDF is time consuming, as it involves a double integration.

Model 3: This model uses the eddy-dissipation approach of Magnusson to simulate combustion. This way of modelling may better represent the limited reaction rate, but it increases the number of dependent equations and can cause convergence problems. This model has also been tried in a two-step version where the hydrocarbon fuel is oxidized in a first step to H_2O and CO, which is modelled by eddy-dissipation. The oxidation of CO to CO_2 as the second step is represented by its proper reaction constant. As this method did not bring any significant improvement, no results are reported here.

Results. The following measured quantities have been taken for comparison:

— axial component of the mean velocity = u
— turbulent kinetic energy = k
— mean temperature = T

Comparisons have been made for each of the three quantities with profiles along the axis and also with profiles in the radial direction at two different axial distances from the burner nozzle (z/D = 20 and z/D = 40).

Temperature: All the comparisons are shown in figs. 3 and 4. All the models produce profiles which correspond qualitatively to the measurements. Deviations of up to 200 K occur. It is surprising that model 1 with the simplest PDF shows an even better agreement than the other models. Only for the peak temperatures do models 2 and 3 produce better values. For the axial distance z/D = 40 it has to be considered that the flame contains some soot. In all the models the additional radiation due to the soot has not been taken into account so that the calculated temperatures are generally too high.

Velocity: The results are shown in figs. 3 and 5. All the models produce similar results, but there is a clear discrepancy between measurement and simulation. The velocity on the

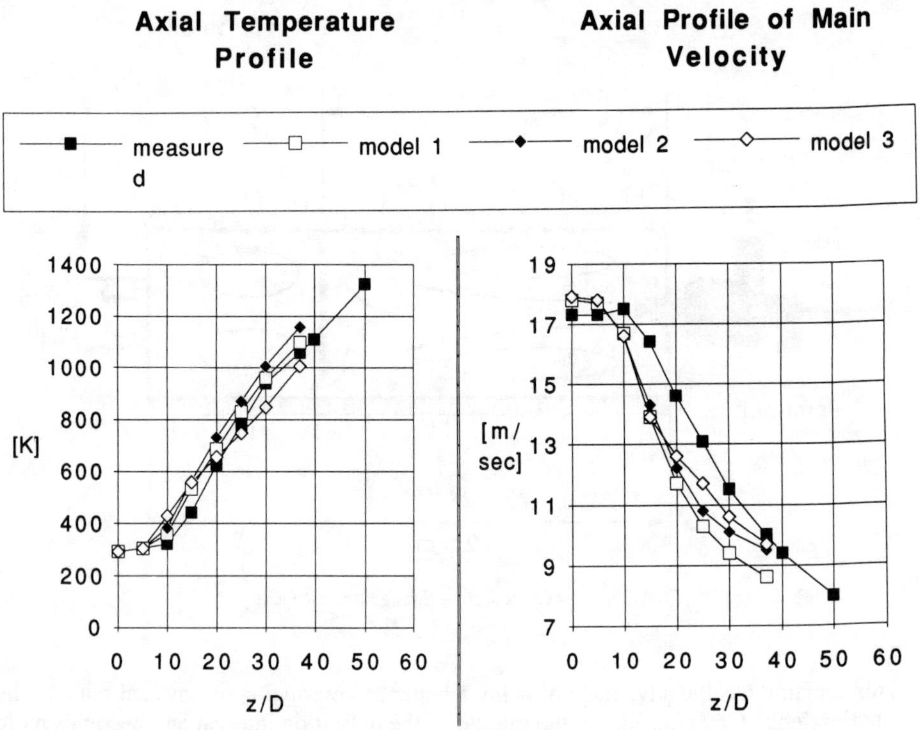

FIGURE 3 Comparison of axial temperature and velocity profiles.

axis remains higher than predicted, which means that the burning jet is less divergent than predicted by the model. This also can be seen clearly in the radial profile at z/D = 20 (see fig. 5). The situation is less clear for the position at z/D = 40.

Turbulent kinetic energy: The turbulent kinetic energy is defined as half the sum of the mean squared velocity fluctuation of the three components. As the model assumes radial symmetry it has only two dimensions. The measurements were also made in only two dimensions, assuming that the turbulent kinetic energy in the tangential direction is negligible. It is not known how valid this assumption is.

From the results (see figs. 6 and 7), it is apparent that the values produced by model 1 and 2 are quite similar, but there is a considerable disagreement with the measurements. The assumed shape of the PDF obviously has only a small influence on the kinetic energy of turbulence. The eddy-dissipation approach (model 3) seems to give a somewhat better agreement, but the improvement is not convincing enough to accept the increased convergence problems due to the additional equation.

For all the models, the turbulent kinetic energy in the near burner region is far too high. Therefore the dissipation rate is also calculated too high, which causes the discrepancy in the jet divergence.

Measurements and comparisons have also been undertaken with a non-burning cold jet. The results are presented in fig. 7. As the distinction between models 1, 2 and 3 only

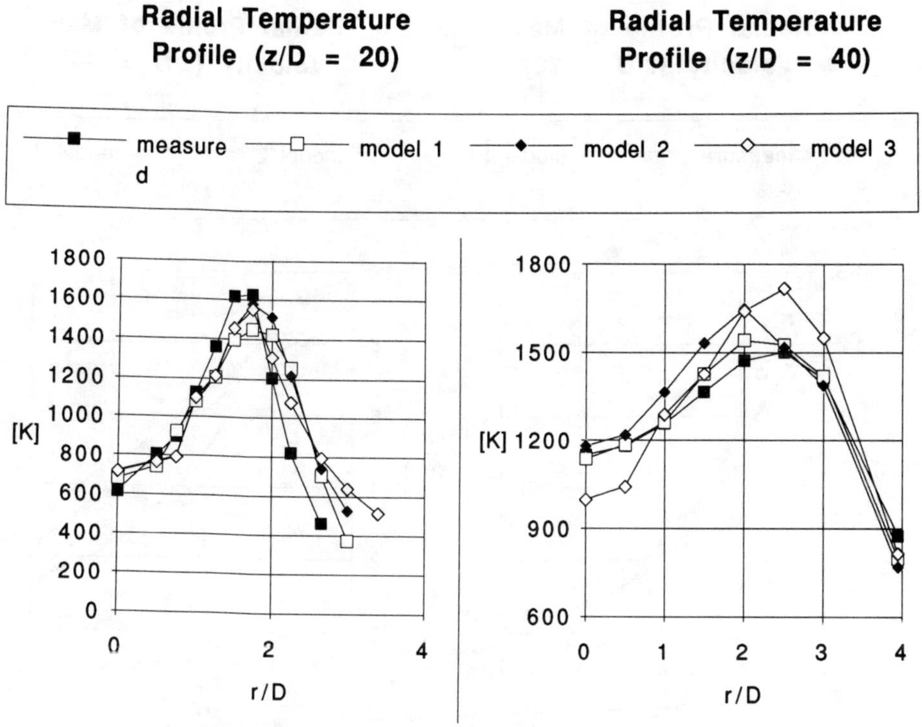

FIGURE 4 Comparison of radial temperature profiles.

concerns the description of combustion, there is only one simulated result representing the k-ε model.

The axial profile of the turbulent kinetic energy shows a much better agreement between measurements and simulation than for the case with combustion. There is still a 30% difference in the maximum value, but the characteristics of the two curves are the same. The dissipation rate of the turbulent kinetic energy, as represented by the gradients, can be seen to be very similar.

When comparing the cold flow with the combustion case, it is realized that the dissipation rate is obviously reduced due to combustion. The axial profile of k (see measurements fig. 6) remains fairly constant in the range z/D = 20 to 50. This effect is not represented by the k-ε model. Another interpretation of this effect may be that the combustion process causes an additional source term for turbulence.

ETH Test Combustor

Flames under consideration. In the test combustor, the flame is enclosed in a cylindrical tube. The flow field, still axisymmetric, is much more complicated than the jet flame and consists of several recirculation zones (see fig. 1 and 2). Also for this situation, a good agreement was found between the simulated flow field and the measurements.

The work with this particular laboratory combustor concentrated on the prediction of NO formation. For the simulations, model no. 2 has been used (infinitely fast combustion with

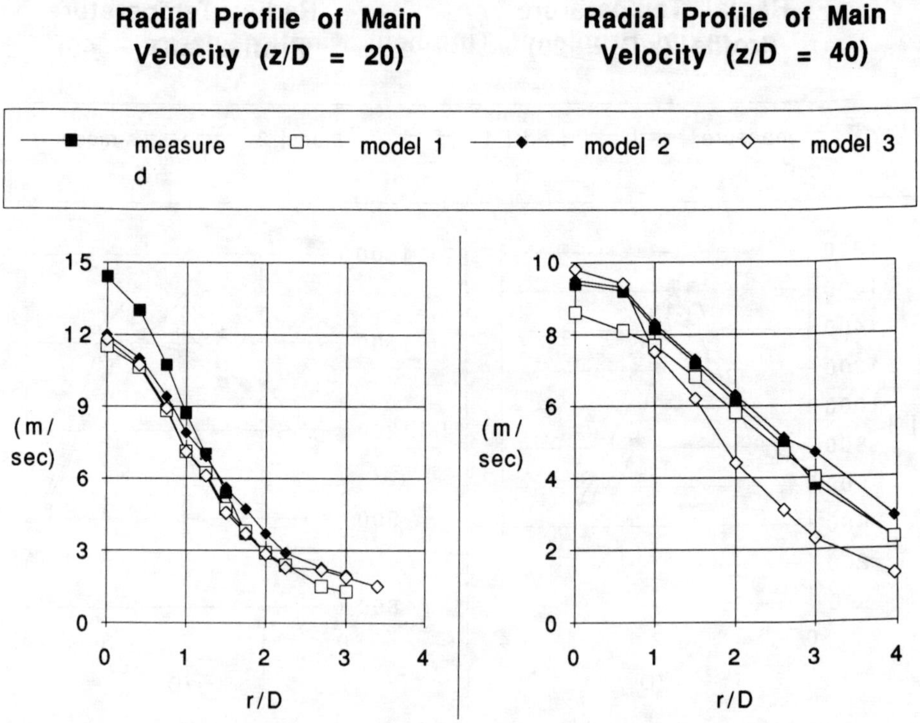

FIGURE 5 Comparison of radial velocity profiles.

TABLE I
SET OF TEST FLAMES FOR COMPARISON WITH SIMULATION

Fuel rate (kW)	Air/ fuel ratio	Staged air supply	Ceramic tube	NO measured (ppm)	NO calculated (ppm)
135	1.3	no	no	9	18
135	2.1	no	no	15	8
220	1.6	no	no	13	8
160	1.6	yes	no	45	72
280	1.4	yes	no	50	71
160	1.3	yes	yes	40	140
160	1.6	yes	yes	50	70
160	2.1	yes	yes	45	40
280	1.4	yes	yes	56	88
235	2.1	no	yes	21	14

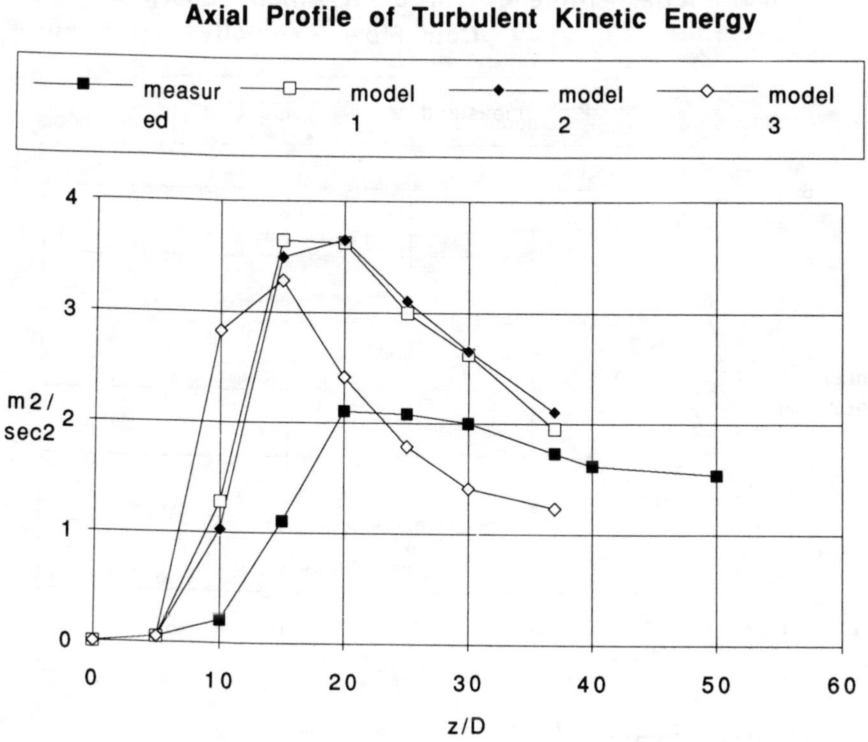

FIGURE 6 Comparison of profiles of turbulent kinetic energy.

a β-function as PDF). For the comparison of simulated NO production with experiments, a set of 10 different flames was investigated. Different fuel rates were combined with different fuel/air ratios. For some flames the air supply was staged, that means that one third of the air is supplied by the inner coaxial tube and the rest by the outer. A ceramic tube was placed within the combustion chamber for certain flames to increase the flame temperature from its uncooled walls. An overview of the 10 different flames is given in table 1.

Results. Figure 8 shows a radial temperature profile of a flame in the ETH test combustor (160 kW, secondary air supply, excess air ratio 1.6, with ceramic tube) at an axial distance of 60 cm from the burner head. For the condition with secondary air, the model does not predict the mixing very well and a much higher peak temperature is predicted than measured. In fig. 9 all the measurements of NO concentrations in the flue gas are plotted and compared with the simulation results. For the situation without a staged air supply, a fairly good agreement was found between the model and the experiment. For flames with a staged air supply there is a clear deviation — the calculated results are noticeably higher than the measurements. Looking at the dependency of the discrepancy between measurements and calculations as a function of the excess air ratio, it is evident that the good agreement is connected to flames without air staging and the disagreement for flames with a staged air supply is obviously increasing with smaller air/fuel ratios. The interpretation

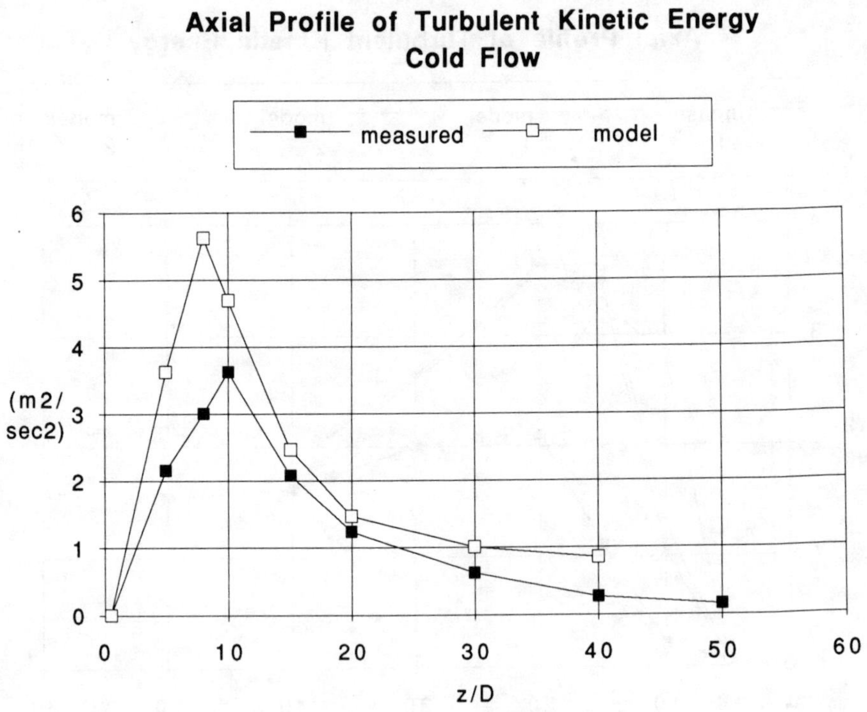

FIGURE 7 Comparison of axial profiles of turbulent kinetic energy for a cold jet.

of this situation is as follows. The staged air supply will cause a fuel-rich zone in the center of the flame as the air flow through the outer concentric tube is much less involved in the small intense recirculation zone behind the bluff body (see fig. 1) and so mixing takes place only further downstream. The disagreement of the calculated NO values can be due to two different phenomena, both connected with the fuel-rich zone.

Firstly, a fuel-rich natural gas flame starts to soot which reduces the flame temperature due to its radiation. This effect could not be taken into account in the model. Due to the too high a temperature, NO formation might have been overestimated. A verification of this effect by measurements of temperature was not possible as the measurable peak value is not significant. The reaction constant for the formation of thermal NO is strongly temperature dependent so that it is mostly produced during the peak values of the fluctuating temperature.

Secondly, in fuel-rich zones with a lack of oxygen, NO, which has already formed, can be reduced in the presence of CH radicals (Myerson, 1957) by the reactions:

$$CH + NO \rightarrow HCO + N$$
$$CH + NO \rightarrow HCN + O$$

The free N radical will react:

$$N + NO \rightarrow N_2 + O.$$

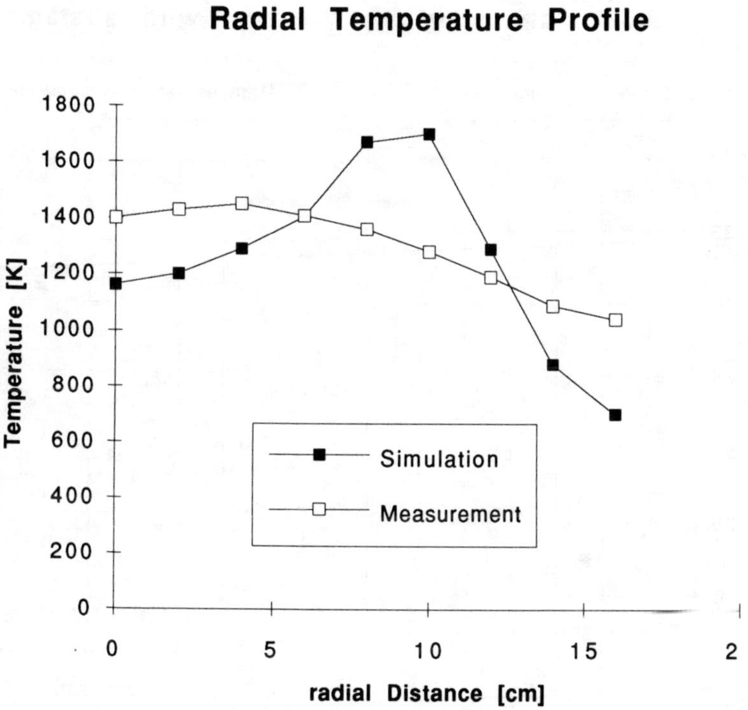

FIGURE 8 Radial temperature profile of a 160 kW flame in the ETH test combustor at an axial distance of 60 cm from the burner head.

Due to the fact that our combustion model does not include a detailed chemistry (no information about CH radical concentration is available) such effects cannot be taken into account.

CONCLUSIONS

As has been shown, this relatively simple flame model is able to predict reasonable values for the flow-field and the temperature distribution in turbulent diffusion flames. The formation of NO by the Zeldovich mechanism can easily be added to this model. It has, however, been found that for fuel-rich flame zones other reactions have a significant influence. As these are based on interim products of the combustion process they cannot be treated by the simple one-step model. Future work will investigate the possibility of including the reduction reaction by correlation to a flow-field variable which is available in the equations. This seems to be the only possibility in order to overcome this problem without giving up the simplicity of the model.

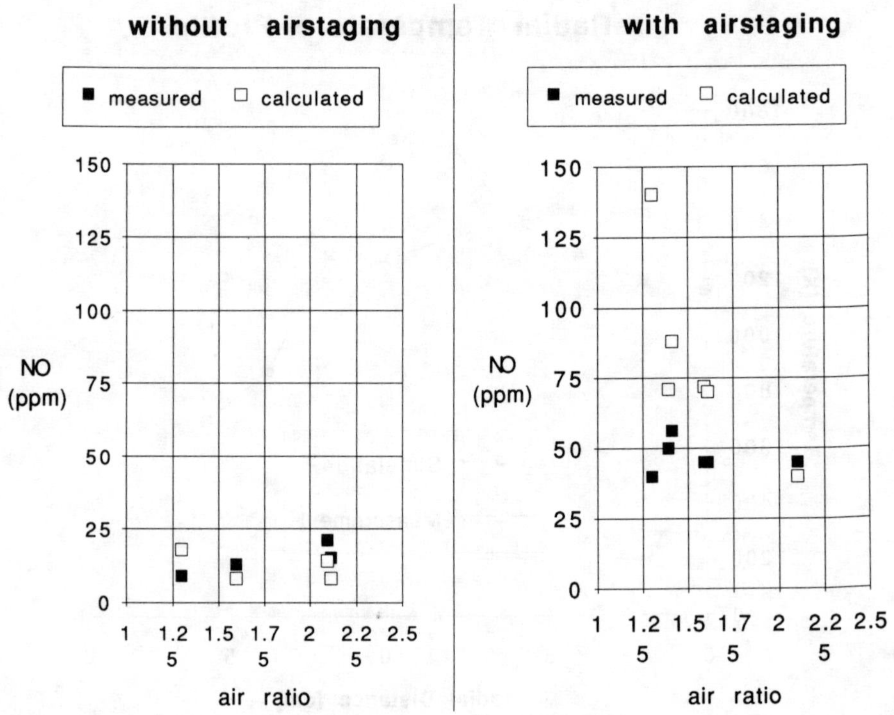

FIGURE 9 Comparison of calculated and measured NO emission without staged air supply (left) and with staged air supply (right).

REFERENCES

Aiulfi, D. (1991). Modeling of technical gas combustion in view of the NO_x formation, Thesis no. 9248 ETH Zürich.

Magnusson, B.F., and Hjertager B.H. (1976). On mathematical modeling of turbulent combustion with special emphasis on soot formation and combustion, *16th Symposium (Int) on Combustion*, The Combustion Institute, p. 719.

Miller J.A., and Bowman, C.T. (1989). Mechanism and modeling of nitrogen chemistry in combustion, *Progress in Energy and Combustion Science*, **15**, p. 287.

Myerson, A.L., Taylor J.P., and Harst, P.L. (1957). *J. Chem. Phys.*, **26**, p. 1309.

Sislian J.P., Jiang L.Y., and Cusworth R.A. (1988). Laser doppler velocimetry investigation of the turbulence structure of axisymmetric diffusion flames, in: *Progress in Energy and Combustion Science*, Vol. 4, p. 99.

Prediction of Nitrogen Oxide in Turbulent Flames: A Two-Parameter PDF Model

R. AGUIRRE *Universidad Nacional Autonoma de Mexico, Mexico*

Abstract — A mayor problem in predicting pollutant formation in flames is the interaction between the turbulence and the combustion chemistry. To obtain a model which allows the prediction of pollutant species, a two-scalar joint probability density function, pdf, was adapted within the k-ε turbulence model. As an intermediate step during the development of such a model, the prediction of nitrogen oxide in the one-dimensional premixed propane/air flame of Robinson (1974) was performed. The pdf approach, proposed by Bray and Moss (1977b), in its expanded version suggested by Jones and Kollmann (1985), was used to evaluate NO as an extra chemical parameter, besides fuel, CO and CO_2 concentrations which were originally considered. This model under predicted NO levels, but showed that this approach can be used, with the correct kinetic information, to predict pollutant species.

INTRODUCTION

Combustion systems which burn fossil fuels are a major cause of the atmospheric pollution since these systems are widely used and the associated pollutants can arise in several ways: as products of incomplete combustion, as impurities in the fuel which are unburned or partially burned, or as by-products of the reaction.

In order to contribute to the development of turbulence/chemistry interaction models which permit the prediction of turbulent gaseous combustion problems, several mathematical approaches have been developed by a research group, directed by Dr. J. J. McGuirk, at Imperial College (Aguirre, 1987). In one of those models, a two-scalar joint probability density function was adopted and used, within the k-ε turbulence model, for both turbulent premixed and turbulent diffusion flame predictions.

The resulting model was tested for one- and two-dimensional flow configurations with both methane and propane combustion. Two different joint-pdf shapes were tested: an exponential shape and the Bray–Moss distribution. The use of the Bray and Moss pdf allowed the expenditure of additional computational effort on the evaluation of NO.

No further references will be made here on the development of the general model, but only the features necessary to understand the approach used for NO prediction will be mentioned.

The extended Zeldovich mechanism of NO formation was adopted assuming chemical equilibrium for oxygen and nitrogen atom concentration as well as zero N_2 recovery from NO decomposition. The chemical rate and equilibrium constants were taken from the kinetic data reported by Baulch *et al.* (1973).

The predicted values of NO were compared with the experimental data of Robinson (1974) and the numerical results of Pope (1981). The outcome of such a comparison is presented here, as well as the way of improving this model's performance.

CHEMICAL MODEL

The evaluation of the mean source term in scalar equations, and the mean density, is performed using the pdf information describing the chemical process involved and whose dimension corresponds to the number of independent scalars in the combustion model.

A premixed flame is generally represented as a three-zone process. In the preheating zone the temperature increase is only produced by heat conduction and very poor chemical activity exists. The reaction zone is characterized by a more rapid increase of temperature and the presence of radicals in concentrations well above their equilibrium values. It is in this zone that CO and H_2 are mainly formed and major consumption of fuel and oxygen takes place. In the third zone, the recombination zone, slow chemical reactions continue and the combustion process can be completed.

Carbon monoxide formation is restricted to the reaction subregion where most fuel and O_2 are consumed and most intermediate species produced. Within the recombination zone, part of the CO formed is oxidized to CO_2 and the CO concentration at the exhaust will be set by the extent of completion of this process. In addition, it may be noted that it is in the post-reaction zone where the major production of nitrogen oxide takes place, thermal NO_x, due to the high temperature conditions; some NO_x is formed early in the reaction zone, prompt NO_x, where nitrogen contained in the fuel reacts with the oxygen present.

In the present model two different chemical models are combined: a combustion model and a NO formation mechanism. The first one is used to describe fuel consumption and carbon monoxide formation, as well as to establish the values of the best possible pdf to be combined with the chosen turbulence model.

The second model, NO formation process, will be introduced after the main chemical species concentrations and flow parameters have been determined. It will be directly affected by the other models outcome, but no inverse effect could be observed.

Combustion Representation

The simplest way of describing a chemical reaction is through a single-step global expression, but no intermediate species representation can be considered in such an approach. Therefore, to improve this chemical representation, without the use of a detailed kinetic mechanism, a two-step reaction scheme was used.

Three chemical reaction groups can be associated with the subregions of a premixed flame: initial pyrolysis of hydrocarbons, fast consumption of fuel and oxygen, and slow reactions of the intermediate species into final products.

Hydrocarbon pyrolysis reactions strongly depend on the chemical composition of the hydrocarbon and the physical conditions of the flame; as their products are hardly present in the exhaust gases, they are not considered here. The consumption of fuel and oxygen is a very fast reaction and, although it might be considered as an instantaneous process, here it will be used as the first step of the chemical scheme, while the slow transformation of an intermediate compound into products will be representing the second chemical step.

The following consecutive two-step mechanism, proposed by Westbrook and Dryer (1981), satisfies the above and describes the formation and destruction of CO:

$$C_3H_8O_2 \leftrightarrow 3CO + 4H_2O \tag{1}$$

and

$$CO + 0.5O_2 \leftrightarrow CO_2, \tag{2}$$

and the kinetic expressions related to these steps are:

$$\dot{r}_{C_3H_8} = -10^{12} \exp\left\{\frac{-30}{RT}\right\} [C_3H_8]^{0.1} [O_2]^{1.65}, \tag{3}$$

and

$$\dot{r}_{CO_2} = \exp\left\{\frac{-40}{RT}\right\} \left[10^{14.6} [CO] [H_2O]^{0.5} [O_2]^{0.25} - 10^{8.69} [CO_2]\right]. \tag{4}$$

NO Formation Mechanism

Nitrogen oxide, NO, was evaluated as the representative species of the nitrogen oxides, NO_x. As reported by Robinson, 95% of these oxides (NO and NO_2) are expected to be in the form of NO in his experiment.

Several reaction mechanisms have been suggested to represent NO formation in flames. For the equivalence ratios used in Robinson's experiments, lean and near stoichiometric, the extended Zeldovich mechanism was assumed to be an adequate description of the formation of thermal NO, this is:

$$O + N_2 \Leftrightarrow NO + N, \tag{5.i}$$

$$N + O_2 \Leftrightarrow NO + O, \tag{5.ii}$$

$$N + OH \Leftrightarrow NO + H. \tag{5.iii}$$

These reaction steps were combined with three further assumptions, suggested by Jones (1979): (1) steady state of nitrogen atom concentration; (2) small reverse rate of the first reaction step (5.i); (3) concentration of oxygen atoms due to the equilibrium of a three body reaction:

$$O + O + M \Leftrightarrow O_2 + M. \tag{5.iv}$$

Therefore, the reaction rate of NO formation can be expressed by the following equation:

$$\dot{r}_{NO} = k_{NO} W_{NO} \rho (C_{O_2}/W_{O_2})^{0.5} \, (C_{N_2}/W_{N_2}), \tag{6}$$

where k_{NO} depends only on the forward reaction rate of the first step (5.i) and the equilibrium constant of the three body reaction (5.iv). Its numerical value was taken from the kinetic data reported by Baulch *et al.* (1973), for high temperature reactions, i.e.,

$$k_{NO} = 4.0917 \times 10^{13} \left(\frac{1}{T}\right)^{0.0675} \exp\left(-\frac{67915}{T}\right). \tag{7}$$

Thus the rate of formation of NO may be uniquely related to the two scalars chosen to describe the main thermochemical events and which therefore determine T, C_{N_2} and C_{O_2}.

TRANSPORT EQUATIONS

To evaluate the characteristics of a flame (flow field, composition and temperature) the following restrictions were imposed: radiation losses are negligible, single gas phase, and low Mach number flames. The conservation principles subject to these restrictions are expressed by the following differential equations:

Continuity

$$\frac{\partial \rho}{\partial t} + \frac{\partial}{\partial x_i}(\rho u_i) = 0. \tag{8}$$

Momentum

$$\frac{\partial \rho u_i}{\partial t} + \frac{\partial}{\partial x_j}(\rho u_j u_i) = \frac{\partial \sigma_{1j}}{\partial x_j} - \frac{\partial p}{\partial x_i} + g_i. \tag{9}$$

Scalars

$$\frac{\partial \rho \phi}{\partial t} + \frac{\partial}{\partial x_j}(\rho u_j \phi) = \frac{\partial}{\partial x_j}\left(\Gamma_\phi \frac{\partial \phi}{\partial x_j}\right) + S_\phi. \tag{10}$$

In the above equation ϕ stands for any scalar: enthalpy, species concentration, temperature, kinetic energy, etc., and Γ_ϕ and S_ϕ are the diffusion coefficient and the source term.

These instantaneous equations were statistically averaged, and written for a stationary one-dimensional flame. To simplify the averaged form of the equations, for the case of flows with density variations, density-weighted averaging was used. Thus, (10) for the mean value of a scalar will be:

$$\frac{\partial}{\partial x}\left(\overline{\rho}\,\widetilde{u}\,\widetilde{\phi}\right) = \frac{\partial}{\partial x}\left(\Gamma_\phi \frac{\partial \widetilde{C}}{\partial x} - \overline{\rho u'' \phi''}\right) + \widetilde{S}_\phi. \tag{11}$$

Turbulence Model

The turbulent transport terms were modeled using a gradient diffusion approximation by introducing a turbulent exchange coefficient; thus for the turbulent scalar fluxes:

$$-\overline{\rho u'' \phi''} = \frac{\mu_t}{\sigma_\phi}\frac{\partial \widetilde{\phi}}{\partial x}, \tag{12}$$

where σ_ϕ is an empirical constant, equal to 0.7, referred to as the turbulent Prandtl number, and μ_t is the turbulent viscosity, evaluated from the kinetic energy and its rate of dissipation (k-ε model; see Jones and Launder, 1972, and Launder and Spalding, 1972):

$$\mu_t = C_\mu \overline{\rho}\,\widetilde{k}^2/\widetilde{\varepsilon}; \tag{13}$$

The transport equations for $\widetilde{k}$ and $\widetilde{\varepsilon}$ are:

$$\frac{\partial}{\partial x}\left(\overline{\rho}\,\widetilde{u}\,\widetilde{k}\right) = \frac{\partial}{\partial x}\left(\frac{\mu_t}{\sigma_k}\frac{\partial \widetilde{k}}{\partial x}\right) + \overline{S}_k, \tag{14}$$

and

$$\frac{\partial}{\partial x}\left(\overline{\rho}\,\widetilde{u}\,\widetilde{\varepsilon}\right) = \frac{\partial}{\partial x}\left(\frac{\mu_t}{\sigma_\varepsilon}\frac{\partial \widetilde{\varepsilon}}{\partial x}\right) + \overline{S}_\varepsilon. \tag{15}$$

It is important to point out here that the scalar parameters in this model are k, ε, $C_{C_3H_8}$, C_{CO_2}, $C''^2_{C_3H_8}$ and $C''^2_{CO_2}$. The source terms of the species concentration equations are mean values of a reaction rate, evaluated from a pdf expression, while the source terms of k and ε are given by the following expressions:

$$\overline{S}_k = \mu_t\left(\frac{\partial \widetilde{u}}{\partial x}\right) - \overline{\rho}\,\widetilde{\varepsilon}, \tag{16}$$

and

$$\overline{S}_\varepsilon = C_1\left(\widetilde{\varepsilon}/\widetilde{k}\right)\mu_t\left(\frac{\partial \widetilde{u}}{\partial x}\right)^2 - C_2\overline{\rho}\left(\widetilde{\varepsilon}^2/\widetilde{k}\right). \tag{17}$$

The values of all constants included in this model are given in Table I.

TABLE I
Constants in the k-ε Model

Constant	Value
C_μ	0.09
C_1	1.44
C_2	1.92
σ_k	1.0
σ_ε	1.3

THE pdf MODEL

The source term of (11), the mean reaction rate, was stochastically evaluated from a two-scalar pdf and the instantaneous reaction rate values, i.e.,

$$\dot{r}\phi = \int_0^1 \int_0^1 \dot{r}_\phi (\mathcal{F}^*, \mathcal{C}^*)\, \mathrm{P}(\mathcal{F}^*, \mathcal{C}^*)\, d\mathcal{F}^*\, d\mathcal{C}^*, \tag{18}$$

where $\mathcal{F}^*$ and $\mathcal{C}^*$ are the normalized concentration of fuel (propane) and carbon dioxide, respectively, and $\mathrm{P}(\mathcal{F}^*, \mathcal{C}^*)$is the pdf, proposed by Bray–Moss (1977b), which has proved to be especially useful for the combustion of hydrocarbons in air where two consecutive steps can be clearly distinguished. This pdf contains five possible different gas states described by the following equation:

$$\mathrm{P}(\mathcal{F}^*, \mathcal{C}^*) = \delta(\mathcal{C})\,[\alpha\delta(1-\mathcal{F}) + \beta\delta(\mathcal{F}) + \gamma\mathrm{F}(\mathcal{F})] + \delta(\mathcal{F})\,[\psi\mathrm{Z}(\mathcal{C}) + \varphi\delta(1-\mathcal{C})], \tag{19}$$

where α, β and φ are the weights of the delta functions representing unburned mixture, all fuel consumed (delay zone) and the final all-burned state, respectively; while γ and ψ are associated with the distribution functions F and Z representing fuel consumption and CO oxidation reactions in progress as occur in an undisturbed laminar flame (see fig. 1).

The distribution functions F and Z were normalized to integrate to unity, therefore

$$\alpha + \beta + \gamma + \psi + \varphi = 1.0$$

has to be satisfied. To evaluate these five parameters and build up the pdf, the following statistical moments of the distribution were used

$$\int_0^1 \int_0^1 \mathrm{P}(\mathcal{F}^*, \mathcal{C}^*)\, d\mathcal{F}^*\, d\mathcal{C}^* = 1.0, \tag{21}$$

$$\frac{1}{\rho}\int_0^1 \int_0^1 \mathcal{F}^* \rho \mathrm{P}(\mathcal{F}^*, \mathcal{C}^*)\, d\mathcal{F}^*\, d\mathcal{C}^* = \widetilde{\mathcal{F}}^*, \tag{22}$$

$$\frac{1}{\rho}\int_0^1 \int_0^1 \mathcal{C}^* \rho \mathrm{P}(\mathcal{F}^*, \mathcal{C}^*)\, d\mathcal{F}^*\, d\mathcal{C}^* = \widetilde{\mathcal{C}}^*, \tag{23}$$

$$\frac{1}{\rho}\int_0^1 \int_0^1 \left(\mathcal{F}^* - \widetilde{\mathcal{F}}^*\right)^2 \rho \mathrm{P}(\mathcal{F}^*, \mathcal{C}^*)\, d\mathcal{F}^*\, d\mathcal{C}^* = \overline{\mathcal{F}}^{*\prime\prime 2}, \tag{24}$$

$$\frac{1}{\rho}\int_0^1 \int_0^1 \left(\mathcal{C}^* - \widetilde{\mathcal{C}}^*\right)^2 \rho \mathrm{P}(\mathcal{F}^*, \mathcal{C}^*)\, d\mathcal{F}^*\, d\mathcal{C}^* = \overline{\mathcal{C}}^{*\prime\prime 2}, \tag{25}$$

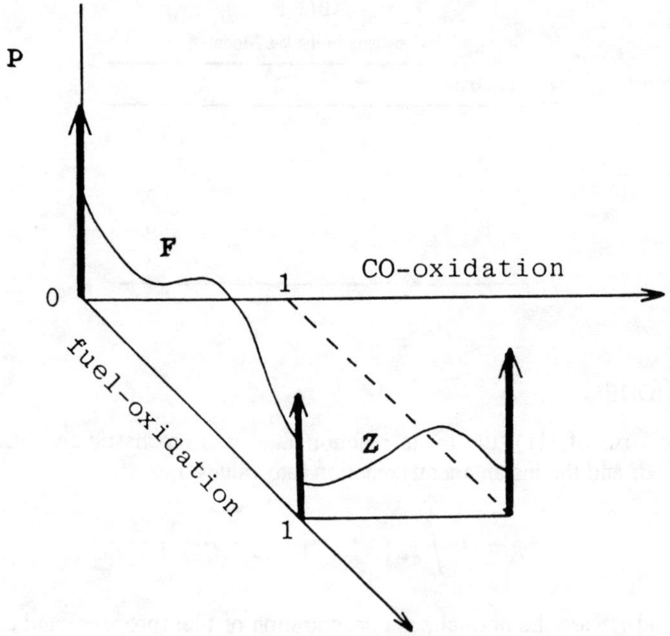

FIGURE 1 Consecutive two-step Bray-Moss pdf.

where, for brevity, $\rho = \rho(\mathcal{F}^*, \mathcal{C}^*)$.

Progress Reaction Functions

The kinetic information introduced into this mode, is represented in the pdf via the reaction progress-variable functions F and Z which have to be specified *a priori* in order to close the pdf description within the Bray–Moss.

These progress-variable functions are closely linked to the description of the flame structure adopted. In the present approach, they were assumed to correspond to laminar structures (inside the reaction zone) and therefore built from reported concentration profiles belonging to laminar flames.

Since the final form of the pdf information used in the transport equations is concerned with integral relationships involving these variable functions, then their detailed shapes are unimportant except at the extremes (the tails) of the flame profile, because small gradients of concentration are typically found at the beginning and at the end of the reaction zone (see Bray and Moss, 1977a).

The incorrect evaluation of these *tails* can strongly bias the reaction rate value representing the laminar reaction zone and therefore the final output of this model. To evaluate this effect two different F functions were tested for the evaluation of fuel concentration, see fig. 2. Shape (a) is quoted by Bray and Moss as a typical laminar flame description whilst shape (b) was obtained from the experimental data of Bechtel *et al.* (1981), albeit at a different equivalence ratio.

To obtain the shapes in fig. 2, the theoretical or experimental progress reaction information was approximated by straight line segments into a normalized space (see fig. 3). The normalized fuel consumption and CO_2 formation variables have been defined as:

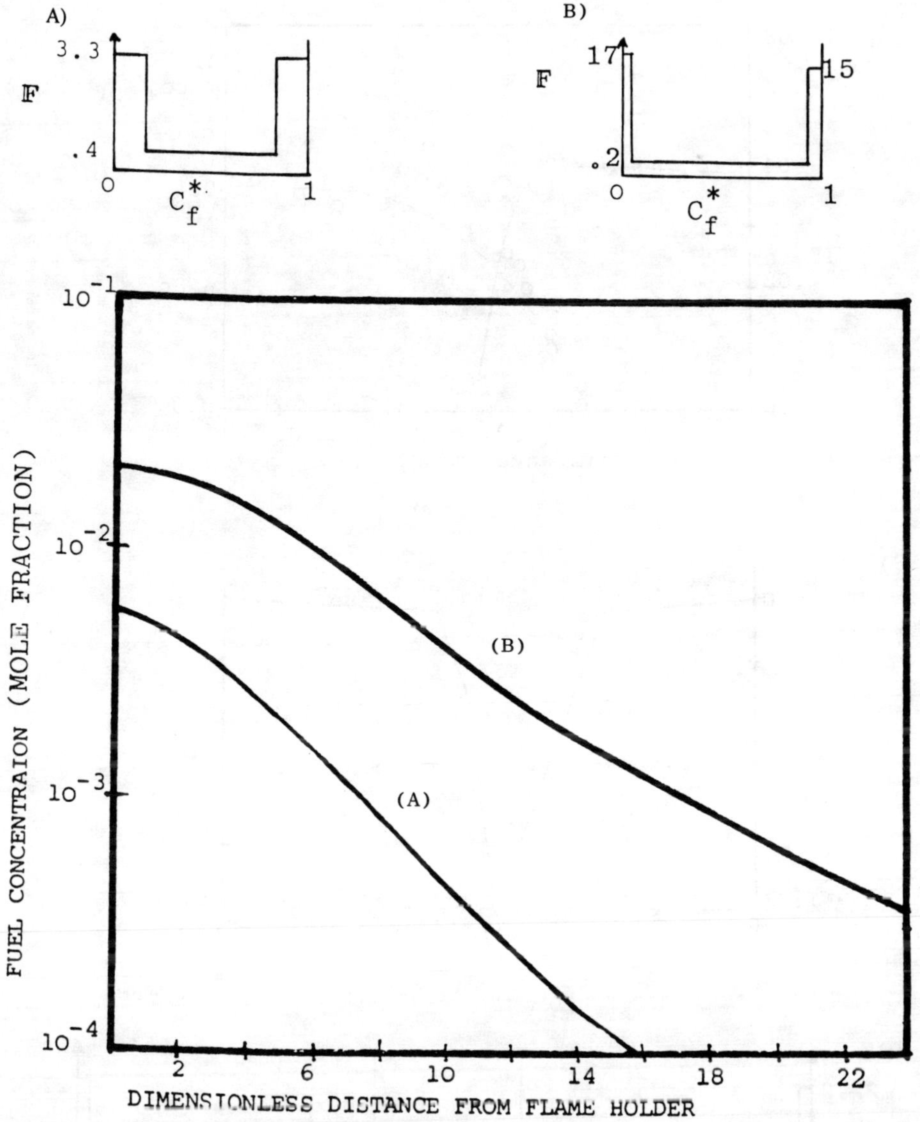

FIGURE 2 Effect of the laminar reaction progress variable function on the turbulent combustion model predictions.

$$\mathcal{F}^* = 1.0 - C_{C_3H_8}/C_{C_3H_8,\max}, \tag{26}$$

and

$$\mathcal{C}^* = C_{CO_2}/C_{CO_2,\max}. \tag{27}$$

Then the progress functions were evaluated from

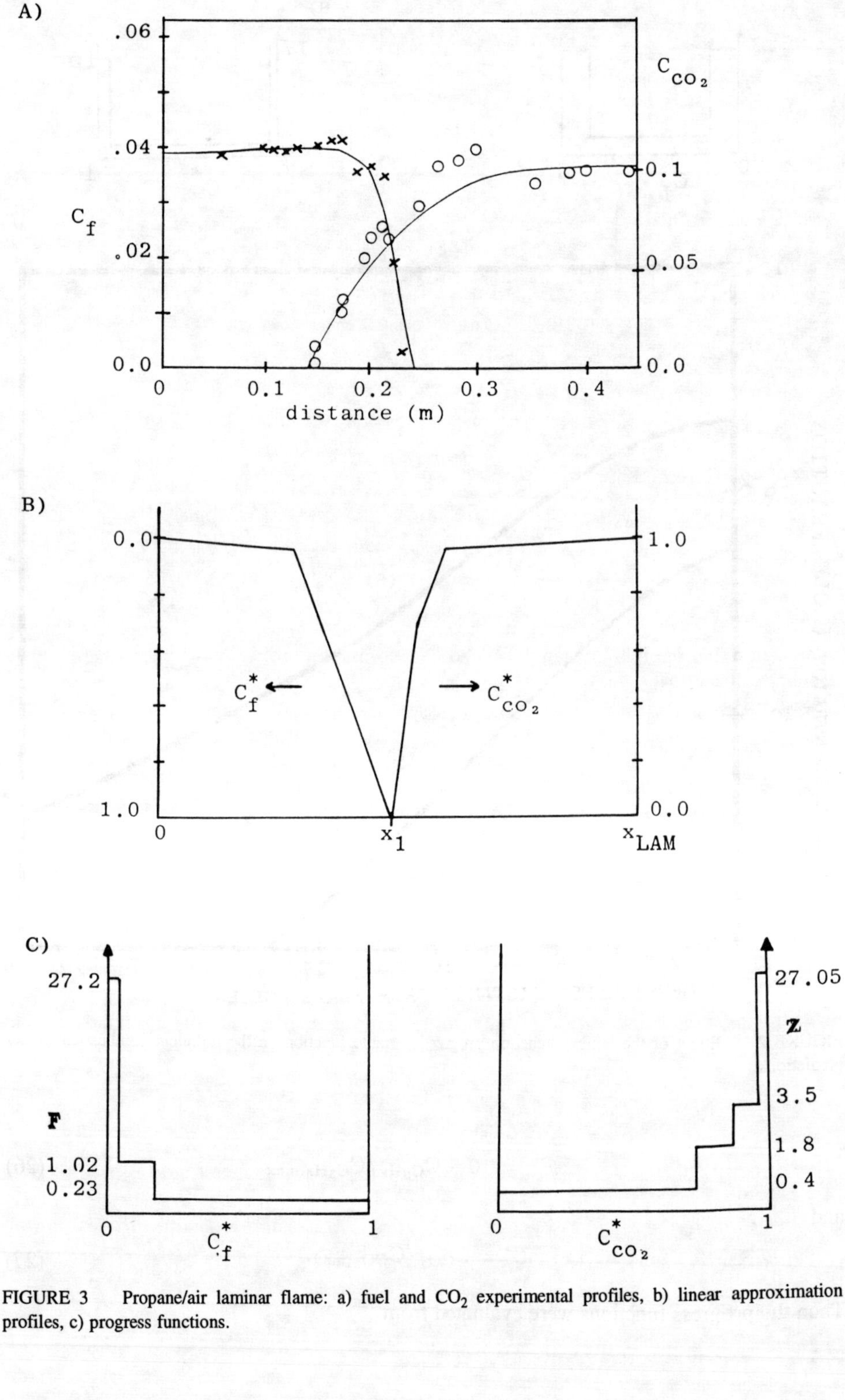

FIGURE 3 Propane/air laminar flame: a) fuel and CO_2 experimental profiles, b) linear approximation profiles, c) progress functions.

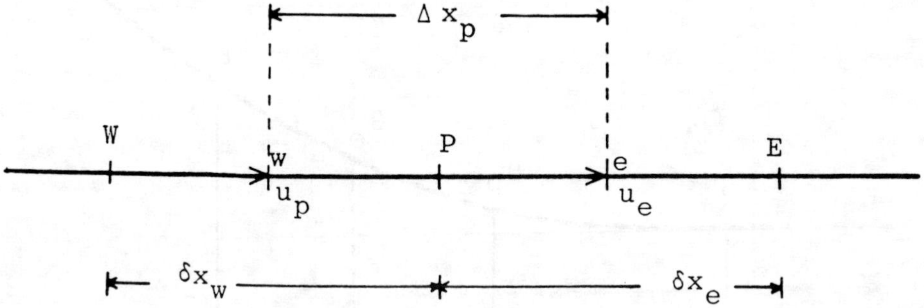

FIGURE 4 Control volume. Computational grid.

$$F = \left(\frac{d\mathcal{F}^*}{d(\alpha/\alpha_1)} \right)^{-1}, \tag{28}$$

where α_1 is the length of fuel consumption, and for the second chemical step

$$Z = \left(\frac{d\mathcal{C}^*}{d(\alpha/\alpha_2)} \right)^{-1}, \tag{29}$$

with $\alpha_2 = \alpha_{lam} - \alpha_1$ as the characteristic length of reaction, see fig. 3. It should be noted in this figure that the fuel-consumption and CO_2 formation profiles have been schematically separated as a further reminder of the consecutive nature of the chemical scheme adopted; the experimental data indicates that for part of the flame zone this is indeed a gross approximation. As mentioned above, the actual experimental profiles have been approximated in a piecewise linear fashion.

NUMERICAL PROCEDURE

All differential equations used in this model, continuity, transport of momentum and scalars (eqs. (11), (14) and (15)), were written in the standard from:

$$\frac{\partial}{\partial x}(\overline{\rho}\widetilde{u}\theta) = \frac{\partial}{\partial x}\left(\Gamma^\theta \frac{\partial \theta}{\partial x}\right) + \widetilde{S}^\theta, \tag{30}$$

which can be reduced to a discretised form by integration over the small volume surrounding each node of a finite difference grid, i.e.,

$$\overline{\rho}_e \widetilde{u}_e \theta_e - \overline{\rho}_w \widetilde{u}_p \theta_w = \Gamma^\theta_e \frac{\partial \theta}{\partial x_e} - \Gamma^\theta_w \frac{\partial \theta}{\partial x_w} + S^\theta_p \Delta x_p. \tag{31}$$

The velocity values do not require interpolation because a staggered grid is used (see fig. 4).

The diffusion coefficients are evaluated at w and e using a central differencing approximation:

$$\Gamma^\theta_e = (\Gamma^\theta_E + \Gamma^\theta_P) / 2, \tag{32}$$

and the same holds for the cell face density:

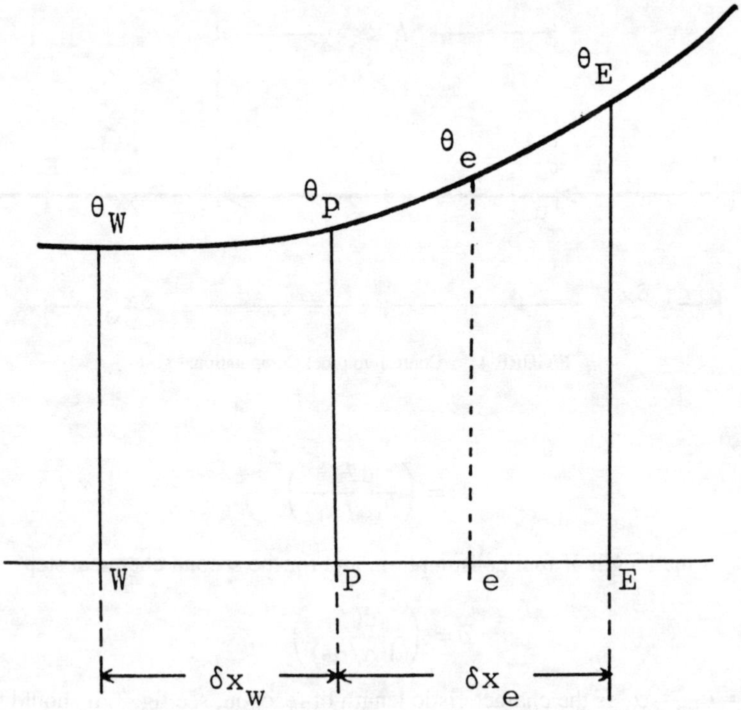

FIGURE 5 Quadratic upstream interpolation.

$$\rho_{\mathrm{e}}^{\theta} = \left(\rho_{\mathrm{E}}^{\theta} + \rho_{\mathrm{P}}^{\theta}\right) / 2. \tag{33}$$

The gradient terms are given by the following interpolation:

$$\left(\frac{\partial \theta}{\partial \mathrm{x}}\right)_{\mathrm{e}} = \left(\theta_{\mathrm{E}} - \theta_{\mathrm{P}}\right) / \delta \mathrm{x}_{\mathrm{e}}. \tag{34}$$

For the interpolation of the convective terms, the QUICK method (quadratic upstream interpolation) was used. This method was proposed by Leonard (1979) and avoids some of the stability problems associated with central differencing, and the numerical diffusion inaccuracies of the upstream differencing technique. Figure 5 presents this parabolic distribution, which was successfully used in these one-dimensional calculations.

Finally, using these discretising techniques, a matrix expression is obtained from (30), i.e.,

$$\mathrm{A}_{\mathrm{P}}\theta_{\mathrm{P}} = \sum_{\mathrm{i=e,w}} \left(\mathrm{A}_i\theta_i + B_i\right), \tag{35}$$

which is conveniently solved by the well know TDMA algorithm (see Patankar, 1980, Leonard *et al.* 1979, and Han *et al.* 1981).

The solution procedure used in the present model is known as the SIMPLE algorithm (Semi Implicit Method for Pressure-Linked Equations); it was proposed by Patankar and Spalding (1972), and establishes the following calculation sequence: i) guess a pressure

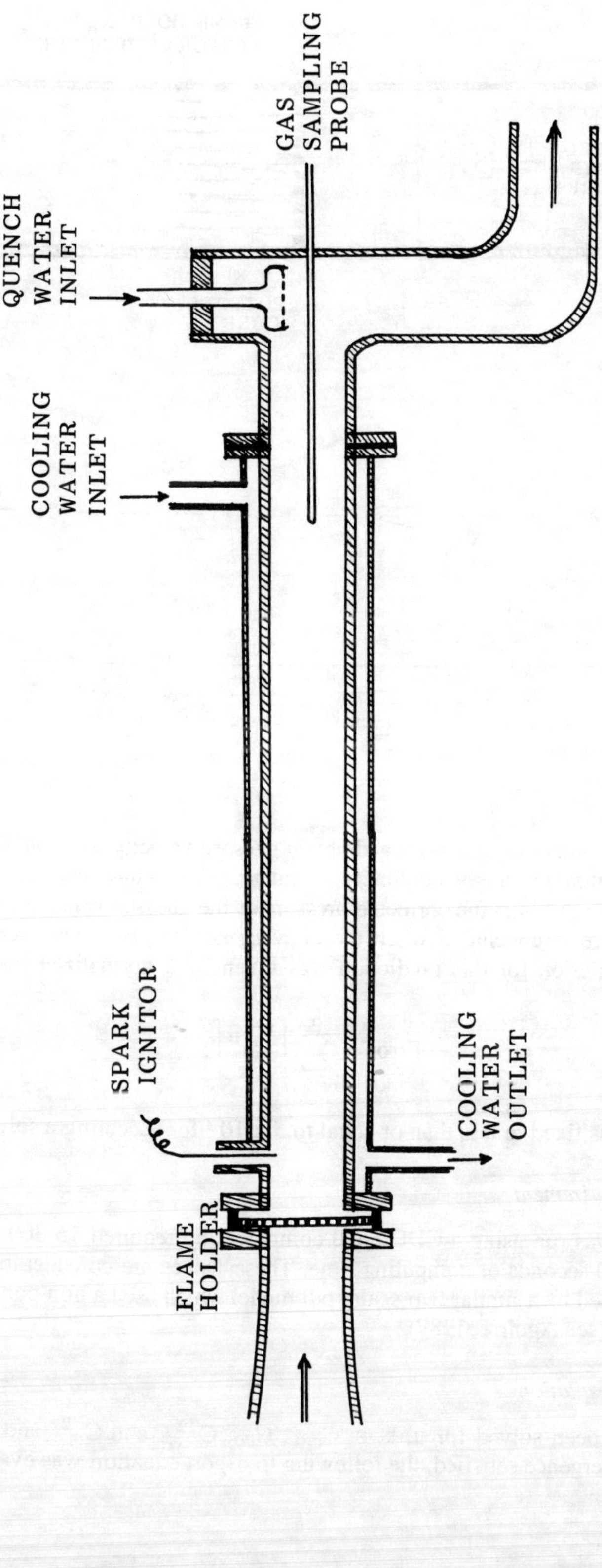

FIGURE 6 Schematic of the turbulent premixed flame setup used by Robinson.

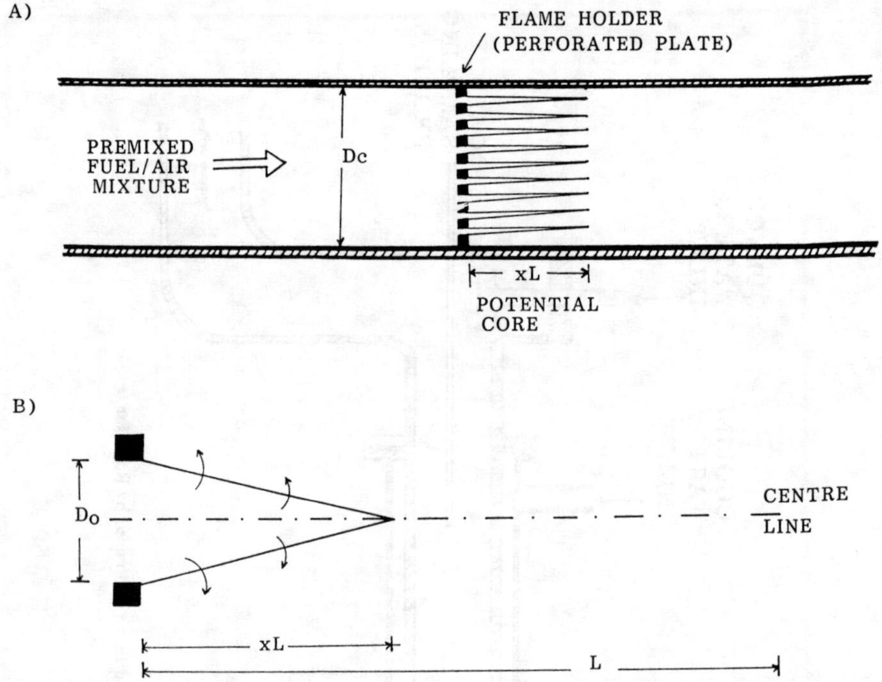

FIGURE 7 Turbulent premixed flame representation: a) combustor schematic representation, b) one-dimensional flow representation.

field; ii) solve momentum equations and obtain guessed velocity components; iii) evaluate pressure correction to satisfy continuity (locally); iv) evaluate corrected pressure and corrected velocity; v) treat the corrected pressure as the guessed value and return to (ii).

This procedure is concluded when the convergence criterion has been reached. The convergence criterion for this prediction was given by a normalized residual error estimator

$$\text{Error} = \sum_{i=1}^{n} \left| \frac{\partial \bar{\rho} \tilde{u}}{\partial x} \right|, \tag{36}$$

whose value was fixed to less than or equal to 5×10^{-4} for accepting a solution.

Computing Requirement

This program was run using a CDC-6600 computer and required 75 300 octal words for storage and 290 seconds of computing time. These values are considerably less than the requirements used by a similar two-scalar pdf model which used a non-consecutive chemical description (see Aguirre, 1987).

NO Transport Equation

Once (30) has been solved for u, k, ε, $C_{C_3H_8}$, C_{CO_2}, $C''^2_{C_3H_8}$ and $C''^2_{CO_2}$ and continuity, and numerical convergence satisfied, the following transport equation was evaluated:

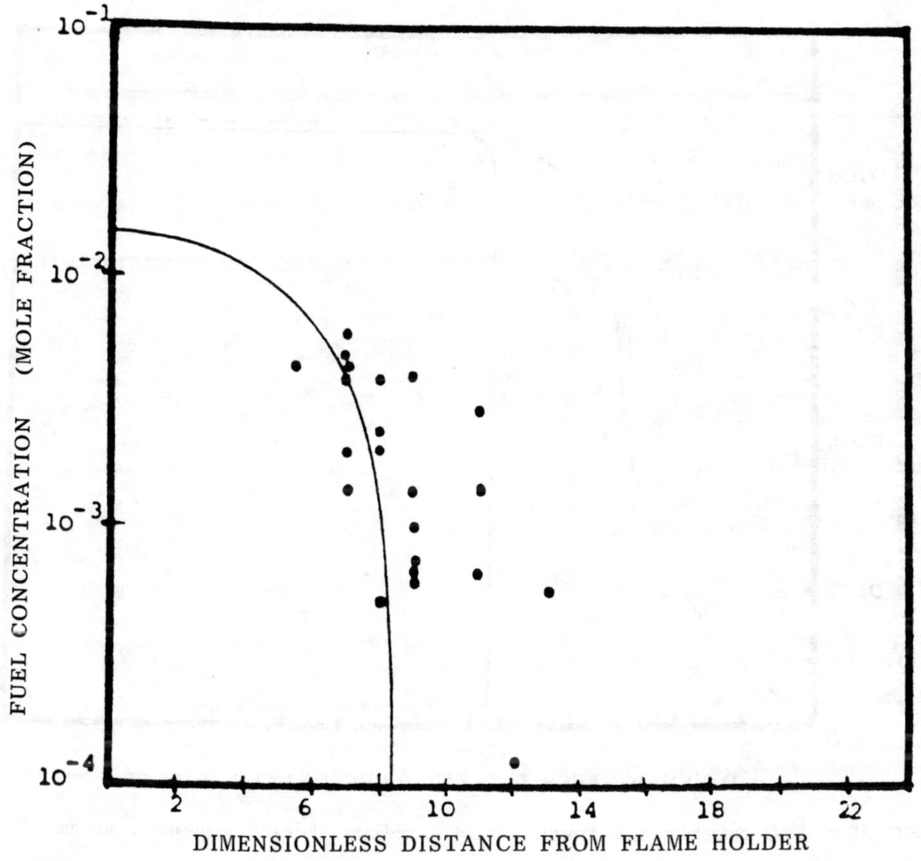

FIGURE 8 Fuel concentration: • experimental and — predicted values (0.73 equivalence ratio and 0.815 solidity ratio).

$$\frac{\partial}{\partial x}(\bar{\rho}\,\tilde{u}\,C_{NO}) = \frac{\partial}{\partial x}\left(\frac{\mu_t}{\sigma_{NO}}\,\frac{\partial C_{NO}}{\partial x}\right) + \tilde{\dot{r}}_{NO}, \tag{37}$$

where the mean reaction rate of NO formation can then be evaluated from the pdf, as follows:

$$\tilde{\dot{r}}_{NO} = \int_0^1\int_0^1 \dot{r}_{NO}(\mathcal{F}^*, \mathcal{C}^*)\,P(\mathcal{F}^*, \mathcal{C}^*)\,d\mathcal{F}^*\,d\mathcal{C}^*, \tag{38}$$

and the instantaneous value of the reaction rate comes from (6) above.

RESULTS

Figure 6 presents the experimental setup used by Robinson to burn out propane, and fig. 7 the one-dimensional turbulent premixed flame representation. The flame holder used in this

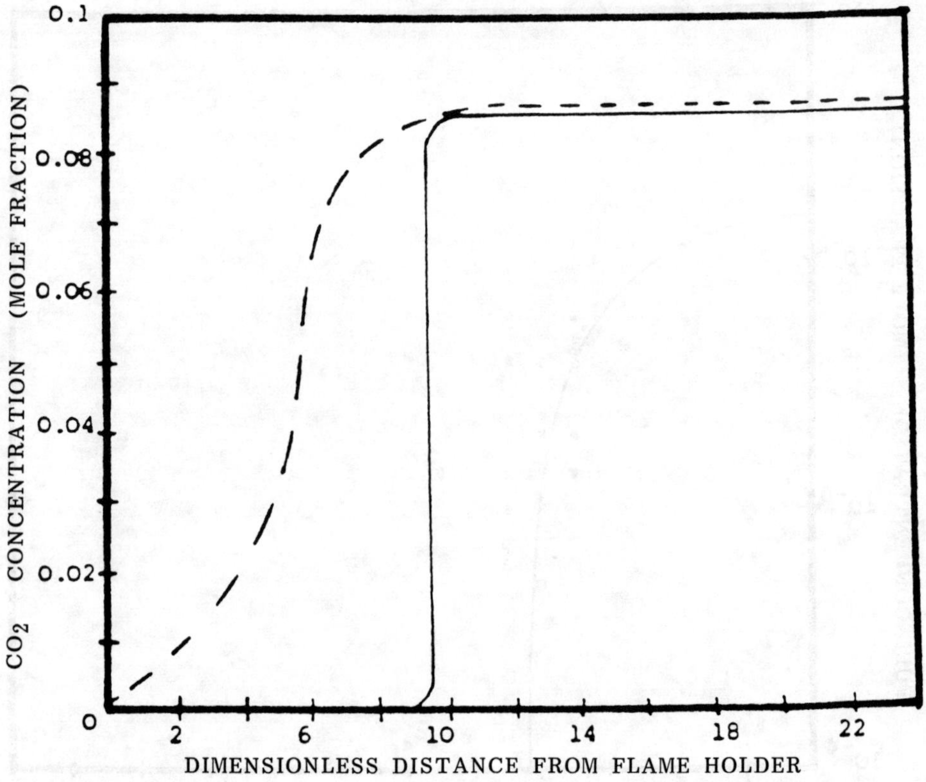

FIGURE 9 CO_2 concentration, --- experimental and — predicted values (0.73 equivalence ratio and 0.815 solidity ratio).

experiment was a perforated plate whose solidity ratio (blocked area/combustor area) was 0.742, 0.815 or 0.892.

Figures 8 and 9 correspond to fuel and CO_2 concentration, respectively. It can be observed that the distance of fuel consumption and final concentration of CO_2 are reasonably well predicted, although a sequential chemistry consideration produced too rapid a decrease and rise of these concentrations.

Figure 10 presents the carbon monoxide concentration; once again a generally good agreement between experiments and numerical predictions is observed, as well as the rapid concentration decrease due to sequential chemistry consideration.

Finally, figs. 11 and 12 correspond to the nitrogen oxide concentrations obtained for two experimental conditions and compared with Robinson's data, fig. 11, and with Robinson's data and Pope's predictions, fig. 12. The observed difference between predicted and experimental data could be directly related to the kinetic constant used in the model or to the exclusion of prompt NO.

In the last figure it can be observed that the NO profile predicted by Pope (1981), presents the same general shape as the present model but its values are closer to the experimental data, which has to be a direct result of the kinetic information adopted, as Pope also omitted prompt NO formation.

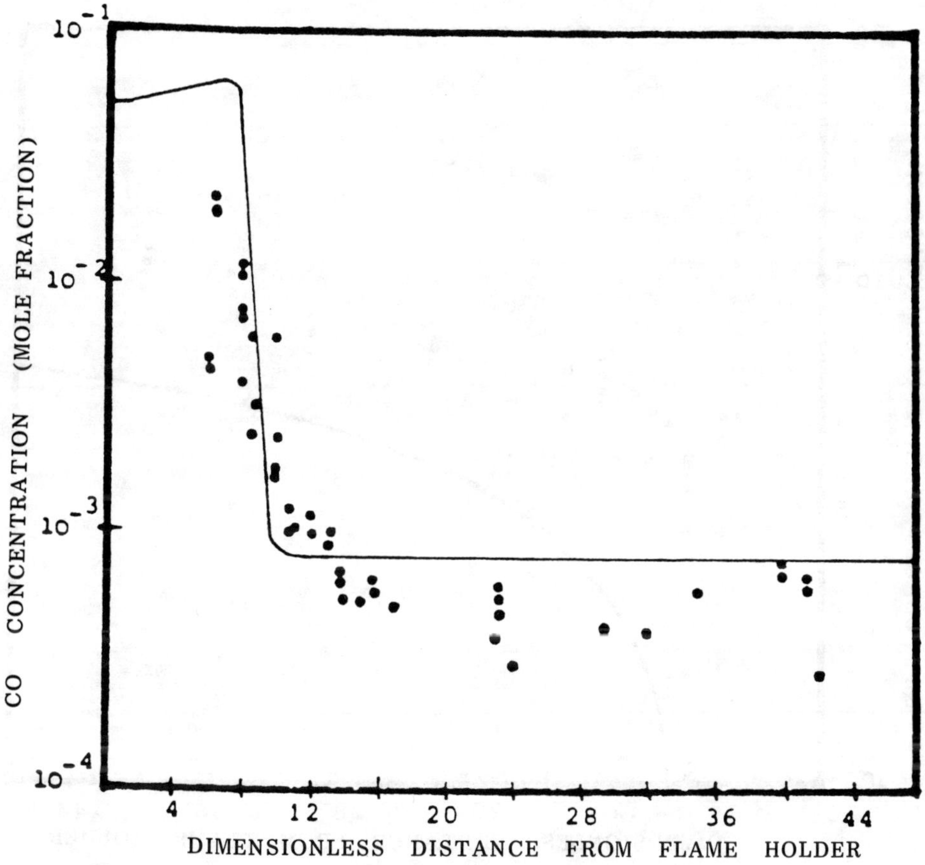

FIGURE 10 CO concentration: • experimental and — predicted values (0.73 equivalence ratio and 0.815 solidity ratio).

NO formation in Pope's model is also based on the extended Zeldovich mechanism with the steady-state assumption for nitrogen atom concentration; but, kinetic information (reaction rate and equilibrium constants) were taken from a different source and, probably, different values of these variables have produced the different output. In Pope's model a two scalar pdf is also used, but it is stochasticaly evaluated using the Monte Carlo method.

CONCLUSIONS

The performance of this model is considered highly satisfactory because of the relatively modest computational effort involved, while the final predicted concentrations compared well with the experimental data — even with the rapid increase and decrease of flame conditions in the reaction zone due to the consecutive approach.

It must be pointed out that this model, from the viewpoint of NO prediction, could be easily modified and improved if better laminar flame descriptions were available for evaluating the progress-variable functions, or a better kinetic expression to represent the consecutive chemical reactions could be found.

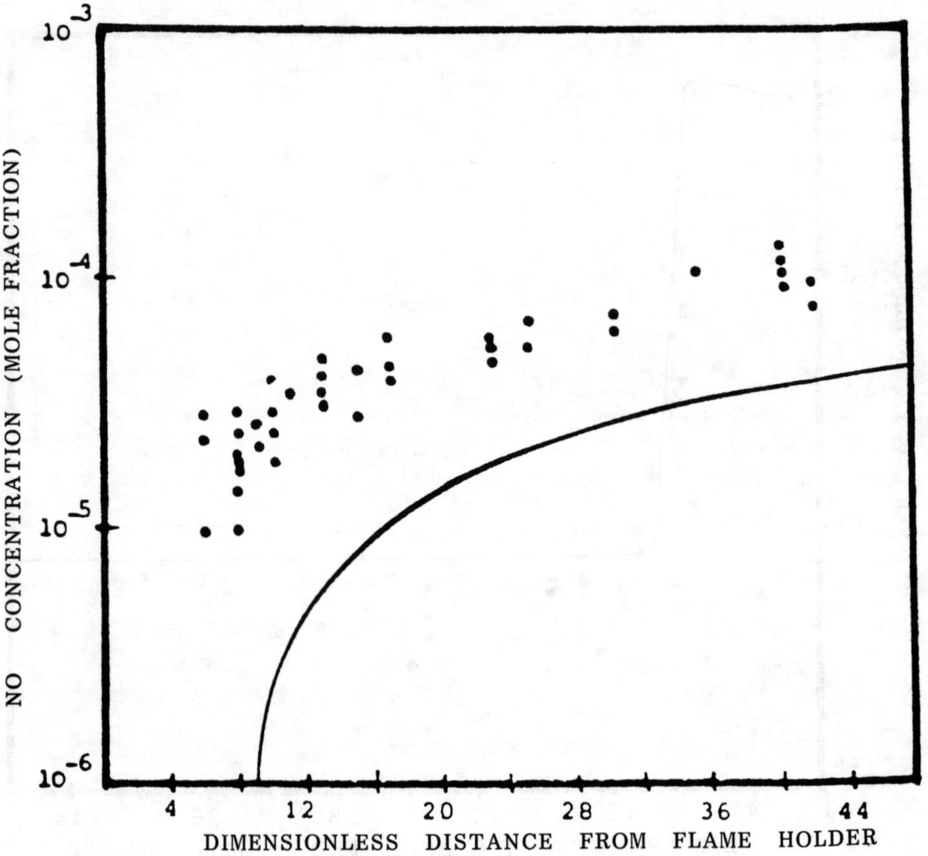

FIGURE 11 NO concentration: • experimental and — predicted values (0.73 equivalence ratio and 0.815 solidity ratio).

Comparisons with all the experimental data of Robinson were performed and similar conclusions could be obtained from all of them. Some figures of those comparisons are included in Aguirre (1987), where temperature profiles were also drawn showing that temperature over-prediction can be reduced from 50 to 40 degrees when moving from a one- to a two-scalar model, mainly due to the CO consideration.

No extensive comparison with Pope's method was carried out because he reports only a few predictions of species concentration; in the figure included here, the NO concentration is better predicted by Pope, but this is probably related to the use, in the present calculations, of a model which assumes that the NO levels are far from equilibrium.

ACKNOWLEDGMENTS

This work was carried out at the Imperial College of Science and Technology, as part of my PhD research program, under the supervision of Dr. J.J. McGuirk, from whom I received many valuable suggestions and constructive criticism.

I would like to express my gratitude to my supervisor, to all members of the fluid mechanics section at Imperial College, to the combustion research group of this College,

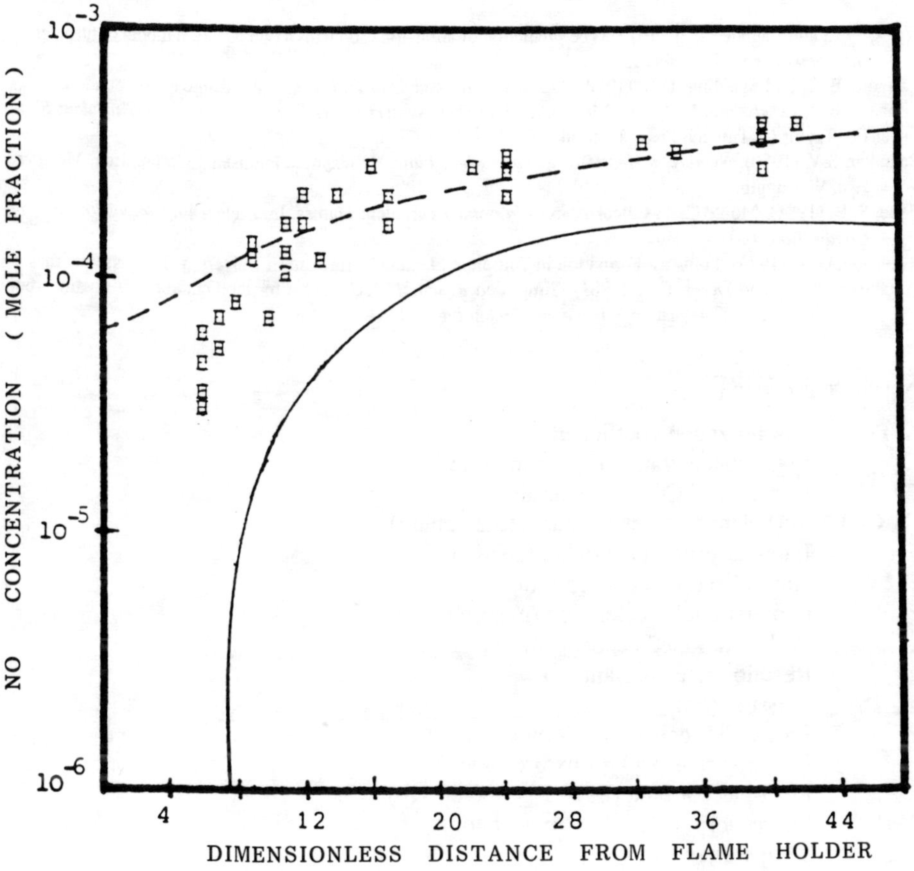

FIGURE 12 NO concentration: ❑ experimental, --- Pope's, and — this model predicted values (0.845 equivalence ratio and 0.815 solidity ratio).

NGTE and Rolls-Royce, and to acknowledge the financial support provided by the National University of Mexico.

REFERENCES

Aguirre, R. (1987). Two-scalar probability density function models for turbulent flames. Imperial College, PhD thesis.

Baulch D.L., Drysdale D.D., Horne D.G., and Lloyd A.C. (1973). *Evaluated Kinetic Data for High Temperature Reactions*. Butterworths Press, London.

Bechtel, J.H., Blint R.J., Dasch C.J., and Weinberger D.A. (1981). Atmospheric pressure premixed hydrocarbon-air flames: theory and experiment. *Combustion and Flame*, **42**, 197.

Bray, K.N.C., and Moss, J.B. (1977a). A unified statistical model of the premixed turbulent flame. *Acta Astronautica*, **4**, 291.

Bray, K.N.C., and Moss, J.B. (1977b). A closure model for the turbulent premixed flame with sequential chemistry. *Combustion and Flame*, **30**, 125.

Han, T., Humprey J.A.C., and Launder, B. (1981). A comparison of hybrid and quadratic-upstream differencing in high Reynolds number elliptic flows. *Computer Methods in Applied Mechanics and Engineering*, **29**, 81.

Jones, W.P., and Kollmann, W. (1985). Multi-scalar pdf-transport equations for turbulent diffusion flames. In: Bradbury L.J.S. (Ed.), *Fifth Symposium on Turbulent Shear Flows*. Springer-Verlag.

Jones, W.P., and Launder, B.E. (1972). The prediction of laminarization with a two-equation model of turbulence. *Int. J. Heat Mass Transfer*, **15**, 301.
Launder, B.E., and Spalding, D.B. (1972). *Mathematical Models of Turbulence*. Academic Press, London.
Leonard, B.P., Leschziner, M.A., and McGuirk, J.J. (1979). *Numerical Methods in Laminar and Turbulent Flow*. Ed. Taylor C., Pentech Press, London.
Patankar, S.V. (1980). *Numerical Heat Transfer and Fluid Flow*. Hemisphere Publishing Corporation, McGraw-Hill. Washington.
Pope, S. B. (1981). Monte Carlo Calculations of Premixed Turbulent Flames. In: *Eighteenth Symposium (Int.) on Combustion*, 1001.
Robinson, G. F. (1974). Pollutant Formation in Turbulent Flames. Northwestern University, Illinois, PhD thesis.
Westbrook C. K., and Dryer, F. L. (1981). Simplified Reaction Mechanisms for the Oxidation of Hydrocarbon Fuels in Flames. *Combustion Science and Technology*, **27**, 31.

NOMENCLATURE

A, D	Discretization coefficient
C	Mass concentration (kg kg-mixture^{-1})
C^*	Normalized CO_2 concentration
C_1, C_2, C_m	Turbulence model constants (see Table I)
F	Laminar progress variable function
$\mathcal{F}^*$	Normalized fuel concentration
g	Gravitational acceleration (m sec^{-2})
k	Turbulent kinetic energy (m^2 sec^{-2})
k_{NO}	Reaction rate constant
p	Pressure (Pa)
P	Probability density function
r	Reaction rate (kg kg-mixture^{-1} sec^{-1})
R	Ideal-gas constant (8314 J kg-mole^{-1} K^{-1})
S	Source term (according to equation)
T	Temperature (K)
t	Time (sec)
u	Axial velocity (m sec^{-1})
u_1	Velocity vector (m sec^{-1})
W	Molecular weight (kg kg-mole^{-1})
x	Axial distance (m)
x_1	Cartesian coordinates (m)
α	Reaction length (m)
Z	Laminar progress variable function

GREEK SYMBOLS

α, β, γ, ψ, φ	Distribution functions
δx	Grid distance (m)
δ	Delta function
ε	Dissipation rate turbulent kinetic energy (m^2 sec^{-3})
Γ	Molecular diffusion coefficient
ϕ	Scalar (according to equation)
μ_t	Turbulent viscosity (kg m^{-1} sec^{-1})
ρ	Density (kg m^{-3})
σ	Prandtl number, $\mu C_p/K$ (dimensionless)
σ_{ij}	Stress tensor (kg m^{-1} sec^{-2})
θ	Surrogate variable (discretization)

OPERATORS

$\widetilde{\ldots}$	Density weighted average
$\ldots''$	Weighted fluctuating component
[...]	Species concentration (kg-mole m^{-3})

Co-Visualization of Flame Structures and Species

D. PROCTOR, I.G. PEARSON, AND M. McLEOD *Commonwealth Scientific and Industrial Research Organisation (CSIRO), Division of Building, Construction and Engineering, Highett, Victoria, Australia*

Abstract — This paper shows how laser-based techniques, such as schlieren and Planar Laser Induced Fluorescence (PLIF) images, can be combined in a novel manner to study how changes to the fluid dynamics alter the emissions and temperatures from gas burners. It will be shown how this results in a cost effective tool to assist in the development of new, more efficient burners and hence lower emissions. The elucidation of specific molecular formation and removal mechanisms is greatly enhanced by using this method. The experimental apparatus and procedure will be discussed along with examples of a series of digitally recorded pictures illustrating the salient features of the technique.

Key Words: Laser diagnostics, LIF, schlieren, NO_2, CH, spectroscopy

INTRODUCTION

Various energy substitution strategies have been put forward as ways of ameliorating the global temperature increase which is scheduled to be making its presence felt early in the 21st century. LPG (propane) and natural gas (methane) have been discussed as possible substitutes for other fuels with higher carbon to hydrogen ratios, such as coal, to reduce the amount of CO_2 emitted to the atmosphere from combustion. The trend has been to produce natural gas burners that are very efficient and emit substantially lower levels of pollutants such as the nitrogen oxides N_2O, NO and NO_2, which in themselves contribute to global warming either directly in the case of N_2O, or through photochemical reactions involving unburnt hydrocarbons, NO_x and UV light from solar radiation to produce O_3. To properly understand these new burners requires specialized tools, usually involving laser diagnostic techniques because of their noninvasive, nonperturbing nature. A laser-based technique is presented in this paper which combines different techniques in a way that has, to our knowledge, not been attempted before.

The effectiveness of many combustion systems depends on the ability to maintain a stable flame at reasonably high combustion efficiency. Both aspects are closely related to the flow characteristics inside the combustor and the degree of mixing obtained between the injected fuel and the air. The mixing involves two important processes. The large-scale structures bring into the mixing layer large amounts of reacting components from the two separated streams. The fine-scale eddies enhance the mixing, at the molecular level between the reactants, which is a necessary condition to initiate the chemical reaction. The understanding of this complex gas-dynamic process requires analysis of the interactions between fluid dynamics, chemical reaction, acoustic waves and heat release of the reactive system (Broadwell and Dimotakis, 1986; Ballal, 1986).

Of particular concern at the moment are diffusion (non-premixed) flames which are the major contributor to combustion energy usage in large industrial boilers and furnaces. The fluid dynamics involved in these types of flames are related to the mixing layer flow, which is governed by large-scale vortices. The roll-up and growth of these vortices are determined by instability forces and mutual interactions (Brown and Roshko, 1974). The large-scale structures entrain flow from the two sides of the shear layer and mix the flow at both large and small scales, leading to the molecular mixing necessary for reaction. The work of Brown and Roshko (1974) led to the understanding of the important role of large coherent structures in mixing layer dynamics and opened up the possibility of controlling the flow in combustors to improve their performance. The development of large-scale vortices in a shear layer is influenced by the reaction through heat release and density gradients. In turn, the vorticity controls the reaction by the turbulent mixing process (Mungal and Dimotakis, 1984). The physical mechanisms involved in this interaction are not fully understood. Although many of the phenomena have been observed in non-combusting flow or postulated on the basis of numerical modeling, experimental results in combusting flows are limited. Previous workers have used several methods to detail the structure of turbulent diffusion flames. Vandsburger *et al.* (1986) used OH Planar Laser Induced Fluorescence (PLIF) and Mie scattering from TiO_2 particles; Chen and Roquemore (1986) have used planar sheet imaging of Mie scatter from TiO_2 particles generated in the flame; Strawa and Cantwell (1985) studied slow-speed acoustically forced diffusion flames using schlieren and photographic imaging. The OH molecule produced by combustion reactions is an important flame radical participating in combustion reactions in the flame front, and also provides an indicator of the main reaction zone. Gutmark *et al.* (1989) and Hanson (1986) claimed that PLIF imaging of OH gives an instantaneous picture of the combustion process inside the flow structures.

OH has been the focus of most laser induced fluorescence (LIF) flame studies, is often present at high concentrations, and is an important intermediate radical. To fully understand the combustion chemistry, however, other species must also be measured.

Cattolica (1987) claimed that NO_2 imaging by planar laser fluorescence techniques will visualize the reaction interface between unburnt reactants and combustion products in a propagating flame. Whilst recognizing that the OH radical also maps the flame front, the NO_2 molecule has a number of characteristics that make it a useful fluorescence tracer for combustion applications. The capability to image this species should prove to be very useful for combustion research, as NO_2 has a wide absorption band and is simpler to image.

This study uses schlieren and molecular species imaging techniques, to compare large-scale structures identified with each, in a laminar diffusion propane-air flame and an oxyacetylene cutting torch. The equipment required is relatively simple and inexpensive. The ability to view and compare transient events in "real-time" leads to a greater understanding of the mixing processes. This results in a cost-effective tool to assist in the development of new, more efficient burners. Conditional phase sampling techniques, as described by Pearson (1989), were used to obtain structural images at the same phase in separate cycles of the acoustically forced diffusion flame which has been shown by Pearson *et al.* (1989) to have a highly reproducible cyclic character. The schlieren technique uses changes of refractive index caused by thermal and/or density gradients to deflect probing light and form light and dark patterns. These patterns represent the large-scale structures in the flame. The NO_2 PLIF technique (similar to the OH PLIF technique developed by Dyer and Crosley (1982) and Kychachoff *et al.* (1982), and recently reviewed by Hanson (1986)) was used in a configuration similar to that used by Muggleton and Proctor (1989), but with a new, novel fluorescence light collection system using part of the schlieren optics.

The experimental apparatus and procedure will be discussed along with examples of a series of digitally recorded pictures illustrating the salient features of the technique.

EXPERIMENTAL APPARATUS AND FACILITIES

The techniques described have been carried out on a number of different burners. The images shown here were obtained from a combustion research facility which allowed optical and other measurements to be made on a diffusion flame formed by a jet of fuel gas and a co-flowing jet of air. In the case of the oxyacetylene flame, the cutting torch was

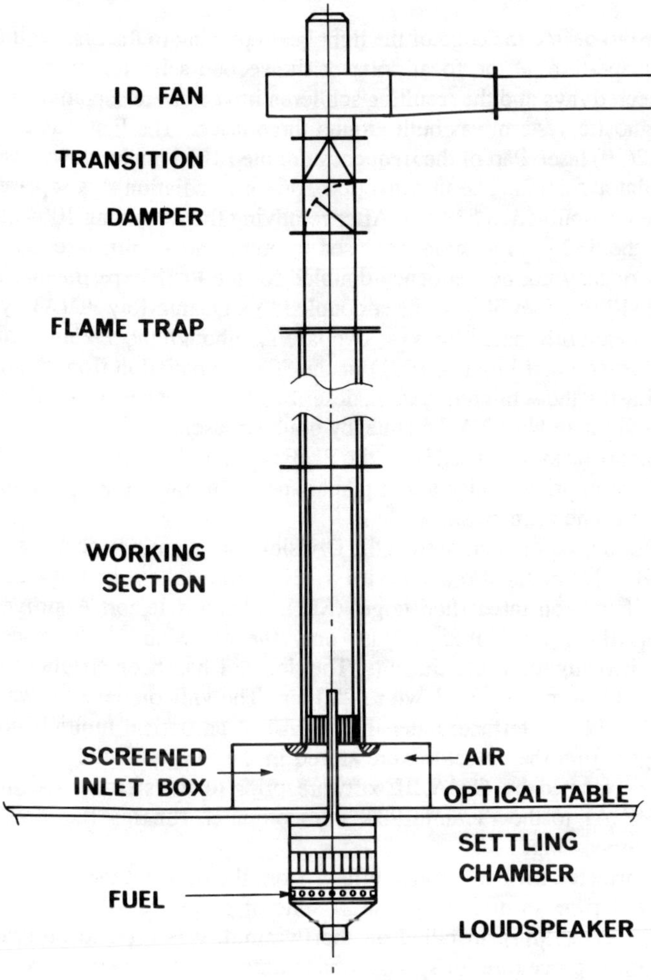

FIGURE 1 Test facility.

positioned on top of the fuel gas supply pipe and pointing downstream. The facility, which is shown in fig. 1, comprised a vertical combustion tunnel around which was mounted an optical table capable of traversing in three mutually orthogonal directions. The tunnel was 248 mm square in cross-section and had 500 mm long transparent quartz glass windows on all sides of the working section. A complete description of the combustion tunnel and facility is given by Pearson and Proctor (1991).

Where perturbation of the fuel supply was required to regularize the flame, sinusoidal signals were normally used for the acoustic driver, which in this case was a 150 mm diameter loudspeaker, driven by a power amplifier and signal generator. This arrangement made possible repetitive measurements at any given phase in the perturbation cycle, enabling events, such as camera gatings or laser pulses, to be synchronized with a given phase of the excitation voltage.

A two-mirror schlieren system was used to visualize the gas flames. The mirrors used were 305 mm diameter with a focal length of 2.4 m and were spaced 5.5 m apart. A circular aperture was used to define the edge of the light beam passing to the first schlieren mirror. Another circular aperture, at the focal point of the second schlieren mirror was used to obstruct the deflected rays and the resulting schlieren images were captured electronically.

The laser diagnostic system was built around three lasers. The first was a Quanta-Ray Nd^{3+}:YAG DCR2(30) laser. Part of the frequency doubled 1064 nm beam was used to pump a dye laser oscillator section. The unconverted 1064 nm radiation was separated off and was again frequency doubled to 532 nm. After removing the remaining 1064 nm radiation from this beam, the 532 nm radiation was used to pump the amplifier section of the dye laser which may or may not be frequency doubled for the PLIF experiments. The second was a Questek 2340 Excimer 308 nm laser coupled to a Quanta-Ray PDL-3 dye laser and WEX. The PLIF beam originated from the dye lasers, although the 532 nm radiation from the Nd^{3+}:YAG (Pearson and Proctor, 1991) or the 308 nm radiation from the excimer has also been used. Lastly, the schlieren system, described fully in Pearson and Proctor (1991), was likewise based on an Nd^{3+}:YAG frequency doubled laser.

In the laser diagnostic system used here, the PLIF beam of the pertinent wavelength light was steered with appropriate optics into a planar sheet (24 mm wide by 1 mm thick) and shone through the flame of interest.

The receiving optics equipment used at the Division for capturing spectra was a Czerny-Turner SPEX 1704 1.0 m spectrometer with a 1200 grooves/mm grating and an EG&G PAR model 1254E silicon intensified target (SIT) vidicon detector. Appropriate narrow bandpass optical filters were used to pass only the emission or fluorescence of the molecules under investigation into the slits. The detector had been specially modified to extend its sensitivity from 350 nm down to 200 nm. The vidicon detector was controlled by an EG&G PAR 1216 interface under the control of an optical multichannel analyser (OMA III). Images from the detector were stored in the OMA III and some preliminary processing was performed by OMA III software prior to transferring the images via a Cleveland 286 PC AT to the Pyramid 90X minicomputer, running the UNIX operating system, for final processing.

Spectra were corrected for background effects from the burners, thermal noise from the detector and irregularities in pixel response. A suite of programs and the "S" package of Becker and Chambers (1984), installed on the Pyramid, was used to process the crude spectrum. The normal procedure for spectra was to obtain a background spectrum with the burner unlit and store the data in an OMA file, then to obtain a spectrum with the burner lit, with the spectrometer settings remaining the same. On the Pyramid, the spectrum file obtained with the burner lit was normalised and background subtracted concurrently using the background file. The resulting spectra were compared with theoretically generated spectra which took into account temperature and instrument optical constants.

THE SCHLIEREN TECHNIQUE

The schlieren technique is one of the oldest, simplest and least expensive procedures for visualizing nonuniformities in transparent media. In the schlieren method, light ray traces deflected from their ideal paths (in the absence of nonuniformities) are selectively intercepted by a spatial filter to display a variation of irradiance in an image of the nonuniformity. The nonuniformity is thereby revealed. A typical system consists of a point light source, collimated by a lens (or mirror), which traverses the nonuniformity of interest, is refocussed by a decollimating lens (or mirror) to image the light source, and finally, is recorded by a camera (or other detector) which images the nonuniformity. A knife edge or small aperture is placed at the second focal point to cut off the deflected rays. In the terminology of Fourier optics, the schlieren apparatus Fourier transforms and spatially filters light transmitted by a nonuniform transparent object and then retransforms the filtered output to recover a spatially filtered image of the object. The system relies on the fact that nonuniformities have nonuniform refractive indices and hence deflect rays different amounts depending on the amount of refractive index gradient at the field of interest. It is unfortunately a linear spatial averaging technique as the rays may be deflected by any inhomogeneities linearly placed between the two collimating lenses (or mirrors).

SPECTRA AND PLIF

Although flow visualization is regarded as a well-developed field, with a long history of important contributions to fluid mechanics and combustion, remarkable advances in measurement capability have occurred in the past few years. Particularly noteworthy has been the development of spectroscopy-based techniques which combine species specificity with the capability for simultaneous, spatially resolved measurements at a large number of points within a plane, in essence providing two-dimensional images of flow field properties.

Laser induced fluorescence is a well-established, sensitive technique for detecting population densities of atoms and molecules in specific quantum states. Although LIF was used initially in studies of spectroscopy and chemical analysis, it is now recognized as a powerful fluid mechanics diagnostic with the potential for species mapping in flow fields. The development of LIF methods has, until recently, been driven by single point measurements, but the underlying concepts carry over directly to multiple point planar imaging (PLIF).

In LIF, a laser source is tuned to excite a specific electronic absorption transition in the species of interest. Following the absorption process, collisional redistribution in the electronically excited state may occur prior to either collisional quenching or radiative de-excitation (fluorescence) of the molecule back to its lower electronic state. The emission, which occurs over a range of wavelengths for different species, is usually collected normal to the plane of illumination and filtered spectrally at the photodetector. For a given species, the variables in the PLIF process are the transition pumped, the detection spectral bandpass, the spectral intensity of the pump laser and, in the case of a narrow bandwidth laser, the location of excitation within the absorption line profile. LIF may be thought of as spatially resolved absorption, with the intersection of the illumination and collection beam paths controlling the spatial resolution of the measurement. In the case of sheet beam illumination and detection with an array detector, the size of the detector pixels, the total detector area and the magnification of the collection optics controls the size of the measurement volume.

Miller and Bowman (1989) have shown that a significant fraction of the nitrogen oxide emissions from some combustion sources indicate that there are large NO_2/NO ratios near the flame zone. Detailed kinetics calculations suggest that NO_2 formation and destruction in flames can occur by the following set of reactions,

$$NO + HO_2 \Leftrightarrow NO_2 + OH,$$
$$NO_2 + H \Leftrightarrow NO + OH,$$
$$NO_2 + O \Leftrightarrow NO + O_2.$$

In the "low" temperature regions of flames, significant HO_2 concentrations are found that can react with NO formed in the "high" temperature regions and transported by diffusion to the low temperature region. The NO_2 removal reactions are rapid, and in the presence of high radical concentrations, NO_2 will be converted rapidly back to NO. It can

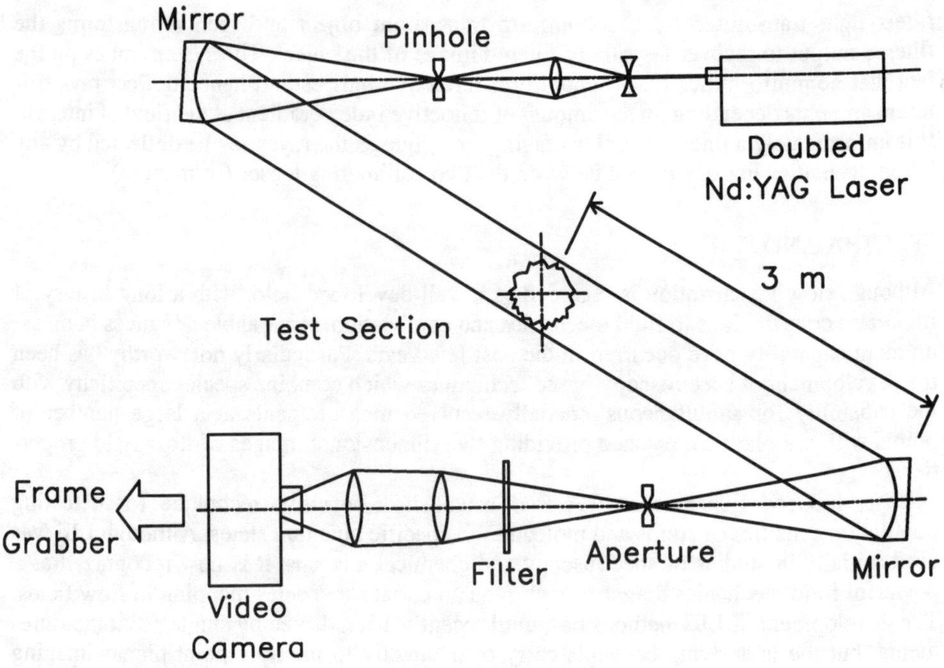

FIGURE 2 Schlieren arrangement.

be seen from the above equations that the use of NO_2 PLIF in effect maps the same regions as OH PLIF.

Keyser *et al.* (1971) have shown that NO_2 fluorescence experiments, with monochromatic excitation at 400–600 nm and monochromatic observation, revealed that the process can be interpreted in terms of a single electronically excited state undergoing vibrational relaxation at gas kinetic rate and electronic quenching about 100 times more slowly than natural fluorescence. An apparent lifetime of about 55 s was reported, independent of excitation frequency, but the possibility of mixing with vibrationally excited ground state makes this an upper limit to the lifetime of the hypothetical unperturbed $^2B_1$ state.

In the above reactions, it is possible to use laser stimulated emission from OH to evaluate temperature, as shown by Rea and Hanson (1988). Alternatively, there are other species which are produced during the oxidation of hydrocarbons, e.g., CH, whose emission spectrum, caused by natural excitation in the flame, can be used to calculate point temperatures. In the case of CH, temperatures were found by comparing the measured spectrum with that calculated using the appropriate molecular constants from Huber and Herzberg (1979), the instrument constants of both the Spex monochrometer and the EG&G PAR vidicon array, and an initial estimate of the temperature. The predicted temperature was obtained when the difference between the normalized theoretical spectrum and the normalized measured spectrum was a minimum.

IMAGING AND CONDITIONAL AVERAGING

Conventional video, by its very nature, is a scanned medium. Consequently, any digitized frame of a video signal represents a time window of at least one frame time. With the 50

Hz CCIR video standard used in Australia, this represents a time difference of 1/25th second (one frame period). That is, the bottom right pixel is 1/25th second behind (in time) the top left pixel. A technique is required that utilizes relatively inexpensive hardware and custom software that allows the digital capture of a video image where every pixel is updated at the same instant in time. The resulting digitized version then represents a "snapshot" of the flow when the video frame was scanned.

Frame grabbing and subsequent digitization of a video image of some fluid flow phenomenon gives advantages not available to other forms of pictorial storing of an image. It allows linearization for the camera response, coordinate transformation, as well as conventional image processing techniques such as contrast enhancement, convolution filtering, edge detection, pseudo colouring, image averaging, subtraction, addition as well as many specialized processes. This allows many image details to be enhanced or even raised from an indistinguishable noisy thermal background of the detector. It also allows image artefacts to be selectively removed. Triggering the laser to pulse at a marginally slower rate than the forcing frequency of the fuel gas supply allows views of rapidly changing events to be slowed down by a "slipping strobe" effect. This is a very valuable technique for studying fluid flow and species formation and removal.

A "PC Vision Plus" frame grabber board from Imaging Technology Inc. in a 33 MHz 80386 IBM AT-compatible desktop computer was used for this work. It provides 512×512 pixels resolution. A General Electric TN2506 CID imaging video camera was connected to the input of the frame grabber. The CID element has 400×287 display elements, but provides a linear continuous video signal which is digitized by the PC Vision Plus card at 12.5 MHz.

The GE CID camera was placed at the schlieren/PLIF image plane to collect the incident light. A pulsed light source (or flash) is necessary for this technique as it relies on the short duration of the flash to "freeze" the motion. The light source for the experiments was one of the three lasers described earlier, which provide light pulses with a duration ranging from 6 to 20 nanoseconds.

The custom software for this technique was written in "C" and "assembly" language. It was compiled and linked with conventional Microsoft compilers, assemblers and linkers. It programs the computer to control the camera and data acquisition processes.

The imaging system (fig. 2) was aligned, adjusted and focussed in the conventional schlieren manner. The camera was then programmed into the "injection inhibit" mode by software in anticipation of an event to be captured. This then interrupted the camera's normal destructive readout of the CID imager, allowing the device to integrate incoming light energy. When the required condition was met, the software triggered a laser pulse which caused the schlieren or PLIF image to fall on the camera's imager surface for a time period of the order of 10 to 80 nanoseconds, depending on the process being viewed. Just prior to the start of the next frame scan, the injection inhibit mode was relaxed and the camera output one field of video representing the transient scene event. The frame grabber captured this field digitally and then interpolation of the data, to allow for interlace artefacts, was initiated. The captured image was then displayed on the external monitor and may then be stored or manipulated appropriately. To mitigate the confounding effect of the chemiluminescence from flames, a narrow bandpass light filter, centred at the 532 nm for schlieren or the appropriate wavelength for PLIF, was placed before the camera.

In the case of spectra capture, the laser used to generate either the schlieren or PLIF images also initiated the detector, coupled to the Spex monochrometer, which was driven by software in the OMA III. Thus, when either a schlieren or PLIF image was being recorded, a spectrum was captured at the same instant in time. In this manner, a point spectrum in a 2D image can be obtained at the same time as the 2D image itself. Again, the

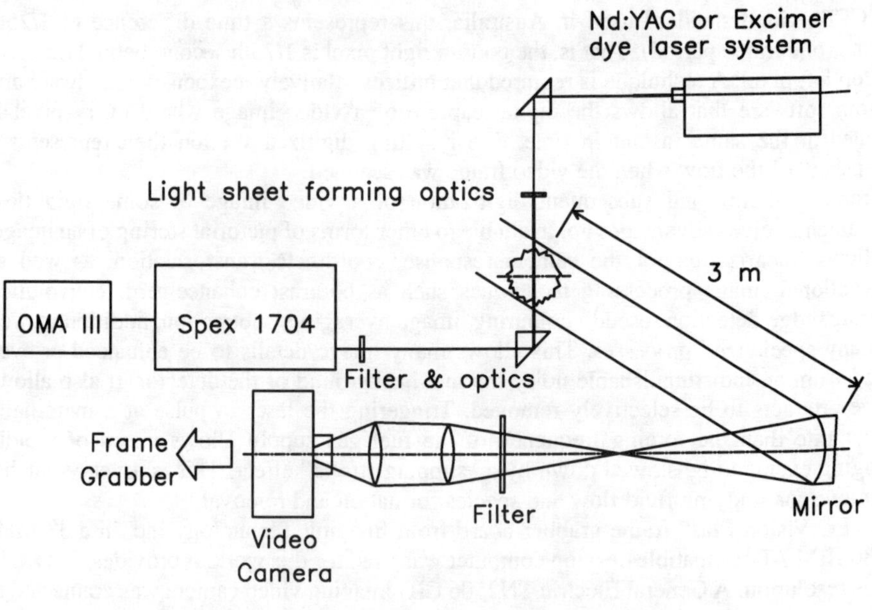

FIGURE 3 PLIF/spectra arrangement.

appropriate narrow bandpass filter was used to pass only the wavelength of interest to the spectrometer.

When conditional sampling was utilized, an image was "grabbed" when a particular phenomenon occurred, hence repetitious flow situations may be studied by removing the noncoherent component of the flow picture. The removal of jitter and clutter reveals features that are otherwise unobservable, as shown in the Results section.

PLIF WITH MIRRORS

In contrast with most other work in this field, the fluorescence emission from the flame of interest was collected with the same mirror that was used for the schlieren setup (figs. 2 and 3). This second mirror imaged the flame (and its emissions) onto the GE video camera as before. In the case of NO_2 PLIF images, the same narrow bandpass filter centred at 532 nm was used to eliminate flame chemiluminescence. However, the electronic gain of the camera had to be raised by 12 dB to allow images to be captured because of the lower light level of the fluorescence emission. Conditional sampling and averaging techniques, as used for the schlieren, were again utilized. A schematic diagram of the laser arrangement and spectrometer setup is presented in fig. 3. The light power input from the Nd^{3+}:YAG laser was 10.6 millijoule per pulse in the 532 nm band. Each picture was taken with just one laser pulse which had a duration of 6 nanoseconds.

RESULTS

A schlieren picture of a representative propane diffusion flame is shown in fig. 4. This picture is a conditional average of 8 separate shots. The fuel flow was perturbed at 14 Hz (the natural frequency of the outer structure; see Cabelli *et al.* (1988) and Pearson *et al.*

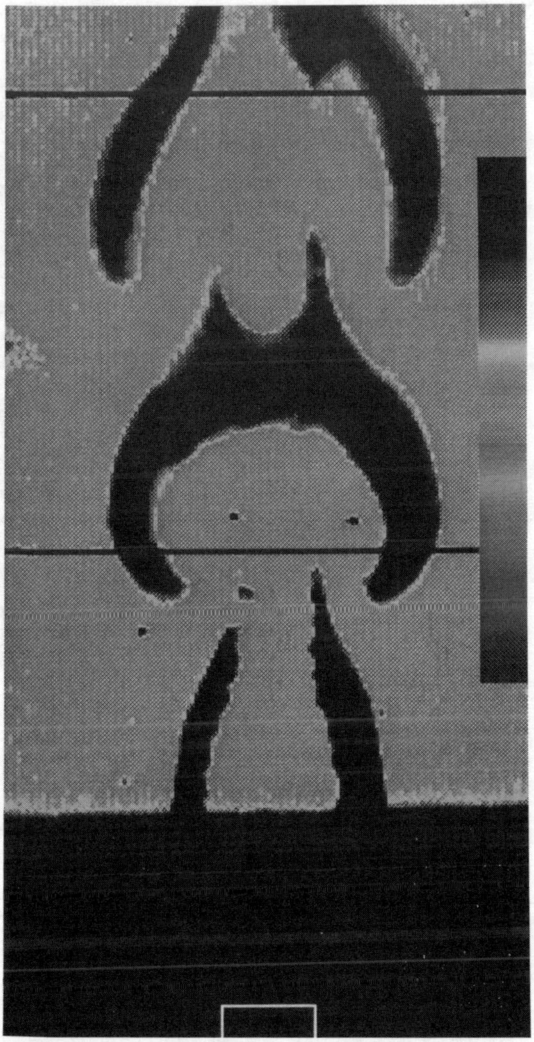

FIGURE 4 Conditionally averaged schlieren image (Re = 180, forcing frequency = 14 Hz, forcing amplitude = 100 mV). SEE COLOR PLATE IV.

(1989)) with 100 mV loudspeaker voltage to regularize the structure formation. The jet Reynolds number was approximately 180 based on jet exit diameter.

An NO_2 PLIF image of the same flame, taken with the same conditional averaging techniques is shown in fig. 5. It is a composite image, built up by averaging 8 shots at each slice location, and then combining the slice images to build a complete NO_2 PLIF image. The scales and positions are identical, so figs. 4 and 5 may be compared directly.

Figure 6, which illustrates the techniques applied to an oxyacetylene cutting torch, results in three different aspects of the same flame. The left picture is the schlieren image, the centre one that of CH at 431 nm, and the right-hand one that of the total emission from the flame. Each picture is exactly the same scale. The cutting torch nozzle is just off the

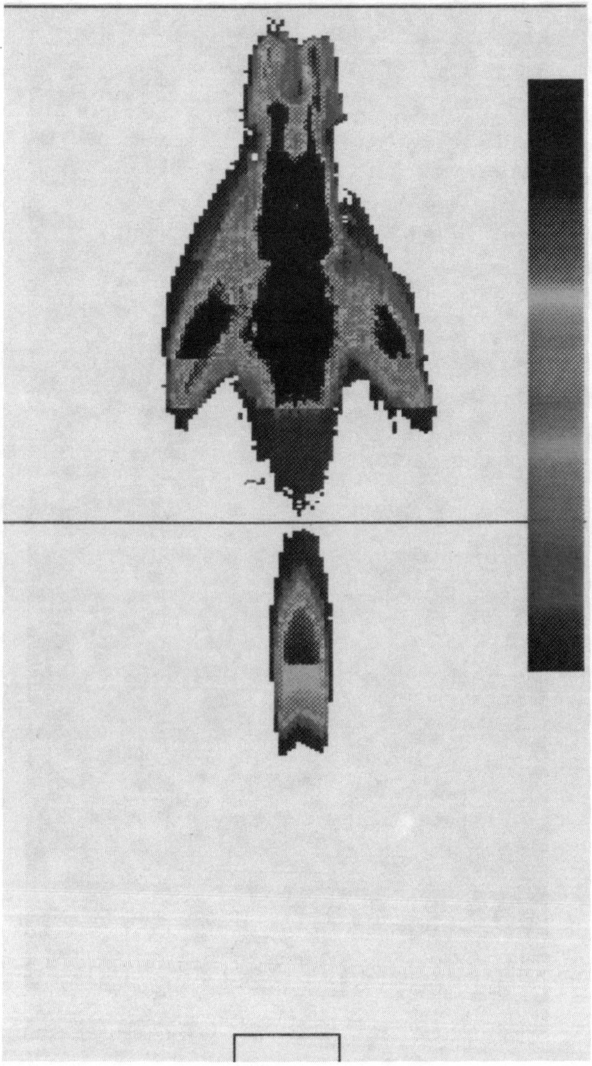

FIGURE 5 Conditionally averaged NO_2 PLIF image (Re = 180, forcing frequency = 14 Hz, forcing amplitude = 100 mV). SEE COLOR PLATE V.

bottom of the picture and the reason for the gap at the bottom of the CH picture is that the filter vignetted the edge of the picture.

The corresponding measured and calculated CH spectra for this flame is shown in fig. 7. Also shown in fig. 7 is the difference between the measured and calculated spectra. The peaks marked X in the difference graph correspond to some obscure spectral lines that are of no consequence to the current results. The other peaks result from detector response. The measured spectrum was taken at the tip of the CH cone and corresponds to a temperature of 2600 ± 50 K.

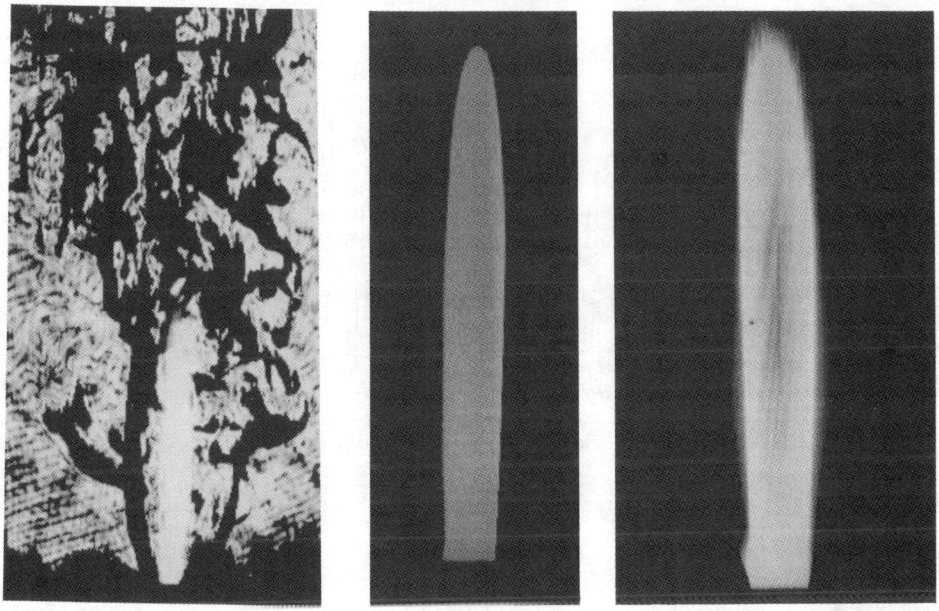

FIGURE 6 Oxyacetylene torch flame images.

DISCUSSION

The advantages of using the same collection optics for both the schlieren method and PLIF technique become obvious as one may conveniently compare images from each technique. It is clear from comparing figs. 5 and 6 that the NO_2 is formed along the inner part of the flame structure and is convected outward into the toroidal vortex shape of the outer shear layer structure.

The schlieren technique is both well documented and relatively easy to operate. To allow conditional phase averaging requires a pulsed light source and digital image capture equipment. On the other hand, the NO_2 PLIF technique is more difficult and requires a high-power pulsed laser source together with digital image capture equipment.

As Hanson (1986) indicates, whilst the line-of-sight schlieren technique is useful, in symmetrical flows, for following the high temperature interface between unburnt reactants and the burned combustion products in the flame, it provides little direct information on the combustion chemistry and hence the possible coupling between chemistry and fluid motion. Further, the spatial resolution of schlieren visualization is limited in highly turbulent flows. The observation of NO_2 (or OH or CH) by PLIF serves as a clear indicator of the chemical reaction front of the flame and the subsequent effect of fluid motion on the state of the burned gases. Each technique is shown to have advantages and disadvantages, and each shows different detail about the large-scale structures inside the combusting flame.

Although the technique has been illustrated using schlieren and NO_2 PLIF imaging, as well as schlieren with CH images and CH spectra, it is not restricted to these species. The experimental arrangement is such that OH PLIF images and/or a third species PLIF image, such as NO, can be generated using an excimer-dye laser system with the appropriate filters and the same collection optics and most of the same input PLIF optics.

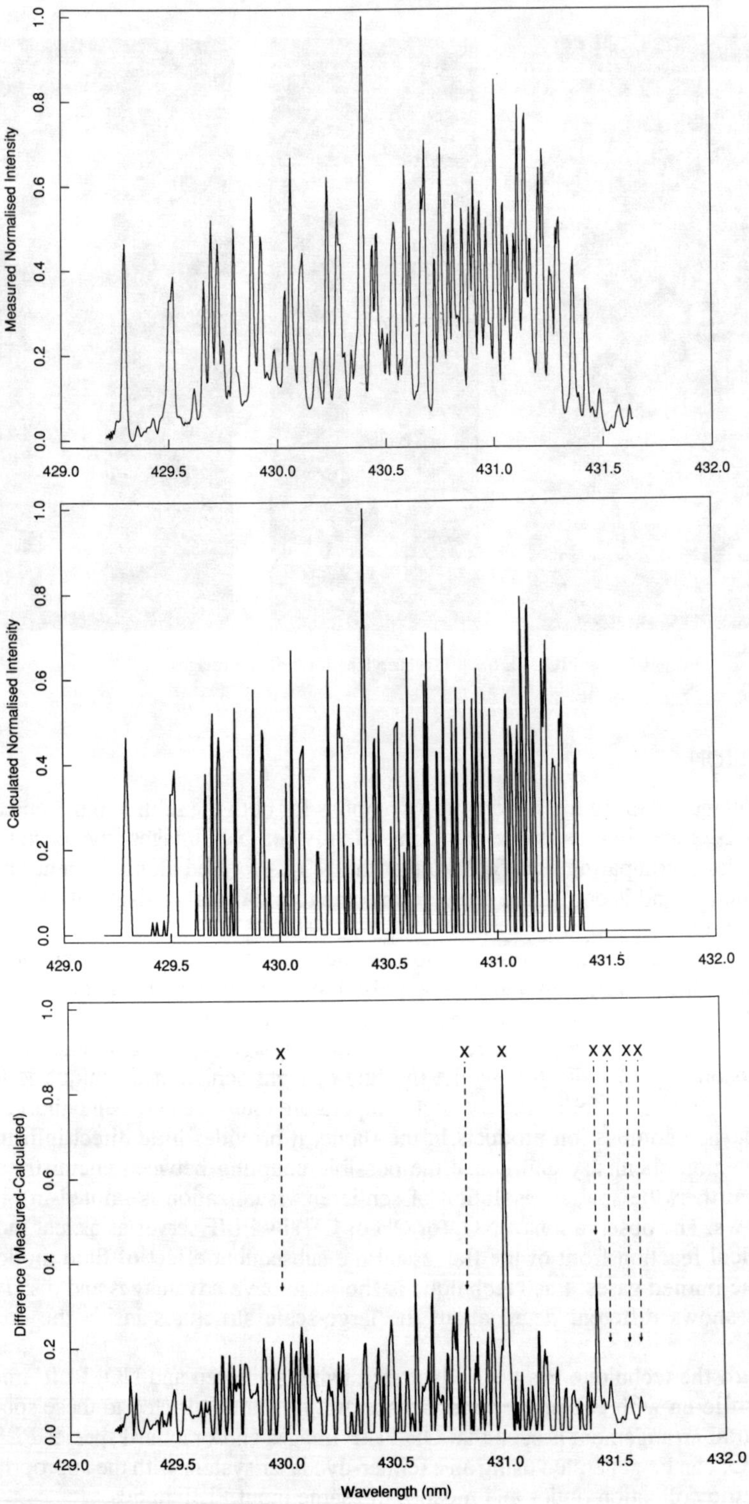

FIGURE 7 CH spectra from oxyacetylene torch flame: measured and predicted.

When one studies the species images in conjunction with the similar schlieren images, explanation of details in both types of image become manifest. These two imaging techniques have a certain synergistic relationship.

REFERENCES

Ballal, D.R. (1986). Studies of turbulent flow-flame interaction. *AIAA J.*, **24**(7), 1148–1154.

Becker, R.A., and Chambers, J.M. (1984). *S: An Interactive Environment for Data Analysis and Graphics.* Wadsworth Advanced Book Program, Belmont, California.

Broadwell, J.E., and Dimotakis, P.E. (1986). Implications of recent experimental results for modeling reaction in turbulent flows. *AIAA J.*, **24**(6), 875–889.

Brown, G.L., and Roshko, A. (1974). On density effects and large structures in turbulent mixing layers. *J. Fluid Mech.*, **64**(4), 775–816.

Cabelli, A., Pearson, I.G., Shepherd, I.C., and Hamilton, N.B. (1988). Fluid dynamics structures in diffusion flames. *Proc. First World Conference on Experimental Heat Transfer, Fluid Mechanics and Thermodynamics*, Dubrovnik, Yugoslavia, 4–9 September 1988, 627–631.

Cattolica, R.J. (1987). Visualization of flame propagation by laser-fluorescence imaging of nitrogen dioxide. *Combust. Sci. & Tech.*, **54**(1–6), 61–67.

Chen, L.D., and Roquemore, W.M. (1986). Visualization of jet flames. *Combust. & Flame*, **66**, 81–86.

Dyer, M.J., and Crosley, D.R. (1982). *Opt. Lett.*, **7**, 382.

Gutmark, E., Parr, T.P., Parr, D.M., and Schadow, K.C. (1989). Planar imaging of vortex dynamics in flames. *J. Heat Trans.*, **111**, 148–155.

Hanson, R.K. (1986). Combustion diagnostics: planar imaging techniques. *Proc. 21st Symposium (International) on Combustion*, 1677–1691.

Huber, K.P., and Herzberg, G. (1979). *Molecular Spectra and Molecular Structure*, IV: *Constants of Diatomic Molecules.* Van Nostrand Reinhold Company, New York.

Keyser, L.F., LeVine, S.Z., and Kaufman, F. (1971). Kinetics and mechanism of NO_2 fluorescence. *J. Chem. Phys.*, **54**(1), 355–363.

Kychachoff, G., Howe, R.D., Hanson, R.K., and McDaniel, J.C. (1982). Quantitative visualization of combustion species in a plane. *Appl. Opt.*, **21**, 3225–3227.

Miller, J.A., and Bowman, C.T. (1989). Mechanism and modeling of nitrogen chemistry in combustion. *Prog. Energy Combust. Sci.*, **15**, 287–338.

Muggleton, S.K., and Proctor, D. (1989). A system for capturing spectra and 2-D pictures. *ACOLS '89*, Adelaide 1989.

Mungal, M.G., and Dimotakis, P.E. (1984). Mixing and combustion with low heat release in a turbulent shear layer. *J. Fluid Mech.*, **148**, 349–382.

Pearson, I.G. (1989). Instantaneous conditional video schlieren. Presented to Workshop on Industrial Fluid Dynamics, Heat Transfer and Wind Engineering, CSIRO, Highett, Melbourne.

Pearson, I.G., Cabelli, A., Shepherd, I.C., and Hamilton, N.B. (1989). Structures in diffusion flames: an acoustic viewpoint. *Proc. Tenth Australasian Fluid Mechanics Conference*, **1**, 5.27–5.30.

Pearson, I.G., and Proctor, D. (1991). Visualization of flame structures. *Proc. Second World Conference on Experimental Heat Transfer, Fluid Mechanics and Thermodynamics*, Dubrovnik, Yugoslavia, 23–28 June, 316–322.

Rea, E.C., and Hanson, R.K. (1988). Rapid laser-wavelength modulation spectroscopy used as a fast temperature measurement technique in hydrocarbon combustion. *Appl. Opt.*, **27**(27), 4454.

Strawa, A.W., and Cantwell, B.J. (1985). Visualization of the structure of a pulsed methane-air diffusion flame. *Phys. of Fluids*, **28**, 2317.

Vandsburger, U., Lewis, G., Seitzman, J.M., Allan, M.G., Bowman, C.T., and Hanson, R.K. (1986). Flame-flow structure in an acoustically driven jet flame. *Proc. Western States Section Combustion Institute Meeting*, Paper No. 86–19.

The Blowout of a Jet Diffusion Flame: The Effects of the Velocity and Composition of the Surrounding Co-Flowing Stream

I. WIERZBA, K. KAR AND G. A. KARIM *The University of Calgary, Department of Mechanical Engineering, Calgary, Alberta, Canada*

Abstract — The blowout limits of a methane diffusion flame in a co-flowing air-fuel or air-diluent stream were determined for a range of surrounding stream velocities, both laminar and turbulent, up to ~1.50 m/s. Methane, ethylene, propane and hydrogen were used as the fuels in the surrounding stream while nitrogen and carbon dioxide were used as diluents. The experimental results show that the velocity of the surrounding stream affects the blowout phenomena significantly. An increase in the stream velocity has a detrimental effect on the blowout limits at very low velocities up to 0.30 m/s (essentially laminar flow) and at velocities higher than 1.50 m/s (turbulent flow). The addition of a fuel to the air stream in most cases enhances the blowout limit of a methane diffusion flame. However, different trends in the variation of the blowout limits with the surrounding fuel concentration were observed, depending on the type of fuel used and on whether the surrounding stream was laminar or turbulent. The addition of nitrogen or carbon dioxide to the air stream results in decreasing the blowout limits. The effect is more severe at the higher velocities.

Key Words: blowout limits, jet diffusion flame, stream velocity

INTRODUCTION

The combustion processes in many practical systems take place through various forms of diffusion flames. Much research effort has been directed over the years towards improving the understanding of the mechanisms of the stability of jet flames in general and their blowout limits in particular.

In some combustion devices the combustion of a fuel jet takes place in a non-quiescent environment that may contain some homogeneously dispersed fuel or a diluent. Examples can be found in engines of the dual fuel compression ignition type, precombustion chamber engines where a flaming jet ignites the bulk of the homogeneously mixed fuel-air charge, prevaporizing gas turbine combustors or in furnaces employing dual fuel systems. The mixing and combustion processes in such devices are very complex. They involve the simultaneous combustion of two fuels varying in their proportions from one point to another, in a combined diffusion and premixed flame mode and most often within a highly turbulent environment. There is a need for a better understanding and control of such processes, especially if alternative fuels, such as natural gas, are to be utilized efficiently, cleanly and economically in conventional combustion devices.

The blowout limit of a jet diffusion flame can be extended substantially when its surroundings contain some fuel premixed with the air (Karim *et al.* 1984; *Karim et al.* 1988). Associated with this approach is the potential of modifying the temperature, velocity and concentration fields of the jet flame for favorable fuel utilization and emissions (Karim and Lapucha, 1988; Kibrya, 1987).

Earlier experiments were conducted at relatively small co-flowing stream velocities that were lower than 0.5 m/s. An increase in the surrounding stream velocity would result in a decrease in the entrainment rate between the jet and the stream and would promote eddy mixing at the jet boundary. The relative contribution of these changes to the jet flame blowout limit is uncertain, requiring further experiments.

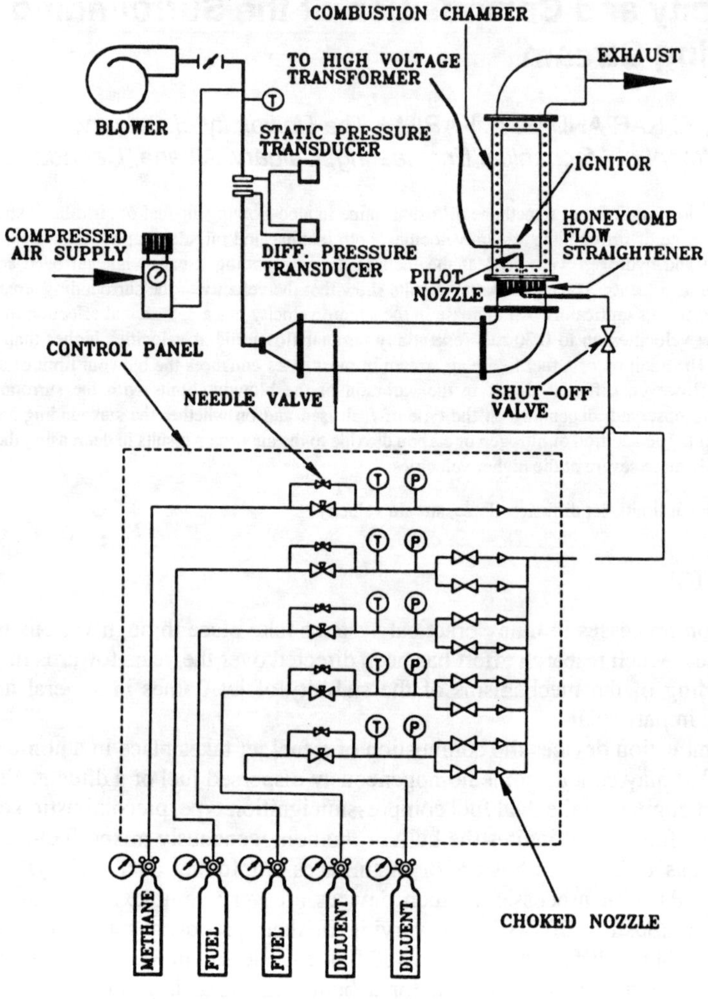

FIGURE 1 Schematic diagram of the apparatus.

The paper describes the results of an experimental investigation that was aimed at establishing the effects of changes in the velocity of the co-flowing stream on the blowout limit of a turbulent jet diffusion flame, as well as the extent of variations in these limits when the surrounding air stream is premixed with small concentrations of various fuels and diluents.

APPARATUS AND EXPERIMENTAL PROCEDURE

The apparatus employed in this study is shown schematically in fig. 1. The blowout limits of a jet diffusion flame were determined in a vertical rectangular (132 × 132 mm^2) stainless steel combustor fitted with long quartz windows for viewing and recording the flame. The fuel jet was discharged vertically along the centre line of the combustor from a sharp edged circular nozzle located at a point 70 mm from the base of the combustor. The work reported here was conducted employing a small 1.5 mm diameter nozzle. The apparatus provided

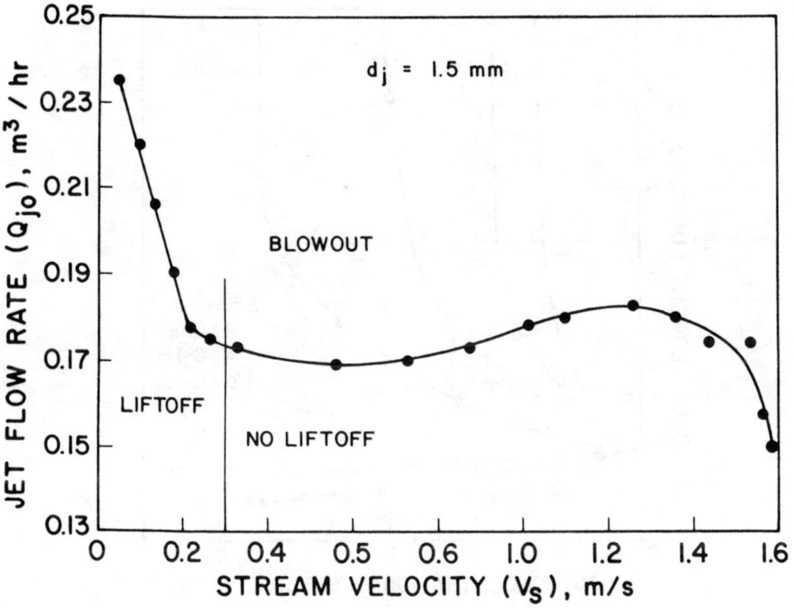

FIGURE 2 The methane diffusion flame blowout limit as a function of the surrounding stream velocity.

for mixing various gaseous fuels or diluents with the surrounding air well upstream of the entrance to the combustor. The mixture of air and fuel was passed through a honeycomb straightening section and fine stainless steel screens before entering the combustor at its base. A relatively uniform velocity distribution was obtained at the inlet section of the combustor. The wire mesh also acted as a flame arrester in case of a flame flashback. The flow rate of air was measured using a sharp edged orifice while the flow rates of fuels and diluents were measured using calibrated choked nozzles.

For any set of operating conditions, the jet discharge rate was increased very slowly until blowout of the flame occurred. The blowout limit was defined as the maximum volumetric flow rate of the jet at which a stable flame can still be maintained and was established visually. Methane was used as the jet fuel while methane, ethylene, propane and hydrogen were employed in the surrounding stream. As diluents, nitrogen and carbon dioxide were used. Much of the testing was carried out with commercial methane, typically composed of 94.7% CH_4, 3.7% C_2H_6, 0.7% C_3H_8, 0.2% C_4H_{10}, 0.4% C_5H_{12} and 0.3% N_2. The stream velocities were varied from 0.20 m/s up to 1.50 m/s at atmospheric pressure and ambient initial temperature conditions.

EXPERIMENTAL RESULTS

The blowout limit of a jet diffusion flame, as shown in fig. 2 is influenced by changes in the co-flowing stream velocity. At small stream velocities, liftoff of the jet flame was observed before flame blowout. Beyond a certain value of the stream velocity the flame remains attached to the nozzle until blowout is encountered. Experiments showed that the value of the limiting velocity is directly proportional to the jet nozzle diameter. For the jet diameter of 1.5 mm this limiting velocity was around 0.30 m/s.

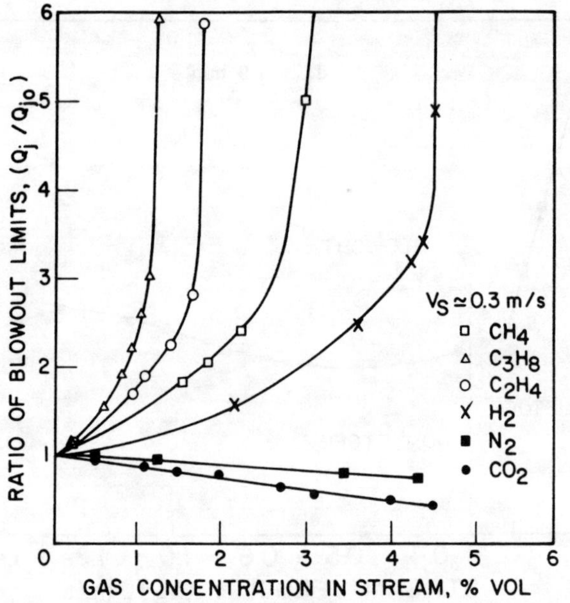

FIGURE 3 Ratio of the blowout limits of a methane diffusion flame in the presence of gas addition to the stream relative to a stream of air, $Q_j/Q_{j,0}$, as a function of the fuel or diluent concentration in the surrounding stream. The stream velocity is 0.30 m/s.

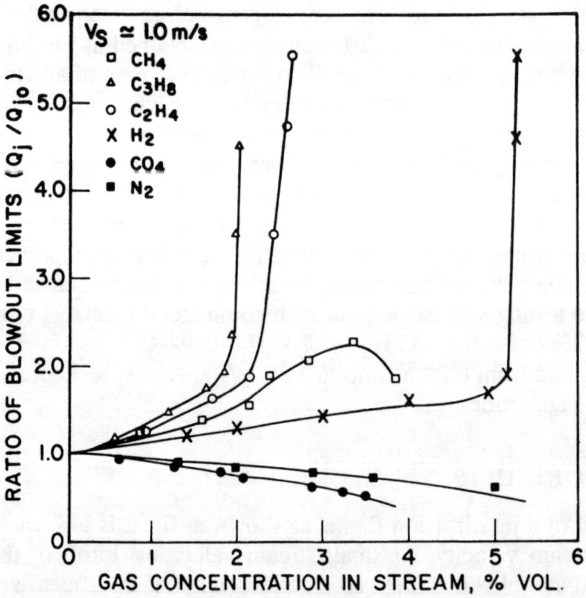

FIGURE 4 Ratio of the blowout limits of a methane diffusion flame in the presence of gas addition to the stream relative to a stream of air, $Q_j/Q_{j,0}$, as a function of the fuel or diluent concentration in the surrounding stream. The stream velocity is 1.00 m/s.

An increase in the stream velocity has a detrimental effect on the blowout limit of the jet flame at very low stream velocities up to 0.30 m/s while laminar flow conditions prevail within the stream. This effect can also be seen at stream velocities higher than 1.50 m/s representing fully turbulent conditions. In between these two regions the effect of the stream velocity on the blowout limit tends to be relatively small. These trends are probably a reflection of the changes in the local mixing and entrainment rates between the jet and surrounding stream with increasing the stream velocity.

It was found that the presence of small amounts of a fuel in the surrounding stream has a significant effect on the jet flame blowout limit. The character of this relationship is dependent not only on the type of fuel being employed, but also on the stream velocity. Observed changes in the blowout limits were evaluated in terms of the ratio of the limit (Q_j) when a fuel was present in the surrounding stream to the corresponding blowout limit in a stream of air (Q_{j0}), under the same operating conditions. A ratio higher than unity represents an improvement in the blowout limit value.

At low stream velocity of 0.30 m/s when the co-flowing stream was essentially laminar, the blowout limits, as shown in fig. 3, are enhanced very significantly for all fuels employed with the increase in the concentration of the fuel in the stream. For any fuel volumetric concentration the extent of improvement observed in the blowout limits varied widely depending on the type of fuel involved. Propane in the stream was the most effective of the fuels tried, followed by ethylene, methane and hydrogen. As expected, the presence of nitrogen or carbon dioxide in the stream influenced the blowout limit adversely. Carbon dioxide was more detrimental than nitrogen.

When the stream velocity was increased to 1.00 m/s, as shown in fig. 4, larger fuel concentrations were required to obtain the same level of improvement to the jet flame blowout limit than in case of the lower stream velocity. The extent of the relative changes in blowout limits also varied with the type of fuel employed. It can be seen that although the apparent order or effectiveness of the fuels used did not change, high concentrations of methane, beyond ~3.5% by volume, in the stream began to affect the blowout limit adversely.

At a higher stream velocity of 1.45 m/s, when the stream became fully turbulent, yet larger fuel concentrations, as shown in fig. 5, were required to obtain the same level of improvement in the blowout limit. Also the adverse influence of the presence of methane at higher stream concentrations was now joined by propane. Increasing the propane concentration beyond 2.0% by volume and that of methane beyond 3.5%, brought about a significant lowering of the jet flame blowout limit.

DISCUSSION

The observed improvement in the blowout limit with the fuel concentration for any stream velocity varied from one fuel system tried to another. The trends, such as those seen in fig. 3 would not correlate any better if the fuel concentrations were plotted on equivalence ratio basis, though the order of the fuel effectiveness would change. As shown in fig. 6, hydrogen appears then to be the most effective, while propane is the least effective.

The maximum value up to which the concentration of a given fuel in the surrounding stream can be increased will be limited by the flashback limit of the fuel in air. This represents the situation when the fuel concentration in the surrounding stream becomes sufficiently high to support flame propagation into the stream independently of the jet flame (Wierzba and Kar, 1991). For the setup and operating conditions employed the flame flashback limit concentrations for the fuels considered are shown in Table I. These values

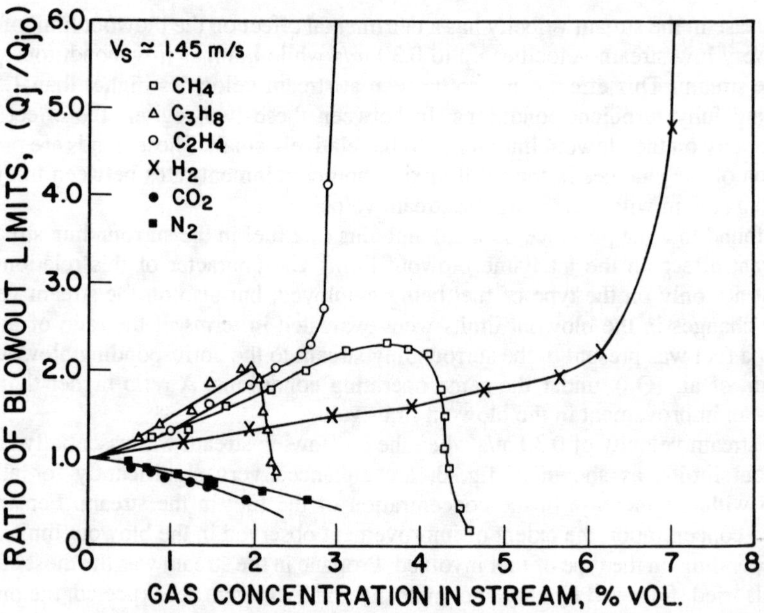

FIGURE 5 Ratio of the blowout limits of a methane diffusion flame in the presence of gas addition to the stream relative to a stream of air, $Q_j/Q_{j,0}$, as a function of the fuel or diluent concentration in the surrounding stream. The stream velocity is 1.45 m/s.

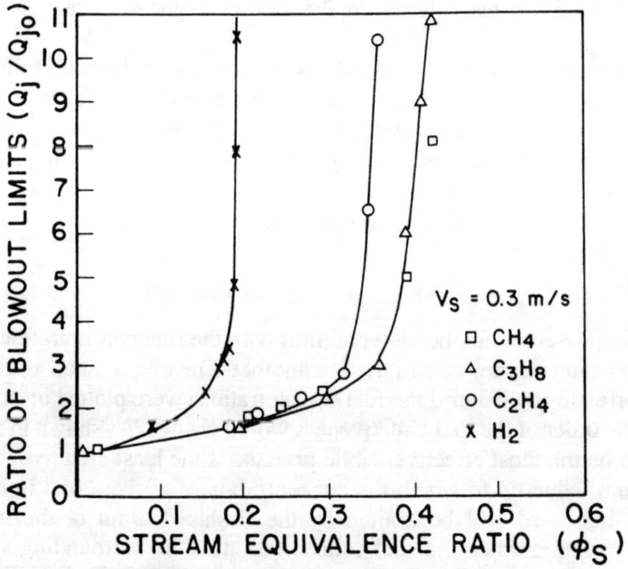

FIGURE 6 Ratio of the blowout limits of a methane diffusion flame in the presence of gas addition to the stream relative to a stream of air, $Q_j/Q_{j,0}$, as a function of the fuel or diluent concentration in the surrounding stream expressed in terms of equivalence ratio.

TABLE I
FLASHBACK CONCENTRATIONS (PERCENT BY VOLUME)

	Stream velocity (m/s)		
Fuel	0.30	1.00	1.45
CH_4	6.10	6.45	6.80
C_2H_4	3.50	3.50	3.60
C_3H_8	2.40	2.75	2.90
H_2	8.30	9.50	9.60

have been shown to be closely represented by the lean flammability limit concentrations for the fuel premixed in air.

The concentration of the fuel in the surrounding stream (C_S) can be considered relative to the corresponding flashback limit concentration ($C_{S,FB}$). A replot of fig. 3 in terms of the relative fuel concentration (i.e., $C_S/C_{S,FB}$) in fig. 7 shows an excellent correlation for all the blowout limit data when the stream was laminar. This correlation was equally effective for all combinations of jet and stream fuels and a range of nozzle diameters tried so long as the stream remains laminar (Kibrya, 1987).

When the stream was turbulent this correlation was also effective as shown in fig. 8, when the relative concentrations were not too high (less than 0.5). At the higher relative concentrations the adverse contribution of methane and propane to the blowout limit of the jet flame was very evident. In the analysis of the mechanism of diffusion flame stabilization (Peters and Williams, 1983; Broadwell, 1984; Pitts, 1989; *Becker et al.* 1967) it is often assumed that the flame is located in regions where the fuels-to-air ratio is optimum, usually close to the stoichiometric value, associated with maximum burning velocity. The physical processes of mixing and entrainment are usually characterized by a mixing time, t_{mix}, and the chemical processes for the oxidation of the fuel and releasing the fuel energy by a characteristic chemical reaction time, t_c. Usually for diffusion flames the chemical time, t_c, is significantly shorter than the mixing time and the mixing processes dictate the flame propagation rates. For a stable diffusion flame to exist, the ratio of the mixing time to the chemical time should be larger than a certain critical value. The flame is blown out when this ratio becomes smaller than the critical value. Accordingly, it is convenient to view the role of the various operating parameters on the flame blowout limits in terms of their contribution to changing the relative values of these two characteristic times.

The main phenomena that bring about the mixing of a turbulent jet with its surroundings are those of entrainment due to momentum differences between the jet and its environment and of the eddy mixing that takes place across and along the boundary of the jet. As the stream velocity is increased relative to that of the jet, the entrainment rate is reduced, while for laminar surrounding stream the mixing rates across the jet boundary will continue to be relatively small. Hence, an increase in the stream velocity will reduce the mixing time and this will lead to a decrease in the jet flame blowout limit.

With further increases in the stream velocity turbulence begins to set in the surrounding stream and the intensity and the corresponding contribution of eddy mixing at the jet boundary will increase markedly. This may compensate for any reduction in the entrainment rate. Hence, as shown in fig. 2, further increases in the stream velocity do not change significantly the blowout limit of the jet diffusion flame. However, when the stream velocity is increased further so that the flow becomes fully turbulent, the intensity of

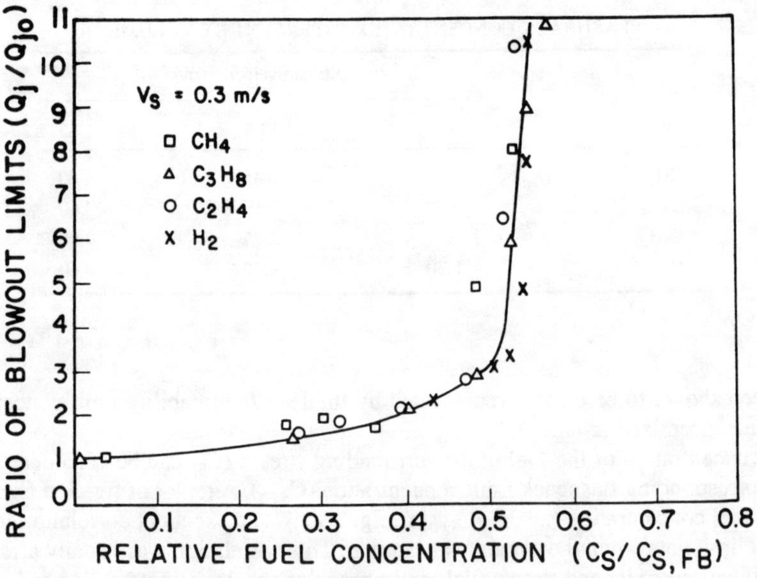

FIGURE 7 Ratio of the blowout limits of a methane diffusion flame as a function of the surrounding fuel concentration expressed as a fraction of the corresponding flame flashback limit ($C_S/C_{S,FB}$). The stream velocity is 0.30 m/s.

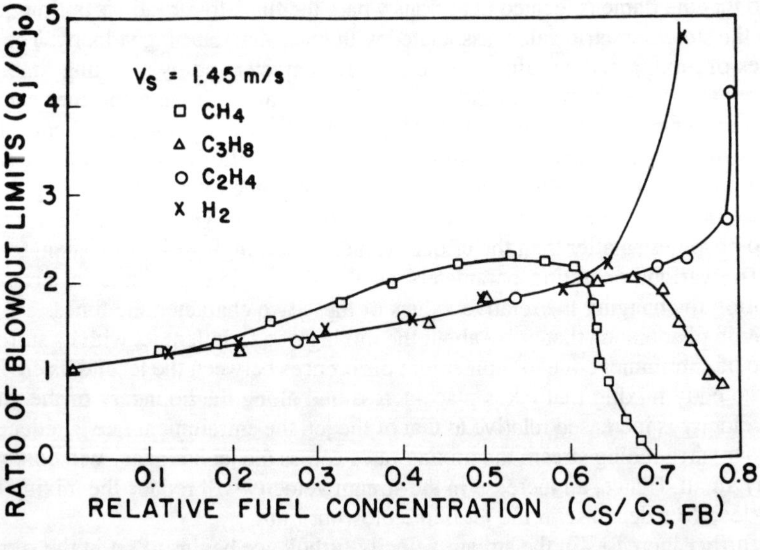

FIGURE 8 Ratio of the blowout limits of a methane diffusion flame as a function of the surrounding fuel concentration expressed as a fraction of the corresponding flame flashback limit ($C_S/C_{S,FB}$). The stream velocity is 1.45 m/s.

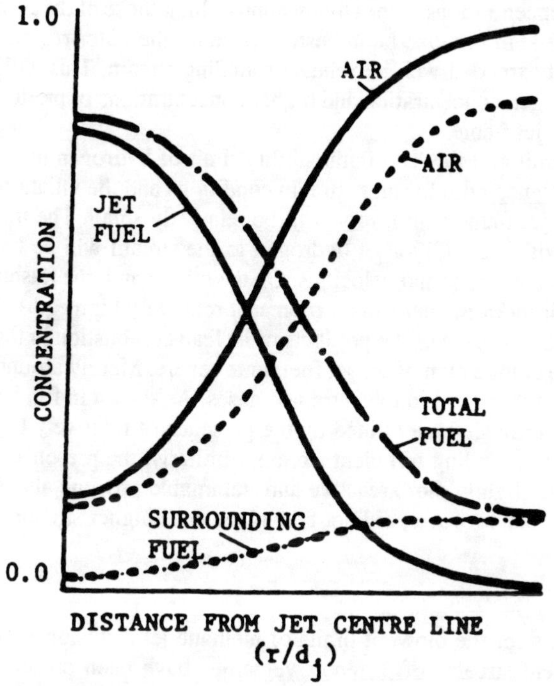

FIGURE 9 Schematic representation of the concentration profiles at a given cross-section within a fuel jet issuing in an atmosphere of air or fuel-air homogeneous mixture.

mixing of the cold surrounding air and interior rich mixtures with the reaction zone increases. The blowout limit then begins to deteriorate significantly.

The addition of a fuel to the surrounding air stream of a fuel jet will change the distribution of the concentrations of the air and the two fuels in both the radial and axial directions. As shown schematically in fig. 9, for any axial plane downstream of the jet nozzle, there will be a richer mixture at any radial location when compared to the case when only air is present in the surrounding stream. Correspondingly, outward shifts in the stoichiometric mixture and lean limit envelopes will be produced radially. The greater the concentration of the fuel in the stream the larger will be such a shift. Also the concentration gradients of the fuel and air in the region of the stoichiometric mixtures will be reduced. The net effect would be an increase in the size of the flame envelope and its shift in the direction of smaller flow velocities. The local burning velocity along the boundaries of the jet flame will be enhanced due to the entrainment of heated air that contains products of combustion that include active reactive species and radicals. As a result, a substantial increase in the combustion rates and the blowout limits of the flame was observed.

For quiescent and relatively slow laminar co-flow surroundings, the increase in the size of the flame envelope, as well as the enhancement to the local burning rate can be shown to be dependent mainly on the relative concentration of the fuel in the stream ($C_S/C_{S,FB}$) for all the fuels examined. Accordingly, the extent of improvement in the jet flame blowout limit is excellently correlated with the corresponding values of the relative concentration of the fuel in the stream, as shown in fig. 7. This trend will not change significantly when the surrounding stream becomes turbulent so long as the relative concentrations are not very large.

At higher fuel concentrations in the stream approaching those of the corresponding flame flashback limit, the lean premixed combustion zone on the outer regions of the jet flame will be formed and extended well into the surrounding stream. This will lead to a further reduction in the oxygen concentration and higher concentrations of products of combustion in the region of the jet flame.

The flashback limit and the lean flammability limit of hydrogen in air are very low in terms of the equivalence ratio for most stream conditions and the vitiation effect of its lean combustion on the jet flame combustion will be relatively small. The improvement in the jet blowout limit with the addition of hydrogen to the stream will be maintained even at relatively high concentrations and velocities. On the other hand, the flashback limit and the lean flammability limit of methane in air represent relatively high equivalence ratios. This will make the vitiation effects of the products of its lean combustion in the outer regions of the jet flame on the combustion of the jet fuel quite severe. Merely around 6.8% of oxygen will be available to the jet fuel in the entrained gases. As shown in fig. 8 the blowout limit of the jet flame is seriously undermined by the presence of relatively high concentrations of methane in the surrounding turbulent stream. Similarly, the presence of relatively high concentration of the slightly more reactive and flammable propane also affects negatively the blowout limit. This, however, did not take place until higher stream velocities.

CONCLUSIONS

Experimental results for the blowout limits of methane jet diffusion flames in co-flowing air-fuel or air-diluent streams of different velocities have been presented. Generally, an increase in the stream velocity has a detrimental effect on the jet blowout limit. The addition of diluents, nitrogen or carbon dioxide, into the stream results in lower values of the blowout limit: the higher the stream velocity, the lower the blowout limit. On the other hand, the addition of a fuel into the surrounding stream in relatively small quantities enhances the blowout limits. The enhancement depends on the stream velocity. At small stream velocities when the flow is essentially laminar, the enhancement is the greatest. All experimental data for the different fuels tried can be correlated very well in terms of the relative fuel concentration ($C_S/C_{S,FB}$). A generally smaller improvement in the blowout limits was observed at higher stream velocities. The addition of the somewhat less reactive fuels such as methane or propane, in comparison to ethylene and hydrogen, and in relatively high concentrations, had an adverse effect on the blowout limits.

ACKNOWLEDGMENT

The financial assistance of the National Sciences and Engineering Research Council of Canada (NSERC) is gratefully acknowledged.

REFERENCES

Becker, H. A., Hottel, H. C., and Williams, G. C. (1967). The Nozzle-Fluid Concentration Field of the Round, Turbulent, Free Jet, *J. Fluid Mech.*, vol. 30, Part 2, pp. 285–303.

Broadwell, J. E., Dahm, W. J. A., and Mungal, M. G. (1984). Blowout of Turbulent Diffusion Flames, *Twentieth Symposium (International) on Combustion*, The Combustion Institute, Ann Arbor, Michigan, pp. 303–310.

Karim, G. A., Kibrya, M., Lapucha, R., and Wierzba, I. (1988). Examination of the Combustion of a Fuel Jet in a Homogeneously Premixed Lean Fuel-Air Stream, *Proc. Int. Fuels & Lubricants Meeting & Exposition*, SAE Paper No. 881662, pp. 1–9.

Karim, G. A., and Lapucha, R. (1988). Flame Flashback for Low Reynolds Number Flows, *AIAA Progress in Astronautics and Aeronautics*, Progress Series (eds. Dr. Allen L. Kuhl *et al.*), vol. 113, pp. 367–383.

Karim, G. A., Wierzba, I. and Hanna, M. (1984). The Blowout Limit of a Jet Diffusion Flame in a Co-Flowing Stream of Lean Gaseous Fuel-Air Mixtures, *Combustion and Flame*, **57**, pp. 283–288.

Kibrya, M. (1987). An Experimental and Analytical Examination of Jet Diffusion Flames in Non-Quiescent Environments of Fuel and Air, Ph.D. Thesis, Department of Mechanical Engineering, The University of Calgary, Alberta, Canada.

Peters, N. and Williams, F. A. (1983). Liftoff Characteristics of Turbulent Diffusion Flames, *AIAA Journal*, **21**, No. 3, pp. 423–429.

Pitts, W. P. (1989). Importance of Isothermal Mixing Processes to the Understanding of Lift-Off and Blowout of Turbulent Jet Diffusion Flames, *Combustion and Flame*, **76**, pp. 197–212.

Wierzba, I. and Kar, K. (1991). Flame Flashback Within Turbulent Streams of Lean Homogeneous Fuel Mixtures and Air, *Proc. of the ASME Energy Resources Conference, Fossil Fuel Combustion*, vol. PD-33, pp. 101–104.

Experimental Studies of NO_x Emissions from Turbulent Propane Jet Diffusion Flames

NILS A. RØKKE AND JOHAN E. HUSTAD *The Foundation for Scientific and Industrial Research at the Norwegian Institute of Technology, Trondheim, Norway*
OTTO K. SØNJU *The Norwegian Institute of Technology, Thermal Energy Division, Trondheim, Norway*

Abstract — In this paper measurements and scaling laws for NO_x-emissions from large flames with up to 9 MW of thermal heat release are presented. These represent much larger flames than reported earlier in the literature. Measurements of oxides of nitrogen have been performed in small vertical laboratory flames (<0.7 MW) with nozzle exit Reynolds and Froude numbers ranging from 13,000 to 110,000 and 17 to 100,000, respectively. This has been achieved by varying the exit gas velocity between 2.2 and 63 m/s and the exit nozzle diameter between 4 and 28 mm. Both axial and radial profiles have been measured above the visible flame region. The measuring equipment includes continuous gas analyzers and water cooled quartz/stainless steel suction probes. Experiments on medium size vertical propane jet diffusion flames have also been performed. The experimental setup was similar to that of the smaller laboratory flames. A nozzle with an exit diameter of 17 mm was used, and an exit Mach number close to 0.9, resulting in a visible flame height of approximately 7.3 m. The maximum heat release was approximately 9 MW. Scaling laws for the global NO_x emission index as a function of the nozzle exit velocity, the nozzle outlet diameter and the Froude number are presented. The results show a 0.2 power law dependence on the exit velocity and a 0.4 power law dependence on the nozzle exit diameter. The results further show that the NO_x emission index is strongly dependent upon the Froude number. These formulas can be employed for engineering purposes and provide guidance for calculating NO_x emissions from turbulent propane jet diffusion flames.

Key Words: Large turbulent diffusion flames, oxides of nitrogen, pollutant emissions, emission scaling laws

1. INTRODUCTION

Emissions of oxides of nitrogen are of major concern due to their contribution to acid rain (as NO_3^-), the photochemical agent forming smog, their impact on the human respiration system and as an indirect greenhouse gas. A recent study of the greenhouse gas emissions from offshore oil and gas production on the Norwegian continental shelf (Lindeberg *et al.*, 1990), concludes that approximately 17% of all the gas burnt offshore is burned in flare systems, while the major source offshore is the gas turbines. This, together with the other pollution aspects of NO_x, was the main objective for performing a study on NO_x emissions from turbulent propane jet diffusion flames.

Offshore flare burners/tips differ a lot in design and the process conditions, such as flow rate, temperature and pressure, which vary over a wide range. The flare tips may be straight tubes, modified burners or Indair/Mardair flare designs which comprise a premixing device called a coanda tulip, which enhances air entrainment, reducing radiation and improving flame stabilization. The heat release from such flames may vary between 10 MW to 5 GW, with flame lengths up to 70 meters.

The structure of large buoyant diffusion flames is still not well understood. Most studies on large diffusion flames are based on experimental derived knowledge focusing on flare geometry, liftoff and thermal radiation (McCaffrey, 1989, Kalghatgi, 1984, Chamberlain, 1986, Sønju and Hustad, 1984, Hustad and Sønju, 1986). Recently Peters and Göttgens (1990) have performed a theoretical study to describe the flame structure of such large buoyant diffusion flames. The theoretical model compares well with experimental results on flame length (Sønju and Hustad, 1984), and it represents the first attempt to describe the importance of both buoyancy and momentum in these large flames. The fact that buoyancy is the dominant force acting on the flame, has implications on flame structure, since fluctuations in temperature and concentration gradients are observed in buoyant driven flames (Ballantyne and Moss, 1977). Temperature fluctuations will have an important influence on the production rates of thermal NO_x, and concentration fluctuations will also effect the prompt NO_x contribution (Fenimore, 1971). A recent study (Williams *et al.*, 1990) has suggested significant NO_x contribution from the prompt mechanism in highly strained flames. Measurements of concentrations of gaseous species in diffusion flames have been performed by various authors, although only a few references can be found on NO_x emissions from unconfined hydrocarbon flames. A recent study (Turns and Lovett, 1989), has made an excellent review of available data and earlier work within this area. Their study shows that only two articles focusing on NO_x emissions in turbulent hydrocarbon free jets have been previously published. Buriko and Kuznetsov (1978) studied the reaction mechanism for NO_x formation in free hydrogen and propane flames. The flames used in their experiments were attached and not lifted from the nozzle. The Reynolds numbers based on the nozzle exit conditions were between 7000 and 17000, meaning that the flames were lying in the transitional flow regime between laminar and turbulent behaviour. Nozzle diameters of 3 and 6 mm were used. Their conclusions were that thermal radiation and residence time control the amount of NO_x formed, together with O radical concentration.

Turns and Lovett (1989) studied propane jet diffusion flames by varying the outlet velocity while keeping the nozzle diameter constant. They used two concentric nozzles in which hydrogen was flowing through the outer nozzle to stabilize the flame. They claimed that this had an insignificant effect on the NO_x emissions, although a hydrogen flame certainly would alter the entrainment of air taking place in the liftoff region. Turns and Lovett (1989) suggest that the emission index is independent of the outlet velocity for laminar to transitional Reynolds number flames, while for fully turbulent flames the velocity becomes an important parameter. The outlet velocity controls the reaction volume and the flame height is nearly constant for high Froude numbers. The heat release rate per unit volume increases with increasing velocity, thus reducing the average residence time. Higher strain rates in flames have been suggested as having the effect of reducing both the thermal and prompt NO_x formation (Drake and Blint, 1989). Bilger and Beck (1975) suggest that the Kolmogorov time scales govern thermal NO_x production for hydrogen flames, thus indicating a reduction of NO_x formation with increasing strain rate (less residence time in the flame). With this hypothesis the emission index should decrease with increasing strain, or by increasing velocity. This is, however, not experimentally experienced in this study. This may be due to the fact that Kolmogorov time scales are too small to govern thermal NO_x production in high velocity flames, due to the small residence time. This is also in agreement with the later work of Bilger (1976), concluding that Kolmogorov time scales are too short to account for the amount of NO_x found in turbulent H_2 flames.

The mechanisms controlling the formation of NO_x and the associated chemical reaction constants are quite well known (Miller and Bowman,1989). The complex interactions between the chemical reactions and fluid mechanics are still not well understood for large buoyant diffusion flames, although some recent advances have been reported (Williams *et al.*, 1990).

The three most relevant compounds in NO_x are NO (Nitric oxide), NO_2 (Nitrogen dioxide) and N_2O (Nitrous oxide). Nitrous oxide is of great concern with regards to the greenhouse effect and ozone depletion but has almost no other pollution effects. In this paper NO_x will be used to denote the sum of NO and NO_2 while the emission of N_2O is not studied in this work.

In the following sections, the experimental apparatus and the results of the present study will be described.

2. EXPERIMENTAL APPARATUS AND METHODS

2.1. Emission Measurement Rig

A schematic drawing of the experimental apparatus is shown in fig. 1. Propane gas, of commercial quality typically 99.85 vol% C_3H_8 (min. 95 vol% C_3H_8, max. 2.5 vol% C_2H_6, max. 2.5 vol% i-C_4H_{10}) was kept in standard gas bottles storing 36 kg of propane. They were connected to a common manifold with independent shutoff valves as shown in fig. 1. The gas pressure was regulated by means of adjustable constant pressure-valves. The rotameter inlet gas pressure was kept at a constant pressure of 2 bar. Rotameters of different sizes were used to control and measure the gas flow. Various shutoff and regulation valves were used in conjunction with the rotameter stand. The test rig was made up of a frame of square steel pipes. The various nozzles were connected to the lower arm and were moved axially and radially to measure emission profiles. Six different nozzle diameters have been used, namely 4, 5, 10, 12, 17 and 28 mm, respectively.

In addition to the lower arm, a movable upper arm with the suction probes was installed. The mast was made telescopic, to make it possible to vary the axial position of the probes over a wide range. Figure 2 shows the suction probe for NO_x measurement. This water-cooled probe was made of quartz. Quartz was selected as a suitable material to prevent catalytic reduction of nitric oxide (NO) and water cooling was required to cool the gas as rapidly as possible. However, to avoid water vapour condensing in the suction probe, hot water was used. England and Houseman(1973), tested several probe materials and designs for NO_x measurements in flames. The water-cooled quartz probe proved to be the best design. Together with NO_x concentrations, the carbon monoxide (CO), the carbon dioxide (CO_2) and the oxygen (O_2) concentration were measured using a water-cooled suction probe made of stainless steel.The stainless steel probe is shown schematically in fig. 3.

2.2. Measuring Equipment

The quartz probe (NO_x) was connected to a Signal Luminox 4000 chemiluminescent analyzer by means of a 6 mm i.d. plastic tube. Water vapour was removed from the sample gas prior to entering the analyzer via a cold trap. This trap was an empty container kept at 0°C by an ice bath. The stainless steel probe was connected to three different analyzers. A Servomex paramagnetic oxygen analyzer measured the O_2 level while the CO and the CO_2 were measured with a Hartmann & Braun nondispersive infrared analyzer.

A platinum-rhodium (Pt/Pt–10% Rh) thermocouple was positioned between the probes. The thermocouple had a bead diameter of 0.2 millimeters, and corrections were made for the radiation contribution. The measured data were continuously logged by a data acquisition system. At the start of the experiment, all the analyzers were calibrated independently with certified calibration gases. A calibration gas mixture of 20.3 ppm NO_2 in synthetic air was used to check the NO to NO_2 converter in the NO_x analyzer. The analyzers were also recalibrated at certain intervals during the measurement period to minimize any instrument drift. The whole experimental setup was surrounded by a fine wire mesh screen, constructed of movable 1×3.5 m (length times height) hinged wire screen panels. This construction succeeded in mitigating the effects of room draft.

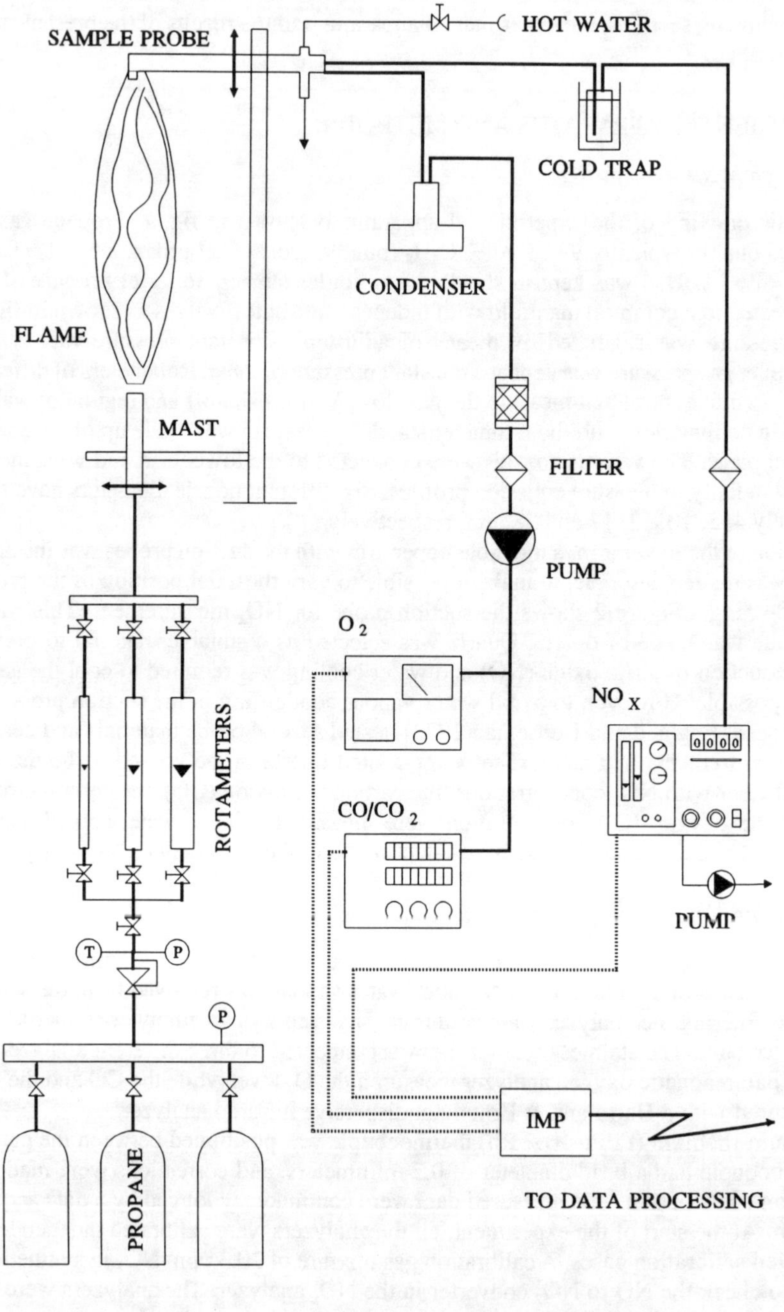

FIGURE 1 Schematic diagram of the experimental apparatus.

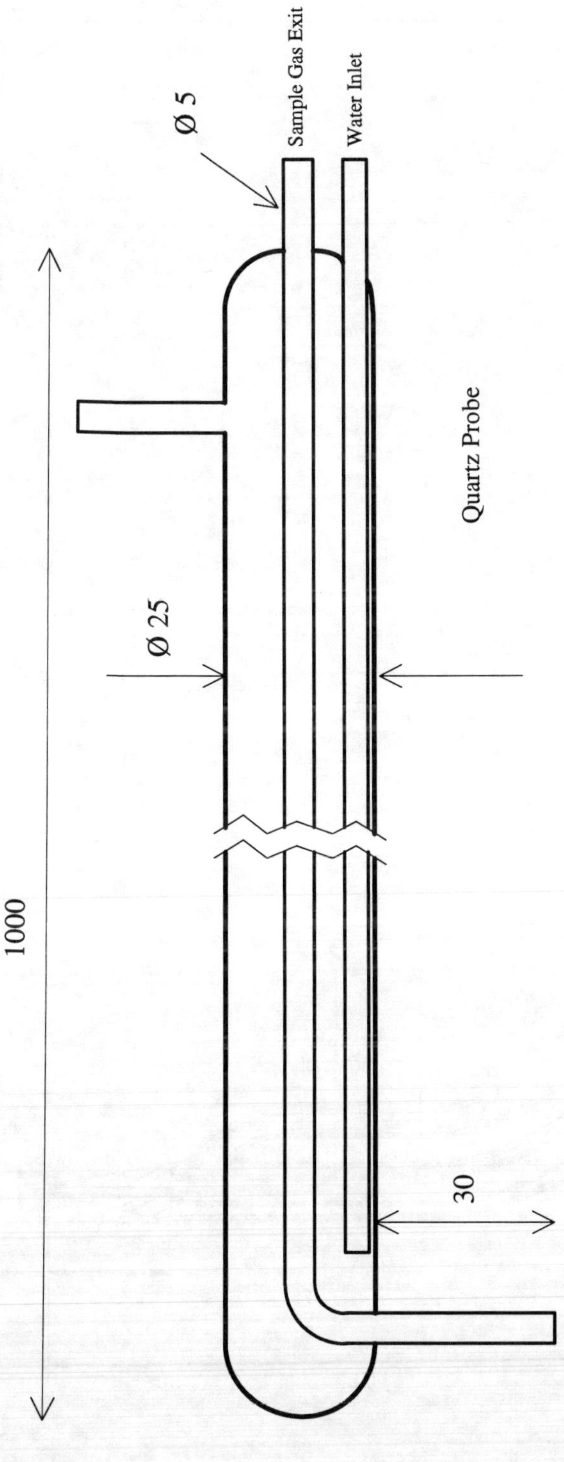

FIGURE 2 Quartz probe used for NO_x measurements.

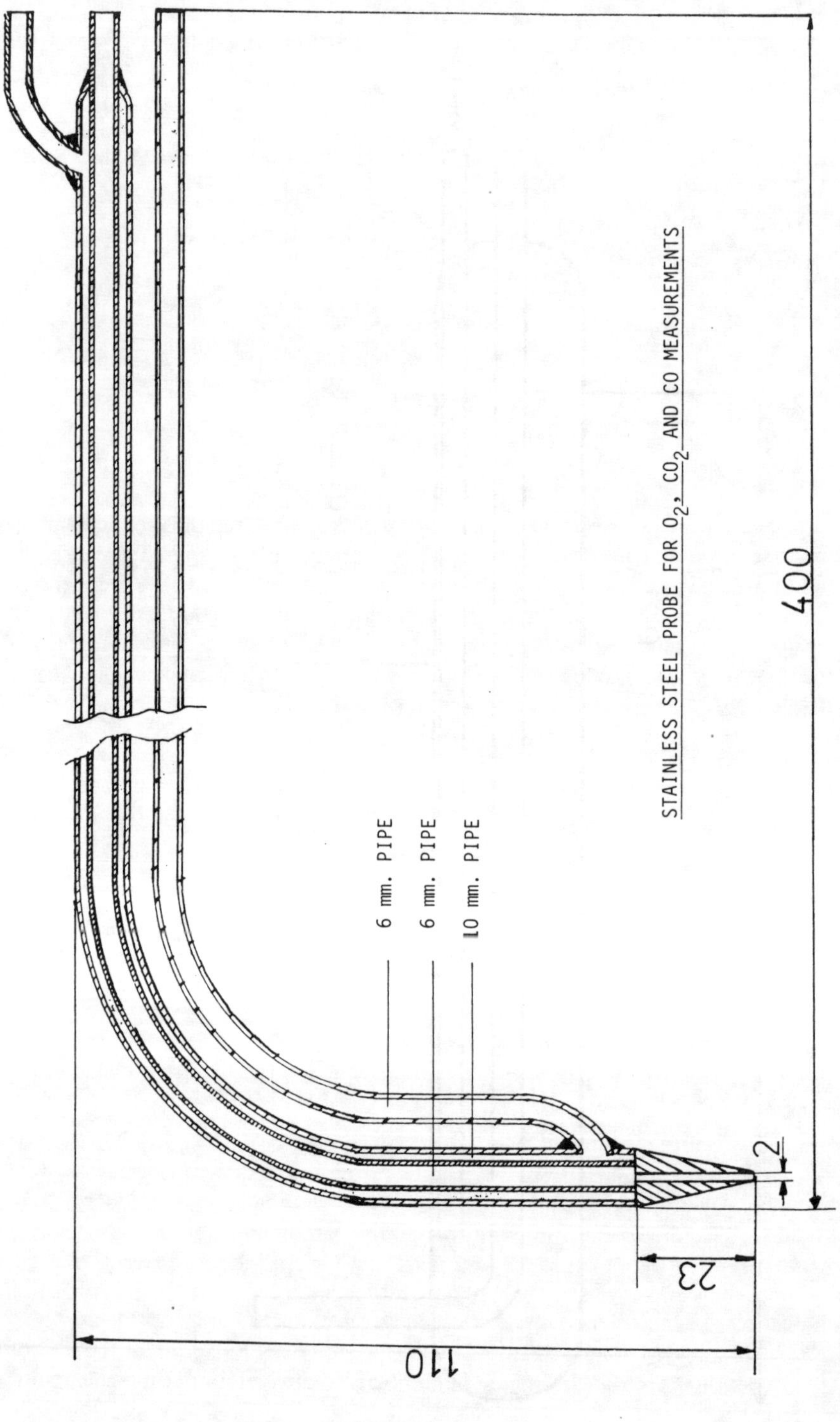

FIGURE 3 Stainless steel probe for O_2, CO_2, and CO measurements.

TABLE I
TEST MATRIX

Nozzle diameter [mm]	Reynolds number interval	Froude number interval	Nozzle exit velocity interval [m/s]
4	15000–60000	6362–101147	15.8–63.0
5	13000–25300	2512–9077	11.1–21.1
10	36000–109112	2300–21100	15.0–45.5
12	43840–60000	1989–3746	15.3–21.0
17	910714	303561	225
28	14700–29500	17.6–70.5	2.2–4.4

2.3. *Experimental Test Matrix*

Varying basic fluid mechanical relations, like the Reynolds number and the Froude number, describe how viscous, momentum and buoyancy forces affect the emissions. Table 1 shows the test conditions used in this study. The nozzle diameters and exit velocities were varied in such a manner that the Froude and Reynolds number could be varied independently.

This does not apply to the medium size flame at 9 MW. This test was only conducted at one value of Reynolds and Froude number and using one nozzle diameter and a fixed nozzle exit velocity.

2.4. *Computational Methods and Assumptions*

The objective was to express emission data for turbulent propane jet diffusion flames on a fuel mass basis in gNO_x/kg fuel. All calculations are based on nozzle exit gas condition, i.e., a temperature of 298 K, and a pressure of 1.013 bar (ambient conditions). Emission indices based on energy input are easy to calculate when the emission is expressed in gNO_x/kg fuel.

Equation (1) was used for calculating a local emission index for NO_x emission expressed in gNO_x/kg fuel:

$$EI_{NO_x} = \frac{n \cdot (NO_x) \cdot MW_{NO_2} \cdot 1000}{(CO_2) \cdot MW_{CnHm}} \left[\frac{g\,NO_x}{kg\,fuel}\right], \tag{1}$$

where n = number of carbon atoms in the hydrocarbon gaseous fuel; (NO_x) = mole fraction of NO_x; MW_{NO_2} = molecular weight of NO_2; (CO_2) = mole fraction of CO_2; MW_{CnHm} = molecular weight of gaseous hydrocarbon fuel.

This equation is based on the assumption that all the carbon atoms in the gaseous fuel are oxidized to CO_2. For propane flames, 3 moles of CO_2 are formed for each mole of propane reacting. The formula can be used for all kinds of hydrocarbon fuels. CO formation will alter the carbon balance, hence it was decided to reject all NO_x measurements with a concentration of CO higher than 1000 ppm. Soot formation will also affect the carbon balance. This effect, however, is considered negligible and thus left out of the calculation procedure.

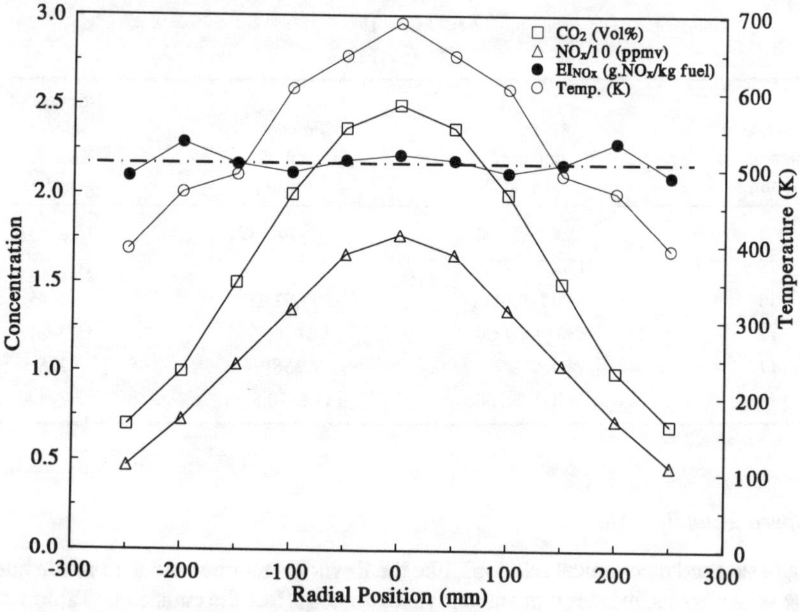

FIGURE 4 Radial temperature and emission profiles, 12 mm nozzle; Re = 60000, $y/d_0 = 207$.

3. RESULTS AND DISCUSSION

Measurements of radial NO_x and CO_2 concentration profiles together with a radial temperature profile are shown in fig. 4. The measurements are taken in the post flame region. This measurement position was calculated using the flame height correlations by Sønju and Hustad (1984):

$$\frac{H}{d_0} = 27 \cdot Fr^{0.2}, \tag{2}$$

where H = total flame height [m]; d_0 = nozzle exit diameter [m]; Fr = Froude number, u_0^2/gd_0; u_0 = nozzle exit velocity [m/s].

Equation (2) has to be modified for small flames (<0.7 MW), using a constant of 40 instead of 27. This is in agreement with the results of Suris *et al.* (1977) for small size flames.

Figure 4 shows the results for NO_x, CO_2 and temperature. The maximum values are at the extended flame centerline.

By calculating an emission index, using Equation (1) at each point in the radial direction, a constant emission index appears, irrespective of the radial position, as can be seen from the figure. The reason for this is the similarity of the profiles of NO_x and CO_2, giving a constant ratio between NO_x and CO_2. This means that the NO_x and CO_2 concentrations are well mixed with the entrained air in the post flame region. It was also attempted to express integrated mean values for the NO_x emission index, based on a flame exit area (flame diameter) and measured velocity profiles. The flame exit area was divided into equal annular areas and the NO_x emission together with the measured velocity in a representative point within each elemental area used to calculate a integrated mean emission index. These results were in good agreement with the calculated values utilizing Equation (1). This area

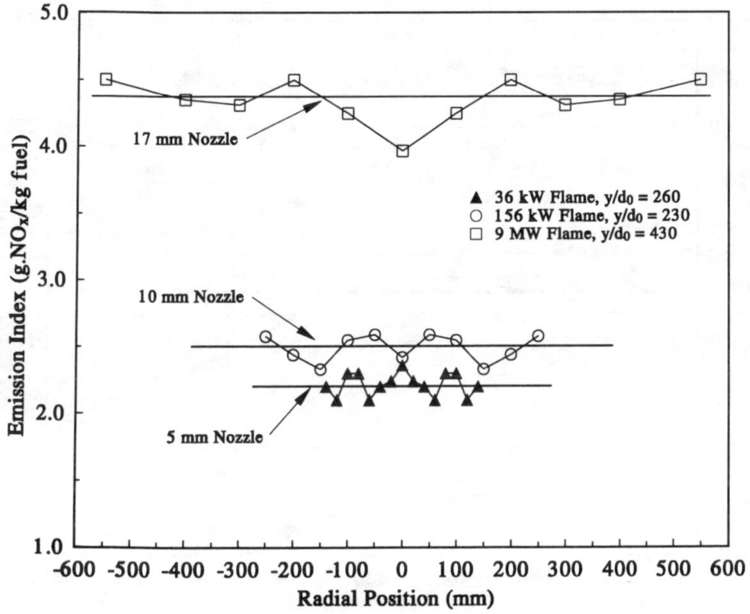

FIGURE 5 Radial emission index profiles for various flame sizes.

method is, however, more cumbersome and needs more measured values, bringing more uncertainty into the calculated emission indices.

Figure 5 shows radial emission index profiles for different heat release rates ranging from a 36 kW flame to a 9 MW flame. These measurements were taken at heights corresponding to Equation (2). This figure shows again the constant radial emission index profile. Deviations of ±5% are observed.

The axial profile in the post flame region is also of great interest, and fig. 6 depicts this profile. All these measurements were taken along the extended flame centerline. As can be seen from the figure, the NO_x concentration together with the CO_2 concentration drop significantly with axial position, while the emission index remains constant. The visible flame height for this flame was $y/d_0 = 240$.

It can be concluded from the emission index invariance with radial and axial position in the post flame region that NO_x is neither destroyed nor formed there. This is as would be expected, due to the relatively low temperature. Both the constant radial and axial emission index profile are in good agreement with the results of Turns and Lovett (1989), although their flames were smaller than the flames studied in this work. The fact that the emission index remains constant in both axial and radial direction in the post flame region implies that only one measurement point is required for each test. A measurement on the centerline was chosen and was, however, performed four times to give an arithmetic mean as the final value.

Figure 7 shows the various gas concentrations as a function of time for the medium size flames. The curves exhibit a kind of intermittent behaviour. The NO_x and CO_2 concentration curves follow each other closely, and large fluctuations in gas concentrations are experienced. This indicates a strong temperature dependence upon the NO_x emissions. This has been verified by plotting the temperature on the same figure.

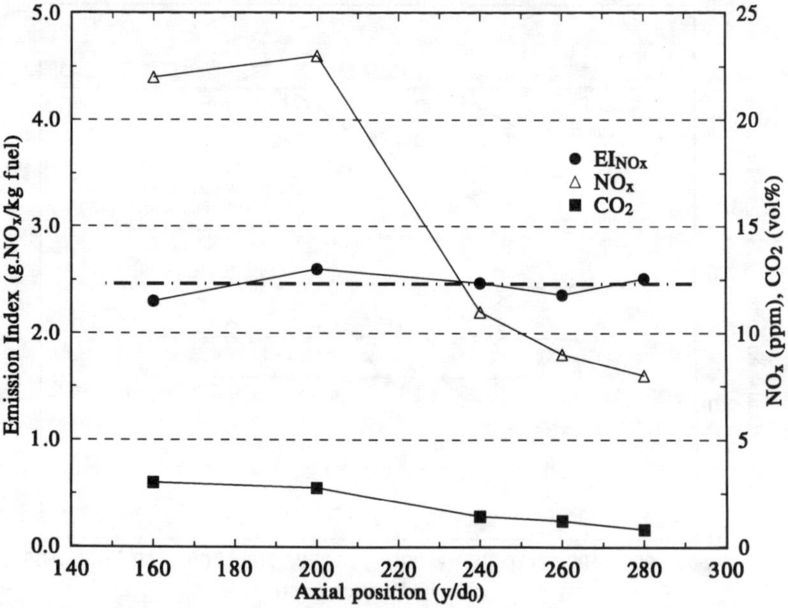

FIGURE 6 Axial CO_2, NO_x, and EI_{NO_x} emission centerline profiles, 5 mm nozzle, Re = 25300.

The question of how the emissions scale with fluid mechanical parameters is also of great interest, and much effort was put into revealing such a dependency. Due to the great variety of nozzle diameters and exit velocities, it was possible to extract some interesting trends.

The work of Williams *et al.* (1990), suggests a nozzle diameter and outlet velocity dependency upon the NO_x emissions. Figure 8 shows a plot of the emission index versus $u_0^{0.2}\ d_0^{0.4}$, with data from the present work together with various earlier studies. The present data and the results of Turns & Lovett and Buriko & Kuznetsov agree well, however, the trend seems different for the data from Turns & Lovett. Buriko and Kuznetsov used a 3 mm nozzle and a Reynolds number of 14,300, Takagi *et al.* used a 2 mm nozzle and a Reynolds number of 13,000 and Turns and Lovett used a 3.86 mm nozzle at various Reynolds numbers. A regression line is drawn through our experimental values. The data give a correlation coefficient of 0.94. The experimental values fit the equation:

$$EI_{NO_x} = 6 \cdot u_0^{0.2} \cdot d_0^{0.4} + 0.85\,[gNO_x/kg\,fuel]\,, \tag{3}$$

or if the Froude number is used in this correlation:

$$EI_{NO_x} = 24 \cdot Fr^{3/5} \cdot (d_0/u_0) + 0.85\,[gNO_x/kg\,fuel]\,. \tag{4}$$

Both the constants in (3) and (4) have the dimension of $[s^{-1}]$.

These equations are basically telling us that the NO_x emission scales as the fuel volume flow to the power of 0.2. The flame volume is thus an important parameter.

The data point from Takagi *et al.* shows a somewhat lower value than the other experiments, but they have only measured NO, and hence not taken the NO_2 concentration into account. The ratio of NO_2/NO_x was measured by Turns & Lovett (1989) and they found this ratio to be dependent on the axial measurement location. A typical value of 0.25 as the

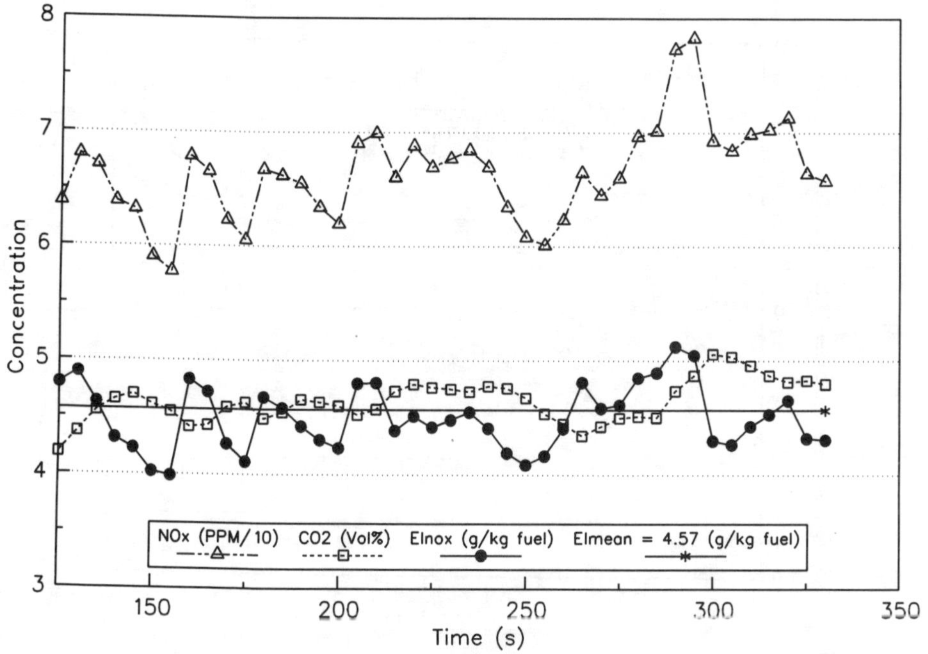

FIGURE 7 Emission indices time history, 9 MW flame, Re = 910000, $x/d_0 = 12$, $y/d_0 = 430$.

ratio of NO_2/NO_x for far downstream positions was found, while this ratio went to zero towards the flame tip.

Their measurements were conducted at a fixed axial position, while varying the radial position and the nozzle exit velocity. A reciprocal relation between the NO_2/NO_x ratio and the nozzle exit velocity was found, while the NO_2/NO_x ratio increased with radial position. The further oxidation of NO to NO_2 is a slow reaction, primarily taking place in the post flame region. The residence time is thus an important parameter for the NO_2/NO_x ratio, together with O radical and O_2 concentration. This explains the results of Turns & Lovett. The present measurements were, however, taken at a fixed nondimensional height at the extended flame centerline and the post region residence time dependency was not discovered from our experiments. Some tests were performed to look at the NO_2/NO_x ratio in these large flames. These tests showed low ratios in the range of 0.05–0.1, which is in good agreement with the results of Turns and Lovett (1989), for measurement positions near the flame tip.

Equation 4 expresses a Froude number dependency upon the emission index. To test how this equation holds for various Froude numbers, the emission index was plotted versus the d_0/u_0 ratio. Figure 9 depicts this Froude number dependency, for Froude numbers ranging from 17.6 to 300,000. It is worth noting that 0.85 is subtracted from the emission index (y-axis), making it possible to get linear correlation curves from (4). The solid lines in figure 9 represent (4), for that particular Froude number. The plotted values are within ±10% of the stated Froude number. The emission index increases with increasing Froude number, which means that buoyancy forces also affect the formation of NO_x. The measured values compare well with the calculated correlation lines, although the amount of data points is limited. The emission index is also increasing with increasing ratio of d_0/u_0. It has

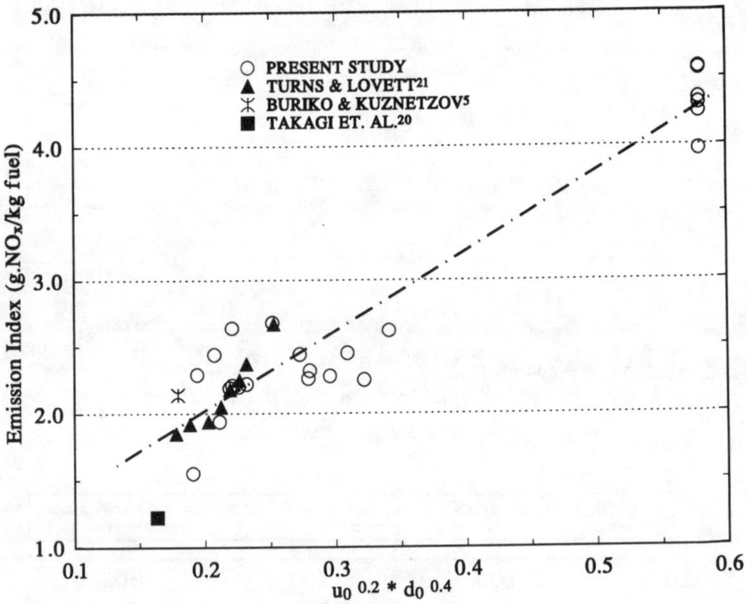

FIGURE 8 Emission index vs. $u_0^{0.2}\, d_0^{0.4}$. Comparison between various studies.

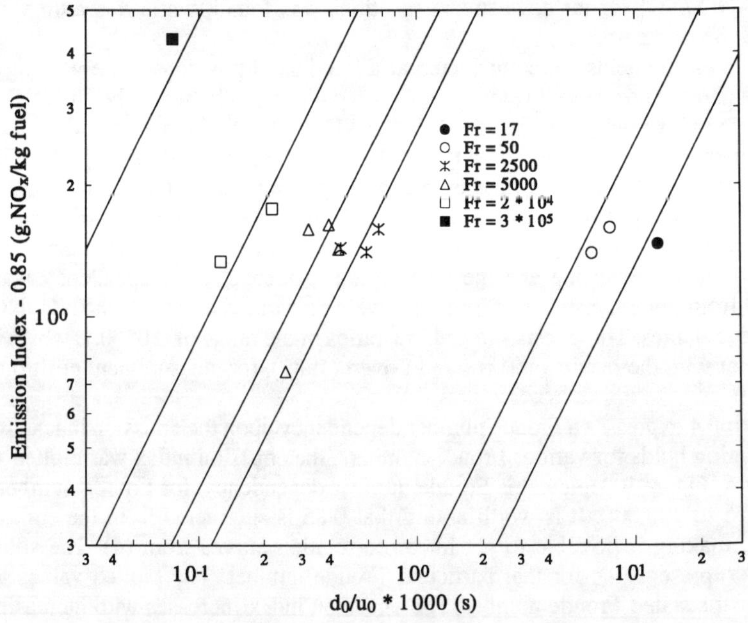

FIGURE 9 Emission index: 0.85 vs. d_0/u_0 ($\log_{10}/\log_{10}$), different Froude numbers.

been shown previously that the flame geometry for these large flames is also dependent on buoyancy (e.g., Hustad and Sønju, 1986).

The sensitivity of (3) was tested by plotting the emission index versus $u_0^{0.2}$ for the experimental results, to see how well they compare with the correlation. These plots show good qualitative agreement, however, the correlation seems to be too sensitive with regards to the nozzle diameter. More data is needed to establish this dependency.

As can be seen from figs. 8 and 9, the range for the suggested correlations is extensive, ranging in Froude numbers from 17.6 to 300,000. More experiments are needed to complete the picture, as the data points are scarce for flames less than 9 MW and greater than 0.7 MW. This will be accomplished in the near future, together with flames greater than 9 MW, meaning near sonic and sonic jets. Some experiments on methane and n-butane together with various mixes of propane/methane/n-butane have also been accomplished and more experiments will be conducted during the fall of 1991/spring 1992.

4. SUMMARY AND CONCLUSIONS

A comprehensive experimental program has been carried out on small (<0.7 MW) and medium size (9 MW) propane jet diffusion flames with the objective of obtaining NO_x emission profiles and indices. These experiments represent the first attempt to obtain emission data for estimating NO_x pollution from offshore flares. The experiments were conducted by varying the nozzle exit velocities from 2.2 to 225 m/s and the nozzle exit diameter from 4 to 28 mm, giving Reynolds numbers in the range of 13,000 to 910,000 and Froude numbers from 17.6 to 303,561.

The experiments show a power law dependence of 0.4 on the nozzle exit diameter and a power law dependence of 0.2 on the jet initial velocity, meaning that the emission index scales with the fuel volume flow to the power of 0.2.

The emission index further shows a Froude number dependence, indicating the NO_x emissions being controlled by the ratio of momentum to buoyancy.

Emission indices ranging from 1.55 to 4.57 gNO_x/kg fuel were measured in the experiments. The experimental results give guidelines by establishing equations for calculating NO_x emissions from small and medium size propane jet diffusion flames up to 9 MW thermal heat release.

5. ACKNOWLEDGMENTS

This work was in part sponsored by the Norwegian State Oil Company (STATOIL), with Kyrre Langnes, Edgar Furuholt and Jan. M. Øverli serving as technical monitors. The authors would also like to thank Yngvil Bjerve and Mona Jacobsen, students at the Norwegian Institute of Technology, for performing some of the experiments.

REFERENCES

Ballantyne, A., and Moss, J.B. (1977). Fine Wire Thermocouple Measurements of Fluctuating Temperature. *Combust. Sci. and Tech.*, **17**, pp. 63–72.

Bilger, R.W. (1976). Turbulent jet diffusion flames. *Prog. Energy Combust. Sci.*, **1**, 87.

Bilger, R.W. (1988). The Structure of Turbulent Nonpremixed Flames. *Twenty-second Symposium (Int.) on Combustion*, The Combustion Institute, Pittsburgh, pp. 475–488.

Bilger, R.W., and Beck, R.E. (1975). Further Experiments on Turbulent Jet Diffusion Flames. *Fifteenth Symposium (Int.) on Combustion*, The Combustion Institute, Pittsburgh, pp. 541–550.

Buriko, Y.Y., and Kuznetsov, V.R. (1978). Possible Mechanism of the Formation of Nitrogen Oxides in Turbulent Diffusional Combustion. *Comb. Expl. Shockwaves*, **14**, pp. 296–302.

Chamberlain, G.A. (1986). The Shell/Thornton Flare Radiation Model. Paper presented at the Fourth International Flare System Seminar, Trondheim, Norway, 35 pp.

Drake, M.C. and Blint, R.T. (1989): Thermal NO_x in Stretched Laminar Opposed-Flow Diffusion Flames with $CO/H_2/N_2$ Fuel. *Combustion and Flame*, **76**, pp. 151–167.

England, C., Houseman, J. and Teixeira, D.P. (1973). Sampling Nitric Oxide from Combustion Gases. *Combustion and Flame*, **20**, pp. 439–442.

Fenimore, C.P. (1971). Formation of nitric oxide in premixed hydrocarbon flames. *Thirteenth Symposium (Int.)* on Combustion. The Combustion Institute, Pittsburgh, pp. 373–380.

Hustad, J.E., and Sønju, O.K. (1986). Radiation and Size Scaling of Large Gas and Gas/Oil Diffusion Flames. *Dynamics of Flames and Reactive Systems, Prog. Astro. Aero.*, **105**, pp. 365–386, AIAA.

Hustad, J.E, Røkke, N.A., and Sønju, O.K. (1991). Heat Transfer to Pipes Submerged in Lifted Buoyant Diffusion Flames. *Second International Conference on Experimental Heat Transfer, Fluid Mechanics and Thermodynamics*, ed. J.F. Keffer, R. K. Shah, and E.N. Ganic, Elsevier, New York, pp. 567–574.

Jeng, S.-M., Chen, L.-D., and Faeth, G.M. (1982). The Structure of Buoyant Methane and Propane Diffusion Flames. *Nineteenth Symposium (Int.)* on Combustion. The Combustion Institute, Pittsburgh, pp. 349–358.

Kalghatgi, G.T. (1984). Liftoff Heights and Visible Lengths of Vertical Turbulent Jet Diffusion Flames in Still Air. *Combust. Sci. and Tech.*, **41**, pp. 17–29.

Lindeberg, E.G.B., Røkke, N.A., and Celius, H.K. (1990). Greenhouse Gas Emissions From Offshore Oil and Gas Production on the Norwegian Continental Shelf. IKU Report 35.3919.00/01/90, IKU, Trondheim, Norway.

McCaffrey, B.J., and Evans, D.D. (1989). Momentum Diffusion Flame Characteristics and the Effects of Water Spray. *Combust. Sci. and Tech.*, **63**, pp. 315–335.

Miller, J.A., and Bowman C.T. (1989). Mechanism and modelling of nitrogen chemistry in combustion. *Prog. Energy Comb. Sci.*, **15**, pp. 287–338

Peters, N., and Göttgens, J. (1990). Scaling of Buoyant Turbulent Jet Diffusion Flames. *Combustion and Flame*, **85**, pp. 206–214.

Suris, A.L., Flankin, E.V., and Shorin, S.N. (1977). Length of free diffusion flames. *Combust. Explos. Shock Waves*, **13**(4), pp. 459–462.

Sønju, O.K., and Hustad, J.E. (1984). An Experimental Study of Turbulent Jet Diffusion Flames. *Dynamics of Flames and Reactive Systems, Prog. Astro. Aero.*, **95**, pp. 320–339, AIAA.

Takagi, T., Ogasawara, M., Fujii, K., and Daizo, M. (1975). A Study on Nitric Oxide Formation in Turbulent Diffusion Flames. *Fifteenth Symposium (Int.)* on Combustion. The Combustion Institute, Pittsburgh, pp. 1051–1059.

Turns, S.R., and Lovett, J.A. (1989). Measurements of Oxides of Nitrogen Emissions from Turbulent Propane Jet Diffusion Flames. *Combust. Sci. and Tech.*, **66**, pp. 233–248.

Williams, F.A., Gordon, A.S., Libby, P.A., Røkke, N.A., Hustad, J.E., and Sønju, O.K. (1990). Emissions From Flares. SINTEF Report, STF15 F90056, SINTEF, Trondheim, NORWAY.

5.

GAS FIRED SYSTEMS

Predictions of Thermal NO_x Emissions from Utility Boiler Furnaces

M. BONVINI and C. PIANA *ANSALDO Componenti Srl, Unità Operativa Legnano, Legnano, Italy*
L. VIGEVANO *Dipartimento di Ingegneria Aerospaziale, Politecnico di Milano, Italy*

Abstract — The present investigation deals with the application of a kinetic model, based on the Zeldovich mechanism, for predicting thermal NO_x emissions from a large utility boiler furnace (320 MWe UP boiler of the Sermide power plant owned by ENEL, Italy). The predictions are obtained by numerical integration of the transport equation for the NO mass fraction, starting from a temperature and velocity field provided by a three-dimensional analysis. The latter has been advantageously decoupled from the calculation of the pollutant concentration and carried out by solving, using a numerical method, the conservation equations of mass, momentum, energy and mixture fraction applied to the volume of the combustion chamber. The thermal radiation is modelled using the discrete transfer method. The simulations presented concern the boiler operating under normal conditions as well as with the burner out of service (BOOS) technique. The computed average values of NO_x concentration at the combustor outlet are compared with experimental measurements: the comparison, although on a restricted basis, shows a satisfactory quantitative agreement when the influence of turbulent fluctuation is neglected. Inclusion in the model of turbulent fluctuations and reverse reactions yields a satisfactory agreement only for the normal operating mode, while for the BOOS mode the emission levels are overpredicted.

INTRODUCTION

The increasing attention given to the ecological impact of the combustion processes, attested by the severe regulations on pollutant emissions presently being issued in many industrialized countries, faces the combustion engineer with new, challenging design problems. Advanced concepts for combustion system design, such as staged combustion or gas reburning, have been proposed, together with the development of low NO_x burners, to drastically reduce the pollutant emissions from utility boiler furnaces.

Experimental tests are required in order to assess the validity of this improved equipment. Such very costly experiments in large-scale furnaces can be usefully supported by numerical predictions, obtained with the aid of the currently available, powerful, computational tools, allowing the combustion processes in a boiler combustion chamber to be simulated.

Fully coupled three-dimensional mathematical modelling, with simultaneous solution of the flow, combustion, heat and mass transfer processes occurring within the furnace, has found increasing application in the field of utility boilers. The first numerical simulation, performed with this kind of model, of a gas-fired furnace with tangential burners was presented by Robinson (1985), who analyzed the influence of excess air and burner configuration on gas temperature and heat transfer distribution. Other simulations of gas-fired furnaces were carried out by Abbas and Lockwood (1986) and Lockwood *et al.* (1988a). Two-phase flow simulation of pulverized coal-fired furnaces was tackled by Boyd and Kent (1986), Goerner and Zinser (1986), and Fiveland and Wessel (1988); calculations of oil-fired combustors were presented by Carvalho and Coelho (1990), De Michele *et al.* (1989), the latter including the simulation of the boiler convective section, but allowing only a weak coupling between flow field and heat transfer predictions. These calculations ended up with boiler performance prediction, through a detailed representation of: a) flow field inside the combustion chamber; b) distribution of the heat flux at the wall.

Due to the lack of experimental data, gathered in large scale plants, the validation of numerical predictions has also been mainly limited to wall temperatures and heat fluxes. A noteworthy exception is represented by the comparison of temperature and oxygen concentration profiles within the furnace, performed by Boyd and Kent (1986) for coal combustion simulation.

Concerning pollutant emissions, very few reliable, mainly proprietary data, exist; furthermore, the still inadequate spatial resolution (dictated by CPU time limitations) of the near-burner field, that can be afforded when simulating large-scale combustors, has prevented the application of pollutant formation models in order to simulate the emissions from utility boilers. Only recently predictions of thermal NO_x emission levels from large-scale boilers have been presented by Carvalho *et al.* (1991) and De Michele *et al.* (1991). The above results, although obtained with rather different mathematical and kinetic models, show a satisfactory agreement with experimental data concerning the overall amount of pollutant emission at the furnace outlet, thus evidencing that a reasonable simulation of emission trends is also possible at a large scale, at least for the thermal NO_x.

The present investigation aims at the simulation of pollutant formation in the combustion chamber of large utility boiler furnaces. The specific goal is to assess the capability of simple kinetic models to reproduce the effects on the global thermal NO_x emissions due to the introduction of advanced combustion systems.

A gas-fired plant is considered; the furnace, equipped with opposite wall burners, is operated in two different modes:

a) the "normal" mode (with all burners performing in air excess), which represents the standard combustion system, and
b) the BOOS (Burner Out Of Service) mode, an advanced combustion system, where the upper rows of burner simply act as overfire air (OFA) ports, while the lower burner rows operate at substoichiometric conditions.

The NO_x concentration measurements, carried out at Sermide power plant by Enel and Ansaldo (Zennaro *et al.* 1991), are compared with the analytical predictions of thermal NO_x concentration obtained by post-processing the temperature and velocity field of a 3D furnace code (CINAR) with an "ad hoc" code, developed by ANSALDO. The results achieved demonstrate that the use of a simple NO_x formation model, in conjunction with an appropriate, fully coupled, three-dimensional simulation of the combustion process, does correctly assess the influence of advanced combustion systems on the emission levels.

DESCRIPTION OF THE MATHEMATICAL MODEL

The mathematical model that accounts for the various coupled physical processes, fluid flow, turbulence, combustion and heat transfer, is formulated in a cartesian reference frame.

The Favre averaged conservation equations of mass, momentum and energy, for a steady, three-dimensional turbulent flow of a gas mixture, can be expressed in the generalized form:

$$\frac{\partial (\overline{\rho} u_j \phi)}{\partial x_j} = \frac{\partial}{\partial x_j}\left(\Gamma_{eff}\frac{\partial \phi}{\partial x_j}\right) + \overline{\rho} S_\phi, \tag{1}$$

where x_j and u_j are, respectively, the spatial coordinates and the velocity components, ϕ is the average transported propriety, $\overline{\rho}$ is the mixture average density, Γ_{eff} is the effective transport coefficient, and S_ϕ is the source term of ϕ-propriety.

A standard "k–ε" turbulence model (Jones and Launder, 1972) is employed as a closure set, consisting of two additional transport equations, having the same form as (1) for the turbulent kinetic energy k and its dissipation rate ε.

The combustion model for a diffusion flame assumes a single step fast global reaction between fuel and oxidant and equality of effective transport coefficients for the gaseous fuel and the oxidant. These assumptions allow the concentrations of the major chemical species to be related to a simple conserved scalar variable, the mixture fraction, which in turn satisfies a transport equation such as (1).

The radiative heat transfer is modelled using the discrete transfer method, developed by Lockwood and Shah (1981), exhibiting the advantages of both reduced computational effort and geometrical flexibility. The gas absorption coefficient is computed with a "mixed grey gas" model, using the weighting coefficients determined by Truelove (1976). A nonuniform wall emissivity may be taken into account.

The temperature field of the ideal-gas mixture is calculated starting from the caloric equation of state, namely:

$$T = \frac{h - Y_{FU} H_{FU}}{\Sigma_i \left(Y_i \overline{C}p_i\right)} + T_0, \tag{2}$$

where h is the specific enthalpy of the mixture, Y_{FU} the fuel mass fraction, H_{FU} the fuel lower heating value, and T_0 a reference temperature used to evaluate the average specific heat of the species, $\overline{C}p_i$.

The turbulent fluctuations, of the temperature and of the major chemical species, are neglected, on the basis of the following considerations: i) the fluctuations are expected to not significantly affect the calculation of average values of temperature and major species in a large combustor, whose temperature field is predominantly governed by the thermal radiation, and ii) hence the heavy additional computational effort, required by a statistical approach, does not seem justified.

The resulting set of seven partial differential equations is solved by using a finite volume/finite difference method, in which the convective terms are discretized with the hybrid central/upwind scheme. Pressure and velocity components are linked and computed by the SIMPLE algorithm (Patankar, 1980); the systems of discretized algebraic linear equations are solved by the Gauss–Seidel line relaxation procedure.

The above mathematical model has been implemented in the CINAR computer code, jointly developed by Imperial College (Lockwood *et al.* 1988b) and ANSALDO Componenti. The code has been recently validated against isothermal velocity measurements in a model boiler (Bonvini *et al.* 1991), while comparison of predicted heat flux distributions with those measured in a large scale plant is currently underway.

THERMAL NO_x FORMATION MODEL

The prediction of the thermal NO_x concentration field is decoupled from that of the major species and the temperature field. This assumption is justified because NO_x is present within the furnaces only in a trace amount, thus not influencing the heat-release kinetics.

An Eulerian viewpoint is adopted and a transport equation is solved for the average mass fraction of nitric oxide, [Picture]: a finite-rate chemistry source term is obtained from the kinetic model as follows.

The formation of NO from atmospheric nitrogen in a lean and near-stoichiometric fuel-air mixture is governed by the extended Zeldovich mechanism (Bowman, 1975):

$$O + N_2 \rightleftharpoons NO + N, \tag{R1}$$

$$N + O_2 \rightleftharpoons NO + O, \tag{R2}$$

$$N + OH \rightleftharpoons NO + H. \tag{R3}$$

Reaction (R1), endothermic and rather slow, represents the rate limiting step, while (R2) and (R3) are exothermic, fast and quickly consuming the N atoms produced by (R1). Consequently, the N atom concentration can be considered in quasi-steady conditions and the NO production rate can be expressed as:

$$\frac{d[NO]}{dt} = \left(1 + \frac{k_{-1}[NO]}{k_2[O_2] + k_3[OH]}\right)^{-1} \left(2k_1[O][N_2] - \frac{2k_{-1}[NO]^2\,(k_{-2}[O] + k_{-3}[H])}{k_2[O_2] + k_3[OH]}\right), \tag{3}$$

where the brackets indicate molar concentrations; the reaction rate constants can be found in Baulch *et al.* (1973).

Equation (3), requiring the knowledge of the radical concentration field, is not compatible with the simple model employed here to compute the temperature field. Additional assumptions are necessary to ensure the compatibility of the pollutant emission model with the combustion model (Sala and Vigevano, 1990). Introducing the hypothesis of partial equilibrium for the reactions

$$O + OH \rightleftharpoons O_2 + H, \tag{R4}$$

$$O_2 + M \rightleftharpoons O + O + M, \tag{R3}$$

equation (3) reduces to:

$$\frac{d[NO]}{dt} = 2k_1[O][N_2]\,\frac{1 - \frac{[NO]^2}{K[O_2][N_2]}}{1 + \frac{k_{-1}[NO]}{K_2[O_2]+k_3[OH]}}, \tag{4}$$

where $K=(k_1/k_{-1})(k_2/k_{-2})$, and the concentration of oxygen atoms is given by:

$$[O] = \frac{K_P}{(RT)^{1/2}}\,[O_2]^{1/2}. \tag{5}$$

For the constant K_P the value reported by Bowman (1975) was used.

It can be argued that super-equilibrium radical concentrations, which have been found to be of little importance for high-pressure combustion, on the basis of the calculations of Jones and Priddin (1979) for gas turbines, cannot, in principle, be neglected at atmospheric conditions (Drake *et al.* 1987).

However, the influence of radical overshoot concentration on NO_x formation decreases with increasing flame temperature. Following this observation, partial equilibrium was assumed by Carvalho *et al.* (1990) in the calculation of NO_x emissions from a high-temperature glass-melting furnace. In the present case, considering that most of the thermal NO_x in large scale diffusion flames is likely to be formed in the high temperature zone, immediately after the flame front, the partial equilibrium hypothesis may also be justified.

Furthermore, considering that: a) (R3) is important only for rich flames; b) the transit time through the combustion chamber is negligible with respect to the time needed for NO to reach equilibrium. Therefore neglecting the reverse reactions in the Zeldovich mechanism, equation (4) takes on the simplified form:

$$\frac{d[NO]}{dt} = \frac{2k_1K_P}{(RT)^{1/2}}\,[O_2]^{1/2}\,[N_2]. \tag{6}$$

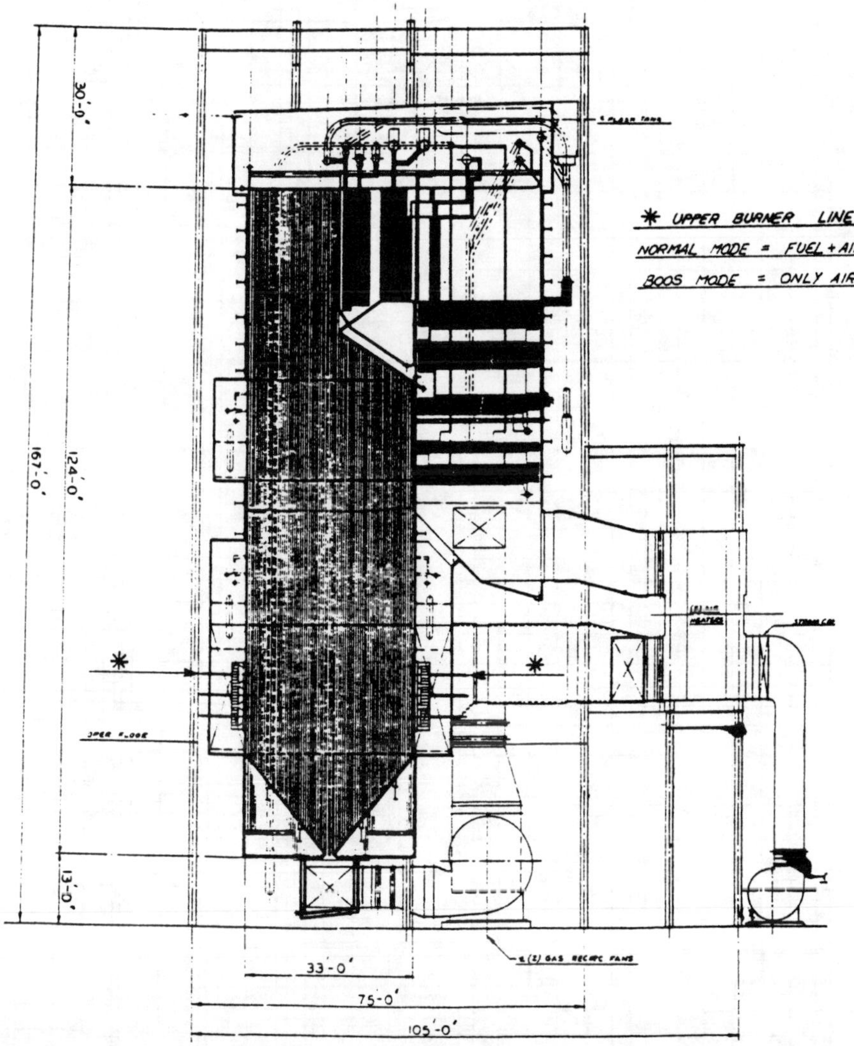

FIGURE 1 Sketch of the 320 MWe utility boiler.

Equation (6) was utilized to determine instantaneous NO production rate, with the value of the rate constant k_1, taken from Monat *et al.* (1979).

The influence of turbulent fluctuations on NO formation is accounted for by a statistical approach (Bilger, 1980). Temperature and major species fluctuations are related to those of the mixture fraction. The average value of the source term in the transport equation for NO mass fraction is then computed by the integral, extended to the mixture fraction space, of the instantaneous production term, weighted with a probability density function (pdf) having an assumed shape. In this work a beta-function is used for the pdf; the last requires the knowledge of the variance of mixture fraction, which, in turn, is modelled with an equation solved with frozen velocity, turbulence and density fields as obtained from the combustion calculations.

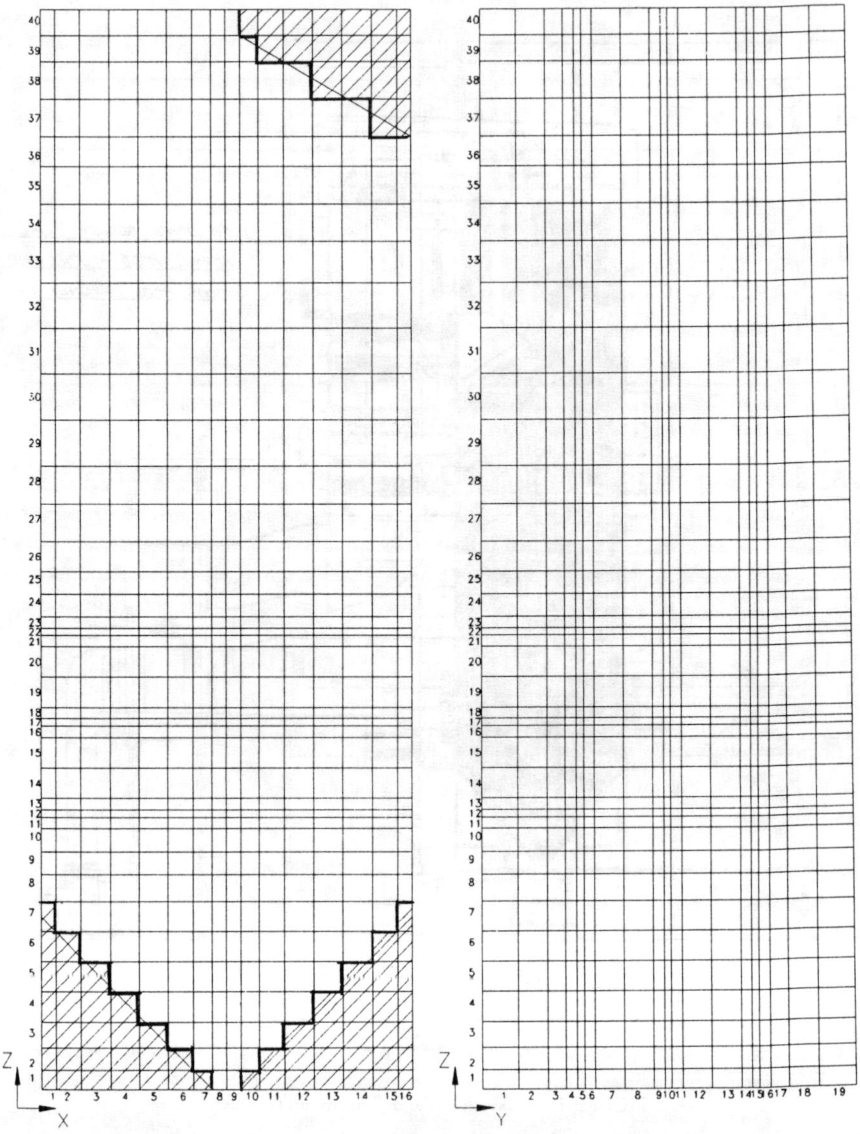

FIGURE 2 Computational mesh used to represent the furnace.

The adopted model, expressed by (6), represents indeed the simplest description of the thermal NO_x formation kinetics. More sophisticated models have been used in recent studies on pollutant emissions from large-scale boilers. Carvalho *et al.* (1991) still assumed partial equilibrium for reactions (R4) and (R5); they neglected the contribution of reaction (R3), but considered the reverse reaction in (R1) and (R2), observing that NO oxidation can take place in regions of the furnace where temperature exceeds 1400 K and oxygen is available. De Michele *et al.* (1991) employed a far more complex NO_x kinetic model, coupled with a finite rate "quasi-global" combustion model, allowing computation of nonequilibrium radical concentrations. The authors can make use of a nonequilibrium

TABLE I
OPERATING CONDITIONS AT 100% THERMAL LOAD

	Normal mode	BOOS mode
Fuel flowrate (t/h)	58	58
Combustion air flowrate	985	717
Overfire air flowrate (t/h)	–	306
Fuel temperature (°C)	30	30
Combustion air temperature (°C)	295	295
Overfire air temperature (°C)	–	295
Excess air (%)	5	5

combustion model, because the calculation of the heat release and species concentration field is weakly coupled with the calculation of flow field and heat transfer.

The scant available data of global furnace emissions do not provide a sufficient basis to give evidence of the really needed sophistication level for a successful prediction of the emissions from a large combustor. In this context, it is valuable to check the predictive capability of a simple model.

DESCRIPTION OF THE MODELLED BOILER

The Sermide plant considered in the present investigation (fig. 1) is a universal pressure (UP) boiler of the Italian Electricity Board (ENEL), rating 320 MWe, designed for oil and natural gas firing. The boiler is equipped with 18 axial flow burners, placed in three rows of 3 burners each on the opposite front and rear walls. The operating conditions at maximum thermal load are summarized in table I. The simulation refers to 100% thermal load and covers two operating modes of the combustion system: a) the normal mode, with all burners operated in balanced, lean conditions at equal thermal load, and b) the BOOS mode, where the upper rows of burners, on both front and rear walls, act as OFA ports, whereas the lower rows operate at substoichiometric conditions.

The model is restricted to the combustion chamber, up to the arch elevation. The overall dimensions of the chamber are approximately $10 \times 10 \times 25$ m^3. The ash hopper and the arch inclined surface are represented in a stepwise fashion, to cope with the cartesian reference frame of the code. Each individual burner is modelled with an array of 9 cells, where the central cell represents the fuel port, while through the surrounding cells air is injected into the furnace. When operating in the BOOS mode the fuel port of the upper burner rows is considered blocked. The cells that simulate the burner throat are sized so as to conserve both mass and momentum fluxes. The above simplifications in modelling the complex geometrical details of the furnace are generally common to large scale simulations and are expected not to practically influence the general features of the prediction.

The furnace walls were assumed to be at a uniform average temperature of 690 K. The emissivity of the wall was set equal to 0.8. The computations were carried out on a mesh with $16 \times 19 \times 40$ cells (fig. 2), covering the whole combustion chamber. Results obtained with this rather coarse mesh are likely to be somewhat mesh-dependent; their interest relies on the fact that the limited spatial resolution of the present computations is typical of "routine" calculations made in an industrial environment. The radiation calculations, that are one of the most time-consuming steps of the overall iterative procedure, are performed

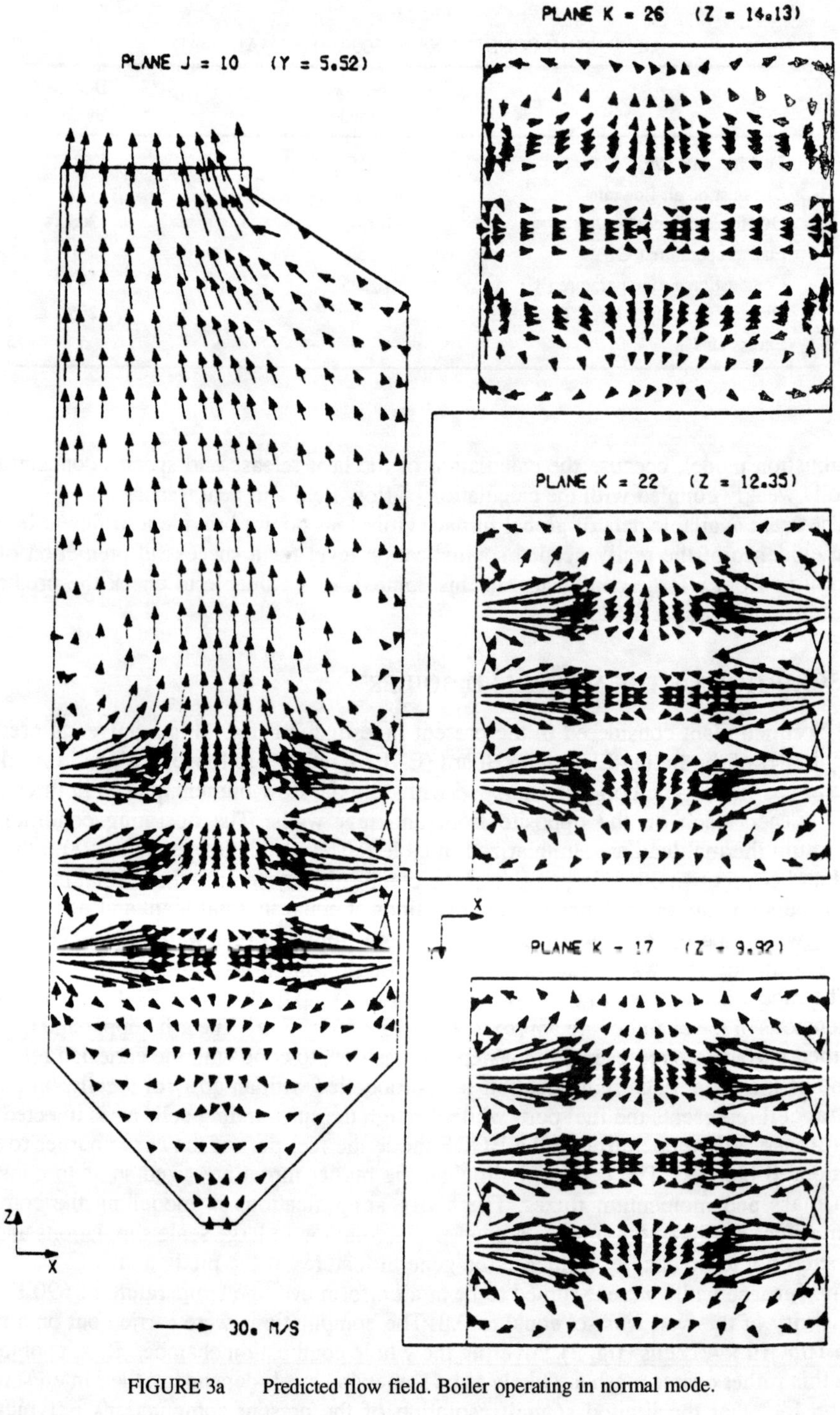

FIGURE 3a Predicted flow field. Boiler operating in normal mode.

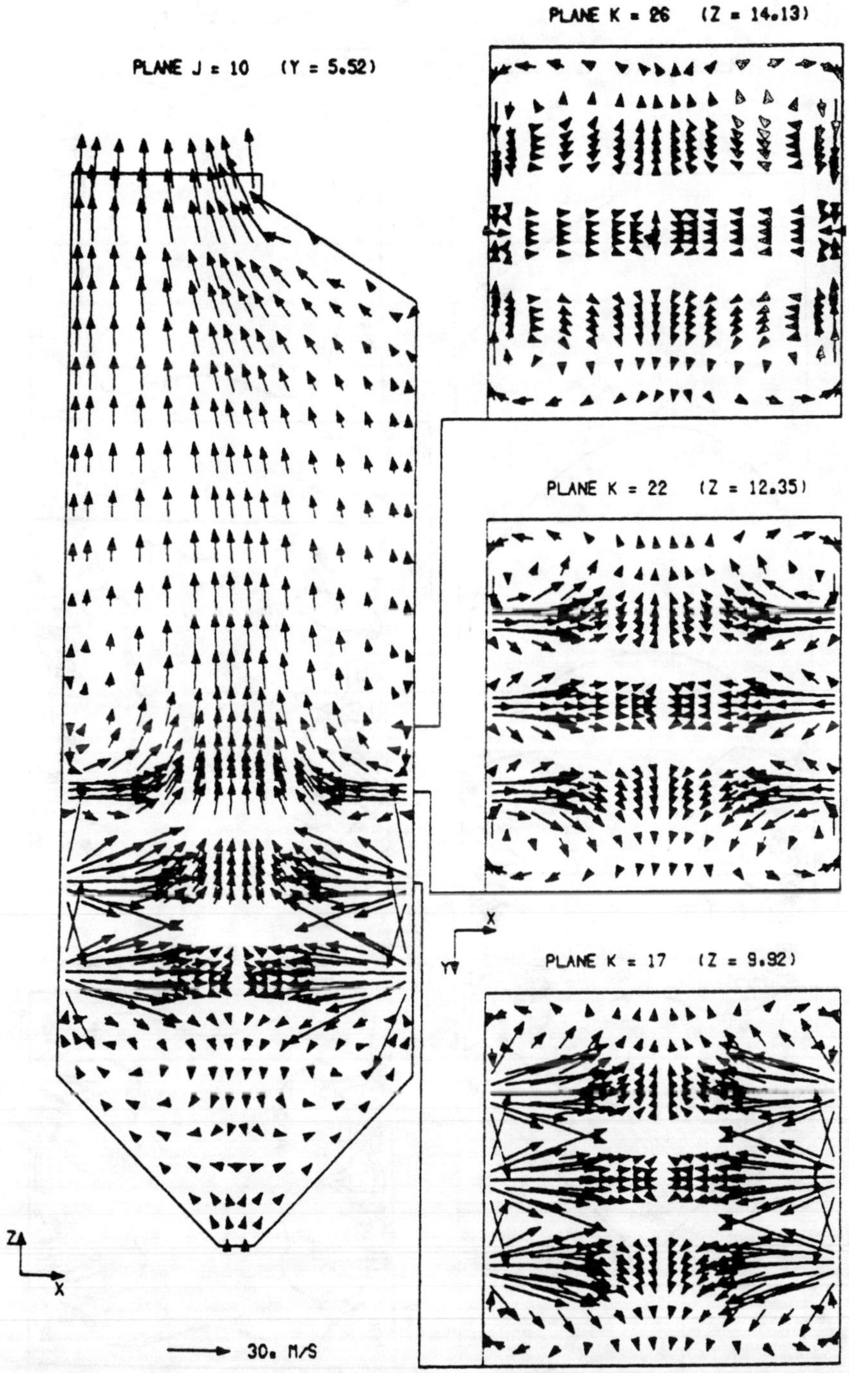

FIGURE 3b Predicted flow field. Boiler operating in BOOS mode.

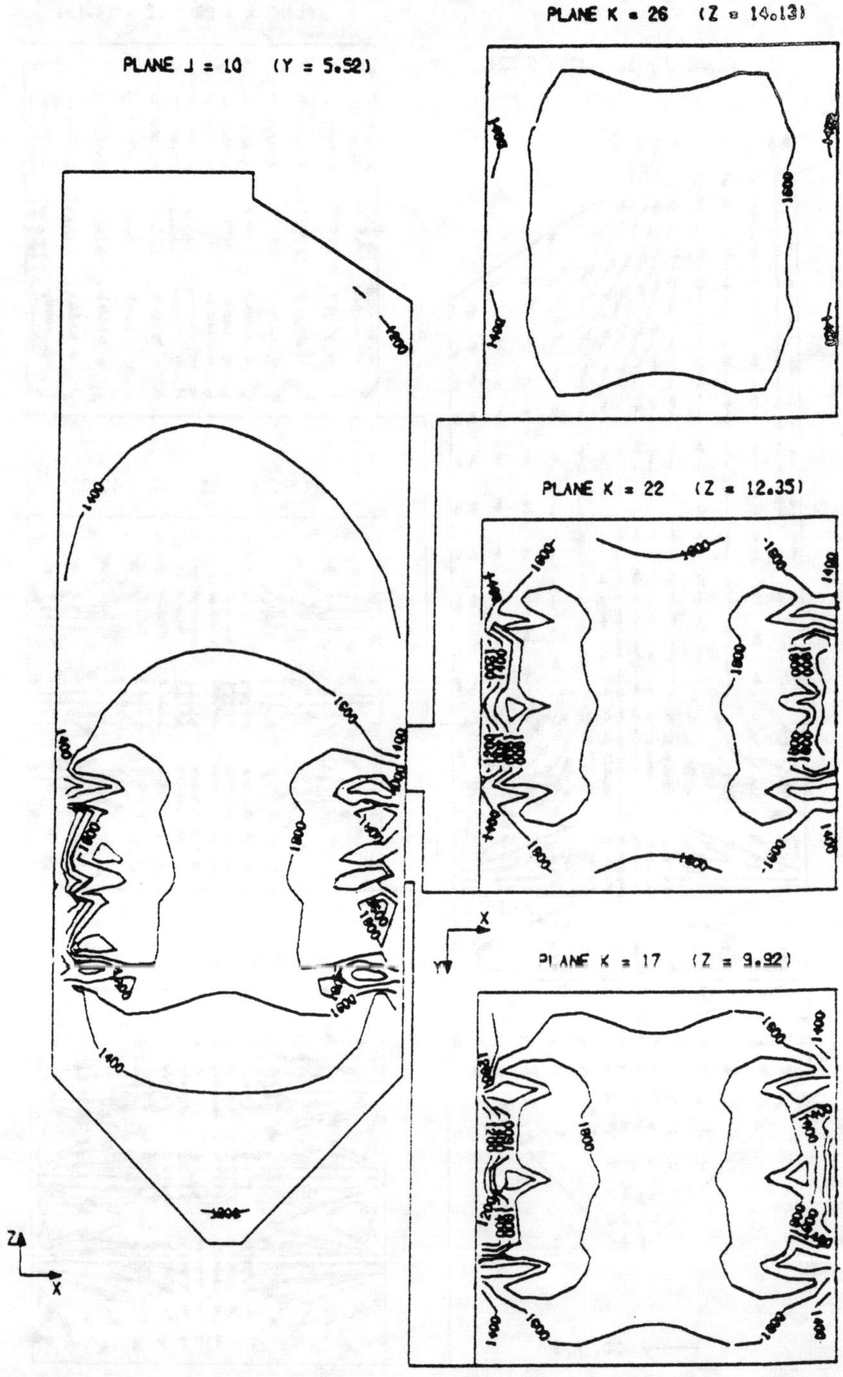

FIGURE 4a Predicted gas temperature distribution (°C). Boiler operating in normal mode.

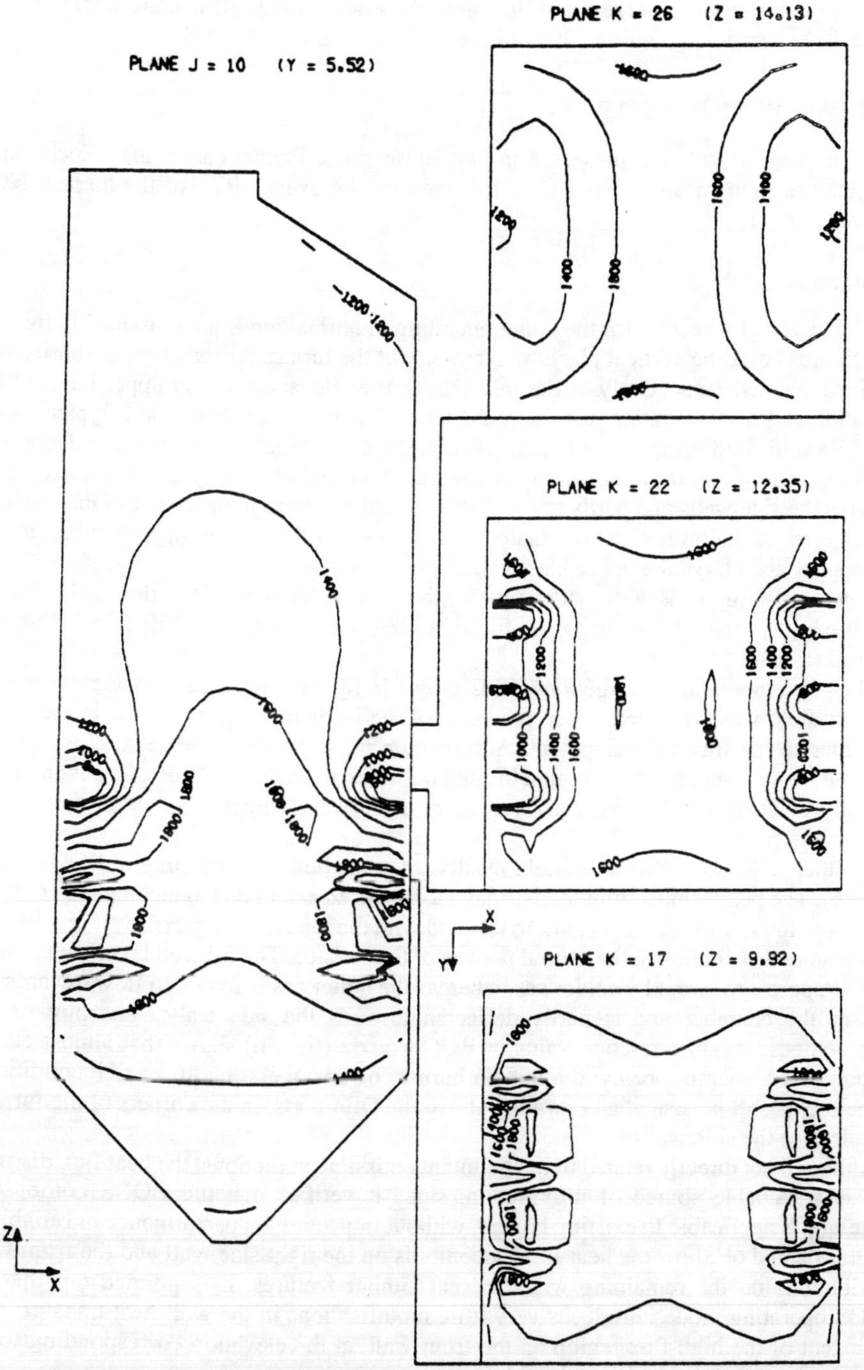

FIGURE 4b Predicted gas temperature distribution (°C). Boiler operating in BOOS mode.

every 15 iterations. Convergence of flow and temperature field calculations is achieved in about 1000 iterations, requiring 40 hours of CPU time on a VAX 8200.

DISCUSSION OF RESULTS

The computed results are presented in two subsections. Predictions of the velocity and temperature fields are discussed first, followed by the examination of the thermal NO_x calculations.

3D Furnace Analysis

Predicted velocity vectors for the boiler operating in normal mode are presented in fig. 3a. The figure shows the vertical plane-of-symmetry of the furnace, together with 3 horizontal sections, located respectively at the mid burner row elevation, at the upper burner row elevation and at an elevation just above the latter. The predicted flow field is typical of an opposite wall fired furnace: the burner jets collide in the middle of the furnace, due to the predominance of axial momentum over the turbulent diffusion; they are deflected partly towards the side walls and partly towards the vertical direction in the center of the furnace.

Large recirculation regions originate in the corners of the combustion chamber, in the ash hopper and above the upper burner rows.

When operating in BOOS mode (fig. 3b) the general features of the flow field are not modified. It is verified that the air jets from the OFA ports penetrate deeply into the hot gas vertical stream.

The gas temperature distribution (°C) is shown in fig. 4a, for the same planes where the velocity field was presented. Steep gradients of the isotherms in front of the burner ports characterize the flame development. A maximum temperature above 1900°C is reached near the flame front and in the region located between the multiple flames. The flame front is easily evidenced in fig. 5a, where contours of the stoichiometric value of the mixture fraction are plotted.

A rather different temperature field results when the boiler is operating in BOOS mode (fig. 4b). The peak temperature regions have a decreased extent due to incomplete combustion; large low-temperature regions, associated with the injection of overfire air, modify the temperature distribution in the central portion of the furnace. The front and rear walls above the OFA ports are swept by colder gas, whereas the hotter gas is forced to flow towards the axis of the chamber and is partly deflected towards the side walls. The plot of the stoichiometric mixture fraction value in BOOS mode (fig. 5b) shows that almost all the portion of the chamber below the middle burner rows is operating under rich conditions. Some fuel is still present at an elevation above the OFA ports, in the corners of the furnace and close to the side walls.

Although not directly related to the pollutant emission predictions, the heat flux distribution is a valuable source of information, since it verifies that the BOOS combustion technique is applicable to existing boilers, without impacting on performance or reliability. Figures 6a and 6b show the heat flux isocontours on the right side wall and the front wall; distributions on the remaining walls present similar features. It is pointed out that the BOOS operating mode introduces very little modifications in the wall heat transfer: only the extent of the high flux region on the front wall, at the elevation corresponding to the upper burner rows, is significantly reduced.

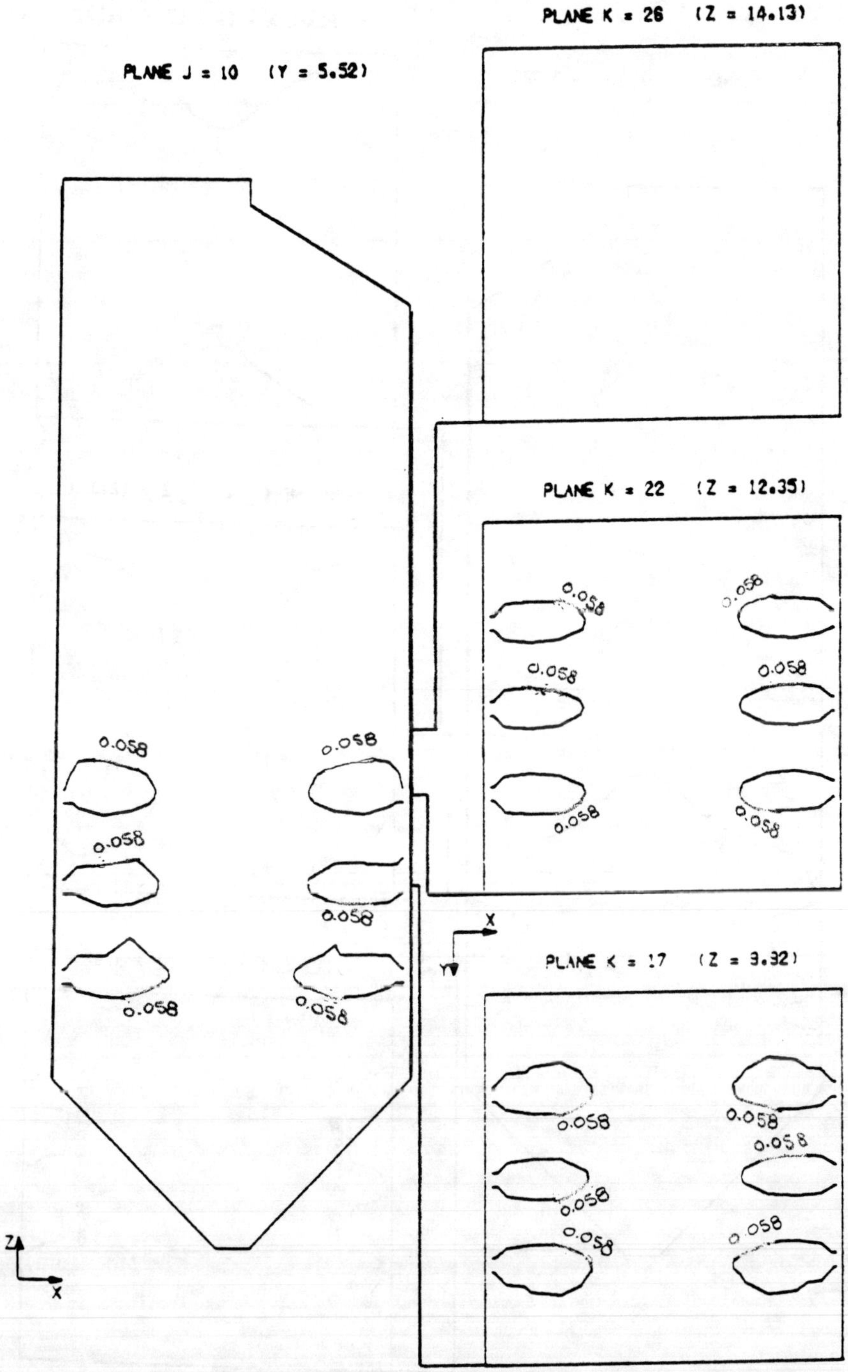

FIGURE 5a Predicted contours of the mixture fraction stoichiometric value. Boiler operating in normal mode.

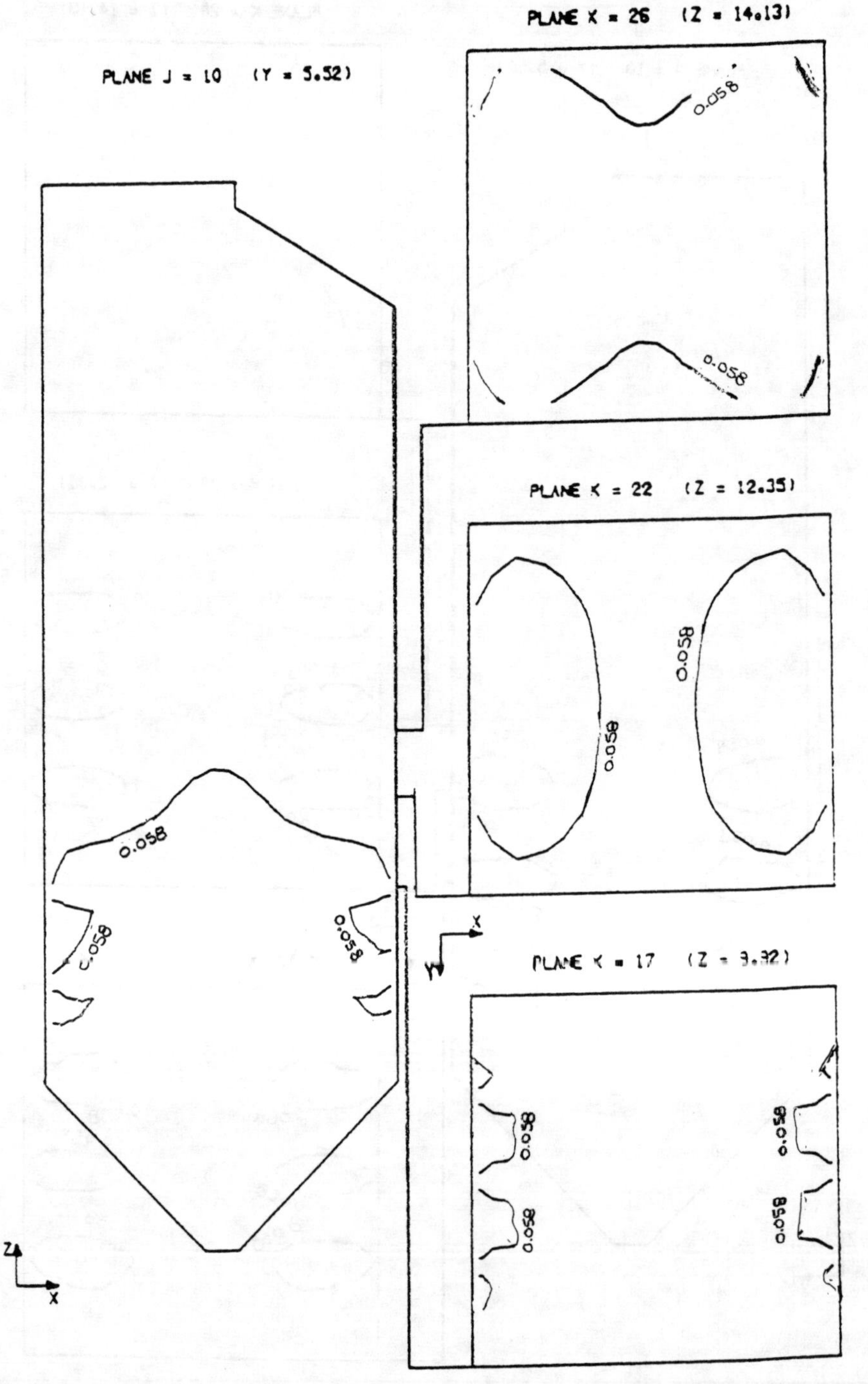

FIGURE 5b Predicted contours of the mixture fraction stoichiometric value. Boiler operating in BOOS mode.

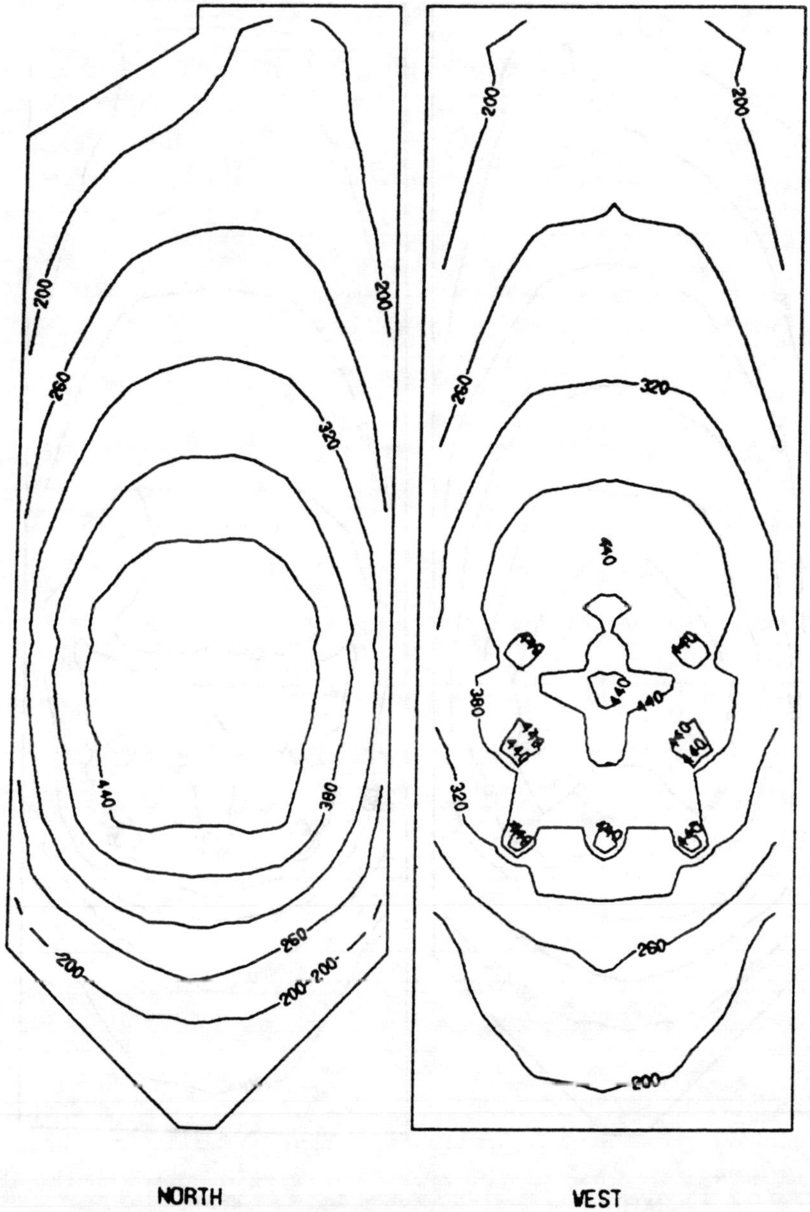

FIGURE 6a Predicted heat flux distribution (kW/m^2) on the right side (north) and the front (west) walls. Boiler operating in normal mode.

Thermal NO_x Predictions

Calculations of the NO mass fraction distribution within the furnace were performed for two cases, corresponding namely to taking into account or neglecting the influence of turbulent fluctuations.

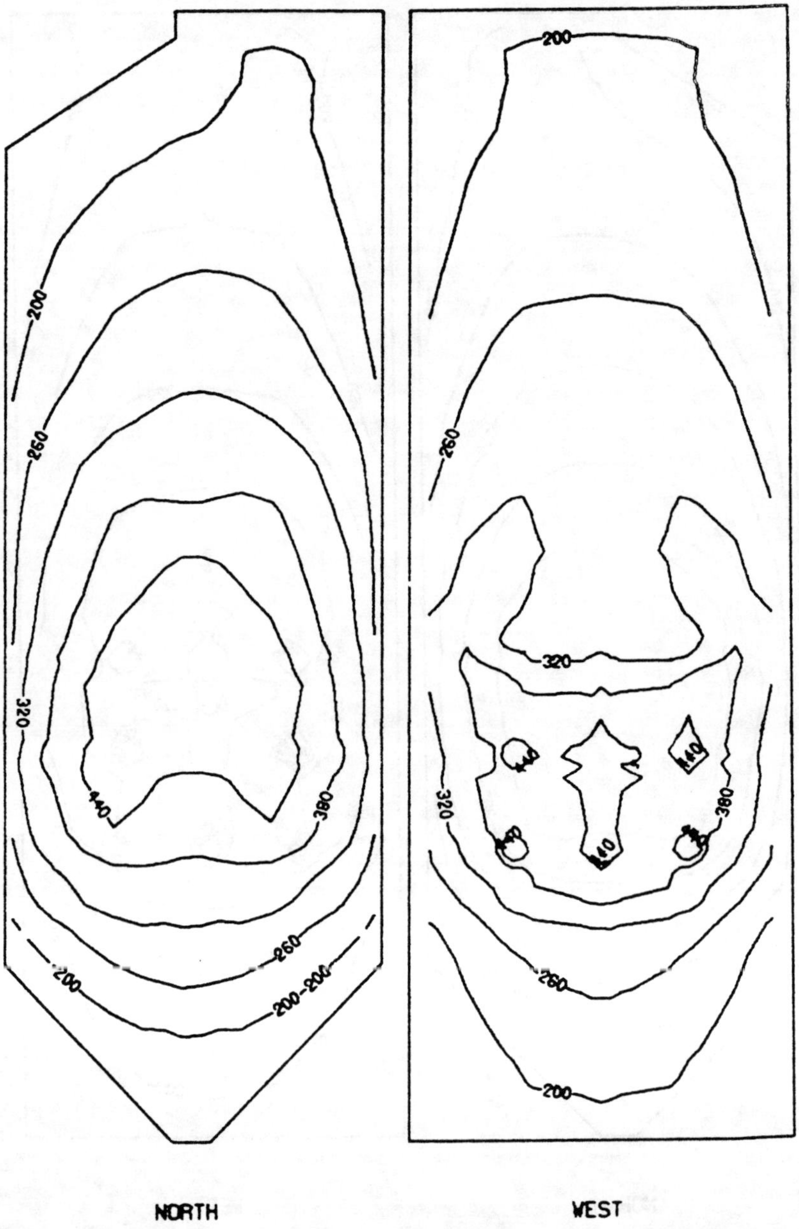

FIGURE 6b Predicted heat flux distribution (kW/m^2) on the right side (north) and the front (west) walls. Boiler operating in BOOS mode.

Predictions neglecting turbulent fluctuations. Isocontours of $Y_{NO} \times 10^4$ for the boiler operating in normal mode are presented in fig. 7a. The thermal NO is mainly produced in the region between the burner jets, where high temperatures occur and sufficient oxygen is available, due to the still incomplete mixing of the combustion air and fuel streams. The

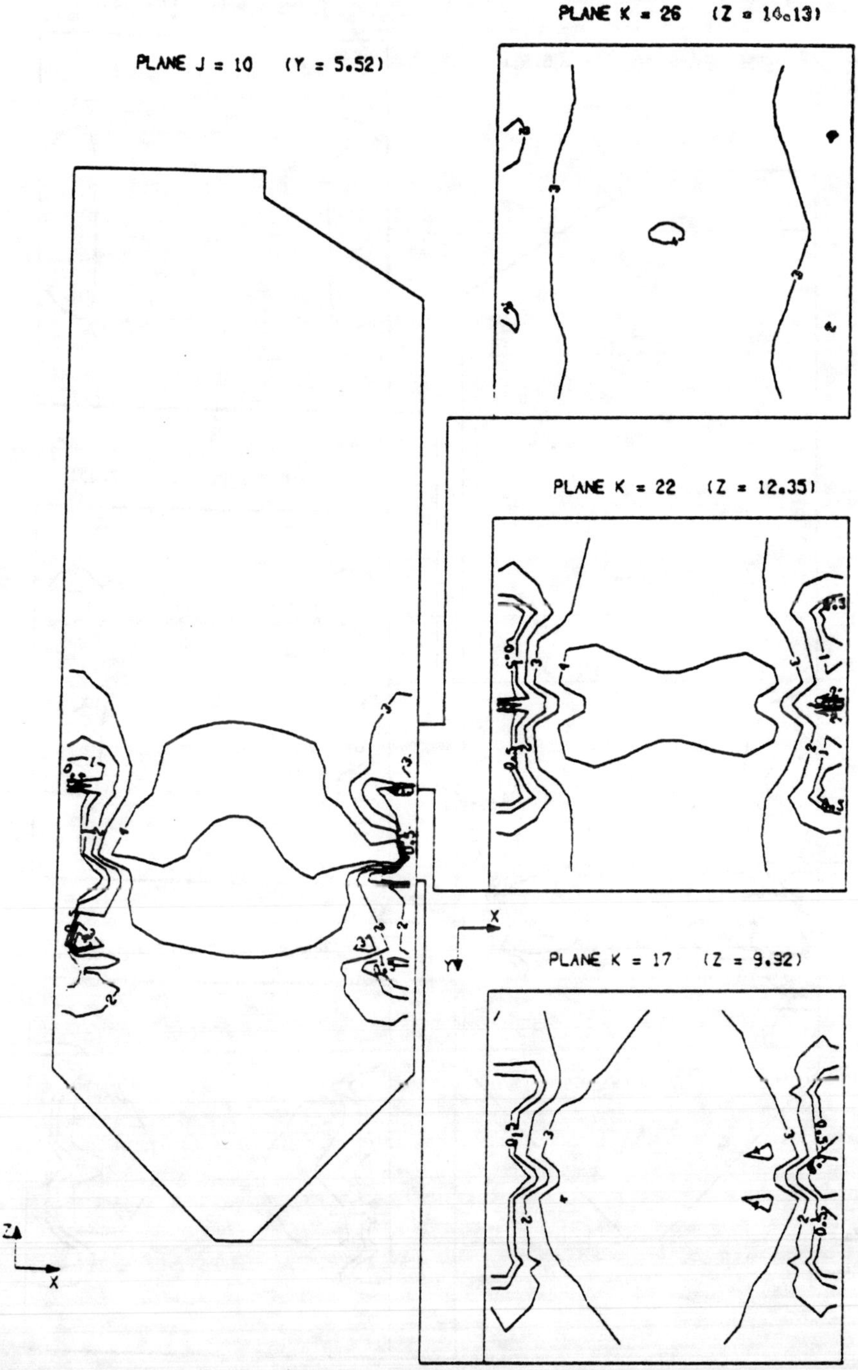

FIGURE 7a Predicted NO mass fraction (× 10^4). Turbulent fluctuations neglected. Boiler operating in normal mode.

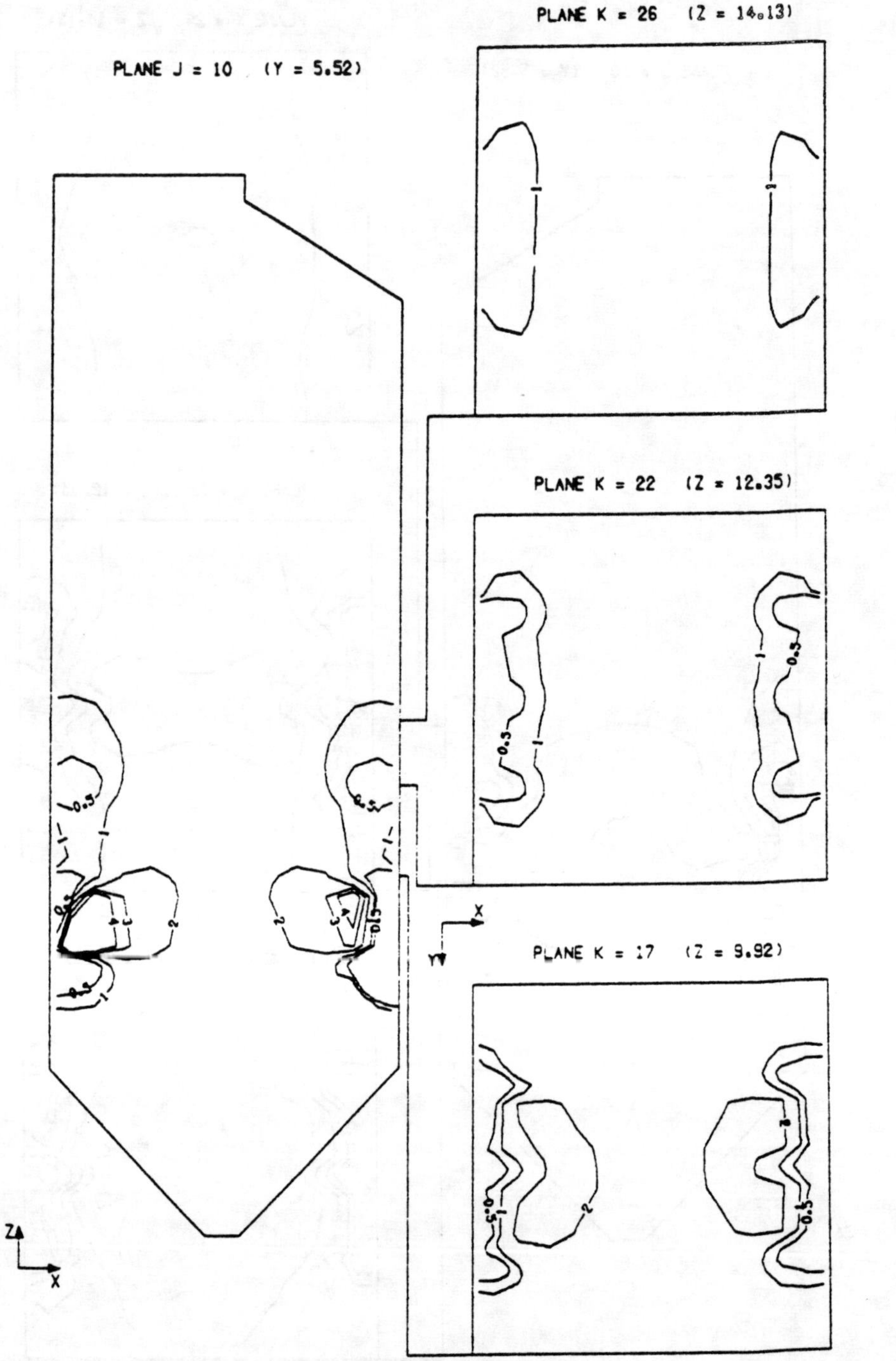

FIGURE 7b Predicted NO mass fraction ($\times 10^4$). Turbulent fluctuations neglected. Boiler operating in BOOS mode.

TABLE II
AVERAGE NO CONCENTRATION AT FURNACE OUTLET

(5% air excess)	Experi-mental data	Predictions neglecting fluctuations	Predictions including fluctuations
NO concentration, normal mode (ppm)	400	321	494
NO concentration, BOOS mode (ppm)	95	130	434
Reduction (%)	76	59	12

NO concentration builds up to its maximum value in the center of the furnace, at about the elevation of the upper burner rows. When the gas temperature decreases the concentration profile becomes flatter, and eventually uniform in the upper section of the furnace.

When operating with the BOOS technique (fig. 7b) the NO mass fraction levels are much lower. It should be noted that, when turbulent fluctuations are neglected, the NO formation is prevented in the lower part of the furnace, where the average mixture composition is substoichiometric. A local increase of NO concentration can be observed between the flames of the two operating burner rows; the concentration is then quickly reduced by diffusion and becomes uniform in the upper section of the furnace.

The predicted average values of the emissions at the furnace outlet, normalized at 3% O_2, are compared in table II with experimental data, obtained for 5% air excess operating conditions. Measurements are actually taken at the outlet of the boiler convective section; the comparison is, however, considered reliable, since the gas temperature in the convective section is low enough to make additional NO formation negligible. The present model, employed without including the influence of turbulent fluctuations, predicts remarkably well the qualitative trend, denoting a large decrease in NO emissions when operating with the BOOS technique. The computed 59% reduction of NO concentration, although somewhat lower than the experimental value, can still be considered in satisfactory quantitative agreement.

Predictions including turbulent fluctuations. The effect of accounting for the turbulent fluctuations, when computing the average NO production term with the present model, results in an increase in the NO concentration levels within the furnace, as expected (Jones, 1975). In normal operating mode the peak values of Y_{NO} are still found in the region between the interacting flames (fig. 8a). A slight asymmetry in the NO concentration profiles above the burners is observed, with higher values occurring towards the front wall; in the upper zone the profiles are levelling out, to become almost uniform at the furnace outlet. The prediction overestimates the measured value by 24% (table II).

This trend is confirmed at different excess air values, as shown in fig. 9a, where the emission levels at the furnace outlet, predicted by both neglecting and including turbulent fluctuations, are compared with the experimental results. In the same figure the effect of accounting for the reverse reactions in the kinetic mechanism, when including turbulent fluctuations is also shown. This effect results in a 5% reduction of the emission levels.

When operating the boiler in BOOS mode (fig. 8b), the effect of the turbulent fluctuations is even more pronounced, inasmuch as the fluctuations allow for nitric oxide formation also in the region of the slightly rich mixture. The oxygen excess in the OFA ports region is also partly contributing to NO production due to temperature fluctuations.

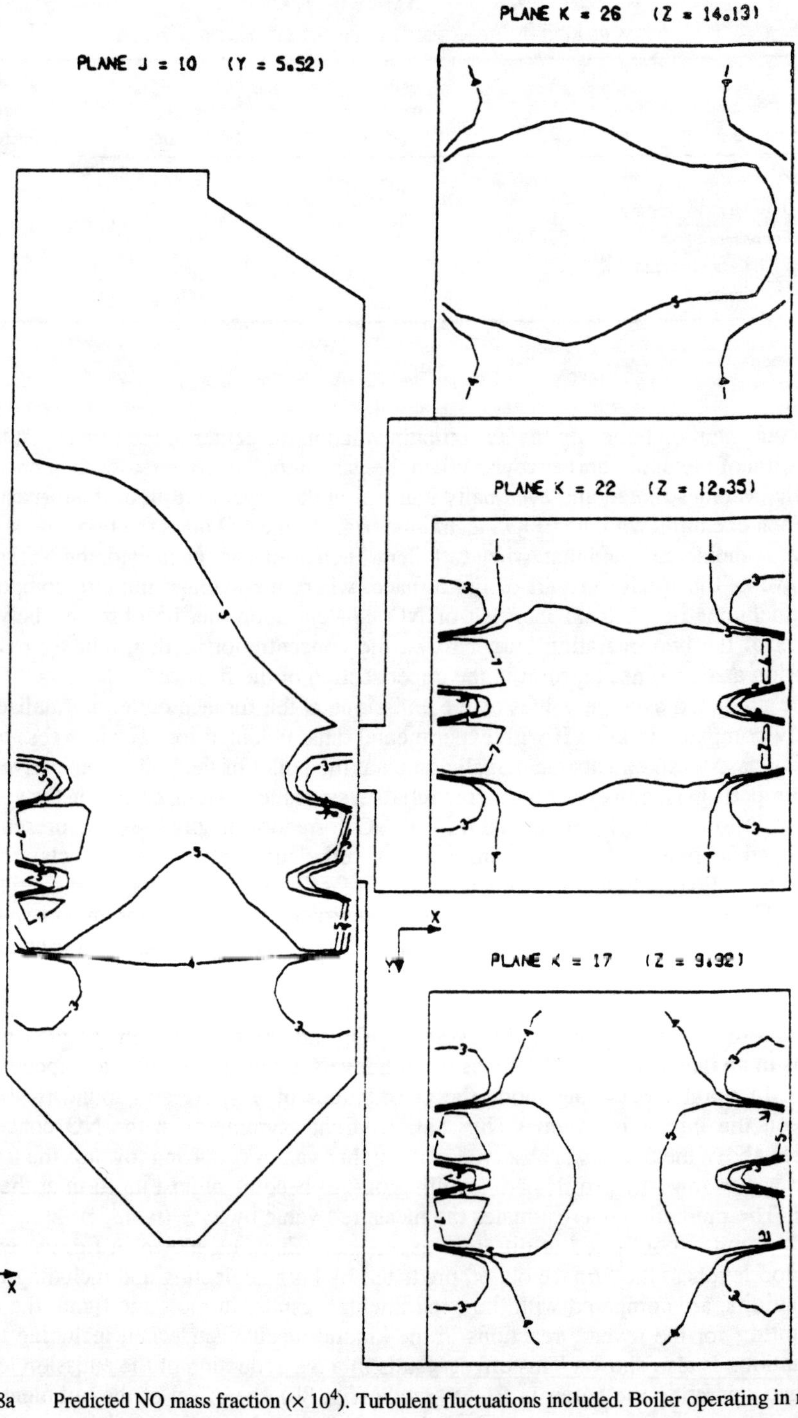

FIGURE 8a Predicted NO mass fraction (× 10^4). Turbulent fluctuations included. Boiler operating in normal mode.

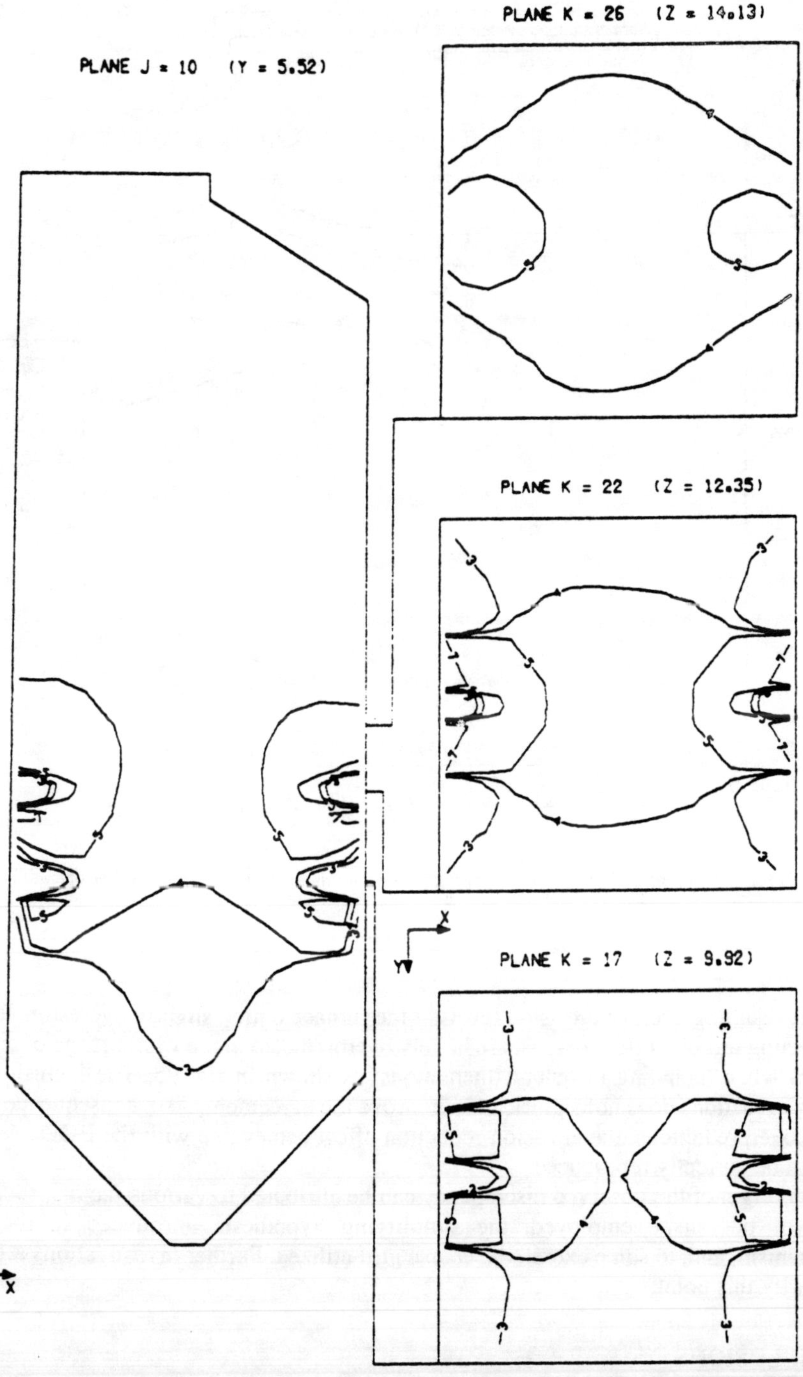

FIGURE 8b Predicted NO mass fraction (× 10^4). Turbulent fluctuations included. Boiler operating in BOOS mode.

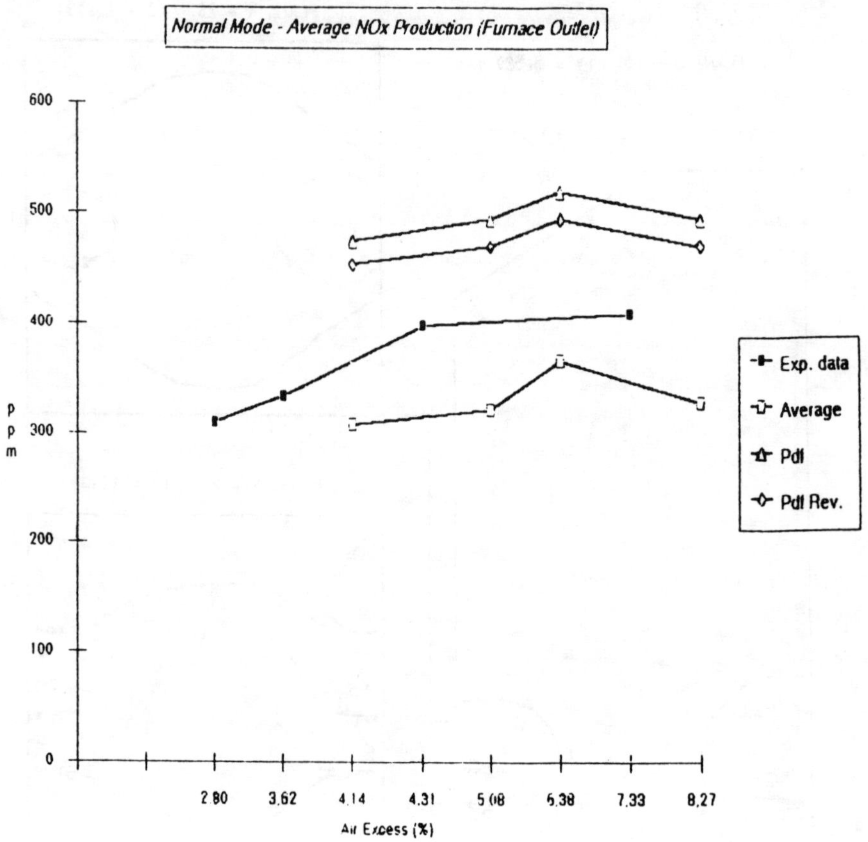

FIGURE 9a Average emission levels at furnace outlet as function of air excess. Boiler operating in normal mode.

The resulting average emission levels at the furnace outlet, slightly overestimated when neglecting turbulent fluctuations, are largely overpredicted in the whole range of air excess values when including turbulent fluctuations, as shown in fig. 9b. The inclusion of the reverse reactions does not significantly improve the agreement. As a consequence of these large overpredictions, the emission reduction effect achievable with the BOOS technique cannot be correctly reproduced.

The origin of the observed discrepancy can be attributed to various factors, such as: the kind of pdf shape employed, the simplifying hypothesis introduced in the kinetic mechanism, and, to some extent, the coarse grid utilized. Further investigations are needed to clarify this point.

CONCLUSIONS

A simple kinetic model of thermal NO formation, coupled with an accurate three-dimensional mathematical model describing the turbulent combustion and radiation processes and yielding a realistic flow field and gas temperature distribution, has been applied to the evaluation of the emission levels from a large scale utility boiler.

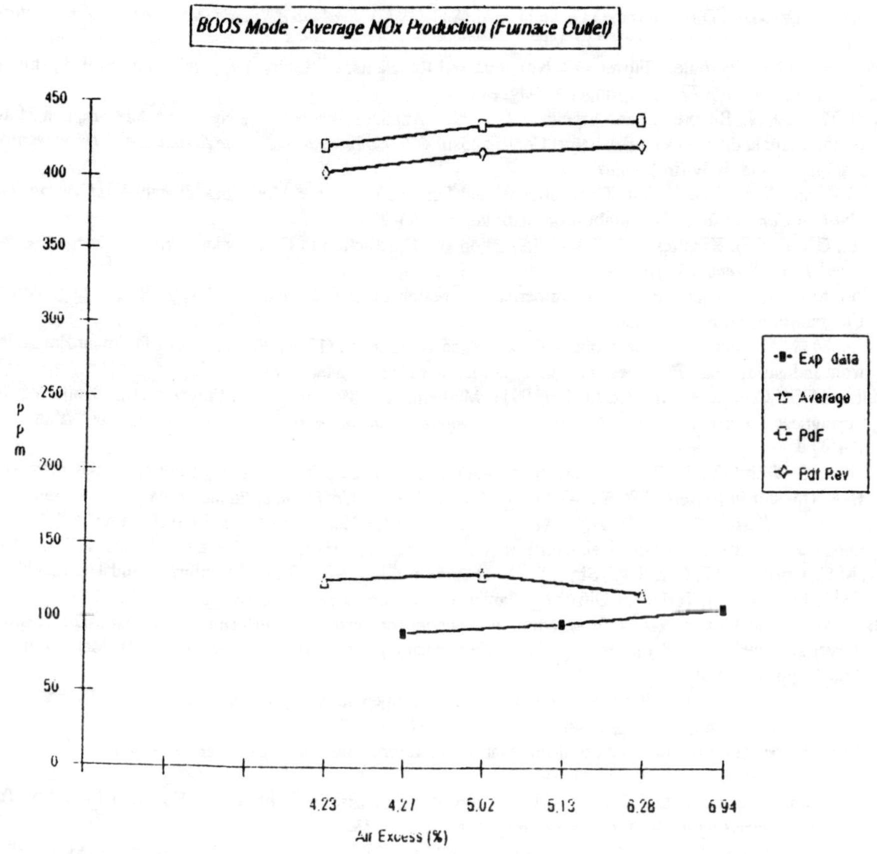

FIGURE 9b Average emission levels at furnace outlet as function of air excess. Boiler operating in BOOS mode.

The NO concentration reduction, experimentally achieved when operating the boiler with the Burner Out Of Service technique, has been predicted under two different hypotheses, namely: a) by neglecting the turbulent fluctuations, and b) by accounting for the turbulent fluctuations.

In the first case, a remarkably good agreement has been obtained, considering the minimum computational effort required by the proposed model. A 59% NO concentration reduction has been computed, compared with the measured 76% reduction.

In the second case, the predictions for the BOOS operating mode considerably overestimate the emission levels, due to the additional NO formation arising from the fluctuations, and fail to yield the expected reduction. This result suggests that a better physical representation of the phenomenon, accounting for the influence of turbulent fluctuations on NO formation, requires an investigation on the influence of the pdf shape and of the grid size employed, together with an improvement of the kinetic model.

REFERENCES

Abbas, A.S., and Lockwood, F.C. (1986). Prediction of Power Station Combustors. *Twenty-First Symposium (Int.) on Combustion*, The Combustion Institute, pp. 285–292.

Baulch, D.L., Drysdale, D.D., Horne, D.G., and Lloyd, A.C. (1973). *Evaluated Kinetic Data for High Temperature Reactions*, Vol. 2, Butterworth, London.

Bilger, R.W. (1980). Turbulent Flows with Nonpremixed Reactants. In: Libby, P.A., and Williams, F.A. (Eds.). *Turbulent Reacting Flows*. Springer Verlag. Berlin.

Bonvini, M., Elia, N., Razore, S., and Vigevano, L. (1991). Analisi Sperimentale e Numerica del Campo di Moto di una Caldaia di Potenza a Bruciatori Contrapposti. *46th Congresso Nazionale Associazione Termotecnica Italiana*, Gaeta, Italy (in Italian).

Boyd, R.K, and Kent, J.H. (1986). Three-dimensional Furnace Computer Modelling. *Twenty-First Symposium (Int.) on Combustion*, The Combustion Institute, pp. 265–274.

Bowman, C.T. (1975). Kinetics of Pollutant Formation and Destruction in Combustion. *Progress in Energy and Combustion Science*, **1**, pp. 33–45.

Carvalho, M.G., and Coelho, P. (1990). Numerical Prediction of an Oil-Fired Water Tube Boiler. *Engineering Computations*, **7**, pp. 227–234.

Carvalho, M.G., Semiao, V.A., Lockwood, F.C., and Papadopoulos, C. (1990). Predictions of Pollutant Emissions from Industrial Glass Furnaces. *Journal of Institute of Energy*, March, pp. 39–47.

Carvalho, M.G., Coelho, P., and Costa, F. (1991). Mathematical Modelling of a Power Station Boiler of the Portuguese Electricity Utility. *2nd European Conference on Industrial Furnaces and Boilers*, Vilamoura, Portugal.

De Michele, G., Ghiribelli, L., Pasini, S. and Tozzi, A. (1989). A 3D Code for Predicting Radiative and Convective Heat Transfer in Boilers. *ASME 26th National Heat Transfer Conference*, Philadelphia, Pennsylvania.

De Michele, G., Pasini, S., and Tozzi, A. (1991). Simulation of Heat Transfer and Combustion in Gas- and Oil-Fired Furnaces. *2nd European Conference on Industrial Furnaces and Boilers*, Vilamoura, Portugal.

Drake, M.C., Correa, S.M., Pitz, R.W., Shyy, W., and Fenimore C.P. (1987). Super-Equilibrium and Thermal Nitric Oxide Formation in Turbulent Diffusion Flames. *Combustion and Flame*, 69, pp. 347–365.

Fiveland, W.A., and Wessel, R.A. (1988). Numerical Model for Predicting Performance of Three-Dimensional Pulverized-Fuel Fired Furnaces. *Journal of Engineering for Gas Turbine and Power*, Transaction of the ASME, pp. 117–126.

Goerner, R., and Zinser, W. (1986). Predictions of Three-Dimensional Flows in Utility Boiler Furnaces and Comparison with Experiments. *ASME 107th Annual Meeting*, Anaheim, California.

Jones, W.P. (1975). The Effect of Temporal Fluctuations in Temperature on Nitric Oxide Formation. *Combustion Science and Technology*, **10**, pp. 93–96.

Jones, W.P., and Launder, B.E. (1972). The Prediction of Laminarisation with a Two-Equation Model of Turbulence. *International Journal of Heat and Mass Transfer*, **15**, p. 301.

Jones, W.P., and Priddin, C.H. (1979). Predictions of the Flow Field and Local Gas Composition in Gas Turbine Combustors. *Seventeenth Symposium (Int.) on Combustion*, The Combustion Institute, pp. 399–409.

Lockwood, F.C., and Shah, N.G. (1981). A New Radiation Solution Method for Incorporation in General Combustion Prediction Procedures. *Eighteenth Symposium (Int.) on Combustion*, The Combustion Institute, pp. 1405–1414.

Lockwood, F.C., Papadopoulos, C., and Abbas, A.S. (1988a). Prediction of a Corner-Fired Power Station Combustor. *Combustion Science and Technology*, 58, pp. 5–23.

Lockwood, F.C., Papadopoulos, C., and Semiao, V.A. (1988b). The CINA Code for Furnace Computation. *1st Eurotherm Seminar on High Temperature Heat Transfer Equipment Modelling and Control*, Mons, Belgium.

Monat, Y.P., Hanson, R.K., and Kruger, C.H. (1979). Shock Tube Determination of the Rate Coefficient for the Reaction $N_2 + O \rightarrow NO + N$. *Seventeenth Symposium (Int.)* on Combustion. The Combustion Institute, pp. 543–552.

Patankar, S.V. (1980). *Numerical Heat Transfer and Fluid Flow*. Hemisphere Publishing Corporation, Washington.

Robinson, G.F. (1985). A Three-Dimensional Analytical Model of a Large Tangentially Fired Furnace. *Journal of the Institute of Energy*, pp. 116–149.

Sala, M., and Vigevano, L. (1990). NO_x Kinetics Modelling for Practical Evaluation of Pollutant Emission from Utility Boiler Furnaces. *Joint Meeting of the Italian and Soviet Section of the Combustion Institute*, Pisa, Italy.

Truelove, J.S. (1976). A Mixed Grey Gas Model for Flame Radiation. U.K. Atomic Energy Authority (Harwell) Report, AERE R8494.

Zennaro, A., Rizzi, R., Zaccarelli, F., and Bianchi G. (1991). Large Industrial Applications of NO_x Control by the Burner Out Of Service (BOOS) Technique. *First Int. Conference on Combustion Technologies for a Clean Environment*, Vilamoura, Portugal.

Measurement and Modelling of NO_x Formation in a Gas Fired Furnace

A. M. A., KENBAR[1,2], S. A. BELTAGUI[1,2], T. RALSTON[2] and N. R. L. MACCALLUM[1]
1–University of Glasgow
2–Heat Transfer and Fluid Flow Service (HTFS), NEL, East Kilbride, Glasgow

Abstract—This paper reports results of NO_x measurements carried out on a gas fired furnace using radial fuel injection. The data consists of a complete mapping of NO_x concentration in the furnace under variable swirl conditions. Swirl produced locally higher NO_x concentrations but little overall increase. Measured stack NO_x variations with burner input conditions indicate that, for this system, operating with a fuel-rich mixture is more effective in suppressing NO_x formation than with excess air operation. The data is needed for testing and developing theoretical models. Theoretical predictions of the overall NO_x formation are made using a well-stirred reactor model based on the Zeldovich mechanism with account for the species concentration fluctuations. The model predictions compare satisfactorily with the measurements.

Key Words: Combustion-generated NO_x, NO_x-Measurement, NO_x-modelling, Furnace, Swirl, Natural-gas, Fuel/air ratio.

NOTATION

$[N_2]$	Nitrogen concentration	mol/cm^3 s
$[O_2]$	Oxygen concentration	mol/cm^3 s
S	Burner swirl number, (ratio of axial flux of tangential momentum to the product of axial momentum flux and burner radius)	—
s	Mixedness factor, unity for perfect mixing	—
T	Flame temperature	K
t	Residence time	s
X	Axial distance along furnace from burner exit	mm
Z	Fluctuation parameter, see Equation 3	—
$\underline{\phi}$	Fuel/air equivalence ratio	—
ϕ	Mean value of ϕ	—
σ	Standard deviation	—

INTRODUCTION

The need to protect the environment from combustion generated NO_x has led to considerable demand to improve burner design. The present paper is part of a study on the effect of burner parameters on the formation of NO_x in gas fired furnaces. These parameters include fuel injection, firing rate, swirl and the mixture fuel/air ratio. These parameters are interrelated and vary from one combustion system to another. Thus it is important to investigate their effects on the level of NO_x emissions. This study makes a significant contribution to the understanding of the main factors involved in the formation of NO_x in these flames. The work also provides a valuable data-base for the assessment and validation of the theoretical models for NO_x prediction.

The present paper concentrates on the effect of swirl and mixture fuel/air ratio on NO_x formation when using a radial-outwards gaseous fuel injector. Swirl is an important feature of burner design since it provides satisfactory flame stability by creating a central reverse-flow zone. The mixture fuel/air ratio is an important parameter in burner operation. Other work to study the effect of the fuel injection modes is reported by Kenbar *et al.* (1991).

In a previous paper (Beltagui *et al.* 1989) an extensive review was made of the NO_x formation mechanisms and methods of abatement. The effect of swirl as observed by previous workers was included and it was realised that it is linked to other parameters, such as fuel injection mode and fuel/air ratio.

The aerodynamics and combustion patterns of the swirling flames in the National Engineering Laboratory (NEL) furnace have already been extensively studied through detailed measurements of the velocity components, gas temperature and species concentrations of CO_2, CO and O_2.

These measurements were carried out in the NEL furnace fired by natural gas through a variable-swirl burner with a quarl. The fuel gas was injected from a central gun radially outward. Four swirl settings were considered, burner swirl numbers ranging from 0.0 to 2.25. The availability of this complete set of data is useful in testing and developing the furnace flow and heat transfer codes, such as the HTFS program PCOC (Stopford, 1989). Part of these measurements and comparisons with modelling using PCOC were reported by Beltagui *et al.* (1991).

The present paper offers two data sets:

The first illustrates complete mapping of NO_x concentrations produced under different swirl levels, but with the same firing conditions, in order to study the effect of swirl on NO_x formation.

The second illustrates the effects of the mixture fuel/air ratio on the overall NO_x emissions as measured in the stack.

The results show that while the effect of swirl on the local NO_x concentration in the near burner zone is significant, it was of secondary importance downstream. The results also demonstrate that increasing excess air above the level required for complete combustion led to an increase in the overall NO_x emission level. However by operating with a fuel-rich mixture, NO_x formation was suppressed significantly.

The paper also presents a simple engineering model which can be used to predict the overall thermal-NO_x formation for furnaces and its variation with input conditions, eg fuel injection mode and equivalence ratio. The model is based on the well-stirred reactor concept with account being taken for local concentration fluctuations. The inputs to the model are limited to the residence time in the reaction zone and the fuel/air mixing parameter. The agreement between model-predicted and measured concentrations was satisfactory.

EXPERIMENTAL PROGRAMME

The aims of the experimental programme were:

a—To produce complete mapping of the local NO_x concentrations within the furnace under a range of swirl intensities. This set of data illustrates the NO_x formation processes in this system. It is also required for the assessment and development of the modelling procedures.

b—To investigate the effect of two burner parameters, namely, swirl and fuel-air ratio on the overall NO_x formation as measured in the stack.

Both sets of results have been analysed and discussed in the light of the temperature, concentrations and velocity measurements reported earlier by Beltagui *et al.* (1991).

The NEL furnace is a model of a cylindrical upshot fired heater. The apparatus is described in detail by Beltagui *et al.* (1988-a). The furnace is fired by natural gas through a variable swirl burner with a quarl. The burner used in this study utilises the moving-block swirl generator of the International Flame Research Foundation (IFRF).

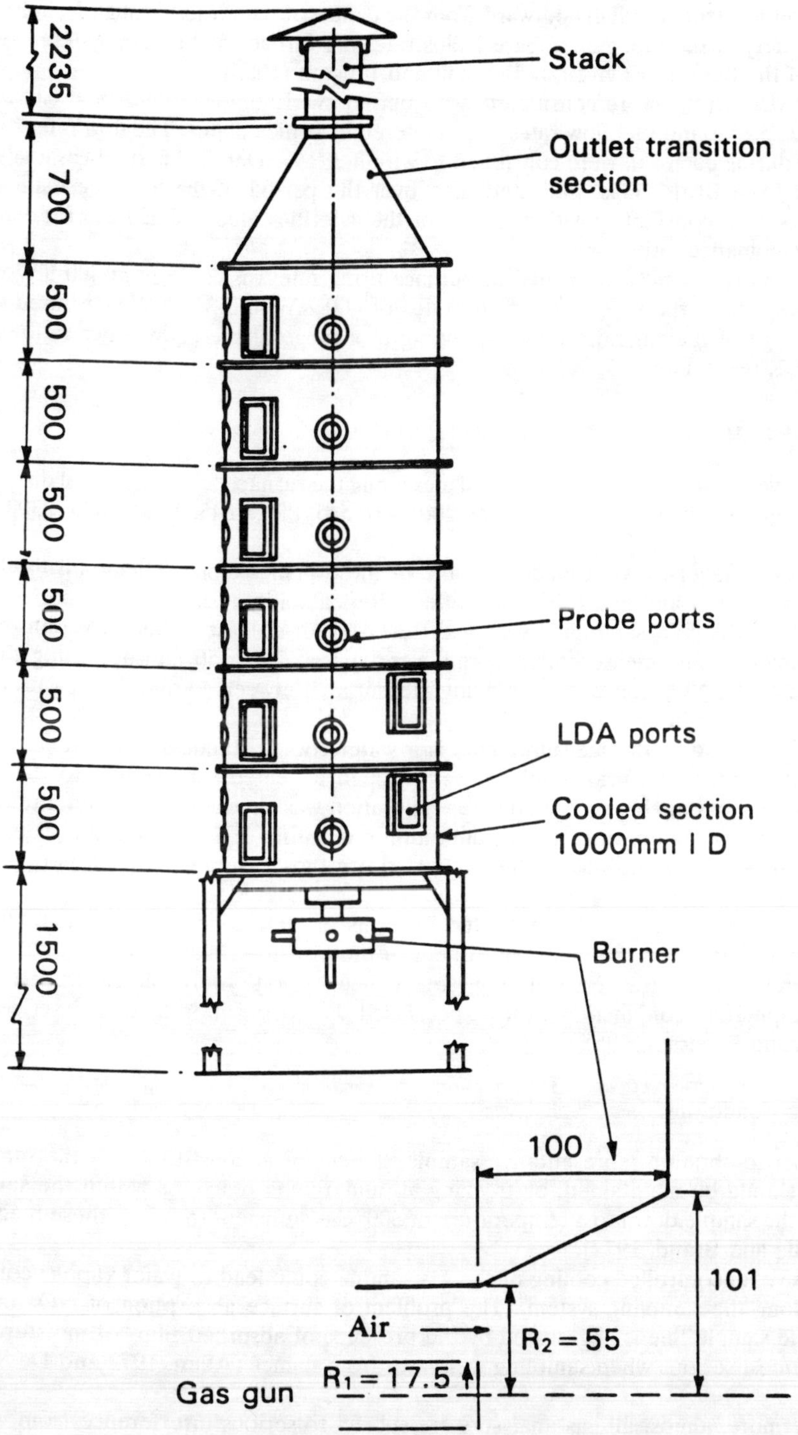

FIGURE 1 Schematic of furnace and burner

The fuel is injected radially outward from the centre of the air jet through 16 holes on the periphery of the fuel gun. Figure 1 illustrates the furnace and burner system. More details of the burner are given by Beltagui and Ralston (1985).

Four swirl settings were considered, with burner swirl number values, S = 0.0, 0.45, 0.9 and 2.25. Air and fuel flow rates were metered by orifice plates. The flow rates of air and fuel during each run were controlled to within ± 1%. Detailed natural gas analysis, obtained from British Gas, indicated that, over the period of the tests the main fuel properties were constant to within ± 2% for the calorific value and ± 1.2% for density and stoichiometric ratio.

In the first set of measurements the furnace firing rate was constant at 400 KW with 5% excess air. In the second set, firing rates of 400 KW and 360 KW were used with a wide range of fuel/air ratios, corresponding to a range of the equivalence ratio, ϕ of 0.71 to 1.1, (excess air 40% to -10%).

MEASUREMENTS AND INSTRUMENTATION

Radial traverses were carried out at 13 planes along the furnace, located at axial distances from the quarl exit of 45, 80, 110, 160, 200, 300, 545, 800, 1045, 1300, 1800, 2300 and 2800 mm.

Additional traverses were made at some of these planes along other radii to check the flow symmetry and probe intrusive effect. Repeatability tests were also performed. In each level the NO_x sampling probe was traversed for at least 1.5 radii to confirm the flow symmetry. The measurement points were spaced 20 to 50 mm according to the gradients of the NO_x concentration being measured. Between 13 and 26 measurement points were made in each plane.

The probe used in the measurements is a water cooled stainless steel probe. After leaving the probe, the gas sample passes through a heated sample line to the NO_x chemiluminescent analyser. The analyser calibration was checked before and after each traverse. The maximum drift in the calibration was within ± 2% of the full scale. The time-mean NO_x concentration values reported are those averaged over a period of 1 minute.

Although all the results are presented in terms of NO_x concentration, it should be appreciated that NO is the main constituent of the measured NO_x value.

The probe radial traverse and the data collection rate were all controlled by a microcomputer in conjunction with a specialised data acquisition system described by Beltagui and Ralston (1985).

Sampling Conditions

In order to obtain a representative sample, the chemical constitution of the sampled mixture should be maintained, by freezing all undergoing reactions within the sample. Cooling the sample down to a temperature of 300°C is sufficient to freeze these reactions (Chedaille and Braud, 1972).

Excessive, uncontrolled cooling of the gas sample could lead to water vapour condensation along the sampling system. The problem of surface absorption of NO_x on the probe and sample line is aggravated by the presence of absorbed films of moisture and hence is most serious when sampling wet gases from flames (Allen, 1973 and De Soete, 1989).

Furthermore, almost all gas analysers are subject to serious interference from water. The use of desiccants has been found to absorb nitrogen dioxide NO_2 to varying degrees. Cold traps also remove NO_2 and water vapour simultaneously (Allen, 1973).

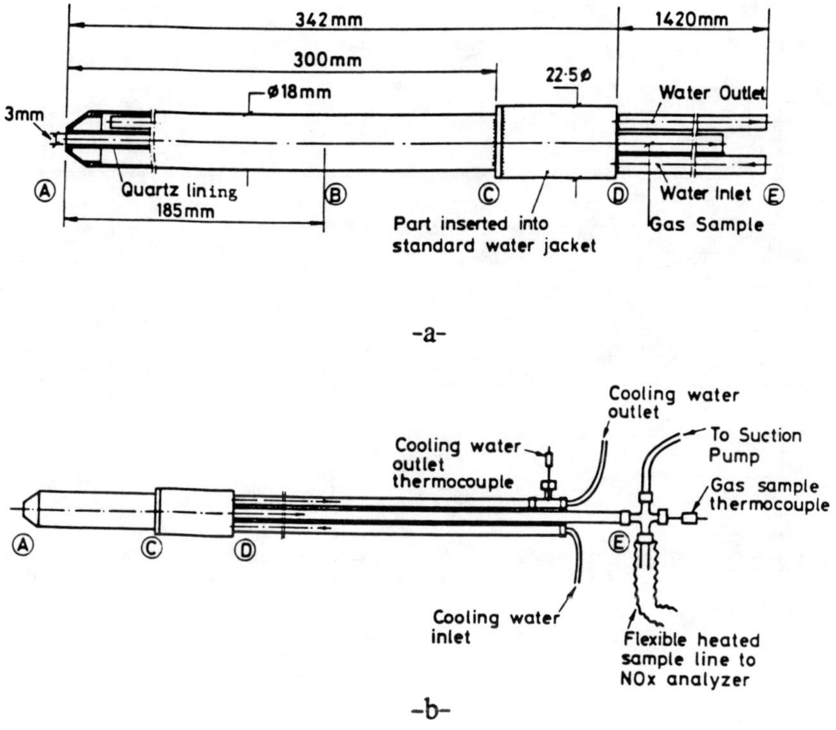

FIGURE 2 NO_X sampling system

The choice of probe material is also important for NO_X sampling to ensure no probe reactions or surface ageing with the sampled gas. A system of water cooled stainless steel probe and a teflon heated sample line was found satisfactory for NO_X measurements by Allen (1973) and Tuttle (1974). Stainless steel probes were also used satisfactorily by Oven *et al.* (1978), Sadakata *et al.* (1980), Ahmad *et al.* (1984) and Mulholland *et al.* (1987).

A water cooled stainless steel probe and a teflon sampling line were therefore used in the current measurements. The receiving part of the probe was lined with quartz to avoid the possibility of ageing especially at the high temperature region.

In order to withdraw a local representative sample, isokinetic sampling is generally recommended. However, in turbulent recirculating flows isokinetic sampling is difficult to achieve due to the large velocity fluctuations especially in the recirculating zones. In the present work, the suction velocity was adjusted to be of similar magnitude to the average flow velocity in the furnace.

Sampling Probe

Figure 2 shows the NO_X sampling system. The tip length A-C is exposed to the flame. The sample tube length D-E passes through a water-cooled probe carrier to convey the sample to the heated sample line. The probe cooling system was designed to control the sample gas temperature at D to 300°C and at E to close to 150°C.

Complete control on the sample temperature under all the operating conditions, without subjecting the probe to excessive heating, has been achieved through the following steps:

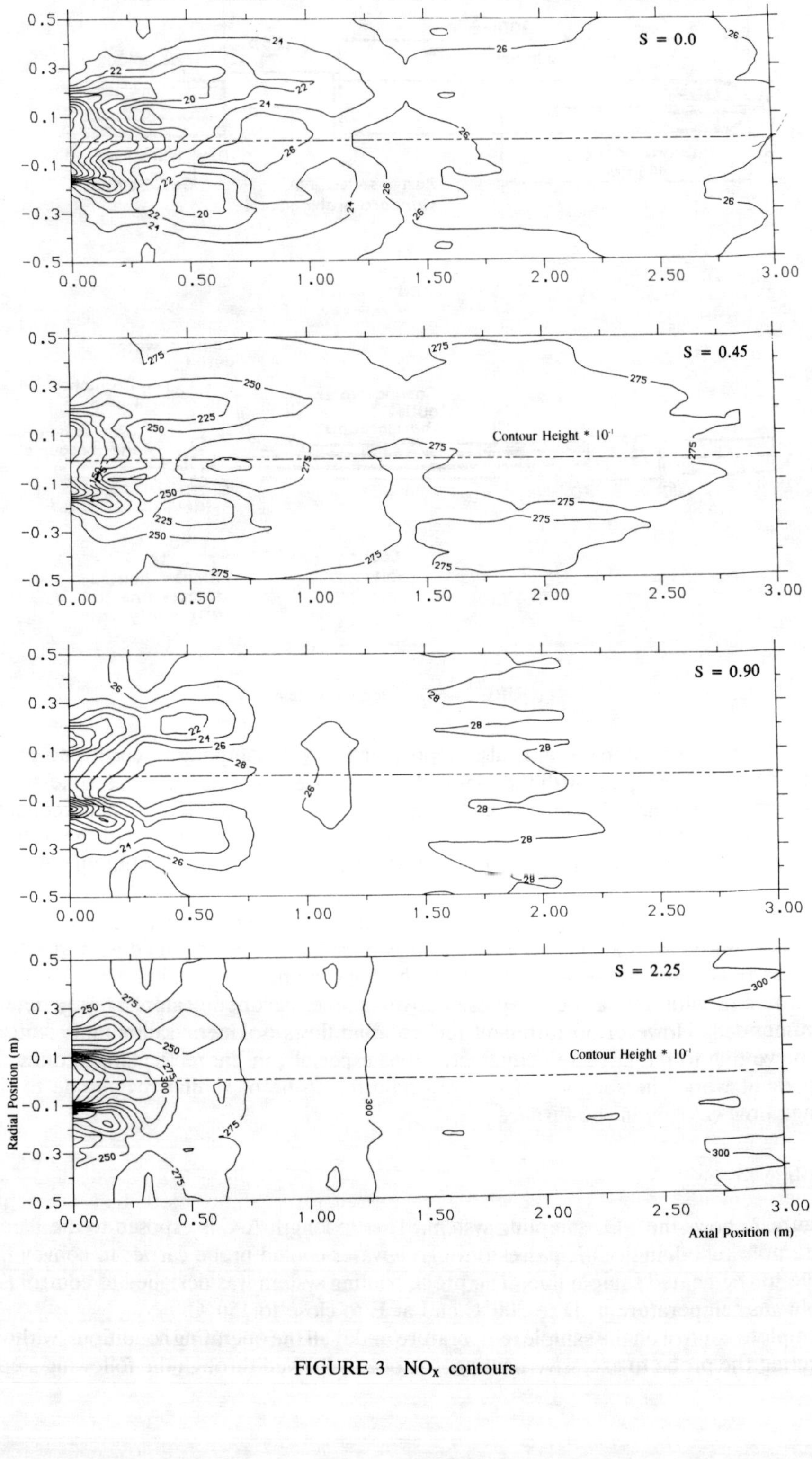

FIGURE 3 NO_X contours

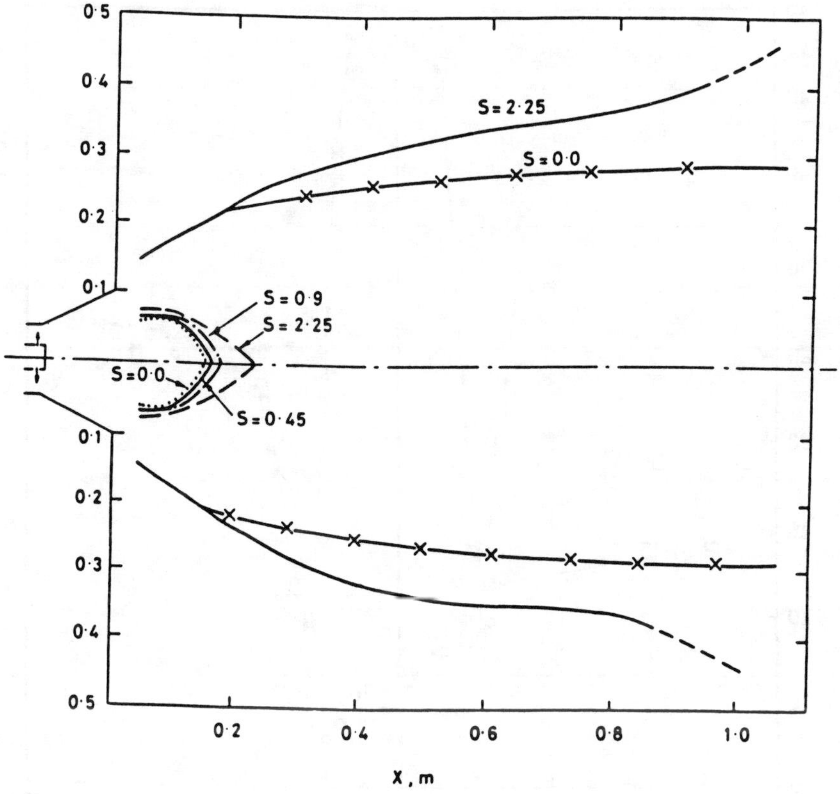

FIGURE 4 Flow boundaries

a—Controlling the cooling water temperature supplied to the probe.

b—Insulating the probe cooling water tubes in contact with the sample tube, length D-E.

c—To compensate for any further heat losses the sample tube D-E was electrically heated

Before the measurements were started, the probe performance was tested in the furnace under a range of operating conditions. Two thermocouples fitted to monitor the sample temperature at points D and E, gave readings within the probe design range.

EXPERIMENTAL RESULTS AND DISCUSSION

Before starting the detailed NO_X measurements and also at the conclusion of the series of measurements, tests were carried out to ensure repeatability of themeasurements. These tests were carried out at axial positions of 200 and 300 mm for all swirl settings. Repeatability of the measurements to within $\pm$ 2 ppm was obtained at all swirl settings.

The effect of probe presence in the flame on the local NO_X concentration was also investigated by traversing the probe through two ports, radially opposite, at the X = 200 mm plane. For all swirl numbers no flame disturbance by the probe was observed. Similar observations were reported by Oven *et al.* (1978) and Amad *et al.* (1984).

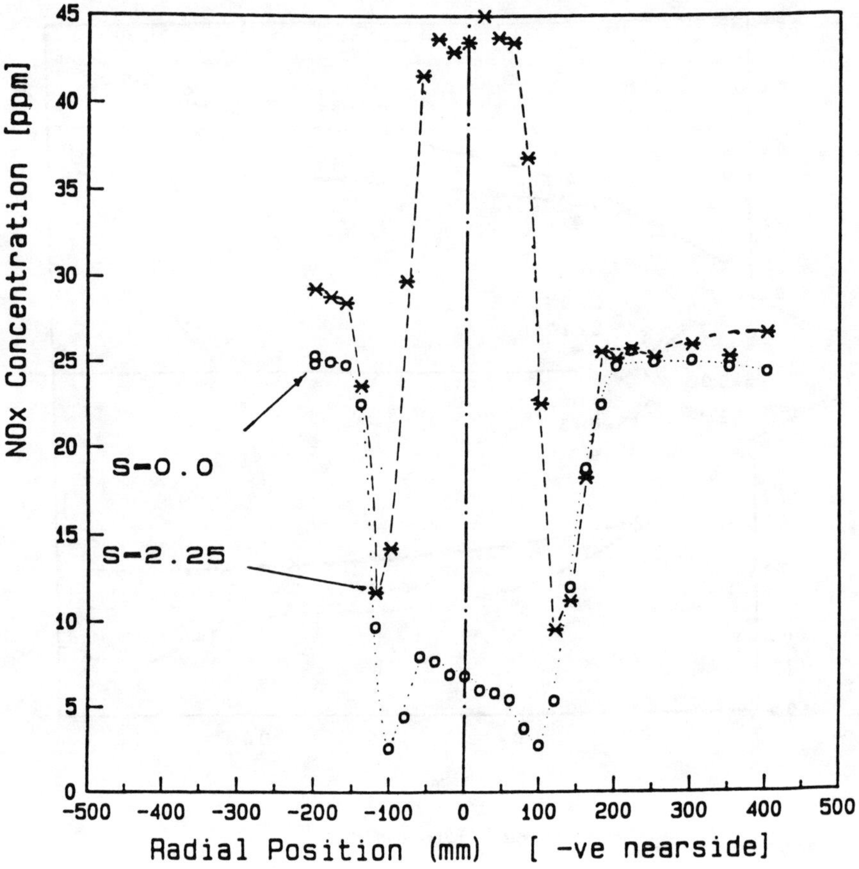

FIGURE 5-a Radial NO_x profile, axial position 45 mm

In-Furnace Measurements

The results of these measurements are presented in Figure 3 as contours of the NO_x concentration field for the four swirl settings surveyed. The above NO_x contours are discussed in relation with the experimental data for the flow and combustion fields reported earlier by Beltagui *et al.* (1991). Figure 4 illustrates the flow boundaries as determined from the flow field measurements, indicating the forward and reverse flow zones.

Some illustrative radial NO_x profiles at three planes near the burner are presented in Figures 5-7, together with the corresponding temperature profiles.

For the present discussion, the flow field can be divided into three zones:

1. The main reaction zone (forward flow zone)
2. The central reverse-flow zone (CRZ).
3. The fully developed flow zone

In general it is seen that most of the NO_x formation takes place in the first zone. Some further reactions occur in the second, leading to uniform concentration distributions in the third zone. The processes occurring in these zones are now discussed.

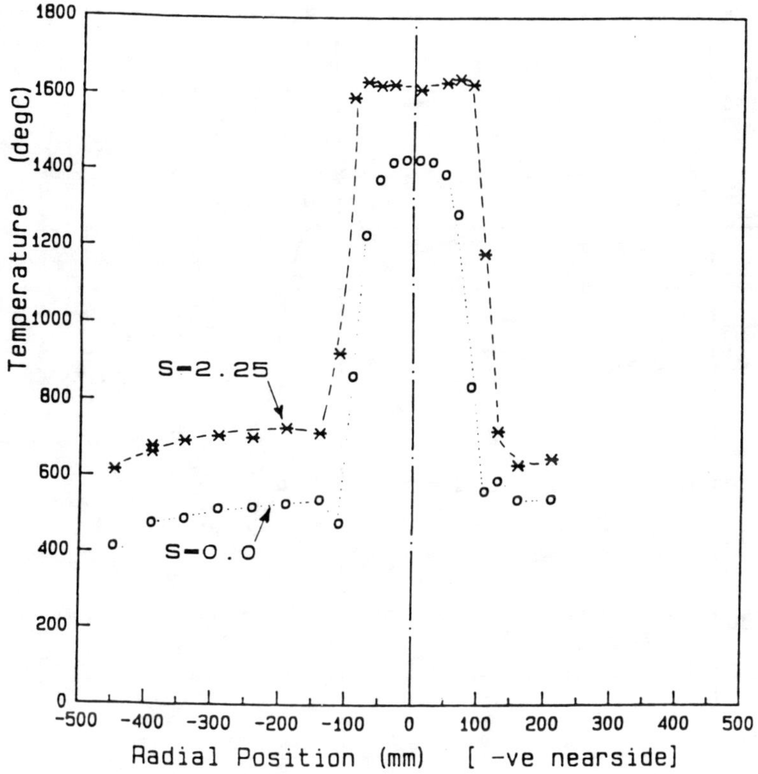

FIGURE 5-b Radial temperature profile, axial position 45 mm

The main reaction zone

This region covers the area containing the forward flow of the fresh mixture discharged from the burner. It is regarded as extending to the point where NO_x concentration becomes uniform across the whole radial profile. The length of the reaction zone decreases from 1.15 m to 0.42 m over the swirl range 0.0 to 2.25. This is in agreement with flame length values estimated from the measured CO concentration. Figures 5-7, show sample profiles of NO_x with those of temperature for three levels near the burner. The figures demonstrate the relation between NO_x concentration and temperature for the cases of swirl numbers 0.0 and 2.25. The main reaction zone is characterised by steep gradients of NO_x concentration in both the radial and axial directions. The rate of NO_x formation follows closely the rate of the main combustion reactions as indicated by the temperature and main species concentrations. The axial distance from the burner to the position of maximum NO_x concentration decreases as swirl is increased. Figure 8 illustrates this further where the minimum NO_x value at any plane is plotted versus the furnace length. The increase in the plane-minimum value of NO_x from one plane to the next indicates the newly formed NO_x. Radial profiles show higher NO_x concentrations on both sides of the minimum value due to mixing with the reverse flow on the boundaries of this region.

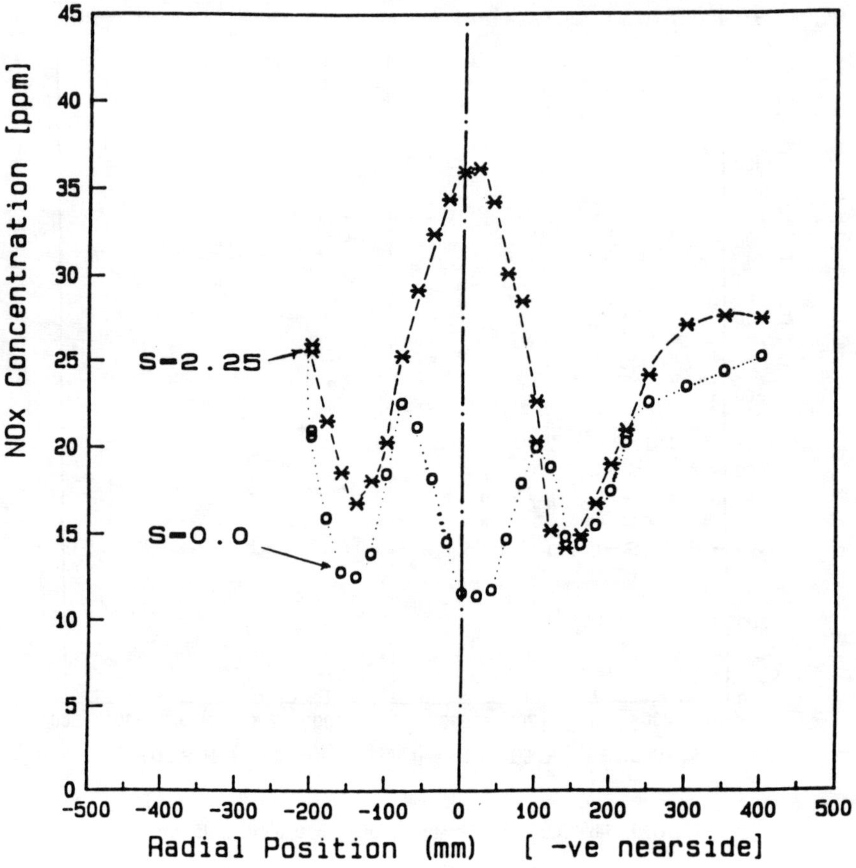

FIGURE 6-a Radial NO_x profile, axial position 200mm

Central reverse-flow zone

The flow entrained into this zone has come from the forward flow described above. Thus reactions here depend on the length of the CRZ relative to that of the reacting zone.

Figure 9 shows the variation of the NO_x concentration at the centreline with axial distance along the furnace.

For low swirl the CRZ is much shorter than the reaction zone, thus the CRZ receives a mixture with reactions still in progress and with low NO_x concentration and low temperature values. In this case it was found that no more NO_x reactions take place within this zone. A decrease of NO_x concentration occurs due to dilution with fresh mixture entrained nearer the burner end, Figure 9.

As swirl is increased the CRZ lengthens and the reaction zone shortens but is still longer than the CRZ. Thus gases with higher NO_x concentrations and at higher temperatures enter the CRZ where further NO_x reactions take place producing peak concentrations near the burner, as shown in Figure 9, with S=2.25.

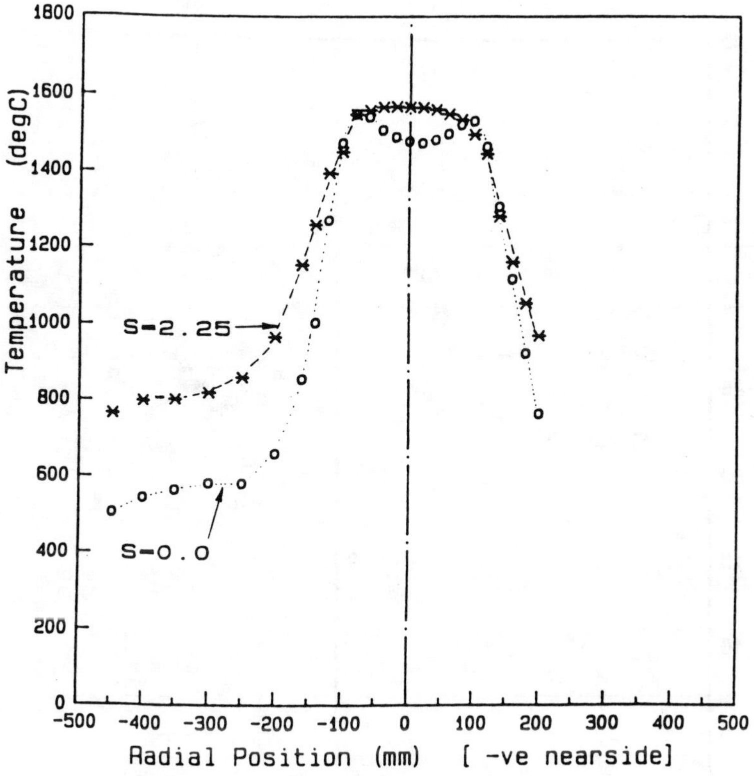

FIGURE 6-b Radial temperature profile, axial position 200 mm

Fully developed-flow zone

This region covers the fully developed forward flow downstream from the reaction zone and the reverse flow in the outer recirculating flow. At each swirl setting constant NO_x concentrations were measured throughout this region and these equal the valuc measured at the stack. This value varied from about 25 ppm to 30 ppm over the range of swirl studied (Figures 5-7). Within the outer reverse-flow zone some dilution of the flow occurs through entrainment from the forward flow giving lower NO_x concentrations at the planes nearer to the burner.

Discussion of the results

In the present system burning natural gas, NO_x is formed by the thermal mechanism. Some additional NO_x may be formed within the flame, as prompt NO_x. The present data was obtained in a lean flame with 5% excess air, thus little or no prompt NO_x is expected (eg Sarofim and Pohl, 1972, Semerjian and Vranos, 1976 and Williams, 1989).

The rate of thermal NO_x reaction is dependent on three parameters, namely temperature, O_2 concentration and residence time. The close relation between the temperature and NO_x profiles presented at Figures 5-7 confirms that NO_x has been formed by the thermal mechanism.

Turning now to the effect of swirl, the experimental results indicate that the overall effect of swirl on the final NO_x emission from the furnace is small. Over the range of

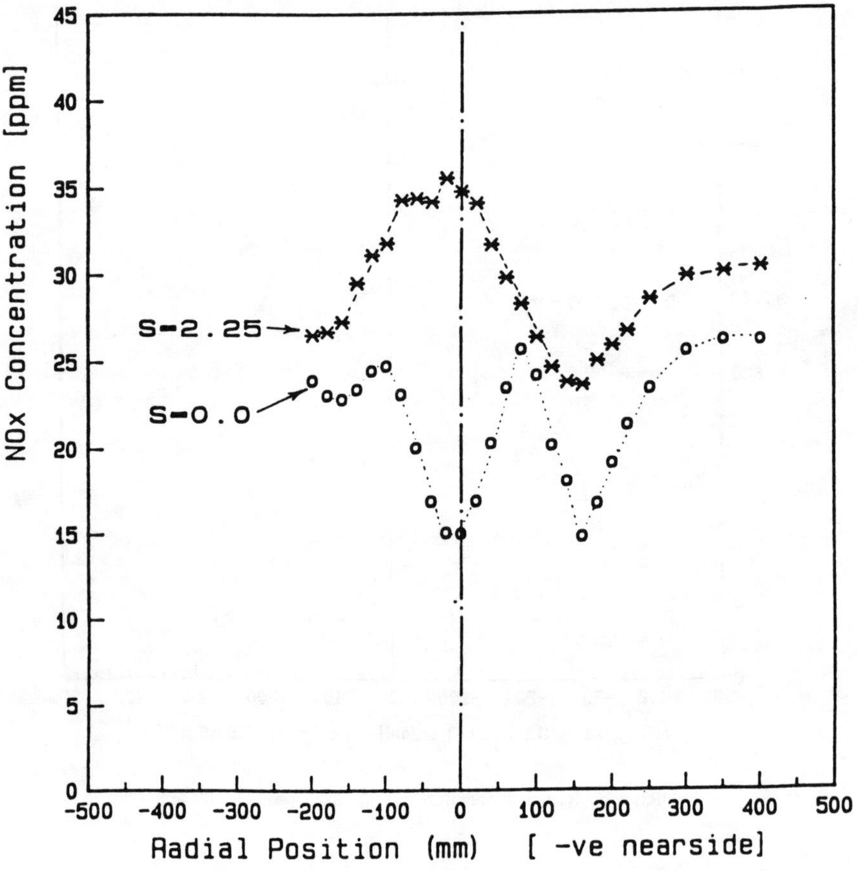

FIGURE 7-a Radial NO_x profile, axial position 300mm

swirl studied, swirl numbers 0 to 2.25, the stack NO_x concentration only varied from 25 to 30 ppm.

The NO_x formation in the forward flow reaction zone, as shown by Figure 8, is nearly the same for all swirls, as measured by the increase of NO_x over the length $X = 0.045$ to $X = 1.15$ m. The value of NO_x concentration near the burner is the result of flow from the CRZ mixing with the fresh flow.

Thus it can be assumed that all the increase in NO_x due to swirl is caused by the reactions taking place within the CRZ. These reactions seem to be most active in the highest swirl case (2.25) as illustrated by Figure 9.

Thus although swirl increases the rate of the main combustion reaction resulting in a shorter flame, it does not contribute to higher NO_x generation within the forward flow. This is due to the higher swirl giving reduced residence time in the shorter high temperature forward flow zone.

It is also realised that the mixing is affected more by the fuel injection system, than by swirl. The mixing pattern in the present system is governed by the radial fuel injection mode. This can be confirmed by the general similarity of the flow patterns obtained for all swirl cases tested, eg even for zero swirl a CRZ is created (Beltagui *et al.* 1989-a).

The use of different fuel injection modes produces different stability ranges and flow patterns, (Beltagui *et al.* 1988-b). The use of central axial fuel injection in this same

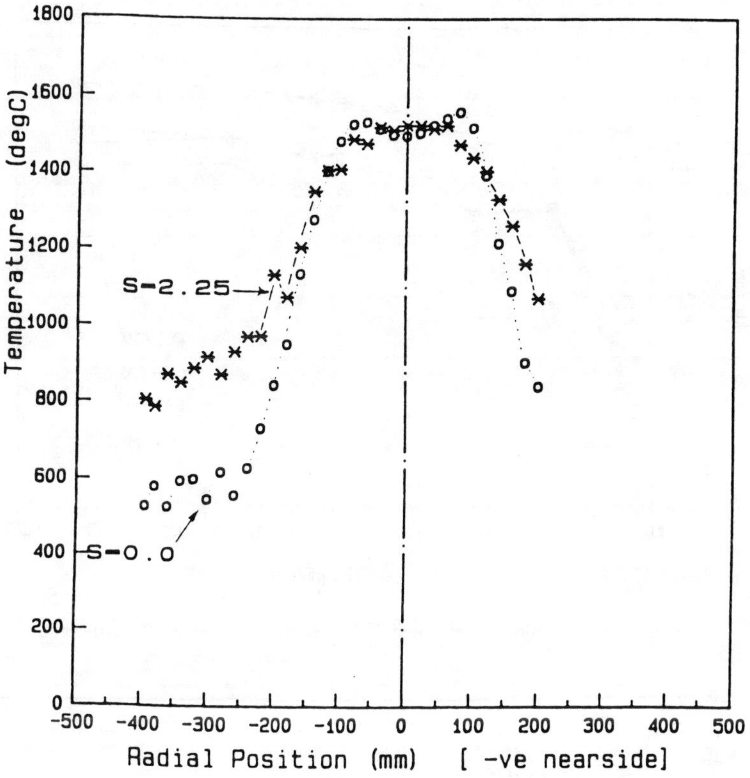

FIGURE 7-b Radial temperature profile, axial position 300 mm

burner system produced significantly lower NO_x emissions for the same swirl although the stability range was seriously reduced (Kenbar *et al.* 1991). Work by Andrews *et al.* (1988) and Al-Shaikhly *et al.* (1989) has also demonstrated the significant effect of fuel injection mode on NO_x formation.

NO_x Variation With Mixture Fuel/Air Ratio

The second set of measurements was aimed at finding the effect of fuel/air ratio upon NO_x formation as measured in the stack. The tests used initially the same input firing rate of 400 KW and covered the same swirl range as for the first set. The fuel equivalence ratio (ϕ) range was 0.71 to 1.10 (40% excess air to 10% deficient air respectively). These tests were also performed at the lower rate of 360 KW. The observed effect of this reduction upon NO_x formation is negligible.

The stack measured NO_x concentrations are given (in mg/Nm^3 corrected to 3% O_2) in Figure 10, for the four swirl settings tested. As discussed above the effect of swirl on the overall NO_x emissions is not significant, especially with the lean mixtures.

Lean mixtures

As the excess air was increased from 5% up to 40% ($\phi = 0.95 - 0.71$), the stack measured NO_x concentration increased in proportion to the excess air for all swirl settings (Figure

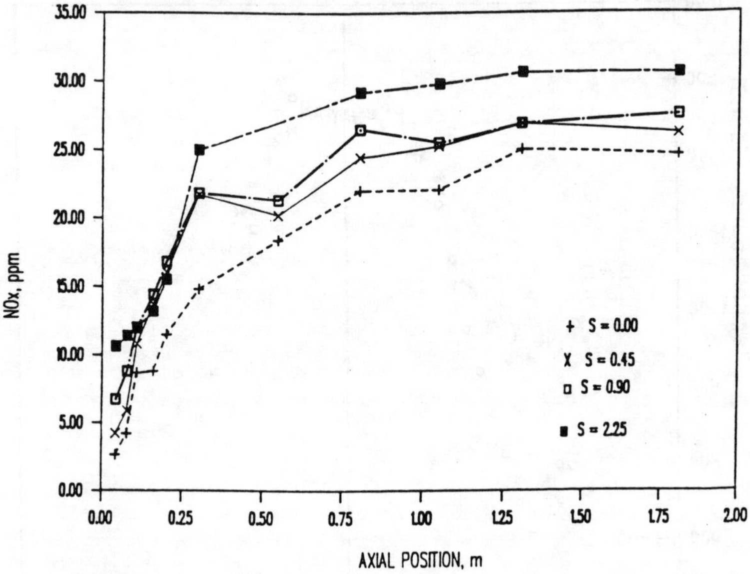

FIGURE 8 Variation of minimum NO_x concentrations along the furnace

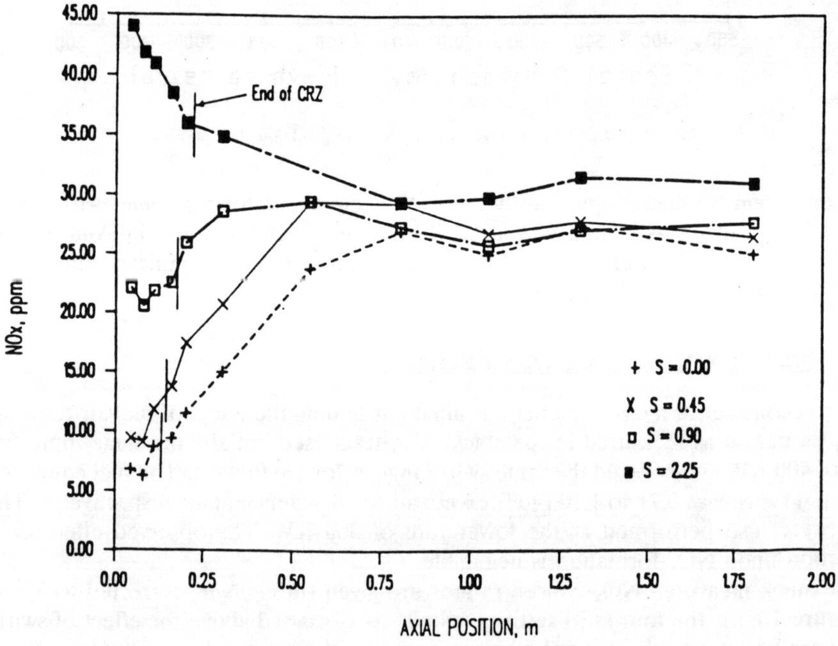

FIGURE 9 Variation of centreline NO_x concentrations along the furnace

10). This indicates that the effect of the increased air upon the local temperature of the NO_x reaction zone is minor, but there is an increase in NO_x due to the increased O_2 concentration.

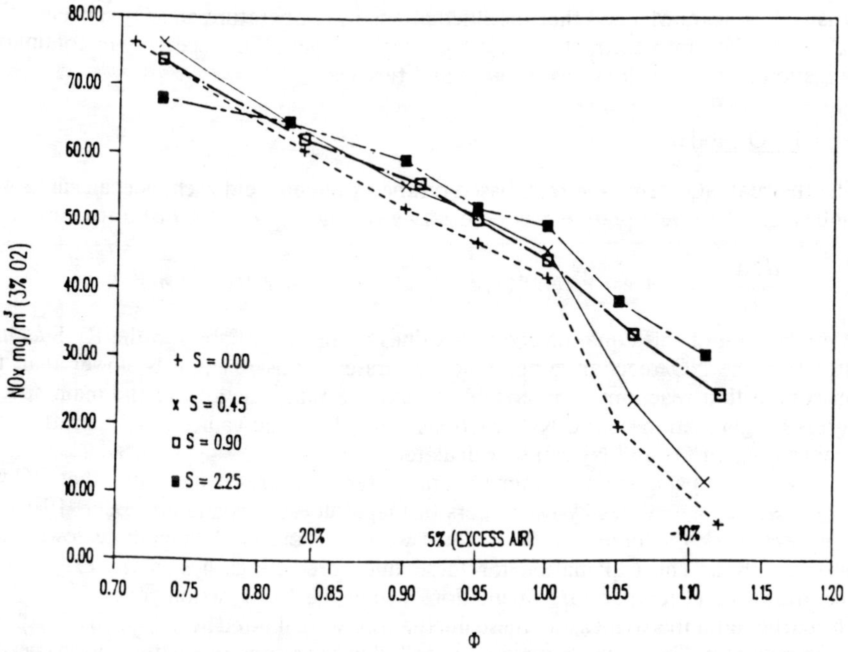

FIGURE 10 Variation of overall NO_X concentration (corrected to 3% O_2) with fuel equivalence ratio.

Rich mixtures

The measured NO_x emissions when burning fuel-rich mixtures with ϕ up to 1.11, are given by Figure 10. These results exhibit a very sharp decrease of NO_x as ϕ was increased, this decrease is more pronounced for the lower swirl settings.

The formation of NO_x occurs in localised regions where combustion products at adiabatic flame temperature and unused oxygen coexist. For rich mixtures the effects of lower temperature and lack of oxygen both work together towards reducing NO_x formation.

Somc of the measured NO_x could be formed within the flame zone by the prompt mechanism although this is unlikely within the relatively modest maximum value of ϕ tested (1.1). The above results indicate that, for the present burner system, air staging is the only effective technique in reducing NO_x formation through combustion control.

MODELLING OF NO_x FORMATION

The aim of this modelling work is to produce a simple engineering model which can be used to predict the overall NO_x formation for furnaces. The model should be able to predict the variation of NO_x formation with input conditions, mainly fuel injection mode and equivalence ratio. The model is based on the well-stirred reactor concept with account being taken for local concentration fluctuations. The inputs to the model are limited to the residence time in the reaction zone and the fuel-air mixing parameter.

In a natural-gas fired furnace system NO_x is formed by fixation of atmospheric nitrogen. Thc main mechanism for this process is known to be the thermal one. This reaction mechanism, assumes post flame NO_x formation and is usually modelled by the Zeldovich equations. The rate of thermal NO_x reaction is dependent on three parameters, namely, temperature, O_2 concentration and residence time within the reaction zone.

It is widely acknowledged that the fluctuations in temperature and O_2 concentration have to be taken into account when calculating thermal NO_x, especially in continuous combustion systems such as gas turbines and furnaces.

Thermal NO Model

The thermal NO formation rate, based on the extended Zeldovich mechanism, is well established. The rate equation was reduced by Beltagui *et al.* (1989-b) to the form:

$$\frac{dNO}{dt} = 1.44 \times 10^{17}[N2][O2]^{1/2}(T)^{-1/2} \exp(-69460/T)\ mole/cm^3 \quad (1)$$

The application of this equation requires values of all the variables in the RHS as they exist within the NO reaction zone. Since the rate of this reaction is slower than the main combustion reactions, it is possible to assume that reactions of the main species progress to completion before NO reactions start. Thus the values of temperature and concentrations of O_2 and N_2 can be calculated.

However when using measured time-averaged temperatures it was found that NO was underpredicted, sometimes by two orders of magnitude (Sadakata and Beer, 1976). In the present work the predicted NO values were one order of magnitude lower than those measured. The explanation for these underpredictions lies in the existence of concentrations and temperature fluctuations within the NO reaction zone.

The earlier attempts to measure these fluctuations were limited by the speed of response of thermocouples. Generally this is too slow to follow the temperature fluctuations beyond 1 kHz (Sadakata and Beer, 1976).

Recent measurements using CARS technique give a much better indication of the range of the temperature fluctuations, for example LaRue *et al.* (1984) measured a 500 K peak to peak range.

These temperature fluctuations are caused by concentration fluctuations. Since the NO reactions take place "immediately" after the main reactions, both temperature and concentration fluctuations have a considerable influence on the NO formation rate.

On the other hand, it is suggested by some investigators that the temperature in the NO reaction zone is the stoichiometric adiabatic flame temperature (Toof, 1986 and Takagi *et al.* 1976). Vranos (1974) assumed that the thermal NO reactions are not faster than the mixing rates so most thermal NO is formed at or near $\phi = 1.0$. This assumption follows from the fact that stoichiometric mixtures have the highest probability of reaction.

When using finite difference solutions of the flow and combustion equations, eg. PCOC (Stopford, 1989), the effect of concentration fluctuations can be taken into consideration through the solution of an extra differential equation (Elghobashi and Pun, 1974). For the present work considering a well-stirred reactor model, a stochastic approach was adopted to estimate the fluctuating components of concentration about the mean value.

Model Assumptions

Assuming the reaction zone to be one well-stirred reactor and using Equation 1, the following analysis is carried out to account for the effect of concentration fluctuations.

On the microscale of mixing, it is assumed that at any location within the well-stirred reactor, the fuel air mixture will have a random distribution of parcels with values of ϕ which have a Gaussian distribution about the mean value $\bar{\phi}$. This assumption agrees with experimental measurements in jet flows as reviewed by Chatwin and Sullivan (1990). The spread of this distribution is dependant upon the value of the standard deviation σ. The value of σ is linked to the mixedness, s, of the mixture by the relation (Appleton and Heywood, 1972 and Fletcher, 1973):

$$\sigma = (1 - s)\bar{\phi} \tag{2}$$

Introducing Z as a non-dimensional parameter of fluctuation gives:

$$Z = (\phi - \bar{\phi})/\sigma \tag{3}$$

The probability density function (p.d.f.) of the distribution function of ϕ is written as:

$$f(\phi) = (2\pi)^{-1/2} \exp\left(-Z^2/2\right) \tag{4}$$

Integrating the area under the above distribution function, leads to the cumulative density function (c.d.f.) of ϕ:

$$F(\phi) = \int_{-\infty}^{\phi} (2\pi)^{-1/2} \exp\left(-Z^2/2\right) dZ \qquad -[\infty < \phi < \infty \tag{5}$$

For standard Gaussian distribution, the value of F(ϕ) is usually between 0 and 1. A stochastic process (Kenbar, 1988) has been implemented to produce the variable Z from Equation 5 using randomly generated values of F (ϕ) using an appropriate random number generator. From Equation 3 above and at known values of $\bar{\phi}$ and σ the stochastic value of ϕ has been predicted. This process gives the Gaussian distribution of ϕ about $\bar{\phi}$.

Of the different parcels, those with ϕ close to unity (ie stoichiometric) have the maximum probability of combustion and thus react first. The products of this reaction will be at the adiabatic stoichiometric flame temperature. The NO reaction will take place effectively at this flame temperature. The required oxygen is assumed to come from the excess air of the unburnt lean parcels with $\phi < 1$. These values are substituted in Equation 1.

Finally a mean value of residence time in the reaction zone was calculated for each swirl case from the length and the average flow velocity in the reaction zone.

Prompt NO_x

Although prompt NO_x was initially defined as NO_x formed in the flame front, it sometimes refers to NO_x resulting from N_2 via reactions other than the Zeldovich mechanism. Detailed models of prompt NO_x are not available yet, because the amounts of prompt NO_x are small compared to other sources of NO_x, and because it seems to occur mainly in very rich flames. It can be important in fuel-rich pockets of diffusion flames. Rate equations were suggested by Fenimore (1970) and De Soete (1974).

In the present analysis both equations were used to estimate prompt NO_x. The results indicate the contribution by this mechanism to be very small, less than 1 ppm.

Model Results

The above model was tested by application to the conditions of the stack measurements to predict the effect of fuel/air ratio (equivalence ratio) on the overall NO_x emission from the furnace. The value for the standard deviation σ can be estimated from the measurements of the fluctuations of concentrations and temperatures in jet flows, as

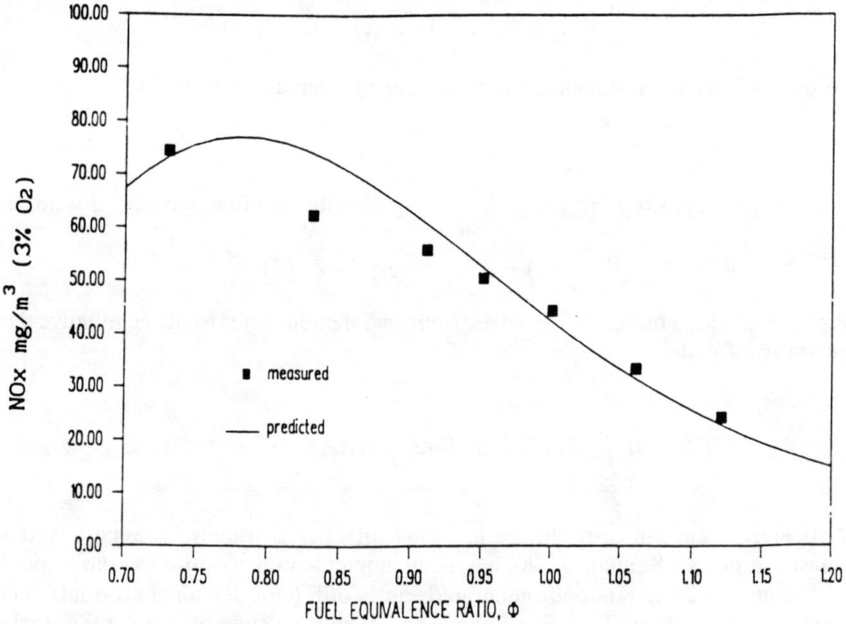

FIGURE 11 Comparison of predicted and measured NO_x concentration, S = 0.90

reported by Becker (1974). His results give an estimated value of 0.70 for the mixing parameter s in Equation 2.

The results for the medium swirl case with S = 0.90 are presented in Figure 11. The agreement between the model predictions and measurements is considered satisfactory and very encouraging considering the simplicity of treating the reaction zone as one well-stirred reactor. As earlier stated, the observed effect of swirl on NO_x formation in the present system is small and therefore could not be used as a reliable test of this model. The model is also being used to model other fuel injection modes. The input data required are only the mixedness of the fuel-air jets and the mean residence time within the reaction zone.

CONCLUSIONS

Two sets of experimental measurements have been carried out in the NEL furnace, when using a variable swirl burner with central radial fuel injection. The first set covers a complete mapping of NO_x concentrations carried out at different swirl settings. In the second set, the effect of fuel/air ratio on the overall NO_x emissions in the stack has been measured.

The reliability of these measurements has been checked by carrying out tests of repeatability and of the probe intrusion effect. NO_x concentration profiles have shown reasonable degree of symmetry. This is important for the meaningful comparison with the axisymmetric numerical predictions.

From the experimental measurements the following conclusions can be drawn:

1— NO_x formation occurred mainly at the high temperature regions downstream of the combustion zone (flame front), hence confirming that NO_x has been formed by thermal reactions which follow the main combustion.

2— For the present burner system, increased swirl did not contribute significantly to increased stack NO_x emissions. At zero swirl, NO_x concentration measured at the stack was about 25 ppm while at the high swirl (S=2.25), NO_x concentration was about 30 ppm.

3— Increased swirl has contributed only to increased local NO_x concentrations in the central recirculation zone.

4— Increasing excess air, from zero up to 40%, led to a slightly more than proportionate increase in the NO_x emissions.

5— At rich flame conditions, increased fuel equivalence ratio resulted in marked reduction of NO_x emission. This finding supports the use of air-staged combustion for NO_x reduction.

Theoretical predictions of the overall NO_x formation under the above test conditions have been made using a well-stirred reactor model based on the extended Zeldovich mechanism. The model takes account of the fluctuations of the concentrations of fuel and oxidant in the NO_x reaction zone. A stochastic analysis has been adopted to calculate the effect of these fluctuations on the NO_x formation rate. The results of these predictions compare satisfactorily with the measurements and confirm the relationship between NO_x formation and equivalence ratio.

ACKNOWLEDGEMENTS

The authors wish to thank colleagues at the National Engineering Laboratory (NEL) and the University of Glasgow for their advice and encouragement. This work was supported in part by the Management Board of NEL.

REFERENCES

Ahmad, N. T., Andrews, G. E., Kowkabi, M. and Sharif, S. F. (1984). Centrifugal mixing forces in enclosed swirl flames. *20th Symp. (Int.) on Combustion,* pp 259-267.

Allen, J. D. (1973). A review of methods of analysis for oxides of Nitrogen. *J. Inst. of Fuel,* **46**, pp. 123-133.

Al-Shaikhly, A. F. A., Andrews, G. E. and Aniagolu, C. O. (1989). Jet shear layer turbulent diffusion flames for ultra-low NO_x emission. Joint meeting of the British and French Sections of The Combustion Institute, Rouen, pp. 5-8.

Andrews, G. E., Abdul Aziz, M. M., Abdul Hussain U. S. *et al.* (1988). High intensity burner with low NO_x emissions. British Flame Days, Furnace Combustion Research and Its Applications, The Institute of Energy, London.

Appelton, J. P. and Heywood, J. B. (1972). The effects of imperfect fuel-air mixing in a burner on NO formation from nitrogen in the air. *14th Symp. (Int.) on Combustion*, pp. 777-786.

Becker, H. A. (1974). Effects of concentration fluctuations in turbulent diffusion flames. *15th Symp. (Int.) on Combustion,* pp 601-615.

Beltagui, S. A. and Ralston, T. (1985). An isothermal model study of aerodynamics and mixing in the flow issuing from a variable-swirl burner. *HTFS Research Symp.,* RS 602, NEL/HTFS 63, also NEL Report DC/355.

Beltagui, S. A., Fuggle, R. N. and Ralston, T. (1988-a). Measurement and prediction of heat transfer in the NEL furnace. I.Mech.E./I.Chem.E./Heat Transfer Society, *2nd UK National Heat Transfer Conference, Glasgow,* pp. 1219-1232, Mechanical Engineering Publications.

Beltagui, S. A., Maccallum, N. R. L. and Ralston, T. (1988-b). The effect of fuel injection modes on combustion. British Flame Days, Imperial College, London.

Beltagui, S. A., Kenbar, A. M. A. and Maccallum, N. R. L. (1989-b). NO_x generation and control in confined swirling flames—review and parametric study. *HTFS Research Symp.,* RS 827, NEL/HTFS 123.

Beltagui, S. A., Fuggle, R. N., Kenbar, A. M. A., Ralston, T., Marriott, N. and Stopford, P. J. (1991). Modelling a gas-fired furnace using the PCOC code. Paper accepted for presentation at the Euroteck Direct 91, Birmingham, 2-4 July.

Chatwin, P. C. and Sullivan, A. N. (1990). A simple and unifying physical interpretation of scalar fluctuation measurements from many turbulent shear flows. *J. Fluid Mech.,* **212**, pp. 533-556.

Cheadaille, J. and Braud, Y. (1972). Industrial Flames. Vol. 1, Measurements in flames. Edward Arnold Publishers, London.

De Soete, G. G. (1974). Overall reaction rates of NO and N_2 formation from fuel nitrogen. *15th Symp. (Int.) on Combustion,* pp 1093-1102.

De Soete, G. G. (1989). N_2O analysis techniques: Gas chromatography with EC detection: sampling techniques. 1st topic Oriental meeting TOTeM1, IFRF, Amsterdam.

Elghobashi, S. E. and Pun, W. M. (1974). A theoretical and experimental study of turbulent diffusion flames in cylindrical furnaces. *15th Symp. (Int.) on Combustion,* pp. 1353-1365.

Fenimore, C. P. (1970). Formation of nitric oxide in premixed hydrocarbon flames. *13th Symp. (Int.) on Combustion,* pp. 373-380.

Fletcher, R. S. (1973). The prediction of nitric oxide formation in combustion systems. *1st European Symp. on Combustion,* Sheffield, pp. 445-450.

Kenbar, A. M. A., Beltagui, S. A. and Maccallum, N. R. L. (1991). Effect of fuel injection modes on the combustion patterns and pollution formation in the NEL furnace. *HTFS Research Symp.,* RS 877, NEL/HTFS 146, pp.183-206.

Kenbar, A. M. A. (1988). MSc thesis, Dept. of Mech. Eng., University of Glasgow.

LaRue, J. C., Samuelsen, G. S. and Seiler, E. T. (1984). Momentum and heat flux in a swirl-stabilized combustor. *20th Symp. (Int.) on Combustion,* pp. 277-285.

Mulholland, J. A. and Hall, R. E. (1987). Fuel oil reburning application for NO_x control to fire tube package boilers. *J. Eng. Gas Turbine Power,* **109,** pp. 207-214.

Oven, M. J., Gouldin, F. C. and McLean, W. J. (1978). Temperature and species concentration measurements in a swirl-stabilized combustor. *17th Symp. (Int.) on Combustion,* pp. 363-374.

Sadakata, M. and Beer, J. M. (1976). Spatial distribution of Nitric Oxide formation rates in a swirling turbulent methane-air flame. *16th Symp. (Int.) on Combustion,* pp. 93-103.

Sadakata, M., Fujioka, Y. and Kunii, D. (1980). Effects of air preheating on emissions of NO, HCN and NH_3 from a tow-staged combustor. *18th Symp. (Int.) on Combustion,* pp. 65-72.

Sarofim, A. F. and Pohl, J. H. (1972). Kinetics of Nitric Oxide formation in premixed laminar flames. *14th Symp. (Int.) on Combustion,* pp. 739-754.

Semerjian, H. and Vranos, A. (1976). NO_x formation in premixed turbulent flames. *16th Symp. (Int.) on Combustion,* pp. 169-179.

Stopford, P. J. (1989). PCOC Users Manual, AERE-R12626.

Takagi, T., Ogasawara, M. and Daizo, M. (1976). NO_x formation from nitrogen in fuel and air during turbulent diffusion combustion. *16th Symp. (Int.) on Combustion,* pp 181-189.

Toof, J. L. (1986). A model for the prediction of thermal, prompt and fuel NO_x emission from combustion turbine. *ASME J. Eng. Gas Turbine and Power,* **108,** pp. 340-347.

Tuttle, J. H., Shisler, R. A. and Mellor, A. M. (1974). Nitrogen Dioxide formation in gas turbine engine: Measurements and measurement methods. *Comb. Sci. and Tech.,* **9,** pp. 261-271.

Vranos, A. (1974). Turbulent mixing and NO_x formation in gas combustors. *Combustion and Flame,* **22,** pp. 253-258.

Williams, A. (1989). Combustion-Generated NO_x. Joint meeting of the British and French Sections of The Combustion Institute, Rouen, pp. A1-A11.

Pulsated O_2/Fuel Flame as a New Technique for Low NO_x Emission

OLIVIER CHARON, DOMINIQUE JOUVAUD, and BERNARD GENIES *Claude-Delorme AIR LIQUIDE Research Center Les Loges en Josas, Jouy en Josas, France*

Abstract—A new primary NO_X reduction method is described. It consists in pulsating one or both fluids from a oxy-gas burner by rotary-plug valve or solenoïd electrovalve. The parameters to be set are C the constant fraction of the pulsated gas flow-rate, P the amplitude of the pulsated fraction of the gas flow-rate, O the duration of the maximum flow-rate, F the duration of the minimum flow-rate and ϕ the dephasing between the two pulsations when both gases are pulsated.

Experiments on 20 kW and 50 kW pilot furnaces have shown that up to 90% NO_X reduction can be achieved without lowering heat transfer when parameters are set as follows (both gases are pulsated): dephasing equal to 1/2, C/P around 0.2 and frequency (1/F+O)) as low as possible.

Further developments are underway to perform test on a 1 MW pilot furnace before industrial trials.

INTRODUCTION

Environmental regulations in all industrialised countries are becoming more and more stringent. For this reason, engineers and researchers continue their effort in developing new combustion techniques and/or modifying established ones to lower pollutant emissions. Among the less desirable species, generally associated with acid rain and the greenhouse effect, are the nitrogen oxides NO_X. The portion of NO_X generated by industry comes mainly from high temperature processes such as glass melting. In such plants, reductions of NO_X emissions can be achieved by two different approaches: primary methods to prevent generation of NO_X, and secondary methods, which are aimed at destroying the undesirable species before their release to the atmosphere. Secondary methods involve the post-treatment of flue gases (SNCR, SCR, steam injection . . .). This type of option is not only capital intensive but it may also demand a continuing effort on equipment maintenance and reagent cost during operation. From both the economical and the environmental perspectives, every effort should thus be made in order to avoid generating the NO_X species in the first place.

Primary methods such as recirculation and staged combustion address the issue of avoiding excessively high flame temperatures by inducing specific turbulence patterns and are generally based on burner design. Charon *et al.* (1990) have shown in a previous publication that a pure oxygen/natural gas flame could, if properly managed, be by itself a primary-type of low-NO_X combustion technique. To further improve the environmental performance of oxygen-based combustion, a new technique has been developed at the AIR LIQUIDE Research Center. The technique is based on the principle of Pulsating Combustion, and results indicated that NO_X reductions between 40 and 90% can be obtained relative to their non-pulsated equivalents. A number of studies on pulsed combustors are mentioned in the scientific literature: Briffa (1982), Putman (1986), Al-Haddad (1989) and Hirose (1978) for instance. These studies are mainly related to air-based flames and their interest focuses on heat transfer research. However very few studies are specifically related to NO_X one such being the GRI work (1987), including a patent (1989), claiming good NO_X performances. The present work deals specifically with **flames involving pure oxygen and natural gas**, the selective pulsation of either one or both of the gaseous streams, and the impact of the pulsation phenomena on the generation of NO_X. In this paper we introduce the principles of pulsating combustion, describe the pulse generating equipment and discuss the influence of operating parameters on NO_x emissions.

PRINCIPLES

The nitrogen for NO_x formation may come from either natural gas (0 to 20%) or oxygen (when produced through non-cryogenic methods). For these experiments, O_2 came from cryogenic sources and was thus 99.999% pure. The N_2 content of the available natural gas varied from 0.5 to 4%.

A typical natural gas/pure oxygen flame produces "prompt" and "thermal" NO_x. Prompt NO_x is generated at an early stage in the flame, induced by high temperature gradients and high oxygen concentrations near the burner tip. Thermal NO_x, on the other hand, is generated further down stream in the flame where the NO_x forming species are exposed to sufficiently high temperatures for a relatively long period of time. Pre-heated air-based flames generate mainly thermal NO_x. An oxygen-based flame will generate more prompt NO_x than a pre-heated air based flame at an equivalent firing rate. It is believed that by pulsating at least one of the gaseous feeds, the fraction of NO_x promptly generated is significantly decreased.

Prompt and thermal NO_x formation is very sensitive to both flame temperature and local O_2 concentration. The maximum flame temperature is obtained for the stoichiometric composition (R=1), since any excess of either oxygen (lean flame, R<1) or natural gas (rich flame, R>1) will dilute the flame. For small O_2 enrichment levels, up to R=0.8, an increase in NO_x formation is observed. Below R=0.8 the dilution effect lowers the flame temperature enough that a decrease in the NO_x concentration in the flue gas, relative to the stoichiometric value, is observed. Consequently, both lean and rich flames generate less NO_x than stoichiometric flame. This is the basic principle behind the use of pulsating combustion for NO_x control. An overall stoichiometric flame, which gives the best combustion efficiency, can be obtained by a succession of periodically alternating lean and rich flames. Such a sequence is obtained by establishing a gas flowrate which varies versus time in at least one of the feed lines. A minimum gas flowrate needs to be maintained in both lines, in order to avoid flame extinction and high levels of CO emissions due to incomplete combustion.

An idealized example of pulsated flowrate follows the square signal shown on figure 1. Characteristics of this signal are:

Q (Nm^3/h)	:	Pulsated gas flowrate
C (Nm^3/h)	:	Minimum flowrate and constant fraction of the gas flowrate
P (Nm^3/h)	:	Amplitude of the pulsated fraction of the gas flowrate
O (s)	:	Duration of the maximum flowrate ($Q = C + P$)
F (s)	:	Duration of the minimum flowrate ($Q = C$)
f (Hz)	:	Pulsation frequency ($1/(O + F)$)
ϕ	:	Dephasing between the two pulsations when both gases are pulsated

When pulsating only one gas, the other gas flowrate is kept constant and is equal to $A^*(C + P/2)$ (A equals 2 or 0.5 if oxygen or natural gas is, respectively, the pulsated gas). Experiments have been run in order to determine the feasibility limits of the technique and the optimum value of parameters C, P, O, F and ϕ.

EXPERIMENTAL SET-UP

Two different furnaces were used which varied in the inner refractory lining material used and in firing rate capability. Furnace no.1, lined with refractory wool, was used for the

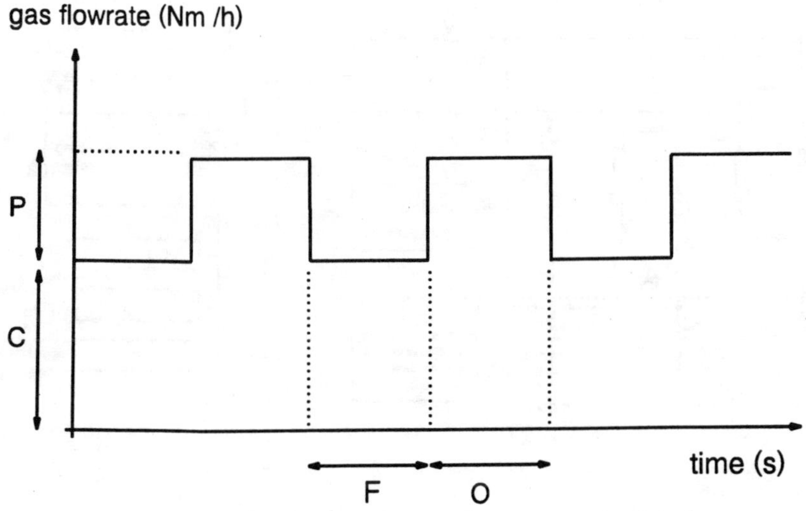

FIGURE 1 Idealized flowrate profile of the pulsated gas and its characteristics.

first experiments at a 20 kW firing rate, and Furnace no.2, lined with alumina brick, was used at 50 kW and a relatively higher flue gas temperature. Gas feed lines were equipped with pulsation generating devices, which consisted of two different types of valves: a) solenoid-type electrovalves and b) rotary valves. Flue gas analysis was provided for the following species: CO, CO_2, NO, NO_2, O_2. Furnace wall temperature was monitored using S-type thermocouples.

Description of furnace no.1 (20 kW)

Furnace no.1 is cylindrical. The combustion chamber is 1.5 m long and the diameter is 0.5 m. The thickness of the 50% alumina refractory wool is 125 mm. The advantage of this type of refractory is its low thermal inertia which allows several experiments to be run during a day. The inconvenience of using this material is that the temperature of the flue gases had to be maintained below 1350°C to prevent refractory damage. A standard AL multi-hole burner was used to develop a 20 kW O_2/natural gas flame. Three S-thermocouples located on the roof of the furnace and a suction pyrometer located immediately after the combustion chamber were used to measure wall and fume temperatures, respectively. The safety of the system is ensured by a UV flame detector which shuts-off all gaseous supplies should the flame be temporarily lost. The middle roof temperature sensor is also interlocked to the safety device tripping the gas flows as the furnace wall temperature exceeds 1400°C.

Description of furnace no.2 (50kW)

Furnace no.2 is also cylindrical. The combustion chamber is 2.0 m long and the diameter is 0.6 m. The thickness of the 96% alumina refractory brick is 230 mm. The advantage of this type of refractory is that it could withstand wall temperature as high as 1800°C. The drawback is its relatively larger thermal inertia. This furnace is also equipped with a standard AL multi-hole burner. Control features and temperature monitoring

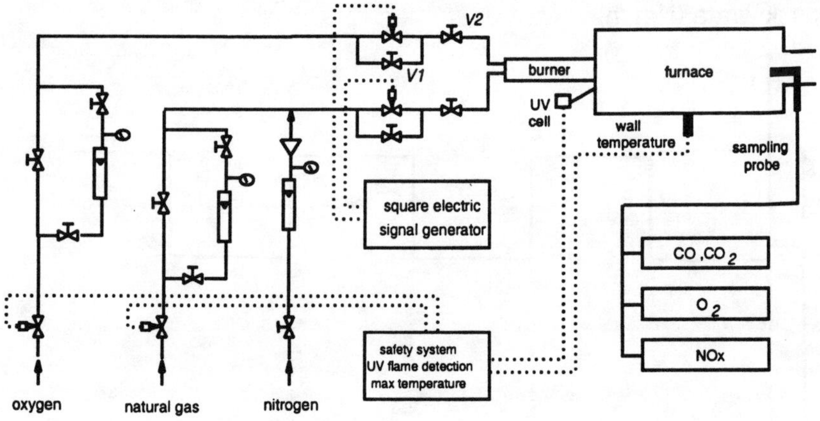

FIGURE 2 Schematic representation of the experimental set-up.

instrumentation are identical to furnace no.1. Furnace no.2 has been used to test the pulsating combustion technique at higher temperature (1500°C).

Analyzers

CO, CO_2, NO_x and O_2 concentrations are measured for all experiments. IR, chemiluminescence and paramagnetic analyzers have been used, respectively.

To draw a flue gas sample, a quenching probe is located at the bottom of the furnace. The quenched sample passes through a condensate trap and then through a $CaCl_2$ bed. The sample is then split into independent streams to each of the analysers. Each analyzer is connected to an N_2 line for purge and zero calibration, and also to a specific standard gas for calibration.

Description of the gas distribution lines

There are 3 different gas lines to the burner: the O_2 line, natural gas line and N_2 line. Figure 2 is a schematic representation of the experimental set up. O_2 and natural gas lines are constituted as follows: First, a safety solenoid interlock will automatically interrupt gas flow in case of emergency. The solenoid is followed by a manual throttle valve, a Rotameter and a pressure gauge. At this point the line is split into two streams:

1) a line for the pulsating fraction of the gas flowrate, equipped with either a solenoid electrovalve which is driven by a square signal, or a rotary-plug valve, and 2) a by-pass line used to establish the constant fraction of the flowrate. The separate lines converge into single lines. Downstream of the rejoining point of the natural gas line, a nitrogen line is used to adjust the N_2 content of the natural gas to the desired value. The content in the natural gas available at the Combustion Research Station varied between 0.5 and 5%. The balance required to attain 7% or 15% (two levels of N_2% were tested) was thus added to the natural gas through the above mentioned line. Further details are given in section **Nitrogen content in natural gas.**

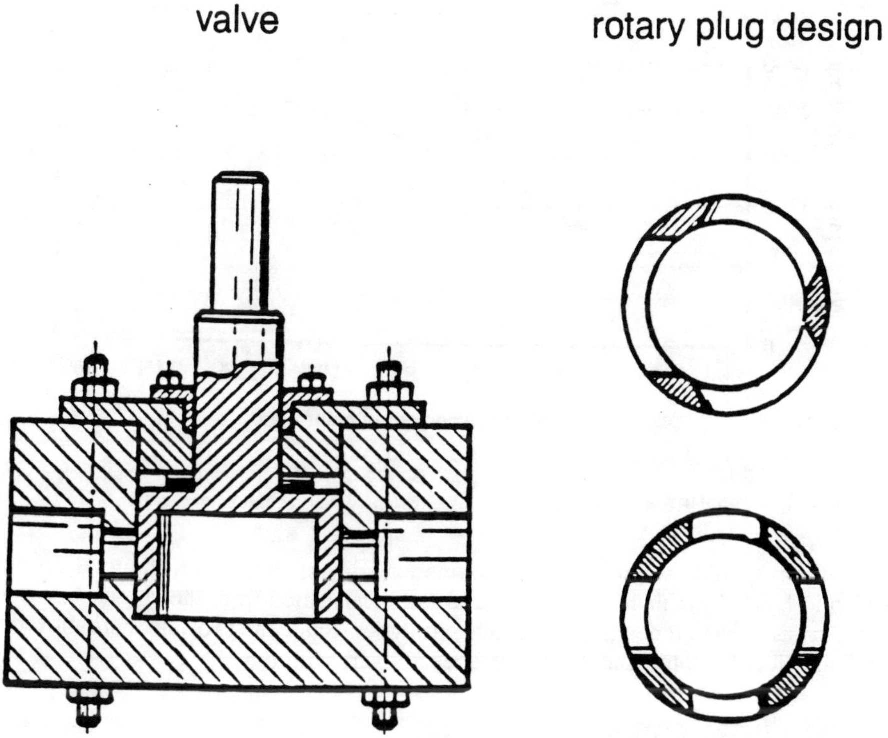

FIGURE 3 Schematic representation of the rotary-plug valve.

Pulsation generators

Commercial solenoid-type electrovalves were first used to generate the pulsating fraction of the flow. Even though they performed satisfactorily throughout the testing period, they are not designed to withstand continuous operation at the frequencies required by the technique. A rotary-plug valve was developed to ensure reliable long-term operation. A schematic representation of the rotary-plug valve is shown in figure 3. The design of the rotary plug determines the O/F ratio and the variable speed electric motor allowed for the setting of the pulsation frequency. This rotary valve responded well to sustained operation. Experiments indicate that for similar pulsating combustion conditions the solenoid-type electrovalve resulted in less NO_x emissions than the rotary-plug valve. This relatively poorer performance of the rotary valve has been associated with plug design considerations, now resolved.

Setting F, O, C, P and Φ

A set of various plugs was fabricated to test different values of the O/F ratio. To set the C/P ratio, valves V1 and V2 are independently adjusted (see figure 2). With the solenoid (or rotary) valve closed and V2 totally open, C is set by opening V1 up to the desired value of the flowmeter. Then the electrovalve is set open (without pulsation) and V2 is manually adjusted until the appropriate (C+P) value on the flowmeter is established. Once C/P set, it is possible to by-pass the flowmeter during the experiment in order to

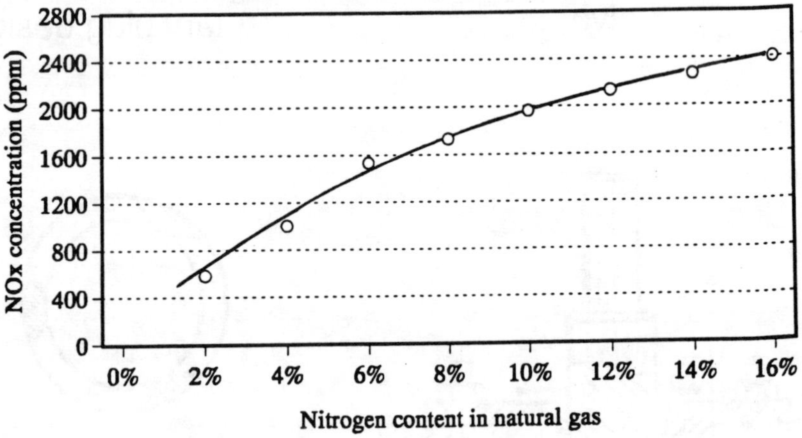

FIGURE 4 NO_x emissions versus nitrogen content of the natural gas.

prevent damage due to pulsation. From one experiment to another, the total flowrate is identical, (C+P) is then constant. C and P are calculated from the chosen C/P ratio. The two-gas pulsation dephasing is mechanically set. Since the two valves are moved by the same electric motor, the dephasing does not vary with time.

EXPERIMENTAL CONDITIONS

In order to test the parameters which characterize the pulsating combustion process, the basic combustion parameters were established: burner firing rate, flue gas temperature, furnace pressure, oxygen excess in flue gas and nitrogen content of natural gas.

Firing rate, flue gas temperature and furnace pressure.

On furnace no.1, the burner firing rate was set to 20kW, yielding flue gases of 1300°C, and on furnace no.2 at 60 kW, with flue gases leaving the furnace at 1500°C. In order to minimize air entrainment into the furnace, experiments were carried out under a slight positive pressure of about 2 mm water gauge.

Nitrogen content in natural gas

A relatively high N_2 content in natural gas (7 to 15%) was selected for the evaluation of pulsating combustion. One reason for this was to supply a source of N_2 for NO_x formation which would be easily measured (non-pulsated NO_x emission levels between 1500 and 2000 ppm). The other was to illustrate relative unfavourable conditions which are however representative of European operations (Groningue natural gas contains 14.7% N_2 before it is diluted to about a 10% range at point of use). A clear relation between the NO_x emission level and the N_2 contained in natural gas is shown in figure 4, at non-pulsating combustion conditions. These results have been obtained at 3% excess O_2 in the dry flue gas. Consequently the work was undertaken in relatively "polluting" conditions.

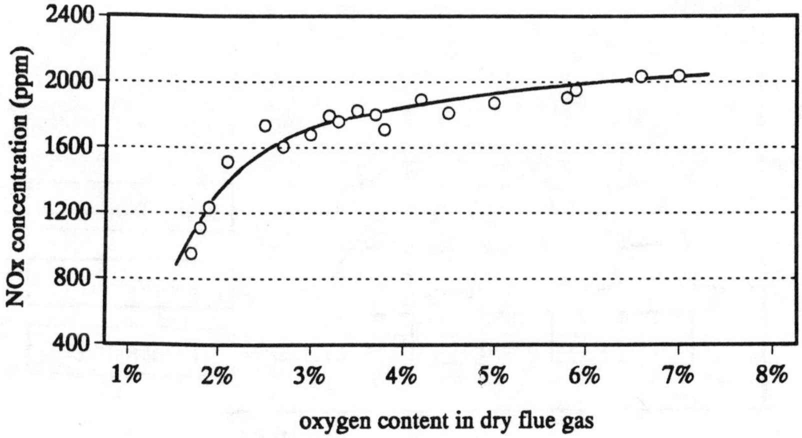

FIGURE 5 NO_x emissions versus oxygen content in the dry flue gas.

Excess oxygen in dry flue gas.

NO_x formation is closely related to the excess $O_2\%$ in the dry flue gas. This was shown in figure 5. In the range between 1% and 4% oxygen in the dry flue gas, NO_x increase drastically. Beyond 4% O_2 the slope is smoother, but NO_x emissions still increase. Above 8%, the impact of lower flame temperature, due to dilution by the extra oxygen, becomes more important than the increased oxygen availability. As flame temperature decreases, so do NO_x emissions. To operate near the "flat" NO_x generating excess O_2 range, it was decided to work at 6% O_2 for all experiments.

EXPERIMENTAL RESULTS

Transmission of the mechanical or electric valve square signal to the gas.

In this section, the performances of the solenoid electrovalve and the rotary-plug valve are compared. Downstream from the valve a pressure sensor was installed to visualize, on an oscilloscope, the gas pressure versus time. A comparison between the 3 signals (idealized square signal, solenoid and rotary-plug valves) is shown in figure 6. Both the solenoid and the rotary-plug valves gave the expected ratio O/F. The rotary valve signal is less "square" than the electrovalve signal. The performance of the rotary valve could be improved on that point by modifying the shape of the rotary plug. Further experiments will be run to evaluate its potential effect on NO_x emissions.

Solenoid electrovalve pulsation

Solenoid electrovalves have been tested on furnace no.1 under the following experimental conditions: firing rate of 20 kW generating flue gas at 1300°C 15% nitrogen in natural gas, 6% oxygen in dry flue gas and O/F equals 1.

Natural gas or oxygen pulsation

All the NO_x results have been expressed on a relative basis as NO_x/NO_x (ref). The reference NO_x (ref) has been defined as the NO_x emissions of the equivalent non-

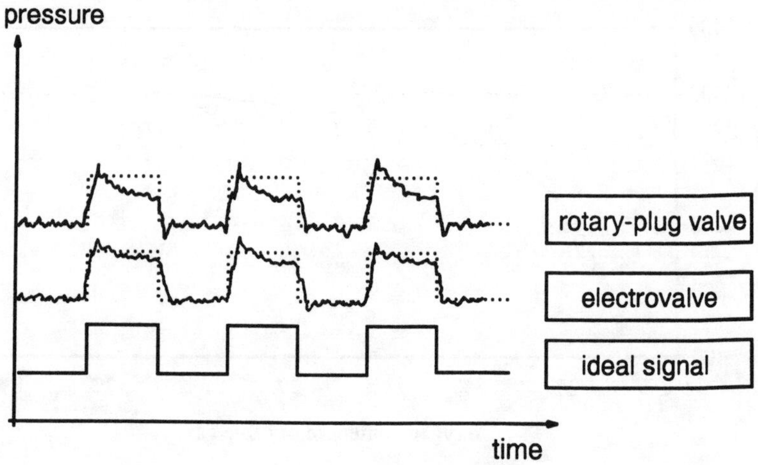

FIGURE 6 Pulsated gas pressure for idealized, solenoid-type electrovalve and rotary-plug valve.

pulsated flame. This reference point was systematically checked after each run in order to take into account effects such as variations in air entrainment. N_2% in natural gas or other experimental fluctuations.

When pulsating natural gas only, the effect of pulsation frequency on NO_x emissions was first investigated. It was observed that when the frequency f decreases, NO_x concentration decreases. Reductions in NO_x emissions of up to 60% of the non-pulsated equivalent were achieved.

The NO_x versus the frequency logarithm curve follows an S-type function. At frequencies below the inflection point, CO begins to form, because of intermittent incomplete combustion between relatively long-interval pulsations. The transient occurrence of CO does not represent a significant drawback of the pulsating combustion technique: in the experimental furnace the flue gas sample has been collected from the bottom of the combustion chamber in a high temperature zone (1300°C) quite close to the end of the flame. This sampling procedure does not provide enough residence time for the gases to allow for the eventual mixing of the rich and lean portions of the product gases. The CO formed by the very rich sequence of the pulsated flame reaches the sampling point before convertion to CO_2 by the excess O_2 available from the following lean sequence. In an industrial furnace, this limitation will not occur. Complete combustion (or post-combustion in this pulsated case) will take place within the combustion chamber itself, resulting perhaps in a relatively longer flame. Should change in flame geometry affect the thermal function of the flame, this low frequency value could be considered as a practical limit of the pulsating combustion technique. Because of the larger furnace-to-flame size ratio in the case of an industrial furnace, it is believed that the impact of the transient occurrence of CO on the heat transfer profile will be negligible. Figure 7 shows NO_x and CO profiles versus the pulsation frequency logarithm.

Similar experiments have been run when pulsating oxygen only. The same conclusion as for the case of natural gas only can be deduced from figure 8: a 60% NO_x reduction is obtained below the same "inflection point" pulsating frequency.

Both natural gas and oxygen pulsation.

Different dephasing between the two gas pulsations have been tested. A dephasing of 1/2 gives the best NO_x reduction. Up to 90% NO_x emission reduction is obtained, as

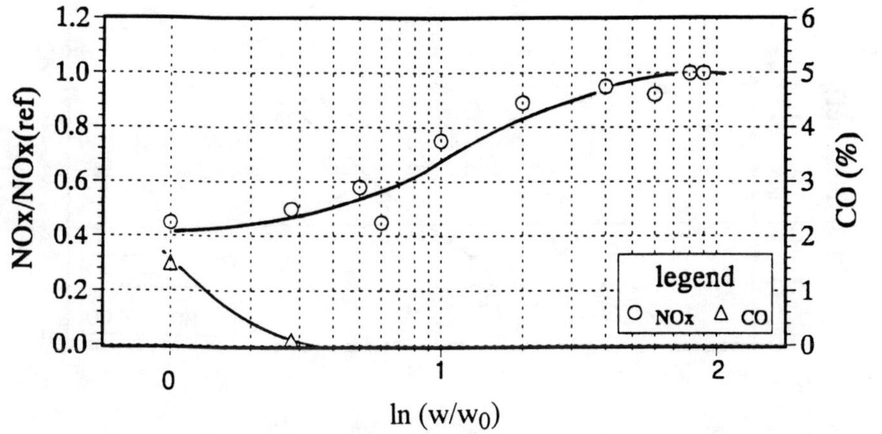

FIGURE 7 Relative NO_x reduction and CO profile from furnace no.1, when pulsating only natural gas. (due to confidentiality reasons, frequency is reported to a relative origin and expressed as w/w_0)

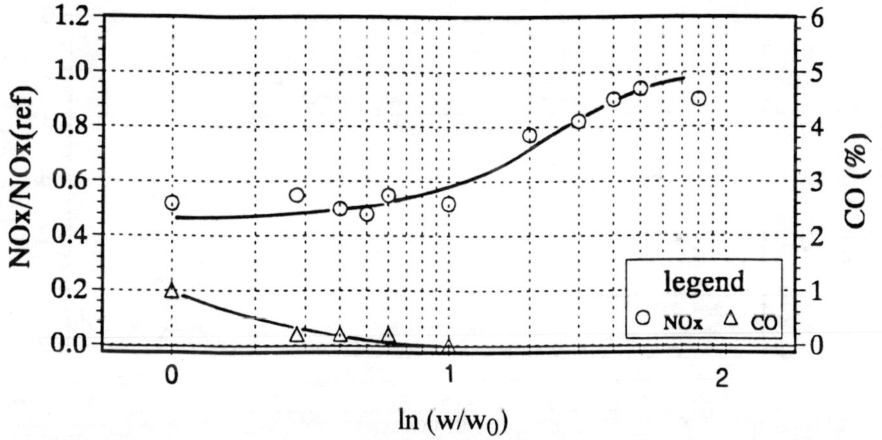

FIGURE 8 Relative NO_x reduction and CO profile from furnace no.1, when pulsating only oxygen. (due to confidentiality reasons, frequency is reported to a relative origin and expressed as w/w_0)

shown on figure 9. The value of the "inflection-point" frequency is of the same order as for the single gas pulsation configurations. Intermittent CO occurrence is likewise observed as pulsating frequency decreases. The inverse relation between NO_x formation and transient CO generation is evidenced throughout this study. At a constant C/P ratio, pulsating both gases with 1/2 dephasing in fact generates successive flames twice as rich and twice as lean, compared with one gas pulsation. As a consequence, NO_x reduction performances are better and transient CO occurrence greater.

Rotary-plug valve pulsation

A second set of experiments was carried out using a rotary-plug valve. These tests were performed in furnace no.2 at a firing rate of 50 kW and maintaining a temperature of 1500°C in the flue gas. With all other parameters kept constant, the higher refractory wall

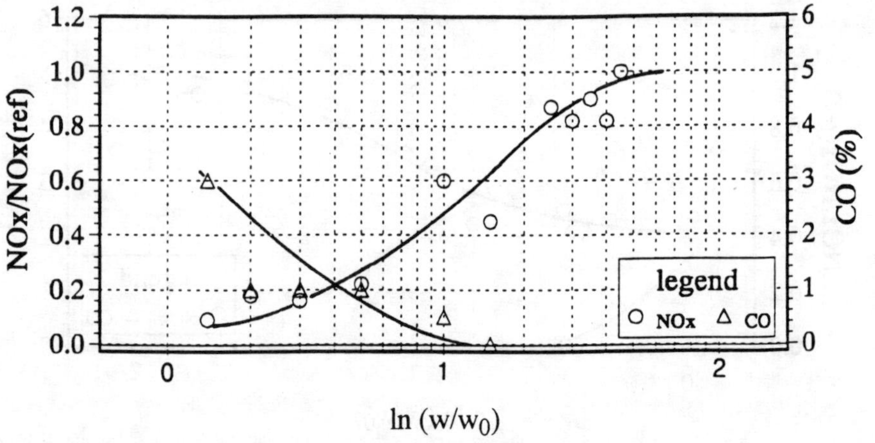

FIGURE 9 Relative NO_x reduction and CO profile from furnace no.1, when pulsating both natural gas and oxygen, dephasing equal to 1/2. (due to confidentiality reasons, frequency is reported to a relative origin and expressed as w/w_0)

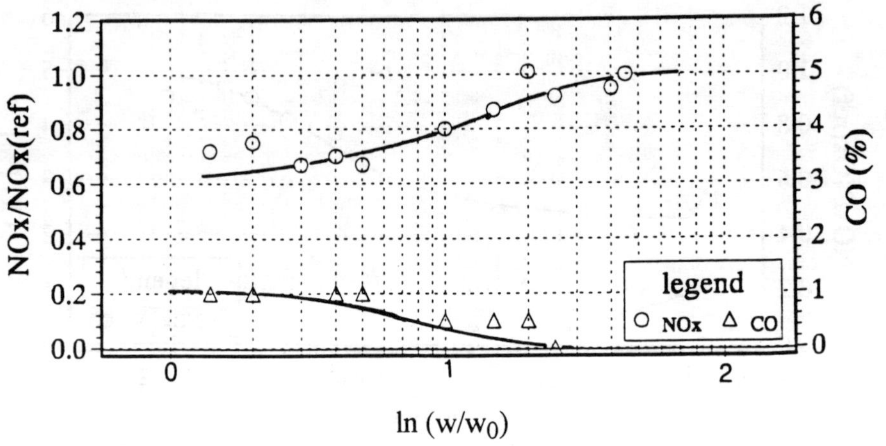

FIGURE 10 Relative NO_x reduction and CO profile from furnace no.2, when pulsating both natural gas and oxygen, dephasing equal to 1/2. (due to confidentiality reasons, frequency is reported to a relative origin and expressed as w/w_0)

and flue gas temperatures resulted in higher levels of NO_x emission. To avoid excessively high NO_x measurements, the N_2 content of natural gas was adjusted to 7% (instead of 15% in furnace no.1).

The results shown in Figure 10 indicate that the same trend prevails when compared with the solenoid valve results. However, a decrease in the NO_x emission reduction performance was observed. Instead of attaining up to 90% reduction, as for the case of the solenoid valve, only a limit of 40% NO_x emission reduction was possible with the rotary-plug valve. Equipment considerations are believed to be responsible for this limitation: the clearance between the rotating plug and body of the valve was too large,in such a manner that it was impossible to set a C/P ratio as low as desired. An improved system will be tested at a firing rate of 1 MW. The impact of the C/P ratio on NO_x emissions is illustrated in figure 11. The rotary-plug valve was used to study the

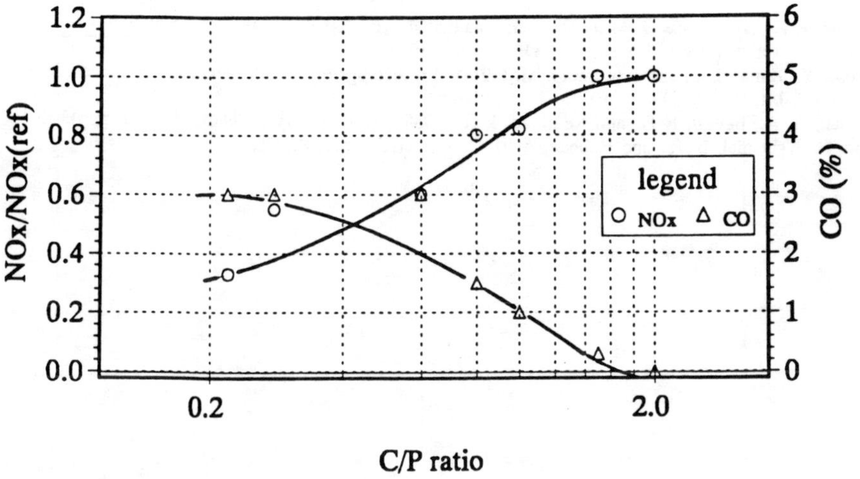

FIGURE 11 Relative NO_x reduction and CO profile from furnace no.2 versus C/P ratio, when pulsating both gas at constant frequency.

dependence of NO_x emissions on the C/P ratio at a constant frequency. After a decrease in NO_x concentration, a constant reduction of 40% was observed. Further decreases were unattainable as C could not be set as low as necessary.

CONCLUSION

Experiments have been run to evaluate the effect of pulsated combustion on NO_x emission reduction in natural gas/pure oxygen flames. Results show the effectiveness of the environmentally friendly technique. Up to 90% NO_x emissions reduction have been obtained, when pulsating both gases with a 1/2 dephasing. This study points out the importance of two main parameters: the pulsating frequency and the C/P (constant to pulsating fractions) ratio. This proprietary technique is being further developed by improving both the equipment as well as the thermal aspects of the operation. Tests on a 1 MW firing rate furnace are being carried out with specially designed rotary valves to study expected NO_x reduction and heat transfer improvement and, once all parameters are set, this technique would be transposed to an industrial unit.

The natural gas/pure oxygen flame is a primary low NO_x technology for very high temperature industrial processes, such as glass melting, for instance. However, when better performances have to be achieved under unfavourable conditions (such as high nitrogen content in natural gas), pulsating combustion is a simple technique which can yield up to 90% NO_x emission reduction. Recent pilot experiments show that these results can be obtained without significant changes in the heat transfer to the charge. The high overall flame temperature and heat transfer efficiency of the pure oxygen flame is maintained. There is no flame dilution or recirculation which may alter the heat transfer. Pulsated combustion represents a technical solution which benefits from all the advantages associated with pure oxygen-based combustion with low NO_x emissions.

REFERENCES

Al-Haddad, A. A., and Al-Binally, N. (1989). *Int. J. Heat and Fluid Flow*, **10**, no.2.

Briffa, E., and Staddon, P. (1982). Symposium on Pulsed Combustion Application, Proceedings vol. 1, Atlanta (Georgia).

Charon, O., et al. (1990). Proceedings of the AFRC Fall International Symposium, October 8-10 1990.

George, P. E., and Corliss, J. M. (1987). GRI annual report GRI 87/0279.
Hamid, A. Abbasi (1989). US Patent 4,846,665, July 1989.
Hirose, Y., Kannari, T., and Kawamoto, M. (1978). Proceedings of the 5th members conference, May 1978, Vol 2, International Flame Research Foundation.
Putman, A. A., Bellest, F. E., and Kentfield, J. A. (1986). *Prog. Energy Combust. Sci.*, **12**, pp 43-79.
Zinn, B. T., Daniel, B. R., and Rabhan, A. B. (1987). GRI report 87/0350.

A New Concept for the Reduction of Pollutant Emissions in Industrial Gas Turbine Plants

UMBERTO DESIDERI, BRUNO FACCHINI, and SERGIO S. STECCO
Department of Energy Engineering, University of Florence, Florence, Italy

Abstract — This paper examines the possibility of reducing pollutant emissions from industrial and ground-based aero-derivative gas turbines by means of a new oxygen enriched combustion system. The emissions considered pollutants are carbon monoxide (CO), unburned hydrocarbons (UHC), nitrogen oxides (NO_x) and smoke, whereas the presence of sulphur dioxide (SO_2) and heavy metals is disregarded because sulphur and metallic compounds are supposed to be removed from or not contained in the fuel. Various thermodynamic parameters, such as specific heat, adiabatic coefficient and diffusion constant are compared when oxygen is used instead of air in a combustion chamber, and a simplified combustion model based on the computation of chemical equilibria of 14 given species is used for predicting the main pollutants. A comparison of the compression power requirements for both pure oxygen and air is also made. The results show that, with suitable oxygen enrichment in gas turbine combustion chambers, a reduction of the major pollutant emissions is theoretically feasible with a combined increase in power production.

Key Words: Gas Turbines; Combustion; Oxygen Enrichment, Pollutant Emissions

INTRODUCTION

New regulations on maximum pollutant emissions are being approved by governments all over the world, and traditional means of pollution reduction are showing an increasing difficulty in meeting requirements of the new laws. Therefore, the authors believe that together with an improvement of the well known techniques, new concepts should also be studied and tested with comparable attention.

Elimination of pollutants such as sulphur dioxide and heavy metallic compounds is possible either through fuel selection itself or by chemical treatment aimed at separating undesired components from the fuel itself. Unburned hydrocarbons and organic compounds in exhaust gases can be reduced by an efficient combustion system, which at the same time helps to reduce carbon monoxide emissions. The presence of carbon monoxide is also due to the dissociation of carbon dioxide. Emissions of nitrogen oxides, where monoxide accounts for the most significant portion, can be avoided only if there is no nitrogen in the air.

The production of nitrogen oxides, NO_x, has been given a great importance owing to their effects on the environment and to the difficulty in eliminating them. Lefebvre (1983) pointed out that techniques used to reduce NO_x are often the reason for an increase in carbon monoxide emissions, so that the combined reduction of both NO_x and CO is much more difficult than reduction of only one.

Oxygen injection was tested in the past as a means to facilitate ignition at high altitudes in aircraft gas turbines (Lefebvre, 1983). The aim of this study is to show how it is possible to reduce emissions to very low levels by using, in the primary zone of the combustion chamber, pure oxygen or a mixture of 90% oxygen and 10% nitrogen which can also be a by-product in industries where nitrogen is produced and used as inert gas.

MAIN ASSUMPTIONS

Owing to the simplified analysis performed in this study (it is a preliminary paper on this subject), it was necessary to make several assumptions for which an explanation is of prime importance.

Reliable computer codes for the calculation of all the pollutant emissions from gas turbine combustors are not yet available and neither the theoretical nor the empirical approach to the problem have provided an adequate answer to pollutant prediction, even though some encouraging results have been published (Wild *et al.* 1987, Romberg, 1983, Mellor, 1983, Swithenbank *et al.* 1983, Jones and Mc Guirk, 1983, Hebrard and Magre, 1983). Moreover, most of the computer codes for modelling gas turbine combustors either require several simplifying assumptions or reveal limitations concerning the geometry of the combustion chamber and the number of chemical species considered.

One of the main advantages in terms of pollutant reduction that ground-based gas turbines have in comparison with aircraft gas turbines is the freedom in the choice of the fuel: if a gaseous fuel, such as natural gas, is used instead of a liquid fuel all the problems connected with injection and the presence of undesired compounds in the fuel are definitely overcome. Expensive techniques for eliminating sulphur or heavy metals are no longer necessary if natural gas is used. A simple premixed injector is sufficient to provide a good mixing of fuel and oxidant and therefore an increase in combustion efficiency.

Arrangements with a single combustion chamber, where it is possible to optimise geometry and flow distribution are not limited by space or weight restrictions.

Fluid dynamics and geometry of the combustor were not dealt with in the present paper, but will be the subject of further studies, while a model (Strehlow, 1985) for the combustion of a hydrocarbon with a mixture of oxygen, nitrogen and an inert species, if present, was tested for different oxygen/nitrogen ratios. Combustion calculations based on the mentioned model provide the equilibrium composition of up to 14 different species, which are in the order: H_2, O_2, H_2O, CO, CO_2, OH, H, O, N_2, N, NO, NO_2, CH_4 and an inert species.

Even though a combustion model based on chemical equilibrium and disregarding fluid dynamics of the gas turbine combustor is not sufficient to predict the combustor's performance and the pollutant emissions, it does, however, provide general indications on the influence of some important parameters, such as temperature, and on the minimum emissions possible.

The same was assumed when computing diffusion coefficients: turbulent mixing of oxygen and fuel may affect local composition of the mixture more than diffusion, but, as far as diffusion flames are concerned, a comparison is nonetheless important for drawing some conclusions.

The problems of producing and storing large quantities of oxygen, either pure or mixed with less than 10% nitrogen, are not dealt with, but the following assumptions were made:

- Oxygen is collected as a by-product in the nitrogen production process, as it is not required to be 100% pure as for laboratory use; it is thus not so expensive as commercial pure oxygen and large quantities are available. It is worth remembering that some types of industries, such as iron and steel factories, where nitrogen is often produced for internal use, often discharge considerable quantities of oxygen into the atmosphere because they do not need it and it is not pure enough to be sold;
- Oxygen is collected in pressurised tanks and less power is then necessary in the gas turbine compressor. Compression power for storage could be also provided by the gas turbine during off-peak hours.

SOME THERMODYNAMIC AND TRANSPORT PROPERTIES OF OXYGEN

Among the thermodynamic properties of a substance, specific heat and the adiabatic coefficient, defined by the c_p/c_v ratio, play a significant role in several calculations concerning work and heat transfer. Therefore it is necessary to use accurate equations to express the dependence of specific heat at constant pressure and at constant volume on temperature and pressure.

Two sets of equations for specific heat at constant pressure and at constant volume as function of the temperature were compared for oxygen and air (Reynolds, 1979, Reid *et al.* 1987). The dependence on pressure is neglected by both Reynolds and Reid.

Reid *et al.* (1987) use a third order polynomial equation to calculate c_p:

$$c_p = A_0 + A_1 T + A_2 T^2 + A_3 T^3 \tag{1}$$

for oxygen nitrogen methane. For air, instead, a weighted average of c_p for single components is used.

Reynolds (1979) bases his calculations on the specific heat at constant volume and the equations for oxygen and nitrogen are as follows:

$$c_v = \sum_{i=1}^{7} B_i T^{i-4} + B_8 e^{B_6/T} \left[\frac{B_9/T}{e^{B_9/T} - 1}\right]. \tag{2}$$

For methane a similar expression is reported by Reynolds (1979):

$$c_v = B_1 + B_2 T^{1/3} + B_3 T^{2/3} + B_4 T + +B_5 e^{B_6/T} \left[\frac{B_6/T}{e^{B_6/T} - 1}\right]. \tag{3}$$

The gas constants are fixed at 259.8 J/kgK for oxygen, 287.1 J/kgK for air and 518.3 for methane.

The specific heat at constant pressure and volume are connected by the relationship:

$$R = c_p - c_v, \tag{4}$$

but due to the different approach used by the two authors there is an error in the determination of the specific heat as computed from equations (1) (2) and (3) and from equation (4). Such error is lower than 2% in the 300 and 700 K temperature range, in all cases considered.

Calculation of the adiabatic coefficient and the sonic velocity defined as:

$$k = c_p/c_v, \tag{5}$$

$$a = \sqrt{kRT}, \tag{6}$$

based on equations (1) and (2) show an even smaller error.

Tables I and II, show some of the thermodynamic parameters such as specific heat, adiabatic coefficients and sonic velocity for oxygen and air with temperature ranging from 300 to 700 K.

Calculations were also made to determine the differences in compression power requirements if oxygen is used instead of air. Table III presents the results for a compression process with an adiabatic efficiency of 0.88 and compression ratios ranging from 8 to 17.

From the results shown in Table III, it is evident that, using oxygen instead of air, a temperature reduction between 15 and 27 K at compressor outlet and, as a consequence, a

TABLE I

SPECIFIC HEAT AT CONSTANT PRESSURE AND VOLUME, ADIABATIC COEFFICIENT AND SONIC VELOCITY FOR OXYGEN, BASED ON EQUATION (1)

T, K	k	c_p, kJ/kgK	c_v, kJ/kgK	a, m/s
300	1.394	0.919	0.659	330
350	1.387	0.931	0.671	355
400	1.380	0.944	0.685	379
450	1.372	0.959	0.699	401
500	1.364	0.973	0.713	421
550	1.357	0.988	0.728	440
600	1.350	1.003	0.743	459
650	1.343	1.018	0.758	476
700	1.337	1.032	0.772	493

TABLE II

SPECIFIC HEAT AT CONSTANT PRESSURE AND VOLUME, ADIABATIC COEFFICIENT AND SONIC VELOCITY FOR AIR, BASED ON EQUATION (1)

T, K	k	c_p, kJ/kgK	c_v, kJ/kgK	a, m/s
300	1.394	1.016	0.729	347
350	1.392	1.019	0.731	374
400	1.390	1.024	0.737	400
450	1.386	1.031	0.744	423
500	1.382	1.039	0.752	445
550	1.377	1.049	0.762	466
600	1.371	1.060	0.773	486
650	1.366	1.072	0.785	505
700	1.360	1.085	0.798	523

notable saving in the work required are possible. It is also important to note that a lower temperature at the combustor inlet has a positive effect on NO_x reduction.

Another aspect of the characteristics of the fluid which was taken into account is the diffusion coefficient, because most of the flames in gas turbine combustors are diffusion flames. As mentioned earlier in this paper, turbulent mixing can affect the local composition of the flame more than the diffusivity, but the diffusion process has a considerable importance.

To calculate diffusion coefficients of the oxygen–methane and the air–methane systems three different methods (Reid *et al.* 1987) were examined:

— Chapman–Enskog method
— Wilke–Lee method
— Fuller method

in the temperature range from 300 to 700 K and with a pressure of 1 MPa.

TABLE III

COMPARISON BETWEEN ISENTROPIC AND REAL FINAL TEMPERATURES AND COMPRESSION POWER WHEN THE FLUID IS OXYGEN AND AIR

	Oxygen				Air				
β	T_{2is}, K	W_{is} kJ/kg	T, K	W, kJ/kg	T_{2is}	W_{is}	T	W	ΔW
8	517.9	207.4	546.8	235.7	530.2	237.6	561.3	270.0	34.3
9	532.9	222.1	563.6	252.3	546.7	254.7	579.9	289.5	37.2
10	546.5	235.4	578.9	267.5	561.7	270.4	396.9	307.3	39.8
11	559.0	247.7	593.0	281.5	575.5	284.9	612.5	323.7	42.2
12	570.6	259.2	606.0	294.5	588.3	298.3	626.9	339.0	44.4
13	581.3	269.9	618.1	306.7	600.3	310.9	640.4	353.3	46.5
14	591.4	280.0	629.4	318.2	611.5	322.6	653.0	366.6	48.5
15	600.9	289.5	640.0	329.0	622.0	333.8	664.9	379.3	50.3
16	609.9	298.5	650.1	339.2	631.9	344.3	676.1	391.2	52.1
17	618.4	307.1	659.6	348.9	641.4	354.3	686.7	402.6	53.7

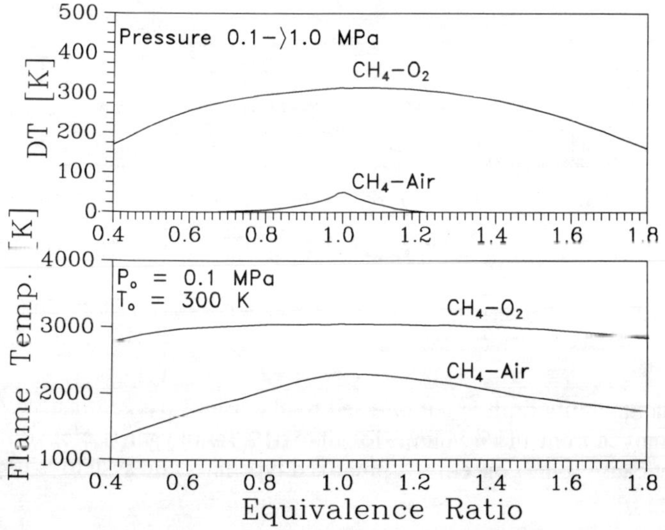

FIGURE 1 Flame temperature for O_2–CH_4 and air–CH_4 combustion at 0.1 MPa and temperature increment when pressure is raised from 0.1 to 1.0 MPa

All the three methods, gave very similar results, even if Fuller's method underestimates the diffusion coefficients in comparison with the Chapman–Enskog and Wilke–Lee methods. The computations gave a difference of less than 5% in the diffusion coefficients of the oxygen–methane and the air–methane systems.

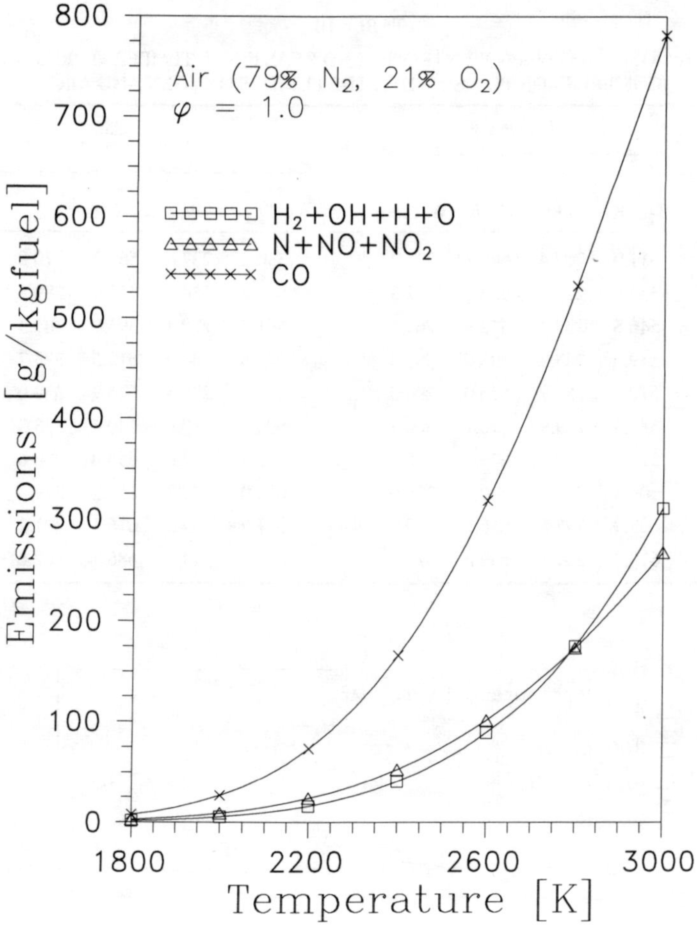

FIGURE 2 Emissions from the combustion of methane with air and $\phi = 1.0$.

The weak flammability limit is not affected by the use of oxygen instead of air, while the rich one is moved from a 15% volume for air to 61% for oxygen (Lefebvre, 1983). That, however, might cause some concern if a premixed type injection system is used.

PREDICTION OF POLLUTANT EMISSIONS

Prediction of major pollutant emissions was performed by using a computer programme by Weinberg (Strehlow, 1984) which makes it possible to compute chemical equilibria of 14 chemical species at given temperature, pressure and reactant compositions. Results obtained with the use of such a programme disregard all the phenomena whose driving process is chemical kinetics or fluid dynamics, but it is interesting to have an estimate of pollutant emissions by means of chemical equilibrium. In fact, chemical equilibrium gives an indication of the minimum pollutant emissions that a gas turbine combustor can produce, as in the real phenomena a significant increase in pollutants is to be expected. Chemical species considered in the programme are H_2, O_2, H_2O, CO, CO_2, OH, H, O, N, N_2, NO, NO_2, CH_4 and an inert species.

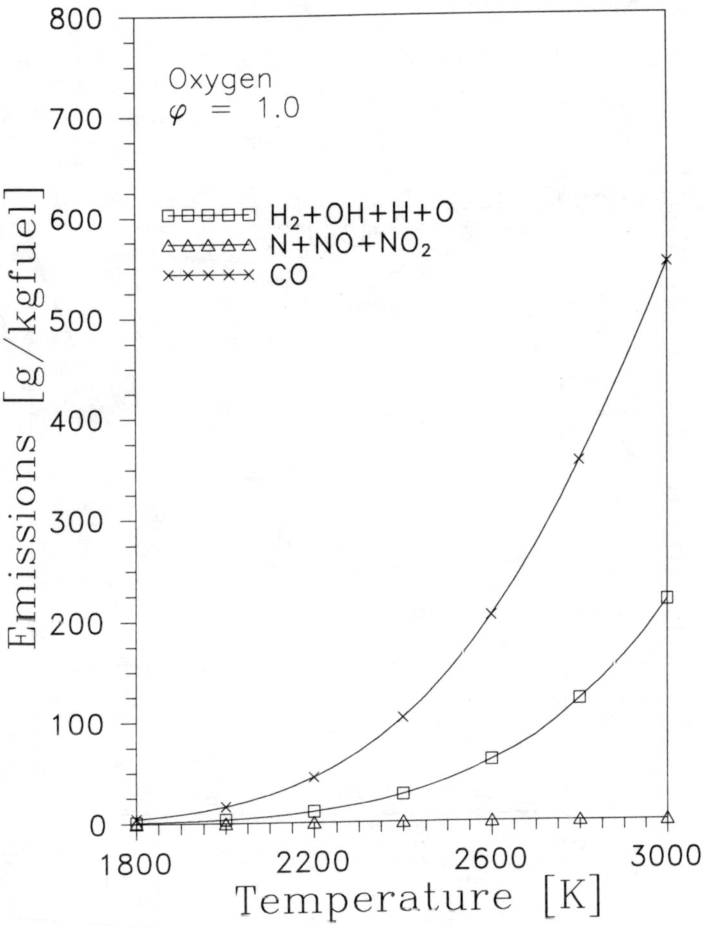

FIGURE 3 Emissions from the combustion of methane with pure oxygen and $\phi = 1.0$.

The programme gives accurate results for the following ranges of temperature, pressure and equivalence ratio:

— $700 < T < 4700$ K

— 5 kPa $< p <$ 1 MPa

— $0.3 < \phi < 2$

In this study pressure was always kept at 1 MPa, because its variations had a very small influence on the pollutant emissions. An analysis was carried out with temperature varying from 1800 to 3000 K which is the range of possible temperatures for the oxygen–methane and air–methane flames (fig. 1, data from Strehlow, 1984) when the equivalence ratio, ϕ, is kept between 0.8 and 1.2, as is usual in gas turbine combustors.

Two different equivalence ratios were considered in this analysis: 0.8 and 1.0. In fact, Lefebvre (1983) states that a reduction in CO emissions is possible with an equivalence ratio of around 0.8, instead of the stoichiometric 1.0 value. It is thus interesting to test not only the 1.0 equivalence ratio but also the 0.8.

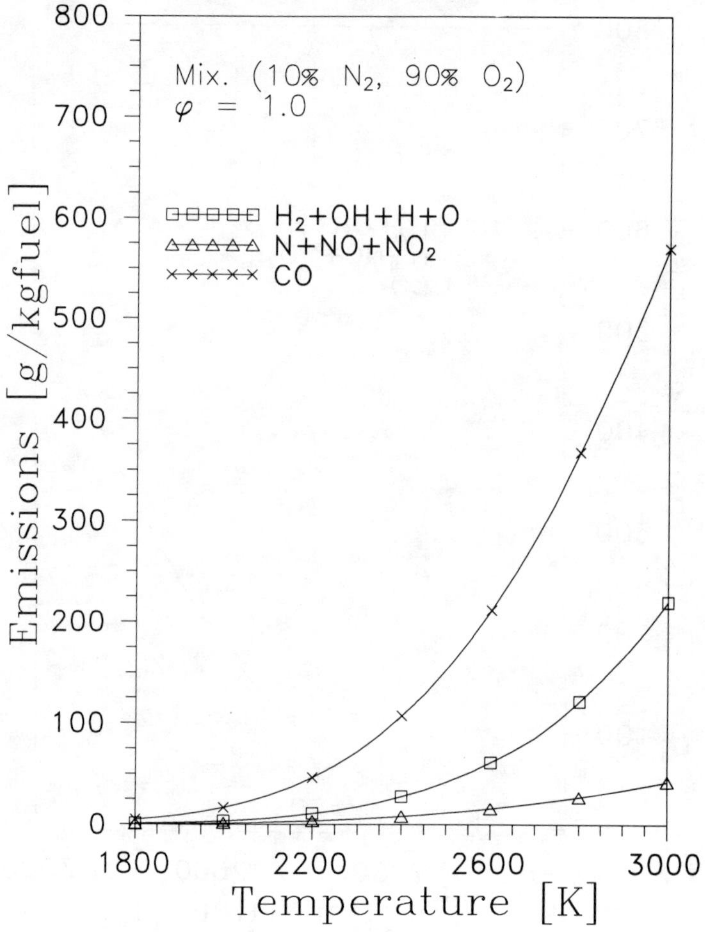

FIGURE 4 Emissions from the combustion of methane with a 90% oxygen–10% nitrogen mixture and ϕ – 1.0.

Three compositions of the oxidant mixture were object of this research: pure oxygen, a mixture with 90% volume oxygen and 10% nitrogen and air with 21% volume oxygen and 79% nitrogen.

The undesired emissions are added in three different groups:

— CO
— Nitrogen compounds: $NO + NO_2 + N$
— Hydrogen compounds: $H_2 + OH + H + O$

and expressed in g/kg_{fuel} in order to simplify the graphs of the results shown in the following part of this paper. It is, however, important to mention that, among the nitrogen compounds NO is definitely the greatest contributor to the total amount of nitrogen emissions, while H_2 has the highest percentage in the hydrogen compounds.

Combustion of methane in air with $\phi = 1.0$ produces emissions as shown in figure 2. There is a considerable amount of CO, due to carbon dioxide dissociation, while nitrogen and hydrogen emissions are almost on the same level.

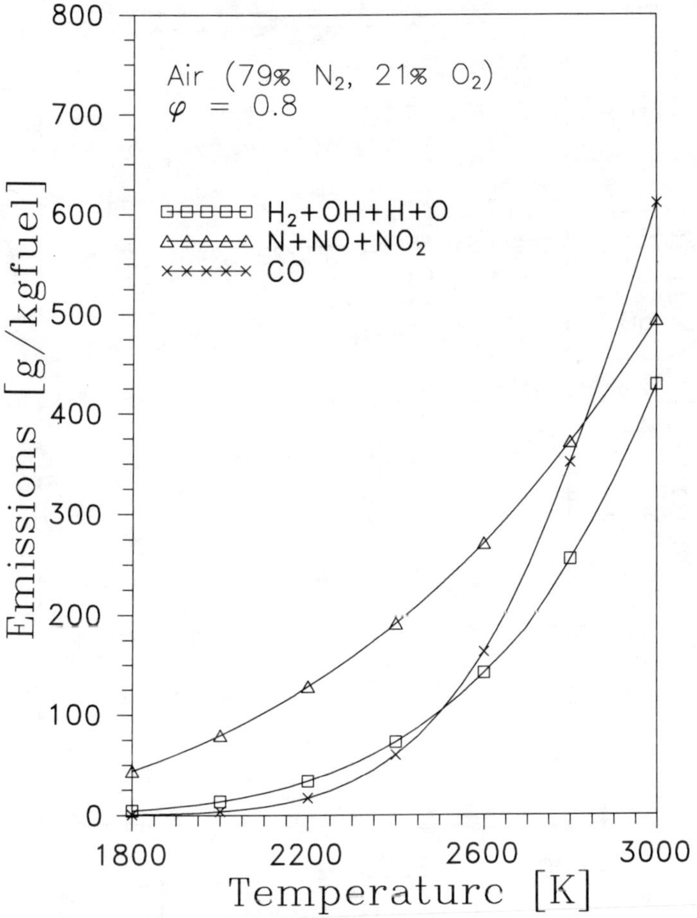

FIGURE 5 Emissions from the combustion of methane with air and $\phi = 0.8$.

Combustion with pure oxygen, decreases the CO emissions by almost 25% and does not produce any nitrogen compounds. A slight decrease is also observable in the hydrogen emissions (fig. 3).

The oxygen–nitrogen mixture with 90% O_2, does not increase CO and hydrogen emissions as compared with the pure oxygen case, while the nitrogen oxides emissions are quite low and almost undetectable for temperatures below 2400 K (fig. 4).

If the equivalence ratio is lowered to 0.8, major changes in the relative amounts of the three groups of emissions are evident. Figure 5 shows the emissions with air as oxidant: CO is lowered to the levels obtained with combustion with pure oxygen and $\phi = 1.0$, but nitrogen emissions are even higher than the CO emissions for most of the temperature range. The absence of nitrogen in the burning mixture (fig. 6), changes the emissions in the same way as when ϕ is equal to 1.0, keeping the sum of hydrogen derived emissions almost constant, and lowering CO emissions by more than 25%.

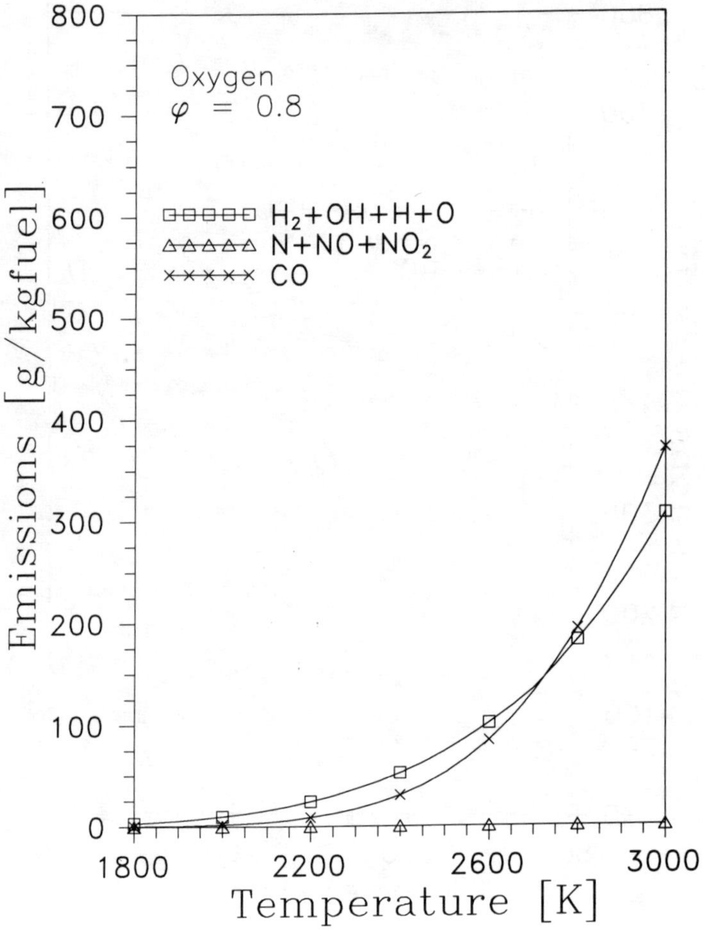

FIGURE 6 Emissions from the combustion of methane with pure oxygen and $\phi = 0.8$.

The oxygen–nitrogen mixture with 90% O_2, gives lower CO emissions and higher NO emissions in comparison with the $\phi = 1.0$ case (fig. 7).

Therefore, from the standpoint of reducing pollutant emissions, using pure oxygen or oxygen enriched air instead of air in gas turbine combustion chambers offers a big theoretical advantage. CO reduction is in the order of 25–30%, while NO_x is practically eliminated at the same temperature levels.

However it must be pointed out that if pure oxygen is used in the primary zone of a combustion chamber instead of air, the adiabatic flame temperature, is much higher as previously shown in fig. 1. As a consequence, the primary zone temperature is very high and mixing the combustion products at such a high temperature with the dilution air would produce an amount of pollutants of the same order of magnitude as common combustion with air. In fact the emissions at the flame temperatures for the stoichiometric combustion of methane with pure oxygen (3350 K) and air (2350 K), are much higher for the combustion with pure oxygen, as can also be seen on figs. 2 and 3.

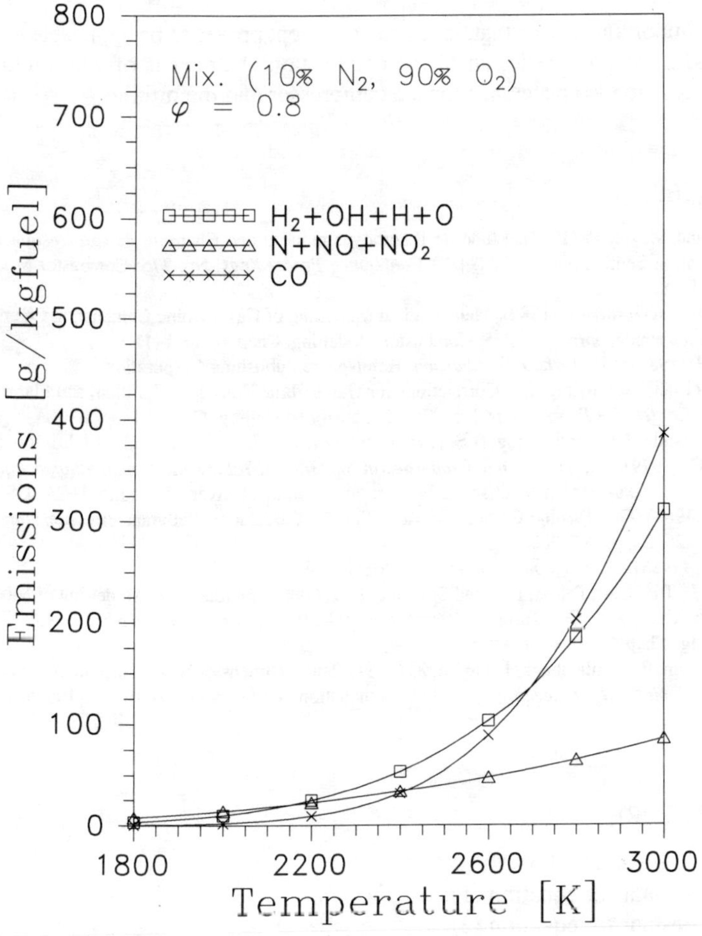

FIGURE 7 Emissions from the combustion of methane with a 90% oxygen–10% nitrogen mixture and $\phi = 0.8$.

It is therefore necessary not only to burn the fuel with pure oxygen in the primary zone but also to perform the first part of the dilution with an oxygen rich mixture, until a temperature as low as 1800 K is reached.

CONCLUSIONS

Combustion of a fuel with oxygen or an oxygen rich mixture in the primary zone of a gas turbine combustor is then a feasible way to reduce pollutant emissions, with the additional advantage of lowering the compression power. However, the use of such a system requires further studies to develop a more accurate model of the oxygen-fuel flame both for chemistry and fluid dynamics and then the combustion process in a real combustion chamber.

Nor can we overlook the technological problem related to the need for large storage tanks for the combustion oxygen and for a separate compression system.

It is also important to note that, if this new concept proves to be applicable to gas turbine combustors, a complete redesign of the combustion chamber is of vital importance, and major changes are also necessary for the compressor and the turbine.

REFERENCES

Hebrard, P., and Magre, P. (1983). Étude de l'Aerodynamique d'une Chambre de Combustion en Vue d'une Modelisation Semi-Empirique. *AGARD Conference Proceedings*, No. 275: Combustor Modelling. Chap. 26, pp. 1–18.

Jones, W.P., and McGuirk, J.J. (1983). Mathematical Modelling of Gas-Turbine Combustion Chambers. *AGARD Conference Proceedings*, No. 275: Combustor Modelling. Chap. 4, pp. 1–11.

Lefebvre, A.H. (1983). *Gas Turbine Combustion*, Hemisphere Publishing Corporation.

Mellor, A.M. (1983). Semi-Empirical Correlations for Gas Turbine Emissions, Ignition, and Flame Stabilisation. *AGARD Conference Proceedings*, No. 275: Combustor Modelling. Chap. 24, pp. 1–13.

Reid, R.C., Prausnitz, J.M., and Poling, B.E. (1987). *Properties of Gases and Liquids*, 4th Edition, McGraw-Hill.

Reynolds, W.C. (1979). *Thermodynamic Properties in SI: Graphs, Tables and Computational Equations for 40 Substances*, Department of Mechanical Engineering, Stanford University, Stanford, CA, USA.

Romberg, G. (1983). Gas Turbine Combustor Modelling for Calculating Pollutant Emission. *Int. J. Heat Mass Transfer*. **26**, 2, pp. 197–209.

Strehlow, R.A. (1985). *Combustion Fundamentals*, McGraw-Hill.

Swithenbank, J., Turan, A., Felton, P.G. and Spalding, D.B. (1983). Fundamental Modelling of Mixing, Evaporation and Kinetics in Gas Turbine Combustors. *AGARD Conference Proceedings*, No. 275: Combustor Modelling. Chap. 2, pp. 1–21.

Wild, P.N., Boysan, F., Swithenbank, J. and Lu, X. (1987). Three-Dimensional Gas Turbine Combustor Modelling. *AGARD Conference Proceedings*, No. 422: Combustion and Fuels in Gas Turbine Engines. Chap. 27, pp. 1–13.

NOMENCLATURE

a	sonic velocity [m/s]
A_i	constant for equation (1)
B_i	constant for equation (2)
c	specific heat [kJ/kgK]
k	adiabatic coefficient
p	pressure [MPa]
R	gas constant [kJ/kgK]
T	temperature [K]
W	power [kW]
α	fuel-air ratio
β	compression ratio
ϕ	equivalence ratio = a/a_{st}

SUBSCRIPTS

is	isentropic
p	constant pressure
st	stoichiometric
v	constant volume

APPENDIX

The constants used in equation (1) are the following (Reid *et al.* 1987):

	O_2	N_2	CH_4
A_0	2.811×10^{1}	3.115×10^{1}	1.925×10^{1}
A_1	-3.680×10^{-6}	-1.357×10^{-2}	5.213×10^{-2}
A_2	1.746×10^{-5}	2.680×10^{-5}	1.197×10^{-5}
A_3	-1.065×10^{-8}	-1.168×10^{-8}	-1.132×10^{-8}

The constants used in equations (2) and (3) are the following (Reynolds, 1979):

	O_2	N_2	CH_4
B_1	-1.29443×10^{6}	-2.18203×10^{5}	1.34761×10^{3}
B_2	5.98232×10^{4}	1.01574×10^{4}	1.35512×10^{2}
B_3	-8.97851×10^{2}	-1.65505×10^{2}	-2.93910×10^{1}
B_4	6.55236×10^{2}	7.43176×10^{2}	2.12775
B_5	-1.13131×10^{-2}	-5.14606×10^{-3}	2.44657×10^{3}
B_6	3.49811×10^{-6}	5.18347×10^{-6}	2009.152
B_7	4.21065×10^{-9}	-1.05922×10^{-9}	
B_8	2.67997×10^{2}	2.8398×10^{2}	
B_9	2239.18105	3353.40610	

Development of New SCR Process for Gas Turbine Cogeneration

K. KOBAYASHI, S. OKAZAKI, and Y. YAMADA *Tokyo Gas Co., Ltd. Production Engineering Development Center, Tsurumi-ku, Yokohama, Japan*
S. MITSUMA *Tokyo Gas Co., Ltd. Air Conditioning Customer's Development and Service Department, Minato-ku, Tokyo, Japan*

Abstract — Cogeneration systems using gas turbines (about 1000 to 4000 kW) and gas engines (about 15 to 4000 kW) are becoming increasingly common for the reason that they make efficient use of energy and are economical. At the same time, regulations concerning atmospheric pollutants in the exhaust gas of such systems have become very stringent. In order to meet the new NO_x regulations, the development of a simple Selective Catalytic Reduction (SCR) process using ammonia in addition to the process of steam or water injection has become essential. We designed a verification process and conducted experiments using ammonia water and ammonium hydrocarbonate aqueous solution in place of liquid ammonia as the reducing agent. We were able to confirm outstanding results with regard to simplicity, economy and ease of handling.

Key Words: NO_x, SCR, ammonia water, ammonium hydrocarbonate aqueous solution, cogeneration, gas turbine.

INTRODUCTION

In Japan, emission standards have been set for gas turbines and gas engines used for cogeneration, independent from automotive emission standards. These standards include NO_x regulations which are made even stricter in large cities such as Tokyo where local governments implement administrative guidance for restriction on top of national standards.

Until now, water or steam injection has been used exclusively for meeting NO_x regulations. However, as NO_x standards become tougher, we expect that these technologies alone will be insufficient to meet the new NO_x standards. A more sophisticated technology for removing NO_x called selective catalytic reduction (SCR) which uses liquid ammonia as a reducing agent is used in applications such as thermal power plants. However, when this technology is applied on a cogeneration scale, problems arise involving the management of liquid ammonia.

In order to meet future NO_x regulations, ammonia water and ammonium hydrocarbonate aqueous solution were used. Both have less legal restrictions and are easier to handle than liquid ammonia as the reducing agent and were used in the research to develop a new SCR process for cogeneration-use gas turbines.

EXPERIMENTAL

Selection of Reducing Agent

For our experiments ammonia water and ammonium hydrocarbonate aqueous solution were selected in place of liquid ammonia as a reducing agent on the basis of economy and ease of handling. Table I compares the respective legal restrictions and handling characteristics of liquid ammonia, ammonia water and ammonium hydrocarbonate aqueous solution.

TABLE I

LEGAL RESTRICTIONS AND HANDLING CHARACTERISTIC OF REDUCING AGENTS

Reducing agent	Liquid ammonia	Ammonia water	Ammonium hydrocarbonate aqueous solution
Legal restrictions	Numerous	Few	None
Handling characteristic	Poisonous flammable	Strong odor	Harmless, odorless, low solubility

Ammonia water, unlike liquid ammonia, has few legal restrictions. Ammonium hydrocarbonate aqueous solution has ever fewer legal restrictions because it is a harmless, odorless substance.

A 25 wt% solution of ammonia water was used with the most widely distributed concentration and a 10 wt% solution of ammonia hydrocarbonate aqueous solution for reasons involving the solubility of this substance.

Test Equipment

In these tests, a 1000 kW class gas turbine (Gaspower 1000 built by Mitsui Engineering and Shipbuilding Co.) was used. This gas turbine is equipped with a NO_x reduction device using water injection. An attempt to obtain the best results possible by combining water injection with the SCR method was undertaken.

Figure 1 shows the flow of the SCR test equipment. The reducing agents, ammonia water and ammonium hydrocarbonate aqueous solution, were pumped from a storage tank and introduced in liquid form into the exhaust gas duct immediately after the gas turbine outlet. The fluids were introduced using a two-fluid nozzle which atomizes with air. This nozzle was chosen on the basis of its atomization characteristics. These nozzles were installed in the exhaust gas duct immediately after the gas turbine outlet. When the 25 wt% ammonia water solution was used as the reducing agent, the solution was diluted with water by a factor of ten in order to satisfy the minimum requirements of nozzles.

Ammonia, generated by evaporation and decomposition of the reducing agents, was made to react with NO_x in a catalyst which was installed between two levels of exhaust boilers in the exhaust gas duct. The SCR process is represented in the following chemical formula:

$$4NO + 4NH_3 + O_2 \rightarrow 4N_2 + 6\ H_2O.$$

The concentration of NO_x was controlled through the feedforward factor by adjusting the quantity of reducing agent introduced and the feedback factor by measuring the NO_x concentration at the catalyst outlet. Adjustments in pump motor revolutions and, consequently, the quantity of reducing agent introduced were made based on these feedback and feedforward factors. Another important consideration was that many cogeneration systems are placed in enclosed spaces such as underground rooms, making the problem of leaking odours troublesome. For this reason, a diaphragm-type pump was chosen to supply the reducing agent. In addition, a water-tight sealpot was installed as a countermeasure to breathing due to temperature fluctuations and to prevent odor leakage.

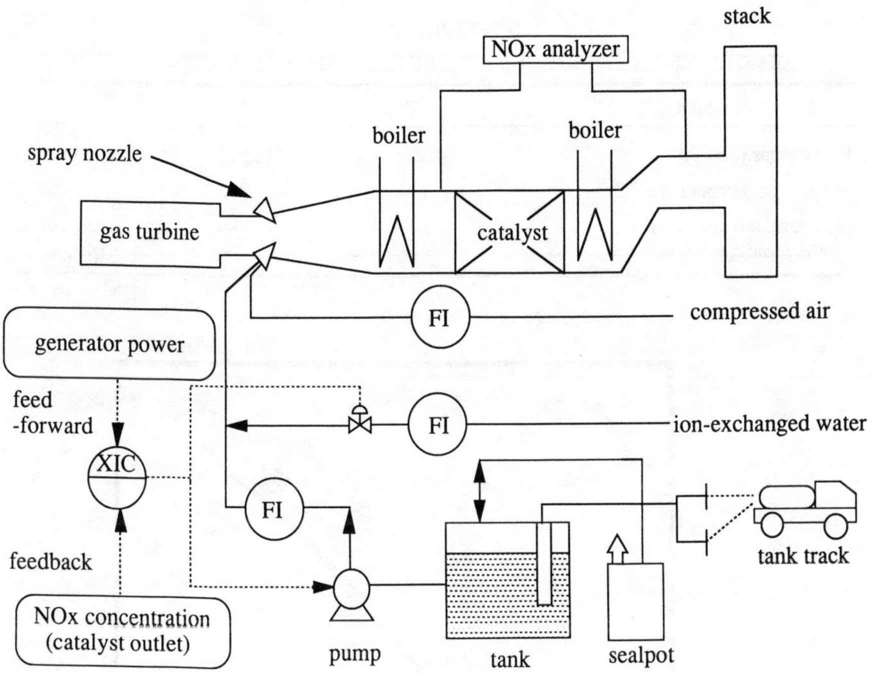

FIGURE 1 Process flow diagram.

Catalyst

The Honeycomb catalyst (TiO_2–V_2O_5–WO_3) which is readily available on the market was used as the catalyst in these experiments. The catalyst which operates at a GHSV of 17000/h was designed so that NO_x concentration after reaction was under 60 ppm (at 0% O_2) and ammonia leakage was controlled to under 10 ppm with consideration given to the problem of odours.

Analysis of Exhaust Gas

NO_x concentration was measured using a chemiluminescence NO_x analyzer and O_2 concentration was measured using a paramagnetic analyzer at the catalyst inlet and outlet. The NO_x reduction rate is defined in the following formula: NO_x reduction rate (%) = [(NO_x concentration at catalyst inlet (ppm)) – (NO_x concentration at catalyst outlet (ppm))]/(NO_x concentration at catalyst inlet (ppm)) × 100. In addition, leaked ammonia was measured using a gas detector tube.

Experimental Conditions

a) gas turbine output: 920 (applies only to cases where ammonia water was used), 690, 460 kW
b) NH_3/NO_x mole ratio: 0.4–1.3
c) W/F (water/fuel ratio during water injection): 0.5

Table II shows the flow volume, temperature and NO_x concentration of exhaust gas at the catalyst inlet for various levels of gas turbine output.

TABLE II
EXPERIMENTAL CONDITIONS WITH RESPECT TO GAS TURBINE OUTPUT

Gas turbine output (kW)	920	690	460
Exhaust gas (Nm^3/h)	14,228	14,579	14,809
Exhaust gas temperature (°C)	340	285	250
NO_x concentration at catalyst inlet (ppm) (at 0% O_2)	130	90	65

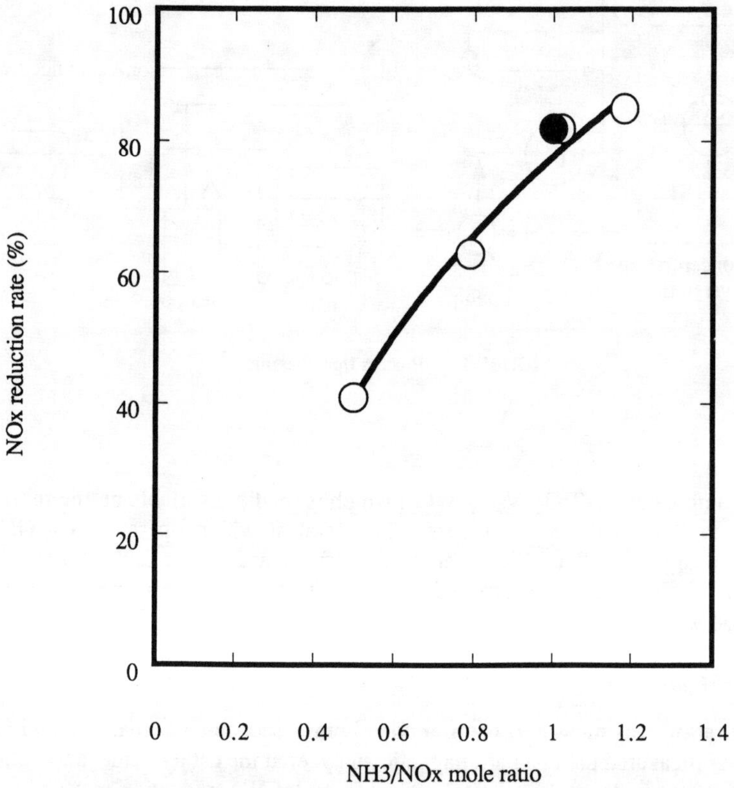

FIGURE 2 Relationship between NH_3/NO_x mole ratio and NO_x reduction rate for a gas turbine output of 920 kW (reducing agent: ○ = ammonia water, ● = designed value).

RESULTS AND DISCUSSION

Efficiency of Reducing Agent Atomization

In order for the ammonia to react efficiently with the NO_x, once the reducing agents are introduced into the duct, they must evaporate or decompose before they reach the catalyst. Furthermore, the ammonia must scatter uniformly throughout the duct. In these experiments, an evaluation was made as to whether or not the ammonia had scattered sufficiently

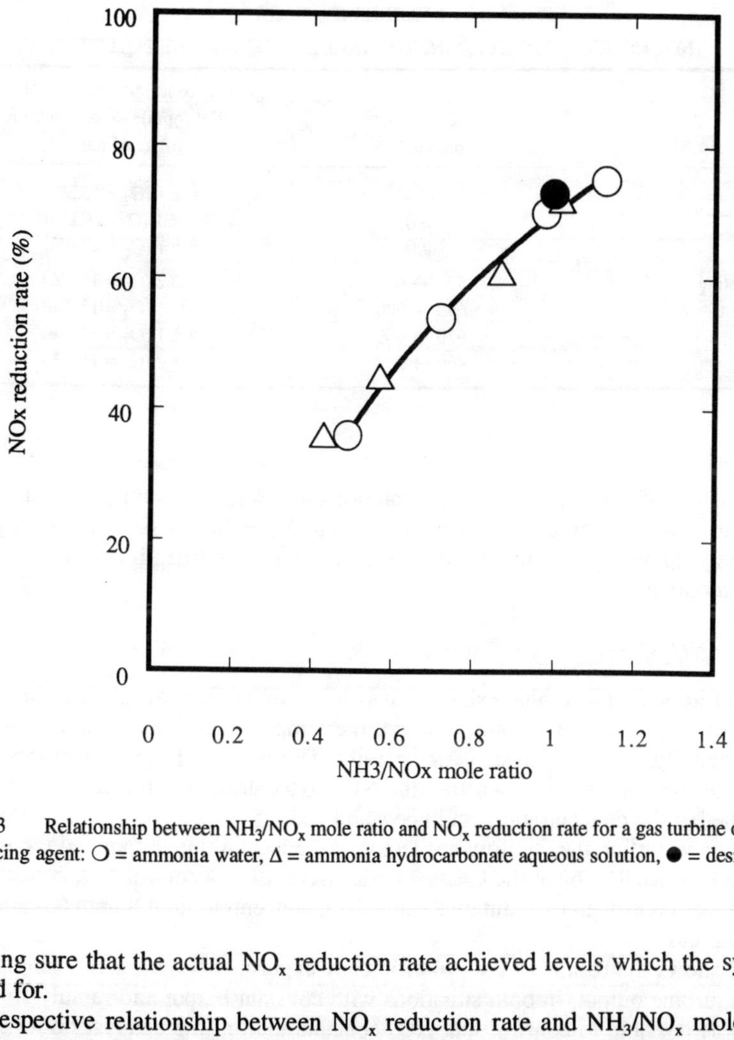

FIGURE 3 Relationship between NH_3/NO_x mole ratio and NO_x reduction rate for a gas turbine output of 690 kW (reducing agent: ○ = ammonia water, Δ = ammonia hydrocarbonate aqueous solution, ● = designed value).

by making sure that the actual NO_x reduction rate achieved levels which the system was designed for.

The respective relationship between NO_x reduction rate and NH_3/NO_x mole ratio for ammonia water and ammonium hydrocarbonate aqueous solution was examined by keeping the gas turbine output constant while adjusting the NH_3/NO_x mole ratio. Figures 2 and 3 represent the relationship between the NH_3/NO_x mole ratio and NO_x reduction rate for gas turbine outputs of 920 and 690 kW, respectively. When ammonia water was used as the reducing agent, NO_x reduction rates of approximately 80% for a gas turbine output of 920 kW and approximately 70% for a gas turbine output of 690 kW were achieved with an NH_3/NO_x mole ratio of 1. These results were close to values that the system was designed for.

When ammonium hydrocarbonate aqueous solution was used as the reducing agent, similar NO_x reduction rates were obtained, regardless of the fact that ammonium hydrocarbonate aqueous solution requires more time to decompose than does ammonia water. In addition, designed values were also roughly achieved using both types of reducing agents when the gas turbine output was 460 kW. In this manner it was possible to confirm that the ammonia generated through the evaporation and decomposition of reducing agents had scattered uniformly by the time it had reached the catalyst, showing that the atomization procedure was an appropriate method of introducing reducing agent.

TABLE III

NO_X CONCENTRATION CONTROL VALUES AT CATALYST OUTLET

	Gas turbine output (kW)	NO_x concentration at catalyst outlet (ppm) (O_2 concentration of actual gas)
Constant output	920	8 ± 1 (O_2 = 15.3%)
	690	8 ± 1 (O_2= 16.1%)
	460	8 ± 1 (O_2 = 17.0%)
Fluctuating output	920 → 690	8 ± 2 (O_2 = 16.1%)
	690 → 920	8 ± 1 (O_2 = 15.3%)
	920 → 460	8 ± 1 (O_2 = 17.0%)
	460 → 920	8 ± 2 (O_2 = 16.8%)

The influence of pressure loss in the catalyst on gas turbine output was a potential problem. However it was confirmed that the actual pressure loss in the catalyst was approximately 37 mm_{aq}, a value below design estimates and negligible in terms of its influence on output.

Efficiency of NO_x Concentration Control

NO_x concentration in gas turbine exhaust gas is in continual flux, even when the output is constant. When gas turbine output is varied, the temperature of the exhaust gas, which influences activity of the catalyst, varies greatly. Through the following process an examination was undertaken as to whether the NO_x concentration at the catalyst outlet was low and whether the concentration could be stabilized.

As mentioned earlier, the feedforward factor of the gas turbine output and the feedback factor of NO_x concentration at the catalyst outlet were used to control NO_x concentration. The system was controlled to maintain a stable NO_x concentration of 8 ppm (O_2 concentration of actual gas).

Table III shows the results of control efficiency experiments during constant and fluctuating gas turbine output. In both situations with constant output and rapidly fluctuating output, it was possible to confirm that NO_x concentration at the catalyst outlet could be consistently controlled to under 50 ppm (at 0% O_2), below the estimated design value. This was notwithstanding the more than one minute lag in NO_x measurement and the change in reaction speed due to fluctuations in the exhaust gas temperature.

Quantity of Leaked Ammonia

The emission of ammonia which did not contribute to the SCR process is a serious problem. Ordinances in Tokyo regulate the concentration of ammonia in exhaust gas to under 50 ppm.

In the system under investigation the worry was that ammonia molecules which had adsorbed in the catalyst would be desorbed with a temperature increase in the catalyst, increasing the ammonia concentration in the exhaust gas. This was most evident when the gas turbine output fluctuated, particularly when it increased. Therefore, during the NO_x concentration control efficiency experiments, the quantity of leaked ammonia which had not undergone the SCR reaction was also confirmed. In addition, an investigation of the

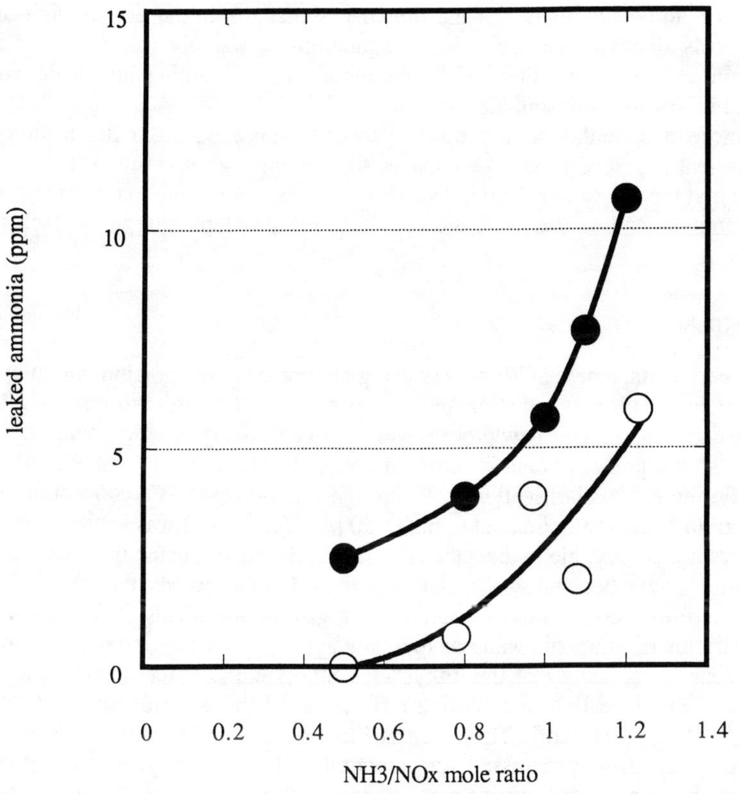

FIGURE 4 Relationship between NH_3/NO_x mole ratio and quantity of leaked ammonia for a gas turbine output of 920 kW (O = actual measured values using a gas detector tube; ● = calculated value using NH_3/NO_x mole ratio and NO_x reduction rate).

relationship between NH_3/NO_x mole ratio and quantity of leaked ammonia during constant gas turbine output in atomization efficiency experiments was undertaken.

In these output fluctuation experiments, a maximum ammonia leakage of 3 ppm was confirmed only when output was increased from 460 to 920 kW. As shown in Figure 4, when output is kept constant and the NH_3/NO_x ratio is increased, an ammonia leakage of 6 ppm was confirmed at the NH_3/NO_x ratio value of 1.3.

Under normal operating conditions where the NH_3/NO_x mole ratio is under 1, it was possible to keep ammonia leakage under 5 ppm.

Through the above results, it was found that ammonia leakage would not be problem because the quantity of leaked ammonia was under 10 ppm in all situations.

Response to Legal Restrictions and Odours

25 wt% ammonia water is a substance which falls under many legal restrictions, making it essential to pay a careful attention to odours and handling upon delivery. However with appropriate preparation, there were no problems such as leaking odours in ordinary use and delivery handling. There were similarly no problems with ammonium hydrocarbonate aqueous solution. Also long-run test results indicated that mechanical troubles caused by these reducing agents are highly unlikely.

The use of liquid ammonia has the drawbacks that the initial cost is high and that personnel costs also rise because of legal regulations. It was concluded that for SCR in cogeneration-scale systems, the use of ammonia water or ammonium hydrocarbonate aqueous solution was more suitable.

Furthermore, despite the fact that running costs are more expensive in situations where SCR equipment is placed in enclosed spaces such as underground rooms, the conclusion drawn was that the use of ammonium hydrocarbonate aqueous solution was more suitable than ammonia water because there are few handling problems due to the lack of legal restrictions.

CONCLUSION

In these experiments, a new SCR process using ammonia water and ammonium hydrocarbonate aqueous solution as reducing agents was developed in order to remove NO_x from gas turbine exhaust gas. This new process was verified as being superior to the conventional process using liquid ammonia in terms of convenience, economy and ease of material handling. Furthermore, clarification that, through this process, NO_x concentration in exhaust gas could consistently be kept to under 50 ppm (at 0% O_2) was achieved.

Nowadays it is impossible to meet the NO_x standards without using this SCR process, as NO_x regulations have become stricter in urban areas. In other words, the process has made it possible to diffuse cogeneration systems using gas turbines. This SCR process using a 7% concentration of ammonia water as reducing agent is presently in operation in the gas turbine cogeneration facility of the Tokyo Gas Shinjyuku District Heating and Cooling Center. This district heating and cooling has surpassed the scale of district heating and cooling of the New York World Trade Center Building area to become the world's largest. To date, operations have progressed very smoothly. Many other systems using ammonia water or ammonium hydrocarbonate aqueous solution are in operation or are being planned.

Flameless Combustion of Premixed Gas Within Porous Radiant Burners Using a Ceramic Fiber Mat as Burner Material

KENICHI ITO, ROMEO M. BALINGCONGAN, and OSAMU FUJITA
Department of Mechanical Engineering, Hokkaido University, Sapporo, Japan

Abstract — "Flameless" combustion of premixed gas within fine ceramic fiber mat was examined experimentally for the development of environmentally acceptable, compact and highly efficient burners. Five distinct combustion stages were observed to occur depending on the air ratio and the mixture velocity, one of which is a stage whereby the flame fully enters the mat and it is not visible above the burner surface. The concentrations of NO_x and CO in the gas products were measured for the different stages and a constant surface temperature map was drawn which showed that the surface temperature during the flameless stage is high enough for radiant heating purposes. Thus, the results of the study show that the use of a fine ceramic fiber mat has great potential in the development of a clean, compact and highly efficient burner.

Key Words: Flameless Combustion, Premixed Gas, Fine Ceramic Fier Mat, Compact Burner, NO_x

INTRODUCTION

Interest in the suppression of NO_x formation during combustion has increased in recent years not only because of the role these gases play in the occurrence of acid rain but, more importantly, because of the hazards these gases pose to the health. These hazards are even more dangerous in the case of small combustors such as heaters, and small furnaces since these are usually used in enclosed spaces. The best way to reduce NO_x emission is to minimize its formation in the flame. Post flame treatment of gas products is difficult and, for small combustors, not practical nor economic.

Since NO_x formation is sensitive to flame temperature, methods of NO_x control that were studied include very lean combustion, EGR, water injection, etc. (Suzuki, 1977). However, significant reduction of NO_x emission is not attained with these methods to satisfy the limits set for NO_x emission.

Flame holding on the combustor requires a minimum temperature which leads to NO_x formation to some extent. Moreover, these methods have lower thermal efficiency and generate other unburnt pollutants. Catalytic combustion, which operates at low temperatures, combines effective suppression of NO_x with combustion efficiency, even for very lean fuel–air mixtures (Ito, 1990). The method, however, imposes an upper temperature limit in order to avoid severe thermal degradation. Besides, catalytic combustion tends to be costly because catalysts are usually made of noble metals.

"Flameless" combustion using inert porous media is another method which effectively suppresses NO_x formation. This method converts the energy released by combustion to radiant energy, thus, lowering the temperature in the reaction zone that results in low NO_x generation. Several studies have been carried out on porous radiant burners. Most, however, have focussed on the mechanism of combustion and used either metallic fibers (Kuwahara, 1988, Shirvill, 1986), spongy porous mediums or ceramic foam (Tong, 1989). Moreover, these studies are primarily intended to elucidate the combustion mechanism and any consideration of the environmental aspects of combustion are merely incidental. Schreiber *et al.* (1983) conducted development studies on combustion of premixed gas

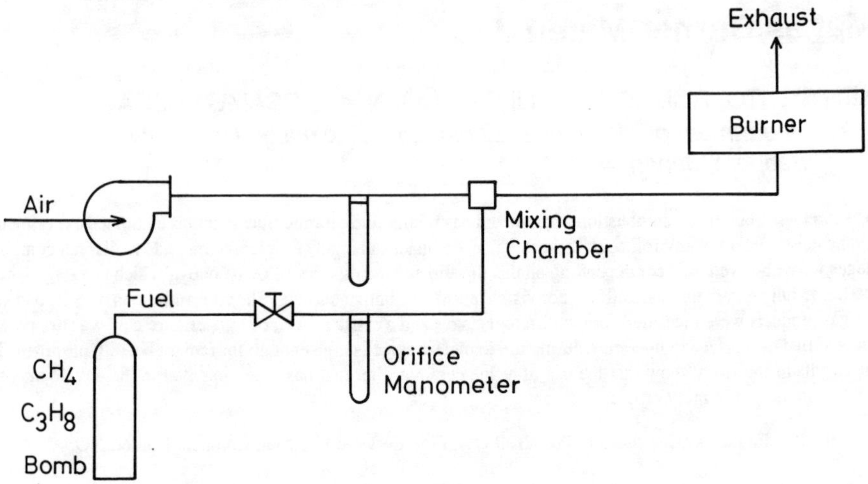

FIGURE 1 Schematic diagram of the experimental setup.

within ceramic fiber mat from the practical standpoint and mentioned the practical superiority of the burner. However, there was no clear definition of the conditions defining the flameless stage so that no comparison between the flameless stage and other combustion stages was presented.

This work investigates the flameless combustion of premixed gas–air using packed fine ceramic fiber mat with the view point of developing an efficient, compact and relatively pollutant free burner. The process was observed to define the combustion stages and measure the axial temperature profiles within the fiber mat and the burner's pollutants emission (NO_x, CO). The burner surface temperature was mapped to obtain information on the system's potential for radiative applications.

EXPERIMENTAL PROCEDURE

Figure 1 shows a schematic diagram of the experimental setup. This consists of an air supply, fuel line, porous burner with detachable fiber mats and sampling line.

Air is supplied through a blower and introduced, along with the fuel gas, into the mixing chamber. The mixture is ignited above the burner and the flame is controlled by varying the air ratio for a given combustion load. The gas products are released to the atmosphere. The air flow rate is controlled by varying the rotary of the blower. The gas flow rate is controlled by a needle valve. The flow rates are measured by orifice manometers.

Figure 2 shows a detailed diagram of the burner. A ceramic fiber mat with diameter of 60 mm and variable thickness is attached at the top of the burner. Premixed gas introduced from beneath burns above and inside the fiber mat. No flashback occurs through the fiber mat. The fuels used were methane and propane and the range of combustion load investigated was from 5 to 30 kcal/cm^2 hr.

The surface temperature of the fiber mat is measured using a Type R and thermocouple. The axial temperature profile within the mat was measured using Type R and Type K thermocouples. The placement of the thermocouples is shown in more detail in Figure 3.

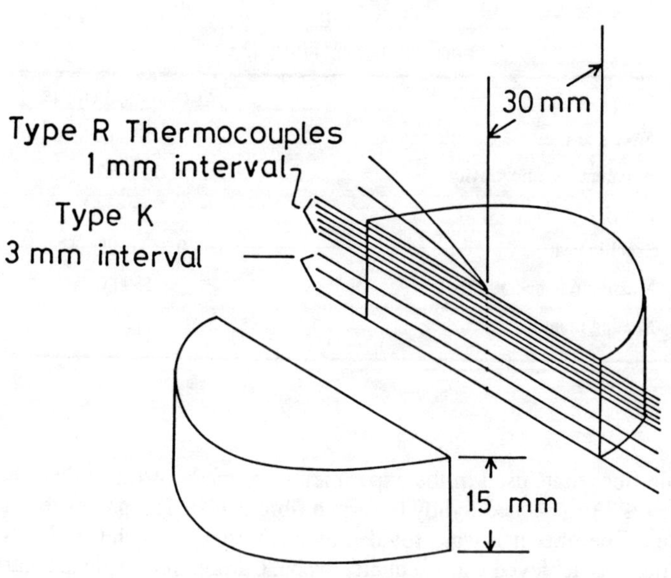

FIGURE 2 Diagram of ceramic fiber mat burner.

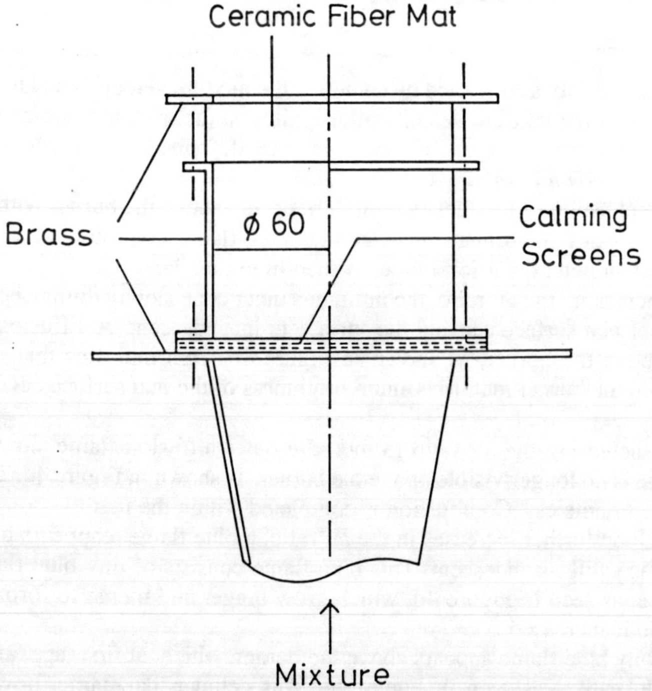

FIGURE 3 Thermocouple placement.

TABLE I
PROPERTIES OF FIBER MAT

Composition	Al_2O_3 49%, SiO_2 51%
Fiber diameter	≤3 mμ
Apparent specific gravity	0.17–0.23
Porosity	54.8%
Specific heat	0.25 kcal/kg °C
Maximum temperature	1300°C
Melting point	1760°C

The ceramic fiber mats used in the experiments are made of fine silica–alumina fibers with a diameter ≤ 3 μm packed tightly to form a fibrous mat. The properties of the mat are listed in Table I. The fiber mat was provided by Nichiasu CeraTech Co., Ltd. of Japan.

Gas sampling was achieved using a quartz probe. Carbon monoxide and carbon dioxide in the products of combustion were analyzed using an infrared spectrophotometer. NO_x was measured by the Saltzman method.

RESULTS AND DISCUSSION

Stages of combustion

The flame characteristics are varied by changing the mixture velocity and air ratio. Increasing the air ratio or the mixture velocity after igniting a fuel-rich mixture causes the flame to undergo five distinct stages. These stages are described below and corresponding photographs are shown in Figure 4.

Stage I. Combustion of the fuel-rich mixture occurs above the burner with the flame no different from that of an ordinary fuel-rich premixed flame, consisting of a luminous blue inner cone and an outer diffusion flame, as seen in Figure 4a.

Stage II. Increasing the air ratio, the luminous inner cone slowly diminishes until it fully enters the fiber mat surface causing the surface to incandescent. A diffusion flame is still observable above the burner, as shown in Figure 4b. This indicates that combustion is initiated within the burner mat. Maximum brightness of the mat surface occurs during this stage.

Stage III. Increasing the air ratio further, the outer diffusion flame slowly diminishes until the flame is no longer visible above the burner, as shown in Figure 4c. Thus, the stage could be called flameless. Combustion is completed within the mat.

Stage IV. Upon further increases in the air ratio, a blue flame reappears over the burner with the surface still incandescent. This blue flame consists of tiny blue flames scattered on the surface, as seen in Figure 4d, which grow bigger and merge to form larger flames as the air ratio is increased.

Stage V. Only blue flame appears above the burner, which, at first, appears flat but later becomes cellular-like, as seen in Figure 4e. This cellular-like flame, however, actually consists of smaller flames attached to the burner surface. The surface is no longer incandescent.

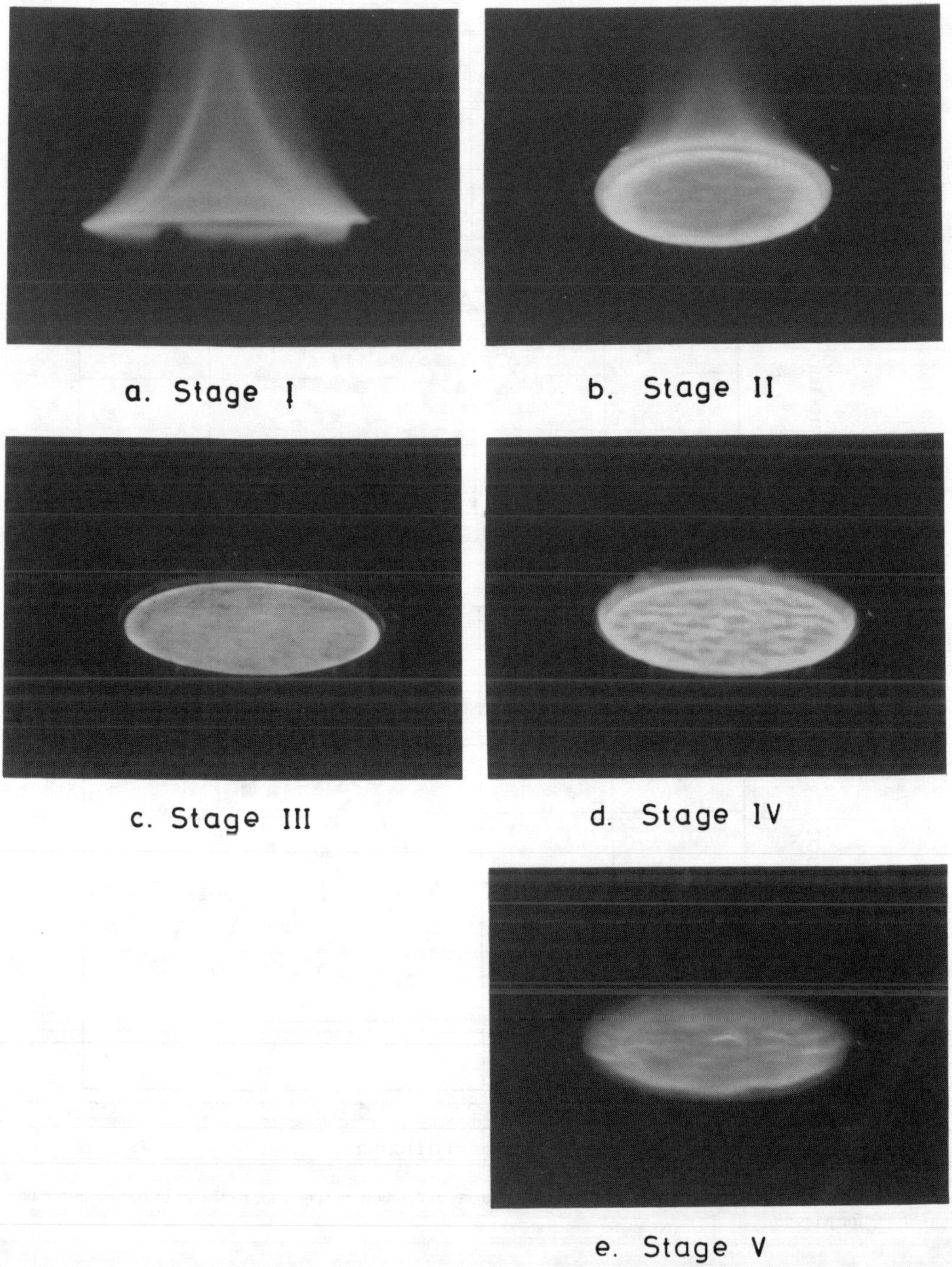

FIGURE 4 Photographs of the five stages of combustion. SEE COLOR PLATE VI.

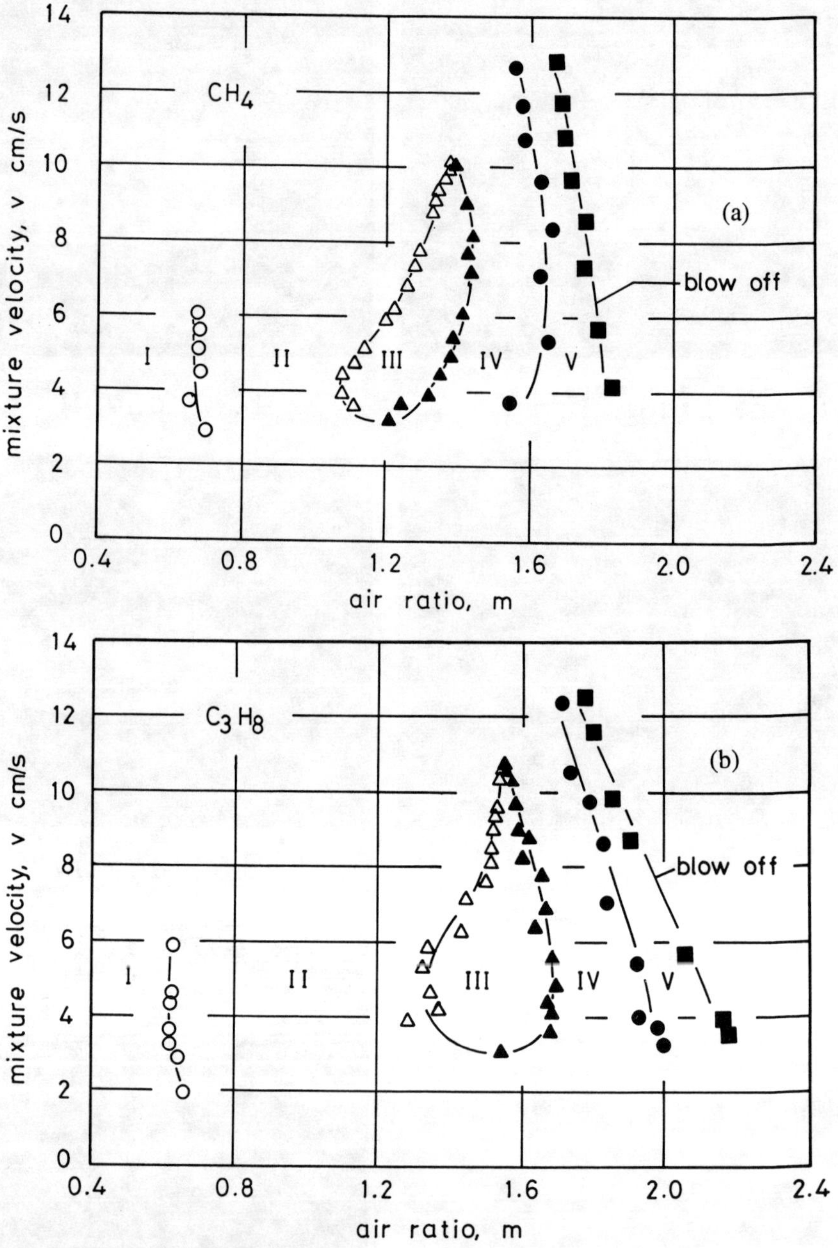

FIGURE 5 Comparison of the combustion stages: a) combustion stages for methane; b) combustion stages for propane.

Extinction occurs when, upon further increases in air ratio, the flames become unattached to the burner and is blown off.

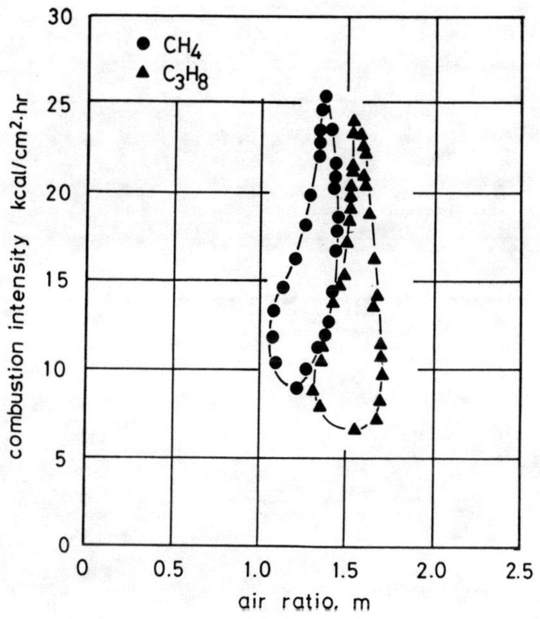

FIGURE 6 Comparison of flameless stages for methane and propane.

Stage III defines the conditions of practical significance to burner application not only because the absence of flame enables more compact designs than is usually possible, but also because high combustion efficiency and minimal pollution are realizable.

The conditions defining the five stages for a 15 mm thick mat are plotted in Figure 5: Figure 5a for methane and Figure 5b for propane. The figure shows that the transition between stages is sensitive to the air ratio of the mixture. Blow off occurs at about an air ratio of 1.90 for methane. Comparing Figures 5a and 5b, the influence of fuel type is seen. Transition conditions differ, occurring at richer conditions when methane is used than when propane is used. Blow off occurs at a leaner air ratio of 2.1 for propane.

Figure 6 focuses on Stage III and compares the plots of the flameless stages for methane and for propane. Although the transitions between stages occur at richer conditions for methane, the width of the plots are almost the same for both fuels. This stage occurs when the air ratio, m, ranges from 1.1 to 1.45 for methane, and from 1.35 to 1.70 for propane.

Figure 7 shows brown lines formed within the mat during combustion. Although these lines do not indicate the precise location of the reaction zone within the mat, they give a good idea of where the reaction region lies. Obviously, a brown line does not appear during Stage I. As the air ratio is increased, the appearance of the line inside the mat means that the reaction region moved from above the mat surface towards the inner part. The deepest penetration occurs at Stage II.

Constant Surface Temperature Map

Before determining the conditions for constant surface temperature, the radial temperature profile on the mat surface was verified. The radial profile was found uniform within 15°C.

The conditions for constant surface temperatures are shown in Figure 8. The shaded area represents the flameless Stage III. The temperature for this stage ranges from 1000 to

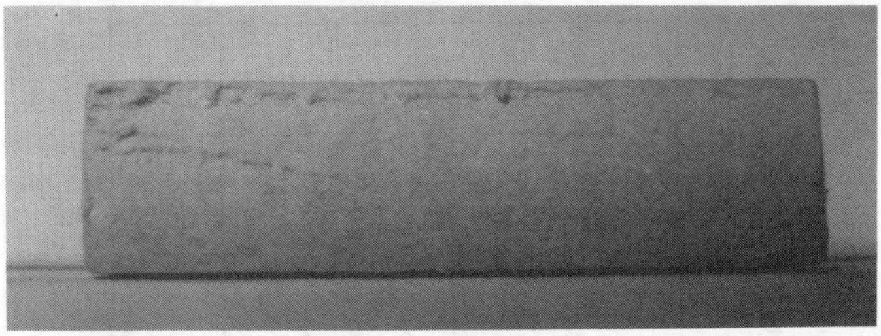

a. Stage I

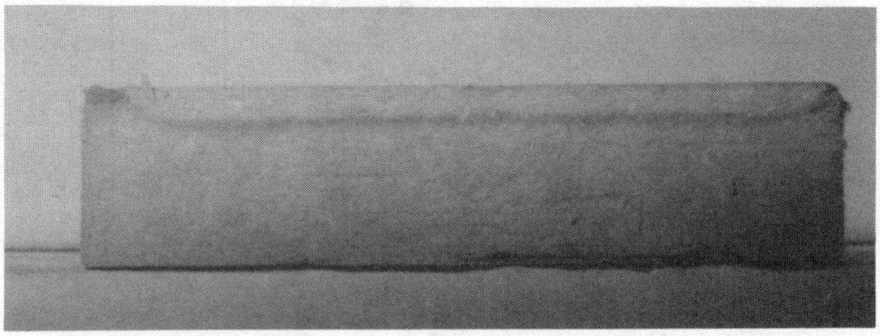

b. Stage II

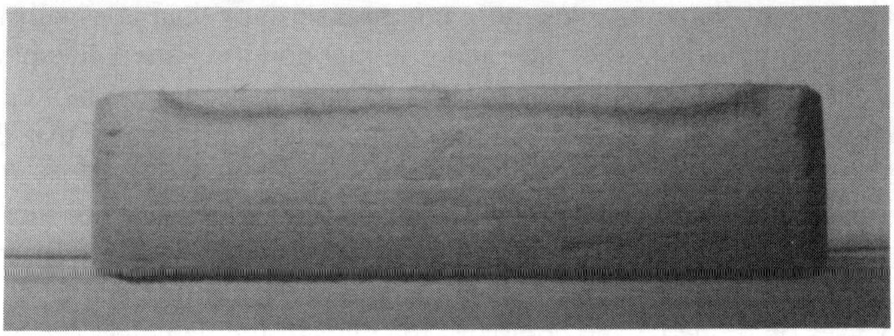

c. Stage III

FIGURE 7 Crosscut of the fiber mat showing the brown line formed during combustion for Stages II and III: a) stage I; b) stage II; c) stage III. SEE COLOR PLATE VII.

1200°C, high enough for good radiant output. Although the maximum surface temperature attainable was not measured to comply with the 1300°C limit recommended by the manufacturer, the figure shows that the maximum occurs at Stage II. This is reasonable because considerable reaction still occurs very near the surface at this stage.

Temperature Profile Within the Fiber Mat

The temperature profiles along the axial distance x for the different stages and for a constant combustion load of 19 kcal/cm^2 hr are plotted in Figure 9, where positive x refers to the distance downstream of the mat surface. Proceeding from Stage I, the figure shows

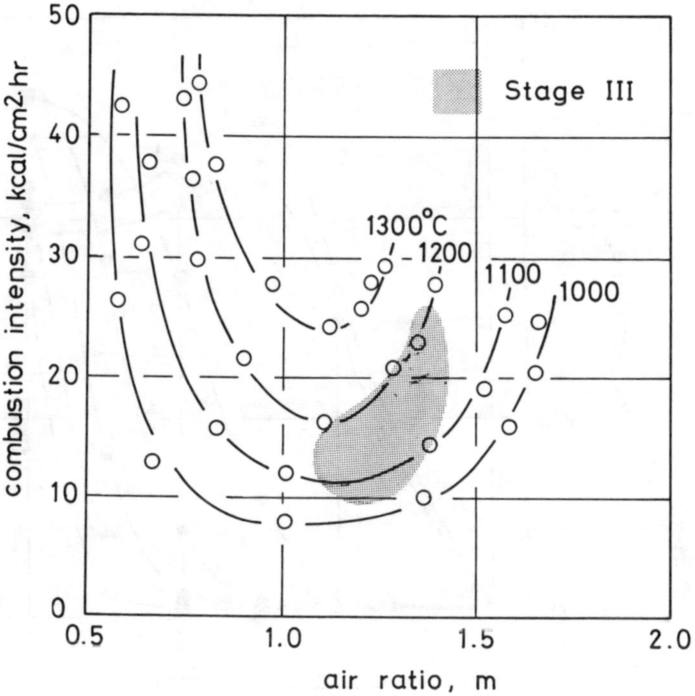

FIGURE 8 Constant surface temperature map.

that the temperature inside the mat rises as the air ratio increases and that the location of the highest temperature within the mat moves deeper into the mat. At m = 0.4, Stage I, the temperature within the mat is generally low. At m = 0.74, Stage II, the location of the maximum temperature has moved inside the mat and a characteristic temperature peak has appeared. At m = 1.31, Stage III, the peak has moved back towards the surface. Although not shown in the figure, the temperature above the mat is lower than at the surface for this condition. These profiles reflect the changes in the location of the brown line noted in the previous section and show that combustion within the mat occurs at a wide range of air ratios.

Products of Combustion

The combustion products are released to the atmosphere and samples were obtained above the burner under conditions corresponding to each stage described earlier. The samples were analyzed for CO_2, CO and NO_x. The results for a combustion load of 19 kcal/cm^2 hr using methane as fuel is discussed below.

Figure 10 shows the axial nitric oxides distribution reduced to $O_2 = 0\%$. Although there is a dependency of the concentration of NO_x on the axial distance caused by air dilution from surrounding air, the rate of concentration change to its mean value is thought to be level. Compared to the self regulation standard for small room heaters set at about 20 ppm (0.005 ml/kJ) by a manufacturer in Japan (Uchida Manufacturing, 1986), the emission level is very low indeed.

Figure 11 shows that NO_x emission of the burner is very sensitive to changes in air ratio, decreasing with air ratio. The sensitivity, however, diminishes at higher air ratio levels and

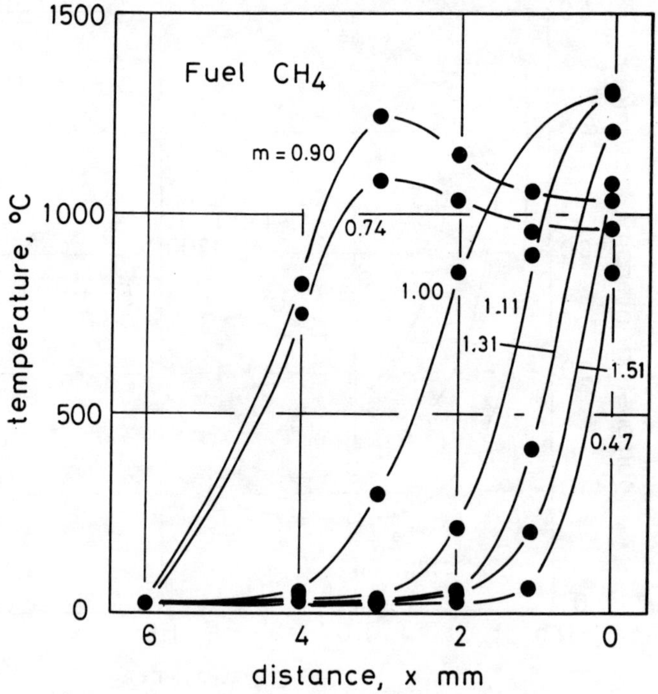

FIGURE 9 Temperature distribution within the fiber mat.

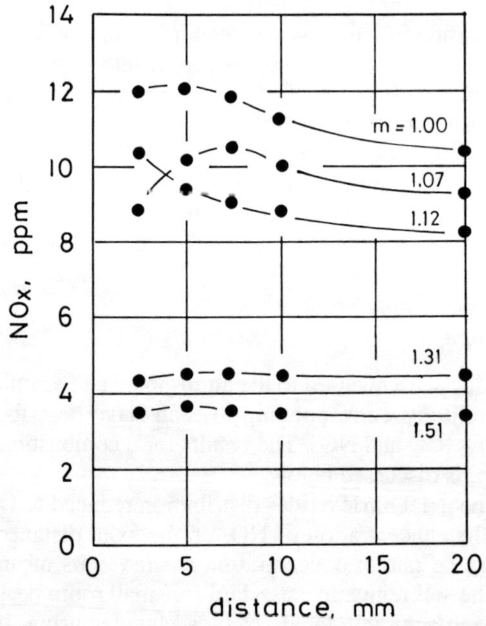

FIGURE 10 Axial distribution of N_x emission (O_2; 0%).

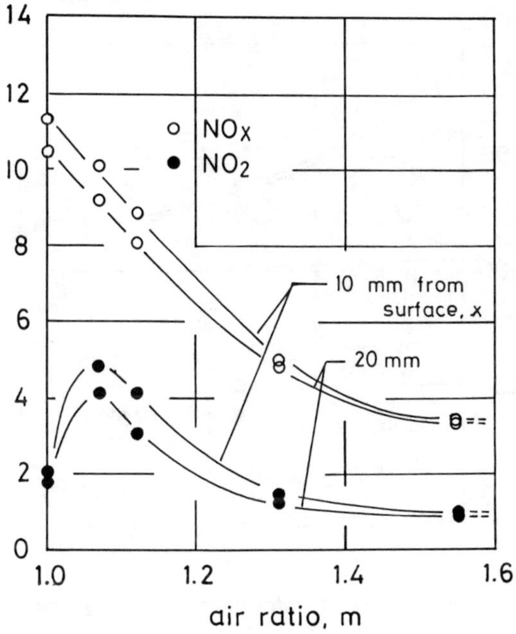

FIGURE 11 Effect of air ratio on NO_x emission (O_2; 0%).

eventually the slope becomes flatter. The NO_x formed at Stage III range is very low at about 5 ppm.

NO_x emissions using flameless combustion compares very favorably with other combustion methods. NO_x emission of about 5 ppm is very much lower than the lowest emission of about 30 ppm for spray, vaporizing and foam combustion (Anzawa, 1990) indicating the practical superiority of flameless combustion over these methods in the development of an environmentally clean burner.

Figure 12 shows the CO and CO_2 emission level as a function of the air ratio. The values are for 10 and 20 mm distance from the surface of the burner. CO emission is very low when m is greater than 1.1. The CO concentration was found to be negligibly small during the flameless stage. The slope of the axial profiles of CO and CO_2 concentrations for m = 1.31, Stage III, is flat, indicating that combustion is practically completed within the fiber mat.

Thermal Degradation

To verify the long term use of the fiber mat, the internal temperature profiles for a new fiber mat and for the same mat after about 100 hours of intermittent use were measured. The results are shown in Figure 13. The profiles show that for the new mat the temperatures are higher than the temperatures at corresponding points after use. The changes may be due to changes in the transport properties of that part of the burner material that has undergone high temperature exposure. However, the difference is small, hence, it can be said that the effect of thermal degradation is very small. In fact, no significant change in the transition between stages were observed.

Figure 14 shows photographs of the fiber mat magnified 360 times using a Scanning Electron Microscope (SEM) both before and after use. Figure 14 shows the SEM image of

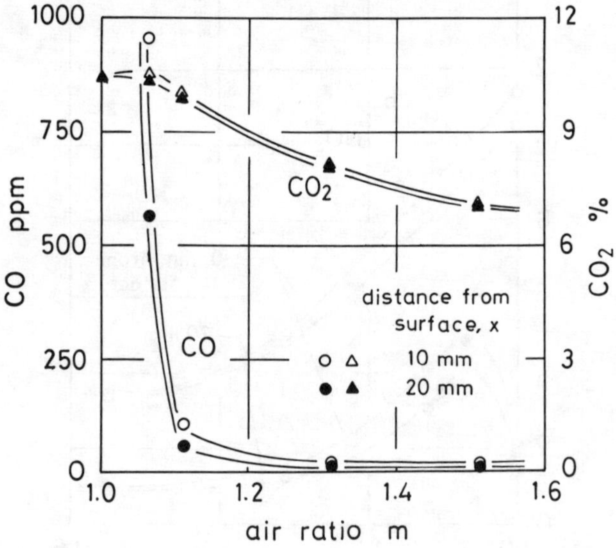

FIGURE 12 Effect of air ratio on CO and CO_2 emission.

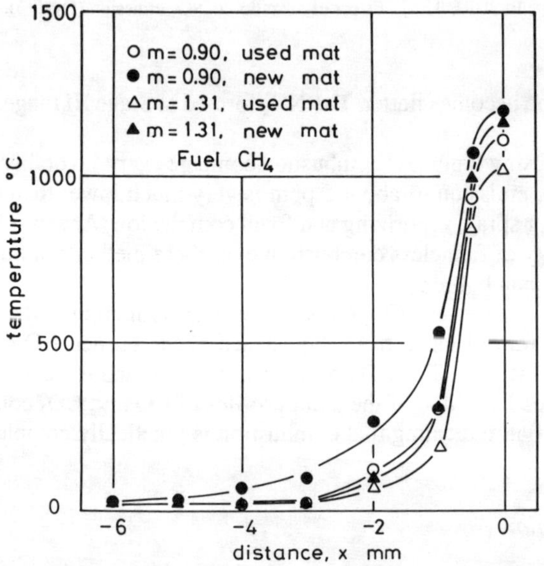

FIGURE 13 Temperature profiles within the mat before use and after 100-hour use.

a fiber mat before use. The fibers' surfaces are smooth. Figure 14b shows the SEM image of a fiber mat after 100 hours of use. The fibers' surfaces appear to have some depositions. Nevertheless, although these depositions may have changed the transport properties of the system leading to the change in the temperature profile, the change in the combustion characteristics were not strongly affected, as mentioned in the previous paragraph.

a. Before use

b. After 100 hours of use

FIGURE 14 SEM photographs of the fiber before use and after 100-hour use: a) before use; b) after 100-hour use.

IMPLICATIONS FOR BURNER DESIGN

For the purpose of developing clean, compact combustion systems, it may seem that the optimum conditions for operating the burner lie in Stage II where the highest surface temperatures occur. However, optimum does not only mean high surface temperature. Optimum conditions should result in high combustion efficiency, effective conversion of the energy released via combustion to radiant energy, and low levels of pollutant emission: conditions that occur within Stage III. No flame is formed above the burner making a compact design feasible. The surface temperature can be limited to the range from 1000 to 1200°C, low enough to limit NO_x emissions to very low levels but still sufficiently high for radiant heating purposes. Combustion is completed within the mat which means the conversion of a greater percentage of the energy released via combustion to radiant heat. CO and NO_x levels are within the allowable standards set by JIS (JIS, 1982) and relatively low compared with an ordinary fan heater using methanol or kerosene as the fuel (Ito, 1987) and is also low compared with thermally stabilized burners using ethane (Strenger, 1988).

CONCLUSIONS

The study was performed to characterize the combustion of premixed methane/air within fibrous ceramic mats. The results show that:

1. Five distinct combustion stages characterized by the visibility of a flame above the burner exist. One of these is a flameless stage where complete combustion is attained within the fiber mat so that no flame is visible above the surface.
2. Conditions defining the combustion stages differ for different fuels.
3. Combustion is completed within Stage III as evidenced by a minimal CO concentration just above the surface and the flat slope of the profile.
4. NO_x emission is greatly influenced by the air ratio but is kept at very low levels, specially during Stage III. The level of NO_x emission is about a third of the recommended regulation for small type burners.
5. Some effects of thermal exposure and the occurrence of reaction within the mat was observed on the surface of the fibers but do not strongly affect the overall combustion characteristics.
6. The mat surface is brightest and reaches maximum temperature in Stage II. However, for the development of a clean, compact and highly efficient combustion system, Stage III offers the optimal conditions: completeness of combustion, more effective conversion of heat released through combustion to radiant energy, absence of a flame that allows for more compact burner designs, and very low NO_x emission.

ACKNOWLEDGMENT

The authors wish to acknowledge the assistance of Mr. K. Yamane and Mr. K. Matsumoto in the performance of the experimental work.

REFERENCES

Anzawa, N., and Adachi, K. (1990). Development of Burning Liquid Fuel in Dispersed Foam Consisting of Numerous Bubbles, In *Proceedings of the Annual Meeting, Japan Furnace Research Committee*, pp. 1–12.

Ito, K., Choi, B.C., and Fujita, O. (1990). The Startup Characteristics of a Catalytic Combustor Using a Methanol Mixture, *JSME International Journal*, Series I, **33**(4), pp. 778–784.

Ito, K. *et al.* (1987). Catalytic Combustion of Lean Methanol Mixtures, In *Proceedings of the Western States Section/Japanese Section of the Combustion Institute, 1987 Fall Meeting*, Honolulu Hawaii, pp. 254–256.

Japan Industrial Standard (1982). Open Type Oil Burning Space Heater, 2

Kuwahara, S. (1989) Combustion Performance of Metal Fiber Burner (in Japanese), In *Nensho Kenkyu*, **81**, pp. 45–57.

Schreiber, S. *et al.* (1983). Industrial Applications of the Radiant Ceramic Fiber Mat, In *IGRC/DO6-83 1983 International Gas Conference*, June 13–16.

Strenger, M.R., and Churchill, S.W. (1988). Formation of NO_x and Burnoff of CO during Thermal Quenching of the Products from Combustion in a Thermally Stabilized Burner, In *Proceedings of the Twenty Second Symposium (International) on Combustion*, Combustion Institute, 817.

Suzuki, S. (Ed) (1977). Nitric Oxides (in Japanese), In the *Publication of the Japan Chemical Society*, **11**, 412.

Tong, T.W., *et al.* (1989). An Experimental Study of Combustion and Heat Transfer in Porous Radiant Burners, In *Proceedings of the 1989 Fall Meeting of the Western States Section of the Combustion Institute*, Livermore, CA.

Uchida Manufacturing Co. Ltd (1986) Corona Roomheater and Small Boiler Line-up, In *Brochure No. 0018601*, 27.

6.

BIOMASS COMBUSTION

A Biomass Fueled Cogeneration Plant Based on an Evaporative Gas Turbine Cycle at the University of Brussels

J. DE RUYCK, K.MANIATIS, M. DISTELMANS, and G. BARON *Department of Fluid Mechanics, Department of Chemical Engineering, Vrije Universiteit Brussels, Belgium*

Abstract — A 0.5 MWe peak-shaving cogeneration plant fueled with biomass is presented. The biomass is gasified and used in a gas turbine power cycle. With the cleaning of the product gas being too critical for direct firing in small gas turbines, external firing is used in combination with a high temperature heat exchanger. Exhaust heat is recovered by heating water for cogeneration purposes and/or water evaporation into the cycle. The fluidized bed gasifier uses wood waste from the particle board and wood working industries. It operates on enriched oxygen-air gasification resulting in a 10 MJ/kg LHV gas. The oxygen enrichment is achieved by Pressure Swing Adsorption at an energetic cost of less than 0.3 kWh/kg for 90% enrichment. A power credit of about 13 MBF/year can be obtained when applying the present cycle in the Brussels University power supply in both peak and off-peak operation. The power credit is sufficient to justify the total demonstration project cost of about 100 MBF. The project aims at demonstrating the viability of biomass gasification in combination with an evaporative gas turbine cycle in the small power range.

Key Words: biomass, gasification, external firing, evaporative cycle, cogeneration

OBJECTIVES

The problems of organic waste disposal and excess agricultural capacity, and the improving economics of biofuel production are strong incentives towards the use of biomass for energy production in Europe. Depending on the availability of organic waste and on the excess agricultural capacity, energy from biomass can cover about 10% of the European energy consumption (Grassi 1990).

Gasification of biomass is recognized as a way to produce biofuel economically (Bridgwater 1990). The product gas can be burned in a gas turbine cycle without cooling it, thus avoiding the problem of tar condensation. At present, developments in the combination of biomass gasification and gas turbines are limited (Larsen 1990) and a 2.5 MWth bio-fueled gas turbine plant is therefore proposed, sited at the VUB University campus in Brussels. The plant will essentially be used in cogeneration mode. The gas turbine combines indirect firing to avoid turbine corrosion and water evaporation to increase efficiency and output power in the power only mode. The project aims at the 0.5 to 5 MW scale biomass market.

PROJECT ORGANISATION

A consortium has been created between the Brussels University VUB (Belgium) and Noel Penny Turbines NPT (UK) for the realisation of the project. Table I summarises the project schedule.

The feasibility study is being financed by the Brussels Executive Government. The other phases are partly funded by the consortium. The project has been submitted to the Commission of the European Communities in 1991 for the remainder of the funding.

TABLE I

PROJECT TIME SCHEDULE

	91	92	93	94	95	96
Fundamentals and feasibility study	—					
Designs and engineering		—	—			
Construction and assembly			—	—		
Commissioning				—		
Demonstration					—	—

BIOMASS GASIFICATION

There are several biomass conversion routes. Gasification is recognised as an efficient process for converting biomass into gaseous and liquid fuels which can be utilised with little or no modification to existing equipment (Buekens *et al.* 1990). Recent work in Europe, Canada and the US has shown that although the technology for the production of liquid fuels (such as methanol) is available, the economics of such a process are prohibitive due to large plant requirements and cheap availability of such products from the oil industry. On the other hand, heat and power production from gaseous fuels from biomass gasification is a recognised commercially available process and several industrial applications have been reported (Ferrero *et al.* 1989). Of the various gasification technologies, the fluidized bed is recognised as the most versatile. Most of the commercial plants are based on various types of fluidized bed systems such as classical bubbling bed and circulating fluidized bed with one or more compartments, their main advantages being relatively easy scale up, operation with a variety of biomass feedstocks and acceptance of feedstocks with a small mean particle diameter.

If air is used as the gasifying agent, the nitrogen in the air significantly dilutes the product gas (which contains large quantities of hydrogen and carbon monoxide) and a low calorific value gas is produced (4 to 6 MJ/Nm^3). It can be substituted for natural gas and oil to fire boilers or can be used instead of gasoline and diesel oil to fuel internal combustion engines, provided that the gas is sufficiently cleaned and cooled. If nitrogen is eliminated from the product gas through steam or oxygen gasification, a medium heating value gas can be produced ranging from 10 to 18 MJ/Nm^3. Recent advances in pressure swing adsorption air separation have made the production of oxygen significantly cheaper, even for small scale applications and energy costs of less than 0.3 kWh/kg for 90% enrichment have been reported (Baron 1990). It is therefore technically feasible and economically viable to consider the fluidized bed gasification processes for power production applications. Other biomass thermochemical conversion processes for power production, such as the direct combustion of pulverised wood in a gas turbine and the utilisation of primary pyrolysis liquid fuels derived from flash pyrolysis in turbines, are still at the R & D level and several problems have to be overcome before they can be used in industrial applications (Hamrick 1985, Andreuccetti *et al.* 1991, Cuevas and Fluxa 1991).

BIOMASS GASIFIER

The VUB gasification plant is illustrated on Figure 1. It combines fluidized bed gasification with oxygen enrichment through pressure swing adsorption. Both processes have been developed and studied for more than a decade at the Department of Chemical Engineering and several lab-scale, as well as pilot scale, plants have been constructed and tested

TABLE II
COMPOSITION OF A 10 MJ/kg LHV PRODUCT GAS FROM WOOD WASTE

Component	Wt%	Vol%
CO	31.8	30.9
CH_4	5.90	10.0
C_2H_4	4.00	3.90
H_2	1.39	18.8
C_2H_6	0.66	0.60
C_3H_6	0.30	0.20
CO_2	51.7	32.9
O_2	3.06	2.60
N_2	1.10	1.12

TABLE III
FLUIDIZATION PARAMETERS

Parameter	Units	
Mean particle diameter (dp)	mm	0.50
Minimum fluid velocity (Umf)	m/s	0.08
Gas velocity (U)	m/s	0.80
Particle terminal velocity (UT)	m/s	5.00
Bubble diameter		
maximum (dBmax)	m	0.50
minimum (dBmin)	m	0.25
Bed height (hb)	m	0.60
Bed diameter (db)	m	0.80

(Maniatis *et al.* 1988, 1990). Wood waste from the particle board and wood working industries will be used. The wood will be fed from a one week capacity silo to the fluidised bed gasifier at a flowrate of 12 t/day. The wood moisture and ash contents are 15 and 1 wt%, respectively. With 90% oxygen enriched air gasification, the product fuel gas will have the composition given in Table II. The heating value of the gas will be about 10 MJ/kg. The temperature in the fluidized bed will have to be maintained at about 750˚C in order to avoid bed sintering and this will be achieved by a heat exchanger in the bed. The freeboard has an expanded volume in relation to the bed in order to increase the residence time of the fly ash and thus eliminate carbon loss. The temperature in the freeboard will drop gradually to about 625˚C at the exit of the gasifier and the fuel gas will then be cleaned of particulates in a cyclone. Part of the nitrogen from the air separation will be recycled in small quantities to the feeding system of the gasifier to eliminate gas leaking from the fluidized bed. The principal design considerations for the fluidized bed are related to efficient fluidization and the feeding of the biomass. The most significant fluidization parameters are summarised in Table III.

It has been shown that efficient fluidization is achieved when $U/U_{mf} > 2.5$. Experience has shown that in practice the above ratio, which offers a measure of solids mixing, should

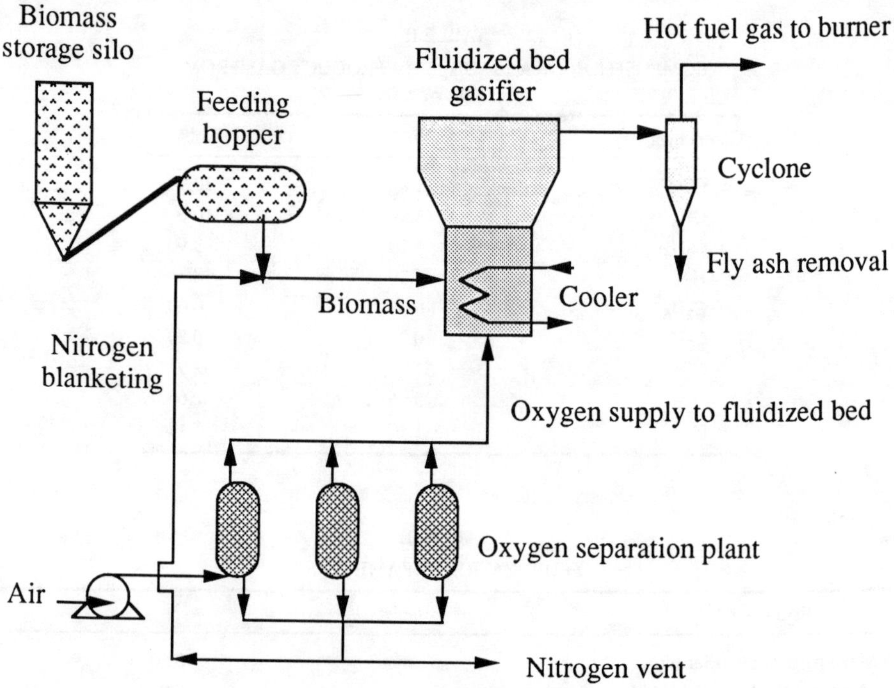

FIGURE 1 Biomass gasification.

be about 10. Especially with feedstocks of low density and small particle diameter (such as biomass residues), which tend to float on the surface of the bed, it is very important to maintain as good a mixing as possible. Otherwise, agglomeration of the bed material will take place due to the creation of low melting alkali metal eutectics from the ash of the biomass and the silica sand. Thus, the mean particle diameter of the feedstock should not be less than 2 mm. In addition, the biomass feeding point should be located close to the distributor in order to allow sufficient residence time for the particles to be completely gasified before they could escape from the bed and thus maximise the carbon conversion.

INDIRECT FIRING OF GAS TURBINES

Synthesis gas from coal gasifiers can be fired directly in large gas turbines after hot gas cleanup. Such turbines, which can withstand the remaining particle erosion, are commercially available. The cleanup systems for coal gas are not suitable for the cleanup of product gas from biomass due to the smaller particle sizes involved however, and the present project will use a small gas turbine. Consequently, the direct firing of the product gas from biomass is presently difficult to achieve.

External firing is considered as an alternative (Figure 2). The external firing reheats the turbine exhaust gas (air) and the energy is recovered in front of the turbine by a high temperature heat exchanger. The efficiency is comparable or even better than the efficiency of a regenerative Brayton cycle (LaHaye 1989). The turbine does not suffer from the corrosive elements in the combustion products, extending the lifetime of the turbine. The key element in external firing is the high temperature gas/gas heat exchanger. Today, high temperature titanium or ceramic heat exchangers operating under pressure are available, but the combustion of a low calorific value gas in such conditions is still to be investigated.

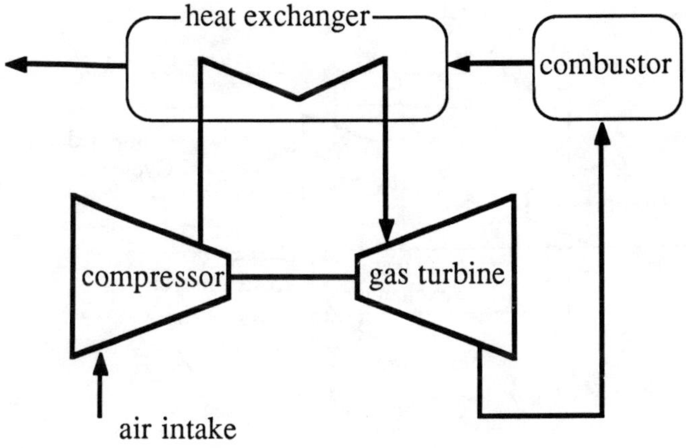

FIGURE 2 Externally fired gas turbine (La Haye 1989).

GAS AND STEAM POWER CYCLES

Both the efficiency and power of a gas turbine cycle can be improved by combining steam and gas. A choice is to be made between three alternatives (Figure 3): combined-cycle plants, steam injected gas turbines (STIG) and evaporation cycles. In combined cycles waste heat is normally recovered in a waste heat steam generator. The steam is expanded in a steam turbine and recovered in a condensor. As an alternative, all the steam can be mixed with the compressed air, superheated in the combustor, expanded in the gas turbine and condensed in the atmosphere (full STIG). Part of the waste heat is released as latent heat in the exhaust vapor. In evaporation cycles, water is injected instead of steam. The idea of evaporation is to reduce availability losses by evaporation at a variable temperature, without use of dual or triple pressure systems.

Comparative studies between the three groups of gas and steam cycles can be found in the literature (Frutschi 1988, El Masri 1988a, Morton 1989). Basically, the three cycles recover the gas turbine exhaust heat in a similar way and they have comparable capabilities. According to Frutschi (1988) the combined cycle is superior through the lower pressure of condensation, and through its capability for high efficiencies at large power to heat ratio's (86% at a ratio of 1.2). For smaller systems, the simplicity of STIG and evaporation is attractive, certainly when compared with simple or dry regenerative cycles. The mixing of steam and gas is limited by the availability of turbosets which can cope with the difference in mass flow rate between compressor and turbine, higher dew point temperatures in the stack (60 to 80°C), and the interdependency of steam and gas cycles. Cogeneration with STIG or evaporation is considered as being less attractive (Frutschi 1988), but it is certainly feasible (e.g., Kolp 1988).

Evaporation cycles have been proposed for a long time. They can yield better efficiencies than simple STIG and combined cycles, but limited results with this cycle are found in the literature. Several arrangements, ranging from a simple recuperative evaporation (Figure 3) to complex systems with different heat exchangers and evaporators are proposed as candidates for future high efficient power plants (El Masri 1988b, Morton 1989, Berta 1990).

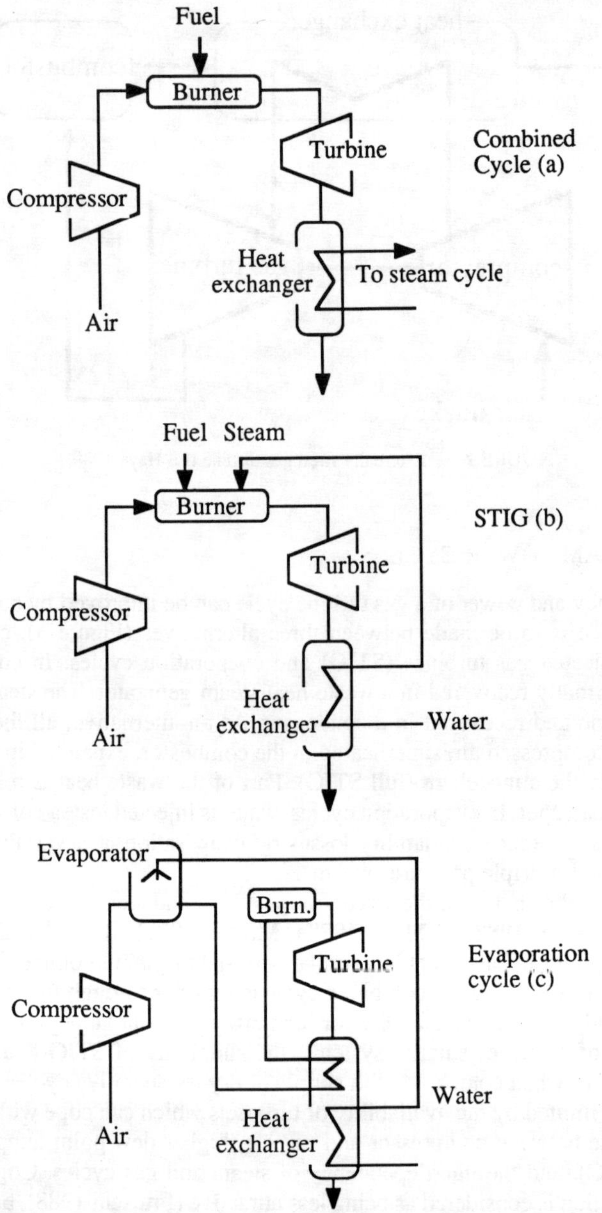

FIGURE 3 Gas and steam power cycles (from Frutschi 1988).

PRESENT POWER CYCLE

The present cycle is illustrated in Figure 4. Between STIG and evaporation, an evaporative cycle has been selected for different reasons:

— there is no need for steam at the University, the heating system being a compressed water system

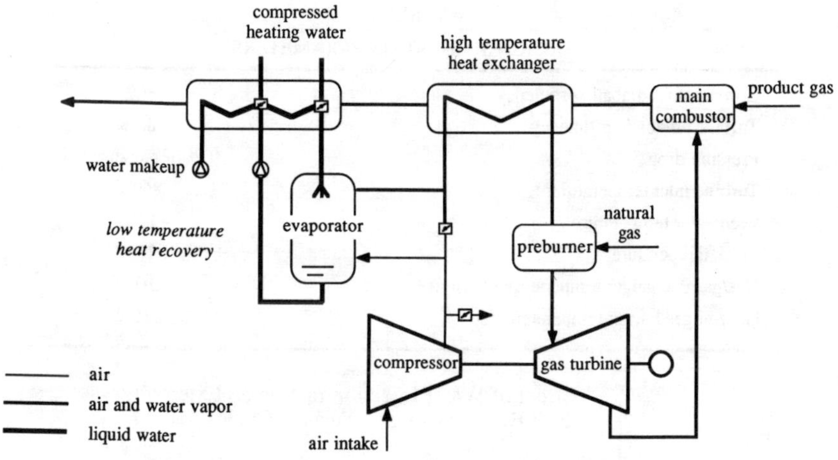

FIGURE 4 VUB peak shaving power plant on biomass.

— the cycle efficiency and output power are slightly better
— experience with evaporative cycles can be gained

The evaporation occurs in the most simple way by direct injection of the compressor exit air in an evaporation tower. Aftercooling of the air before entering the evaporator has been considered (El Masri 1988a) but no incentive was found to add this extra component in the present combination, as will be discussed below. The gas and steam mixture is further heated by external firing of the product gas in a high temperature heat exchanger. Extra burning of some natural gas next to the external firing is included for different reasons:
— lower temperatures in the heat exchanger
— no availability loss due to gasifier outages, by switching to natural gas
— essential for power control

The hot turbine exhaust air enters the product gas combustor, which provides the major part of the energy input. About ⅔ of this energy is recovered in the gas/gas heat exchanger, which is a major piece of equipment. Next to the gas/gas exchanger, the exhaust air enters a liquid/gas exchanger which heats water for the evaporation and/or the University heating system. The water pressure and temperature levels in these two water circuits are similar, and both heat recoveries can occur in the same heat exchanger by switching from one water circuit to the other. The remaining exhaust heat is used to preheat the feedwater.

CYCLE BEHAVIOUR

Simulation Package

The cycle is simulated in a simulation package for energy systems which is under development at the Fluid Mechanics department. The background of this package is LABVIEW, a commercially available object-oriented software using the dataflow principle. Its graphical interface allows easy problem solving and debugging in the design phase as well as in the demonstration phase of the project. Future plant control and measurement facilities can easily be implemented. A description of this simulation package can be found in Distelmans and De Ruyck 1992).

TABLE IV
DESIGN SIMULATION PARAMETERS

Compressor adiabatic efficiency	80%
Turbine adiabatic efficiency	85%
Pressure drop	10% of pressure ratio
Turbine inlet temperature	860°C
Feedwater temperature	15°C
Stack temperature	≥120°C
Gas/gas exchanger temperature difference	≥100°C
Natural gas burner temperature rise	≥60°C

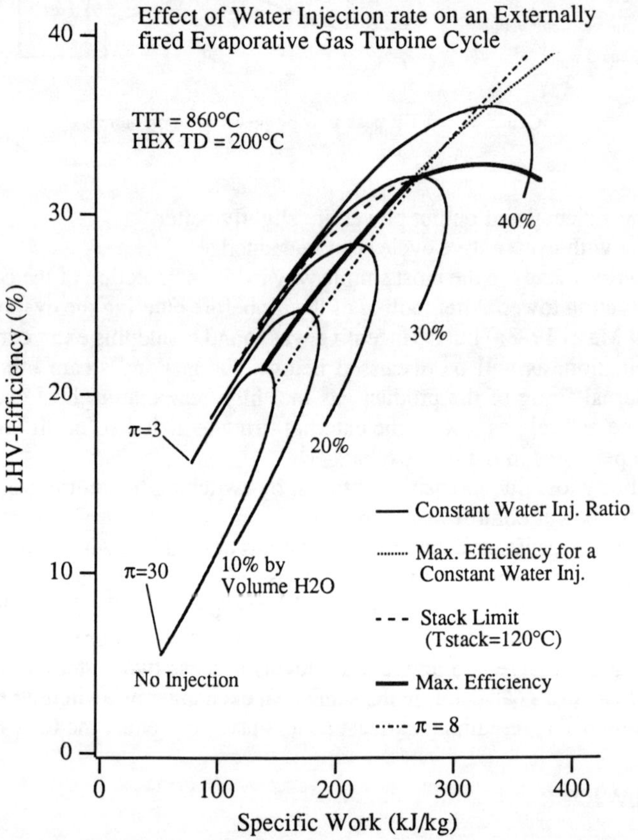

FIGURE 5 Mixed air and water vapor cycle, effect of water injection.

Parameters and Constraints

The main parameters influencing the specific power and the efficiency are the turbine inlet temperature, the compressor pressure ratio π, the amount of water injected, the temperature difference on the hot side of the black box heat recovery system (HexTD) and the stack temperature. The inlet temperature for small gas turbines is in the range 800 to 860°C and its effect will not be discussed. The stack temperature is imposed as 120°C (the dewpoint

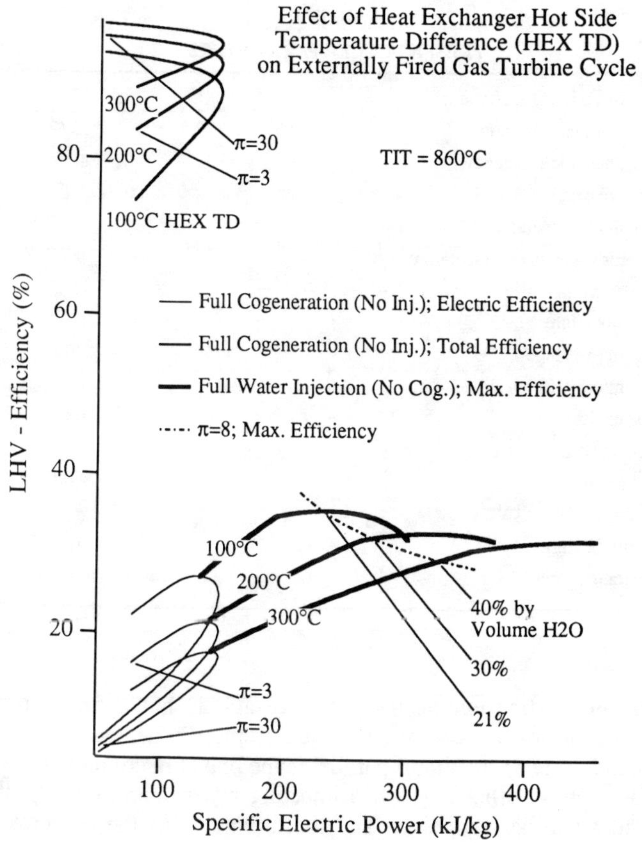

FIGURE 6 Mixed air and water vapor cycle, effect of temperature difference (Hex TD) at the hot side of the heat recovery system.

may reach 80°C). A minimum of 100°C difference at the hot side of the recovery system is required. The constraints for the cycle behaviour are summarised in Table IV.

Effect of Water Injection in the Power Mode

The cycle behaviour is shown in Figures 5 and 6. Both figures analyse the first law LHV efficiency against the specific electric power. Figure 5 shows the effect of water injection when operating in the power mode, at a fixed turbine inlet temperature of 860 and 200°C temperature difference at the hot side of the recovery system. The compressor pressure ratio is varied from 3 to 30, for different values of water injection. The water injection is expressed as the volume fraction of water vapor in the turbine.

The quantity of water which can be evaporated is limited by the minimum stack temperature of 120°C, provided no second law violation occurred. This stack temperature constraint shows a maximum efficiency, which is the overall maximum that may be attained by such a cycle within the given constraints. To reach this maximum, a gas turbine with a pressure ratio much higher than the original dry gas turbine would, however, be required. The thick solid line in Figure 5 joins the maximum efficiencies, taking the stack limiting temperature into account.

TABLE V
PRESENT CYCLE PARAMETERS

Compressor adiabatic efficiency	80%
Compressor pressure ratio	7.7
Turbine adiabatic efficiency	85%
Turbine pressure ratio	7.0
Turbine inlet temperature	860°C
Evaporator water inlet temperature	160°C
Feedwater temperature	15°C
Stack temperature	≥110°C
Gas/liquid exchange temperature difference	≥30°C
Gas/gas exchanger temperature difference	≥100°C
Natural gas burner temperature rise	≥60°C
University heating temperature range	110/180°C

Starting from an existing engine the pressure ratio is not a free parameter, and a maximum efficiency is not necessarily achieved. Fortunately the lines of constant pressure ratio (e.g., $\pi = 8$ in Figure 5) are almost parallel to the line of maximum efficiency, and gas turbines therefore behave efficiently when injecting water. Unfortunately, the excess of water injected between the compressor and the turbine shifts the compressor operation towards the surge limit, limiting the maximum amount of water injection. This maximum depends on the turboset characteristics which are unknown in the present stage of the project. This maximum will also be eventually limited by the stack limiting temperature. The corresponding maximum efficiency increase when injecting water is 10% whereas the specific power almost doubles. As can be seen from Figure 5, the maximum efficiency at a pressure ratio of 8 is not much lower than the absolute maximum allowed by the stack limit.

Effect of Hex Temperature Difference in Power Mode

The lower part of Figure 6 repeats the dry gas turbine and the maximum efficiency characteristics for different temperature differences at the hot side of the recovery system. The turbine inlet temperature remaining the same, an increase in the Hex temperature difference corresponds to higher firing temperatures. The power output of the dry gas turbine cycle is not dependent on the Hex temperature difference. The drop in electric efficiency is due to the higher heat content in the exhaust gases, which can be used for cogeneration or evaporation purposes. When using the exhaust heat for evaporation purposes, the higher firing temperature allows more water evaporation, resulting in better electric efficiencies and a strong increase in specific power. The strong imbalance by injecting much water however limits this potential to an unknown extent. A line of constant pressure ratio of 8 (while water injection is maximized until the stack temperature limit is reached) has been plotted to allow the choice of heat exchanger temperature difference for a given gas turbine. It should be observed that at the lower limit of 100°C difference, the cycle efficiency would reach an absolute maximum. Hex pressure losses, however, become stringent at such low temperature differences.

Effect of Hex Temperature Difference in Cogeneration Mode

Figure 6 shows the electric and total first law efficiencies for different temperature differences in the heat exchanger, when using the exhaust heat for cogeneration purposes. High firing temperatures and high pressure ratios yield the best global efficiencies, but at the cost of lower (high valued) power output. For the general application of the plant under development, a thermoeconomic analysis is to be applied to decide on the power to heat ratio. As will be shown below, the present project yields better economics at the highest firing temperatures.

VUB COGENERATION PLANT

A gas turbine has been sought for the practical realisation of the plant at the university campus at Etterbeek, near Brussels. The thermal power is limited to 2.5 MWth due to local constraints at the University campus. NPT will provide a 500 kWe gas turbine running at an inlet temperature of about 860°C and a pressure ratio of 7 (these numbers are still subject to change). Compressor modifications will be necessary to adapt the mass flow rates, since large amounts of water will be injected. For the purposes of the project, the compressor adaptation may be simulated by bleeding compressor air before entering the evaporator. Although the air bleeding is an inefficient solution, it will allow the plant to be tested in different configurations, with and without important water injections. The pinch-point difference in the low temperatures will be kept above 30°C and a minimum of 100°C difference seems to be required at the high temperature end. The university heating system runs on compressed water where the temperature ranges from 110 to 180°C, which is a constraint for the cogeneration of heat. The parameters for the present application are summarised in Table V. The same constraints have been used to compute a corresponding simple gas turbine cycle, a dry externally fired cycle, and a single pressure STIG cycle. Results are summarised in Table VI for power production and Table VII for combined heat and power production.

Power Generation Without Cogeneration

Present cycle. The compressor air enters the evaporator at a temperature of 306°C (Figure 4). The water to be evaporated enters at 160°C, close to the boiling point, and leaves the evaporator at about 110°C. The humid air leaves the evaporator at about 125°C with 23.5% water vapor by total volume. The corresponding water consumption is 1.55 ton/hr. According to cooling tower simulations, the lowest water temperature at the exit of the evaporator, without aftercooling the compressor, air is 98°C. This temperature is in the range of the stack temperature limit of 110°C and aftercooling will therefore not be effective for a further reduction of the stack temperature. A water temperature of 110°C at the exit of the evaporator sustains the evaporation at a sufficient rate, such as to limit the size of the evaporator. The humid air temperature is raised from 125 to 800°C by the gas/gas heat exchanger, recovering about ⅔ of the biomass energy. The preburner yields about 10% of the input energy and raises the temperature to 860°C. The air leaves the gas turbine at 483°C and is reheated to 900°C by the product gas combustor.

The temperatures in the heat recovery system are indicated in Figure 7. In this figure a comparison is made between the present cycle and corresponding STIG temperature profiles, which will be discussed below. The product gas combustor exhaust is cooled in

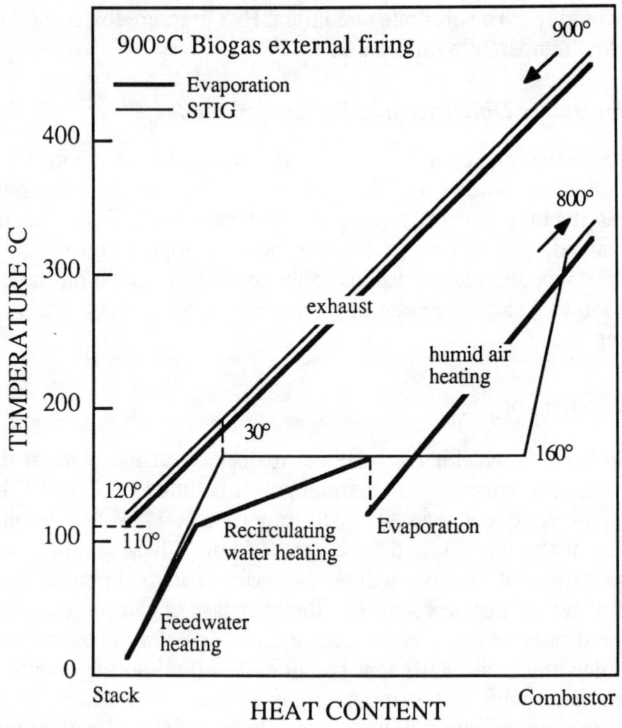

FIGURE 7 Heat recovery temperatures in the case of external firing at 900°C. Comparison between evaporation and STIG.

the gas/gas heat exchanger from 900 to 291°C, raising the temperature of the humid air from 125 to 800°C. In the liquid/gas heat recovery, the exhaust temperature is further decreased from 291 to 159°C, raising 22.7 ton/hr of recirculating water for the evaporator. The exhaust gas temperature is further decreased by preheating the feedwater from 15 to 110°C, reducing the stack temperature to the imposed limit of about 110°C. As can be seen in Figure 7 the stack energy content is determined by the stack limit temperature of 110°C, and not by the pinch-point temperature difference limit of 30°C.

Comparison with other cycles. Based on the same simulation parameters, comparisons are made with the following options in the absence of cogeneration:

— natural gas fired simple gas turbine cycle
— natural gas fired simple STIG cycle
— natural gas fired evaporation cycle
— product gas externally fired dry cycle
— product gas externally fired STIG cycle
— product gas externally fired evaporation cycle (present cycle)

The external firing is considered at two firing temperatures: 900 and 1000°C. Table VI summarises the overall output data. Figure 7 compares the heat recovery temperatures between the present cycle and the corresponding STIG cycle. Figure 8 makes a similar comparison when using direct natural gas firing instead of product gas external firing. Switching from direct natural gas firing to product gas dry external firing yields an efficiency increase of 0.8%, but the power drops to 478 kW. The efficiency increase is

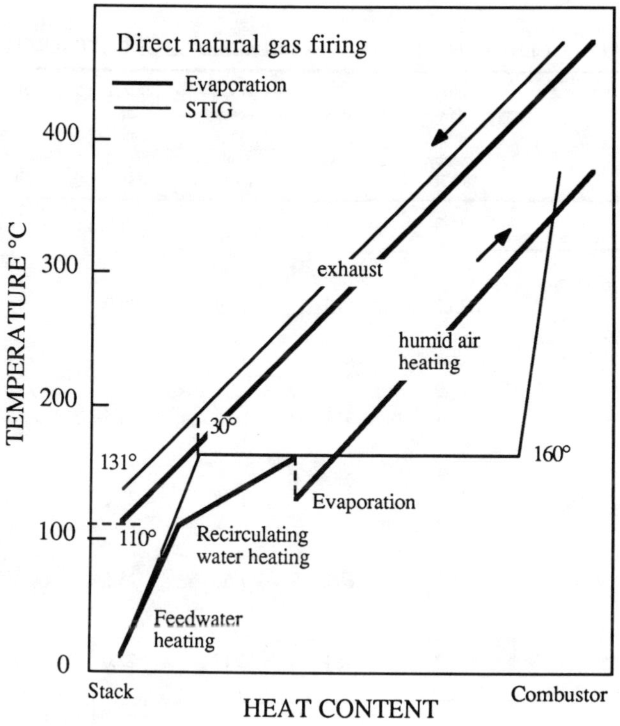

FIGURE 8 Heat recovery temperatures in the case of natural gas firing. Comparison between evaporation and STIG.

limited although the cycle is recuperative, which is due to the gas/gas heat exchanger loss and to a nonoptimal pressure ratio. The power drop is typical for regenerative dry gas turbine cycles. Inclusion of steam raises the cycle efficiency from 24.3% to 35.7% for STIG and to 36.3% for evaporation, when using direct fired natural gas. The power increases by 31 and 34%, respectively. Inclusion of steam injection in the product gas case raises the efficiency from 25.1 to 35.4 for STIG and 35.7% for evaporation, with power increases of 43.7 and 44.5%. According to these numbers, the efficiency and power output in the natural gas and product gas cases are very comparable when including steam. Hence, a penalty in power output is found when firing product gas in a dry cycle whereas no penalty is found when mixing gas and steam.

In both direct and externally fired cases, the benefit of switching from STIG to evaporation is limited. As can be seen from Figure 7, the effect of starting the evaporation at 110°C instead of 160°C pulls down the exhaust temperature profile, reducing the energy content to the stack. This effect is however limited by the minimum stack temperature of 110°C, and by the proximity of the boiling and stack temperatures. These limiting effects are more pronounced for the product gas case: as can be seen when comparing Figures 7 and 8, the shift in exhaust heat content is more important in the direct fired case, the STIG stack temperature being 131 vs. 120°C for external firing, giving a maximum shift of 21°C for STIG vs. 10°C for evaporation.

Since the stack temperature limit is easily reached when evaporating, there is no incentive to aftercool the compressor air or to include more complex evaporation systems at such low pressure ratios. At higher pressure ratios and corresponding higher boiling tempera

TABLE VI

OVERALL CYCLE OUTPUT PARAMETERS IN ABSENCE OF COGENERATION

	First law efficiency % LHV	Electric power kW	Water consumption ton/hr	Stack air humidity % total vol	Stack temperature °C	Avoided costs[a] BF/hr
Natural gas-fired simple Brayton gas turbine cycle	24.3	517	–	5.5	474	46
Natural gas-fired simple STIG cycle	35.7	679	1.19	20.9	131	637
Natural gas-fired evaporation cycle	36.3	693	1.29	22.2	110	672
Biogas externally fired dry cycle fired at 900°C	25.1	478	–	5.5	452	1340
Biogas externally fired STIG cycle fired at 900°C	35.4	687	1.48	23.0	120	1853
Biogas externally fired evaporation cycle fired at 900°C	35.7	691	1.51	23.5	110	1863
Biogas externally fired evaporation cycle fired at 1000°C	32.2	781	2.15	32.0	110	2036

[a]Power valued at 3.58 BF/kWh. Natural gas savings at 0.94 BF/kWh. Natural gas consumption at 0.84 BF/kWh. Biogas at 0.1 BF/kWh. Water at 150 BF/ton. 1 USD ≈ 30 BF.

tures, the benefit of evaporation (with aftercooling) should be more pronounced since the exhaust temperature profile is pulled up by the pinch-point in the STIG case. A further analysis of this effect is found in the literature (Berta 1990, El Masri 1988a, El Masri 1988b, Frutschi 1988, Morton 1989).

Product gas firing temperature. Since the stack limit temperature is easily reached, the evaporator parameters are not very relevant in the cycle behaviour. The only important parameter left is the temperature at which the product gas is fired. Increasing the firing temperature to 1000 instead of 900°C results in an efficiency drop of 3.5% (Table VI), but the output power increases up to 781 kW, which is an increase of 63% when compared with the simple gas turbine cycle. This increase comes from the higher energy content in the exhaust gas which raises more water vapor for injection (32% vapor in the exhaust). Although the efficiency is lower, the cost savings associated with the peak shaving can justify extra external firing of product gas. Obviously the imbalance in mass flow rates between compressor and turbine becomes very important and the limits of this imbalance are to be determined.

Cogeneration

Cogeneration can be achieved by reducing the water evaporation and by using the saved heat in the University heating system. The heating water pressure is 12 bar and its temperature ranges from 110 to 180°C. These properties being very similar to the properties

TABLE VII
OVERALL CYCLE OUTPUT PARAMETERS IN COGENERATION MODE

	First law efficiency %LHV	Power output kW	Heat output kW	Water consumption ton/hr	Stack temperature °C	Avoided costs[a] BF/hr
Natural gas-fired simple Brayton gas turbine cycle	79.0	517	1175	–	140	1156
Biogas externally fired dry cycle fired at 900°C	84.3	478	1124	–	140	2402
Biogas externally fired dry cycle fired at 1000°C	88.6	478	1593	–	140	2801
Biogas externally fired evaporation cycle fired at 900°C	61.9	585	585	0.74	116	2222
Biogas externally fired evaporation cycle fired at 1000°C	55.8	660	660	1.27	110	2373

[a]Power valued at 3.58 BF/kWh. Natural gas savings at 0.94 BF/kWh. Natural gas consumption at 0.84 BF/kWh. Biogas at 0.1 BF/kWh. Water at 150 BF/ton. 1 USD ≈ 30 BF.

of the recirculating water in the evaporator, both heat recoveries can occur in the same heat exchanger by switching from one water circuit to the other.

In theory all the exhaust heat can be recovered for heating purposes and the cycle would run as a dry externally fired cycle. The compressor exit air would enter the gas/gas exchanger at 306°C instead of 125°C. The main combustor exhaust gas would be cooled from 452 to 140°C, raising 13.8 ton/hr of water for heating purposes. The global first law efficiency would reach 84.3%, but the power output decrease from 691 to 478 kW. Since electric power is valued much higher than the thermal power, this dry mode is not necessarily attractive. In addition, the limited imbalance in mass flow through compressor and turbine will probably put a limit on the available cogenerated heat.

Table VII compares different options with cogeneration. The first two lines compare a dry external firing with the corresponding simple gas turbine cycle. Since the simple gas turbine cycle is always dry its electric power output is obviously constant and a maximum of waste heat can always be used, reaching an efficiency of 79%. The dry externally fired cycle again suffers from a lower electric power output, but the global efficiency is better than with direct firing. Firing at higher temperatures (1000°C) yields even better efficiencies. In the same table outputs are given at a heat to power ratio of 1, giving fair efficiencies of 61.9% when firing at 900°C and 55.8% when firing at 1000°C. Here again, extra firing of the product gas may increase both heat and power outputs at the cost of lower efficiency, but with higher cost savings.

ECONOMICS

The economic feasibility of the project is based on the avoided costs for the university, as follows. The power demand rises from 950 kW to 2,350 kW between 6 AM and 10 AM.

The peak power demand is already limited to 2,500 kW by shutting down some machinery between 10 AM and 3 PM. The demand decreases slowly from 2,500 to 1,000 kW between 3 PM and 12 PM.

Tarification

The mean rate of off-peak power was 2.38 BF/kWh variable cost in 1990 (1 USD ≈ 30 BF). The monthly fixed cost is normally 133 BF per kW peak power demand, regardless of the power consumption. The corresponding mean power rate was 2.58 BF/kWh. From November to February an additional cost of 666 BF per kW is counted for peak demand between 10.30 AM and 12.30 PM, and between 5.30 PM and 7.30 PM. The total power rate during these 175 hours is 17.6 BF/kWh and represents about 20% of the total power bill. These 175 hours will further be denoted as peak hours. Maximum auto-power is to be achieved during such hours: 781 kW when firing product gas at 1000°C in an evaporation cycle without cogeneration (see Table VII).

The natural gas rate was 236 BF/GJ (0.85 BF/kWh). The cost of wood waste has been rated at 10% of the natural gas cost. The exact number will depend on negotiation, and may even be negative due to the avoided disposal costs. The global cost of demineralised water is estimated at 150 BF/ton. The variable cost of the oxygen enrichment can be kept very low when using pressure swing adsorption and enrichment below 90%.

Analysis

The off-peak avoided costs of the different cycle options were already indicated in Tables VI and VII. A detailed analysis is given in the Appendix. Five options are compared:

— Natural gas dry cycle with cogeneration
— Natural gas Evaporation cycle without cogeneration
— Natural gas Evaporation cycle with cogeneration
— Product gas fired Evaporation cycle without cogeneration
— Product gas fired Evaporation cycle with cogeneration

The dry cycle can always work in cogeneration mode since the power output is constant. The evaporation cycles work in power mode during peak hours since the extra power output is valued 15 times higher than heat output. In off-peak hours, cogeneration and non-cogeneration modes are to be compared since the ratio between power and heat values is uncertain.

Due to a different tarification, the total savings are to be reduced by an auto-production penalty of about 1 MBF/year. Missing one peak in auto-production cancels the avoided fixed cost for the current month. The maximum risk of missing at least one peak each month is subtracted from the savings to find the minimum savings. The availability of the plant is essential to avoid these penalties. Switching to natural gas, bypassing of the evaporator and bypassing of the recovery unit are therefore part of the demonstration project. The availability of a 500 kW emergency-only diesel unit reduces the risk considerably.

The results show that cogeneration during off-peak hours is to be considered in any case. The evaporation yields extra savings mainly through the extra power during peak hours. The external firing of product gas triples the savings, due to higher power output and lower fuel costs when compared with the direct firing of natural gas.

A maximum cost saving of 15 MBF/year is envisaged with the considered 500 kW gas turbine, which is enough to justify the project. Upscaling to a few MW should furthermore improve the economics. A further cost analysis will be published in the near future.

CONCLUSIONS

The present project combines the gasification of biomass, external firing and water evaporation into one gas turbine cycle. The externally fired evaporation cycle yields a higher efficiency and power output than dry and/or direct fired cycles. A substantial power credit can be obtained when applying the present cycle in the Brussels University power supply in both peak and off-peak operation, encouraging a further economic analysis of the proposed cycle.

ACKNOWLEDGMENTS

The present work has been financed by the Brussels Executive Government, contract no. 90/1057.

REFERENCES

Andreuccetti, P., Barbucci, C., and Trebbi, G. (1991). Electricity production by an advanced turbo generator fired with biomass. Presented at the *6th European Conference on Biomass for Energy*, Industry and Environment, Athens.

Baron, G. (1990). Pressure swing adsorption: Latest developments, economic aspects and modelling. In Proceedings Gas Separation Technology, Eds. E.F. Vansant and R. Dewolfs, Elsevier, Amsterdam.

Berta, G.L., Durelli, E., and Prato, A.P. (1990). A special arrangement of hybrid gas turbines. *AICHE 25th IECEC Conference Proceedings*, **5**, p. 495.

Bridgwater, A.V. (1990). Economics of primary fuels production from biomass. *AICHE 25th IECEC Conference Proceedings*, **5**, p. 107.

Buekens, A., Bridgwater, A., Ferrero, G., and Maniatis, K. (1990). Commercial and Marketing Aspects of Gasifiers, Commission of the European Communities, EUR 12736, Brussels.

Cuevas, A., and Fluxa, J. (1991). Combustion tests on gas turbines and power generation equipment. Presented at the *6th European Conference on Biomass for Energy, Industry and Environment*, Athens.

Distelmans, M., and De Ruyck, J. (1992). Labcycle: An integrated simulation package for energy systems based on Labview and Oracle, *ASME ECOS'92 Conference Proceedings*, pp. 93–100, 1992.

El Masri, M.A. (1988a). A flexible, efficiency gas-turbine cogeneration cycle with a novel dual-mode heat recovery system. *ASME COGEN II Conference Proceedings*, 1988.

El Masri, M.A. (1988b). A modified, high-efficiency, recuperated gas turbine cycle. *ASME Journal of Engineering for Gas Turbines and Power*, **110**, April p. 232.

Ferrero, G.L., Maniatis, K., Buekens, A., and Bridgwater, A.V. (1989). *Pyrolysis and Gasification*. Elsevier Applied Science, 1989.

Frutschi, H.U., and Plancherel, A. (1988). Comparison of combined cycles with steam injection and evaporisation cycles. *ASME COGEN II Conference Proceedings*.

Grassi, G., and Bridgwater, A.V. (1990). European research in biomass energy. *AICHE 25th IECEC Conference Proceedings*, **5**, p. 126.

Hamrick, J.T. (1985). Development of wood as an alternative fuel for gas turbines. In *Proceedings of the 1985 Biomass Thermochemical Conversion Contractors Meeting*, Minneapolis, MN.

LaHaye, P.G., and Zabolotny, E. (1989). Externally fired combined cycle. *ASME COGEN III Conference Proceedings 1989*, p. 263.

Larson, E.D., and Svenningsson, P. (1990). Development of biomass gasification systems for gas turbine power generation. *Proceedings of the Energy from Biomass and Wastes XIV Conference*, Lake Buena Vista, Florida.

Kolp, D.A., and Moeller, D.J. (1989). World's first full STIG LM5000 Installed at Simpson Paper Company. *ASME J. of Gas Turb. and Power*, **111**, April p. 200.

Maniatis, K. (1990). Fluidized bed gasification of agricultural residues. In *Proceedings International Symposium on Application & Management of Energy in Agriculture*, New Delhi.

Maniatis, K., Bridgwater, A.V., and Buekens, A. (1988). Fluidized bed gasification of wood. In *Research in Thermochemical Biomass Conversion*, Eds. A.V. Bridgwater and J.L. Kuester, Elsevier.

Morton, T.R. (1989). Perspective for advanced high efficiency cycles using gas turbines. *EPRI 1989 Conference on Technologies for Producing Electricity in the Twenty-First Century*.

APPENDIX

PLANT AVOIDED COST ANALYSIS

Approximate 1990 tarification parameters	
Mean off-peak power cost	2.38 BF/kWh
Fixed peaktime cost	666 BF/kW/month
Fixed off-peak cost	133 BF/kW/month
Peaktime months	4 months/year
Off-peaktime months	10 months/year
Raw biomass price	23 BF/GJ
Natural gas price	236 BF/GJ
Raw biomass price	0.08 BF/kWh
Natural gas price	0.85 BF/kWh
Water cost	150 BF/ton
Heating system efficiency	90%

Auto-production parameters	Evap. cycle	Dry cycle	Evap. cycle	Evap. cycle	Evap. cycle
Firing	Interior	Interior	Exterior	Exterior	Exterior
Fuel	Nat. gas	Nat. gas	Nat. gas	Biogas	Biogas
Cogeneration	No	Yes	Yes	No	Yes
Peak production time (hr/year)	175	175	175	175	175
Peak power autoproduction (kW)	603	517	693	781	781
Peak heat autoproduction (kW)	0	1175	0	0	0
Peak biomass consumption (kW)	0	0	0	2168	2168
Peak natural gas consumption (kW)	1900	2123	1900	260	260
Peak water consumption (ton/hr)	1.29	0.00	1.29	2.15	2.15
Off-peak production time (hr/year)	6500	6500	6500	6500	6500
Off-peak power production (kW)	693	517	517	781	478
Off-peak heat production (kW)	0	1175	1175	0	1593
Off-peak biomass consumption (kW)	0	0	0	2168	2097
Off-peak natural gas consumption (kW)	1900	2123	2123	260	241
Off-peak water consumption (ton/hr)	1.29	0.00	0.00	2.15	0.00

	Results				
Peak kWh valued at (BF/kWh)	17.60	17.60	17.60	17.60	17.60
Peak avoided fixed costs (MBF/year)	1.846	1.377	1.846	2.081	2.081
Peak avoided kWh costs (MBF/year)	0.289	0.215	0.280	0.325	0.325
Peak biomass consumption (MBF/year)	0.000	0.000	0.000	–0.031	–0.031
Peak gas consumption (MBF/year)	–0.282	–0.316	–0.282	–0.039	–0.039
Peak water consumption (MBF/year)	–0.034	0.000	–0.034	–0.056	–0.056
Total peak savings (MBF/year)	**1.818**	**1.277**	**1.818**	**2.279**	**2.279**
Off-peak kWh valued at (BF/kWh)	2.58	2.58	2.58	2.58	2.58
Off-peak avoided fixed costs (MBF/year)	0.922	0.688	0.688	1.039	0.636
Off-peak avoided kWh costs (MBF/year)	10.721	7.998	7.998	12.082	7.395
Off-peak avoided gas (MBF/year)	0.000	7.210	7.210	0.000	9.775
Off-peak biomass consumption (MBF/year)	0.000	0.000	0.000	–1.167	–1.129
Off-peak gas consumption (MBF/year)	–19.493	–11.724	–11.724	–1.436	–1.331
Off-peak water consumption (MBF/year)	–1.258	0.000	0.000	–2.096	0.000
Total off-peak savings (MBF/year)	**–0.108**	**4.171**	**4.171**	**8.422**	**15.346**
Autoproduction penalties (MBF/year)	–1.000	–1.000	–1.000	–1.000	–1.000
Estimated maintenance costs (MBF/year)	–1.000	–1.000	–1.000	–1.500	–1.500
Maximum savings (MBF/year)	**–0.289**	**3.448**	**3.990**	**8.201**	**15.125**
Autoproduction risk (MBF/year)	–2.768	–2.065	–2.534	–3.119	–2.716
Minimum savings (MBF/year)	**–3.057**	**1.383**	**1.456**	**5.082**	**12.409**

Clean Combustion and Excess Air Factors

E. MOERMAN and K. KRISHNA PRASAD *Woodburning Stove Group, Faculty of Physics, Eindhoven University of Technology, Eindhoven, The Netherlands*

Abstract — In this paper a model based on simple elementary reactions is presented which enables the prediction of the CO/CO_2 ratios resulting from a woodfire in a so-called downdraft stove. When the fuel composition and the reaction rates are known the excess air factor is the sole parameter determining the combustion quality. Comparison with experimental data shows a good qualitative agreement. Improvements can be made by adding reactions, taking another temperature distribution and more detailed reaction mechanisms. In the downdraft stove the fuelbed has an important influence on the excess air factor. The airflow through the fuelbed depends as (a exp{–b M}) on the fuelbed mass and was experimentally verified for a cold fuelbed for both wood and charcoal. For a weight up to 0.06 kg it also holds for burning charcoal. However for higher fuelbed weights the temperature evolution at the air inlet must probably be taken into account.

Key Words: airflow, biomass, carbon monoxide, combustion, downdraft, stove

INTRODUCTION

Cooking tasks in most of the developing countries are still carried out on open fires or their close relatives using one form or another of the biomass, the principal one of which is wood. Many energy studies in the past 15 years or so have clearly established that cooking accounts for as much as 80% of the total energy used in rural areas (see for example Goldcmberg *et al.* 1988). The same reference also establishes how inefficient the traditional open fires are: the households using the latter can consume as much as 8 times that of a simple kerosene wick stove in India. In view of the ongoing deforestation, most developing countries and international assistance agencies have launched programs to introduce improved cookstoves.

There is a second dimension to this problem. The traditional stoves emit immense amounts of pollutants such as carbon monoxide and unburnt products of pyrolysis — which are poisonous, affect the respiratory systems and are carcinogenic — into ill-ventilated kitchens (Smith, 1987). Many an improved cookstove attempts to overcome this problem by providing a chimney to vent the pollutants out of the kitchen. Such a procedure is not only expensive in its first cost but also quite difficult to maintain. It is this problem that was addressed in this paper by attempting to develop combustion systems that can burn biomass fuel cleanly.

In the past three years the group has been investigating the process of downdraft combustion as a means of burning biomass cleanly. The paper starts out with a brief description of the early work with this mode of combustion that establishes the role of excess air on the cleanliness of combustion. This is followed by a mathematical model based on global chemical reactions that attempts to capture the role of excess air on the quality of combustion. Since the earlier experiments of Khan *et al.* (1989) indicate that fuel bed flow resistance is an important component in determining the air flow in a naturally aspirated system, a novel method of measuring this resistance in cold fuel bed experiments is described.

In the entire work we use CO/CO_2 ratio as a marker for clean combustion since it has been shown in earlier investigations that a high CO/CO_2 ratio is always accompanied by large amounts of other unburnt compounds of pyrolysis (van Hoek 1989, Sangen 1983).

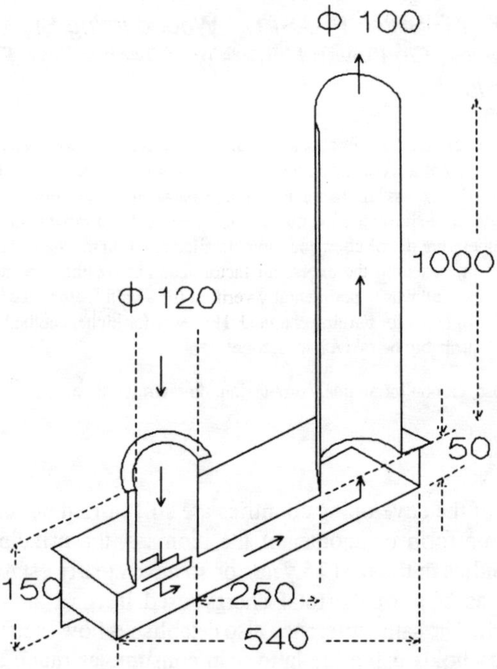

FIGURE 1 The downdraft stove used in our experiments. Dimensions are in mm.

DESCRIPTION OF A DOWNDRAFT STOVE

In a downdraft stove the fuel and the airflow are in the same direction (see figure 1). In order to provide the draft needed to force the airflow downwards a chimney is needed. The hot combustion products passing through the chimney provide the necessary draft with a chimney length of about 0.5 m. Since the flue gas is clean it can be emitted into the kitchen without a health hazard; thus a longer chimney is not needed. A more detailed description of the stove can be found, for instance, in Khan (1990). Experiments performed by Khan indicate that it is possible to obtain very low CO/CO_2 ratios (<0.1%) while smoke and odor in the flue gas are absent. In figure 2 the CO/CO_2 ratio as a function of the excess air factor for a downdraft stove is compared with those for a conventional stove (Dirks (1989)). This graph not only shows the superiority of the downdraft stove as far as combustion quality is concerned but also shows the importance of the air supply for the downdraft stove. Since the air has to pass the fuel bed and the fuel bed can vary in height and structure it is important to know its resistance to flow.

A MATHEMATICAL MODEL OF WOOD COMBUSTION

Experiments have shown that operating conditions and the design of the downdraft stove determine the combustion quality. In a real cooking situation operating conditions are too cumbersome for practical use which is why it was necessary to make the operating conditions less of a constraint. To help develop designs less sensitive to these operating conditions, and therefore better suited for practical use, a qualitative picture of the combustion in the first instance was required. The model presented in this section was developed with this idea in mind.

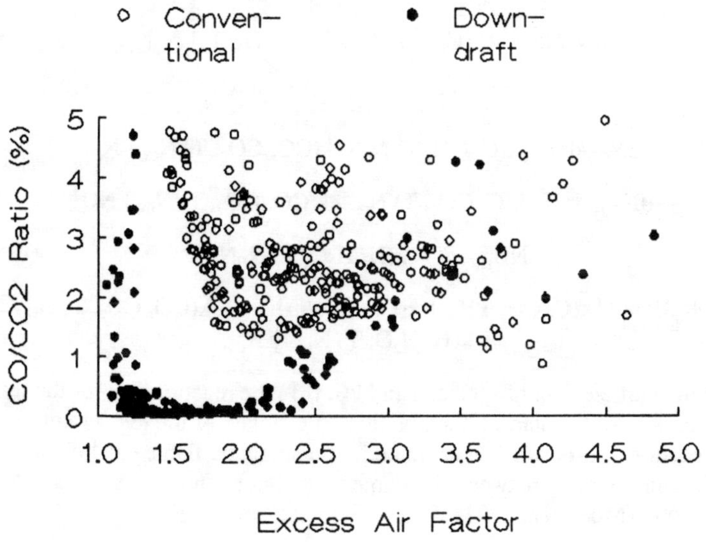

FIGURE 2 Combustion quality (CO/CO_2 ratio) as a function of the excess air factor in a conventional stove and in a downdraft stove.

The detailed evolution equations for the fuel and for the reaction products are often solved numerically in order to predict the flue gas composition. However this is a painful task when the combustion of wood is involved. According to Emmons and Atreya (1983) over 200 hydrocarbons can occur as a result of the pyrolysis occurring in a wood fire. In essence the same method of solving the evolution equations numerically was used in this instance. However a few crude approximations are made in the hope of grasping the essential phenomena occurring with a simple model. It was assumed that the fuel is partly atomized (in the fuel carbon and hydrogen occur as atoms and oxygen as O_2). Further assumptions are that the flame is pre-mixed, there is only one reaction temperature and that an ignition temperature or a flame front is not taken into account. The only reactions considered are:

$$C + \tfrac{1}{2}O_2 \rightarrow CO, \tag{1}$$

$$C + O_2 \rightarrow CO_2, \tag{2}$$

$$CO + \tfrac{1}{2}O_2 \rightarrow CO_2, \tag{3}$$

$$2H + \tfrac{1}{2}O_2 \rightarrow H_2O. \tag{4}$$

Dissociation effects have been ignored by noting that even the thermal equilibrium is not established in the system. The reactions given here are somewhat arbitrary because not all reaction mechanisms are known. Especially the reaction mechanisms and reaction rates of reaction 2 are subject of discussion (see Laurendeau, 1978 and Yang, 1984). Reaction 2 is of major importance when the CO/CO_2 ratio is the subject of study. Again a crude approximation is made and only reactions 1 to 4 are considered. The reaction $CO_2 + C \rightarrow CO$ is neglected since the neglected mechanisms of reaction 2 have a far more important influence on the predicted CO/CO_2 ratios. The reactions 1 to 4 lead to the following evolution equations.

$$dN_C/dt = -\{k(C, CO, T) + k(C, CO_2, T)\}\, N_C N_{O_2}, \tag{5}$$

$$dN_H/dt = -k(H, H_2O, T)\, N_H N_{O_2}, \tag{6}$$

$$dN_{CO}/dt = \{k(C, CO, T)\, N_C - k(CO, CO_2, T)\, N_{CO}\}\, N_{O_2}, \tag{7}$$

$$dN_{CO_2}/dt = \{k(C, CO_2, T)\, N_C + k(CO, CO_2, T)\, N_{CO}\}\, NO_2, \tag{8}$$

$$dNH_{H_2O}/dt = k(H, H_2O, T)\, N_H N_{O_2}/2, \tag{9}$$

$$dN_{O_2}/dt = \{\{k(C, CO, T)/2 + k(C, CO_2, T)\}\, N_C + k(CO, CO_2, T)\, N_{CO}/2 + k(H, H_2O, T)\, N_H/4\}\, N_{O_2}, \tag{10}$$

where N_i is the particle flux of species i and k(i, j, T) the reaction rate for the reaction i $\rightarrow$ j at temperature T. At the start of a simulation an estimate of the temperature is needed at which the reaction will occur. The estimate for an adiabatic flame can be obtained. In that case the following relation between the flame temperature and the excess air factor can be derived for wood (Moerman, 1991).

$$\lambda = 1940/\Delta T - 0.218, \tag{11}$$

where λ is the excess air factor (the ratio of the air used and the stoichiometric amount of air), and ΔT, the temperature difference between flame and surroundings. Experiments performed by Moerman (1991) result in figure 3. Each line gives the behavior for the period between two consecutive fuel charges when steady-state burning is reached. The coefficients of these lines differ from the theoretical ones in equation 11. When the experimentally found coefficients are averaged the following equation results:

$$\lambda = 14800/\Delta T - 19.4. \tag{12}$$

Assuming the reaction rate to be described by the Arrhenius factor the reaction rate can be written as:

$$k(i, j, T) = k(i, j, T_0)^{\alpha}, \tag{13}$$

with $\alpha = T_0/T$. In the literature (Lewis and von Elbe, 1961) the ratios of some of these reaction rates can be found and assume to be obtained at a certain equivalent temperature $\Delta T = 726$ K (Moerman, 1991) for a stoichiometric mixture and leaving only the unknown absolute reaction rates and one unknown reaction rate which can be crudely fitted to experimental data. In this simulation the following reaction rates were used: $k(C, CO, T_0) = 10^{-21}\ s^{-1}$, $k(C, CO_2, T_0) = 1.5 \times 10^{-20}\ s^{-1}$, $k(CO, CO_2, T_0) = 5.7 \times 10^{-20}\ s^{-1}$ and $k(H, H_2O, T_0) = 4 \times 10^{-20}\ s^{-1}$ ($T_0 = 726$ K) and the time step Δt for the simulation is 2×10^{-4} s. The initial conditions for the N_i are obtained from the burning rate and the air flow in the simulation. Since the oxygen present in wood (White Fir) accounts for the reaction with hydrogen the excess air factor is taken as:

$$\lambda = \{O_2\}/\{C\}. \tag{14}$$

The error introduced due to this is only a few percent. With λ determined, the reaction temperature T_1 and the appropriate reaction rates can be calculated using equations 12 and 13, respectively. The result of this simulation is plotted in figure 4 together with experimental data. On observing the carbon that is left at the end of the simulation it can be seen that part of the fuel is not burned and therefore has not taken part in heating the mixture to the

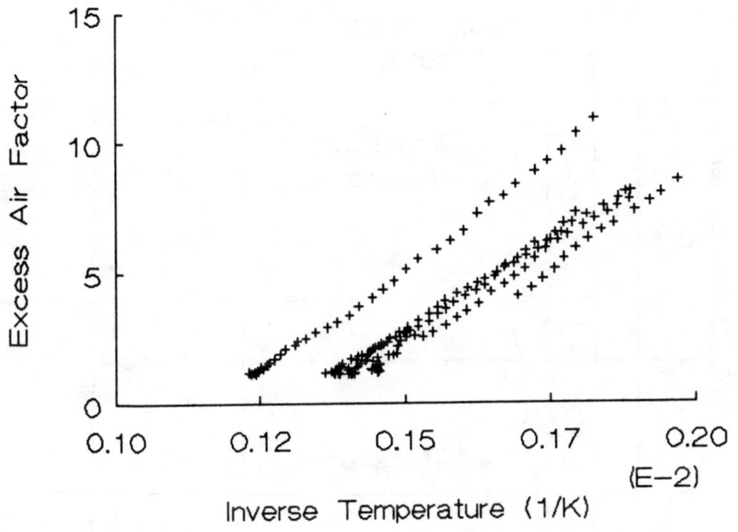

FIGURE 3 Experimental relation between the excess air factor and the adiabatic flame temperature in a downdraft stove.

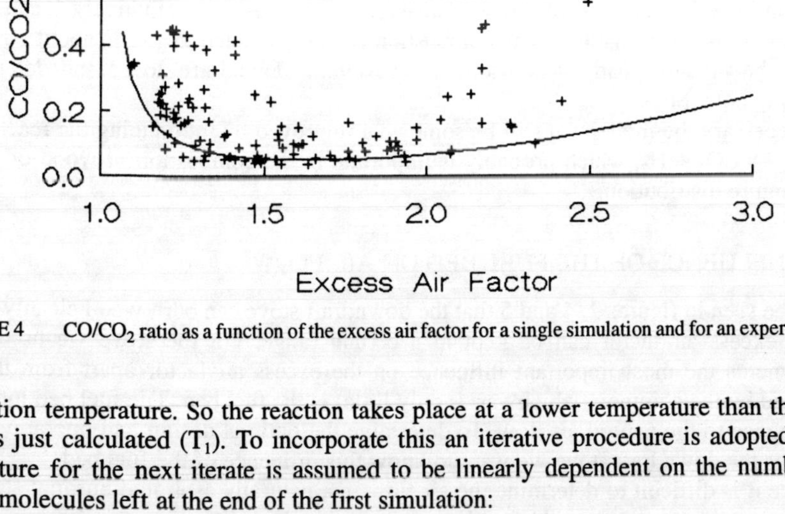

FIGURE 4 CO/CO_2 ratio as a function of the excess air factor for a single simulation and for an experiment.

combustion temperature. So the reaction takes place at a lower temperature than the one that was just calculated (T_1). To incorporate this an iterative procedure is adopted. The temperature for the next iterate is assumed to be linearly dependent on the number of oxygen molecules left at the end of the first simulation:

$$T_2 = \frac{T_1}{[C] + [H]/4 - [O]/32} \left([O_2] - N_{O_2}\right). \tag{15}$$

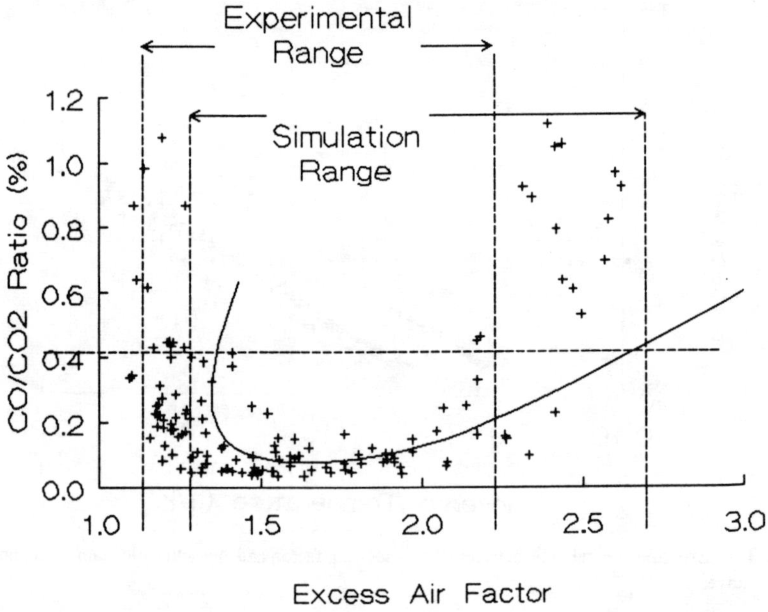

FIGURE 5 The combustion quality versus excess air factor for an experiment and according to the iterative simulation.

where [i] denotes the fraction of element i according to the ultimate analysis. By setting $k(C, CO_2, T_0) = 1.9 \times 10^{-20}\ s^{-1}$, this unknown parameter is adjusted to the new situation. The result of this simulation is presented in figure 5 together with experimentally obtained data. Naturally this procedure should be repeated until it converges to a certain temperature, however, it does not converge rapidly enough so has been stopped after the second iteration. The simulation results agree with experimental results qualitatively. It can be seen in figure 5 that the range for clean combustion can be predicted to within about 10%. This makes the model already very useful for designing downdraft stoves and determining operating conditions.

If necessary the model can still be somewhat improved by introducing the reaction $CO + H_2O \rightarrow CO_2 + H_2$, which probably is important in the combustion of wood or a better temperature distribution.

THE INFLUENCE OF THE FUEL BED ON AIR FLOW

It can be seen in figures 2, 4 and 5 that the downdraft stove can burn wood cleanly so long as the excess air factor can be kept in a certain range. For the stove we used in our experiments the most important influence on the excess air factor apart from the draft provided by the chimney (which varies only little) is the fuel bed. The fuel bed introduces a resistance to the natural air flow. To determine the optimal design and operation conditions for the downdraft stove we want to know this influence of the fuel bed.

Since it is difficult to determine the air flow in a naturally aspirated burning stove first a few experiments with a cold fuel bed were undertaken. For this case we can derive a relation between the mass M of the fuel bed (which is easier to measure than the height of the fuel bed) and the air flow ϕ through the fuel bed. It is assumed that the woodblocks used are of uniform shape and size with a thickness Δh. A single layer of randomly distributed blocks reduces the free grate area by a factor x. Each additional layer reduces the free grate area by a further factor x. Thus the air flow through the fuel bed made up of multiple layers can be written as:

$$\phi = v\, A\, \rho_{air}, \tag{16}$$

where v, the velocity for an incompressible fluid, can be written as:

$$v = (2\Delta p/\rho_{air})^{1/2}, \tag{17}$$

and A is the effective area of flow given by:

$$A = (1 - x)^{M/\rho A_{fb}\Delta h}, \tag{18}$$

where A_{fb} is the total free area of the grate and ρ the bulk density of the fuel on the fuel bed. Thus,

$$\phi = (2\Delta p/\rho_{air})^{1/2}(1 - x)^{M/(\rho A_{fb}\Delta h)}\, \rho_{air}. \tag{19}$$

Experimental Procedure

Equation 19 can be verified experimentally. The normal burning situation is simulated by placing a bunsen burner at the bottom of the chimney (see figure 1) to create a draft. In order to determine the air flow through the fuel bed the composition of the gas is needed. Natural gas ("Gronings aardgas") was used with its composition according to the gas company being 0.89 vol% CO_2, 0.01% O_2, 14.35% N_2, 81.3% methane, 2.85% ethane, 0.37% propane, 0.14% butane, 0.04% pentane and 0.05% hexane. The remaining gases are negligible. The burner provided a power of 945 W. Since the composition of the gas is known and the flow through the bunsen burner is measured with a flow meter the total number of C atoms present in the flue gas can be calculated. Using the composition of the fuel given above a calculation shows that 1 mole of fuel (gas) is equivalent to 0.9006 mole C atoms, 3.4784 mole H atoms and 0.009 mole O_2 molecules. To burn this completely,

$$0.9006 + \tfrac{1}{4}\, 3.4784 - 0.009 = 1.7612 \text{ mole } O_2 \tag{20}$$

is needed, this being the stoichiometric amount of air. The fuel is assumed to be completely burned, therefore the sum of the measured concentrations CO and CO_2 in the flue gas has to equal the flux of C-atoms.

$$[CO] + [CO_2] \equiv 0.9006\phi \text{ (mole/s)}, \tag{21}$$

where ϕ is the fuel flow in mole/s. To burn the fuel completely 1.76 ϕ mole of O_2 per second is needed. The amount of oxygen left in the flue gas is $[O_2]/([CO]+[CO_2])\ 0.90\ \phi$ (mole/s). This gives a total oxygen flow of:

$$\phi(1.76 + [O_2]\, 0.90/([CO] + [CO_2])) \tag{22}$$

The air flow (kg/s) is now easily calculated from the oxygen flow.

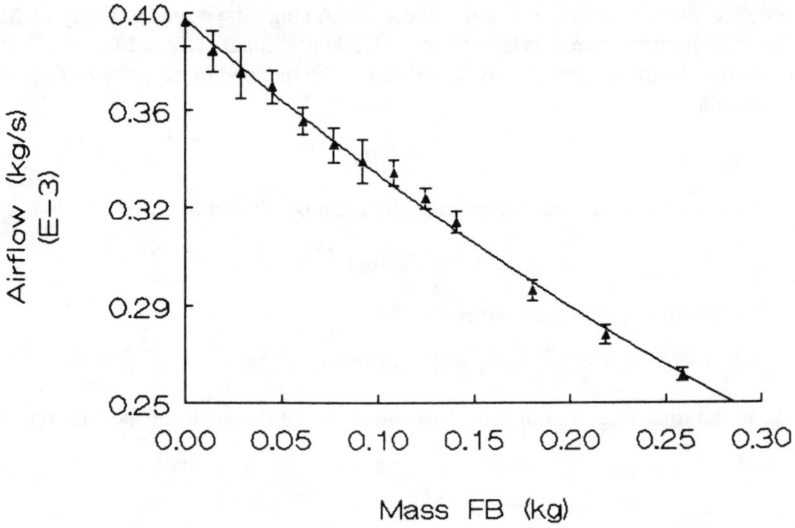

FIGURE 6 Air flow through a cold stove as a function of the weight of the fuel (charcoal) on the fuelbed

Cold Experiments

The experiment is started when steady state is reached. During the experiment, fuel is added without disturbing the fuel bed. In figure 6 the air flow is seen, as a function of the fuel bed weight for an experiment with charcoal; an exponential ($a \cdot \exp\{-bM\}$) is fitted to the data as is suggested by equation 19 with constant pressure difference Δp. Although the shape and sizes of the charcoal pieces were certainly not uniform, the air flow can be described by the exponential very well. The data can also be described by a linear relation between the air flow and the inverse fuel bed weight ($c/\{M + d\}$). The difference between that and the exponential is virtually impossible to see on the scale of the graph in figure 6. The experiment was also carried out for wood blocks (White Fir $20 \times 20 \times 30$ mm). The fitted parameters are given in table I.

In the experiments for wood, the wood blocks were numbered and in each experiment thrown on the fuel bed in the same order. The constant a represents the air flow through the system if the fuel bed is empty and it varies in these experiments by about 10% in case of wood, and slightly over 10% in case of charcoal. The constant b varies substantially for wood, while it is reasonably constant for charcoal.

TABLE I
PARAMETERS FITTED TO EXPERIMENTS

White fir		Charcoal	
a (10^{-3} kg/s)	b (kg^{-1})	a (10^{-3} kg/s)	b (kg^{-1})
3.73	1.88	3.33	2.40
3.83	1.92	2.96	2.46
3.47	1.54	2.89	2.47
3.47	1.40		

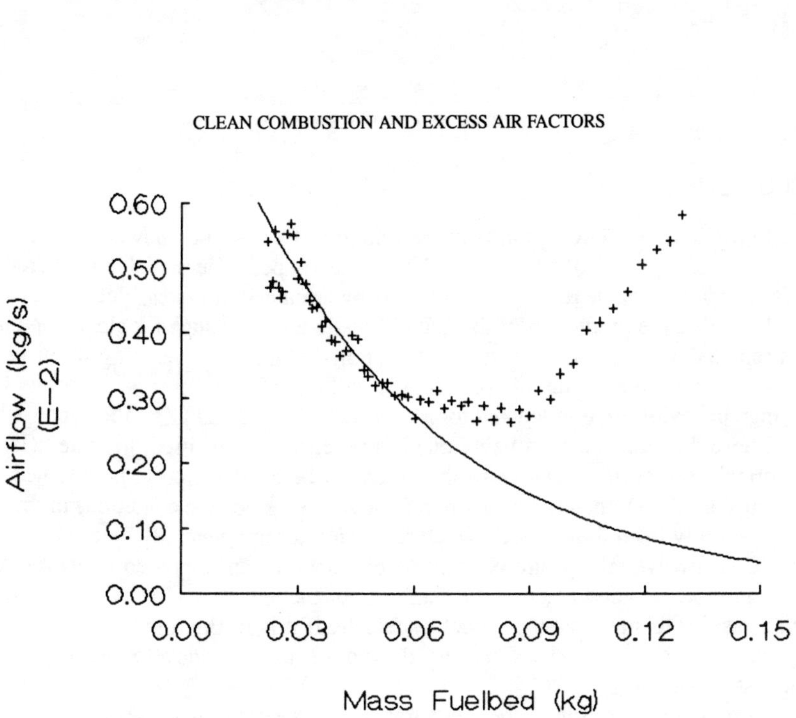

FIGURE 7 Air flow through a cold stove burning charcoal versus fuelbed weight.

Burning Experiment

The above results justify making an attempt to determine the air flow through the system in a charcoal burning stove. In this case the charcoal on the fuel bed is burning and the hot combustion products passing through the chimney provide the necessary draft. The weight of the stove, the temperature at the top and at the bottom of the chimney and the CO, CO_2, and O_2 contents of the flue gas are measured every 10 seconds. The composition of the fuel is taken as 90.2% C, 2.4% H and 2.9% oxygen as given by Rose and Cooper (1977). The stoichiometric amount of air can now be calculated in the same way as for natural gas. To burn 1 kg of char the stoichiometric amount of air required is:

$$(0.902/12 + 0.024/4 - 0.029/32)\ 10^3 = 80.3 \text{ mole } O_2 \equiv 2.57 \text{ kg } O_2 \qquad (23)$$

From the burning rate we can determine the flux of C atoms and from that we can derive the air flow through the system in the same way as was done for the bunsen burner. The air flow can now again be given as a function of the fuel bed weight for a stove burning charcoal (see figure 7). In this case the draft provided by the chimney was certainly not constant as is necessary to fit the exponential to the data. For a weight up to 0.06 kg on the fuel bed, however, the curve is virtually exponential. The strange fact that the air flow is very high when a charge has just been added (so a high fuel bed weight) can be explained as follows. The temperature of the air entering the stove is influenced by the burning char. When a fresh charge is added it shields the air from the fire, resulting in a large density difference between the flue gas in the chimney and the entering air. When the newly added fuel itself starts burning (in this experiment a few minutes after charging) the air is heated by the fire, resulting in a smaller density difference and therefore a smaller air flow. This seems to be confirmed by experiments. However more experiments will be carried out to check the details of equation 19.

CONCLUSIONS

In order to predict the CO/CO_2 ratios in the flue gas of a so-called downdraft stove for a biomass fuel under given conditions a model was developed. The model is very crude and simple (a simulation can be run on an PC-XT taking around 10 minutes), yet it gives a quite reasonable qualitative prediction of the CO/CO_2 ratio and the range for clean combustion can be predicted with an error of about 10%. For the aim of developing clean burning woodstoves with usable operating conditions it will do fine. For excess air factors below 1.3 a strange behavior is exhibited and for λ exceeding 2.0 the CO/CO_2 ratio is consistently too low. These deviations are probably due to the reaction rates used; the rate $k(H, H_2O, T_0)$ is probably not correct relative to the others. A better description is obtained if the reaction rates are determined in the downdraft stove itself since the conditions in this stove differ substantially from those under which reaction rates are generally determined. Other improvements involve taking into account other reactions, an improved temperature distribution and more detailed reaction mechanisms for the reactions already considered. All in all it seems fair to state that the model, especially if it is further improved, can be used to design naturally aspirated, clean-burning, downdraft stoves for developing countries (for which it was meant).

In the present configuration of the downdraft stove, the air flow passes through the fuel bed. This fuel bed presents a resistance to the passing air flow. For a cold fuel bed the air flow through the system can, to quite a good degree, be taken as exponentially dependent on the mass of the fuel bed.

A burning fuel bed, however, poses problems in that it is difficult to obtain useful data with the equipment we have. However we managed to obtain the air flow as a function of the fuel bed weight for burning charcoal. This experiment clearly shows that the exponential dependence can only be used for a fuel bed weight up to 0.06 kg (with all fuel burning). For higher fuel bed weights, phenomena occur that need to be looked at more closely and a number of new experiments undertaken. Some preliminary experiments seem to suggest that the temperature of the air at the air inlet has a large influence on the airflow. Thus the influence of the fuel bed on the air flow is not yet completely known but further well-planned experiments to throw some light on the matter are being planned.

ACKNOWLEDGMENT

The authors wish to thank A.M. Dirks for performing the experiments used on the conventional stove. We would also like to thank O. van Dongen for the performance of the experiments to determine the influence of the fuel bed on the air flow.

REFERENCES

Dirks, A.M. (1989). *Tests on the Experimental Stove.* Eindhoven University of Technology, Eindhoven, The Netherlands.

Emmons, H.W., and Atreya, A (1983). The science of wood combustion. In K. Krishna Prasad and P. Verhaart, Eds., *Wood Heat for Cooking*, Indian Institute of Science, Bangalore, India, Chap. 1, pp. 5–14.

Goldemberg, J, Johansson, T.B., Reddy, A.K.N., and Williams, R.H. (1988). *Energy for a Sustainable World.* Wiley Eastern, New Delhi.

Khan, A.M., and Hasan, R. (1990). *Clean Combustion of Wood Part II.* WSG, Eindhoven University of Technology, Eindhoven, The Netherlands.

Khan, A.M., Hasan R, Schutte, E., Krishna Prasad, K., and Verhaart, P. (1989). Application of downdraft combustion to woodburning devices. *Proceedings of the Ninth Heat Transfer Conference, Jerusalem.*

Laurendeau, N.M. (1978) Heterogeneous kinetics of coal char gasification and combustion. *Prog. Energy and Combust. Sci.*, **4**, pp. 221–270.

Lewis, B. and von Elbe, G. (1961). *Combustion, Flames and Explosions of Gases*, 2nd ed. Academic Press, London.

Moerman, E. (1991). Theory of clean combustion. In K. Krishna Prasad, E. Schutte, and E. Moerman, Eds., *Woodcombustion Studies Part II*, WSG, Eindhoven University of Technology, Eindhoven, The Netherlands, Chap. 3, pp. 68–89.

Rose, J.W., and Cooper, J.R. (1977). *Technical Data on Fuel.* The British National Committee World Energy Conference, London.

Sangen, E. (1983). A survey of test results on woodstoves. In K. Krishna Prasad and E. Sangen, Eds., *Technical Aspects of Woodburning Stoves, WSG*, Eindhoven University of Technology, Eindhoven, The Netherlands, Chap. 7, pp. 199–259.

Smith, K.R. (1987). *Biofuels, Air Pollution and Health. A Global Review*. Plenum Press, New York.

van Hoek, R. (1989). *Woodstoves, the Possibilities of Clean Combustion.* Polytechnic College, Breda, The Netherlands.

Yang, R.T. (1984). Etch-decoration electron microscopy studies of the gas/carbon reactions. In P. A. Thrower Ed., *Chemistry and Physics of Carbon*, Marcel Dekker Inc., New York, Vol. 19, Chap. 3, pp. 163–210.

Gasification of Pine and Cork Suspensions in a Cyclone Reactor

L. ARAÚJO *Nat. Lab. of Eng. and Ind. Tech., Lisboa*
J. CAMPOS *Dept. of Mech. Eng., Fac. of Science and Tech., Univ. of Coimbra, Coimbra, Portugal*
A. PORTUGAL *Dept. of Chem. Eng.-Fac. of Science and Tech., Univ. of Coimbra, Coimbra, Portugal*

Abstract — This communication concerns the influence of equivalence ratio, temperature and residence time, on the gasification levels of pine and cork particles and emissions of pollutants from a cyclone reactor. The structure of the solid residues, the collection efficiency of the cyclone, as well as the chemical composition of the exhaust gases at the most significant regimes, are analysed and discussed. The results obtained show mass loss increasing with the equivalence ratio and reducing particle size. Increasing heating rates also raises the gasification levels. The microscopic analysis of the solid residues led us to define three intervals of heating rate, as a function of the structural degradation of the particles, and confirmed the existence of a structural explosion of the smallest particles at the highest heating rates. This microscopic analysis, coupled with mass balances of the solid fuel, suggests increasing internal swirl and collection efficiency with the expansion phenomena associated with the burned gases. The equilibrium composition of the exhaust gases was calculated with the numerical code THOR, using different equations of state for real gases and solid species. The minimum value of Gibbs free energy was calculated for free pressure and temperature conditions. Standard gas chromatography was used for gas analysis of CO_2, CO, CH_4, O_2 and N_2. A chemiluminescence analyser was used for the determination of NO and NO_2. The measured values are in good agreement with the calculated results and show low concentrations of CO and CO_2. No significant values of NO_x in the exhaust gas were found.

Key Words: pine and cork; cyclone reactor; gasification; structural degradation; collection efficiency, pollutants

INTRODUCTION

The gasification and combustion of wood dusts are nowadays interesting sources of heat production for industrial purposes. The majority of the studies published so far, have dealt with large samples, and their application to small particles is often inadequate. This lack of information concerning the ignition, gasification and combustion of micron-sized wood particles, led us to study the thermochemical behaviour of these dust materials, based on the dependence of the mass depletion rate on environmental gas temperature and particle size, for different concentrations of solid particles in atmospheric air (Araújo and Campos, 1987, 1988, 1989; Araújo 1990).

The recent EEC directives on emissions of pollutants (CO, NO_x and SO_x concentrations) and particulate emissions encourage the development of gasification or combustion systems with high levels of burnout and of solid residues collection efficiency. Cyclone combustors, well known as effective means of reducing particulate emission, seem to be one of the most well-adapted reactors for flash pyrolysis and gasification of biomass particles, producing a large yield of gases, while tar and char are minimized by fast heating (Cook and Ford, 1991).

In this communication a study of the influence of equivalence ratio, temperature and residence time on the gasification levels and final structure of pine and cork particles, in a cyclone reactor, is presented. Gasification and combustion was carried out in a cyclone provided with two closed collector pockets where stationary vortexes are formed allowing the separation of the particles from the transport flow. The water cooled vortex collector pockets also allow the quenching of reacting particles, immediately after their trajectory inside the reactor, for further study of the effects of heating rate on their structure. Gas sampling was carried out during the tests by means of a probe tube placed in the exhaust pipe.

Three working regimes of the cyclone reactor were selected, related to the global equivalence ratio r (ratio wood concentration/air, for stoichiometry $r = 1$):

— $1.0 < r < 1.3$, equivalence ratio near to the best combustion conditions,
— $1.4 < r < 1.8$, transition to rich mixtures,
— $3.0 < r < 4.2$, high wood solid concentrations high mass depletion rates are expected.

In a previous communication (Araújo, 1990) the ignition temperature of wood and cork particles was studied (Fig. 1), correlating the experimental results with those calculated from theoretical formulations of Tognotti *et al.* (1985) Chen *et al.* (1981) and Bandyophadyay and Bhaduri (1972).

These results show a mean ignition temperature of 700 ± 100 K, indicating the working temperature level of a cyclone reactor.

This study presents:

— the theoretical analysis of the mass depletion rates of pine and cork samples, as a function of temperature, and the equilibrium calculations of the composition of reaction products, assuming the isobaric adiabatic combustion conditions;
— the cyclone reactor and the experimental procedures;
— the experimental results and theoretical correlations and conclusions.

THEORETICAL ANALYSIS AND PREDICTED VALUES

Mass Depletion Theory and Results

The prediction of the mass depletion rate of wood-dust particles as a function of particle size, environmental temperature and heating rate, was based on two formulations developed by Horton (1979) and Howard (1981). The single reaction model considers the gasification of a particle as a unique decomposition process, following a first-order Arrhenius reaction, where the total volatile production dm_v/dt can be expressed as:

$$dm_V/dt = k_A \exp(-E/RT)\,(m_{vf} - m_v), \tag{1}$$

where m_v is the cumulative volatile mass produced until the instant t, and m_{vf} is the final value ($t = \infty$). Assuming a constant heating rate $H = dT/dt$ and $(E/RT) > 1$ (Howard, 1981), the total mass of volatiles yet to be released can be expressed, as a function of T, as:

$$(m_{vf} - m_v)/m_{vf} = \exp[(-k_A RT^2/HE)\exp(-E/RT)] \tag{2}$$

The multiple reactions model proposed by Anthony *et al.* (1975) describes the gasification process as a series of first order independent reactions, occurring in different time intervals, each one referred to a specific component or group of components with characteristic mass loss rates. It is assumed that the Arrhenius constants of all components or groups have the same value, equal to the one of the global reaction ($K_A = K_{Ai}$), and that the number of the individual reactions is large enough to allow E to be expressed as a continuous distribution function f(E) (Araújo, 1990). When the volatile production is clearly characterized by the existence of a small number of peaks, corresponding to a few dominant groups of components, the formulation can be simplified considering only a limited number of first order reactions, each with constant values of E and k_A. The total mass of volatiles to be released is expressed as:

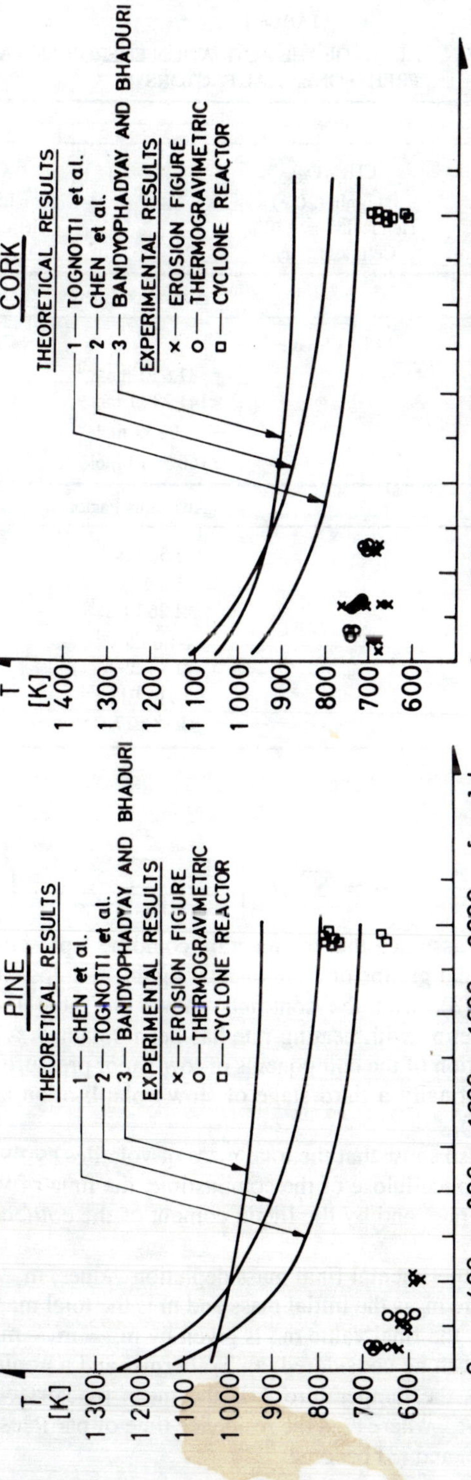

FIGURE 1 Ignition temperature of pine and cork particles.

TABLE I
GLOBAL COMPOSITION, VALUES OF THE ACTIVATION ENERGY AND ARRHENIUS PREEXPONENTIAL FACTORS

	Pine		Cork
Global chemical composition	$CH_{1.46}O_{0.60}$		$CH_{1.40}O_{0.40}$
Components	Lignin (20%) Hemicellulose (30%) Cellulose (50%)		Lignin (40%) Hemicellulose (30%) Cellulose (30%)
		Activation Energy	
Global	125.4 kJ mole^{-1}		142.12 kJ mole^{-1}
Lignin		>37.6 kJ mole^{-1} <141.28 kJ mole^{-1}	
Hemicellulose		123.72 kJ mole^{-1}	
Cellulose		>166.36 kJ mole^{-1}	
		Arrhenius Factor	
Global		1.5 E9 s^{-1}	
Lignin		>0.93 s^{-1} <1.26 E8 s^{-1}	
Hemicellulose		>1.6 E8 s^{-1} <1.44 E9 s^{-1}	
Cellulose		>3.916 E11 s^{-1} <1.18 E17 s^{-1}	

$$(m_{vf} - m_v)/m_{vf} = \frac{m_{vfi} - m_{vi}}{m_{vfi}} = \sum \exp\left[-\frac{KA_iRT^2}{HE_i} \sum \exp\left(-\frac{E_i}{RT}\right)\right]. \quad (3)$$

In the case of wood, we assumed that the reaction could be represented by the decomposition of the three principal groups of components: lignin, cellulose and hemicellulose. Using equations (2) and (3), with the constants shown in Table I, the characteristic evolution of the mass depletion with heating-rate has been obtained: a short initial stage, corresponding to the evolution of the components of low vapor pressure, a second stage of fast yield of volatiles and finally a third stage of slow volatilisation and char oxidation (Araújo and Campos, 1989).

The theoretical results also show that the total mass of volatiles evolved depends on the content in cellulose and hemicellulose of the combustible, the time of volatilisation being determined by the heating rate, and by the lignin content of the combustible (Araújo and Campos, 1989).

In order to predict the experimental final mass depletion values, m_v was assumed to be given by $m_v = m_0 - m$, where m_0 is the initial mass and m is the total mass of solid residues at the instant t. When $t = \infty$ the final value m_{vf} is given by $m_{vf} = m_0 - m_f$.

Two physical situations can be considered: an isothermal and a nonisothermal process. For an isothermal analysis the temperature T is the mean reactor temperature and the heating rate is $H = (T - T_0)/t_{re}$, where t^{re} is the residence time of particles inside the reactor. Consequently, equations (2) and (3) become:

$$(m_0 - m_f)/m_0 = \exp\left[(-k_A RT^2/HE) \exp(-E/RT)\right], \quad (4)$$

$$(m_0 - m_f)/m_0 = \sum \frac{m_{0i} - m_{fi}}{m_{0i}} = \sum \exp\left[-\frac{KA_iRT^2}{HE_i} \exp\left(\frac{E_i}{RT}\right)\right], \qquad (5)$$

and the results obtained are shown in Fig. 6. In the nonisothermal analysis the numeric solutions were obtained, as a function of the equivalence ratio and residence time, assuming that the instantaneous temperature of the reacting particles could be calculated using a polynomial regression representative of the axial evolution of the temperature along the reactor wall (vd. Fig. 7).

The results show increasing mass depletion rates with temperature, heating rate and residence time, in good agreement with Howard (1981) and Simmons and Ragland (1986) (Araújo and Campos, 1989). The numerical results are very sensitive to the adopted value for the Arrhenius factor and are not adequate to predict gasification results at low temperatures (500–600 K), as referred by Tang (1967).

Composition of Reaction Products

The emission of the most common pollutant gases resulting from the gasification and combustion of the wood materials was made using the computer code THOR (IEPG Reps, 1989). Assuming the chemical and thermodynamic equilibrium, the final composition (solid and gas components) is calculated for the minimum value of Gibbs free energy using different equations of state (Campos, 1991).

It is possible with this code to consider up to m atomic species ($m < 20$) and to form n chemical components with these atomic species ($n < 40$).The selection of the chemical components depends on the equivalence ratio of the mixture, related to the stoichiometry, and they are those which are expected to have significant concentrations in final product's composition. The mass balance yields a linear system involving m atomic species and m–n equilibrium reactions. The m–n equilibrium equations are determined by the method of Lagrange multipliers or the equilibrium constants (Brinkley, 1947). In order to determine the chemical concentration of the n components, for imposed P and T conditions, a chemical affinity method can be used, proposed by Heuzé *et al.* (1985) or a interactive method optimizing the relative concentration inside the group related to the same atomic specie, for the minimum value of global Gibbs free energy $G = \Sigma\, x_i\mu_i$, being the Gibbs free energy of each component $\mu_i = G_i^0(T) + R\,T \ln P + R\,T \ln(x_i/x)$, with x_i the mole fraction. The values of $G_i^0(T)$ can be obtained from JANAF Thermochemical Tables, 1971, and from polynomial expressions of Gordon and McBride (1971) (Campos, 1991). The solution of the composition problem simultaneously involves:

- the thermodynamic equilibrium, obtained with the mass and species balance, and the equilibrium condition $G = G_{min}(P, T, x_i)$, previously described;
- the thermal equation of state (EOS) based on statistical mechanics and intermolecular forces, developed for detonation fluids and associated computer programs (BKW EOS, Mader Code, 1979; Quatuor Code, Heuzé *et al.* 1985)
- the solid species can be calculated based in Cowen and Fickett (Fickett and Davies, 1979) and Debye physical models (Heuzé, 1989);
- the energetic equation of state, related to the internal energy $U = \Sigma\, x_i\, u_i(T) + \Delta u$, $u_i(T)$ calculated from JANAF Thermochemical Tables and from polynomial expressions of Gordon and McBride.

The isobaric adiabatic combustion condition has been chosen because it gives the upper limit of concentration of combustion products.

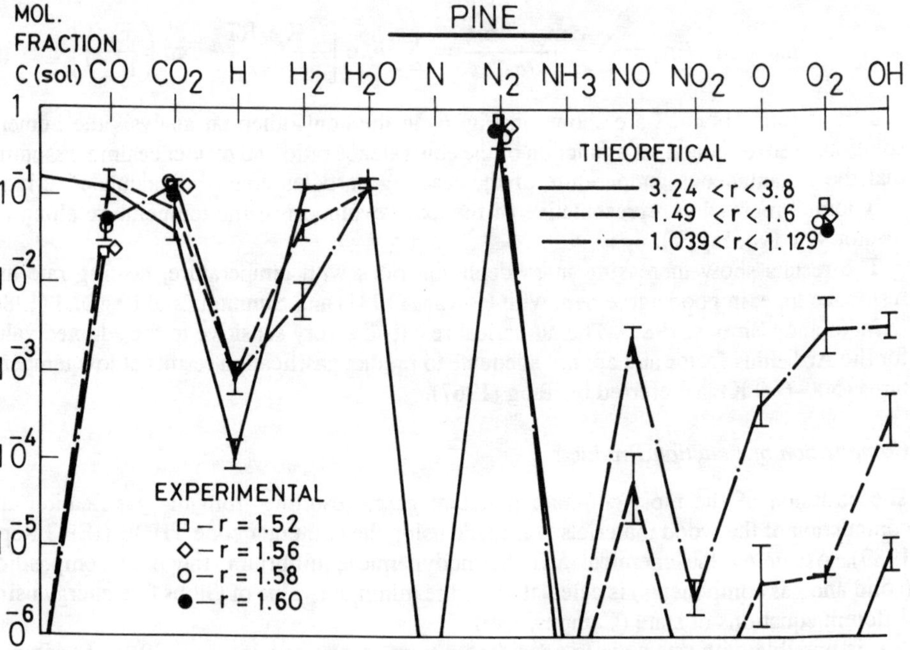

FIGURE 2 Solid and gas composition from pine reactions.

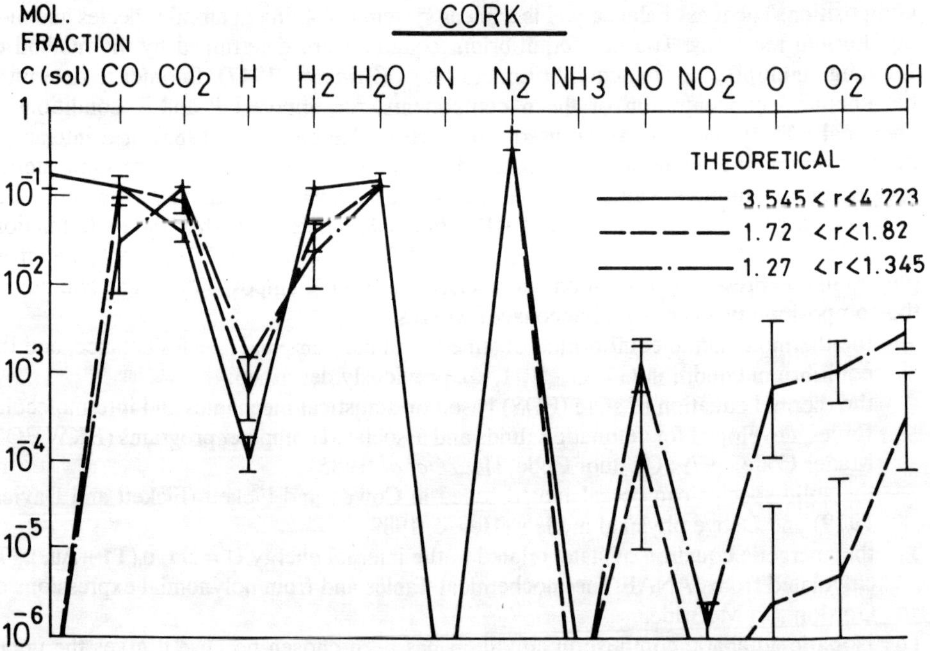

FIGURE 3 Solid and gas composition from cork reactions.

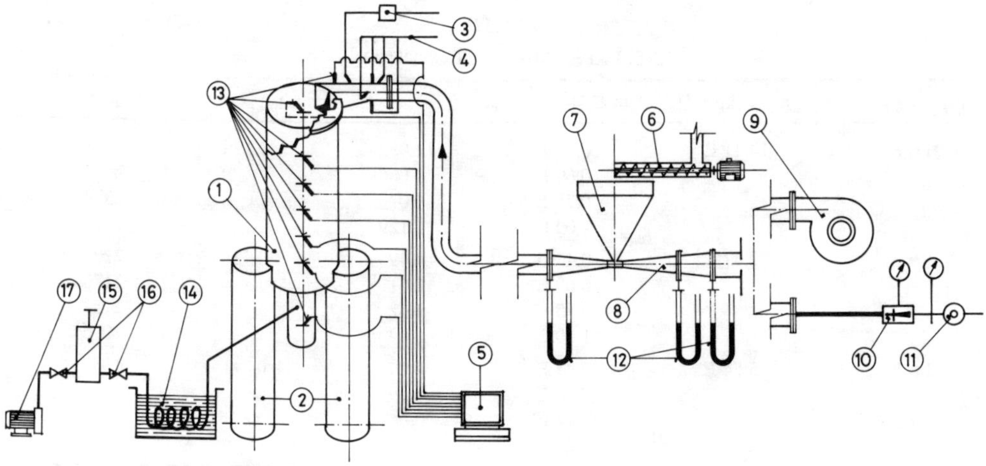

FIGURE 4 The cyclone reactor experimental apparatus.

The results show the equilibrium concentrations, in the final products composition, of CO_2, CO, H_2O, OH, N_2, O_2, H_2, NO, NO_2, H, N, O, NH_3 and CH_4 gases and two kinds of solid carbon (graphite and diamant) (vd. Figs. 2 and 3).

The presented results have been calculated for a perfect gas equation of state but no significant differences were found between these results and those obtained with BKW, H9 and H12 EOS. The obtained results show combustion temperature T_{comb} decreasing (vd. Table II) and solid carbon and CO concentrations increasing with equivalence ratio (vd. Figs. 2 and 3). The calculated NO_x concentrations have significant values only for mixtures of $1 < r < 1.7$. This situation is in good agreement with other bibliographical results.

EXPERIMENTAL APPARATUS AND PROCEDURES

The experimental apparatus (Fig. 4) has a particle's feeding system composed of a hopper provided with a calibrated grid that retains all particles larger than 2000 μm, a screw feeder, a venturi tube, a radial ventilator or, alternatively, a sonic meter.

The mass flow of solid particles in the pneumatic feeding circuit is a function of the air velocity in the throat of the venturi tube.

The gasification and combustion reactor is a cyclone, provided with two water-cooled, closed, vortex collector pockets (vd. McGregor *et al.* 1987). This characteristic design of

TABLE II
PINE, EXPERIMENTAL CONDITIONS

m_{air}, [kg s^{-1}]	m_{pine} [kg s^{-1}]	m C_3H_8 [kg s^{-1}]	r	t_{re} [s^{-1}]	T_{comb} [K]
91.6 E^{-3}	14.11 E^{-3}	0.43 E^{-3}	1.032	0.196	2377
		0.60 E^{-3}	1.066		2380
		0.75 E^{-3}	1.07		2379
		1.00 E^{-3}	1.129		2373
56.9 E^{-3}	12.77E^{-3}	0.43 E^{-3}	1.49	0.32	2204
		0.60 E^{-3}	1.51		2170
		0.75 E^{-3}	1.52		2137
		1.00 E^{-3}	1.62		2096
11.8 E^{-3}	5.54 E^{-3}	0.43 E^{-3}	3.24	0.711	1837
		0.60 E^{-3}	3.39		1818
		0.75 E^{-3}	3.6		1800
		1.00 E^{-3}	3.8		1778

TABLE III
CORK, EXPERIMENTAL CONDITIONS

m_{air} [kg s^{-1}]	m cork [kg s^{-1}]	mC_3H_8 [kg s^{-1}]	r	t_{re} [s^{-1}]	T_{comb} [K]
91.6 E^{-3}	13.53 E^{-}	0.43 E^{-3}	1.27	0.196	2379
		0.60 E^{-3}	1.282		2369
		0.75 E^{-3}	1.294		2356
		1.00 E^{-3}	1.345		2337
56.9 E^{-3}	10.18	0.43 E^{-3}	1.72	0.32	2254
		0.60 E^{-3}	1.729		2221
		0.75 E^{-3}	1.732		2006
		1.00 E^{-3}	1.83		2001
11.8 E^{-3}	4.87 E^{-3}	0.43 E^{-3}	3.545	0.711	1835
		0.60 E^{-3}	3.73		1817
		0.75 E^{-3}	4.006		1798
		1.00 E^{-3}	4.223		1776

the reactor permits the entrainement of the particles from the main flow into the side pocket, where stationary vortexes are formed. Gravitational and friction forces against the pocket walls facilitate deposition. The water-cooled vortex collector pocket also allow the reacting particles to quench, immediately after their trajectory inside the reactor, for further study of the effects of heating rate and temperature on their structure.

The cyclone was also designed in order to ensure the separation of all particles larger than d = 50 μm, and the complete combustion of particles of d = 2000 μm (Araújo, 1990).

The reactor is heated at the inlet section by a lean propane/air flame. The temperature of the flow and particles, in the traject from the inlet to the gas exhaust sections of the reactor, can be measured by means of eight k type thermocouples. The output of the thermocouples is amplified and digitized in order to obtain both numerical and graphic records of the local instantaneous temperatures during gasification (Fig. 5). Sampling of the exhaust gases was carried out during the trial tests through a probe tube connected to the exhaust pipe. Visualization of the flame inside the cyclone is possible, through a glass window placed in the upper part of the reactor.

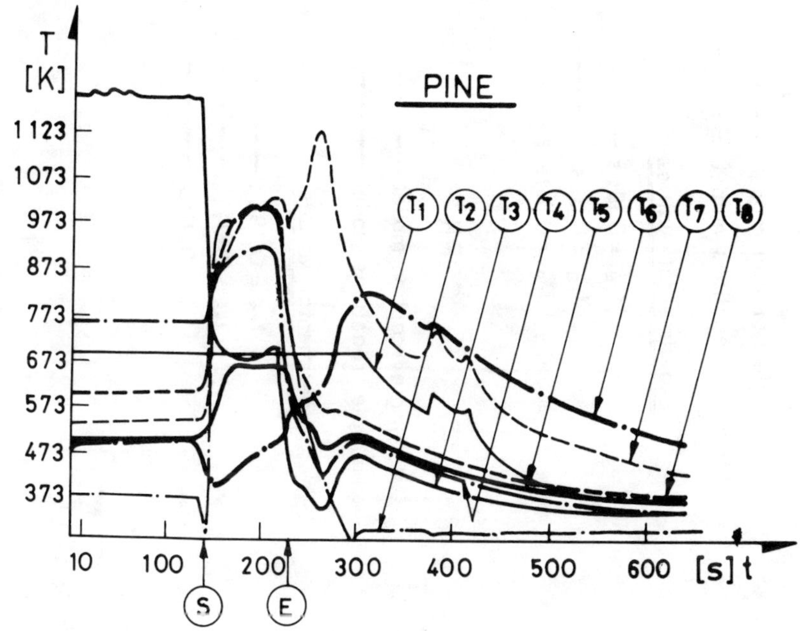

FIGURE 5 Experimental graphic records.

Prior to the gasification and combustion experiments, the system was calibrated in order to select the global equivalence ratios (pine/propane/air) and the corresponding particle's residence time that would be the object of our tests (vd. Tables II and III).

The experiments began with the warming-up of the reactor, by the air/propane lean flame attached to the inlet section of the cyclone, until the thermal stabilisation of the reactor was obtained. A sample of dust particles was then introduced into the hopper, injected in the pneumatic circuit and transported to the reactor.

The suspension of the pine or cork particles, in air and products of the propane/ air combustion, was injected into the reactor for a period of time always larger than its thermal relaxation time (Fig. 5).

Samples of the exhaust gases were collected in pressurized aluminium bags for further analysis by gas chromatography and chemiluminescence.

The analyses were carried out using a Hewlett-Packard chromatographer, model 5890 HP, with a thermal conductivity detector. The CO_2 peak was detected together with CO, N_2, O_2 and CH_4, by using a system of two columns, and a switching valve. The first peak to appear was the CO_2 peak after the sample passed through a Porapak N column 1/8" in diameter. The commuting valve was then activated and the remaining gases were detected in a second column, Molecular Sieve 5X 1/8" in diameter. For NO_x analysis a Thermo Electron 10 A, chemiluminescence automatic analyser was used.

The remaining unburnt particles and char, that were quenched in the vortex pockets, were collected after each experiment in order to measure the mass depletion of the sample, and to analyse the effect of temperature and residence time on their physical structure.

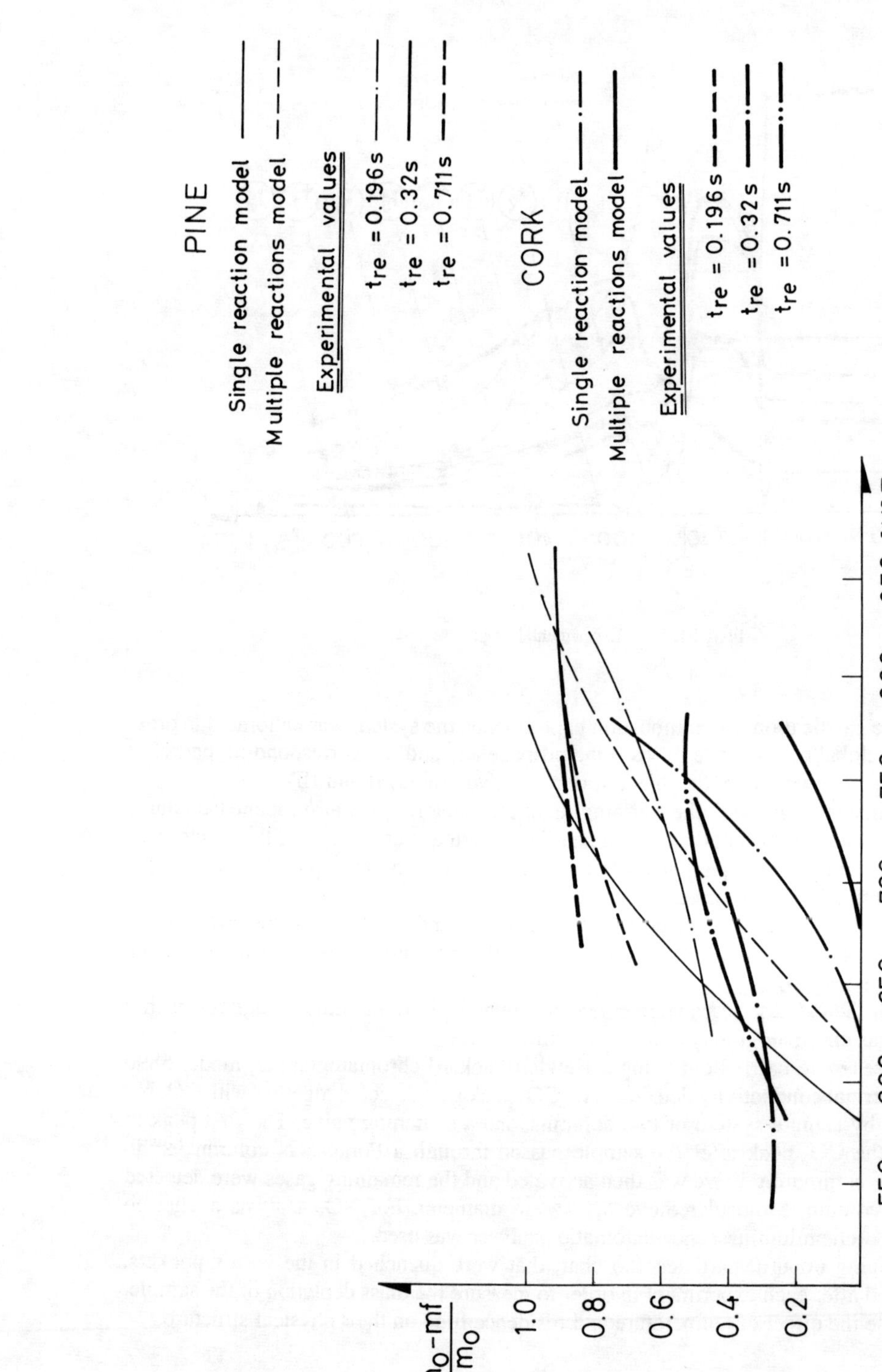

FIGURE 6 Evolution of the mass loss of pine and cork samples (isothermal model), as a function of temperature and residence time.

EXPERIMENTAL RESULTS AND THEORETICAL CORRELATIONS

Mass Depletion Results

In the analysis of the mass depletion results we have considered the physical situations of isothermal and nonisothermal gasification processes.

For an isothermal analysis, considering the flow velocity inside the reactor as almost constant and mass conservation for the global system, the global temperature of the reactor, Tp_{mm}, was expressed as the arithmetic mean of all local Tp_{mi}'s (Araújo, 1990). Furthermore we assumed that the solid particle's radiation is the fundamental heat transfer process inside the reactor and, consequently, the local temperatures measured by the thermocouples and the temperature of the particles had the same value.

The evolution of the nondimensional mass depletion of the samples, as a function of the temperature, residence time, and particles concentration in air was obtained using equations (4) and (5), with $T = Tp_{mm}$. The values of E and k_A for pine, cork, cellulose, hemicellulose and lignin are shown in Tables I and II. Consequently we correlated the experimental results of mass depletion with those predicted by the single and multiple reaction models.

The obtained results, presented in Fig. 6, show the experimental values of the ratio $m_0 - m_f/m_0$ increasing with growing values of Tp_{mm}, heating rate and residence time, in a good agreement with that Howard (1981) and Simmons and Ragland (1986). The mass loss levels are, for the same conditions, always higher in the case of pine. As mentioned before, the theoretical models are unable to predict reasonable values in the low range of global temperatures (500–600 K).

This is in agreement with previous results (Tang, 1967) and reflects the fact that, at this temperature level, higher values of k_A should be considered. In a nonisothermal analysis the correlation of the mass depletion experimental results with that predicted by the two theoretical models was made assuming the evolution of the temperature of the particles, during their traject inside the cyclone, as polynomial regressions of the previously calculated Tp_{mi} (Araújo, 1990). The results obtained (vd. Fig. 7), show that the gasification levels of both materials increase with the global equivalence ratio. For the same heating rate, the gasification of pine dust is initiated before cork dust, being gradual in time.

Solid Residues Analysis

Collecting efficiency. The collecting efficiency of the cyclone reactor was computed by coupling the mass balances of the solid fuel introduced in the reactor and the residues collected after each trial (vd. Figs. 6 and 7). The microscopic analysis of residues permits the evaluation of its size and suggests an increasing internal swirl and collection efficiency with the expansion phenomena associated with burned gas.The presence of a large number of very small particles (Fig. 8) among the solid residues, indicates that the collecting efficiency of the cyclone is increased for particle diameters up to 10 μm.

Structural analysis. The microscopic analysis of the collected ashes, char and partially burnt particles (Fig. 9) showed that pine wood is formed of packed parallel fibers, in agreement with the structural model adopted by most of the authors. Cork has a special cellular structure that includes a high percentage of void spaces. This analysis allowed us to establish, for short residence times and severe heating rates, three levels of structural degradation of the particles as a function of the global heating rate.

When $dT/dt < 3300\ K\ s^{-1}$, the volume and the superficial structure of the particles remains almost unchanged, with the formation of small pores or local gasification spots. For $3300\ K\ s^{-1} < dT/dt < 3750\ K\ s^{-1}$ there is a severe shrinking of the particles, with the

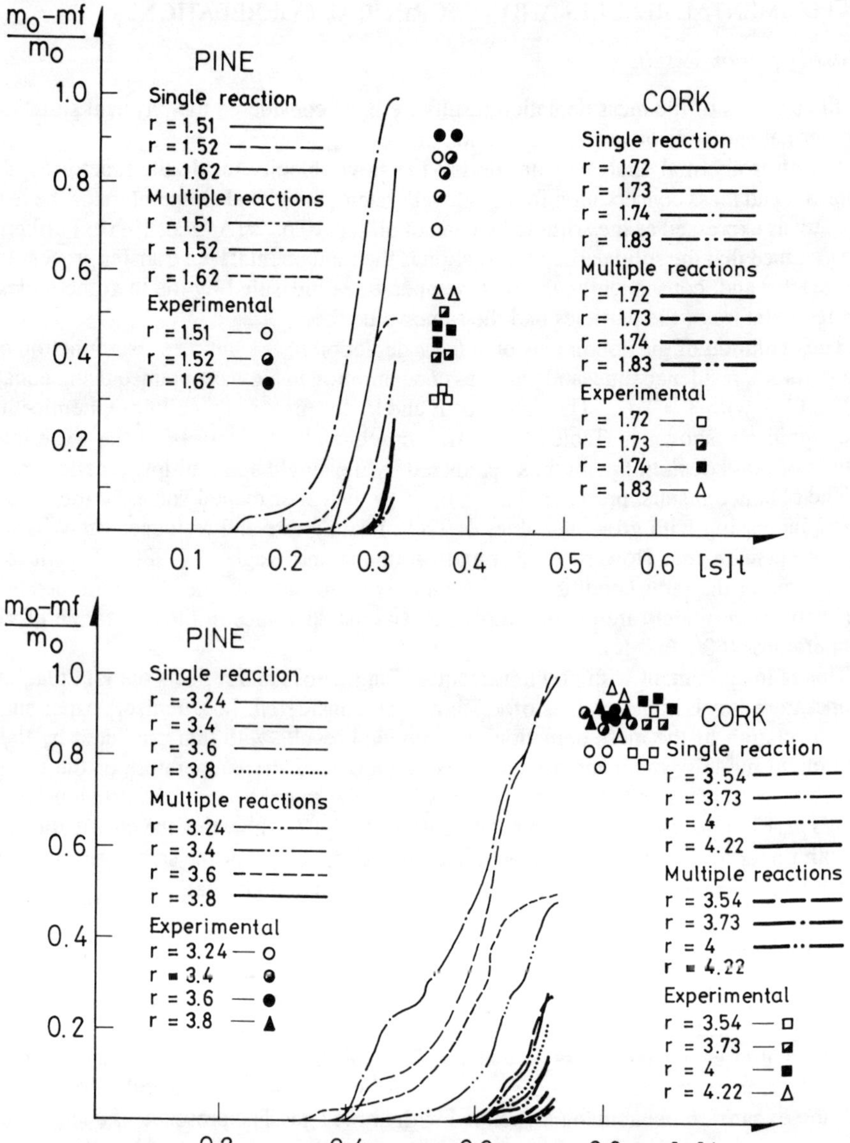

FIGURE 7 Evolution of the mass loss of pine and cork samples (nonisothermal model), as a function of equivalence ratio and residence time.

formation of deep pores and cracks in the structure of the pine particles, the wall cells of the cork particles burst, and a characteristic hexagonal pattern is formed on its surface.

For 3750 K s^{-1} < dT/dt < 3850 K s^{-1}, there is in both cases a structural explosion of the smallest particles, and gasification is almost complete.

Analysis of the exhaust gases. For the study of the composition of the exhaust gases and the emission of pollutants we have chosen only one working cyclone regime (m_{air} = 56.9

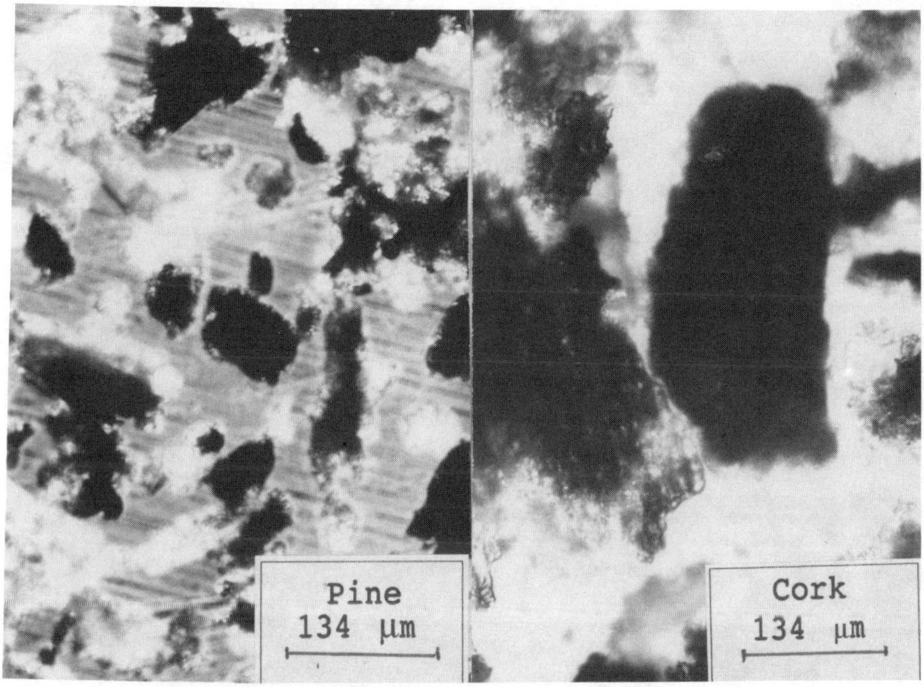

FIGURE 8 Photos of pine and cork collected particles.

E^{-3} kg s^{-1}; m_{pine} = 12.97 E^{-3} kg s^{-1}), because it is the transition regime between stoichiometric and rich mixtures for which significant values of NO_x were expected (vd. Fig. 2).

For this regime, several global equivalence ratios and global temperatures, depending on the propane mass flow, were tried (vd. Table IV). The results (Table IV) show a slight increase of CO_2 (10% to 12%), CO (2% to 6%) and CH_4 (0.1% to 0.6%) with the equivalence ratio. No significant variation of NO_x contents was detected (20 ppm).

Comparing theoretical and measured values (vd. Fig. 2) a good agreement for N_2, CO and CO_2 concentrations has been found.

The experimental results for O_2 and CH_4 are higher than the ones predicted. Inversely, the calculated values for NO_x are higher than measured values. This can be explained by the difference between theoretically assumed conditions (isobaric adiabatic combustion) and real operating conditions (combustion and gasification).

CONCLUSIONS

The theoretical and experimental mass depletion results show mass loss increasing with the equivalence ratio and reducing particle size. Increasing heating rates also raise the gasification levels. The predicted values from the theoretical models, assuming isothermal conditions, are in good agreement with the experimental data. The results obtained by the multiple reaction's model are closer to the experimental results. Gasification of pine dust samples is more complete and homogeneous. The more important deviation between

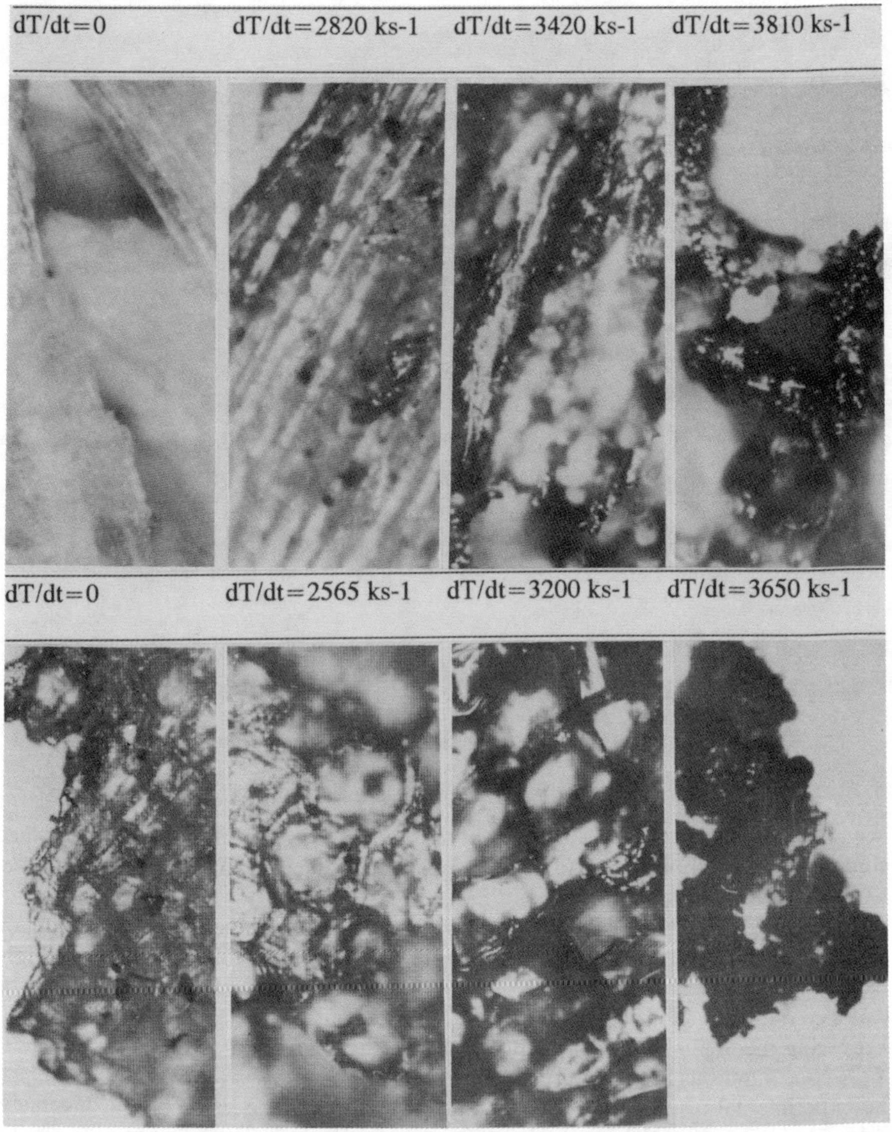

FIGURE 9 Thermal degradation of pine and cork particles.

theoretical and experimental cork results can be explained by the lack of a variable representing the special structure of this material.

Microscopic analysis of the solid residues, coupled with mass balances of the solid fuel introduced into the reactor and collected in the side pockets, seems to indicate that increasing the swirl of the internal flow increases the collecting efficiency of the cyclone. From the analysis of collected particles, after gasification, three levels of structural degradation were established.

The equilibrium composition of the exhaust gases was predicted using the numerical code THOR, using different equations of state for real gases and solid species.

TABLE IV
ANALYSIS OF THE EXHAUST GAS FROM PINE REACTIONS

Pine: experimental conditions:
$m_{air} = 56.9\ E^{-3}\ [kg\ s^{-1}]$; $m_{pine} = 12.97\ E^{-3}\ [kg\ s^{-1}]$

$\dot{r}$	mC_3H_8 [kg s^{-1}]	CO_2	CO	CH_4 [mole fraction]	O_2	N_2	NO_x [ppm]
1.52	0.75 E^{-3}	0.102	0.026	0.0	0.97	0.788	20
1.56	0.84 E^{-3}	0.101	0.026	0.001	0.094	0.788	20
1.58	0.86 E^{-3}	0.117	0.058	0.006	0.063	0.758	20
1.60	0.94 E^{-3}	0.120	0.062	0.007	0.059	0.748	20

For gas analysis of CO_2, CO, CH_4, O_2 and N_2, standard gas chromatography was used. A chemiluminescence analyser was utilised for NO_x. The measured values are in good agreement with the calculated results and show low concentrations of CO and CO_2. No significant values of NO_x in the exhaust gas were found.

REFERENCES

Anthony, D.B., Howard, J.B., Hottel, H.C., and Meissner, H.P. (1975). Rapid devolatilisation of pulverized coal. *15th Symposium International on Combustion*, pp. 1303–1317. The Combustion Institute.

Araújo, L. (1990). Inflamação e Gasificação de Partículas de Pinho e de Cortiça, Doctoral Thesis, Mech. Eng. Dept., Fac. of Sci. and Tech., University of Coimbra, Portugal.

Araújo, L., and Campos, J. (1987). Gasification and combustion of white pine and cork dusts. Dynamics of reactive systems. Part II: Heterogeneous combustion and applications. *Progress in Astronautics and Aeronautics*, **113**, pp. 303–319.

Araújo, L., and Campos, J. (1988). Ignition and gasification of pine and cork dusts. *1st European Conference on Industrial Furnaces and Boilers*, vol. 1, pp. 83–94, Lisbon, Portugal.

Araújo, L., and Campos, J. (1989). Ignition and gasification of pine and cork dusts in a cyclone. *12th International Colloquium on Dynamics of Explosions and Reactive Systems*. Michigan University. USA.

Bandyophadyay, S. and Bhaduri, D. (1972). Prediction of the ignition temperature of a single coal particle. *Combustion and Flame*, **18**, pp. 411–415.

Brinkley, S. R. (1947). Chemical equilibria. *J. Chem. Phys.*, No. 15, pp. 107.

Campos, J. (1991). Thermodynamic calculation of solid and gas combustion pollutants using different equations of state. *First International Conference on Combustion Technologies for a Clean Environment*. Vilamoura, Portugal.

Chen, M.R., Fan, L.S., and Essenhigh, R.H. (1981). Prediction and measurement of ignition temperatures of coal particles, *Colloque International Berthelot-Vieille-Mallard-Le Chatellier*, Tome 1, pp. 349.

Cook, M. J. and Ford, J. N. (1991). Development in particulate control for industrial boilers. *Proceedings of the 2nd International Conference on Industrial Furnaces and Boilers*, vol. 2. Vilamoura, Portugal.

Déglise, X., and Lédé, J. (1980). Valorisation energétique de la biomasse par voie thermique. *Entropie*, **94**, pp. 24–37.

Fickett, W., and Davis, W.C. (1979). *Detonation*. University of California Press.

Gordon, S., and McBride, B.J. (1971). Computer program for calculation of complex chemical equilibrium compositions, rocket performance, incident and reflected shocks and Chapman–Jouguet detonations. Report NASA, SP 273, NASA Lewis Research Center.

Heuzé, O. (1989). Cálculo numérico das propriedades das misturas gasosas em equilibrio termodinâmico. Universidade de Coimbra, Portugal.

Heuzé, O., Bauer, P, Presles, H.N., and Brochet, C. (1985). The equations of state of detonation products and their incorporation into the quatuor code. *Proceedings of the Eighth Symposium (International) on Detonation*, pp. 762–769, Albuquerque Conventional Center, New Mexico.

Heuzé, O., Presles, H.N., and Bauer, P. (1985). Computation of chemical equilibria. *J. Chem. Phys.*, **83**(9), pp. 4734–4735.

Horton, D. M. (1979). Pulverized coal combustion and gasification. *Theory and Applications for Continuous Flow Processes.* Plenum Press.

Howard, J. B. (1981). *Chemistry of Coal Utilisation*, 2nd suppl. vol. John Wiley and Sons.

IEPG. Campos, J., Luz, P. and Marques, C. (1989, 1990). BKW KHT H9 equation of state calculations. Report of Progress of Technological Area 25, Collaborative Technical Programme 1, Portugal.

JANAF (1971). *Thermochemical Tables*, 2nd ed. National Bureau of Standards, Washington, DC.

Mader, C. L. (1979). *Numerical Modeling of Detonations.* University of California Press, Berkeley, USA.

McGregor, S.A., Claypole, T.C., Mohamed Ali, A., and Cooper, R. (1985). Aerodynamics of cyclone combusters for burning solid fuels. *Preprints of the 10th International Colloquium of Dynamics of Explosions and Reactive Systems.* Berkeley.

Simmons, W.W., and Ragland, K.W. (1986). Burning rate of millimeter sized particles in a furnace. *Combustion Science and Technology*, **4**, p. 293.

Tanaka, K. (1983). Detonation properties of condensed explosives computed using the Kihara-Hikita-Tanaka equation of state. Report from National Chemical Laboratory for Industry, Ibaraki, Japan.

Tang, W. K. (1967). US Department of Agriculture and Forest Services.

Tognotti, L., Malotti, A., Petrarca, L., and Zanelli, S. (1985). Measurement of ignition temperature of coal particles using a thermogravimetric technique. *Combustion Science and Technology*, **44**, pp. 15–28.

NOMENCLATURE

E	Activation energy
E_0	Mean activation energy
$f(E)$	Continuous distribution function of the activation energy
G	Gibbs free energy
k_A	Arrhenius preexponential factor
m	Mass
P	Pressure
R	Perfect gases constant
r	Equivalence ratio
T	Temperature
Tp_m	Mean local temperature
Tp_{mm}	Mean global temperature of the reactor
t	Time
t_{re}	Residence time
x	Mole fraction
u	Molar internal energy
μ	Molar Gibbs free energy

SUBSCRIPTS

0	Initial
f	Final
i	Of the component or group
p	Of the particle
v	Of volatiles

Combustion of Coffee Lignocellulose Waste

G. ANTOLÍN, E. VELASCO, R. IRUSTA, and J.J. SEGOVIA *Chemical Engineering Department, University of Valladolid, Valladolid, Spain*

Abstract — This paper presents the determination of the composition of three samples of lignocellulose waste from coffee, as well as its particle sizes and heating power. The results obtained demonstrate how, if its moisture content is reduced, its heating power would make it competitive as a fuel with those traditionally used. The analysis of the drying conditions for lignocellulosic waste from coffee, is also presented here, along with the need for using two successive stages (pressing and forced drying with hot air), both having been observed. Thermograms in inert and oxidizing atmospheres are determined, and combustion tests in an Atmospheric Fluidized Bed Combustor (AFBC) and a grate furnace are performed, demonstrating the viability of the use of this product as an alternative fuel with accompanying energy savings. An incineration plant is proposed along with its elements. The use of coffee lignocellulose waste includes a possible economic saving and the lowering of the atmospheric contamination due to SO_2 thanks to its lower sulphur content compared with conventional fuels.

Key Words: Biomass, Coffee, Waste, Combustion, Environment

INTRODUCTION

One of the solutions for the promotion of a clean environment and alleviation of an energy problem for some industries might be by the use of wastes generated through the manufacturing process. Their utilization minimizes the energy component coming from petroleum and the environmental contamination by SO_2. This solution could be adopted in the case of the instant coffee industry in Spain and Europe, which generates about 12 and 121 Gg every year, respectively, of lignocellulose waste.

The instant coffee industry generates, approximately, some 55% lignocellulose waste which is not usually utilized later, fundamentally due to its high moisture (85–90%) and its lignocellulose character. Eliminating the water through conventional drying is very expensive, and its possible transformation by means of the wet method is not considered viable because of its lignocellulose character.

This coffee waste, normally discharged to dumping locations, causes the contamination of the nearest springs with their lixiviate waters. Also, it frequently causes fires in these dumping places, mainly in summer, that could propagate to the nearest agricultural or forested areas.

The purpose of this paper is to characterize the lignocellulose residue of coffee, arising from the production of instant coffee, to reach a decision as to its possible use as a fuel. Also presented in this document is the experience undertaken to reduce, by means of pressing and forced drying with air, the initial water content to 30%, which would make its combustion profitable (Antolín *et al.* 1990). A thermic analysis is also performed and divers samples have been burned in an atmospheric fluidized bed combustor (AFBC) and in an industrial boiler equipped with a grate of solids.

MATERIALS AND METHODS

Three samples of coffee waste resulting from the manufacturing process of the company SEDA in Palencia (Spain) have been used in this work.

Characterization of Waste

The parameters evaluated in the characterization of the residues are: proximate analysis, elemental analysis, granulometric distribution and higher heating power with different moisture contents. The analyses of the waste characteristics were performed in the laboratories of CEDER in Soria (Spain), using the following methodology (Antolín and Irusta, 1989):

Moisture. In a drying chamber at 105°C for 24 hours. Under these conditions the water content of the samples disappears completely, while no losses in volatiles have been observed. The moisture results are shown in the tables in wet basis.

Proximate analyses. Submitting the samples to specific gradients of temperature in both inert and oxidizing atmospheres, during which the content of volatiles and fixed carbon could be respectively determined. The final waste matter obtained corresponds to the ash content of the samples.

Higher heating power. This was determined in a "Parr" brand, completely automatic, calorimeter.

Granulometric distribution. Performed with the dry sample, given that moisture provokes the compacting of the residues and prevents its classification by sizes. The test was performed using a conventional wobble screener equipped with spans of screen to 8, 4, 2, 1, 0.5, 0.125 and 0.0625 mm, respectively.

Elemental analysis. In an elemental analyzer, "LECO" brand. The principal components of the organic material were analyzed, and the oxygen content was obtained by difference.

Drying Waste

The drying of the coffee residue has been performed in two successive stages: first, part of the initial water content was eliminated through pressing, thus reducing its moisture to 50%; and then, by means of forced drying with air, a moisture of 30% was achieved.

Drying by pressing: This was achieved using a bench press (BMZ-10) whose characteristics are given in the table I. This press was equipped with two cylindrical screens, 30 cm in diameter and 15 cm in height, which have lateral outlets, for the water from the pressing placed at quincunx, of 1 mm and 1 cm diameter, respectively. A metallic screen located below the press-cake facilitates the release of the water through pressing from the inferior section.

The pressing of the waste has been optimized by varying the following parameters: size of the screen mesh, press-cake height, piston pressure, pressing time and initial moisture. The experimental design applied to this model was a complete factorial development, so as to optimize it to the maximum. Noise parameters were not calculated, as their influence on the results are not thought to be great. However, in case they were so, the robustness zones, in which the variations in the noise signals do not influence the final results of the process, must be found.

The pressing study was performed at two levels for each parameter (see table II).

Forced drying: Forced drying of the previously pressed waste was performed in a fluidized bed constructed using methacrylate. The distributor was in the form of bubble caps, to which slightly hot air was borne.

The characteristics of the equipment and the operating conditions were as follows:

— Regulated air flow for fluidization and drying.
— Set of distributing plates with different openings.
— Adjustable height of drying press-cake.
— Temperature of air entering approximately 35°C.
— Air moisture depends on the environmental conditions.

TABLE I
CHARACTERISTICS OF THE BENCH PRESS

Capacity	10,000 kg
Pitch	375 × 990 mm^2
Manometer	BGP-105 (0 to 70,000 kPa)
Hydraulic pump	P-39 M
Hydraulic cylinder	RC-106
Outstroke	15 cm
Volume	2,240 cm^3
Useful area	14.5 cm^2

TABLE II
PARAMETER LEVELS OF PRESSING

Parameter	Level 1	Level 2
Pressure press-cake (kg/cm^2)	14	8
Time pressing (minutes)	15	5
Height press-cake (cm)	15	8
Span of screen (mm)	10	1
%Initial moisture	72.69	79.58

Thermograms

In order to determine the evolution of the waste with temperature, thermograms were performed, using a Perkin-Elmer apparatus, with a heating rate of 20°C/min in a range from 20°C to 940°C, and in atmospheres of nitrogen and dynamic air.

Combustion

The coffee waste combustion, with a 30% moisture content, has been carried out in two boilers; one an AFBC plant and the other a grate boiler for solids.

In the AFBC (figure 1), which has been designed by this research team and installed in CEDER, the combustion was performed with a mixture of 90% sand, which acts as an inert in the interior of the burner, facilitating the fluidization and combustion processes (Antolín *et al.* 1991).

In order to corroborate the results of the combustion test in the AFBC furnace, several tests were performed in the industrial furnace of the firm IDEA, which has a double chamber, one for solid material and another for natural gas.

RESULTS

The results obtained in the characterization of the three samples analyzed are very similar, indicating that it is to be expected that other tests will yield results which do not differ significantly from those shown in the tables III–VI.

FIGURE 1 Atmospheric fluidized bed combustion pilot plant. SEE COLOR PLATE VIII.

TABLE III

PROXIMATE ANALYSIS

Component	% Mass No. 1	% Mass No. 2	% Mass No. 3
Moisture	83.28	78.72	69.54
Volatile (db)	83.12	81.61	81.93
Fixed carbon (db)	16.48	17.64	17.74
Ash (db)	0.40	0.75	0.33

TABLE IV

ELEMENTAL ANALYSIS

Chemical element	% Mass No. 1	% Mass No. 2	% Mass No. 3
C	54.93	56.20	54.85
H	7.07	7.10	6.88
N	2.28	2.42	2.56
S	0.18	0.17	0.19
O	35.54	34.11	35.62

TABLE V

HIGHER HEATING POWER (kJ/kg)

% Moisture	Sample No. 1	Sample No. 2	Sample No. 3
0	23.876	24.298	23.826
25	17.907	18.225	17.869
50	11.938	12.151	11.913
75	5.969	6.074	5.956

TABLE VI

GRANULOMETRIC DISTRIBUTION

Size (mm)	% Mass No. 1	% Mass No. 2	% Mass No. 3
0.063	0.09	0.26	0.01
0.063–0.125	1.66	1.92	1.15
0.125–0.250	2.01	2.16	2.48
0.250–0.50	4.69	5.18	6.68
0.50–1	13.87	12.58	14.01
1–2	44.54	36.78	31.86
2–4	33.14	38.40	42.02
4–8	–	2.72	1.28
8–16	–	–	0.51

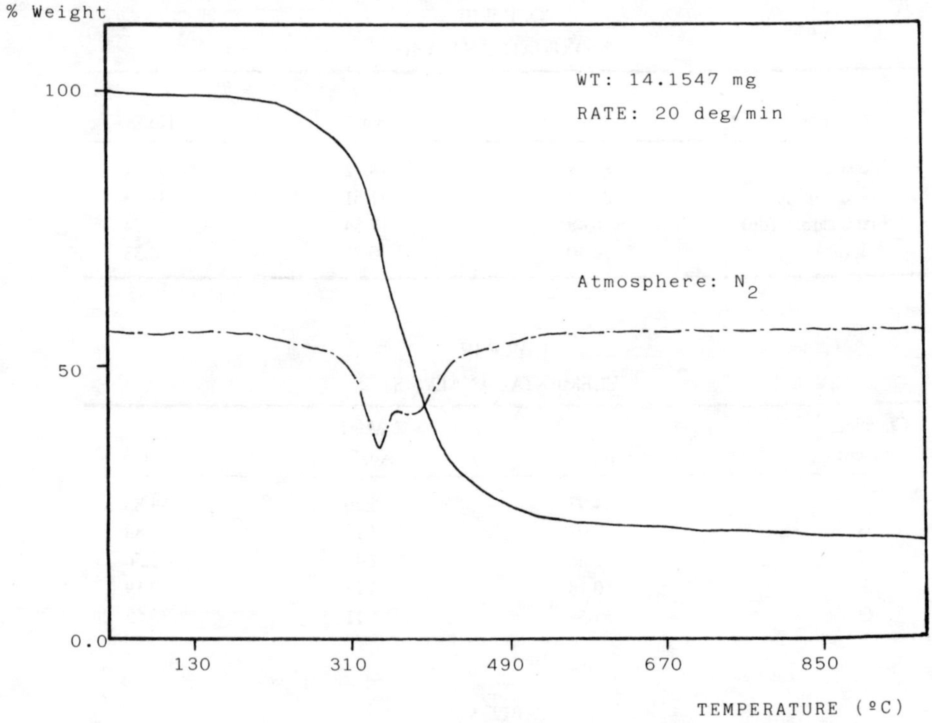

FIGURE 2 Thermogram in inert atmosphere.

Drying by Pressing

The moisture content was reduced to 50% by means of a complete factorial design of the factors which influence the waste pressing. The optimal operating conditions were deduced by following the conditions below:

— Screen size: Acts as a betterment parameter.
— Press-cake height: 80 mm (best lesser parameter).
— Piston pressure: 14 kg/cm^2 (best major parameter).
— Pressing time: 15 min (best major parameter).
— Initial moisture: Noise factor in the process.

The treatment of the water of the pressing was completed in two consecutive stages: First a flocculation to eliminate the organic material in suspension with a flocculent (zetag 66) followed by an airing of the water, if it is to be discharged afterwards into a river, in accordance with legal requirements.

Forced Drying

The waste previously pressed was crumbled, and the fraction of thick matter, which would produce segregation in the fluidized bed, eliminated: drying tests were then performed, varying different parameters. The results shown in Table VII correspond to waste pressed samples with an initial moisture of 50.8%.

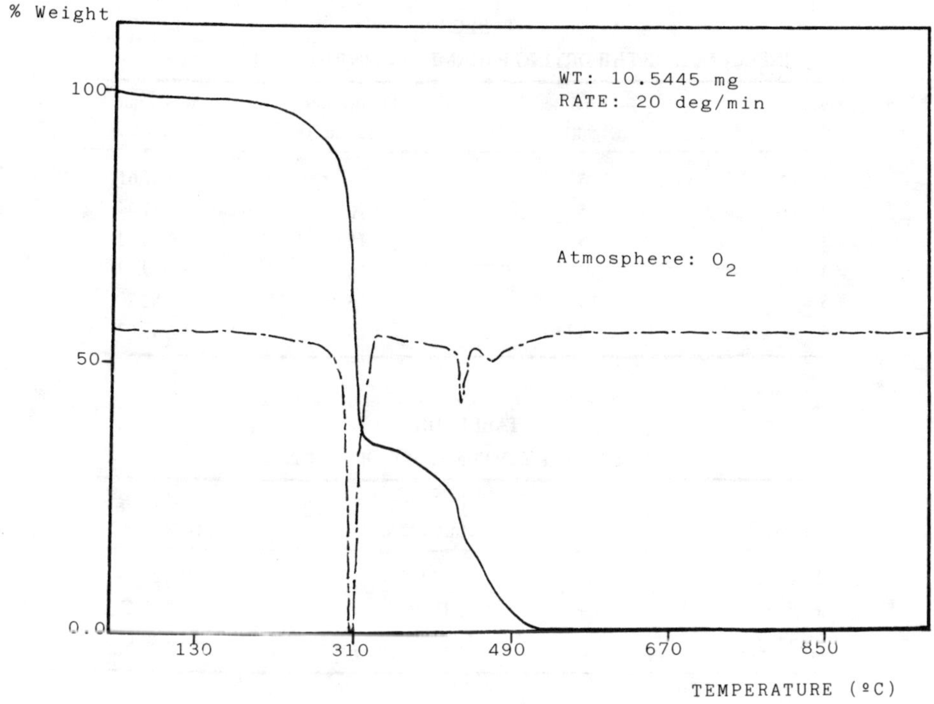

FIGURE 3 Thermogram in oxidizing atmosphere.

Thermograms

Figures 2 and 3 present, respectively, the thermograms in inert and oxidizing atmospheres obtained with microsamples of coffee waste.

Similar results were obtained with macrosamples in a "Thermolyne" brand muffle furnace.

Combustion

The combustion of the coffee waste with a 30% moisture content in the AFBC demonstrates a similar behaviour to that of other lignocellulose fuels tested. A faster combustion was observed for this waste due to its high volatile content and its great specific surface. In the top of the boiler the temperature was close to 400°C.

The test in the industrial furnace of the SEDA factory demonstrated yields of almost 50%, and it was observed that the volatiles released upon waste feeding were burned as they rose inside the furnace in which natural gas was being used.

The fusibility temperature, the elemental analysis of the components and the analysis of the ashes are given in the tables VIII, IX and X.

DISCUSSION

The elevated volatile content of the coffee waste in the proximate analysis is notable, as well as its low ash content. This will impose some special characteristics on the waste combustion process (Burnley, 1988), given that the escape of unburned volatiles should be

TABLE VII

INFLUENCE OF THE DRYING PARAMETERS IN THE FLUIDIZED BED

Test no.	Initial bed height (cm)	Fluidization time (min)	% Final moisture
1	5	5	47.61
2	5	10	41.44
3	5	15	45.73
4	10	10	49.58
5	15	5	52.70
6	15	15	49.03

TABLE VIII

FUSIBILITY TEMPERATURE OF ASHES

Fusibility	Reducing atmosphere	Oxidating atmosphere
Deformation	1,120°C	1,120°C
Fusion	1,165°C	1,165°C
Fluidity	1,180°C	1,180°C

avoided, especially since the granulometric nature of the waste is so low that it permits a very rapid release of the gases. This is avoided by the introduction of secondary air from above the feed zone of the coffee waste, as the volatiles released in the lower grate zone should be burned inside the furnace.

From the elemental analysis a low sulphur content can be observed, below 2% in dry basis, which will limit the release of SO_2 into the atmosphere through the combustion gases.

The heating power, both higher and lower, are elevated. They vary in the different samples received, possibly due to the origin of the samples, the class to which they correspond, or to the decaffeinization process prior to extraction to which they might have been subjected. In the experimental results it can be seen that, for a moisture of approximately 35%, the higher heating power of waste of about 14,630 J/g is obtained, which is comparable with that of other lignocellulosic wastes normally utilized in combustion.

The method of initial elimination of the water of the residue, by pressing, has demonstrated the ease, speed and low cost with which it is possible to reduce the moisture to nearly 50%.

Forced drying with air has shown that it is convenient to perform this stage intermittently, since the elimination of superficial water is very rapid, thus converting the diffusion of the water by capillary action towards the surface of the waste particle into the controlling stage of the process. This fact makes it logical to think about the necessity of using thermic drying which could take advantage of the residual heat from the fumes in the boiler.

The thermogravimetric analysis of the waste confirms its high degree of volatile material (≈82%), as well as the fact that only the ashes remain at temperatures above 550°C and that all the volatiles (light and heavy) are eliminated at 490°C.

The fusibility temperatures observed in the ash's behaviour are very high, which indicates that slag is not generated as the furnace work will be carried out at lower temperatures

TABLE IX
ELEMENTAL ANALYSIS OF ASHES

Element	Concentration (ppm)
Ba	48
Be	5
Co	36
Cr	1,375
Cu	4,900
La	38
Mo	42
Ni	99
Sr	950
U	13
W	25
Y	6.4
Zn	1,575
Zr	89

TABLE X
COMPONENTS ANALYSIS OF ASHES

Component	% concentration
Al_2O_3	2.40
CaO	20.00
Fe_2O_3	4.20
MgO	5.90
MnO	0.58
Na_2O	0.73
P_2O_5	7.60
TiO_2	0.13
S (total)	0.90
SiO_2	1.60
K_2O	1.70

than in these tests. However, the existence of hot points in some zones of the furnace, which might cause slag, must not be neglected.

The range of superior sizes in the granulometric distributions varies with the samples, which could be due to the grinding stage, or the characteristics of the original products processed.

PLANT FOR COFFEE WASTE COMBUSTION

The specifications needed for the design of the incineration plant for the combustion of coffee waste from the manufacturing process in the company SEDA of Spain are:

— Moist material: 4,000 kg/h.
— Initial moisture waste: 85% water.
— Moisture after pressing: 55% water.
— Moisture after drying: 25% water.
— Dried material: 800 kg/h.
— Heating power value at 25%: 18,000 kJ/kg.
— Maximum release of heat: 14.4 10^6 kJ/h.
— Temperature of flue-gas boiler: 400°C.
— Temperature of flue-gas drying: 200°C.

Installation Proposed

A diagram of the proposed pilot plant for the combustion of coffee waste, with the specifications indicated previously, is shown in figure 4. This plant will be installed next year in the SEDA factory of Palencia (Spain).

This plant has the following elements: A silo for coffee waste; a screw press; a catch basin; a drying drum; a cyclone; a double stoker for feeding the coffee waste to the incinerator and a boiler with a double chamber — one for the coffee waste and another for the natural gas.

The spent waste, obtained in the drying drum, is fed by a screw conveyor to the incinerator.

The hot fumes from the boiler flow will be transported, by means of a special exhauster which operates at 400°C, to the inlet of the drying drum and thereafter to the cyclone, where dust will be separated and another exhauster will transport the gases to the chimney.

The flue-pipe between the exhauster and inlet box is "T-shaped," with one branch leading to the chimney, the other a bypass in case of break down in the system. This allows the boiler to operate with the natural gas burners and the incinerator with waste without the necessity of drying drum.

The industrial process proposed to eliminate the water of pressing consists of a flocculation aimed at eliminating the organic material in suspension. The flocs will be recycled to the pressing stage, and the resultant water will be also recycled to the extraction process of the soluble coffee. Owing to its better economics, this process seems to be very convenient.

CONCLUSIONS

The higher heating power of the lignocellulose wastes of coffee indicates the necessity to reduce, by close to 25%, the initial moisture content in order to make them profitable as a fuel.

The advantages of this residue as a fuel include its low sulphur content, with the resultant low dew point in the vapours, as well as the lack of transportation cost due to the fact that this expense is included in the price of the initial product.

It should also be pointed out that the advantages of its use as a fuel include possible economic and energy saving, elimination of the risk of fires in the spillways during the summer season, and the lowering of SO_2 atmospheric contamination due to its low sulphur content.

One inconvenience existing in its utilization is the need to eliminate the high degree of initial moisture.

The biological degradation of stored waste makes it obligatory to perform the pressing process no later than 72 hours after the generation of the material; otherwise, operational problems will be caused by the rupture of the lignocellulose structures.

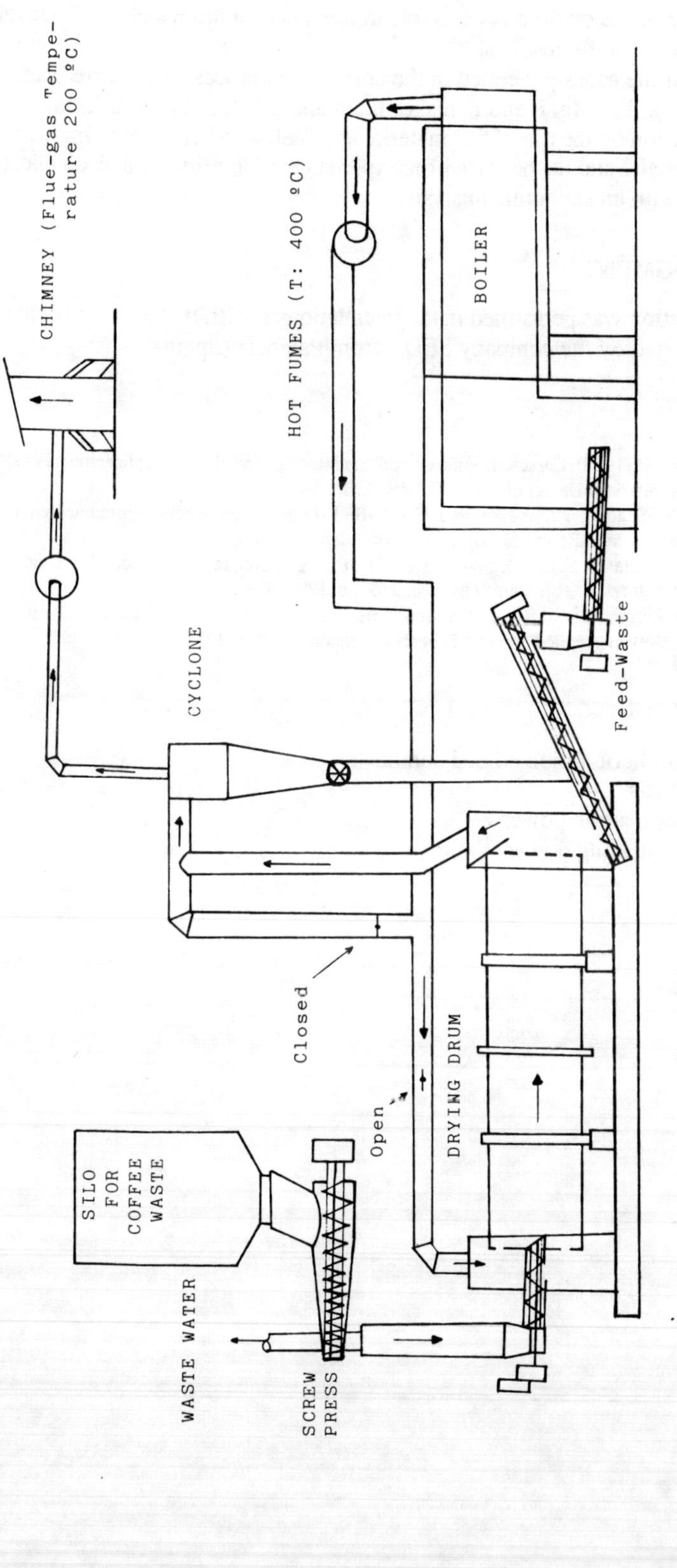

FIGURE 4 Combustion pilot plant proposed.

For the drying of the waste it is advisable to use a drying drum and to take advantage of the residual heat of the fumes.

An analysis of the ashes generated in the combustion process was carried out, and their high content of Ca, Fe, Mg, P and K makes them apt to be used as fertilizers.

The final decision on the use of this material as a fuel, whether alone or mixed with other traditional materials, should be based on experience in the drying and combustion processes together with an economic analysis.

ACKNOWLEDGMENTS

The characterization was performed in the installations of CEDER at Soria (Spain) and the work was supported by the company SEDA from Palencia (Spain).

REFERENCES

Antolín, G. and Irusta, R.(1989). Caracterización de combustibles lignocelulósicos: aplicación a la paja de cereal. Publication Secretary of University of Valladolid, ISBN 84-7762-072-5.

Antolín, G., Irusta, R., Segovia, J.J. and Velasco, E. (1990). Estudio del aprovechamiento energético del residuo lignocelulósico del café. Report SEDA industry of Palencia (Spain).

Antolín, G., Irusta, R., Velasco, E. and Segovia, J.J. (1991). Planta de combustión en lecho fluidizado atmosférico para residuo agroforestal. *Ingeniería Química*, **269**, pp. 128–132.

Burnley, S.J. (1988). Refuse derived fuel combustion trials. *Energy Recovery Through Waste Combustion*. eds. A. Brown *et al.*, Elsevier Applied Science Publisher, London, pp. 298–301.

NOMENCLATURE

AFBC	Atmospheric fluidized bed combustor
db	Dry basis
HHP	Higher heating power
LHP	Lower heating power

7.

FLUIDISED BED COMBUSTION

Mathematical Model of a Circulating Fluidized Bed with Chemical Reaction

F.R. STEWARD *Energy Conversion Engineering Group, University of New Brunswick, Fredericton, New Brunswick, Canada*

Abstract — A mathematical model has been developed for a gas/solid reaction within a circulating fluidized bed based on certain assumptions with respect to the flow of gas and solid particles. The analysis is applied specifically to the reaction between sulfur dioxide and particles of lime. A number of cases have been considered based on the mechanism by which sulfur dioxide is introduced into the bed and the mechanism of the gas solid reaction. Diagrams are presented which allow the prediction of the fraction of sulfur dioxide absorbed within the bed for various operating parameters and characteristics of the fluidized bed. The model is compared with data obtained from the Chatham Unit operated by the New Brunswick Electric Power Commission.

Key Words: fluidized bed, sulfur dioxide removal

DESCRIPTION OF A CIRCULATING FLUIDIZED BED AS A CHEMICAL REACTOR FOR SULFUR DIOXIDE CAPTURE

A typical circulating fluidized bed is shown in Figure 1. Solid material enters the base of the fluidized bed as fresh feed and solids return. The gas feed is distributed over the base of the bed and supplies fluid support of the solid particles. The solids and gas flow up the bed with the solids at a somewhat lower overall upward velocity than the gas. This effect produces a solids concentration profile within the bed which results in a higher concentration of solids at the base of the bed than at the top.

The stream leaving the fluidized bed containing both solids and gas enters a cyclone which separates a high percentage of the solids from the gas. The exhaust gas leaves the system after a single pass with a low concentration of solids.

The solids from the cyclone enter a return leg which feeds back into the base of the bed. If insufficient solids leave with the exhaust gas to satisfy the solids balance within the bed, the usual case, a fraction of the solids are removed from the return line. This fraction is usually small compared to the fraction returned to the base of the fluidized bed. Therefore, on the average the solids circulate through the fluidized bed a large number of times before leaving the system.

For the particular case when the circulating fluidized bed is used for the combustion of coal in the presence of limestone the following observations have been made.

Experimental data [1–4] indicate that the temperature distribution within the bed is reasonably uniform both along and across the bed. This can be attributed to the relatively small amount of coal in the fresh feed compared to recirculating solids and the significant carbon content in the recirculating solids which will react throughout the entire bed.

The limestone in the fresh feed is heated within the bed and reacts rapidly to form lime. The lime in turn reacts with the sulfur dioxide formed from the combustion of the sulfur in the coal. The amount of fresh limestone in the feed is also small compared to the amount of recycled solids.

Finally the overall reaction rate between the sulfur dioxide generated by the combustion of the coal and the lime in the solid particles is relatively slow. Therefore, the particles recirculate many times and continue to react over a long period of time relative to the contact time of the gas and solids on a single pass.

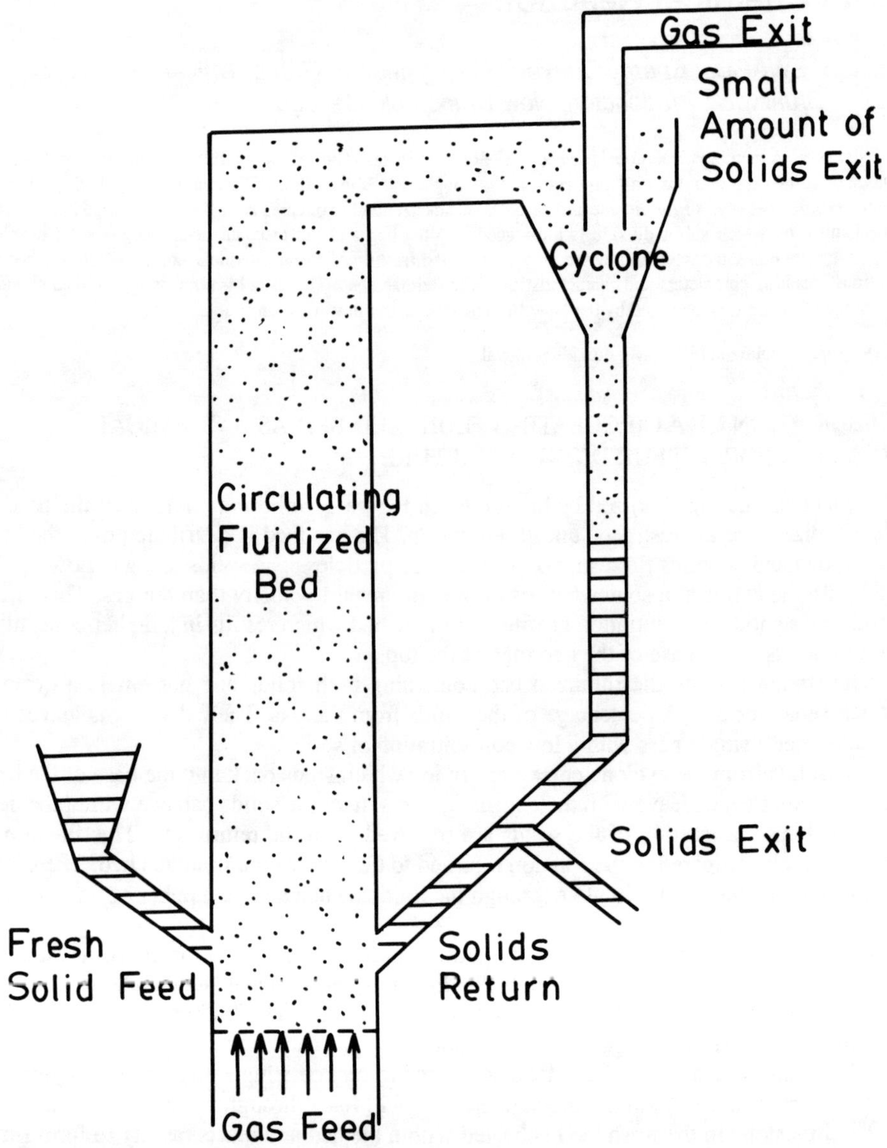

FIGURE 1 A circulating fluidized bed.

In view of the above it would seem reasonable to treat a circulating fluidized bed, in so far as the reaction between sulfur dioxide and lime particles is concerned, as a chemical reactor with the gas in plug flow, the solids completely mixed and the temperature uniform throughout the bed. The concentration of solids along the bed is given by an exponential relation as indicated by experimental evidence [1–4].

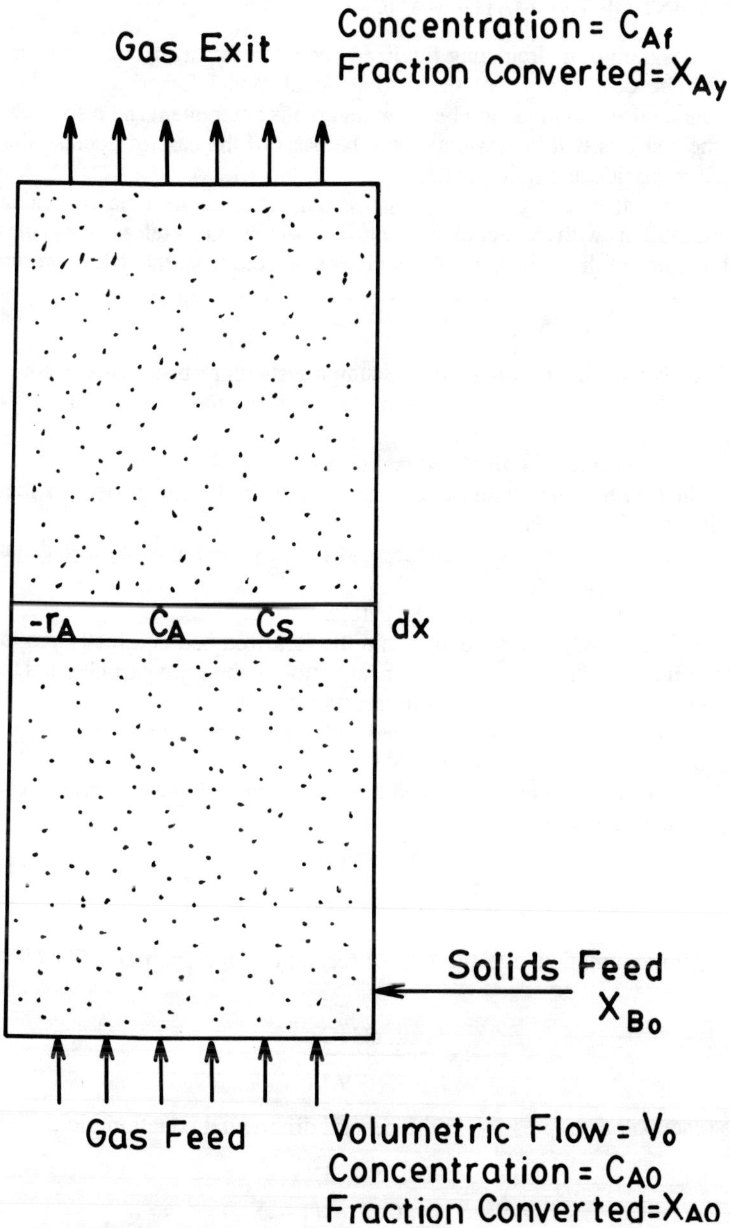

FIGURE 2 Mathematical representation of a circulating fluidized bed.

This paper develops a number of mathematical models assuming the above to be the physical situation. It is believed that these models are useful in qualitatively considering the operating parameters of such a circulating fluidized bed in so far as sulfur dioxide capture is concerned.

DEVELOPMENT OF THE MATHEMATICAL MODEL

A schematic diagram of a circulating fluidized bed treated as a chemical reactor is shown in Figure 2. The development of the mathematical model for such a fluidized bed is completely general for any reaction between a gaseous component and a solid component. However, the analysis will be developed on the basis of the case of interest, the reaction between sulfur dioxide and lime particles.

The mathematical analysis has been carried out for three separate cases that involve different assumptions with respect to the manner in which SO_2 is released into the fluidized bed. Combinations of these three cases can also be derived from the results presented here.

Model 1

The first case involves the following five assumptions with respect to the manner in which the circulating fluidized bed operates and the nature of the chemical reaction which occurs within the bed.

1. The gas flow through the fluidized bed is in plug flow.
2. The solids within the fluidized bed are well mixed with respect to their overall residence time within the bed.
3. The concentration of solids within the fluidized bed is exponential with distance along the bed.
4. The fluidized bed is at a uniform temperature.
5. The reaction of SO_2 at any point within the fluidized bed is directly proportional to the local concentration of SO_2, the local concentration of the solid particles and the relative activity of the solid particles compared to fresh solid particles.
6. The SO_2 enters the base of the fluidized bed at a concentration which is equivalent to its total release within the combustion process.

The differential equation for a chemical reactor shown in Figure 2 and based on the six above assumptions is given by

$$\frac{\mathrm{A}dx}{\mathrm{C_{AO}V_0}} = \frac{\mathrm{DX_A}}{-\mathrm{r_A}}. \tag{1}$$

According to assumption five the rate of reaction of the gas per unit volume can be written as

$$-\mathrm{r_A} = \frac{\mathrm{k_1 a C_S C_A}}{\mathrm{M_B}}. \tag{2}$$

Using this reaction rate expression in the general differential equation yields

$$\frac{\mathrm{A}dx}{\mathrm{C_{AO}V_0}} = \frac{\mathrm{M_B dX_A}}{\mathrm{k_1 a C_S C_{AO}(1 - X_A)}}. \tag{3}$$

Since it is assumed that the concentration of solids varies exponentially along the length of the bed the local concentration of solids is given by

$$\mathrm{C_s = C_{SO} e^{-k_s x}}. \tag{4}$$

Thus the differential equation becomes

$$\frac{\mathrm{A}dx}{\mathrm{V_0}} = \frac{\mathrm{M_B dX_A}}{\mathrm{k_1 a C_{SO} e^{-k_s x}(1 - X_A)}}, \tag{5}$$

with the boundary condition

$$x = 0, \quad C_A = C_{AO}, \quad X_A = X_{AO}.$$

Rearrangement and integration yields

$$\int_0^L \frac{Ak_1aC_{SO}e - \exp k_s x}{V_0}\,dx = \int_{X_{AO}}^{X_{Af}} \frac{dX_A}{(1 - X_A)}, \tag{6}$$

which gives upon integration

$$\frac{1 - X_{Af}}{1 - X_{AO}} = \frac{C_{Af}}{C_{AO}} = \exp - \frac{k_1 C_{SO} V (1 - e^{-k_s L}) a}{M_B k_s L V_0}. \tag{7}$$

Let $\overline{X}_B$ be the mean fraction of solid reactant B, lime, converted in the solids leaving the system. Given the feed rate of gaseous reactant to the system, A_f and the feed rate of solid reactant to the system, B_f, stoichiometry for the overall reaction within the fluidized bed system yields

$$A_f(X_{Af} - X_{AO}) = \frac{B_f}{b}\left(\overline{X}_B - X_{BO}\right). \tag{8}$$

The relative reactivity of the solids to that of fresh solids would normally be a function of the fraction of B reacted in the solids X_B. For the assumption of well mixed solids in the bed the mean relative reactivity for the entire bed would be a function of the outlet fraction of B converted. Thus,

$$a = f\left(\overline{X}_B\right). \tag{9}$$

Equations (7)–(9) constitute a unique solution to the fraction of gas phase reactant converted, X_{Af}. This solution would also give the mean fraction of solid converted, $\overline{X}_B$.

The solution to the above three equations depends on the mechanism of the reaction between the gas and the solid. This mechanism defines the relationship (9) from which a particular solution can be obtained.

Model 2

The second case involves the same first five assumptions as the previous model. However, the six assumption concerning the manner in which SO_2 enters the system is replaced by the following:

6. The SO_2 is generated throughout the fluidized bed at a rate which is directly proportional to the local concentration of solid particles.

The differential equation for the chemical reactor shown in Figure 2 based on the above assumption is given by

$$\frac{A\,dx}{V_0} = \frac{dC_A}{-r_A + K_s C_s}. \tag{10}$$

Using the same reaction rate expression for the reaction of SO_2, equation (2), and the same solids concentration distribution within the bed, equation (4), yields the differential equation

$$\frac{A}{V_0} C_{SO}\, e^{-k_s x}\, dx = \frac{dC_A}{\left(\frac{-k_1 a C_A}{M_B} + K_s\right)}. \tag{11}$$

The boundary condition at the inlet of the bed would be

$$x = 0 \quad \text{and} \quad C_A = 0.$$

since no reactant gas enters with the gaseous feed.

Applying the above inlet boundary condition and integrating over the limits yields.

$$\frac{-k_1 AVC_{SO}}{M_B V_0 k_s L}(1 - e^{-k_s L}) = \ln\left[\frac{\frac{-k_1 a C_{Af}}{M_B} + K_s}{K_s}\right]. \tag{12}$$

The reaction coefficient K_S must satisfy the mass balance for the gaseous reacting component A in that all the A generated by the solid fuel is released within the fluidized bed.

$$\int_0^L K_s C_s A dx = \frac{F_C X_s}{M_s}. \tag{13}$$

Substitution of the solids concentration distribution, equation (4), into equation (13) and integrating yields the reaction coefficient K_s in known quantities:

$$K_s = \frac{F_C X_s k_s L}{VM_s C_{SO}(1 - e^{-k_s L})}. \tag{14}$$

Substitution of this relation into equation (12) and solving for C_{Af} yields

$$C_{Af} = \frac{M_B F_C X_s k_s L}{k_1 a V C_{SO}(1 - e^{-k_s L})}\left(1 - e^{\frac{-k_1 a V C_{SO}\left(1 - e^{-k_s L}\right)}{M_B V_0 k_s L}}\right). \tag{15}$$

The concentration of reactant gas can be related to the fraction of A reacted

$$\frac{F_C X_s}{M_s}(1 - X_{Af}) = C_{Af} V_0. \tag{16}$$

Substitution of the above for C_{Af} into relation (15) yields the final result for the fraction of A reacted in the bed

$$(1 - X_{af}) = \frac{1 - e^{\frac{-k_1 V C_{SO}(1 - e^{-k_s L}) a}{M_B V_0 k_s L}}}{\frac{k_1 V C_{SO}(1 - e^{k_s L}) a}{M_B V_0 k_s L}}. \tag{17}$$

The overall mass balance (8) and the relation for the relative activity of the solid (9) would be the same for this model as Model 1 so that equations (8), (9) and (17) constitute a solution for Model 2 for a given mechanism of the reaction.

Model 3

The third case involves the first five assumptions as the previous two models. However, the six assumption concerning the manner in which SO_2 enters the system is replaced by the following:

6. The SO_2 is generated throughout the fluidized bed at a uniform rate per unit volume.

The differential equation for the chemical reactor shown in Figure 2 based on the above assumption is given by

$$\frac{\mathrm{Adx}}{\mathrm{V_0}} = \frac{\mathrm{dC_A}}{-\mathrm{r_A} + \mathrm{K_v}}. \tag{18}$$

Using the same reaction rate expression for the reaction of SO_2, equation (2), and the same solids concentration distribution within the bed, equation (4), yields the differential equation

$$\frac{\mathrm{dC_A}}{\mathrm{dx}} + \frac{\mathrm{k_1 a C_{SO}} e^{-\mathrm{k_s x}}\mathrm{A}}{\mathrm{V_0 M_B}}\mathrm{C_A} = \frac{\mathrm{A}}{\mathrm{V_0}}\mathrm{K_v} \tag{19}$$

along with the boundary condition at the inlet of the bed

$$\mathrm{x} = 0, \quad \mathrm{C_A} = 0$$

yields a differential equation in terms of C_A.

The solution to this differential equation with the above boundary condition is given by

$$\mathrm{C_{Af}} = \exp\left(\frac{\mathrm{k_1 V C_{SO} a} e^{-\mathrm{k_s} L}}{\mathrm{M_B V_0 k_s L}}\right)\int_0^L \frac{\mathrm{AK_v}}{\mathrm{V_0}} e^{\frac{-\mathrm{k_1 C_{SO} V a} e^{-\mathrm{k_s x}}}{\mathrm{M_B V_0 k_s L}}}\,\mathrm{dx}. \tag{20}$$

The reaction coefficient K_v must satisfy the mass balance for the gaseous reactant component A in that all the A generated within the reactor entered the reactor with the solid feed:

$$\int_0^1 \mathrm{K_v A dx} = \mathrm{K_v V} = \frac{\mathrm{F_C X_s}}{\mathrm{M_s}} = \mathrm{C'_{AO} V_0}. \tag{21}$$

Substitution of (21) into (20) yields the fraction of A reacted within the bed.

$$\frac{\mathrm{C_{AF}}}{\mathrm{C_{AO}}} = 1 - \mathrm{X_{Af}} = e^{\frac{\mathrm{k_1 V C_{SO} a} e^{-\mathrm{k_s} L}}{\mathrm{M_B V_0 k_s L}}}\left\{\int_0^1 e^{\frac{-\mathrm{k_1 V C_{SO} a} e^{-\mathrm{k_s L}(\mathrm{x/L})}}{\mathrm{M_B V_0 k_s L}}}\, d\left(\frac{\mathrm{X}}{\mathrm{L}}\right)\right\}. \tag{22}$$

Again the overall mass balance, (8) and the relation for the relative activity of the solids, (9), would be the same for this model so that equations (8), (9) and (23) constitute a solution for Model 3 for a given mechanism of the reaction.

Mechanism of the Reaction

The solution to the three models given above are independent of the reaction mechanism except for the assumption that the reaction rate per unit volume of bed is first order with respect to the gaseous reactant A and directly proportional to the local concentration of solid particles. Other assumptions could have been made but the integrated expressions giving the fraction of A reacted in the bed depend on these assumptions.

In order to obtain a specific result for each of the above models it is necessary to postulate the mechanism of the reaction within the solid particles so that an expression for the relative activity of the solids, equation (9), can be obtained. Several typical reaction mechanisms are considered below and the results presented.

Case 1

The case which has been given the most consideration here is based on the following assumptions which seem to be reasonably well verified for the reaction between sulfur dioxide and calcined limestone particles used in fluidized bed combustion units.

1. There is a certain fraction of solid reactant B available for reaction in a fresh particle. The remainder of reactant B within the particle does not react.

2. The rate of reaction is directly proportional to the amount of solid reactant B available for reaction within a particle divided by the total amount of B available for reaction in a fresh particle.

The above assumptions give the reaction rate for a single particle as

$$-r_B = \frac{4kv\pi r^3\rho_B}{3M_B}\frac{(X_{B\,max} - X_B)}{X_{B\,max}}C_A. \tag{23}$$

The reaction rate per unit volume of bed is given by equation (2). The reaction rate per particle and the reaction rate per unit volume of bed are related according to the expression

$$-r_A = \frac{-r_B n_B}{b}. \tag{24}$$

The local concentration of solid particles is given by

$$n_B = \frac{C_S}{4/3\pi r^3\rho_B}. \tag{25}$$

Substitution of equation (2), (23) and (25) into (24) yields

$$k_1 = \frac{k_v}{b}, \tag{26}$$

and

$$b = \frac{\text{g moles B}}{\text{g moles A}}. \tag{27}$$

Since the solids within the bed are assumed to be well mixed, assumption 2, the reaction rate per unit volume throughout the entire bed is based on the relative reactivity of the solid particles leaving the bed. Thus

$$a = \frac{X_{B\,max} - \overline{X}_B}{X_{B\,max}}. \tag{28}$$

The solution to all three models becomes specific based on the fraction of A reacted in the bed, the overall stoichiometric balance and the relative reactivity of the solids.

For the various models the solutions are given by

— Model 1: equations (7), (8) and (28)
— Model 2: equations (17), (8) and (28)
— Model 3: equations (22), (8) and (28)

From the point of view of numerical evaluation $\overline{X}_B$ can be eliminated from equation (35) and equation (8) to yield

$$a = 1 - \frac{bX_{Af}}{\frac{B_f}{A_f} X_{B\,max}}. \tag{29}$$

Equation (29) must be solved with the particular model equation for X_{Af} by trial and error.

Results of Models

The solution for each of the above models is given in terms of three dimensionless groups: the reaction parameter:

$$R = \frac{k_v V C_{SO}}{bM_B\, V_0 k_s L}. \tag{30}$$

the calcium/sulfur parameter:

$$S = X_{B\,max}\, Ca/S. \tag{31}$$

and the solids distribution parameter:

$$K = k_s L. \tag{32}$$

The solution is presented graphically in Figures 3–5 for Models 1, 2 and 3. Each diagram gives the fraction of SO_2 removed within the fluidized bed vs. the calcium/sulfur parameter for various values of the reaction parameter for a fixed value of the solids distribution parameter.

In each case the percent SO_2 removed increases with an increase in the calcium/sulfur parameter for a given reaction parameter. Also the percent SO_2 removed increases with an increase in the reaction parameter for a given calcium/sulfur parameter.

For a given set of the three parameters Model 1 gives a somewhat higher SO_2 removal than Model 2. Both Model 1 and Model 2 give a significantly higher SO_2 removal than Model 3 for the same three reaction parameters, see Figure 6. These results are consistent with the physical situation.

Notice that the calcium/sulfur parameter is directly proportional to the amount of calcium in the limestone *available* for reaction and inversely proportional to the sulfur in the fuel. It is a particular feature of the reaction between limestone and SO_2 that only part of the calcium is available for reaction.

The reaction rate parameter is directly proportional to the volume of the bed, the concentration of solids in the bed, and the reaction rate coefficient. It is inversely proportional to the volumetric flow rate. This is a typical reaction rate parameter which gives a higher conversion for a higher value.

The fraction of SO_2 removal decreases with an increase in the solids distribution parameter for a given set of values for the calcium/sulfur and reaction parameters. This is due to a reduction in total solids within the bed for an increase in this parameter.

Case 2

A common assumption for a gas solid reaction is the shrinking core model for a reaction mechanism. This case assumes that the reaction occurs at a defined interface with pure unreacted material in the core surrounded by pure product. For the particular reaction

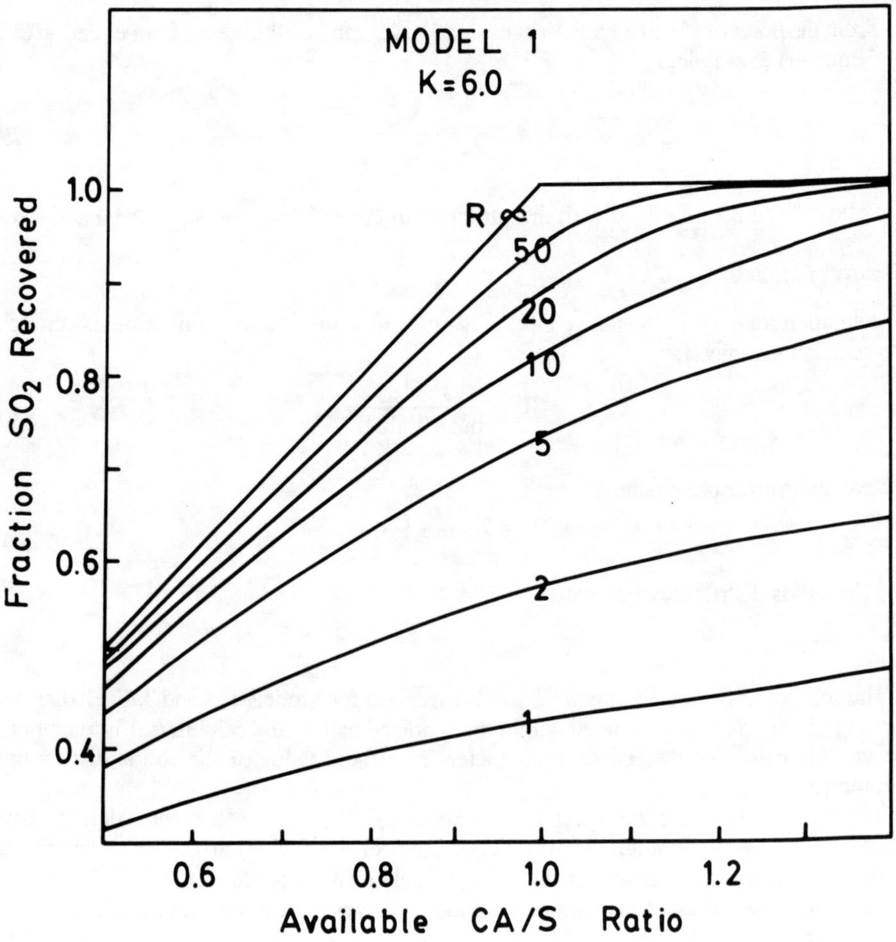

FIGURE 3 Fraction SO_2 recovered vs. available Ca/S ratio for various reaction rate parameters, Model 1, first order solid kinetics.

between SO_2 and CaO this mechanism is modified to allow for the unavailable CaO which does not participate in the reaction.

The shrinking core model can have various mechanisms controlling the rate of reaction. For this case it is assumed that the reaction is controlled by the gaseous reactant diffusion through the product layer.

The above assumptions give a rate of reaction for the gaseous component A in a single particle as

$$-r_{AS} = \frac{4\pi \mathcal{D}_e C_A}{\left(\frac{1}{r_c} - \frac{1}{r}\right)}. \tag{33}$$

Using the equation giving the local concentration of particles in the bed, (25), the reaction rate coefficient per unit volume can be determined by

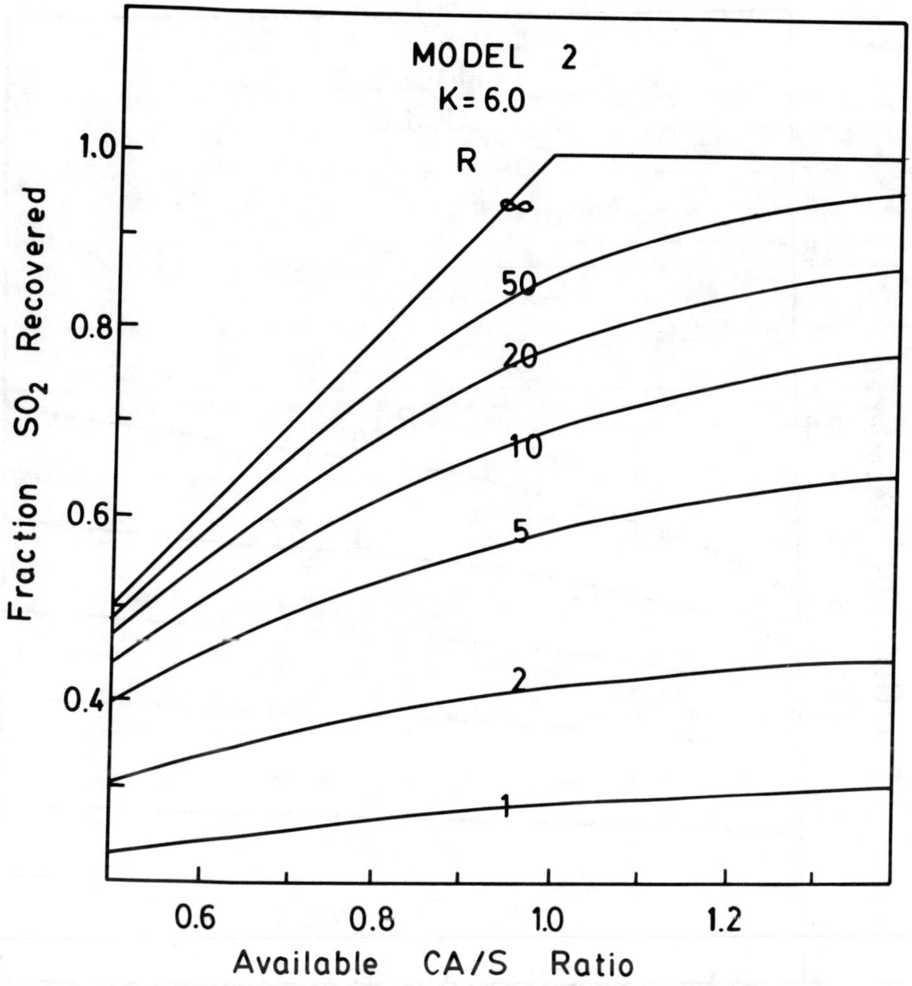

FIGURE 4 Fraction SO_2 removed vs. available Ca/S ratio for various reaction rate parameters, Model 2, first-order solid kinetics.

$$\frac{k_1 a C_s C_A}{M_B} = \frac{4\pi \mathcal{D}_e C_A C_s}{\frac{4}{3}\left(\frac{1}{r_c} - \frac{1}{r}\right) \rho_B \pi r^3}. \tag{34}$$

Define the relative solid activity for this case as

$$a = \frac{\frac{1}{r_{ref}} - \frac{1}{r}}{\frac{1}{r_c} - \frac{1}{r}}. \tag{35}$$

The above two equations yield

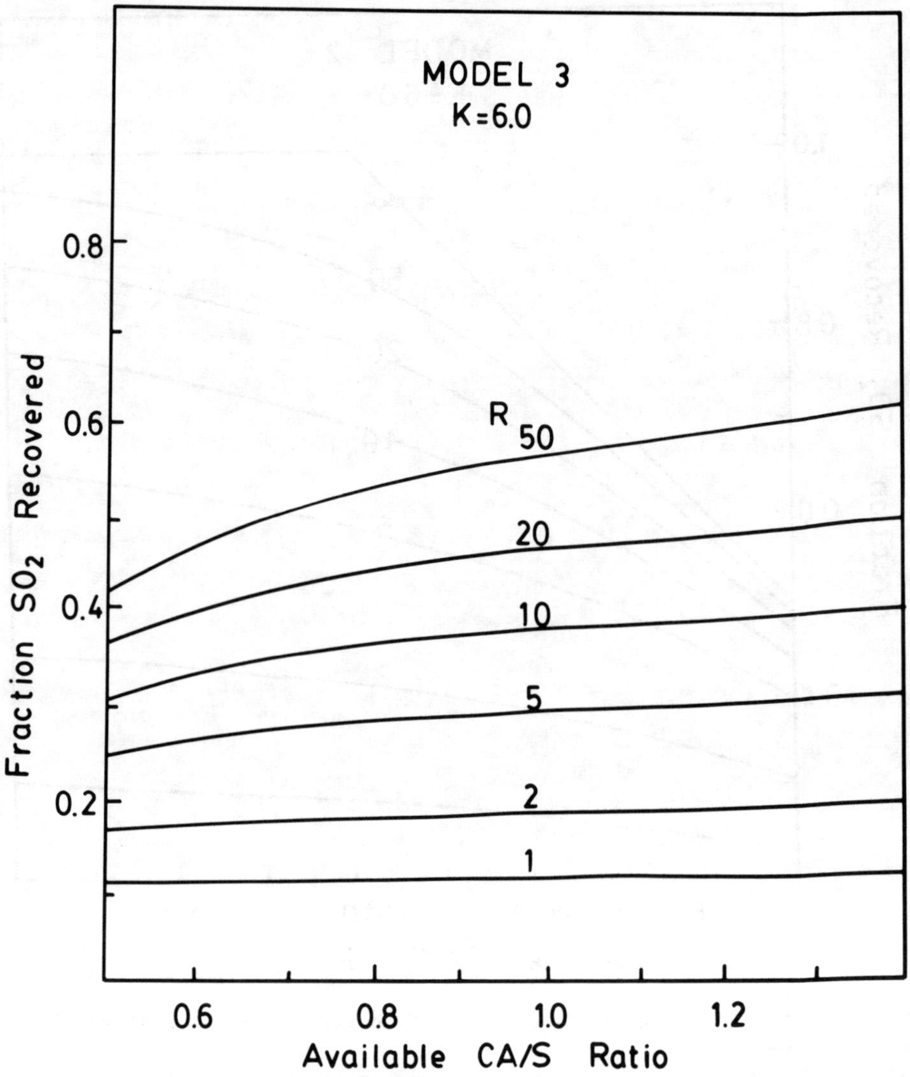

FIGURE 5 Fraction SO_2 removed vs. available Ca/S ratio for various reaction rate parameters, Model 3, first-order solid kinetics.

$$k_1 = \frac{3\mathcal{D}_e M_B}{\rho_B r^2 \left(\frac{r}{r_{ref}} - 1\right)}. \tag{36}$$

The r_{ref} must be less than R to prevent the reaction rate from going to infinity for fresh solid entering the system.

Since

$$\frac{\overline{X}_B}{X_{Bmax}} = 1 - \left(\frac{r_c}{r}\right)^3. \tag{37}$$

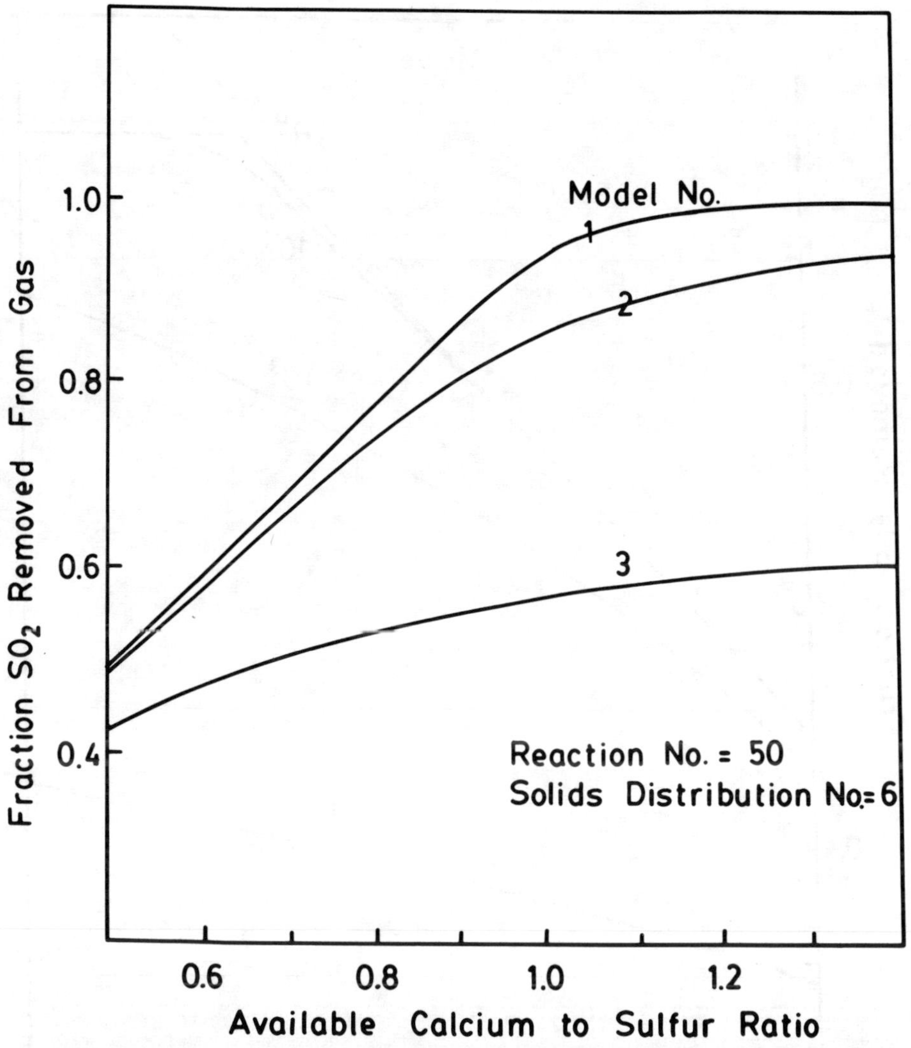

FIGURE 6 Fraction SO_2 removed vs. available Ca/S ratio for the three models, first-order solid kinetics.

The solution for this case with the assumptions of Model 1 is given by

$$\frac{1-X_{Af}}{1-X_{AO}} = \exp\frac{-3\mathcal{D}_e C_{SO} V(1-e^{-k_s L})a}{\rho_B R^2 k_s L V_0 \left(\frac{r}{r_{ref}}-1\right)}, \tag{38}$$

along with equations (35), (39) and the total mass balance (8).

The solution can be obtained in terms of the three parameters

$$R' = \frac{\mathcal{D}_e C_{SO} V(1-e^{-k_s L})}{\rho_B R^2 k_s L V \left(\frac{r}{r_{ref}}-1\right)}, \tag{39}$$

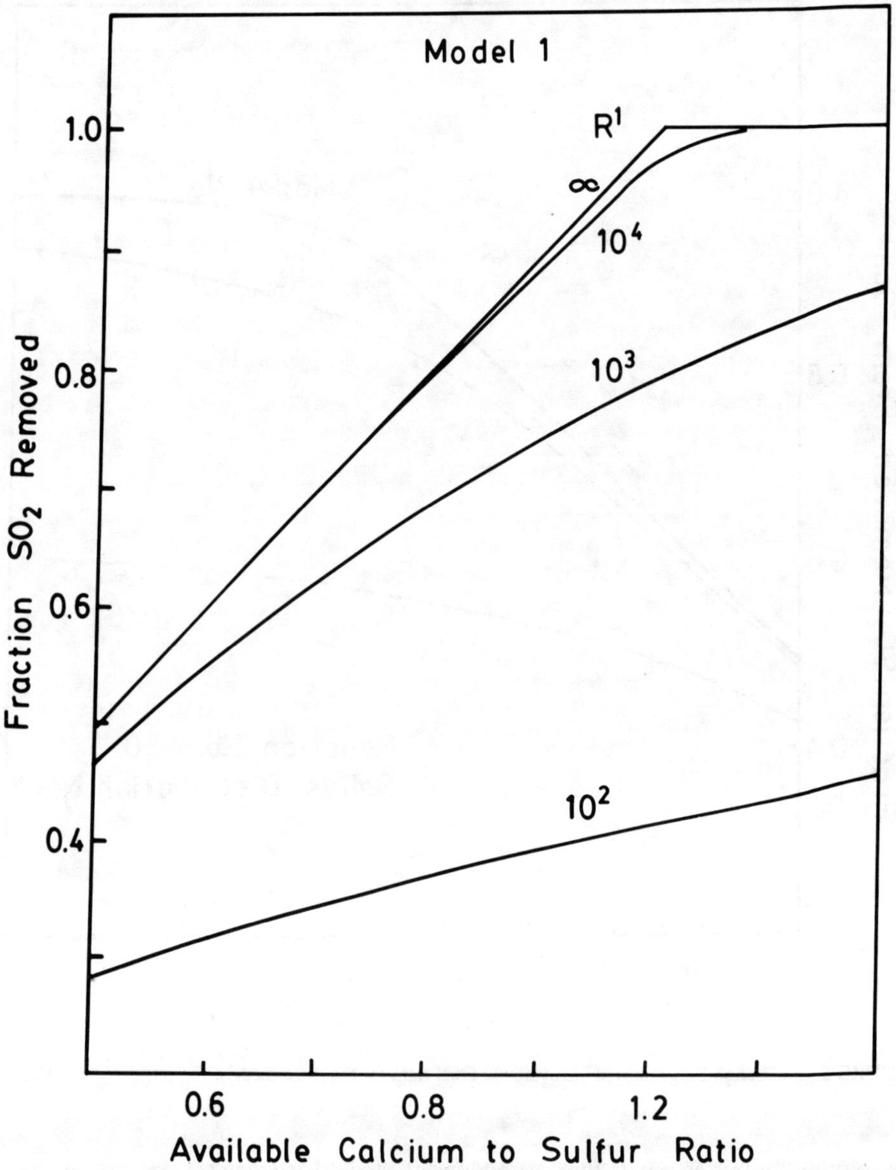

FIGURE 7 Fraction SO_2 removed vs. available Ca/S ratio various reaction rate parameters, Model 1, shrinking core mechanism.

$$S = X_{Bmax}\, Ca/S, \tag{40}$$

$$K = k_s L. \tag{41}$$

The solution to Case 2 for Model 1 is presented graphically in Figure 7. The form of the solution is similar to that of Model 1 with the solid reaction mechanism of Case 1.

Solid Side Analysis

The solutions given above were obtained for a reaction rate expression based on the reaction per unit volume within the gas phase. In order to get a specific solution a reaction mechanism for the specific reaction between the gas and the solid must be postulated. A solid side analysis also gives additional information about the nature of the solid reaction with respect to time. Each particle within the fluidized bed is reacting at a certain specific rate depending on the amount of time it has spent within the bed. The overall reaction rate is given by the integration over all such particles. This in turn gives a separate expression for the mean overall conversion of the solid particles which pass through the bed.

The general expression for the mean conversion of the solid particles is given by the expression

$$\overline{X}_B = \int_0^\infty X_B(t)\,E(t)dt. \qquad (42)$$

The solution to this equation for given residence times distribution E(t) gives the mean residence time the particles must spend in the bed since $\overline{X}_B$ is fixed by the fraction of A removed from the gas X_{Af} and the stoichiometry of a given set of feed conditions between reactant in the gas and reactant in the solid particles, equation (8).

Considering a specific reaction mechanism, Case 1, equation (42) can be solved. The reaction rate of a single particle for this mechanism is given by equation (23). Setting the rate of reaction to the change in fraction B reacted gives

$$-r_B = \frac{4}{3}\pi R^3 \frac{\rho_B}{M_B}\frac{dX_B}{dt} = \frac{4}{3}\frac{k_v \pi R^3 \rho_B}{M_B}\frac{(X_{Bmax} - X_B}{X_{Bmax}} C_A, \qquad (23)$$

and

$$\frac{dX_B}{dt} = k_v \frac{(X_{Bmax} - X_B)}{X_{Bmax}} C_A. \qquad (43)$$

This can readily be solved to yield

$$X_B = X_{Bmax} - (X_{Bmax} - X_{BO})\, e^{\frac{-k_v C_A t}{X_{Bmax}}}. \qquad (44)$$

Assumption 2 for the circulating fluidized bed model states that the solids are well mixed within the bed due to the high fraction of recirculation. The residence times distribution for a well mixed material in a reactor is given by

$$E(t) = \frac{e^{-t/\tau}}{\tau}. \qquad (45)$$

The value of C_A in equation (43) represents the local concentration of the gas for the reacting solid. Within the fluidized bed this concentration is variable throughout the bed and an appropriate mean value must be determined. The reacting mean gas concentration for the solid particles is determined by integrating the solids concentration and reacting gas concentration over the bed.

$$C_{Amean} = \frac{\int_0^L C_A C_s\,dx}{\int_0^L C_s\,dx}. \tag{46}$$

Substituting for the solid concentration distribution within the bed equation (4), yields

$$C_{Amean} = \frac{\int_0^L C_A C_{SO} e^{-k_s x}\,dx}{\int_0^L C_{SO} e^{-k_s x}\,dx}, \tag{47}$$

and

$$C_{Amean} = \frac{k_s L}{(1 - e^{-k_s L})} \int_0^{1.0} C_A\, e^{-k_s L y}\,dy. \tag{48}$$

For each of the four models C_A is given as a function of relative distance along the bed y = X/L.

The mean residence time of the solid material within the bed can now be found by substituting equations (44) and (45) into equation (42) while using equation (48) to determine the appropriate C_{Amean}. As an example consider Model 1. The value for C_A as a function of y can be determined from equation (7) as

$$C_A = C_{AO}\, e^{-Ra(1-e^{-k_s L y})}. \tag{49}$$

Substitution of (45) and (44) into equation (42) with $X_{B0} = O$ yields

$$\overline{X}_B = \frac{X_{Bmax}}{\tau} \int_0^{\infty} (1 - e^{-At}) e^{-t/\tau}\,dt, \tag{50}$$

where

$$A = \frac{k_v C_{Amean}}{X_{Bmax}}, \tag{51}$$

and

$$\overline{X}_B = \frac{X_{Bmax}}{\tau}\left[\tau - \frac{1}{A + \frac{1}{\tau}}\right], \tag{52}$$

or solving for τ yields

$$\tau = \frac{1}{A}\left(\frac{1}{1 - \frac{\overline{X}_B}{X_{Bmax}}} - 1\right), \tag{53}$$

where C_{Amean} must be determined by numerical integration of equation (50).

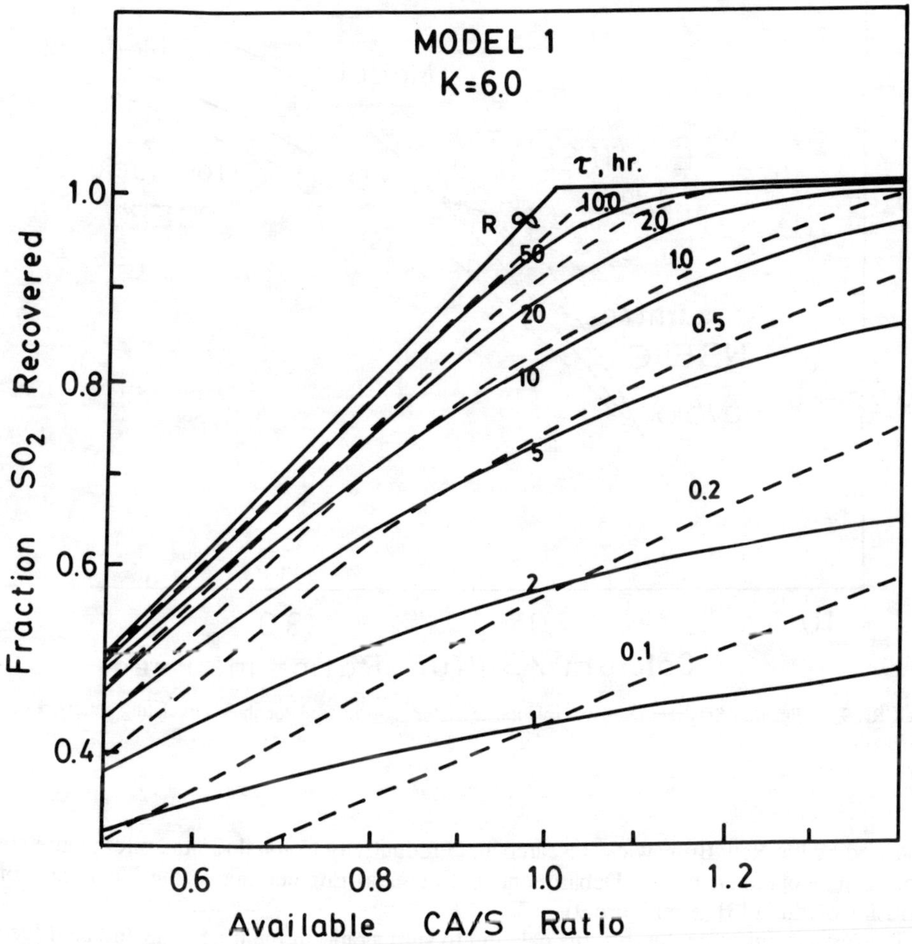

FIGURE 8 Fraction SO_2 removed vs. available Ca/S ratio various mean residence times for solid particles, Model 1, first-order solid kinetics.

This now represents a solution for any of the given models for any given reaction mechanism of the solid gas reaction. The solution includes the mean residence time of the reaction particles.

The fraction A converted as a function of the calcium/sulfur parameter for various mean residence times for the solid particles is given in Figure 8 along with the curves presented previously, Figure 3, for various values of the reaction rate parameter. For a given calcium/sulfur parameter the fraction of SO_2 removed increases with an increase in solids mean residence time. An increase in mean residence time obviously makes better use of the solid material.

Comparison with Plant and Pilot Plant Data

The model for the gas side analysis is compared with data from the Chatham Plant operated by the New Brunswick Electric Power Commission [1], pilot plant data from a unit

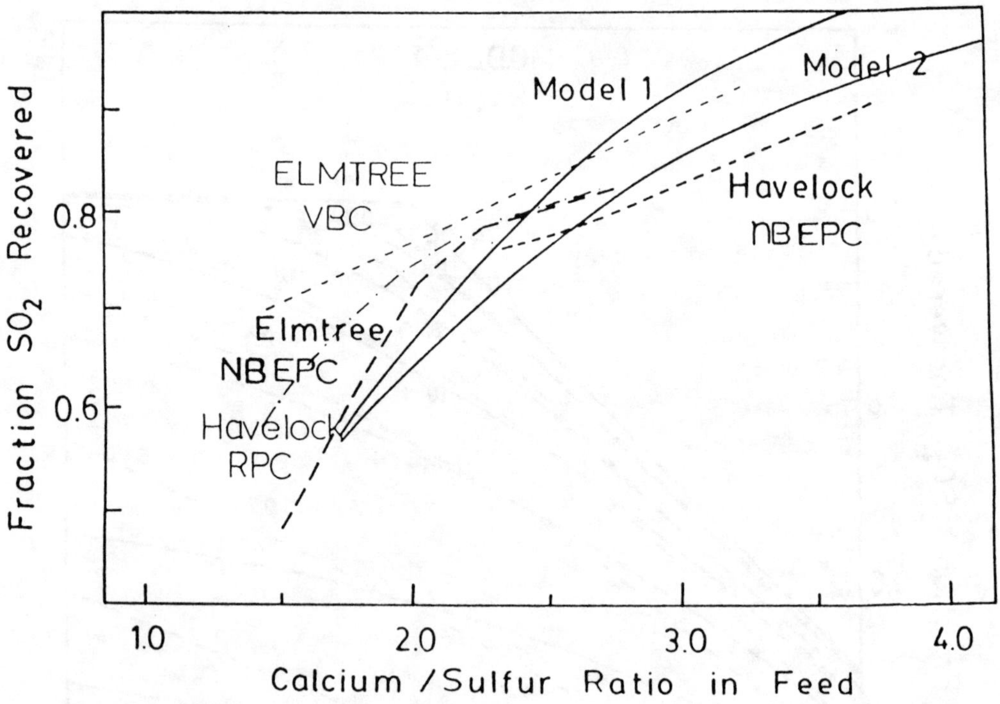

FIGURE 9 Fraction SO_2 recovered vs. calcium to sulfur ratio in feed for three circulating fluidized bed combustors.

operated by the New Brunswick Research and Productivity Council [2] and pilot plant data from a unit operated by the Department of Chemical Engineering at the University of British Columbia [3] (see Figure 9).

The percent sulfur capture vs. the calcium to sulfur ratio in the limestone and coal feed is presented for the three units with three types of limestones. In all cases the temperature was approximately 850°C and the bed operating conditions were such that the reaction rate parameter R was between 30 and 60.

Curves for Model 1 and Model 2 are also presented in Figure 9 based on the following assumptions:

1. One third of the calcium in the limestone is available for reaction with the sulfur dioxide. This is based on laboratory studies carried out on Havelock limestone [5] and plant data [1].

2. The reaction parameter, R, for the circulating fluidized bed is 50 based on a reaction coefficient determined in the laboratory for Havelock limestone [5] and the operating conditions during the tests conducted in [1–3].

3. The solid distribution parameter for the circulating fluidized bed during these tests was 6 (see [1–3]). This is the actual value for Chatham. The values for the UBC and RPC units were 7.3 and 4.3 respectively.

For the above assumptions it is apparent that Models 1 and 2 give a reasonable range for the prediction of the sulfur capture for the three units. It should be pointed out, however, that all three units were operating with conditions which allow the limestone to remain in the circulating fluidized bed for an extended period of time, several hours. This allows a high percentage of the calcium which is available to react to do so. Therefore, kinetic limitations are not significant under these operating conditions. This is predicted by the model, but the kinetic part of the model is only marginally relevant for these cases.

CONCLUSIONS

1. A mathematical model for the gas/solid reaction within a circulating fluidized bed has been developed based on the principles of chemical reactor design.
2. The model is applied specifically to the reaction of sulfur dioxide capture by lime particles.
3. A number of cases have been considered based on the mechanism by which SO_2 is introduced into the bed and the mechanism of the gas solid reaction.
4. The model is compared with data from an operating plant and two pilot plants which fired high sulfur coal.

REFERENCES

1. The New Brunswick Electric Power Commission and Energy Mines and Resources, Sorbent Optimization for FDC Technology, Chatham Circulating Fluidized Bed Demonstration Project, May 1989.
2. Salib, P. (1989). Investigation of Sulfur Capture During Circulating Fluidized Bed Combustion of Eastern Canadian Coals, New Brunswick Research and Productivity Council, Report No. MDP/88/17, March 20.
3. Grace, J.R., Brereton, CMH, LIM, C.J., Legros, R., Zhao, J., Senior, R.C., Wu, R.L, Muir, J.R., and Ergman, R. (1989). Circulating Fluidized Bed Combustion of Western Canadian Fuels, Department of Chem. Eng., University of British Columbia, August.
4. Couturier, M.F., Doucette, B., and Poolpol, S. (1989). *A Study of Gas Concentration, Solids Loading and Temperature Profiles within the CFB Combustor*, Energy Conversion Engineering Group, University of New Brunswick, November.
5. Richie, C.D. (1989). Sulphation of Limestone in a Circulating Transport Reactor, MSc Thesis, Dept. of Chem. Eng., University of New Brunswick, Fredericton, N. B.

NOMENCLATURE

a	The relative reactivity of the solids compared to that of fresh solids, dimensionless
A	Cross-sectional area of the reactor, m^2
A_f	Molal flow rate of gas A to bed, g moles A/s
b	Stoichiometric coefficient for reaction, (g moles B)/(g moles A)
B_f	Molal flow rate of solid B to bed, g moles B/s
C_A	Local concentration of gas phase reactant, SO_2, g moles A/m^3
C_{AO}	Inlet concentration of reacting gas, SO_2, g moles A/m^3
C'_{AO}	Concentration of A equivalent to the inlet if all of A generated within the bed was in the gas feed, g moles A/m^3
C_S	Local concentration of solids, CaO, gB/m^3
C_{SO}	Concentration of solids at the base of fluidized bed, gB/m^3
E(t)	Residence times distribution of particles in the bed, s^{-1}
F_c	Feed rate of the solid material containing A, g/s
k_1	Reaction rate coefficient for fresh solids for first order reaction, m^3/g moles B s
k_s	Solids distribution coefficient within bed, m^{-1}
k_v	Reaction rate coefficient within a particle, m^3/g moles A s
K	Solids distribution parameter, k_sL, dimensionless
K_s	Reaction rate coefficient for the generation of SO_2 per unit volume within bed for solids, g moles A/g B s

K_v	Reaction rate coefficient for the generation for SO_2 per unit volume with bed in gas phase, g moles A/m^3 s
L	Length of bed, m
M_B	Molecular weight of solid reactant, CaO, gB/g moles B
M_s	Molecular weight of gas phase reactant, SO_2, gA/g moles A
n_B	Concentration of solid particles per unit volume in bed, m^{-3}
r	Radius of particle, m
r_{ref}	Reference radius of particle in shrinking core model, m
$-r_A$	Reaction rate per unit volume of bed, g moles A/m^3 s
$-r_{AS}$	Reaction rate of gas phase component in a single solid particle, g moles A/s
$-r_B$	Reaction rate of a single particle, g moles B/s
r_c	Reaction interface in a particle, m
R	Reaction parameter for first order model, $(k_vVC_{SO})/(bM_BV_0k_SL)$, dimensionless
R′	Reaction parameter for shrinking core model, $(\mathcal{D}_eC_{SO}V(1 - e^{-k_SL})/(\rho_BR^2k_SLV_o)r/r_{ref} - 1))$, dimensionless
S	Calcium/sulfur parameter, X_{Bmax} Ca/S
t	Time, s
V_o	Volumetric flow rate of gas, m^3/s
V	Volume of bed, m^3
x	Distance along bed, m
X_A	Fraction of reacting gas converted, dimensionless
X_{Ao}	Fraction of reacting gas converted in feed, dimensionless
X_{Af}	Fraction of reacting gas converted in exit gas, dimensionless
$\underline{X}_B$	Fraction of reacting solids converted, dimensionless
X_B	Mean fraction of reacting solids converted in exit particles, dimensionless
X_{BO}	Fraction of reacting solids converted in feed, dimensionless
X_{Bmax}	Maximum fraction of reacting solids that can be converted, dimensionless
X_s	Mass fraction of material which generates gas in solid feed, gA/g
y	Dimensionless length along bed = x/L
$\mathcal{D}_e$	Diffusivity of gas through reaction product layer of particle, m^2/s
ρ_B	Density of solid particles, gB/m^3
τ	mean residence time of particles in bed, s

One-Dimensional Fluidized Bed Furnace Model

R. LEITHNER and J. WANG *Institut für Wärme und Brennstofftechnik, Technischen Universität Braunschweig, Germany*

Abstract — Based on the fact that the solid particles in a fluidized bed show a particle size distribution, a one-dimensional model of a circulating fluidized bed has been developed. Using this model, one can determine the continuous porosity distribution and the continuous pressure drop throughout the fluidized bed dependent only on known parameters, such as particle size distribution, volumetric flow, etc. This model simultaneously allows for the calculation of the recirculating solid mass flow.

Key Words: Circulating Fluidized Bed, One-Dimensional Model, Solid-Gas Flow, Porosity Profile, Pressure Profile, Solid Mass Flow

INTRODUCTION

Fluidized bed technology can be utilized in a wide variety of processes, such as the combustion or gasification of coal, dry exhaust-gas cleaning, drying, calcination and roasting, as well as ore-reduction processes. The modelling of two-phase flow is extremely significant in terms of the thermal design of fluidized bed combustors, and crucial for calculating combustion kinetics and kinetics of pollutant formation and reduction. Previous studies of fluidized bed models generally used a mean particle diameter; only in a few instances has the particle size distribution been mentioned. It is acknowledged, however, that there is a sorting effect in a fluidized bed as well. With this fact in mind, one can develop a simple one-dimensional model. This model can be used to calculate very plausible porosity and pressure distributions in a fluidized bed. This simultaneously facilitates the determination of circulating solid mass flow.

OUR MODEL CONCEPT

Particle Size Distribution

The solid particles in fluidized beds (see figures 1 and 2) show a specific particle size distribution. If one displays the weight of the particle fractions with certain particle sizes (diameters), one gets the particle size density distribution curve (see figure 2). The integral of this curve from the diameter 0 to d_p and, at the same time the curve of the passage fraction in sieve analysis, is called the particle size cumulative curve (see figure 2 as well). For such particle size cumulative curves Rosin, Rammler and Sperling found first the following simple formula (see e.g., Brandt (1981)):

$$D = 1 - \frac{1}{e^{(d_p/k)^n}}, \tag{1}$$

in which D is the passage fraction, which has a particle size smaller than d_p, k and n are constants and are determined by the particle size cumulative distribution curve (sieve analysis). The particle size density distribution curve can be established using the derivation of the particle size cumulative distribution curve:

$$V_d = \frac{dD}{dd_p} = \frac{n}{k}\left(\frac{d_p}{k}\right)^{n-1} e^{-(d_p/k)^n}. \tag{2}$$

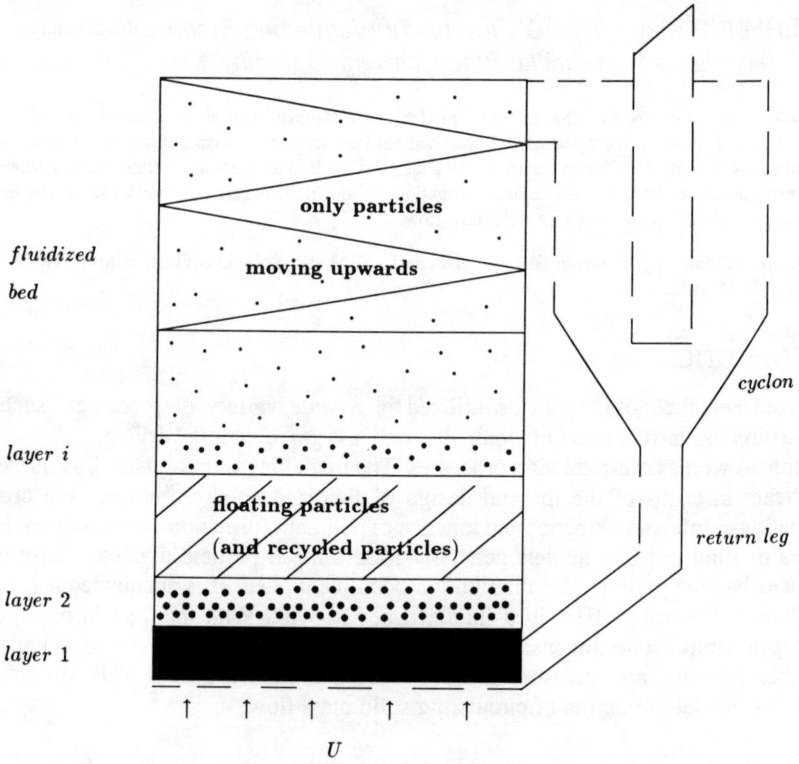

FIGURE 1 Schematic diagram of a CFBC.

By setting the particle size density distribution curve derivative to zero ($dV_d/dd_p = 0$), the maximum of the particle size density distribution curve (and at the same time the inclination changing point in the particle size cumulative curve) will be given by:

$$d_{pmax} = k\sqrt{\frac{n-1}{n}}. \tag{3}$$

Fluidized Beds with Recirculation of Particles in the Uppermost Layer (Without Cross-Sectional Extension)

A single particle in an upward gas flow in a channel falls, floats, or flies along with the gas flow, depending on whether the gas velocity is lower, equal to, or higher than the particle terminal velocity.

Consider a fixed bed composed of solid particles in a channel which is allowed to flow throughout. A certain volumetric gas flow would have so much force on the solid particles in the fixed bed that the particles would begin to loosen until they cease to touch each other. From that point on, called the minimum fluidization, one can refer to a fluidized bed. With increasing gas flow the distances between the particles increase as well, giving more free cross-sectional area and by this keeping the actual gas velocity constant. This expansionary behavior can be described by the porosity, i.e., the gas volume relative to the total volume. Due to different particle sizes, some of the particles with smaller diameters, and thereby a lower terminal velocity than the gas velocity between the particles, would move upwards

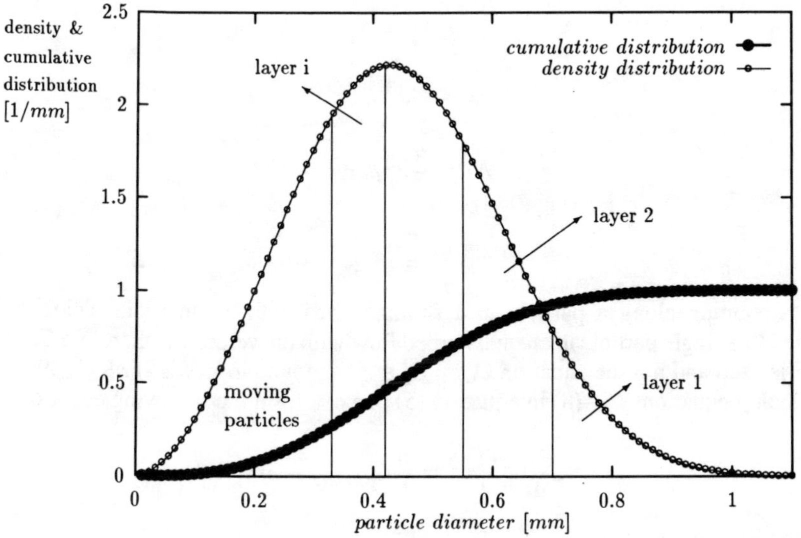

FIGURE 2 Particle size distribution.

even at the minimum fluidizing condition. The coarser particles stay down and form a stationary fluidized bed with the porosity at minimum fluidization. The displaced particles cause the porosity in the upper section at minimum fluidization to be still approximately 1.

If one increases the volumetric gas flow, and thereby the superficial gas velocity, more particles are carried upwards. The particles continue to move upwards until they reach a certain height where the high porosity and therefore the reduced velocity between the particles causes them to fall or to float. In a steady state, the fluidized bed forms a continuous vertical distribution of porosity with high concentrations of solid particles (low porosity) and larger particle diameter in the bottom section, and lower solid concentrations (higher porosity) and smaller particle diameters in the top sections.

Depending on flow conditions such as volumetric gas flow, the geometry of the fluidized bed, and locations of air- and recirculatory gas supply, particle size distribution, the total mass of solids found in the bed consisting of bed material, ash and fuel, the material properties of the solids and gas, one can calculate the porosity distribution and the pressure drop caused by the solid particles, as a function of the fluidized bed height. Simultaneously, the recirculating solid particles flow can be calculated.

Let us begin in the bottom layer with the porosity at minimum fluidization ε_L. Using the superficial gas velocity U and the porosity ε, one can calculate the mean gas velocity u_f between the particles:

$$u_f = \frac{U}{\varepsilon}. \tag{4}$$

Using the gas velocity u_f, one can establish a particle diameter d_p, whose terminal velocity is identical with this gas velocity. This particle diameter can be labelled the sinking particle diameter d_{ps}. It can be determined as follows.

We initially consider a single spherical particle in a vertically upwards gas flow. Weight G and buoyancy A_f, as well as the friction force W grip the particle. If the gas velocity achieves the terminal velocity of the particle, these three forces are in equilibrium:

$$G = W + A_f, \tag{5}$$

$$G = \frac{\pi}{6} d_p^3 \rho_s g, \tag{6}$$

$$A_f = \frac{\pi}{6} d_p^3 \rho_f g, \tag{7}$$

$$W = C_\omega \frac{\pi}{4} d_p^2 \frac{\rho_f}{2} u_f^2. \tag{8}$$

(The streaming along a particle in a fluidized bed with a superficial velocity U is simplified to a single particle in an undisturbed flow with the velocity u_f therefore $C_{\omega,\varepsilon}$ equal to C_ω/ε^2 is used and not the equation $C_{\omega,\varepsilon} = C_\omega\, \varepsilon^{-4.65}$ as found in Lewis *et al.* (1949).

By placing equations (6)–(8) in equation (5) one can form the following equation:

$$C_\omega \frac{\pi}{4} d_p^2 \frac{\rho_f}{2} u_f^2 = \frac{\pi}{6} d_p^3\, g\, (\rho_s - \rho_f), \tag{9}$$

which can be rearranged as

$$Ar = 0.75 C_\omega\, Re_s^2, \tag{10}$$

in which Ar is the Archimedes number, Re_s the Reynolds number formed with the terminal velocity and C_ω the resistance coefficient of an individual spherical particle.

The resistance coefficient C_ω can be described in the Reynolds number range up to 104 (which is also the interesting range for the fluidized bed) using the following approximation [6]:

$$C_\omega = \frac{24}{Re} + \frac{5.48}{Re^{0.573}} + 0.36. \tag{11}$$

Equations (10) and (11) allow the sinking particle diameter to be implicitly determined.

Only a small quantity of particles, whose diameters are larger than the sinking particle diameter would not be carried upwards at a given porosity ε_1 and remain in this layer. The mass fraction M_{G1} and the bed height h_1 (see figures 1 and 2, layer 1) of the remaining particles can be thus expressed:

$$M_{G1} = \frac{MG}{e^{(d_{ps}/k)^n}}, \tag{12}$$

$$h_1 = \frac{M_{G1}}{(1 - \varepsilon_1)\rho_s A}. \tag{13}$$

If the first calculated fluidized bed height h_1 with a porosity of ε_1 is still smaller than the sinking particle diameter, this layer does not really exist. This indicates that for this particle size distribution this superficial gas velocity is too rapid to form an actual, existing layer with the porosity ε_1. One has to start with a higher porosity at the lowest layer, until a layer larger than the sinking particle diameter exists. The porosity in upper layers can not be lower than that of the lower layers. Allow us to examine the next layer, beginning with a higher porosity. Due to the higher porosity and reduced gas velocity between particles, a corresponding smaller sinking particle diameter d_{ps2} will be obtained. In this second layer with a porosity of ε_2, only particles with diameters between d_{ps1} and d_{ps2} can float. The mass fraction M_{G2} and bed height h_2 is thus established:

$$M_{G2} = MG\left(\frac{1}{e^{(d_{ps2}/k)^n}} - \frac{1}{e^{(d_{ps1}/k)^n}}\right), \tag{14}$$

$$h_2 = \frac{M_{G2}}{(1-\varepsilon_2)\rho_s A}. \tag{15}$$

As previously stated, the porosity ε_2 must be sufficiently high so that the bed height h_2 can exist. The remaining smaller particles "fly" further about, and some settle in a certain level where they float.

Similarly, one can establish the succeeding bed heights step by step, until the porosity in the last such layer achieves a porosity of ε_{ob} in its uppermost layer, formed by the remaining particles:

$$M_R = M_G - \Sigma M_{Gi}, \tag{16}$$

$$\varepsilon_{ob} = 1 - \frac{M_R}{(H - \Sigma h_i)\rho_s A}. \tag{17}$$

Figure 3 shows the porosity distribution throughout the fluidized bed layers, with different superficial gas velocities as a parameter. One can see that the porosity distribution at a slower superficial velocity is characterised by a zone of high porosity in the upper portion, and a zone with low porosity in the lower part of the fluidized bed. As the superficial gas velocity increases, the lower fluidized bed then expands and the height with low porosity and coarser particles shrinks.

The relationship between the porosity and the pressure drop caused by solid particles in a layer is known as the following:

$$\Delta p_i = (1-\varepsilon_i)(\rho_s - \rho_f)gh_i. \tag{18}$$

This allows one to show the pressure gradient in the fluidized bed layer as being dependent on the superficial gas velocity, as in figure 4. From beyond a certain superficial gas velocity, the pressure gradient would become a straight line, therefore, the fluidized bed would become homogeneous.

This model was simplified, so that the fluidized bed cross-sectional areas would not be altered, and the particles that fly along are immediately separated in the cyclone, and then directly returned to the fluidized bed, to the uppermost layer. By further development of this model, circulating fluidized beds with solid particles recirculation into lower layers and cross-sectional change could also be examined.

Fluidized Beds with Particles Recirculation into the Bottom Layer (Circulating Fluidized Bed) and Cross-Sectional Extension

In a circulating fluidized bed furnace, solid particles would be carried with the gas flow out of the fluidized bed into a cyclone. The particles separated in the cyclone are returned to the fluidized bed, achieving a complete burning, and reaching a temperature as equally distributed as possible throughout the entire furnace.

The bottom section of a circulating fluidized bed is generally, from the gas distributor onwards, hopper-shaped, so that the applicable gas flow for firing would suffice even at the minimum load, for example, 30% part load, to cause the solid particles in the furnace to fluidize. The small bed area has another favorable effect: The fuel and the additional limestone is equally distributed throughout the cross-section of the furnace, without forming strands.

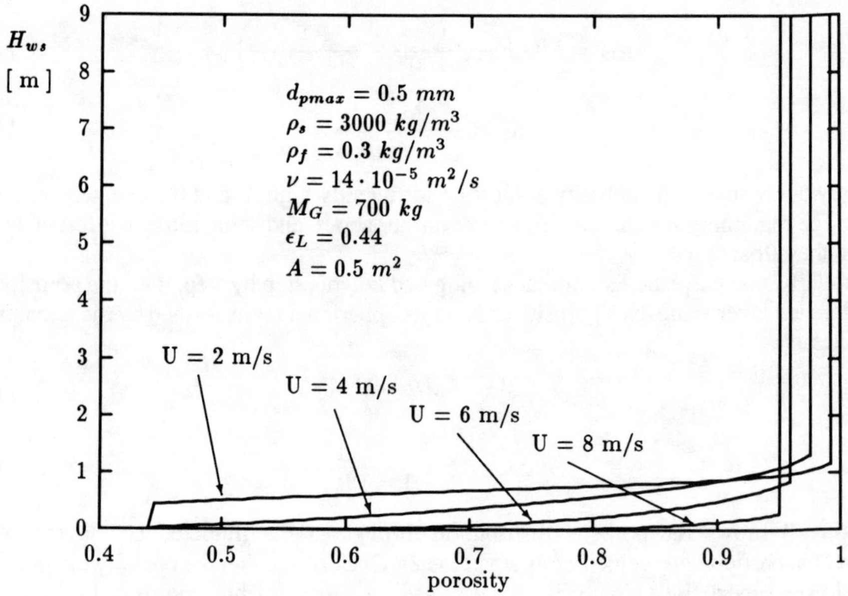

FIGURE 3 Porosity profile depending on the gas velocity.

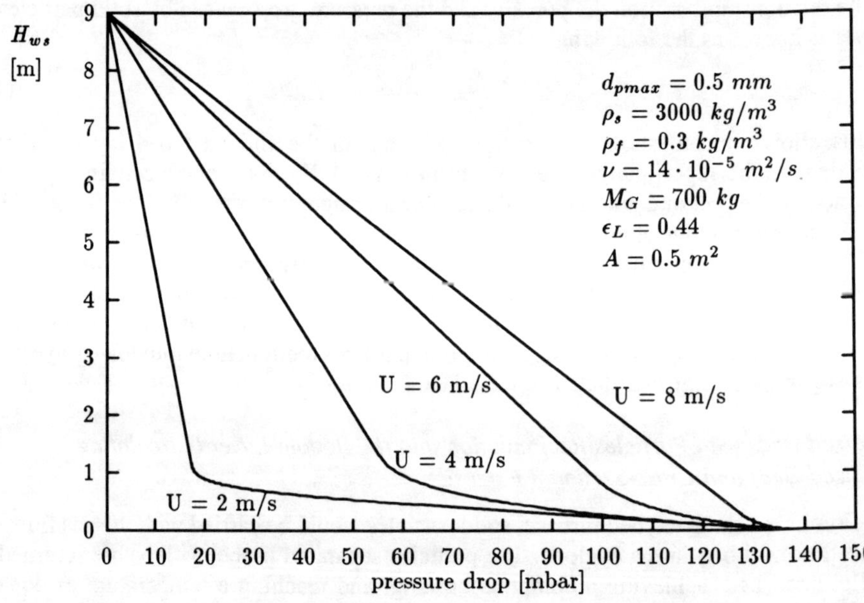

FIGURE 4 Pressure drop depending on the gas velocity.

The principle already illustrated in the model for fluidized beds with particle recirculation can be applied to cross-sectional enlargement as well. The fluidized bed without cross-sectional extension would, with increasing height, also increase constantly in porosity. This is not always the case when the fluidized bed cross-section increases. In such instances, the porosity in the bottom portion due to cross-sectional extension causing a decreasing gas velocity can decrease as well. The gas velocity would decrease despite decreasing porosity, if the cross-section increases more strongly. The expansion of the fluidized bed is dependent on the gas velocity, the particle size distribution, the cross-section over the height, and the recirculating solid mass flow as well.

The recirculating solid mass flow in a circulating fluidized bed can be thus calculated: A distinction is made between circulating and floating (noncirculating) particles. All particles whose terminal velocity u_s is slower than the mean gas velocity U/ε_{ob} would fly along, therefore circulating. In the cyclone all particles will be separated from the gas flow. In the return leg from the cyclone to the fluidized bed, the particles form a dense layer with a porosity of $\varepsilon_r \approx 0.44$. As long as the height h_r of this layer in the return leg is so small that its pressure $\rho_s\, g h_r(1 - \varepsilon_r)$ is smaller than the pressure difference in the fluidized bed, $\Delta p_{\omega s}$, recirculation will not start, i.e., a minimum quantity of fine particles is necessary to cause recirculation. An opposite direction flow through the return leg is discouraged by a siphon, or by a non-return flap.

If the layer height were to increase so significantly in the return leg that a pressure difference would remain, circulation starts. This additional height h_{r2} is, considering its solid mass, so minimal that it can be neglected.

If one distinguishes the particles as being either circulating or floating (noncirculating), the solids mass in the layer i, M_{Gi}, is composed of circulating particles M_{upi} and the floating particles M_{spi}:

$$M_{Gi} = M_{spi} + M_{upi}. \tag{19}$$

The volume of layer i, V_i, is equal to the product of cross-section A_i and the layer height h_i:

$$V_i = h_i A_i. \tag{20}$$

The entire porosity of the layer ε_{Gi} can be thus defined:

$$\varepsilon_{Gi} = \frac{V_i - M_{Gi}/\rho_s}{V_i}, \tag{21}$$

and rearranged to arrive at

$$\varepsilon_{Gi} = \varepsilon_{spi} + \varepsilon_{upi} - 1, \tag{22}$$

with

$$\varepsilon_{spi} = 1 - \frac{M_{spi}/\rho_s}{V_i}, \tag{23}$$

$$\varepsilon_{upi} = 1 - \frac{M_{upi}/\rho_s}{V_i}. \tag{24}$$

ε_{spi} is the floating particle formed porosity, ε_{upi} is the circulating particle formed porosity, ε_{Gi} is the entire particle formed porosity of the layer i. Each porosity refers to the volume Vi and the total cross-section A_i.

The calculation of the circulation solid mass flow is iterated from the beginning with the calculation of the porosity distribution throughout the fluidized bed height. A prerequisite is that the recirculating solid mass flow for each layer in a stationary circulating condition must be the same:

$$\dot{m} = A_i\left(1-\varepsilon_{upi}\right)\rho_s\left(\frac{U_i}{\varepsilon_{Gi}} - u_s\right), \tag{25}$$

where U_i/ε_{Gi} is the real mean velocity of the gas phase between the particles and u_s is the terminal velocity of the circulating particles calculated for a mean particle diameter. It is assumed that the real velocity of the solid phase is the difference between the real mean gas velocity and the mean particle terminal velocity. Should U_i/ε_{Gi} change, then ε_{upi} must also change, because u_s (in a stationary condition) and ρ_s are constant. In a fluidized bed with cross-sectional change, A_i will also change with the height.

With an iterative, given recirculating solid mass flow $\dot{m}$, one can calculate the porosity ε_{upi} formed by circulating particles for a layer i using the equation:

$$\varepsilon_{upi} = 1 - \frac{\dot{m}}{A_i\rho_s\left(\frac{U_i}{\varepsilon_{Gi}} - u_s\right)}. \tag{26}$$

From the equation (22) one can establish the porosity ε_{spi} formed by floating particles:

$$\varepsilon_{spi} = 1 + \varepsilon_{Gi} - \varepsilon_{upi}, \tag{27}$$

where ε_{Gi} is calculated according to equation (22), or in the case of cross-sectional change still iteratively. Together with the floating solid mass of a layer i M_{spi}, one arrives at the equation (23) to establish the layer height h_i

$$h_i = \frac{M_{spi}}{(1-\varepsilon_{spi})\rho_s A_i}. \tag{28}$$

From equation (24) one can arrive at the following for the circulating solid mass in a layer M_{upi}:

$$M_{upi} = A_i(1-\varepsilon_{upi})\rho_s h_i. \tag{29}$$

For cross-sectional change of the fluidized bed, a height of h_i in a layer would be selected and the total porosity iteratively determined by use of the following equation, by eliminating ε_{upi} from (22) and (25):

$$u_s\varepsilon_{Gi}^2 - \left(u_s\varepsilon_{spi} + U_i + \frac{\dot{m}}{A_i\rho_s}\right)\varepsilon_{Gi} + \varepsilon_{spi}U_i = 0. \tag{30}$$

ε_{spi} is calculated from equation (23).

For each given, or afterwards, newly calculated circulating solid mass flow, one can reach a porosity distribution throughout the fluidized bed. The circulating solid mass flow $\dot{m}$ can be calculated only with consideration of the upper layer, which is solely composed of circulating particles.

$$\dot{m} = A\left(1-\varepsilon_{ob}\right)\rho_s\left(\frac{U_{ob}}{\varepsilon_{ob}} - u_s\right). \tag{31}$$

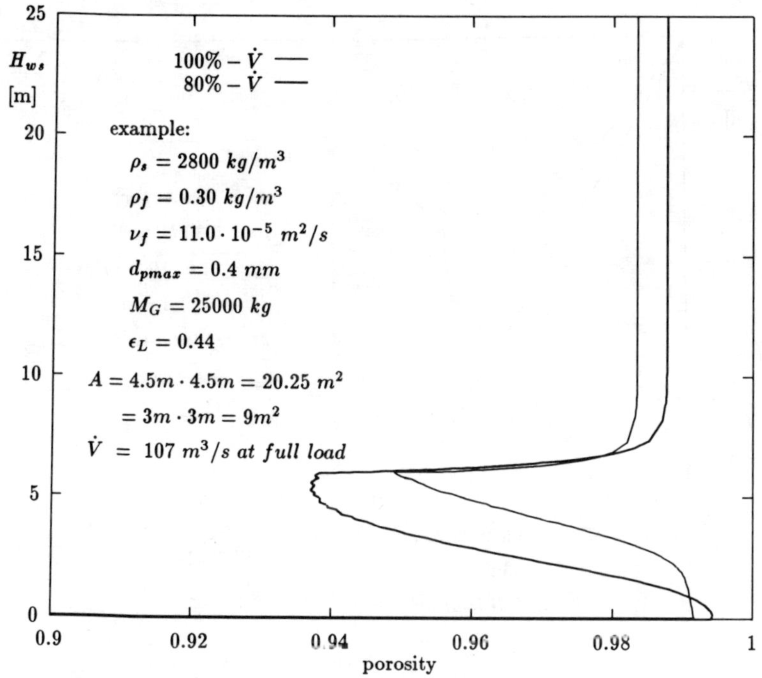

FIGURE 5 Porosity profile depending on the gas velocity.

After each loop, the present solid mass flow is replaced by the newly calculated solid mass flow. It would need to be repeated until the new value achieved by using equation (31) is sufficiently close to the solid mass flow of the previous iteration step.

Two extreme cases can arise in the fluidized bed:

— there would not be a sufficient height h_r in the return leg, i.e., the filling of the solid particles in the return leg would be too low, or in some cases, the particle size distribution would not be suitable for a circulating fluidized bed, or the downcomer is too large, then, only fine particles would be separated and a stationary fluidized bed would be the result.

— the return leg would be so filled up that h_r would be greater than H, therefore, the fluidized bed would be overfull or the downçomer too small.

Figure 5 shows the porosity distribution throughout the fluidized bed height dependent on the volumetric gas flow for an operating condition. The example cited here is of a selected fluidized bed unit with a height of 25 m. The top straight portion of the furnace has a cross-sectional area of A = 4.5 m × 4.5 m = 20.25 m². Along the lower 6 m, the cross-sectional area is reduced to A = 3 m × 3 m = 9 m² forming a hopper. The whole volumetric flow through the gas distributor $\dot{V}$ amounts to 107 m³/s at full load.

One can see that the porosity in relationship to the particle size distribution, the volumetric gas flow, and the form of the hopper-shaped lower portion could become smaller until the upper end of the hopper. In the upper portion with vertical walls, porosity distribution is the same as in the fluidized bed without the cross-sectional extension.

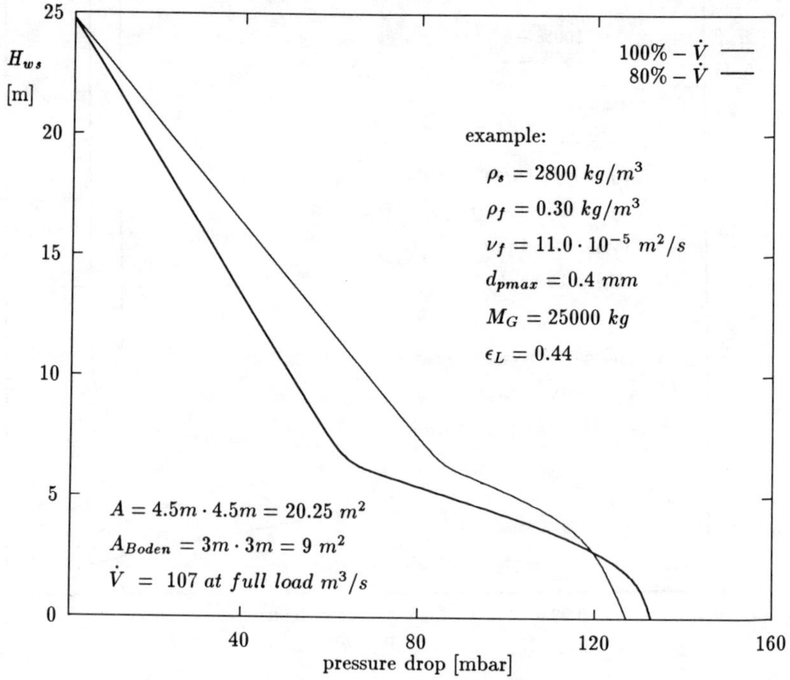

FIGURE 6 Pressure drop depending on the gas velocity.

Circulating Fluidized Beds with Cross-Sectional Extension and Secondary Air Supply

The same equations previously developed can be applied to circulating fluidized beds with cross-sectional extension and secondary air supply as well. Due to secondary air supply, a sudden change in local porosity can be expected.

Figure 6 displays the pressure drop distribution throughout the fluidized bed, dependent on the volumetric gas flow, as referred to in figure 5. Based on this diagram, one recognizes that at different volumetric gas flow, a differing total pressure drop results throughout the gas distributor, although the total solid mass in the fluidized bed remains constant. Only the solids derived pressure drop between the gas distributor and the furnace outlet of a fluidized bed without cross-sectional extension would not change with the volumetric flow, or the superficial gas velocity, but only with the solid mass in the fluidized bed, while the pressure drop in a fluidized bed with cross-sectional extension, is not only dependent on the solid mass, but also on the applied volumetric gas flow. This is illustrated in figure 7, which shows two similar fluidized bed units, both furnaces having a total height of 2 m. In the upper, straight portion of the furnace, the cross-sectional area is A = 0.3 m × 0.3 m = 0.09 m^2. Along the lower 0.3 m the cross-sectional area is reduced to A = 0.1 m × 0.1 m = 0.01 m^2. The solid mass in each furnace is 33.2 kg. For the first fluidized bed, the volumetric gas flow through the gas distributor is so great, that it has a homogenizing effect on the porosity in the entire fluidized bed, while the volumetric gas flow for the second fluidized bed is provided, so that the minimum fluidization will just be attained. The total pressure drop in the homogeneous fluidized bed reaches 39.2 mbar, while the total pressure drop at minimum fluidization attains 53.3 mbar. Both pressure drops are different, although both

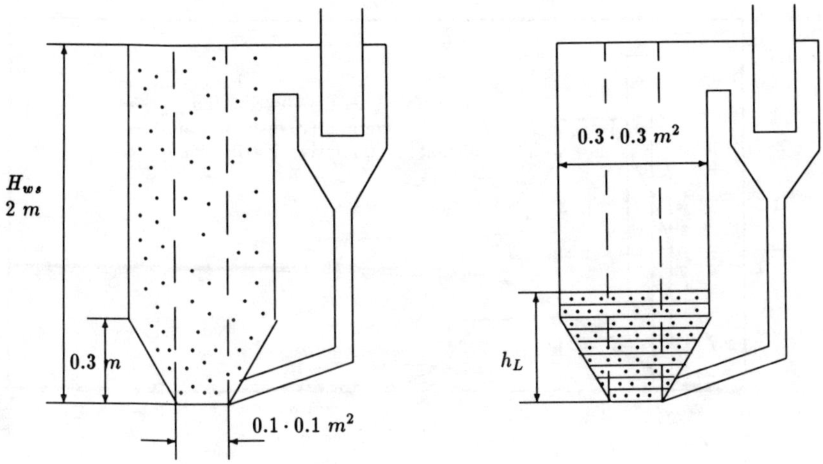

homogeneous porosity in total CFBC

$\rho_s = 2000\ kg/m^3$

$M_G = 33.2\ kg$

$=> \ \epsilon = 0.9$

$\Delta p = \rho_s(1-\epsilon)\cdot g \cdot H_{ws}$

$\Delta p_{H_{ws}} \approx 39.2\ mbar$

at mimimum fluidizing conditions

$\rho_s = 2000\ kg/m^3$

$M_G = 33.2\ kg$

$\epsilon_L = 0.44$

$=> \ h_L = 0.485\ m$

$\Delta p = \rho_s(1-\epsilon_L)\cdot g \cdot h_L$

$\Delta p_L \approx 53.3\ mbar$

$$\Delta p_{H_{ws}} < \Delta p_L$$

FIGURE 7 Pressure drop in the CFBC at part and full load.

fluidized bed units and the total masses are identical. The difference is due to the fact that the pressure difference Δp in a fluidized bed with a variable cross-sectional area cannot be calculated by the difference between the weight-force and the buoyancy divided by the bottom cross-section, but by

$$\sum_{i=1}^{n} \rho_s(1-\varepsilon_i)gh_i$$

and the remaining weight would be taken by the inclined hopper walls.

COMPARISON WITH MEASUREMENTS

In the first test case the cold CFB pilot plant at IWBT with a rectangular cross-section of $0.2 \times 0.3\ m^2$ and a height of ca. 9 m was used. Air was used as the fluid gas and quartz sand as bed material, with a particle size distribution according to figure 8 was used. The pressure drop over the CFB was determined by virtue of U-shaped tube pressure gauges. According to the following relationship:

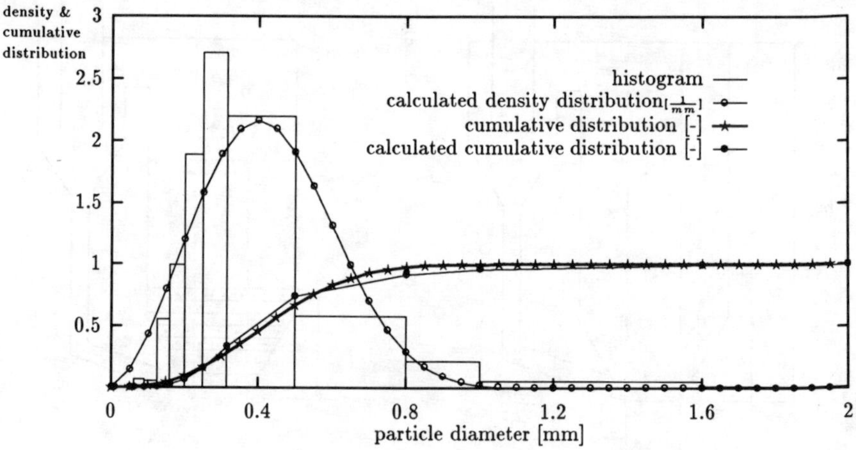

FIGURE 8 Particle size analysis of the solid material used.

$$\dot{m}_s = \Delta h_{fr}(1 - \varepsilon_{fr})\rho_s A_{fr}/\Delta\tau, \tag{32}$$

with Δh_{fr} as the measured height of repose ε_{fr}, as the porosity of repose measured in the pretest, A_{fr} as the cross-section in the return leg tube and $\Delta\tau$ as the collection time for the circulating solid mass flow can be determined.

In addition to the particle size analysis, figure 8 shows the calculated curve suited to the particle size analysis of the solid used. In this figure the histogram, the calculated density distribution, the cumulative distribution and the calculated cumulative distribution are plotted against the particle size.

Figure 9 shows the comparison between the pressure drop calculated according to the model and the measurement carried out in the cold circulating fluidized bed test rig. The pressure difference during the measurement shook reasonably. This makes the comparison between the calculation and the measurement more difficult to analyze. The arrows in figure 9 show approximately the amount of shaking in the experiment.

Figure 10 plots the calculated porosity distribution against the fluidized bed height. In figures 9 and 10 it is clear that there are floating particles only between the gas distributor level and the 0.4 m level and one finds only circulating solid particles above the 0.4 m level, for a certain superficial gas velocity applied.

The total solid mass in the riser amounts to ca. 20 kg. For certain operating conditions the external circulating solid mass flow is equal to 1.8 kg/s, according to the model calculation and it is ca. 1 kg/s, according to the measurement. The difference in the external circulating solid mass flow between the model prediction and the measurement maybe due to the small rectangular cross-section of the riser (only 0.06 m^2) and so the wall effect on the gas flow cannot be easily neglected. This means that the solid particles which can fall near the wall are not available to the external recirculation because of decreasing gas velocity.

The second fluidized bed represents the circulating fluidized bed furnace pilot plant at the power station in Niederaussem. Figure 11 shows the geometry of the combustor of the CAFB pilot plant. The most important operating data can be found on the right side of figure 11. The fluidized bed furnace has a total height of ca. 9.5 m. The cross-section of the

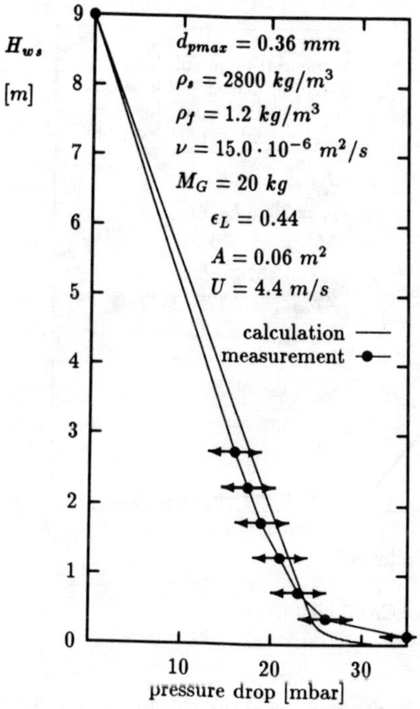

FIGURE 9 Pressure drop in the fluidized bed.

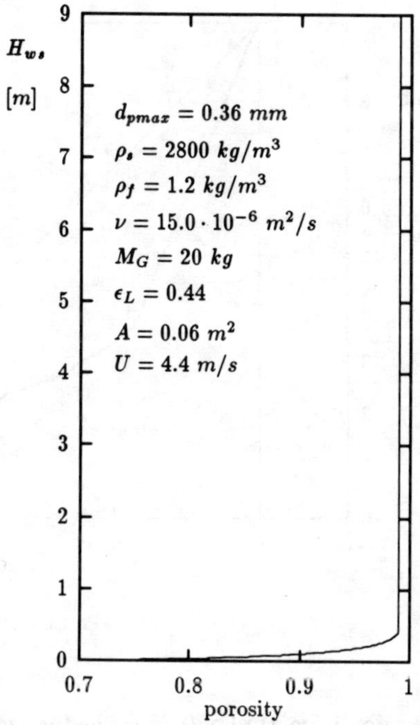

FIGURE 10 Porosity distribution in the fluidized bed.

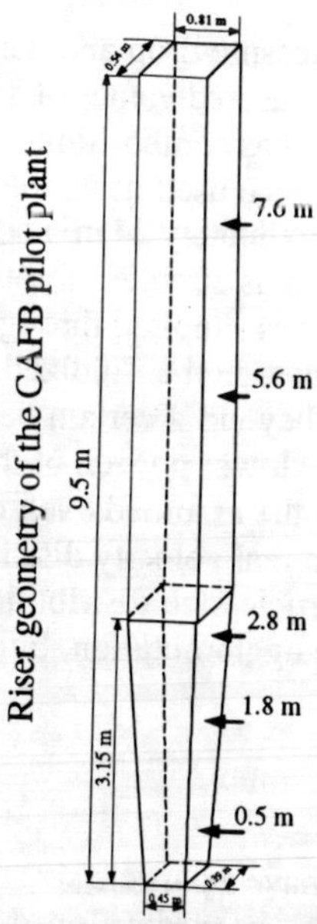

The most important operating dada during the test

operating data	*full load*	*part load*	*unit*
primary air	856	465	m_N^3/h
secondary-1	111	174	m_N^3/h
secondary-2	121	123	m_N^3/h
secondary-3	73	97	m_N^3/h
secondary-4	58	26	m_N^3/h
secondary-5	75	25	m_N^0/h
seal air for coal	221	222	m_N^3/h
pressure in 0.2 m	70	70	*mbar*
pressure in 4.0 m	30	15	*mbar*
pressure in 9.2 m	-2	-2	*mbar*
temperature	789	734	*°C*
solid mass	ca. 280	ca. 280	*kg*

FIGURE 11 CAFB pilot plant.

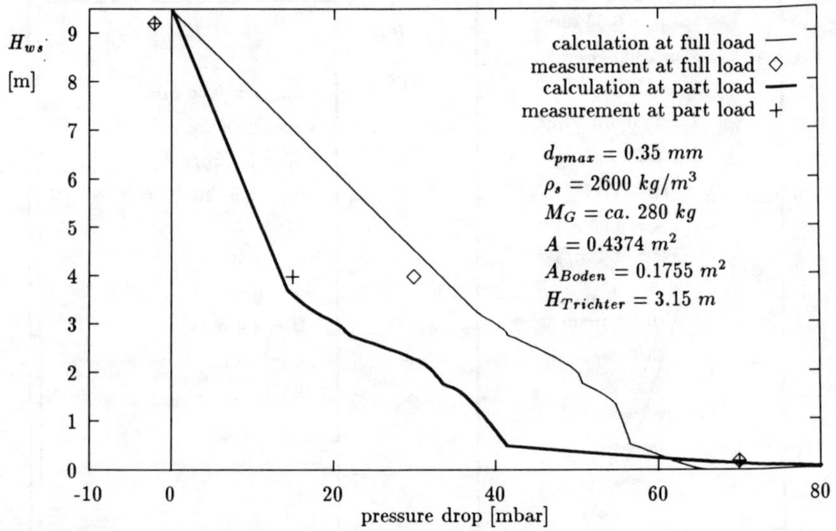

FIGURE 12 Pressure drop in the CAFB pilot plant.

upper portion is ca. 0.81×0.54 m^2 and is reduced in the hopper along the lower, ca. 3.15 m to ca. 0.45×0.39 m^2. The thicker arrows represent secondary air supply.

Figure 12 shows the comparison between the model prediction and the measurement carried out in the CAFB pilot plant. At full load as well as at part load the predicted curves of the pressure drop correspond well to the measurements.

Figure 13 shows a comparison between the calculated pressure drop and the calculated recirculating solid mass flow based on this model and the measured values of an industrial unit. A temperature of approximately 850°C is used, and flue gas compositions are uniform throughout the fluidized bed furnace. The fluidized bed furnace used has a total height of 24.8 m. The cross-sectional area of the upper portion is approximately 24 m^2 and is reduced in the hopper along the lower 6 m to 9 m^2. The primary air is supplied through the gas distributor and partly at a height of 0.5 m. The secondary air is supplied through openings at two levels, at a height of 2.5 m and 5 m. The total solid mass in the fluidized bed can be 17 t. The diagram suggests that the calculated result from beyond a certain height would agree well with the measured values. Deviations occur in the lower portion of the fluidized bed. This is due to effects on the secondary air supply and the assumed even distribution of the gas velocities in the cross-sections deviating from the real velocity distribution, and on the other hand, that the model assumptions that the particle size distribution deviates from the real one and that the combustion process, particle agglomeration, bubble formation, and internal solid particles recirculation are neglected.

CONCLUSIONS

Based on the fact that the solid particles in the fluidized bed have a particle size distribution, a one-dimensional model of a circulating fluidized bed has been developed. Using this model, one can determine the continuous porosity distribution and the continuous pressure drop throughout the fluidized bed dependent only on known parameters such as particle size distribution, volumetric flow, etc. This simultaneously allows for calculating the recirculating solid mass flow.

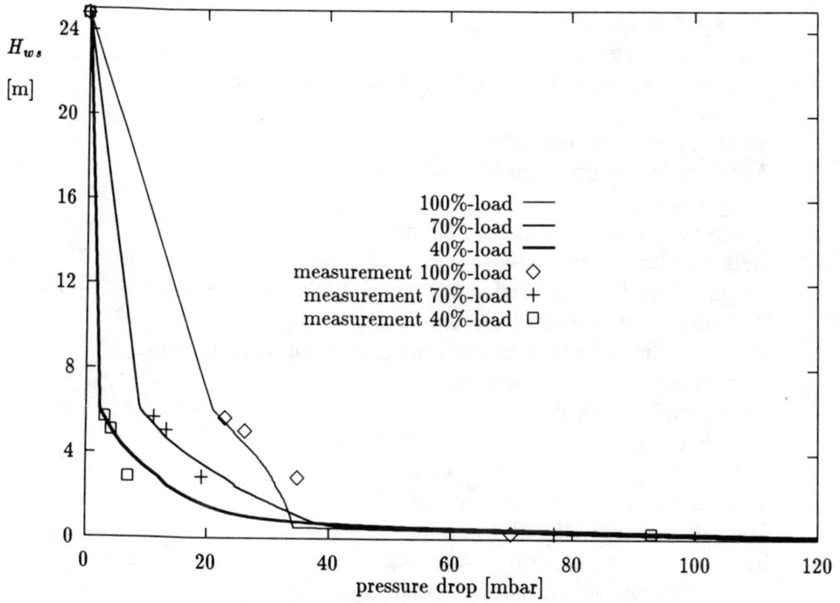

FIGURE 13 Pressure drop depending on the load.

It is assumed in this model, that the solid particles in the fluidized bed are spherical, and ideally streamed by the gas flow with a mean velocity. There are no variable parameters or coefficients necessary for this model.

The particle size distribution need not be one according to Rosin, Rammler and Sperling, but it can be established by using other functions, which better approximate the measurements.

This model is suitable for calculations of stationary and circulating fluidized beds with or without cross-sectional extension as well as secondary air supply in the upper areas.

REFERENCES

Leithner, R. (1989). Einfluß unterschiedlicher Wirbelschichtfeuerungssysteme auf Auslegung, Konstruktion und Betriebsweise der Dampferzeuger, *VGB Kraftwerkstechnik 69*, 7, pp. 679 ff.

Leithner, R., und Wang, J. (1990). Eindimensionales Wirbelschichtmodell, BWK Bd. 42, Nr. 7/8, S. 436 ff.

Leithner, R., und Wang, J. (1990). Eindimensionales Wirbelschichtfeuerungsmodell, Internationale VGB-Konferenz "Wirbelschichtsysteme 1990," am 14/15 Nov., Essen.

Lewis, W.K., Gilliland, E.R., and Bauer, W.C. (1949). Characteristics of Fluidized Particles, *Industrial and Engineering Chemistry*, **41**, No. 6,, p. 1104 ff.

Brandt, F. (1981). Brennstoffe und Verbrennungsrechnung, FDBR, Fachbuchreihe Band 1, Vulkan-Verlag, Essen, S. 38/39.

Molerus, O. (1982). *Fluid-Feststoff-Strömungen*, Springer-Verlag, Berlin/Heidelberg/New York.

NOMENCLATURE

A	Cross-sectional area of the fluidized bed (indices: 1, 2, 3, …, i, n corresponding to layers), m^2
A_f	Buoyancy, N
Ar	Archimedes number = $[(\rho_s - \rho_f)d_p^3 g]/\rho_f \nu^2$, –

D	Passage fraction, –
d_p	Particle diameter, m
d_{pmax}	Particle diameter at the maximum of the particle size density distribution, m
d_{ps}	Sinking particle diameter, m
g	Acceleration of gravity, m/s^2
G	Weight force, N
H	Entire height of the fluidized bed, m
h_i	Height of layer i, m
h_L	Height of the fluidized bed at minimum fluidization, m
h_r	Bed height in the return leg = $h_{r1} + h_{r2}$, m
h_{r1}	Height of the layer r1 in the return leg, with the same pressure drop as in the fluidized bed, m
h_{r2}	Layer height above h_{r1}, m
k	Constant of the particle size distribution, –
$\dot{m}$	Recirculating solid mass flow, kg/s
M_G	Total solid mass in the fluidized bed, kg
M_{Gi}	Total solid mass in the *i*th layer, kg
M_R	Solid mass in the uppermost layer, kg
M_{spi}	Mass of suspended particles in the *i*th layer, kg
M_{upi}	Mass of circulating particles in the *i*th layer, kg
n	Constant of the particle size distribution, –
Δp_H	Pressure drop in entire homogeneous fluidized bed, N/m^2
Δp_i	Pressure drop in the *i*th layer, N/m^2
Δp_L	Pressure drop at minimum fluidization, N/m^2
Δp_{ws}	Pressure drop in the entire fluidized bed, N/m^2
Re	Reynolds number = ud_p/ν, –
Re_s	Reynolds number based on terminal velocity = $u_s d_p/\nu$, –
U	Superficial gas velocity (indices: 1, 2, 3, …, n, ob corresponding to layers), m/s
u_f	Mean gas velocity between the particles, m/s
u_s	Terminal velocity of the circulating particles formed for a mean particle diameter, m/s
V_d	Particle size density distribution, 1/m
V	Volume (indices: 1, 2, 3, …, i, n corresponding to layers), m^3
$\dot{V}$	Gas volume flow in operating state, m^3/s
W	Drag, N
ε	Porosity, –
ε_G	Total porosity (indices: 1, 2, 3, …, i, n corresponding to layer), –
ε_i	Porosity in the *i*th layer, –
ε_L	Porosity at minimum fluidization, –
ε_{ob}	Porosity in the uppermost layer, –
ε_r	Porosity in return leg, –
ε_{sp}	Floating particles formed porosity (indices: 1, 2, 3, …, i, n, corresponding to layers), –
ε_{up}	Circulating particles formed porosity (indices: 1, 2, 3, …, i, n, corresponding to layers), –
ν	Kinematic viscosity of the gas, m^2/s
ρ_f	Gas density, kg/m^3
ρ_s	Particle density, kg/m^3

Mathematical Simulation of a Circulating Fluidised Bed Combustor

P. C. SARAIVA, J. L. T. AZEVEDO and M. G. CARVALHO *Instituto Superior Técnico, Mechanical Engineering Department, Lisbon, Portugal*

Abstract—In this paper a mathematical model to describe a Circulating Atmospheric Fluidised Bed Combustor (CAFBC) is presented. For the fast section of the bed, momentum and energy balance equations are used to predict temperature and velocity profiles for gas and particles. The model performs mass balances for the chemical gas species (O_2, H_2O, CO_2, CO and SO_2) with consideration on the last being given for retention by limestone particles. A bubbling bed model is considered to simulate the bottom of the CAFBC. The model is applied to typical conditions of a boiler and the results show the expected trends.

NOMENCLATURE

A	—Area (m^2)
C	—Concentration (moles)
C_p	—Specific heat Capacity (J/KgK)
d	—Particle diameter (m)
D_g	—Diffusion coefficient (m^2/s)
H	—Bottom region height (m); Heating Value (J/Kg)
h	—Convection heat transfer (W/m^2K)
k_c, k_d	—Kinetic rate; Diffusion rate (s^{-1})
m	—Particle mass (Kg)
P	—Pressure (Pa)
T	—Temperature (K)
u	—Velocity (m/s)
$\dot{m}$	—Rate of combustion (Kg/m^2s)
W	—Mass flux (Kg/m^2s)

Greek

δ	—Density (Kg/m^3)
ε	—Voidage; Emissivity
μ	—Viscosity (Kg/ms)
ρ	—Bulk density (fuel and limestone) (Kg/m^3)
ρ_i	—Density of a chemical specie ($Kg_i/m^{3}{}_{air}$)
σ	—Stefan Boltzman constant (W/m^2K^4)

Subscripts

o	Initial conditions
f	—fuel particles
g	—gas
l	—limestone particles
mf	—minimum fluidization
p	—Particle
rad	—radiation
s	—inert particles
v	—volatiles
w	—wall

INTRODUCTION

Circulating fluidised bed boilers have a number of advantages over conventional bubbling fluidised bed boilers and became an important topic of research. The empirical design of boiler furnaces practiced for many years promoted studies to understand the processes occurring in such equipment. Nowadays, due to restrictive pollution regulation as well as the necessity of high performance and the development of powerful computers, research programs were created.

Investigations on high-velocity fluidization have been conducted by many investigators (see e.g., Yerushalmi and Avidam, (1985), Kwauk *et al.,* (1987), Rhodes and Geldart, (1987)). These models allowed the identification of the circulating fluidised bed structure, with a more dense zone at the bottom and a fast bed above, with higher particle concentrations near the wall region. Kunni and Levenspiel (1990), developed a model for the hydrodynamics which considers three different phases, (dilute particles entrained, ascending and descending clusters), with mass transfer between them. The key parameter in their model is the decay constant for the fall off of bulk density of solids with height in the freeboard. Berruti and Kalogerakis (1989), modelled the fast bed hydrodynamics using a core-annulus theory which is applied to small diameter risers ($d < 0.3$ m) used in catalytic processes.

Weiβ *et al.* (1987) developed a mathematical model which describes coal combustion in a CAFBC, based on mass and energy balances for a compartment and considers eight chemical species. However, no details are given about the solids distribution in the reactor and no special treatment is given to the bottom region. At the bottom of a circulating fluidised bed there is a dense phase which has particle concentrations typical of a bubbling fluidised bed (see e.g. Kwauk *et al.,* (1987), Kunni and Levenspiel, (1990)). A compilation of all relevant work on the modelling of circulating fluidised beds and conventional bubbling beds is presented by Manno and Rietema (1990).

The present paper proposes a model for a Circulating Atmospheric Fluidised Bed which includes hydrodynamics of the fast bed as well as for a bubbling bed. Combustion, heat transfer and SO_2 retention are modelled throughout the reactor (for details see Saraiva (1992)). After this introduction the mathematical model for the fast and bubbling bed are described, followed by the model of SO_2 retention by the limestone particles. The model is applied to operating conditions of a typical boiler and the results are discussed. In the last section the conclusions of the present work and future work for the model are outlined.

MATHEMATICAL MODEL

Fast Bed

The model used for the fast bed region of the combustor was adopted from Chung and Carlson (1981) and extended to a two-zone core-annulus model. In the core, particle velocities are obtained by integration of the equations of motion. When only particle drag and gravitational forces are considered the particle momentum equation can be written as:

$$u_p \frac{du_p}{dx} = \frac{1}{\tau_p}(u_g - u_p) - g \tag{1}$$

where τ_p is the particle's relaxation time expressed as:

$$\tau_p = \frac{24\delta_p dp^2}{18\mu\, C_D\, Re_p} \tag{2}$$

The drag coefficient is calculated using the well known correlation

$$C_D = \frac{24}{Re_p}(1 + 0,15 Re_p{}^{0,687})\ \varepsilon^{-4.67} \qquad 1 < Re_p < 10^3 \tag{3}$$

The gas phase velocity is evaluated by integration of the momentum equation:

$$u_g \frac{du_g}{dx} = \frac{1}{\tau_f}(u_g - u_f)\frac{\rho_f}{\rho_g} + \frac{1}{\tau_s}(u_g - u_s)\frac{\rho_s}{\rho_g} - (u_p - u_g)\,S_{burn} \tag{4}$$

where momentum exchange between the gas and particles phase is considered and including the source of momentum exchange due to coal combustion, S_{burn}.

For combustion calculations the energy equations for both phases are used to predict the temperature profiles. The evolution of coal particles is assumed to be in three steps. In the first one, coal particles are dried (in this step temperature is assumed constant and equal to 100° C) and heated until the devolatisation temperature. Then devolatisation starts and this process is kinetically controlled. When devolatisation is complete, char combustion is considered to be controlled by both kinetics and oxygen diffusion to the particle's surface. The energy equation for a particle is based on mass and energy balances and can be written as:

$$Cp_f\, m_f \frac{dT_f}{dt} - A_f\, \dot{m}_{v,c}\,(H_{v,c} - Q_v) + A_f\, h\,(T_g - T_f) + A_f\, \sigma\, \varepsilon_f\,(T_g^4 - T_f^4) + H_{H_2O}\frac{d\dot{m}_{H_2O}}{dt} \tag{5}$$

where Q_v represents the fraction of the particle heat of combustion which is transferred to the particle and $\dot{m}_v$ is the mass flow rate released:

$$\dot{m}_v = \frac{\delta_f}{3}\left[1 - \left(\frac{R_c}{R_f}\right)^3\right] R_f A_v \exp\left(-\frac{E_v}{RT_f}\right) \tag{6}$$

for volatile combustion. For char consumption the mass flow rate by combustion, $\dot{m}_c$, is expressed by:

$$\dot{m}_c = \pi\, k_r\, \delta_f\, d_f^2 P_{O_2} \tag{7}$$

where

$$\frac{1}{k_r} = \frac{1}{k_c} + \frac{1}{k_d} \tag{8}$$

k_c and k_d represent the kinetic and diffusion rates. The main assumptions done are: i) the particles are spherical ii) a one-step reaction for combustion iii) the particle density remains constant iv) particles are sufficiently separated from each other that the single-particle combustion analysis is valid for each and v) the temperature of the particle is uniform.

The energy equation for inert particles is similar to equation (5) but without considering the drying, volatilization and combustion steps ($\dot{m}_{v,c} = 0$).

$$C_{ps}\, m_s \frac{dT_s}{dt} = A_s\, h(T_g - T_s) + A_s\, \sigma\, \varepsilon_s\,(T_g^4 - T_s^4) \tag{9}$$

To predict gas phase temperature, an energy balance is performed on a small compartment of height Δx, leading to:

$$\frac{d}{dx}(\rho_g U_g \, Cp_g T_g + \rho_f \, U_f \, (Cp_f \, T_f + H_{v,c} - Q_v) + \rho_s \, U_s \, Cp_s \, T_s) = \frac{4h}{D}(T_W - T_g) + S_{rad} + Q_{rad} \tag{10}$$

where D is the combustor diameter. The radiative heat fluxes to the combustor walls as well as radiative sources for the gas phase, are evaluated by the Discrete Transfer Method. The overall emissivity is calculated considering the gas as a mixture of grey gases and particles obeying to the Geometrical Optics Theory. The convection coefficient, h, is predicted according to the model of Mahalingam and Kolar (1991) for circulating fluidised beds.

The mass concentrations of fuel and chemical species are expressed as a function of the mass combustion rate:

$$\frac{d}{dx}(\rho_f \, u_f) = -\frac{3\dot{m}_{v,c}\rho_f}{\delta_f R_f} - a(\rho_f \, u_f) \tag{11a}$$

$$\frac{d}{dx}(\rho_s \, u_s) = -a(\rho_s \, u_s) \tag{11b}$$

$$\frac{d}{dx}(\rho_{O_2} \, u_g) = -C_1\dot{m}_{v,c} - \frac{M_{O_2}}{2}\frac{d[CO]}{dt} \tag{11c}$$

$$\frac{d}{dx}(\rho_{CO_2} \, u_g) = C_2\frac{M_{CO_2}}{M_C}\dot{m}_{v,c} + M_{CO_2}\frac{d[CO]}{dt} \tag{11d}$$

$$\frac{d}{dx}(\rho_{CO} \, u_g) = C_3\frac{M_{CO}}{M_C}\dot{m}_{v,c} - M_{CO}\frac{d[CO]}{dt} \tag{11e}$$

$$\frac{d}{dx}(\rho_{H_2O} \, u_g) = N_P\frac{dm_{H_2O}}{dt} + C_4\frac{M_{H_2O}}{2M_H}\dot{m}_v \tag{11f}$$

$$\frac{d}{dx}(\rho_{SO_2} \, u_g) = C_5\dot{m}_{v,c} \quad C_6K_r \tag{11g}$$

where the constants C_1 to C_5 are evaluated from stoichiometry. The last term of equation (11a) and (11b) represents the radial diffusion of particles from the core to the annulus. The value of the constant "a" was taken from Rhodes and Geldart (1987). The CO oxidation rate used in equation (11d) was taken from Howard et al. (1973) correlation and the last term of equation (11g) represents the SO_2 retention by limestone.

The gas density is related with temperature to a given pressure by the equation of state:

$$\rho_g = P/R_g T_g \tag{12}$$

Bubbling Bed

The model adopted for the bubbling bed is a "Level II" model classified as a "learning model" by Van Swaaij (1985). The gas flow is divided into two phases—emulsion and bubble. Mass transfer between the bubble and particle phases is evaluated by a mass coefficient k_{be}, proposed by Sit and Grace (1981), considering diffusive and throughflow effects:

$$K_{be} = \frac{\left[2U_{mf} + 12\left(\frac{D_g \varepsilon_{mf} U_b}{D_b}\right)^{1/2}\right]}{D_b} \tag{13}$$

where D_b is the bubble diameter predicted by a correlation established by Mori and Wen (1975). For a heterogeneous reaction, with negligible particle concentration in the bubble phase, it is assumed that no reaction occurs in the lean phase. Assuming uniform concentration in the particulate phase C_p, the gas concentration in the bubble phase C_b can be expressed along the height as:

$$C_b = C_p + (C_o - C_p)\exp\left(-\frac{K_{be}\varepsilon_b x}{\beta U_g}\right) \tag{14}$$

Considering a heterogenous first order reaction with rate k_r in the particulate phase, an overall balance is:

$$U_g\ C_o = \beta\ U_g\ C_{bH} + (1-\beta)\ U_g\ C_p + k_R\ (1-\varepsilon_b)\ C_p H \tag{15}$$

The overall rate of carbon consumption may then be expressed as

$$k_R = \sum_i \pi\ d_{fi}^2\ k_{ri}\ F_i\ C_{char} \tag{16}$$

Carvalho *et al.* (1991) established the reaction rate as:

$$\frac{1}{k_r} = \frac{1}{k_c} + \frac{d_f}{\phi\ Sh'\ D'_m} \tag{17}$$

where Sh' is the Sherwood number:

$$\frac{Sh'}{\varepsilon} = \left[4 + 0.576\left(\frac{U_o\ d_f}{D'_m \varepsilon}\right)^{0.78} + 1.28\left(\frac{U_o\ d_f}{D'_m \varepsilon}\right) + 0.141\left(\frac{d_f}{d_s}\right)\left(\frac{U_o\ d_f}{D'_m \varepsilon}\right)^3\right]^{1/2} \tag{18}$$

and the kinetic rate k_c is given by La Nauze (1985);

$$k_c = 0.513\ T_f \exp(-9160/T_f) \tag{19}$$

Equations (14) and (15) are solved together to calculate C_{bH} ($x = H$) and C_p. The parameter β of equations (14) and (15), as well as bubble fraction, ε_b, must be calculated from bed hydrodynamics. A modified two phase theory is used (e.g. Azevedo (1991)) considering the through flow in bubbles (n=3) and the gas velocity in the emulsion is assumed to be larger than the minimum fluidization velocity by a K_e factor correlated for ($\varepsilon_b < .3$) as:

$$K_e = 1 + .25(U_b \varepsilon_{mf}/U_{mf} + 2)\varepsilon_b \qquad (\varepsilon_b > .04) \tag{20}$$

The particulate phase voidage is assumed to remain equal to the incipient fluidization voidage and the bubble velocity is obtained according to Davidson and Harrison (1963):

$$U_b = U_o - U_{mf} + 0.711\sqrt{g\ D_b} \tag{21}$$

Considering the above mentioned modified two-phase theory the bubble void fraction is given by:

$$\varepsilon_b = \frac{U_o - K_e U_{mf}}{U_b + U_{mf}(n + 1 - K_e)} \tag{22}$$

The fraction of flow within the bubbles, β, is predicted by:

$$\beta = 1 - \frac{U_{mf}\, K_e}{U_o}(1 - \varepsilon_b) \tag{23}$$

The objective of the dense region model is to predict the char concentration in the bed for each class of particles considered and also the mass flow rate due to splashing. Wen and Chen (1982) correlated the entrainment flux of solids at the bed surface as:

$$E_o = 3 \times 10^{-9} D_b (U_o - U_{mf}) \frac{\pi D^2}{4} \frac{\rho_g^{2.5}\; g^{0.5}}{\mu_g^{2.5}} \tag{24}$$

so the mass flux for each size of particles is a function of carbon concentration and mass fraction of each diameter:

$$E_i = E_o \cdot F_i \cdot C_{char} \tag{25}$$

To predict the mass fraction of each class (F_i) a mass balance for each size is performed:

$$\begin{bmatrix}\text{Falling flux}\\ \text{from fast bed}\end{bmatrix} - \begin{bmatrix}\text{Ejecting flux from}\\ \text{bubbling bed}\end{bmatrix} + \begin{bmatrix}\text{Flux from } i+1 \text{ size}\\ \text{by combustion}\end{bmatrix} -$$

$$\begin{bmatrix}\text{Flux from } i-1 \text{ size}\\ \text{by combustion}\end{bmatrix} - \begin{bmatrix}\text{Fraction of combustion}\\ \text{particles burnt}\end{bmatrix} = 0$$

which can be written as:

$$W_{fall_i} - E_i + W_{i+1} - W_i - W_{burn_i} = 0 \tag{26}$$

The first term of the above equation is the particle flux of the falling particles related to equations (1) and (11a). The mass flux duc to combustion is directly proportional to the reaction rate (see equations (15) and (16)):

$$W_{burn_i} = \pi\, d_{fi}^2\, C_{char} k_{ri}\, C_p O_2\, F_i\, H\, (1 - \varepsilon_b) \tag{27}$$

It is assumed that, from the initial number N of particles of ith size, N_{i-1} will change to i-1th size and N_i will remain with its original dimensions. The corresponding mass flux from a given size to the smaller one is:

$$W_{i-1} = \frac{H\, \delta_{bed}\, C_{char}}{t_r} \frac{(d_{fi}^3 - d_f^3)}{(d_{fi}^3 - d_{fi-1}^3)} \left(\frac{d_{fi-1}}{d_{fi}}\right)^3 \cdot F_i \tag{28}$$

For the smallest class W_i = O and for the largest $W_{i+1} = 0$. In equation (28) d_f is the particle diameter after the combustion process during a mean particle residence time (t_r):

$$d_f = \left(d_{fi}^3 - \frac{6}{\pi} \frac{k_r t_r}{d_f\, C_{char}}\right)^{1/3} \tag{29}$$

Equation (26) can be re-written in the form:

$$A_i\ F_i\ +\ B_i\ F_{i+1} = C_i \tag{30}$$

forming a set of algebraic linear-equations. The char concentration (C_{char}) is calculated from an overall mass balance of the bubbling bed. The bed height is a result of the mass balances of the limestone particles in all the combustor. Since bulk density in the fast bed is calculated, the total mass of solids in the upper region of the combustor is a result of the integration over the total height. Knowing the total amount of solids in the combustor, the dense region height is then calculated.

SO_2 Retention

During coal combustion, the sulfur compounds are oxidized and the resultant sulfur dioxide is reduced by calcium oxide particles, (produced by the limestone's calcination), forming calcium sulfate according to the reaction:

$$SO_2 + Ca\ O + \frac{1}{2}\ O_2 \rightarrow Ca\ SO_4$$

The reaction rate of a limestone particle can be expressed as (Borgwardt, (1970)):

$$k_l = \frac{\pi}{6}\ d_s^3\ k_{vl}\ C_{SO_2} \tag{31}$$

In equation (31), k_{vl} represents the overall volumetric reaction rate constant and C_{SO_2} is the SO_2 concentration in the combustion gases. The overall volumetric reaction rate is calculated by:

$$k_{vl} = 490\ \exp\ (-17500/R\ T_l)S_g\ \lambda_l \tag{32}$$

where S_g is the specific surface area correlated with calcination temperature (Borgwardt et al (1971)) given by:

$$S_g = -38.4\ T + 5.6 \times 10^4 \qquad T \geq 1253K \tag{33a}$$

$$S_g = 35.9\ T + -3.67 \times 10^4 \qquad T < 1253K \tag{33b}$$

and λ_l is the limestone's reactivity which is a function of the fractional conversion of CaO, temperature and particle size.

NUMERICAL PROCEDURE

The set of non-linear differential equations governing momentum and energy in the fast bed are solved using the Runge-Kutta method. Equations (11a) to (11g) are manipulated by the introduction of equation (11a) into equations (11c) to (11g). By direct integration, the gaseous species are calculated in each compartment. For SO_2 retention, the reaction rate is evaluated at all the compartments and mass balances are performed.

In the bubbling bed, the set of algebraic linear equations written in the form of equation (30) is solved directly assuming the char concentration in the bed. Thus an overall mass balance is performed to predict the new char concentration and an iterative procedure is established.

The initial conditions for the fast bed are taken from the feed conditions and from the solids splashing from the bubbling bed. For the bubbling bed calculation, the flux of

TABLE I
Operating Conditions

Conditions	Mass Flow
Air	$\lambda = 1.2$
Fuel (Kgh^{-1})	480
Additives (Kgh^{-1})	15.3

TABLE II
Coal Composition

	Coal
Lower Heating Value (MJ/Kg)	9.85
H_2O (%) (raw)	53.60
Ash (%) (raw)	3.2
Volatil Matter (%)	49.40
C(%) (waf)	66.30
H(%) (waf)	4.76
S(%) (waf)	0.46
N(%) (waf)	0.97
O(%) (waf)	27.20

incoming particles is the sum of the falling flux of solids and the downflow of particles near the walls of the fast bed.

The coal particles are considered in a discrete number of sizes which are tracked along the fast bed height. Each particle size is partially burned out in each passage in the combustor and its diameter is thus reduced. The recirculated particles are considered in the size classes closer to the diameter at the end of the previous passage. The size classes include diameters down to the size which are completely burned in one passage. The procedure considered allows the consideration of only two kinds of coal particles along the combustor, the ones in the first passage which are still releasing moisture and volatiles and the recirculated ones with char combustion.

DISCUSSION OF RESULTS

The model described in the previous section was applied to a typical circulating fluidised bed boiler. The overall dimensions are 9,4 m high, and a cross-sectional area varying from 0,16 m^2 (at the distributor) to 0,42 m^2 (at 3,1 m height and above). The combustion air is supplied through the distributor (primary air) and at the secondary air inlets at 0,7 m, 2,7 m, 5,5 m and 7,1 m.

Most of the energy released by the combustion of air and fuel is radiated to the walls covered by water tubes. It is assumed that the boiler's walls have tubes at uniform temperature. In the present work the bubbling bed temperature was fixed at 1123 K. The operating conditions and coal composition used are shown in tables 1 and 2, respectively. The results of the bubbling bed section are not analysed in detail in the present work. This module was introduced in the calculation procedure to allow for the calculation of the gas and reacting particles' initial conditions for the fast bed.

The outlet gas concentrations from the bed are shown as the initial conditions for the fast bed section after dilution with the secondary air (24.8% of the total). The downflow

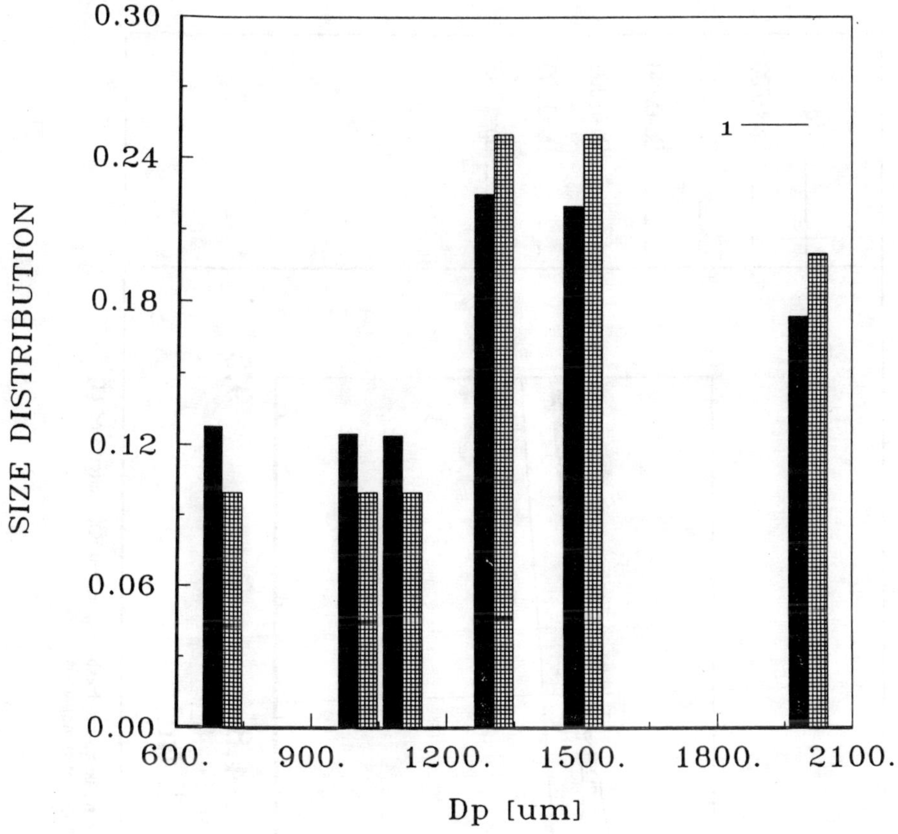

FIGURE 1 ■—Bed; □—Feed

solids near the walls (and calculated from equations (11a) and (11b)), together with the falling particles, is the total feed rate to the bubbling bed.

The char in the bed was found to be about 2.8% of the total weight. The particle size distribution of the feed coal and the resultant in the bed are represented in Figure 1. The projected solids have a particle size distribution similar to the one existing in the bed and the initial particle velocity was taken as twice the bubble velocity.

The predicted temperature profiles along the fast bed of the combustor are presented in Figure 2. The inlet temperature of feed, splashed and recirculated coal particles is about 373 K, 1100 K and 700 K, respectively. The feed coal particles are dried and heated in the initial zone of the fast bed leading to a slight decrease in the gas temperature until coal particles ignite. The volatiles released from the coal particles falling in the bed were assumed to be burned instantaneously at the bed surface. For the splashed and recirculated particles, only char combustion is considered since drying and devolatisation are complete in the first passage in the combustor and during the residence time in the bed. Due to these particles, the gas temperature remains almost constant along the whole combustor, which is typical for CFBC. The figure also shows that the drying time is not negligible and takes about 20-25% of the particle's residence time in the first passage on the combustor.

Figure 3 represents particles' and gas velocities. The gas velocity evolution is a direct consequence of the gas temperature and gas/particle interactions. The drift velocity is

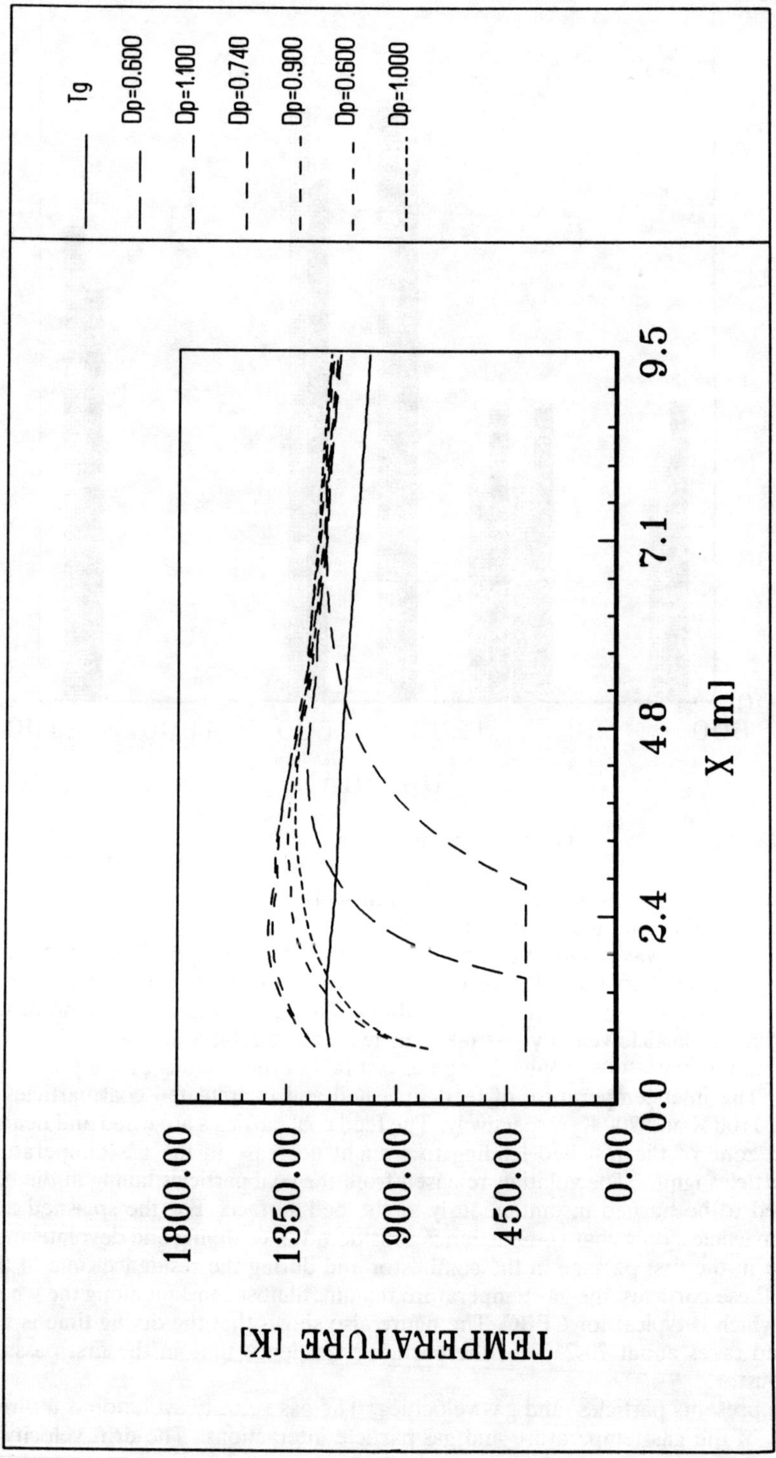

FIGURE 2 Temperature profiles. The dashed lines are representing fresh (T_{fo} = 100 K), recirculated (T_{fo} = 770 K) and projected (T_{fo} = 1123 K) particles for different dimensions.

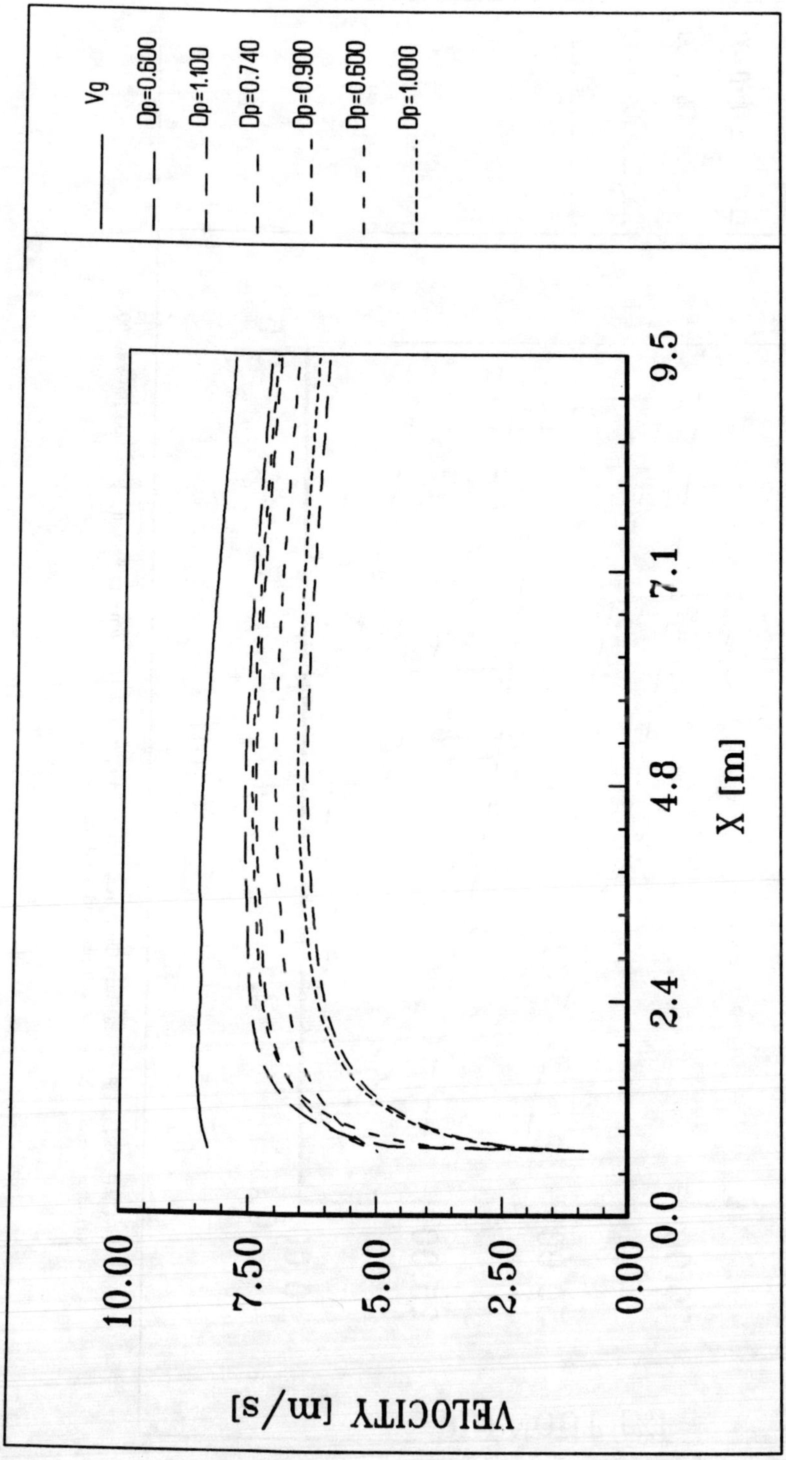

FIGURE 3 Velocity profiles. The dashed lines are representing fresh (d_f = 600 μm and 1100 μm and 900 μm).

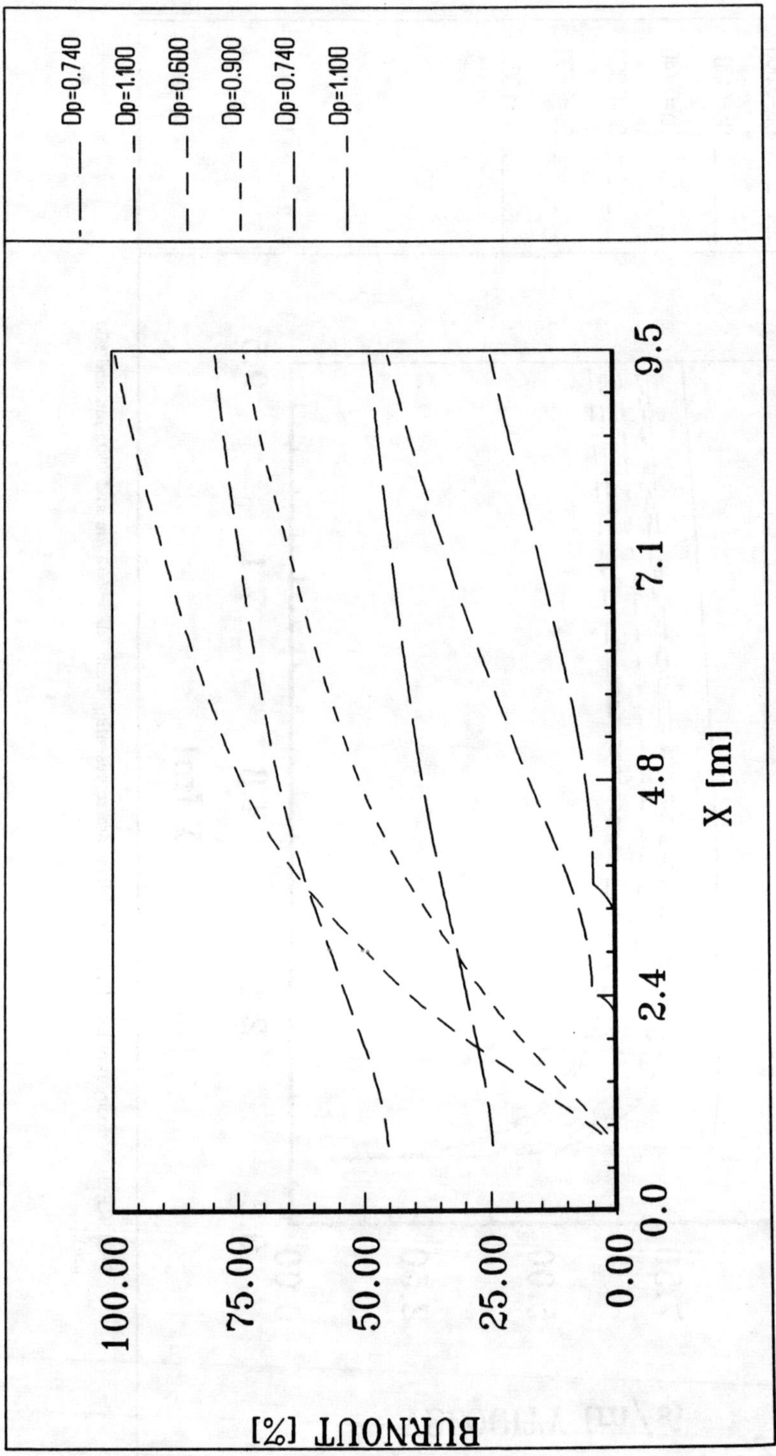

FIGURE 4 Coal Particles Burn-Out: fresh (d_f = 740 μm and 1100 μm), projected (d_f = 600 μm and 900 μm) and recirculated (d_f = 740 μm and 1100 μm)

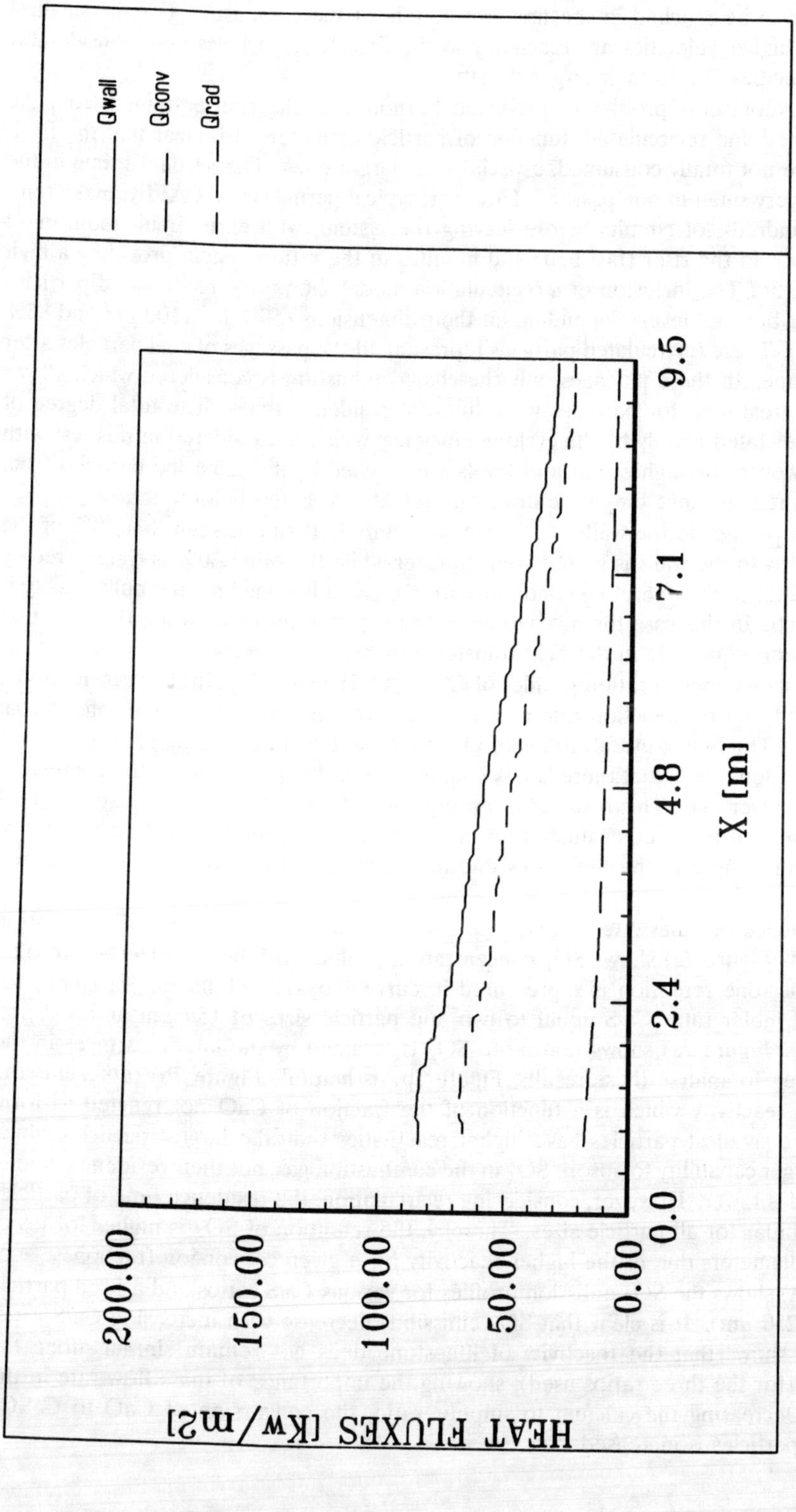

FIGURE 5 Heat Flux to the Walls; Radiative Heat Flux to (about 70% of the total); Convective Heat Flux (about 30% of the total)

higher for the feed particles leading to a high initial acceleration allowing the dynamic equilibrium to be reached in the first two meters of the combustor. For the smallest diameters, higher velocities are reached and the limestone particles have considerably smaller velocities due to their larger density.

Figure 4 represents profiles of predicted burnout, for the coal entrained from the feed, splashed and recirculated, function of particle diameter. It is clear that the fresh particles are not totally consumed, especially the larger ones. The residence time in the fast bed is very small in one passage. However, typical particles in a CAFBC make tens, or even hundreds, of circuits before leaving the system, with each circuit requiring 2 to 20 seconds in the riser (fast bed) and minutes in the return system providing a high residence time. The inclusion of a recirculation model shows that recirculated particles have higher burnout levels depending on their dimensions (50% for 1100 μm and 85% for 740 μm). These recirculated particles represent all the passages of coal particles after the initial one. In these passages only the char combustion is considered which allows a common treatment for particles with different residence times. The total degree of burnout is dictated mainly by the cyclone efficiency which is considered in this test with $\eta = 1$. However, the highest burnout levels are reached by the projected particles from the bottom region, since they have dried and released volatiles before.

Heat flux profiles to the walls are shown in Figure 5. It can be seen that 70% of the total heat flux to the walls is by radiation transfer while the other 30% is transferred by convection. Since the operating conditions are typical of low loading, the bulk density is not very high. In this case the gas/particle mixture is not dense, allowing the radiation to play an important role in the heat transfer process.

Figure 6 shows concentration profiles of CO_2, CO, H_2O and O_2. In the present model it is assumed that the reaction rate is directly proportional to the reaction rate of coal combustion. The high concentrations of O_2 (10% for 20% of excess air) and low levels of CO_2 emissions show that there is less combustion in the upper part of the combustor than in the lower one. This result is a consequence of the value "a" in equations (11a) and (11b) since it has been evaluated for different operating conditions. The content of water vapour is high in the flue gases due to the high moisture content in the lignite considered.

The influence of limestone particle size and Ca/S molar ratio on the SO_2 retention was studied. Figure 7a) shows SO_2 concentration profiles with height. The SO_2 profile without limestone retention is represented in curve 1 in the combustor (fast bed) for the case of molar ratio Ca/S equal to two and particle sizes of 150 μm and 500 μm, respectively. Figure 7a) shows that more SO_2 is retained by the larger particles in the first passage. To analyse these results, Figure 7b), is helpful. Figure 7b) represents the limestone's reactivity which is a function of the fraction of CaO not reacted to form $CaSO_4$. The smallest particles have higher reactivities than the largest particles which means a larger capability to absorb SO_2 in the combustion gas, but their residence time in the fast bed is lower. However, considering recirculation, the residence time of particles becomes similar for all particle sizes. Thereby, the retention of SO_2 is higher for lower limestone diameters due to the higher reactivity for a given conversion fraction.

Figure 8a) shows the SO_2 emission profiles for various Ca/S ratios and a fixed particle size (d_s = 250 μm). It is clear that SO_2 emissions decrease with increasing Ca/S ratio. Figure 8b) shows that the reactivity of limestone does not remain similar along the combustor (for the three ratios used), showing the importance of mass flow rate in the fast bed. Decreasing the calcium to sulphur ratio, the conversion of CaO to CaSO4 within the particles is increased

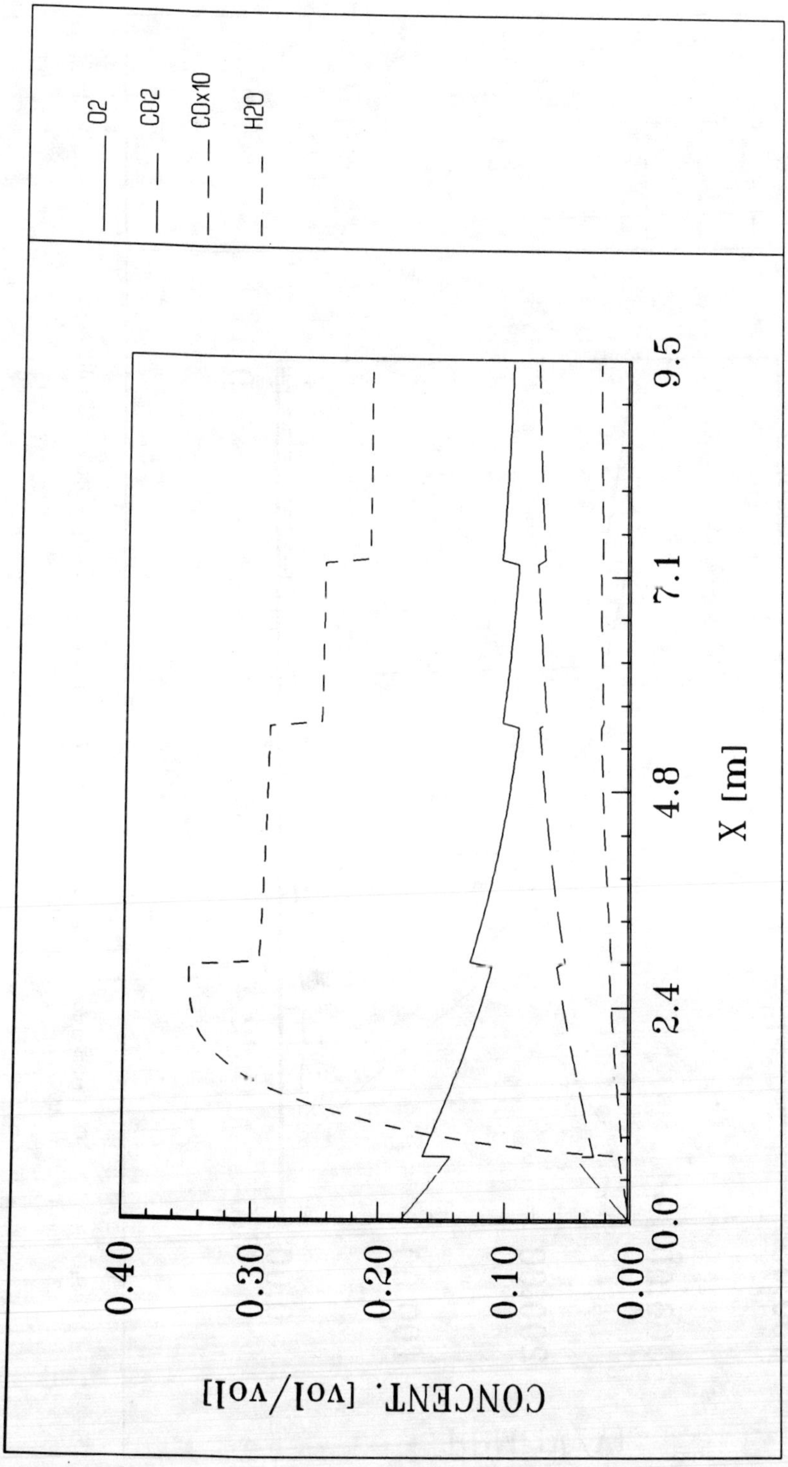

FIGURE 6 O_2, CO_2, CO and H_2O profiles along the combustor with air staging.

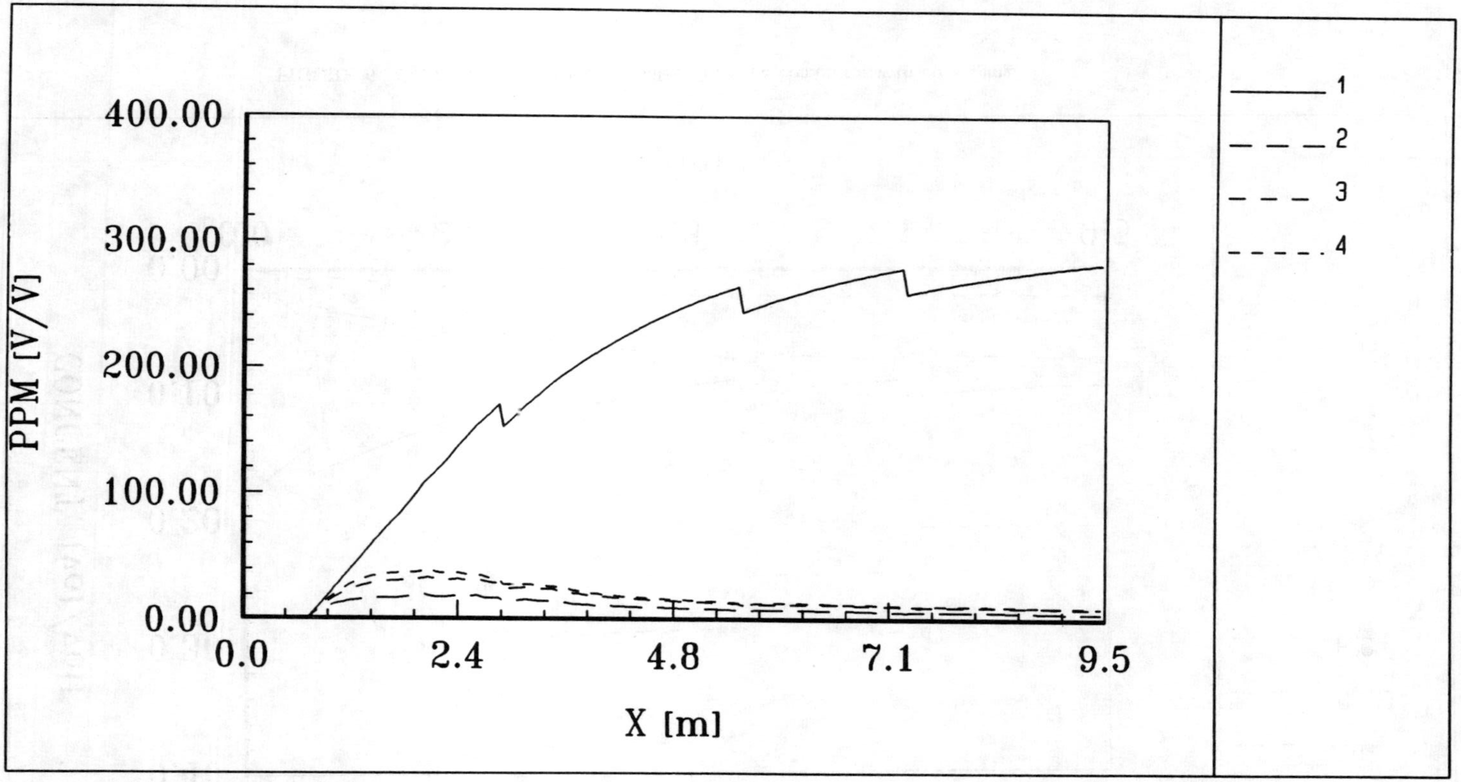

FIGURE 7a) SO_2 Emission for the Case of Ca/S = 2:1—Without retention; 2 - d_1 = 150 μm; 3 - d_1 = 250 μm; 4 - d_1 = 500 μm.

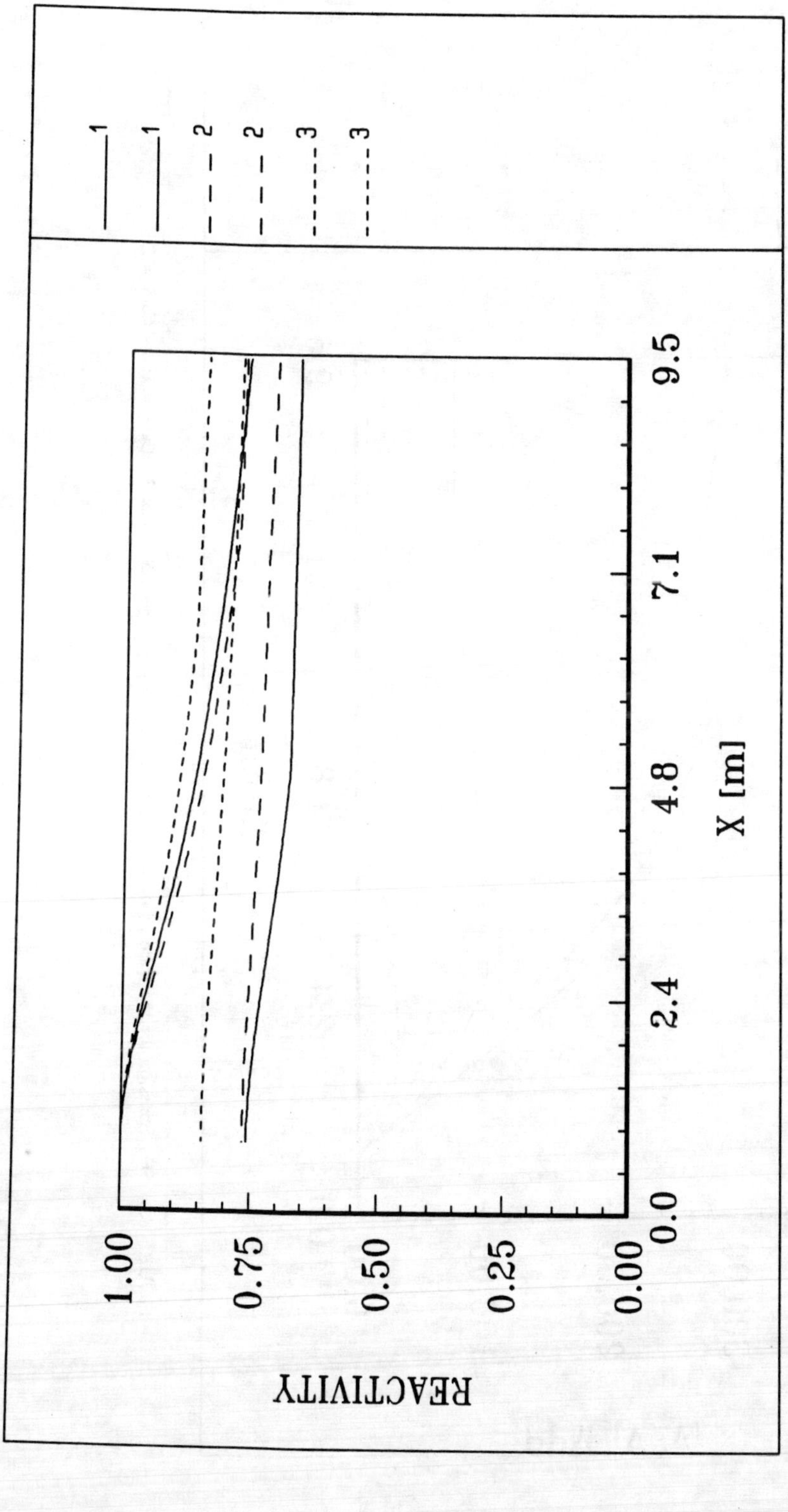

FIGURE 7b) Reactivity of Limestone Particles for the Case of Ca/S = 2; 1 - d1 = 500 μm; 2 - d_1 = 250 μm; 3 - d_1 = 150 μm.

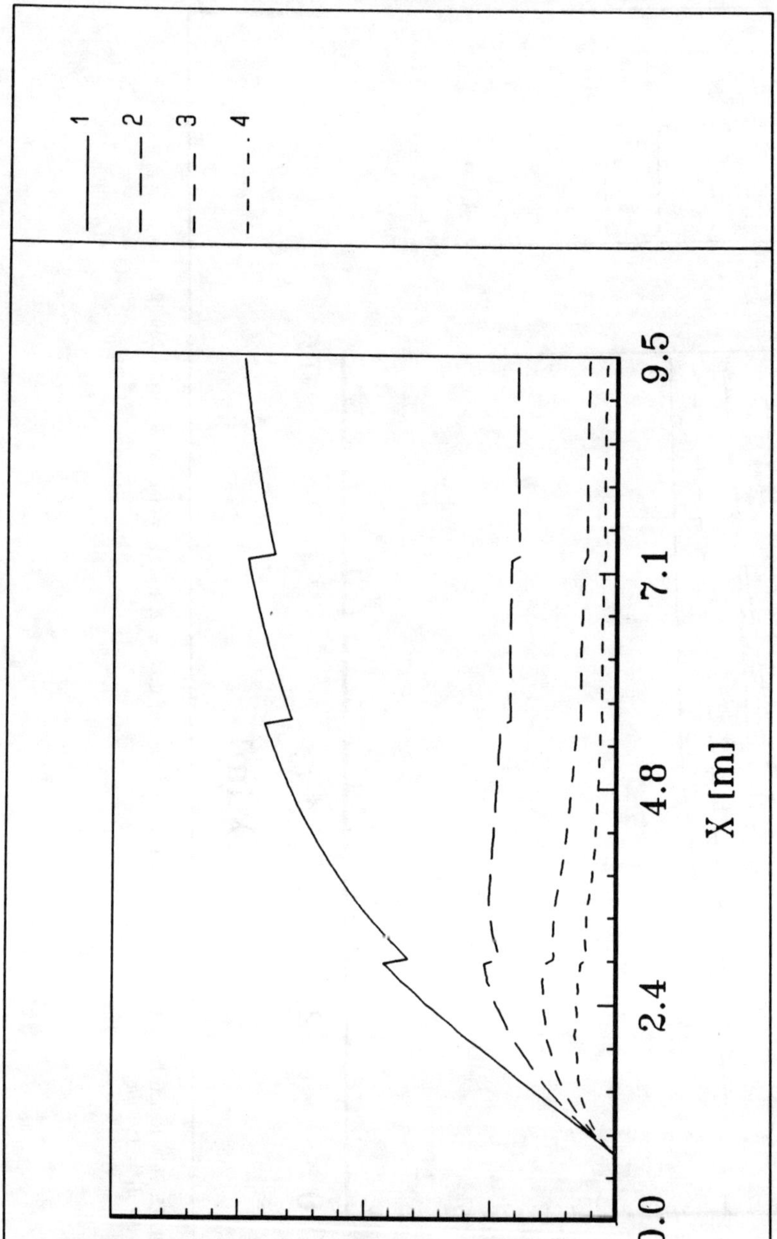

FIGURE 8a) SO_2 Emission for the Case of $d_1 = 250\ \mu m$: 1 - Ca/S = 0; 2 = Ca/S = 0.5; 3 - Ca/S = 1; 4 - Ca/S = 2.

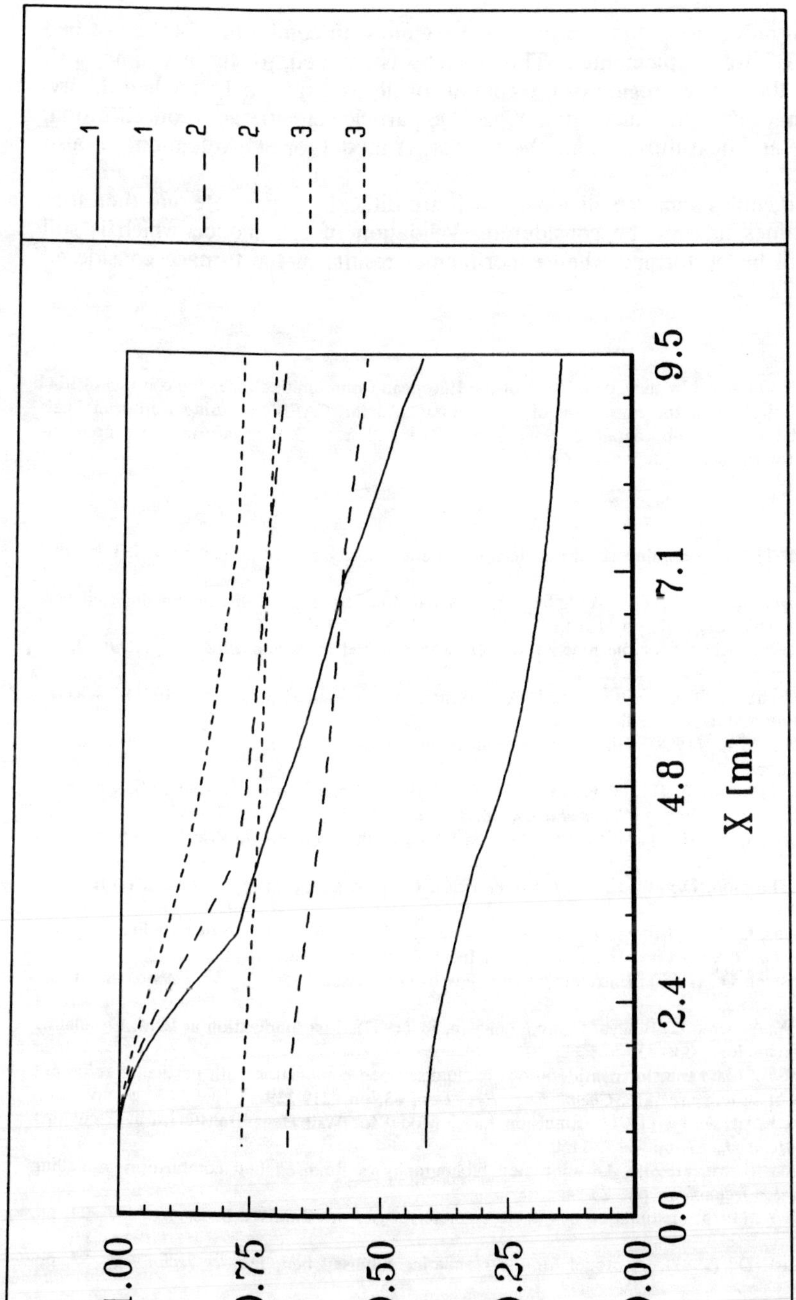

FIGURE 8b) Reactivity of Limestone Particles for the Case of $d_1 = 250\ \mu m$: 1 - Ca/S = 0.5; 2 - Ca/S = 1; 3 - Ca/S = 2.

SUMMARY

- A numerical model to simulate a two zone regime with combustion in the fast bed of the CAFBC was implemented. This model was coupled, for the first time, with a model for the bottom region of the combustor derived from bubbling bed theory.
- The model allows for the calculation of gas and particle velocities, gas concentration, temperature and heat fluxes, along the reactor. A model for SO_2 retention was also included.
- The model includes some coefficients which are difficult to calculate and therefore empirical values have to be considered. Validation of this model, which is still required, will be performed when experimental results of the furnace considered become available.

ACKNOWLEDGEMENTS

This work is currently supported by the Commission of the European Communities under the contract 0030-C (MB) entitled "Minimization of the Formation of Air Pollutants in the CAFBC by using European Fuels and Additives" of the JOULE subprogram of Solid Fuels. The author P. C. Saraiva wants to express his gratitude to JNICT for the scholarship given for Msc.

REFERENCES

Azevedo, J. L. T., (1991). On the influence of bubble fraction and velocity on bed permeability, IST internal report.

Berruti, F. and Kalogerakis, N., (1989). Modelling the internal flow structure of circulating fluidised bed, *The Can. J. of Chem. Eng.,* **67,** pp. 1010-1014.

Borgwardt, R. H., (1970). Kinetics of the reaction of SO_2 with calcined limestone, *Env. Sci. Techn.,* **4,** pp. 49-53.

Borgwardt, R. H., Drehmel, D. C., Kittleman, T. A., Mayfield, O. R. and Owen, J. S., (1971). Selected studies on alkaline additives for sulphur dioxide control, EPA-RTP.

Breault, R. W., Mathur, V. K., (1989). High velocity fluidised bed hidrodynamic modelling, *Ind. Eng. Chem. Res.,* **28,** pp. 684-688.

Carvalho, J. R. F. G., Pinto, M. F. R. and Pinho, C. M. C. T, (1991). Mass transfer around carbon particles burning in fluidised beds, *Trans. IChem. E.,* **69,** pp. 63-70.

Chung, P. M. and Carlson, L., (1981). Vertical combustion for particulate refuse, *Numerical Heat Transfer,* **4,** pp. 101-122.

Davidson, J. F., and Harrison, D., (1963). Fluidised Particles, Cambridge, Cambridge University Press, pp. 21.

Howard, J. B., Williams, G. C. and Fine, D. H., (1973). Kinetics of Carbon Oxidation in Post Flame Gases, *14th Symp. (Int.) on Combustion,* The Combustion Institute, pp. 975-986.

Kunni, D. and Levenspiel, O., (1990). Entrainment of solids from fluidised beds, *Powder Technology,* **61,** pp. 193-206.

Kwauk, M., Ningde, W., Youchu, L., Bingyu, C., and Zhiyuan, S., (1987). Fast fluidization at ICM, Circulating Fluidised Bed Technology, pp. 33-45.

La Nauze, R. D., (1985). Mass transfer considerations in fluidised bed combustion with particular reference to the influence of system pressure, *Chem. Eng. Res. Des.,* **63, pp. 219-229.**

Mahalimgam, M. and Kolar, A. K., (1991). Emulsion Layer Model for Wall Heat Transfer in a Circulating Fluidised Bed, *AIChE J.,* **37,** pp. 1139-1150.

Manno, V. P., Rietema, S. M., (1990). An annotated bibliography of fluidised bed combustion modelling information, *Powder Technology,* pp. 23-34.

Mori, S. and Wen, C. Y., (1975). Estimation of bubble diameter in gaseous fluidised beds, *AIChE J.,* **21,** pp. 109-115.

Rhodes, M. J., Geldart, D., (1987). A model for the circulating fluidised bed, *Powder Technology,* **53,** pp. 155-162.

Saraiva, P. C. (1992). Modelação de Combustão, Transferência de Calor e Emissão de Poluentes em Caldeiras de Leito Fluidizado Circulante, Msc Thesis (in Portuguese).

Sit, S. P. and Grace, J. R., (1981). Effect of bubble interaction on interphase mass transfer in fluidised beds, *Chemical Eng. Science,* **36,** pp. 327-335.

Van Swaaij, J. R., (1985). Chemical Reactors, Fluidization, Academic Press, pp. 595-629.

Weiβ, V., Schöler, J. and Fett, F. N., (1988). Mathematical modelling of coal combustion in a circulating fluidised bed reacting, Circulating Fluidised Bed Technology II, pp. 289-298.

Wen, C. Y. and Chen, L. H., (1982). Fluidised bed freeboard phenomena: Entrainment and elutriation, AIChE J., 28, pp. 117-1287.

Yates, J. G., (1983). Fundamentals of fluidised bed chemical processes, Butterworths.

Yerushalmi, J. and Avidam, A., (1985). High Velocity Fluidization, Fluidization, Academic Press, pp. 225-291.

Combustion of a High Ash and Sulfur-Containing Lignite and Its Pollution Characteristics in a Circulating Fluidized Bed Combustor

ÖZKAN GÖKSEL and DOGU GÜLSEN *Department of Chemical Engineering, Gazi University, Ankara, Turkey*
DURMAZ ALI *Department of Mechanical Engineering, Gazi University, Ankara, Turkey*

Abstract — In this study a 50 kW circulating fluidized bed combustor was constructed for the combustion of high ash containing lignites. Beypazarı lignite which contains 45.3% ash and 3.72% sulfur was mixed with limestone and was burnt in this combustor. Emission characteristics were investigated at different Ca/S mole ratios. It was shown that SO_2 removal efficiencies reached values of 98%. At an average bed temperature of 850°C, NO_x emission levels were found to be around 300 ppm. Temperature variation along the combustor (of 1.8 m in height) was found to be about 140 and 90°C for two different feed point locations of 0.17 and 0.52 m above the distributor, respectively.

Key Words: circulating fluidized bed, combustion, desulfurization

INTRODUCTION

One of the major advantages of circulating fluidized bed combustors is their efficiency for the combustion of lignites with low calorific value and high ash content. It was shown in the literature (Plass *et al.* 1986; Kelly and Layden, 1987) that combustion efficiencies reaching 99% are possible in pilot-scale circulating fluidized bed combustors. These combustors operate at high gas velocities and generally with an excess air level of about 20–30%. Hirama *et al.* (1984) investigated the effect of particle size on combustion efficiency in a fluidized bed combustor and reported that the combustion efficiency of larger particles is higher.

Oxides of sulfur produced in burning the coal may be retained in solid form by reaction with particles of limestone or dolomite which is fed to the fluidized bed combustor together with the solid fuel. The porous alkaline solid, CaO, which is produced by the calcination of limestone reacts with SO_2.

$$CaO + SO_2 + \tfrac{1}{2}O_2 \rightarrow CaSO_4.$$

Since molar volume of the reaction product $CaSO_4$ is about three times greater than the molar volume of CaO, formation of $CaSO_4$ causes pore mouth closure and the reaction stops before all the CaO is consumed by this reaction (Dogu, 1981; Marsh and Ulrichson, 1985). As a result, Ca/S mole ratio is usually chosen between 2 and 4 in a classical fluidized bed combustor (Vogel *et al.* 1978; Uysal *et al.* 1983). On the other hand, high SO_2 retention efficiencies were obtained for Ca/S mole ratios of less than two in a circulating fluidized bed combustor (Goblirch *et al.* 1987; Plass *et al.* 1986).

It was shown in the literature that (Furusawa *et al.* 1978) rather low NO_x emissions are obtained by staged combustion in a fluidized bed combustor. By use of primary and secondary air injected at different locations in a circulating fluidized bed combustor, its temperature and combustion atmosphere is well-regulated and generally low NO_x emissions of about 150–350 ppm are reported (Lucat and Tondeur, 1988; Plass *et al.* 1986).

TABLE I
SOME SPECIFICATIONS OF THE CIRCULATING FLUIDIZED BED COMBUSTOR USED IN THIS STUDY

Combustor height	1.80 m
Bed cross-sectional area	0.0123 m^2
Minimum fluidization velocity	0.30 m/s
Terminal velocity	1.20 m/s
Operation velocity	1.75 m/s
Pressure drop (bed)	14.4 cm H_2O
Pressure drop (distributor)	35.0 cm H_2O
Orifice diameter (distributor)	0.001 m
Number of orifices	600
Coal feed rate	15.6 kg/h
Power	50 kW
Particle diameter of coal	0.9 mm
Feeding system	Screw feeder with an L-valve
Recycling system	Fluidized bed (0.10 × 0.14 × 0.10 m)
Recycle feed rate	100 kg/m^2 h

EXPERIMENTAL WORK

A schematic diagram of the circulating fluidized bed combustor used in this study is shown in Figure 1. The combustion chamber has a diameter of 0.125 m and its height is 1.80 m. Beypazarì lignite, which is crushed and sieved to 0.9 mm average particle size, is fed to the combustor at different feed locations using a screw feeder. Particles leaving the combustor are collected by a cyclone and recycled back to the combustor by use of a fluidized bed feeding unit. Some specifications of the circulating fluidized bed combustor are summarized in Table I. Details of the design parameters are discussed by Özkan (1989).

In most of the experiments 20% excess air was used. Coal feed was started after the system was preheated to about 650°C with LPG. Chemical composition of the Beypazarì lignite used in this work is given in Table II. Composition of the flue gas was monitored by making analysis every five minutes during combustion. For this, Enerac-80 and Adosdublex NVT analysis equipment were used. Ash samples collected from the fluidized bed recycling unit were also analyzed to determine the combustion efficiencies.

RESULTS AND DISCUSSION

Typical temperature profiles obtained in the combustor at steady state are given in Figures 2 and 3. When the coal is fed to the combustor at a position of 17 cm above the distributor plate, approximately 140°C decrease of temperature was observed along the combustor. The maximum temperature measured was 870°C around the feed point. On the other hand, when the feeding was carried out at 52 cm above the distributor, the maximum temperature was detected at around 85 cm above the distributor. In this case, about 90–100°C lower temperatures were measured both at the bottom (distributor) and at the top of the combustor. As it is shown in these figures, combustion was carried out at rather low temperatures.

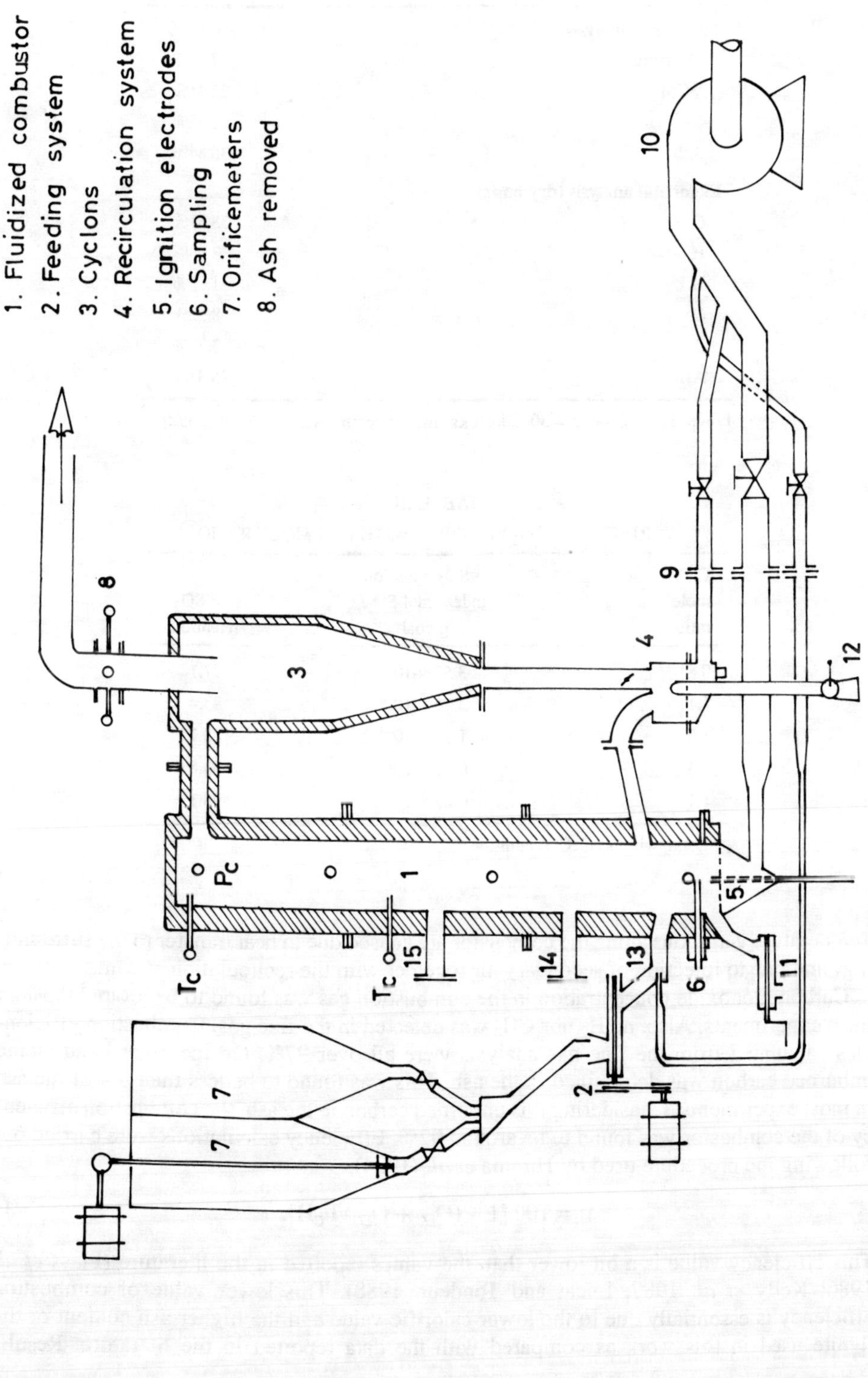

FIGURE 1 Schematic diagram of the experimental setup.

TABLE II

CHEMICAL ANALYSIS OF THE BEYPAZARÌ LIGNITE

Proximate analysis	
Moisture	12.40%
VCM	26.41%
C	22.35%
Ash	38.84%
Elemental analysis (dry basis)	
C	38.31%
H	3.03%
N	1.11%
O	8.52%
S	3.72%
Ash	45.31%

Lower heating value = 3002 kcal/kg; higher heating value = 3207 kcal/kg.

TABLE III

VARIATION IN SO_2 REMOVAL WITH Ca/S MOLE RATIO

Ca/S mole ratio	SO_2 emission index, mol SO_2/ g coal	% SO_2 retention
0.8*	3.52×10^{-4}	77.1
1.1	2.55×10^{-4}	83.4
1.2	1.73×10^{-4}	88.7
1.3	0.95×10^{-4}	93.8
1.4	0.29×10^{-4}	98.1

*Lignite with no $CaCO_3$ addition.

Temperature variations along the combustor are caused due to heat transfer to the surroundings and due to injection of secondary air together with the recirculation stream.

Carbon monoxide concentration in the combustion gas was found to be around 0.3% in most experiments. Also, no H_2 nor CH_4 was detected in the flue gas. Combustion efficiencies calculated from the flue gas analysis were all over 97%. On the other hand, some unburned carbon was determined in the ash. This was found to be less than 6% of the ash in most experiments. Considering this unburned carbon in the ash, the combustion efficiency of the combustor was found to be around 92%. Efficiency calculations were carried out following the procedure used by Hirama *et al.* (1984).

$$\eta = 100\,[1 - (Q_{LC} + Q_{LG})/F_C H]. \tag{1}$$

This efficiency value is a bit lower than the values reported in the literature (Plass *et al.* 1986; Kelly *et al.* 1987; Lucat and Tondeur, 1988). This lower value of combustion efficiency is essentially due to the lower calorific value and the higher ash content of the lignite used in this work as compared with the data reported in the literature. Results

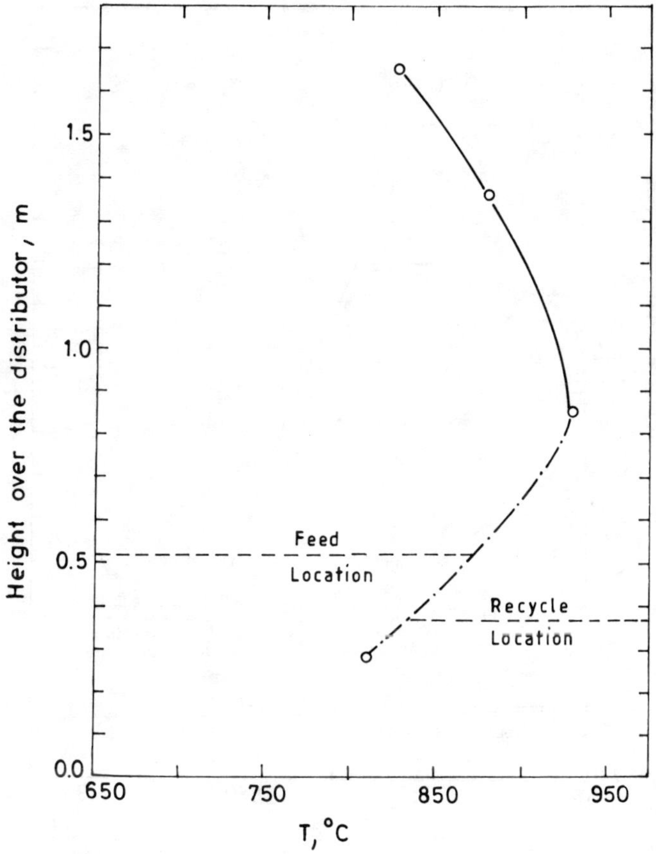

FIGURE 2 Temperature profile in the circulating fluidized bed combustor.

indicated that low quality lignites would be successfully burnt in a circulating fluidized bed combustor with much higher combustion efficiencies than conventional fixed bed systems.

The circulating fluidized bed combustor used in this work can be considered as a small-scale pilot unit. In a larger unit, the hydrodynamics of the system might be altered, which in turn might alter the temperature distribution in the combustor. Using the data base obtained in this study, a larger unit should be designed and results should be compared.

SO_2 Emission Characteristics

Chemical analysis of the Beypazarì lignite showed a rather high calcium content. The molar ratio of Ca to S of this lignite is 0.8. Experiments were also conducted for the addition of limestone to the lignite. Results of the experiments conducted at different Ca/S mole ratios and at combustion temperatures around 850°C are shown in Table III and Figure 4. It was shown that about 77% of the sulfur is retained in the ash without addition of limestone. This high value is partially due to the good dispersion of $CaCO_3$ in the lignite structure in rather fine grains. This allows for very efficient utilization of Ca and very high conversion of CaO to $CaSO_4$. Similar results were reported in the literature for lignites containing high calcium (Dogu and Togay, 1979; Özsaygì, 1981).

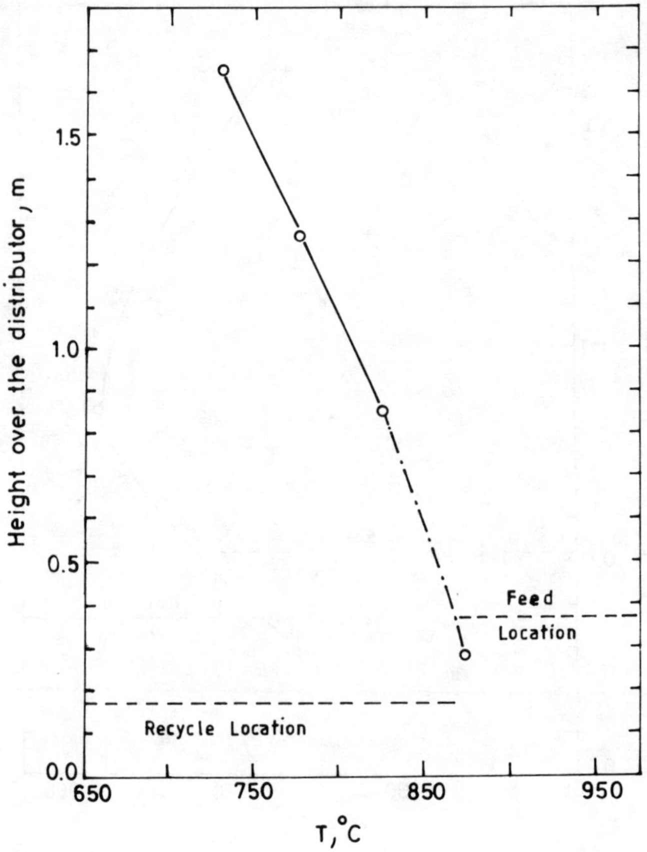

FIGURE 3 Temperature profile in the circulating fluidized bed combustor.

In the present study, SO_2 retention in the ash was increased to 98.1% at a Ca/S mole ratio of 1.4. This is a rather high value and shows the efficiency of a circulating fluidized bed combustor retention of SO_2 by limestone addition. In classical fluidized bed combustors this ratio should generally be 2–4 to achieve over 90% SO_2 removal. Plass *et al.* (1986) and Lucat and Tondeur (1988) also reported higher SO_2 removal in circulating fluidized beds. This higher SO_2 removal efficiency may be partially due to attrition of the $CaSO_4$ layer by CaO particles and exposition of the active CaO surface to SO_2 due to the high gas velocities in circulating beds.

NO_x Emission Characteristics

The NO_x emissions of the circulating fluidized bed combustor used in this study were found to be less than 300 ppm in the temperature range studies. NO_x emissions detected at different bed temperatures are summarized in Table IV and Figure 5. Below 800°C NO_x emissions are rather low.

Over 800°C some increase in NO_x was observed, but still these values are much less than the values reported for conventional fluidized bed combustors (Dogu and Togay, 1979; Gibbs and Hedley, 1978). Such low values are also reported in the literature for circulating

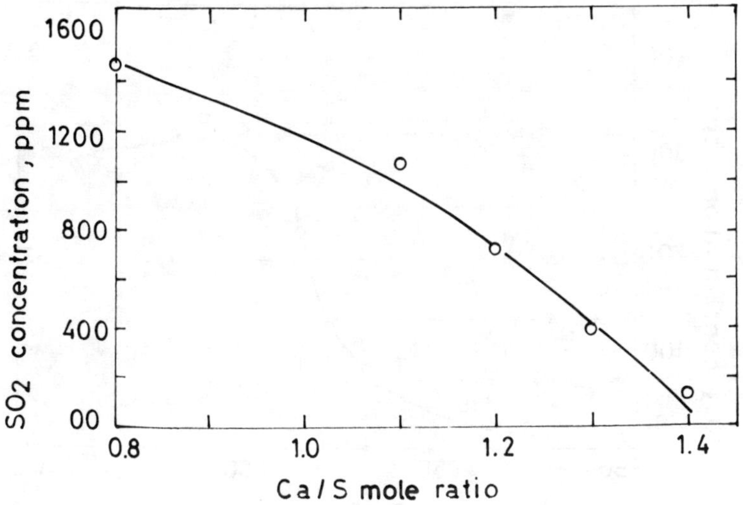

FIGURE 4 Effect of Ca/S ratio on SO_2 emissions.

TABLE IV

VARIATION OF NO_x EMISSIONS WITH AVERAGE BED TEMPERATURE

Mean bed temperature, °C	NO_x emission, ppm	NO_x emission index, $E \times 10^4$, mol/g coal
705	31	0.074
740	42	0.101
770	23	0.055
785	66	0.160
790	80	0.190
810	123	0.30
825	218	0.52
845	340	0.82
875	333	0.80
890	320	0.77
895	275	0.66
910	277	0.67
925	286	0.69

fluidized bed combustors (Plass *et al.* 1986; Lucat and Tondeur, 1988). Lower bed temperatures and the possibility of using secondary air and staged combustion lowers NO_x emissions.

CONCLUSION

It was shown that lignites with low calorific value and high ash content can be successfully burned in a circulating fluidized bed combustor with rather high combustion efficiencies.

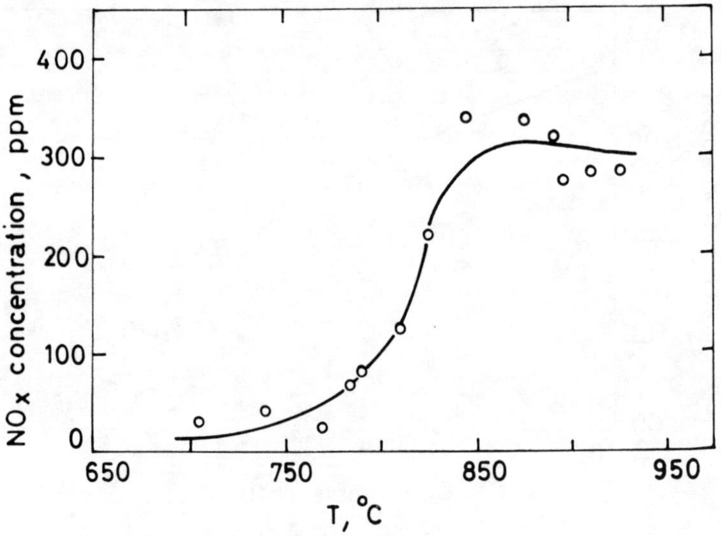

FIGURE 5 Effect of bed temperature on NO_x emissions.

Sulfur dioxide emissions can be considerably reduced by the addition of limestone. SO_2 retention efficiency of a circulating fluidized bed combustor is higher than a conventional fluidized bed combustor. Also by carrying out combustion below 850°C rather low NO_x emissions were obtained.

ACKNOWLEDGMENT

The NATO TU Air Pollution SfS Grant is gratefully acknowledged.

NOMENCLATURE

F_C	Coal feed rate, kg/h
H	Higher heating value, kcal/Kg
Q_{LC}	Heating value of combustibles in ash, kcal/h
Q_{LG}	Heating values of combustibles in flue gas, kcal/h.
η	Combustion efficiency

REFERENCES

Dogu, T. (1981). The Importance of Pore Structure and Diffusion in the Kinetics of Gas–Solid Non-Catalytic Reactions; Reaction of Calcined Limestone with SO_2. *Chem. Eng. J.*, **21**, 213.

Dogu, T., and Togay M. (1979). Effect of Air to Fuel Ratio on the Reaction of SO_2 and NO_x Emissions in a Fluidized Bed Combustor. *Technologie Des Lits Fluidises et Disperses Applications Industrielles*, Compiegne, 1–5.

Furusuwa, T., Honda, T., Tatano, J., and Kunii D. (1978). Abatement of Nitric Oxide Emission in Fluidized Bed Combustion of Coal. *J. of Chem. Eng. of Japan*, **11**(5), 377–383.

Gibbs, B.M., and Hedley A.B. (1978). Combustion of Large Coal Particles in a Fluidized Bed, *Fluidization*, Cambridge Univ. Press., 235.

Goblirch, G., Molen, V.R.H., Wilson, K., and Hajicek, D. (1980). Atmospheric Fluidized Bed Combustion Testing of North Dakota Lignite. *Sixth International Conf. on Fluidized Bed Combustion*, III, Georgia, USA, 9–11.

Hirama, T., Hosada H., Nishizaki H., Chiba, T., and Kobayashi H. (1984). Effects of the Size Distribution of Coal on Fluidized Bed Combustion, *Int. Chem. Eng.*, **24**(3), 502–510.
Kelly, A.G., and Layden B. (1987). Fluidized Bed Combustion of Coal with High-Ash Deformation Temperature. *J. of Inst. of Energy*, **29**, 29–31.
Lucat, P., and Tondeur G. (1988). *Circulating Fluidized Bed Boilers*, Stein Industrie, Seminar, Ankara, May 26.
Marsh, D.W., and Ulrichson, D.L. (1985). Rate and Diffusional Study of the Reaction of Calcium Oxide with Sulfur Dioxide. *Chem. Eng. Sci.*, **40**, 423–433.
Özkan, G. (1989). Design and Combustion Studies of a Pilot Circulating Fluidized Bed Combustor. MS Thesis, Gazi University, Ankara.
Özsaygì, R. (1981) Fluidized Combustion of Beypazarì Lignite in a Pilot Fluidized Bed Combustor and its Pollution Characteristics. MS Thesis, METU, Ankara.
Plass, L., Bierbach H., and Gummel P. (1986). *Experience with Combustion in Circulating Fluidized Beds*. 2, Lurgi GmbH Gervinusstraße, 17/19, Frankfurt.
Uysal, B.Z., Mutel, T., Tunçman, N., and Dogu, T. (1983). Fluidized Bed Coal Combustion Emission and Heat Transfer Characteristics:, Fluidization, *AIChE Symp. Ser.*, 435–442.
Vogel, G.J. (1973). Reduction of Emission of Sulfur Oxides and Nitrogen Oxides by Addition of Limestone or Dolomite During the Combustion of Coal in Fluidized Beds, *Proc. III, Int. Conf. on FBC*, EPA 650/2-73-053.

Status and Environmental Impact of PFBC Plants

SVEN A. JANSSON *ABB Carbon AB, Finspong, Sweden*

Abstract — The world's first commercial PFBC power plant, at Stockholm Energi's Värtan power plant in Stockholm, Sweden, has been taken over by the customer, following a successful availability test during a 600 hour period in late March and early April 1991. When the summer revision is over and the heating season starts again on 15 September, Stockholm Energi will thus assume full responsibility for the operation of this combined heat and power plant. Performance testing at the Värtan PFBC plant has shown that strict environmental regulations can be met. In addition, even lower emission levels for NO_x have been demonstrated at full load, i.e., 20 mg/MJ fuel input, or 57 mg/Nm^3 (calculated as NO_2). Nitrous oxide emissions have been below 10 ppm. The Escatron and Tidd PFBC plants in Spain and the US are in the trial operation phase. These repowered plants use high-sulfur Spanish black lignite and Eastern US high-sulfur coal respectively, as fuel. With typically 90% desulfurization in the fluidized bed and some 50–80% reduction of the NO_x emissions compared to existing power plants that use these fuels, the environmental impact of repowering with PFBC is very favorable.

INTRODUCTION

PFBC plants are clean, efficient and compact. They combine the advantages of fluidized bed units, i.e., fuel flexibility and reduced emissions of sulfur dioxide, nitrogen oxides, and other pollutants, with the benefits of combined cycle plants. As a result, PFBC plants have a high efficiency for generation of electricity, typically 40 to 44% (LHV) with today's PFBC technology, but with a potential for improvements to 50%. They also provide a high power to heat ratio in combined heat and power applications.

Figure 1 shows schematically the PFBC cycle and the important systems. Pressurization increases the amount of air and coal that can be supplied to a fluidized bed, at similar fluidizing velocities. A higher bed level can also be used. This results in an increased contact time between solid and gaseous phases in the bed, which in turn leads to a greatly improved combustion efficiency as well as very efficient desulfurization within the fluidized bed, where dolomite or limestone functions as a sulfur sorbent.

Nitrogen oxide formation is low due to the modest bed temperature of approximately 850 to 880°C, and also because there is a breakdown of nitrogen oxides as well as hydrocarbon compounds on the surface of the bed solids.

The compactness of a PFBC plant compared to a conventional pulverized coal power plant can be seen in Figure 2. This comparison is valid for plants designed for use with Spanish high-sulfur black lignite.

Fifteen years after intensive development of PFBC began within the ABB (then ASEA) group, there are now three PFBC plants in existence in the world, and they all use ABB's PFBC technology. This paper describes briefly these plants and their status. It also discusses the environmental characteristics of PFBC plants and result obtained to date.

THE VÄRTAN PFBC PLANT

Figure 3 shows an aerial view of the Värtan power station in Stockholm with the new PFBC plant. Following a turnkey order from Stockholm Energi which owns the Värtan plant and is responsible for supplying Stockholm with heat and power, ABB Carbon has built, commissioned and supervised the operation of the new 135 MWe/224 MW heat PFBC plant during one and a half heating seasons.

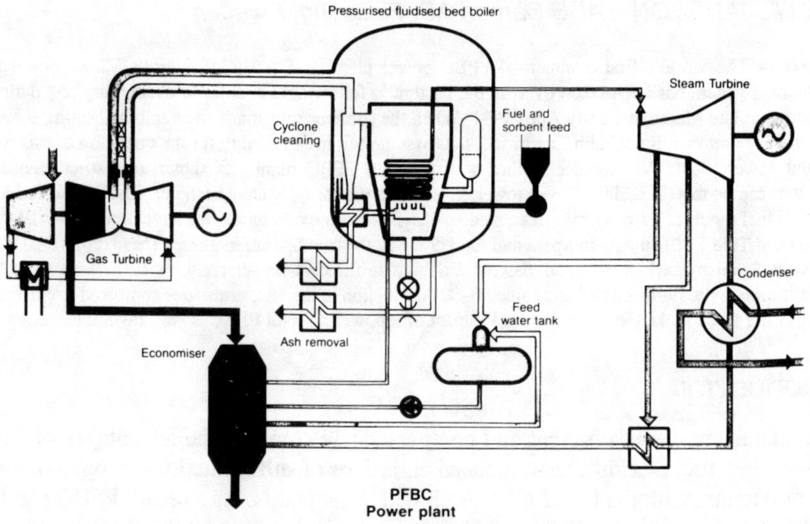

FIGURE 1 The PFBC cycle.

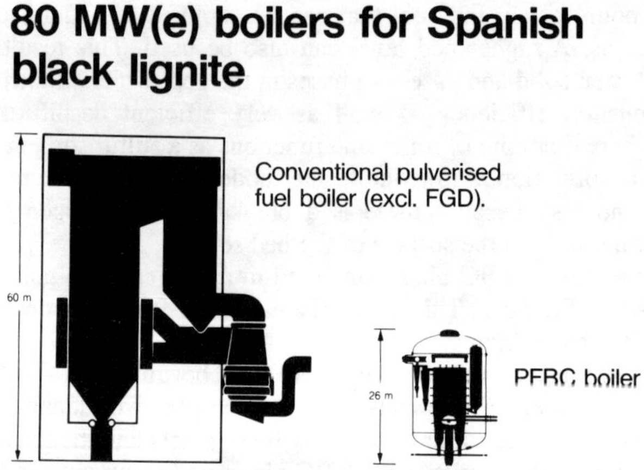

FIGURE 2 Size comparison.

Totally, some 4800 boiler operation hours have been achieved, with 40% of those on one of the two P200 PFBC modules and the rest on the other. A P200 module consists of an ABB STAL GT35P gas turbine and a P200 PFBC boiler, which basically is a pressure vessel which contains a fluidized bed boiler as well as cyclones for primary cleanup of the combustion gases after the fluidized bed, storage vessels for bed material, etc.

The Värtan PFBC plant has two of these P200 modules. The steam which is generated and superheated within the PFBC boilers is fed to a common steam turbine. A bag-house is used for final cleanup of the combustion gases which after passing an economizer are directed to an existing stack.

Coal and dolomite, or limestone, are delivered to a receiving station in the harbor, adjacent to the Värtan plant. Storage of coal, which is delivered by boat, is underground in

FIGURE 3 The Värtan plant.

a huge rock cavern which was previously used for storing heavy fuel oil. This coal store can hold 110,000 tons of coal or approximately half the amount needed for operation during a heating season. The coal consumption is about 200,000 tons per year.

Fuel is withdrawn from the underground coal store and is subsequently crushed and mixed with dolomite or limestone, and with water, to a coal-sorbent-water paste. This paste is pumped to the PFBC plant and injected into the fluidized beds.

Ashes are withdrawn from the fluidized beds and from the cyclones. After depressurization and recovery of their sensible heat, these ashes are transported to ash silos in the harbor area. The intention is to use these products, which are mixtures of partly sulphated dolomite or limestone and coal ash, for large scale construction purposes. Experiments have shown that simple mixing of these rest products with water results in the formation of a concrete-like product, which can replace natural rock in road beds, etc.

The location of the plant, not far from the very center of Stockholm, the fact that housing developments exist just outside the boundary of the site, and a general tightening of the environmental legislation during the period of permitting discussions, resulted in very strict emission limits being established.

Sulfur emissions were limited to 30 mg/MJ thermal input, which corresponds to 60 mg/MJ or 170 mg/Nm^3 of SO_2. Nitrogen oxides (calculated as NO_2) were limited to 50 mg/MJ, which corresponds to 140 mg/Nm^3. These are annual average numbers. Particulate emissions were limited to 5 mg/MJ (14 mg/Nm^3) and 10 mg/MJ (28 mg/Nm^3), respectively, as inspection and monthly average values.

An annual emission limit for mercury of 8 kg was also introduced. Similarly, it was the Swedish government that decided that the coal storage was to be underground. Special demands were also placed on the appearance of the buildings, including the fuel preparation building in the harbor next to the plant.

A contractual condition for takeover of the plant by the customer was a successful 600 hour availability test with the operating conditions determined by Stockholm Energi. This availability test was carried out during the period from 17 March to 11 April 1991. Each morning the operating profile, i.e., power and heat output from the plant, was specified for the coming 24 hour period. The test simulated a full year's expected operation and major load changes, compressed into the 600 hour test period.

The availability of the whole plant during the test was 88.7% and the availability of the "PFBC islands" was 93.9%. These are, of course, very good availability numbers. The major causes for unavailability were a steam turbine trip, and also a cyclone pluggage in one module during a scheduled shutdown of the other, while the steam turbine was out of service.

The reason for the steam turbine trip was water in the gland leakage system, when extremely low load had been requested. The turbine was inspected and restarted again after 50 hours. Altogether, the whole plant was out of service for 25 hours during the availability test period.

Performance testing has shown the following emissions data at full load:

SO_2	<60 mg/MJ
NO_x	20 mg/MJ, or 57 mg/Nm^3 (as NO_2)
CO	<10 ppm
N_xO	<10 ppm
Particulates	<1 mg/MJ

THE ESCATRON PFBC PLANT

The Escatron PFBC plant, Figure 4, has been built by a consortium of ABB Carbon and Babcock & Wilcox Espanola. ENDESA has been responsible for the supply of some systems. Here one P200 module is used to repower an existing steam turbine. The nominal output of the repowered plant is 76 MWe.

Due to the very special nature of the fuel at Escatron, a number of components and systems are substantially different from those in the Värtan plant. One big difference relates to fuel preparation, which here is dry.

A significant effect of coal quality and especially the high contents of ash and sulfur is that the particulate loading in the combustion gases that leave the fluidized bed is very high. Accordingly, the cyclones and the ash removal systems must be able to handle much higher solids mass flows at Escatron than at Värtan.

First coal-fire at Escatron was at the end of November 1990. Commissioning and trial operation has since proceeded relatively well considering the special features of the project. Through August 1991, a total of about 1000 hours of operation on coal had been achieved, with uninterrupted periods of 110, 150 and 250 hours. The longest full load operation period has been 220 hours. This was an acceptance test before takeover.

Environmental performance has been as expected. In-bed desulfurization well over 90% has been reached with less than expected sorbent usage. Nitrogen oxide emissions have also essentially been as expected. The continued work at Escatron will aim to improve the availability of the plant through optimization of the performance of the different systems.

FIGURE 4 The Escatron plant.

THE TIDD PFBC PLANT

The Tidd plant (see Figure 5) has been supplied by ASEA Babcock. AEP is responsible for architect engineering and construction management as well as for certain systems. Following first coal-fire in November 1990, the Tidd plant has now operated over 600 hours on coal, with a longest continuous period of 90 hours. The plant, which also uses on P200 module, operates on a good quality, high-sulfur coal. The coal is crushed and then mixed with water into a coal-water paste which is pumped into the combustor. The sorbent is crushed and fed dry with the aid of lock-hoppers.

Some of the experiences from the early commissioning and trial operation at Värtan have been applied at Tidd (and also at Escatron when applicable). As a result it was possible to go from first coal fire to full bed height operation at Tidd in half the time required at Värtan.

Work continues at Tidd to check out and optimize function of the different systems and to improve the plant's availability, in preparation for takeover.

MORE PFBC PLANTS WILL BE BUILT

It has become widely recognized that PFBC is the leading clean coal technology for power generation, due to the high thermal efficiency, which translates to reductions in fuel use and emissions of carbon dioxide. The potential for further improvements is also very important and naturally the same is true for the promise of a competitive price level, due to the compactness of PFBC plants.

FIGURE 5 The Tidd plant.

ABB Carbon's licensee in Japan, IHI, has received an order for repowering one unit at the Wakamatsu power plant in Japan with PFBC. AEP has been selected for partial funding of a larger PFBC repowering project, under the US Clean Coal Technology Demonstration Program (the Tidd PFBC plant was also funded partly under that program). It seems likely that the intended repowering project will instead be a new PFBC plant, using ABB Carbon's larger P800 module, which would give an output of over 360 MWe. PFBC plants based on the P800 module are also planned for several other countries, including Spain.

The US Clean Coal Technology Demonstration Program will probably also fund the first constructing of a PFBC plant of the circulating fluidized bed type.

ENVIRONMENTAL IMPACT

The strength of PFBC as well as other clean coal technologies that use combined cycle technology is that fuel use and carbon dioxide emissions are much lower than for the current generation of coal-fired power plants.

A comparison of what currently exists and what can be achieved with PFBC is very enlightening. The conventional coal-fired plants around the world today range in efficiency from perhaps 25 to 40% depending on age and local conditions. The average efficiency is 34%. Those units which have the lowest efficiency are often old and serve as peaking or stand-by units. In some countries they are used for base-load. The average age of the plants in operation is 20 years.

Most of these plants have no cleaning at all, other than perhaps devices for some sort of particulate cleanup. The emissions of sulfur dioxide and nitrogen oxides per unit of fuel used or electricity produced are typically at least 10 times higher than what is considered acceptable in newer plants.

Low emissions of acid rain precursors can be achieved also with conventional power plants, if these are equipped with scrubbers and de-NO_x equipment. It is, basically, only a matter of cost.

A comparison of fuel consumption for the production of electricity in power plants with different efficiencies shows some interesting numbers. If the fuel consumption index is 100 for plants with the average world efficiency, 34%, the same index for a PFBC plant with today's technology is 77. Future PFBC plants, with efficiencies that approach 50%, will have a fuel consumption index of 68. The emissions of carbon dioxide will follow the fuel consumption index.

The major issue is how quickly the world's power plants can be improved to benefit from these very large potentials for reduction in emissions of acid rain precursors, as well as carbon dioxide.

In view of the expected continued population growth and the needs for improvement in standard of living, it seems obvious that while efforts are made to find ways of utilizing renewable and sustainable energy resources for the future, steps must also be taken to improve the means for utilization of the world's most abundant fossil energy resource, which is coal.

Depending on the age and location of existing plants they should be upgraded, repowered or replaced. PFBC is a strong contender for repowering and replacement of old plants on sites which are often cramped.

PFBC has also the potential for becoming the dominant clean coal technology for the next several decades.

8.

INCINERATORS

Solid Waste Incineration in Rotary Kilns

D. W. PERSHING, J. S. LIGHTY & G. D. SILCOX *University of Utah, Salt Lake City, Utah, USA*

M. P. HEAP & W. D. OWENS, *Reaction Engin. International, Salt Lake City, Utah, USA*

Abstract—Rotary kilns are used to dispose of many solid wastes and sludges and to thermally treat contaminated soils. In this communication the fates of hydrocarbon and metal species are examined with a view toward optimization of new kiln designs and maximizing existing unit throughout while minimizing pollutant emissions.

Initially, process fundamentals are considered to characterize the controlling phenomena. Pilot- and large-scale data are then examined to define practical system complexities. Finally, techniques for data scale-up and performance prediction are summarized.

Temperature is clearly the most important parameter with respect to the fate of both metal and hydrocarbon species; hence, heat transfer is often rate limiting. High temperatures favor hydrocarbon evolution, but can also enhance the formation of toxic metal fumes. Both the solid composition and the moisture content can significantly influence the time at temperature required for hydrocarbon destruction and metal vaporization.

Improving bed mixing helps contaminant release but can also aggravate puffing tendencies with batch charging. Full-scale performance predictions currently require a combination of small scale data and computer modeling. Future work needs to focus on verification of large-scale predictions for complex mixtures and sludges so that expensive trial burns can be minimized.

INTRODUCTION

The disposal of all types of waste and remediation of hazardous waste sites has become a major issue worldwide and no single treatment option provides a complete answer to the problem. Waste minimization and recycling must play a part in the final solution, but in themselves, they cannot solve the problem facing our modern society. Thermal treatment systems, such as incineration, are attractive methods of waste disposal for those materials that cannot be recycled. Thermal treatment reduces the bulk of material that must be landfilled and there is the potential to recover the energy content of the waste which may help defray the cost of disposal.

Recently Seeker (1991) characterized the overall problem of waste disposal (see Figure 1) and noted that virtually every human activity generates some type of waste material. The United States generates approximately 10^{12} lbs of waste in one year. Even if extensive waste minimization plans are implemented, this amount is projected to increase by at least 1% annually. Waste incineration, as opposed to open burning, is practiced in many industrialized nations. Tens of thousands of tons of municipal waste are burned daily in the United States and the practice of municipal waste combustion is even more common in Japan and Europe.

Properly designed and operated thermal treatment devices do not pose any major threats to the environment; however, this view is not held by many environmentalists. There are legitimate concerns, but none that cannot be solved by the application of sound combustion engineering practices. Many concerns are common to all combustion sources, such as, pollutant emissions which are the by-products of the combustion process and are the result of inefficiencies in the combustion process or the presence of heteroatoms in the fuel or waste. The quality of any solid residue is also of concern, particularly the level of hydrocarbons and the leachability of metal compounds. Although incineration has met with considerable public resistance in some states, such as Arizona, there are others who recognize the need for permanent, environmentally-safe disposal of waste materials and have granted permits to thermal treatment facilities fitted with advanced pollution control systems.

In 1988 there were more than a dozen fixed hazardous waste incinerators in the United States, the majority of which are based on rotary-kiln technology (Tillman, 1991a). There are numerous other facilities in Sweden, Finland, Germany, Switzerland, and Denmark. The U.S. EPA has estimated that there could be in excess of 2,000 existing waste sites requiring some form of clean-up and transportable incinerators are used quite often for this purpose when a permanent installation cannot be justified. The projected increase in the generation of hazardous waste, the reduction of landfill disposal options, and the required clean-up of existing landfill sites have all increased the need to understand incineration processes more thoroughly since, to date, incineration and thermal treatment have proven to be one of the few permanent solutions for many waste disposal problems minimizing future "cradle to grave" liabilities.

A typical incinerator system consists of two parts: the primary combustor and the secondary combustor. In the primary combustor—a rotary kiln, fluidized bed, multiple hearth, or other type of incinerator—the hydrocarbon contaminants are volatilized from the solid. Initial decomposition also takes place and some oxidation may occur. The temperature and time in the primary combustor are such that the solid exiting the combustor should be free of organics and can be delisted. In the secondary combustor or afterburner, the hazardous off-gas is decomposed and oxidized to CO_2, H_2O, and halogenated compounds which can be removed in subsequent downstream scrubbing systems. In contrast to hydrocarbons, metal contaminants may remain in the solid phase or be vaporized and recondense as a fume which can be difficult to collect downstream. This review focuses on the primary combustor, in particular on rotary kiln incinerators because they are among the most flexible and wide-spread incineration systems. The fates of both hydrocarbons and metals are examined with a view toward the specification and design of optimized rotary kiln systems for various solid wastes.

Overview Of Rotary Kiln Incineration

Figure 2 illustrates the complex chemical and physical processes which occur when a solid is incinerated in a rotary kiln (Cundy, *et al.* 1989). The solid waste enters the kiln via a continuous screw-feeding mechanism (as in the case of contaminated soil) or in discreet barrels. The solid is heated by conduction, convection, and radiation from the primary flame, bulk gases, and refractory walls. In the case of a sludge or high moisture solid, water is initially driven off. Hydrocarbon contaminants are desorbed and diffuse through the solid bed to the free stream where they may ignite (if the local concentration is sufficiently high and appropriate radicals are available) or they may be transported into the free stream. With high Btu waste, bed burning and/or gas-phase burning immediately adjacent to the bed can provide significant thermal feedback to the solid material. Thus, the primary processes in the thermal treatment of a contaminated solid are:

(1) Bed heating (including moisture removal)
(2) Contaminant desorption
(3) Intra-particle diffusion
(4) Inter-particle diffusion
(5) Bulk transport and possible chemical reaction

It is important to understand which of these steps is rate controlling in a specific, practical application in order to accurately scale laboratory test results and provide overall design guidance. In both new system design and evaluation of existing equipment suitability for a specific waste problem, the critical issue is the minimum time/temperature profile required to achieve a certain level of contaminant evolution. The required solids' temperature and associated residence time then determine the auxiliary fuel firing rate, solids through-put, and kiln operating parameters.

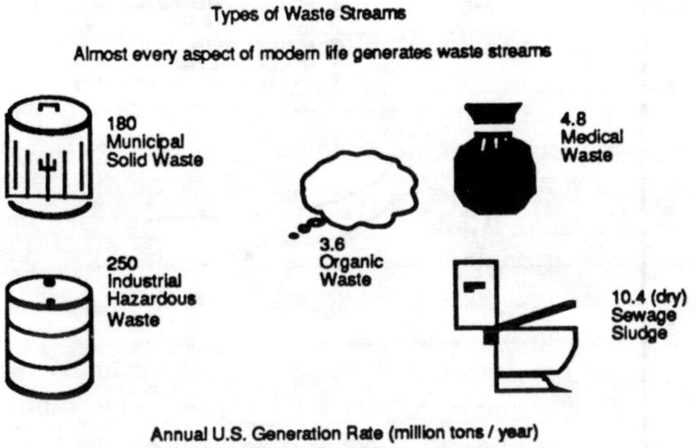

FIGURE 1 Types of waste streams.

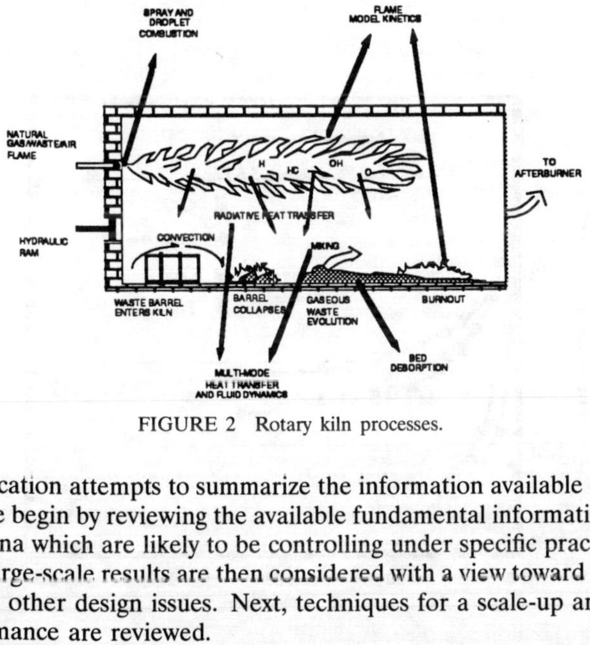

FIGURE 2 Rotary kiln processes.

This communication attempts to summarize the information available on these overall design issues. We begin by reviewing the available fundamental information to characterize the phenomena which are likely to be controlling under specific practical conditions. Pilot-scale and large-scale results are then considered with a view toward practical system complexities and other design issues. Next, techniques for a scale-up and prediction of full-scale performance are reviewed.

Although considerable effort has been directed toward understanding the thermal treatment of solids with hydrocarbon contaminants, relatively little attention has been given toward metal emissions which are potentially equally toxic. The final section of this review summarizes current understanding of metal vaporization and reaction in a rotary kiln environment.

FUNDAMENTALS OF HYDROCARBON DESORPTION

Problem Identification

The effectiveness of a rotary kiln incineration system is directly influenced by the ability

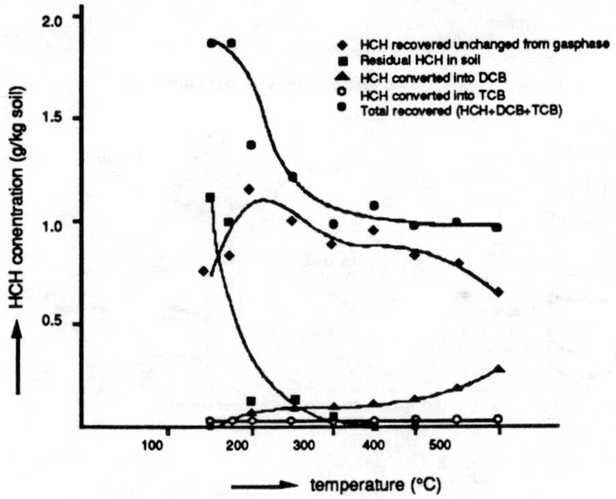

FIGURE 3 Recovery of HCH and HCH derived decomposition products as a function of the oven temperature.

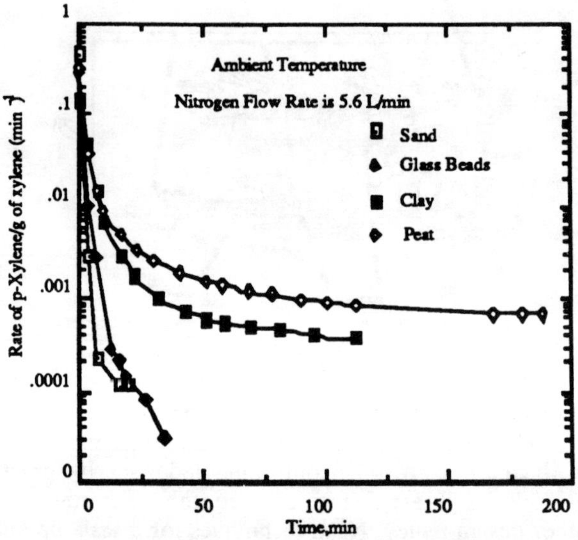

FIGURE 4 Comparison of the desorption rates of porous clay and peat versus nonporous sand and glass beads.

of the kiln to drive the hydrocarbon contaminants from the solid phase into the vapor phase where they can subsequently be destroyed. One of the earliest studies on the desorption of hydrocarbons from contaminated solids was conducted by deLeer and his associates (1985) using an indirectly-heated laboratory reactor with a nominal 30 minutes residence time. Figure 3 shows typical results from their work (Borkent-Verhage, *et al.* 1986) on the recovery of hexachlorocyclohexane (HCH) from humic soil. These data show the influence of thermal environment. When the contaminant was evolved from the sand, with low humic content in an air atmosphere, the HCH evaporated and partially

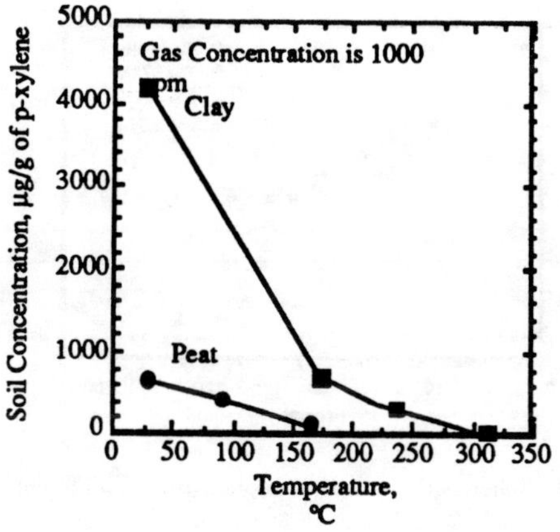

FIGURE 5 Adsorption of peat and clay as a function of temperature for a constant gas-phase concentration.

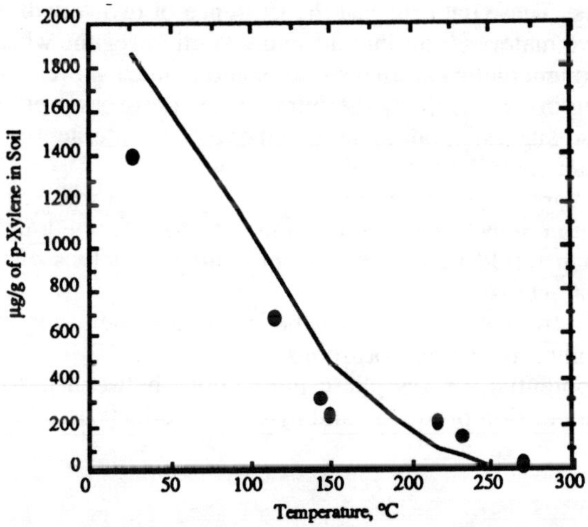

FIGURE 6 Model predictions of concentration of p-xylene remaining on the soil after desorption for one hour compared with experimental results (symbols).

decomposed into other products of incomplete combustion (PIC's) including significant quantities of dechlorobenzene (DCB) and small amounts of chlorinated phenols (which are apparently formed from chlorine which reacts with phenols from the pyrolysis of the humic soil). Since chlorinated phenols are potentially precursors of dioxin, they must of course be subsequently removed. As the temperature increased, the residual HCH in the soil decreased.

Lighty and coworkers conducted similar experiments on the desorption of p-xylene from a variety of materials. Figure 4 summarizes results from their early work (Lighty, *et al.* 1988a) with 0.5 weight percent p-xylene. These data are reported as a normalized evolution rate where the amount of p-xylene evolved has been normalized with respect to the initial amount present. With the nonporous glass beads or sand the evolution rate was initially high and fell off rapidly. With the more porous and reactive peat and clay

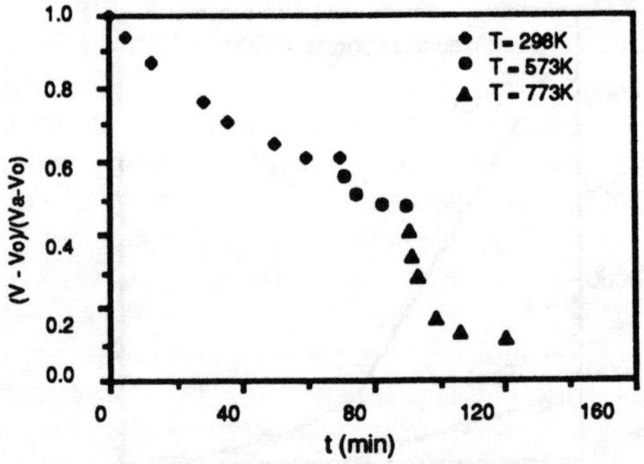

FIGURE 7 Spherocarb-toluene desorption run for different temperatures.

the rate decreased quickly and then, as time progressed, the rate of decline slowed. In addition, with the latter two materials the overall evolution rate was only 80% at 120 minutes while with the sand and glass beads 99.9% of the p-xylene had been evolved prior to 60 minutes. These data suggest the existence of two very different processes for the porous, reactive materials: an initial rapid evolution regime where the contaminant is evolving from nonenergetic external surfaces and a much slower second regime where the rate controlling process is likely the intraparticle desorption process. The existence of the long tail also suggests that contaminant-substrate interactions may be significant when trying to clean contaminated soils to low levels.

Thus, the early work on soil decontamination suggested three key ideas:

- Complete evolution of hydrocarbon contaminants may require longer times and higher temperatures than would be expected from a simple analysis of vapor pressure and boiling point characteristics.
- Contaminant-substrate interactions may be important when trying to remove the final few parts per million of the hydrocarbons.
- There exists a potential for gas phase interactions between desorbed hydrocarbons (and their decomposition products) and pyrolysis products associated with the soils or solids.

Particle Results

Adsorption equilibria. Based on the initial, early data, a number of investigators have proceeded with detailed characterizations of the related desorption phenomena. Lighty, *et al.* (1988a) developed a particle characterization reactor which allowed preheated desorbent gas to be passed through a thin bed of soil at temperatures up to 1200°C. The exhaust gas was analyzed using GC/MS and GC/FID techniques to characterize the detailed evolution rates. These investigators also used the reactor to characterize the process fundamentals by studying the equivalent adsorption processes. Figure 5 illustrates typical adsorption data for p-xylene using peat and clay soils.

In this instance, the gas-phase concentration was 1000 ppm and the equilibrium soil concentrations were measured for temperatures from 25 to 310°C. Even 100°C above the boiling point of p-xylene (139°C) the clay soil still retained more than 200 μg/g of the hydrocarbon with a gas phase concentration of only 1000 ppm. Desorption experiments with the same two materials indicated that after 110 minutes at 160°C, 100 μg/g was left on the soil in the case of peat and 240 μg/g was left with the clay.

To evaluate the overall importance of desorption equilibrium, Lighty and coworkers (1990a) developed a thin bed model for their reactor and predicted soil desorption rates based on their own measurements of adsorption equilibrium. These predictions are compared with their measured desorption rates for p-xylene evolution as a function of temperature in Figure 6 (Lighty, *et al.* 1990a). Here the concentration of p-xylene remaining in the soil after 60 minutes was integrated over the depth of the bed (1.27 cm). The desorption results were in general agreement with the model predictions based on adsorption equilibria measurements although for temperatures below approximately 150°C the model predicted slightly higher concentrations than are actually present. The steep dependence on temperature is consistent the work of Borkent-Verhage, *et al.* (1986) illustrated in Figure 3.

More recently Wu, *et al.* (1991) and Tognotti, *et al.* (1991) have conducted detailed studies of the adsorption/desorption characteristics for organic compounds on soils. Wu, *et al.* derived mass transfer parameters for a variety of chlorinated hydrocarbons. A conventional desorption model developed by Owens, *et al.* (1991b) suggests that the controlling process is likely the intraparticle desorption process which is highly dependent on temperature; however, the desorption and particle diffusion steps were lumped together and further work is needed to elucidate the relative importance of these mechanisms.

Figure 7 summarizes data from Tognotti and coworkers at MIT (1991) for the transient desorption of toluene from a Spherocarb particle originally exposed to a saturated nitrogen toluene stream at 25°C. These data were obtained in an electrodynamic thermogravimetric analyzer with laser heating. After one hour of exposure to a stream of dry nitrogen at 25°C, 60% of the amount originally absorbed on the particle remained trapped in the porous solid. Laser heating of the particle to 300°C resulted in a slight decrease in this amount to about 50%, and most of the organic finally desorbed upon heating to 500°C. Since the boiling point of toluene is only 110.8°C, these data again clearly confirm the importance of hydrocarbon substrate binding for trace levels of organic contamination and the clear presence of equilibrium constraints.

Impact of moisture. The results discussed previously have largely neglected the influence of water; however, in the incineration of actual hazardous solid waste materials or the thermal treatment of contaminated soils, natural moisture should play a significant role. Figure 8 summarizes test results from Lighty, *et al.* (1988b) which were conducted in a particle characterization reactor to quantify the influence of moisture on the desorption of p-xylene. Two different cases were investigated relative to the dry soil conditions. First, the clay was dried, contaminated with p-xylene, and then 10% water by weight was added. Alternatively, soil containing 10% natural moisture was contaminated. Figure 8 shows the results of these experiments as a function of desorption time. The amount of p-xylene remaining in the dried, then contaminated soil was the highest. Adding moisture after decontamination actually increased the rate of desorption and reduced the overall residual hydrocarbon level, probably due to "steam distillation effects" during the period of water evaporation. The naturally wet soil produced the lowest final concentration. These results suggest that in addition to steam distillation, water probably replaced some of the sites that would have been occupied by contaminant. In this later case, less p-xylene was able to physically absorb onto the soil and, hence, the contaminant evolved from the soil more easily. All three cases again show clear indication of asymptotic equilibrium levels which are probably associated with binding between organic hydrocarbon and the inorganic clay matrix.

Packed Bed Results

The results discussed previously were obtained in either thin bed or single particle reactors where heat and mass transfer limitations were minimized; however, in any large-scale

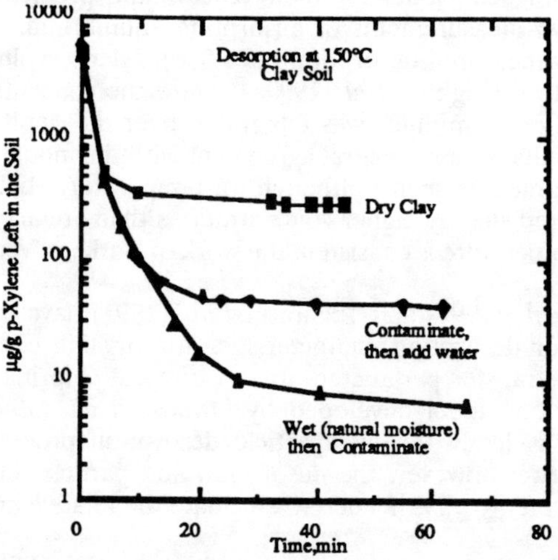

FIGURE 8 Data taken in the particle-characterization reactor showing the effects of moisture on the desorption of p-xylene on clay at 150° C.

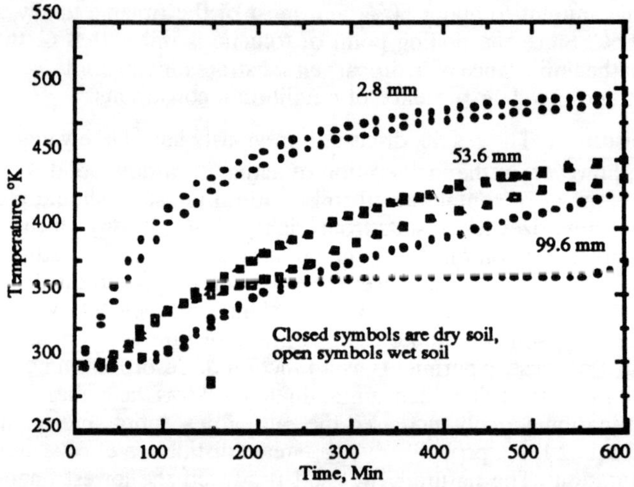

FIGURE 9 Temperature distributions for west (5% by weight and dry clay soil at 523° K desorption.

system, energy must be transported from the solid surface through a packed or slumping bed and the evolved contaminants must diffuse through the bed back to the surface. To investigate these effects Lighty and coworkers (1988a) constructed a bed characterization reactor, a one-dimensional experiment which heats a bed of solid waste from the top. The bed is continuously weighed and the exhaust is sampled using GC/FID and GC/MS techniques. The bed also contains thermocouples which are used to characterize the transport of energy downward from the top. Figure 9 illustrates typical temperature measurements at different vertical locations within the bed. In this instance a 10.2 cm bed of clay soil with 5% moisture was desorbed with a radiant heater temperature of 523 K. Data from similar experiments with dry soil are also shown for comparison. With the dry soil, the temperatures rose steadily at the three axial locations as the thermal wave propagated down through the bed. It is important to note that even with this relatively thin bed (total thickness of 4 inches) the thermal delay was significant. For the wet soil the delay was greatly aggravated by the energy required for the heat of vaporization of water. The bottom temperature rose to the evaporation temperature (approximately 370 K) and then the temperature profile was flat until approximately 550 minutes when it began to rise again. Thus, the presence of the moisture has a dramatic impact on the thermal history.

Surprisingly, however, when the tests were repeated with 1% p-xylene contamination, the average p-xylene remaining in the soil was found to be 240 μg/g for the dry clay and less than 5 μg/g for the soil containing the moisture. Thus, even though the moisture depresses the temperature profile throughout the bed, this effect is more than compensated by the steam distillation benefits associated with the water.

To characterize contaminant evolution for the complex multicomponent hydrocarbon mixtures typically found in contaminated soils, Lighty, *et al.* (1990b) studied actual contaminated soils from a field site where a coal gasification plant once stood. The complex organic contaminants were quantified using on-line, short column chromatography in conjunction with mass spectrometry (GC/MS) McClennen, *et al.* (1990). Figure 10a (Lighty, *et al.* 1990b) shows typical results from these experiments at a desorption temperature of 500°C with a 2.5 cm thick bed. Two different tests are shown to indicate overall reproducibility of the results. The thermal distribution data for the bed were essentially identical for test one and test two, however the hydrocarbon data (Figure 10a) indicate that while the trends reproduce nicely, the total hydrocarbons evolved for test one were 27% higher than for test two. This difference was believed to result from differences in the homogeneity of the soil from sample to sample. At the conclusion of both runs the concentration of PNA's left in the soil was below 0.1 μm/g.

Figure 10b shows the corresponding detailed GC/MS data. The naphthalene peak (molecular weight 128 and boiling point 218°C) was much higher than the other components as expected based on the initial distribution of hydrocarbons in the soil. The bimodal distribution of total hydrocarbons indicated in Figure 10a was clearly due to the evolution of phenanthrene and anthracene (boiling point 340°C, mass 178). This second peak, which occurred at 40 minutes, was after the evaporation of water had been completed and the bottom of the bed was continuing to rise in temperature; the first peak appears to be associated with steam stripping and organics from the top portions of the bed. These data illustrate the complexity of multi-component hydrocarbon evolution in an actual field system.

Closure—Hydrocarbon Fundamentals

In summary the fundamental hydrocarbon evolution results indicate several general characteristics which must be considered when designing a large-scale system:

- The bulk of the hydrocarbon contaminants can generally be relatively easily evolved at temperatures only slightly above the boiling point of the specific hydrocarbons.
- Depending on the substrate composition and porosity, the final trace amounts of hydrocarbon contamination may be difficult to remove because of binding between the hydrocarbons and the inorganic matrix and/or substantial intraparticle mass transport limitations. This behavior dominates the time and temperature required to ensure complete cleanup and can best be estimated from adsorption data for the specific system of interest (not from boiling point information).

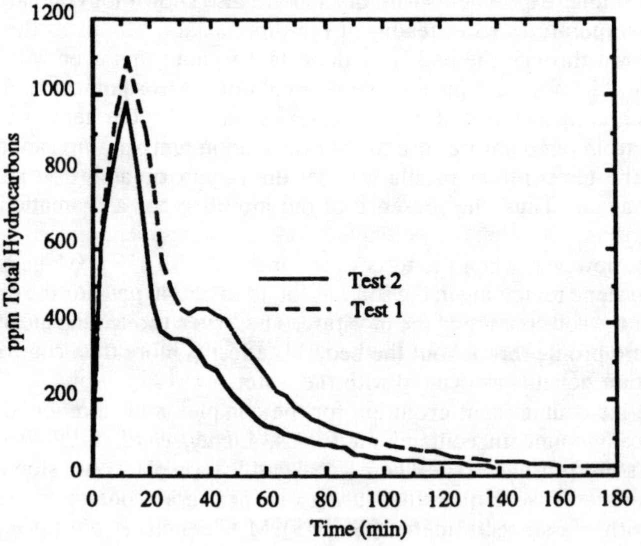

FIGURE 10a Total hydrocarbon data for soil A at 400°C in the BCR—Test 1 and 2.

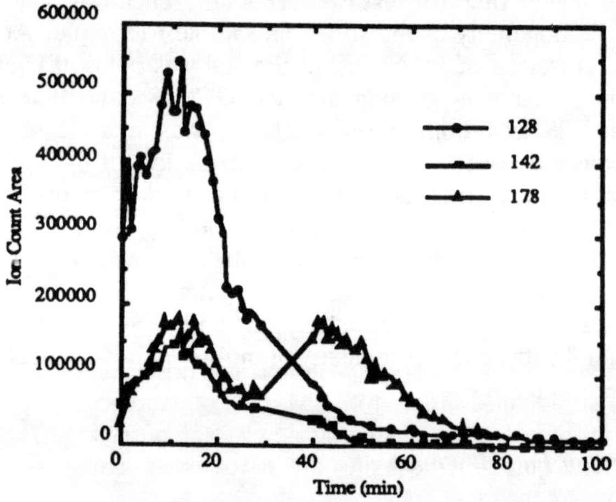

FIGURE 10b MINITMASS data for soil A desorbed at 400°C in the BCR.

- Complex mixtures of hydrocarbons can be expected to evolve selectively in relative correspondence to their volatility in association with the specific solid matrix.
- The influence of moisture on the contaminant evolution is complex. Typical moisture levels can significantly reduce the heating rate of the solid phase because of the large latent heat of vaporization of water; however, this effect may be partially or totally mitigated by the beneficial stripping and the convective flux which the associated steam produces.

- The overall hydrocarbon evolution process is likely to be heat-transfer limited in any full-scale system; hence, characteristic particle and bed sizes have an important influence on the evolution rate.

PILOT-SCALE STUDIES

The previous sections have focused on generic, fundamental experimental and theoretical results that are generally applicable to a wide variety of different types of thermal destruction systems; in this section we turn our attention to the specific case of a rotary kiln incinerator and examine the available pilot-scale data to characterize the controlling phenomena.

Much of the publicly available pilot-scale data comes from 24 to 30 inch diameter rotary kiln simulators located at the Energy and Environmental Research Corporation, Irvine, California; the University of Utah, Salt Lake City, Utah; and the EPA Combustion Laboratory at Research Triangle Park, North Carolina. Large, pilot-scale test facilities are also available at various industrial locations including IT (Tulsa), DB (Germany), and AAB Raymond (Chicago), as well as the EPA test facility in Pine Bluff, Arkansas.

In a typical commercial kiln, a control volume of solid charge moves progressively from the entrance to the exit in a predetermined residence time based on the rotation speed, angle, and length of the kiln. In the rotary kiln simulators, the charge does not move axially. Rather, the simulator is designed to follow the fate of a control volume of solid by replacing the variable of distance with time; thus, it is possible to directly characterize the evolution of hazardous components as they "pass through" a kiln. In the simulators, a batch of solid is loaded into the kiln and combusted for a predetermined residence time, corresponding to the time required for the solid to flow from the entrance to the exit of the commercial unit being simulated.

Figure 11 illustrates the details of a typical rotary kiln simulator (from the University of Utah). This is a 73 kW natural gas fired facility constructed in three refractory-lined steel sections with the core of the main section being 61 cm in diameter and 61 cm long. The burner and the main section rotate from 0.5 to 2.0 rpm while the exhaust section remains stationary. Solid-phase residence time in the kiln can be varied from a few minutes to six hours depending upon the experiments. The variable swirl burner fires axial and radial gas and can produce kiln temperatures up to 1650°C. After the kiln, the gas passes through a flue where the continuous gas measurements are taken using two probes. A quartz lined stainless steel probe is used for hydrocarbon analysis including total hydrocarbons, GC/FID, and GC/MS. A 304 stainless steel probe is used for oxygen, CO, CO_2, and NO_x. After the sampling flue, the gas enters a refractory-lined afterburner which contains two opposed natural gas fired burners, each rated at 73 kW.

Pilot-scale data are ideal to illustrate the complexities in actual, full-scale kiln systems which must be considered when designing new systems or evaluating the suitability of existing units for new wastes. Five issues are particularly significant:

- Thermal environment
- Waste composition
- Transient charging
- Bed stirring (mixing)
- Moisture

We will examine these issues by means of specific examples.

Thermal Environment

The fundamental results presented in the previous section demonstrated the importance of particle temperature. Pilot-scale testing has confirmed this conclusion and demonstrated

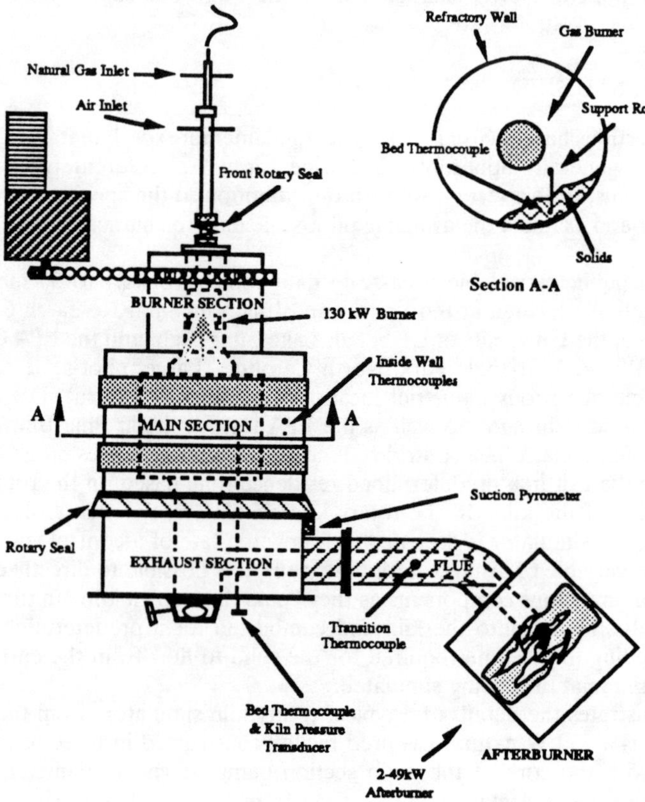

FIGURE 11 Top View of rotary kiln simulator.

that heat transfer is often the rate limiting step in the thermal treatment process Figure 12 shows typical pilot-scale results obtained by Owens, *et al.* (1991a) showing the effect of kiln wall temperature on the heat up of a bed of sorbent material. At this operating condition (rotation rate of 0.5 rpm and 8% kiln fill fraction), slightly more than 10 minutes was required for the solid to reach its asymptotic temperature value. (The lines represent model predictions which will be discussed in a subsequent section.)

Figure 13 shows the comparable results for toluene desorption (Owens, *et al.* 1991b) and demonstrates the strong influence of kiln temperature on the characteristic desorption time. These data suggest that the toluene can be largely removed in approximately five minutes at the high temperature, but nearly twenty five minutes are required to clean up the same sample at 189°C.

Waste Composition

Pilot-scale testing is often used to help evaluate the suitability of incineration for remediation of both contaminated sites and the disposal of hazardous solid waste. Tillman, *et al.* (1991b) report the results of a study which was conducted for the Bogg Creek Farms CERCLA (Superfund) site in New Jersey. Treatability studies were conducted by

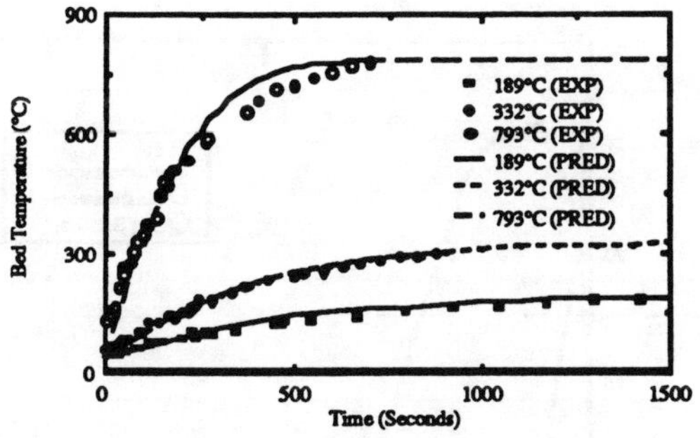

FIGURE 12 Influences of kiln operating temperature on sorbent bed temperature at constant rotation rate of 0.5 rpm and constant fill fraction of 8%.

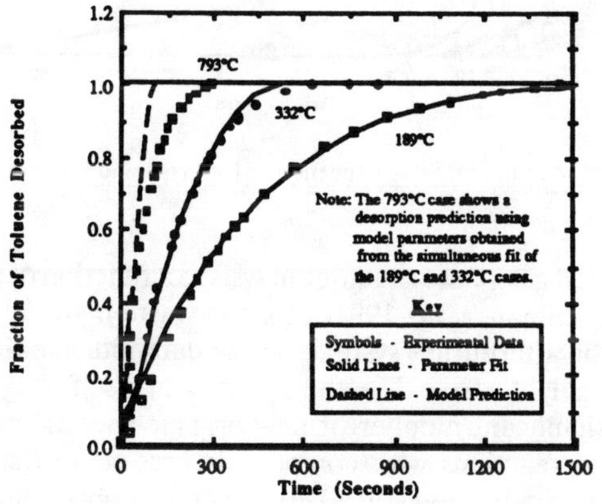

FIGURE 13 Simultaneous parameter fit at 0.5 rpm and 8% fill fraction with kiln wall temperatures ranging from 189°C to 793°C.

Secker and coworkers at EER to define the conditions required for cleaning the soil. Figure 14 (Tillman, *et al.* 1991b) shows the effect of thermal treatment on the carbon content of samples from different site locations. The initial contamination varied greatly and even materials from the same general site locations had variations of 2.5 to 25% in carbon content. This figure illustrates the complexity of the desorption process in a rotary kiln environment. Upon heat treatment at 1460°F the waste samples and the soil lost carbon very rapidly; in the first five minutes their carbon content was reduced to less than 1% and at 30 minutes it was below 0.1%. However, the Bogg sediment carbon content increased initially and subsequently fell, but even after 60 minutes the carbon content was still 0.6%. This delay is associated with the high moisture content of the Bogg sediment. The initial apparent increase in carbon is simply due to the loss of water (without an associated carbon loss). However, even during the initial drying regime (approximately 30 minutes) there was some carbon evolution which is believed to be associated with steam stripping as will be discussed subsequently. After thermal treatment for 60 minutes all hazardous organics were removed from the waste samples to below the detection limit of approximately 600 ppb. This removal occurred even at temperatures as low as 1000°F.

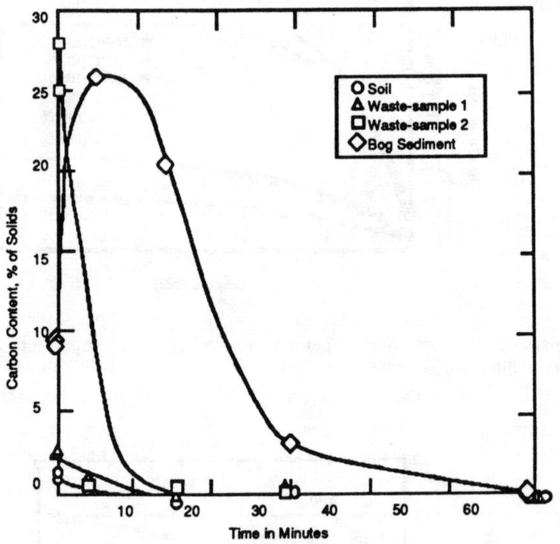

FIGURE 14 Carbon content of the solids as a function of feed type and residence time with kiln temperature at 1460° F.

The results of detailed gas-phase organic analysis as a function of temperature are shown in Figure 15 and 16 (Tillman, *et al.* 1991b) for the waste material. These measurements were taken prior to the afterburner system. These data indicate that the primary organic constituents were largely destroyed in the rotary kiln itself. Figure 16 does, however, demonstrate that a significant number of new organic species are produced in the kiln zone via cracking and reactions with compounds present in the waste sample. These products of incomplete combustion (PICs) are generally polyaeromatic hydrocarbons such as pyrene, phenanthrene, and chrysene. Both the remaining original organic constituents and the PICs would need to be destroyed in a conventional downstream afterburner.

Overall these results demonstrate (1) the utility of pilot-scale testing for characterizing the thermal conditions required to ensure complete soil cleanup, (2) the complexity of actual waste testing, and (3) the potential of PIC generation under realistic combustion conditions.

Transient Charging

Another complexity of rotary kiln incineration occurs since drums are often charged in a batch-wise fashion and transient emissions of organics can result. Linak, *et al.* (1987) have conducted a series of experimental studies in a pilot-scale rotary kiln at EPA, RTP which focused on characterizing the conditions under which transient puffs are produced and on determining the hazardous constituents present in the exhaust gases during such transient upset conditions. The formulation of a mathematical model (Wendt and Linak, 1988) allowed some of the fundamental physical phenomena which are important to the generation of these puffs to be delineated. In a recent paper (Lemieux, *et al.* 1990) these investigators have completed this work by exploring the potential of operating changes for minimizing transient emissions. Figure 17 (Lemieux, *et al.* 1990) summarizes the influence of temperature and kiln rotation rate on potential transient emissions. These

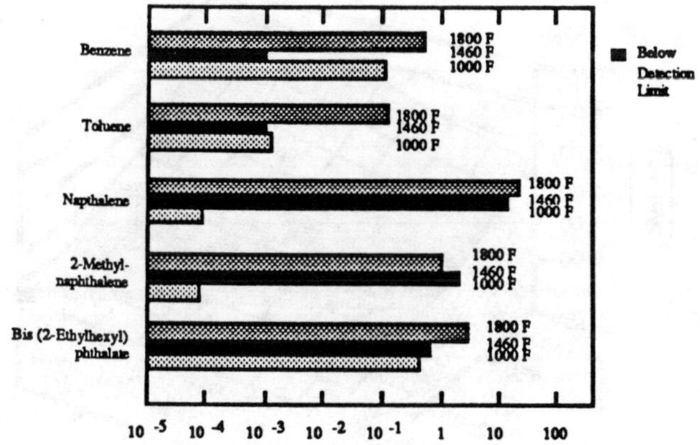

FIGURE 15 Fate of organics in kiln feed, measured in the kiln off-gas, as a function of temperature.

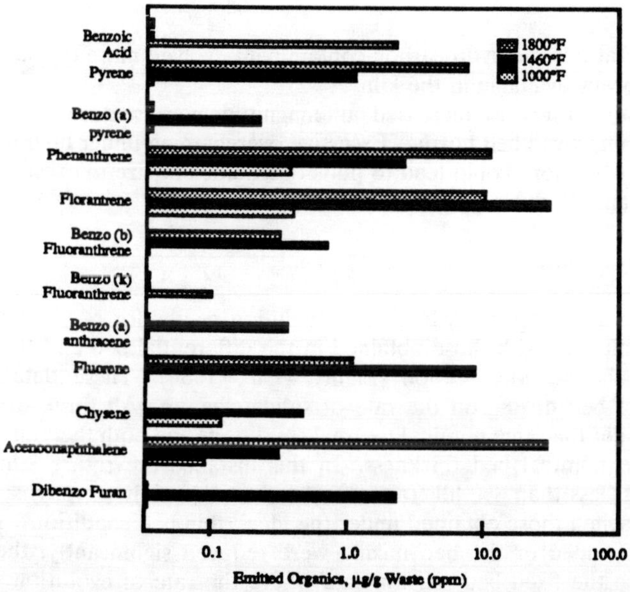

FIGURE 16 Organics measured in the kiln off-gas but not present in the waste stream, as a function of kiln temperature.

results are plotted in terms of a puff magnitude ratio, based on the measured particulate filter weight and normalized to a standard operating condition of 1250 K, 0.5 rpm, and 100 grams of toluene waste added to the sorbent. Their results indicate that transient puffs can be minimized by operating the rotary kiln at low rotation speeds and/or low temperatures. The puff magnitude increased with increasing temperature because at

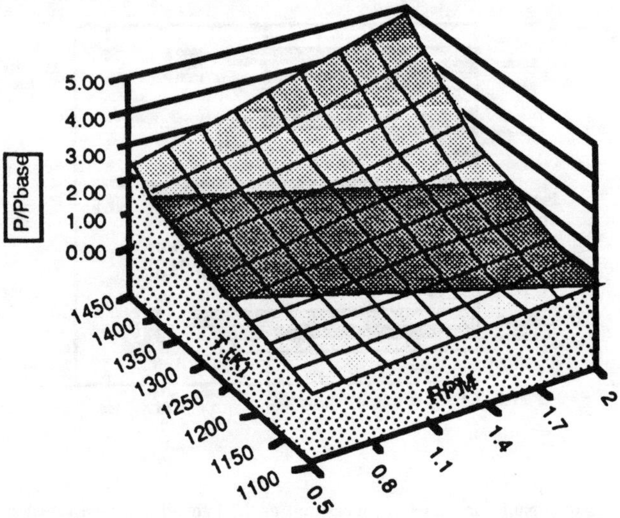

FIGURE 17 Measured effect of Tk and RPM on pull magnitude normalized to base case conditions.

higher temperatures the hydrocarbon contaminant evolved more quickly and consumed the oxygen locally available in the kiln.

Higher rotation rates also increased puff magnitude because devolatilization rates were enhanced by improved bed mixing. Excessive overcharging under high temperature and rapid mixing conditions could lead to puffs of a sufficient size to exceed the capacity of the downstream afterburner system.

Influence of Bed Mixing

Figure 18 compares typical hydrocarbon evolution results for p-xylene obtained in the rotary kiln simulator with those obtained in the differential bed and thick bed reactors discussed in the previous section (Lighty, *et al.* 1989). These data emphasize the importance of bed mixing on the rate of solids cleanup. All three experiments were conducted under the same nominal thermal conditions and both the kiln and bed reactor had the same nominal bed thickness. In this instance the rotary kiln was relatively lightly loaded (less than 8% fill fraction) and rotated rapidly (1 rpm) so the pilot-scale results approached those obtained under the ideal, thin bed conditions. If the kiln were more heavily loaded or the bed mixing were reduced significantly, the results would then approach the fixed bed reactor data where the rate of evolution was completely heat-transfer limited. To investigate the role of bed mixing with typical kiln constraints Owens, *et al.* (1991a,b) conducted detailed studies on the influence of kiln rotation rate and fill fraction.

Figure 19 shows the effect of rotation rate on bed temperatures for a 332°C wall temperature and a constant fill fraction of 8%. Increasing the rotation rate from 0.1 to 0.9 rpm reduced the time required for the bed to reach 95% of asymptotic temperature from 22 min to 12.5 min. At this temperature (332°C), the controlling heat-transfer mechanism, as shown in the inset, is wall-to-solid convection. Thus, it is reasonable that rotation rate should have a strong influence on bed heat up. Similar data were also

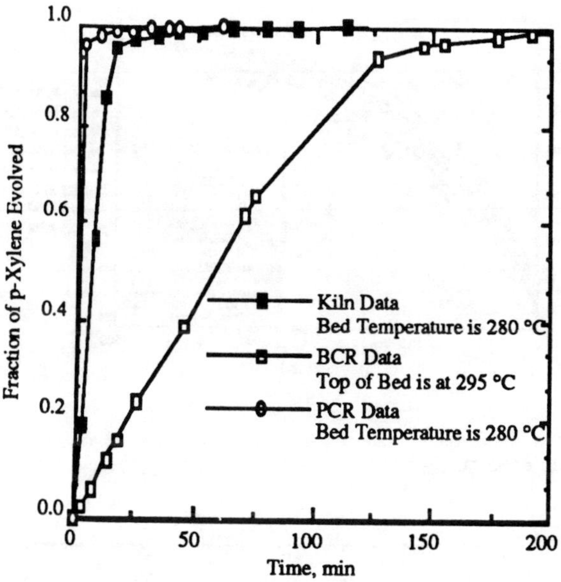

FIGURE 18 Comparison of results from the three reactors at a desorption temperature of ~280°C.

obtained at the higher temperature (793°C) where radiation was the dominant mode of heat transfer and, as expected, the effect of rotation rate was less significant.

Figure 20 shows the influence of rotation rate on toluene evolution. Higher kiln rotation settings resulted in faster toluene desorption rates; however, again the dependence on rotation rate was relatively small. Owens and coworkers also obtained data regarding the influence of fill fraction on both bed temperature and toluene evolution rates. These results indicated that fill fraction did not have a dramatic influence on rotation rate unless the solid contained significant moisture.

Moisture Effects

The fundamental studies discussed previously indicated that moisture has a complex effect on the rate of decontamination because it can dramatically alter the heating rate of the solid phase as well as enhance the hydrocarbon evolution. Figure 21 summarizes a series of tests by Lighty, *et al.* (1991) at a constant wall temperature of 390°C. Figure 21 shows the solid temperature (measured with the thermocouple positioned within the slumping bed) as a function of time after charging for moisture contents of 10, 20, and 30%. The data indicate that the presence of moisture significantly increases the time required for the solid to reach kiln temperature; with 30% moisture the delay associated with the moisture evolution (approximately 800 seconds) far exceeds the required heat-up time (approximately 250 seconds). The dotted lines show model predictions for the measured wall temperature and the agreement is good, except in the 30% moisture case where the model predicted a faster bed heat-up than was experimentally observed. This appears to be because the 30% moisture case was sufficient to drop the overall wall temperature; the model predictions with a 340°C wall agree much more favorably with the experimental data. Figure 22 illustrates the hydrocarbon evolution results which are initially somewhat surprising in light of the bed temperature measurements.

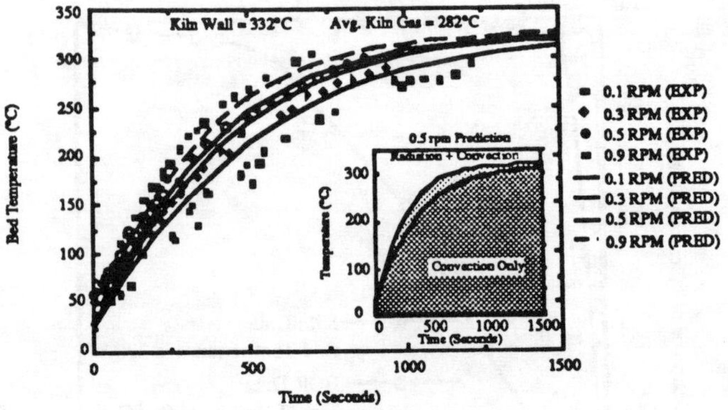

FIGURE 19 Influence of kiln revolution rates on sorbent bed temperature at constant fill fraction of 8% and kiln wall temperature of 332°C.

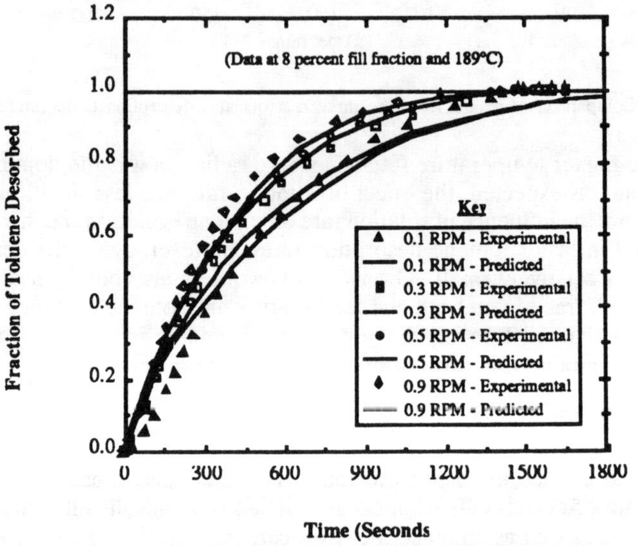

FIGURE 20 Predicted effect of rpm at 189°C using model parameter fit at 0.5 rpm and 8% fill fraction.

As Figure 22 indicates, the toluene actually evolved *faster* with the low moisture levels than with the dry solid in spite of the less favorable thermal profile. This must be the result of steam stripping during the low temperature water vaporization process; a significant fraction of the hydrocarbon is actually evolved prior to the solid phase being heated above 100°C. Ultimately, at the high, 30% moisture condition, the negative impact of the thermal profile becomes dominant and the toluene evolution curve is below that of the dry sorbent material. These data suggest that, as in the fundamental studies, the influence of moisture is a trade-off between a negative impact on the solid heating and beneficial steam stripping.

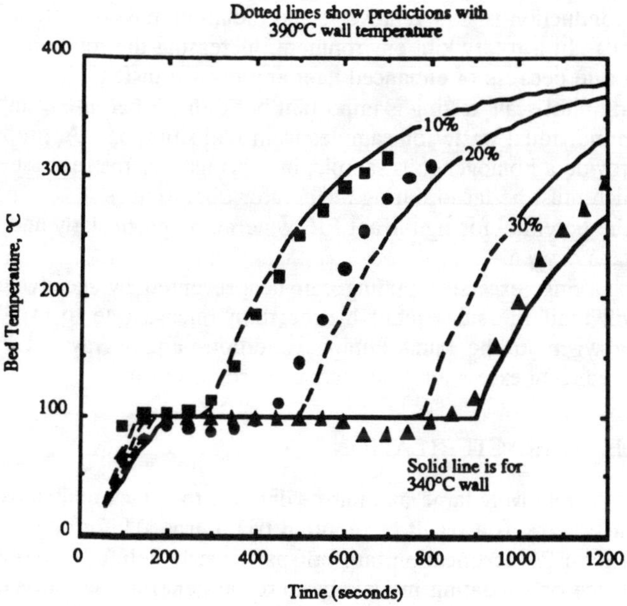

FIGURE 21 Predictions of bed temperature and waste evolution for 10, 20, and 30% water concentration.

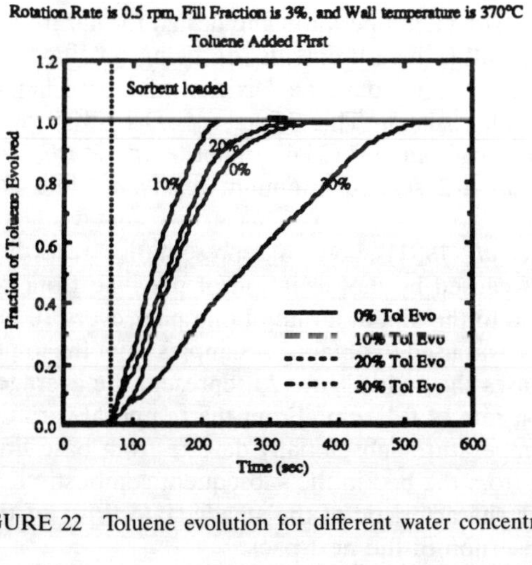

FIGURE 22 Toluene evolution for different water concentration.

Closure

In summary, the pilot-scale experimental results suggest several general ideas which should be incorporated into full-scale design considerations:

- Solids temperature is by far the most important parameter in defining the overall rate of decontamination because it controls the rate of desorption.

- Bed mixing is important because it controls heat transfer which will usually be overall rate limiting in a practical system. Static beds produce inherently poor performance because bed conduction is slow relative to the radiation and convection in a kiln (or a fluidized bed). In a rotary kiln environment, increasing the rotation rate increases the evolution rate because of enhanced heat and mass transfer.
- Laboratory and pilot-scale testing is important but difficult because of inherent variations between individual waste/soil samples from real situations. Appropriate sample mixing can provide a homogeneous sample, but this ignores the real issue of inherent variability which must be faced during incinerator operation.
- There exists the potential for significant PIC generation, particularly under conditions of limited excess oxygen.
- With batch charging, excessive puffing can be prevented by specifying a charging rate which will limit the subsequent hydrocarbon release rate to avoid consuming all available oxygen in the kiln. Puffing conditions are aggravated by increasing temperature, reducing excess air, and increasing rotation rate.

LARGE-SCALE CHARACTERIZATION

There now exists a relatively large amount of data on the stack emissions from rotary kiln incineration systems as a result of required trial burns. Unfortunately, these data, although useful for EPA certification, are not particularly helpful in terms of defining system performance or validating models because, in general, the emissions are at or below the detection limits and internal measurements (prior to the afterburner) are almost always lacking. However, Professors Lester, Cundy, Sterling, and their coworkers at LSU are conducting a pioneering study (Lester, *et al.* 1991) involving flame zone probing of a full-scale rotary kiln. The overall goal of the program is to provide a better understanding of the fundamental processes that occur during rotary kiln incineration with a view toward improving kiln operation and design for hazardous waste destruction. Gas samples and temperature measurements are being obtained during the operation of the incinerator facility at the Louisiana Division of Dow Chemical Company USA located in Plaquemine, Louisiana. This 3.2 meter ID by 10.7 meter long kiln has a design capacity of 17 mw with an 800°C kiln outlet temperature. The 7 MW secondary combustion chamber has a 2 second minimum residence time and a 1000°C outlet temperature.

Figure 23 (Lester, *et al.* 1991) shows a representative recording of the CO_2 and temperature transients caused by the evolution of p-xylene from contaminated plastic packs which were added to the kiln at a rate of one pack every 10 minutes. A 7.8 meter long water cooled probe was used to obtain gas samples from the upper and lower regions of the kiln. The responses shown in Figure 23 represent the average of 6 packs fed into the kiln with a rotation rate of 0.1 rpm. From the temporal variation shown it is clear that both CO_2 and temperature substantially deviate from base line conditions due to the p-xylene evolution from the bed in the subsequent combustion. They then return to a baseline value that is characteristic of the steady state firing of the natural gas/air in the kiln prior to the insertion of the next pack.

These data also indicate that there are significant differences in both the temperature and composition of the upper and lower regions of the kiln. Previous work by Cundy, *et al.* (1989) also indicated this effect during CCl_4 incineration. These major composition and temperature variations appear to be the result of a non-uniform auxiliary burner pattern on one side of the kiln and large leaks associated with the drum injection door on the other.

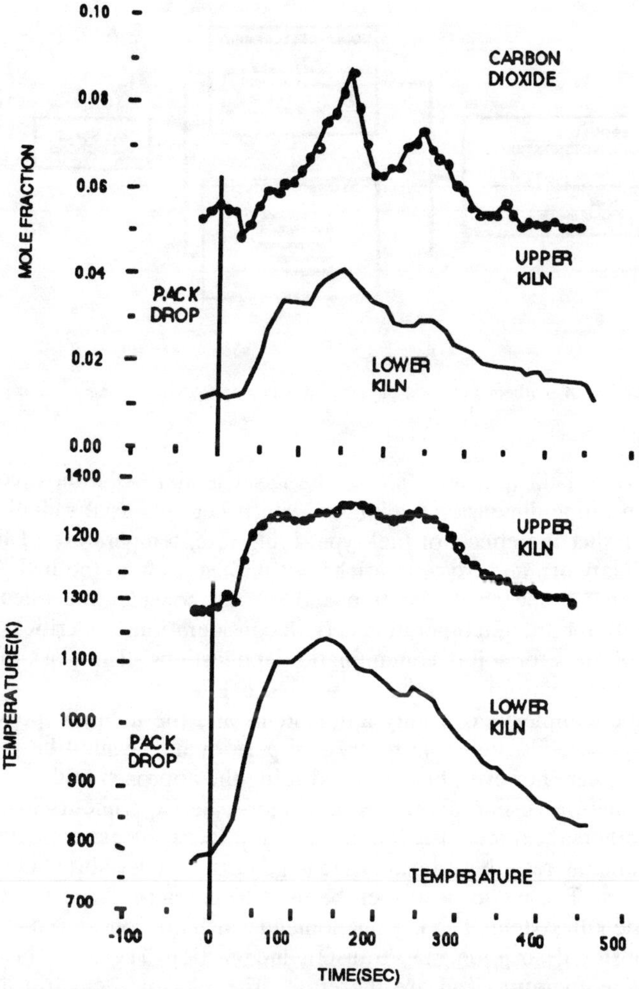

FIGURE 23 Temperature and species distribution after pack charging.

ROTARY KILN MODELING

Analysis Concept

Because of the industrial importance of kilns, numerous investigators have developed heat transfer models. Recent models include those of Jenkins and Moles (1981) and Gorog, Brimacombe, and Adams (1981, 1982, 1983). Jenkins and Moles focus on predicting gas and refractory temperatures using a modification of the method of Hottel and Sarofim (1967). Their results show encouraging agreement with cement kiln measurements but they do not include burden temperature predictions in their calculations.

In a series of three papers, Gorog *et al.* examine the radiative heat transfer between the freeboard gas and the interior surfaces (1981), the overall heat transfer to the solid burden,

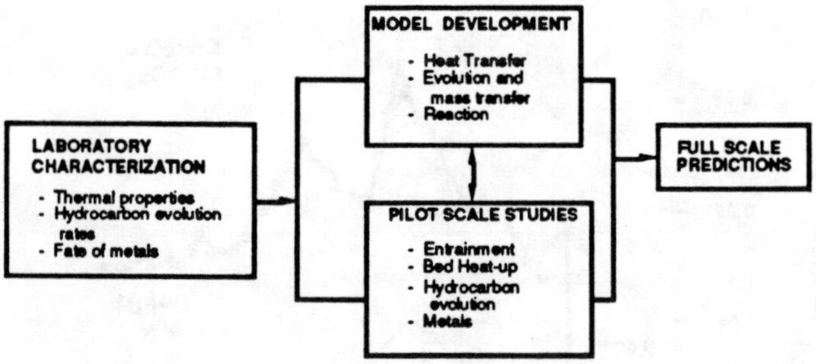

FIGURE 24 Overview of mathematical and physical modeling approach showing relationship to bench and pilot-scale testing.

and the flame zone heat transfer (1983). The latter model is the most comprehensive and is based on a one-dimensional approach which neglects axial radiation. They use the model to predict the effects of fuel type, firing rate, temperature of the secondary air, percent primary air, and oxygen enrichment on heat flow to the bed. They assume an isothermal bed in the axial direction and do not consider the effects of burden moisture content on burden temperatures. Both considerations are critical to the design and operation of kilns for soil decontamination applications. The work of Gorog *et al.* formed the starting point for the model discussed below.

Because of the complexity of rotary kiln systems and the multicomponent nature of most contaminated solids, full-scale rotary kiln performance cannot be predicted from first principles. It can, however, be estimated using the approach outlined in Figure 24. This approach involves using laboratory and pilot-scale experiments in an interactive fashion with mathematical models. The models help focus the experiments on essential measurements and the resulting data are used to guide model development and to provide model parameters. The models can then be used to predict full-scale performance for the specific waste/kiln system. The key phenomena which are considered in the modeling include heat transfer, desorption, mass transfer, and reaction. The latter three phenomena are sensitive to temperature and are therefore driven by the heat transfer. The heat transfer is only affected by mass transfer and desorption if the amount of hydrocarbons entering the kiln with the feed is significant relative to the fuel input.

Model Formulation

The authors have developed an overall rotary kiln simulation based on a heat transfer model described in Silcox and Pershing (1990), Silcox, *et al.* (1991), and Owens, *et al.* (1991a). The heat transfer modeling concept is shown schematically in Figure 25.

The gas phase is represented by a series of completely mixed zones. The resulting model is simply a one-dimensional, long furnace model as described by Hottel and Sarofim (1967) including wall convection as recommended by Tscheng and Watkinson (1979). The flame length and flame heat release rate are specified by the user. The movement of the solids is represented by simple plug flow which allows for temperature and concentration gradients in the axial direction, but which considers the bed to be completely mixed in the radial direction. The solids flow direction can be parallel or

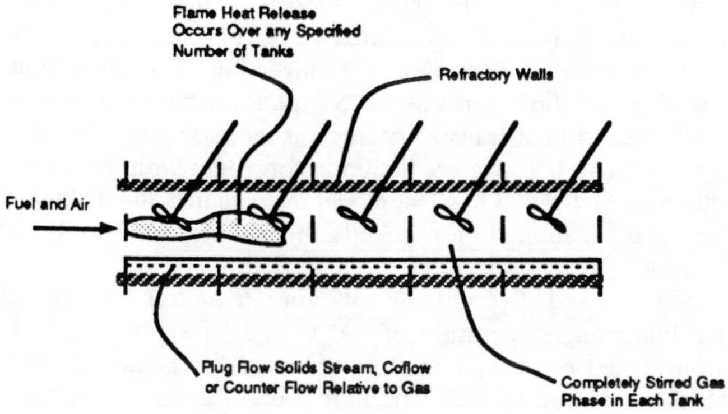

FIGURE 25 Schematic of one-dimensional kiln heat transfer model.

counter to the gas and the kiln can be directly or indirectly fired. The vaporization of moisture and/or hydrocarbons from the bed are included in the energy balances.

The overall simulation includes heat transfer, energy balances, transient behavior, volatile hydrocarbon emissions, and hydrocarbon removal at low levels. Hydrocarbon evolution rates at high solids contamination levels are controlled by the rate of heat transfer to the bed. At low contamination levels, a detailed mass transfer model (Owens, *et al.* 1991b) accounts for the following phenomenon: (1) bed slumping, (2) diffusion within the bed, (3) advection of contaminant from the bed, and (4) the generation of vapor phase contaminant within the bed. The latter is described by a simple first order expression with a rate constant that varies in an Arrhenius fashion with temperature, and with an activation energy that varies linearly with the adsorbed species concentration. At the small, pilot-scale conditions examined by Owens, *et al.* (1991b), the generation term, in conjunction with heat transfer predictions, completely described the observed toluene evolution rates.

Comparison With Pilot-Scale Results

The heat and mass transfer models described have been compared with the pilot-scale data of Owens, *et al.* (1991a,b). These results were all obtained in the small, pilot-scale rotary kiln shown in Figure 11. The effect of kiln-wall temperature on dry solid temperature profiles is shown in Figure 12 for an 8% fill fraction at 0.5 rpm. The data are closely predicted by the model for both radiation-dominated (790°C) and wall-to-solid convection-dominated (190, 330°C) systems without the adjustment of any parameters.

The fraction of toluene evolved from a slumping bed as a function of time is shown in Figure 13 for three temperatures at 0.5 rpm, a fill fraction of 8%, and an initial toluene concentration of 0.25 weight percent (Owens, *et al.* 1991b). The data were obtained at kiln wall temperatures of 190, 330, and 790°C. The model was fitted to the two lower temperatures as shown by the solid lines. The high temperature data was not included in the fit because at 790°C the toluene chars and the resulting carbon leaves the bed more slowly than would be expected based on the lower temperature results and their correlation. The dashed line shows the ability of the lower temperature results to predict the 790°C data. Measured bed temperatures were used in the fitting procedure, however, the calculated thermal profiles are accurate enough (Figure 12) that the prediction of other results, both thermal and desorptive, are satisfactory.

Figure 19 shows the effects of rotation rate on dry solids temperatures at a kiln wall temperature of 330°C with an 8% fill fraction. Increasing the rotation rate from 0.1 to 0.9 rpm decreases the time to reach temperature by a factor of 1.8. (As shown in the inset in Figure 19, the heat transfer is dominated by convection. The convection only curve was obtained by setting radiation terms in the complete model to approximately zero.) The calculated wall-to-solid heat-transfer coefficient increases by a factor of 1.9 when increasing speed from 0.1 to 0.9; this prediction is consistent with the data as discussed above (a time difference factor of 1.8). The model overpredicts the bed temperature for the slower rotation rates because, at these speeds, the bed may not have been well-mixed as assumed in the model.

The predicted and measured effects of kiln rotation rate on toluene removal are shown in Figure 20 for a kiln wall temperature of 190°C and an 8% fill fraction. The solid lines are predictions based on the parameters obtained for Figure 13. Measured bed temperature profiles were used to make the model predictions. As shown, higher kiln rotation settings resulted in faster toluene desorption rates and this is in good agreement with the model predictions. The unequal spacing of the model predictions for this kiln temperature is due to the variability in the experimental thermal profiles.

Full-Scale Predictions

Shown in Figure 26 are the toluene evolution predictions of Owens, *et al.* (1991b) for a matched set of pilot- and full-scale rotary kiln conditions. Four different cases were evaluated by changing the inner wall temperatures of the kilns. The full-scale conditions were for a 3.6 by 18 m kiln, operating at 0.5 rpm, 5.5% fill fraction, 5 tons/hr, and 0.25 weight percent toluene. The pilot-scale kiln was 0.6 by 0.6 m and operated at the same conditions.

The results from both pilot- and full-scale predictions show an increase in inner wall temperature causes faster toluene desorption rates. At pilot-scale, an increase in temperature by a factor of ≈ 3 (300 to 1000°C) reduces bed clean-up time by a factor of ≈ 4.3 (260 to 60 seconds). Full-scale prediction trends are similar to the pilot-scale results: an increase in temperature by a factor of ≈ 3 (300 to 1000°C) reduces bed clean-up time by a factor of ≈ 5.3 (3700 to 700 seconds).

However, the most important conclusion from Figure 26 pertains to the difference in desorption times between pilot- and full-scale kilns for the *same* given set of operating conditions. Removal rates for pilot-scale predictions are considerably faster than predictions at full-scale conditions due to inherent physical differences in bed size for both scales. The average bed depth at the same fill fraction of 5.46% is 0.043 m (1.7″) for pilot-scale and 0.25 m (10″) for full-scale. The different bed thicknesses greatly affect thermal and mass transport thus effecting overall toluene desorption rates.

These results suggest that considerable caution must be exercised when using pilot-scale desorption results to predict full-scale performance. Pilot-scale data are extremely useful for calibrating a model which can subsequently be used to predict behavior of large-scale systems; however, direct extrapolation of small-scale results can lead to considerable overprediction of actual kiln performance as illustrated in Figure 26.

FATE OF METALS

The previous sections have focused on the fate of hydrocarbon species during rotary kiln incineration; however, there is also much concern about potential emissions of the trace toxic metals that can be found in both combustible and non-combustible fractions of solid wastes (Korzun and Heck, 1990; Pacyna, 1987; Oppelt, 1987).

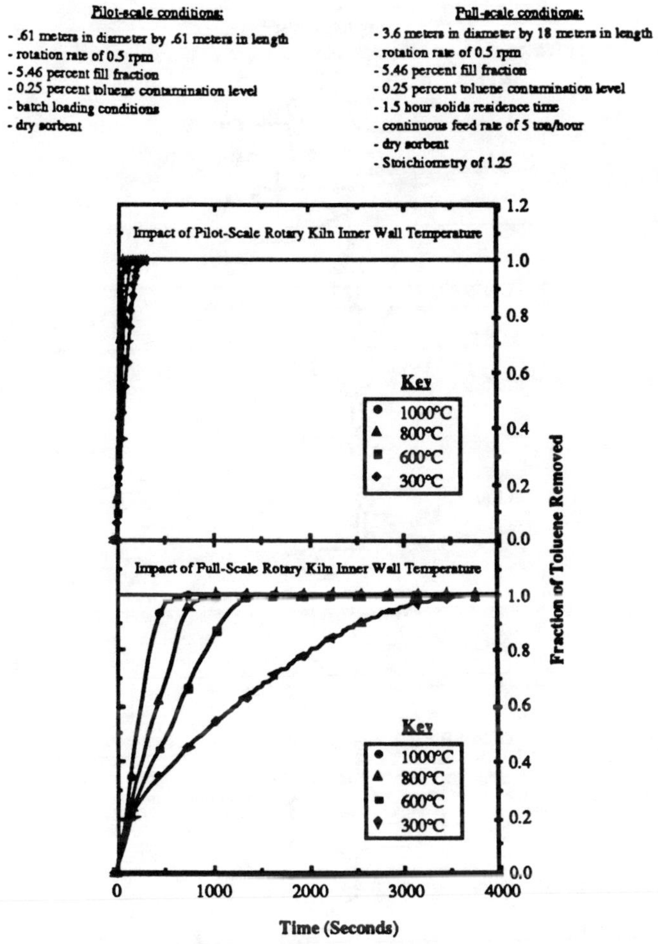

FIGURE 26 Pilot- and full-scale desorption predictions at different kiln inner wall operating temperatures.

Since waste combustion systems do not destroy metals, the ultimate fate of the metals is of concern. Metals that enter the combustion system with the waste exit by any of several pathways (Seeker, 1991). Most of the metals in a waste pass through the incinerator unchanged and are found in the residual ash or bottom ash. Other metals will become entrained in the flue gases and exit the combustion zones as fly ash. Fly ash can be caught subsequently in heat recovery and flue gas cleaning equipment. Finally, in some instances, a toxic heavy metal fume may be formed as a result of vaporization of volatile metal species and subsequent homogeneous nucleation of the supersaturated metal vapors (Germani and Zoller, 1988; Lee, K.C. 1988; Quann and Sarofim, 1982). Further condensation of these metals takes place on this fume due to its high surface area-to-volume ratio leading to highly toxic emissions of concentrated fine particulates which are difficult to capture and have tremendous bioavailability (Denison, 1988; Ulrich, 1978). Minimization of metal vaporization from its host substrate can help limit stack

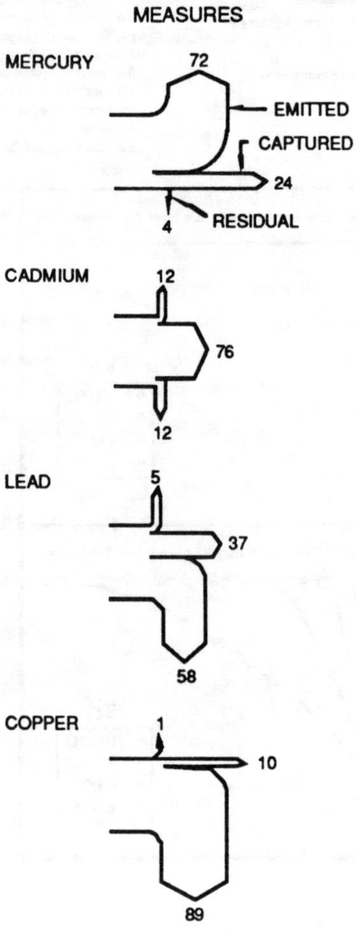

FIGURE 27 Example of the partitioning of four metals in a municipal waste combustion facility.

emissions of toxic heavy metals. The safe design and operation of incineration systems that handle metal bearing wastes are dependent on an understanding of the rates and mechanisms involved in the vaporization of the metal contaminants (Penner, *et al.* 1988; Vogg, 1987).

Some metals can escape the combustion system and flue gas cleaning device and hence, be emitted into the air. The division of the metal is sometimes referred to as the partitioning of the metal since an individual metal will partition itself into the various emissions streams. An example of the partitioning that can occur in a waste combustion process for four metals is shown in Figure 27 based upon detailed studies conducted by Brunner and Monch (Seeker, 1991) on a large municipal waste combustion facility.

Three exit pathways were examined and compared to the amount of each metal fed: residual bottom ash, captured fly ash and emissions. Each metal partitioned itself very differently through the combustion system. Mercury was largely emitted while copper largely stayed with the bottom ash residue. Cadmium and lead were more evenly distributed into the three emissions streams. The following sections will examine the parameters which control this partitioning.

Thermal History

The effect of temperature on the vaporization of metals has been identified as being of tremendous importance in determining the volatility of metal species (Barton, 1990a). A small change in temperature has an exponential influence on the vapor pressure of most toxic metals of interest. For example, Figure 28 shows the enrichment of metals in the fly ash relative to the waste feed concentration after the thermal treatment of a soil from a Superfund site in a pilot-scale rotary kiln (Tillman, *et al.* 1991a). At low temperatures, the metals in the fly ash were of similar concentration as in the untreated soil. After high temperature treatment, many of the metals were highly concentrated in the fly ash. Since the organic removal from the soil takes substantially longer at the lower treatment temperatures, there is a trade-off between organic removal and destruction and metal release as a function of temperature. This tradeoff suggests that an optimum temperature exists, sufficient to remove the organics without significantly releasing the metals.

Gas-Phase Environment

The impact of chlorine on metal partitioning to the fly ash has recently been confirmed in pilot-scale rotary kiln tests by Waterland and associates at the EPA facility in Arkansas and in laboratory tests by Lighty and Eddings at the University of Utah. Figure 29 (Lighty, *et al.* 1990b) shows results from laboratory experiments with PbO. These data demonstrate the influence of gas environment on the fraction of lead evolved from glass beads at 540°C. Losses of lead attributed to loading the reactor were low, on the order of 5% and the experiments were reproducible. Of the gas-environments studies, hydrogen chloride (HCl) had the most significant fraction of the lead oxide reacting to a more volatile species; 82% of the lead partitioned into the gas phase. The product formed was white, in contrast to the initially yellow PbO. Oxygen, argon, and carbon dioxide all behaved similarly where only 10% of the species was volatilized, including the 5% lost as a result of the experimental technique. The beads in these experiments were still yellow supporting the fact that little of the PbO reacted and left the solid. Sulfur dioxide at 540°C exhibited a slightly lower ratio of final to initial concentration, 81% indicating that this environment formed a species which was slightly more volatile than the parent PbO. The beads from this experiment were white in color verifying that some reaction had occurred. The sulfur dioxide experiment was repeated at 20°C and no reaction occurred as evidenced by an absence of color change.

Impact of Substrate Composition

Most attempts to predict the fate of metal species have assumed thermodynamic equilibrium when calculating the partitioning between the solid and gas phases (Tillman and Leone, 1990; Barton, *et al.* 1990a; Lee, C.C., 1988). Although there has been some success using this method, there is experimental evidence that equilibrium calculations might overpredict the amount of metal vaporized (Barton, *et al.* 1990b). Identification of the conditions under which the equilibrium assumption may be used and quantitation of the true vaporization rates for non-complying metals is required for accurate predictions of the fate of metals.

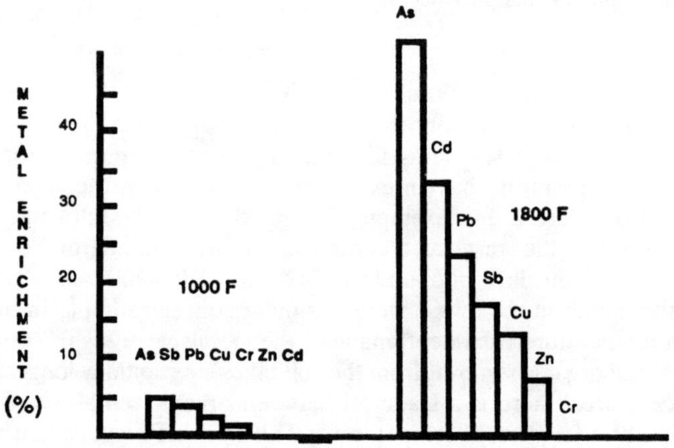

FIGURE 28 Impact of the treatment temperature on the enrichment of metals in the fly ash after the thermal treatment of soils from a superfund site.

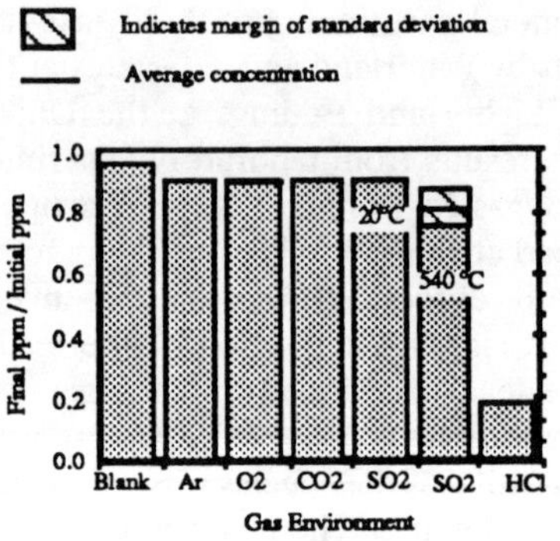

FIGURE 29 Retained fraction of lead as a function of gas environment for argon, oxygen, carbon dioxide, and hydrogen chloride. Metals reactor at 540°C unless specified.

Figure 30 illustrates the results of experiments by Eddings and Lighty (1991) where multiple metals were contaminated on a commercial, clay sorbent with a narrow sized distribution (1–2 mm) particles.

The contamination method involved mixing an aqueous solution of the metals with the sorbent particles and subsequently carefully drying the resultant mixture at 105°C for 24 hours. The particles were then exposed to a simulated combustion gas (including HCl and moisture) for thirty minutes at 538°C. The resulting data are compared to equilibrium predictions. The predictions for lead and copper differed notably from the experimental results. The amount of lead that vaporized was very low; however, equilibrium would predict complete vaporization. The copper results, however, exceeded the level of vaporization predicted by the equilibrium model. The data suggest that either the predicted equilibrium reactions are very slow due to some type of inhibition, or that the resulting species volatility is affected by some other factor. Comparison of Figures 29 and 30 suggests one possible reason for the discrepancy between the predicted and

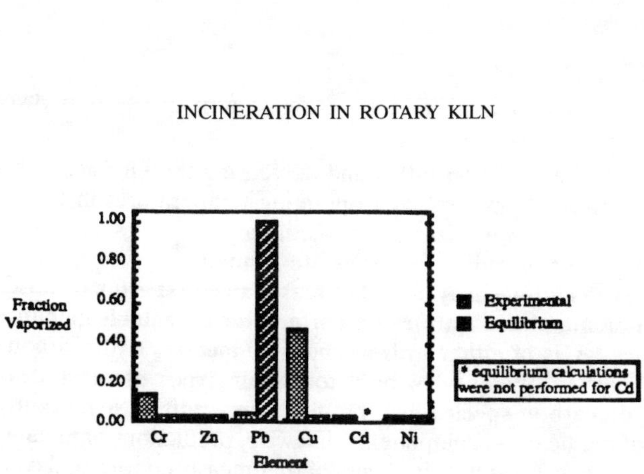

FIGURE 30 Experimental values vs. equilibrium predictions for a thirty minute exposure of contaminated clay sorbent to simulated combustion gas at 538°C.

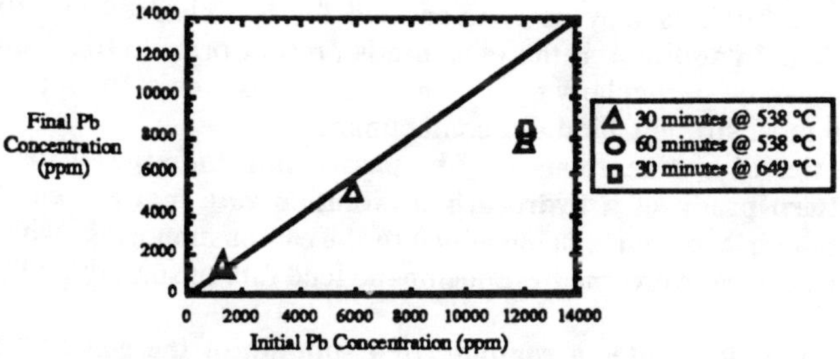

FIGURE 31 Effect of initial contamination level on the vaporization behavior of $PBCl_2$ from a highly porous silica substrate in a gaseous nitrogen environment.

measured volatilization in the case of lead; when the lead was deposited on the glass beads, it was found to be relatively volatile if chloride species were present, but when the lead was deposited onto the clay sorbent particles, little volatility was observed.

To evaluate the impact of substrate and contaminant concentration on relative metal vaporization, Eddings and Lighty (1991) performed experiments with $PbCl_2$ contaminated silica particles for 30 and 60 minutes at two different temperatures. Figure 31 summarizes these results which are plotted as final vs. initial concentration of lead on the silica particles. In all cases, significant vaporization was noted only at high initial concentrations of lead. Increasing either the time or the temperature had little affect on the vaporization behavior as the data demonstrate. The authors speculate that the critical coverage level is associated with a monolayer covering of the most active adsorption sites. Equilibrium calculations for these experimental conditions would predict complete vaporization of the lead chloride at all concentrations; however, the equilibria do not consider interactions with the substrate. Further work is need to verify the authors hypothesis; however, it appears that, like hydrocarbon contamination, interactions between the contaminant and the solid phase can be significant.

CONCLUSIONS

The available information on the thermal treatment of solid wastes and contaminated soils in rotary kilns suggests the following overall conclusions:

1. **Temperature is the most important parameter with respect to the fate of both hydrocarbons and metals.** Heat transfer is, therefore, likely the rate limiting step in many actual systems. Hydrocarbon evolution rates are enhanced by increasing

bed mixing (via higher rotation rates) and decreasing the fill fraction because both enhance bed heat-up. However, high operating temperatures in the kiln can also significantly increase the vaporization of volatile metal species, particularly if, in the presence of HCl, volatile chloride species are formed.

2. **Vapor pressure calculations may provide a reasonable first estimate for the evolution of bulk contaminants; however, they do not provide acceptable guidance regarding the fate of trace levels of either hydrocarbons or metals.** Hydrocarbon and metals both show a marked tendency to bind to certain types of substrates. Complete removal of hydrocarbon species may require temperatures significantly above the boiling point of the heaviest components; likewise, predictions of metal vaporization using equilibrium may be greatly in excess of that measured for small concentrations.
3. **Large kilns may be characterized by highly non-uniform temperature and species concentrations due to their inherent three-dimensional geometry.** The burners may be all on one side of the kiln and charging devices often result in large air in-leakage.
4. **The impact of increasing moisture is complex because it depresses the overall thermal profile, but this may be compensated for by enhanced evolution due to steam stripping.** In addition, if the moisture is present prior to the contamination, hydrocarbons can be particularly easily removed because moisture appears to block sites likely to most strongly bind the contaminants.
5. **Large quantities of hydrocarbons may be passed into the afterburner if the kiln charging pattern produces a hydrocarbon evolution rate that exceeds the rate at which stoichiometric oxygen is being added to the combustion zone.** This is a design problem which can be overcome by reducing the feed rate or smoothing the evolution peaks.
6. **A kiln model that integrates a computerized solution of the governing equations with bench- and pilot-scale experimental data on the specific waste/solid substrate of interest can be used to predict full-scale performance.** Caution must be exercised when extrapolating laboratory and pilot-scale results directly to full-scale conditions because without proper scaling the characteristic resistances can change.

ACKNOWLEDGEMENTS

The authors gratefully acknowledge the assistance of their academic associates Eric Eddings and Brian Keyes at the University of Utah, Vic Cundy and Chris Leger at LSU, Thomas Lester at the University of Kentucky, Adel Sarofim at MIT, and Jost Wendt at the University of Arizona. They also gratefully acknowledge the support of their industrial colleagues Randy Seeker and Bob Barton of EER, Blake Beckstrom of Ford, Bacon, and Davis, Martha Choroszy Marshal of Ciba Geigy, and David Tillman at Ebasco. The funding for much of the work reported herein was provided by the NSF, EPA, GRI, and the Advanced Combustion Engineering Research Center Affiliates and Associates.

REFERENCES

Barton, R. G., and W. R. Seeker (1990a). Affect of System Operation on Toxic Air Emissions. Proceedings of the American Flame Research Committee's 1990 Fall International Symposium. Oct. 8, San Francisco, CA.

Barton, R. G., W. D. Clark, and W. R. Seeker (1990b). Fate of Metals in Waste Combustion Systems. *Comb. Sci. & Tech.* **74,** pp 327-342.

Borkent-Verhage, C., C. Cheng, L. deGalan, and Ed W. B. deLeer (1986). Thermal Cleaning of Soil Contaminated with g-Hexachlorocyclohexane. *Contaminated Soil, Eds.* J. W. Assink and W. J. van den Brink Martinus Nijhoff Publishers, Dordrecht, Netherlands.

Cundy, V. A., et al. (1989). Rotary Kiln Incineration I. An In Depth Study. *JAPCA.* **39,** 63.

deLeer, Ed W. B., et al. (1985). Thermal Cleaning of Soil Contaminated with Cyanide Wastes from Former Coal Gasification Plants. Conference on New Frontiers for Hazardous Waste Management.

Denison, R. A. (1988). The Hazards of Municipal Incinerator Ash and Fundamental Objectives of Ash Management. *AIChE Symposium Series.* **84,** No. 265, 148-153.

Eddings, E. G. and J. S. Lighty (1992). Fundamental Studies of Metal Behavior During Solids Incineration. Submitted to *Comb. Sci. and Tech.* **85,** 375-390, 1992.

Germani, M. S., and W. H. Zoller (1988). Vapor-Phase Concentrations of Arsenic, Selenium, Bromine, Iodine, and Mercury in the Stack of a Coal-Fired Power Plant. *Environ. Sci. Technol.* **22,** 1079-1085.

Gorog, J. P., J. K. Brimacombe, and T. N. Adams (1981). Radiative Heat Transfer in Rotary Kilns. *Metallurgical Transactions B.* **12B,** 55.

Gorog, J. P., T. N. Adams, and J. K. (1982). Brimacombe, Regenerative heat transfer in rotary kilns, *Metallurgical Transactions B.* **13B,** 153.

Gorog, J. P., T. N. Adams, and J. K. (1983). Brimacombe, Heat transfer from flames in a rotary kiln, *Metallurgical Transactions B.* **14B,** 411.

Hottel, H. C. and Sarofim, A. F. (1967). Radiative Transfer, McGraw-Hill, New York.

Jenkins, B. C., and F. D. Moles (1981). Modeling of Heat Transfer From a Large Enclosed Flame in a Rotary Kiln. *Trans. IChemE.* **59,** 17.

Korzun, E. A., and H. H. Heck (1990). Sources and Fates of Lead and Cadmium in Municipal Solid Waste. *J. Air Waste Manage. Assoc.* **40,** No. 9, 1221.

Lee, C. C. (1988). A Model Analysis of Metal Partitioning. *JAPCA.* **38,** No. 7, 941-945.

Lee, K. C. (1988). Research Areas for Improved Incineration System Performance. *JAPCA.* **38,** No. 12, 1542.

Lemieux, P. M., W. P. Linak, J. A. McSorley, J. O .L. Wendt, and J.E. Dunn (1990). Minimization of Transient Emissions from Rotary Kiln Incinerators. *Combust. Sci. and Tech.* **74,** 311.

Lester, T. W., et al. (1991). Rotary-Kiln Incineration Comparison and Scaling of Field-Scale and Pilot-Scale Contaminant Evolution Rates from Sorbent Beds. *Environ. Sci. Technol.* **25,** 1142.

Lighty, J. S., D. W. Pershing, V. A. Cundy, and D. G. Linz (1988a). Characterization of Thermal Desorption Phenomena for the Cleanup of Contaminated Soil. *Nuclear Chem. Waste Management.* **8,** 225.

Lighty, J. S., D. W. Pershing, V. A. Cundy, and G. D. Silcox (June 19—24, 1988b). 81st Annual Meeting of Air Pollution Control Association. Dallas, TX, Air Pollution Control Association: Pittsburgh, PA, Paper 88-17.5.

Lighty, J. S., G. D. Silcox, D. W. Pershing, VA. Cundy, and D. G. Linz (1989). Fundamental Experiments on Thermal Desorption of Contaminants from Soils, *Environ. Progress.* **8,** 57.

Lighty, J. S., G. D. Silcox, D. W. Pershing, V. A. Cundy, and D. G. Linz (1990a). Fundamentals for the Thermal Remediation of Contaminated Soils. Particle and Bed Desorption Models. *Environ. Sci. and Tech.* **24,** 750-57.

Lighty, J. S., et al. (1990b). Rate Limiting Processes in the Rotary-Kiln Incineration of Contaminated Solids. *Comb. Sci. & Technol.* **74,** 31.

Lighty, J. S., et al. (1991). Toluene and Water Evolution from a Clay Sorbent in a Rotary Kiln Simulator. Second International Congress on Toxic Combustion By-Products: Formation and Control. March 26—29, Salt Lake City, UT.

Linak, W. P., J. A. McSorley, J. O. L. Wendt, and J. E. Dunn (1987). On the occurrence of transient puffs in a rotary-kiln incinerator simulator: II Contained liquid wastes on sorbent. *JAPCA.* **37,** 934-942.

McClennen, W. H., et al. (1990). Fast Repetitive GC/MS Analysis of Thermally Desorbed Polycyclic Aromatic Hydrocarbons (PAHs) From Contaminated Soils. *Comb. Sci. & Technol.* **74,** 297.

Oppelt, E. T. (1987). Incineration of Hazardous Waste: A Critical Review, *JAPCA.* **37,** 558.

Owens, W. D., Silcox, G. D., Lighty, J. S., Deng, X., Pershing, D. W., Cundy, V. A., Leger, C., and Jakway, A. L. (1991a). Thermal Analysis of Rotary Kiln Incineration: Comparison of Theory and Experiment. Accepted to *Combustion and Flame.* **86,** 101.

Owens, W. D., Silcox, G. D., Lighty, J. S., Deng, X., Pershing, D. W., Cundy, V. A., Leger, C., and Jakway, A. L. (1991b). The Desorption of Toluene From a Montmorillonite Clay Adsorbent in a Rotary Kiln Environment. *Jrn. of the Airwaste Mng. Assoc.* **42,** 681, 1992.

Pacyna, J. M. (1987). Atmospheric Emissions of Arsenic, Cadmium, Lead and Mercury from High Temperature Processes in Power Generation and Industry. Lead, Mercury, Cadmium and Arsenic in the Environment. edited by T. C. Hutchinson and K. M. Meema, John Wiley & Sons Ltd.

Penner, S. S., D. P. Y. Chang, R. Goulard, and T. Lester (1988). Waste Incineration and Energy Recovery. *Energy.* **13, 12,** 845-851.

Quann, R. J., and A. F. Sarofim (1982). Vaporization of Refractory Oxides During Pulverized Coal Combustion. Nineteenth Symposium (International) on Combustion. The Combustion Institute, pp. 1429-1440.

Seeker, W. R. (1991). Twenty-Third Symposium (International) on Combustion. The Combustion Institute.

Silcox, G. D., and Pershing, D. W. (1990). The Effects of Rotary Kiln Operating Conditions and Design on Burden Heating Rate as Determined by a Mathematical Model of Rotary Kiln Heat Transfer. *J. Air Waste Manage. Assoc.* **40,** 337.

Silcox, G. D., Larsen, F. S., and Pershing, D. W. (1991). Mathematical and Physical Modeling of Rotary Kilns With Applications to Scaling and Design. Accepted to CS& T.

Tillman, D. A., and C. Leone (1990). Control of Trace Metals in Fly Ash at the Tacoma, Washington Multifuels Incinerator. Proceedings of the American Flame Research Committee's 1990 Fall International Symposium. Oct. 8, San Francisco, CA.

Tillman, D. A. (1991a). The Combustion of Solid Fuels and Wastes, Academic Press, San Diego, CA.

Tillman, D. A., W. R. Seeker, D. W. Pershing, and K. DiAntonio (1991b). Developing Incineration Process Designs and Remediation Projects from Treatability Studies. Remediation. Summer, 251.

Tognotti, L., M. Flytzani-Stephanoporilos, A. F. Sarofim, H. Kopsinis, and M. Storekides (1991). Study of Adsorption-Desorption of Contaminants on Single Soil Particles Using the Electrodynamic Thermogravimetric Analyzer. *Environ. Sci. Technol.* **25,** 104.

Tscheng, S. H. and A. P. Watkinson (1979). Convective Heat Transfer in a Rotary Kiln, *Can. J. of Chem. Eng.* **57:**433.

Ulrich, G. D. (1978). An Investigation of the Mechanisms of Fly-Ash Formation in Coal-Fired Utility Boilers. Quarterly Report. February 1-April 30, FE-2205-11, U. S. Department of Energy.

Vogg, H. (1987). Behavior of (Heavy) Metals in the Incineration of Municipal Wastes. *International Chemical Engineering.* **27, 2,** 177-182.

Wendt, J. O. L., and W. P. Linak (1988). Mechanisms Governing Transients from the Batch Incineration of Liquid Wastes in Rotary Kilns. *Combust Sci. Tech.* **61,** 169.

Wu, Y-p G., J-i. Dong, and J. W. Bozelli (1991). Mass Transfer of Hazardous Organic Compounds in Soil Matrices Experiment and Model. Presented at Second International Congress on Toxic Combustion By-Products: Formation and Control. Salt Lake City, UT, March.

Adsorption of Mercuric Chloride Vapours from Incinerator Flue Gases on Calcium Hydroxide Particles

A. LANCIA*, D. MUSMARRA**, F. PEPE*, G. VOLPICELLI* *Dipartimento di Ingegneria Chimica, Universitá di Napoli "Federico II"; **Istituto di Ricerche sulla Combustione, C.N.R. P.le V.Tecchio, Naples, Italy

Abstract—The adsorption of mercuric chloride vapours on a fixed bed of calcium hydroxide fine particles has been experimentally studied. The study has been conducted at laboratory scale using simulated flue gases, constituted of mercuric chloride vapours in nitrogen, which have been in contact with a $Ca(OH)_2$ fixed bed. The effect of the bed temperature, the inlet $HgCl_2$ concentration, and the relative gas-solid velocity on the $HgCl_2$ removal efficiency has been studied. The experimental results indicated that high removal efficiencies of up to 95% can be obtained. The bed temperature has been found to be the most relevant parameter; particularly as the experimental results show that the lower the bed temperature, the higher the removal efficiency. Moreover, the removal efficiency increases when the $HgCl_2$ concentration in the inlet gas increases and when the relative gas-solid velocity decreases. A model based on a simplified expression of the Freundlich adsorption isotherm has been proposed, which is able to describe quite accurately the dependence of $HgCl_2$ removal efficiency on the parameters investigated.

Key Words— Mercuric chloride; adsorption; municipal solid wastes; incineration.

SYMBOLS

a	specific contact surface (cm^{-1})
a_I	activity of I species (-)
C	gas phase $HgCl_2$ equilibrium concentration at the gas-solid interface as defined in equation (15) ($\mu g/Nm^3$)
c	$HgCl_2$ concentration in the gas phase ($\mu g/Nm^3$)
c_{Hgtot}	total mercury concentration ($\mu g/Nm^3$)
c_I	I species concentration ($\mu g/Nm^3$)
c_{out}	$HgCl_2$ outlet concentration to the reactor ($\mu g/Nm^3$)
c_o	$HgCl_2$ inlet concentration to the reactor ($\mu g/Nm^3$)
c^*	gas phase $HgCl_2$ equilibrium concentration at the gas-solid interface as defined in equation (11) ($\mu g/Nm^3$)
K	thermodynamic equilibrium constant (-)
K_{ads}	thermodynamic equilibrium constant for adsorption reaction (21) (-)
k_d	overall mass transfer coefficient (cm/s)
L	bed thickness (cm)
M	molecular weight (g/mol)
N	dimensionless group defined in equation (19) (-)
p	total pressure (Pa)
p_s	vapour pressure (Pa)
R	gas constant (J/mol K)
r	$HgCl_2$ adsorption rate ($\mu g/Nm^3$ s)
T	absolute temperature (K)
t	time (s)
u_o	superficial velocity of the gas in the bed (cm/s)
z	axial coordinate along the reactor (cm)

Greek symbols

α	constant in the Freundlich equation

β exponent in the Freundlich equation
ε void fraction of the bed (-)
η $HgCl_2$ removal efficiency (-)
ρ_p particle density (g/l)
ω adsorbate loading (g/g solid)

INTRODUCTION

In recent years the increasing concern about the environment has pointed out that flue gases from Municipal Solid Waste (MSW) incineration have to be treated in order to reduce the emissions of heavy metals together with the other contaminants.

The toxicity and the relatively high concentration of mercury in the flue gas can result in severe environmental damage. In the lakes surrounding the mercury emission sources, elevated levels of this pollutant have been found, probably as a consequence of deposition from the air. High concentrations of methylmercury have been found in tissues of some kinds of comestible lake fishes, making them unsuitable for consumption, since the methylmercury is highly poisonous to man and higher animals (Roberts, 1981).

The mercury content in MSW usually lies in the range 1-2g/t; most of it comes from exhausted batteries, while other sources are pesticides, chemicals, etc. About 80% of this is released in the flue gases, resulting in a mercury concentration ranging between 1000 and 2000 μg/Nm^3, while the remaining 20% is found as metallic mercury adsorbed on the fly ash.

Air quality standards of the E.E.C. set the maximum emission of mercury for stationary sources at 100 μg/Nm^3. It is therefore of great importance to reduce mercury emissions by developing processes capable of decreasing the amount of this metal discharged into the atmosphere.

The thermodynamic analysis shows that in the combustion process of refuse, all mercury compounds are transformed to elemental mercury, since heating mercury compounds above 700°C leads to thermal decomposition that gives metallic mercury. Moreover, during the cooling of the gas, up to 200°C, a high fraction of the mercury is oxidized by alogenated compounds (Hall *et al*. 1990), NO_2 and H_2S. In particular, in the flue gases, the mercury is found as $HgCl_2$ due to the relatively high amount of hydrogen chloride.

Flue gases from MSW incinerators are usually treated with $Ca(OH)_2$ spray drying technology to remove acid gases (Lago, 1988), and it has been observed that in several cases this treatment is beneficial for mercury removal, since mercuric chloride is partly adsorbed on to the solids captured in electrostatic precipitators and/or bag filters. The mechanism behind this retention is not well understood and needs to be studied further.

In this paper the attention is focused on the phenomenon of $HgCl_2$ vapour adsorption on to solid particles of calcium hydroxide. The study has been performed in an apparatus at laboratory scale in which simulated flue gases ($HgCl_2$ vapours in N_2) have been contacted with a $Ca(OH)_2$ fixed bed. Experimental runs have been conducted at different temperatures, $HgCl_2$ inlet concentrations to the fixed bed reactor, and gas-solid relative velocities. The results, expressed as $HgCl_2$ removal efficiency, have been correlated using a model which takes into account the thermodynamic equilibrium of the adsorption process and a diffusional resistance for $HgCl_2$ transport from gas to solid surface.

THERMODYNAMIC ANALYSIS

The thermodynamic analysis has been performed with the aim of differentiating the most relevant mercury compounds in the flue gases from MSW incinerators. The species considered in this analysis are limited to elemental mercury and mercuric oxide (Hg, HgO)

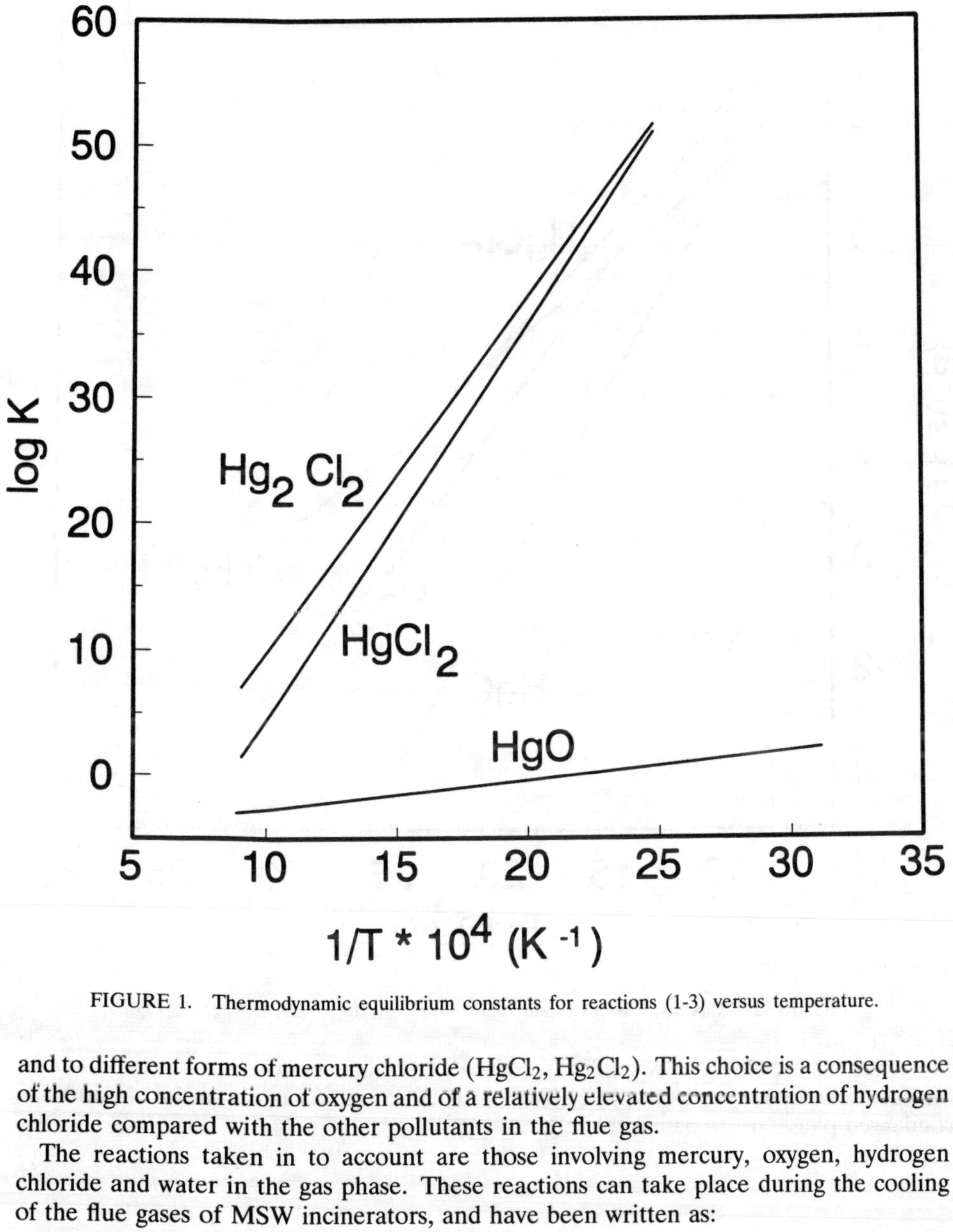

FIGURE 1. Thermodynamic equilibrium constants for reactions (1-3) versus temperature.

and to different forms of mercury chloride ($HgCl_2$, Hg_2Cl_2). This choice is a consequence of the high concentration of oxygen and of a relatively elevated concentration of hydrogen chloride compared with the other pollutants in the flue gas.

The reactions taken in to account are those involving mercury, oxygen, hydrogen chloride and water in the gas phase. These reactions can take place during the cooling of the flue gases of MSW incinerators, and have been written as:

$$2Hg_{(g)} + O_{2(g)} = 2HgO_{(g)} \quad (1)$$

$$2Hg_{(g)} + 4HCl_{(g)} + O_{2(g)} = 2HgCl_{2(g)} + 2H_2O_{(g)} \quad (2)$$

$$4Hg_{(g)} + 4HCl_{(g)} + O_{2(g)} = 2Hg_2Cl_{2(g)} + 2H_2O_{(g)} \quad (3)$$

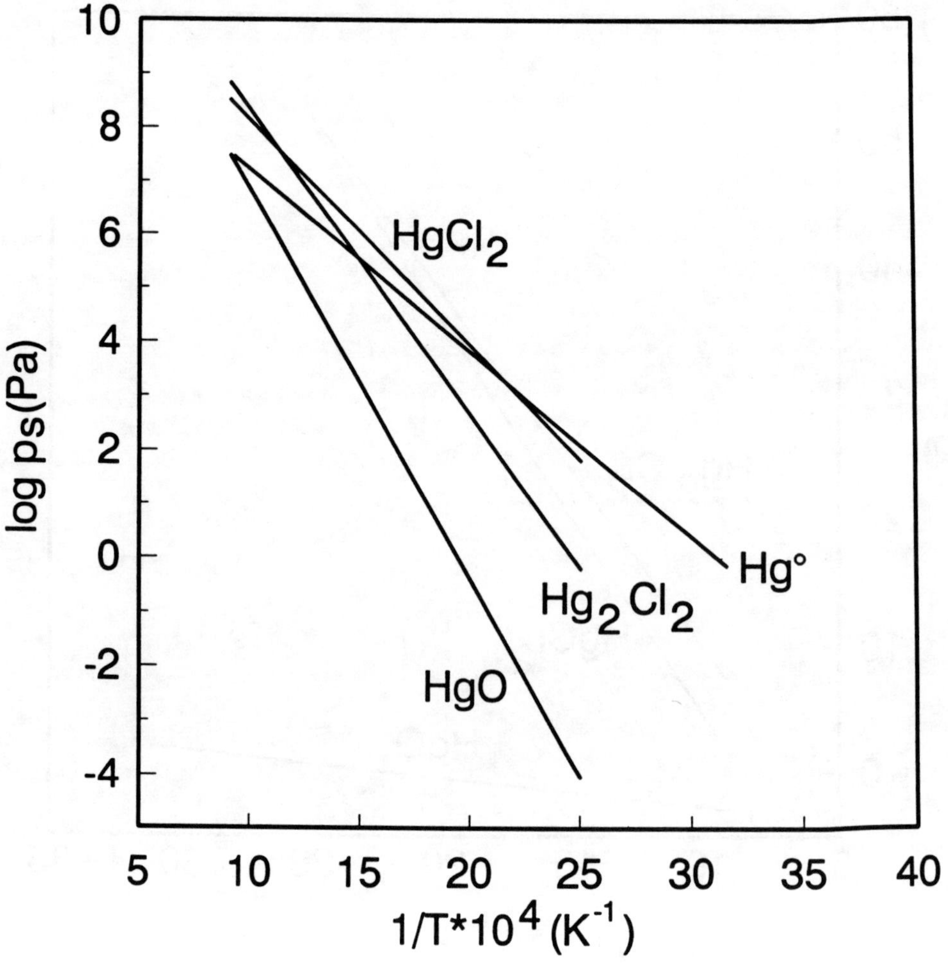

FIGURE 2. Vapour pressures versus temperature for Hg^0, HgO, $HgCl_2$, Hg_2Cl_2.

Moreover, as the flue gas is gradually cooled in the incinerator, the condensation of mercury compounds can take place. Therefore the equilibria between the gas and the condensed phase are to be considered:

$$Hg_{(g)} = Hg_{(l)} \tag{4}$$

$$HgO_{(g)} = HgO_{(s)} \tag{5}$$

$$HgCl_{2(g)} = HgCl_{2(s)} \tag{6}$$

$$Hg_2Cl_{2(g)} = Hg_2Cl_{2(s)} \tag{7}$$

Figure 1 reports the thermodynamic equilibrium constants of the reactions (1-3) as a function of the temperature. This figure was obtained by calculating the values of ΔG^0 for

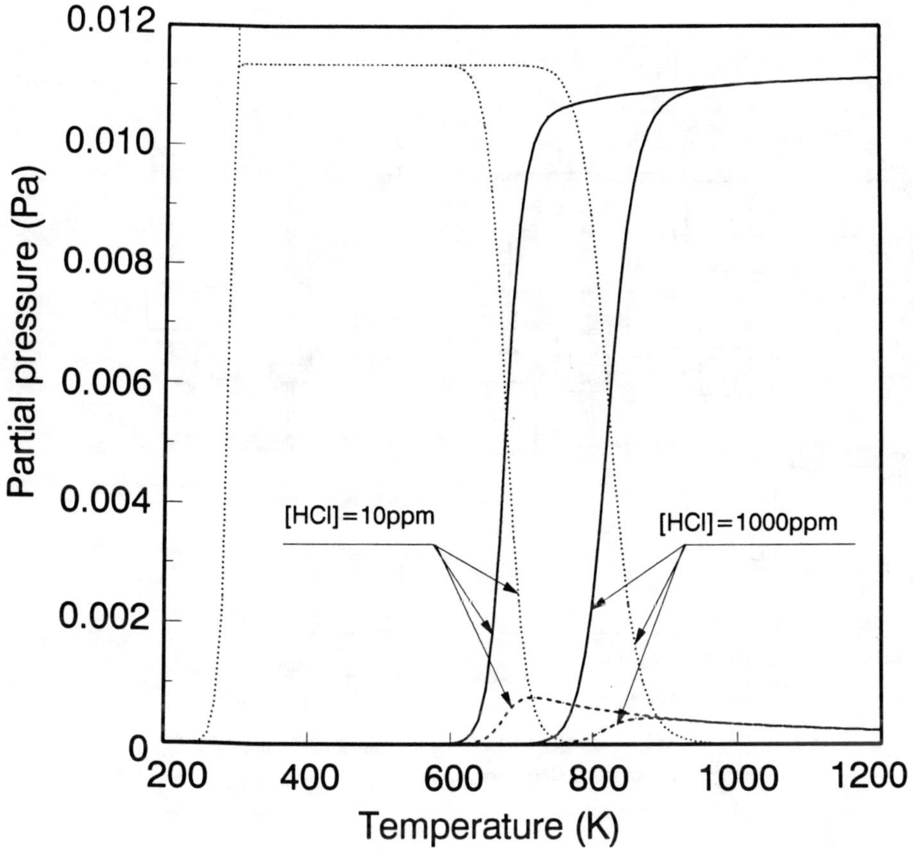

FIGURE 3. Theoretical calculations of gas phase mercury compound concentrations versus temperature. Total mercury concentration 1000μg/Nm^3; oxygen concentration 10%; water concentration 10%; hydrogen chloride concentration 10 and 1000 p.p.m.

Hg^0 : – – – –

HgO: – – –.

$HgCl_2$:........

different values of temperature from data tabulated in standard thermodynamic tables, such as those edited by Stull and Prophet (1971). On the other hand, figure 2 shows vapour pressures p_s as a function of temperature for the different mercury compounds considered, as calculated from data reported by Weast (1977) and by Anonymous (1978).

The thermodynamic analysis has been performed at different temperatures fixing the following values of concentrations in the system:

—1000 μg/Nm^3 total mercury concentration;

—10 p.p.m. and 1000 p.p.m. as different levels for hydrogen chloride concentration;

—10% oxygen concentration;

—10% water concentration. These concentration values typically occur in the flue gases from MSW incinerators.

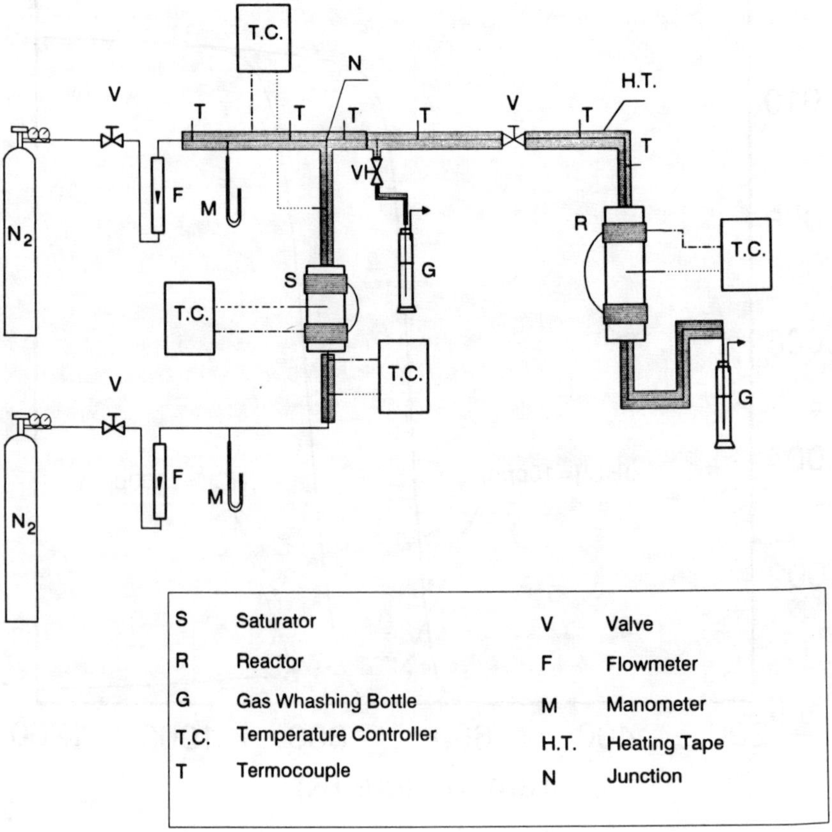

FIGURE 4. Sketch of the experimental apparatus.

The equations considered in this analysis are the thermodynamic equilibrium equations of the reactions (1-3) and the following mercury balance equation:

$$c_{Hg_{tot}} = c_{Hg_{(g)}} + c_{HgO_{(g)}} + c_{HgCl_{2(g)}} + 2c_{Hg_2Cl_{2(g)}} \tag{8}$$

where c_I is the gas phase concentration of the I species, with I = Hg, HgO, $HgCl_2$, Hg_2Cl_2, and $c_{Hg\ tot}$ is the total concentration of mercury.

The thermodynamic equilibrium equations of reactions (1-3), together with equation (8), allow the calculation of the concentrations of all species considered at different values of temperature when only the gas phase is present. On the other hand, when a condensed species occurs it is necessary to take into account its phase equilibrium equation; moreover equation (8) has to be modified considering the presence of the condensed phase too:

$$c_{Hg_{tot}} = c_{Hg_{(g)}} + c_{HgO_{(g)}} + c_{HgCl_{2(g)}} + 2c_{Hg_2Cl_{2(g)}} + (\text{cond.phase}) \tag{9}$$

Figure 3 reports the results of the analysis, obtained at different temperatures for the two levels of hydrogen chloride concentration. For the examined system the $HgCl_2$

condensation occurs at about 290K. Two different temperature windows can be identified: at the lower temperature the mercury is present in the gas phase only as $HgCl_2$, while at the higher the mercury is present practically only in the elemental form. Between the two temperature windows the HgO concentration reaches its maximum; however, its concentration is one order of magnitude lower than the Hg^0 concentration. The amplitude of these zones is slightly affected by the hydrogen chloride concentration. Moreover, the results indicate that in the whole range of temperature the Hg_2Cl_2 concentration is negligible. This analysis suggests that it is reasonable to assume that in the flue gases reaching the spray drying tower, most of the mercury present is in the form of $HgCl_2$ vapours.

EXPERIMENTAL APPARATUS AND TECHNIQUE

The experimental apparatus (fig.4) can be considered divided into two sections: in the first the simulated flue gases are produced at the required temperature and $HgCl_2$ concentration; in the second the adsorption of mercuric chloride vapours on $Ca(OH)_2$ is realized. The simulated flue gases are obtained by sublimation of reagent grade solid $HgCl_2$ into a stream of high purity nitrogen from a cylinder. Solid $HgCl_2$ is held in an AISI 316 stainless steel cylindrical saturator (S) which is kept at a selected temperature by a heating tape driven by a P.I.D. temperature controller. In the junction (N), the outlet stream from S, that is saturated by $HgCl_2$ vapours, is diluted with clean nitrogen in order to obtain the required $HgCl_2$ inlet concentration to the adsorption reactor, in a range between 800 and 2200 $\mu g/Nm^3$. To avoid condensation of $HgCl_2$ vapours in cold points, all pipes are kept hot by heating tapes. The mercuric chloride vapours adsorption is conducted in an AISI 316 stainless steel vessel (R) 26mm I.D. and 250mm high in which $Ca(OH)_2$ particles with 90 μm average diameter are used as the adsorbing material. In order to operate with a bed deep enough to avoid channelling and with a small reactive surface, the $Ca(OH)_2$ particles have been dispersed in inert glass beads of the same size. In all runs performed the bed thickness has been 4 mm, the $Ca(OH)_2$ content has been 12.5mg and a downward gas flow has been realized with the relative gas-solid velocity ranging between 5 and 13cm/s. A P.I.D. temperature controller provided with the heating tapes is used to keep the selected temperature in the adsorption reactor at a fixed value between 100° and 250°C.

$HgCl_2$ vapours upstream and downstream of the adsorption reactor have been collected by absorption into a 12% HNO_3 aqueous solution (Feldman, 1974) by means of gas washing bottles (G), and the collected mercury was determined by an atomic absorption spectrophotometer (I.L., model Video 12).

A check of this procedure has been realized by leaching the bed material after each run in concentrated HNO_3 to directly measure the mercuric chloride adsorbed on the bed. A difference not larger than 8% has been found in the mercury material balance, confirming the reliability of the experimental technique. Results of experimental runs conducted with the same operating conditions but different sampling times up to three hours showed that the $HgCl_2$ amount collected in the bottles, in the range of time explored, is proportional to time. This enables an evaluation of the mercuric chloride removal efficiency to be made as follows:

$$\eta = \frac{c_o - c_{out}}{c_o} \tag{10}$$

where η is the $HgCl_2$ removal efficiency and c_o and c_{out} are, respectively, the inlet and the outlet $HgCl_2$ concentrations to the reactor.

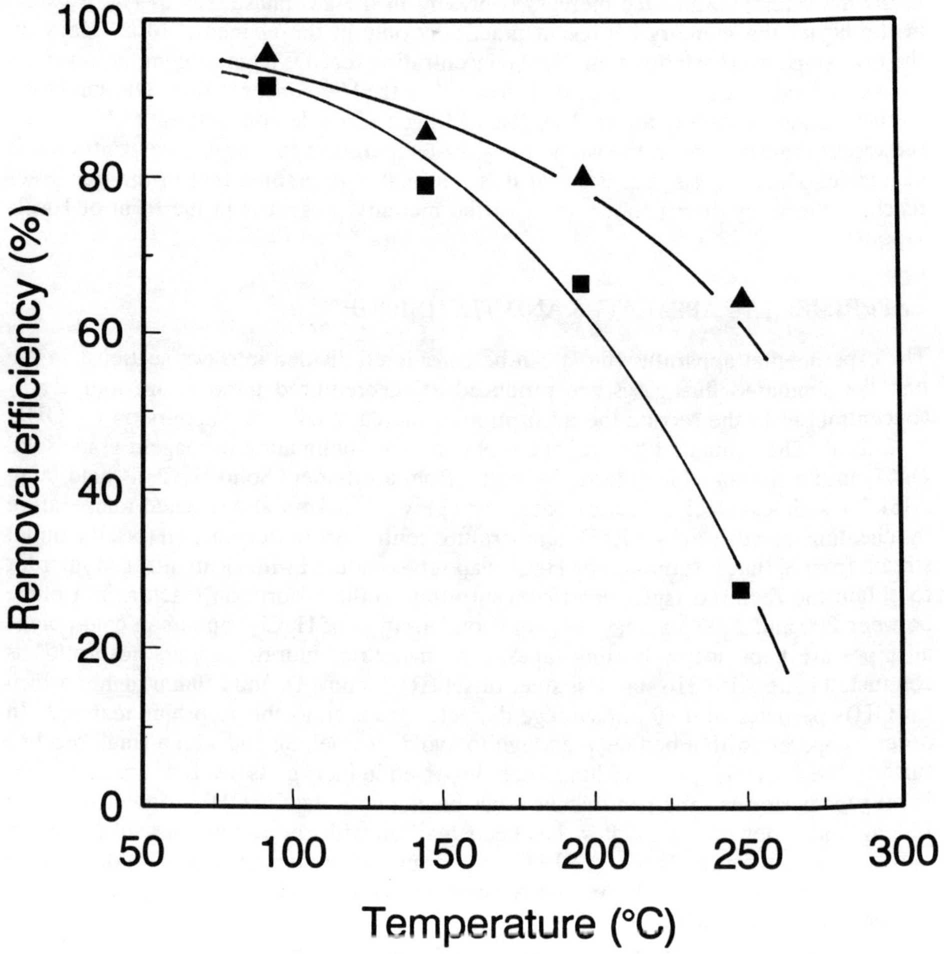

FIGURE 5. $HgCl_2$ removal efficiency versus temperature. Gas-solid relative velocity: 8.4 cm/s; ■800μ/Nm^3 $HgCl_2$ gas inlet concentration: $HgCl_2$ gas inlet concentration; ▲ 1600μg/Nm^3; —— proposed model (eq.20)

RESULTS AND DISCUSSION

The main results of the study are reported in figures 5 to 7. In these figures the $HgCl_2$ removal efficiency is reported as a function of the operating parameters investigated.

Figure 5 reports the removal efficiency versus the bed temperature for runs conducted with a constant superficial velocity of the gas u_0=8.4 cm/s and for two different $HgCl_2$ inlet concentrations to the reactor c_0=800 and 1600 μg/Nm^3.

In the figure it appears that high mercuric chloride removal efficiency (about 95%) can be obtained by operating at a relatively low temperature. Figure 5 also shows that the lower the bed temperature, the higher the mercuric chloride removal efficiency. This behaviour is particularly remarkable for the lowest $HgCl_2$ inlet concentration.

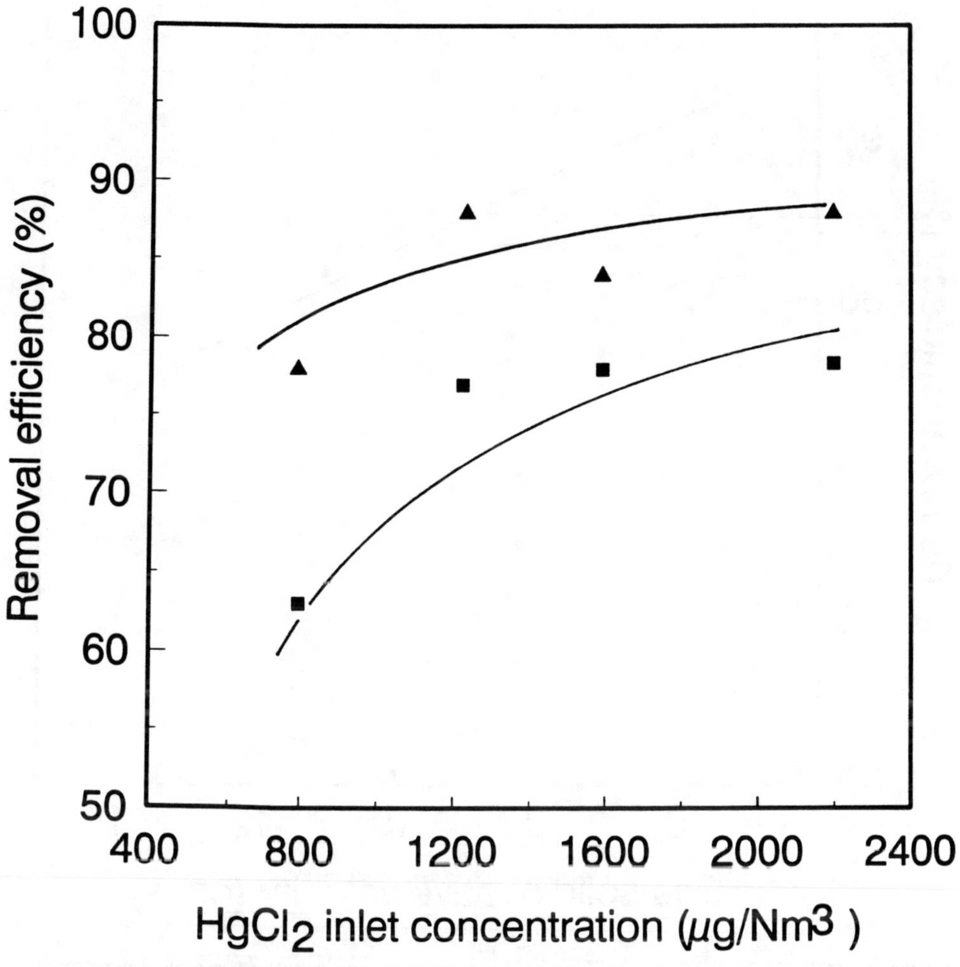

FIGURE 6. $HgCl_2$ removal efficiency versus inlet concentration to the reactor. Gas-solid relative velocity: 8.4cm/s; ▲ 150°C bed temperature; ■ 200°C; —— proposed model (eq. 20)

Figure 6 reports the removal efficiency versus the $HgCl_2$ inlet gas concentration for experimental runs conducted at u_0=8.4 cm/s and with bed temperatures of T=150° and 200°C. This figure shows that increasing the $HgCl_2$ inlet gas concentration increases the removal efficiency. This effect is higher for the lower $HgCl_2$ inlet concentration.

Figure 7 reports the removal efficiency versus the superficial velocity of the gas in the bed, for runs conducted with inlet concentration c_0=1600 $\mu g/Nm^3$ and with bed temperatures T=150° and 200°C. This figure shows that an almost linear dependence exists between the removal efficiency and the gas-solid relative velocity.

The experimental results that have been obtained can be used to find the best operating parameters for $HgCl_2$ removal. Furthermore, a model can be developed based on the following considerations: the dependence of $HgCl_2$ removal efficiency on temperature may be justified considering that a thermodynamic limit is reached at the gas-solid interface, and no kinetic control is present; the influence of $HgCl_2$ inlet concentration

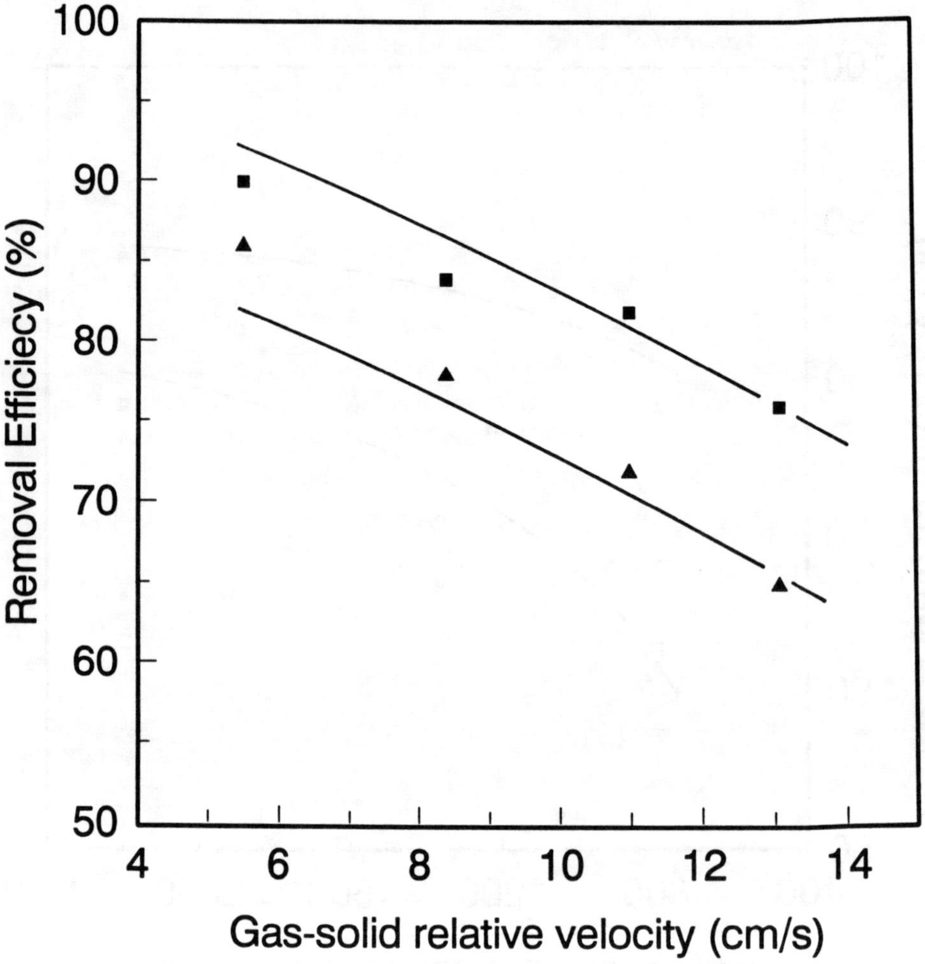

FIGURE 7. $HgCl_2$ removal efficiency versus gas superficial velocity in the bed. $HgCl_2$ gas inlet concentration: 1600μg/Nm^3; ■ 150°C bed temperature; ▲ 200°C bed temperature; —— proposed model (eq. 20)

on the removal efficiency can be explained considering a process which is not linear in concentration. As a consequence the following hypotheses have been assumed:

i) the mechanism for $HgCl_2$ transport from the gas to the solid includes diffusion through the gas film around the particles and diffusion into the pores to internal adsorption sites;
ii) the process of physical adsorption of $HgCl_2$ vapours on $Ca(OH)_2$ particles is practically instantaneous, and thermodynamic equilibrium is assumed to exist at the gas-solid interface between the concentration in the gas phase and the adsorbate concentration on the particles.

Therefore the $HgCl_2$ adsorption rate can be expressed as follows:

$$r = k_d a(c - c^*) \tag{11}$$

where r is the $HgCl_2$ adsorption rate, $k_d a$ is the product of the overall mass transfer coefficient and the specific gas-solid contact surface, while c and c^* are the $HgCl_2$ gas concentrations in the bulk and at the gas-solid interface, respectively.

A material balance on $HgCl_2$ in the gas phase for a section dz of the bed, where z is the axial coordinate along the bed, has to be considered (McCabe *et al.* 1985). Neglecting the accumulation in the gas phase in comparison with the accumulation on the solid, the following equation is obtained:

$$(1-\varepsilon)\rho_P \frac{\partial \omega}{\partial t} = -u_o \frac{\partial c}{\partial z} \tag{12}$$

where ε is the external void fraction of the bed, ρ_p is the particle density, ω is the adsorbate loading, t is the time, and the superficial velocity of the gas is assumed to be constant along the reactor. Considering equation (11), a balance for the adsorbed $HgCl_2$ can be written as follows:

$$(1-\varepsilon)\rho_P \frac{\partial \omega}{\partial t} = k_d a(c - c^*) \tag{13}$$

Generally the equilibrium gas phase concentration of the adsorbed species c^* and the adsorbate loading at a given temperature can be correlated by the Freundlich isotherm (Freundlich, 1926):

$$\omega = \propto (c^*)^\beta \tag{14}$$

where $\propto$ is the Freundlich adsorption equilibrium constant and β is an empirical exponent. Isotherms characterized by $\beta < 1$ are called favourable, because a relatively high solid loading can be obtained at low concentrations in the gas. On the contrary, an isotherm is called unfavourable when $\beta > 1$, because of the relatively low solid loadings that can be obtained.

The experimental results indicate that the $HgCl_2$ removal efficiency increases when $HgCl_2$ inlet concentration to the reactor increases. This dependence can be explained by assuming an unfavourable isotherm for $HgCl_2$ adsorption on $Ca(OH)_2$.

In order to obtain an analytical solution for an unfavourable isotherm, the limiting case of $\beta \to \infty$ has been considered. Therefore the mass transfer rate can be expressed as:

$$(1-\varepsilon)\rho_s \frac{\partial \omega}{\partial t} = k_d a(c - C) \tag{15}$$

where C is the gas-phase concentration of $HgCl_2$ which at a given temperature is in equilibrium with the solid, regardless of the adsorbate loading.

Substituting equation (15) in equation (12) one obtains:

$$u_o \frac{dc}{dz} = -k_d a(c - C) \tag{16}$$

this is an ordinary differential equation in which c is a function of the axial position along the reactor and is independent of time. This is in agreement with the experimental results showing that the $HgCl_2$ removal efficiency is independent of time. The initial condition for equation (16) is:

$$z = 0 \qquad c = c_o \tag{17}$$

The outlet concentration value calculated by solving equation (16) is:

$$c_{out} = (c_o - C)e^{-N} + C \tag{18}$$

where N is defined as:

$$N = \frac{k_d a L}{u_o} \tag{19}$$

and L is the bed thickness.

Therefore the $HgCl_2$ removal efficiency is:

$$\eta = \frac{c_o - c_{out}}{c_o} = 1 - (1 - \frac{C}{c_o})e^{-N} - \frac{C}{c_o} \tag{20}$$

The equilibrium concentration C can be expressed in terms of the thermodynamic equilibrium constant of the adsorption reaction:

$$HgCl_2 + (\sigma) = (HgCl_2 - \sigma) \tag{21}$$

where (σ) is used to indicate the active sites and $(HgCl_2 - \sigma)$ to indicate the adsorbed complex.

The equilibrium constant of reaction (21) K_{ads}, which is a function of the absolute temperature, can be related to the activities of the different species through the equation:

$$K_{ads} = \frac{a_{(HgCl_2} - \sigma)}{a_{HgCl_2} a_{(\sigma)}} \tag{22}$$

where a_I is the activity of the I species, with $I = HgCl_2$ (σ) and $(HgCl_2$ - $\sigma)$

Making the assumption that for solids $a_I = 1$, and assuming an ideal gas behaviour for $HgCl_2$, from equation (22):

$$C = \frac{Mp}{K_{ads} RT} \tag{23}$$

where M is the molecular weight of $HgCl_2$, p is the total pressure, and R the gas constant.

A regression analysis performed on the experimental results enables the dependence of $k_d a$ and K_{ads} on the temperature to be evaluated. An equation of the Arrhenius form can be used to describe the dependence on $k_d a$ on temperature as:

$$K_d a = 25.6 e^{-\frac{320}{T}} \tag{24}$$

On the other hand for K_{ads}:

$$\ln K_{ads} = \frac{\Delta G^0_{ads}}{RT} = \frac{\Delta H^0_{ads} - T\Delta S^0_{ads}}{RT} \tag{25}$$

and neglecting the dependence of ΔH°_{ads} and ΔS°_{ads} on temperature the parameters obtained from the regression analysis are:

$$\Delta H^\circ_{ads} = 34 \text{kJ/mol} \tag{26}$$

$$\Delta S^\circ_{ads} = 70 \text{J/molk} \tag{27}$$

The curves reported in figures 5, 6, 7 have been obtained by equation (20). The model results correctly describe the dependence of mercuric chloride removal efficiency on the operating parameters investigated, showing that the assumption of $\beta \gg 1$ in the

Freundlich isotherm is in agreement with the physical behaviour of this system. This result is close to that reported by Otani *et al.* (1986) for mercury adsorption on various soils.

CONCLUSIONS

The high toxicity of mercury compounds constitutes a severe environmental problem; therefore an improvement in developing technologies capable of reducing the mercuric chloride vapour emissions into the atmosphere is necessary.

Spray drying technology with $Ca(OH)_2$ aqueous suspensions has been used for a long time to remove acid gas from flue gases downstream of M.S.W. incinerators. This treatment has been found to be beneficial for mercury compounds removal. This is due to the adsorption of mercury compounds on the solids captured in electrostatic precipitators and/or bag filters. In this work the $Ca(OH)_2$ adsorption capacity for $HgCl_2$ removal has been studied with experiments conducted at laboratory scale. The dependence of $HgCl_2$ removal efficiency on temperature, concentration, and gas-solid relative velocity has been considered. The experimental results indicate that it is convenient to operate at the lowest possible temperature compatible with the need to avoid condensation. The experimental results, furthermore, indicate that removal efficiency increases when concentration increases and gas-solid relative velocity decreases.

The dependence of $HgCl_2$ removal efficiency on the parameters investigated is described quite accurately by a model which is based on a simplified expression of the Freundlich adsorption isotherm for the unfavourable case. According to this simplification, at a given temperature the equilibrium concentration of $HgCl_2$ in the gas phase does not depend on the adsorbate loading on the solid.

For the future development of the work it will be of primary importance to obtain adsorption isotherms at different temperatures for the system $HgCl_2$-$Ca(OH)_2$ in order to take into account the saturation of the solid.

REFERENCES

Anonymous, (1978). Pressions de vapeur du mercure et de ses chlorures, étude bibliographique., *Inf. Chim.*, **182,** 227.

Feldman, C. (1974). Preservation of diluted mercury solutions., *Anal. Chem.* **46,** 99.

Freundlich, A. (1926). Colloid and Capillary *Chemistry,* Dutton, N.Y.

Hall, B., Schager, P., Lindqvist, O. (1991). Chemical reactions of mercury in combustion flue gases, Water, Air and Soil Pollution **56,** 3-14.

Lago, A. (1988). L'incenerimento dei rifiuti senza danni per l'ambiente: il controllo delle emissioni., *Simposio sul Controllo dell'Inquinamento Industriale,* Milano 6/10/1988.

Mc Cabe, W. L., Smith, J. C., Harriot, P. (1985). Unit Operations of Chemical Engineering, 4th ed., McGraw Hill, N.Y., pp. 698.

Otani, Y., Kanaoka, C., Matsui, S., Emi, H. (1986). Adsorption of mercury vapor on particles., *Environ. Sci. Technol.* **20,** 735.

Roberts, R. H. (1981). *Food Safety,* Wiley & Sons, N.Y., pp. 148-169.

Stull, D. R., Prophet, H., Eds. (1971). *JANAF Thermochemical Tables,* 2nd ed., Nat. Stand. Ref. Data, Ser., Nat. Bur. Stand. (U.S.A.).

Weast, R. C., Ed., (1977). *CRC Handbook of Physics and Chemistry,* 58th ed., CRC Press, Cleveland.

Thermal-Catalytic Incineration of Waste Gases

TERESA BANASZAK, RYSZARD MILLER, and ANITA ZIEBA *Technical University of Wroclaw, Wroclaw, Poland*

Abstract — Numerous waste gases consist of a low concentration of hydrocarbons in air. Therefore, in catalytic incineration it is necessary to heat up the waste gases to the temperature required. As a rule this heating up is achieved by the combustion of a high calorific value fuel. The combustion of a high calorific fuel is not only the thermal energy source but it also creates the high chemical activity of the combustion zone. In this zone the initiation of oxidation processes and the oxidation of toxic components of the waste gases is observed. An effective utilization of the chemical activity of the combustion zone can be assured through the direct contact of the waste gases and the flame and post-flame zones. Such a new approach to the incineration of the waste gases in catalytic incineration will be called "thermal-catalytic incineration."

Key Words: combustion, waste gas, thermal incineration, catalytic incineration

INTRODUCTION

Catalytic incineration is one of the frequently used methods of neutralizing the waste gases. The performance of a catalytic incinerator is affected by several factors including: operating temperature, space velocity (reciprocal of residence time), composition and concentration of the waste gases and catalyst properties.

The process temperature is the most important of the above parameters.

The operating temperature for a particular destruction efficiency is dependent on the concentration and composition of the waste gases in the emission stream and the type of catalyst used. Maintaining a specific operating temperature of a catalytic bed is necessary to initiate and support the catalytic incineration reaction. Too low a process temperature may lead to the creation of equally toxic intermediate products as a result of the oxidation of the waste gas, while too high a temperature may destroy the catalytic bed.

A lot of gases usually contain hydrocarbons at a low concentration in the air. Therefore, it is necessary in catalytic incineration to heat up the waste gases to the required temperature. As a rule this heating is achieved by the combustion of the supplementary fuel, usually a high calorific value gaseous fuel.

The quantities used of this fuel are small which is a consequence of commonly used heat recuperation. Burning of supplementary fuel takes place in various burners and combustion chambers. Usually, one does not pay any attention to the organization of the burning process and particularly to the process of the mixing of the combustion gases with the waste gases.These processes should ensure quick heating and an evening of the temperature of the combustion gases and large volumes of the waste gases across the catalytic incinerator chamber. This can only be achieved by intensive mixing ensuring a homogenous mixture of waste gases and combustion gases before the catalytic bed.

A proper organization of the process of heating the waste gases facilitating efficient use of such small quantities of fuel poses a complex and important problem. This is not only because of the economy, but also because of the possibility of improving the catalytic bed operating conditions.

The combustion of supplementary fuel is not only a source of thermal energy but also creates the high chemical activity of the combustion zone.

Initiation of oxidation processes, as well as the oxidation of toxic components of the waste gases, are observed in this zone (Banaszak *et al.* 1987, Miller and Zembrzuski 1986).

An effective utilization of the chemical activity of the combustion zone can be assured through the direct contact of the waste gases and the flame and post-flame zones in the mixing process. Such a new approach to the incineration of waste gases in catalytic incineration will be called "thermal-catalytic incineration."

It has been suggested that the combustion of supplementary value fuel should be used in the preliminary stage of the destruction of the waste gases in catalytic incineration. A comparison between the efficiency of the waste gas neutralization, with and without contact with flames, has been carried out.

EXPERIMENT

Apparatus

The experimental installation used is shown in Fig. 1. It consists of a flow quartz reactor (1) 0.3 m in length and 0.03 m in diameter. A thermal incineration chamber (2) is located in the bottom part of the reactor while the catalytic incineration chamber (3) in its top part. Within the thermal incineration chamber there are burners (4). Hydrogen burners perform the flame heating of the waste gases. The catalytic incineration chamber is filled with the catalyst. Outside the reactor a kanthal heating coil (5) is installed for heating up the waste gases without the need for a flame. The temperature of each incineration chamber is controlled separately using thermocouples (6). Standard measuring instruments and techniques were employed. The waste gas samples for the chromatographic analyses were collected at the reactor inlet (A), before the catalytic bed (B), and at the reactor outlet (C).

Procedure

The study was carried out in two series of measurements. In the first series catalytic incineration was used, where the waste gas was heated up to the required temperature with electric heaters.

In the second series, thermal-catalytic incineration was investigated where the waste gas was flame-heated to the required temperature. In this series the burners were supplied with a constant amount of hydrogen, and the additional gain in temperature was achieved by electric heating.

All experiments were carried out at temperatures ranging from 100–475°C, maintaining the steady-state temperature of the catalytic bed and the thermal incineration chamber and a space velocity of 20,000 hr^{-1}. Toluene, ethyl acetate and ethyl alcohol were selected from among the organic compounds as toxic components of the waste gases.

The catalyst examined was a commercial catalyst containing Pt supported on γ-Al_2O_3. The catalyst takes the form of balls $3–4 \times 10^{-3}$ m in diameter. The experimental procedure consisted of measuring the reactor inlet (A), outlet (C) and before catalyst bed (B) concentrations at a series of constant temperatures and gas flow rates for each gas.

RESULTS

The analysis of the operational efficiency of the thermal-catalytic incineration installation was carried out for two methods of waste gas heating. The first was flameless heating, with electric heaters used to heat up the waste gas. The second was flame heating where the waste gas was heated through the direct contact with the hydrogen flame and its post-flame zone.

In order to evaluate the incineration efficiency of the thermal-catalytic installation it was subdivided into parts where thermal and catalytic incineration takes place. Thermal in-

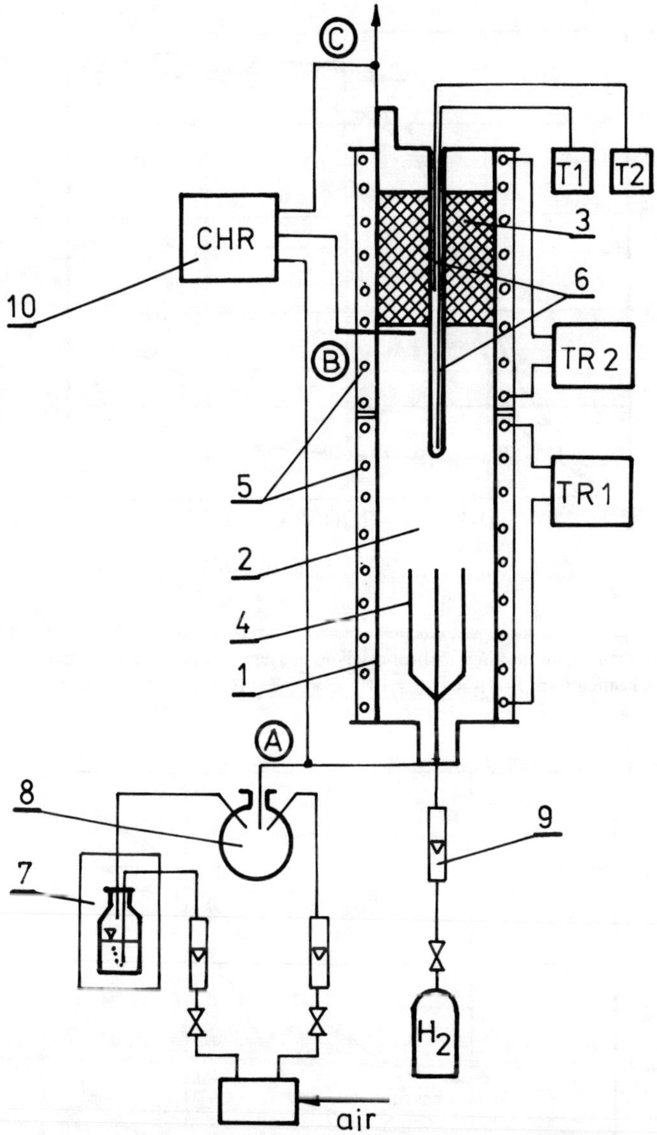

FIGURE 1 Scheme of experimental installation: 1 — reactor, 2 — thermal chamber, 3 — catalytic chamber, 4 — burners, 5 — heating coil, 6 — thermocouples, 7 — toxic component feeder, 8 — mixer, 9 — rotameter, 10 — gas chromatograph.

cineration takes place in a chamber below the catalytic bed. It encompasses the flame zone from the burner outlets and the post-flame zone up to the bottom layer of the catalytic bed. The catalytic incineration takes place during the electric heating of the waste gas. The thermal-catalytic incineration takes place in the whole installation including the flame zone, the post-flame zone, as well as the catalytic bed.

A chromatographic analysis of the waste gas in the selected measuring points (A, B, C) was used to determine the conversion of its toxic components to CO_2 and H_2O in the area

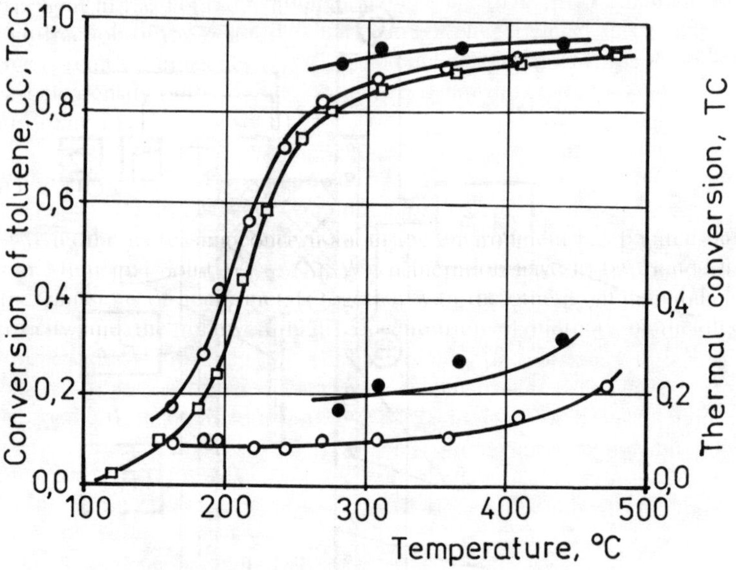

FIGURE 2 Conversion of toluene as a function of temperature of the catalytic bed: ❐ = catalytic conversion (CC) of toluene without flame heating, thermal-catalytic conversion (TCC) and thermal conversion (TC) of toluene with flame heating for: ○ – $H_2 = 13.9 \times 10^{-7}$ m^3 s^{-1}, ● – $H_2 = 27.8 \times 10^{-7}$ m^3 s^{-1}.

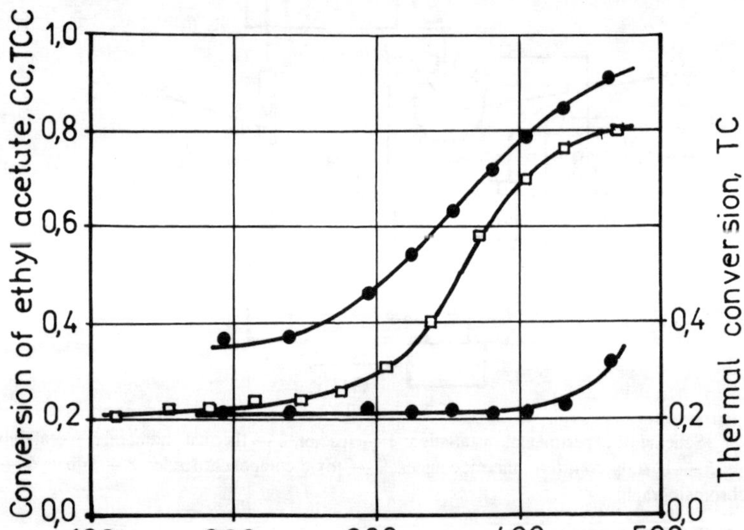

FIGURE 3 Conversion of ethyl acetate as a function of temperature of the catalytic bed: ❐ = catalytic conversion (CC) of ethyl acetate without flame heating, thermal-catalytic conversion (TCC) and thermal conversion (TC) of ethyl acetate with flame heating for: ● – $H_2 = 27.8 \times 10^{-7}$ m^3 s^{-1}.

of catalytic, thermal and thermal-catalytic incineration. Thus obtained efficiency characteristics of the thermal, catalytic, and thermal-catalytic incineration were used to evaluate the efficiency of the incineration installation.

Figures 2 and 3 show the conversion of toluene and ethyl acetate with respect to the catalytic bed temperature.

The flame heating of the waste gas in the thermal-catalytic system results in a higher neutralization efficiency as compared to the catalytic incineration. The difference is some 10 to 20% and depends on the amount of burnt highly calorific fuel. Increasing the amount of this fuel causes an increase in the percentage of thermal incineration during the thermal-catalytic waste gas neutralization.

The percentage of the thermal incineration amounts to about 10% which creates better conditions for the operation of the catalytic bed. Similar results were obtained for ethyl acetate (Fig. 3). Ethyl acetate is difficult to neutralize in a catalytic installation. Initially, up to the temperature of 270°C, the conversion does not change and is about 20%. Then it rises slowly, reaching 80% at the temperature of 470°C. In the thermal-catalytic installation with flame heating, much higher conversion is achieved (10 to 20%) across the whole range of temperatures. Also, thermal heating constitutes a significant part (about 20%) as the initial conversion of ethyl acetate and depends on the amount of burnt fuel.

The neutralization (oxidation) of toxic additives in the waste gas usually proceeds as a complex, multistage reaction with a number of secondary organic waste products. It is therefore important to organize the neutralization process so as to limit the creation of the products of incomplete oxidation, especially when they are toxic. A comparison of the capabilities of the catalytic and thermal-catalytic processes to limit the creation of incomplete combustion products was conducted using the waste gas containing ethyl alcohol (Figs. 4 and 5). Acetaldehyde is a product of the partial oxidation of ethyl alcohol. In order to evaluate the efficiency of the neutralization of the waste gas an additional parameter was introduced, called the process selectivity, towards the complete oxidation of the substrate (Straszko, 1979).

The selectivity was defined as the ratio of the amount of ethyl alcohol oxidating to CO_2 and H_2O to the amount of ethyl alcohol oxidized to CO_2 + H_2O and CH_3CHO in the installation. Figure 4 shows the selectivity of ethyl alcohol oxidation as a function of the catalytic bed temperature. Utilization of flame heating in the process of neutralization of ethyl alcohol significantly affects the oxidation processes. The selectivity of the oxidation reaction in the thermal-catalytic system is higher than the selectivity of the reaction in the catalytic process. The difference depends on the amount of fuel. The advantageous influence of the flame heating of the waste gas can also be seen in the conversion of ethyl alcohol to acetaldehyde (Fig. 5). Initially, an increase in temperature causes an increase of acetaldehyde up to a maximum. Thereafter production of acetaldehyde decreases until it stops completely at a temperature of 375–475°C. At the same time one can see that the amount of acetaldehyde created decreases with the increase in the amount of the highly calorific fuel — the maximum shifts towards lower temperatures.

Utilization of the supplementary fuel burning in the waste gas heating also results in its preliminary destruction, ranging from 10 to 30% (Fig. 6). The basic results indicate that the efficiency of the thermal-catalytic incineration is higher than the efficiency of the catalytic incineration without flame heating of the waste gas. Similar results have been observed in a prototypic experimental installation with a capacity of 1000 $m^3\ h^{-1}$ of the waste gas.

DISCUSSION

An analysis of the efficiency of the thermal-catalytic incineration system was based on the results obtained for two methods of waste gas heating. This research confirmed the in-

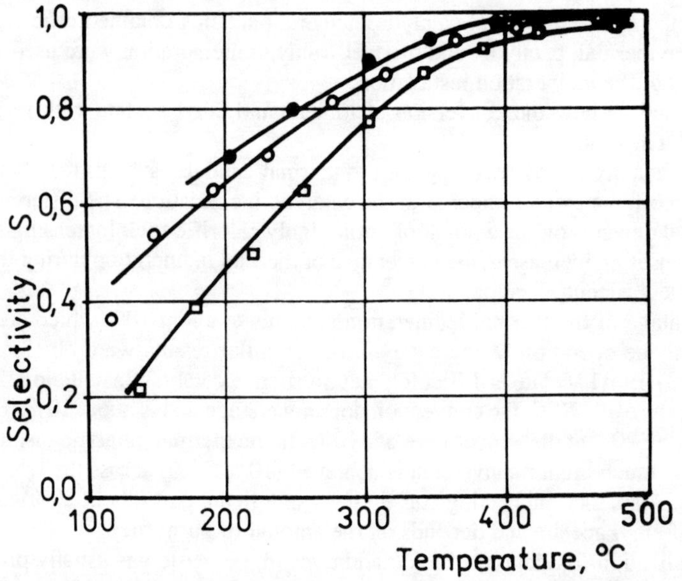

FIGURE 4 Selectivity (S) of complete oxidation of ethyl alcohol as a function of temperature of the catalytic bed: ❐ = without flame heating, with flame heating: ○ – $H_2 = 13.9 \times 10^{-7}\ m^3\ s^{-1}$, ● – $H_2 = 27.8 \times 10^{-7}\ m^3\ s^{-1}$.

fluence of the waste gas heating method on the efficiency of the waste gas neutralization. Then, when burning supplementary fuel is necessary, it is important how this heating is accomplished. Burning this fuel ensures not only the waste gas heating but also partial oxidation of the toxic components. The preliminary destruction phase occurring in the thermal combustion chamber, where the gas is heated to the required catalytic bed temperature, is between 10 and 20%. It is important to perform the process of oxidation to CO_2 and H_2O, e.g., without forming new toxic compounds as products of incomplete oxidation (CO, CH_3CHO). An examination of ethyl alcohol indicates that it is possible to limit the amount of toxic acetaldehyde with a simultaneous increase in the level of oxidation to the final products. Not only the amount of acetaldehyde decreases but also the operating temperature of the catalytic bed is lowered.

An evaluation of the incineration systems with different methods of waste gas heating was based on the temperature required for 90% neutralization efficiency. Table I contains the results obtained for the examined toxic components of the waste gas. It follows from the results presented that the utilization of the chemical activity of the flame zone and the post-flame zone makes it possible to lower the catalytic bed operating temperature from 50 to about 100°C.

What could cause such a strong effect?

Using flame heating causes the following changes in the waste gas:

— as a result of thermal incineration occurring in the flame as well as the post-flame zones the concentration of the toxic component is lowered,

— the flame and its post-flame zone is a chemically active area and a source of radicals which initiate and propagate the oxidation processes by:

a) particle excitation to higher energy levels in the reactions of the type:

$$C_nH_m + M^* = C_n H_m^* + M, \qquad (1)$$

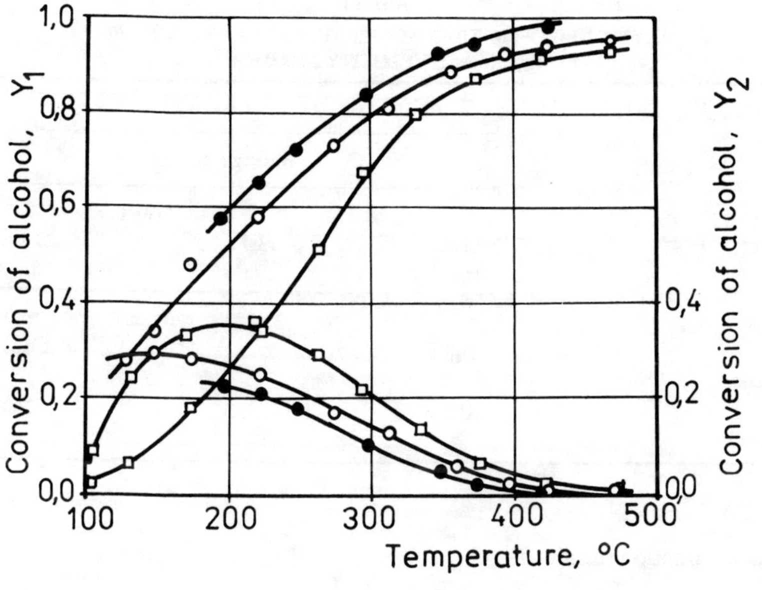

FIGURE 5 Conversion of ethyl alcohol to final oxidation product $CO_2 + H_2O$ (Y_1) and to acetaldehyde (Y_2) as a function of temperature of catalytic bed: ❐ = without flame heating, with flame heating: ○ – $H_2 = 13.9 \times 10^{-7}\ s^{-1}$, ● – $H_2 = 27.8 \times 10^{-7}\ m^3\ s^{-1}$.

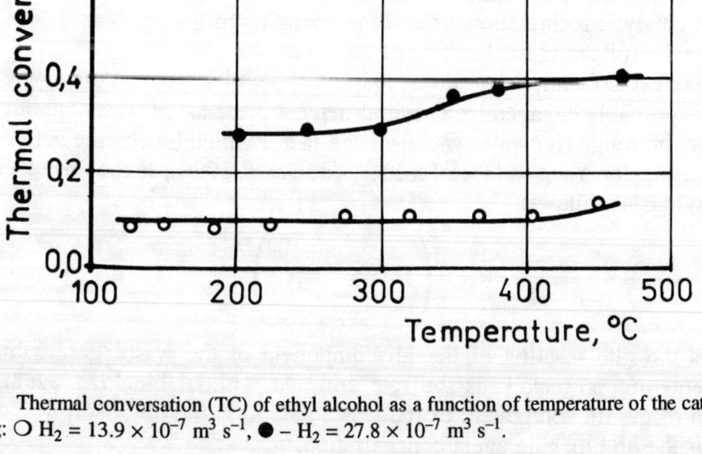

FIGURE 6 Thermal conversation (TC) of ethyl alcohol as a function of temperature of the catalytic bed with flame heating: ○ $H_2 = 13.9 \times 10^{-7}\ m^3\ s^{-1}$, ● – $H_2 = 27.8 \times 10^{-7}\ m^3\ s^{-1}$.

TABLE I
CATALYTIC BED OPERATING TEMPERATURE FOR 90% CONVERSION $T_{90\%}$, SPACE VELOCITY 20,000 HR^{-1}

Waste gas	$T_{90\%}$, °C: Waste gas heating, without flame	with flame: 13.9×10^{-7} m^3 s^{-1}	with flame: 27.8×10^{-7} m^3 s^{-1}
Toluene	410	380	280
Ethyl acetate	Doesn't reach		460
Ethyl alcohol	330	290	260

b) radical initiation reactions:

$$C_nH_m + (O, OH, H) = C_nH_{m-1} + (OH, H_2O, H_2), \tag{2}$$

$$C_nH_m + O + R_1 + R_2, \tag{3}$$

c) thermal decomposition reactions:

$$C_nH_m + M = C_nH_{m-1} + H + M, \tag{4}$$

— as a result of the fuel burning, water vapour appears in the waste gas.

In order to verify whether the described changes in the waste gas affect the process of catalytic incineration, the kinetic parameters of the process of catalytic oxidation with or without flame heating were determined based on the kinetic-diffusion model (Musialik-Piotrowska, 1986). A comparison of the kinetic parameters contained in Table II indicates that, for the process occurring with flame heating, both the preexponential factor (A) and the activation energy (E) are different. This means that the process of thermal incineration, preceding the catalytic incineration, affects the operation of the catalyst.

We propose the following explanation of these phenomena.

The preexponential factor, proportional to the number of active centers on the surface of the catalyst, is probably changed by a several-percent presence of water vapour created as a result of fuel burning. The water vapour is the factor which blocks the active centers of the catalyst (Tsong-Jen Yang and Lunsford 1987, Liu *et al.* 1990), so the formula expressing the reaction rate is as follows:

$$r_i = K_i C_i \left(1 + \sum_j K_j C_j\right)^{-1},$$

where r_i is the rate reaction of the *i*th component of the waste gas oxidation when blocking agents are present, k_i is the rate constant without blocking agent, C_i is the concentration of the *i*th component of the waste gas, K_j is the *j*th adsorption equilibrium constant, C_j is the *j*th blocking agent concentration.

TABLE II
KINETIC PARAMETER

Waste gas	Temperature range °C	Kinetic parameter: Waste gas heating: without flame A, s^{-1}	E, $Jmol^{-1}$	with flame A, s^{-1}	E, $Jmol^{-1}$	Fuel (hydrogen) m^3s^{-1}
Toluene	130–230	1.40×10^{13}	83100	7.40×10^{11}	54400	21.4×10^{-7}
	280–475			1.44×10^{8}	15400	21.4×10^{-7}
		1.12×10^{9}	43300	1.05×10^{8}	10000	27.8×10^{-7}
Ethyl acetate	280–478	1.70×10^{10}	51150	8.60×10^{10}	57250	27.8×10^{-7}
Ethyl	130–380	4.94×10^{9}	32600	2.65×10^{9}	27350	13.9×10^{-7}
alcohol				1.02×10^{10}	33950	27.8×10^{-7}

The toxic waste gas components in direct contact with the hydrogen flame and its active post-flame zone reach the higher energy level as a result of excitation (reaction 1), or radical reactions (2)–(4). Highly reactive hydrocarbon radicals, easily oxidate to final oxidation products, which would explain the decrease in the activation energy in the catalytic process.

Additionally, the kinetic parameters may be influenced by the homogenous oxidation process occurring among the catalyst grains. It should be noted, that in this experiment diffusion hydrogen flames were placed in a stream of the waste gas where high excess oxygen is present. Such flames mainly generate hydroxyl and oxygen radicals which much more slowly initiate the process of oxidation of the hydrocarbons contained in the waste gas. Applying a rich flame so as to enforce the generation of high concentrations of hydrogen radicals should activate the process even more.

CONCLUSIONS

The presented data permits one to draw the following conclusions:

1. When supplementary fuel is necessary for heating up the waste gas it is important how this heating is achieved. Proper heating requires a direct contact of the waste gas with a large area of the heating flame and its post-flame zone.
2. The initial destruction of the waste gas during flame heating is 10–20%, which improves the operating conditions of the catalytic bed.
3. The efficiency of thermal-catalytic incineration with flame heating of the waste gas is better than heating without it. The difference is 10 to 20% and depends on the on the operating conditions of the catalytic bed.
4. The described thermal-catalytic incineration is an energy-efficient technology for the neutralization of the waste gas. This technology requires specific features in the burner construction in order to produce a fuel-rich post-flame zone thereby improving the oxidation of the hydrocarbon components of the waste gas.

REFERENCES

Banaszak, T., Miller, R., and Zembrzuski, M. (1987). The influence of flame-generated free radicals on the thermal oxidation of waste gases. *JAPCA*, **37**, No. 12, pp. 1434–1438.

Liu, W., Eliyas, A., and Petrov, L. (1990). Influence of feed water vapour on the selective oxidation of ethylene over silver catalyst. *Applied Catalysis*, **61**, pp. 265–174.

Miller, R., and Zembrzuski, M. (1986). Thermische Nachverbrennung von Abgas in einer Hilfsflamme. *Staube Reinhaltung der Luft*, **46**, No. 11, pp. 463–466.

Musialik-Piotrowska, A. (1986).Catalytic combustion of toluene, n-heptane and ethanol in air. *Ochrona Powietrza*, **20**, No. 5, pp. 120–123 (in Polish).

Straszko, J. (1979). *Modelowanie kinetyki procesow kontaktowych*. (Ed.) Prace Naukowe Politechniki Szczecinskiej, 124, 172 (in Polish).

Tsong-Jen Yang, and Lunsford, J.H. (1987). Partial oxidation of methane to formaldehyde over molybdenum oxide on silica. *Journal of Catalysis*, **103**, pp. 55–64.

9.

INTERNAL COMBUSTION ENGINES

...following ... can be ... concentration. As a consequence the following hypotheses have been assumed:

i) the mechanism for HgCl transport from the gas to the ... through the gas film around the particles and diffusion ... to the adsorption sites;

ii) the process of physical adsorption ... vapours on $Ca(OH)_2$ particles is practically instantaneous and thermodynamic equilibrium is assumed ... the gas-solid interface between the concentration in the gas phase and the adsorbed concentration on the particles.

Therefore the $HgCl_2$ adsorption rate can be expressed as follows:

$$... = k_g ... \qquad (1)$$

Fluid-Dynamic Investigation and Optical Characterization of Particulate to Reduce Diesel Emissions

F. E. CORCIONE, B. M. VAGLIECO, G. VALENTINO *Istituto Motori CNR Napoli-Italy*

Abstract—This paper reports recent studies on air motion in a new combustion chamber geometry that reduces the exhaust emission and an on-line optical characterization of the particulate emitted at the exhaust for a development of an after treatment device. A direct injection (d.i.), heavy-duty, single cylinder diesel engine with an optical access, operating at 1000 rpm under motored conditions, was used.

Laser Doppler Anemometry (LDA) measurements of cycle resolved instantaneous tangential velocity, performed in different locations of the conventional straight sided and new generation of re-entrant combustion chamber, were reported. The mean motion and turbulence intensity were computed, during the compression stroke, using a filtering procedure of LDA data.

In addition, simultaneous multiwavelength scattering and extinction measurement from the near ultra-violet (u.v.) to visible were carried out at the exhaust of a d.i. diesel engine operating at 2200 and 3000 rpm.

Size, concentration, optical properties and mass of particulate were evaluated versus different engine parameters. A good agreement was observed on comparing the data with those obtained by conventional techniques.

Key Words: Engine, Fluid-dynamic, Soot, Optical diagnostic

INTRODUCTION

The superior fuel economy of diesel engines over the gasoline one has resulted in an increasing penetration in its automotive applications both in trucks and duty passengers cars. However, the use of the diesel in such vehicles might be ruled out if it cannot meet the severe regulations for exhaust particulate emissions.

Available control strategies are categorized into those which attempt to limit particulate formation in the engine and those which treat the exhaust stream.

At present, two particular methods are used for the reduction of exhaust emissions: the first one regards the optimization of the combustion system in terms of fluid-dynamic behaviour and interaction between the air motion and fuel injection (air-fuel mixing) (Zelenka *et al.* 1990, Corcione *et al.* 1991); the second one regards the development of after treatment techniques based on particulate filtering, with and without the presence of catalytic activity (Zalenka *et al.* 1990).

Gas motion within the engine cylinder is one of the major factors that controls the fuel air-mixing and combustion process in diesel engines (Heywood, 1988). In fact, recent studies (Mayer, 1987a, Mayer, 1987b, Ceronetti *et al.* 1990) have shown that the geometry of the combustion chamber plays an important role in the development of strategies and ways toward the "clean" diesel engine.

This paper analyzes the modifications of combustion system that were undertaken to reduce significantly the smoke emissions from a single cylinder d.i. diesel engine, representative of the heavy duty family.

Improvement in smoke emissions were achieved using a new generation of re-entrant combustion bowls resulting from their better air utilization with respect to conventional omega-shaped ones. The mean motion and turbulence intensity were measured by Laser Doppler Anemometry (LDA) for both chambers.

In addition, multiwavelength scattering and extinction measurements, in the spectral range from 200-500nm were carried out in the exhaust of the engine. Using the simultaneous spectral scattering and extinction measurement, a complete comprehension of the chemical and physical structure of the particulate was possible. In fact sizes, concentrations and optical properties were evaluated. Optical measurement attracted interest because it provided superior spatial and temporal resolution compared with sampling techniques.

Historically, the majority of studies on diesel particulates was carried out using intrusive techniques (Lipkea *et al.* 1978, Williams *et al.* 1988). These techniques were time consuming, not in real time, required long period of analysis and the data obtained may be influenced by sampling methods.

In the last decade optical measurements were carried out in exhausts. They were made by nephelometry or transmissiometry or opacimetry, generally at a single wavelength in the visible range (Lipkea *et al.* 1978) or in the infrared by photoacustic and absorption measurements (Roessler *et al.* 1981 and Krempl *et al.* 1985). These techniques permitted the determination of particulate mass. No simultaneous information concerning sizes, concentrations and optical properties was obtained.

EXPERIMENTAL APPARATUS AND PROCEDURES

Two Ruggerini RP 170 single cylinder, four-stroke, air-cooled, direct injection diesel engines of 100 mm bore and 95 mm stroke, with compression ratio of 17.2:1, were used. The first engine was modified to allow LDV measurements while the second one was used for engine tests and particulate characterization.

Figure 1 shows the engine with optical access and the two combustion chamber geometries tested. They are a conventional straight-sided and a re-entrant bowl with the same volume and squish area, respectively.

LDV System

The measurements of the tangential velocity were performed on the engine with optical access in the head and a shrouded intake valve allowing it to run with a swirl ratio of 1.0 at the end of intake stroke, as measured on a steady state air flow test rig. Figure 2 shows the experimental set-up for LDV measurements. It comprises a 4 watt Argon-ion laser that utilizes the two strong lines at 488 and 514.5 nm. The optical system was able to work both as a three-beam and a four-beam system, depending on the application. The signal quality of the individual component was improved by introducing an optical between beam laser output and standard optics. A single mode polarization optical was used. It had an attenuation of about 30 dB/Km and was suitable for a wavelength band from 450 to 650 nm. The optical terminated into two plugs suitable to link the two manipulators mounted directly on the laser output and on the standard optics.

The doppler signal was processed by a Burst Spectrum Analyzer (BSA). It performed a spectrum analysis in order to extract the doppler frequency.

An optical shaft encoder, connected directly to the engine shaft, drove a gating circuit enabling the BSA for the measurements. It drove the input gate during the measurement range with an angular resolution of 0.6 crank angle degree. More details concerning the LDV system, engine with optical access and data reduction were reported in Corcione and Valentino (1991).

Instantaneous tangential velocity was measured under motoring conditions at 1000 rpm during the last 60 degrees of the compression stroke using a crank angle window of 0.6 degree. The cycle, having a number of validated measurements lower than 60, was refused. This was because the cycle-resolved analysis rule required valid measurements to occur at least every 1-2 degrees. For each measurement point about 200 validated cycles were computed.

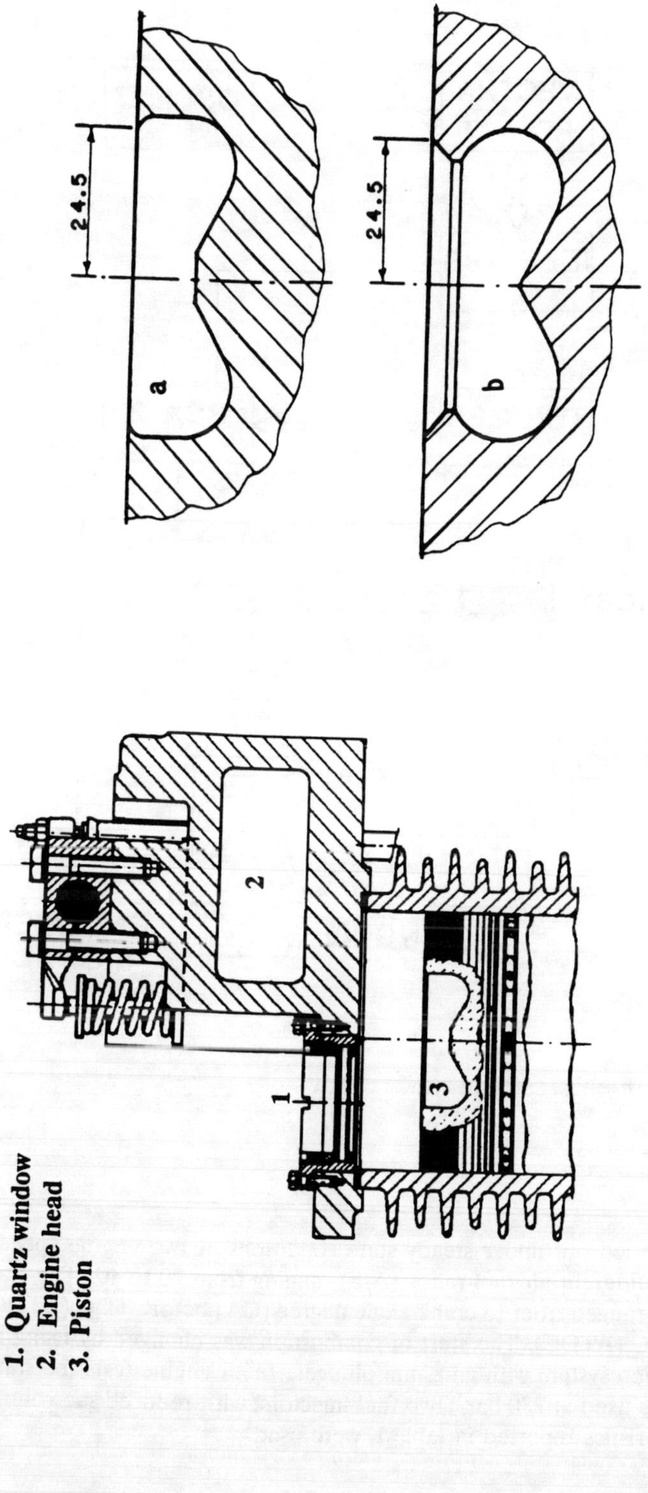

FIGURE 1 Engine with optical access and combustion chamber geometries tested: (a) straight sided and (b) re-entrant

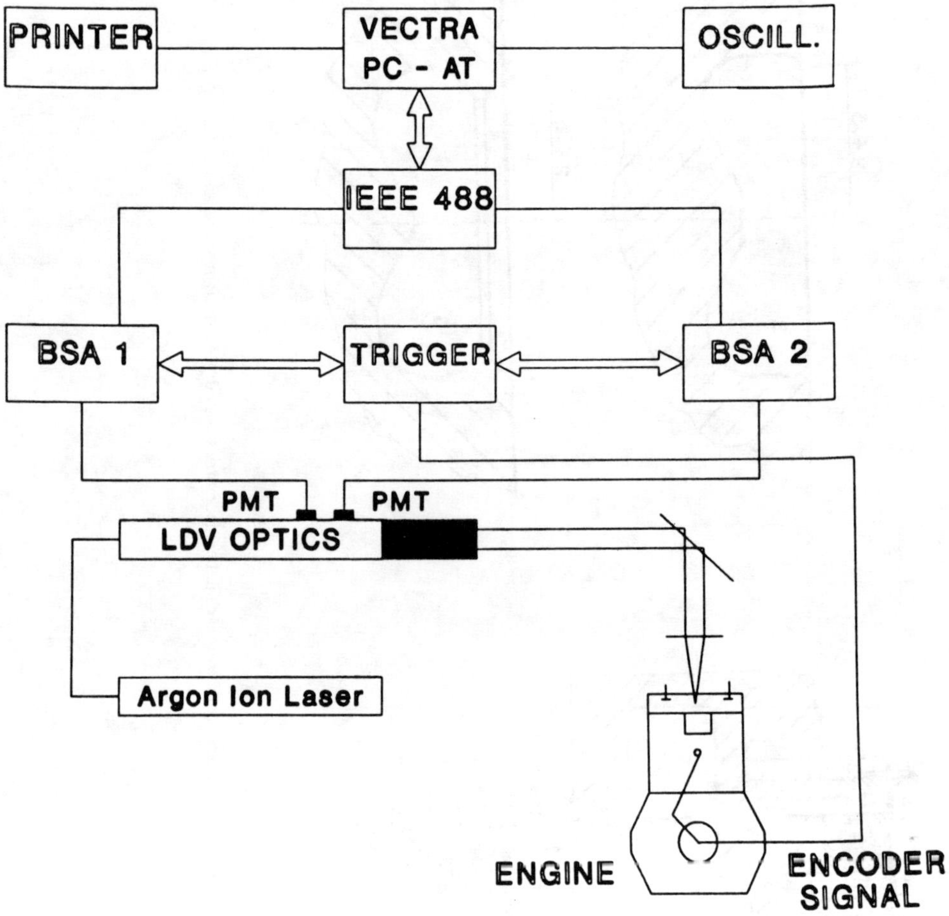

FIGURE 2 Schematic diagram of LDV system

Engine Tests

The engine tests were carried out under steady state conditions at two engine speeds, 2200 and 3000 rpm, with different air-fuel ratios (A/F) ranging from 20 to 40. The start of combustion (SOC) was ranged from 15 crank angle degree (CA) before TDC (BTDC) to 6 CA degree after TDC (ATDC). The start of combustion was changed by using an external Bosch fuel injection system with an 8 mm plunger. In all engine tests the static fuel injection pressure was fixed at 230 bar. Two fuel injectors, with reduced sac volume (sacless) and the characteristics reported in table I, were used.

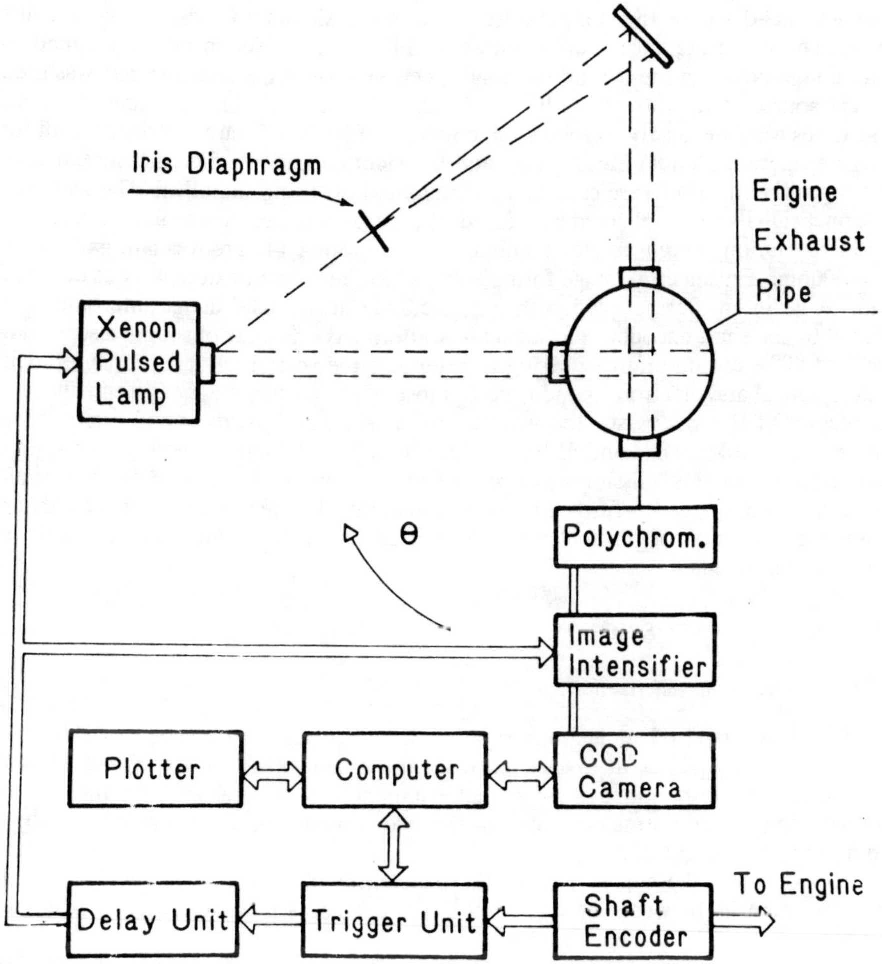

FIGURE 3 Experimental set-up for particulate measurements

TABLE I
Nozzles descriptions

Holes number	Diameter[mm]	Spray angle
4	0.30	160
5	0.22	150

The smoke emission was measured by a Bosch smokemeter, while the total gaseous HC and NO_x emissions were measured using a heated line (190°C) connected with flame FID and chemiluminescence analyzers, respectively.

Apparatus for On-line Particulate Measurements

Polychromatic light scattering and extinction measurements were carried out at the undiluted exhaust stream of the engine. The test section as located 1.5 m downstream the engine head where the temperature values were similar to those of the exhaust vehicle. The experimental set-up is shown in Figure 3. A Xenon lamp, designed to generate high power in the u.v. and a broad spectrum covering u.v. to infrared, was used as a light source. The pulse power had a rise time of 5 μs and a duration time of 1.2 ms. These times were necessary to follow the emission of the gas from the exhaust both for an engine operating under steady state and transient conditions. The measurement of scattering and extinction were carried out simultaneously using an optical. The scattered and extinct signals were collected by a fused silica lens on the entrance slit of a luminous and low resolution polychromator equipped with a grating 147 grooves/mm and blazed at $\lambda = 300$nm. The spectral image formed on the polychromator exit plane was detected on an image intensifier coupled with a photodiode array. The image intensifier was triggered by the shaft encoder. The absolute scattering coefficients of soot at a scattering angle θ of 90° were determined taking as reference the scattering of a propane flux in identical optical and electronic conditions to those of the particulate measurements. The cross-section of the gas versus the wavelength, at room temperature and atmospheric pressure, was well-known (Landolf-Bornstein, 1962). The extinction coefficients required no calibration since transmission was a relative measurement. All the signals were stored on a personal computer for further numerical analysis. The temperature of the exhaust gas and the collection of particulates were carried out in the same section as for the optical measurements.

RESULTS AND DISCUSSION

Fluid-mechanical Characterization

LDV data were acquired under motoring conditions at an engine speed of 1000 rpm in three horizontal planes at a distance from the engine head of 4, 7.5 and 11 mm, respectively. The instantaneous tangential component was measured on the last 60 degrees during the compression stroke in the two combustion chambers: the straight-sided and the re-entrant ones.

Figure 4a reports the comparison of the iso-contours of the tangential velocity and turbulence intensity between the two combustion chambers measured at 15 CA degree BTDC.

The swirl and the turbulence assume a different spatial distribution inside the two chambers. The re-entrant bowl produced a higher tangential velocity close to the wall and moved the high velocity zones toward the bottom of the bowl. In particular, on the periphery of the re-entrant bowl, the velocities were up to 50% higher than conventional one. The zones at higher turbulence intensity expanded toward the bottom and close to the wall.

Same behaviour was observed at 5 CA degree BTDC as shown in Figure 4b. The re-entrant bowl increased the swirl level at the periphery up to 40% and the turbulence intensity up to 25%. This spatial distribution can have a positive effect on the mixing between the air and the fuel that was injected during the entire injection period. This conditioned the choice of the nozzle spray angle in order to inject the fuel in to the zone at higher velocity. It was evident that the l/d ratio (length to diameter ratio of the nozzle) had to be appropriate to limit the effects of possible underpenetration of the jets (Yoshikawa *et al.* 1989).

Effect of Engine Parameters

Engine tests were carried out on re-entrant bowl and the straight-side one equipped with appropriate nozzles. Performance and exhaust emissions in terms of brake specific fuel

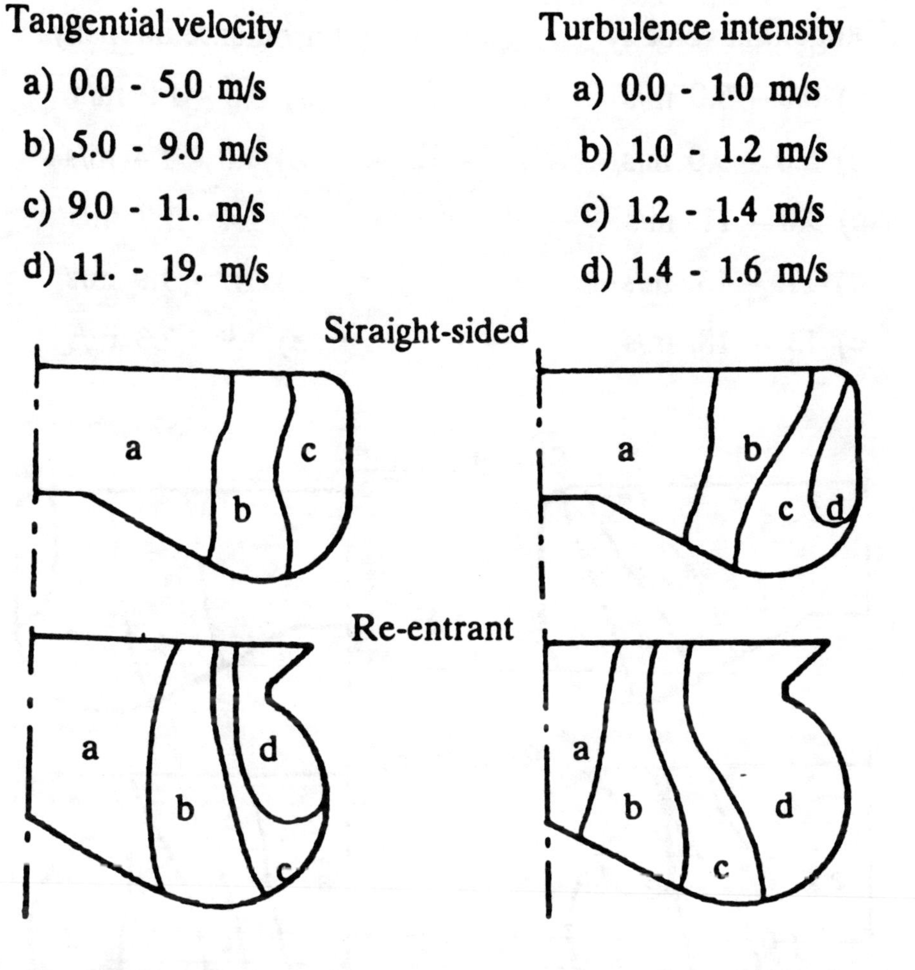

FIGURE 4a Spatial distribution inside combustion bowls of tangential velocity and turbulence at 100 rpm and 15 crank angle degrees BTDC

consumption (BSFC), brake medium effective pressure (BMEP), smoke, EIHC (hydrocarbons emission index) and $EINO_x$ (nitrogen oxides emission index) were measured at 2200 and 3000 rpm versus SOC.

Figure 5 shows BSFC, BMEP, Smoke, EIHC and $EINO_x$ for the two combustion chambers at A/F=30:1. The air fuel ratio chosen is an illustrative engine operating condition. A high reduction in smoke emission was obtained for both engine speeds using a re-entrant bowl equipped with the 5/0.22/150 sacless injector, the engine performance remaining unchanged in terms of BSFC and BMEP. As light worsening in NO_x emissions was observed before the TDC while the EIHC showed low values at higher engine speed for all the SOC.

The smoke reduction depended on the better air-fuel mixing obtained with the re-entrant bowl that had higher swirl and turbulence levels with respect to the straight side bowl, as was observed by LDA measurements. In addition, the use of 5/0.22/150

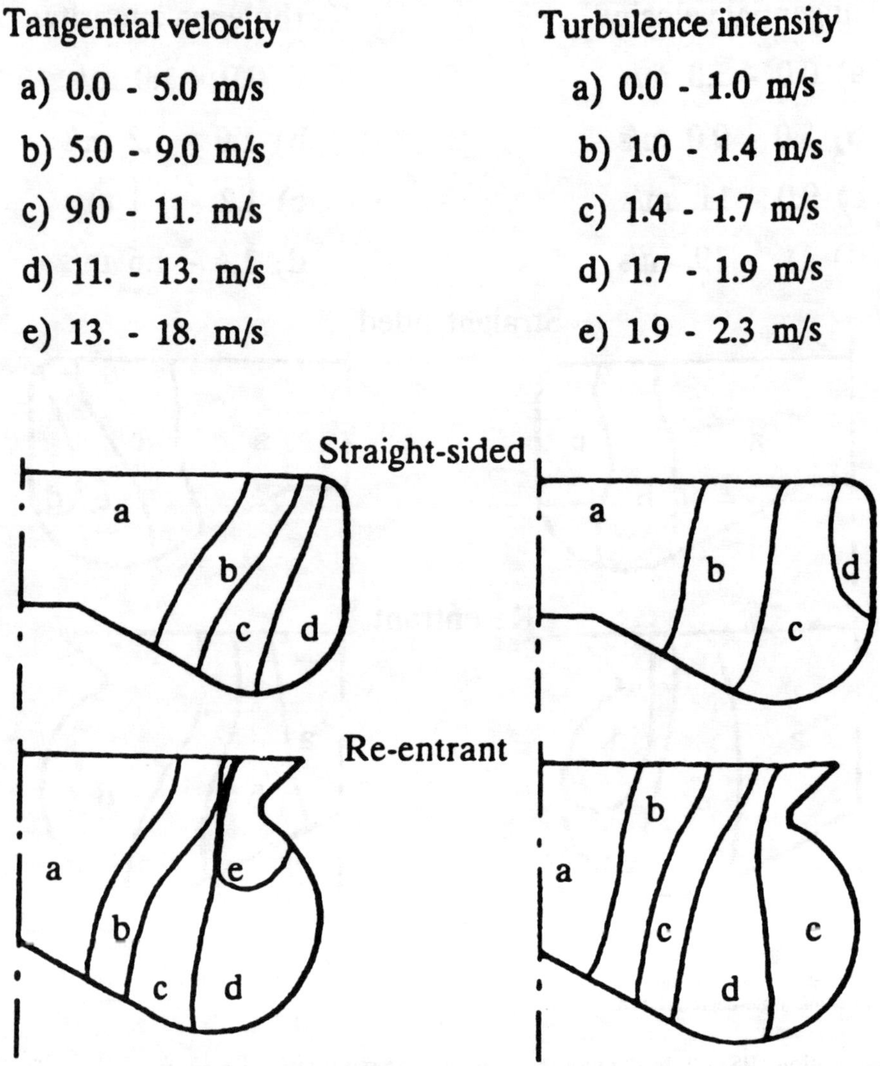

FIGURE 4b Spatial distribution inside combustion bowls of tangential velocity and turbulence at 1000 rpm and 5 crank angle degrees BTDC

sacless nozzle permitted a better fuel atomization giving a positive effect both on low soot formation and EIHC (Yoshikawa *et al.* 1989).

Particulate characterization

A review of the literature indicated that most of the previous research work on particulates in the exhaust of diesel engines was limited to measurements of their size distribution and mass concentration by microscopy and chemical sampling. They were time consuming, not real time and not able to characterize the engine in transient operating conditions

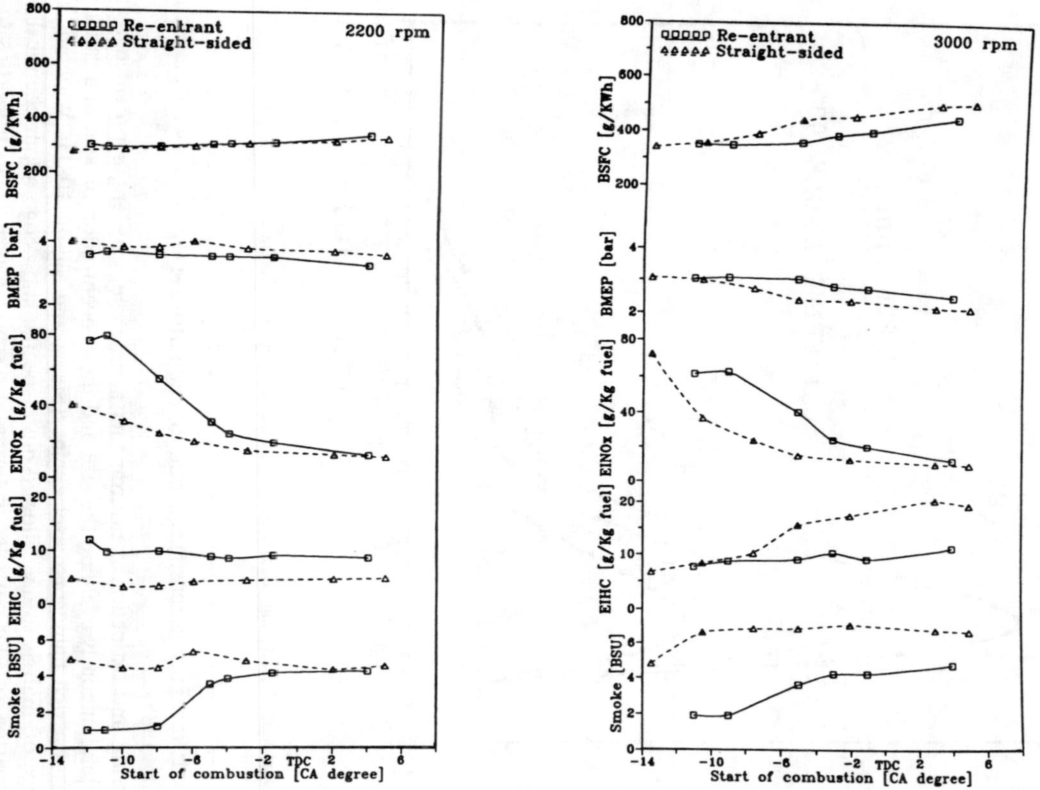

FIGURE 5 Effect of combustion chamber geometries on engine performance and exhaust emissions versus SOC at A/F = 30:1 and engine speeds of 2200 and 3000 rpm.

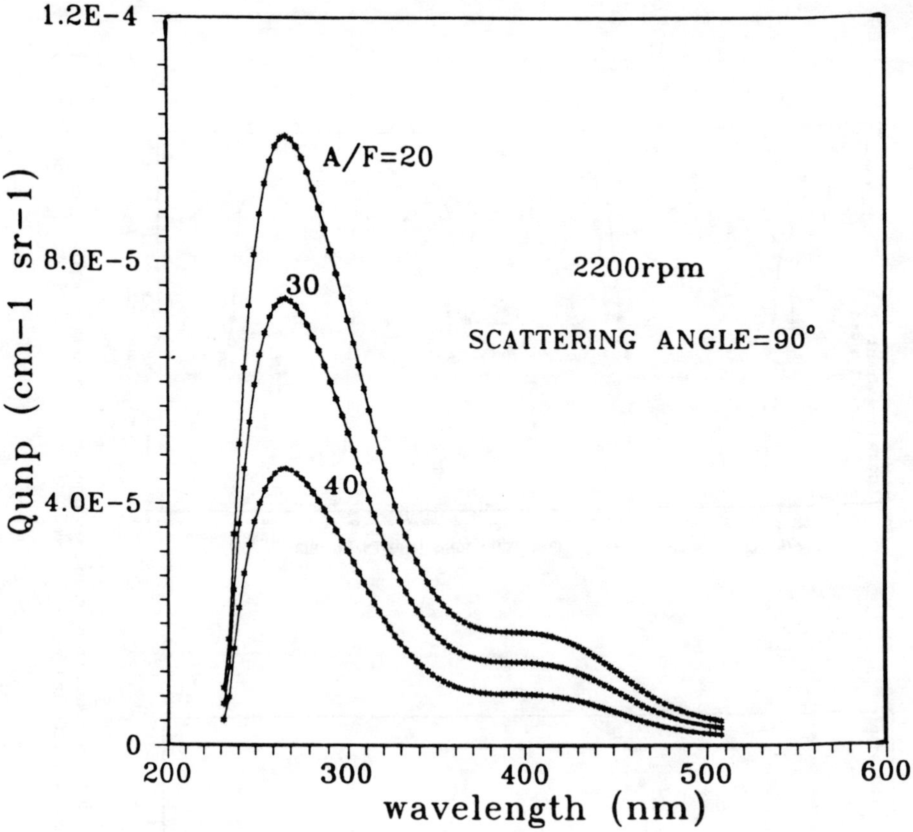

FIGURE 6 Unpolarized scattering coefficients versus wavelength for different air-fuel ratio (A/F) at engine speed of 2200 rpm.

(Lipkea *et al.* 1978). For this reason real time and nonintrusive techniques were developed (Krempl *et al.* 1985).

Light scattering and extinction measurements at a single wavelength provided a fast method to simultaneously evaluate the size and the concentration of soot in a radiative system such as a rich flame or diesel engine (Siegla and Smith, 1981). Widening the spectral range of analysis to u.v. allowed more specific knowledge about chemical and physical properties of soot particles in terms of dispersion of optical properties to be obtained (Vaglieco *et al.* 1990). Considering the previous experience on rich flames we investigated the exhaust at different engine operative conditions by polychromatic scattering and extinction.

Scattering and extinction measurements in the near u.v. were particularly promising because carbonaceous material and aromatic molecules exhibit an absorption band between 200 to 350 nm. On the other hand, data in the visible permitted us to evaluate dimensions, concentration and mass of the particulate.

In Figures 6 and 7 the spectral behaviour of scattering and extinction coefficients at different air fuel ratios and 2200 rpm are reported. The data showed a sharp peak around 280-300 nm, probably due to π-electron characteristic of graphitic-like material. In the

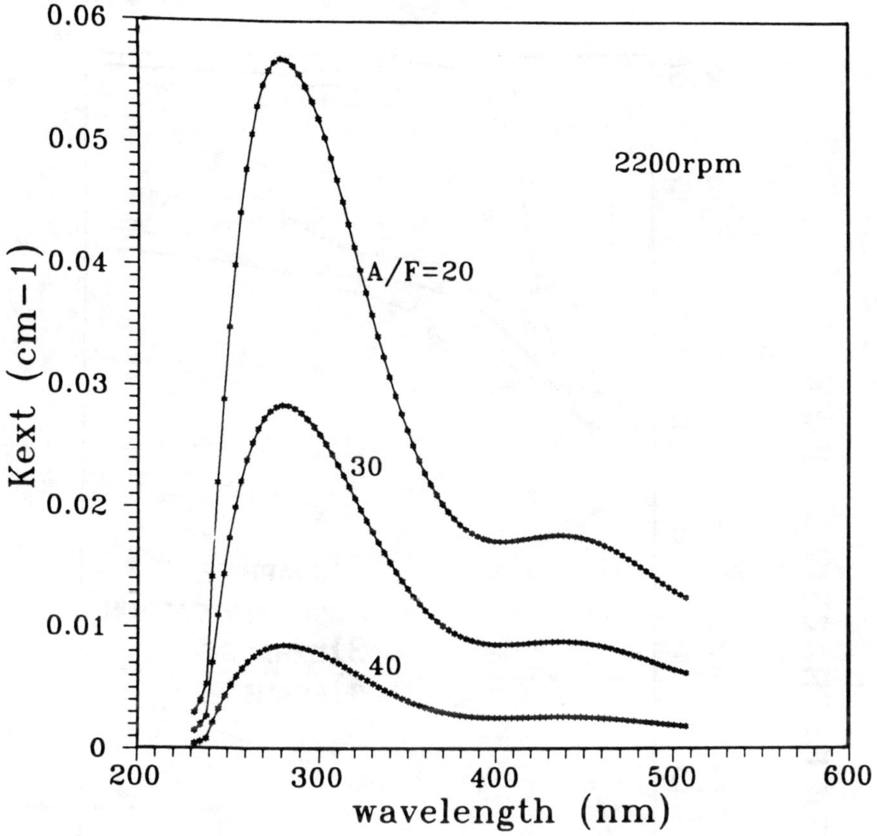

FIGURE 7 Extinction coefficients versus wavelength for different air-fuel ratio (A/F) at engine speed of 2200 rpm.

visible band a decrease which was more rapid for the scattering than the extinction was observed. This behaviour was different than that predicted by the Rayleigh theory. It was probably due to dispersion of the optical properties of the particulate.

Evaluation of particulate properties from scattering and extinction measurements was based on Rayleigh theory for isotropic spheres assuming the diameters were less than the wavelength (Kerker, 1969). For a monodisperse system of spherical particles the scattering factor, Q_{unp} is given by:

$$\mathrm{Qunp}(\theta) = \mathrm{N}\alpha^6 \left| \frac{\mathrm{m}^2 - 1}{\mathrm{m}^2 + 2} \right|^2 \frac{1 + \cos\theta^2}{2} \frac{\lambda^2}{4\pi^2} \tag{1}$$

and the extinction coefficient, K_{ext}

$$K_{ext} = -\mathrm{N}\alpha^3 \mathrm{Im} \left\{ \frac{\mathrm{m}^2 - 1}{\mathrm{m}^2 + 2} \right\} \frac{\lambda^2}{\pi^2} \tag{2}$$

where $a = \pi D/\lambda$ and m=n-ik are the size parameter and the particle refractive index, respectively. N is the concentration of particles (cm-3), D the diameter of the particle

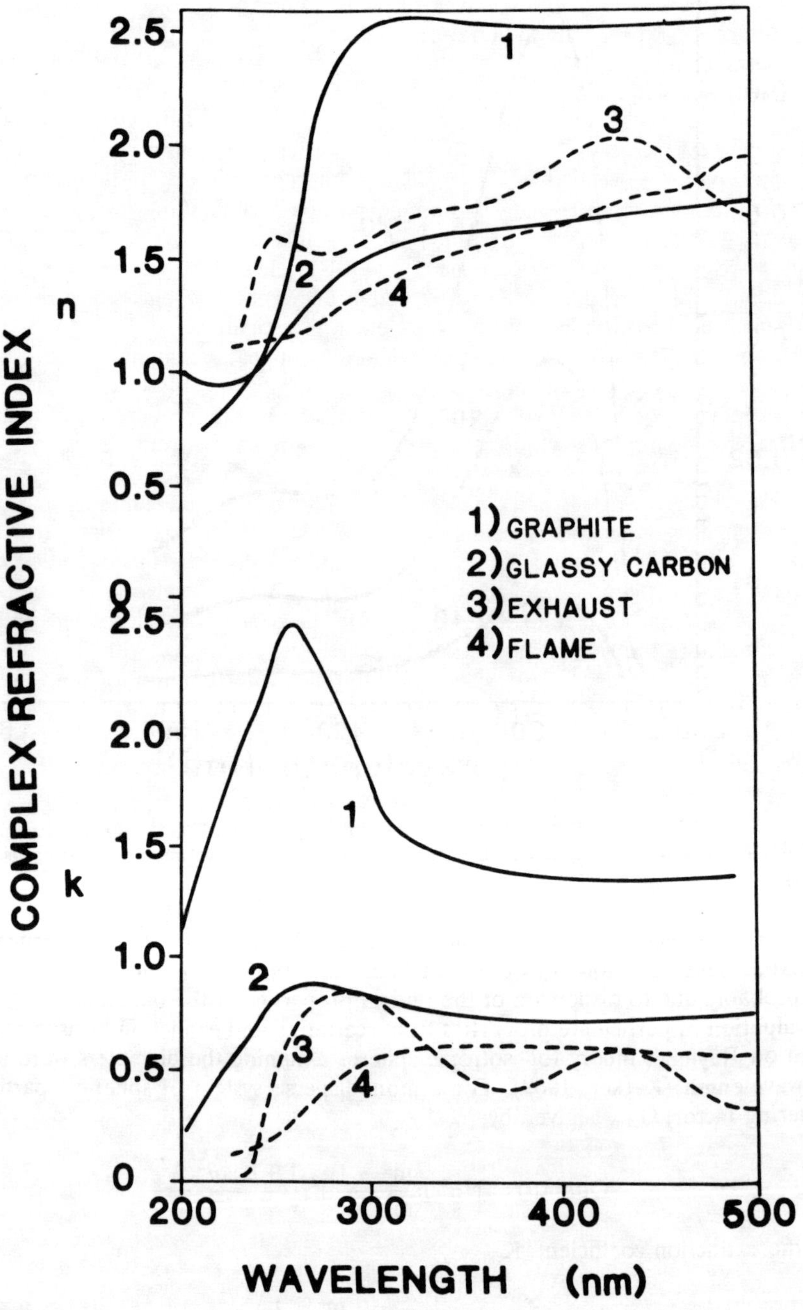

FIGURE 8 Particulate refractive index compared with graphite, glassy carbon and soot flame data.

(nm), λ the wavelength (nm) and θ the scattering angle. Combining equations 1 and 2 we obtained D, N and the particulate mass. The uncertainty concerning these values was relative to the choice of the complex refractive index m for the soot particles. In fact different values were found in literature (Dalzell and Sarofim, 1969) (Chippett and Gray, 1978), (Pluchino *et al.* 1980) (Lee and Tien, 1981), (Rosseler *et al.* 1981). Estimating the uncertainty in the determination of N, D and mass we chose the refractive index 1.9-0.55 (Lee and Tien, 1981, Bengtsson 1988 and Vaglieco *et al.* 1990), for a soot particle not containing hydrocarbons, with 1.7-0.33 for a particle containing a part of hydrocarbon absorbed on it (Rosseler *et al.* 1981). These different values of m did not influence the D value comparable with data of primary particles, constituting the agglomerate and observed by transmission electron microscopy (Lipkea *et al.* 1978). The D values range from 25-35nm, and the N range from 10E9-10E11 cm-3. The value of particulate mass was comparable with that obtained by weight of particulate collected at the exhaust in the same operative condition as for the optical investigation.

An apparent contrast between size obtained by optical techniques and those by TEM was observed relative to the agglomeration. In fact the spectral behaviour of scattering and extinction coefficients shown in Fig. 6 and 7 are different from the behaviour of spectra obtained in the agglomeration region and simulated with Mie theory for spheres of hundreds of nanometers (Vaglieco *et al.* 1988). This disagreement might be due to an agglomeration that takes place on the filter during the sampling and extraction of particulate. A complete chemical and physical characterization of particulate also required a study of the optical properties. In this regard, by the spectral measurement of scattering and extinction we evaluated the n and k values, numerically solving the system of equations 1 and 2. Details were reported in the paper of Vaglieco *et al.* (1990). The Figure 8 shows the optical properties of particulate compared with the data of other carbonaceous material. There was a big difference with graphite for both the value of n and k.

The particulate resonance in the u.v. was typical of graphitic-like material, but it was much enlarged with respect to graphite and more similar to glassy carbon. In the visible, a large dispersion was observed for the particulate that was similar to the soot of the flame. In conclusion, the behaviour of particulate was between that of glassy carbon and soot of the flame. This was probably due to both a disorder in space orientation of the graphitic layer and a turbostatic structure. Furthermore, the narrower and higher value of k of the particulate with respect to soot of flame might depend on a different value of temperature, that ranges from 100 to 200° C in the exhaust with respect to the flame.

CONCLUSIONS

An LDA technique was successfully used for fluid-mechanical characterization of the combustion chamber of a single cylinder d.i. diesel engine operating at 1000 rpm under motored conditions. Two combustion chamber geometries have been investigated with the aim of evaluating their fluidynamic behaviour.

The re-entrant bowl modified the swirl and turbulence at the end of compression stroke. It increased the swirl up to 50% at 15 CA degrees before TDC and expanded the zones at higher turbulence with respect to the straight-sided one. This behaviour improved the air-fuel mixing, thereby reducing the smoke emissions up to 50% with respect to the straight-sided one.

Low values of HC emission were obtained using a sacless nozzle 5/0.22/150.

To meet the stringent regulations relative to particulate diesel, additional exhaust system modifications were required. On-line particulate characterization using an optical technique was carried out at the exhaust operative conditions, whose temperature was similar to that at the vehicle exhaust. Size, concentration and mass of particulate were evaluated. The mean dimension of particles of soot obtained by the optical technique was comparable with the data of primary particles of agglomerate measured by TEM.

Optical properties of particulate obtained by polychromatic scattering and extinction measurements showed that it is a "poorly crystalline" arrangement with aromatic character and therefore the nomenclature of "graphitic-like" to describe soot was not correct.

REFERENCES

Bengtsson, P. E. and Alden, M. (1989). Application of a Pulsed Laser for Soot Measurements in Premixed Flames. *Appl. Phys. B.* **48,** 155-164.

Ceronetti, G. F., Imarisio, R., Stroppiana, B. (1990). Control Strategies and Design Modifications for Exhaust Emission Reduction in Direct Injection Diesel Engines. FISITA 905074.

Chippet, S., Gray, W. A. (1978). The size and optical properties of soot particles. *Comb. Flame* **31,** 149-159.

Corcione, F. E., Vaglieco, B. M., Valentino, G. (1991). In-Cylinder Fluid Motion and Emissions of a Conventional and Re-entrant Diesel Combustion System. SAE Paper No. 911842.

Corcione, F. E., and Valentino, G. (1991). Characterization of Fluid-Dynamic Behaviour of Diesel Engines by LDA Technique, Accepted at Fourth International Conference on Laser Anemometry, August 5-9, 1991. Cleveland, USA.

Dalzell, W. H., Sarofim, A. F. (1969). Optical constant of Soot and their Application to Heat Flux Calculations. *J. Heat Transfer* **91,** 100-104.

Kamimoto, T., Kobayashi, H. (1991) Combustion Processes in Diesel Engines, *Prog. Energy Comb. Sci.* **17,** 163.

Kerker, M. (1969). The scattering of light and other electromagnetic radiation. Academic Press, New York. Chap. 3, 37-91.

Krempl, P., Schindler, W., Schieler, E., (1985). Dynamic Measurements of Particulate Mass Emission on Light-Duty Diesel Engines under Steady State and Transient Conditions. SAE Paper No. 850269.

Lipkea, W. H., Johnson, J. H., Vuk, C. T. (1978). The Physical and Chemical Character of Diesel Particulate Emissions-Measurement Techniques and Fundamental Considerations. SAE Paper 780108.

Landolt-Bornestein (Ed.) (1962). Zahlenwerte und Funktionen aus Physik, Chemie, Astronomie, Geophysic und Technik, II Band, 8 Tiel, Optische Konstanten Springer, Berlin, Heidelberg, Chap. 6 p. 881.

Mayer, K. P. (1987). Medium and Heavy Duty Diesel Engine-Development Trends. International Internal Combustion Engine Exbition, April 7-12, Shanghai.

Mayer, K. P. (1987). Light Duty High Speed DI Diesel Engine-Development Trends. International Internal Combustion Engine Exbition, April 7-12, Shanghai.

Pluchino, A. B., Goldberg, S. S., Dowling, J. M., Randall, C. M. (1980). Refractive-index measurements of single micron-sized carbon particles. *Appl. Opt.* **19,** 3370-3372.

Roesseler, D. M., Faxvog, F. R., Stevenson, R., Smith, G. W. (1981). Optical Properties and Morphology of Particulate Carbon: Variation with air fuel ratio.

Siegla, D. C. and Smith, G. W (Ed.) (1981). Particulate Carbon Formation During Combustion. Plenum Press-New York.

Vaglieco, B. M., D'Alessio, A., Beretta, F. (1990). In Situ Evaluation of Soot Refractive Index in the UV-Visible from the Measurement of the Scattering and Extinction Coefficients in Rich Flames *Comb. Flame* **79:**259-271.

Vaglieco, B. M., Beretta, F., D'Alessio, A., De Natale P. (1988). Characterization of early formation and late agglomeration of soot particles through spectral measurements of the scattering and extinction coefficients. Tenth Task Leaders Meeting International Energy Agency, Amalfi, 291-295.

Williams, P. T., Abbass, M. K., Tam, L. P., Andrews, G. E., Ng, K. L., Bartle, K. D. (1988). A Comparison of Exhaust Pipe, Dilution Tunnel and Roadside Diesel Particulate SOF and Gaseous Hydrocarbon Emissions, SAE Paper No. 880351.

Yoshikawa, S., Furosawa, R., Arai, M., Hiroyasu, H. (1989). Optimizing Spray Behaviour to Improve Engine Performance and to Reduce Exhaust Emissions in a Small D.I. Diesel Engine", SAE Paper No. 890469.

Zelenka, W., Kriegler, P. L., Herzog, W. P. Cartellieri. (1990). Ways Toward the Clean Heavy-Duty Diesel", SAE paper No. 900602.

Some Measures Towards Enhancing the Performance of Gas-Fueled Diesel Engines at Light Load

G.A. KARIM *Department of Mechanical Engineering, The University of Calgary, Calgary, Alberta, Canada*

Abstract — The dual fuel engine is a ready means for utilizing gaseous fuel resources efficiently in diesel engines after appropriate conversion. These converted engines can provide an effective method for producing power while reducing exhaust emissions, especially in relation to exhaust particulates and oxides of nitrogen. More efficient and increased power output relative to the corresponding diesel operation can be achieved with dual fuel engines at high load. The light load performance, especially with high gas to diesel utilization ratios, remains relatively inferior. Poor fuel utilization efficiencies and relatively high unburnt hydrocarbons and carbon monoxide exhaust concentrations are readily encountered at light loads. This trend has necessitated usually the resorting back to diesel operation at idling and very light load conditions. The paper describes the combustion phenomena that bring about these limitations at light load. It then sets out some guidelines that can be followed to permit improved performance and exhaust emissions of diesel engines converted to dual fuel operation when idling and at light load.

Key Words: dual fuel engines, conversion, high and light load, reducing exhaust emissions, increased power, improved performance

INTRODUCTION

The dual fuel engine is a conventional diesel engine in which much of the energy release, and hence power, comes about from the combustion of a gaseous fuel such as natural gas/methane. Through continuing the usual timed cylinder injection of the diesel liquid fuel, ignition and the remaining part of the energy release are obtained. The introduction of the gaseous fuel is usually made through fumigation which involves its controlled continuous aspiration into the engine air by appropriate carburation equipment. More recently, increased efforts are being directed to develop intermittent electronically timed injection of the gaseous fuel into the air. This is done either just ahead of the intake valve or occasionally, directly into the engine cylinder.

The dual fuel engine represents a very attractive and economic means for utilizing gaseous fuel resources for the production of power while using the efficient conventional diesel engine. Usually the main aim in dual fueling, largely due to economic reasons, is to minimize the use of diesel fuel injection required for initiating ignition. Its replacement by the usually more abundant and often cheaper gaseous fuel is maximized throughout the whole load range. This is accomplished without undermining the operation and performance of the engine as a "straight diesel." There is a need to select optimally the liquid fuel quantity injected in relation to the type and extent of gaseous fuel supply for any specific engine and operating conditions so as to achieve the best of performance over the whole load range. There is a continuing research effort aimed at achieving this goal.

Dual fuel operation can have higher output with superior specific energy consumption than the corresponding "straight diesel" operation. This can be achieved while displaying superior exhaust emission characteristics, quieter and smoother operation, reduced thermal loading of engine parts and improved low ambient temperature operation in comparison to diesel operation (Karim, 1987).

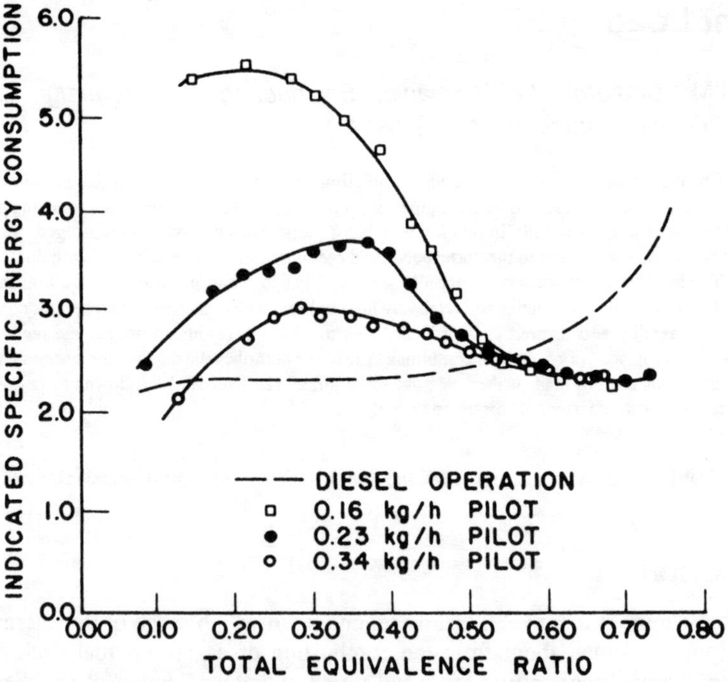

FIGURE 1 Variations of the indicated specific energy consumption with total equivalence ratio of an engine fueled with methane for a range of pilot quantities. The corresponding values for diesel operation are also shown.

The dual fuel engine can also have superior operational characteristics to those of the corresponding spark ignition engine. For example, it accepts a wider spectrum of fuels at high compression ratio without requiring wide changes in spark timing with load and fuel, nor exhibiting wide cyclic variations. The ability of the dual fuel engine to operate over a wide range of charge mixture strength permits mixture quality control without excessive exhaust emissions or resorting to throttling and its associated deficiencies.

The provision of an adequate and consistently timed liquid fuel pilot ignition in dual fuel engines permits the utilization of almost any gaseous fuel or vapor, albeit with varying degrees of success. Various gaseous fuel mixtures such as natural gases, biogases, producer gases, liquified petroleum gases, "low btu" gases, etc. have been employed successfully in dual fuel engines.

A serious problem associated with the dual fuel engine is the relatively poor light load performance associated with low output efficiency and inferior exhaust emission characteristics, as shown typically in Figures 1 and 2. This limitation arises primarily from a flame originating from an ignition region cannot propagate fast enough to consume the mixture due to its being either too lean and/or well below the effective flammability limit under the prevailing conditions. Figure 3 shows typically the variation in the unburnt fuel in the exhaust for a relatively small pilot quantity usage representing around 4% of the total thermal energy release. The need to provide mixtures richer than a limiting threshold of charge mixture strength for methane conversion is evident.

Another problem is the onset of autoignition and knock which imposes a limit to the maximum load that can be achieved for any engine with any gaseous fuel. The operational

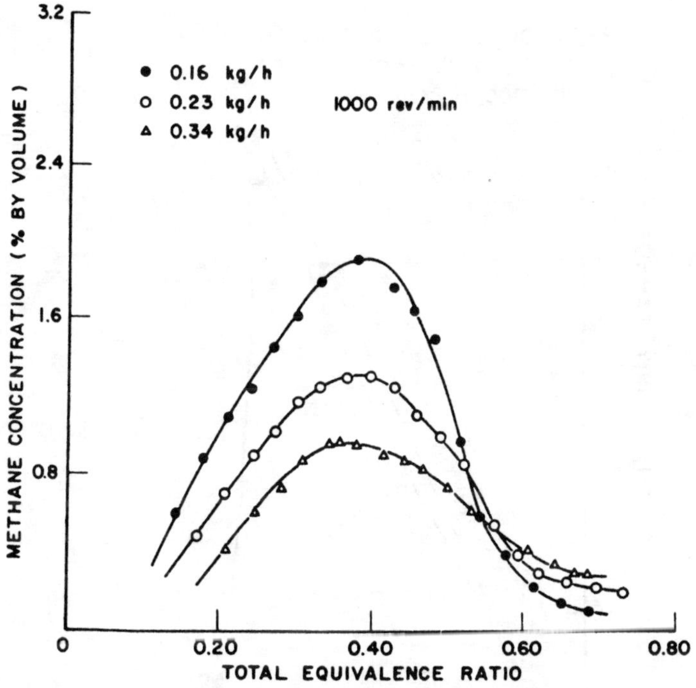

FIGURE 2 Typical variations of the concentrations of methane in the exhaust of a dual fuel engine fueled with methane with total equivalence ratio for a range of pilot liquid fuel quantities.

and design requirements for the avoidance of knock tend to be different from those needed to improve light load operation of dual fuel engines, Karim *et al.* (1967).

The present contribution considers the light load operation of dual fuel engines in general and suggests some measures that can enhance light load performance, including the reduction of the exhaust concentrations of the unburned methane and carbon monoxide.

LIGHT LOAD DUAL FUEL ENGINE PERFORMANCE

The injection of a small quantity of pilot liquid diesel fuel is done so that its autoignition will provide ignition centres from which turbulent flames travel subsequently. These would then consume the lean homogeneous gaseous fuel/air mixture. Changes in the magnitude of pilot liquid fuel, especially at light load when very lean homogeneous mixtures are involved should influence the combustion process very significantly. Changes in the size of the initial combustion zone, the number of ignition centres and the associated thermal energy released would follow almost proportionally to the mass of the pilot. The presence of a gaseous fuel with the engine air during the relatively long compression period will affect, sometimes markedly, the processes of preignition and subsequent combustion of the pilot and the whole cylinder charge.

The extent of deterioration in engine performance at light load and idling depends on a number of factors. These include the size of the pilot employed, the gaseous fuel used, operating conditions and the type of engine being used. With some fuels and engine combinations, there may be occasions when engine idling or even light load operation may

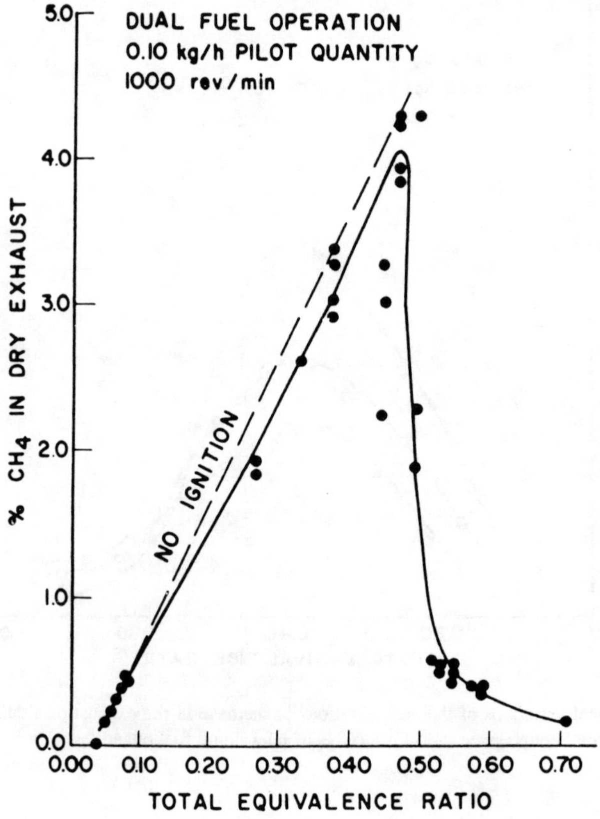

FIGURE 3 Variations of the concentrations of ambient methane in the exhaust of a dual fuel engine fueled with methane with the total equivalence ratio for a relatively small pilot quantity. The corresponding concentrations when there is no combustion is also shown.

become totally impaired. The ignition delay of the pilot, reckoned on the basis of cylinder pressure changes, increases considerably with the introduction of the gaseous fuel with the air, but decreases later as further gas is added to the air, as shown typically in Figure 4.

The increase in the delay period cannot be accounted for by the slight reduction of the partial pressure of oxygen with the small addition of the gaseous fuel. The associated small reduction in the temperature of the charge at around top dead centre due to the slightly higher overall specific heat of the charge cannot also account for the observed increase in the delay.

The homogeneously mixed gaseous fuel with the air will undergo chemical reactions prior to diesel injection during the relatively long compression stroke. Even lean mixtures with air of the not so reactive fuel methane can undergo significant reaction activity. There is a gradual build up of the concentrations of reactive species, such as radicals and partial oxidation products, that becomes quite rapid as the top dead centre portion is approached. Figure 5 shows typically the calculated concentrations of formaldehyde, CH_2O and CO at various stages of the compression process for a wide range of equivalence ratio using a detailed chemical kinetic model, Karim (1991). Accordingly, the homogeneously dispersed gaseous fuel and its partial oxidation products within the pilot spray region can participate

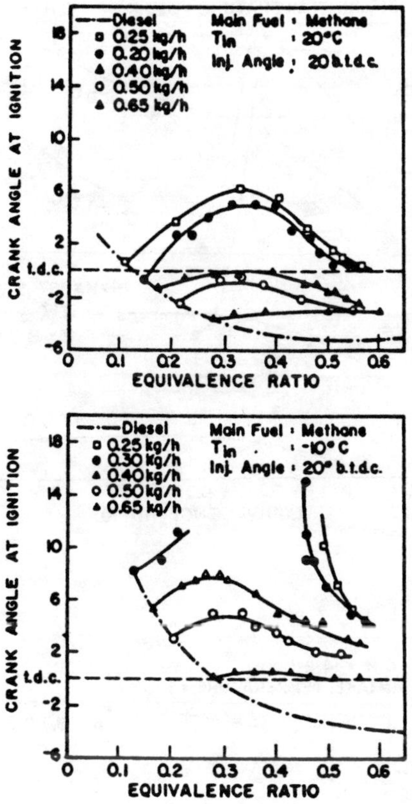

FIGURE 4 Typical variations in the point of ignition with the total equivalence ratio of a dual fuel engine fueled with methane for a range of pilot quantities at two different intake temperatures.

actively in the preignition chemical processes of the high hydrocarbon pilot fuel. This may contribute significantly to the observed variations in the delay period. The competition for active radicals between the two fuel systems can, especially at very lean mixtures, affect the preignition processes of the diesel pilot adversely. As can be seen in Figure 5, when a considerable amount of the gaseous fuel is added, significant amounts of species and associated energy are produced earlier in the compression process. These would aid the ignition processes following pilot fuel injection and contribute towards not only halting the increase in the ignition delay, but also helping in its reduction. These changes, also will have a major effect in establishing the extent and nature of exhaust emissions from dual fuel engines at light load.

At very light load, when very lean gaseous fuel/air mixtures are being used, significant proportions of the low concentrations of the gaseous fuel will not burn completely. This, as shown in Figure 6, can be a significant source for exhaust fuel emissions when very small pilots are used. This is despite the presence of much excess air and pilot ignition. The small amount of the gaseous fuel present in the air, even at the high temperatures encountered near the end of compression and the ignition of the diesel pilot releasing energy and raising the effective temperature locally, cannot permit turbulent flame propagation. A flame starting from any pilot ignition centre will be quenched quickly leaving some of the

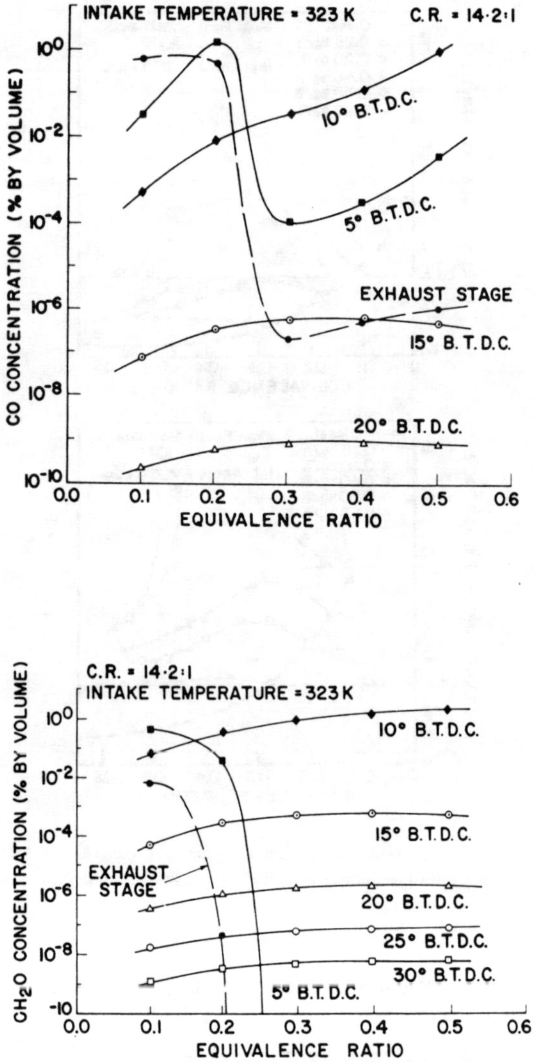

FIGURE 5 Calculated concentrations of carbon monoxide and formaldehyde during different stages of the compression process while using a detailed chemical kinetic model in an engine fueled with methane for a wide range of equivalence ratios. The calculated concentrations surviving to the exhaust stage are also shown.

gaseous fuel unoxidized and survives to the exhaust stage. For the same pilot fuel quantity, only through using a sufficiently rich gaseous fuel air mixture that flames can propagate throughout the charge. There is a need to reach an effective flammability limit within the mixture environment to permit widespread turbulent flame propagation subsequent to pilot ignition. This low gaseous fuel utilization at very lean mixtures leads also to a significant increase in the concentrations of the partial oxidation products, notably carbon monoxide. The exhaust levels obtained will be beyond those normally observed in the corresponding diesel operation. Hence, there has been a tendency to retain virtually exclusive diesel operation for idling and very low load operation in dual fuel applications.

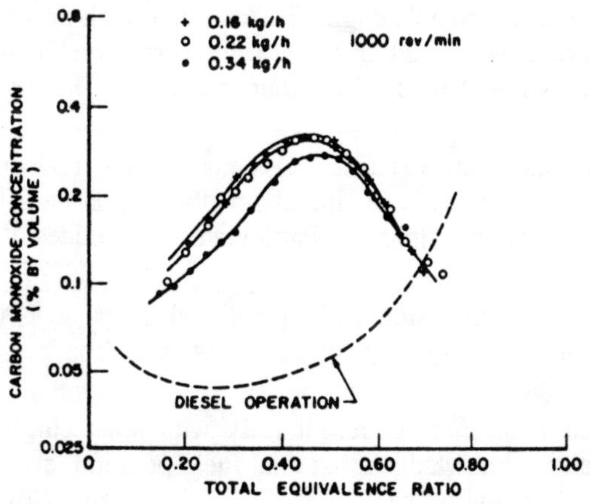

FIGURE 6 Typical variations of the concentrations of carbon monoxide in the exhaust gas of a dual fuel engine fueled with methane, with total equivalence ratio for a range of pilot liquid fuel quantities. The corresponding values for diesel operation is also shown.

When the cylinder charge is too lean to support combustion, the flame fronts propagating from the various ignition centres will not extend to all regions of the cylinder. Some pre-ignition reactions may proceed within the charge but the net effect is to leave a sizeable fraction of the homogeneously dispersed gaseous fuel unchanged to emerge in the exhaust. Some partial oxidation products will be produced as a result of the preignition reactions and in regions of the fringes of the main combustion zone where flame propagation and combustion are quenched. Thus, mainly the fuel within these combustion zones will be oxidized. When the gaseous fuel concentration exceeds a limiting range, these turbulent flame fronts will sweep throughout the cylinder. Faster energy release rates, higher rates of pressure rise and improved power output and efficiency will then ensue.

The poor fuel conversion at very lean mixture concentrations, increases the specific energy consumption based on the pilot diesel fuel and the gaseous fuel supplied. The associated power output is also lower than those obtained under the corresponding straight diesel operation. This situation becomes especially severe at low intake temperature operation, such as under cold ambient temperature conditions or when using liquified natural gas. Light load operation can also be unacceptably poor when using low heating value gaseous fuel mixtures having excessive amounts of diluents such as nitrogen, carbon dioxide and water vapor. Examples of such fuels can be found with certain natural gases, industrially produced gases and biogases.

SOME DESIRED GOALS OF HIGH LOAD OPERATION

The conversion of conventional diesel engines so as to be able to operate as dual fuel engines satisfactorily at idling and light load needs to satisfy the following main goals:

i. Avoid whenever possible the need to use relatively large quantities of the liquid pilot and minimize its use in relation to that of the gaseous fuel.

ii. Retain throughout the option of being able to operate the engine wholly as a diesel whenever desired without undermining normal engine performance as a diesel.Often the change over from dual fuel to diesel operation needs to be carried out without interruption to engine load nor producing fluctuations in the speed.

iii. Maintain throughout acceptable engine performance with power output and efficiency values comparable or exceeding those of the corresponding diesel operation. This is to be achieved without reduced combustion stability nor increased cyclic variations.

iv. Reduce to a minimum the exhaust emissions of dual fuel engines, especially in relation to unburnt hydrocarbons, mainly methane and partial oxidation products which are mainly carbon monoxide. Particulates and oxides of nitrogen tend to be insignificantly low.

v. Maintain throughout conversion and operational costs at a reasonably low level without undermining in any way the safety of operation nor the expected life of trouble free operation.

These objectives cannot be met at present fully with most current engine conversion practices. It is normally acknowledged that dual fuel operation, even on methane, has not been optimized fully. The environmental benefits arising from its low emissions are yet to be fully recognized and demonstrated. There is a need for continued research and development to achieve these highly desirable features. Meanwhile, some of the main measures that can be pursued are being outlined in this contribution.

SOME MEASURES RELATING TO THE PILOT FUEL SYSTEM

The pilot liquid quantity is one of the most important variables that can have a controlling influence on the performance of dual fuel engines at high load. Increasing the relative size of the pilot quantity will increase the conversion of the gaseous fuel to complete combustion products, as shown typically in Figures 1 and 2. A larger pilot may provide, in principle, a greater multitude of ignition centres with larger reaction zones within the overall very lean fuel gas/air mixture. A larger pilot can also provide a greater energy release with the production of higher pressures and local temperatures. Flame propagation within the charge will cover relatively shorter paths.

The overall ignition delay period will also be reduced somewhat (Karim, 1987). This would permit improved and less patchy flame propagation. There is a reduction in the likelihood of some of the liquid fuel from the main spray to be dissipated and mixed with the charge to produce fuel concentrations that are much too low for rapid ignition and combustion. Mixtures of diesel fuel vapor and air tend to be more flammable than those involving methane. The presence of some diesel vapor within the charge will modify the local composition and lower the effective lean flammability limit of the charge. Engine performance will be affected favourably.

Normally, the pilot is introduced into the charge through the use of the conventional diesel injection system without significant modifications made to it. The quantity injected is often very small. Occasionally, it may be so small as to be outside the operational range of the injector when the engine is operating as a diesel. Conventional engines also are not operated normally with variable injection timing. Accordingly, the options available to improve light load performance with small pilot quantities without changes to the fuel injection system tend to be limited. A notable factor is to ensure that the injector does not get overheated when small liquid fuel quantities are injected while significant energy release through gas combustion is proceeding within the cylinder. There is also a need to ensure that the fuel injection system is well matched to the engine when operating as a diesel even at light load, Karim *et al.* (1988). There is evidence to show that the lowest injector nozzle opening pressure needs to be used while still providing satisfactory engine operation on liquid fuel alone, Karim (1987).

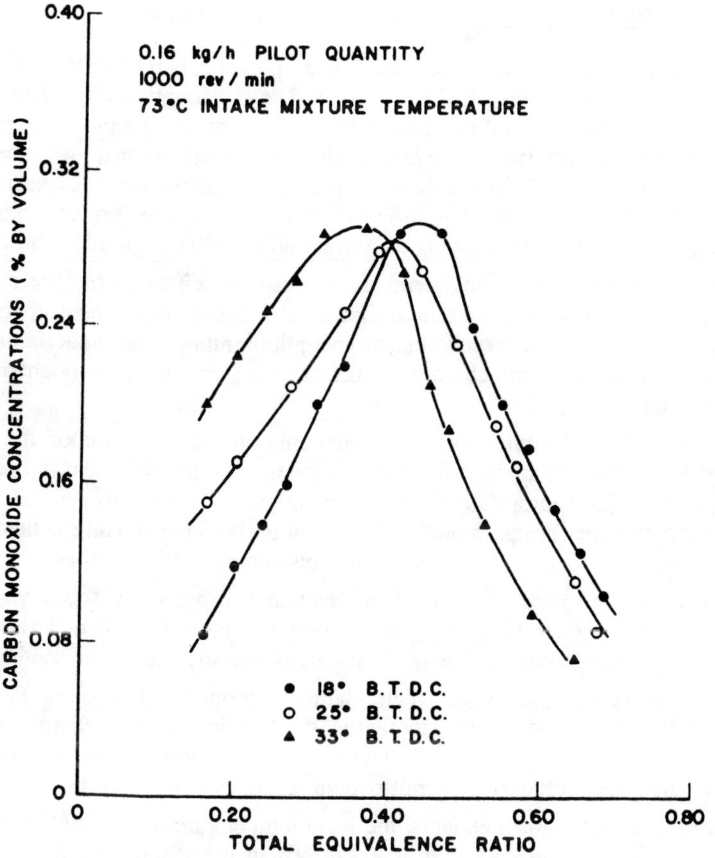

FIGURE 7 Typical variations of the concentrations of carbon monoxide in the exhaust gas with the total equivalence ratio for an engine fueled with methane for a range of pilot injection timing at a constant small pilot quantity of 0.16 kg/h.

With the advent of more versatile diesel fuel injection aided by electronic controls, it may be possible to vary the injection characteristics so as to obtain increased performance benefits at light load when dual fuel operation is used. The injection timing of the pilot needs to be varied from that of diesel operation. Advancing the injection timing of the pilot fuel beyond that of diesel operation increases the residence time and the activity of the partial oxidation reactions and reduces the flammability limit boundary. The injection characteristics, including the timing of the pilot, as shown typically in Figure 7, need not remain fixed over the whole load range, but can be optimized continuously in relation to the gas quantity used. The optimum pilot atomization characteristics for a dual fuel engine may not be associated necessarily with optimum diesel operation. Hence, for a dedicated dual fuel operation, there is a need for a thorough evaluation of the optimum injection characteristics, including injection pressure, number of spraying jets involved, location of the jet in the cylinder, size of jet, type of nozzle being used, etc. The physical and chemical characteristics of the pilot fuel also need to be looked at thoroughly for such a dedicated operation.

SOME MEASURES RELATING TO THE FUEL/AIR CHARGE

The chamber volume, even near the top dead centre position, is relatively very large when compared to the size of the diesel fuel pilot region. The consumption of much of the charge and the release of the associated energy need to be accomplished rapidly and completely following ignition via turbulent flame propagation. Any measures that lower the effective lean flammability limit of the cylinder charge and promote rapid subsequent flame propagation in the presence of pilot ignition will improve engine performance during idling and light load operation. The following measures can contribute towards achieving this:

i. Adopting measures that would tend to increase the effective fuel/air ratio of the charge, such as by some preferential throttling of the air component of the cylinder charge. Care needs to be taken to ensure that pilot ignition is not undermined by such throttling nor adverse turbocharger performance is permitted in turbocharged engine applications.

ii. The introduction of some favourable fuel mixture stratification of the charge is another less tried and more difficult measure to accomplish. It can be achieved in principle, through appropriate charge preparation, turbulent mixing and/or suitable engine and chamber design modifications. A slightly richer mixture is then produced in the vicinity of the pilot fuel regions than elsewhere in the cylinder.

iii. Small amounts of various auxiliary fuels, such as hydrogen or gasoline vapor, can be selectively added to the natural gas to enhance its combustion characteristics. This is achieved at the expense of increased complexity and the use of yet another fuel.

iv. Resorting to some exhaust gas recirculation without much cooling increases the initial charge temperature and increases the seeding of the charge with active products. Avoiding the presence of diluents, such as carbon dioxide, nitrogen or water vapor in the gaseous fuel will permit overall leaner mixture operation.

v. Measures that will tend to enhance the local temperature of the charge at light load will be helpful in improving light load performance and reducing the extent of un-reacted gaseous fuel surviving through to the exhaust stage. As shown typically in Figure 8, the slight preheating of the intake gas/air mixture provides for reduced exhaust emission at very light load. Other similar measures may include operation with higher water jacket temperatures and/or adopting engine designs that would reduce heat transfer from the engine cylinder. In turbocharged engine applications, excess air may be directed into the compressor intake to increase the temperature of the charge further.

vi. Reduction in the engine operational speed would permit longer time around top dead centre for flame propagation from the ignition centres to proceed within the lean fuel/air charge.

SOME MEASURES FOR CONTROLLING GASEOUS FUEL INTRODUCTION

Much of the current practice of dual fuel conversion of diesel engines introduces the natural gas into the incoming air outside the cylinder. This procedure while employing various carburation systems tends to be simple, economically attractive and does not require a high pressure gas supply. The effective compression ratio of the engine can be reduced by the presence of the gaseous fuel with the air due to the increase in the specific heat of the charge in comparison to that of air operation. It is also difficult to achieve low exhaust emissions of unburnt or partially oxidized fuel especially with two stroke engine applications or with engines having a significant valve overlap period. This fumigation type of fuel introduction

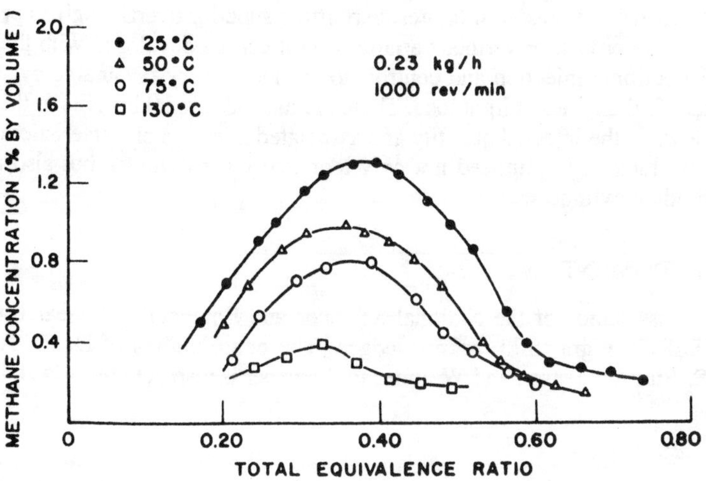

FIGURE 8 Typical variations of the concentrations of methane in the exhaust gas with total equivalence ratio for an engine fueled with methane for a range of intake mixture temperatures.

does not lend itself well to charge stratification while it permits reactions to proceed unimpeded during compression. This reaction activity may influence the preignition processes of the pilot uncontrollably. With richer mixtures and certain fuels, it may contribute to the onset on autoignition and knock.

Increased efforts are being directed recently at adopting timed port natural gas injection (e.g., Beck *et al.* 1989; Wakenell *et al.* 1987). Much of the mixing with the air then takes place within the cylinder. This procedure, which can provide improved controls of the gas injection into the air, has features that would contribute towards enhanced light load performance of dual fuel engines. It would be especially attractive for two-stroke applications and may permit in principle suitable charge stratification. This may be neither easy nor cheap to achieve in practice without further research effort.

An attractive approach that requires further research and development, especially for medium and small size dual fuel engines is the direct injection of natural gas in the form of a gas jet from a high pressure supply into the cylinder after valve closure and well into the compression stroke. This approach requires not only a high pressure gas supply but also relatively sophisticated and expensive gas injectors with electronic controls (e.g., Beck *et al.* 1989; Wakenell *et al.* 1987). During diesel operation, the gas injector remains idle while being exposed to high temperature combustion gases without the benefit of injected gas cooling.

High pressure gas injection into the cylinder permits some controlled charge stratification. This approach requires further research and development efforts especially since a lack of uniform mixing of the injected fuel with the air, within the short time available, will lead to patchy flame propagation and poor engine performance and exhaust emissions.

The potential future advancement in gas injection controls, with appropriate microelectronics may permit significant improvement to the light load operation of multicylinder dual fuel engines. Selective controls of the gas injection among the individual cylinders can be made also. This will permit the injection of a gaseous fuel at any cycle into some of the cylinders while others operate on pilot diesel only. Thus, for the same overall lean gas fuel/air ratio, those cylinders operating on the dual fuel mode would be operating at much richer gas/air mixtures, permitting superior overall engine operation and emissions. This approach or various variations on it can be developed with the advent of appropriate electronic injection and controls to produce a very favourable overall performance of dual fuel engines at light load. There are also additional benefits to the light load performance when the injected quantity and associated injection characteristics of both the gas and liquid fuels are optimized not only identically per cylinder, but also among the various individual cylinders.

ACKNOWLEDGMENT

The financial assistance of the National Sciences and Engineering Research Council of Canada (NSERC) is gratefully acknowledged. The contributions of Dr. D. Azzouz and Messrs. K.S. Burn, L. Jensen and W. Jones to the paper is also acknowledged.

REFERENCES

Beck, N.J., Johnson, W.P., George, A.F., Pekison, P.W., Van der Lee, B., and Klopp, G. (1989). Electronic Fuel Injection for Dual Fuel Diesel-Methane, SAE Technical Paper No. 891652.

Karim, G.A. (1987). The Dual Fuel Engine, in: *Automotive Engine Alternatives*, ed. Robert L. Evans, Plenum Press, New York/London, pp. 83–104.

Karim, G.A. (1991). An Examination of the Role of Formaldehyde in the Ignition Processes of a Dual Fuel Engine, SAE paper presented at the Fuels and Lubricants Meeting of the SAE, SAE Paper No. 912367, pp. 85–92.

Karim, G.A., Jones, W., and Raine, R.R. (1989). An Examination of the Ignition Delay Period in Dual Fuel Engines, SAE International Fuels & Lubricants Meeting, Baltimore, Maryland, SAE Paper No. 892140, pp. 97–103.

Karim, G.A., Klat, S.R., and Moore, N.P.W. (1967). Knock in Dual-Fuel Engines, *Proc. of The Institution of Mechanical Engineers*, **181**, pp. 453–466.

Karim, G.A., Raine, R.R., and Jones, W. (1988). An Examination of Cyclic Variations in a Dual Fuel Engine, *Proc. of the International Fuels and Lubricants Meeting and Exposition of the SAE*, Portland, Oregon, SAE Paper No. 881661, pp. 1–9.

Wakenell, J.F., O'Neal, G.B., and Baker, Q.A. (1987). High Pressure Late Cycle Direct Injection of Natural Gas in A Rail Medium Speed Diesel Engine, SAE Technical Paper No. 872041.

On Simulation of the Inflammation Period in Spark Ignition Engines

H. MÜLLER and J. KREIS *Technische Universitat Braunschweig, Institut fur Verbrennungskraftmaschinen und Flugtriebwerke, Braunschweig, Germany.*

W.-D. WAGNER and B. ERATH *Mercedes-Benz AG, PKW-Motoren-Konstruktion, Stuttgart, Germany.*

Abstract—A simulation model has been evolved which performs the duration of the inflammation period t_{EP}. The principal parameters of this model are environmental condition, air supply, mixture composition with including residual gas, and engine speed. Due to the intention of the model to be a design tool rather than a precise but complicated model for fundamental research the algorithms used are fast and according to physical properties. These algorithms have been fitted to experimental results obtained by a spark ignition engine. The prediction of t_{EP} at reference conditions has been formulated based on combination of mixture density at a particular crankangle and the ratio of integral length scale and turbulence intensity at ignition time. Deviation flame speed relations. The influence of engine speed has been assumed to result in changed turbulence intensity which influences t_{EP}. The in mixture composition from reference conditions has been performed by laminar performance of the model is good, on most conditions the difference between predicted and analyzed value of t_{EP} is better than 4 deg CA.

Key Words: spark ignition engine, energy conversion, ignition delay inflammation period, exhaust emission, simulation model.

INTRODUCTION

The improvement of new generations of homogeneous charge spark ignition engines with regard to environmental aspects is based on sufficient simulation of the energy conversion process. Such kind of simulation supports proper engine adjustment for highest efficiency and therefore minimizes the emission of CO_2 as well as other gaseous pollutants.

Detailed mathematical models are available to study multi-dimensional engineflow (e.g. Henriot *et al.*, 1989; Ahmadi-Befrui *et al.*, 1989) or complex chemical reaction mechanisms (Westbrook *et al.*, 1988). Models for predicting the combustion process in turbulent engine flow are still under development (Tang-Wei Kuo and Reitz, 1989). The detailed models, however, are very expensive in terms of computation time and cost, which is why these models are primarily tools for fundamental research.

A drawback of most models, even the very detailed ones, is an empirical assumption of the inflammation period of the combustion process. For best simulation results, the inflammation period and the main conversion period of combustion should be handled separately. This is due to the different dependency of these periods on changes in engine operating conditions. The effect is shown in figure 1 left by the influence of a varied (relative) air-fuel-ratio λ on these two periods (Müller *et al.*, 1986). It is preferable to divide the combustion process into inflammation period, from ignition up to 1% of converted energy, main conversion period, between 1% and about 90% of converted energy, and a post burning period, of minor interest from 90% to the end of energy conversion.

The duration of the period is able to be counted using an absolute time scale, named t_{EP}, and in crankangle, named α_{EP}. Both scales are linked by engine speed. Figure 1 shows the rate of typical energy conversion at the top left and the sum of energy conversion at the middle left. A change in air-fuel-ratio λ at ignition time α_{ig} for best efficiency, results in clear effects in the relative sum of energy conversion only, left bottom. With increasing λ, ignition time for best efficiency becomes earlier and the duration of inflammation period α_{EP} increases. On the other hand, the main conversion period remains nearly unaffected by variation of λ. In a first approach the duration of inflammation period indicates a laminar flame behaviour while the main conversion period seems to be influenced by turbulent transport properties only. The influence of λ can be seen more clearly in the right top diagram of figure 1 right. The appearance of definite converted energy portions CE, counted in crankangle, has been connected to lines with this diagram.

The combustion process in spark ignition engines is accompanied by the so-called cycle-by-cycle variations. These cbc variations are evident in the pressure development and can be determined quantitatively from the energy conversion process. This has been achieved by the standard deviation of constant portions of converted energy from 200 subsequent cycles. Increasing the air-fuel-ratio results in a lengthened α_{EP} and a nearly constant duration of main conversion period, as mentioned in top diagram of fig. 1 right. An additional effect of increasing λ is a progressive rise in cbc variations as shown in the bottom diagram of fig. 1 right.

In this paper a design tool will be presented which gives information about the duration of the inflammation period on an engineering level. An accurate determination of the inflammation period duration is essential for good simulation results of the whole energy conversion process. An inaccuracy of t_{EP} determination increases the error in simulation of the main conversion period. On the other hand, the estimated schedule of the inflammation period is of minor importance due to the small amount of released energy. In simulating t_{EP} the relevant physical properties have been considered qualitatively. This has been done with regard to good negotiability to different engines and combustion chamber shapes. Nevertheless, attention has been primarily laid on developing a model for a type of engines similar to the spark ignition engine family from Mercedes-Benz.

The ability to dispence with details of the local turbulent flow field and the local chemical reaction mechanism results in a very fast algorithm which is essential to perform a task like determining an ignition time map, for instance. Determination and validation of the proposed model has been worked out using a M103 Mercedes-Benz 6 cylinder engine with fuel injection and a displacement of 2962 cm^3.

DURATION OF THE INFLAMMATION PERIOD

As pointed out, only the main operating parameters of the engine have to be considered in simulating t_{EP}. These parameters are principally environmental condition, air supply, mixture composition, ignition timing and engine speed. Environmental condition includes air pressure, temperature and humidity.

Influence of Air Supply

An increase of supplied air leads to shortened α_{EP}. The value of α_{EP}, derived from measurement, has been marked by symbol in Figure 2. The air supply is varied between minimum at idle load and high air supply near full load. Engine speed, air-fuel-ratio, and ignition timing have been held constant. The value of supplied air is represented by the density of the air-fuel-mixture at the particular position of -23 deg CA, nominated as the reference position. The density of the mixture contains the masses of air, fuel, gaseous water and residual gas, which is determined by use of the PROMO code. PROMO uses one-dimensional, instationary scheme for computing the gas exchange process. This code is frequently used in the German automobile industry. In the proposed model the gaseous water and the internal and external residual gases are mentioned as residual gas.

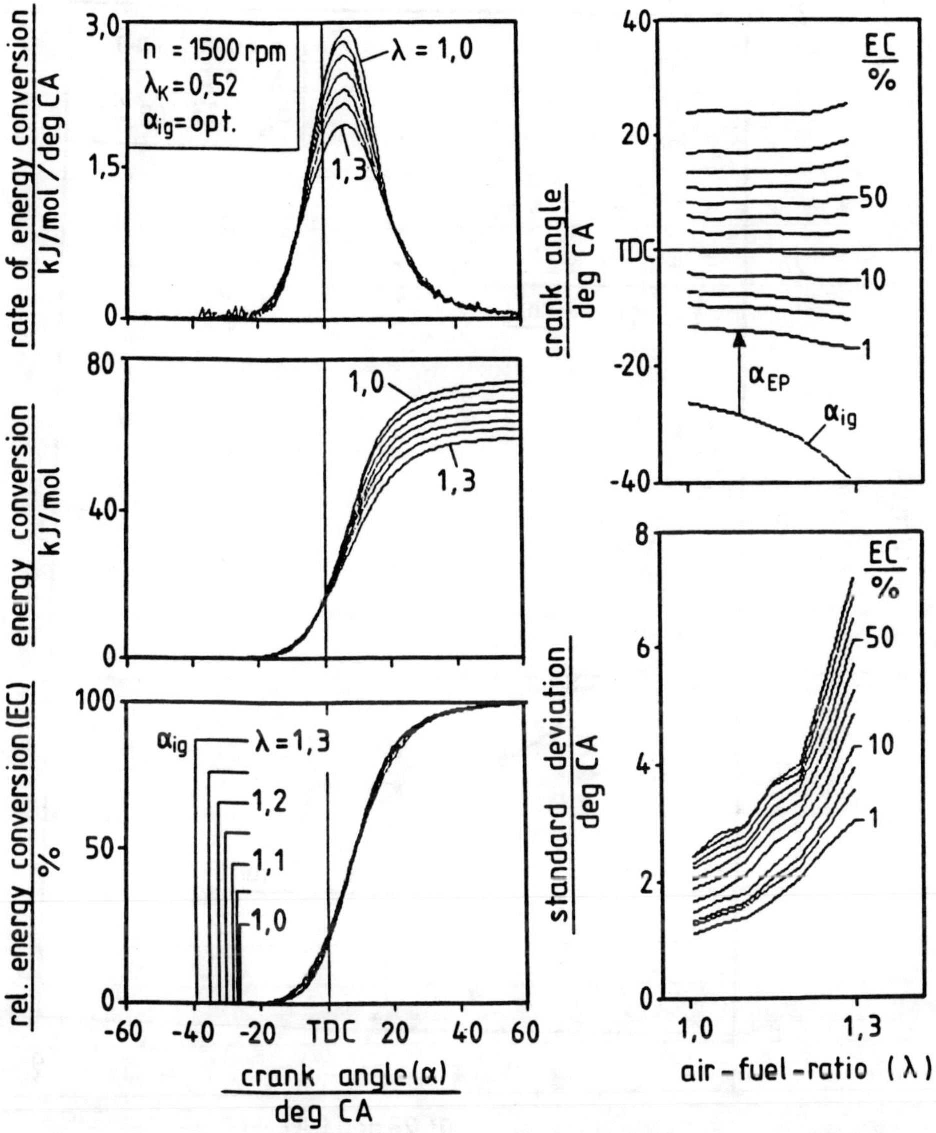

FIGURE 1: Variation of (relative) air-fuel-ratio λ in a spark ignition engine. Influence on inflammation period, from α_{ig} up to 1% EC, and main conversion period, from 1% up to 90% EC, in 200 subsequent cycles. Scatter of energy conversion position characterised by standard deviation.

The assumption that laminar flame behaviour alone has the main influence on t_{EP} when air supply increases doesn't agree well with frequently used formulae of laminar flame speed S_L (Metghalchi and Keck, 1982; Rhodes and Keck, 1985). With assumed constant temperature these formulaes of laminar flame speed predict lowered S_L at increased air supply, which in fact yields enlarged t_{EP}. This disagreement with experimental results from spark ignition engines is caused by an additional influence of the flame kernel development process within the inflammation period.

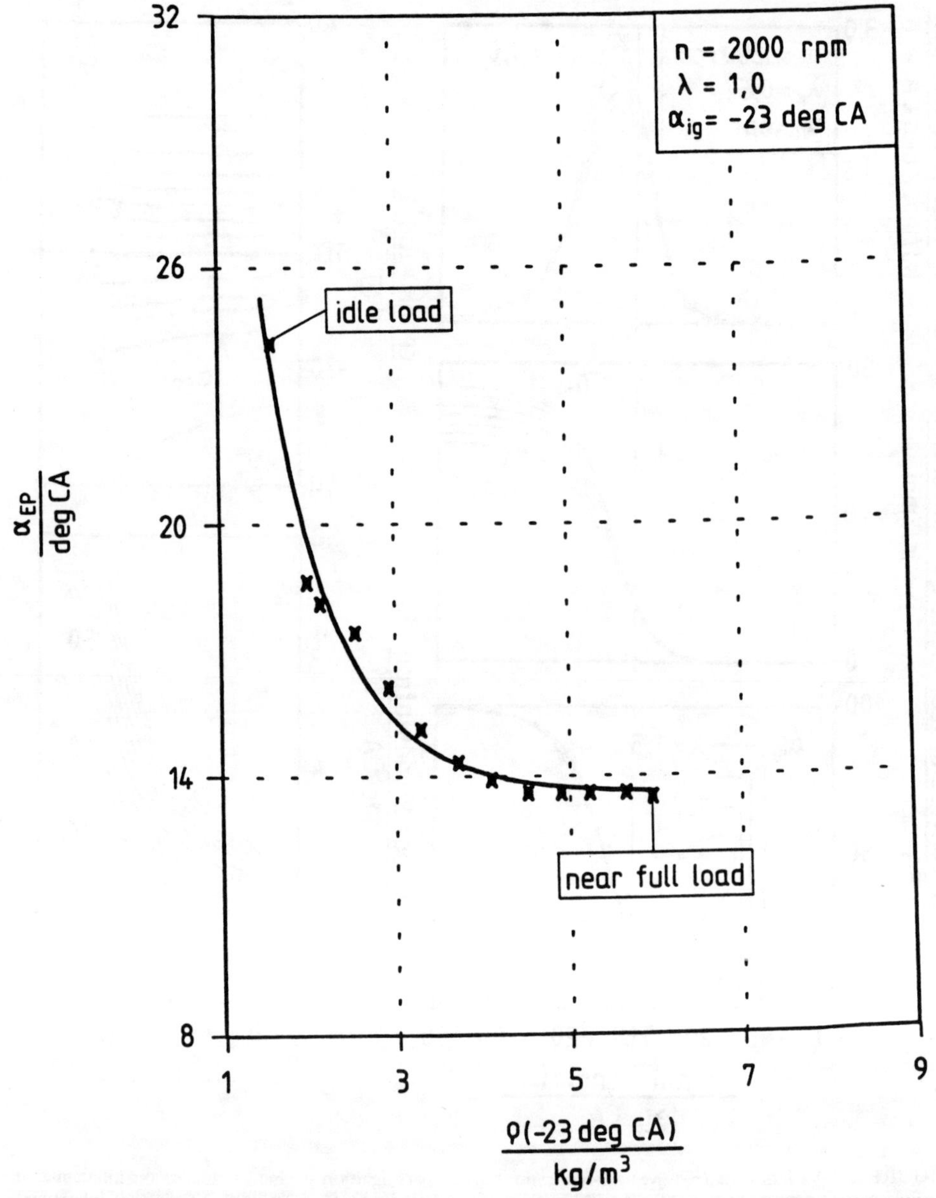

FIGURE 2: Duration of inflammation period α_{EP} for reference conditions. Air supply represented by mixture density at particular crankangle.

With scope to selfignition processes of hydrocarbon mixtures, Schmidt (1967) derived an expression of a mean reaction rate dc_K/dt with pressure p and temperature T:

$$dc_K/dt \sim p^n \cdot d/(e^{b/T} \cdot T^{1/2}). \tag{1}$$

The constant d, and exponents b and n have been empirically ascertained as characteristic

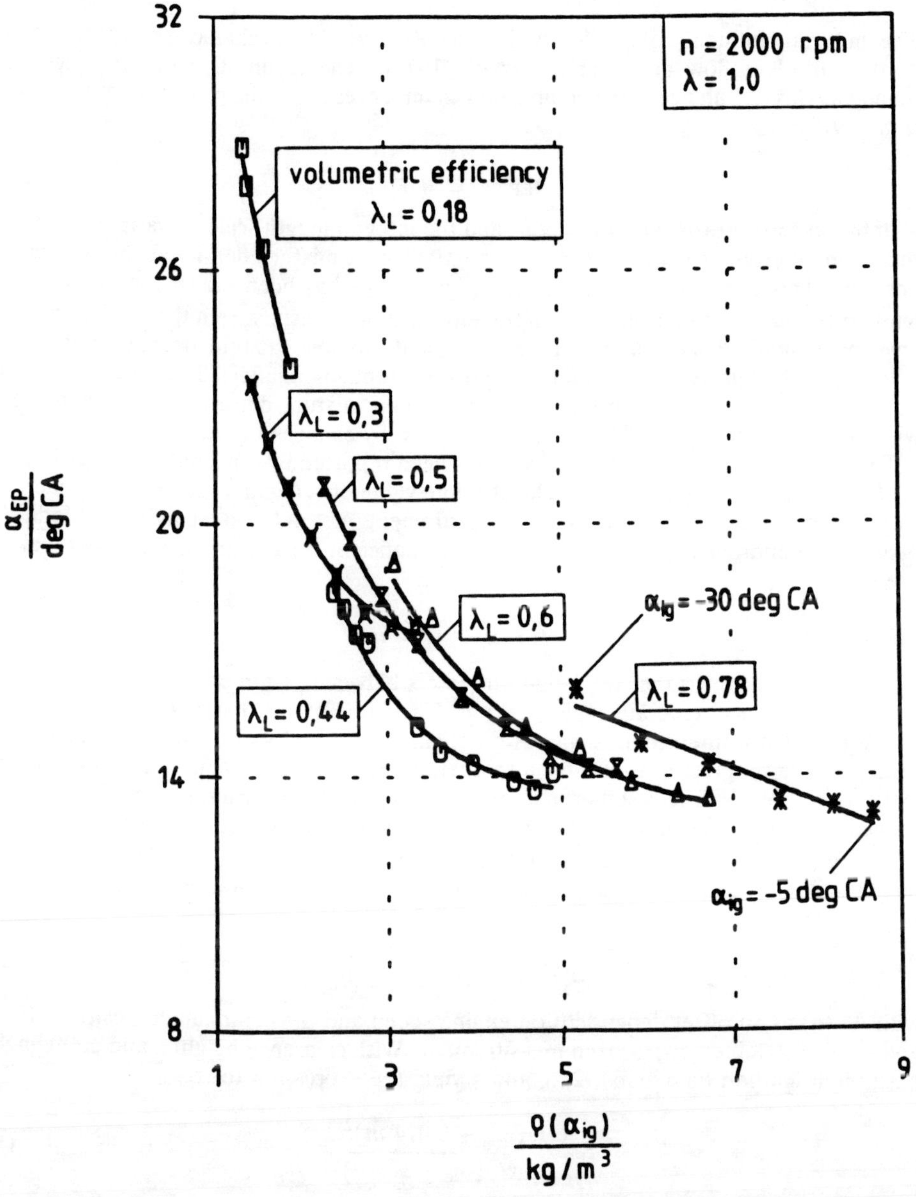

FIGURE 3: Mixture density at ignition time as a link of varied volumetric efficiency and ignition timing. Influence on duration of inflammation period at reference engine speed and mixture composition.

values of fuel type and air-fuel-ratio within a particular temperature and pressure range. Eq.1 yields an increase in mean reaction rate with pressure, which in turn results in a decrease in t_{EP} with increasing mixture density. A best fit curve of the same type as eq.1 has been drawn in fig.2, which is used as a calibration curve for this engine. Fig.2 gives the impression that the mechanism of flame kernel development is of paramount importance at low load conditions.

Influence of Ignition Timing

The important influence of ignition timing on t_{EP} is given by the characteristic parameters of the turbulent flow at α_{ig} (Hires *et al.,* 1978). The assumption of isotropic and homogeneous turbulence in the combustion chamber results in the proportionality (Hires *et al.,* 1978):

$$t_{EP} \sim (L/u')^{2/3} \tag{2}$$

with the so called turbulent intensity u' and the integral length scale L. Variation of L/u′ results in changes of t_{EP} even for a constant value of mixture density at the reference position. This effect is shown in figure 3, where α_{EP} has been drawn for variation of ignition timing α_{ig} at constant volumetric efficiency λ_L. Results from the measurements have been marked as symbols. Early α_{ig} yields to low mixture density at ignition crankangle though density at reference position remains unaffected due to a constant air supply. It is obvious from fig.3, that at constant mixture density, α_{EP} is lengthened with increasing air supply. This influence is given by changes in L/u′.

The progress of integral length scale L has been reported by some authors (e.g. Fraser *et al.,* 1986; Dinsdale *et al.,* 1988). The absolute value of L depends on many parameters such as combustion chamber shape, measurement method and analysis method. A quantitative approach that is considered to be sufficient for the proposed model makes use of:

$$L \sim x + 0.014\ m \tag{3}$$

where (x + 0.014 m) is the free height in meters between piston crown and combustion chamber top at a given crankangle position. The value of x is 0 at TDC.

The turbulence intensity u′ depends on engine speed and varies with crankangle, for instance. The progress of u′ has been frequently reported (e.g. Dent and Salama, 1975, Witze *et al.,* 1984). The results for u′ are scattered as demonstrated for a simple disk-shaped combustion chambers in figure 4. Reasons for the great differences in u′, even at constant value of mean piston speed c_m, are the same as mentioned above at integral length scale L. A quantitative approach of turbulence intensity evolution with crankangle is given by:

$$(u')^{2/3} \sim \cos(\alpha + \Delta\alpha(n)) \tag{4}$$

with $\Delta\alpha$(n) as an offset dependent on engine speed and α as crankangle position. Eq 3 and 4 are ascertained by t_{EP} from measurement. With clearance height x and crankangle position at ignition time, eqns. 2, 3, and 4 yield the expression for t_{EP}:

$$t_{EP} \sim \frac{(X_{ig} + 0.014\ m)^{2/3}}{\cos(\alpha_{ig} + \Delta\alpha(n))}. \tag{5}$$

As an example, the progress of L/u′ has been drawn by the use of eq.5 in figure 5. The minimum of L/u′ indicates a convenient minimum for t_{EP} with α_{ig} = -8 deg CA at an engine speed of 2000 rpm.

The ignition timing for minimized t_{EP} depends on $\Delta\alpha$(n) which shifted the minimum of L/u′ to an earlier crankangle with increasing engine speed, figure 6. The lower diagram of fig.6 shows this shift with engine speed in α_{ig} for minimum t_{EP} at distinct conditions. The curve is derived from eq.5 with the best fit approximation for $\Delta\alpha$(n) which is shown in the top diagram of fig.6. Values of $\Delta\alpha$(n), which have been derived from experiment, have been marked with symbols.

Author		Measurement method	Chamber geometry	Flow spin	Analyze method
Winsor and Patterson (1973)	○	HW	Disk	none	
Witze (1977)	□	HW	Disk	none	Ensemble-averaged
Cole and Swords (1978)	◇	LDA	Disk	none	Cycle-resolved
Witze et.al. (1984)	▲	LDA	Disk	with	Ensemble-averaged
Ikegami et.al. (1987)	∘	LDA	Disk	none	Cycle-resolved
Tomita and Hamamoto(1988)	◆	LDA	Disk	with	Cycle-resolved

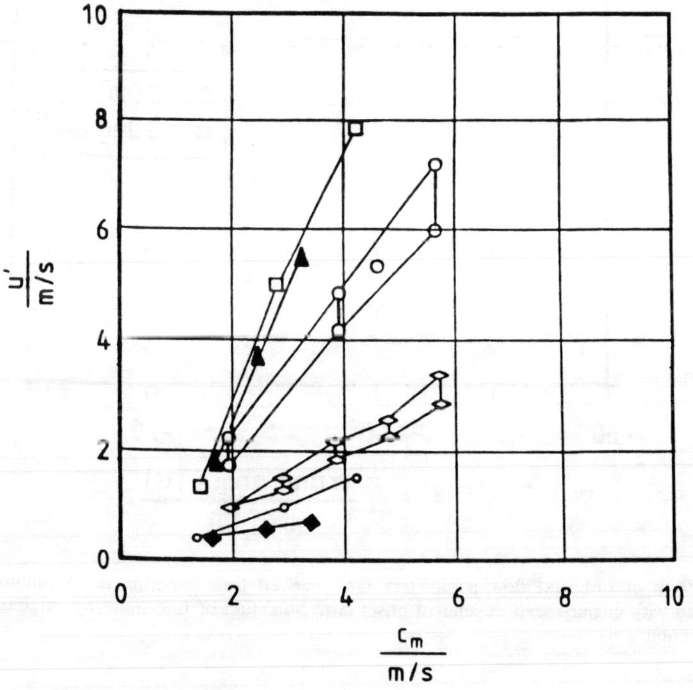

FIGURE 4: Results of turbulence intensivity u′ as presented by some authors derived from diverse engines with disk shape combustion chamber. Good comparison is given by use of mean piston velocity c_m as a representative of engine speed. Results depend on engine type, measurement and analyze method.

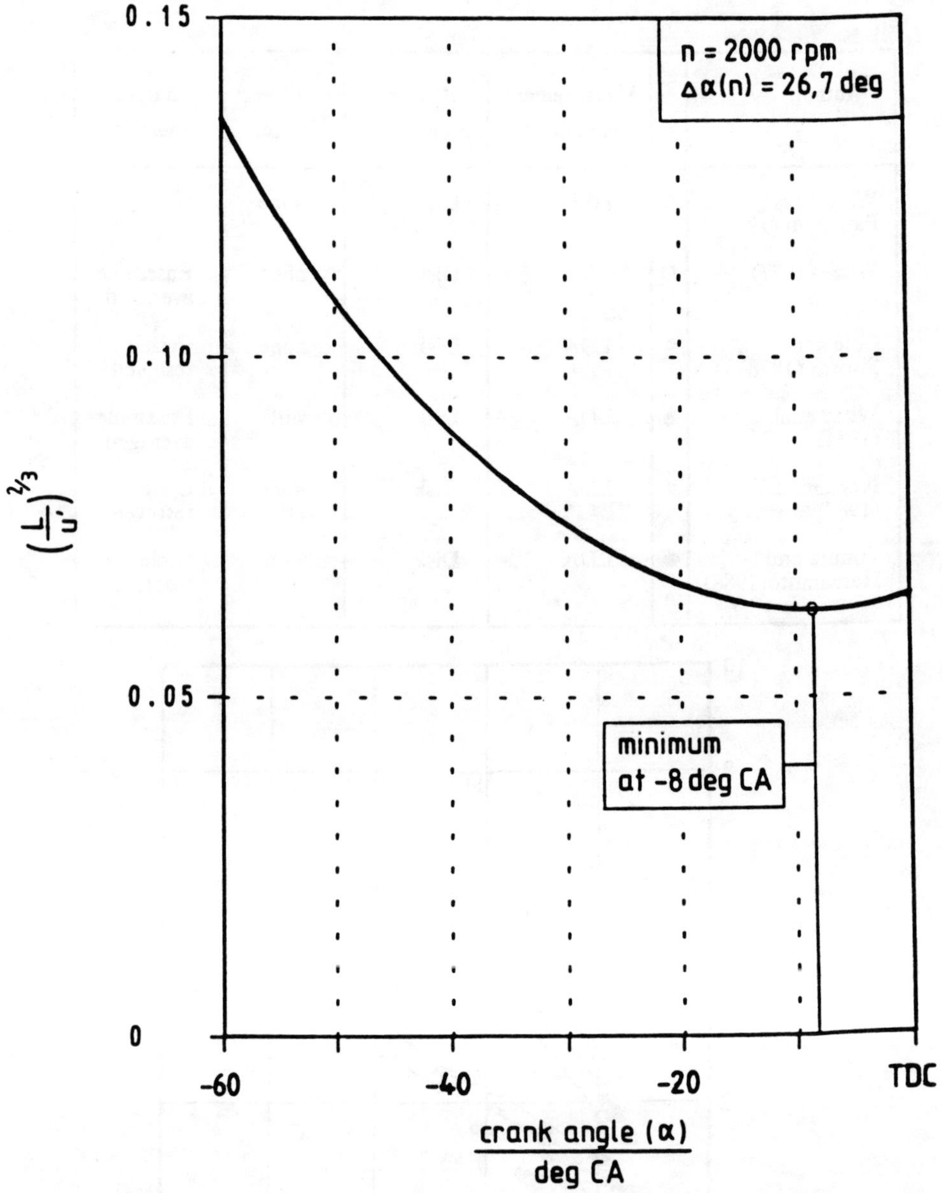

FIGURE 5: Rate of turbulent flow parameters L/u′, derived from experiment. Minimum of L/u′ has been determined with engine-speed dependent offset $\Delta\alpha$. Minimum of L/u′ indicates minimize duration of inflammation period.

To predict t_{EP} with respect to air supply and ignition timing a combination of mixture density and characteristic turbulent parameters is necessary. At reference conditions, that is an engine speed of 2000 rpm and an air-fuel-ratio of 1.0, the expression for t_{EP} in deg. CA becomes:

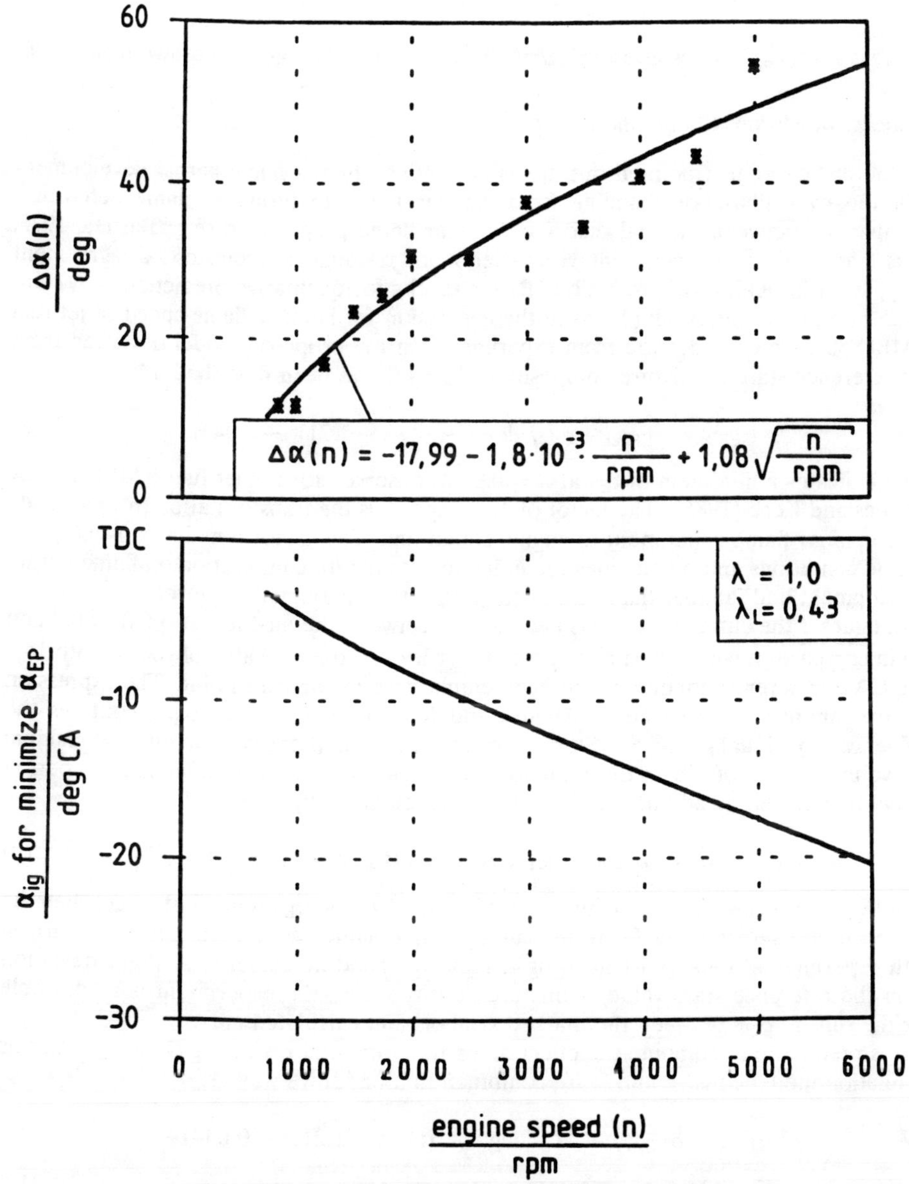

FIGURE 6: Offset $\Delta\alpha(\mathrm{n})$ least square fitted from experimental results. The curve of $\Delta\alpha(\mathrm{n})$ leads to predicted values of ignition timing for minimize duration of inflammation period.

$$\alpha_{\mathrm{EP,ref}} = K(\rho_{(-23\ \mathrm{deg\ CA})}.\frac{(X + 0.014m)^{2/3}}{\cos(\alpha_{\mathrm{ig}} + \Delta\ \alpha(\mathrm{n}))} \tag{6}$$

with $K(\rho_{(-23\ \mathrm{deg\ CA})})$ in (deg CA) $\cdot$ $m^{3/2}$ as a formulation of mixture density influence:

$$K(\rho_{(-23\text{ deg CA})}) = 193 + 1112.\, e^{-1.27.(\rho\ -23\text{ deg CA})} \tag{7}$$

whereas $\rho((_{-23\text{ deg CA}}))$ is given in kg/m^3. The curve in fig.2 has been drawn from eq. 6.

Influence of Mixture Composition

As the preceding text pointed out, t_{EP} is determined by the flame kernel development followed by a phase of changing from laminar flame to turbulent flame behaviour. Comparison between thermal theory of laminar flame propagation (e.g. Bartlaä, 1975, Glassman, 1987) and experiment with varied λ and residual gas indicates a disagreement in t_{EP} even in qualitative prediction. Better results for qualitative prediction of t_{EP} are given by Rhodes and Keck (1985) in the expression for laminar flame speed of jet fuel RMFD 303, which is derived from experiment. In the proposed model deviation from the reference state in mixture composition, $\lambda = 1.0$, has been described as:

$$\alpha_{EP,\lambda} = \alpha_{EP,ref} + 47.7\text{ deg CA}.(1/\lambda - 1.21)^2 - 0.044). \tag{8}$$

Eq.8 indicates a minimum of t_{EP} at $\lambda = 0.83$ as observed for the jet fuel RMFD 303 by Rhodes and Keck (1985). The factor of 47.7 deg CA is the transformation from a steady state laminar flame experiment to engine conditions. This factor remains constant at all engine conditions and it combines the influence of mixture composition on flame kernel development and laminar flame character within the inflammation period.

In figure 7 the curve of eq.8 has been drawn between the air-fuel-ratio (λ) of 0.7 and 1.3 in comparison with experimental results for lean mixtures. Value of $\alpha_{EP,ref}$ with 14.0 deg CA at reference conditions has been emphasized as the base point. The expression for the laminar flame speed of Rhodes and Keck (1985), part of eq.8, contains the influence of residual gas on S_L. Starting at reference conditions with an internal residual gas volume $V_{Res,0}$ of 9% in the total volume V_{tot}, the duration of inflammation period would be affected by additional exhaust gas recirculation V_{EGR} as:

$$\alpha_{EP,\lambda} = \alpha_{EP,ref} + 196.5\text{ deg CA}.((V_{Res}/V_{tot})^2 - (V_{Res,0}/V_{tot})^2_{ref}). \tag{9}$$

with $V_{Res} = V_{Res,0} + V_{EGR}$. The influence of residual gas on α_{EP} is much stronger than the influence of λ shown in fig.7 with the curve of eq.9, named as residual gas. Comparison with experimental data, marked in fig.7, indicates good agreement. A slight deviation from the reference state value in mixture density within the experiment is responsible for the shift in α_{EP} between the marked symbols and calculated curves.

It is preferable to combine the effect of air-fuel-ratio and residual gas on α_{EP}. In this simulation model eqns. 8 and 9 are combined in an additive way as:

$$\alpha_{EP,\lambda} = \alpha_{EP,ref} + 47.7\text{ deg CA}.((1/\lambda - 1.21)^2 - 0.044)+$$
$$196.5\text{ deg CA}.((V_{RES}/V_{tot})^2 - (V_{Res,0}/V_{tot})^2_{ref}). \tag{10}$$

Influence of Engine Speed

The progress of turbulence intensity u', fig.4, depends on engine speed if the crankangle is held at a reference position. High engine speed yields high values of u'. Only a few authors report a nonlinear increase of u' with engine speed (e.g. Bopp *et al.*, 1986).

Figure 8 bottom shows the decrease in total time of t_{EP} with increasing engine speed from experimental results of the M103 engine. In figure 8 top the progress of exponent

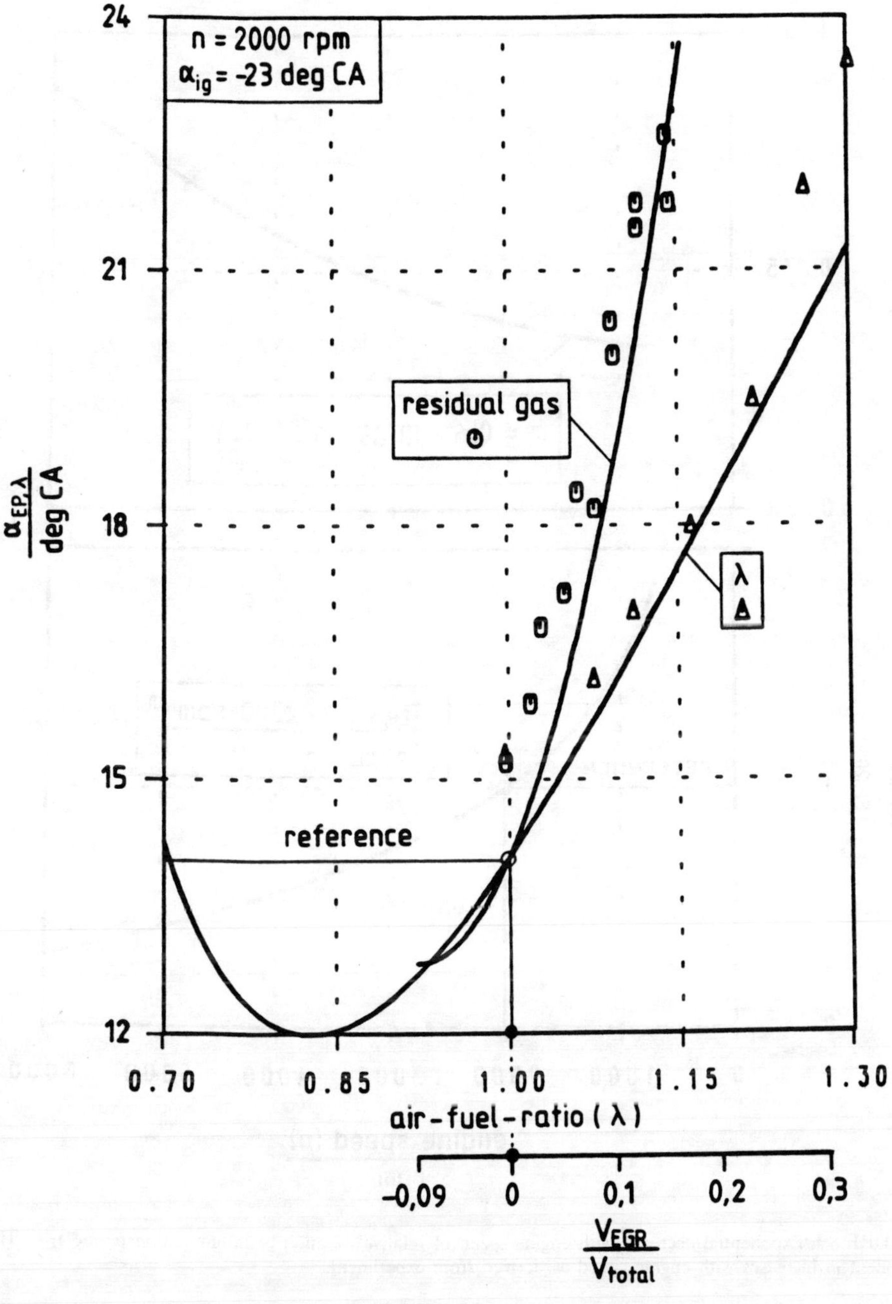

FIGURE 7: Variation of air-fuel-ratio and exhaust gas recirculation (EGR) influence duration of inflammation period α_{EP} as shifted from reference conditions. Comparison between simulated and experimental results.

m has been drawn as derived from experiment. It is markable that in fig.8 top m is not constant but increases with engine speed. The value of m = 0.47 proposed by Bopp *et*

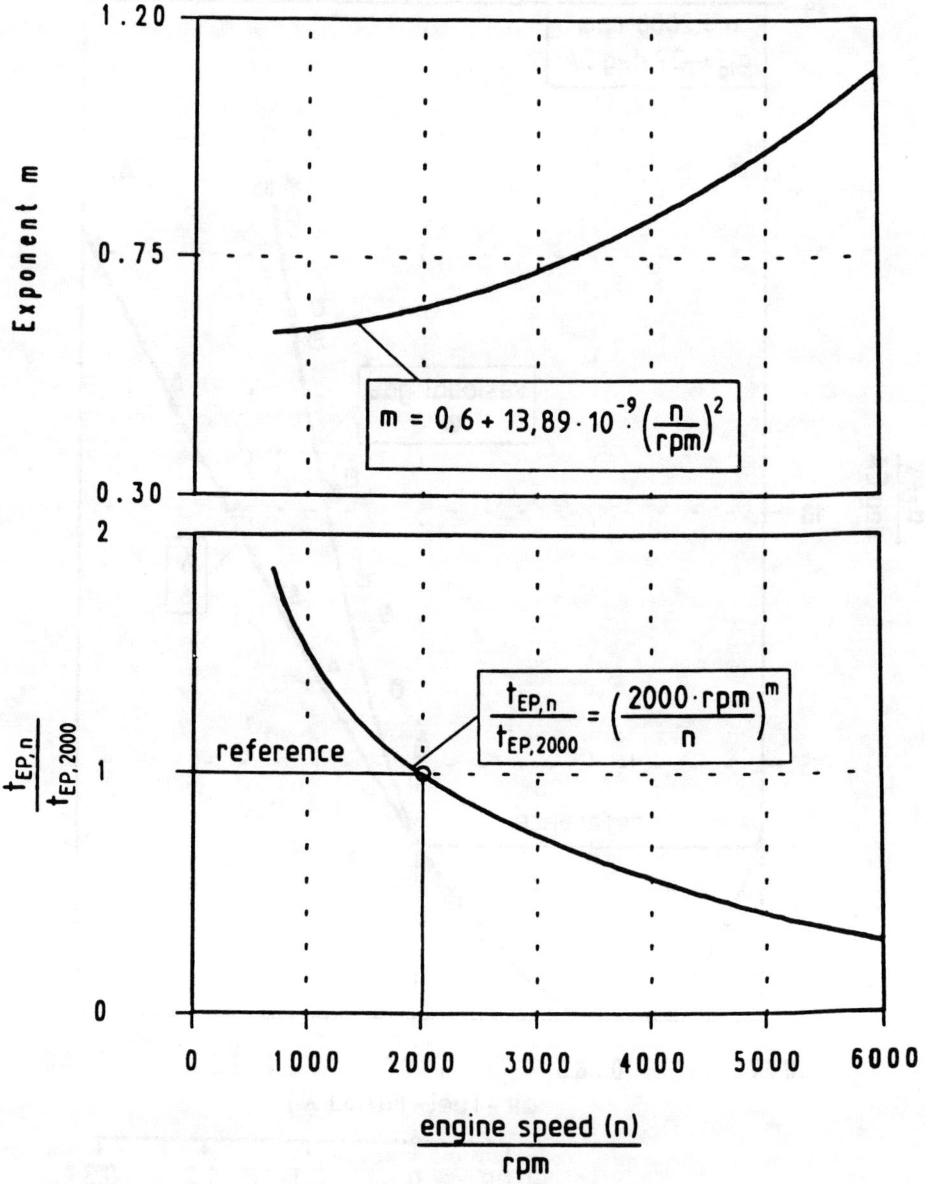

FIGURE 8: Exponential decrease with engine speed of relative duration of inflammation period t_{EP}. The exponent m increases with engine speed as derived from experiment.

al. (1986) at minimum speed of 300 rpm has not been reached with M103 even as m = 0.59 at maximum speed of 2000 rpm, but the results of Bopp *et al.* (1986) confirm an increase of m with engine speed. The frequently used direct proportionality of m = 1.0 has only been reached at approximately 5400 rpm in this engine. It is not ascertained that the proposed curve of m is caused only by u' but this assumption is easy to use in the simulation model. Consideration of engine speed influence results in the expression:

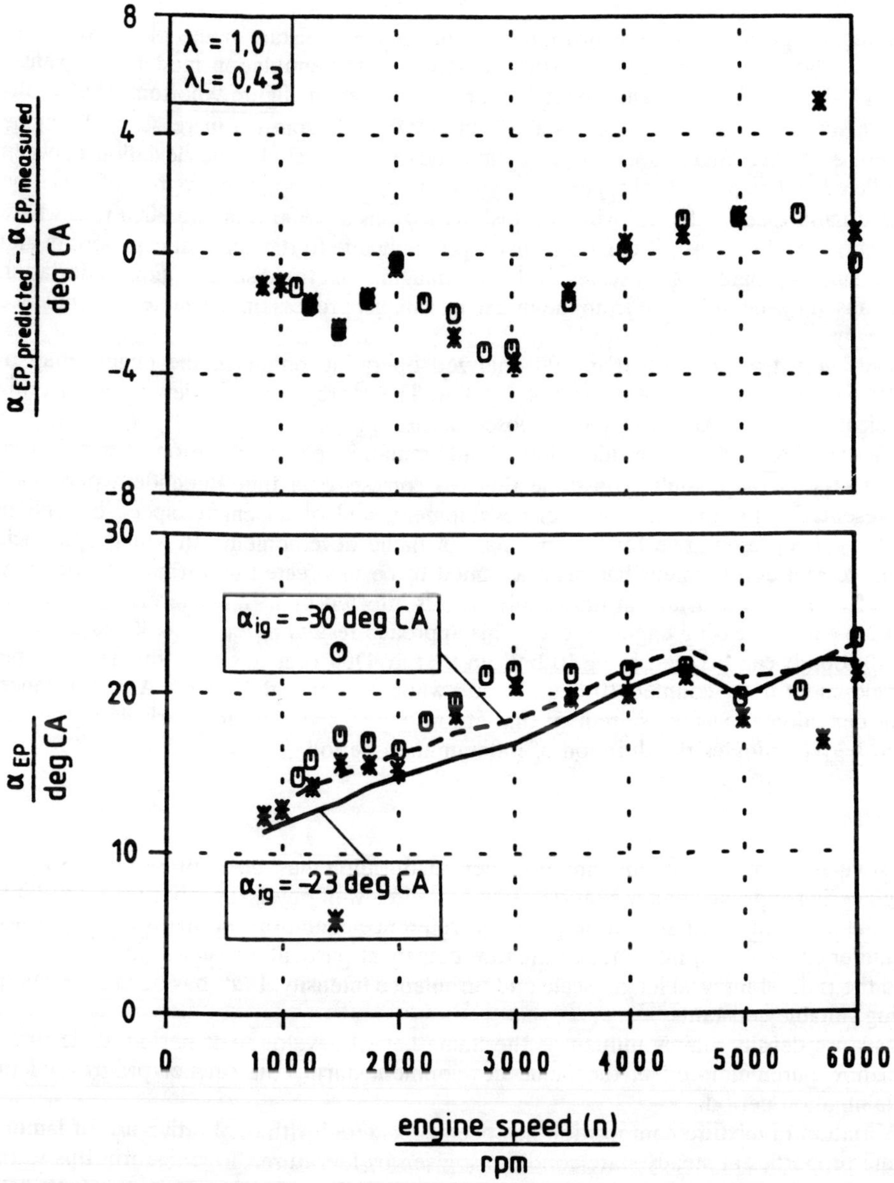

FIGURE 9: Comparison of predicted and experimental results for duration of inflammation period α_{EP}. Performed at two engine speed variations at constant operating conditions.

$$t_{EP,\lambda,n} = t_{EP,\lambda}.(2000/\eta)^{m} \tag{11}$$

with engine speed η in rpm and exponent m from fig.8.

RESULTS

An engine speed variation at two ignition timings and constant values of λ and λ_L has been analyzed as an example for the capability of the simulation model. The values of α_{EP} derived from measurement have been marked in figure 9 bottom, the results of the simulation have been delineated out. Until 4500 rpm t_{EP} increases with engine speed, at higher engine speed α_{EP} remains nearly unaffected. The deviation between predicted and measured values of α_{EP} is shown in figure 9 top. The progress of deviation with engine speed is typical with two distinct regions at 1500 rpm and 3000 rpm where deviation reaches nearly 4 deg CA. This is probably due to the curve of exponent m and the assumed progress of u', respectively. We imagine that the distinct regions indeviation are caused by the influence from the intake or exhaust process on the value of turbulence intensity.

Only very few of more than 700 analyzed operating points in the engine map of M103 yield higher a deviation than 4 deg CA. Therefore, from the view of engineering application, the capability of the proposed model is good.

Figure 10 top allows consideration of inflammation period duration changed from α_{EP} to t_{EP}. It is possible to assume that t_{EP} consists of a time dependent part $t_{EP,c'}$ representative for the flame kernel development, and of an engine speed dependent part $t_{EP,\nu'}$ representative for the first part of flame development. In a first approach, flame kernel development has been assumed to be unaffected by turbulence intensity, therefore $t_{EP,c}$ is constant in time while $t_{EP,\nu}$ is affected by u' and therefore shortened in time with increasing engine speed. This approach results in $t_{EP,c} = 0.36$ ms drawn in fig.10 top. It can be seen in fig.10 bottom that at 4700 rpm this time for flame kernel development $\alpha_{EP,c}$ counted 10 deg CA and whole α_{EP} lasted 22 deg CA, for instance. This depiction allows the statement that at low engine speed up to nearly 4500 rpm the time $t_{EP,c}$ dominates the duration of inflammation period.

SUMMARY

A simulation model of inflammation period duration has been presented using an engineering approach under consideration of the relevant physical properties. In order to get the duration of inflammation period at reference conditions in mixture composition λ and engine speed η influence of mixture density at particular crankangle $\rho_{(-23deg CA)}$, and the ratio of integral length scale and turbulence intensity, L/u', has been determined using suitable constants.

Mixture density mainly influences the flame kernel development period while turbulent flow parameters effect the flame development during the further progress of the inflammation period.

Variation in mixture composition has been considered with qualitative use of laminar flame properties at steady state conditions, given by literature. To transform this to the duration of flame kernel development and subsequent flame development, constants have been fixed for air-fuel-ratio influence and for residual gas influence. Leaner air-fuel-ratio or increase in EGR lengthens the total inflammation period.

The influence of engine speed results in an increase in turbulence intensity which results in a shortened duration of inflammation period. Turbulence intensity has been assumed to affect the flame development only, the flame kernel growth.

Comparison with experiment at M103 engine reveals good simulation results within the whole engine map. Variations in air supply, air-fuel-ratio, ignition timing and engine speed have been worked out. In most cases the difference between predicted and analyzed duration of inflammation period is better than 4 deg CA.

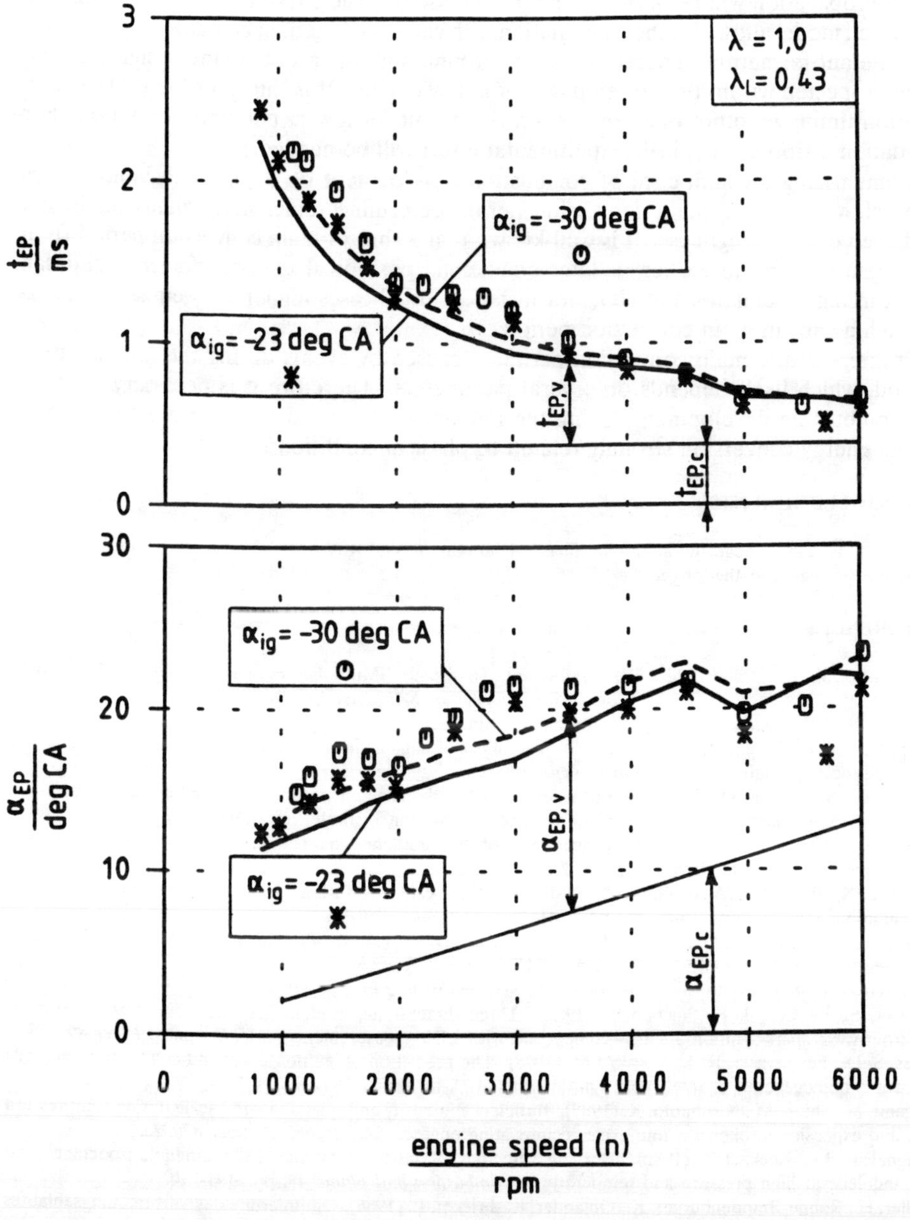

FIGURE 10: Comparison of predicted and experimental results for duration of inflammation period. Results counted in time t_{EP} and crankangle α_{EP} for clear presentation of time constant part $t_{EP,c}$ and $\alpha_{EP,c'}$, respectively. $t_{EP,v}$ is the engine speed dependent variable part of t_{EP}.

The capability of the proposed model is sufficient not only as a fast design tool but reveals further details in the dependence of the inflammation period from engine operating conditions. For similar engines and configurations this tool should allow good simulation results of the whole energy conversion process, with assistance of a subsequent model of flame propagation in the main conversion period. Such a model of flame propagation will be presented later. The use of basic physical properties within the proposed model allows for the combination with chemical kinetic submodels for prediction of pollutant formation. Therefore, the environmental aspects of engine design are able to be respected within the design phase of a new engine. It is thus possible to design an ignition timing map not only for best efficiency but for lowest pollutant formation. For validation purpose only, little experimental effort will be needed.

Summarizing the influence of combustion development on engine behaviour, it has been shown that the main conversion period determines efficiency, exhaust emission, and even engine roughness. It is well-known that a shift of main conversion period from best ignition time to earlier or later crankangle position decreases efficiency and can result in higher emissions of NO_x, for instance. In all cases higher CO_2 emission results by such a shift in main conversion period.

Progress of the main conversion period is caused by events during the inflammation period, which itself depends on several parameters. Therefore it is necessary to study the combustion development during the inflammation period and to evaluate this part of the energy conversion strongly related to physical conditions.

ACKNOWLEDGEMENTS

We gratefully acknowledge the additional financial support of VOLKSWAGENSTIFTUNG, which promotes continuous research in this project.

REFERENCES

Ahmadi-Befrui, B., Brandstätter, W., Kratochwill, H. (1989). Multi-dimensional calculation of the flow processes in a loop-scavenged two-stroke cycle engine. *SAE Technical Paper 890841.*

Bartlmä, F. (1975). Gasdynamik der Verbrennung. Springer-Verlag Wien, pp. 52-55.

Bopp, S., Vafidis, C., Whitelaw, J. H. (1986). The effect of engine speed on the TDC flowfield in a motored reciprocating engine. *SAE Technical Paper 860023.*

Cole, J.B., Swords, M.D. (1978). Optical studies of turbulence in an internal combustion engine. *17. Symposium (International) on Combustion,* The Combustion Institute, pp. 1295-1303.

Dent, J. C., Salama, N. S. (1975). The measurement of the turbulence characteristics in an internal combustion engine cylinder. *SAE Technical Paper 750886.*

Dinsdale, S., Roughton, A., Collings, N. (1988). Length scale and turbulence intensity measurements in a motored internal combustion engine. *SAE Technical Paper 880380.*

Fraser, R. A., Felton, P. G., Bracco, F. V., Santavicca, D. A. (1986). Preliminary turbulence length scale measurements in a motored IC engine. *SAE Technical Paper 860021.*

Glassman, I. (1987). Combustion. Academic Press New York, 2.Ed., pp.119-125.

Henriot, S., Le Coz, J. F., Pinchon, P. (1989). Three dimensional modelling of flow and turbulence in a four-valve spark ignition engine—comparison with LDV measurements. *SAE Technical Paper 890843.*

Hires, S. D., Tabaczynski, R. J., Novak, J.M. (1978). The prediction of ignition delay and combustion intervals for a homogeneous charge, spark ignition engine. *SAE Technical Paper 780232.*

Ikegami, M., Shioji, M., Nishimoto, K. (1987). Turbulence intensity and spatial integral scale during compression and expansion stroke in a four-cycle reciprocating engine. *SAE Technical Paper 870372.*

Metghalchi, M., Keck, J.C. (1982). Burning velocities of mixtures of air with methanol, isooctane, and indolene at high pressure and temperature. *Combustion and Flame* **48,** pp. 191-210.

Müller, H., Rohde-Brandenburger, K., Obländer, K., Fricker, L. (1986). Optimierung des Verbrennungsablaufes eines Mehrzylinder-Ottomotors im dynamischen Betrieb. *XXI FISITA Congress (865015),* pp. 1.103-1.110.

Rhodes, D. B., Keck, J. C. (1985). Laminar burning speed measurements of indolene-air-diluent mixtures at high pressures and temperatures. *SAE Technical Paper* 850047.

Schmidt, F. A. F. (1967). Verbrennungskraftmaschinen, Springer-Verlag, Berlin, 4.Ed., pp.546-551.

Tang-Wei Kuo, Reitz, R. D. (1989). Computation of premixed-charge combustion in pancake and pent-roof engines. *SAE Technical Paper 890670.*

Tomita, E., Hamamoto, Y. (1988). The effect of turbulence on combustion in cylinder of a spark-ignition engine—evaluation of entrainment model. *SAE Technical Paper 880128.*

Westbrook, C. K., Warnatz, J., Pitz, W. J. (1988). A detailed chemical kinetic reaction mechanism for the oxidation of iso-octane and n-heptane over an extended temperature range and its application to analysis of engine knock. *22.Symposium (International) on Combustion,* The Combustion Institute, pp. 893-901.

Winsor, R. E., Patterson, D. J. (1973). Mixture turbulence—a key to cyclic combustion variation. *SAE Technical Paper 730086.*

Witze, P. O. (1977). Measurements of the spatial distribution and engine speed dependence of turbulent air motion in an IC. *SAE Technical Paper 770220.*

Witze, P. O., Martin, J. K., Borgnakke, C. (1984). Fluid motion during flame propagation in a spark ignition engine. *SAE Technical Paper 840377.*

10.

REACTION KINETICS

Effect of Fuel Ratio on Coal Pyrolysis and Nitrogen Emission Behaviour

MASANOBU HASATANI, HITOKI MATSUDA, and YONG CHEN *Department of Chemical Engineering, Nagoya University, Furo-cho, Chikusa-ku, Nagoya, Japan*

Abstract — The present study is concerned with the pyrolysis of ten kinds of coals of different rank in which of fuel ratio ranges relatively widely from 0.9 to 6.6. The coal pyrolysis experiment was carried out using Thermogravimetric Analysis (TGA) in an atmosphere of nitrogen under the following conditions: coal particle size 37–2000 μm, heating rate 2–2000 K/min. The pyrolysis parameters as the reaction order, the activation energy and the frequency factor were determined, and then these were correlated with the ratio of percent carbon to percent volatile matter, referred to as the fuel ratio. From the results, it was recognized that the experimental data were well reproduced with respect to TG curves by means of the determined parameters. The behaviours of volatile-N released from original coals of different ranks were investigated during coal pyrolysis. The results show that volatile-N starts to be released from coal when the temperature reaches about 673–773 K, which is about 50–100°C higher than the temperature where volatile-C and volatile-H start to be released; The conversion of volatile-N increases with a decrease of fuel ratio at same heating temperature.

Key Words: Coal Pyrolysis, Fuel Ratio, Reaction Kinetic Parameter, Volatile-N

INTRODUCTION

Recently there has been a considerable interest in the coal utilization approach involving coal pyrolysis, coal gasification and coal combustion, from the viewpoint of securing a substitute for petroleum. In general, thermal decomposition (pyrolysis) of coal plays a significant role in producing coke and in the carbonization process, and is also an important preliminary step in combustion and gasification. Therefore, it is necessary to study the mechanism of coal pyrolysis for designing a coal combustor and also for controlling combustion conditions with higher efficiency and with lower air pollution. To this end, numerous researches on coal pyrolysis so far have been undertaken from various angles in order to investigate the coal pyrolysis characteristics and to determine and derive reaction rate expression of coal in the process of pyrolysis by means of varied methods.

Kobayashi *et al.* (1976) studied the devolatilization of lignite and bituminous coal at high temperatures of 1000–2100 K. Anthony *et al.* (1974) conducted a devolatilization experiment of involving lignite and bituminous coal, and derived a multi-reaction model by applying a Gaussian distribution for the activation energy density. Juntgen and van Heek (1979) concluded that the devolatilization temperature shifted to a high temperature range as the heating rate increased. Devanathan and Sanean (1986) derived a theoretical model for nonplastic coal particles by considering both heat and mass transfer resistances as well as the reaction kinetics of coal pyrolysis. Solomon *et al.* (1990), recently, proposed a network-model for coal pyrolysis. Furthermore, there are review articles on coal devolatilization by Anthony *et al.* (1976) and Saxena (1990). However, it is rather difficult to derive and express a general rate equation for coal pyrolysis with consistency by using common parameters, since coal is much complicated in composition and structure, which differ from one another according to the kind of coal.

The present research is concerned with pyrolysis of coals of various ranks by Thermogravimetric Analysis (TGA). Such parameters as the reaction order n, the activation energy E and the frequency factor A were determined. The relation between the kinetic

TABLE I
PROPERTIES OF VARIOUS COALS

Coal	Location	Proximate analysis (dry basis, wt%)				Ultimate analysis (daf, wt%)			
		F.C.	V.M.	Ash	Fuel ratio	C	H	N	S+O (diff.)
Kaixi (A)	China	63.0	25.4	11.6	2.5	80.1	4.1	1.2	14.6
Sangei (B)	China	68.2	22.6	9.2	3.0	81.0	4.0	1.1	13.9
Kaipin (C)	China	55.1	32.0	13.0	1.7	76.8	4.6	1.0	17.6
Jinjin (D)	China	68.8	21.6	9.6	3.2	79.7	4.2	1.1	15.0
Datong (E)	China	61.1	31.4	7.5	2.0	80.2	4.8	0.8	14.2
Loy Yang (F)	Australia	46.3	51.3	2.4	0.9	66.1	4.3	0.7	28.9
Newlands (G)	Australia	59.3	26.5	14.2	2.2	73.0	4.1	1.2	21.7
Blair Athol (H)	Australia	63.4	28.8	7.8	2.2	76.1	4.4	1.5	18.0
Kangra (I)	South Africa	68.8	18.3	12.9	3.8	82.3	3.9	2.0	11.8
Hongay (J)	Vietnam	64.3	9.8	25.9	6.6	74.8	2.3	0.9	22.0

parameters thus obtained and the fuel ratio: the ratio of the amount of fixed carbon to volatile, was discussed.

It is very important to make clear the nitrogen emission behavior at various temperatures, because 75% of NO is generated from fuel-N involving 80% volatile-N and 20% char-N. The emission behavior of nitrogen from coal in the process of coal pyrolysis was also investigated by measuring the residual amount of the nitrogen.

EXPERIMENTS

Samples

Kaixi, Sangei, Kaipin, Jinjin, Datong (China), Loy Yang, Newlands, Blair Athol (Australia), Kangra (South Africa), and Hongay (Vietnam) were employed. The proximate and ultimate analyses of these coals are shown in Table I.

Prior to the coal pyrolysis experiment, the sample coals were pulverized to a particle size ranging from 37 to 2000 μm, and they were well dried in a drier at 378 K for 24 hours.

Experiment and Procedure

Figure 1 shows a schematic diagram of the thermobalance used for coal pyrolysis. The pyrolysis experiment was carried out in an atmosphere of nitrogen and in the temperature range 573–1173 K. The specimen coal particles of about 10 mg were loaded on the sample holder and were heated at constant heating rates from 2 to 200 K/min.

Determination of Parameters

Up to the present, there have been different reaction models presented for coal pyrolysis: the first-order reaction model, the nth-order reaction model by Wiser (1968), Skilar *et al.* (1969) and Juntgen (1984), the multiple-reaction model by Pitt (1962), and the two competing reaction model by Anthony (1974) and Kobayashi (1976).

In the present study, nth-order reaction model is adopted for describing the rate expression of coal pyrolysis as follows:

$$dx/dt = dx/dT\,(dT/dt) = k\,(1 - x)^n, \tag{1}$$

$$k = A\exp(-E/RT). \tag{2}$$

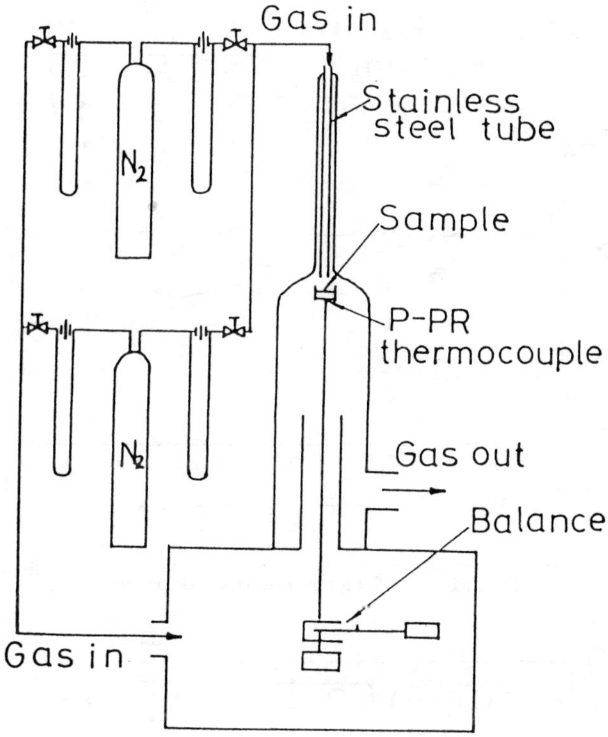

FIGURE 1 Thermobalance employed.

In the determination of the kinetic parameters by use of the above equations, the non-isothermal analysis method proposed by Ito *et al.* (1983) was used.

Taking the logarithm of the equation resulting from the combining of (1) and (2) and rearranging the result equation, the following equation is obtained:

$$\ln((1 - x)^{-n} (dx/dt)) = \ln A - E/RT. \tag{3}$$

The values of dx/dt and x can be obtained directly from the measured thermogravimetric curves. In determining the kinetic parameters in (3), the reaction order, n, is first assumed. By using this n-value, the calculated results of the left term in (3) are plotted against the reciprocal of absolute temperature. The n-value is determined as the value which gives a straight line of $\ln(1 - x)^{-n}$ (dx/dt) vs. 1/T (see Figure 2). The activation energy E and the frequency factor A are determined by the slope of the straight line and the intercept of the straight line to the ordinate, respectively.

RESULTS AND DISCUSSIONS

Kinetic Parameters

In order to determine the kinetic parameters, it is important to establish the experimental condition where both heat and mass transfer resistances in coal particles are negligible. For that reason, the pyrolysis experiments of coals were carried out by changing the heating rates as well as particle sizes until a unique TG curve could be obtained irrespective of the

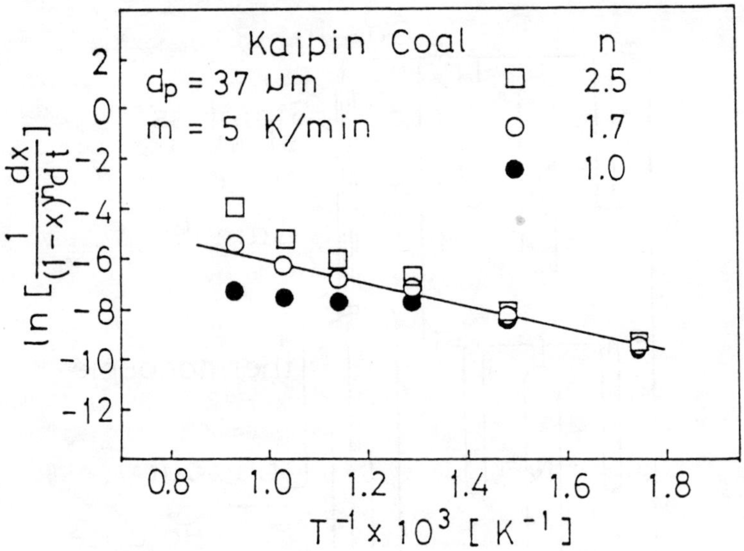

FIGURE 2 Plot for estimating kinetic parameters.

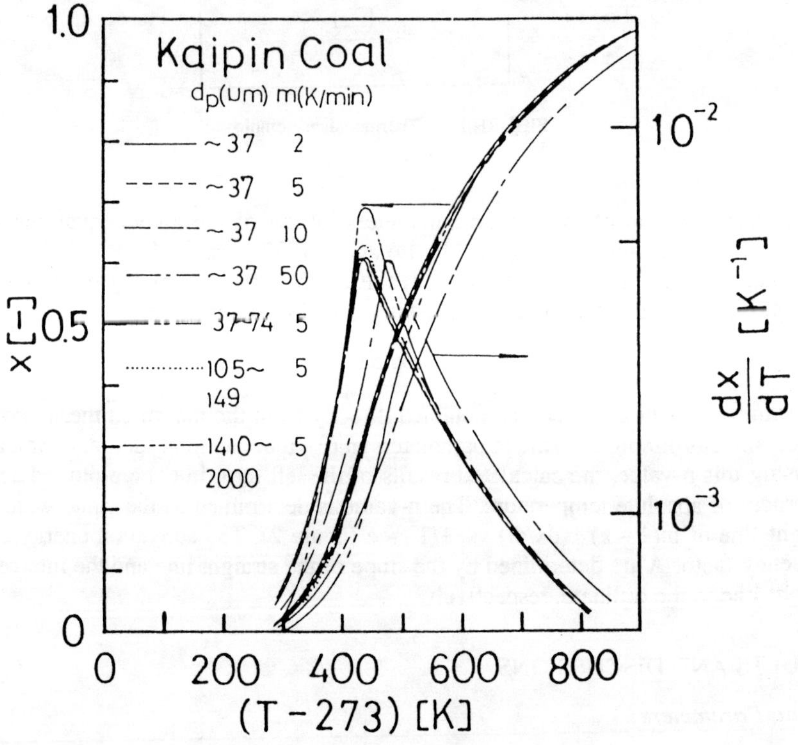

FIGURE 3 TG and DTG curves.

TABLE II

KINETIC PARAMETERS

Coal	Reaction order n	Activation energy E (kJ mol^{-1})	Frequency factor A (s^{-1})
Kaixi (A)	1.8	33.3	0.10
Sangei (B)	1.5	35.6	0.12
Kaipin (C)	1.7	37.9	0.24
Jinjin (D)	1.8	44.4	0.62
Datong (E)	2.0	43.5	0.53
Loy Yang (F)	1.9	33.1	0.19
Newlands (G)	1.6	40.6	0.27
Blair Athol (H)	1.5	48.7	0.66
Kangra (I)	1.5	36.5	0.10
Hongay (J)	1.6	47.1	0.22

experimental condition. A typical example of this examination is shown in Figure 3. From this figure, it can be seen that the TG curves obtained under several experimental conditions converge to a constant TG curve which is obtained with a particle size below 149 μm in diameter and a heating rate below 10 K/min.

Taking this preliminary experimental result into consideration, TG experiments were made with a coal particle size under 37mm in diameter at a heating rate of 5 K/min. The kinetic parameters of various coals thus obtained are shown in Table II.

Relationship Between Reaction Parameters and Fuel Ratio

As it is well known, the fuel ratio is a kind of characteristic value of coals. Therefore, an attempt was made to investigate the reaction parameters obtained from various coals by adopting the fuel ratio as a parameter, so as to characterize the quantitative feature of coal pyrolysis. From Table II, it is found that the reaction order, n, lies within the range of 1.5 and 2.0, almost independent of the fuel ratio. This result coincides fairly well with the result obtained by Juntgen (1984) presenting n = 1.3–2.0. On the other hand, it is seen from Figure 4 that there is a certain relation between reaction rate constant k and fuel ratio FR. It seems that this relation is represented by a straight line obtained by means of the least-squares method. The evaluated rate constant k is expressed by (4)

$$k = (-1878T^{-0.84} - 0.3FR). \qquad (4)$$

Validity of Kinetic Parameters

In the preceeding section, the reaction kinetic parameters for the pyrolysis of different coals was determined. In order to confirm the validity of these obtained parameters, the fundamental equation (1) was arranged by substituting k in (4). The average value of n = 1.75 was adopted as the reaction order for a relatively wide range of fuel ratios.

$$dx/dT = (1/m)(-1878T^{-0.84} - 0.3FR)(1 - x)^{1.75}. \qquad (5)$$

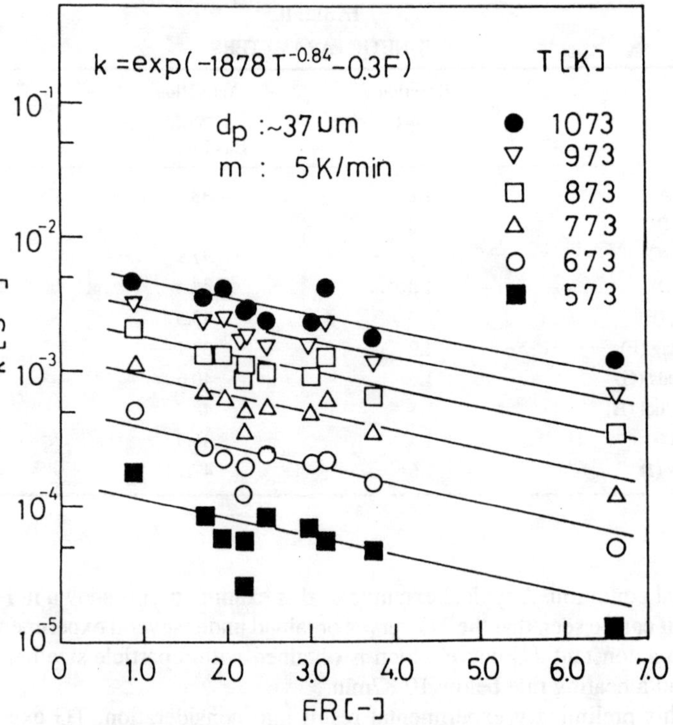

FIGURE 4 Reaction rate constant vs. fuel ratio.

When m is chosen as m = 5 K/min, where all the kinetic parameters were obtained by this heating rate, the integration of (5) with respect to T yields

$$x = 1 - (1 + T^{1.84} \exp(-1878T^{-0.84})/175.33 \exp(0.3FR)))^{(-1.33)}) \tag{6}$$

The relation between conversion and temperature for the heating rate of 5 K/min can be calculated using (6) for coals for which the fuel ratio is known.

Figure 5 represents the comparison of x vs. T between the experimental results and the calculated ones obtained from (6). From this figure, it can be noted that the experimental results are fairly well reproduced by (6).

Effect of Heating Temperature on the Emission of Volatile-N

Figures 6a–c show the relation between conversion of volatile-N, -C, -H and temperature. It is found that for each coal, the volatile-N starts to be released from coal when the temperature reaches about 673–773 K, which is about 50–100°C higher than the temperature where volatile-C and -H start to be released. According to the results shown in the figures, the emission behaviours of volatile-N depend on the temperature, and are summarized: (i) The emission rate of volatile-N is the highest when the temperature is below 873 K; (ii) After the completion of the first stage of the emission of volatile-N, the emission rate of the remaining N is reduced; (iii) However, with a further rise in temperature, the emission rate of volatile-N becomes high again.

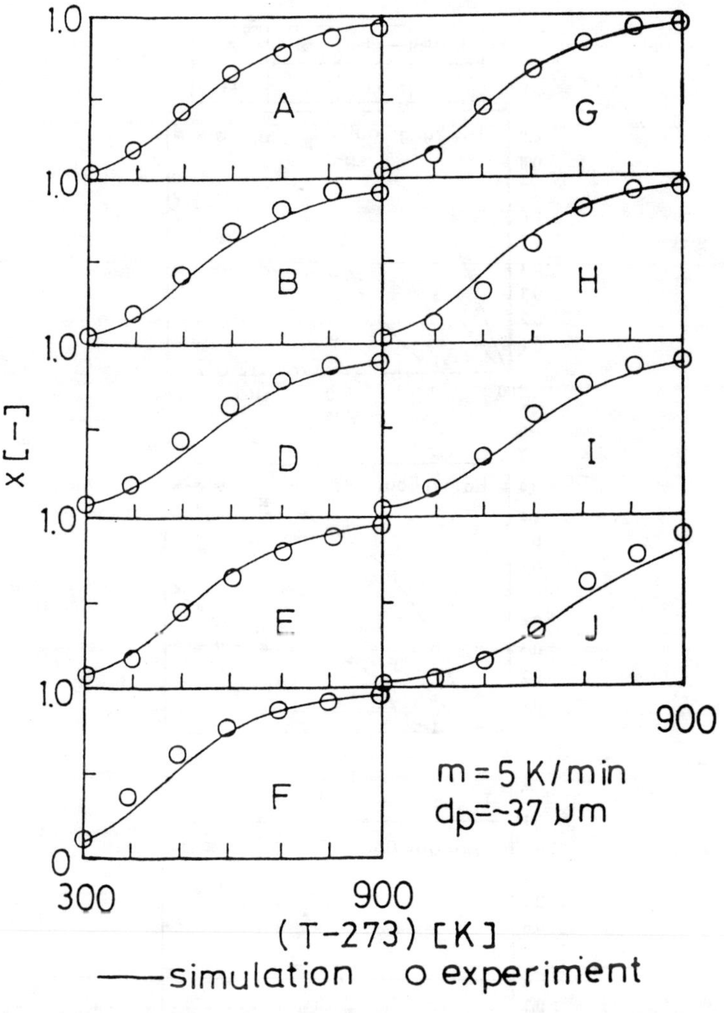

FIGURE 5 Comparison between the simulated result and the experimental data.

Relation Between Conversion of Volatile-N and Fuel Ratio

Figure 7 shows the relation between conversion of volatile-N and fuel ratio. From this figure, it is seen that the volatile-N conversion decreases as the fuel ratio increases in the temperature range of 773–1273 K. In this study, the volatile-N conversion X_N can be correlated with the fuel ratio FR as

$$X_N = \exp[A + B(FR)], \tag{7}$$

with the constants A and B expressed as follows:

$$A_1 = -8.22 + 8.75 \times 10^{-3}\,T \qquad (T < 873\ K), \tag{8a}$$

$$A_2 = -0.89 + 3.85 \times 10^{-4}\,T \qquad (873\ K < T < 1173\ K), \tag{8b}$$

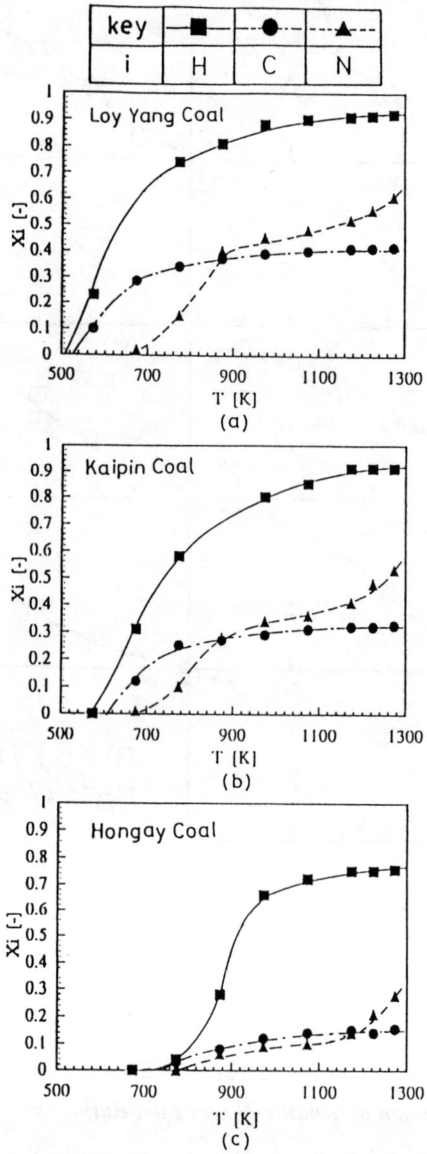

FIGURE 6 Relation between the conversion of C, H and N released from coals and the temperature.

$$A_3 = -1.05 + 5.30 \times 10^{-4}\,T \qquad (T > 1173\ K), \tag{8c}$$

$$B = -\exp(1.48 - 2.7 \times 10^{-3}\,T). \tag{9}$$

Figure 8 represents the comparison of X_N vs. T between the experimental results and the calculated ones obtained from (7), (8) and (9). Figure 9 shows the comparison between the experimental data presented by Pohl and Sarofim (1977) and the calculated ones obtained

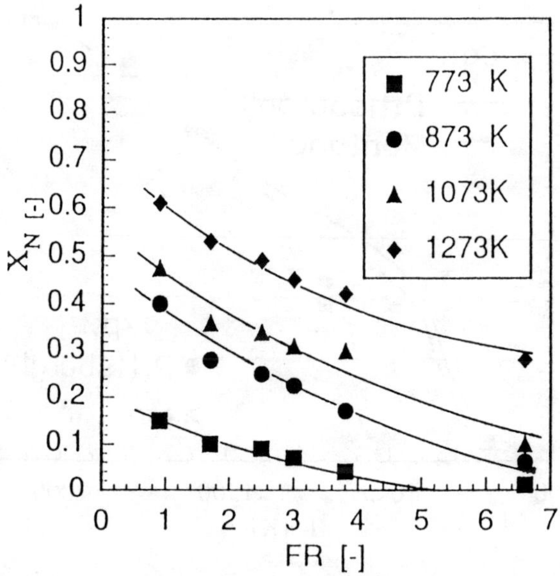

FIGURE 7 Relation between the conversion of volatile-N and the fuel ratio.

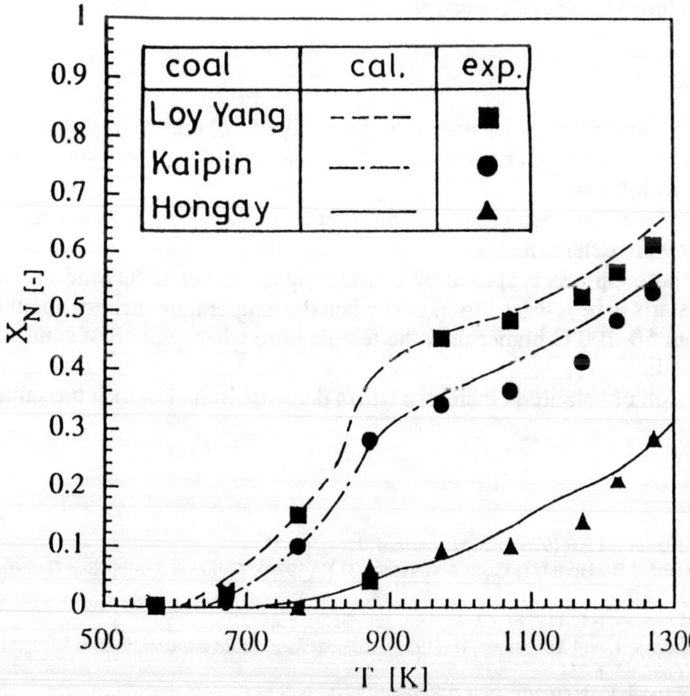

FIGURE 8 Comparison between the calculated results and the experimental results on X_N vs. T.

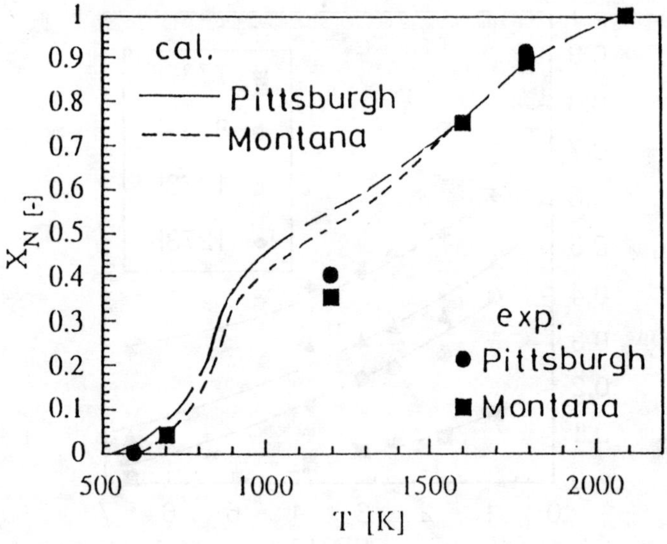

FIGURE 9 Comparison between the calculated results and the experimental data by Pohl *et al.* (1977).

by (7), (8) and (9). From these figures, it is recognized that the experimental results are fairly well reproduced by the calculated results.

CONCLUSIONS

The pyrolysis characteristics and emission behaviours of volatile-N of different kinds of coal were investigated in terms of fuel ratio. The results obtained from the present study are summarized as follows.

1. Pyrolysis reaction parameters such as the reaction order, the activation energy and the frequency factor were determined.
2. The rate expression was proposed by considering the effect of fuel ratio as well.
3. Volatile-N starts to be released from coal when the temperature reaches about 673–773 K, which is about 50–100°C higher than the temperature where volatile-C and volatile-H start to be released.
4. The conversion of volatile-N increases with a decrease in fuel ratio at the same heating temperature.

REFERENCES

Anthony, D.B., and Howard, J.B. (1976). *AIChE Journal*, **22**(4), p.625.

Anthony, D.B., Howard, J.B., Hottel, H.C., and Meissner, H.P. (1974). *Fifteenth Symposium (International) on Combustion*, p. 1303.

Devanathan, N., and Sanean, S.C. (1986). *Chem. Eng. Sci.*, **41**(9), p.2442.

Ito, N., Obata, K., Hakuta, T. and Yoshitome, H. (1983). *Kagaku Kogaku Ronbunshu*, **9**, p.434.

Juntgen, H. (1984). *Fuel*, **63**, p.731.

Juntgen, H., and Van Heek, K.H. (1979). *Fuel Processing Tech.*, **2**, p.261.

Kobayashi, H., Howard, J.B., and Sarofim, A.F. (1976). *Sixteenth Symposium (International) on Combustion*, p.411.

Niksa, S. (1984). *Twentieth Symposium (Intern.) on Combustion*, p. 1445.

Pohl, J.H., and Sarofim, A.F. (1977). *16th Symp. (Int) Combust.*, p. 491.

Pitt, G.J. (1962). *Fuel*, **41**, p. 267.
Skilar, M.G., Shustikov, V.I., and Virozub, I.V. (1969). *Intern. Chem. Eng.*, **9**, p. 595.
Solomon, P.R., Hamblen, D.G., Yu, Z.Z., and Serio, M.A. (1990). *Fuel*, **69**, p.754.
Saxena, S.C. (1990). *Prog. Energy Combust. Sci.*, **16**, p.55.
Wiser, W.H. (1968). *Fuel*, **47**, p.475

NOMENCLATURE

A	Frequency factor (s^{-1})
E	Activation energy (kJ mol^{-1})
FR	Fuel ratio (–)
k	Rate constant (s^{-1})
m	Heating rate (K s^{-1})
n	Reaction order (–)
R	Gas constant (kJ K^{-1} mol^{-1})
T	Absolute temperature (K)
t	Time lapse (s)
x	Conversion (–)

Chemical and Thermochemical Properties of Heavy Molecular Weight Hydrocarbons Evolved During Rapid Heating of Coal of Varying Rank Characteristics

J. D. FREIHAUT and W. M. PROSCIA *United Technologies Research Center East Hartford, Connecticut 06108, USA*
J. C. MACKIE *University of Sydney, Sydney, Australia*

Abstract—The evolution phenomenology, yields, chemical, and thermochemical properties of heavy hydrocarbons, tars, evolved from a wide range of coal types are presented. Coals varying in rank characteristics from lignite to anthracite were rapidly heated to varying extents of devolatilization, the tars isolated from the char stream and subjected to chemical and thermochemical analysis. Systematic variations in the yields, chemical characteristics, thermophysical and thermochemical properties are observed for a given extent of tar evolution from coals of varying rank. Systematic variations are also observed for a given extent of tar evolution from coals of varying rank. Systematic variations are also observed for a given coal but with changing extents of devolatilization. Despite significant variation in the molecular types, yields and temperature-sensitivity of heavy hydrocarbons with coal type, underlying invariances are observed in the evolution sequence of parent coal nitrogen, volatility, functional group distributions, and calorific values of evolved tar species as a function of extent of tar evolution.

Key Words: Devolatilization, Pyrolysis, Fuel Bound Nitrogen, Calorific Values, Volatility, Heating Condition Effects

Symbol	*Definition*
ATP	attached tar precursor
DTP	detached tar precursor
LG	light gas species
T	tar species
k	global rate process

Subscript	*Definition*
char2	char species formed by high-temperature secondary reactions after tar evolution
chem	chemical pyrolysis process
conv	convective transport
det	detachment process
g	gaseous state
i,j,k	specific chemical component
l	liquid state
m	mass transport
ned	nonequilibrium desorption
phy	physical bond-breaking process
vap	vaporization
1	low-temperature devolatilization/pyrolysis-formed light gases (H_2O, CO_2, CO, CH_4, C_2H_6 and higher alkanes, C_2H_4)
2	high temperature pyrolysis light gases (CO, H_2, HCN, C_2H_2, C_2H_4)

Superscript	*Definition*
′	low-temperature secondary reactions of tars
″	high-temperature secondary reactions of tars

INTRODUCTION

On a single particle scale, the combustion of coal appears to occur in a sequential process—volatiles evolution and combustion followed by residual char burnout, as indicated by Seeker, *et al.* (1981) and McLean, *et al.* (1981). This sequential combustion phenomenology does not appear to vary greatly either with combustion intensity, particle heating rate conditions or coal type (Seeker *et al.* 1981; McLean, *et al.* 1981; Smith, 1982; Peters, 1965). In moderate, $10^2 - 10^3$°C/sec, to rapid, 10^4°C/sec or greater, heating rate conditions, to final particle temperatures of 800°C or greater, many coal types will volatilize 50% or more of the parent coal mass (Pohl, 1976; Tyler, 1980). For a fuel introduced as an array of particles, approximately half the particulate-introduced mass will react in a homogeneous combustion process, with the remainder reacting in heterogeneous processes. The combustor is in actuality interacting with two fuel types. Such phenomenology compounds the difficulties involved in burner design modifications to minimize potential pollution formation while maximizing carbon conversion. Substantial changes in the chemical nature of the volatiles with coal type adds complexity to burner design considerations (Freihaut, *et al.* 1989).

In particular, fuel bound nitrogen-related (FBN) NO_x formation is known to be dependent on the so-called "volatile nitrogen" fraction of the parent coal nitrogen (Seeker, *et al.* 1981; Song, *et al.* 1982; Chen, *et al.* 1982). The FBN to NO_x, transformations are intimately linked to the tar and light-gas volatiles flame surrounding the coal particle source. This, in turn, implies the potential influence of micro-scale mixing, turbulence intensity and frequency distributions, established by the combustion system, on ultimate NO_x yields (Wendt, 1977). Similar arguments would apply to soot formation processes. A rational approach to low pollutant burner design requires some knowledge of the combustion properties and yields of tars as a function of coal rank. Since heavy hydrocarbons, tar yields, can dominate the volatiles evolution of bituminous coals, pollutant/carbon conversion chemistry is coupled to these species (Freihaut, *et al.* 1982; Solomon, 1978; Blair, *et al.* 1976).

EXPERIMENTAL

Due to the chemical complexity and heat transfer sensitivity of the rapid devolatilization process, a multi-reactor approach was employed. The three reactors included a calibrated heated grid device, an entrained flow reactor, and a flash lamp reactor. Nominal heating rates for 50 – 100 micron particles in each of these reactors are 1000°C/sec, 5000°C/sec and 10,000 – 100,000°C/sec, respectively. Maximum particle temperatures achieved in the respective reactors are 1000°C, 1100°C and 1600°C, with maximum residence times at temperature on the order of seconds in the heated grid device, hundreds of msecs in the entrained flow device, and 10 msecs in the flash lamp device. Ambient pressures from near vacuum to 1 atmosphere of pressure are employed, depending on the reactor configuration. Particle sizes investigated ranged from 10 microns to 700 microns.

UTRC Entrained Flow Reactor

Conventional entrained flow reactors tend to integrate the tar evolution and secondary reaction processes due to the high temperatures required to achieve a reasonable volatile yield in the residence times available. These conventional reactors are "isothermal" in that the entrainment gas is preheated to match the wall temperature in the main sections of the reactor, that is, exclusive of sample introduction and extraction regions. Depending on the peak gas temperature, the tars escaping the particle may undergo cracking reactions in the hot entrainment gas. At peak gas or particle temperatures

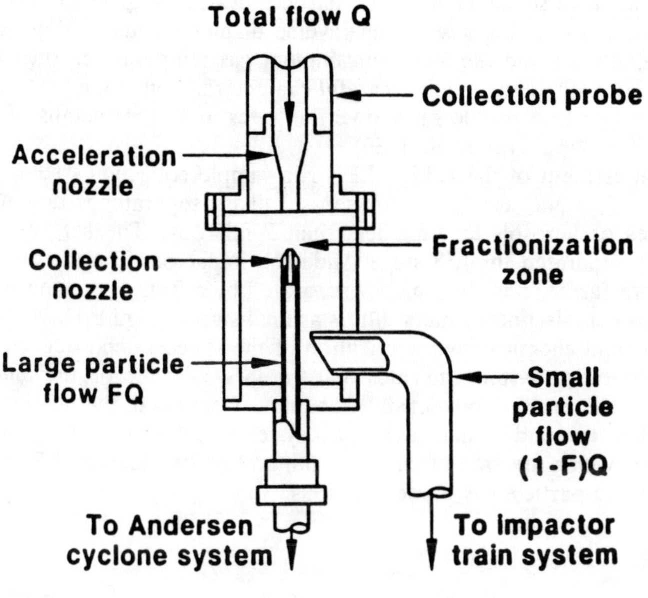

FIGURE 1a. Char and Aerosol/Fines Separation: Entrained Flow Reactor

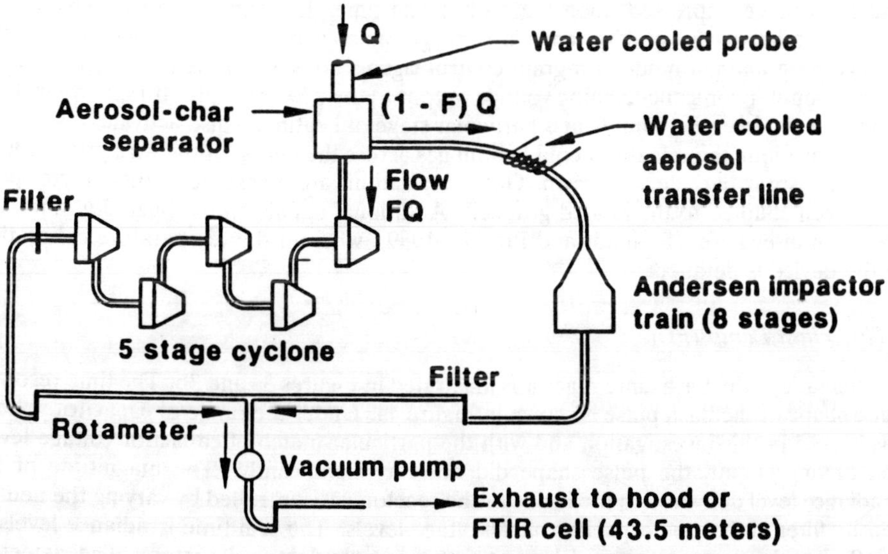

FIGURE 1b. Aerodynamic Separation of Char and Tar/Fines Streams

below 700°C, the EFR will maximize tar yields while minimizing secondary reactions of evolved tars. The entrainment gas enters the reactor at room temperature and is heated by the hot walls as it flows through the reactor. The resulting gas profiles are non-isothermal. However, the peak gas temperature is typically two hundred degrees below the nearly axially isothermal, reactor wall temperature. The irradiance of the reactor walls provides an inverse dependence of particle heating rate on particle diameter, while loss to the surrounding gas has an inverse diameter squared dependence. The evolved tars rapidly assume the local entrainment gas temperature. Interpretation of results obtained in the reactor requires careful characterization of the heat transfer field. Detailed gas temperature profiles, radiative flux rates and other details of the reactor are given elsewhere (Freihaut, *et al.* 1989).

An additional element of the UTRC EFR is a sample collection system designed to isolate tar from char and ash fines. An aerosol phase separator is designed to "pull off" all particles or aerosols that are less than 2 microns. The aerosols are further aerodynamically separated in an 8 stage Andersen impactor. The inertially separated char particles are further aerodynamically separated by a 5 stage cyclone system. Both separation trains contain porous metal filters as final stages (Figure 1). A detailed mass fraction analysis and chemical characterization of the species deposited on the various stages in the separation system indicates that the tars collected on the final filters are representative of the evolved, condensible tar mass. Combined, the heating conditions, aerosol separation train and on-line gas analysis system enables one to generate significant quantities of tars which are free of char and mineral matter fines and have undergone a minimum of extra-particle gas phase reactions.

UTRC Heated Grid Reactor

Figure 2 schematically displays the UTRC heated grid system with its associated microcomputer-based, power programming and data acquisition system. Provisions are made to routinely monitor two thermocouples, a pressure transducer, delivered current and the voltage impressed upon the grid in real time. Electrical power is delivered to the electrodes of the grid system via a programmable power supply operated in the voltage-programmed mode. Program control signals are sent to the power supply via a microcomputer controlled analog voltage output to the power supply. If rapid quenching is desired to "freeze" a sample at a particular stage of heating, it also activates a solenoid valve controlling the release of cold helium gas across the length of the wire grid through a helium "spray bar" device shown. Gas compositions are measured in the 1 meter path length cell coupled to the heated grid cell. Additional details on the heated grid system are given elsewhere, (Freihaut and Proscia, 1989), wherein the temperature calibration of the device is detailed.

UTRC Flash Lamp (FL)

A schematic of the flash lamp reactor is illustrated in Figures 3a and 3b. The time resolved pulse shape of the flash pulse is shown in Figure 4a. Under the range of capacitor voltage levels used in this investigation and with the particular match of capacitor voltage levels and driving circuit, the pulse shape did not vary significantly. The magnitude of the irradiance level delivered to the inside of the reactor was controlled by varying the neutral density filter and/or the capacitor bank voltage levels. The real time irradiance levels of the flash pulse were monitored by fast response pyroelectric detectors. The detectors employed were calibrated with NBS irradiance standard lamps and a NBS traceable radiometer system used in conjunction with a chopped Argon ion laser. Pyroelectric detectors have a nearly constant responsivity with respect to wavelength from the UV to IR spectral regimes.

Because of the possibility of inducing photochemical reactions with the UV component of the flash pulse, the reactors employed were pyrex, as were the neutral density filters. The delivered radiation is characterized by wavelengths ranging from 0.4 to 2.0 microns with peak intensities between 0.8 and 1.1 microns (Figure 4b).

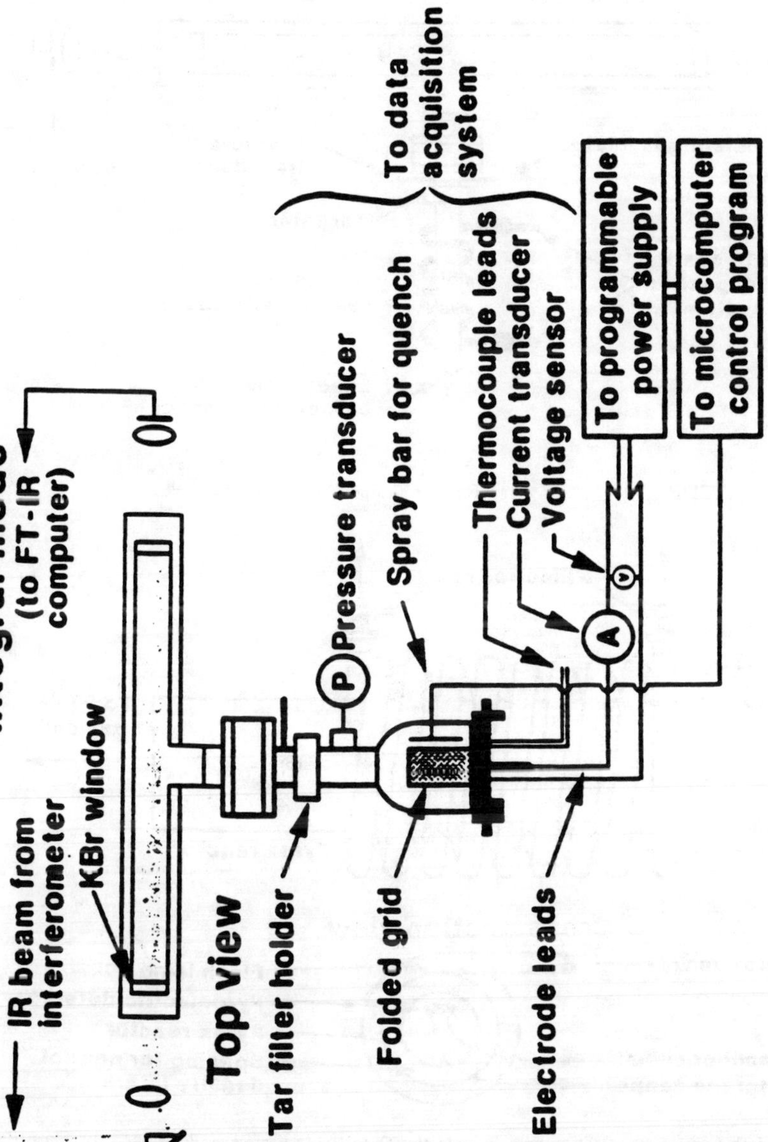

FIGURE 2. Heated Grid Control and Data Acquisition System

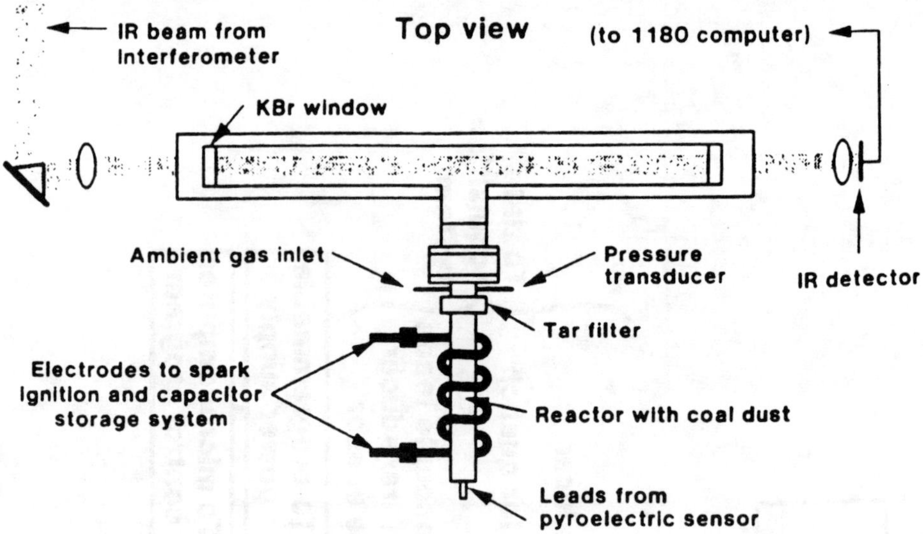

FIGURE 3a. Flash Lamp Reactor Coupled to FT-IR Gas Cell

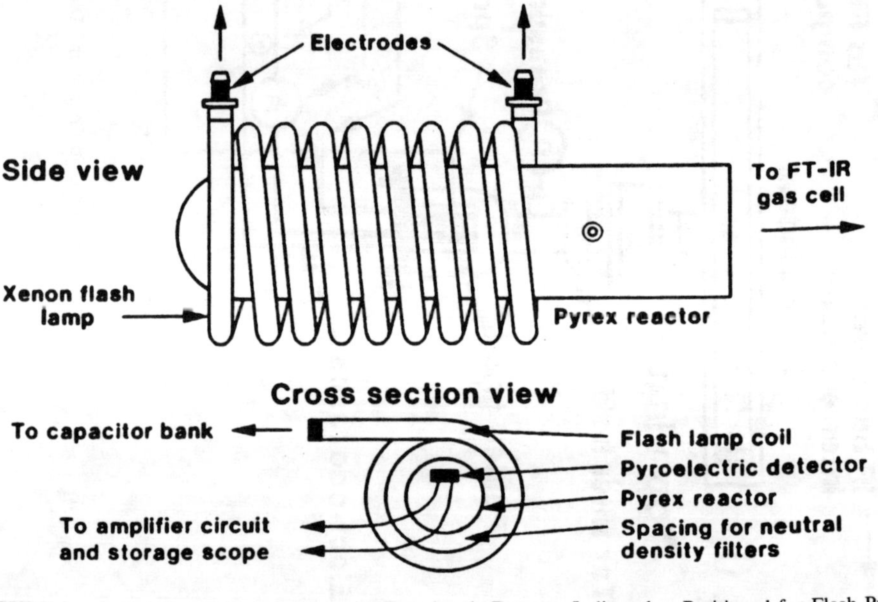

FIGURE 3b. Flash Lamp Cross Section with Pyroelectric Detector Indicated as Positioned for Flash Pulse Calibration

Reactor Design Rationale Relative to Bituminous Coal Devolatilization

The multi-reactor approach to examining devolatilization is necessitated by the inherent chemical nature of coal. The complex chemical nature of coal by molecular group type and bonding type distributions provides a nearly unlimited set of primary thermal

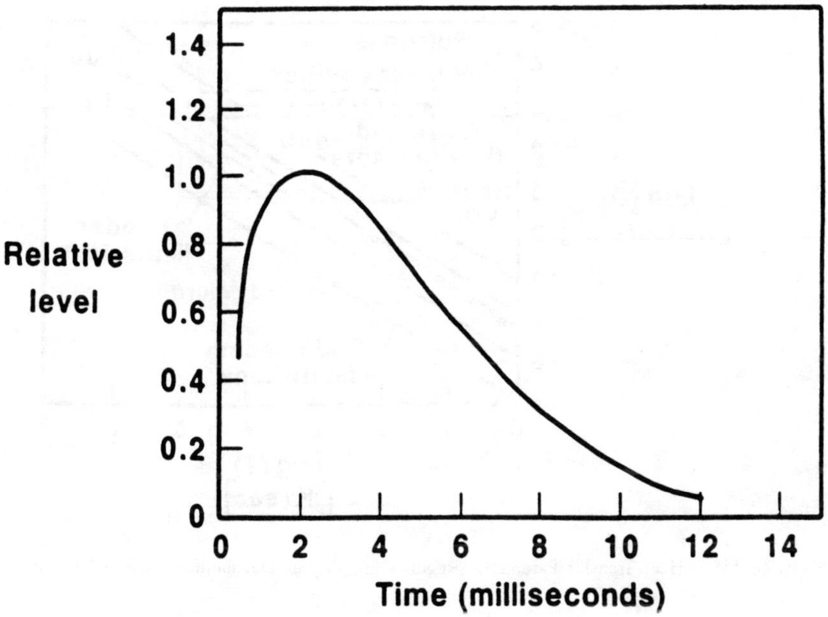

FIGURE 4a. Pulse Shape of Helical Flash Lamp

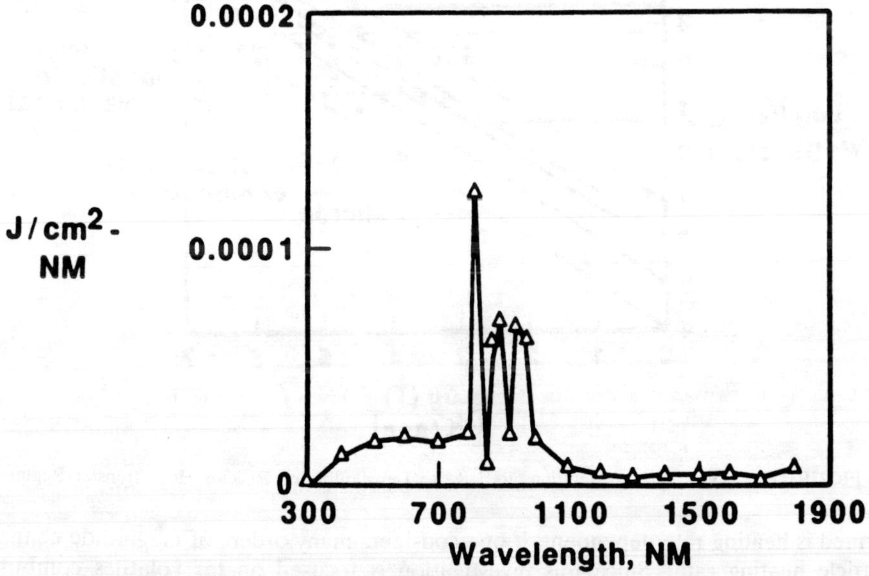

FIGURE 4b. Wavelength Distribution of Delivered Flash Pulse

decomposition pathways. A "primary" decomposition pathway may be arbitrarily defined as a devolatilization process whereby volatiles formed within the parent coal particle undergo minimal intra- or extra-particle secondary decomposition reactions. For a hydrocarbon substance as complex as coal the mix of "primary," intraparticle volatiles

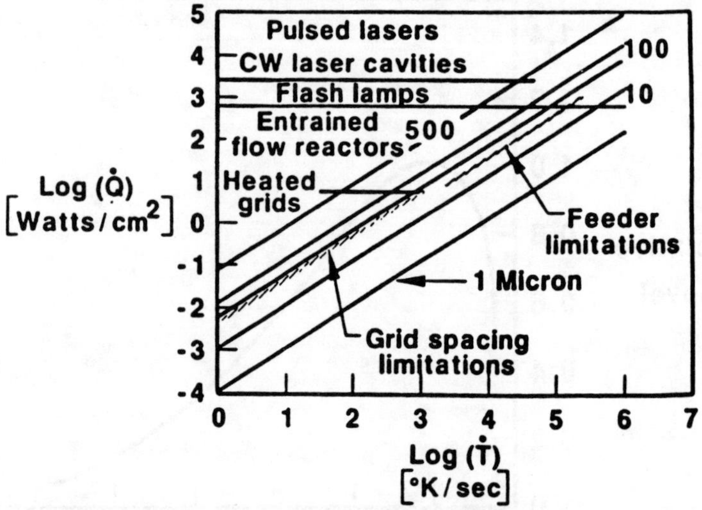

FIGURE 5a. Heat Transfer Parametric Regimes for Various Devolatilization Reactor Types

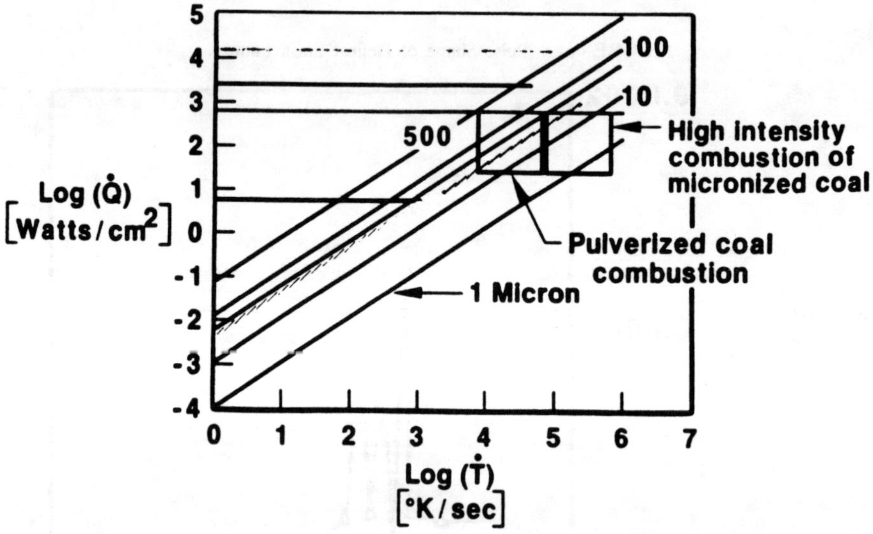

FIGURE 5b. PFC Firing Conditions Relative to Devolatilization Reactor Heat Transfer Regimes

formed is heating rate dependent, if one considers many orders of magnitude change in particle heating rate. Since this investigation is focused on tar volatiles combustion properties related to pulverized coal combustion conditions (pfc), imposed particle heating rates are restricted to 1000° C/sec or greater.

By subjecting samples of the same coal to the different heat transfer fields one is able to experimentally focus on a particular phase of the devolatilization process (Figures 5a and 6). Figure 5b indicates the portion of heat transfer parameter space applicable to pfc firing. The HG and FL reactors allow particle heating to be accomplished in a thermally quenching environment, once the evolved tar volatiles have traversed the particle or

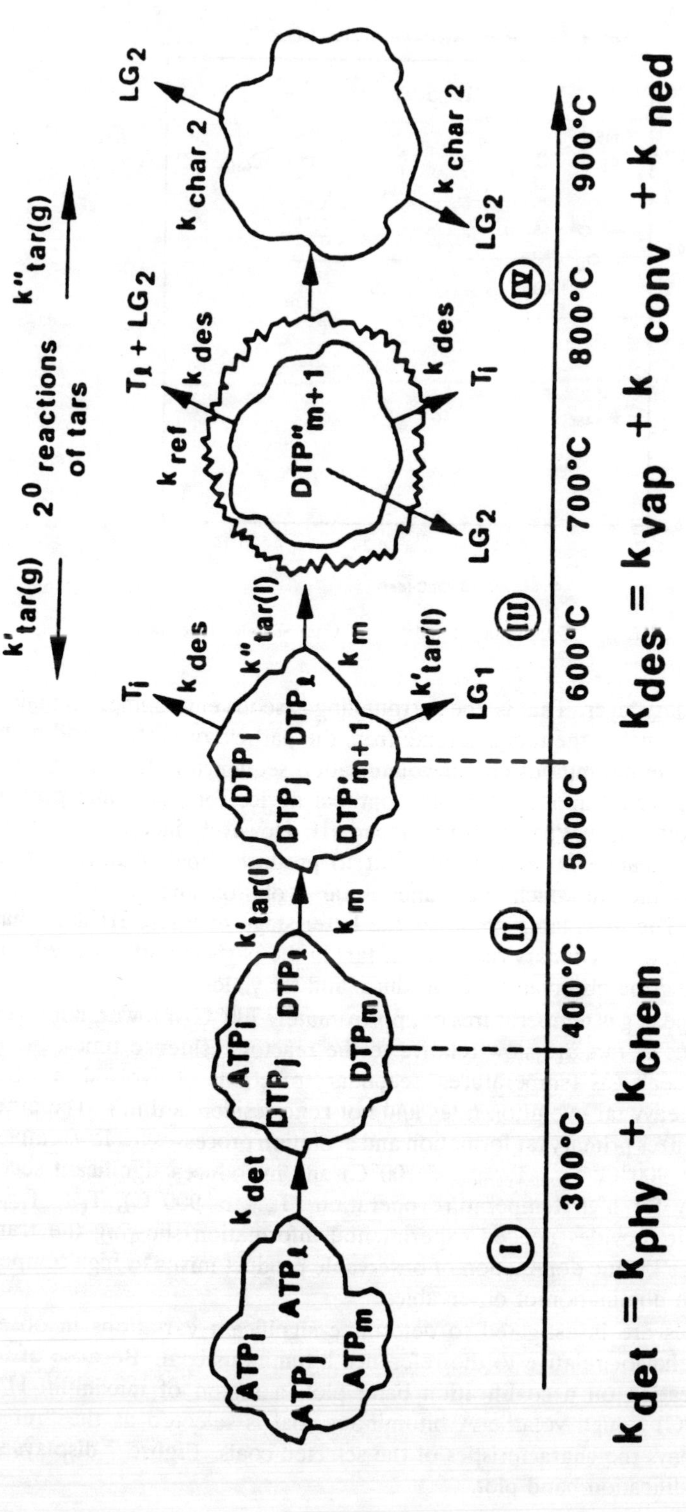

FIGURE 6. Devolatilization Sequence of Bituminous Coals

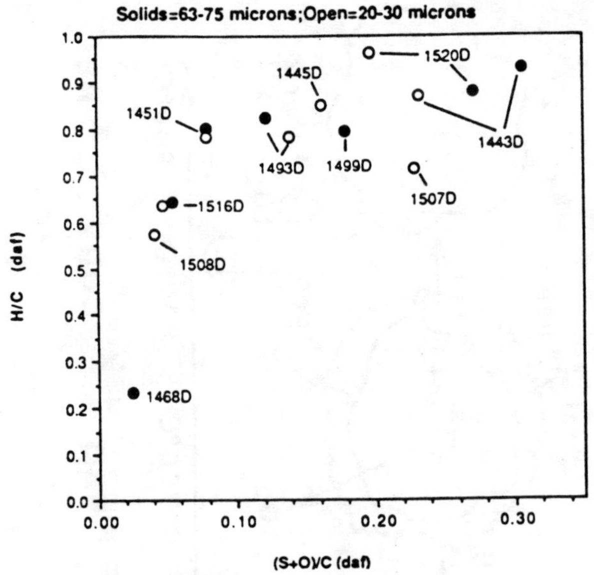

FIGURE 7. Coalification Band Plot of Coal Samples Investigated

particle/grid boundary layer. That is, the surrounding gaseous environment remains cool and immediately quenches the tars emerging from the particle or particle/grid boundary layer. Relative to the bituminous coal devolatilization sequence noted in Figure 6, the heated grid device would appear to be the optimal device for examining primary tar formation and evolution in Phases I and II and III. However, in the case of the FL reactor, imposed radiative fluxes are sufficient to produce rapid heating of particles to temperature regimes in which simultaneous tar evolution and secondary reaction chemistry occurs. The flash lamp amplifies the latter stage of Phase III and Phase IV and the influence of k'' secondary reactions of tars in the particle and particle boundary layer in determining the observable gas products and tar yields.

In the EFR for peak gas temperatures of approximately 700° C or lower, homogeneous secondary reactions of tars are slow relative to the reactor residence times. At higher reactor wall and peak gas temperatures, secondary reactions of evolved tars become competitive with heavy tar evolution rates and hot zone residence times. The entrained flow reactor maximizes primary tar formation and evolution processes for low temperature operation ($T_{wall} < 900°C$, T_{gas}, $T_{particle} < 700°C$) and introduces significant secondary reaction chemistry for high temperature operation ($T_{wall} > 900°C$), T_{gas}, $T_{particle} > 700°C$). As such it provides needed experimental information showing the transition from primary phase kinetic domination of observable product mixes to high temperature secondary reaction domination of observables.

A range of coals are investigated to determine significant variations in observable devolatilization behavior relative to the reference bituminous coal. Because of its high tar yields and location on a coalification band plot, a region of maximum H/C and minimized (S+O/C) a high volatile A bituminous coal is selected as the "reference" coal. Table I displays the characteristics of the selected coals. Figure 7 displays sample positions on a coalification band plot.

TABLE I
Elemental Composition of PSOC Coals on a Dry, Ash-Free Basis
(High Temperature)

PSOC no.	Rank	% C	% H	% N	% O	% S	Seam Location
		20-30 μm Particles					
1443D	Lignite	71.42	5.17	1.35	21.01	1.06	Lower Wilcox, TX
1507D	Lignite	72.49	4.32	1.09	20.22	1.88	Beulah, ND
1520D	Sub C	73.67	5.90	1.11	18.22	1.10	Smith Roland, WY
1445D	HVC bit	76.72	5.44	1.28	15.65	0.91	Blue, NM
1451D	HVA bit	83.98	5.48	1.67	7.41	1.46	Pitt. No. 8
1493D	HVA bit	78.79	5.16	1.48	10.15	4.42	Ill. No. 6
1508D	MVB	89.71	4.29	1.10	4.27	0.63	Pocahontas No. 3
1516D	LVB	88.88	4.71	1.49	4.30	1.25	Lower Kittaning
		63-75 μm Particles					
					% S + O		
1443D	Lignite	66.45	5.16	1.28	27.11		Lower Wilcox, TX
1520D	Sub C	69.03	5.04	0.98	24.96		Smith Roland, WY
1499D	HVA	79.90	5.49	1.57	13.04		Colorado B, CO
1493D	HVA	75.52	5.02	1.46	18.01		Illinois #6, IL
1451D	HVA	83.95	5.62	1.62	8.81		Pittsburgh #8, PA
1516D	LVB	87.40	4.69	1.55	6.36		Lower Kittaning, PA
1468D	Anthracite	94.34	1.86	0.74	3.06		Buck Mountain, PA
		% Tar Yield (daf) 1 atmos. He, 10 sec					
		1443D	17.6				
		1520D	19.3				
		1499D	22.2				
		1493D	22.6				
		1451D	31.04				
		1516D	10.8				
		1468D	2.5				

RESULTS

Fuel Bound Nitrogen Evolution in Tar Volatiles

Previous investigations indicated that large fractions of the fuel bound nitrogen in bituminous coals are volatilized as a heteroatom in the tar component of the total volatiles (Freihaut, *et al.* 1982; Solomon, 1978; Blair, *et al.* 1976). The fraction of parent coal nitrogen evolved with the tar is in direct proportion to the mass fraction of parent coal mass evolved as tar for high volatile bituminous coals in the moderate heating rate, thermal quenching environment characteristic of heated grids (Freihaut, *et al.* 1982). Since tar mass dominates the initial mass release of bituminous coals, the FBN release is nearly proportional to the total mass fraction of ash free coal volatilized.

In the rapid heating, hot gas environment of high temperature entrained flow reactors, lignite coals appear to display a lag in nitrogen release relative to bituminous coals, on a total volatiles release basis (Pohl, 1976). Although lignite coals are known to produce less

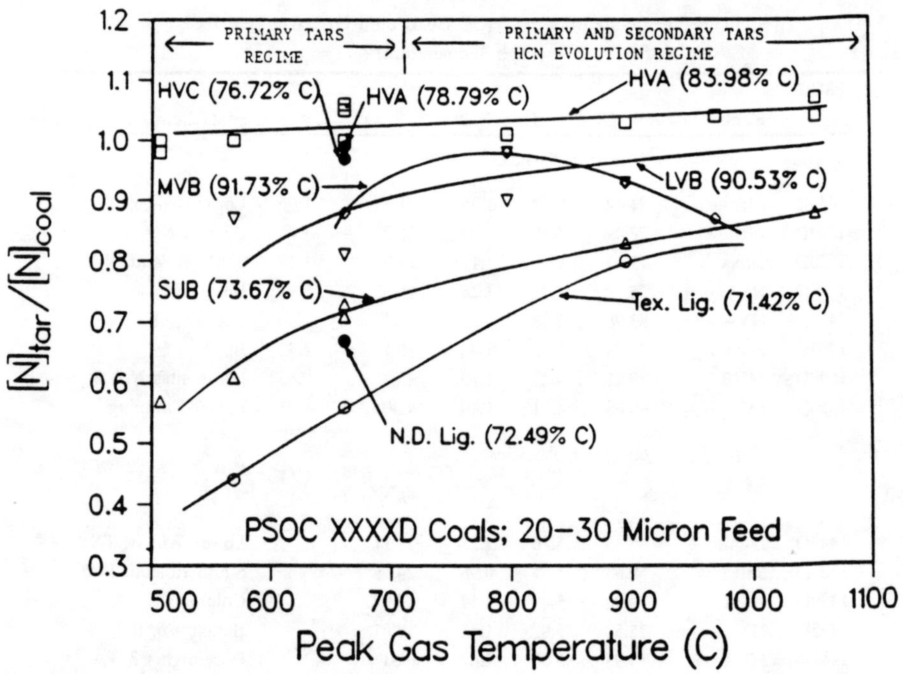

FIGURE 8. Nitrogen Concentrations of Evolved Tars as a Function of Coal Type and Peak Particle/Gas Temperature in Entrained Flow Reactor

tar mass in moderate to rapid heating conditions than bituminous coals (Freihaut, *et al.* 1989; Wei-Chun Xu, 1987), it is not clear if the differences in tar yield alone account for differences in early FBN volatilization. A number of investigations of tar evolution from low rank coals indicated such tar species differ significantly from the average structural characteristics of the parent coal (Tyler, 1980; Nelson, 1986; Calkins, 1986; Suuberg, 1982).

Figure 8 indicates the nitrogen mass fraction concentration of the captured tar species relative to that of the parent coal as a function of peak particle or gas temperature and coal type. As indicated, the captured low rank coal tar volatiles contain significantly less nitrogen on a mass fraction basis than the parent coal during the initial, primary tar evolution phases. Bituminous coal tars display a constant FBN concentration, at parity to that of the generating, parent coal, irrespective of extent of tar evolution. Table II displays the detailed FBN evolution for the reference coal, PSOC 1451D, as a function of peak gas (particle), temperature. Figure 9 displays the results graphically. The results indicate that the FBN evolution behavior of low rank coals differs from that of bituminous coals both with respect to tar yield potential and intrinsic concentrations in the tar. That is, not only are tar yields substantially lower than bituminous coals as a fraction of total volatiles but also molecular types are evolved that do not contain nitrogen concentrations as great as that of the parent coal average.

Volatility and Functional Group Characteristics of Primary Tars

Tars evolved at conditions in which yields are maximized while secondary reactions are minimized were examined by standard KBr pellet techniques using an FT-IR instrument.

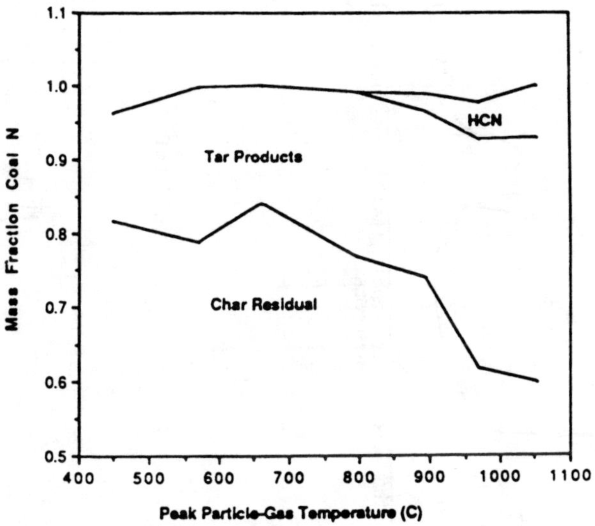

FIGURE 9. FBN Distribution of Reference Coal (PSOC 1451D) in Devolatilization Products

TABLE IIa
COAL NITROGEN DISTRIBUTION IN DEVOLATILIZATION PRODUCTS
PSOC 1451D, 20-30 microns

Peak Gas Temp. (°C)	F_{char} (char / coal-N)	F (char / coal-N)	F (HCN / coal-N)	F (NH_3 / coal-N)	F (Σproducts / coal-N)
450	0.816	0.145	ND	ND	0.961
570	0.787	0.211	ND	ND	0.998
660	0.840	0.159	ND	ND	0.999
796	0.768	0.222	0.0007	ND	0.991
895	0.739	0.224	0.0241	ND	0.989
970	0.618	0.308	0.0506	ND	0.977
1053	0.598	0.330	0.0720	ND	1.000

ND—None Detectable

TABLE IIb
TAR YIELDS AS A FUNCTION OF REACTOR CONDITIONS
PSOC 1451D, 20-30 microns

Reactor Wall Temp. (°C)	Max. Gas Temp. (°C)	Residence Time (msec)	TAR (DAF) Mass Fraction
668	507	580	0.14
704	569	545	0.20
825	660	515	0.16
939	796	450	0.22
1058	895	410	0.21
1163	969	355	0.31
1241	1053	335	0.30

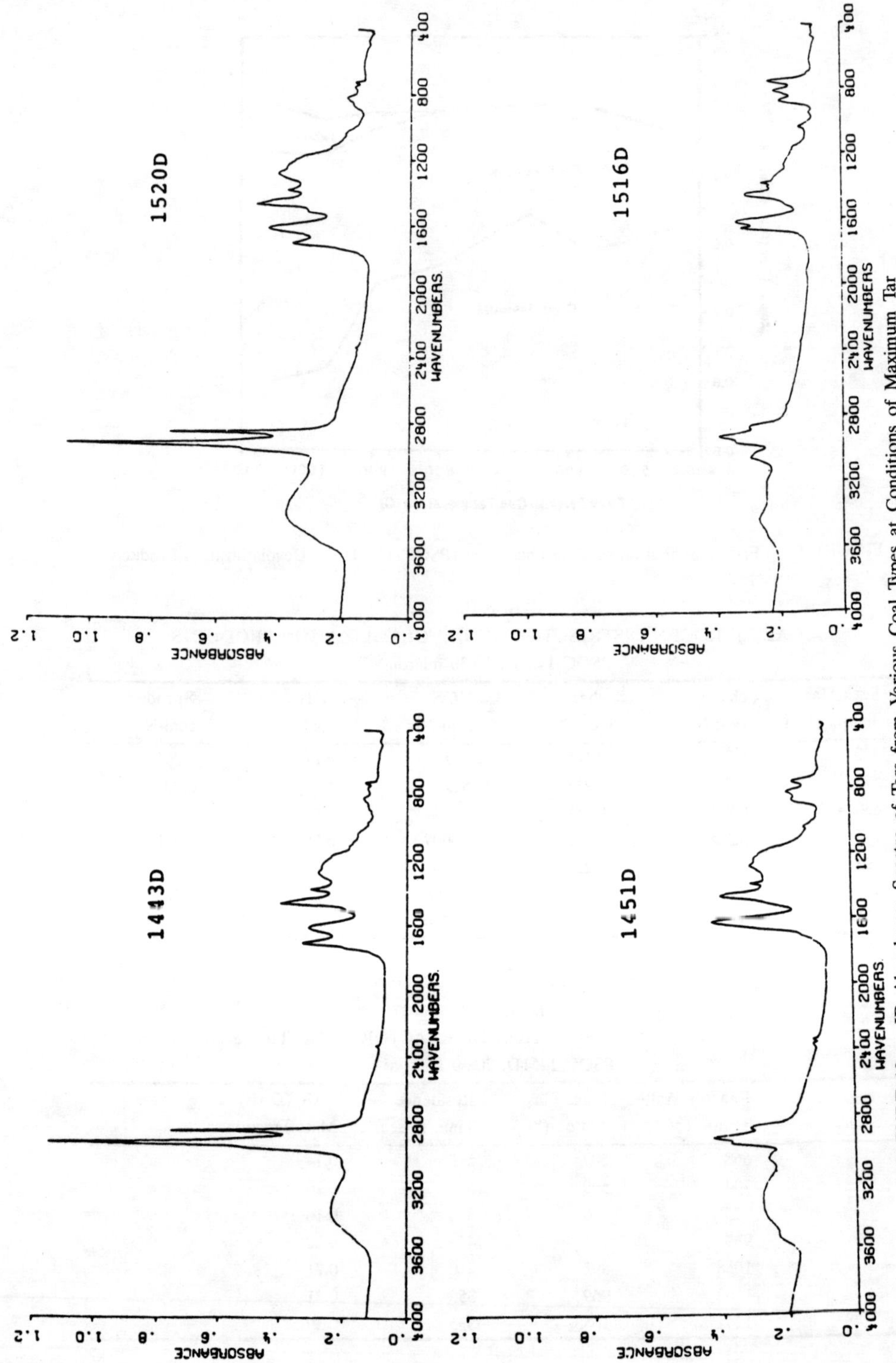

FIGURE 10. IR Absorbance Spectra of Tars from Various Coal Types at Conditions of Maximum Tar Yield and Minimum Secondary Reactions (Peak Particle Temp. @660)

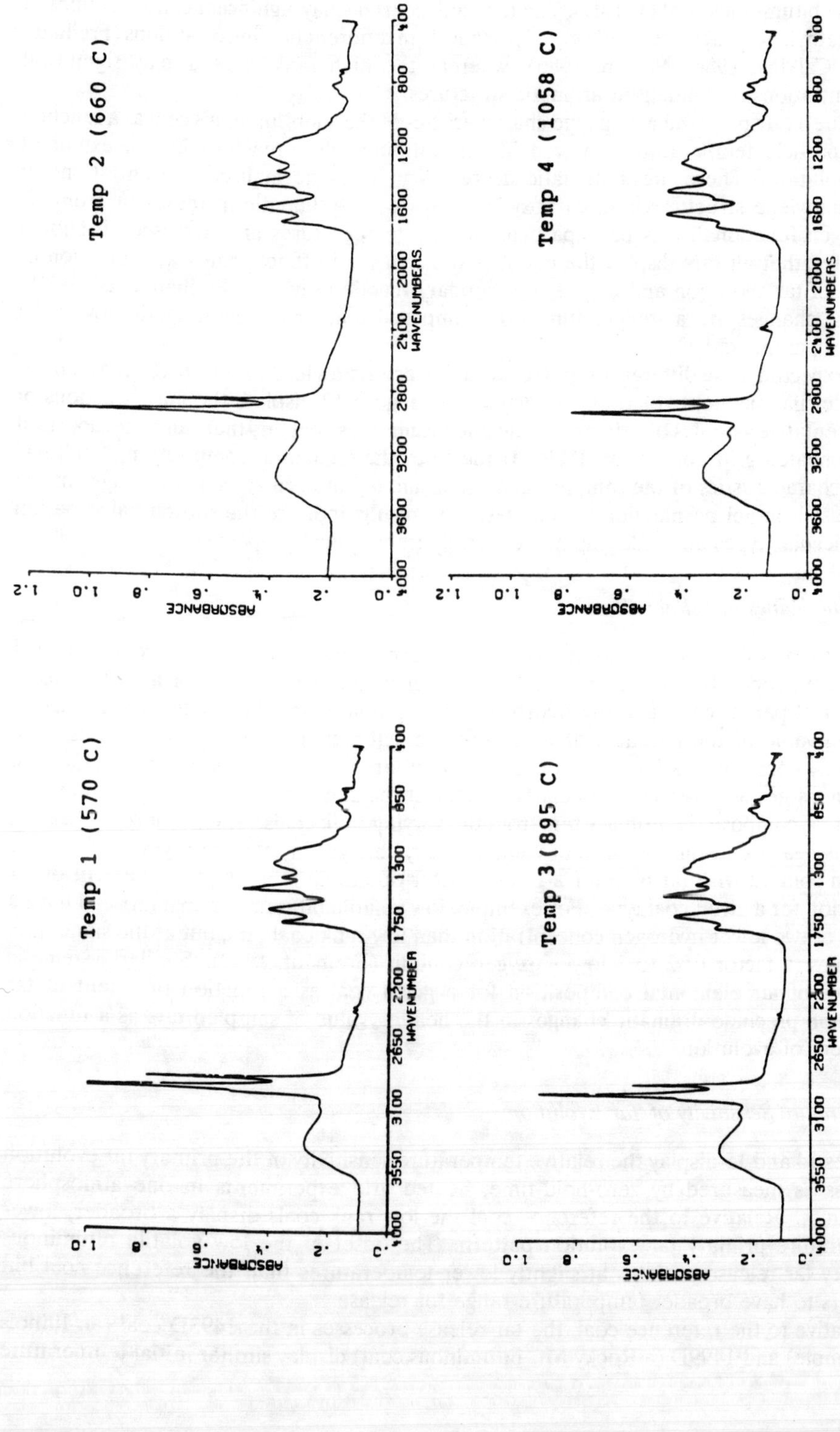

FIGURE 11. IR Absorbance Spectra of Subbituminous Coal Tars as a Function of Peak Particle/Gas Temperature in Entrained Flow Reactor

Representative absorbance spectra are indicated in Figure 10 for a lignite (1443D), a subbituminous coal (1520D), the reference bituminous coal (1451D), and a low volatile bituminous coal (1516D). The low rank coals display significant concentrations of polymethylene structures, as observed previously in different heating conditions (Freihaut, 1989; Calkins, 1986; Nelson, 1986), whereas the high rank coals display significant concentrations of condensed aromatic structures.

Figure 11 displays the group type characteristics of the subbituminous coal as a function peak particle temperature achieved in the entrained flow reactor, that is, extent of tar evolution. There are systematic decreases in mass normalized concentrations of polymethylene structures and carboxylic groups with systematic increases in aromatic hydrogen functionality as peak particle and gas temperatures are increased. It should be noted that all tars display these systematic changes in functionality as a function of extent of tar evolution and degree of secondary reaction severity (Freihaut *et al.* 1989), but the changes are a steeper function of imposed reactor temperature the lower the rank of parent coal.

As expected these differences in structural characteristics lead to corresponding changes in molecular size and volatility characteristics. Figure 12 displays the mass fractions of representative samples of primary lignite and bituminous coal tars that can be revaporized in the heated grid apparatus. Table III indicates the elemental composition, "melting" point characteristics of the samples, and the estimated molecular weight averages of the tars based on gel permeation techniques. The results indicate the substantial variation in molecular types and resulting associated thermophysical properties.

Calorific Values of Primary Tars

The calorific values of the parent coals and corresponding primary tars were determined using a microbomb calorimeter. Figure 13 displays the measured calorific values for a number of parent coals and their corresponding primary tars. The systematic change in heating value of the parent coal with rank characteristic is immediately obvious. The relatively constant heating value of the primary tars irrespective of parent coal source is perhaps not as obvious, particularly in light of the substantial variations in molecular types that compose the primary tars from the varying rank coals. The primary tar heating value is nearly constant because of compensatory changes in total hydrogen, carbon and oxygen content with tar type for a given reactor operating temperature or extent of tar evolution for a given coal type. For example, low volatile bituminous coal tars can have a factor of two lower hydrogen concentration than low rank coal tars, but at the same time they have a factor of 2 to 3 lower oxygen content (Freihaut, 1989). Similar systematic changes of tar elemental composition for a given coal as a function of extent of tar evolution preclude dramatic changes in the heating value of sampled tars as a function of extent of evolution.

Temperature Sensitivity of Tar Evolution

Figures 14 and 15 display the relative temperature sensitivity of the primary tar evolution process as measured by zero-hold-time, heated grid experiments in one atmosphere of helium. Relative to the reference coal the low rank coals display a broader, lower temperature primary tar evolution pattern. The onset of the low volatile bituminous primary tar release initiates at slightly lower temperatures than the reference coal but appears to have broader temperature range for release.

Relative to the reference coal, the tar release processes in the 1493D coal (an Illinois #6 sample) and 1499D (a Rocky Mt. bituminous coal) display similar initial temperature

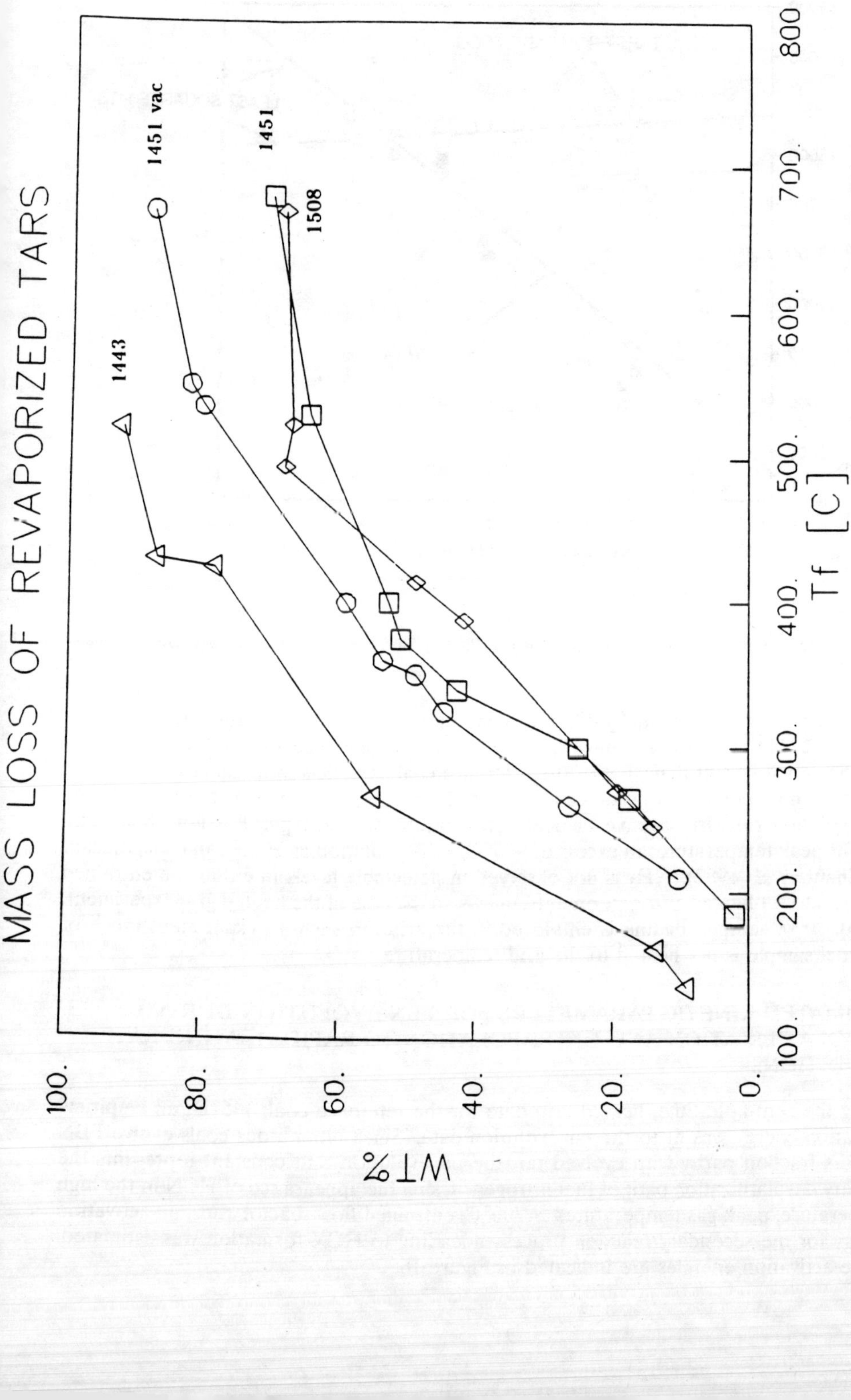

FIGURE 12. Relative Volatility of Lignite and Bituminous Coal Tars as a Function of Heated Grid Vaporization Temperature

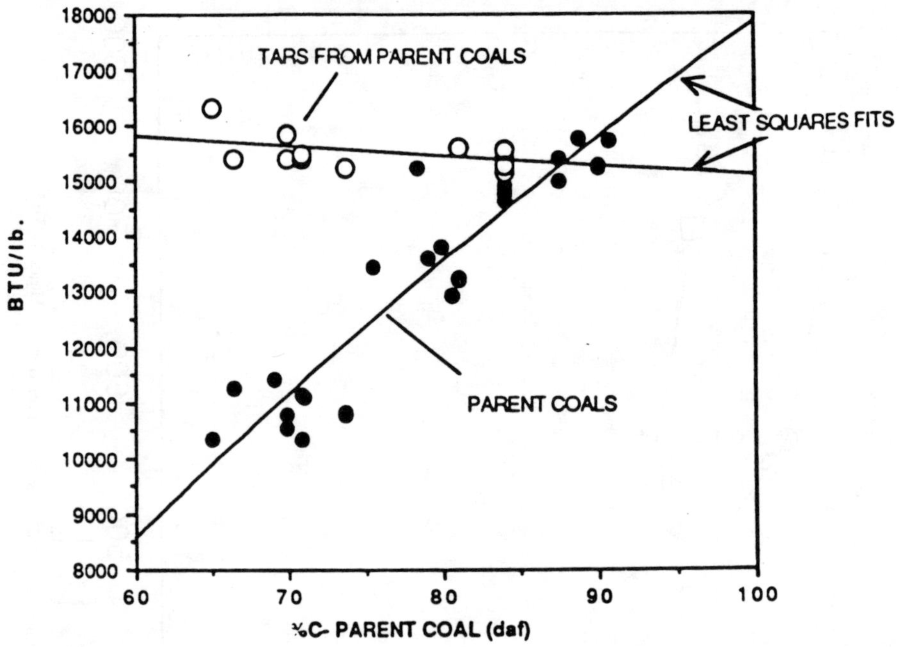

FIGURE 13. Calorific Values of Parent Coals and Primary Tars Plotted wrt Parent Coal Rank Parameter C% (daf)

dependencies, but substantially different temperature sensitivities at higher temperatures (Fig. 15). Ultimate yield behavior, as displayed by long hold time tar yields, is also significantly different than that of the reference coal as indicated in Table I.

It should be noted that no detectable HCN or NH_3 is observed in the zero hold time tar evolution runs irrespective of peak temperature or coal type. For long hold time runs at peak temperatures in excess of 900°C, FBN evolution as HCN can be significant (Freihaut, *et al.* 1982). NH_3 is not observed in detectable levels in either the entrained flow reactor or heated grid experiments, unless, in the case of the heated grid experiment, non-disperse sample loading is employed in the grid. In such an experiment much of the coal sample is not heated to the grid temperature.

ESTIMATED KINETIC PARAMETERS FOR FBN EVOLUTION DURING DEVOLATILIZATION AND EXTRAPOLATION TO RAPID HEATING RATE CONDITIONS

Using the zero-hold-time, heated grid data for the reference coal, 1451D, an empirical activation energy was fit to the tar evolution data. Since bituminous coals evolve FBN in mass fraction parity with evolved tar, the tar evolution rate constant represents the primary devolatilization path of fuel nitrogen. Using the appearance of HCN in the high temperature, peak gas temperatures > 700°C, entrained flow reactor runs, an activation energy for the secondary reaction processes leading to HCN formation was estimated. These activation energies are indicated in Figure 16.

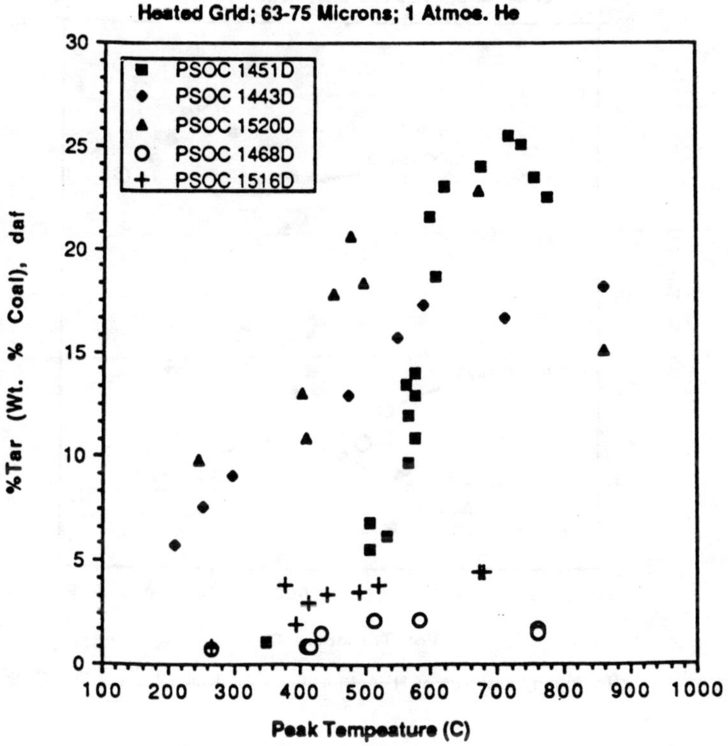

FIGURE 14. Tar Evolution Sensitivity of Non-HVA Bituminous Coals Relative to "Reference" Coal as Observed in HG Apparatus

Extrapolation to High Intensity Heating Conditions

The kinetic parameters imply that a process in which the coal particle mass can be heated to temperatures in excess of 1000°C in the order of tens of milliseconds or less would result in simultaneous tar evolution and secondary cracking reactions for a bituminous coal. Rapid heating to such temperature regimes would result in competitive tar evolution and high temperature cracking processes at the particle surface or particle-gas boundary layer. Table IV indicates the estimated peak temperatures of small particles of the reference bituminous coal subjected to flash lamp, pulse heating. As noted above, the ambient argon or helium gas remains relatively cold in the radiant pulse heating, and thereby rapidly quench any evolved tars that escape the particle-gas boundary layer. Table V displays the measured product distributions as a function of given irradiance levels.

The sensitivity of the observable product mix to the heat transfer related parameters is apparent. Both the tar evolution and secondary cracking reactions of tars from bituminous coals are rapid at high particle temperatures. Consequently, the mass transfer resistance to heavy hydrocarbon evolution would be expected to result in significant changes in observable product distributions under intense heating conditions. As the product distributions indicate, at the highest irradiance levels significant quantities of secondary decomposition product gases—HCN, C_2H_2 and CO—are observed. At the lowest flux levels, only tar and low temperature decomposition gases are observed.

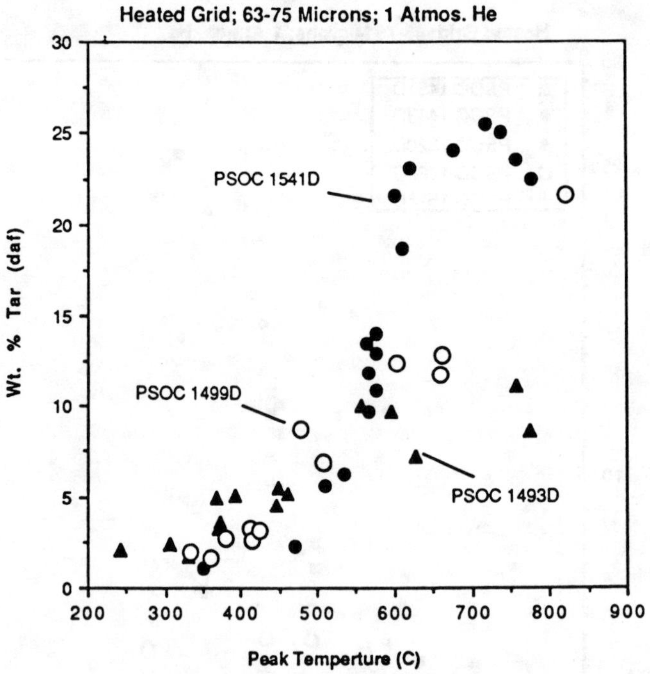

FIGURE 15. Tar Evolution Sensitivity of HVA Bituminous Coals as Observed in HG Apparatus

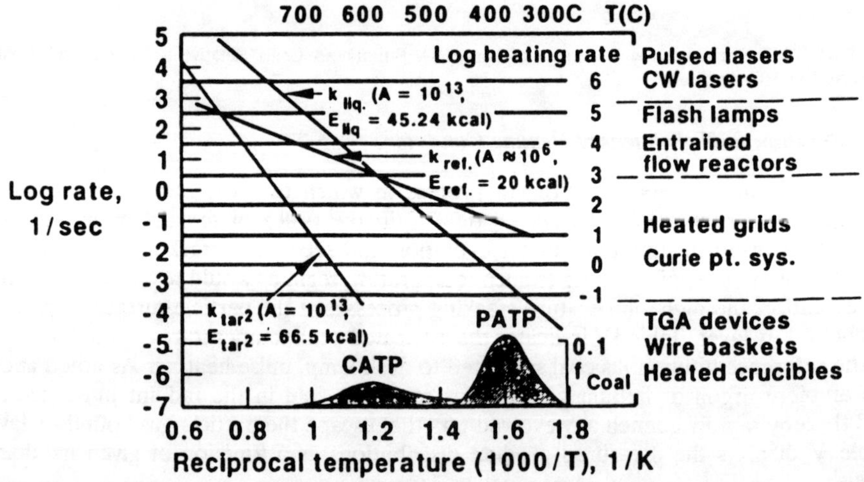

FIGURE 16. Rate Constant Plots of FBN Evolution as Primary Tar and Homogeneous Secondary Reactions of Tars to Form HCN

Significant quantities of tars are observed at all low net flux levels, provided particle temperatures in excess of 500°C are obtained during the transient heating. Significant tars are evolved at all transient conditions, but in the presence of ambient pressure a large fraction of the tars that are formed and evolved in the low pressure conditions are not evolved or extractable with room temperature THF. The results indicate that

TABLE III
THERMOPHYSICAL CHARACTERISTICS OF ENTRAINED FLOW REACTOR TARS

EFR Run Conditions: 825°C Wall Temperature
660°C Peak Gas Temperature
0.515 Sec Average Particle Residence Time

A.) Tar Characteristics

I.D.	ASTM Rank	Yield (Wt %)	% C	Elemental Composition % H	%N	% (S+O)
PSOC 1443D (Texas)	Lig	0.24	80.27	8.88	0.76	10.09
PSOC 1451D (PA)	HVA Bit	0.24	84.46	5.93	1.76	7.83
PSOC 1508D (WV)	MVB	0.14	89.18	5.18	0.96	4.65

B.) Tar Melting Point Determinations

	Sublimation (°C)	Initial Softening (°C)	Completed Melting (°C)
PSOC 1443D (Texas)	NA*	68-70	76-78
PSOC 1451D (PA)	91-95	110-115	140-150
PSOC 1508D (WV)	85-90	90-95	135-140

* NA = Not Applicable

C.) SEC MOMENTS

	M_n	M_w
PSOC 1443D	212	414
PSOC 1451D	185	259
PSOC 1508D	168	207

the ambient pressure, coupled to the rapid rise in particle temperature, results in high temperature cracking of the less volatile tar fraction before it can escape to the quenching ambient gas.

SUMMARY OF TAR EVOLUTION PHENOMENOLOGY

The following invariant phenomenology is observed irrespective of changes in parent coal type.

— the tars evolve in a "distillation-like" fashion, i.e., relatively aliphatic-rich, lower molecular weight species are released early in the tar evolution process with more aromatic, heavy molecular weight species evolved in the latter stages of tar evolution;
— due to systematic changes in the chemical structure of primary tars with coal rank characteristics, the mass normalized heating values of primary tars are relatively constant;
— the mass fraction of parent coal nitrogen evolved during primary devolatilization is entirely contained in the heavy hydrocarbon fraction of the volatiles;
— for HV bituminous coals and higher ranks, the mass fraction of parent coal nitrogen evolved in the tar species is in parity with the mass fraction of parent coal mass evolved as tar on a daf basis;
— the gas phase, secondary reactions of primary tars which produce HCN initiate at gas temperatures of approximately 700°C;

TABLE IV
CHARACTERISTICS OF DELIVERED FLASH LAMP PULSES

Ambient Conditions	Capacitor Bank Voltage (kV)	Delivered % NDF	Maximum* Average Irradiance (w/cm^2)	Time** Averaging Particle Temp. (°C)	Maximum Interval (msec)	Average*** Particle Temp. (°C)	Heating Rate (°C/sec)
Vacuum	1.5(0.5)‡	30	301	412	35.4	516	5.1×10^4
1 atm. Ar	1.5	30	301	399	19.5	471	4.7×10^4
1 atm. He	1.5	30	301	-	0.0	229	5.0×10^4
Vacuum	1.8(0.7)	30	478	516	54.1	740	7.3×10^4
1 atm. Ar	1.8	30	478	490	29.6	663	7.8×10^4
1 atm. He	1.8	30	478	314	3.6	328	6.3×10^4
Vacuum	1.5(1.0)	60	620	583	66.0	909	9.0×10^4
1 atm. Ar	1.5	60	620	550	35.0	808	9.1×10^4
1 atm. He	1.5	60	620	361	6.2	399	8.3×10^4
Vacuum	1.5(1.5)	90	909	710	81.5	1241	1.1×10^5
1 atm. Ar	1.5	90	909	660	41.8	1080	1.2×10^5
1 atm. He	1.5	90	909	445	8.7	532	1.0×10^5
Vacuum	2.2(2.2)	60	1650	953	103.0	2056	1.9×10^5
1 atm. Ar	2.2	60	1650	872	51.4	1718	1.9×10^5
1 atm. He	2.2	60	1650	620	11.2	829	1.8×10^5

* See Figure 4a for time-resolved pulse shape.

** Average over time interval for which particle temperature exceeds 300°C. See next column for averaging interval in msec.

***Average heating rate calculated from ($T_{\text{particle maximum}}$ - 20°C/(t_{maximum} - 0 sec); that is, linear.

‡ Numbers in parentheses represent relative flux delivered.

Note: Particle temperature calculations are for an inert particle with diameter 50μm, absorptivity 0.7, and heat capacity given by the Merrick model. The vacuum condition calculations allowed for conductive losses through a gaseous tar cloud surrounding the particle.

— in high temperature secondary reaction regimes, HCN is the principal, nitrogen-containing light gas evolved.

The following variant phenomenology is observed with respect to changes in coal rank characteristics:

— the yields of primary tars;

— the temperature-sensitivity of the evolution of primary tars;

— the chemical nature of the primary tars varies systematically for high vitrinite coals from primarily polymethylene for low rank coals to primarily condensed aromatics for high rank coals;

— the fuel nitrogen concentration of low rank primary coal tars is significantly lower than that observed for the parent coals on a daf basis;

— as a result of variance in fuel bound nitrogen evolution with primary tars the global kinetics of nitrogen evolution varies with coal type;

— the thermophysical (softening temperatures) and volatility characteristics (transient vaporization temperatures and extent of revaporization) vary systematically with rank of the parent coal, with lower rank coals producing primary tars having lower softening temperatures, lower vaporization temperatures and greater mass fraction re-vaporization.

For heating rates in excess of 10^4C/sec to particle temperatures in excess of 900°C, high temperature secondary cracking reactions of tars are kinetically competitive with tar evolution processes in determining observable product distributions.

TABLE V
FLASH LAMP DEVOLATILIZATION PRODUCT DISTRIBUTIONS:
PSOC 1451 D AT VARYING INCIDENT FLUX AND AMBIENT ATMOSPHERES

VOLATILES RELEASE
PSOC 1451 (HVA BIT) Flash Lamp Apparatus
270/325 Mesh Particles

Relative Flux Level: 0.7 (1.8 kV, 30% NDF)

Product	Vacuum	1 atm. Ar	1 atm. He
Char	59.0	70.0	90
Tar	31.0	19.0	6
CH_4	0.13	0.16	–
C_2H_4	–	–	–
C_2H_2	–	–	–
C_6H_6	–	–	–
HCN	–	–	–
CO	0.12	0.15	–
CO_2	0.63	0.12	–

Relative Flux Level: 1.0 (1.5 kV, 30% NDF)

Product	Vacuum	1 atm. Ar	1 atm. He
Char	57.0	59.0	78
Tar	37.0	24.0	12
CH_4	0.25	0.29	–
C_2H_4	0.09	0.12	–
C_2H_2	–	0.36	–
C_6H_6	–	–	–
HCN	–	0.47	–
CO	0.23	0.69	–
CO_2	0.28	0.12	–

Relative Flux Level: 1.5 (1.5 kV, 90% NDF)

Product	Vacuum	1 atm. Ar	1 atm. He
Char	59.0	63.0	76
Tar	27.0	18.0	19.0
CH_4	0.82	0.34	0.25
C_2H_4	0.51	0.09	0.06
C_2H_2	0.79	1.16	1.01
C_6H_6	0.09	–	–
HCN	0.32	0.65	0.41
CO	1.43	3.01	1.77
CO_2	0.34	–	0.13

TABLE V (continued)
FLASH LAMP DEVOLATILIZATION PRODUCT DISTRIBUTIONS:
PSOC 1451 D AT VARYING INCIDENT FLUX AND AMBIENT ATMOSPHERES

VOLATILES RELEASE
PSOC 1451 (HVA BIT) Flash Lamp Apparatus
270/325 Mesh Particles

Relative Flux Level: 2.2 (2.2 kV, 60% NDF)

Product	Vacuum	1 atm. Ar	1 atm. He
Char	53.0	67.0	70.0
Tar	25.0	22.0	17.0
CH_4	0.66	0.38	0.75
C_2H_4	0.01	0.14	0.23
C_2H_2	1.41	1.90	1.48
C_6H_6	–	–	0.17
HCN	0.56	1.39	1.11
CO	1.68	5.35	4.44
CO_2	–	0.25	0.14

REFERENCES

Blair, D. W., Wendt, J. D. L., and Bartok, W. (1976). Evolution of Nitrogen and Other Species During Controlled Pyrolysis of Coal. *Sixteenth Symposium (Intern.) on Combustion,* The Combustion Institute, Pittsburgh, 475.

Calkins, W. H. and Tyler, R. J. (1986). Coal Flash Pyrolysis. Polymethylene Compounds in Low Temperature Flash Pyrolysis Tars. *Fuel,* **63,** 1119.

Chen, S. L., Heap, M. P., Pershing, D. W. and Martin, S. B. (1982). Fate of Coal Nitrogen During Combustion. *Fuel,* **61,** 1218.

Chen, S. L., Heap, M. P., Pershing, D. W. and Martin, D. B. (1982). Influence of Coal Composition in the Fate of Volatile and Char Nitrogen During Combustion. *Nineteenth Symposium (Intern.) on Combustion,* The Combustion Institute, Pittsburgh, 1271.

Freihaut, J. D., Proscia, W. M. and Seery, D. J. (1989). Chemical Characteristics of Tars Produced in a Novel Low Severity, Entrained-Flow Reactor. *Energy and Fuels,* **3,** 692.

Freihaut, J. D., Zabielski, M. F., and Seery, D. J. (1982). A Parametric Investigation of Tar Release in Coal Devolatilization. *Nineteenth Symposium (Intern.) on Combustion,* The Combustion Institute, Pittsburgh, 1159.

Freihaut, J. D. and Proscia, W. M. (1989). Tar Evolution in Heated-Grid Apparatus. *Energy and Fuels,* **3,** 625.

McLean, W. J., Hardesty, D. R. and Pohl, J. H. (1981). Direct Observations of Devolatilizing Pulverized Coal Particles in a Combustion Environment. *Eighteenth Symposium (Intern.) on Combustion,* The Combustion Institute, Pittsburgh, 1239.

Nelson, P. F. and Tyler, R. J. (1986). Formation of Light Gases and Aromatic Species During the Rapid Pyrolysis of Coal. *Twenty-First Symposium (Intern.) on Combustion,* The Combustion Institute, Pittsburgh, 427.

Peters, W. and Bertling, H. (1965). Kinetics of the Rapid Degasification of Coals. *Fuel,* **44,** 317.

Pohl, W. and Sarofim, A. (1976). Devolatilization and Oxidation of Coal Nitrogen. *Sixteenth Symposium (Intern.) on Combustion,* The Combustion Institute, Pittsburgh, 491.

Seeker, W. R., Samulson, G. S., Heap, M. P. and Trollinger, J. D. (1981). The Thermal Decomposition of Pulverized Coal Particles. *Eighteenth Symposium (Intern.) on Combustion,* The Combustion Institute, Pittsburgh, 1213.

Smith, I. W. (1982). The Combustion Rates of Coal Chars: A Review. *Nineteenth Symposium (Intern.) on Combustion,* The Combustion Institute, Pittsburgh, 1045.

Solomon, P. R. and Colket, M. B. (1978). Evolution of Fuel Nitrogen in Coal Devolatilization. *Fuel,* **57,** 749.

Song, Y. H., Pohl, J., Beer, J. and Sarofim, A. (1982). Nitric Oxide Formation During Pulverized Coal Combustion. *Combustion Science and Technology,* **28,** 31.

Suuberg, E. M. and Scelza, S. T. (1982). Gas Evolution During Rapid, Low-Temperature Pyrolysis of a North Dakota Lignite. *Fuel,* **61,** 198.

Tyler, R. J. (1980). Flash Pyrolysis of Coals. Devolatilization of Bituminous Coals in a Small Fluidized-Bed Reactor. *Fuel,* **59,** 218.

Wei-Chun Xu and Akira, Tomita (1987). Effect of Coal Type on the Flash Pyrolysis of Various Coals. *Fuel,* **66,** 626.

Wendt, J. D. L., and Pershing, D. W. (1977). Physical Mechanisms Governing the Oxidation of Volatile Fuel Nitrogen in Pulverized Coal Flames. *Combustion Science and Technology,* **16,** 111.

Wei, C. (1979). [illegible] ... Dewetting of Bitumen ... [illegible]
[illegible]
Van Olphen, [illegible] ... Various Gases and ... [illegible]
Wendt, [illegible] and [illegible] (1979). Physical Mechanisms Governing the Formation of [illegible] ... Colloid and Interface Science ... 18 [illegible]

Determination of Rapid Heating Volatile Matter Contents as a Routine Test

J. R. GIBBINS, C. K. MAN, K. J. PENDLEBURY *Mechanical Engineering Department, Imperial College, London*

Abstract—Better characterisation of coal volatile yields under rapid heating conditions could improve fuel matching for existing plant, leading to improved overall performance, and also assist in the development of advanced clean combustion technologies. The relative merits of entrained flow and captive sample techniques for routine coal characterisation under rapid heating are discussed. Although the former can apply conditions more representative of PF combustion, volatile yield measurements are likely to be easier and more accurate in the latter. A version of the well known wire-mesh captive sample reactor is described which dispenses with fine thermocouples for temperature measurement and is therefore easier to use on a routine basis. Results obtained in this apparatus for a variety of coals are presented. These show that the R factor, the ratio between yields under rapid heating conditions and the proximate VM content, can vary significantly between coals. A brief comparison with results from drop tube reactors indicates approximately similar trends.

Key Words: coal, devolatilisation, pyrolysis, rapid

INTRODUCTION

Background

It is well known that the amount of volatiles liberated from a coal under rapid heating conditions can exceed the volatile matter measured in the standard proximate analyses by 50% or more. There are several reasons why this should be so: the peak temperature in the devolatilisation process will often be different, and previous work at Imperial College has also shown that, even for the same peak temperature, yields differ significantly, with roughly equal effects for differences in heating rate and particle dispersion respectively (Gibbins, 1990). While the standard proximate analyses do give reasonable supporting data for coke production and combustion of lump coal in fixed beds (the main coal-consuming processes when these tests were introduced), both the current main consumer for coal, PF combustion, and proposed advanced technologies (gasification, MHD) would ideally use a more appropriate test to characterise the volatile-forming potential of a coal. More accurate rapid heating volatile release data could thus improve fuel matching for existing plant, leading to improved overall performance, and also assist in the development of advanced clean combustion technologies. In addition to overall volatile release, the partition of fuel nitrogen between volatiles and char under realistic volatile release conditions would also be of relevance for low-NO_x burner performance. Currently a variety of test methods are used in research laboratories to measure volatile release under rapid heating conditions. These can be categorised as entrained sample (e.g. drop tube, free jet in pre-heated gas stream) or captive sample (e.g. Curie-point analyzer, wire-mesh). In this paper the relative merits of these two types of apparatus will be discussed, and a development of the wire-mesh reactor described which appears to be close to technical feasibility for use as a routine test.

Entrained flow tests

While tests in this category more closely match the conditions encountered in real combustion processes, since the individual particles are dispersed in a gas stream during heating, they also suffer from the problem of having to recapture the particles for analysis.

In general, complete sample recovery is not possible and hence the measurement of volatile release in entrained flow tests usually relies on the use of ash or mineral element tracer techniques. Complete sample capture and hence direct weight loss measurements were reported by Kobayashi (1976, 1977) at MIT, who used a drop tube reactor with a water-quenched sampling probe, and Kimber and Gray (1967) at BCURA, but most workers do not appear to have found this approach feasible. Estimates of devolatilisation using the ash tracer technique rely on proximate analyses being undertaken of the coal feed and the product char, with the volatiles yield (% dry coal) being given by:

$$V = 100x(1 - \frac{\text{ash content of coal(\% dry coal))}}{\text{(ash content of char (\% dry char))}} \qquad (1)$$

Rather than collect the large amounts of sample (1 gram or more) required for the standard crucible-based proximate ash content tests, many workers find it convenient to use a thermobalance. This can use as little as 10 mg of sample per determination, and gives results which are acceptably close to the standard tests in most cases. The validity of the ash tracer technique relies on the ash in the coal being conserved. Of course, if ash analyses are not undertaken on the actual sample being devolatilised, considerable errors can result owing to differential ash enrichment of the various size fractions during grinding (Littlejohn, 1967). Even if all possible precautions are taken, volatilisation of the mineral matter is likely at elevated temperatures (>1500 deg.C) (Kobayashi, 1977). At lower temperatures systematic losses of ash in handling have also been reported (Badzioch, 1967). The inevitable tendency for the denser mineral particles to separate, even when using a narrow particle size range, may have been responsible. The normal experimental uncertainty in determining the ash content of the coal and char samples can also lead to large errors if the ash content of the coal is low (say < 7% dry coal), particularly at low extents of devolatilisation. In an attempt to overcome the uncertainties involved in ash tracer techniques, Fletcher (1989) also undertook elemental analysis of the ash and used titanium, silicon and aluminium as additional tracers for devolatilisation experiments in a laminar jet reactor. The simultaneous determination of four (generally different) extents of weight loss allowed statistical analysis of the results to determine an average yield, but 90% confidence limits for the scatter still ranged from +/− 2 % to +/− 13%. Since the element tracer technique would still be susceptible to ash volatilisation (the maximum temperature reported by Fletcher in this study was only 980 deg.C) and (though perhaps to a lesser extent) to ash losses in handling and errors in analysis, the extra complication would not seem to be warranted in most cases.

Given the uncertainty associated with indirect weight loss determinations using tracer techniques, it is fortunate that an alternative exists to estimate the relative extents of weight loss during devolatilisation in entrained flow systems, allowing studies of the devolatilisation kinetics to be conducted with a reasonable degree of accuracy. In early experiments using a drop tube reactor (Badzioch, 1967; Badzioch and Hawksley, 1970) it was found that, although the tracer-derived weight losses for different extents of devolatilisation showed considerable scatter, there was a linear correlation between the underlying tracer weight loss trend and the measured loss in volatile matter of the coal—the well-known Q factor defined as:

$$Q = \frac{\text{weight loss of sample (\% daf coal)}}{\text{prox. VM of coal} - \text{prox. VM of char}} \qquad (2)$$

where both the VM content of the coal and the char are expressed as a percentage of the original daf coal. The Q factor is thus a multiplier representing the greater efficiency of volatile production/release compared to the standard proximate analysis.

The principle was extended by Essenhigh (1981) and Gray (1974), who restated the Q factor as its reciprocal, the 'capture factor' for volatiles, although the Q factor has remained the more commonly-used parameter. It should be noted that the Q factor is not an intrinsic property of a coal, but depends on the devolatilisation conditions used. Thus, for example, Kimber and Gray (1967) devolatilising a bituminous coal in a drop tube reactor found a variation in Q factor of 1.3 to 1.95 as the peak temperature, and hence also the heating rate, was raised (from 1050 K to 2170 K).

Given the complicated nature of pyrolysis reactions and volatile transport, it is perhaps surprising that, for a given set of devolatilisation conditions, there is such a good linear relationship between the weight loss and the change in the residual volatile content of the char. Because the relatively imprecise ash tracer technique was used to get weight loss measurements for comparison in the original BCURA study, small variations from linearity might have been concealed, and there appears to have been no subsequent systematic examination presented in the literature which would allow the linearity hypothesis to be tested further. It was therefore decided to make a preliminary test of the linearity in a series of trials using the wire-mesh apparatus, which allows direct weighing to establish volatile yields with a reproducibility of about +/− 1%. Samples of Gedling (UK bituminous) coal were first heated at about 4000 K/s to temperatures between 300 deg.C and 1100 deg.C and then rapidly cooled (500-1000 K/s), giving volatile yields between 0 and about 50%. After weighing to establish the initial volatile yield, the same wire-mesh sample holder was then replaced in the apparatus, heated to 950 deg.C and held there for 10 s to give a second weight loss, analogous to the proximate analysis of char captured in an entrained flow reactor. Unfortunately a fairly high heating rate (1000 K/s) had to be used for the second heating stage, in contrast to 10-20 K/s in the proximate test; as will be discussed below this was limited by the use of a non-contact IR pyrometer for temperature control. Previous work (Gibbins, 1988) has shown, however, that in the wire-mesh reactor there is generally very close linear correspondence between yields at 1000 K/s and lower heating rates, down to 1 K/s. The wire-mesh apparatus also fails to reproduce the secondary cracking reactions in the bed of coal in the proximate VM test crucible, but it might be expected that for a partially-pyrolysed char sample these will be low, since melting and fusion of the bed is unlikely and the less-thermally-stable volatile products will already have been liberated. As shown in Fig. 1, a close linear correlation was in fact observed between initial weight loss and the change in char volatile content, as measured by the second heating step, confirming the validity of the Q factor analysis. In the light of this encouraging preliminary result, it is planned to extend the work at a later stage to include higher heating rates and temperatures in the first step and the use of a TGA for devolatilisation of the char, in line with common practice.

It is convenient at this stage to discuss the use made of the Q factor in reporting the results of devolatilisation experiments. In the experience of the authors, it appears that the Q factor sometimes is mistakenly employed in practice as the ratio between the volatile yield that will actually be obtained under rapid heating conditions and the proximate VM content. However this ratio, which is generally different, was also defined by Kimber and Gray (1967), at the same time as they introduced the Q factor, as the R factor:

$$R = \frac{\text{weight loss of coal (\% daf coal)}}{\text{prox. VM content of coal (\% daf coal)}} \tag{3}$$

The R factor will clearly vary with the extent of devolatilisation, unlike the Q factor, and will generally be lower. For practical purposes, one would like to know the R factor under PF combustion conditions to allow the 'true' volatile-producing potential of different coals to be known. It might be thought that with the fairly intense devolatilisation conditions

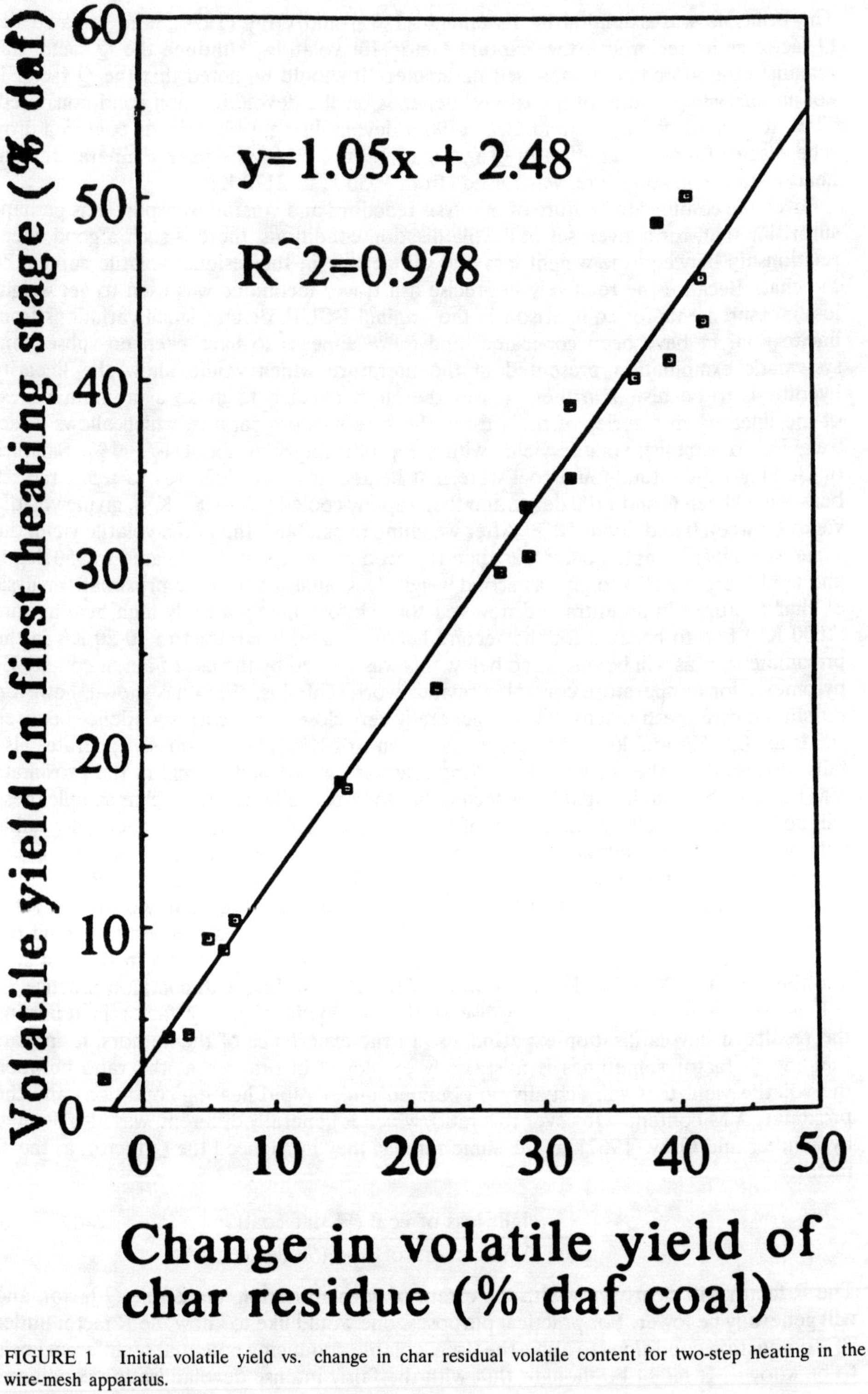

FIGURE 1 Initial volatile yield vs. change in char residual volatile content for two-step heating in the wire-mesh apparatus.

(> 10,000 K/s heating to approx. 1500 deg.C) in PF combustion the residual proximate VM content of the chars would be zero, making the R factor under these conditions equal to the Q factor. However, even though the heating rates and final temperatures used in proximate VM tests are lower, the soak time at peak temperature is much longer (minutes instead of milliseconds) and while it appears that the relatively rapid processes involved in primary volatile (mainly tar) release can be enhanced by intense heating, the much slower processes involved in the removal of heteroatoms from the solidified char matrix still need a significant length of time (or extremely high temperatures) to approach completion. Thus Kimber and Gray (1967), using a drop tube, still found a residual char proximate VM content corresponding to 3.8% of the daf coal (out of 38.4%), after pyrolysis at a peak temperature of 1720 K for 55 ms had produced 60% weight loss. The corresponding Q factor is 1.75 but the R factor only 1.57. Even for peak temperatures of 2170 K the char still contained between about 1% and 2% VM (% daf coal) for residence times of 45 ms. Given the likely slow rate of release of the remaining volatiles from the char, it can therefore be assumed that heterogeneous char combustion will begin before this residual material can be released, and the majority of it will be burnt in situ.

Captive sample techniques

Captive sample techniques, in contrast to entrained flow techniques, allow relatively accurate measurement of volatile release, although special techniques are required to obtain the necessary high heating rates and sample dispersion. High heating rates (of the order of 1000 K/s) can be obtained in commercially available pyroprobe units attached to gas chromatographs or mass spectrometers. These use Curie-point heating of fine wires or direct electrical resistance heating of platinum filaments, but although this technique enables very detailed analysis of the more-volatile products, the size of coal sample used (of the order of μg) is usually too small to allow the weight loss to be determined. The particles also often have to be 'stuck' onto the heating element by deposition from a wet suspension (e.g. in acetone), which makes the retention of all of the sample unlikely. The problems of sample retention can be overcome, for Curie-point heating, by placing the sample inside a small tube (e.g. Xu, 1987). In this case, however, the likelihood that the sample distribution will result in at least several overlying layers of particles, and the relatively tortuous path for volatile release out of the tube, makes secondary char deposition from cracking reactions a serious possibility, especially at elevated temperatures. The latitude for varying the peak temperature in Curie-point heating is also limited by the materials available, while only a narrow range of heating rates can be applied unless the heating system is extended to such a point that the fundamental advantage of Curie-point heating, the self-limiting peak temperature at the Curie-point of the metal used for the sample holder, is essentially negated.

To date, however, the most widely used captive sample research apparatus for rapid heating devolatilisation studies has been the wire-mesh reactor, in which a mono-layer of coal particles is held between two layers of electrically-heated wire-mesh. Originally invented by Loisin and Chauvin (1964) at CERCHAR in France, the apparatus has subsequently been used at a number of research institutions in Europe (e.g. Arendt, 1981; Desypris, 1982; Gibbins, 1988, 1990; Johnson, 1988; Juntgen, 1968; Lowenthal, 1986), N. America (e.g. Anthony, 1974, 1976; Bautista, 1986; Fong, 1986; Franklin, 1981; Freihaut, 1982, 1983; Gavalas, 1980; Glass, 1988; Jain, 1980; Niksa, 1982, 1984; Solomon, 1978; Stangeby, 1981; Suuberg, 1985; Unger, 1984) and Australia (Hamilton, 1979, 1981).

Although the basic principle of the wire-mesh apparatus has not changed, a range of different approaches have been adopted to the related problems of controlling the heating rate of the sample holder and monitoring its temperature. To date, temperature measurements during heating have always been undertaken with one or more thermocouples. A common problem with the use of thermocouples is the conflicting requirements to place the thermocouple junction as close to the mesh as possible, to get more accurate temperature measurements, and to avoid placing the thermocouple leads in contact with the mesh in such a way that they act as voltage probes and pick up interference (easily very large compared to the thermal emf) from the heating current. The most satisfactory approach to the problem seems to be the use of alternating current (AC) for heating, which allows the interference to be averaged out, (e.g. Loisin, 1964; Arendt, 1981; Gibbins, 1988; Glass, 1988). If direct current (DC) is used, the heating current can be switched off briefly while reading the thermocouples (Hamilton, 1979) or the two thermocouple wires can be welded to a single strand of the mesh, running perpendicular to the voltage gradient across the sample holder and hence at the same potential (Freihaut, 1982, 1983). Other methods, in which the thermocouple is insulated from the mesh by the coal sample itself, or the actual junction offset from the surface of the mesh, cannot be regarded as reliable. In these cases it is often impossible to detect the interference, however, since the deviation occurs when the current is first applied, when the output of the thermocouple would be expected to change anyway. Even when the thermocouple can be positioned appropriately, the fine size of the wires used (typically 0.025 or 0.05 mm), make attachment difficult. Usually spot welding is involved; the aid of a microscope may be necessary. Perhaps the simplest form of thermocouple attachment method is that developed at Imperial College (Gibbins, 1988, 1989), in which the thermocouple wires are simply threaded through the mesh about 1 mm apart and on approximately the same equipotential line with respect to the heating current (within about 1 mm, residual interference is suppressed by the AC-based temperature control/logging system). This involves no special equipment, is very quick to attach and remove for weighing, and has the advantage that the thermocouple junction is unambiguously located on the sample holder itself. Even then, however, a reasonable level of manual dexterity and good close-up vision are required and several potential operators, principally older staff, have experienced problems, particularly when using the finer thermocouple wire. Irrespective of the other merits or demerits of the wire-mesh apparatus it was therefore felt that, unless a easier method of temperature measurements could be devised, while it might be valuable as a research tool the wire-mesh apparatus could not be considered for use as a routine test.

EXPERIMENTAL

Apparatus

The particular problem of temperature measurement has been addressed as part of an interdepartmental project at Imperial College, funded by the Royal Society and the UK SERC, to extend the operating range of the wire-mesh apparatus. The key to a more acceptable test has proved to be the use of non-contact infra-red pyrometry for temperature measurement, together with advances in computer-controlled power regulation. This has both avoided the operational difficulties previously encountered due to the use of fine thermocouples for temperature measurement and enabled the temperature range to be extended without the use of exotic-metal thermocouples and with a reduced measurement lag at very high heating rates. The containment vessel for the new apparatus is shown in Fig. 2, and a diagram of the heating circuit is given in Fig. 3. Further development work on the mechanical design, facilitated by the possibility of non-contact temperature measurement, has also led to a more robust system package

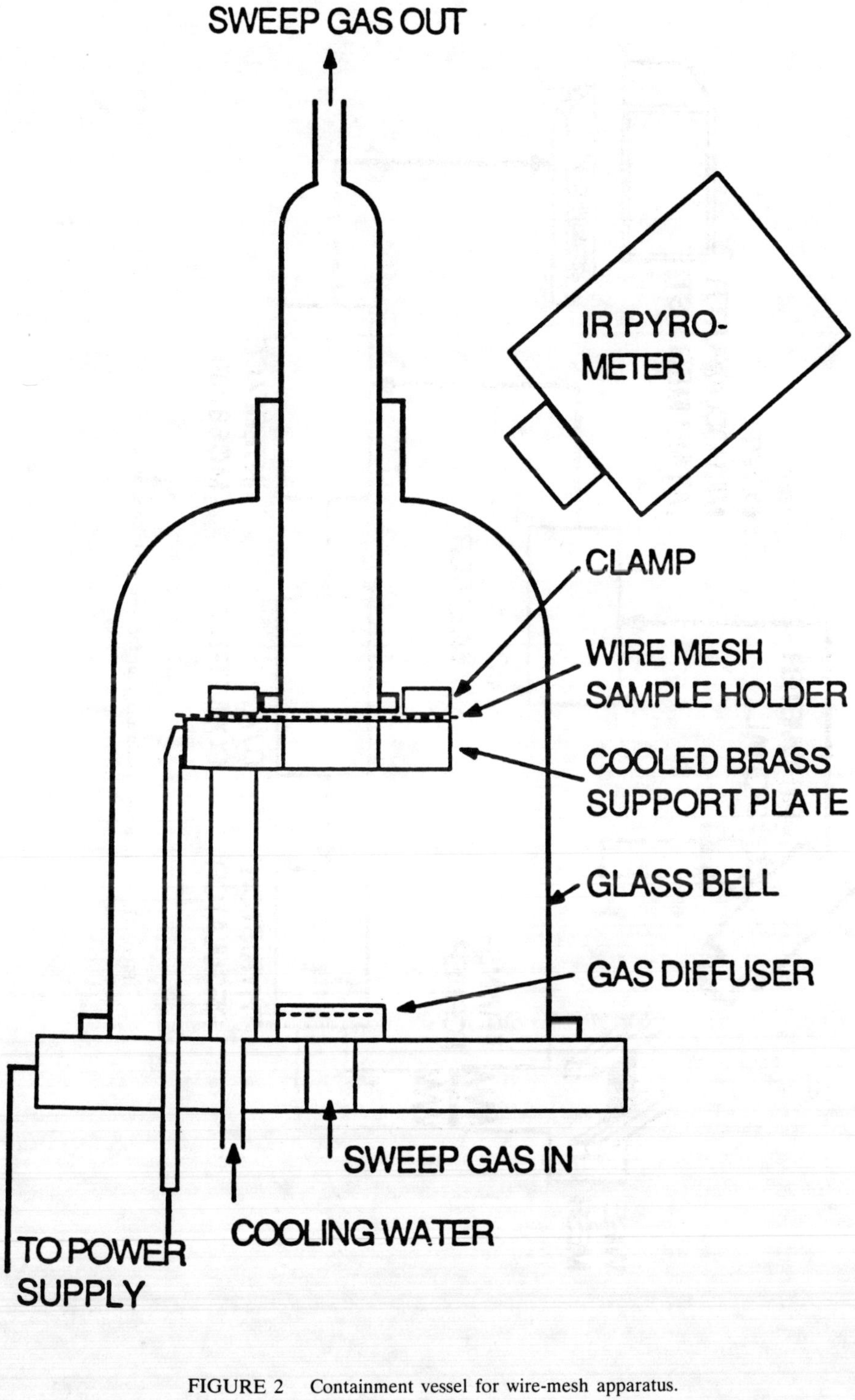

FIGURE 2 Containment vessel for wire-mesh apparatus.

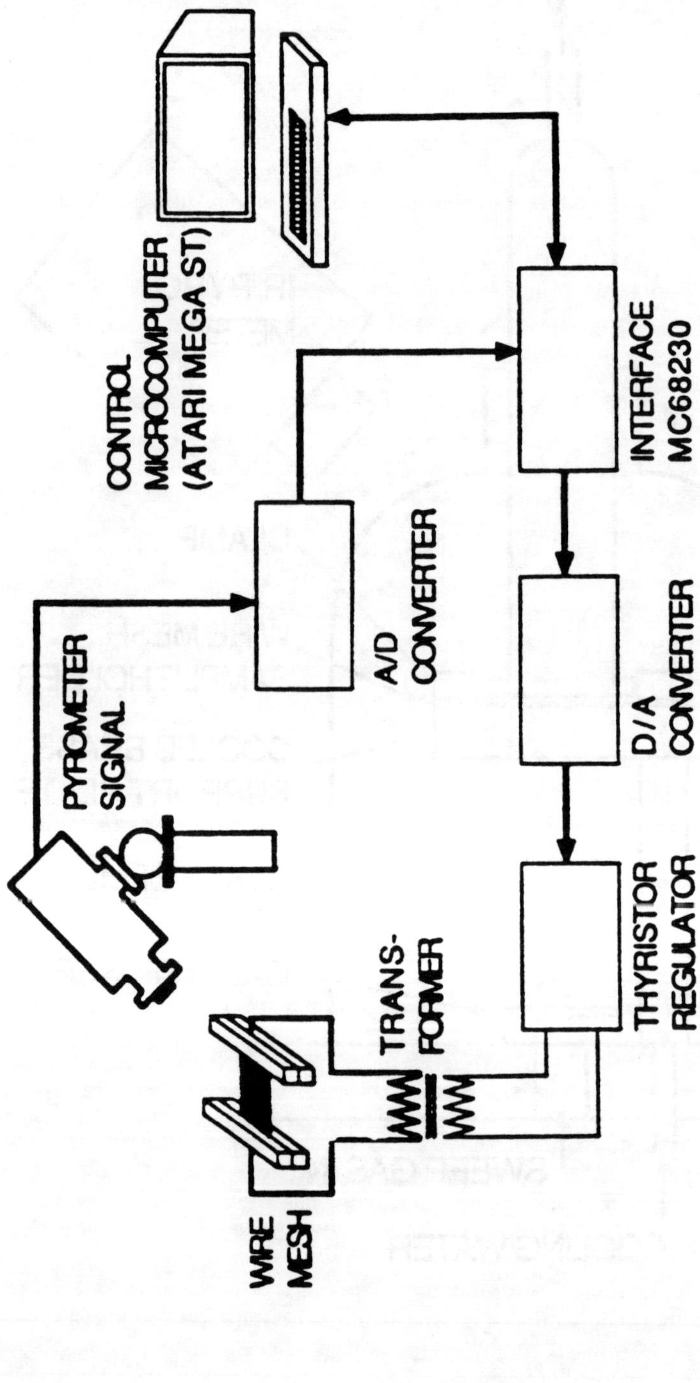

FIGURE 3 Heating system diagram for wire-mesh apparatus.

and made it possible to replace many special-purpose components with off-the-shelf equipment. The two-colour pyrometer used (Ircon type SR) has a measuring spot area of about 4 mm diameter, and is trained on the centre of the area of mesh containing the coal sample. The instrument uses a thin-film detector working at wavelengths of 0.7-1.08 μm and 1.08 μm. Linearisation of the two signals can be accomplished within the instrument, including compensation for the non-ideal ratio between intensities at the two wavelengths due to the non-grey behaviour of the mesh material and selective absorption in the silica quartz/pyrex glassware. The compensation is calibrated at steady state using thermocouples; typically less than 20 K difference is observed between the thermocouple readings and the pyrometer readings under subsequent transient conditions. The minimum time constant for the standard linearisation circuitry is 1 ms, which is adequate for heating rates up to about 10,000 K/s. This is also the maximum heating rate that can reasonably be applied with alternating current at 50 Hz, since power adjustments can be made only at half-wave intervals (i.e. at 10 ms time intervals or every 100 K at 10,000 K/s). For higher heating rates, for which DC heating is planned, modifications to allow the sensor outputs to be recorded individually will be required. It is envisaged, however, that AC heating at normal mains frequency would be used in a wire-mesh apparatus designed as a routine test instrument since this greatly simplifies (and reduces the cost of) the heating system. A similar techno-economic limitation on the peak temperature used in a standard instrument is also likely to apply. The peak temperature is limited mainly by the high-temperature strength of the mesh material. This gives a maximum operating temperature of about 1100 deg.C for stainless steel. Higher temperatures (1500 deg.C and above) can be obtained by the use of more exotic mesh materials (e.g. tungsten, platinum) but these are expensive and/or difficult to obtain in a wide range of mesh sizes. Particularly since it is most convenient to use a fresh sample holder for each experiment, the use of stainless steel mesh, which is cheap and readily obtainable, is likely to be the best option. Fortunately, as will be discussed below, while these limitations mean that the absolute volatile yields will be slightly lower than at more typical PF combustion conditions (e.g. peak temperatures around 1500 deg.C and heating rates > 10,000 K/s) it appears likely that the measured yields will still be proportional to those under more intense conditions. This is probably adequate for a routine characterisation test, since in practice a wide range of conditions will apply with some (e.g. entrained flow gasification or MHD plants) being well in excess of the norm. A complete description of an earlier version of this apparatus is available (Gibbins, 1989), and a more detailed account of the new apparatus will be given elsewhere.

Experimental method

The coal sample (typically 7 mg) to be tested is first loaded between the layers of a pre-weighed wire-mesh sample holder, and then a suction line is applied to draw the coal into place and remove any remaining fines. Routinely a 32 μm aperture stainless steel mesh is used, allowing a minimum particle size of about 38 μm. The coal sample may be either a narrower size cut or, if the greatest possible yield is required on grinding, a sieved 38-150 μm sample. After the sample holder has been reweighed to establish the sample weight, it is stretched between the electrodes in the apparatus and clamped in place. The suction line is reapplied to obtain an even distribution of the coal within a 15 mm circle at the centre of the sample holder. This is particularly important in order to ensure even thermal loading of the mesh and hence uniform heating of the coal and accurate temperature measurements. The apparatus is sealed and evacuated and refilled with high-purity helium three times to remove all traces of air. A constant flow of helium (at a velocity of approx. 0.1 m/s) is then set to sweep upwards through the

TABLE I.
Wire-mesh volatile yields and R factors for coal sample set A.
Heating rate: 4000 K/s approx. Peak temperature: 1100 deg.C
Isothermal hold time: 10 s Sweep gas: helium

Coal	Proximate VM (%daf)	Wire-mesh vol's (%daf)	R factor
Daw Mill	42.7	54.9	1.29
El Cerrejon	40.4	48.6	1.20
Hem Heath	40.9	49.9	1.22
Kellingley	39.0	47.9	1.23
Lounge	44.8	55.0	1.23
Markham Stavely	39.4	49.9	1.27
Maryport	36.0	47.7	1.33
Rietspruit	31.8	41.5	1.30

sample holder to remove the liberated volatiles. The desired heating sequence, which can be specified within very wide limits using the proprietary control software, is then initiated. When using only the IR pyrometer, feedback control cannot be used below about 700 deg.C, since the intensity of the radiation emitted is too low for temperature measurements. In this case, a constant power, which can be set to give a heating rate of anywhere between 500 K/s and 5000 K/s, is used to give pure feed-forward control up to 700 deg.C, where feedback control takes over. The heating rate for feed-forward control has been found, by test measurements using thermocouples, to be reproducible to within better than a factor of 2. Previous work (Gibbins, 1988) has shown that at these low temperatures the effect of such heating-rate variations will be negligible, and any larger variations that might occur in practice are readily identifiable from the time that elapses before the pyrometer signal is registered. Typically heating rates of 1000 or 5000 K/s to peak temperatures of 950 deg.C (in line with standard proximate VM tests) or 1100 deg.C (the maximum temperature with stainless steel sample holders) have been used in most runs to date, with a hold time of 10 seconds at peak temperature. This has been found to be long enough for the bulk of the rapid volatile-forming processes to run to completion.

RESULTS

Results for pyrolysis of two sets of coal samples in the wire-mesh apparatus are shown in Tables I and II (note that coal samples with the same name in sets A and B are not identical). As can be seen, *R* factors well in excess of unity were obtained in all cases, even for Table II where a peak temperature only slightly in excess of the temperature for the proximate analyses was used. There appears to be no obvious explanation for the variation in *R* factor between the different coals. Since at least some of the variation will be due to the differing behaviour of the coal in the proximate test (e.g. extent of melting in the crucible) then it is likely that a more thorough investigation of the artifacts introduced in the proximate test would be required to resolve the issues involved. The variation in *R* factor serves to illustrate, however, the inadequacy of the proximate VM except as little more than a qualitative indicator of behaviour during rapid heating. Both sets of coal samples have also been devolatilised by others in drop tube reactors. The results will be discussed in detail elsewhere, but it was found that, although the peak temperatures used in the drop tube experiments were considerably higher than in the wire-mesh experiments (1450 deg.C vs. 1100 deg.C in Table I and 1350 deg.C vs. 950

TABLE II.
Wire-mesh volatile yields and R factors for coal sample set B.
Heating rate: 4000 K/s approx. Peak temperature: 950 deg.C
Isothermal hold time: 10 s Sweep gas: helium

Coal	Proximate VM (%daf)	Wire-mesh vol's (%daf)	R factor
Bentinck	39.0	54.2	1.39
Bilsthorpe	40.0	51.1	1.28
Daw Mill	43.6	51.2	1.17
Gedling	41.3	47.9	1.16
Hem Heath	38.2	47.2	1.24
Kellingley	39.2	50.6	1.29
Markham Stavely	38.1	48.3	1.27
Riccall	38.9	44.2	1.14
Selby	38.0	48.6	1.28
Westside	36.8	48.0	1.31

deg.C in Table II), the R factors were only greater by averages of 10% (s.d. 7%) and 6.5% (s.d. 8%, i.e. some drop tube R factors were lower) of the wire-mesh R values respectively.

The small differences may in part be due to the greater time at peak temperature in the wire-mesh reactor, 10 s compared to a total residence time of 150 ms in both sets of drop tube experiments, although the latter was considered an adequate time at the temperatures used for the bulk of pyrolysis reactions to run to completion. Further work to explore the reasons for the different trends between the two types of reactor is currently in progress, but the similarity of the R factors obtained suggests that the wire-mesh reactor might be an acceptable alternative to the drop tube reactor for routine rapid heating characterisation of coals. Future work to measure the amount (and later the rates) of volatile production at elevated temperatures in the wire-mesh apparatus is also planned. Using tungsten mesh, peak temperatures in excess of 1500 deg.C are anticipated. Although the techniques involved are not likely to be applicable in a standard instrument, for reasons already discussed, comparison with results for more severe condition will help to determine whether or not the preferred limits of 10,000 K/s heating to 1000-1100 deg.C are adequate to obtain the correct characteristic trends for different coals. The possibility of extending the sample size will also be explored, to allow the residual nitrogen content of the char from a single run to be measured. To achieve the minimum char sample size of 15-20 mg required for micro elemental analysis, about twice the quantity of coal would have to be heated. While this is technically feasible using a larger sample holder, the currents involved will also rise. At present the apparatus can be powered from a single-phase 240V, 13 amp socket, and ways of improving the 'efficiency' of the system to retain this desirable feature will be investigated.

CONCLUSIONS

A wire-mesh apparatus with non-contact temperature measurement using a two-colour IR pyrometer has been developed. With this modification, the wire-mesh apparatus becomes a potential candidate for use as an instrument for the routine determination of the volatile content of coals under rapid heating conditions. As an alternative to entrained flow instruments, the wire-mesh apparatus has the advantage of giving easier and potentially more accurate volatile yield measurements, but for practical reasons the heating conditions used in a standard instrument are likely to be less intense.

Using the new wire-mesh reactor, supporting evidence for the invariance of Q factor with the extent of devolatilisation has been obtained. This helps to confirm this approach for improving the accuracy of kinetic measurements in entrained flow studies. Volatile yields and R factors (the ratio of volatiles obtained to proximate VM) for coal devolatilisation experiments in the new wire-mesh reactor are reported. The relatively wide range of R factors obtained confirms the inadequacy of the proximate VM yield for characterisation of a coal's rapid heating behaviour. Despite the less-intense heating regime used in the wire-mesh experiments, R factors obtained are quite close to those based on yields in drop tube reactors.

ACKNOWLEDGEMENTS

The authors would like to thank the Royal Society (Grant 10383) and the SERC (Grant GR/F 89817) for financial support for the work, and the members of the Steering Committee for the SERC project for their help and advice. Coal samples were kindly provided by British Coal, Coal Research Establishment and NEI International Combustion, Derby. The contributions of collaborative workers at Imperial College, R. Kandiyoti, F. C. Lockwood and J. Williamson, are also gratefully acknowledged.

REFERENCES

Anthony, D. B., Howard, J. B., Hottel, H. C. and Meissner, H. P. (1974). Apparatus for determining high pressure coal-hydrogen reaction kinetics under rapid heating conditions. *Rev. Sci. Instrum.* **45,** 992-995.

Anthony, D. B., Howard, J. B., Hottel, H. C. and Meissner, H. P. (1976). Rapid devolatilization and hydrogasification of bituminous coal. *Fuel* **55,** 121-128.

Arendt, P. and van Heek, K-H. (1981). Comparative investigations of coal pyrolysis under inert gas and hydrogen at low and high heating rates and pressures up to 10 MPa. *Fuel* **60,** 779-787.

Badzioch, S. (1967). Thermal decomposition. Ch. 4 in Combustion of Pulverised Coal, Field, M.A., Gill, D. W., Morgan, B. B. and Hawksley, P. G. W., BCURA, 1967, 155-174.

Badzioch, S. and Hawksley, P. G. W. (1970). Kinetics of thermal decomposition of pulverized coal particles. *Ind. Eng. Chem. PDD* **9(4),** 521-530.

Bautista, J. R., Russel, W. B. and Saville, D. A. (1986). Time-resolved pyrolysis product distributions of softening coals. *Ind. Eng. Chem. Fundam.* **25,** 536-544.

Desypris, J., Murdoch, P. and Williams, A. (1982). Investigation of the flash pyrolysis of some coals. *Fuel* **61,** 807-816.

Essenhigh, R. H. (1981). Fundamentals of coal combustion. Ch. 19 in Chemistry of Coal Utilization, 2nd Supp., Elliott, M. A. (Ed.), Wiley, NY, 1153-1312.

Fletcher, T. H. (1989). Time-resolved particle temperature and mass loss measurements of a bituminous coal during devolatilisation. *Combustion and Flame* **78,** 223-236.

Fong, W. S., Peters, W. A. and Howard, J. B. (1986). Kinetics of generation and destruction of pyridine extractables in a rapidly pyrolysing bituminous coal. *Fuel* **65,** 251-254.

Franklin, H. D., Peters, W. A., Coriello, F. and Howard, J. B. (1981). Effects of calcium minerals on the rapid pyrolysis of a bituminous coal. *Ind. Eng. Chem. PDD* **20,** 670-674.

Freihaut, J. D., Zabielski, M. F. and Seery, D. J. (1982). Reaction induced temperature deviations during coal devolatilisation in a heated grid. *ACS DFC Prepr.* **27(2),** 89-98.

Freihaut, J. D. and Seery, D. J. (1983). Effects of pyrolysis conditions on coal devolatilisation. *ACS DFC Prepr.* **28(4),** 265-277.

Gavalas, G. R. and Wilks, K. A. (1980). Intraparticle mass transfer in coal pyrolysis. *AIChE Jnl.* **26,** 201-212.

Gibbins, J. R. (1988). Investigation of primary coal pyrolysis processes using a variable heating rate wire-mesh apparatus. PhD thesis, Imperial College, London.

Gibbins, J. R., King, R. A. V., Wood, R. J. and Kandiyoti, R. (1989). A variable heating rate wire-mesh pyrolysis apparatus. *Rev. Sci. Instrum.* **60(6),** 1129-1139.

Gibbins, J. R., Khogali, K. and Kandiyoti, R. (1990). Relationship between results from conventional proximate analyses and pyrolysis yields under rapid-heating conditions. *Fuel Proc. Tech.* **24(1),** 3-8.

Glass, M. W., and Zygourakis, K. (1988). Computerized microreactor and video microscopy system for coal pyrolysis. *Rev. Sci. Instrum.* **59,** 580-587.

Gray, D., Cogoli, J. G. and Essenhigh, R. H. (1974). Problems in pulverized coal and char combustion. In Coal Gasification, *Am. Chem. Soc. Adv. Chem. Ser.* **131,** 72-91.

Hamilton, L. H., Ayling, A. B. and Shiboaka, M. (1979). A new experimental device for pyrolysing coal particles under controlled conditions over a wide range of heating rates. *Fuel* **58,** 873-876.

Hamilton, L. H. (1981). Char morphology and behaviour of Australian vitrinites of various ranks pyrolysed at various heating rates. *Fuel* **60,** 909-913.

Jain, R. (1980). Mathematical modeling and experimental studies of thermal reactions of coal. Ph.D Thesis, CIT, Pasadena, Calif.

Johnson, G. R., Murdoch, P. and Williams, A. (1988). A study of the mechanism of the rapid pyrolysis of single particles of coal. *Fuel* **67,** 834-842.

Juntgen, H. and van Heek, K. H. (1968). Gas release from coal as a function of the rate of heating. *Fuel* **47,** 103-117.

Kimber, G. M. and Gray, M. D. (1967). Rapid devolatilization of small coal particles. *Combust. Flame* **11,** 360-362.

Kobayashi, H. (1976). Devolatilization of pulverised coal at high temperatures. PhD thesis, MIT.

Kobayashi, H., Howard, J. B. and Sarofim, A. F. (1977). Coal devolatilization at high temperatures. *Proc. 16th Symp. (Int.) Combustion,* Combustion Inst., Pittsburgh, 411.

Littlejohn, R. F. (1967). Mineral matter and ash distribution in pulverised fuel. Appendix B in Combustion of Pulverised Coal, Field, M. A., Gill, D. W., Morgan, B. B. and Hawksley, P. G. W., BCURA, 1967, 250.

Loison, R. and Chauvin, R. (1964). Rapid pyrolysis of coal. *Chim. Ind.* **91(3),** 269-275.

Lowenthal, G., Wanzl, W. and van Heek, K. H. (1986). Kinetics of swelling and plasticity of coal during rapid pressurized pyrolysis and hydropyrolysis, *Fuel* **65,** 346-353.

Niksa, S. J., Russel, W. B. and Saville, D. A. (1982). Captive sample reactor for kinetic studies of coal pyrolysis and hydropyrolysis on short time scales, *Fuel* **61,** 1207-1212.

Niksa, S. J., Heyd, L. E., Russel, W. B. and Saville, D. A. (1984). On the role of heating rate in rapid coal devolatilization. *20th Symp. (Int.) on Combustion,* The Combustion Institute, 1445-1453.

Solomon, P. R. and Colket, M. B. (1978). Evolution of fuel nitrogen in coal devolatilization. *Fuel* **57,** 749-755.

Stangeby, P. C. and Sears, P. L. (1981). Rapid pyrolysis and hydropyrolysis of Canadian coals, *Fuel* **60,** 131-135.

Suuberg, E. M., Unger, P. E. and Lilly, W. D. (1985). Experimental study on mass transfer from pyrolysing coal particles, *Fuel* **64,** 956-962.

Unger, P. E. and Suuberg E. M. (1984). Molecular weight distributions of tars produced by flash pyrolysis of coals. *Fuel* **63,** 606-611.

Xu, W. C. and Tomita, A. (1987). Effect of coal type on the flash pyrolysis of various coals. Fuel 66, 627-631.

Kinetics and Mechanism of the HNCO de-NO_x Process at High Temperatures

C.H. WU* *Chemistry Division, Naval Research Laboratory, Washington, DC 20375-5000, USA*

YISHENG HE and M.C. LIN *Department of Chemistry, Emory University, Atlanta, GA 30322, USA*

Abstract — The reaction kinetics of the HNCO-RAPRENO$_x$ process have been studied at temperatures between 1700 and 2500 K in a shock tube equipped with two CW CO probing lasers. The rates of disappearance of NO and formation of CO and H_2O products were measured by laser resonance absorption. The reaction was found to be controlled primarily by the rate of the HNCO decomposition reaction which produces the reactive, reducing radical, NH. Addition of NO_2 to the system was found to enhance the rates of HNCO disappearance and CO and H_2O production. This effect could be attributed to the enhanced production of NH, OH and NCO by the O + HNCO reaction via different product channels.

Key Words: Kinetic Modeling, HNCO–RAPRENO$_x$, Shock Waves

INTRODUCTION

Isocyanic acid, HNCO, has been proposed as a potential agent for the rapid reduction of NO_x (RAPRENO$_x$) in exhaust gas streams (Perry and Siebers, 1986). Both HNCO and NO_x are also known to be important early products of nitramine propellant decomposition reactions (Bogg, 1981). A thorough understanding of the kinetics and mechanism of the reactions of HNCO with NO_x is therefore of vital importance to our ability in predicting as well as improving the performance of these practical systems.

In this series of studies to elucidate the high-temperature chemistry of HNCO reactions, we have employed a shock tube equipped with two frequency-stabilized CW CO lasers for simultaneous CO and NO or H_2O product formation measurements. In our recent study of the HNCO decomposition reaction in the temperature range of 2000–2600 K (Wu *et al.* 1990), we have shown that the values of the rate constant for HNCO $\rightarrow$ NH + CO evaluated from the production of CO agrees quantitatively with those reported by Hanson, Bowman and coworkers, who employed entirely different product diagnostic methods (Mertens *et al.* 1989).

In the present study, we have examined the effect of NO on the formation of CO and H_2O in the decomposition of HNCO above 1700 K. In order to understand the important effect of O atoms, we have also investigated the reactions with added NO_2. These new experimental results are reported herein.

EXPERIMENTAL

The shock tube-CO laser resonance absorption apparatus used in the present study has been described in detail previously (Hsu *et al.* 1982, 1984). The 6.35-cm I.D. stainless steel shock tube, which has a combined leak/degas rate of 10^{-4} torr/min under normal operation

---

*ONT/NRL postdoctoral associate (1988–89); present address: US Patent and Trade Mark Office, Department of Commerce, Washington, DC 20231.

conditions, has been interfaced with two frequency-stabilized CW CO lasers for simultaneous detection of CO and NO or H_2O using different rovibrational lines of the CO laser. The procedure for calibrations and concentration measurements of these three molecules has been reviewed previously by Hsu and Lin (1984). Two liquid N_2 cooled InSb detectors were used to detect the absorptions of CO and NO or H_2O. The absorption signals were transferred to an IBM PC for analysis and storage, after proper amplification and digitization. The computer also converted the absorption signals directly to concentration-time profiles using prior calibration data for all three species. The temperature and pressure of the shocked gas were calculated using the NASA/Lewis equilibrium program (Choudhury and Lin, 1990).

Four different mixtures of HNCO and NO, highly diluted with Ar, were employed in the present study. The compositions of three mixtures given by the HNCO:NO:Ar ratio are: (A) 0.0030:0.010:0.987, (B) 0.010:0.0050:0.985, and (C) 0.0050:0.0054:0.9896. In the fourth mixture, prepared separately with a small amount of NO_2 added to test the effect of O atoms on CO and H_2O production, the following composition, HNCO:NO:NO_2:Ar = 0.0050:0.0050:0.000517:0.9895, was used.

HNCO was prepared by heating solid cyanuric acid (Aldrich) at about 650 K; the gaseous product was dried with P_2O_5 before trap-to-trap distillation from 195 K (dry ice bath) to 150 K (n-propanol slush bath), with continuous pumping to remove CO_2 impurity. An FTIR analysis of the distillate collected at 150 K was found to be quite pure and free from CO_2 absorption peaks.

Both NO and NO_2 were obtained from Matheson Gas Products, Inc. Since commercial sources of NO invariably contain a trace amount of NO_2, which has an expected deleterious effect on the present kinetic study, it was carefully purified by using a clean silica gel trap maintained at 195 K to remove the NO_2 contaminant. NO_2 was purified by extensive pumping of an appropriate amount of sample condensed at 195 K to remove noncondensable impurities including NO. A high-purity Ar (99.9995% pure) obtained from Matheson was used for dilution without further purification.

RESULTS

Four different reaction mixtures, one of them contained 517 ppm NO_2, were shocked in the temperature range 1700–2500 K. In the absence of NO_2, no detectable products (CO and H_2O) could be measured below 2000 K. Accordingly, most of the measurements were made above that temperature. The conditions used in the present experiment are listed in Table I.

Figures 1–3 summarize several typical concentration profiles of CO, H_2O and NO measured; they are compared with calculated profiles using the mechanism presented in Table II. The mechanism is a simplified version of the one employed recently by Hanson and coworkers to model their measured NH and OH radical production profiles in the shock-induced reactions of HNCO with NO, O_2 and N_2O (Mertens *et al.* 1991, 1992).

The results of our kinetic modeling, using the CHEMKIN program developed by Sandia National Laboratory (Kee *et al.* 1980), indicate that almost all of our measured concentration profiles for CO, H_2O and NO, with and without added NO_2, could be modeled satisfactorily with the rate constants recently employed by Mertens *et al.* (1991). Figures 1 (a) and (b) compare the experimentally measured CO profiles with those of the calculated ones. The agreement between the observed and modeled values appears to be quite good. Similar good agreements are noted for H_2O and NO profiles, with and without added NO_2, as illustrated in Figures 2 and 3, respectively. The dependence of the product yields on key individual processes will be discussed later.

TABLE I

EXPERIMENTAL CONDITIONS AND EVALUATED RATE CONSTANTS[A]

Mixture[b]	T/K	P/atm	$k_1/10^7$
A	1974	0.552	0.44
A	2014	0.509	0.45
A	2073	0.588	1.2
A	2099	0.491	1.2
A	2113	0.483	2.0
A	2183	0.451	3.3
A	2327	0.409	9.8
A	2330	0.414	8.5
A	2444	0.338	26.0
A	2461	0.375	21.0
A	2479	9.356	30.0
A	2482	0.343	29.0
B	1707	0.500	
B	1960	0.496	
B	2069	0.485	
B	2106	0.498	
B	2177	0.465	
B	2237	0.437	
B	2249	0.397	
B	2289	0.430	
B	2389	0.390	
B	2492	0.339	
B	2509	0.344	
C	1970	0.489	0.42
C	2160	0.455	2.2
C	2260	0.408	5.9
C	2319	0.413	8.9
C	2353	0.391	12.0
C	2368	0.385	15.0
C	2383	0.381	14.0
C	2447	0.368	23.0
C	2452	0.329	25.0
C	2452	0.342	23.0
C	2452	0.349	26.0
C	2464	0.350	25.0
C	2500	0.347	34.0
D	1823	0.584	
D	2091	0.496	
D	2147	0.467	
D	2215	0.435	
D	2227	0.445	
D	2367	0.386	
D	2383	0.387	
D	2482	0.341	
D	2517	0.342	

[a]Rate constants are given in units of cm^3/mol s. [b]The compositions of the reaction mixtures are: (A) HNCO:NO:Ar = 0.0030:0.010:0.987; (B) HNCO:NO:Ar = 0.0100:0.0050:0.985; (C) 0.0050:0.0054:0.9896; (D) HNCO:NO.NO_2:Ar = 0.0050:0.0050:0.000517:0.9895.

TABLE II

REACTION MECHANISM AND RATE CONSTANTS USED IN KINETIC MODELING[a]

	Reaction	A	B	E_a/R
1.	$HNCO + M = NH + CO + M$	8.40E15	0.0	42,600[b]
2.	$HNCO + M = H + HCO + M$	5.0E15	0.0	60,000
3.	$NH + HNCO = NH_2 + NCO$	2.3E13	0.0	12,000
4.	$H + HNCO = NH_2 + NCO$	1.1E14	0.0	6,400
5.	$H + HNCO = H_2 + NCO$	1.1E7	2.0	8,000
6.	$NH_2 + HNCO = NH_3 + NCO$	5.0E12	0.0	3,500
7.	$N + HNCO = NH + NCO$	4.0E13	0.0	18,000
8.	$NH + NH = 2H + N_2$	5.1E13	0.0	0
9.	$NH + N = N_2 + H$	6.3E11	0.5	0
10.	$NH_3 + M = NH_2 + H + M$	2.5E16	0.0	47,030
11.	$NH_2 + M = NH + H + M$	3.2E23	–2.0	46,000
12.	$NH_2 + N = N_2 + 2H$	7.2E13	0.0	0
13.	$2NH_2 = NH_3 + NH$	5.0E13	0.0	5,000
14.	$2NH_2 = N_2H_2 + H_2$	4.0E13	0.0	12,000
15.	$N_2H_3 + M = N_2H_2 + H + M$	1.0E16	0.0	25,000
16.	$N_2H_3 + M = NH_2 + NH + M$	1.0E16	0.0	21,000
17.	$H + H + M = H_2 + M$	1.0E18	–1.0	0
18.	$N_2H + M = N_2 + H + M$	1.0E14	0.0	1,500
19.	$N_2H_2 + M = N_2H + H + M$	1.0E16	0.0	25,000
20.	$NH + M = N + H + M$	2.6E14	0.0	38,000
21.	$NH + NH_2 = N_2H_2 + H$	1.0E15	–0.5	0
22.	$N + H_2 = NH + H$	1.9E14	0.0	13,000
23.	$NH_2 + H = NH + H_2$	4.0E13	0.0	1,840
24.	$NCO + M = N + CO + M$	6.3E16	–0.5	24,150
25.	$NCO + N = N_2 + CO$	2.0E13	0.0	0
26.	$NCO + H = NH + CO$	1.1E14	0.0	1,000
27.	$N_2H + O = NH + NO$	1.0E13	0.0	0[c]
28.	$NH + NO = N_2O + H$	1.4E14	0.0	6,400
29.	$NH + NO = N_2 + OH$	3.4E13	0.0	6,400
30.	$NCO + NO = N_2O + CO$	1.0E13	0.0	–200
31.	$NH_2 + NO = N_2H + OH$	6.4E15	–1.25	0
32.	$NH_2 + NO = N_2 + H_2O$	6.2E15	–1.25	0
33.	$NH_2 + NO = N_2O + H_2$	5.0E13	0.0	12,400
34.	$N_2O + M = N_2 + O + M$	6.9E23	–2.5	32,710
35.	$N_2O + O = N_2 + O_2$	1.0E14	0.0	14,100
36.	$N_2O + O = 2NO$	6.9E13	0.0	13,400
37.	$N_2O + H = N_2 + OH$	4.4E14	0.0	9,500
38.	$NH + OH = HNO + H$	1.0E12	0.5	1,000
39.	$NH + OH = N + H_2O$	5.0E11	0.5	1,000

TABLE II, cont'd

Reaction	A	B	E_a/R
40. $H + OH + M = H_2O + M$	1.6E22	−2.0	0
41. $NH_3 + OH = NH_2 + H_2O$	2.0E6	2.04	285
42. $NH + O = NO + H$	4.7E13	0.0	0
43. $NH + O = N + OH$	3.1E13	0.0	0
44. $NH_2 + O = HNO + H$	6.6E14	−0.5	0
45. $NH_2 + O = NH + OH$	6.8E12	0.0	0
46. $NH_2 + OH = NH + H_2O$	4.0E6	2.0	500
47. $CO + OH = CO_2 + H$	1.5E7	1.3	380
48. $CO + O + M = CO_2 + M$	6.2E14	0.0	1,500
49. $CO + N_2O = CO_2 + N_2$	4.1E14	0.0	23,150
50. $H + O_2 = O + OH$	9.1E13	0.0	7,400
51. $H_2 + O = H + OH$	5.0E4	2.67	3,167
52. $O + NO = N + O_2$	3.8E9	1.0	20,800
53. $NH_3 + H = NH_2 + H_2$	6.5E5	2.39	5,120
54. $NH_3 + O = NH_2 + OH$	2.1E12	0.0	4,530
55. $HNO + M = H + NO + M$	1.5E16	0.0	24,500
56. $NO_2 + H = NO + OH$	3.5E14	0.0	740
57. $NO_2 + O = NO + O_2$	1.0E13	0.0	300
58. $NO_2 + M = NO + O + M$	6.8E14	0.0	26,800[d]
59. $HNCO + O = NH + CO_2$	4.8E12	0.0	8,300[f]
60. $HNCO + O = HNO + CO$	1.5E8	1.6	22,150[g]
61. $HNCO + O = OH + NCO$	4.7E13	0.0	15,600[f]
62. $HNCO + OH = H_2O + NCO$	2.7E12	0.0	2,790
63. $NO_2 + NO_2 = 2NO + O_2$	2.0E12	0.0	13,500
64. $O_2 + M = 2O + M$	1.8E18	−1.0	59,380
65. $NO_2 + NH = HNO + NO$	1.0E11	0.5	2,000
66. $NO_2 + NH_2 = N_2O + H_2O$	2.0E20	−3.0	0[e]
67. $NH + O_2 = HNO + O$	3.6E13	0.0	9,000
68. $NH + O_2 = NO + OH$	7.6E10	0.0	770
69. $H_2 + OH = H + H_2$	8.9E13	0.0	4,400
70. $2OH = H_2O + O$	6.0E8	1.3	0
71. $N_2 + O = NO + N$	1.8E14	0.0	38,370
72. $NO + H = N + OH$	2.6E14	0.0	25,370
73. $N_2H + NH = N_2 + NH_2$	2.0E11	0.5	1,000
74. $N_2H + OH = N_2 + H_2O$	3.0E13	0.0	0

[a]Rate constants are given in the form of $k = AT^B \exp(-E_a/RT)$, in units of cm^3, mol and sec. The mechanism and all rate constants given except those noted are based on those of Merten *et al.* (1991) from which the original references can be found. [b]Wu *et al.* (1990). [c]Estimated. [d]C.-Y. Lin, H.-T. Wang, and M.C. Lin, unpublished results. [e]Hanson and Salimian (1984). [f]This work. [g]He *et al.* (1992).

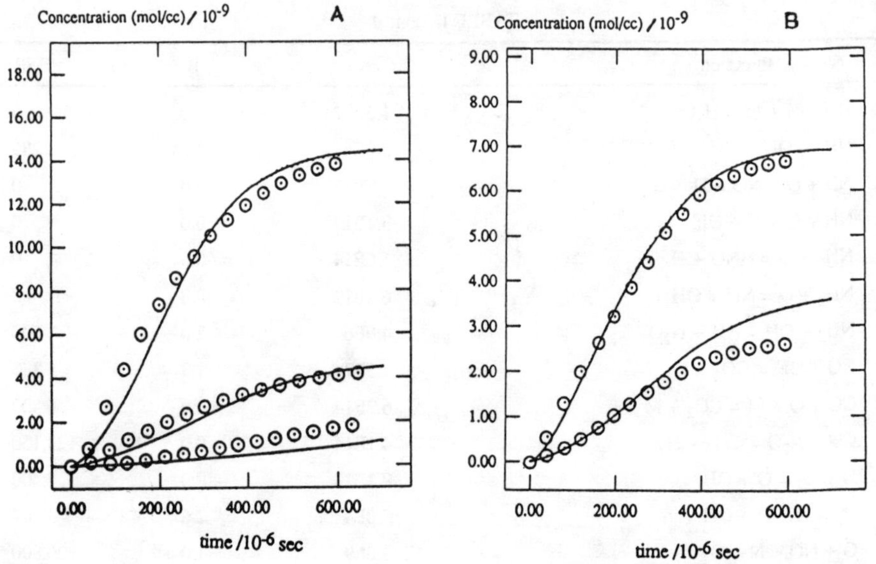

FIGURE 1 Concentration profiles of CO formed in the pyrolysis of HNCO and NO with and without added NO_2: (a) upper curve: mixture (B), T = 2492 K, P = 0.339 atm; middle curve: mixture (A), T = 2461 K, P = 0.375 atm; lower curve: mixture (A), T = 2073 K, P = 0.588 atm. (b) upper curve: mixture (D), T = 2482 K, P = 0.341 atm; lower curve: mixture (C), T = 2452 K, P = 0.349 atm. Solid curves: kinetically modeled values using the mechanism and rate constants given in Table II.

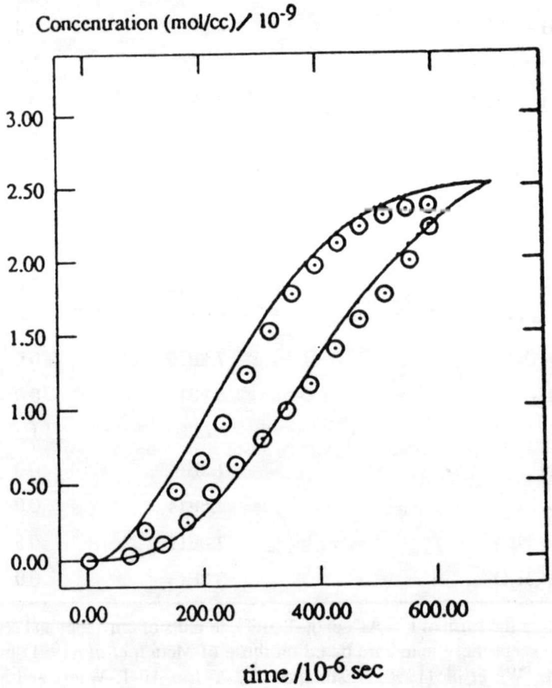

FIGURE 2 Concentration profiles of H_2O formed in the pyrolysis of HNCO and NO with and without added NO_2, upper curve: mixture (D), T = 2517 K, P = 0.342 atm., lower curve: mixture (C), T = 2452 K, P = 0.342 atm. Solid curves: kinetically modeled values using the mechanism and rate constants given in Table II.

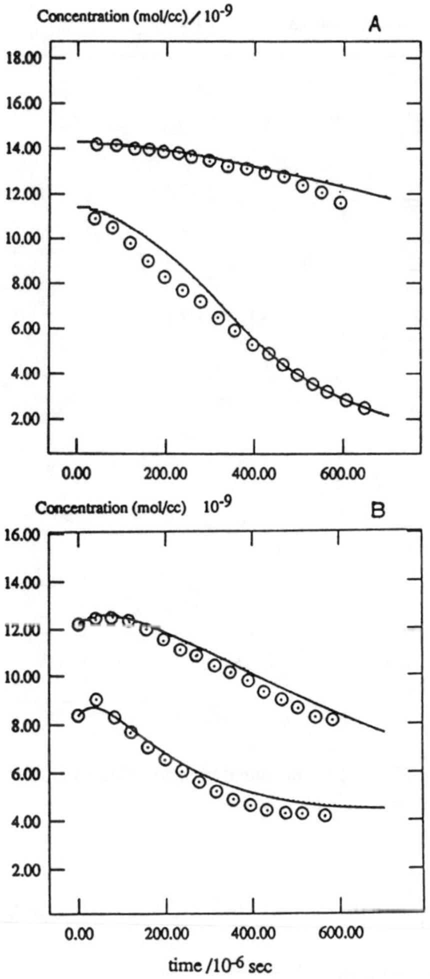

FIGURE 3 Concentration profiles of NO measured in the pyrolysis of HNCO and NO with and without added NO_2: (a) Mixture (B), upper curve: T = 2069 K, P = 0.485 atm; lower curve: T = 2289 K, P = 0.430 atm. (b) Mixture (D), upper curve: T = 2215 K, P = 0.435 atm; lower curve: T = 2482 K, P = 0.341 atm. Solid curves: kinetically modeled values using the mechanism and rate constants given in Table II.

In addition to the direct concentration measurements for CO, H_2O and NO, we also monitored the visible emission profiles of NO_2 using a photomultiplier with an appropriate bandpass filter (546 ± 15 nm). From the observed emission traces and the initial concentration of NO_2, we are able to evaluate the time-dependent NO_2 concentrations as shown in Figure 4. These concentration profiles can be reasonably accounted for by kinetically modeled values as illustrated by the figure.

DISCUSSION

Our measured CO, H_2O, NO and NO_2 concentration profiles could be reasonably accounted for by the mechanism listed in Table II. The reduction of NO was found to be controlled

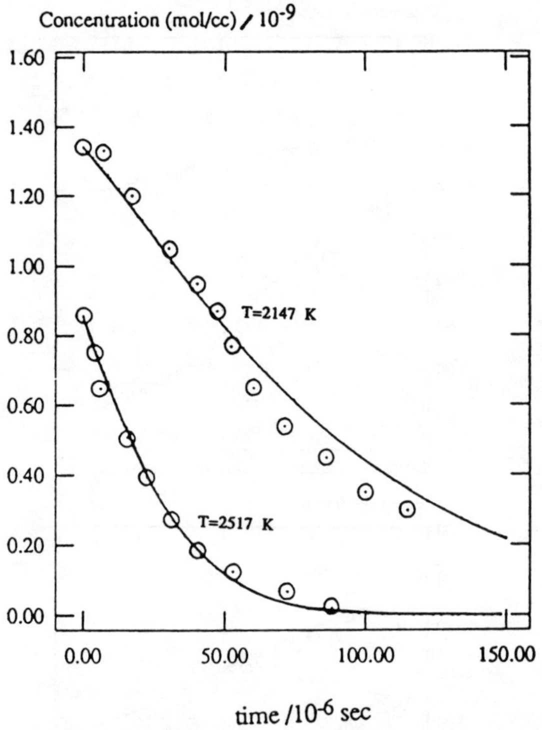

FIGURE 4 Concentration profiles of NO_2 measured by monitoring its emission at 546 ± 15 nm for mixture (D). Solid curves are calculated values.

primarily by the rate-limiting process, reaction (1), which generates the key reactive species in the system, NH:

$$HNCO + M \rightarrow NH + CO + M. \tag{1}$$

Since there are no observable products (CO and H_2O) formed at temperatures below 2000 K, the direct bimolecular initiation reaction, NO + HNCO → products, can be ruled out under the conditions employed.

For reaction (1), we have previously determined its second-order decomposition rate constant over the temperature range 2120–2570 K by kinetic modeling of the CO formed in the reaction (Wu *et al.* 1990). The modeled values of k_1 were found to agree very closely with those reported by Hanson, Bowman and coworkers (Mertens *et al.* 1989) using entirely different diagnostic methods. Combination of these two sets of data gave rise to

$$k_1 = 8.40 \times 10^{15} \exp(-42{,}630/T) \text{ cm}^3/\text{mol s} \tag{I}$$

covering the broad temperature range, 1830–3340 K (Wu *et al.* 1990).

In view of the fact that the rate of CO production in the present HNCO–NO system controlled mainly by reaction (1), we have also attempted to fit our CO profiles by adjusting k_1. The values of k_1 thus obtained with mixtures (A) and (C) are summarized in Table I and compared with the values given by (1) in Figure 5. The present results agree closely with those calculated by (I).

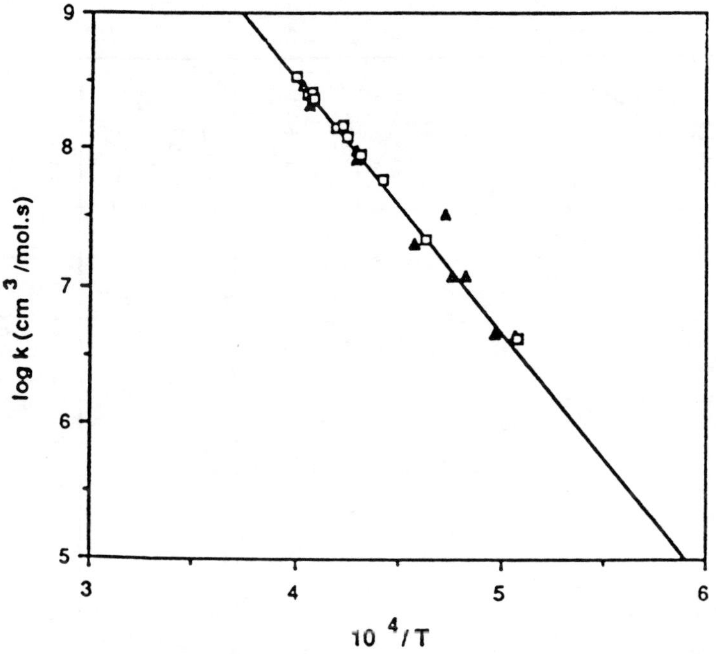

FIGURE 5 Arrhenius plot of the rate constant for HNCO + M → NH + CO + M (k_1) obtained from mixtures A (triangles) and C (squares). Solid curve is the value predicted by (I).

In the presence of NO_2, the rate of CO production was found to be enhanced noticeably, apparently by the increased concentration of O atoms produced by the thermal decomposition of NO_2 in reaction (58). The decay of NO_2 shown in Fig. 4 could be reasonably accounted for by the model. The decay rate was controlled mainly by the rate of its rapid decomposition reaction under the conditions employed. Because of the low NO_2 concentration present in the system, the O atom is expected to attack mainly HNCO via the two key reaction paths (Mertens *et al.* 1992; He *et al.* 1992):

$$O + HNCO \rightarrow NH + CO_2, \quad (59)$$

$$O + HNCO \rightarrow OH + NCO. \quad (61)$$

Reaction (59), the dominant channel, generates more NH, which in turn reacts with O, NO and HNCO to produce more reactive chain carriers such as H, OH and NH_2 (see Table II).

In order to identify the key processes responsible for NO reduction by HNCO, we carried out sensitivity analysis using the SENKIN program recently developed by Sandia National Laboratory (Lutz *et al.* 1987). The results of the analysis for mixture (D) at 2517 K and mixture (B) at 2387 K are shown, respectively, in Figures 6 and 7 for CO production and NO removal with and without added NO_2. In the presence of NO_2, the major processes responsible for the removal of NO are:

$$HNCO + M \rightarrow NH + CO + M, \quad (1)$$

$$O + HNCO \rightarrow OH + NCO, \quad (61)$$

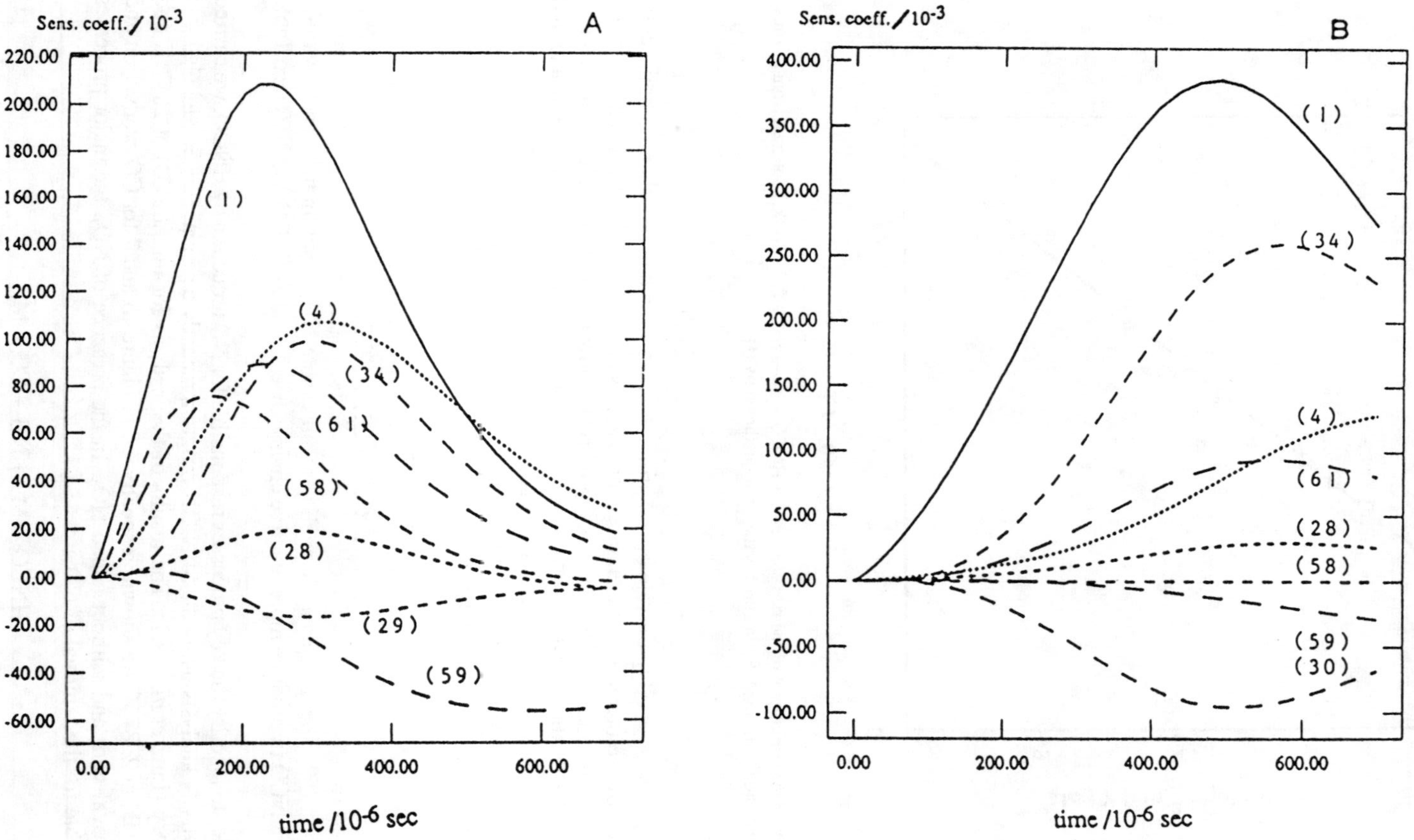

FIGURE 6 Normalized sensitivity coefficients for CO: (a) mixture (D), T = 2517 K, P = 0.342 atm; (b) mixture (B), T = 2289 K, P = 0.430 atm. Key reactions are given in parentheses.

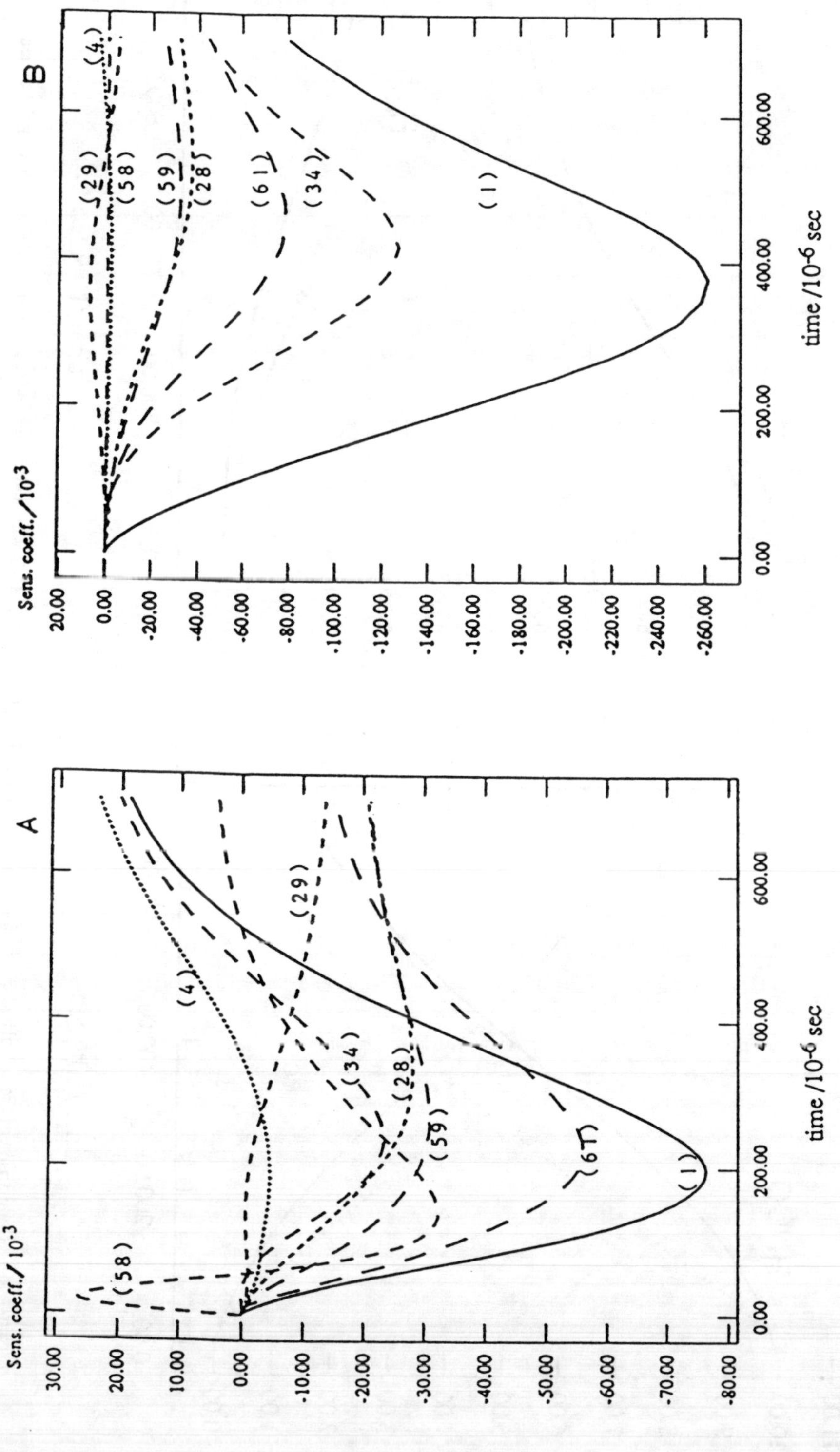

FIGURE 7 Normalized sensitivity coefficients for NO: (a) mixture (D), T = 2517 K, P = 0.342 atm; (b). mixture (B), T = 2289 K, P = 0.430 atm, Key reactions are given in parentheses.

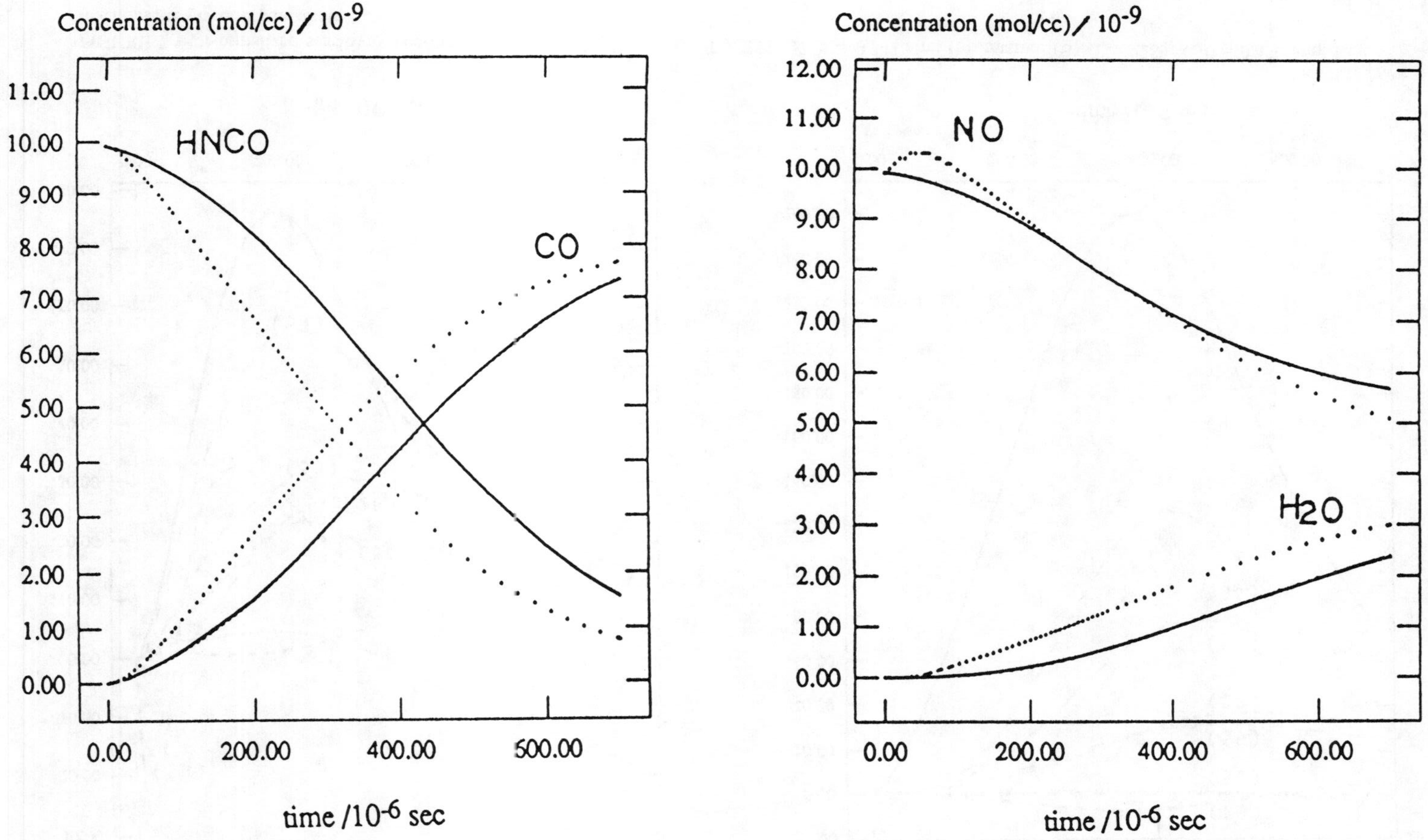

FIGURE 8 Comparison of the calculated concentration profiles for HNCO, CO, NO and H_2O with and without added 517 ppm NO_2 as given by the dotted and solid curves, respectively, using mixtures (D) anc (C) as examples. T = 2367 K, P = 0.386 atm.

$$O + HNCO \rightarrow NH + CO_2, \quad (59)$$

$$NO_2 + M \rightarrow NO + O + M \text{ (late stage)}, \quad (58)$$

$$NH + NO \rightarrow N_2O + H. \quad (28)$$

Those responsible for the enhanced NO production are

$$NO_2 + M \rightarrow NO + O + M \text{ (early stage)}, \quad (58)$$

$$O + NH \rightarrow NO + H, \quad (42)$$

and to smaller extents,

$$H + NO_2 \rightarrow NO + OH, \quad (56)$$

$$NH + OH \rightarrow HNO + H. \quad (38)$$

For the production of CO, the following reactions dominate:

$$HNCO + M \rightarrow NH + CO + M, \quad (1)$$

$$N_2O + M \rightarrow N_2 + O + M, \quad (4)$$

$$O + HNCO \rightarrow NH + CO_2, \quad (59)$$

$$O + HNCO \rightarrow OH + NCO. \quad (61)$$

In the absence of NO_2, the results presented in Figures 6 and 7 reveal that reactions (1), (34), (4) and (61) affect CO production most strongly, while (1), (34), (61), (59) and (28) affect the removal of NO quite significantly.

The effects of the 517 ppm NO_2 on the destruction of HNCO and the production of CO and H_2O are quite noticeable, as illustrated in Figure 8. These effects evidently result from reactions (59) and (61), which generate additional NH, OH and NCO radicals.

These results, together with those of Hanson, Bowman and coworkers (Mertens *et al.* 1991), clearly indicate that the reduction of NO by HNCO is controlled primarily by the amount of NH and NH_2 radicals present in the system. Additionally, the O atom also plays an important role in enhancing the reduction rate through the exothermic process, O + HNCO $\rightarrow$ NH + CO_2, which generates additional NH radicals.

In view of the high thermal stabilities of both NO and HNCO, and of the efficacy of NO_2 as a radical source as illustrated in the present study, experimentally observed HNCO de-NO$_x$ effects at low temperatures (Perry and Siebers, 1986; Caton and Siebers, 1989; Lyon and Cole, 1990; Wicke *et al.* 1989) must have resulted from the participation of NO_2, which can be readily formed by the well-known reaction, $2NO + O_2 \rightarrow 2NO_2$, with negligible activation energy.

CONCLUSIONS

We have studied the kinetics of CO and H_2O production and of NO destruction in shock-heated mixtures of HNCO and NO between 1700 and 2500 K in an attempt to elucidate the mechanism of the HNCO–RAPRENO$_x$ process (Perry and Siebers, 1986). The reaction between HNCO and NO was found to be slow below 2000 K and the rates of stable product formation at temperatures above 2000 K were found to depend very strongly on the rate of the HNCO decomposition reaction and on the small amount of added NO_2, which provides an additional pool of O atoms.

We are able to quantitatively model the formation of CO and H_2O and the disappearance of NO using the kinetic data recently measured by Wu *et al.* (1990) and by Hanson, Bowman and coworkers (Mertens *et al.* 1989, 1991) on several key processes related to the HNCO–NO system.

ACKNOWLEDGMENT

The support we have received from the Office of Naval Technology (to CHW) and the Office of Naval Research (to Emory University through Project No. N00014-89-J-1949) are gratefully acknowledged.

REFERENCES

Bogg, T.L. (1981). Fundamentals of Solid-Propellant Combustion, eds. K.K. Kuo and M. Sommerfield, *Prog. in Aeronautics and Astronautics*, **90**, 121.

Caton, J.A., and Siebers, D.L. (1989). Comparison of Nitric Oxide Removal by Cyanuric Acid and by Ammonia, *Combust. Sci. and Tech.*, **65**, 277.

Choudhury, T.K., and Lin, M.C. (1989). Pyrolysis Methyl Nitrite and 1,3,5-Trioxane Mixtures in Shock Waves: Kinetic Modeling of the H + CH_2O Reaction Rate. *Combust. Sci. and Tech.*, **64**, 19.

Hanson, R.K., and Salimian, S. (1984). *Combustion Chemistry*, ed. W.A. Gardiner Jr., p. 361, Springer-Verlag.

He, Y., Lin, M.C., Wu, C.H., and Melius, C.F. (1992). The Reaction of HNCO with NO_2 in Shock Waves. *24th Symp. (Intern.) on Combust.*, The Combustion Institute, p. 711.

Hsu, D.S.Y., Burks, G.L., Beebe, M.D., and Lin, M.C. (1984). Thermal Decomposition of Methyl Nitrite in Shock Waves Studied by Laser Probing. *Int. J. Chem. Kinet.*, **16**, 1139.

Hsu, D.S.Y., and Lin, M.C. (1984). Infrared Laser Probing of Combustion and Energetic Molecular Decomposition Reaction in Shock Waves, SPIE 482, *Applications of Laser Chemistry and Diagnostics*, p. 89.

Hsu, D.S.Y., Shaub, W.M., Blackburn, M., and Lin, M.C. (1982). Thermal Decomposition of Formic Acid at High Temperatures in Shock Waves. *19th Symp. (Intern) on Combust.* The Combustion Institute, p.89.

Kee, R.J., Miller, J.A., and Jefferson, T.H.(1980). CHEMKIN: A General-Purpose, Problem: Independent, Transportable, Fortran Chemical Kinetics Code Package, Sandia National Laboratory Report No. SAND 80-8003.

Lutz, A.E., Kee, R.J., and Miller, J.A. (1987). SENKIN: A Fortran Program for Predicting Homogeneous Gas Phase Chemical Kinetics with Sensitivity Analysis, Sandia National Laboratory Report No. SAND 87-8215.

Lyon, R.K., and Cole, J.A. (1990). A Reexamination of the $RAPRENO_x$ Process, *Combust. Flame*, **82**, 435.

Mertens, J.D., Chang, A.Y., Hanson, R.K., and Bowman, C.T. (1989). Reaction Kinetics of NH in the Shock Tube Pyrolysis of HNCO. *Int. J. Chem. Kinet.*, **21**, 1049.

Mertens, J.D., Chang, A.Y., Hanson, R.K., and Bowman, C.T. (1991). A Shock Tube Study of the Reactions of NH with NO, O and O_2, *Int. J. Chem. Kinet.*, **23**, 173.

Mertens, J.D., Chang, A.Y., Hanson, R.K., and Bowman, C.T. (1992). A Shock Tube Study of Reactions of Atomic Oxygen with Isocyanic Acid, *Int. J. Chem. Kinet.*, **24**, 279.

Perry, R.A., and Siebers, D.L. (1986). Rapid Reduction of Nitrogen Oxides in Exhaust Gas Streams, *Nature*, **324**, 657.

Tully, F.P., Perry, R.A., Thorne, L.R., and Allendorf, M.D. (1988). Free-Radical Oxidation of Isocyanic Acid, *22nd Symp. (Intern.) on Combustion*, The Combustion Institute, p. 1101.

Wicke, B.G., Grady, K.A. and Ratcliffe, J.W. (1989). Cyanuric Acid + Nitric Oxide Reaction at 700°C and the Effects of Oxygen, *Combust. Flame*, **78**, 249.

Wu, C.H., Wang, H.-T., Lin, M.C., and Fifer, R.A. (1990). Kinetics of CO and H Atom Production from the Decomposition of HNCO in Shock Waves. *J. Phys. Chem.*, **94**, 3344.

Reduction of Nitric Oxide by Hydrogen Sulfide at High Temperatures

M. OYA and K. TSUCHIYA *National Institute for Resources and Environment, Tsukuba, Ibaraki, Japan*
T. ASABA *Department of Resources Engineering, Waseda University, Shinjuku, Tokyo, Japan*

Abstract — NO was reduced by H_2S much faster than by H_2 behind incident shock waves in the temperature range between 1600 and 3500 K. In the system of NO and H_2S the elementary reaction, NO + S = N + SO, was found to be the key reaction of the system and the rate constant was determined in the separate system, i.e., that of NO and COS which produced atomic S directly. By adopting this rate constant and the existing rate constants of the other elementary reactions, the simulative computations on the NO and H_2S systems heated in the shock waves were performed. The results of the simulation were found to be consistent with the experiments.

Key Words: nitric oxide, hydrogen sulfide, high temperatures, chemical kinetics

INTRODUCTION

The high temperature reaction system of nitric oxide and hydrogen sulfide, which are known to be the pollutant substances, have been investigated through use of the shock tube.

In the previous papers (Ando and Asaba 1976; Koshi *et al.* 1974), it had been ascertained that the elementary step, H + NO = OH + N, was the rate-controlling one in the reduction system of NO and H_2, at least above 2000 K. In this paper H_2S has been used to reduce NO in place of H_2 to examine its reducibility and to clarify the reaction mechanism of the system. While the interactions between N-containing species and S-containing ones have attracted attention in connection with the furnace combustion of the fossil fuels (Hills *et al.* 1987; Levy, 1982; Tseregounis and Smith 1983), there has been little accumulation of knowledge on the reaction kinetics of them. The investigation of the system of NO and H_2S may be the first step in their understanding.

EXPERIMENTAL

Experiments were carried out through the use of a shock tube of stainless steel of 50 mm diameter and 4 m long, which was used in two ways, i.e., as the conventional shock tube and as the single-pulse type.

Through the use of the conventional shock tube two groups of gaseous mixtures were heated in incident shock waves: One group of the mixtures was of NO, H_2S and Ar, of which the initial concentrations were $(2–106) \times 10^{-9}$, $(2–106) \times 10^{-9}$ and $(2–4) \times 10^{-6}$ mol/cm^3, respectively. This group of mixtures was used to obtain the overall rate constant and to determine the reaction mechanism of the system. In these experiments the variation of the concentrations of NO and H_2S were monitored by the IR emission of 5.3 and 2.7 mμ, respectively and that of atomic S was monitored by the UV recombination emission of 3705 A. Here it was necessary to measure the IR emission of 3.2 mμ (H_2O) simultaneously, since the emission of 5.3 mμ contained that of H_2O produced during the reaction period. The other group of mixtures was NO, COS and Ar, of which the initial concentrations were $(8.2–17) \times 10^{-8}$, $(2.0–2.2) \times 10^{-8}$ and $(1.9–2.0) \times 10^{-6}$ mol/cm^3, respectively. This group of mixtures was used to obtain the rate constant of the elementary reaction, S + NO = N + SO,

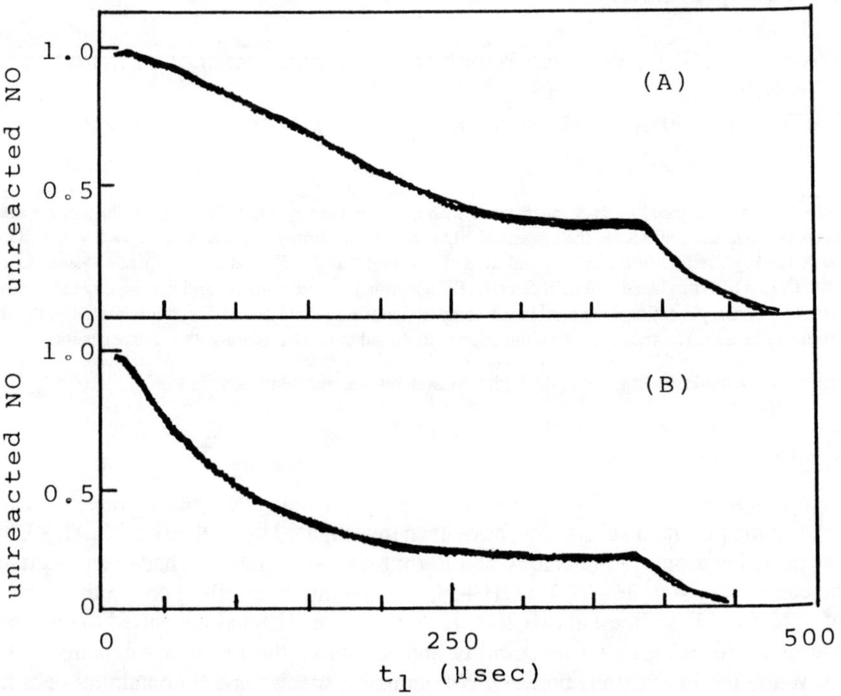

FIGURE 1 Examples of oscillograms of NO reduction curves in incident shock waves: A) NO, H_2 and Ar system, $(NO)_0$, $(H_2)_0$ and $(Ar)_0$ were 9.2×10^{-7}, 9.2×10^{-7}, and 9.0×10^{-5} mol/cm^3, respectively; temperature was 2800 K; B) NO, H_2S and Ar system $(NO)_0$, $(H_2S)_0$ and $(Ar)_0$ were 5×10^{-8}, 5×10^{-8}, and 4.9×10^{-6} mol/cm^3, respectively; temperature was 2890 K.

which was proved to be the key reaction in the high temperature system of NO and H_2S (Oya *et al.* 1987). In these experiments the concentrations of NO and S were monitored in the same way as that described above, but since both COS and CO, which was produced in the course of the decomposition reaction of COS, had the emission at 5.3 mμ, the emission of COS at 9.5 mμ and of CO at 4.7 mμ needed to be measured simultaneously.

By use of the single-pulse type shock tube two gaseous mixtures of NO, H_2S, Kr and Ar, of which the molar ratios were 4.67/5/1/89.33, and 4.67/3/1/91.33 were heated in the reflected shock waves. The initial concentrations of the component of the mixtures were; for NO (3.5–5.3) $\times 10^{-7}$, for H_2S (2.2–5.5) $\times 10^{-7}$ mol/cm^3. Here Kr was added as the internal standard. The gaseous mixtures were heated in the shock waves for about 1.5–2.0 msec and then cooled during about 1.5 msec to the room temperature by the rarefaction waves. The pressure profiles were monitored by the piezo-gauge fixed at the end plate of the shock tube. Shock-heated gaseous samples were drawn out of the outlet installed 10 mm from the end plate of the shock tube, and were analyzed by the use of the mass spectrograph combined with the gas chromatograph. The mass signals of NO, H_2S, SO_2, N_2, O_2 and N_2O were obtained.

The experiments were carried out in the temperature range between 1600–3600 K. The sample gases NO, H_2S, Kr and COS were of a purity of 99.99% and that of Ar of 99.999%.

RESULTS AND DISCUSSION

Comparison of the Reaction Rates of the Reduction of NO by H_2 with that by H_2S

In previous papers (Ando and Asaba 1976; Koshi *et al.* 1974) it was found that decreasing curves of NO in the system of NO and H_2 showed the reverse sigmoid shape, which suggested the reaction was of an autocatalytic nature, as shown in Fig. 1A, and they could be expressed under the experimental conditions as:

$$-d(NO)/dt = k_{overall}(H_2)_0^k\,[(NO)_0 - (NO)]^1\,(NO)_0^m(Ar)^n \quad ((NO)/(NO)_0 > 0.7) \qquad (1)$$

where the exponents of the equation were determined roundly as k = 0.35, 1 = 0.5, m = 1 and n = 0, here subscript 0 means initial condition. On the other hand the decreasing curves of NO in the system of NO and H_2S were found to be quite similar to those in the NO and H_2 system as shown in Fig. 1B. They were expressed in a similar form, i.e.,

$$-d(NO)/dt = k'_{overall}\,(H_2S)_0^{k'}\,[(NO)_0 - (NO)]^{1'}\,(NO)^{m'}\,(Ar)^{n'} \quad ((NO)/(NO)_0 > 0.6), \qquad (2)$$

but the exponents of the equation were somewhat different from the system of NO and H_2 and determined as k′ = 1, 1′ = 0.3, m′ = 0.5 and n′ = 0.3.

The Arrhenius plot of the overall rate constants of both systems are shown in Fig. 2, where the most probable expressions of them were:

$$5.4 \times 10^{10} \exp(-50\ \text{kcal/RT})\ (\text{mol, cm}^3\text{, sec unit})$$

and

$$4.7 \times 10^{12} \exp(-40\ \text{kcal/RT})\ (\text{mol, cm}^3\text{, sec unit}),$$

respectively. It is clear that H_2S is much more effective as a reducing reagent for NO than H_2.

The Rate Constant of Elementary Reaction NO + S = N + SO,k_{2-1}

It was found in the previous investigation (Oya *et al.* 1987) that the system of NO and H_2S above 2200 K could be approximated to the system in which the decomposition products of H_2S in chemical equilibrium reacted with NO, since H_2S decomposed very fast comparing the decreasing rate of NO. Then it was concluded that the elementary reaction, NO + S = N + SO,k_{2-1}, was the most influential one in the system for reduction of NO by H_2S. Consequently, it was most important to precisely determine the rate constant, k_{2-1}, in order to understand the system.

Because of the coexistence of H and S atoms which were to react with NO directly, the system of NO and H_2S was too complicated and was considered not to be suitable to determine the rate constant, $k_{2\text{-}1}$. In this study it was determined directly from the concentration of NO and atomic S in the simple system of NO and COS. In this case COS was decomposed in the early stage of the reaction to produce S atoms and CO. The details of the procedure of the experiment, are described elsewhere (Oya *et al.* 1991). The Arrhenius plot of the reaction rate is shown in Fig. 3 and is expressed as follows:

$$k_{2\text{-}1} = 7.94 \times 10^{12} \exp[(-28.2 \pm 3.0)\ \text{kcal/RT}]\ \text{cm}^3\ \text{mol}^{-1}\ \text{sec}^{-1}. \qquad (3)$$

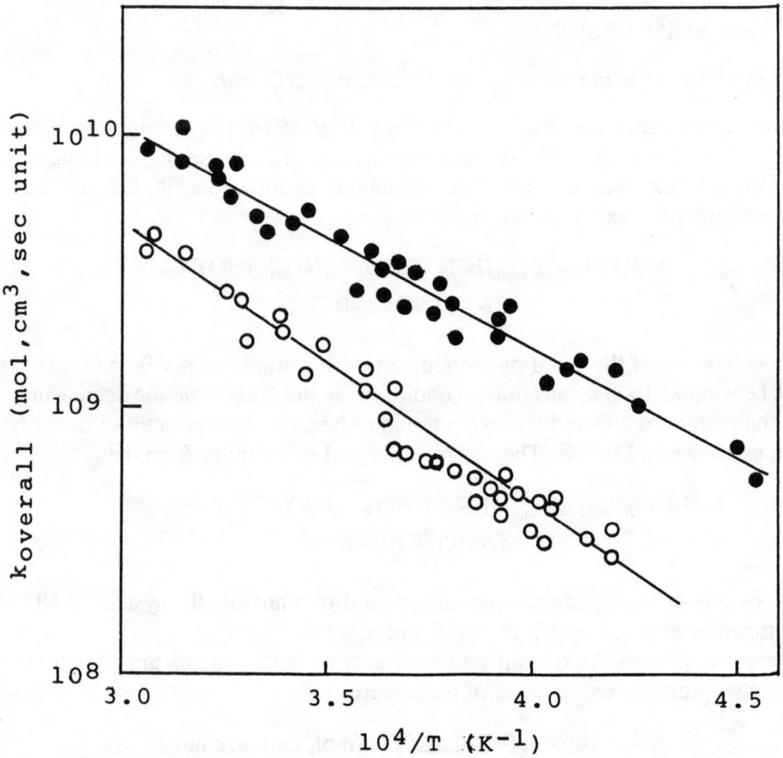

FIGURE 2 Arrhenius plots of the overall rate constant of the NO reduction reaction: ● = NO, H_2S and Ar system; ○ = NO, H_2 and Ar system.

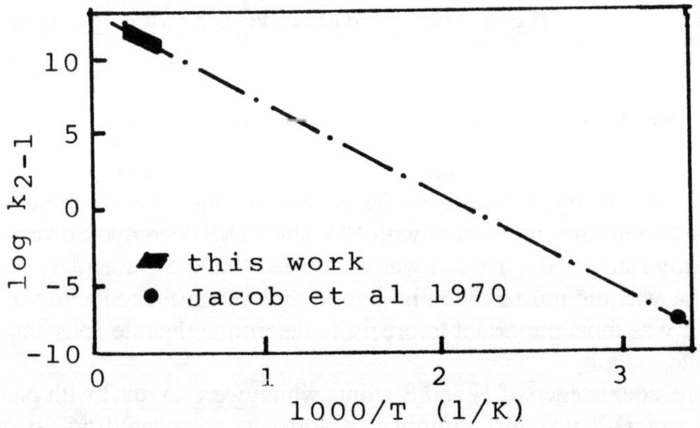

FIGURE 3 Arrhenius plot of the rate constant of the elementary reaction, NO + S = N + SO,k_{2-1}. The line was obtained by connecting the present data with that determined on the reverse reaction of k_{2-1} at room temperature by Jacob and Winkler (1970).

TABLE I

REACTION SCHEME AND RATE PARAMETERS, $k = AT^B \exp(-E/RT)$, cm^3, mol, sec, unit

	H_2S – NO – Ar	log A	B	E	ΔH_{298}	Reference
1–1)	$H_2S + M = HS + H + M$	14.67		82.5	91.6	Roth *et al.* (1982)
1–2)	$H + H_2S = H_2 + HS$	13.03		3.0	–12.6	Roth *et al.* (1982)
1–3)	$S + H_2S = HS + HS$	13.85		7.2	7.4	Baulch *et al.* (1976)
1–4)	$HS + H = H_2 + S$	13.18		0	–20.0	Baulch *et al.* (1976)
1–5)	$HS + S = S_2 + H$	13.5		0	–18.3	Baulch *et al.* (19)
1–6)	$HS + M = H + S + M$	14.79		49.4	84.2	Bowman *et al.* (1976)*
1–7)	$H + H + M = H_2 + M$	17.81	–1	0	–100.2	Baulch *et al.* (1976)
1–8)	$S + S + M = S_2 + M$	15.0		0	–102.5	Baulch *et al.* (1976)
2–1)	$S + NO = N + SO$	12.9		28.2	26.3	This work
3–1)	$H_2 + NO = HNO + H$	13.6		57.7	54.3	Koshi *et al.* (1974)
3–2)	$H + NO = OH + N$	14.13		49.2	48.7	Flower *et al.* (1977)
3–3)	$H + NO + M = HNO + M$	15.46		–0.6	–49.7	Clyne *et al.* (1961)
3–4)	$N + NO = N_2 + O$	13.2		0	–75.0	Baulch *et al.* (1973)
3–5)	$O + NO = N + O_2$	9.18	1	38.6	31.8	Baulch *et al.* (1973)
3–6)	$N + N + M = N_2 + M$	14.48		0	–0.9	Baulch *et al.* (1973)
4–1)	$O + H_2 = OH + H$	10.26	1	8.9	2.0	Baulch *et al.* (1972)
4–2)	$H + O_2 = OH + H$	14.35		16.8	16.9	Baulch *et al.* (1972)
4–3)	$OH + H_2 = H_2O + H$	13.34		5.2	–15.1	Baulch *et al.* (1972)
4–4)	$H + OH + M = H_2O + M$	21.92	–2	0	–119.3	Baulch *et al.* (1972)
5–1)	$SO_2 + M = SO + O + M$	16.18		112.0	132.0	Plach *et al.* (1984)
5–2)	$SO + M = S + O + M$	15.00		111.0	124.6	Plach *et al.* (1984)
5–3)	$SO_2 + O = SO + O_2$	12.62		19.5	13.0	Baulch *et al.* (1976)
5–4)	$SO + SO = SO_2 + S$	12.54		3.5	–7.5	Baulch *et al.* (1976)
5–5)	$S + O_2 = SO + O$	12.13		0.0	–7.5	Baulch *et al.* (1976)

Reverse reactions were included in calculations. *$k_{1\text{-}6}$ given in the table was calculated from the value of the reverse reaction obtained by Bowman and Dodge (1976) which was different from their original value.

Simulation of the Reaction System of NO, H_2S and Ar in the Incident Shock Waves

Adopting the rate constants of the reasonable elementary reactions, which are shown in Table I, the simulative computations were carried out on the reaction system of NO, H_2S and Ar heated in the incident shock waves.

Two examples of the simulations are shown in Fig. 4. Figure 4A is an example of the case where the concentrations of the reactants, NO and H_2S, were low enough that the temperature change in the course of the shock heating was considered to be negligible. When the concentrations of the reactants became high, however, there might be a temperature change in the reacting shock waves due to the endo- and exotherm in each elementary reaction, i.e., in the early stage of the reactions the temperature dropped through the thermal decomposition of H_2S and then the temperature rose gradually due to the oxidation reactions of the decomposition products of H_2S, i.e., H, S and HS, by NO. Figure 4B shows an example of this case, in which the concentrations of the reactants were so high that it was necessary to take account of the temperature variation of the systems. Two kinds of calculations were performed; assuming constant pressure and constant volume in the shocked gas mixtures. The actual conditions might lie between them. Lastly, it should be

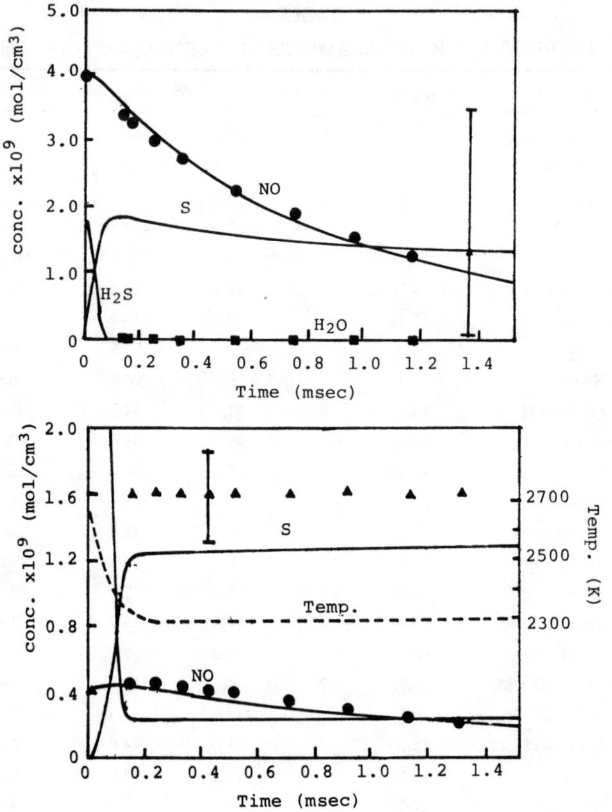

FIGURE 4 Examples of the simulative computation of the system of NO, H_2S and Ar heated in the incident shock waves: A) the case that the concentration of reactants was sufficiently low that the temperature change of the system was negligible: ● = NO, ■ = H_2O, T = 3, 458 K; $(NO)_0 = 3.98 \times 10^{-9}$, $(H_2S)_0 = 1.99 \times 10^{-9}$ mol/cm^3; B) the case that the concentration of the reactants were fairly high that the temperature change must have been considered: ● = NO, ▲ = S; lines: simulation; T_0 = 2687 K; $(NO)_0 = 4.2 \times 10^{-9}$, $(H_2S)_0 = 1.05 \times 10^{-7}$ mol/cm^3.

commented, that since there exists ambiguities in the rate constants of some elementary reactions, e.g., NO + H_2S = HNO + HS ? and NO + HS = HNO+ S ?, they were neglected in the present calculations. The simulations show that the sensitivities of these side reactions are expected to be quite low.

Simulation of the Composition of the Reaction Products of the System of NO, H_2S, Ar and Kr in the Reflected Shock Waves

Aside from the unreacted reactants NO and H_2S, and the products N_2, SO_2, O_2 and N_2O, which were analyzed quantitatively by considering the balance of species, other products, i.e., H_2, H_2O, and solid S, which should have been but could not be detected, could be estimated.

The experimental results of the system of NO/H_2S/Kr/Ar = 4.67/3/1/91.33 are shown in Fig. 5. In the case of another gaseous mixture, NO/H_2S/Kr/Ar = 4.67/5/1/89.33, the

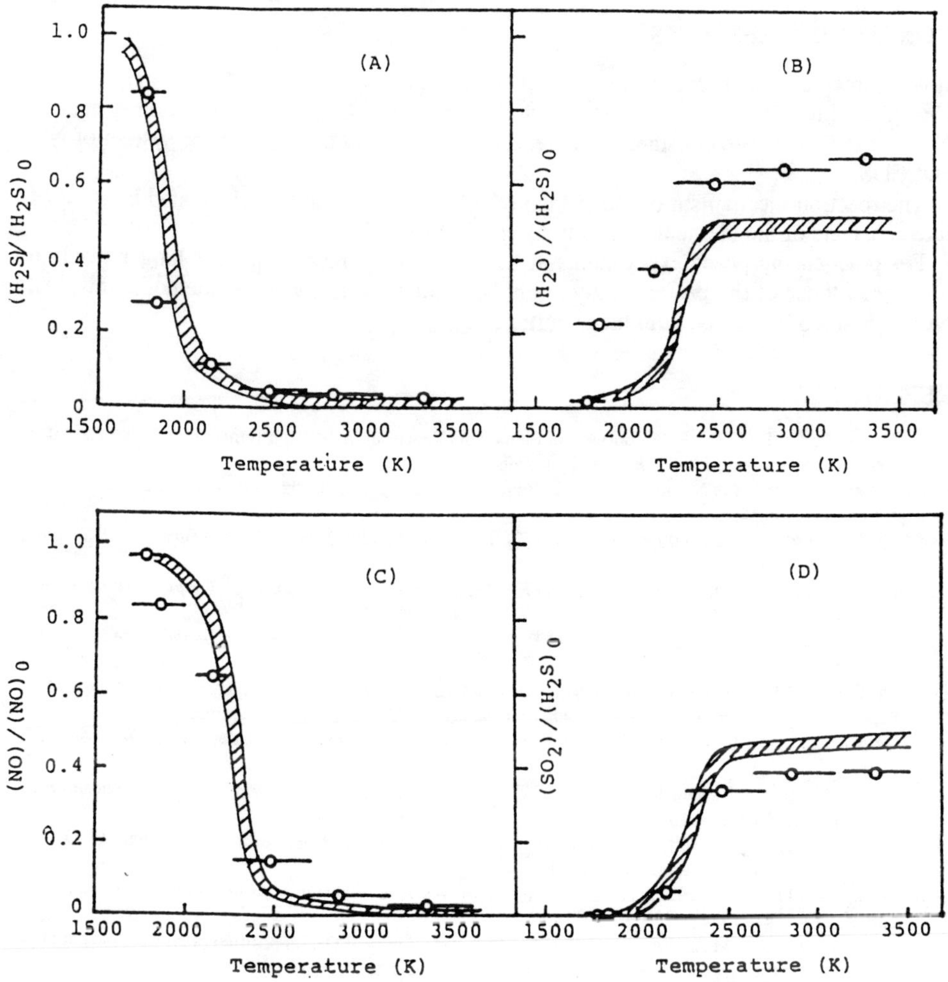

FIGURE 5 The distribution of the unreacted reactants and products of the system of $NO/H_2S/Kr/Ar$ = 4.67/3/1/91.33 in single-pulse shock tube experiments: A) $(H_2S)/(H_2S)_0$; B) $(H_2O)/(H_2S)_0$; C) $(NO)/(NO)_0$; D) $(SO_2)/(H_2S)_0$; –O– = experiments; hatched = simulative calculations.

experimental results were similar. The oscillograms of pressure records often showed that after the sudden pressure rise of the systems by the reflected shock waves the gradual pressure rise appeared due to the boundary layer effect. Accordingly the uncertainties in the estimations of the temperature of the system, which are plotted in the figure, were fairly large. The simulative calculations were performed by adopting the rate constants of the elementary reactions cited in Table I, here endo- and exothermicities of each elementary reactions were considered assuming constant volume in the system. In the cooling stage the temperatures were estimated from the pressure profiles of the oscillograms assuming the adiabatic expansion. The results of the simulations are also shown in Fig. 5. Coincidence of the simulative calculations with the experiments is said to be satisfactory.

CONCLUDING REMARKS

It was shown that the reduction reaction of NO by H_2S was much faster than by H_2 and this is attributed to the elementary reaction, S + NO = N + SO(k_{2-1}), in the reaction system of NO and H_2S. The rate constant of the reaction, k_{2-1}, was determined in the system of NO and COS.

The reaction mechanism of the system of NO and H_2S could be elucidated by a set of rate constants of the elementary reactions cited in Table I.

For practical purposes experiments at the lower temperatures and for longer reaction times than those of the present study might be required to test whether the pollutants, i.e., NO, H_2S and SO_2, can be eliminated sufficiently.

REFERENCES

Ando, H., and Asaba, T. (1976). Rate Constants of Elementary Reactions in the High Temperature System of Nitric Oxide and Hydrogen., *Int. J. Chem. Kin.*, **8**, 259.

Ando, H., and Asaba, T. (1977). Reaction of Hydrogen and Nitric Oxide in Shock Waves, *J. Fac. Eng.*, Univ. Tokyo (B) **34**, 2251.

Baulch, D.L., Drysdale, D.D., and Horne, D.G.(1972). *Evaluated Kinetic Data for High Temperature Reactions*. Vol. 1, Butterworths.

Baulch, D.L., Drysdale, D.D., and Horne, D.G.(1973). *Evaluated Kinetic Data for High Temperature Reactions*. Vol. 2, Butterworths.

Baulch, D.L., Drysdale, D.D. and Horne, D.G.(1976). *Evaluated Kinetic Data for High Temperature Reactions*. Vol. 3, Butterworths.

Bowman, C.T., and Dodge, L.G.(1976). Kinetics of the Thermal Decomposition of Hydrogen Sulfide behind Shock Waves, *Proc. of the 16th (Int) Symp. on Combustion*, The Combustion Institute, p. 971.

Clyne, M.A.A., and Thrush, B.A. (1961). Reaction of Hydrogen Atoms with Nitric Oxide, *Trans. Farad. Soc.*, **57**, 1305.

Flower, W.L., Hanson, R.K., and Kruger, C.H. (1977). *Experimental Study of Nitric Oxide Decomposition by Reaction with Hydrogen*, **15**, 115.

Hills, A.J., Cicerone, R.J., Calvert, J.G., and Birks, J.W. (1987). Kinetics of the Reactions of S_2 with O, O_2, O_3, N_2O, NO and NO_2, *J. Phys. Chem.*, **91**, 1199.

Jacob, A., and Winkler, C.A. (1970). The Rate Constant of Reaction of Nitrogen Atoms with Sulfur Monoxide, *Can. J. Chem.*, **48**, 1774.

Koshi, M., Ando, H., Oya, M., and Asaba, T. (1974). Shock Tube Study of Decomposition of Nitric Oxide at High Temperatures, *Proc. of the 15th (Int) Symp. on Combustion*, The Combustion Institute, p. 809.

Levy, A. (1982). Unresolved Problems in SO_x, NO_x, Soot Control in Combustion, *19th Symp. (International) on Combustion*, The Combustion Institute, p. 1223.

Oya, M., Tsuchiya, K., and Asaba, T. (1987). High Temperature Reaction of Nitric Oxide and Hydrogen Sulfide in Shock Waves, *Proc. of the 16th Int. Symp. on Shock Tubes and Waves*, The Combustion Institute, p. 421.

Oya, M., Tsuchiya, K., and Asaba, T. (1991). The Rate Constant of the Elementary Reaction, S + NO = SO + N, *18th Int. Symp. on Shock Waves*, forthcoming.

Plach, H.J., and Troe, J. (1984). UV Absorption Study of the Dissociation of SO_2 and SO in Shock Waves, *Int. J. Chem. Kinet.*, **16**, 1531.

Roth, P., Lohr, R. and Barner, U. (1982). Thermal Decomposition of Hydrogen Sulfide at Low Concentrations, *Comb. & Flame*, **45**, 273.

Tseregounis, S.I., and Smith, O.I. (1983). Enhancement of Fuel Nitrogen Oxidation, *Comb. Sci. & Tech.*, **30**, 231.

A Shock Tube Study of the High-Temperature Oxidation of Soot by Nitric Oxide

P. CADMAN and R.J. DENNING *University of Wales, Aberystwyth, Dyfed, Wales, UK*

Abstract — The high temperature oxidation of soot by nitric oxide has been studied in a shock tube at temperatures in the range of 2000 to 3200 K and total pressures near 6 atm. The surface oxidation rate of soot by nitric oxide was found to be

$$\omega = 4.9 P_{NO} \exp(-126 \text{ kJ mol}^{-1}/RT) \text{g cm}^{-2} \text{ atm}^{-1} \text{ s}^{-1},$$

where P_{NO} (atm) is the partial pressure of NO. The rate of reaction was found to be slower than the oxidation rate of soot with oxygen. Unlike the oxygen case where a maximum in the rate occurs around 2200 K maybe due to diffusion control, no maximum was reached in the rate of the NO reaction, suggesting that the surface complex is produced by a nondissociative mechanism in which more active sites are available. The values of the surface collision reaction probability was also calculated and found to be given by

$$\alpha = 4.08 \exp(-16340/T).$$

This agreed very well with the best fit expression given by Roth from measurement of the gaseous species in this reaction. The behaviour of soot/argon mixtures at temperatures greater than 2900 K was also investigated. A slow decay in the soot load was observed at the lower temperatures, becoming significant only at temperatures greater than 3500 K. This loss of soot was attributed to reaction into smaller species. The activation energy for this sublimation was found to be about 216 kJ mol^{-1}.

Key Words: Soot, Oxidation, Rate, Nitric Oxide, Shock Tube

INTRODUCTION

The reduction of nitric oxides by carbonaceous particles is recognised as a potential means of controlling NO emissions from combustion systems (Merryman *et al.* 1979). The mechanism and kinetic parameters for the reduction of NO-carbon particles, however, is not well understood. Nitric oxide formation in combustion systems is from two main sources, oxidation of fuel nitrogen and the thermal production of NO from the oxidation of N_2 (Tang and Churchill 1981). Control of NO emissions from fuel nitrogen is presently achieved by controlling the combustion parameters so that any NO produced in the early stages of combustion is destroyed in the later stages (Wendt 1979). It has been noted that NO reduction in coal flames is greater than in gaseous of liquid fuel flames, suggesting that carbonaceous particles are responsible for the enhanced reduction (Levy *et al.* 1981). Low carbon fly ash was found to reduce NO emissions significantly (Merryman *et al.* 1979).

The main kinetic studies of the reduction of NO by carbon (Levy *et al.* 1981, Cheng *et al.* 1989, Song *et al.* (1981) at flame temperatures report very different activation energies. Levy *et al.* (1981), using a laminar flow furnace, found an activation energy of 145 kJ mol^{-1} while Cheng *et al.*(1989), using a single pulse shock tube to study the reaction, report an activation energy of 80 kJ mol^{-1}. Both groups studied the reaction by monitoring the gaseous species involved, either the removal of NO or the production of CO, CO_2. At low temperatures an O-atom balance was not achieved, believed to be caused by a chemisorbed oxygen layer on the soot surface The major product was CO but a small amount of CO_2 was also produced (Cheng *et al.* 1989). Song *et al.* (1981) using the same technique, report an activation energy of 137 kJ mol^{-1} and a rate of reaction two orders of magnitude slower than the rate of reaction of O_2 with coal char. Beer *et al.* (1978) found the activation energy changed with temperature over the temperature range 673 to 1223 K, with its value being greatest at temperatures greater than 923 K. Values between 82 kJ mol^{-1} at temperatures

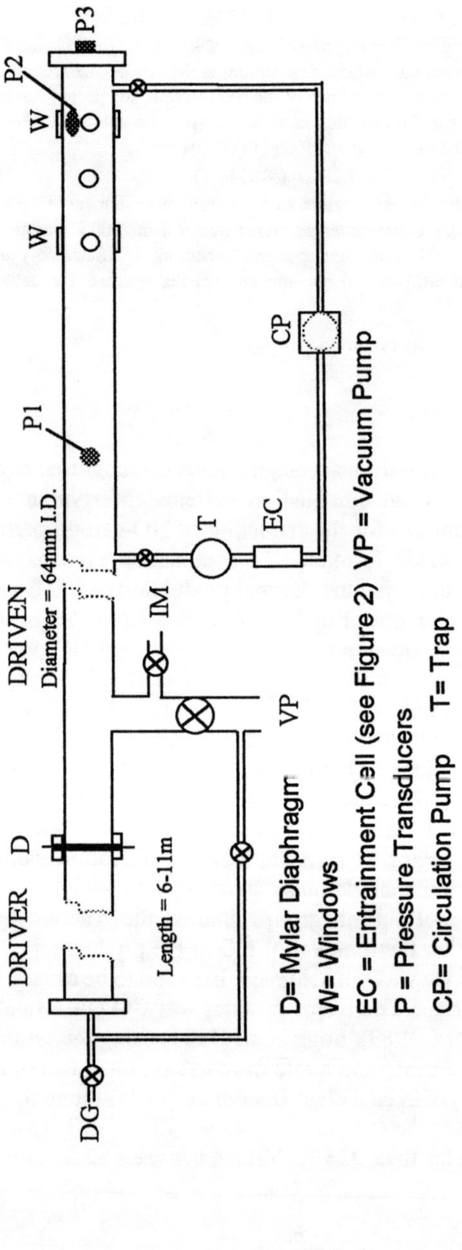

FIGURE 1 Experimental arrangement of shock tube. The three window port sites are at 10-cm intervals (only the windows 10-cm from the end of the tube were used).

below 923 K and 304 kJ mol^{-1} at higher temperatures are reported. Beer *et al.* (1981) found an activation energy of 145 kJ mol^{-1} for the NO/char reaction. Watts (1958) found an activation energy for the first order reaction of NO/charcoal of between 34–58 kJ mol^{-1} around 580–723 K. Shelef *et al.* (1969) found that various carbons reacted with NO slower than with O_2. Edwards (1972) found an activation energy of 87 kJ mol^{-1} for the reaction of NO with a carbon black. Myerson (1978) found an activation energy of 142 kJ mol^{-1} for a carbonaceous product reaction with NO.

Soot oxidation has been studied at pressures near 6 atm in the temperature range 2000–3200 K. Nitric oxide concentrations of 5, 10 and 15% in argon were used. A soot aerosol in argon was also investigated to determine the reactivity of the soot in an inert atmosphere.

EXPERIMENTAL METHOD

The experimental apparatus is shown in Figs. 1 and 2. It is essentially similar to that used earlier (Cadman and Dennings 1990). The 6.4 cm internal diameter shock tube has a driver section of 1.83 m and a driven section of 4.60 m. Four observation ports were situated 10 cm from the end plate, two used for laser absorption measurements and two for emission or scattering measurements. Three piezoelectric pressure transducers (PCB gauges) at 70 and 10 cm from the end plate and one in the end plate were used to monitor the shock pressure and determine the shock wave velocities. Both the incident and reflected velocities were used to calculate the temperature behind the reflected shock wave using the ChemKin program. The maximum available reaction time in the reflected wave was 1 msec.

Soot aerosols were prepared from lampblack soot by circulating the test gas mixture through an entrainment cell shown in Fig. 2. The cell consists of a soot reservoir with four jets in the base and bypass loop. The test gas was pumped through the soot, entraining the smaller particles. Larger aggregates of soot were trapped in a spherical reservoir (with the inlet and outlet at 45°) before the aerosol enters the shock tube. The flow rate through the entrainment cell was regulated by the valve in the bypass loop. Flow rates of 5 1/min through the cell were used. A mean particle radius of the soot entering the tube of 20 nm was determined by electron microscopy. The soot aerosol was circulated for 40 sec. through the final metre of the driven section and left to settle to remove any larger particles still in the gas, prior to the shock initiation.

Laser absorption measurements of the soot aerosols were made using an argon ion laser at 488 nm. The transmission of the laser through the soot was detected by a photodiode amplified and digitized for later analysis. An eight bit digitizer with 2 kbytes memory per channel, sampling rate of 1 MHz and 8 channels was used to record the pressure, absorption and emission records. Light emission from one window was collected using an optical fibre bundle with a field of view of 1 mm and a divergence of 2°. A beam-splitter was used to divide the collected light into two paths for two colour pyrometry measurements of the soot particle surface temperature. Filters at 441.6 and 546.6 nm and photomultiplier tubes were used to determine the ratio of emission at these wavelengths. The surface temperature could then be calculated from Planck's theory. At higher temperatures it is believed (Levy *et al.* 1981) that the NO reaction is absorption controlled, while at lower temperatures it is believed to be desorption controlled, as has been postulated for the O_2/soot reaction. The formation of a layer of chemisorbed oxygen at low temperatures on a bed of graphite was found to effectively stop the graphite/NO reaction while at higher temperatures an appreciable reaction was observed. In this paper the high temperature reaction of NO with soot has been studied. The temperature range was selected such that the reaction times would be less than 1 msec.

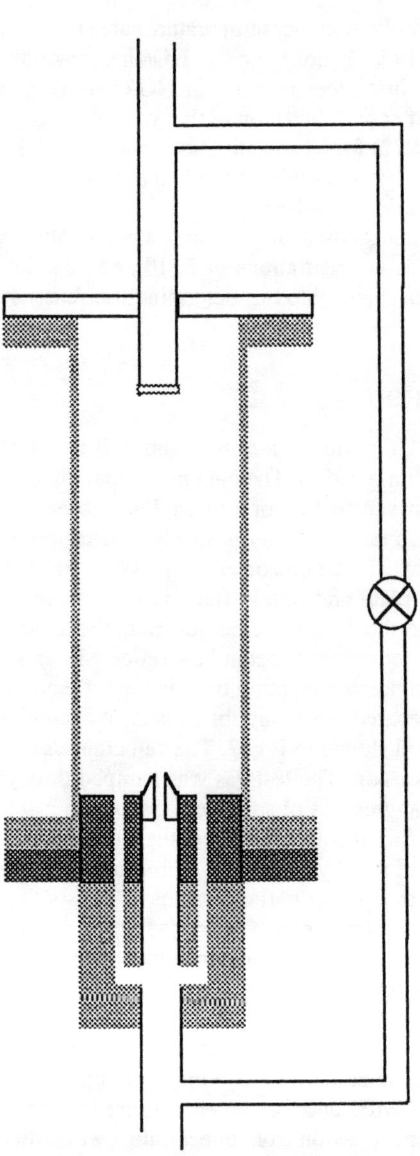

FIGURE 2 Entrainment cell. Soot is placed in cell and entrained by circulating test gas. Three jets in base of cell help to loosen aggregated soot under vacuum. A fine wire gauge over outlet trap stops large aggregates from being circulated.

RESULTS AND DISCUSSION

Typical time resolved laser absorption and mission records are shown in Fig. 3.

Immediately after the arrival of the reflected shock front the soot particles are heated to the gas temperature. Maximum heatup times of 5 μsec were measured although the typical time was less than 2 μsec. As the soot aerosol is compressed, a sudden decrease in laser

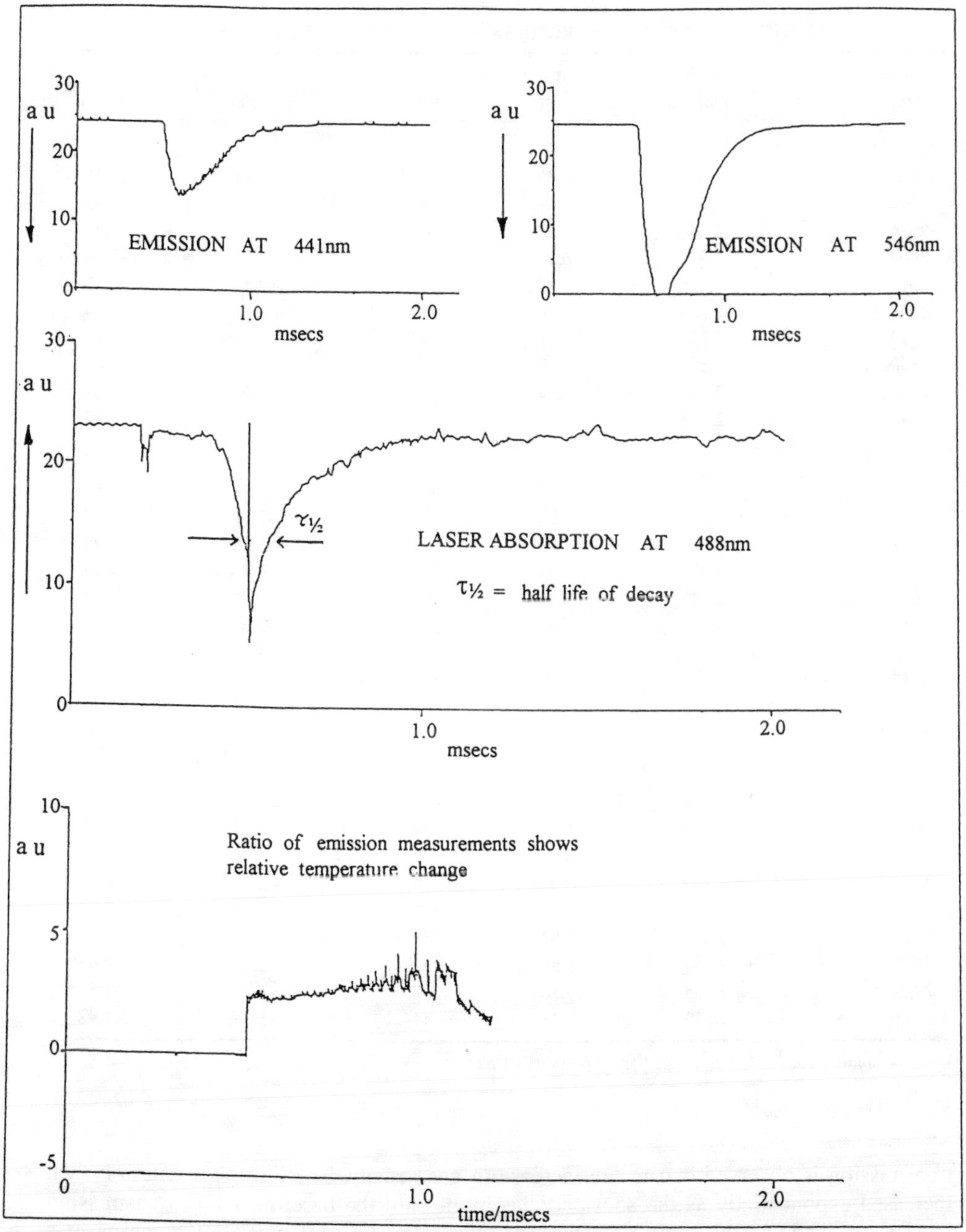

FIGURE 3 Transient signals. Typical emission and laser absorption profiles. Top records are emissions at 441.6 and 546.6 nm. Middle profile is laser absorption at 488 nm (τ is absorption half-life). Bottom profile is ratio of two emissions. Temperature = 2950 K. Pressure = 6 atm. Horizontal scale is 2.2 msec. Vertical scale is digitised volts (arbitrary).

TABLE I
TYPICAL EXPERIMENTAL RESULTS FOR OXIDATION OF SOOT BY NO

Temperature, K	$\rho \times 10^4$, mol cm^{-3}	Pressure, bar	$\tau_{1/2}$, μsec	log(ω)	log (ω/P_{NO})
5% NO					
2222	3.511	6.47	822	−2.84	−2.35
2270	3.791	6.63	630	−2.73	−2.25
2376	3.388	6.22	507	−2.63	−2.12
2462	3.010	6.19	438	−2.57	−2.06
2517	3.125	5.85	329	−2.44	−1.91
2647	2.780	5.93	219	−2.27	−1.74
2713	2.720	5.51	247	−2.26	−1.76
2830	2.562	6.02	164	−2.14	−1.62
2976	2.387	5.74	137	−2.06	−1.52
3146	2.213	5.05	123	−2.02	−1.42
10% NO					
2061	4.203	6.74	849	−2.85	−2.68
2111	3.882	6.27	520	−2.64	−2.44
2245	3.504	6.08	384	−2.51	−2.29
2347	3.214	5.63	247	−2.32	−2.07
2474	2.822	5.71	198	−2.22	−1.98
2580	2.567	5.47	137	−2.06	−1.80
2675	2.400	5.37	137	−2.06	−1.79
2825	2.227	5.18	116	−1.99	−1.70
2953	2.078	5.01	82	−1.84	−1.54
3101	2.730	6.52	41	−1.54	−1.35
15% NO					
1915	4.325	6.45	644	−2.73	−2.72
1994	4.029	6.13	343	−2.45	−2.42
2094	3.648	6.34	240	−2.31	−2.28
2111	3.690	5.93	233	−2.29	−2.24
1288	3.410	5.48	246	−2.32	−2.23
2430	3.648	5.81	151	−2.10	−2.04
2464	2.958	5.56	151	−2.10	−2.03
2758	2.303	5.38	69	−1.76	−1.67
3143	1.920	5.12	41	−1.54	−1.42

Units of ω are g cm^{-2} s^{-1}; P_{NO} is the partial pressure of NO.

transmission is observed due to the increase in gas density. The laser transmission then increased exponentially as the soot particles burnt until the baseline 100% transmission level is reached. At the same time the emission from the soot particles increases to a maximum and then decreases again as the soot particles burn away. The ratio of the emission at 441.6 and 546.6 nm shows a jump at the shock front and then remains constant until the soot particles have burnt out. This constant level indicates the surface temperature of the soot particles remains approximately constant during combustion. This is different to that found during coal particle combustion in shock waves (Lester *et al.* 1980).

Soot oxidation rates were calculated as before (Cadman *et al.* 1990) from the decay of the laser absorption curves. In calculating the oxidation rate it was assumed the particles

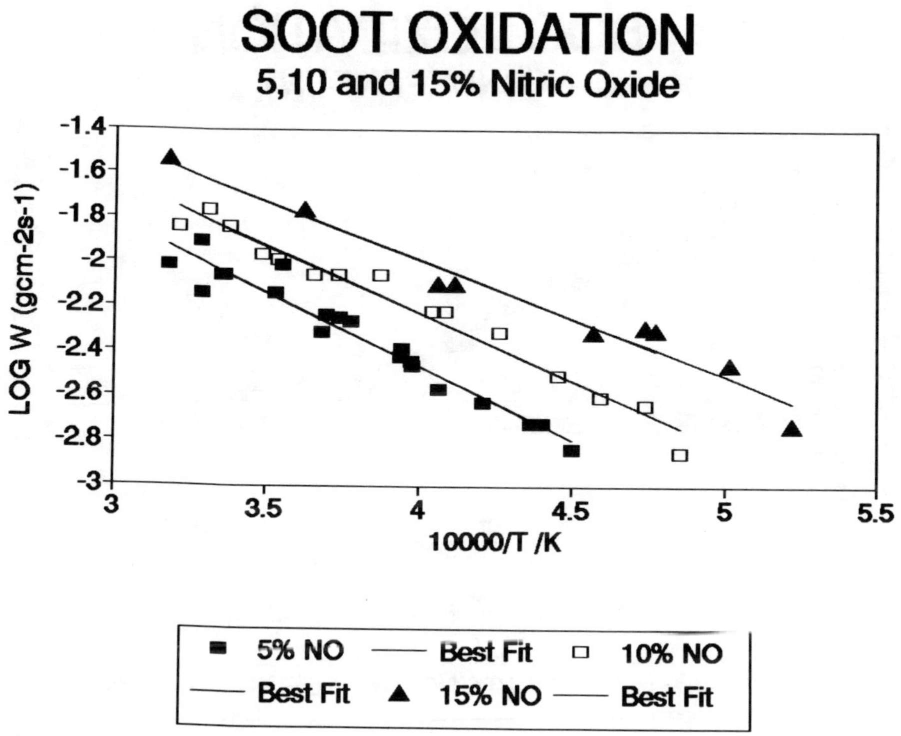

FIGURE 4 Soot oxidation: 5, 10 and 15% nitric oxide.

were spherical (based on the electron micrographs) and that any soot aggregates left were broken to single particles by the incident shock front and the particles maintained a constant density as they burn. The overall oxidation rate, ω, can then be expressed as:

$$\omega = 0.33R_m\, \rho/\tau_{1/2} \quad \text{g cm}^{-2}\ \text{s}^{-1},$$

where R_m is the mean soot particle radius, ρ is the density of the soot particle (1.8 g cm-3) and $\tau_{1/2}$ is the half life of the laser absorption curve (Park and Appleton 1972). Typical results are shown in Table I.

No induction time was observed, indicating the soot particles begin to oxidize immediately and start to decrease in size from the onset of oxidation.

Arrhenius plots for the three concentrations of nitric oxide are shown on Fig. 4. and the rate coefficients used in these plots are shown in Table II.

Increasing the nitric oxide concentration increases the rate of soot oxidation. The increase in rate was found to be first order in nitric oxide. If ω is divided by the pressure of NO this gives the nitric oxide independent results shown on Fig. 5 confirming this first order dependence.

The final rate of expression derived from this data is

$$\omega = 4.9 \times P_{NO} \times \exp(-126\ \text{kJ/RT}) \quad \text{g cm}^{-2}\ \text{atm}^{-1}\ \text{s}^{-1}.$$

This rate constant is in reasonable agreement with that found by Levy *et al.* (1981) who reported a rate of $\omega = 2.8 \times P_{NO} \exp(145\ \text{kJ/RT})$ g cm^{-2} atm^{-1} s^{-1}, although the latter has a

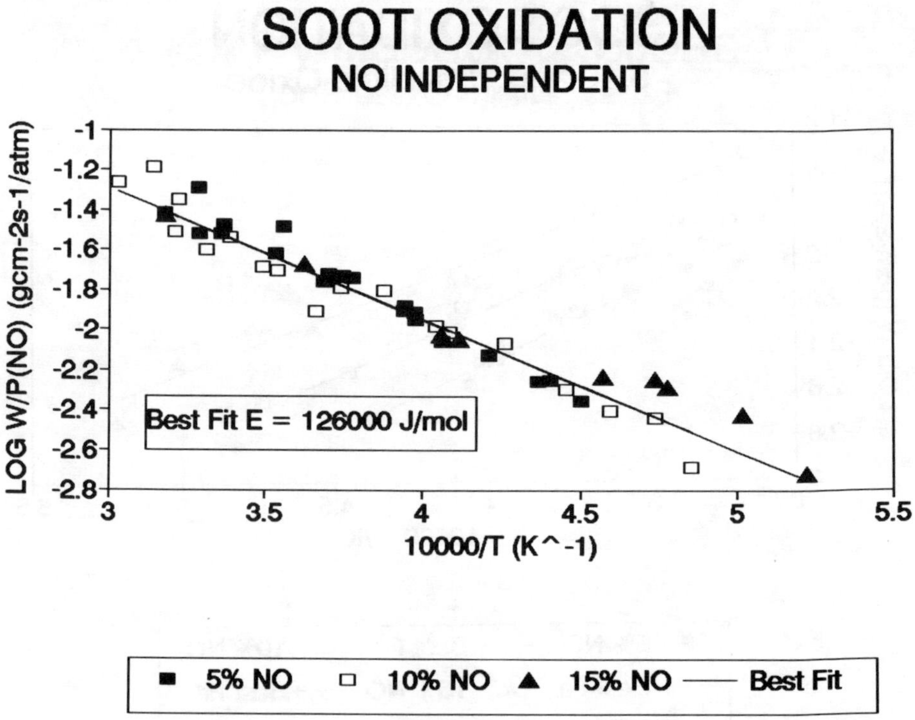

FIGURE 5 Soot oxidation: NO-independent.

TABLE II

ARRHENIUS PARAMETERS FOR OXIDATION OF SOOT BY NO

%NO	log(A)	E_A
5	1.50	126
10	1.54	115
10	1.27	100
Argon	11.50	216
5% O_2	2.09	102

Units of A are g cm^{-2} s^{-1}; units of E_a are kJ mol^{-1}.

smaller preexponential factor, but not with those of Song (1981). The rate found by Cheng *et al.* (1989), however, if extrapolated to these temperatures is also considerably slower than those determined in this work. Their results may be influenced by the formation of a chemisorbed layer of oxygen on the soot surface resulting in apparently faster rates of reaction at low temperatures. This would tend to decrease the activation energy. It should be noted our results are in reasonable agreement at temperatures near 2000 K, the maximum of their study. Roth *et al.* (1989, 1991, 1992) has also been studying the oxidation of soot aerosols with nitric oxide in a shock tube. They primarily measured the consumption

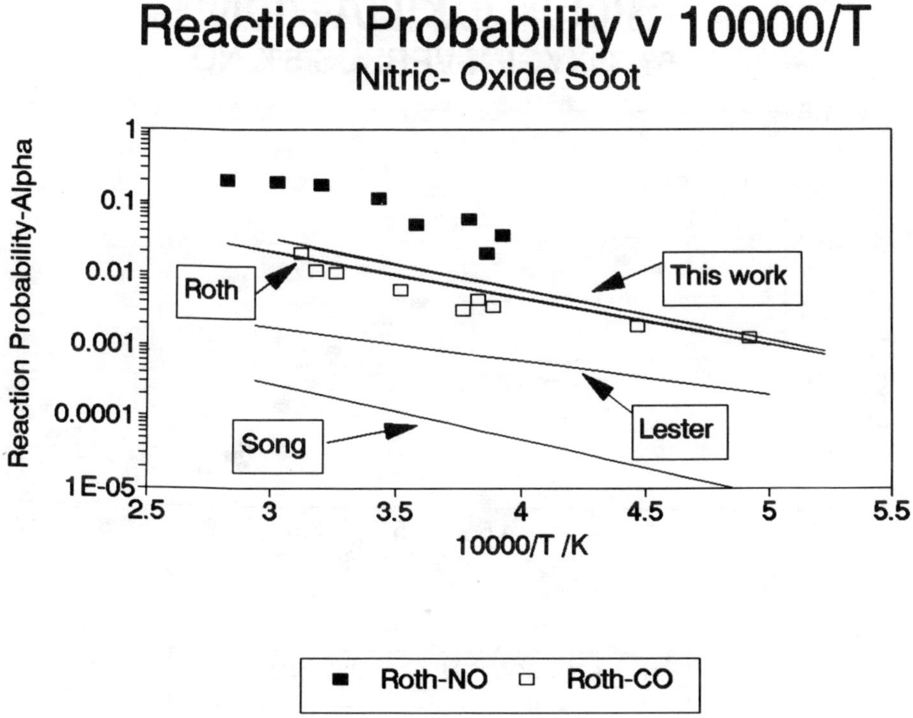

FIGURE 6 Reaction probability vs. 10000/T: nitric oxide soot.

of nitric oxide and the appearance of products using laser-diode infrared spectroscopy, although they also monitored the carbon loss by laser absorption in some experiments. Their experimental technique is described in an earlier publication. The rates of change of gaseous reactants and products are normally measured in units such as mol cm^{-3} s^{-1} (or similar ones) while it is more convenient to express the carbon loss in wt. per surface area per time. In order to compare measurements in the two different systems common units are necessary. One way of unifying these different results is to use the collisional efficiency α, for gas-surface collisions where α is given by

$$\alpha = \text{(rate of reactive collisions with surface } \omega)/(\text{rate of all collisions } z),$$

where $z = (3.51 \times 10^{22} P)/[T \cdot M]^{1/2}$ collisions/cm^2/s, M is the molar mass in g, T is the temperature in K, P is the pressure of gas in Torr. For NO,

$$z = 6.41 \times 10^{21}\, PT^{-1/2},$$

α has a maximum value of 1 for production of CO and 0.5 for CO_2 formation.

The results shown in Fig. 6 can be expressed as

$$\alpha = 4.08 \exp(135820/RT) \quad \text{or} \quad 4.08 \exp(16340/T).$$

The results of Roth (1991, 1992) at about 1 atm total pressure, are also shown for comparison. Individual results for NO and CO are shown along with the best line given as $\alpha = 1.82$ (exp 15000/T). The values of NO consumption are too high due to decomposition

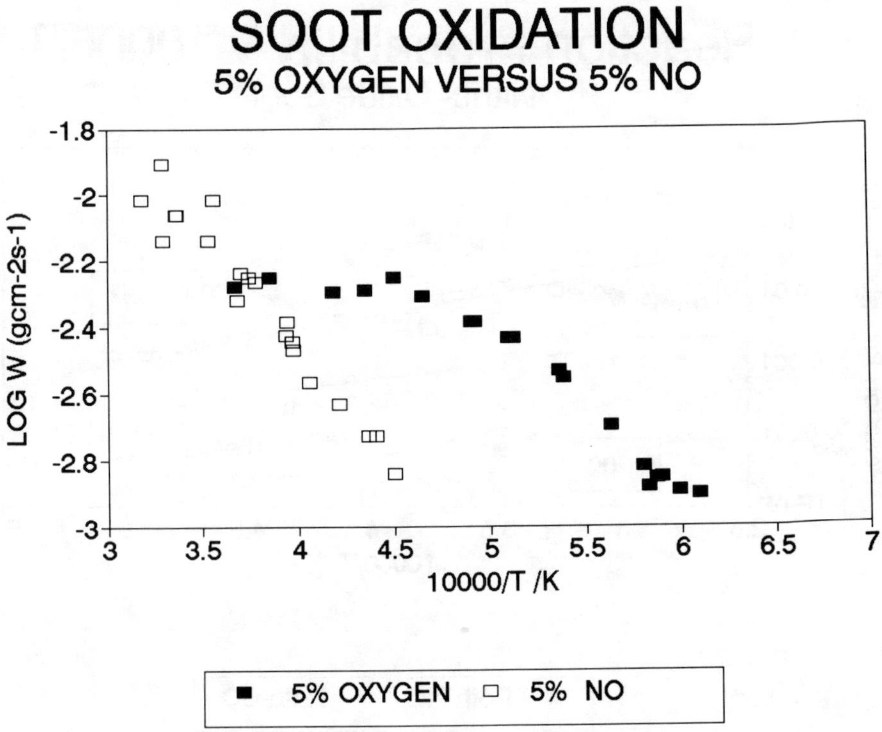

FIGURE 7 Soot oxidation: 5% oxygen vs. 5% NO.

of the reactant and are reduced by modelling this decomposition reaction to give agreement between both sets of results (Roth 1991, 1992). The values of a calculated from the results of Lester (1989) and Song (1981) are also given.

As with the values of ω, the values calculated from the work of Lester (1989) and Song (1981) do not agree with this work. The values of Roth (1991, 1992) are though in good agreement.

The observed rate of oxidation of soot by NO is considerably slower than that by O_2 at temperatures up to 2500 K as shown in Fig. 7.

Oxidation by O_2 is observed to reach a maximum rate at about 2100 K and at lower temperatures to have an order of 0.5–0.6 with respect to O_2 whereas the observed rate with nitric oxide continues to increase over the temperature range studied. The rate with NO is about a factor of 4× slower than with O_2 at 2200 K, but by 2600 K the rates are now equal as the rate with O_2 has levelled off. It has been proposed that there are two types of site for soot oxidation by molecular oxygen (Park and Appleton 1973, Nagle and Strickland-Constable 1962), possibly edge and planar sites of graphite-like structures in soot. Each type has a different oxidation rate and thermal rearrangement rate to the other kind. Nagle (1962) have developed a mechanism and empirical formula to model these processes in graphite. At low oxygen concentration and low temperatures there is good agreement between this theory and experiment. At higher temperatures or high oxygen there appears to be a large discrepancy between the observed rates and theory (Cadman and Denning 1990, Lee *et al.* 1962). Tesner (1969) found that the surface of soot became uniform with respect to energy and therefore the reaction rate does not depend upon the nature of the reactive site. A possible alternative explanation for the observed behaviour of oxygen oxidation is that at high temperatures the rate is diffusion controlled. The rate of reaction is also limited by the availability of reactive sites. Oxygen requiring two adjacent sites to form the surface complex. Two mechanisms for the adsorption of nitric oxide on a carbon surface have been proposed (Levy *et al.* 1981, Cheng *et al.* 1989). The first is a duel site mechanism (DSD) where the NO is adsorbed on the surface in a similar way to oxygen. Nitrogen is then released and the oxygen remains bound to the surface. The other is a nondissociative mechanism where the oxygen atom is bound to the surface but the nitrogen is free. This only requires one site and to produce one molecule of nitrogen requires two adjacent NO molecules. This mechanism, however, places no constraints on the adsorption of NO on the surface.The diffusion limit would therefore be expected to be greater than for the DSD mechanism. As the DSD mechanism is similar to that proposed for oxygen above, a similar diffusion controlled rate would be expected. The reaction has further constraints to the oxygen case, requiring nitrogen to be adsorbed on adjacent sites before it is desorbed as N_2. As no limit in the rate of reaction was observed, the nondissociative mechanism is favoured for the initial adsorption of NO on to the soot surface. Adsorbed oxygen is then released as most probably CO (Roth and Brandt 1991, 1992) as shown below:

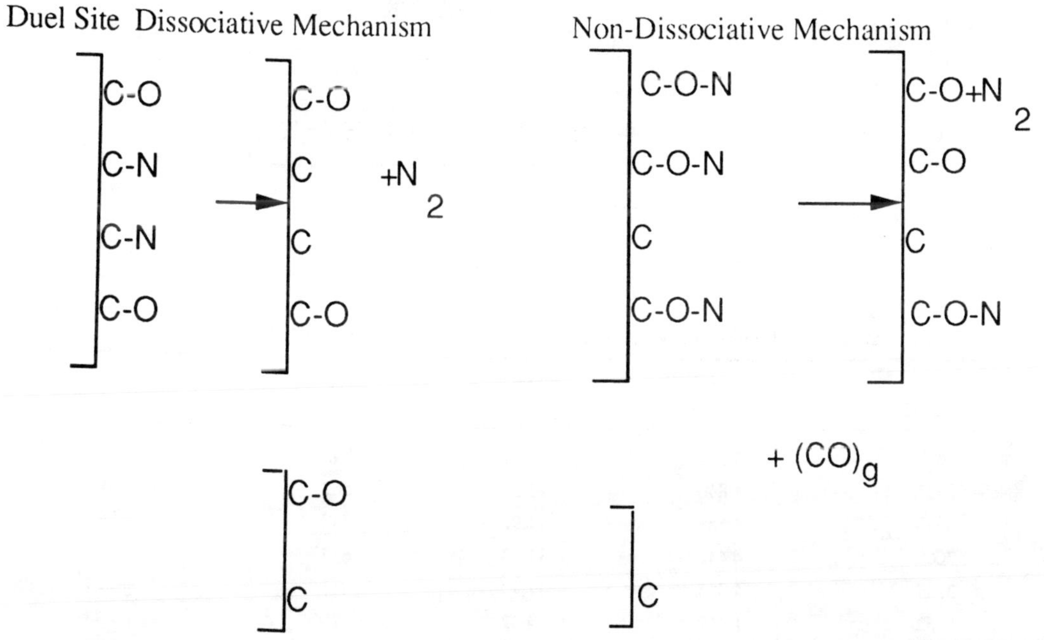

Soot absorption was observed to decrease even in the absence of oxygen at temperatures above 2799 K. A series of experiments were conducted in the presence of argon only. These experiments are shown in Table III and Fig. 8.

This decrease in soot absorption must be due to the soot particles breaking down to smaller species which have less absorption. At lower temperatures used in the oxidation study no significant reaction was found in argon showing that this reaction was not due to just adsorbed oxygen reacting but a genuine different reaction. The activation energy found is 216 kJ mol^{-1} which is considerably less than the enthalpy change for the formation of C

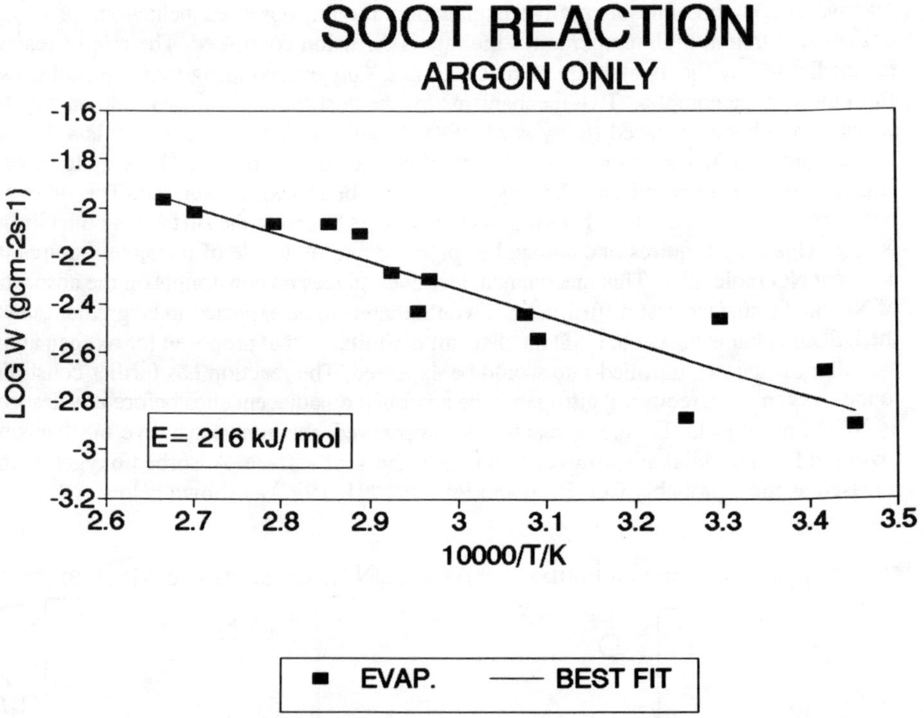

FIGURE 8 Soot reaction: argon only.

TABLE III
EXPERIMENTAL RESULTS FOR EVAPORATION OF SOOT

Temperature, K	$\rho \times 10^4$, mol cm^{-3}	Pressure, bar	$\tau_{1/2}$ μsec	log(ω)
2897	4.731	11.61	945	–2.90
2927	4.840	11.96	562	–2.67
3033	4.555	11.90	342	–2.45
3069	4.442	11.23	877	–2.87
3237	4.055	11.13	411	–2.54
3252	1.719	4.32	329	–2.44
3371	3.676	13.32	233	–2.29
3386	1.685	4.59	315	–2.42
3421	3.359	9.32	219	–2.27
3462	1.523	4.12	151	–2.10
3504	1.591	4.16	137	–2.06
3583	1.538	4.32	137	–2.06
3703	1.548	4.41	123	–2.02
3752	3.064	9.11	110	–1.97

Units of ω are g cm^{-2} s^{-1}.

atoms, C_2 or C_3 suggesting that the species formed is much bigger than these ones. Recently it has been found that fullerenes (McKinnon 1992) are produced in flames from soot at temperatures above 2500 K and the production of these cannot be ruled out. The rate was found to be independent of pressure.

CONCLUSION

The rate of oxidation of soot by NO has been determined across the temperature range 2000–3200 K at high pressures. The reaction was found to be slower than that with oxygen at 2200 K but similar at 2700 K. although both had an activation energy of 126 kJ mol. While the oxygen rate reached a maximum at around 2200 K, NO shows no such maximum. The order w.r.t. NO is 1 different to the value of 0.56 found with O_2. It is suggested that nondissociative adsorption of NO at a single site probably occurs which can then react with adjacent adsorbed NO to produce CO and N_2. This reaction can contribute to the production of a cleaner environment. Soot also reacts in some way in argon with an activation energy of 216 kJ mol^{-1}.

ACKNOWLEDGMENTS

We would like to thank the SERC for their support.

REFERENCES

Beer, J.M., Sarofim, A.F., Chan, L.K., and Sprouse, A.M. (1978). *5th Int. Conf. on Fluidised Bed Combustion*, 577, Mitre Corp.

Cadman, P., and Denning, R.J. (1990). *17th Int. Symp. on Shock Waves*, ed. Y. Kim, 751.

Cheng, M.T., Kirsch, M.J., and Lester, T.W. (1989). *Comb. and Flame*, **77**, 213.

Edwards, H.W. (1972). *Amer. Inst. Ch. E.*, **66**, no. 126, 91.

Lee, K.B., Thring, M.W., and Beer, J.M. (1962). *Comb. and Flame*, **6**, 137.

Lester, T.W., Seeker, W.R., and Merklin, J.F.,(1980). *Rev. Sci. Inst.*, **51**, 1523.

Levy, A., Chan, L.K., Sarofim, A.F., and Beer, J.M. (1981). *18th Symp. on Combustion*, 111, The Combustion Institute.

McKinnon, J.T. (1992). *Comb. and Flame*, **88**, 102.

Merryman, E.L., Miller, S.E., and Levy, A. (1979). *Comb. Sci. and Techn.*, **20**, 161.

Myerson, A.L. (1978). *Environ. Sci. and Techn.*, **12**, 213.

Nagle, J., and Strickland-Constable, R.F. (1962). *Proc. 5th Carbon Conf.*, **1**, 154.

Park, C., and Appleton, J.P. (1973). *Comb. and Flame*, **20**, 369.

Roth, P., and Brandt, O. (1989). *Comb. and Flame*, **77**, 69.

Roth, P., and von Gersum, S. (1991). Personal communication.

Roth, P., and von Gersum, S. (1992). *24th Symp. on Combustion*, Sydney

Song, Y.H., Beer, J.M., and Sarofim, A.F. (1981). *Comb. Sci. and Techn.*, **25**, 237.

Tang, S., and Churchill, S.W. (1981). *18th Symp. on Combustion*, 73, The Combustion Institute.

Tesner, P.A., and Tsybulevskii, A.M. (1969). *Comb. Explos. and Shock Waves* **3**, 163.

Watts, H. (1958). *Trans. Farad. Soc.*, **54**, 93.

Wendt, J.O.L., Pershing, D.W., Lee, J.W., and Glass, J.W. (1979). *17th Symp. on Combustion*, 77, The Combustion Institute.

On the Cool Flames of Methane

MARCEL VANPEE *Chemical Engineering Department, University of Massachusetts, Amherst, MA 01003*

Abstract—Methane is an attractive alternate fuel. It burns cleanly and is resistant to "knock." For many years methane was believed not to give rise to the phenomenon of cool flame, a process basic to the problem of "knock" in internal combustion engine. This paper shows that methane exhibits the same combustion events as those observed in higher hydrocarbons: slow combustion; production of hydrogen peroxide; multistage, low temperature ignition; single and multiple cool flames; regions of negative temperature coefficient; and high temperature ignition. The ignition diagrams are normal, but the cool-flames regions are found at higher temperatures and are much narrower than for higher hydrocarbons. This is the reason why the methane cool flames escaped experimental observation for such a long time. These findings give a rationale for the high, but not absolute, resistance of methane to "knock."

Key Words—Methane, Cool Flames, Multi-Stage Ignition, Knock, Hydrogen Peroxide, Negative Temperature Coefficient

INTRODUCTION

Due to the rising cost of petroleum and the need for a cleaner environment, there is a recurrence of interest in alternative fuels. Among these, methane is especially attractive since it burns cleanly and in addition is very resistant to "knock." In accord with this last observation, it was believed for a long time that methane could not give rise to the phenomenon of cool flame, a process which is recognized as basic in the problem of "knock" in internal combustion engines. In 1956, Vanpee reported experimental observations, for the first time, of the cool flames of methane and the determination of the pressure/temperature region where these flames occur for the methane/oxygen mixture of 67 mole percent methane. In 1965 Egret et al. confirmed these findings and discussed the mechanism of oxidation of these new cool flames. In view of current interest in alternative fuels we restate in this paper our first findings on the cool flames of methane. In addition, we present further information on these exceptional cool flames with respect to the following: (1) the effect of the reactor size on the cool flame domain in the pressure/temperature plane, (2) the determination of the pressure/concentration domain for the temperature of 515°C where the cool flame region exhibits its maximum extension, (3) the results of hydrogen peroxide sampling during the cool flame process.

In this paper, it will be shown that methane exhibits the same combustion events as those observed in its higher homologues, namely: slow combustion, multistage low temperature ignition, single and multiple cool flames, regions of negative temperature rate coefficient and high temperature ignition.

EXPERIMENTAL

Figure 1 shows the experimental apparatus. It consists of a Pyrex reaction vessel of cylindrical shape (A), a reservoir (B) to prepare the reactant mixtures, a mercury manometer (M), a mercury pump (C) and two reservoirs for oxygen and methane gas respectively. Vacuum up to 10-4 torr pressure can be realized in all parts of the apparatus with a Speedivac oil pump. The reaction is followed by recording photographically the pressure change during combustion with a glass spoon gauge.

Before each run the furnace (F) is brought to a temperature, T, and the reactor and differential manometer are evacuated. A methane-oxygen mixture of concentration, C,

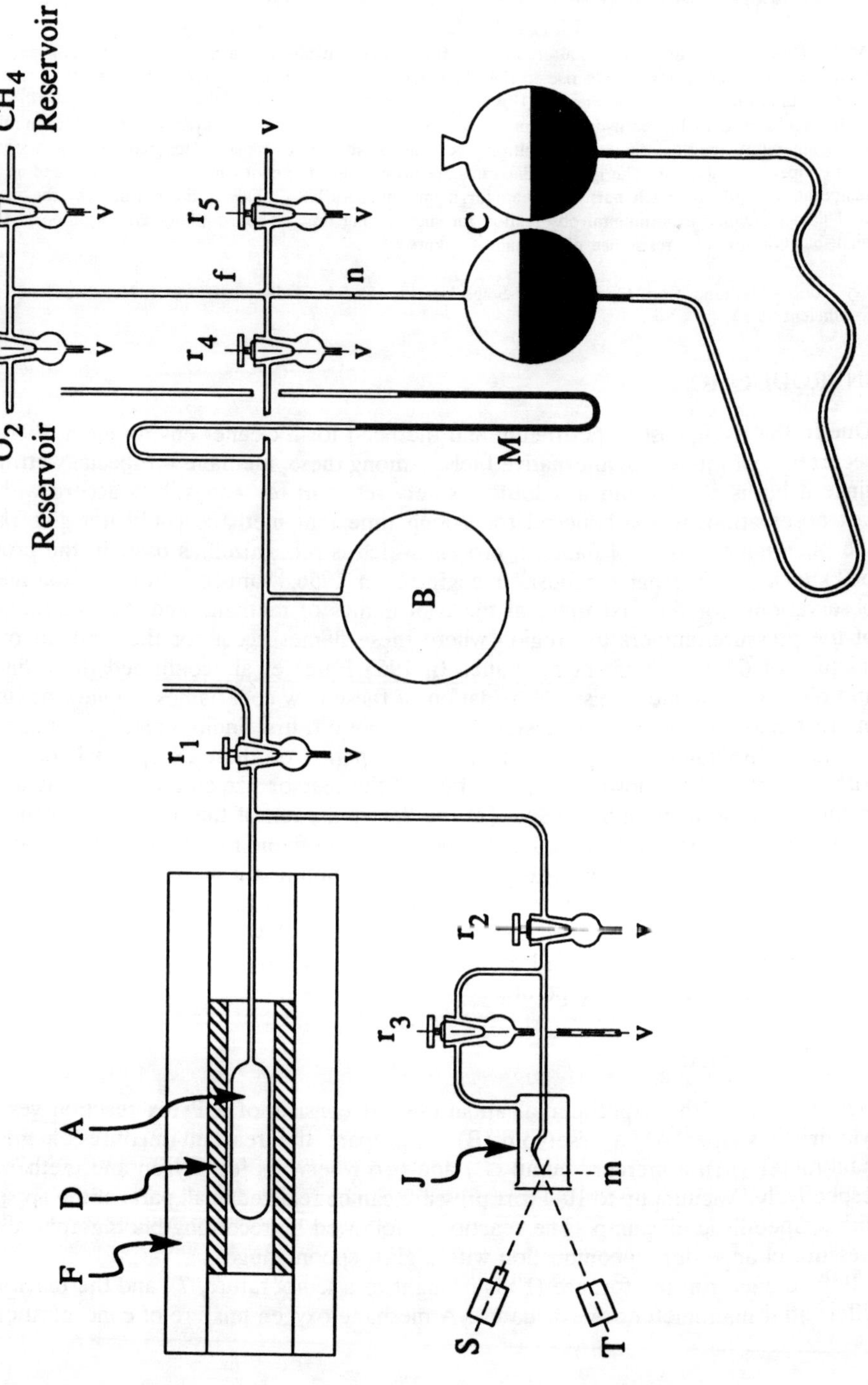

FIGURE 1 Apparatus A: pyrex reactor, D: cast-iron mantel, F: electrical furnace, J: differential manometer, S: light source, T: recording drum, B: mixer, M: mercury manometer, C: mercury pump, V: vacuum

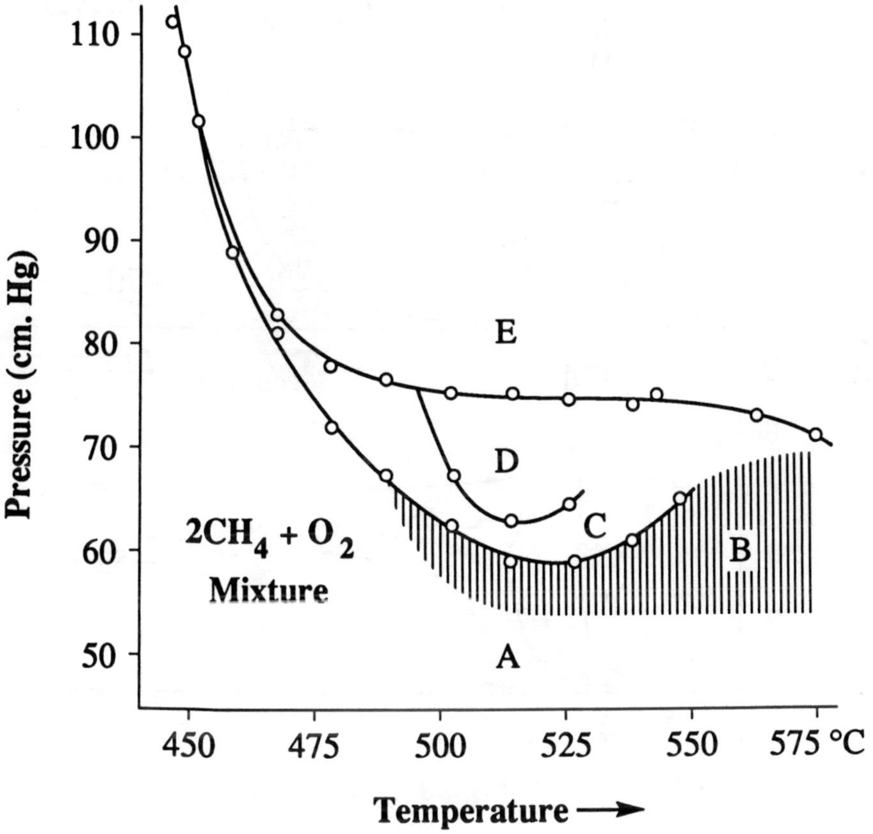

FIGURE 2 Pressure temperature combustion diagram of the $2CH_4 + O_2$ mixture. Reaction vessel pyrex, interior diameter 4.4 cm, length 15 cm. A: slow combustion, B: luminescence, C: one cool flame, D: two cool flames, E: two-stage ignition.

is prepared in (B) and the pressure is adjusted so that after expansion into the reaction vessel the pressure is P. The admission time is of the order of 1 to 2 seconds. Stopcock r1 is closed 5 seconds after the complete admission of the gases. The change of pressure is then monitored until the reaction comes to an end. During the reaction one also visually observes the light emission through one side opening in the furnace. After completion of the run, one reads the pressure, P, in the reservoir (B). This is the initial pressure of the experiment. A number of records and visual observations with different values of the parameters, T, P and C are then made to permit the construction of the combustion diagrams with their various zones of reactions.

The change of pressure is only recorded in the regions of slow combustion and of cool flames. In the zones of explosions only visual observation are made. Stopcock r2 then remains closed in order to avoid breaking the spoon gauge.

RESULTS

Three series of experiments are reported. Figures 2 and 4 give the different regions of combustion for the $2CH_4 + O_2$ mixture as a function of pressure and temperature. The

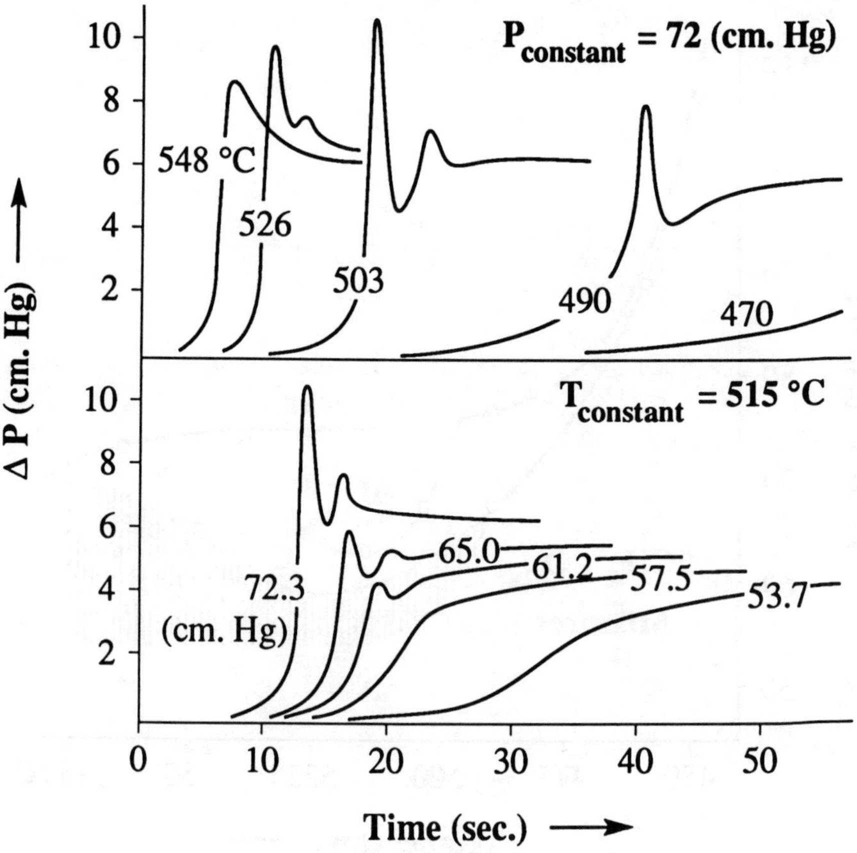

FIGURE 3 Pressure records of single and double cool flames.

first series is that from our early work and is reported here for comparison. It used a cylindrical reactor of 15 cm length and 4.4 cm internal diameter in contrast with the second series which used a reactor of 11 cm length and 2.6 cm diameter. The third series gives the pressure-concentration diagram for the temperature of 515°C where the cool flame region exhibits its maximum extension. This series was performed with a reactor of 13 cm length and 4.3 cm diameter. The pyrex reaction vessels were treated with concentrated nitric acid and then rinsed several times with distilled water.

1. Pressure/Temperature Diagrams

The two series of experiments performed are reported in the diagrams shown in Figures 2 and 4. These diagrams show the following diverse zones of combustion:

(A) a zone of slow combustion, non-luminous;
(B) a zone of luminescence, pale blue;
(C) a zone of single, cool flame, blue;
(D) a zone of double cool flames; and
(E) a zone of two-stage ignition.

The lower limits of the cool flame regions are determined on the basis of the pressure records. Figure 3 illustrates records of single and double, cool flames. The flames were

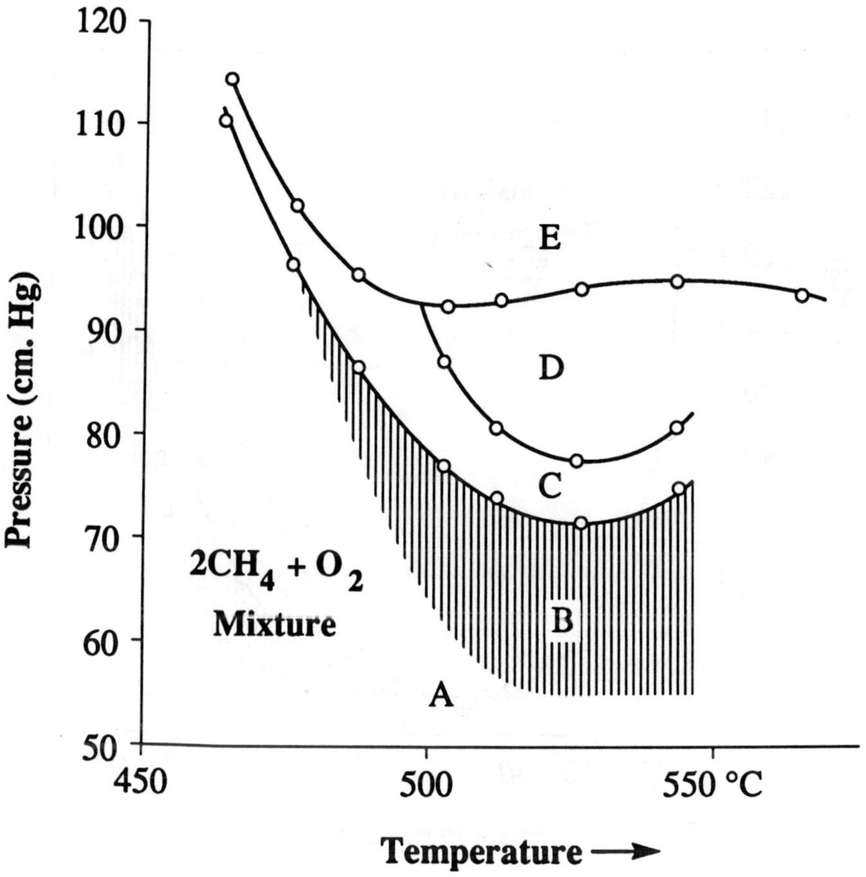

FIGURE 4 Pressure-temperature combustion diagram of the $2CH_4 + O_2$ mixture. Reaction vessel pyrex, interior diameter 2.6 cm, length 11 cm. A, B, C, D, E as in Figure 2.

sharper when the pressure was higher and the temperature lower. For instance, at the temperature of 478°C and the pressure of 750 torr (first series of experiments in the large reactor), the propagation of the cool flame was clearly visible in the reactor. At the higher temperatures (500°C) in the zone of luminescence, one only observed an increase in the luminous intensity during the pressure oscillation. At still higher temperatures (540°C) the pressure records do not distinguish the flames, because the reaction became so fast that the thermal effect of the slow combustion was superimposed on that of the cool flames.

The lower bound of the region of luminescence coincides very closely with that of the duration of the slow combustion. The two-stage ignition region has only been investigated in the vicinity of its boundary by visual observation. In that region, the explosive reaction was not too violent and the flames of second stage did not always fill the reactor. The flames were bright and yellow.

As can be seen from Figures 2 and 4, the size of the reactor did not significantly change the extension of the cool flame region. However, reducing the volume displaces the diagram toward the high pressure region which, in turn, increases the sharpness of the cool flames. Furthermore, in the small reactor, the two-stage ignition boundary

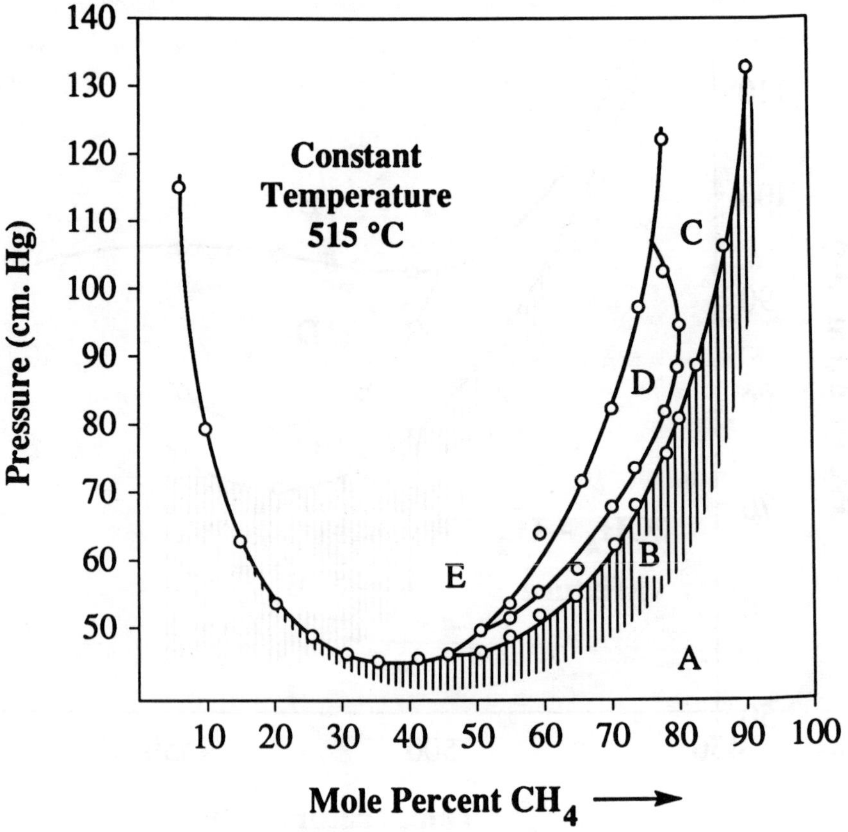

FIGURE 5 Pressure-concentration combustion diagram of CH_4/O_2 mixtures at the constant temperature of 515°C. Reaction vessel pyrex, interior diameter 4.3 cm, length 13 cm. A, B, C, D, E as in Figure 2.

presents a minimum and a maximum at 500°C and 540°C respectively, showing that the combustion has a negative reaction temperature coefficient in that region.

2. *Pressure//Concentration Diagram*

Figure 5 gives the diverse regions of combustion as a function of pressure and concentration for the constant temperature of 515°C. The diagram comprises the same zones, A, B, C, D and E as previously described. It will be seen that the cool flame zone is entirely situated in the region where the concentration of methane is above 40 percent, whereas the zone of luminescence starts at a methane concentration of 20 percent. One sees, furthermore, that the region of double flames is a closed region.

As previously mentioned, the two-stage ignition region has only been explored in the vicinity of its lower limit. The flames were always very bright yellow. The explosions had their maximum violence between the concentrations of 10 and 35 percent methane, at which point the flame propagates in the external tubes, resulting in eventual breaking of the apparatus. It should be emphasized that the pressure/concentration diagram at 515°C shows that the double, cool flame region is extremely narrow, comprising a band

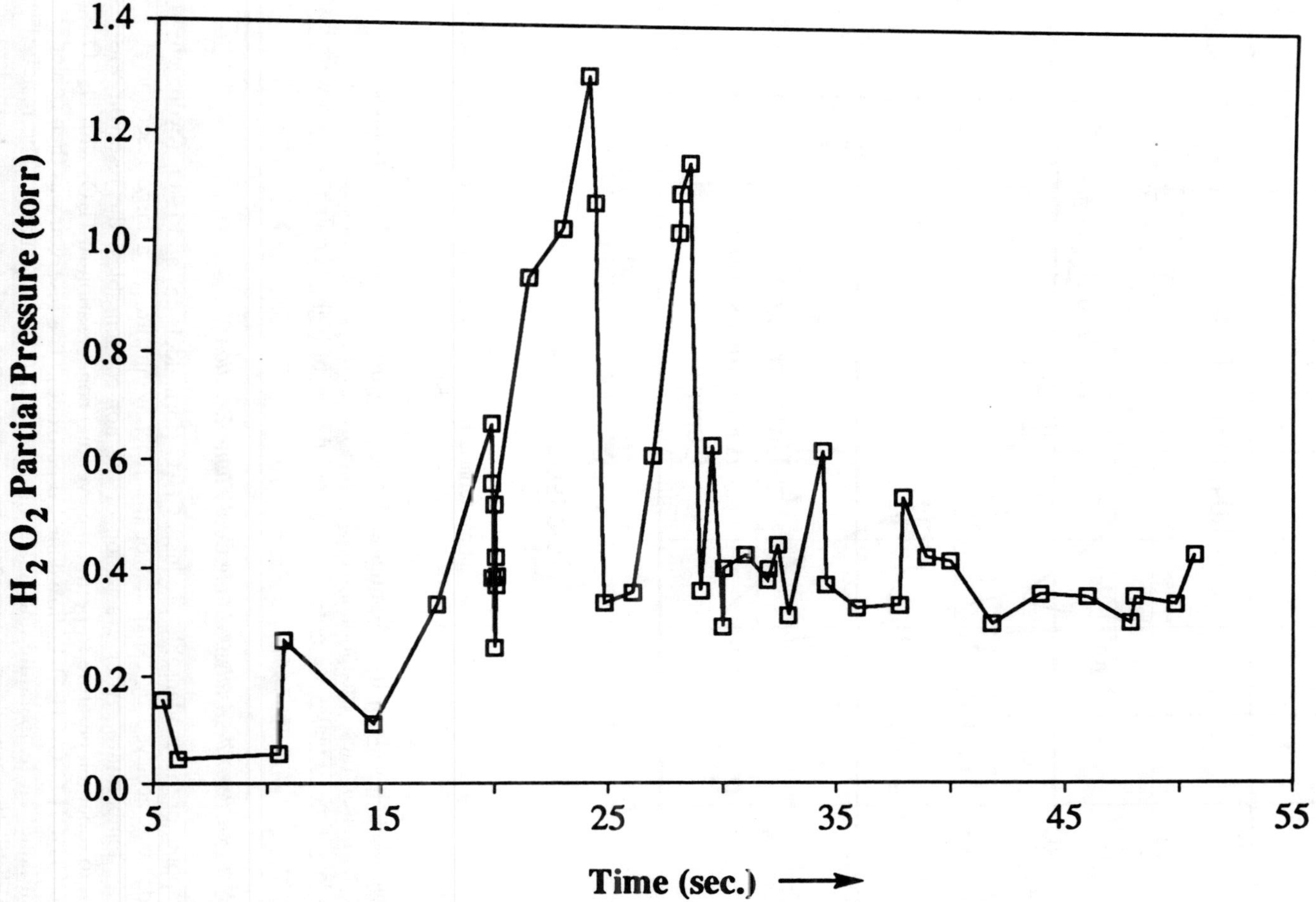

FIGURE 6 Hydrogen peroxide partial pressure during cool flame combustion in $2CH_4 + O_2$ mixture at 489°C and 723 torr pressure. Reaction vessel pyrex, interior diameter 4.5 cm., length 13 cm.

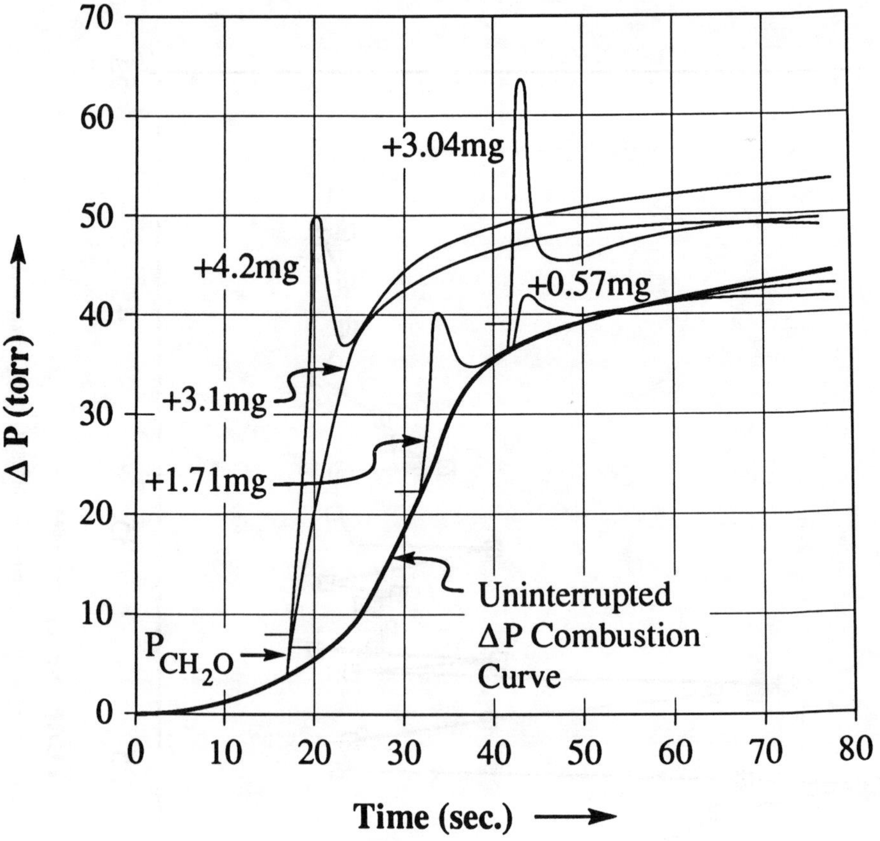

FIGURE 7 Effect of adding small amounts of paraformaldehyde at different stages of the reaction $2CH_4 + O_2$ at 500 torr and 498°C (Spherical quartz reaction vessel 1,340ml).

0.1 mole fraction wide at its maximum and extending from 500 to 1100 torr pressure. This is the reason these exceptionally cool flames escaped experimental observation for such a long time. Finally, we note that no cool flames of higher order than two were observed.

3. *Hydrogen Peroxide Analysis during Cool Flame Reaction*

The trapping of the condensable products CH_2O (formaldehyde) and H_2O_2 (hydrogen peroxide) was achieved by expansion of the reacting mixture in a large reservoir at liquid air temperature. After evacuation of the non-condensables, the reservoir was brought to room temperature to permit the eventual condensation of H_2O_2 with H_2CO to form $CH_2OH\text{-}O\text{-}O\text{-}CH_2OH$ (dioxydimethyl peroxide). The reservoir was then cooled again at liquid air temperature and finally brought to 0°C and evacuated. The residue of this distillation was then decomposed by passage over a platinum wire heated to about 1000°C. The products of decomposition H_2, CO, CO_2, and H_2O were separated by low temperature distillation and determined by direct P, V, T measurements. The results of 47 samplings at time intervals elapsing over the first 50 seconds in the double cool flame region are reported in Table I. The total amount of condensable material decomposed over the 47 runs gives: 196.00E-6 mole H_2, 150.65E-6 mole CO, 64.62E-6 mole CO_2 and 93.86E-6 mole H_2O, representing 9.142 mgr of material. This material contains the elements H, O and C in the proportions 0.495, 0.320 and 0.184, respectively. This compares favorably with the theoretical values of 0.5, 0.333 and 0.166 for the dioxydimethyl peroxide. If it is assumed that this compound is formed by condensation of two molecules of formaldehyde with one molecule of hydrogen peroxide, one can estimate the partial pressure of H_2O_2 present during the cool flame reaction. These partial pressures are reported in the last column of Table I. The analytical results of hydrogen peroxide sampling are illustrated graphically in Figure 6. They show that this compound builds up progressively during the phase preceding the cool flame formation and then decays abruptly with the passage of the cool flame. On the diagram, one notes two definite pressure peaks, which very likely correspond to the passage to the cool flames.

CONCLUSION

The combustion diagrams of the methane-oxygen mixtures are normal and are with respect to their shape, similar to those of higher hydrocarbons. Taking as an example the classical diagram obtained by Newitt and Thornes (1937) for the combustion of a 1/1 propane/oxygen mixture, one observes the same zones of combustion as with the methane. However, the zones are more extended and are at lower temperatures and pressures. In addition, the two-stage ignition peninsula is more pronounced and extends deeper into the cool flame region than in the case of methane. Qualitatively, there is no essential difference between the ignition diagrams of methane and those of its higher homologous. It must be noted, however, that the temperature for which the cool flames present their maximum extension is notably higher for methane than for other hydrocarbons. This temperature is 340°C for n-hexane (Freling, 1955), 350°C for n-pentane (Lucquin et al., 1955), 315°C for n-butane (Bardwell, 1955), 310°C for propane, (Newitt and Thornes, 1937) 320°C for ethane (Knox and Norrish, 1954) and 515°C for methane.

The notable difference of temperature for the first member of the hydrocarbon series and the lack of any noticeable trend in temperatures for the higher members indicate that the mechanism of cool methane flame formation differs from that of its higher homologues.

Two compounds play an important, if not essential, role in the mechanism of formation of the methane cool flames: formaldehyde and hydrogen peroxide.

Vanpee and Grard, (1955) have shown that one could produce artificially the phenomenon of methane cool flame by adding small quantities of formaldehyde during the course of the slow reaction. This feature is illustrated in the series of experiments shown in Figure 7, where formaldehyde was injected into the reaction vessel by the introduction and subsequent bursting of a tiny thin walled pyrex bulb containing paraformaldehyde. On the other hand, formaldehyde itself can give rise to chemiluminescent phenomenon (Vanpee, 1956) when reacting in rich mixtures close to the explosion limits. It is noteworthy that, under these conditions, hydrogen peroxide could be detected in an intermediate of reaction. This fact is illustrated in Figure 8 which represents the combustion of a rich mixture of formaldehyde and oxygen at two different pressures, one above the explosion limit and the other below it, in the zone on chemiluminescence. The progress of the reaction is followed by its thermal effect and the hydrogen peroxide is analyzed according to the method described above. It will be seen that hydrogen peroxide builds up progressively. In the zone of luminescence it decays slowly after all the oxygen has been consumed. In the explosion region it disappears catastrophically (after this event).

TABLE I
Hydrogen Peroxide Analysis During Methane Cool Flame Combustion in $2CH_4 + O_2$ mixture at 489 °C.

TIME sec	H_2 MOL*10^6	CO MOL*10^6	CO_2 MOL*10^6	H_2O MOL*10^6	$C_2O_4H_8$ MOL*10^6	H_2O_2 Torr
5.40	0.41	0.37	0.51	1.71	0.69	0.15
6.00	0.48	0.35	0.10	0.23	0.20	0.05
10.40	0.84	0.56	0.09	0.08	0.24	0.05
10.60	0.89	0.66	0.60	3.50	1.17	0.26
14.60	0.95	0.71	0.37	0.33	0.47	0.10
17.40	2.31	1.91	1.19	1.22	1.41	0.32
19.80	4.79	4.12	1.98	2.91	2.81	0.63
19.80	4.15	3.46	1.74	2.34	2.38	0.53
19.80	2.94	2.19	1.47	1.21	1.64	0.37
19.90	2.86	1.94	0.84	0.27	1.08	0.24
20.00	3.39	2.77	1.31	1.55	1.81	0.41
20.00	3.06	2.08	1.47	1.10	1.58	0.36
20.00	3.08	2.12	1.56	1.17	1.65	0.37
20.00	4.43	3.18	1.52	2.35	2.20	0.49
21.50	6.76	5.49	2.91	4.39	3.98	0.89
22.90	6.65	5.38	3.38	5.40	4.36	0.98
23.90	8.88	6.23	4.86	6.58	5.58	1.25
24.30	8.31	5.02	4.30	4.71	4.59	1.03
24.80	3.12	2.46	0.64	1.71	1.42	0.32
26.00	4.43	4.46	0.04	0.22	1.48	0.33
26.90	6.16	4.04	2.00	1.61	2.58	0.58
28.00	7.58	5.48	3.94	5.31	4.66	1.05
28.00	5.39	3.79	4.07	6.58	4.41	0.99
28.40	11.73	8.33	3.52	2.55	4.87	1.09
29.00	3.59	2.75	0.75	1.68	1.57	0.35
29.00	3.04	2.29	0.84	1.94	1.51	0.34
29.50	4.67	2.57	2.46	3.59	2.70	0.61
30.00	4.64	3.66	0.64	1.06	1.69	0.38
30.00	3.36	3.10	0.34	0.43	1.24	0.28
31.00	4.94	3.53	0.93	1.11	1.80	0.41
31.90	4.66	3.56	0.67	0.71	1.61	0.36
32.00	2.98	2.20	1.20	2.37	1.74	0.39
32.40	2.79	2.08	1.40	2.86	1.88	0.42
32.80	3.62	3.64	0.17	0.30	1.30	0.29
34.40	6.36	5.10	1.53	1.31	2.62	0.59
34.50	2.87	1.91	1.18	2.09	1.58	0.36
35.80	3.93	4.18	0.00	0.22	1.37	0.31
37.80	4.12	4.21	0.01	0.19	1.38	0.31
37.90	6.23	5.04	0.94	0.82	2.23	0.50
38.80	4.66	3.40	1.02	1.05	1.79	0.40
40.00	4.81	3.14	1.12	1.03	1.76	0.39
41.80	3.57	3.80	0.01	0.15	1.24	0.28
44.00	3.08	2.11	1.03	1.87	1.54	0.34
46.00	2.91	2.17	0.94	1.86	1.50	0.34
47.80	3.75	3.82	0.00	0.16	1.25	0.28
47.90	3.18	2.19	0.97	1.48	1.46	0.33
50.00	2.66	1.97	0.94	1.62	1.39	0.31

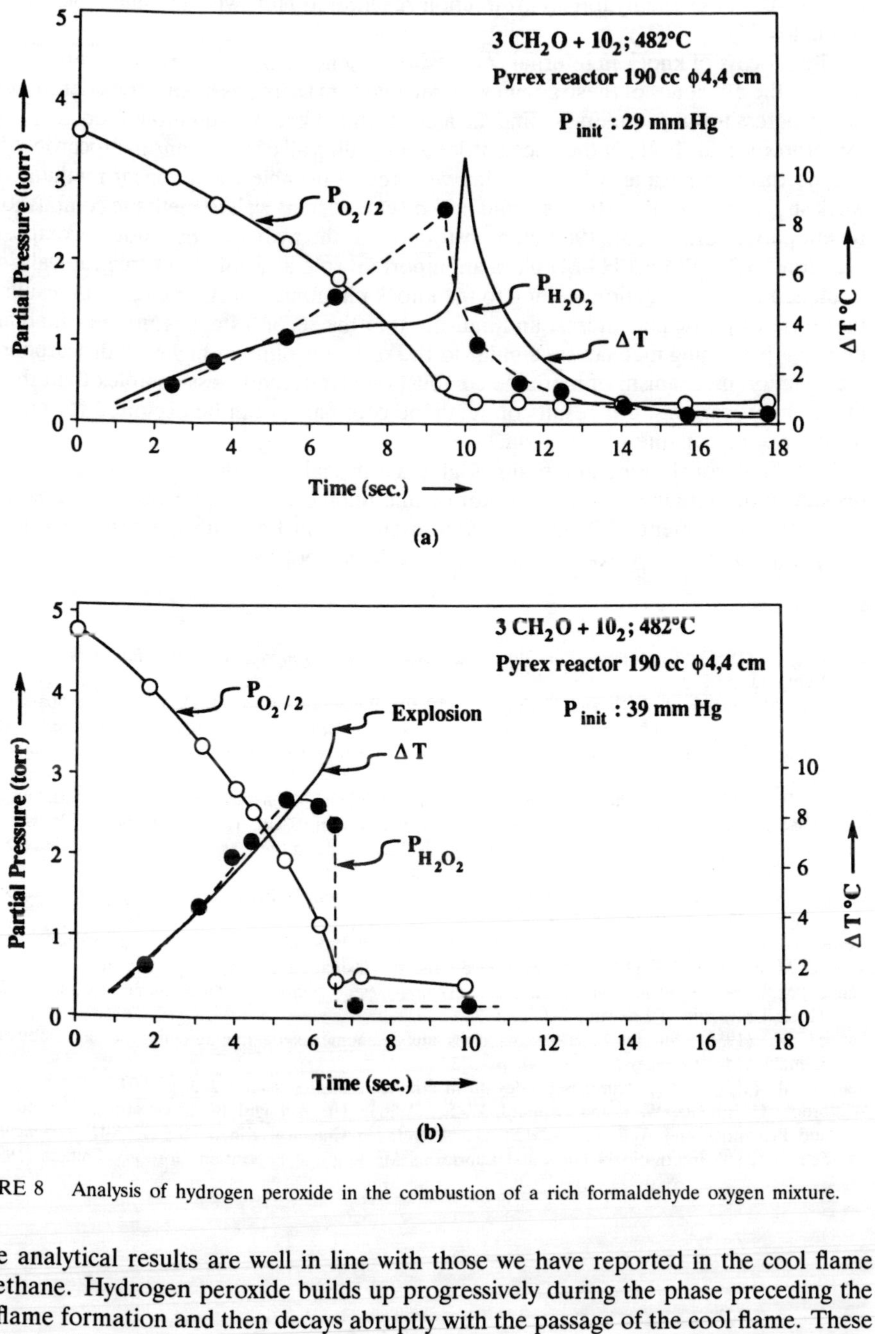

FIGURE 8 Analysis of hydrogen peroxide in the combustion of a rich formaldehyde oxygen mixture.

These analytical results are well in line with those we have reported in the cool flame of methane. Hydrogen peroxide builds up progressively during the phase preceding the cool flame formation and then decays abruptly with the passage of the cool flame. These experimental observations tend well to indicate that in the combustion of methane, formaldehyde is a precursor to the formation of hydrogen peroxide. This would be the reason why this latter compound, that is essential for the cool flame mechanism, builds up at higher temperatures (500°C) than in the combustion of higher hydrocarbons, where alternative paths for the formation of H_2O_2 are available at lower temperatures

(310°-340°) (e.g. the hydrogen abstraction reaction in HO_2 with the fuel molecule HO_2 + Fuel = H_2O_2 + R.)

The process of knock in internal combustion engines is of great practical importance as it limits the efficiency of these engines. In recent years there has been a considerable effort and progress towards understanding the mechanism of knock. Numerous modeling studies (Westbrook et al. 1991) of the knock process, including all of the alcane hydrocarbons from ethane through hexane, where studied with the noticeable exception of methane. Our work suggests that these studies could now be extended as well to methane combustion. A recent paper (Chun et al. 1989) demonstrated that the hydrogen peroxide decomposition reaction (H_2O_2+M=2OH+M) plays an important role as a source of free radicals in the branched thermal reaction leading to the knock phenomenon. Our analytical results are, therefore of particular interest and indicate that the second stage ignition of methane is due to a branching mechanism similar to the one prevailing in higher hydrocarbons. As the reaction mechanism of methane combustion is inherently less complex than those of higher hydrocarbons, our results on methane cool flames can be exploited for modeling these important flame phenomena.

As a final conclusion, our findings give a rationale for the high, but not absolute, resistance of methane to knock (motor octane number: 105). Our results indeed are in qualitative agreement with this observation and fall in line with the restricted domain, compared to higher hydrocarbons, where methane cool flames are observed.

REFERENCES

Bardwell, J. (1955). Cool flames in Butane Oxidation. *Fifth Symposium (International) on Combustion,* Reinhold, p. 529.

Chun, K. M., Heywood, J. B. and Keck, J. C. (1989). Prediction of Knock Occurrence in Spark-Ignition Engine *Twenty-Second Symposium (International) on Combustion,* The Combustion Institute, p. 455.

Egret, J., Sochet, L. R. and Lucquin, M., (1965). La combustion menagee du methane a basse temperature. *Bull. Soc. Chim.,* p. 2205.

Freling, E. (1955). Recherches sur l'oxydation et l'auto-inflammation des hydrocarbures paraffiniques en phase gazeuse; Etude particuliere des melanges hexane normal-air. Thesis University of Paris.

Knox, J. H. and Norrish, R. J. W. (1954). Cool Flame Phenomena in the Oxidation of Ethane. *Trans. Faraday Soc.,* **50,** p. 928.

Lucquin, M., Ben-Ain, R. and Laffitte, P. (1955). Sur les flammes froide des melanges de pentane et d' óxygene. *Comptes Rendus* **241,** p. 949.

Newitt, D. M. and Thornes, L. S. (1937). The Oxidation of Propane. *J. Chem. Soc.,* p. 1656.

Vanpee, M. and Grard, F (1955). Formaldehyde and the Oxidation of Methane. *Fuel,* **36,** p. 433.

Vanpee, M. (1956). Contribution a l'etude des mechanismes de combustion du methane et du formaldehyde. Thesis University of Louvain, Belgium.

Vanpee, M., (1956). Sur les reactions associees aux phenomenes de luminescence de la combustion du formaldehyde. *Comptes Rendus,* **242,** p. 373.

Vanpee, M., (1956) Sur les flammes froides du m éthane. *Comptes Rendus* **243,** p. 804.

Westbrook, C. K., Pitz, W. J. and Leppard, W. R., (1991). The Autoignition Chemistry of Paraffinic Fuels and Pro-knock and Anti-knock Additives: A Detailed Chemical Kinetic Study. *SAE Technical Paper Series 912314,* International Fuels and Lubricants Meeting and Exposition, Toronto, Canada (1991).

Experimental and Theoretical Studies on Soot Formation in an Ethylene Jet Flame

A. COPPALLE and D. JOYEUX *U.R.A. C.N.R.S. 230 C.O.R.I.A. Faculté des Sciences de Rouen 76134 Mont Saint Aignan Cédex, FRANCE*

Abstract—Soot formation was studied in any ethylene turbulent diffusion flame. A small probe was built to carried out pyrometric measurements in the flame. Instantaneous and local values of temperature and soot volume fraction were determined from emission-absorption measurements, and that made it possible to follow the fluctuations of temperature and soot volume fraction at the same time. Numerical simulation of the jet flame was performed and comparison with the experimental results is shown.

1. INTRODUCTION

Soot formation is difficult to predict in many combustion systems. But it is an important phenomenon which has to be better understood, especially in turbulent flows. Very often such particle emissions are responsible for damage and are to be eliminated.

A lot of experimental and theoretical work on laminar flames, carried out a few years ago, has given the fundamental mechanisms of formation and oxidation of soot. But the important point is that these chemical processes are not fast compared to gas reactions. So in turbulent combustion, there is a competition between these slow reactions and turbulent mixing. Up to now, the theoretical means available to calculate soot formation are those which were developed for gas combustion. For example, the flamelet concept (Peters, 1986) has been used by Moss (1990) giving interesting results, but with some new assumptions and more difficulties than in the gas combustion case. The problem is that when soot formation occurs radiation increases and cannot be neglected; the modelling with mixture fraction concept thus becomes more difficult.

In order to deal with this subject, experimental works were carried out in a turbulent diffusion flame, which was produced by an ethylene jet in still air. A small probe was built to carry out pyrometric measurements in the flame. Instantaneous and local values of temperatures T and soot volume fractions fv were determined from emission-absorption measurements, and that made it possible to follow the fluctuations of T and fv at the same time.

The mean and the r.m.s. values measured across the jet are presented. The temporal correlation coefficient was calculated from temperature values, enabling us to determine the integral scales in the jet. The probability distribution functions are also presented for the temperatures and soot volume fractions. Because these two parameters were determined simultaneously, correlation coefficients were calculated and are shown.

As was noted above, the soot formation and destruction models have to be improved. A numerical simulation of the jet flame was performed using the Genmix procedure (Patankar *et al.*, 1970). The fast chemistry model and pdf beta function were used for gas combustion and radiation was taken into account in the energy equation. For soot formation a single step process was chosen as it had already been proposed in other work on acetylene flames (Abbas *et al.*, 1985). The soot oxidation was calculated using two different processes and comparison with the experimental results are shown.

2. OPTICAL PROBE AND EXPERIMENTAL METHOD

The diffusion flame was produced by jet of pure ethylene in still air. The diameter of the ethylene exit was 9 mm, and the gas speed was 6.3 m/s. In order to carry out local optical

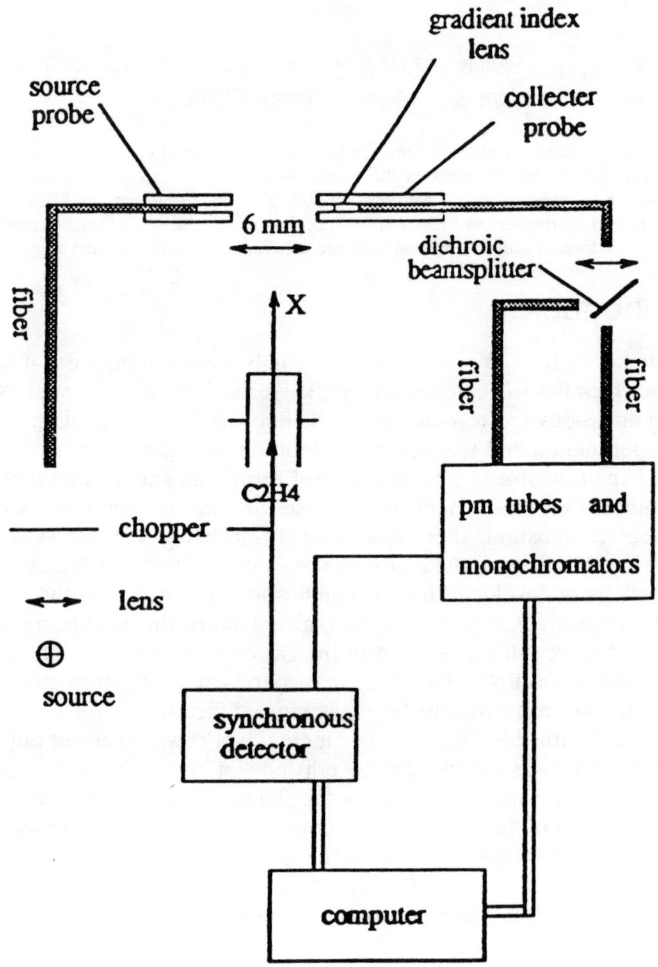

FIGURE 1: Experimental optical system

measurements in this turbulent flame, a small probe (6 mm external diameter) was built, and is shown in figure 1. Water cooling was used and no sooting up was detected during the first fifteen minutes of each experiment.

The pyrometric device allowed the temperature and the soot volume fraction within the flame to be measured locally and instantaneously. The chopper gave the two sequences which are necessary: during the emission sequence the source was blocked by the chopper and then the transmitted signal of the source was measured. A 2 KHz sampling rate was found to be necessary. For this frequency the acquisition time was faster than the fluctuations observed in the flame. In order to avoid any doubt, the same measurements were carried out with a higher sampling rate of 10 KHz. These gave us the confirmation that 2 KHz is sufficient.

With the fibre in the reception probe, the optical signal was routed onto a plate beamsplitter and thereby divided into two parts to be detected by the photomultiplyers fitted with interference filters ($\lambda 1 = 700$ nm; $\lambda 2 = 900$ nm). The signal was then shared and processed by a computerized measuring chain.

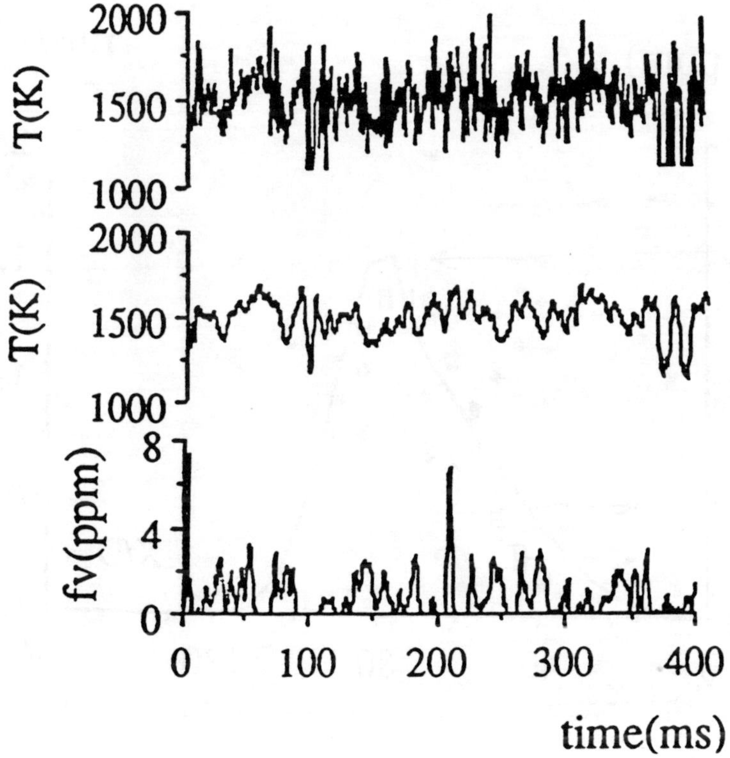

FIGURE 2: Temperature with and without filter and soot volume fraction, signal at 2kHz sample

To determine the flame temperature accurately the so-called "2 colour pyrometry" method was used. From Wien's law, we have

$$\frac{S_1^\lambda}{S_2^\lambda} = \frac{C_1^\lambda}{C_2^\lambda}\frac{\varepsilon_1^\lambda}{\varepsilon_2^\lambda}\left(\frac{\lambda_1}{\lambda_2}\right)^5 \exp\left[\frac{C_2}{T}\left(\frac{1}{\lambda_1}-\frac{1}{\lambda_2}\right)\right] \tag{1}$$

in which $S\lambda$ is the measured signal at the wavelength λ, $C\lambda 1/C\lambda 2$ a constant given by calibration with a black body source, and C_2 is the Planck's second constant. In our case, the space between the probes was small and the emissivity of this flame portion was in the optically thin limit. So we didn't need the soot volume fraction to calculate the emissivity ratio $\varepsilon\lambda_1/\varepsilon\lambda_2$, but an inverse power law for the spectal absorption coefficient $K_{o\lambda}$ of soot was assumed (Lee *et al.,* 1981) with a complex refractive index value of $n*$ = 1.57-0.56 i (Dazell *et al.,* 1969).

The soot volume fraction was provided by the absorption measurements given at λ_1 = 700 nm. For this wavelength $Ko\lambda_1$ is greater than at 900 nm, and is related to fv by

$$f_v = -\frac{1}{k_{0\lambda}L} 1n\,(1-\varepsilon_{\lambda\;1}) \tag{2}$$

where L is the optical spatial resolution of the device. In fact the geometrical width between the two probes was 6 mm, but photographs of the probes taken in the flame have shown that the optical width was smaller and it was assumed to be 5 mm.

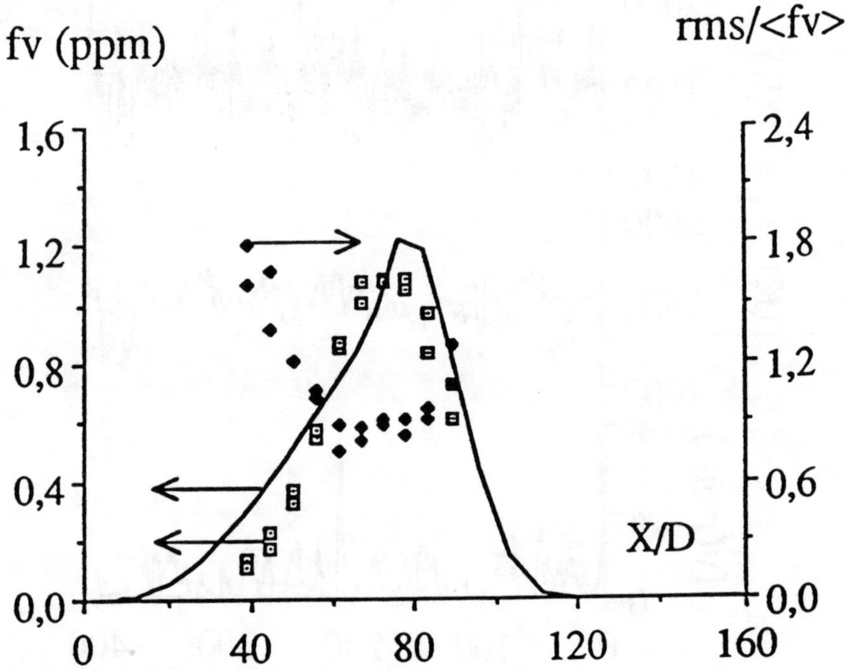

FIGURE 3: Rms and mean values of soot volume fraction on the axis experimental point —calculated value

Our optical device provided us with the means to calculate and to follow the fluctuations of ninepoint T and fv in the flame at the same time. Due to some noise both in the signals and also provided by the acquisition process, it was necessary to filter the temperature and the soot concentration results using the method of Stavitzky *et al.* (1964). The effect of the procedure is shown in figure 2 which shows the result of an experiment. We can see that the numerical filter was correctly chosen in order to avoid a loss of information in the experimental values. It must be pointed out that the characteristic fluctuating time is short, a few milliseconds.

3. EXPERIMENTAL RESULTS AND DISCUSSION

To carry out pyrometric measurements, sufficient optical signals are necessary, particularly for temperature determination. Only measurements around the flame axis were carried out, and near the edge of the flame no measurements were possible due to the intermittency of the jet.

a) *Mean values*

Figure 3 shows the soot volume fraction and the root mean squares of the fluctuations obtained along the axis. The decrease in the soot concentration, $(x/D > 70)$ corresponds to the zone where the oxidation process is great. In the case of ethylene fuel, the soot burnout is sufficiently strong to avoid smoke emissions. The fluctuation intensities of the soot concentration are high (100%) compared to the usual values which are obtained

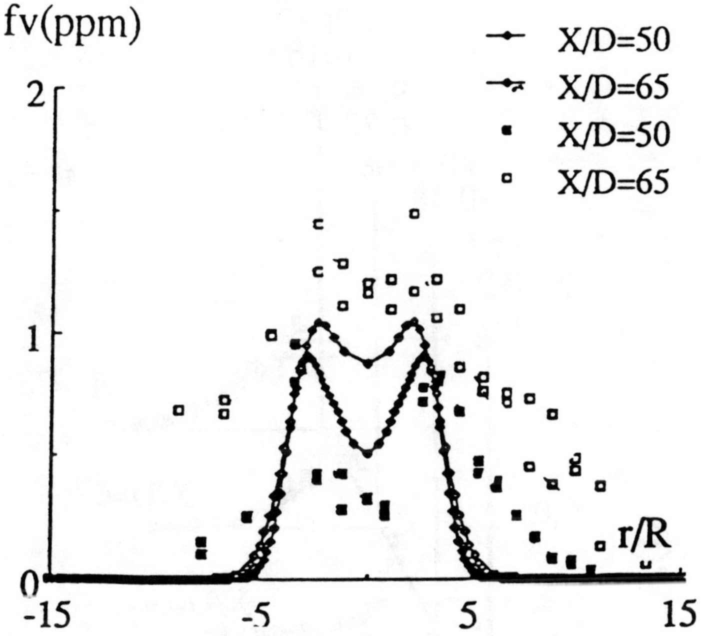

FIGURE 4: Radial variations of the mean soot volume fraction experimental point at X/D = 50 experimental point at X/D = 65 —calculated value

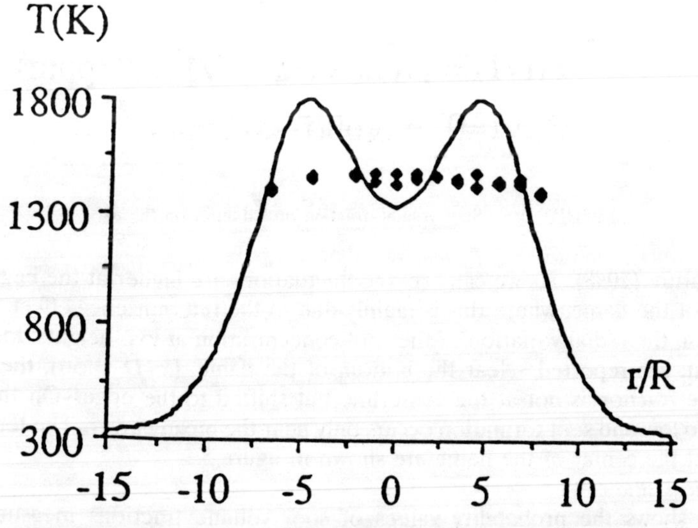

FIGURE 5: Radial variations of the mean temperature at X/D = 65 experimental point —calculated value

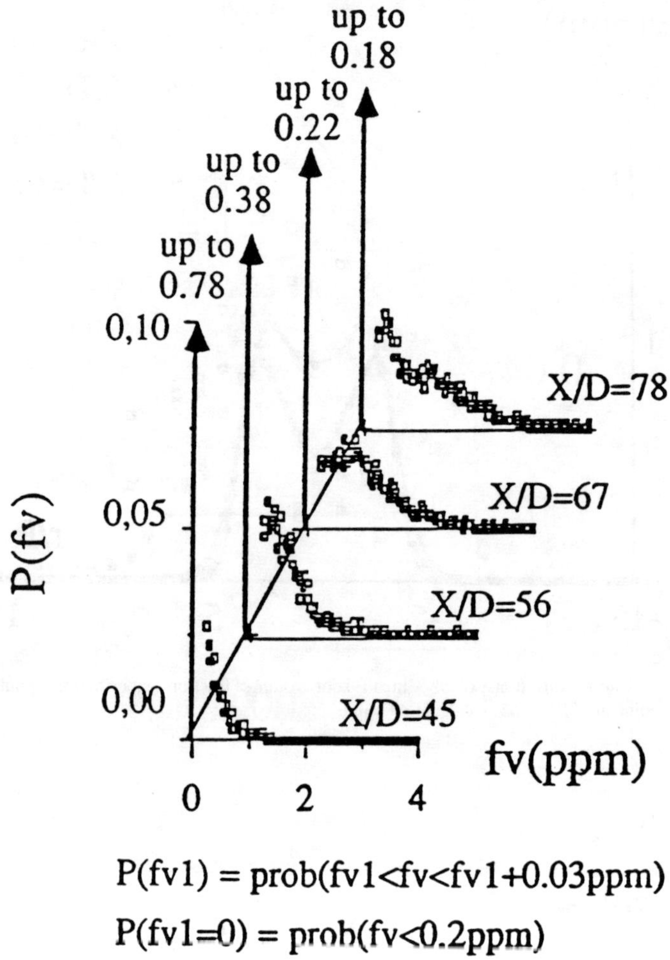

FIGURE 6: Soot volume fraction probabilities on the axis

for temperature (20%). As we can see, the fluctuations are higher at the beginning and at the end of the flame, where this is mainly due to the intermittency effect.

In figure 4, the radial variation of the soot concentration at two heights, located from the fuel exit, are reported. Near the bottom of the flame, ($x/D = 50$), the maximum soot volume fraction is not at the centerline but shifted to the edge. On the axis, the mixing is too low and soot formation occurs only near the mixing layer. The temperatures measured at the center of the flame are shown in figure 5.

b) *Statistical studies*

Figure 6 shows the probability values of soot volume fractions measured on the centerline and at four heights in the flame. The very low soot concentrations are difficult to measure, so a threshold of 0.2 ppm was considered. This value corresponded to a noise level of nearly one percent on the non-absorbed signal. As is shown in figure 6, the probability of low soot concentration values (0.2 ppm) is high at the beginning of

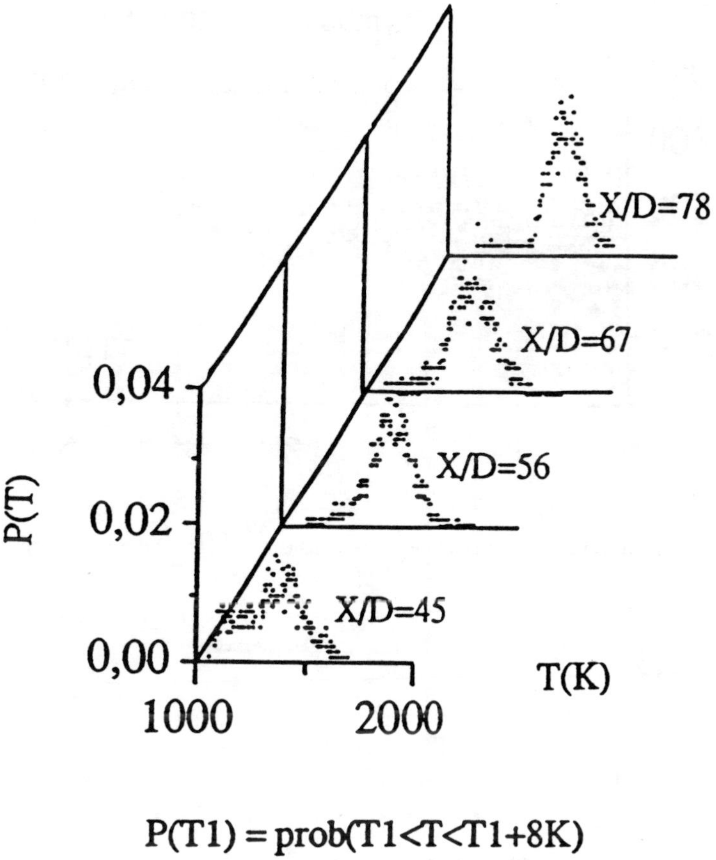

FIGURE 7: Temperature probabilities on the axis

the jet and then it decreases due to the intense mixing occurring at higher places in the flame. These strong fluctuations have been mentioned earlier when the rms values were given (see figure 3).

Figure 7 shows the probability values of temperatures measured at the same location. Except near the beginning of the jet, the probability distribution has a gaussian shape (skewness and flatness approximately equal to zero and three respectively. These results are different to those reported in figure 6. An important point is that at the same time a high temperature value is possible but with no presence of soot. As explained above, the experimental process allowed us to determine at the same time the temperature T and the soot concentration fv. So the correlation between these two quantities was studied. Figure 8 shows such a correlation graph determined at $x/D = 65$ and on the axis. The spread in the experimental results is large and is nearly uniform. The correlation is low between instantaneous values of T and fv at this height in the flame.

No upper limit for soot concentration appears. The only conclusion that might be made is that soot exists only in the temperature range of 1 200-1 800 K. On the same graph the mean values are also reported in order to look at the reality of a relationship between T and fv. Figure 9 is another presentation and it is more easier to analyse. We

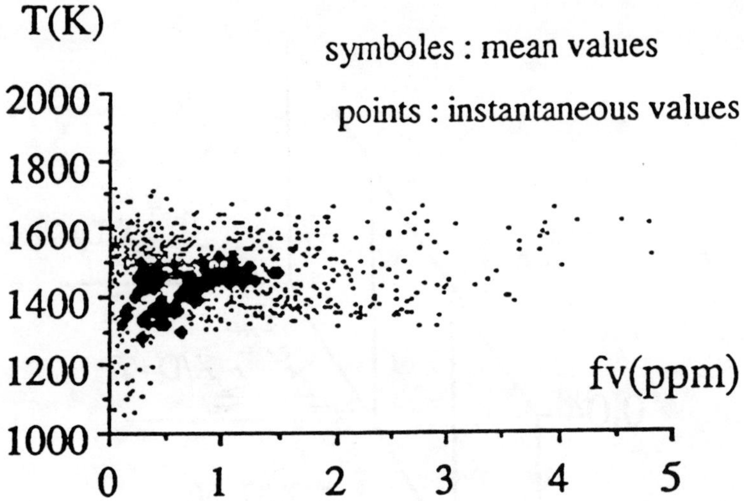

FIGURE 8: Instantaneous temperature versus instantaneous soot volume fraction on the axis.

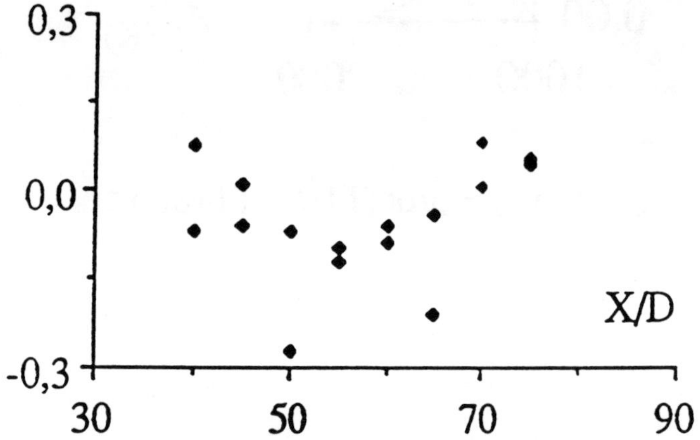

FIGURE 9: Coefficient correlation between temperature and soot volume fraction along the axis.

can see that the remarks which have been made above, may be applied also in the edges of the jet. If the temperature T is linked to the mixture fraction 8, as is assumed using the flamelet approach, and, if it is still true for the quantity of soot present in this flame (Kent *et al.,* 1987), the relationship is no longer simple according to our experimental correlations. It is well known that the radiation destroys the simple relation between T and f (Bilger, 1976); but as expected with finite chemical rate, T and fv may also depend on two or more parameters (Peters, 1984; Borghi, 1988; Tien *et al.,* 1989).

In order to avoid these difficulties, Moss (1990) has recently proposed to link not the soot quantity but the soot production rate with the mixture fraction. This problem remains a difficult one and we will come back to it in the following section.

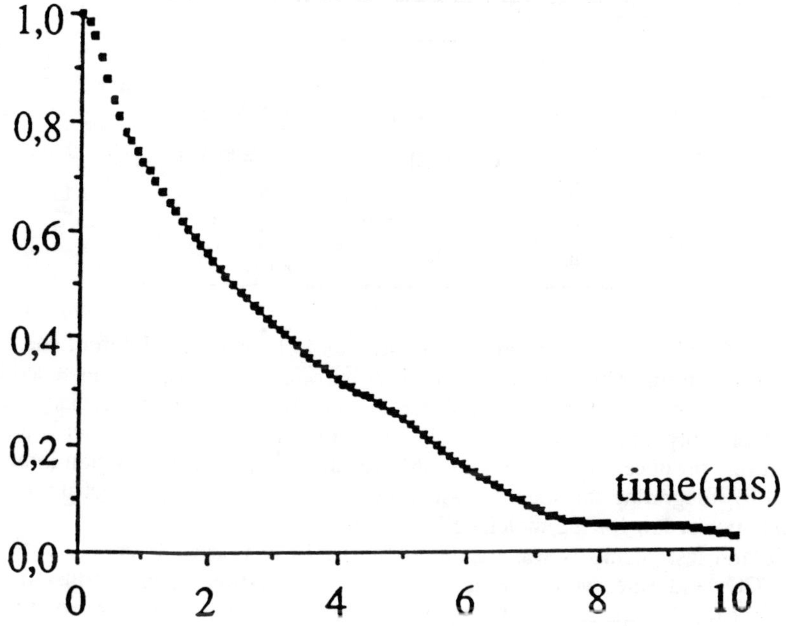

FIGURE 10: Temperature temporal autocorrelation on the axis at $X/D = 55$

Figure 10 shows the temporal auto-correlation coefficient *R(t)*, calculated from the temperature result and taken at ninepoint $x/D = 55$. This value was obtained with a 10 KHz sampling rate in order to increase the accuracy of the measurement. From this curve, we have determined the integral scale and it is equal to 3 ms. On the center line of the jet and for x/D in the 55-75 range, the values lie between 2 and 6 ms. If we take the Taylor assumption on turbulence, the length integral scale Λ may be calculated. At $x/D - 55$, the mean speed (provided by a numerical simulation of the jet) is 5 m/s; this gives Λ equal to 2.5 cm. This last result has also provided us with an argument to confirm that the spatial resolution (6 mm) of our experimental device was well chosen compared to the characteristic scale in the jet.

4. NUMERICAL SIMULATION OF THE JET FLAME

a) *Flame field predictions*

The numerical predictions were made using the procedure of Patankar and Spalding (1970). The mass, momentum and energy equations were solved. Continuity is obtained through the transformation to streamline co-ordinates. The $K - \varepsilon$ turbulence model (Launder *et al.*, 1972) gives the turbulence kinetic energy k and the turbulent dissipation rate, which allow us to calculate the effective turbulent exchange coefficient. The jet is a two-dimensional flow, and the usual boundary layer assumptions were made. The general equations are of the kind given below:

$$\rho u \frac{\partial \phi}{\partial x} + \rho v \frac{\partial \phi}{\partial r} = \frac{1}{r} \frac{\partial}{\partial r} \left(r \frac{\mu}{\sigma \phi} \frac{\partial \phi}{\partial r} \right) + S_\phi \tag{3}$$

TABLE I

ϕ	σ_ϕ	S_ϕ
u	1.0	$(\rho_\infty - \rho)a_g$
k	1.0	$\mu\left(\frac{\partial u}{\partial}r\right)^2 - \rho\varepsilon + \frac{\mu a_g \rho_\infty}{\rho^2}\frac{d\rho}{dx}$
e	1.3	$\left(C_{\varepsilon 2}\mu\left(\frac{\partial u}{\partial r}\right)^2 - C_{\varepsilon 2}\rho\varepsilon + \frac{\mu a_g \rho_\infty}{\rho^2}\frac{d\rho}{dx}\right)\varepsilon k$
f	0.7	0
h	0.7	$-k_0(T) f v\sigma(T^4 - T_\infty^4)$
g	0.7	$Cg^1\mu\left(\frac{\partial f}{\partial}r\right)^2 - C_{g2}\rho g\frac{\varepsilon}{k}$

where $\mu = C\mu 1k^2/\varepsilon$ is the turbulent viscosity. Table I shows the different coefficients and the source terms which have been used. A buoyancy term has been included in the momentum equation and in the turbulence kinetic energy equation. Radiation effects are great in sooty flames. In the enthalpy equation heat loss by radiation is assumed to be entirely produced by the soot and it is calculated in the optically thin limit; K_0 is provided by integrating the spectral assumption coefficient over the spectrum and with an inverse power law for the wavelength dependence.

Simple and fast chemistry was assumed (Bilger, 1976, Lockwood *et al.,* 1975). Only air, fuel, CO_2 and water were considered. A balance equation for the mixture fraction f and another for the square of its fluctuations were solved. A presumed mixture fraction probability density function was required to predict the mean concentration of the species involved. The Beta function was considered and the mean temperature was calculated from the enthalpy value.

b) *Soot formation and destruction*

There has been a great deal of research on soot formation in laminar flames (Harris, 1985; Megaridis, 1989; Aayus, 1981; Bockhorn, 1984; Bockhorn, 1988; Moss, 1988; Kent, 1982; Kent, 1990...). Each study seeks a basic law which could account for the main characteristics observed in premixed or diffusion flames. The oxidation process is more difficult to study and less data is available. (Nagle *et al.,* 1961; Lee *et al.,* 1962; Neoh *et al.,* 1981; Garo *et al.,* 1990). The important point is that these chemical processes are not fast. The chemical characteristic times are of similar magnitude to those of the turbulent mixing. In order to calculate the average formation rate, the flamelet concept (Moss, 1990) was applied but a great deal of work remains to be done to resolve certain questions, such as oxidation or radiative effects on combustion for example.

In this work, we only intended to predict soot with a simple model. A single step process was chosen for the formation (Khan *et al.,* 1974; Abbas *et al.,* 1985)

$$\left[\frac{dm_s}{dt}\right]_f = C_f P_{fu}\phi^3 \exp\left(-\frac{2000}{T}\right) \tag{4}$$

where P_{fu} is the fuel partial pressure, ϕ is the local equivalent ratio and C_f is taken as 0.01 with Kg/m^3 Pa. Two mechanisms were considered for the soot burnout. The first was proposed by Nagle (1961).

$$\left[\frac{dm_s}{dt}\right]_{b,k} = \frac{720}{\rho_s d} m_s \left[\left(\frac{k_a P_{O2}}{1 + K_z P_{O2}}\right) x + K_b P_{O_2}(1 - x)\right] \tag{5}$$

where

$$x = \frac{K_b P_{O2}}{K_b P_{O2} + K_t}$$

d_s is taken as 300 nm and PO_2 in atm. The different constants K are given by Nagle. If the turbulent mixing is supposed to control the soot burnout, Magnussen's law (1976, 1978) is applied and gives the time average value directly

$$\left[\frac{dm_s}{dt}\right] b, t = 4\frac{\varepsilon}{K}\left[ms, \frac{\overline{mo_2}}{r}\left(\frac{\overline{m_s r_s}}{\overline{m_s r_s} + \overline{m_{fu} r}}\right)\right] \tag{6}$$

where $\bar{m_{fu}}$ and $\bar{m_{02}}$ are the time average fuel and oxygen mass fractions, r and r_s are stoichiometric coefficients for fuel and soot, K is the turbulence kinetic energy and its dissipation rate.

In order to average these source terms, instantaneous values of each variable were replaced by their time average. This straightforward insertion is a crude procedure for time averaging, but we have to remember that firstly, $m_s(f)$ is actually unknown, as explained above, and secondly the radiation, which is important in this flame, destroys the link between T and f. This method is also a way of estimating the relative importance of soot formation and destruction in a jet diffusion flame.

The soot mass gives the volume fraction directly, thus allowing the radiative effect in the enthalpy equation to be determined.

c) *Results*

The numerical results have already been reported in figs 3-5. As we can see in figs 3-4, good agreement with the experimental data is obtained for the soot volume fraction. At these observed heights in the flame, the formation rate is still great, particularly at the center of the jet, and as expected soot oxidation has an efficiency which is higher near the mixing layer. The first chemical burnout rate (see above) is a controlling step in the upstream region of the jet while turbulence mixing controls the burnout in the downstream region. Figure 5 shows the comparison between experimental and calculated temperature values. The agreement is also reasonable but not as good as that for the soot. It must be pointed out that the radiation is very high and without taking it into account, the calculated temperature would be 25% higher with a more pronounced peak outside the centreline.

5. CONCLUSION

The optical probe allowed us to determine mean and fluctuating values of temperature T and soot concentration fv in a turbulent jet flame diffusion. The mean values of these two quantities have similar evolutions in the jet. But the behavior of the r.m.s. values and the probabily density function were different. Low correlation between T and fv was found in the flame centre.

All the measurements were carried out in a flame in which buoyancy effects were significant. We intend to do similar measurements in a jet flame with a higher initial speed in order to compare results obtained with different Reynolds numbers and different turbulence levels.

REFERENCES

Abbas A. S., Lockwood F. C. (1985). *Journal of the Institute of Energy,* pp 112-115.
Bilger R. W. (1976). *Progress in Energy and Combustion Science,* **1** No. 2/3, pp 87-110.
Bockhorn H., Felting F., Hedcbrick A., Wannemacher G. (1984). *20 Symposium on Combustion,* pp 979-988.

Borghi R. (1988). *Progress in Energy and Combustion Science,* **14** No. 4, pp 245-292.
Chen J. Y., Kollmann W., Dibble R. W. (1989). *Combustion Science and Technology,* **64,** pp 315-346.
Dalzell W. A., Sarofim A. F. (1969). *Journal of Heat Transfer.* February, 91, pp 100-104.
Garo A., Prado G., Lahaye J. (1990). *Combustion and Flame.* **79,** pp 226-233.
Harris S. J., Weiner A. (1985). *Ann. Rev. Phys. Chem.,* **36,** pp 31-52.
Haynes B. S., Wagner N. G. (1981). *Progress in Energy and Combustion Science,* **7,** pp 229-279.
Kent J. H., Wagner H. G. (1982). *Combustion and Flame,* **47,** pp 53-65.
Kent T. H., Honnery D. (1987). *Combustion Science and Technology,* **54,** pp 383-397.
Kent J. H., Honnery D. R. (1990). *Combustion and Flame,* **79,** pp 287-298.
Khan I. M., Greeves G. (1979). *Heat Transfer in Flames.* N.H. Afgan and J. M. Beer (eds), Scripta back 10.
Launder B. E., Morse A., Rodi W. Spalding B. B. (1972). *Nasa Free Shear Flows Conference,* Virginia, NASA report number sp-311.
Lee K. B., Thring M. W., Beer T. M. (1962). *Combustion and Flame,* **6,** p 137.
Lee S. C., Tien C. C. (1981). *18 Symposium on Combustion,* pp 1159-1166.
Lockwood, F. C., Naguile A. S. (1975). *Combustion and Flame,* **24,** pp 109-124.
Magnussen B., Hjertager B. A. (1976). *16 Symposium on Combustion,* pp 719-729.
Magnussen B., Hjertager B. H., Glenn J. G., Bkaduri D. (1978). *17 Symposium on Combustion,* pp 1383-1393.
Megaridis C. M., Dobbins R. A. (1989). *Combustion Science and Technology,* **66,** pp 1-16.
Moss J. B., Stewart C. D., Syed K. J. (1988). *22 Symposium on Combustion,* pp 413-423.
Nagle J., Strickland R. F. (1961). Proc. Fifth Conference on Carbon, Pergamon, p 154.
Neoh K. G., Howard J. B., Sarofim A. F. (1981). *Particulate Carbon Formation during Combustion,* Plenum Press.
Peters N. (1984). *Progress in Energy and Combustion Science,* **10** No. 3, pp 319-341.
Peters N. (1986). *21, Symposium on Combustion* pp 1231-1250.
Stavitsky A., Coolay J. E. (1964). *Analytical Chemistry,* **36,** No. 8.
Syed K. J., Steward C. D., Moss J. B. (1990). *23 Symposium on Combustion*
Patankar S. V., Spalding D. B. T. (1970). *Heat and mass transfer in boundary layers,* Intertext books, London.
Wieschnawky H., Bockhorn H., Vetting F (1988). *22 Symposium on Combustion,* pp 343-352.

Ignition of Fine Particles in Shock-Heated Environment

SUSUMU KOBAYASHI and TAKASHI ADACHI *Department of Mechanical Engineering, Saitama Institute of Technology, 1690 Fusaiji, Okabe-machi, Ohsato-gun, Saitama 369-02, Japan*
TATEYUKI SUZUKI *Department of Mechanical Systems Engineering, Toyama Prefectural University, 5180 Kurokawa, Kosugi-machi, Imizu-gun, Toyama 939-03, Japan*

Abstract — An experiment on ignition of fine particles was performed by means of a shock tube. The particles used were pulverized coal, food powder (cornstarch) and super-fine metallic particle (tungsten). They were dispersed in an atmosphere by a pneumatic injection device specially designed for this purpose. The temperature behind the reflected shock ranged from about 600 to 1100 K. Loci of ignited particles were photographed together with direct shadowgraphs of dispersed particles. Direct photographs of ignited tungsten particles were also taken. The ignition mechanism for these particles was suggested. The ignition delay time was investigated and its definition was examined. The effect of degree of particle dispersion on the ignition delay time was investigated. The ignition probability was also estimated.

Key Words: ignition, dispersive media, ignition delay time, ignition probability, ignition mechanism, shock tube

1. INTRODUCTION

The remarkable development of modern technology has made it possible to deal with exceedingly fine particles. In such an environment as a grain silo or high-technology factory, fine particles of inflammable material are apt to drift in the atmosphere. Consequently, a serious hazard may occur through a dust explosion triggered by various energy sources such as mechanical or electrostatic sparks, mechanical heating and hot surfaces (Palmer, 1973; Bartknecht, 1989).

It is vitally important to secure an accurate knowledge of the ignition mechanism and ignition delay time, not only for a discussion of safety in the environment mentioned above, but also for a fundamental investigation on combustion efficiency in order to realize clean combustion. The ignition mechanism has been a question at issue and its elucidation will be of great benefit in the enhancement of combustion efficiency. The ignition delay time can, if only suitably defined, be a measure of inflammability in the environment concerned.

For this purpose, there are two typical ways of igniting fine particles in a hot atmosphere. One is to insert particles into a preheated atmosphere, the other to disperse them behind a reflected shock front. In the former case, the particles are radiated from a hot vessel wall and it is difficult to account for or evaluate this influence correctly. In the latter case, on the other hand, the atmosphere surrounding the particles is instantaneously heated while maintaining its stagnancy, with the heat transfer to particles being limited to conduction and convection.

Thus, because of these advantages, shock tubes have been used successfully in many ignition studies of combustible (mainly coal) particles. One of the early experiments was carried out by Nettleton and Stirling (1967, 1971). They measured the ignition delay times for graphite and anthracene, as well as various coal particles, and tried to identify the ignition mechanism. Hwang and Pillay (1978) confirmed the existence of a minimum gas temperature for ignition and concluded that the effect of particle size on the ignition delay

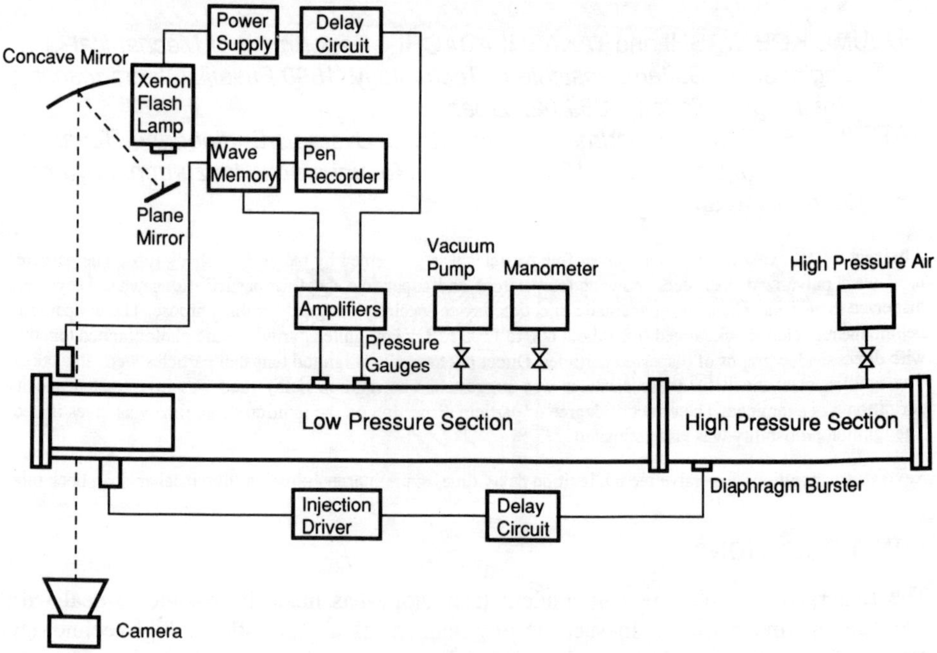

FIGURE 1 Schematic diagram of experimental setup.

was small for the particles they had used. Kobayashi *et al.* (1988) examined the difference of particle dispersion method over the ignition delay time.

In the present paper, the ignition delay times of pulverized coal (mean diameter $d_m = 340$ and 100 μm), cornstarch powder (25 μm) and super-fine metallic particles (0.89 μm) of tungsten have been investigated by means of a shock tube. For the investigation on the ignition mechanism, a shadowgraph of dispersed coal particles and direct image of ignited particles were recorded on the same image plane through a CCD camera. Direct photographs of ignited tungsten particles were also taken.

2. EXPERIMENTAL APPARATUS AND METHOD

The experimental setup is the same as that used in our previous research (Suzuki *et al.*, 1990) and shown in Fig. 1. The shock tube is composed of a 78 mm diameter driver section 2.95 m long and a 50 × 50 mm rectangular driven section 4550 mm long. At the end of the driven section is a test section with a pair of 50 × 200 mm optical windows on both sides. A shock wave is generated by bursting a diaphragm (Mylar) between the driver and driven sections. High temperature air compressed by the shock wave reflected at the end of test section (i.e., end-plate) was utilized to ignite dispersed particles.

The injection device for dispersing particles in the atmosphere is similar to the pneumatic dust injector as used by Sichel *et al.* (1985) and its detail is shown in Fig. 2. It was installed at the center of the test section floor 100 mm away from the end-plate. The diameter of the aperture is 8 mm. The upper surface of the particles placed loosely on the bed of sintered

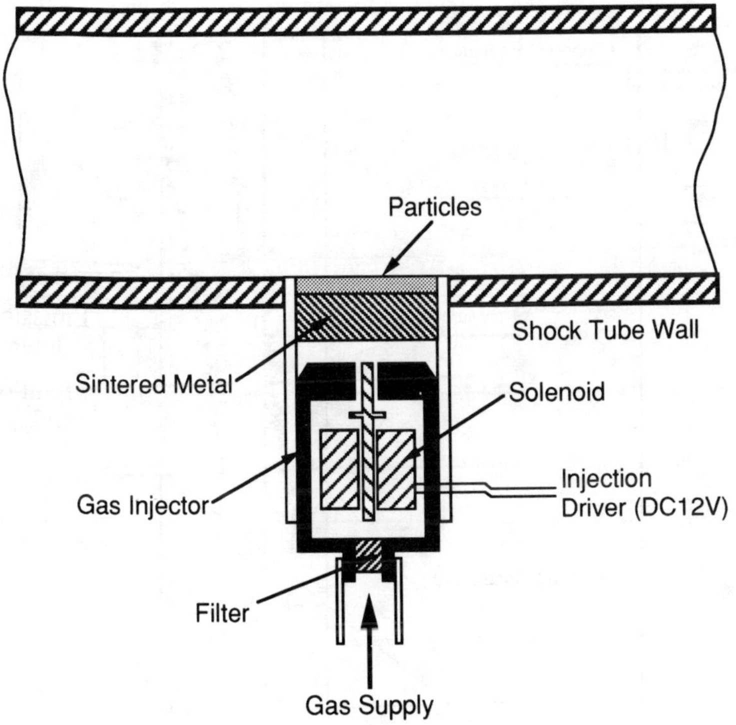

FIGURE 2 Particle dispersion device.

metal was flattened so as to be flush with the shock tube floor. By regulating a digital delay timer, a diaphragm burster could be operated at any desired instant after a solenoid-controlled valve of the injection device was opened. Thus a various *dispersity* (degree of particle dispersion) in shock-heated air could be realized. The required temperature behind the reflected shock was achieved by presetting the pressure ratio of driver to driven section. The pressure in the driver section was fixed at 3.5 atm (3.55×10^2 kPa) and the low pressure in the driven section was realized by a vacuum pump.

The incident shock velocity was estimated by measuring the time interval of the passage of the incident shock between two pressure gauges (Kistler model 601A) mounted on the wall ahead of the test section. The temperature, T_5, behind a reflected shock was evaluated from the normal shock relations with the incident shock Mach number thus measured.

For photographing an image of dispersed particles, the pressure gauge nearest to the test section was used to trigger a single-spark xenon light source, whose half pulse width is 0.7 μs, via a digital delay timer.

The light emission due to ignition was monitored by a photomultiplier tube (Hamamatsu Photonics Co., R269, spectrum sensitivity characteristic 185–650 nm, the maximum sensitivity wavelength 420 nm) set perpendicular to the optical window, and 30 mm away from the end-plate of the shock tube. Together with the pressure gauge outputs, the photomultiplier tube output was recorded by a pen-recorder via a digital wave memory (Iwatsu Electric Co., DM-7100).

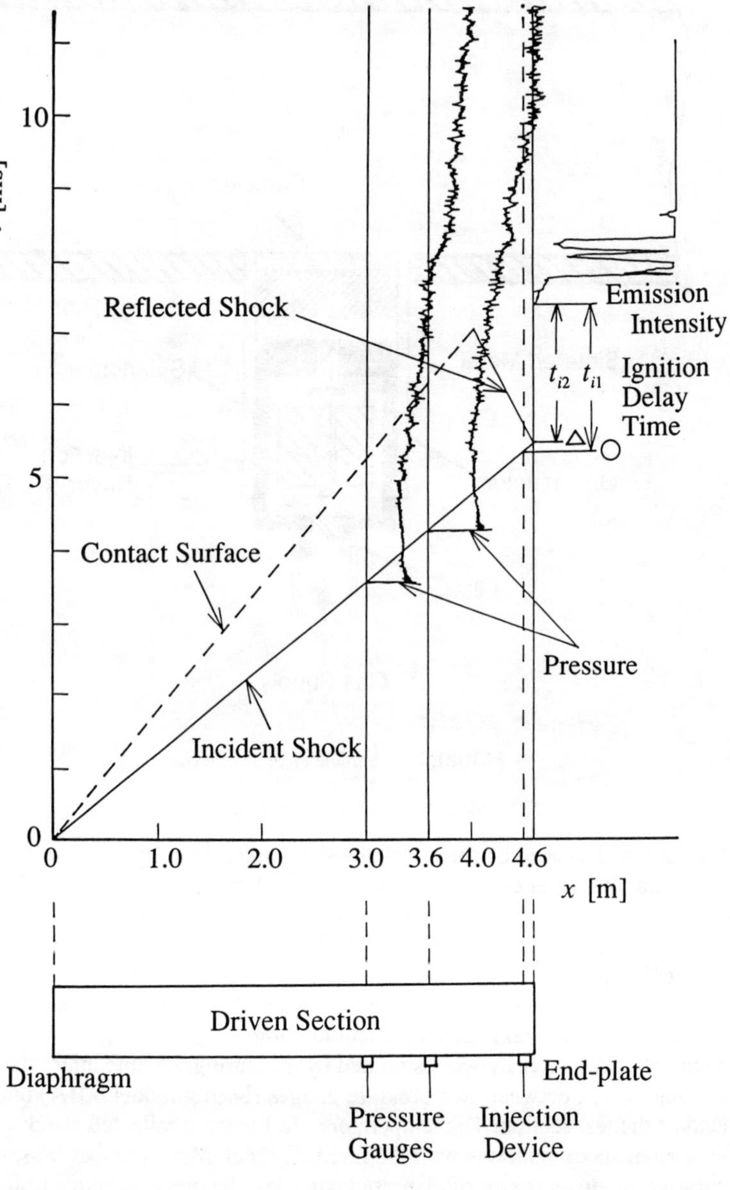

FIGURE 3 Definition of ignition delay time t_i and x–t diagram.

In the present paper, particles are assumed to be ignited when the photomultiplier tube output exceeds a certain value. The traditional ignition delay time t_{i1} is defined as the time interval between the passage of the incident shock at the aperture of the injection device and the onset of the photomultiplier tube output leading to immediate ignition. Another

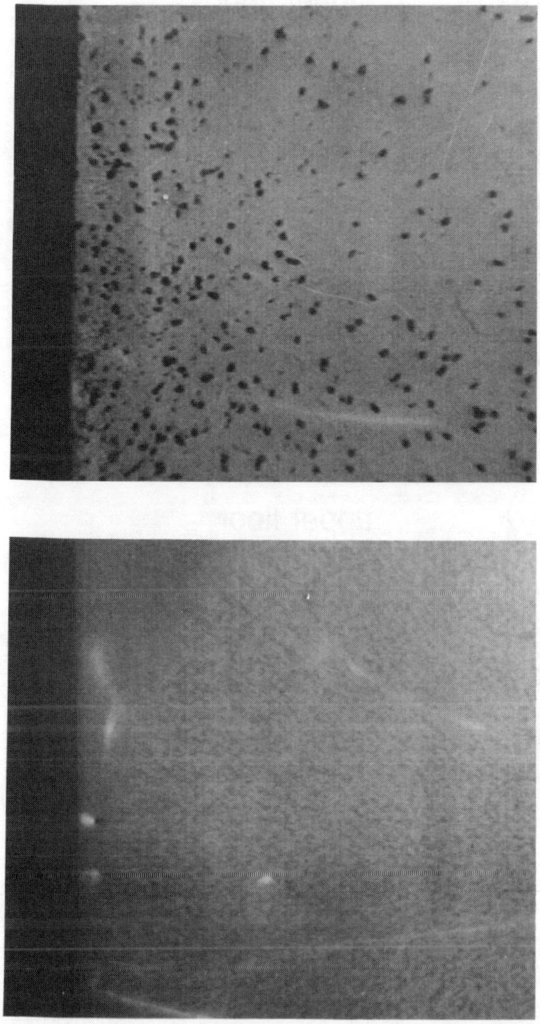

FIGURE 4 Loci of ignited particles and shadowgraphs of dispersed coal particles. The black belt on the left is the end-plate of the shock tube: (a) d_m = 340 μm, the estimated temperature behind reflected shock T_5 = 960 K, the time interval between the impact of shock on the end-plate and the flash t_f = 2.8 ms, (b) d_m = 100 μm, T_5 = 985 K, t_f = 3.0 ms.

ignition delay time t_{i2} is defined as the time interval measured from the shock incidence on the end-plate. The latter ignition delay time is physically significant since the atmosphere behind the incident shock is not so high as to initiate a chemical reaction making it possible to count out the time interval between the passage of the shock at the aperture and the incidence on the end-plate from the ignition traditional delay time. This is schematically

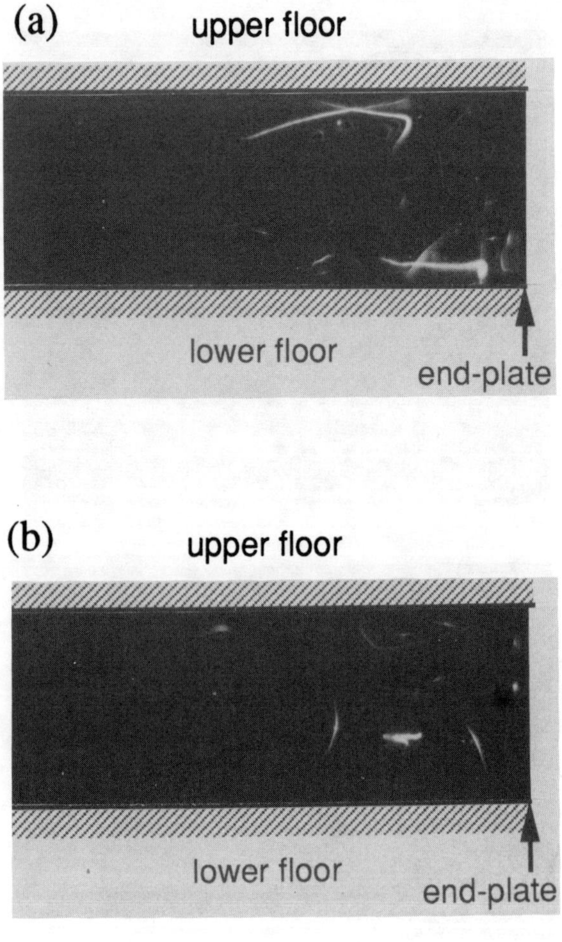

FIGURE 5 Direct photographs of ignited tungsten particles. (a) $T_5 = 1000$ K, (b) $T_5 = 830$ K. SEE COLOR PLATE IX.

illustrated on the x–t diagram in Fig. 3, where x and t denote the spatial coordinate along the shock tube axis and the time, respectively.

3. RESULTS AND DISCUSSION

The loci of ignited coal particles and the direct shadows of dispersed particles were photographed on the same films (Fig. 4). During a single run of the experiment, the shutter was kept open and thus the *loci* of ignited particles were recorded. On the other hand, the duration of a single spark was so short that dispersed particles remained stationary. Some ignited particles were entrained by the drag force due to the flow behind the incident shock and their own inertia, while some were almost at rest. The ignited particles at rest do not

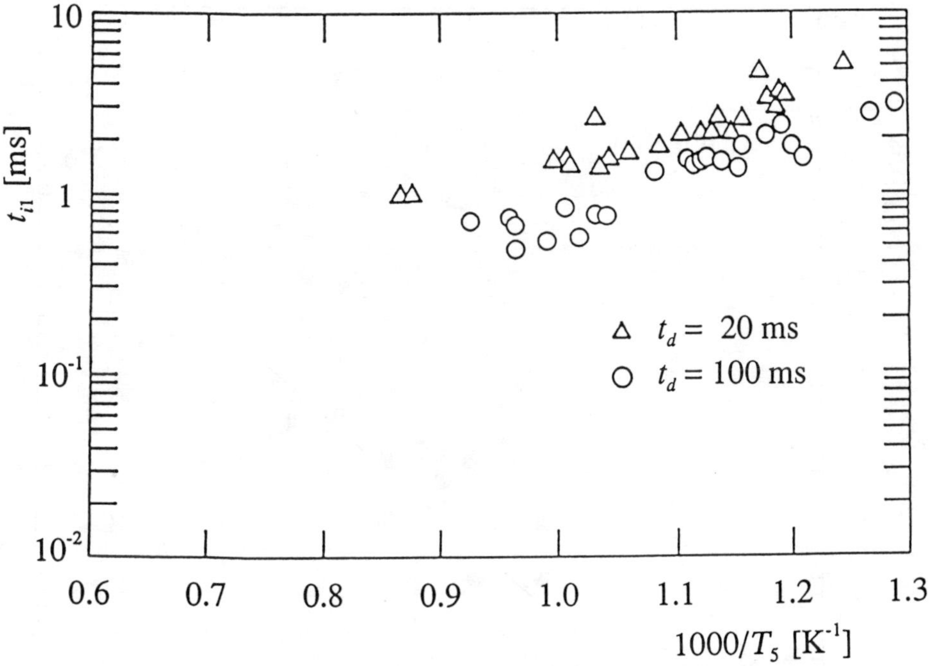

FIGURE 6 Effect of dispersity on ignition delay time. The time interval between the operation of solenoid-controlled valve and the passage of incident shock at the aperture of injection device is denoted by t_d.

necessarily mean they were not in motion, because it is possible that the duration of the light emission was too short. The figures show that of all the particles several were ignited at most. This indicates that coal particles ignite as individuals, not as a group. It must be noted that the temperature T_5 behind a reflected shock was relatively low, below 1000 K, in the present experiment.

Figure 5 is a direct color photograph of ignited tungsten particles. More particles were ignited than for pulverized coal. Considering that the mean diameter of tungsten particles is 0.89 μm, the width of the orange light locus is exceedingly large. It is possible that either an agglomerate of tungsten or group of tungsten particles was ignited.

In Fig. 6, the ignition delay times for different dispersities are plotted, where t_d is the time interval between the operation of the solenoid-controlled valve and the passage of the incident shock wave at the aperture of the injection device. The ignition delay time for a "better-dispersed" particle cloud (t_d = 100 ms) is evidently smaller. This is physically explained in that the heat transfer from the hot atmosphere to the particles is better for a sparser cloud because the particles act as heat sinks and cool down the temperature of the environment around them. The present results are in contradiction to those by Borisov *et al.* (1967). However, in the present experiment, the mass of dispersed particles was of the same order as the air compressed by the reflected shock in the test section, so that the temperature around the particles was sensitive to dispersity.

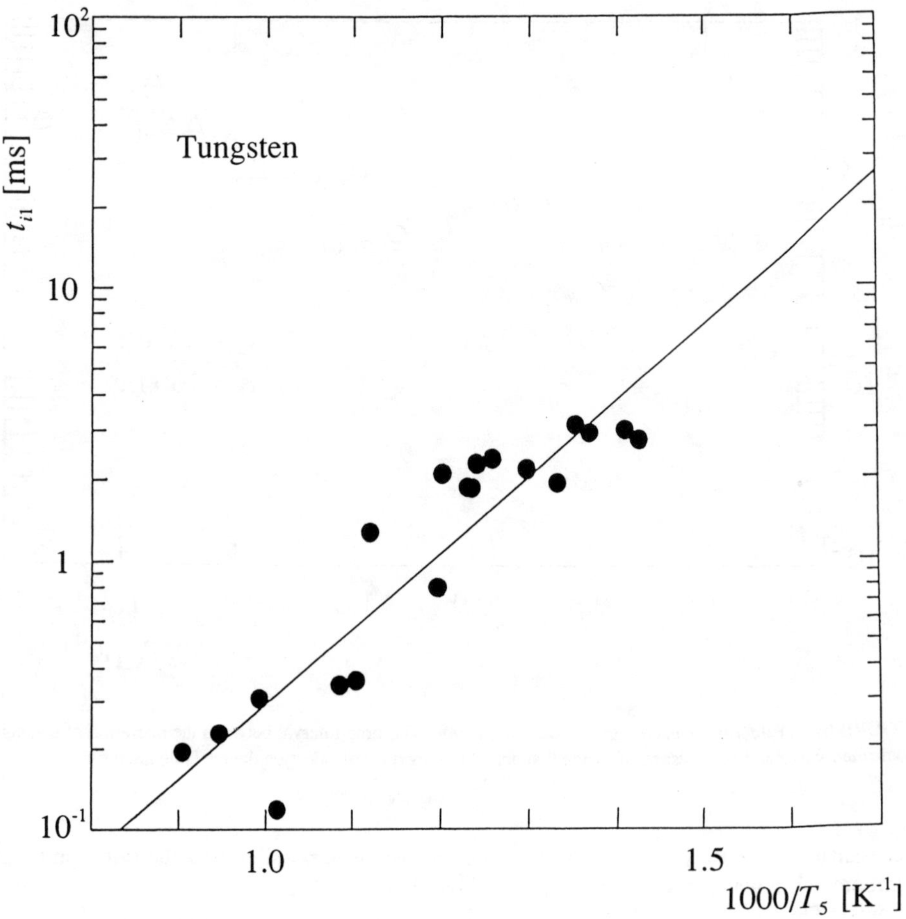

FIGURE 7 Ignition delay time of tungsten particles. The solid line shows the least-squares plot.

The ignition delay time of tungsten particles is plotted in Fig. 7. As compared with pulverized coal, the slope is steeper and so the activation energy is higher. On the other hand, the value of the ignition delay time itself is by far larger than that for coal particles.

Figure 8 compares two different definitions of ignition delay time. Circles denote the traditional definition t_{i1}, while triangles denote a new delay time t_{i2} measured from the instant the shock is incident on the end-plate (see also Fig. 3). The two lines are corresponding to a least-squares plot. Namely, the new delay time is shorter than the traditional one by the time elapsed during which the incident shock proceeds from the aperture of injection device to the end-plate. As the temperature behind the incident shock is lower than T_5 by a few hundred degrees Kelvin, the chemical reaction rate in such an atmosphere is much slower than that behind the reflected shock. Therefore, this cool environment hardly contributes to a substantial ignition delay time. From a physicochemical point of view, such a superfluous time should be counted out from the delay time. In the present experiment, this superfluous time is approximately 0.1 ms, so that the difference between both ignition delay times is not essential. However, it is noteworthy that the ignition delay time is

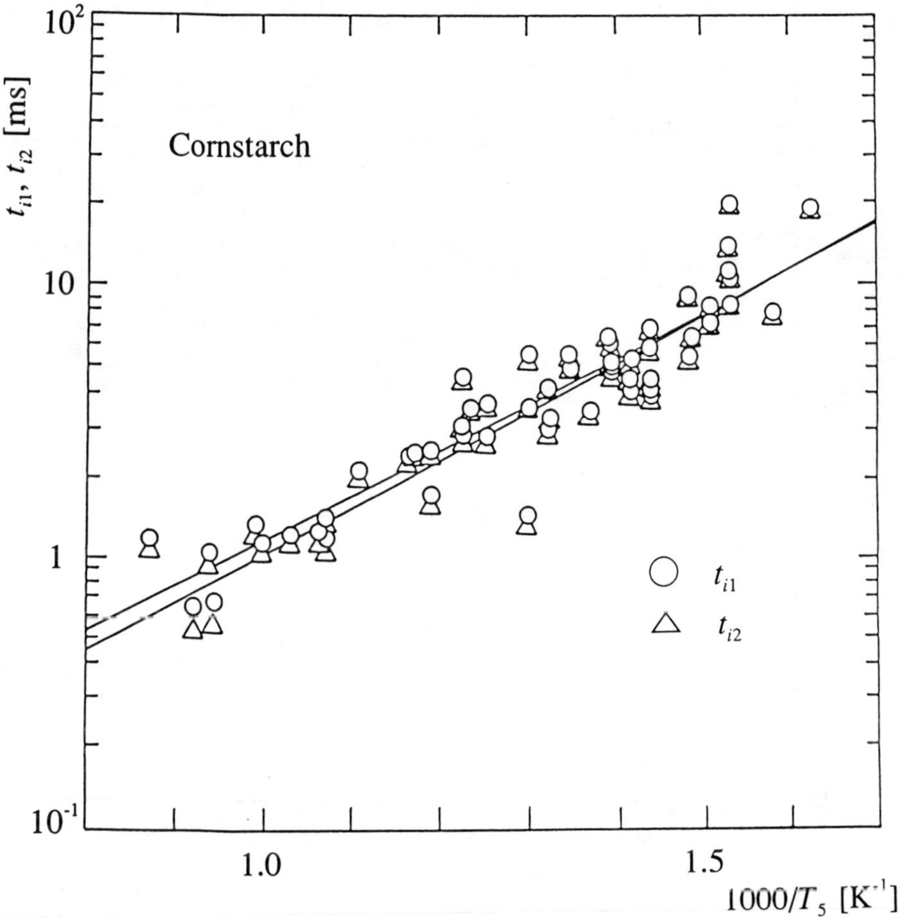

FIGURE 8 Comparison of two definitions of ignition delay time (cornstarch).

dependent on the experimental apparatus used, and this fact makes it difficult to compare the data obtained by different research workers.

The ignition probability of cornstarch is shown in Fig. 9. As expected, it increases monotonously from 0 to 1 between T_5 = 590 and 710 K. According to the US Bureau of Mines (Palmer, 1973), the minimum ignition temperature of cornstarch was 660 K, however, the present result was around 600 K.

4. CONCLUSION

An experiment on the ignition of fine particles was performed by means of a shock tube. The loci of ignited coal particles, as well as their state of dispersion., were photographed on the identical image plane via a CCD camera. Direct photographs of ignited tungsten particles were also taken. The ignition delay time for several kinds of particle was investigated and its definition was examined. The ignition probability was estimated for

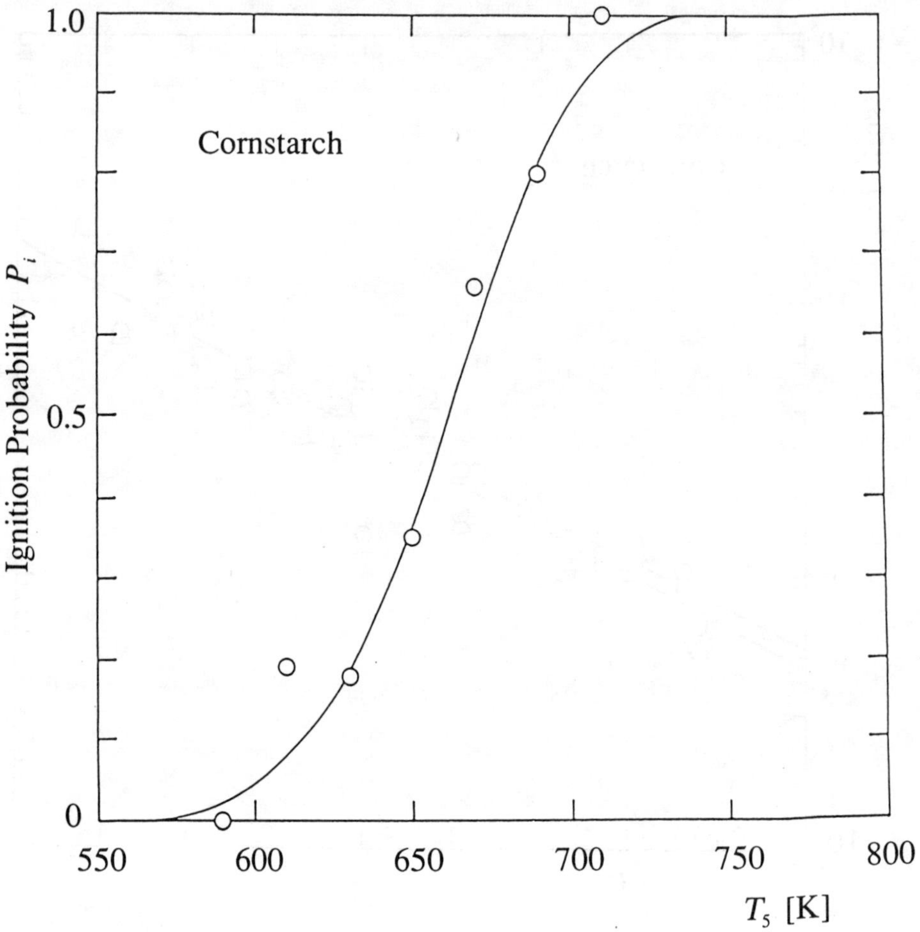

FIGURE 9 Ignition probability of cornstarch.

cornstarch. The following conclusions are obtained within the restricted conditions of the present experiment. First, coal particles ignite and burn as individuals rather than as a group, whereas for much smaller particles of tungsten, it is possible they ignite as group. Second, the ignition delay time depends on the dispersity of the particles. Third, the ignition probability of cornstarch increases monotonously from 0 to 1 between $T_5 = 590$ and 710 K.

REFERENCES

Bartknecht, W. (1989). *Dust Explosions*, Springer-Verlag, Berlin.

Borisov, A.A., Kozenko, V.P., Lyubimov, A.V., and Kogarko, S.M. (1967). Ignition of fuel powders behind shock waves, *Combustion, Explosion, and Shock Waves*, **3**, 192 (English translation).

Hwang, C.C., and Pillay, S. (1978). Ignition delay of coal particles in shock-heated air. *Combust. Sci. Tech.*, **17**, 241.

Kobayashi, S., Adachi, T. and Suzuki, T. (1988). The ignition of coal particle clouds in shock-heated air. *Proc. Int. Symp. Space Tech. Sci.*, **16**, 749.

Nettleton, M.A., and Stirling, R. (1967). The ignition of clouds of particles in shock-heated oxygen, *Proc. Roy. Soc. Lond.*, **A300**, 62.

Nettleton, M.A., and Stirling, R. (1971). The combustion of clouds of coal particles in shock-heated mixtures of oxygen and nitrogen, *Proc. Roy. Soc. Lond.*, **A322**, 207.

Palmer, K. N. (1973). *Dust Explosions and Fires*, Chapman & Hall, London.

Sichel, M., Baek, S.W., Kauffman, C.W., Maker, B., Nicholls, J.A., and Wolanski, P. (1985). The shock wave ignition of dusts, *AIAA J.*, **23**, 1374.

Suzuki, T., Adachi, T., and Kobayashi, S. (1990). An experimental study on the ignition of pulverized coal particles in a shock tube, *Trans. Japan Soc. Aeron. Space Sci.*, **32**, 147.

Effect of the Air/Fuel Ratio Fluctuation on the Formation of Nitrogen Oxides

K. HASE, Y. KORI, and K. OHGI *Research and Development Institute, Tokyo Gas Co., Shibaura, Minato-ku, Tokyo, Japan*

Abstract — The relationship between the air/fuel ratio fluctuation and NO_x formation was investigated in a combustor with a multiport burner. By using a laser absorption method, the air/fuel ratio fluctuation was measured under various mixing conditions. The fluctuation of the burned gas temperature was measured by a fine thermocouple. From the measured fluctuation of the air/fuel ratio, the probability density distribution of the air/fuel ratio was obtained to quantify the degree of the fluctuations. Using the model proposed by Heywood *et al.*, the effect of the air/fuel ratio fluctuation on NO_x formation was investigated by numerical simulation. The simulation results using the measured probability density distribution of the air/fuel ratio gives a reasonable prediction of the effect of the fluctuation on NO_x formation.

INTRODUCTION

Driven by the growing concern for environmental protection, combustion systems must operate with greatly reduced emission levels of nitrogen oxides (NO_x), carbon monoxide and hydrocarbons. At present, emission levels of carbon monoxides and hydrocarbons from most combustion systems using city gas as a fuel are kept within satisfactory levels, however, the NO_x emission has yet to be reduced to a satisfactory level. One promising method for reducing the NO_x emission is lean premixed combustion, and this appears to be very attractive and has already been employed in some combustion systems. However, it is well known that if the fuel and air are not sufficiently mixed, the NO_x emission level is not as low as expected, despite lean premixed combustion. The effect of mixing fuel and air on NO_x formation has been investigated by several researchers (Pompei and Heywood (1972), Appleton and Heywood (1973), Anderson (1973), Semerjian and Ball (1978), Lyons (1982)). In recent work on the effect on NO_x formation in a premixed combustor, Lyons concluded that in lean premixed combustors, highly uniform spatial fuel/air distribution was necessary to achieve a low NO_x emission. However, no mention was made by Lyons on the effects of temporal fluctuation of the air/fuel ratio on NO_x formation. When mixing of fuel and air is not perfect, there is a temporal fluctuation of the air/fuel ratio, in addition to a spatial nonuniformity of the fuel/air distribution. Semerjian and Ball (1978) has suggested that in very lean flames, the increased level of NO_x emission compared with that of premixed flames can be attributed to the large statistical fluctuations in the local equivalence ratio. Pompei and Heywood have tried to estimate the magnitude of the equivalence fluctuation by measuring the change of O_2 concentration in the burner. Semerjian has also estimated the magnitude of the fluctuation using the same method. However, this method does not directly measure the fluctuation. In the quantitative estimation of the effect of equivalence fluctuation on NO_x formation, a more precise estimation of the fluctuation is desired. The objective of this research is to quantitatively measure the equivalence ratio fluctuation in the mixture and investigate the relation between its magnitude and NO_x formation.

EXPERIMENTAL APPARATUS

Figure 1 shows a schematic diagram of the experimental system. The lower part of the combustor functions as a mixer, to which the fuel gas is injected perpendicularly to the air

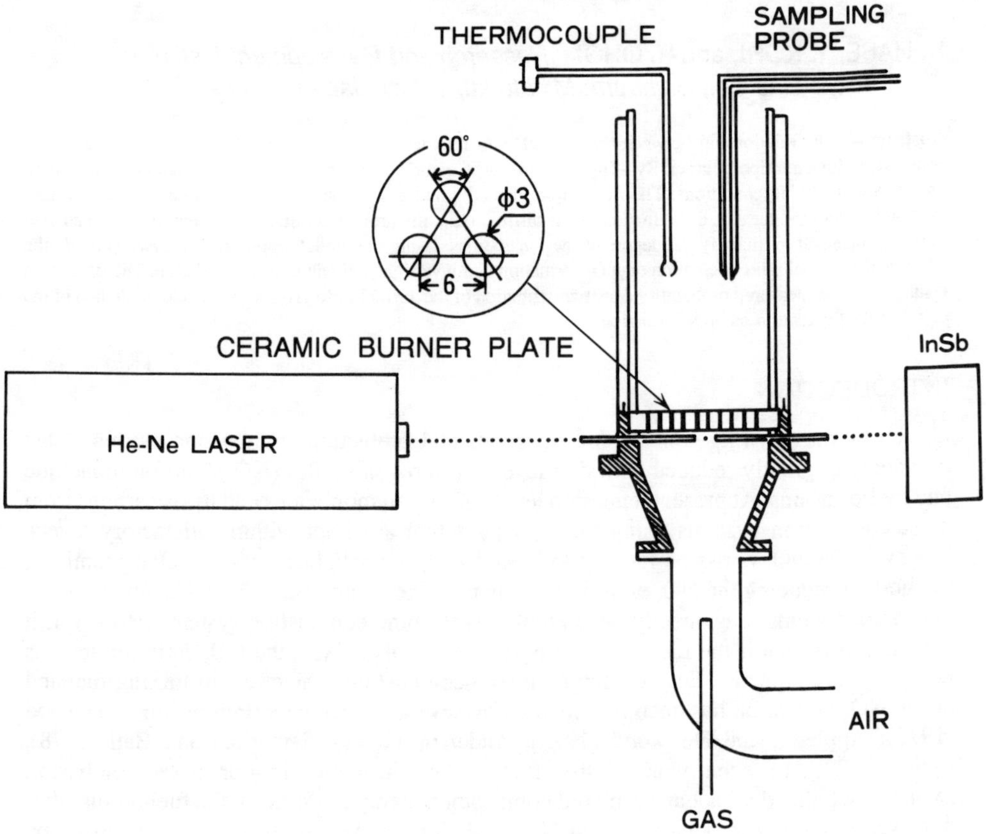

FIGURE 1 Schematic of the experimental system.

flow from a nozzle. The nozzle has 12 holes of diameter 2 mm. The ceramic burner plate is also shown schematically in Figure 1. A ceramic cylinder is mounted above the burner plate, which forms a combustion chamber. The ceramic cylinder is covered with glass wool to keep heat losses as low as possible. A city gas (CH_4 88.5%, C_2H_6 4.6%, C_3H_8 5.4%, C_4H_{10} 1.5%) is used as the fuel. The flow rates of the fuel gas and air supplied to the burner were regulated by orifice flow meters. The burned gas was sampled by a water-cooled stainless probe at atmospheric pressure. The sampled gases were conveyed to a cooler through an electrically heated Teflon tube to remove water, then supplied to a NO_2-to-NO converter. The tube was heated above 110°C to prevent condensation of water. After flowing through the converter, the sampled gases were sent to analyzers. The NO_x levels were measured with a chemiluminescence analyzer and the concentrations of CO, CO_2, O_2, H_2, N_2, CH_4 were measured by using gas chromatography. Since the conversion of NO to NO_2 may occur in the sampling probe, measurements were conducted only in the NO_x mode.

In the temperature fluctuation measurements, a 50 μm diameter silica coated thermocouple (Pt/Pt–13% Rh) was used. The signal from the thermocouple fed into a computer through a digital volt meter and the thermal inertia effects on the signal were compensated by the computer software program. The time constant of the thermocouple was determined by the alternating current heating method (Kunugi and Jinno, 1959). This compensation allows temperature fluctuations up to 2 kHz to be measured.

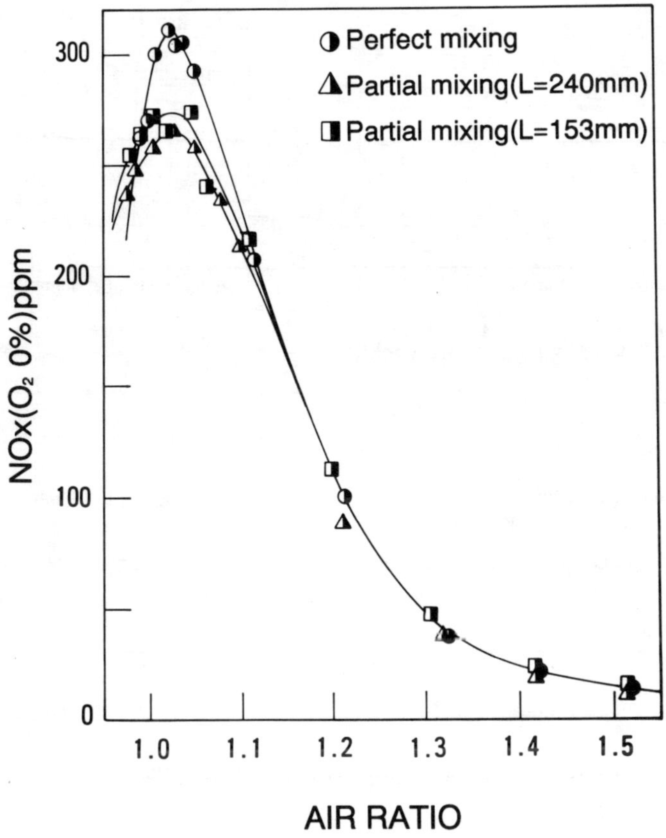

FIGURE 2 Variations of NO_x formation with air/fuel ratio in perfectly premixed and partially premixed flames.

The fluctuation of the fuel concentration in the mixture was measured by a laser absorption method without igniting the fuel. In these measurements, because CH_4 is the major component of city gas in Tokyo, a He–Ne laser light of 3.392 μm wavelength was used, as this particular wavelength is strongly absorbed by CH_4. Since the amount of absorbed light is equal to the integrated value along the entire light pass in the mixture, the light guiding tube was constructed so that the light pass in the mixture could be changed. The two stainless steel tubes with mounted magnesium fluoride windows were used for the light guiding tubes. The outer diameter of the tube was 5 mm and the inner diameter was 3 mm. The firing rate of 18.6 kW (16000 kcal/h) was kept constant and the air/fuel ratio was adjusted by changing the air flow rate. In the case of perfect mixing, the air and fuel were mixed by means of a venturi mixer and supplied to the burner through a 2 m long tube. In the case of partial mixing, gas was injected from the nozzle described previously. The distance L between the nozzle and the burner was set equal to 153 and 240 mm, respectively.

RESULTS AND DISCUSSION

The burned gases were sampled 100 mm downstream of the center of the burner. The measured NO_x are plotted against the local air/fuel ratio (λc) in Figure 2. The local air/fuel

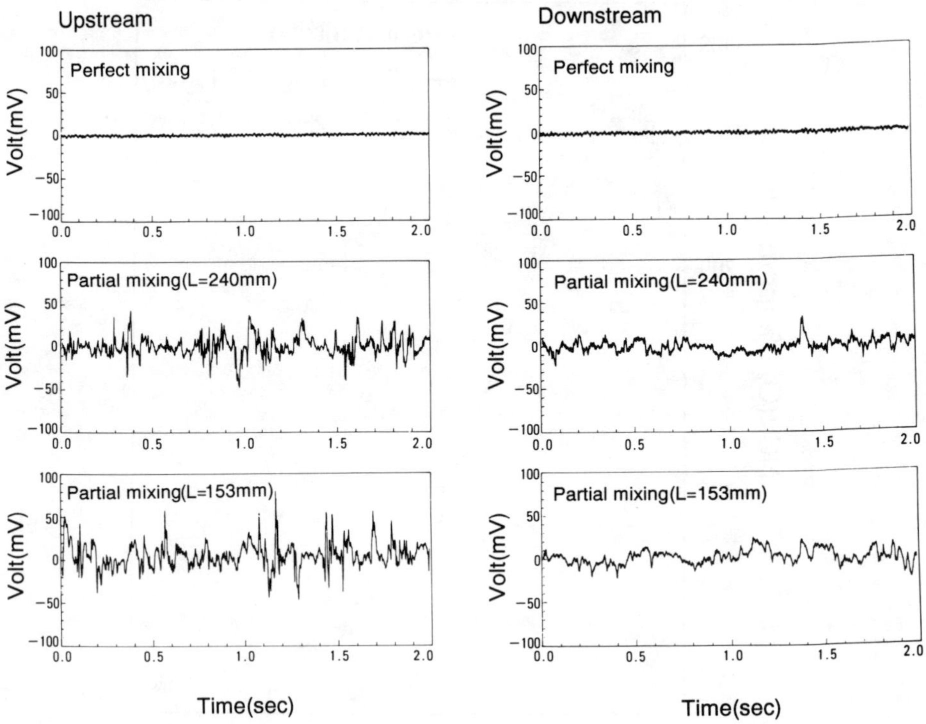

FIGURE 3 Light sensor outputs indicating air/fuel ratio fluctuations.

ratios were calculated from the burned gas composition. In the nearly stoichiometric air/fuel ratios, the NO_x levels of the two cases of the partially mixing flames (L = 240 and 153 mm) are lower than those of the perfect mixing case. A possible reason for this difference may be attributed to the existence of the temporal air/fuel ratio fluctuations. However, in the leaner side above $\lambda c = 1.2$, there is virtually no difference in the NO_x levels of the perfectly mixing flames, L = 153 and 240 mm. In Figure 2, the effect of the air/fuel ratio fluctuation on the NO_x formation was more evident near the stoichiometric air/fuel ratio. However it was not clearly observable in the lean air/fuel ratio.

To quantitatively assess the effect of the air/fuel fluctuation on NO_x formation, it is necessary to measure the degree of the fluctuation. In the premixed mixture this was measured by monitoring the CH_4 concentration fluctuation by laser absorption. First, the relationship between the air/fuel ratio and the output from the light sensor was determined with various light pass lengths. As the light pass length was shortened, the output became less sensitive to changes in air/fuel ratio. The light pass length of 8 mm was therefore chosen for the measurement.

The measurement was conducted at two points, 10 mm upstream and 10 mm downstream from the center of the noncombusting burner surfaces. The typical signals in upstream and downstream sides for perfect mixing, L = 153 and 240 mm, are shown in Figure 3. In the perfect mixing case, fluctuations were smaller than the strong fluctuations observed for L = 153 and 240 mm. A similar level of fluctuations to the small fluctuations of the perfect mixing example was observed only in the air flow. Therefore, the small fluctuations of perfect mixing probably arise due to electrical noise and not from the

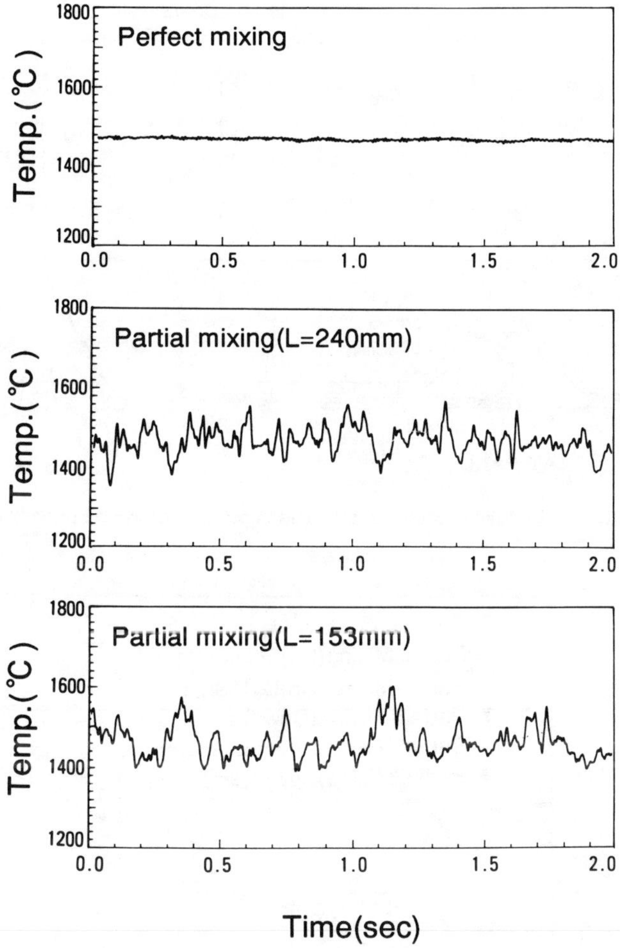

FIGURE 4 Burned gas temperature fluctuations.

air/fuel ratio fluctuation. Between the results of L = 153 and 240 mm, the difference is not apparent. Comparing the signals upstream and downstream, the signals attenuate as they pass through the ceramic burner.

Figure 4 shows the measured temperature fluctuations for a flame formed by an air/fuel ratio of 1.4. The temperatures were measured 100 mm downstream from the center of the burner. The effect of the radiation loss on the measured temperatures was not corrected. Similar to the fuel concentration fluctuations, temperature fluctuations are clearly present in the flames when L = 153 and 240 mm; this is in contrast to the small fluctuations in the perfectly premixed flame.

To quantify the degree of the fluctuations, the sensor outputs were converted into the air/fuel ratio and the probability density distributions were calculated. Figure 5 shows the probability density distributions of the air/fuel ratio in the upstream and downstream sides for a mean air/fuel ratio of 1.4. In the perfect mixing case, the air/fuel ratio is concentrated around 1.4. The small variance in the ratio may be mainly attributed to electrical noise in the light sensor. In the case of partial mixing, the air/fuel ratio varies from 1.1 to 1.7

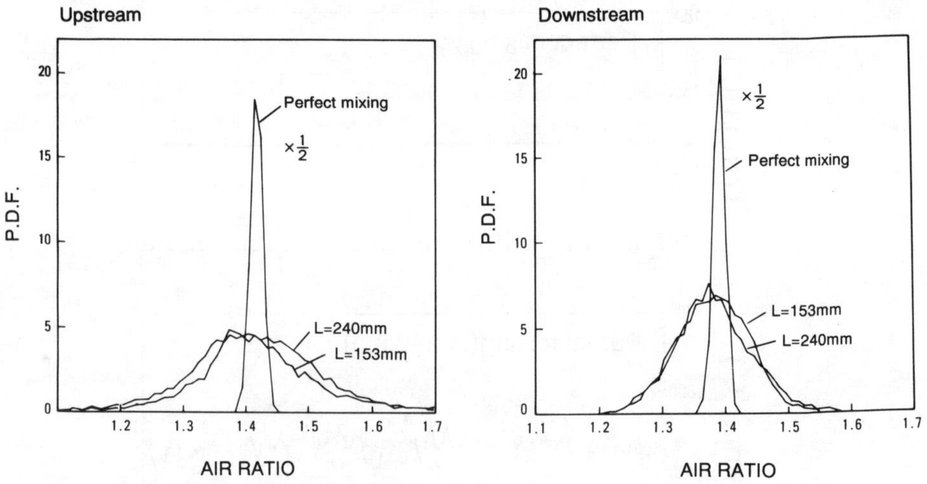

FIGURE 5 Probability density distributions in perfectly mixed and partially mixed gases.

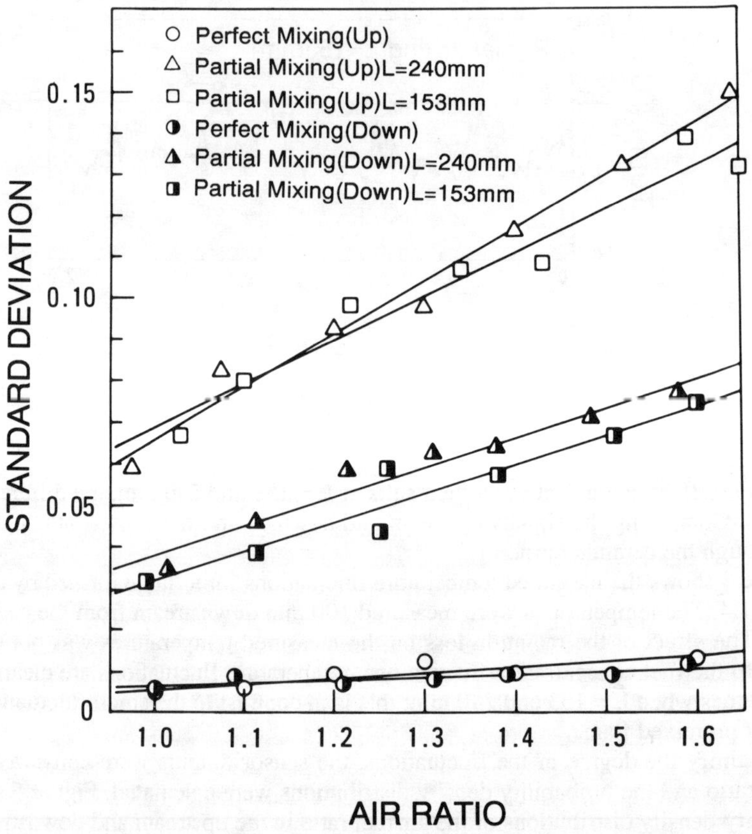

FIGURE 6 Variation of standard deviations with air/fuel ratio calculated from air/fuel ratio fluctuations.

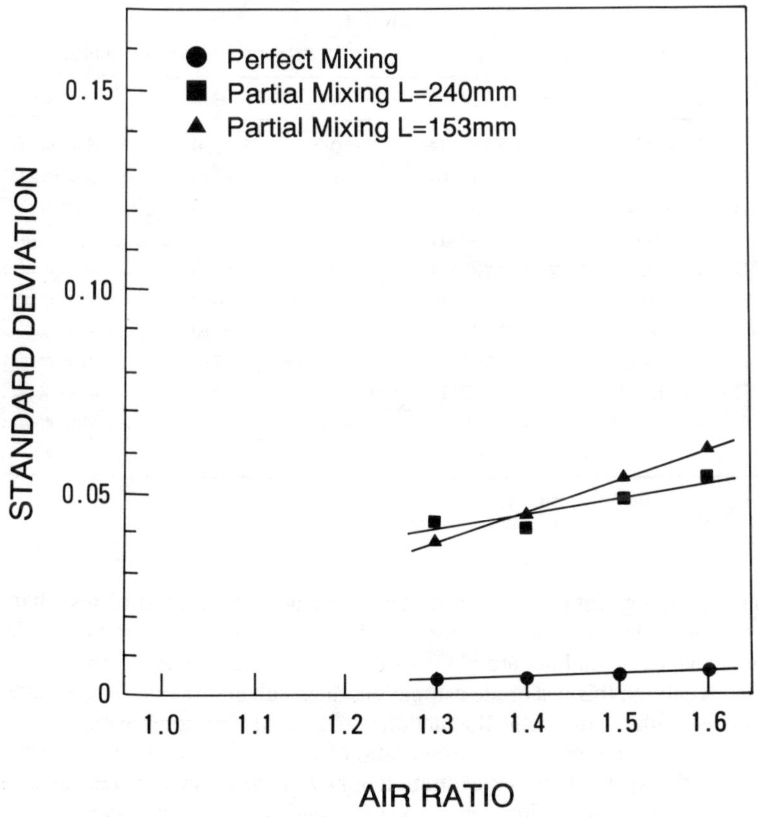

FIGURE 7 Variation of standard deviation with air/fuel ratio calculated from burned gas temperature fluctuations.

upstream, and from 1.2 to 1.55 downstream. From the distributions for each mixing condition, the standard deviations were calculated. Figure 6 shows the variation of the standard deviations with the air/fuel ratio. It is clear that the standard deviations decrease after the mixture passes the burner. The difference of the results between L = 153 and 240 mm is not clearly distinguishable. In the perfect mixing case, the standard deviations are small. As described above, the values can be attributed to the noise in the light sensor.

An estimation of the probability density distributions of the air/fuel ratio from the temperature fluctuations of the burned gases has also been tried. In the preliminary measurements of the temperature distribution in this combustion chamber, temperature profiles in the radial direction were found to be almost flat near the center of the combustion chamber. In the direction of the burned gas flow, for the flame with an air/fuel ratio of 1.4, only a 15°C decline in 100 mm was observed. Therefore, it can be assumed that the thermal state in this chamber is adiabatic. Due to radiation losses from the thermocouple, the measured temperatures are lower than the actual values, which should be close to the adiabatic flame temperature. Thus, the measured mean temperatures were corrected to the adiabatic flame temperature for the measured local air/fuel ratio. By referring to the relationship between the adiabatic flame temperature and the air/fuel ratio, the measured temperature fluctuations were converted into air/fuel ratio fluctuations, and the probability density distribution of the air/fuel ratio was obtained. The standard deviations in the air/fuel ratio distribution are shown in Figure 7. The standard deviations were obtained only for

TABLE I

CONSIDERED ELEMENTARY REACTIONS AND REACTION RATE COEFFICIENTS

No.	Reaction	A	n	E	Reference
1	$N_2 + O = NO + N$	1.84E+14	0	76250	Glarborg (1986)
2	$N + O_2 = NO + O$	6.40E+09	1	6280	Glarborg (1986)
3	$N + OH = NO + H$	3.80E+13	0	0	Glarborg (1986)
4	$CO + OH = CO_2 + H$	5.60E+11	0	1080	Mori *et al.* (1976)
5	$H + O_2 = OH + O$	2.20E+14	0	16600	Mori *et al.* (1976)
6	$P + H_2 = OH + H$	1.74E+13	0	9450	Mori *et al.* (1976)
7	$H_2 + OH = H_2O + H$	2.19E+13	0	5150	Mori *et al.* (1976)
8	$OH + OH = H_2 + O$	5.75E+12	0	780	Mori *et al.* (1976)
9	$H + H + M = H_2 + M$	3.29E+18	–1.2	0	Mori *et al.* (1976)
10	$O + O + M = H_2 + M$	2.62E+16	–0.84	0	Mori *et al.* (1976)
11	$OH + H + M = H_2O + M$	1.50E+16	0	0	Mori *et al.* (1976)

$k = A \cdot T^n \exp(-E/RT)$ (mol, cal, cm^3, K, s).

air/fuel ratios ranging from 1.3 to 1.6, because, for an air/fuel ratio of less than 1.3, the temperature fluctuations could not be measured due to the high temperature of the burned gases. The standard deviations are close to those calculated from the fuel concentration fluctuations. Although this is a crude estimation, the result shows that temperature fluctuations associated with air/fuel ratio fluctuations do occur in this combustor.

The effect of fluctuations of the air/fuel ratio of the mixture on the NO_x formation was estimated numerically. First, the concentration of NO_x in the flame of perfect mixing, which is formed in the post-combustion zone in various air/fuel ratios, was calculated by using the simple reaction scheme shown in Table I. Only the Zeldovich NO formation mechanism was considered. In this calculation, the flow in the chamber is assumed to be one-dimensional and adiabatic. As initial conditions, the equilibrium concentrations of the components, except NO and N, in the adiabatic flame temperatures of each air/fuel ratio were used, and the initial concentration of NO and N were set equal to zero. Although the residence time changes with the air/fuel ratio, it was set to 25 msec over the entire range of the air/fuel ratio.

To estimate the effect of the air/fuel ratio fluctuation on the NO formation, the model proposed by Heywood *et al.* (1972) was used. In this model, the burned gas is assumed to be composed of a large number of fluid eddies having a Gaussian distribution of air/fuel ratio about the mean value. The NO formed in the burned gas in the presence of air/fuel ratio fluctuations is calculated from equations (1) and (2). The standard deviation σ in equation (2), which represents the degree of air/fuel ratio fluctuation, was expressed as a function of the ratio. The function approximates the data in Figure 6.

$$\overline{[NO]} = \int_0^\infty [NO](\lambda c, t) \cdot f(\lambda c) \cdot d\lambda c, \tag{1}$$

$$f(\lambda c) = 1/\sigma\sqrt{2\pi} \cdot \exp\left[-\left(\lambda c - \overline{\lambda c}\right)^2 / 2\sigma^2\right], \tag{2}$$

where $[NO](\lambda c,t)$ is the NO formed in the burned gas at the adiabatic flame temperature for λc with a residence time of t sec, and σ is the standard deviation.

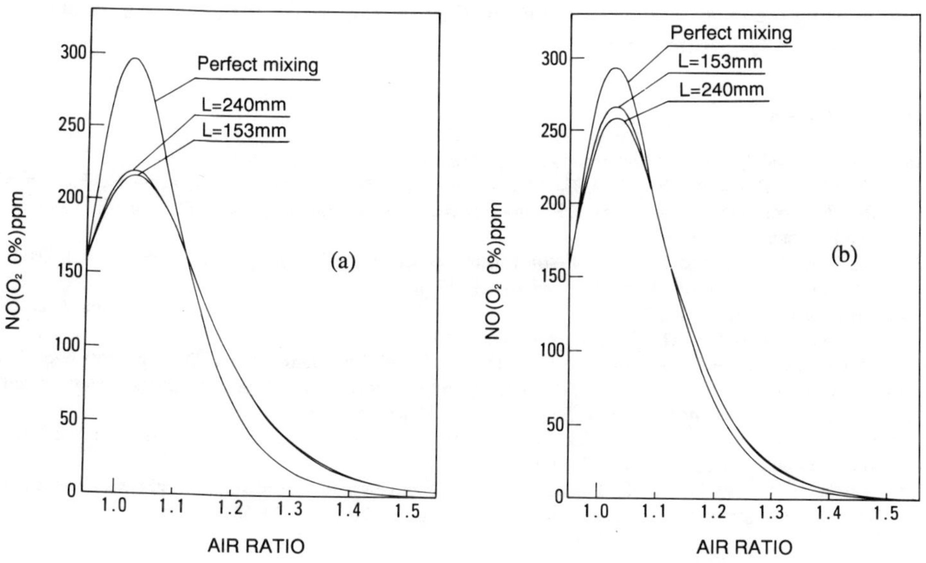

FIGURE 8 Predicted NO levels using the standard deviation measured from air/fuel ratio fluctuation upstream (a) and downstream (b).

Figures 8a and 8b show the predicted NO levels with the measured standard deviations upstream and downstream, respectively. In Figure 8a, clear differences in the NO formation can be observed for perfect mixing and L = 153 and 240 mm. Near the stoichiometric air/fuel ratio, the NO levels in the flames of perfect mixing are higher than those for L = 153 and 240 mm. However, in the leaner air/fuel ratios, more NO is formed in flames of L = 153 and 240 mm than for perfect mixing. The difference in NO levels between L = 153 and 240 mm is small.

In Figure 8b, the difference between perfect mixing and L = 153 and 240 mm is smaller than that in Figure 8a. The reason, as described above, is because fluctuations in the air/fuel ratio are attenuated on passing through the multiport burner. Near the stoichiometric air/fuel ratio, the predicted NO for L = 153 mm shows higher NO values than for L = 240 mm. This is because air/fuel fluctuations for L = 153 mm are slightly less than those for L = 240 mm.

From the results of the simulation, it can be concluded that if the air/fuel ratio fluctuates, NO_x formation decreases near the stoichiometric air/fuel ratio, while NO_x formation increases for the lean air/fuel ratio. Comparing the predicted NO_x with the experimental results, the predicted effect of air/fuel fluctuation on NO_x formation using the measured standard deviation in the downstream side agrees well with the experimental results.

CONCLUSION

Laser absorption was used to quantitatively measure the degree of air/fuel ratio fluctuations in the mixture. The degree of fluctuation was obtained as a standard deviation for the probability density distribution of the air/fuel ratio. The standard deviation is almost halved by passing the mixture through the multiport burner. Standard deviations were also obtained from the burned gas temperature fluctuations. The deviations from the temperatures were close, but slightly less than those from air/fuel ratio fluctuations. Numerical estimation by using the measured air/fuel ratio fluctuations can predict the effect of air/fuel ratio fluctuations on NO_x formation.

REFERENCES

Anderson, D.N. (1973). *Effect of Premixing on Nitric Oxide Formation*, NASA TMX-68220.

Appleton, J.P., and Heywood, J.B. (1973). The Effects of Imperfect Fuel/Air Mixing in a Burner on NO Formation from Nitrogen in the Air and the Fuel, *Fourteenth Symposium (International) on Combustion*, The Combustion Institute, 777–786.

Glarborg, P., Miller, J.A., and Kee, R.J. (1986). Kinetic Modeling and Sensitivity Analysis of Nitrogen Oxide Formation in Well-Stirred Reactors, *Combust. and Flame*, **65**, 172–202.

Knugi, M., and Jinno, H. (1959). Measurement of Fluctuating Flame Temperature, *Seventh Symposium (International) on Combustion*, The Combustion Institute, 942–948.

Lyons, V.J. (1982). Fuel/Air Nonuniformity: Effect on Nitric Oxide Emissions. *AIAA J.*, **20**, 5, pp. 660–665.

Mori, Y., Ootake, K., and Nakamura, K. (1976). A Study on NO Formation in Combustion Reaction. *Transactions of The Japan Society of Mechanical Engineers*, **42**, 357, pp. 1511–1524.

Pompei, F., and Heywood, J.B. (1972). The Role of Mixing in Burner-Generated Carbon Monoxide and Nitric Oxide. *Combust. and Flame*, **19**, 407–418.

Semerjian, H.G., and Ball, I.C. (1978). Pollutant Emissions From Partially Mixed Turbulent Flames, *Seventeenth Symposium (International) on Combustion*, The Combustion Institute, 679–687.

Control of Stoichiometric Propellant Supply in an H_2/O_2-Spinning Reserve System

S. BEER, O.J. HAIDN, and H. WILLMS *German Aerospace Research Establishment (DLR), Langer Grund, Hardthausen, Germany*

Abstract — The concept of a H_2/O_2-spinning reserve system for conventional power plants and its control system for stoichiometric propellant supply is described. The H_2/O_2 based spinning reserve concept is able to reduce, in comparison with conventional methods, the consumption of fuel and the output of exhaust gases as well. The control concept is based on the on-line metering of unburnt propellants (H_2 and O_2) by a calcia-stabilized ZrO_2-sensor. Experiments have been carried out to study the time dependant behaviour of the sensor against H_2 and O_2 in superheated steam. The results show that the suggested control concept which is based on the ZrO_2-sensor is sufficient for a proper operation of the steam generator.

Key Words: hydrogen/oxygen spinning reserve system, control of stoichiometric supply, zirconia sensor, on-line hydrogen/oxygen concentration monitoring

INTRODUCTION

The continuous adaption of the generation of electricity to the actual power consumption within a network is an important task in the field of electric power supply. Since every unexpected rise in power consumption has to be compensated almost immediately, UCPTE (Union for Coordination of Production and Transport of Electrical Energy) and DVG (Deutsche Verbundgesellschaft) have settled requirements for electric power utilities. The negative frequency deviations within the UCPTE network and their cumulative duration per year are listed in Table I.

According to those requirements every company has to be able to activate a spinning reserve of about 5% of its total momentary power capacity; the reserve of a power plant unit should amount to at least 5% of its nominal capacity, half of it being activated within 5 seconds, the remainder within the following 25 seconds. These requirements are even more stringent for isolated, smaller networks.

METHODS FOR SPINNING RESERVE

Today's conventional method, i.e., live steam throttling, provides a reserve of steam in the boiler. Opening the control valve yields an increase in steam mass flow until a new equilibrium pressure has been established in the steam generating unit, typically after some 200 s. As the HP turbine is loaded by a high mass flow only in the first seconds — the middle and low pressure turbines follow with a time constant corresponding to the accumulation time of the intermediate superheater — the resulting power increase corresponds only to the contribution of the high pressure turbine to the overall output of the turbogenerator set. This means that the degree of throttling for fulfilling the requirements of UCPTE or DVG has to be around twice the desired power increase in the first 5 s. Thus provision of, for instance, 5% of the momentary power plant output as spinning reserve in the first 5 s requires that the live steam flow has to be throttled down by 10% (see e.g., Falgenhauer and Kurten, 1985).

The costs of a spinning reserve system based on the live steam throttling concept are made up of the specific excess fuel consumption due to throttling, typically 0.04 percentage points per degree of throttling, and in particular from the capital costs of the power plant

TABLE I

CUMULATIVE DURATION OF NEGATIVE FREQUENCY DEVIATIONS FROM 50 Hz NOMINAL VALUE IN THE UCPTE NETWORK PER YEAR

Negative frequency deviation [mHz]	Duration of deviation [min/year]
0–25	192.000
25–50	42.300
50–75	13.600
75–100	1.900
100–150	350
>150	20

capacity which has to be foreseen as the throttling reserve. Besides, the fact that the turbogenerator set does not work at its optimum working place which leads to a somewhat lower specific turbine efficiency, energy losses due to the irreversible throttling process cause, for the same electric output, a higher fuel consumption. As a result live steam throttling results in an unnecessary high output of exhaust gases and dust and lowers the life time of the denitrification catalyst.

Sternfeld *et al.* (1989) have shown that the generation of steam by the reaction of hydrogen and oxygen, subsequently injecting preheated feed water, is a way of creating a spinning reserve system which represents an alternative to live steam throttling. Figure 1 shows a flow diagram of a fossil-fired power plant with H_2/O_2-steam supply for spinning reserve.

The energetics section of DLR (see e.g., Beer (1991), Sternfeld (1984) and Sternfeld and Wolfmüller (1985)) has been engaged for a couple of years in the development of a steam generator which is able to produce steam at 40 bar, 530°C and a thermal power of 40 MW within only one second. A schematic of this experimental version of a steam generator is shown in Figure 2. Downstream of the fuel injection head is the combustion chamber which is followed by the feed water injection area and the vaporization zone.

In order to achieve an increase in the electrical output of the power plant of 12 MW, and at the same time to meet the DVG requirements, the H_2/O_2-steam generator has to operate within the first 5 s at a thermal output (referred to hydrogen input) of around 70 MW. During the following 25 s, this is reduced linearly to a steady-state value of approximately 43 MW. Figure 3 shows the necessary thermal output capacity of the H_2/O_2-steam generator to fulfill the above-mentioned requirements for the electrical output.

STOICHIOMETRIC PROPELLANT SUPPLY

While the reaction of H_2 and O_2 is a common process in space technology, the control of this process for application in a spinning reserve system is not an easy task. Temperature, pressure and steam mass flow have to be adjusted according to the requirements of the power plant. Special attention to the stoichiometry of the propellant supply is essential for such a spinning reserve system: hydrogen residues in the steam cause decarburization of steel and reduce the life time of the turbine blades; unburnt hydrogen in combination with surplus oxygen yields a combustible atmosphere in the condenser; the remaining oxygen causes rust formation in the entire flow system. Additionally, both gases are non-conden-

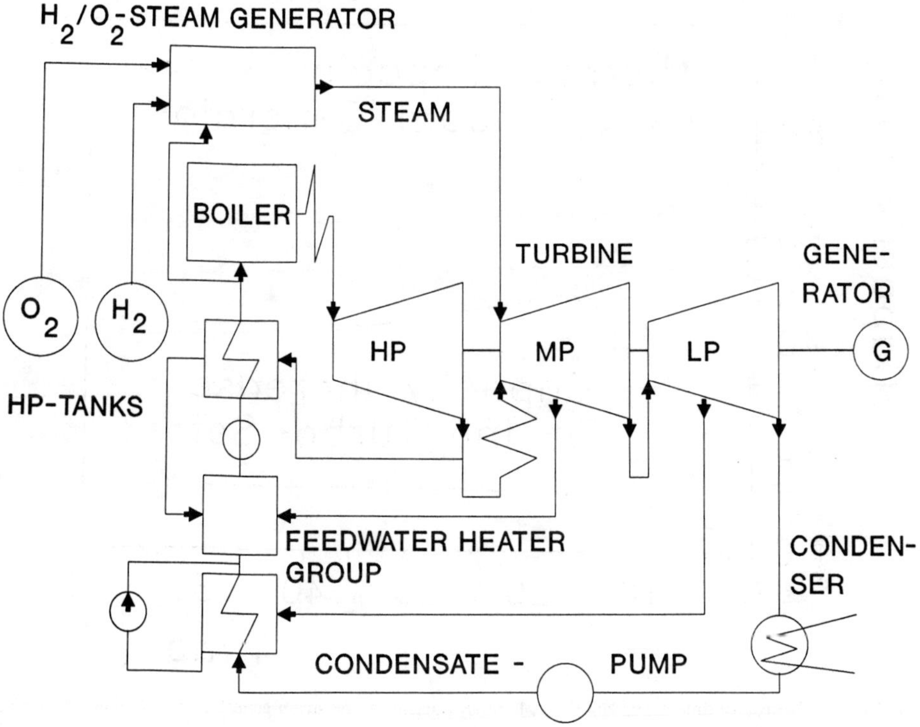

FIGURE 1 Flow diagram of a fossil-fired power plant with an H_2/O_2-steam supply for spinning reserve.

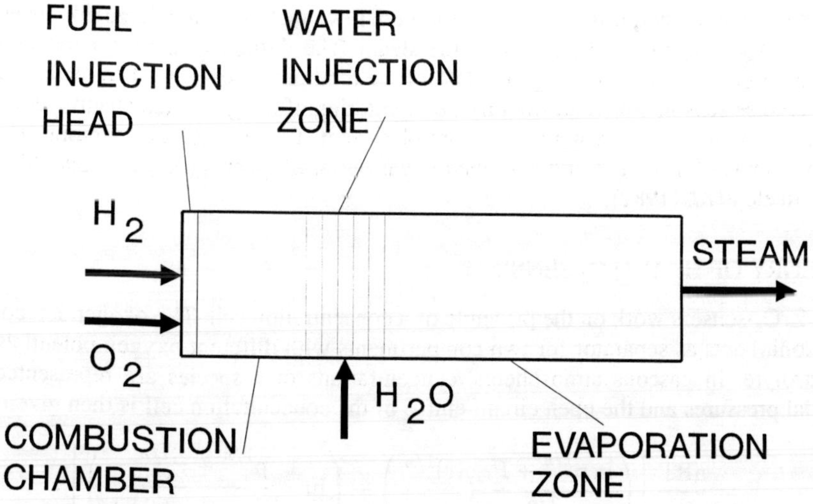

FIGURE 2 Schematic of the H_2/O_2-steam generator.

sable and therefore lower the efficiency of the condenser. The only way to overcome that problem is the precise control of the stoichiometric supply of the propellants.

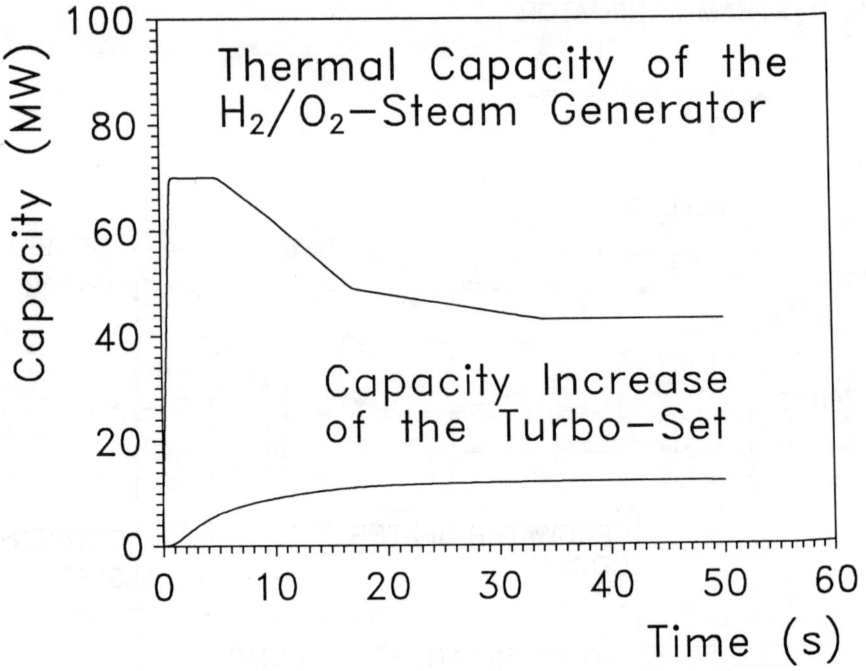

FIGURE 3 Necessary time-dependant thermal output capacity of the steam generator and resultant capacity increase of the turbo set.

The inherent measuring error of common flow meters is too large for meeting the requirements regarding the purity of the steam. The desired maximum concentration of oxygen is 0.3%, that of hydrogen 0.01% of the total mass flow rate. The readout of a fast and precise sensor, which allows on-line monitoring of the generated steam quality, could be used as input for the automatic control system. Calcia-stabilized zirconia is a well-known sensor for oxygen and hydrogen in various atmospheres (see, e.g., Schottky (1935) and Steele *et al.* (1981)).

THEORY OF H_2 AND O_2 SENSING

The ZrO_2-sensors work on the principle of a concentration cell. The oxygen ion conductor (zirconia) acts as separator for two compartments with different oxygen potentials and as electrolyte. In gaseous atmospheres, concentrations of a species are represented by its partial pressures and the open circuit emf E of the concentration cell is then given by:

$$E = \frac{nRT}{4F}\left[\left(\frac{\ln p_-^{1/n} + P_{O_2}(r)^{1/n}}{p_-^{1/n} + P_{O_2}^{1/n}}\right) + \left(\ln \frac{p_+^{1/n} + P_{O_2}^{1/n}}{p_+^{1/n} + P_{O_2}(r)^{1/n}}\right)\right], \tag{1}$$

where R, T, and F have their usual significance, p_- and p_+ are concentration measures for the number of electrons and holes in the electrolyte, and $PO_2(r)$ and PO_2 are the partial pressures of oxygen in the reference gas and the steam, respectively. Taking into account that

$$p_- < P_{O2} < P_{O2}(r) < p_+ \tag{2}$$

yields the well-known Nernst relation:

$$E = \frac{RT}{4F} \ln \frac{P_{O_2}(r)}{P_{O_2}}. \tag{3}$$

Knowing the partial pressure P_{O2} and the temperature T, the unknown partial pressure P_{O2} is obtained. It has to be mentioned that (3) holds only as long as (2) is fulfilled.

When P_{O2} is of the same order of magnitude as p_-, (1) has to be rewritten as:

$$E = \frac{nRT}{4F} \left[\ln \frac{P_{O_2}(r)^{1/n}}{p_-^{1/n} + P_{O_2}^{1/n}} \right]. \tag{4}$$

If there is no oxygen present in the steam ($P_{O2} = 0$), the output of the probe depends only on the partial pressure of oxygen in the reference gas $P_{O2}(r)$ and the number of free electrons in the electrolyte p_-:

$$E(P_{O_2} = 0) = E_0 = \frac{RT}{4F} \ln \frac{P_{O_2}(r)}{p_-}. \tag{5}$$

For the probe in use this value is $E_0 = 405$ mV. When there is hydrogen in the steam at the anode, the emf E is given by the following equation:

$$E = E_0 + \frac{RT}{2F} \ln \frac{P_{H_2}}{P_{H_2O}}. \tag{6}$$

Unfortunately the probe indicates the presence of H_2 only, even if there are H_2 and O_2 present in the steam. The hydrogen in the steam reacts preferably with the activated oxygen at the three-phase contact of the sensor. This reaction reduces the number of molecules at the anode and induces an additional oxygen ion transport through the electrolyte resulting in a higher emf signal.

The presence of both gases in the steam is caused by small geometrical inaccuracies of the coaxial H_2/O_2 injection elements and a coupling of thermodynamic and aerodynamic effects which yield small fluid elements with nonstoichiometric mixtures of propellants. These fluid elements cannot fully decompose in the steam generator because its residence time is too small (typically 10–20 ms). Furthermore the persistence of unburnt propellants is promoted by quenching through injection of preheated water and rarefication due to the evaporation of the injected water.

EXPERIMENTAL APPARATUS

Sensor Element

Figure 4 shows a sketch of the sensor with the gas/steam mixture supply. The temperature of the ZrO_2-sensor element is kept constant at 830°C by electrical heating. There are two reasons why the sensitivity of the ZrO_2 probe increases when it is operated at higher temperatures. First, the emf E is directly proportional to the temperature and second, the electrolyte becomes an exclusive oxygen ion conductor at temperatures above approximately 400°C. Heat losses from hot sensor elements are reduced by thermal insulation. The sensor itself is coated on both sides with a layer of porous platinum. The reference side of the sensor element is fed with hot air with a known partial pressure of oxygen $P_{O2}(r)$.

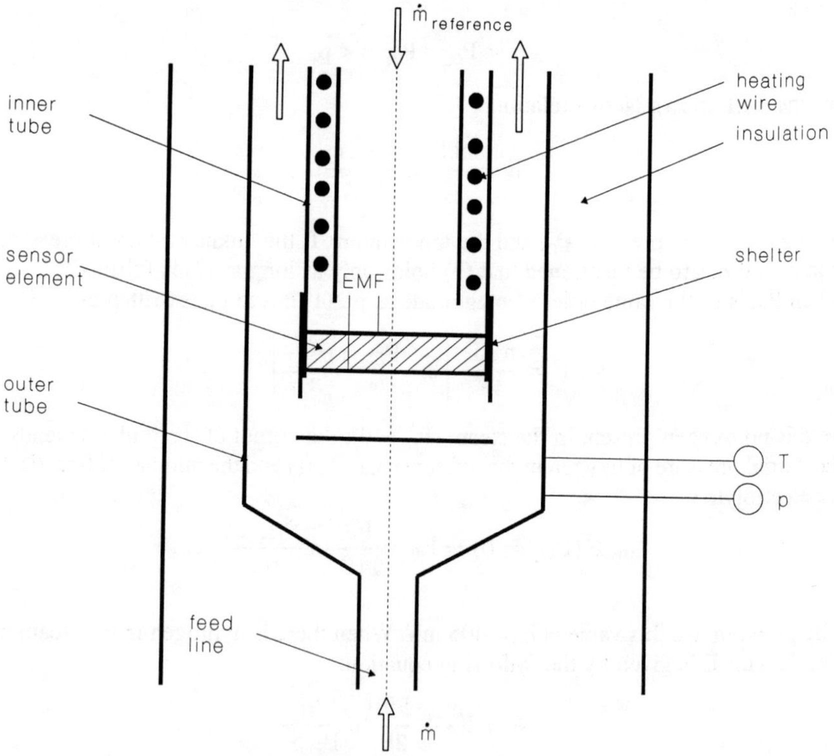

FIGURE 4 Schematic of ZrO_2 sensor.

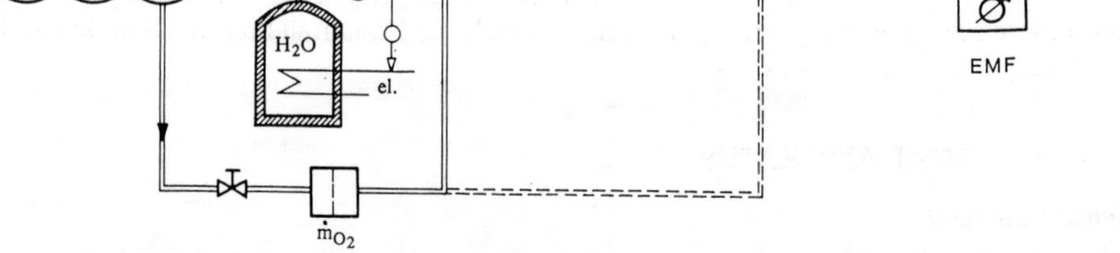

FIGURE 5 Schematic of calibration setup.

The steam side of the sensor is sheltered against direct contact of water droplets which can cause damage to the hot sensor element. The tube containing the sensor element and the heating device is a commercially available Westinghouse oxygen concentration probe. The outer tube is especially designed to meet the following conditions: Improvement of the preheating behaviour and minimizing the overall time constant of the probe by reducing the gas volume in front of the sensor element.

Sensor Calibration

The unknown constant E_0 in (6) has to be determined by calibration of the probe. This constant is a function of the material properties, the operating temperature and the geometry of the sensor element. It is useful to repeat the calibration procedure from time to time because impurities in the gases which pass through the sensor can diffuse into the sensor element and slightly alter the material properties causing a measurement failure. Figure 5 shows a sketch of an apparatus designed to calibrate the sensor for measuring oxygen or hydrogen in super-heated steam. The apparatus allows concentrations of hydrogen between 0.008 and 0.16% and 0.1 to 2.0% for oxygen with a maximum failure of ±4%. The total mass flow of steam can be varied between 0 and 5.31×10^4 kg/s. The steam temperature in front of the sensor element can be varied between 100 and 600°C at ambient pressure. The steam is generated by boiling demineralized water and superheated by electrical heating. The water contains dissolved gases like O_2, N_2 and CO_2. They are removed by boiling before the calibration procedure starts. Experiments have shown that the results for E_0 are the same for pure steam, nitrogen or argon.

Calibration Results

Figure 6 shows the emf E over concentrations of hydrogen and oxygen in pure steam. The equations to describe the data points are derived from a least squares fit method to give

$$E = 66 - 19.54 \ln C_{O2}, \tag{7}$$

$$E = 900 + 47.77 \ln C_{H2}. \tag{8}$$

These forms of equation can only be used if the temperature of the sensor element is kept constant during the calibration process and the measurements. The sensitivity of the sensor element for oxygen is

$$\frac{dE}{dC_{O_2}} = -\frac{19.54}{C_{O_2}}, \tag{9}$$

and for hydrogen

$$\frac{dE}{dC_{H_2}} = \frac{47.77}{C_{H_2}}. \tag{10}$$

The derivative of (7) and (8) with respect to the concentration yields the sensitivity of the probe to changes in the concentration of oxygen and hydrogen. An analysis of (9) and (10) shows that:

— the sensitivity of the probe increases for both gases with decreasing concentration;
— the sensitivity against hydrogen concentration changes is about 2.5 times higher than that for changes in oxygen concentration;
— changes in hydrogen and oxygen have opposite signs.

Time Constants

Time constants are measured by imposing a concentration step of hydrogen or oxygen on the sensor. Figure 7 shows the results of these tests for a hydrogen step function of 0.35% and an oxygen step function of 2.0%. For both cases the time constant is less than 100 ms.

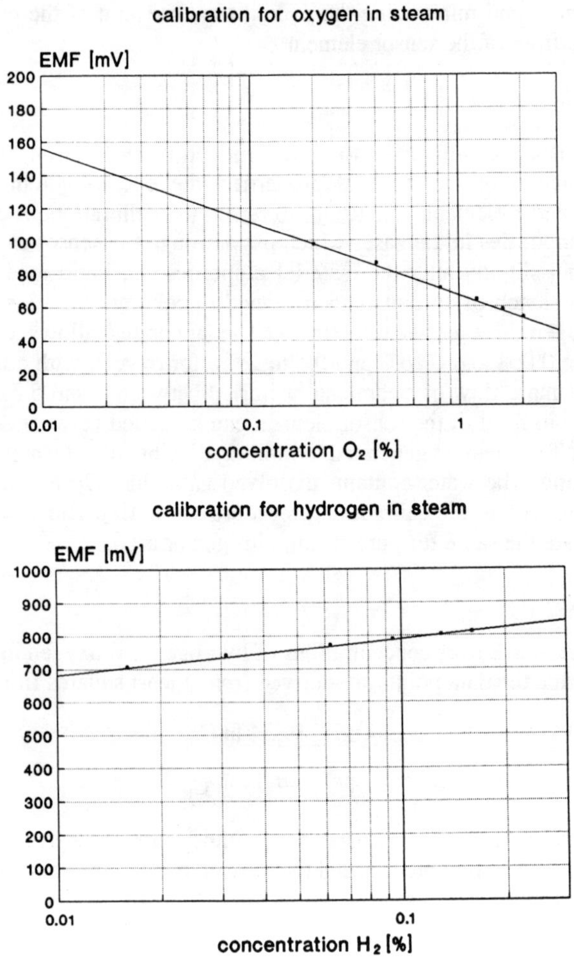

FIGURE 6 emf E for oxygen and hydrogen.

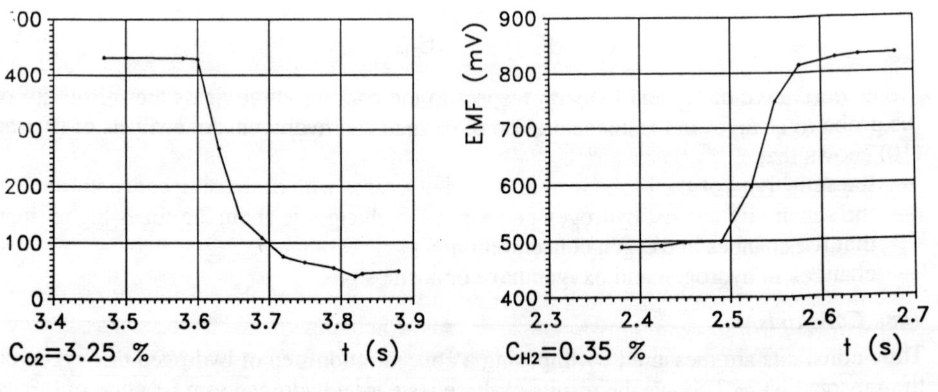

FIGURE 7 emf E vs. concentration step functions of hydrogen and oxygen.

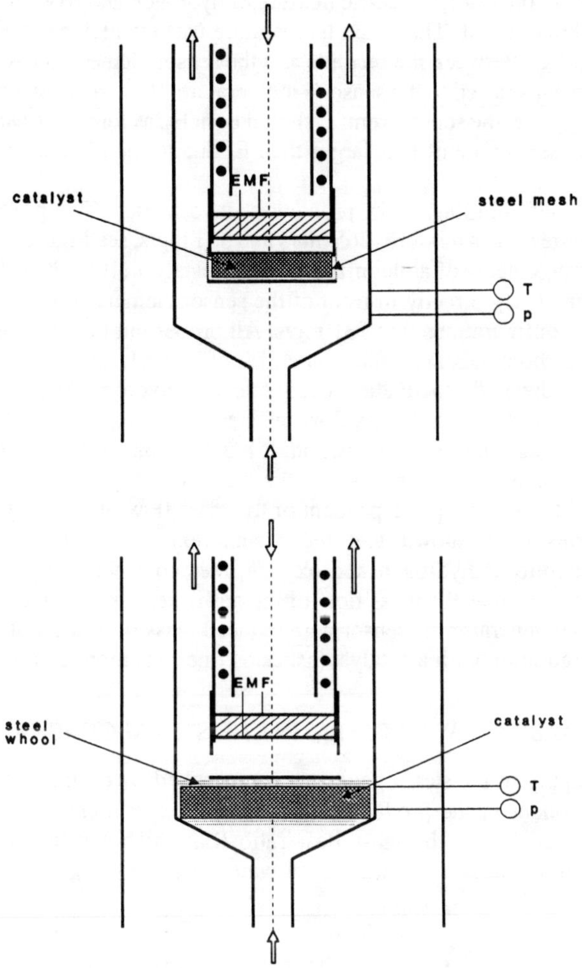

FIGURE 8 Schematic of the two sensor configurations with catalyst.

Theoretically, a time constant can be derived by dividing the mass of steam contained in the outer tube of the sensor in front of the sensor element (see Figure 4) by the total mass flow of steam through the sensor. The measured and the calculated values for the time constant are in fair agreement and they also agree with the literature (see, e.g., Pfeifer and Wertheimer (1985)). Therefore it can be assumed that this time constant is caused by gas exchange in the sensor.

Measurement of Concentration in the Presence of Hydrogen and Oxygen

As already mentioned both components are always present in the steam, even if the mass flow ratio R_{of} of oxygen to hydrogen is exactly stoichiometric. When the sensor is used as an on line measuring device for R_{of} in the steam generator, it has to detect only the surplus component. Therefore the sensor's behaviour in a mixture of steam with H_2 and O_2 has to be investigated.

For this investigation a stoichiometric mixture of hydrogen and oxygen was added to the steam in the calibration unit. The steam temperature was maintained at 500°C, the sensor temperature at 830°C. Between the feedline and the sensor element, the temperature of the steam/gas mixture was raised to the sensor's temperature. If the reaction between hydrogen and oxygen is complete, the sensor's emf E should be at E_0, as for pure steam. If the reaction is incomplete, the sensor's emf E is larger than E_0 due to the remaining hydrogen in the steam.

Four different configurations of the probe were investigated. Two probes with different mass flows of the steam gas mixture (265 mg/s and 531 mg/s, see Figure 4), and two probes with different arrangements of a platinized asbestos catalyst (10% Pt) with 1 g of catalyst in the feed line and 0.1 g directly in front of the sensor element (see Figure 8). The mass flow for the later configurations was 531 mg/s. All probes but the one with the mass flow of 531 mg/s and without a catalyst show an emf E of E_0 fairly well. This is due to the fact that the catalyst reduces the activation energy for the hydrogen/oxygen reaction in the steam. In the probe with reduced mass flow the steam/gas mixture was heated to a higher temperature promoting the chemical reaction. It must be considered that the time constant of the sensor is changed to higher values if the mass flow through the sensor is reduced. The calibration of the sensor is independent of the mass flow of the mixture.

The experiments have shown that the assumption if the steam contains a non-stoichiometric mixture of hydrogen and oxygen, the probe will only detect the surplus component holds as long as the mass flow of the gas is kept small enough. For measuring the R_{of} in the steam generator the sensor with reduced mass flow without a catalyst is used because the configuration with a catalyst (asbestos fibers) is more complicated to handle.

CONTROL OF MASS FLOW RATIO R_{OF} IN THE STEAM GENERATOR

The propellant supply in the steam generator is controlled by on-line measurement of the surplus gas component in the produced steam. Three experiments were conducted with three different set points for the mass flow ratio R_{of}, which can be calculated from the concentrations and the mass flow ratio R_{wg} of water and gases to give

$$R_{of} = \begin{cases} \frac{7.94 + C_{O_2}(R_{wg}+1)}{1 - C_{O_2}(R_{wg}+1)} & \text{for } R_{of} \geq 7.94 \\ \frac{1 - C_{H_2}(R_{wg}+1)}{1/7.94 + C_{H_2}(R_{wg}+1)} & \text{for } R_{of} \leq 7.94 \end{cases} \tag{11}$$

where mass flow ratio R_{wg} depends on the set points for the steam temperature and pressure, on the temperature of the injected water and on the thermal efficiency of the steam generator. This efficiency depends only on the concentration of the unburnt hydrogen in the steam. The heat losses from the system are negligible for steady state operation because the steam generator is regeneratively cooled.

In Table II the relevant experimental set points for the control system R_{of}, steam temperature and pressure and the measured R_{wg} in the stationary state are given.

Figure 9 shows the measured emf E of the steam generator over time for a given set point of R_{of} = 7.90($\{C_{H2}\}$ = 0.013%) and the calculated R_{of} according to (7), (8) and (11). Beginning with the fourth second a gas sample was taken over 10 seconds. This sample was analyzed using a gas chromatograph for comparison. During the first 3 seconds the emf E cannot be used as an indicator because of residual air inside the sensor.

To compare the concentration measured with the sensor to the gas chromatographic analysis, a time averaged concentration value has been calculated according to

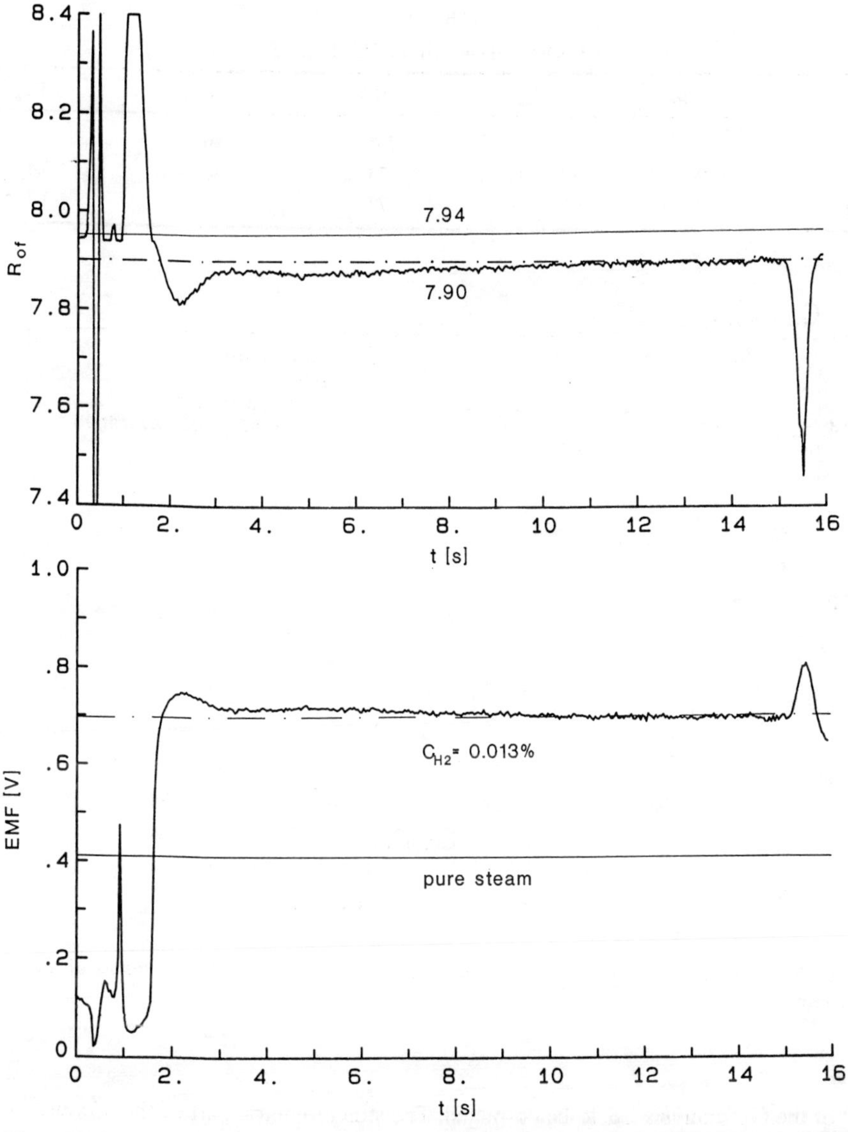

FIGURE 9 R_{of} and emf E of the H_2/O_2-steam generator.

$$C_{O_2,av} = \frac{1}{t_2 - t_1} \int_{t_1}^{t_2} C_{O_2}(t)\,dt \text{ and } C_{H_2,av} = \frac{1}{t_2 - t_1} \int_{t_1}^{t_2} C_{H_2}(t)\,dt. \quad (12)$$

The time interval for the integration of (12) is the same as the time interval during which the steam probe is taken.

In the gas sample there is, in addition to the surplus component, a stoichiometric mixture of hydrogen and oxygen due to incomplete reaction in the steam generator. Therefore the

TABLE II

SET POINTS FOR THE EXPERIMENTS

R_{of}	C_{H2}	C_{O2}	T [K]	P [bar]	R_{wg}
7.94	0	0	773	40	3.55
7.98	0	0.1	773	40	3.4
7.90	0.013	0	773	40	3.5

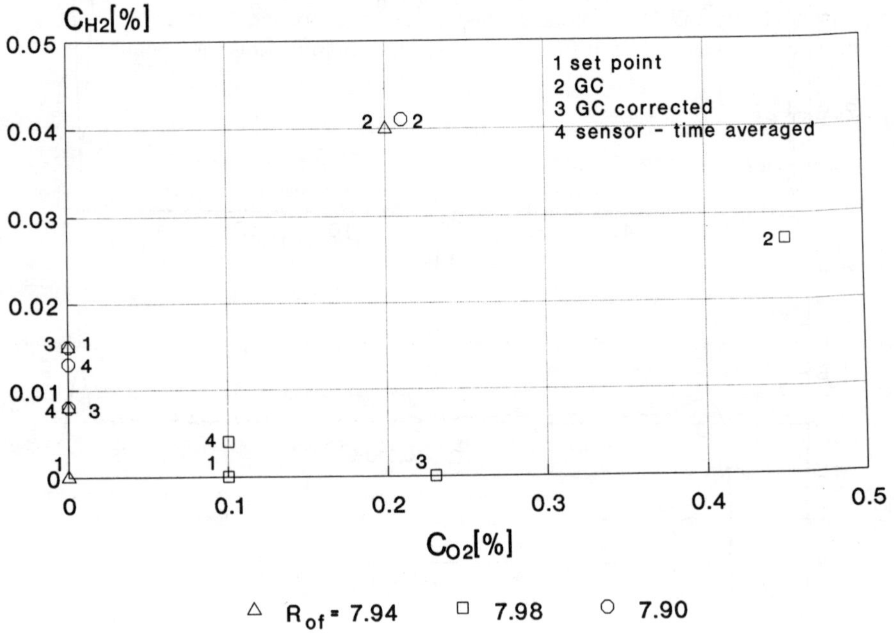

FIGURE 10 Comparison of the gaschromatic analysis of the gas sample with the results of the sensor measurements.

result of the GC analysis has to be corrected. The stoichiometric part of the mixture in the sample is defined by $CO_2 = 7.94 \times CH_2$. The corrected result yields the surplus component in the sample. Figure 10 shows a comparison for the experiments defined in Table II. Number 1 corresponds to the concentration set point for the control system, number 2 to the result of the GC analysis which is the real content of hydrogen and oxygen in the steam, number 3 to the corrected result of the GC analysis and number 4 to the time averaged result of the probe signal calculated using (12).

From inspecting Figure 10 one can see that the gas concentrations measured in the steam sample (numbers 2) are far away from the set points (1s) because of incomplete reaction in the steam generator. However the corrected values (3s) are in good agreement with the set points. This means that the control system based on the sensor is working properly although there is an incomplete reaction. The time averaged values (4s) are very close to the set points (1s). If the system were to work correctly, the results for the cases 1, 3 and 4 should fall on one point.

CONCLUSION AND PERSPECTIVES

The experiments with the ZrO_2-sensor and the steam generator have shown that this sensor can be successfully used as an in situ measuring device for direct control of the stoichiometric propellant supply in a H_2/O_2-steam generator and that this steam generator is a promising method for a spinning reserve system. This method reduces fuel consumption and exhaust gas output. A demonstration of the successful operation of the H_2/O_2-spinning reserve concept in a commercial power plant is planned for early 1993.

Once the hydrogen prices become low hydrogen from solar energy this kind of steam generator can even be used in peak load power plants. Electrization of water and burning the propellants H_2 and O_2 to produce energy means: Store the sun's energy in gaseous hydrogen and oxygen and set it free again to produce nothing but water again.

REFERENCES

Arbeitsgemeinschaft H_2/O_2 Sofortreserve-Dampferzeuger, Abschlußbericht Phase B: Nachweis der technischen und wirtschaftlichen Durchführbarkeit einer H_2/O_2 Sofortreserve-Dampferzeugungsanlage im industrieähnlichen Demonstrationsmaßstab, Stuttgart, 1988.

Beddies, G. (1990). Eichung einer ZrO_2-Sonde zur Bestimmung von H_2- und O_2-Konzentrationen in überhitztem Dampf, interner Bericht IB 64390/11, Lampoldshausen.

Beer, S. (1991). Theoretical Investigation of the Injection and Evaporation of Water in a Hydrogen/Oxygen Steam Generator, ESA-TT-1252.

Falgenhauer, G., and Kürten, H. (1985). Zweckmäßigkeit von Maßnahmen zur schnellen Leistungssteigerung in Dampfkraftwerken, VGB-Kraftwerkstechnik 65, no. 4.

Pfeifer, J.L., and Wertheimer, H.P. (1985). Heated oxygen sensor based on thick-film titania, *Inst. Mech. E. Conf. Publ.*, **12**, pp. 357–363.

Schottky, W. (1935). Über stromliefernde Prozesse im Konzentrationsgefälle fester Elektrolyte, *Veröffentlichungen aus den Siemens-Werken*, XIV, 2.

Steele, B.C.H., Drennan, J., Slotwinski, R.K., Bonanos, N., and Butler, E.P. (1981). Factors Influencing the Performance of Zirconia-Based Oxygen Monitors. In: *Advances in Ceramics*, Vol. 3, Am. Cer. Soc., Columbus.

Sternfeld, H.J. (1984). A Hydrogen/Oxygen Steam Generator for the Peak-Load Management of Steam Power Plants, Hydrogen Energy Progress V, *Proc. of the 5th WHEC*, Toronto, Pergamon Press, Vol. 4.

Sternfeld, H.J., Kusterer, H., Kunzmann, R., and Schippers, K. (1989). H_2/O_2 Direktdampferzeugung: Zusammenwirken mit Dampfturbinen und Dampfnetzen, VGB Kraftwerkstechnik 69, 7.

Sternfeld, H.J., and Wolfmüller, K. (1985). Spinning Reserve for Power Plants from Stored Hydrogen, Hydrogen Energy Progress VI, *Proc. of the 6th WHEC*, Vienna, Pergamon Press.

Weissbart, J., and Ruka, R. (1962). A Solid Electrolyte Fuel Cell, *J. of Electrochemical Society*, **109**, No. 8.

LIST OF SYMBOLS

C_{H2}	$= m_{H2}/(\dot{m}_g + \dot{m}_w)$
C_{O2}	$= \dot{m}_{O2}/(\dot{m}_g + \dot{m}_w)$
E	emf
F	Faraday coefficient
P_{O2}	Partial O_2 pressure in the steam
P_{H2}	Partial H_2 pressure in the steam
$P_{O2}(r)$	Partial O_2 pressure in the air
p_+	Partial pressure of holes in ZrO_2
p_-	Partial pressure of e^- in ZrO_2
R	Gas coefficient
R_{of}	$= \dot{m}_{O2}/(\dot{m}_{H2})$
R_{wg}	$= \dot{m}_w/\dot{m}_g$
T	Temperature

t	Time
t_1	Start of gas sampling
t_2	End of gas sampling

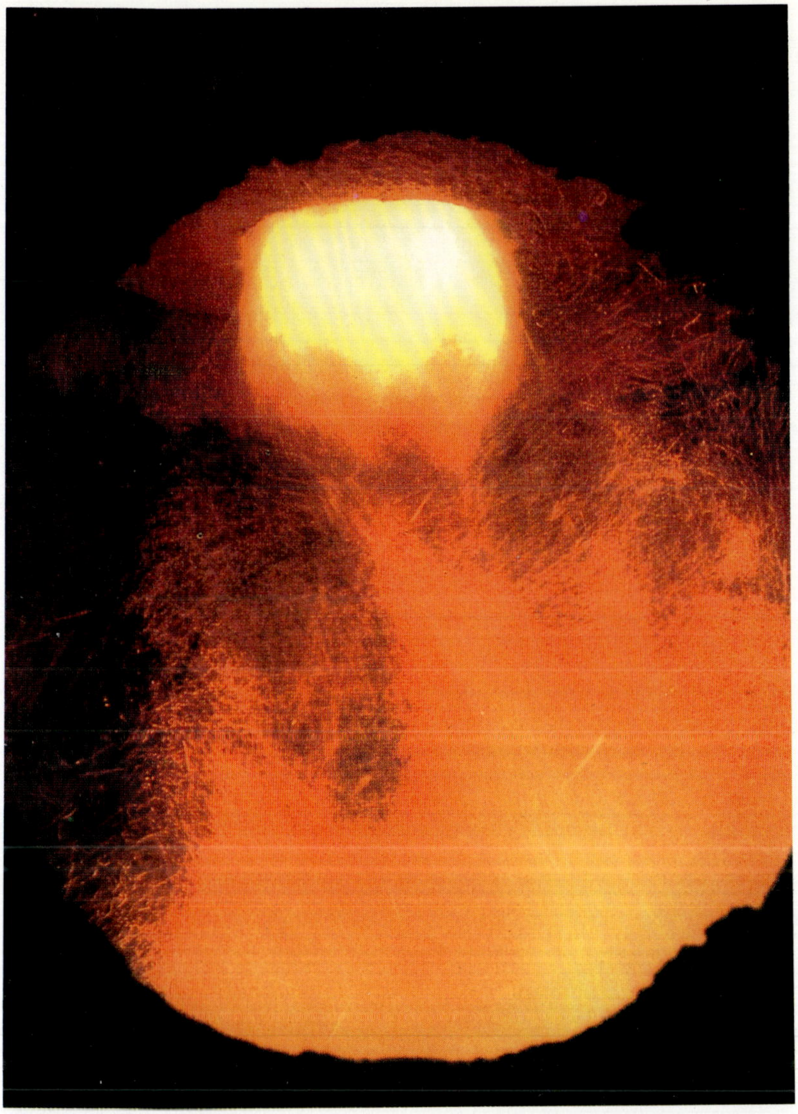

COLOR PLATE I (see p. 140) A photograph of flame 1 (SAO burner).

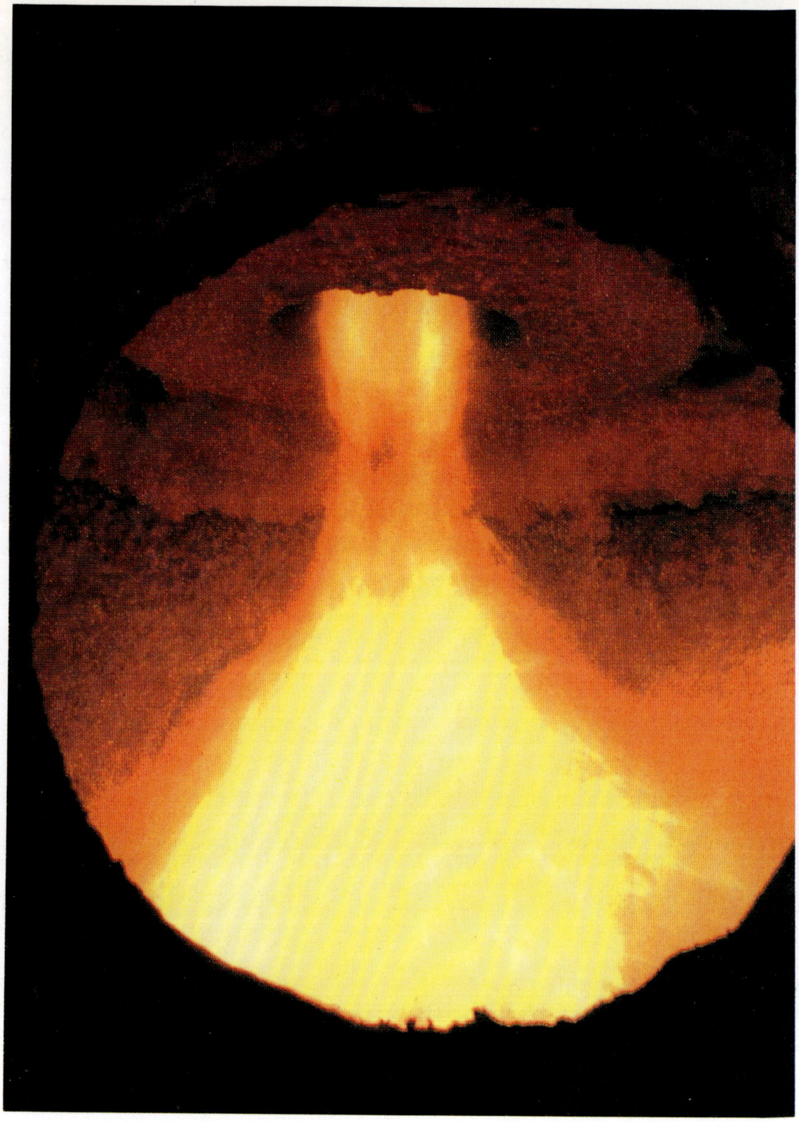

COLOR PLATE II (see p. 141) A photograph of flame 2 (SCO burner).

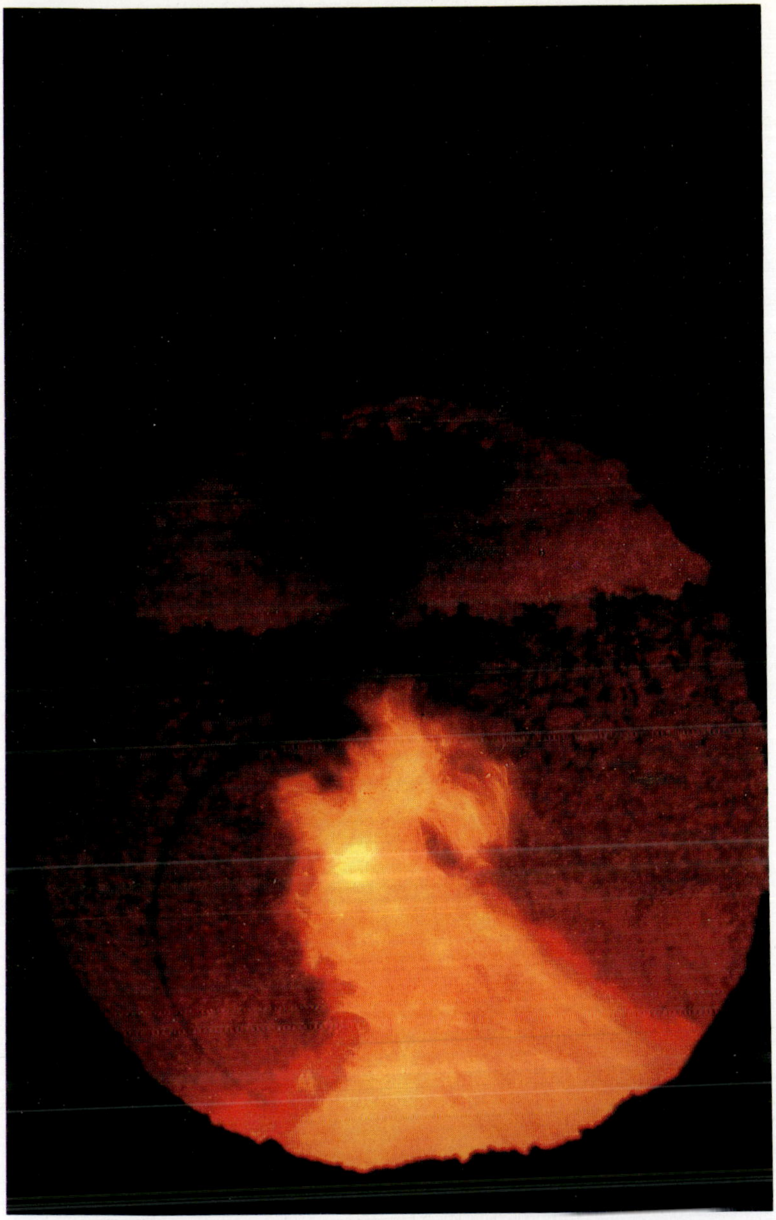

COLOR PLATE III (see p. 142) A photograph of a typical unstable flame.

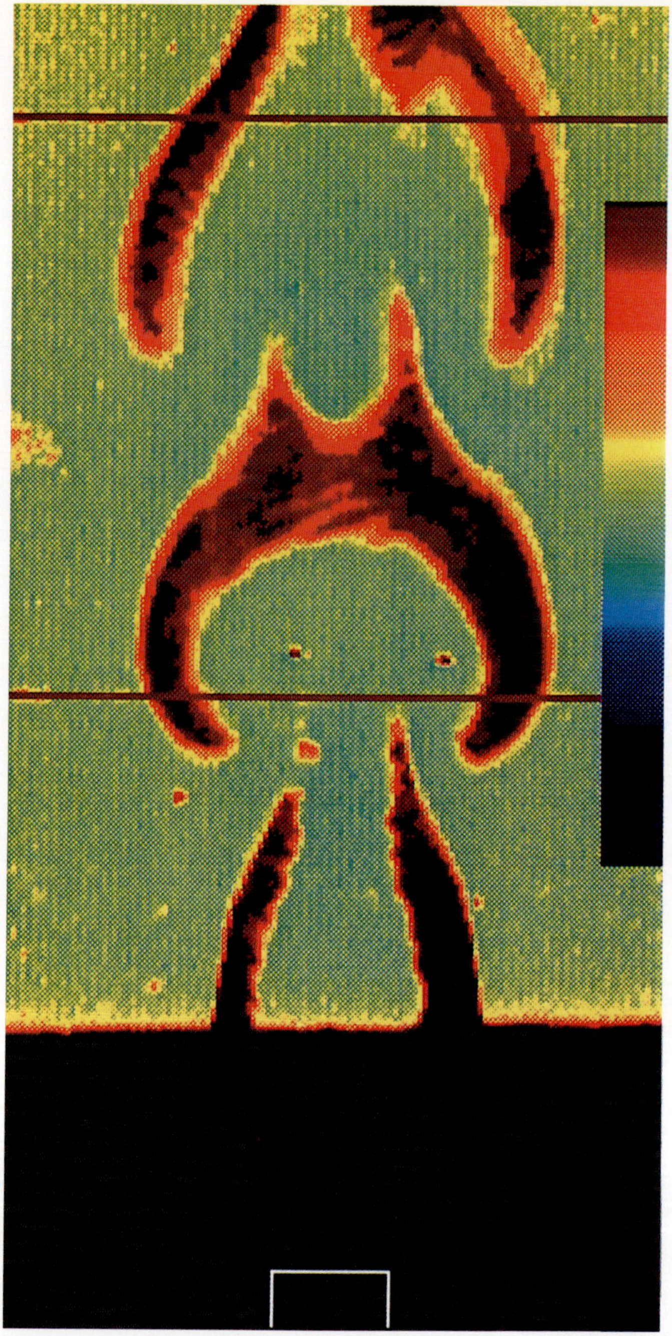

COLOR PLATE IV (see p. 317) Conditionally averaged schlieren image (Re = 180, forcing frequency = 14 Hz, forcing amplitude = 100mV).

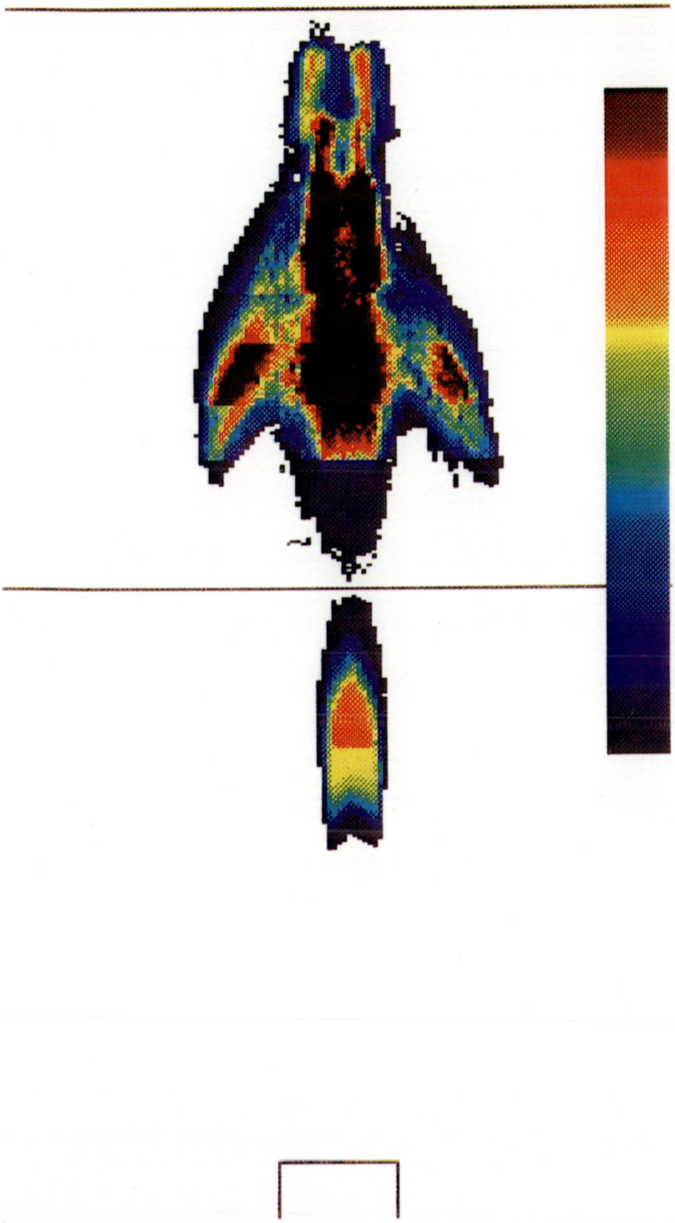

COLOR PLATE V (see p. 318) Conditionally averaged NO_2 PLIF image (Re = 180, forcing frequency = 14 Hz, forcing amplitude = 100 mV).

a. Stage I

b. Stage II

c. Stage III

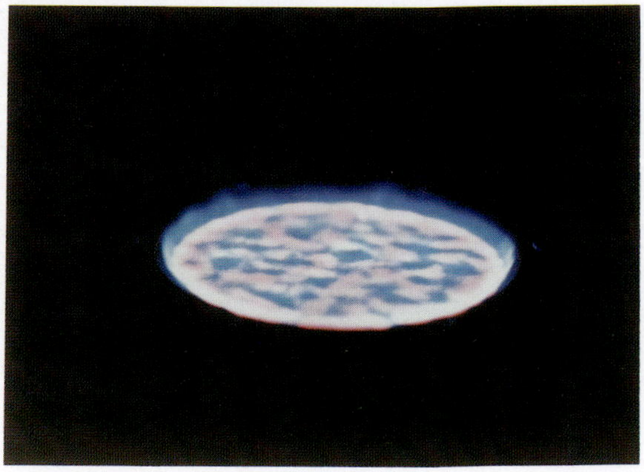

d. Stage IV

e. Stage V

COLOR PLATE VI (see p. 433) Photographs of the five stages of combustion.

a. Stage I

b. Stage II

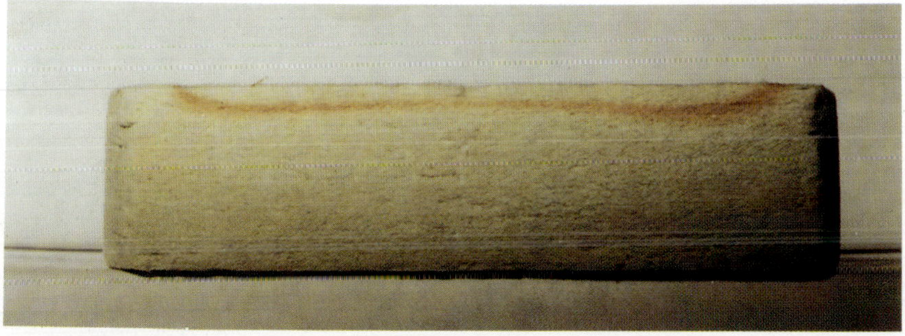

c. Stage III

COLOR PLATE VII (see p. 436) Crosscut of the fiber mat showing the brown line formed during combustion for Stages II and III: a) stage I; b) stage II; c) stage III.

COLOR PLATE VIII (see p. 498) Atmospheric fluidized bed combustion pilot plant.

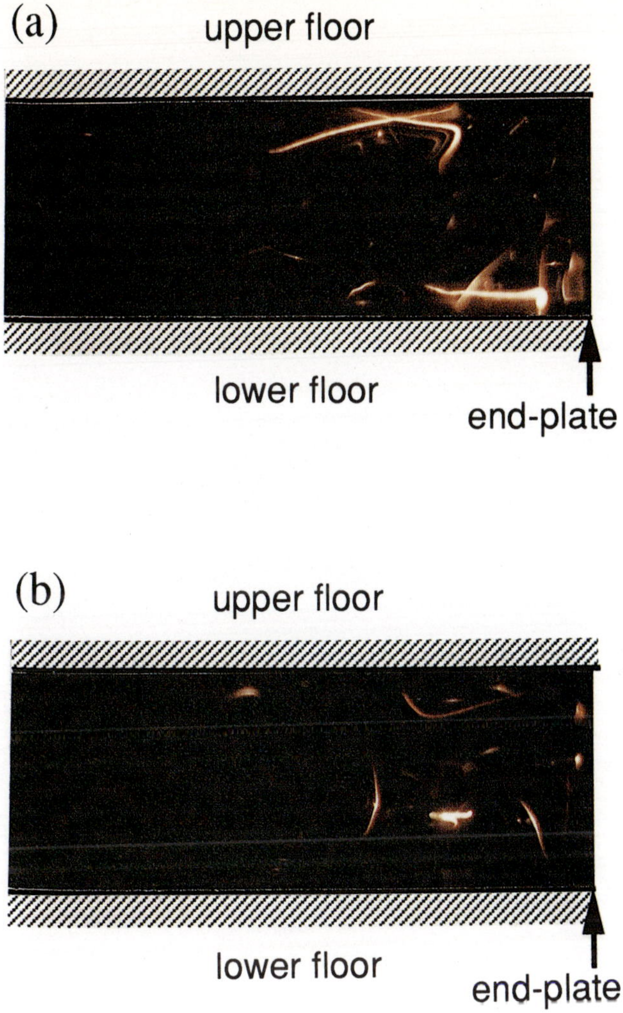

COLOR PLATE IX (see p. 808) Direct photographs of ignited tungsten particles: a) $T_5 = 1000$ K, (b) $T_5 = 830$ K.